Ergonomics for Children

Designing products and places for toddlers to teens

Ergonomics for Children

Designing products and places for toddlers to teens

Rani Lueder

Valerie J. Berg Rice

CRC Press is an imprint of the
Taylor & Francis Group, an informa business

CRC Press
Taylor & Francis Group
6000 Broken Sound Parkway NW, Suite 300
Boca Raton, FL 33487-2742

First issued in paperback 2019

ISBN-13: 978-0-415-30474-0 (hbk)
ISBN-13: 978-0-367-37644-4 (pbk)

Visit the Taylor & Francis Web site at
http://www.taylorandfrancis.com

and the CRC Press Web site at
http://www.crcpress.com

FOREWORD

For many centuries children have been seen as an adjunct to adults, either as helpers, earners, or little people to be humanized. The last hundred years have seen a gradual change, in schools, in families, and in the environment in which children live. At last they are seen, in general, as children. They have unique requirements, problems, needs, and desires.

Children's lives impinge on society in ways which differ from the influences of adults. Yet adults are responsible for the environment which children experience. So it is important that adults so design the environment of children that the children themselves grow, benefit, and enjoy their developing lives. The world of a child is broad, just as is that of an adult. That is why this book covers so many areas. It matches the world in which the child lives.

Research in human factors has not always matched the reality of the world we live in, however. It has tended to concentrate on adults, on military requirements, on adult workplaces, and on (male) college graduates as well as on the interface between the public and the world around us. In spite of the large population of children, school (the workplace of the young) has a relatively small body of ergonomics studies and the ergonomics of play even less. Yet the psychology of childhood is a large field, with huge implications for ergonomics.

This book is a first attempt to put the lives of children in the context of the society in which they live, to give a comprehensive analysis with explanations, reasons, and design recommendations for the betterment of their lives. Anyone who has responsibility for the welfare of children should become familiar with the contents of this book, to recognize the complexity of their task and to give themselves the background to cooperate in systematic and holistic approaches to their problems. This is not a book just for the affluent West, but for children everywhere. If we are to have a better future world, we must use our knowledge to make it better for our children.

Nigel Corlett, Phd, FREng., FErgs
Emeritus Professor
University of Nottingham

ACKNOWLEDGMENTS

Writing this book about ergonomics for children has been exciting, inspiring, and terrifying: Exciting, because editing and writing this book expanded our world and crystallized our thinking; inspiring while reviewing the vast research literature on children that cuts across disciplines and the momentous contributions from nations and communities around the world, and terrifying at times as we tried to find the words that might adequately represent these efforts and help our most vulnerable citizens at their different developmental stages and life circumstances.

We are grateful to many people along the way; it would be difficult to name all of them. First, thanks to our families for not giving up on us (especially Kiran, Elia, Vince, and Chuck) as we were consumed by this vortex of information. We can not thank Ursy Potter enough, the brilliant professional photographer who fortunately (for us) is Rani's mother. Her wonderful photos of children around the world fill most of the chapters and made all the difference.

We are also grateful to our contributors; this book was a team effort in every respect and we can not express our thanks enough for their patience with us as they each continued to make (up to 20) revisions while we struggled to create a common vision for the book. They did not know what they were in for, yet they stuck by us with humor and faith all the time. We owe a special thanks to Cheryl Bennett for her participation and suggestions in the early stage of this book. Our gratitude to our editor Cindy Carelli for her help, belief in the book, and patience. We are indebted to Randy Burling, Taylor & Francis's wonderful production manager, who (when we thought all was lost) stepped in and volunteered long hours to help the publication better reflect its intended vision. Our thanks to Nigel Corlett for the foreword and for inspiring generations of ergonomists with his invaluable contributions. Finally, we thank the children of this world for being such astonishing and brilliant creatures—they are our future and remind us that all things are possible.

EDITORS

Rani Lueder, MSIE, CPE, is President of Humanics ErgoSystems, Inc., an ergonomics consulting firm in Encino, California, where, since 1982, she consults, performs research, and provides expert testimony in occupational ergonomics and the design and evaluation of products and places for adults, children, and people with disabilities. She also teaches human factors and ergonomics in product design at Art Center College of Design in Pasadena. She has been a consultant for corporations, governments, and universities in seven countries and served as an expert witness on a range of cases related to the design of products, built environment, occupational health, and accommodating people with disabilities. She has performed numerous large-scale evaluations for organizations and served on retainer for numerous organizations including Waseda University, the Institute for Human Posture Research, and other organizations in Japan. Her service-related activities include participating in ergonomics standards committees and chairing a Human Factors and Ergonomics Society task force on the human factors and ergonomics content and services on their Web site for the short- and long term. Previously, she edited and coauthored two books: *Hard Facts about Soft Machines: The Ergonomics of Seating* (with Kageyu Noro: Taylor & Francis) and *The Ergonomics Payoff: Designing the Electronic Office* (Holt, Rinehart and Winston). Her Web site, www.humanics-es.com, provides extensive content on a range of topics, including ergonomics for children.

Valerie J. Berg Rice, PhD, CPE, OTR/L, FOTA, has graduate degrees in occupational therapy, health care administration, and industrial engineering and operations research (human factors/ergonomics option). She is a board-certified ergonomist, a registered and licensed occupational therapist, and a fellow of the American Occupational Therapy Association. Dr. Rice (COL, R), completed 25 years of active duty in the Army. She has provided patient treatment to children with developmental, physical, and cognitive injuries and illnesses, and helped children and their parents address design issues in their homes, schools, and programs of learning. She has presented and published extensively on medical human factors, cognitive ergonomics (learning, attention deficit and hyperactivity disorder, stress), industrial ergonomics (injury prevention, safety, and physically demanding tasks), user evaluation, and macroergonomics. She served on the Board of Directors for the Human Factors and Ergonomics Society and the Board of Certification in Professional Ergonomics, and is President of the Board of Directors for the Foundation for Professional Ergonomics. She previously edited the textbook, *Ergonomics in Health Care and Rehabilitation* (Butterworth), and has authored and presented over 100 professional papers. Her general ergonomics Web site at www.genergo.com provides additional background on her ergonomics consulting services and related child ergonomics activities. She is the chief of the Army Research Laboratory Army Medical Department Field Element at Ft. Sam Houston, San Antonio, Texas.

CONTRIBUTORS

Gary L. Allen
Department of Psychology
University of South Carolina
Columbia, South Carolina

Dennis R. Ankrum
Ankrum Associates
Oak Park, Illinois

Mary Benbow
Research Clinician
Private Practice
La Jolla, California

Cheryl L. Bennett
Youth Ergonomics Strategies
Livermore, California
and
Lawrence Livermore National Laboratory
Livermore, California

Melissa Beran
Interek-RAM
Oak Brook, Illinois

Barbara H. Boucher
TherExtras.com
San Antonio, Texas

Tina Brown
Interek-RAM
Oak Brook, Illinois

Cindy Burt
University of California at Los Angeles
Los Angeles, California

Hsin-Yu (Ariel) Chiang
Department of Occupational Therapy
Fu-jen Catholic University
Taiwan

Knut Inge Fostervold
Department of Psychology
University of Oslo
Oslo, Norway
and
Faculty of Health and Social Work
Lillehammer University College
Lillehammer, Norway

David Gulland
Hassell
Perth, Western Australia

Libby Hanna
Hanna Research and Consulting
Seattle, Washington

Alan Hedge
Department of Design and Environmental Analysis
Cornell University
Ithaca, New York

Hal W. Hendrick
Hendrick & Associates
Englewood, Colorado

Karen Jacobs
Sargent College of Health and Rehabilitation Sciences
Boston University
Boston, Massachusetts

Michael J. Kalsher
Rensselaer Polytechnic Institute
Troy, New York

Jeff Kennedy
Jeff Kennedy Associates, Inc.
Somerville, Massachusetts

Rani Lueder
Humanics ErgoSystems, Inc.
Encino, California

Lorraine E. Maxwell
Department of Design and Environmental Analysis
Cornell University
Ithaca, New York

Beverley Norris
Institute for Occupational Ergonomics
University of Nottingham
Nottingham, United Kingdom

Jake Pauls
Consulting Services in Building Use and Safety
Silver Spring, Maryland

Jeff Phillips
Department of Education and Training Western Australia
Perth, Western Australia

Clare Pollock
School of Psychology
Curtin University of Technology
Perth, Washington

Marjorie Prager
Jeff Kennedy Associates, Inc.
Somerville, Massachusetts

Valerie J. Berg Rice
General Ergonomics
Universal City, Texas
and
USA Research Laboratory
Army Medical Department Center & School Field Element
Ft. Sam Houston, San Antonio Texas

Stuart A. Smith
Institute for Occupational Ergonomics
University of Nottingham
Nottingham, United Kingdom

Robin Springer
Computer Talk, Inc.
Encino, California

Leon Straker
School of Physiotherapy
Curtin University of Technology
Perth, Western Australia

Brenda A. Torres
S.C. Johnson & Son, Inc.
Racine, Wisconsin

Nancy L. Vause
USA Center for Health Promotion & Preventive Medicine – Pacific
Camp Zama, Japan

Alison G. Vredenburgh
Vredenburgh & Associates, Inc.
Carlsbad, California

Thomas R. Waters
National Institute for Occupational Safety and Health
Cincinnati, Ohio

David Werner
HealthWrights
Palo Alto, California

Michael S. Wogalter
North Carolina State University
Raleigh, North Carolina

Ilene B. Zackowitz
Vredenburgh & Associates, Inc.
Carlsbad, California

TABLE OF CONTENTS

Section C
Injuries, Health Disorders, and Disabilities

Section D
Children and Product Design

Section E
Children at Home

Section F
Children and School

Section G
Children and Public Places

SECTION A

Introduction

CHAPTER 1

INTRODUCTION

RANI LUEDER AND VALERIE RICE, WITH CHERYL L. BENNETT*

This book is an attempt to provide a practical user's manual about ergonomics† and children for professionals who design products and places for and work and play with children. They include ergonomists, product designers, manufacturers, technology specialists, educators, rehabilitation therapists, architects, city planners, attorneys, and even parents.

As such, it is a different sort of book than others in the field. It cuts across a wide swath of disciplines such as ergonomics, psychology, medicine, rehabilitation, exercise physiology, optometry, education, architecture, urban planning, law, and others.

Children are clearly not "little adults," but how do they differ, and how do such differences affect the design of products and places that they use? How can we better help them face new and unique challenges, such as when using new technologies? The questions were simple, but the answers were not.

The process of trying to make sense of this vast and sometimes contradictory array of information was often daunting, but we have learned much along the way. While we mourn for the material we were not able to include in this version, we are grateful to our contributors for their struggle with us through this wonderful and creative experience. We thank you for opening this book. We hope it helps you (our readers) practice your important efforts.

Table 1.1 briefly summarizes the chapters to come.

* Our thanks to Cheryl Bennett for her initial participation in the book.

† Ergonomics is an applied research discipline concerned with the fit between people, the things they do, the objects they use, and the environments they work, travel, and play in (Ergonomics Society).

Table 1.1 Chapter Topics in this Book

Section A ***Introduction***	Chapter 1 Introduction Rani Lueder and Valerie Rice, with Cheryl L. Bennett		
Section B ***Child abilities and health***	Chapter 2 Developmental stages of children Tina Brown and Melissa Beran	 (Ursy Potter Photography)	Each child is unique, yet children also undergo universal developmental stages that are affected by their life-experiences and culture. We can predict when, how, and why children do the things they do based on the cognitive, physical, social, emotional, and language dimensions of each developmental stage.
	Chapter 3 Child anthropometry Beverley Norris and Stuart A. Smith	Anthropometry is the scientific measurement of human body sizes, shapes, and physical capabilities. Anthropometric data helps us evaluate the fit between children and the products and environments they use. An understanding of this fit is critical to ensure that children can use (and enjoy) products intended for them. At the same time, it protects them from harm by ensuring that hazards are properly guarded or placed out of reach.	 (Valerie Rice)
	Chapter 4 Visual ergonomics for children Knut Inge Fostervold and Dennis R. Ankrum	 (Ursy Potter Photography)	Children are exposed to very different visual environments than previous generations. Yet these technologies are often used without regard for the corresponding implications for children's comfort, health, and long-term well-being. Further, children's visual environments may adversely affect their postures and postural risk. This chapter reviews what we know about the ergonomic implications related to vision and to the use of new technologies.
	Chapter 5 Hearing ergonomics for children: Sound advice Nancy L. Vause	Loud and distracting noise can damage our hearing, potentially hindering our ability to learn and fully experience our lives. This chapter reviews the literature on the impact of noise on hearing, the nature of hearing loss, and its effect on learning in the classroom. It also provides guidance for designing effective classroom environments that will reduce sound distractions and promote hearing.	 (Ursy Potter Photography)

Table 1.1 Chapter Topics in this Book (Continued)

	Chapter 6 Physical development in children and adolescents and age-related risks Rani Lueder and Valerie Rice	Children today must face different demands than with preceding generations. In general, children are also taller, heavier, and less fit. Many children (particularly girls) experience puberty earlier than previous generations, increasing the potential for musculoskeletal pain and disorders. Children's injury risks are also different than those adults commonly experience. This chapter reviews how children's musculoskeletal systems develop and the corresponding implications for developing back pain and soft tissue disorders.	 (Ursy Potter Photography)
	Chapter 7 Physical education and exercise for children Barbara H. Boucher	 (Ursy Potter Photography)	Children benefit mentally, physically, and socially from physical activity. Exercise is essential for proper development of bones, muscles, and joints. It promotes health, improves alertness, self-esteem, and outlook. User-centered physical education programs incorporate principles of child development, build on environmental influences, and consider the child as a whole. This chapter reviews how children develop physiologically, cognitively, and socially between the ages of 5 and 15, and the implications for effective PE programs.
Section C ***Injuries, health disorders, and disabilities***	Chapter 8 Children and injuries Valerie Rice and Rani Lueder	 (Valerie Rice)	A child's "work" is to develop physically, emotionally, socially and cognitively. Eliminating all risks would also eliminate challenges essential to children's growth and development. This chapter focuses on preventing injuries that are most common among children: falls, burns, poison, choking, and drowning. It includes design and safety considerations for products, places, and tasks to promote effective environments while reducing the potential for injuries.
	Chapter 9 Assistive technologies for children Robin Springer	Assistive technologies enable people with disabilities to participate in activities of daily living, helping to ensure equal opportunities. These items range from high tech to everyday products that include wheelchairs, adapted vans, communication devices, and modified computers. This chapter aims to help the reader understand a broad range of assistive technologies and to evaluate their appropriateness for the user. It also reviews considerations for using them effectively.	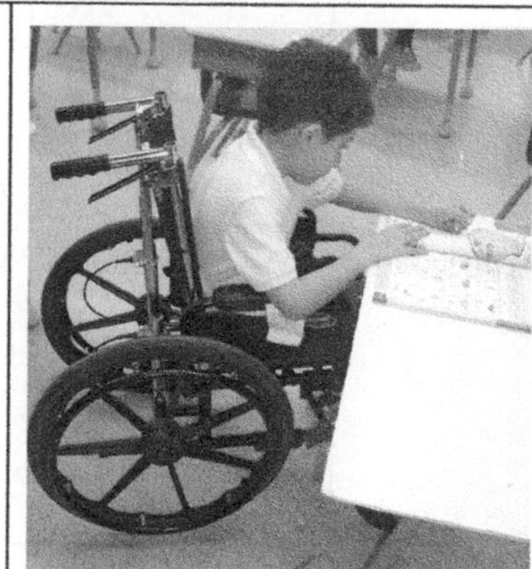 (Valerie Rice)

(continued)

Table 1.1 Chapter Topics in this Book (Continued)

	Chapter 10 Meeting the needs of disabled children in developing countries David Werner	(Ursy Potter Photography)	Helping disabled children in developing villages requires working with them to create solutions that provide a "goodness of fit" with their life circumstances. Adapting simple, locally made equipment while involving the disabled child and their family helps them fit in their community. Even rehabilitation exercises can isolate a child from their community and emphasize how they differ from their peers. Work-based therapeutic activities can help them build on strengths rather than limitations.
Section D ***Children and product design***	Chapter 11 Designing products for children Valerie Rice and Rani Lueder	This chapter reviews the ergonomic considerations for designing products for children to encourage their proper use, safety, and fun. It also summarizes the research and provides guidelines for evaluating a range of products, from baby cribs and strollers to youth recreation.	 (Ursy Potter Photography)
	Chapter 12 Children's play with toys Brenda A. Torres	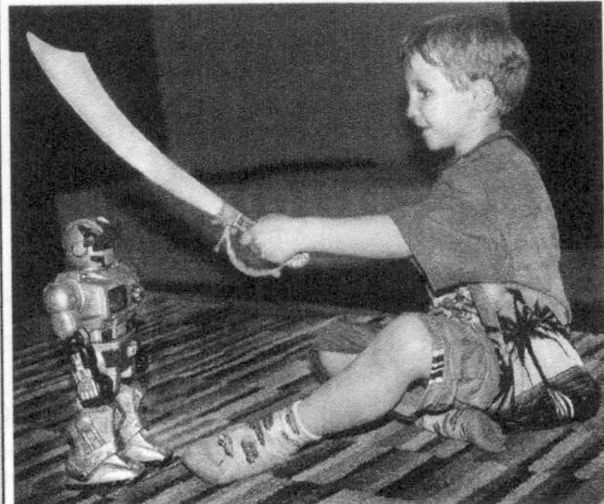 (Ursy Potter Photography)	Today, the focus on ergonomics in toy design has broadened beyond safety and comfort to designing toys to fit the child user. Today's consumers also expect products to be easy to use, functional, pleasurable, and to contribute to growth and development. This requires an understanding of children's developmental stages.
	Chapter 13 Bookbags for children Karen Jacobs, Renee Lockhart, Hsin-Yu (Ariel) Chiang, and Mary O'Hara	Backpacks are a practical way to transport schoolwork. Even so, research indicates that the design of the book bag and how it is used influence the risk of developing musculoskeletal injuries.	 (Valerie Rice)

Table 1.1 Chapter Topics in this Book (Continued)

	Chapter 14 Warnings: Hazard control methods for caregivers and children Michael J. Kalsher and Michael S. Wogalter	 (Ursy Potter Photography)	Young children do not have the cognitive abilities to recognize and avoid risks. This chapter reviews the range of issues surrounding how best to inform children and caregivers of hazards. It considers the roles of manufacturers and caregivers in protecting children by designing out the hazard, guarding against the hazard, and warnings.
Section E ***Children at home***	Chapter 15 Stairways for children Jake Pauls	Climbing and descending stairs requires a combination of strength, balance, timing, and equilibrium. It takes coordinated effort to avoid missteps, falls, and injury. This chapter focuses on design factors that contribute to ease-of-use and safety for children on stairs. Important considerations include stairway visibility, step dimensions, and handholds.	 (Ursy Potter Photography)
	Chapter 16 Child use of technology at home Cheryl L. Bennett	New technologies are increasingly common in the home. These must do more than accommodate children's physical and mental abilities; it is important for child users to understand basic principles of posture, body mechanics, and the risks and benefits associated with using computers and other electronic devices. Imparting an awareness of the importance of position, posture, and comfort to children at an early age can establish habits that will empower them throughout their lives.	 (Ursy Potter Photography)
	Chapter 17 Children in vehicles Rani Lueder	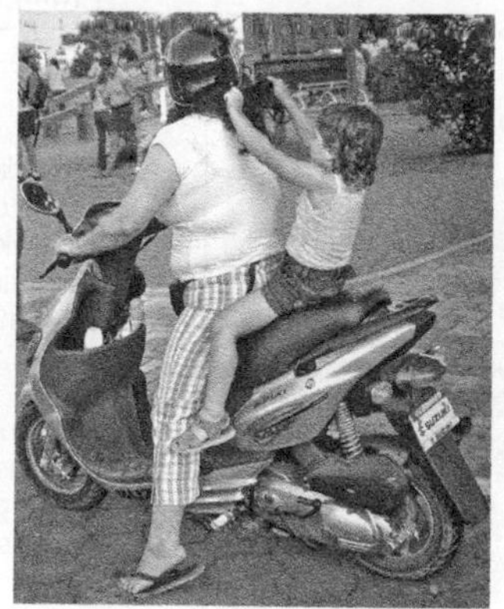 (Ursy Potter Photography)	Motor vehicle accidents are the leading cause of death and severe injuries in children at every age after their first birthday. Children's injuries are also more severe than those of adults; their small size and developing bones and muscles make them more susceptible to injury in car crashes if not properly restrained. Many of these deaths could have been avoided, such as by proper design and use of child restraints that are appropriate for the child's age, positioning in the vehicle, and many other factors.

(continued)

Table 1.1 Chapter Topics in this Book (Continued)

	Chapter 18 Preventing musculoskeletal disorders for youth working on farms Thomas R. Waters	Farms are one of the most hazardous places for anyone to work. Children and adolescents on family farms begin helping at very young ages, often performing physically demanding jobs designed for adults. These jobs may exceed children's capabilities and lead to acute or chronic musculoskeletal disorders. The jobs include lifting and moving materials, operating farm equipment, and performing other tasks that require strength and coordination. This chapter describes the ergonomic risks for children that are associated with farming tasks and provides guidelines for avoiding hazards.	 (Ursy Potter Photography)
Section F ***Children and school***	Chapter 19 Preschool and daycare design Lorraine E. Maxwell	 (Ursy Potter Photography)	Physical preschool environments play a critical role on children's cognitive, social, physical, and emotional development. Stimulating and well-organized childcare settings help children develop their vocabularies, attention and memory skills, and social interactions with peers.
	Chapter 20 Children and handwriting ergonomics Cindy Burt and Mary Benbow	Many of us take children's ability to learn to write for granted. Yet handwriting is physically and intellectually demanding. Postural instability, paper and pencil positioning, and limited gripping ability are correlated with poor handwriting performance. Creating effective child environments for writing requires more than simply supplying a place for children to copy letters. Effective learning environments must be directive, supportive, and intriguing for children as they develop this new way to communicate. Children need appropriate writing tasks and tools and must be developmentally ready to write.	 (Ursy Potter Photography)
	Chapter 21 School furniture for children Alan Hedge and Rani Lueder	 (Ursy Potter Photography)	Research indicates that many schoolchildren sit in furniture that does not fit them properly. Schoolchildren who sit in awkward postures for long durations can experience musculoskeletal symptoms that worsen with time. Yet common assumptions about what is ergonomically "proper" for adults may not be appropriate for children. This chapter reviews ergonomic design considerations for classroom furniture and summarizes worldwide ergonomics research into the design of comfortable school furniture.

Table 1.1 Chapter Topics in this Book (Continued)

	Chapter 22 Child-friendly user interfaces in the digital world Libby Hanna	Children, like adults, need ergonomic "user-friendly" interfaces in the broad range of electronic media they use on a regular basis. Ergonomic design guidelines can set the stage for children's initial interactions with a product, enabling them to use products intuitively and fluidly. The products should be easy to use and geared to their particular developmental stages in hand–eye coordination and cognitive skills. Children also need media content that offers opportunities for growth. Ergonomic guidelines can stimulate and nurture development by providing electronic media that enrich children's lives.	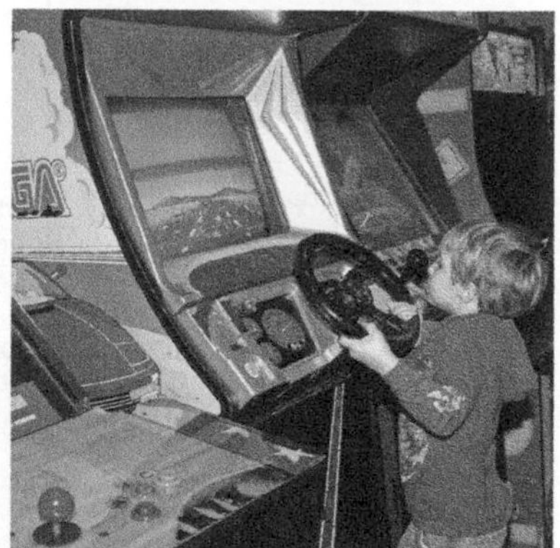 (Ursy Potter Photography)
	Chapter 23 Information and communication technology in schools Clare Pollock and Leon Straker	This chapter reviews the ergonomic implications of using information and communication technologies in schools. It also provides guidelines for implementing and using these technologies to promote child learning and well-being.	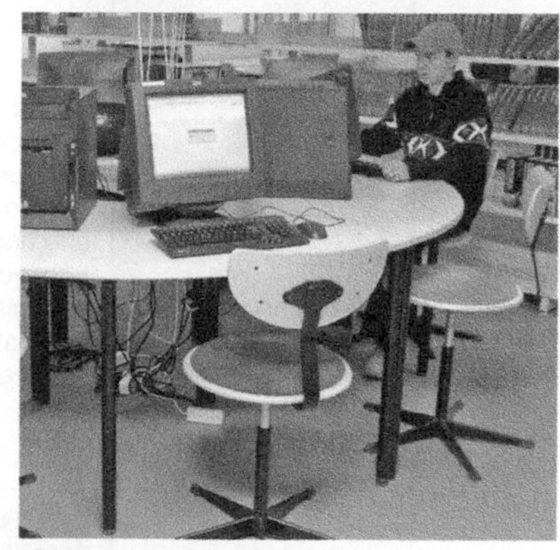 (Courtesy of SIS-USA)
	Chapter 24 Rethinking school design: New directions in child learning David Gulland and Jeff Phillips	 (Courtesy of SIS-USA)	The quality of learning environments can enhance learning outcomes, as learners respond positively to stimulating spaces. Poor school designs create barriers to learning by physically isolating students from each other and hindering the sense of belonging, ownership, or engagement with the space. This chapter compares traditional schools with innovative new schools that reflect a paradigm shift on environmental influences on learning.
Section G ***Children and public places***	Chapter 25 Designing cities and neighborhoods for children Rani Lueder	This chapter focuses on the design of public places to help children be safe and to use them effectively.	 (Ursy Potter Photography)

(continued)

Table 1.1 Chapter Topics in this Book (Continued)

Chapter 26 Children and wayfinding Gary L. Allen, Rani Lueder, and Valerie Rice	Children are explorers by nature, and their wayfinding skills improve dramatically with age and practice. Further, individual children vary greatly in their wayfinding abilities. This chapter identifies wayfinding strategies and reviews methods to enhance children's wayfinding skills. It also provides guidelines for design that will help children orient themselves.	 (Ursy Potter Photography)
Chapter 27 Designing museum experiences for children Jeff Kennedy and Marjorie Prager	The last 25 years has seen a boom in museum experiences for children. Museums of all kinds—science centers, art and history museums, zoos, aquariums, and nature centers—offer experiences that engage and entertain children. This chapter reviews ergonomic implications for accommodating children and their caregivers in these environments.	 (Ursy Potter Photography)
Chapter 28 Playground safety and ergonomics Alison G. Vredenburgh and Ilene B. Zackowitz	 (Ursy Potter Photography)	Playgrounds enable children to develop physical and social skills in fun and stimulating environments. Playgrounds provide children with "work" activities that help them develop skills such as eye–hand coordination and balance. The fit between the child user and their environments is critical in playgrounds. Children need challenge in safe environments. Improper playground design or maintenance can contribute to injuries. This chapter focuses on safety issues that help reduce children's exposure to danger and reviews common playground hazards, child user requirements, and design guidelines.

SECTION B

Child Abilities and Health

CHAPTER 2

DEVELOPMENTAL STAGES OF CHILDREN

TINA BROWN AND MELISSA BERAN

TABLE OF CONTENTS

Each child is unique, yet children also exhibit consistent similarities as they grow and develop. We can predict when, how, and why children do the things they do based on the cognitive, physical, social, emotional, and language dimensions of each developmental stage. This chapter describes these developmental stages as children mature from infancy to their 14th year.

Of course, such developmental timelines may not apply to a particular child. Children evolve through universal patterns, but these commonalities occur against a backdrop of each child's culture, heritage, and individual differences.

INFANCY TO TODDLERHOOD (BIRTH TO 2 YEARS)

COGNITIVE DEVELOPMENT

Jean Piaget (1896–1980) was a leading authority on child development. Through his work, he recognized that young children think and reason very differently from older children and adults. He described how mental processes change as children progress through four qualitatively different stages of cognitive development. These theories underlie much of the content in this chapter.

During Piaget's *sensorimotor stage* (birth to 2 years), children explore the world through their senses and motor skills (Figure 2.1). They smell, look at, touch, taste, and listen as they explore new objects. These interactions begin accidentally, but become purposeful and goal-oriented as they learn about their world.

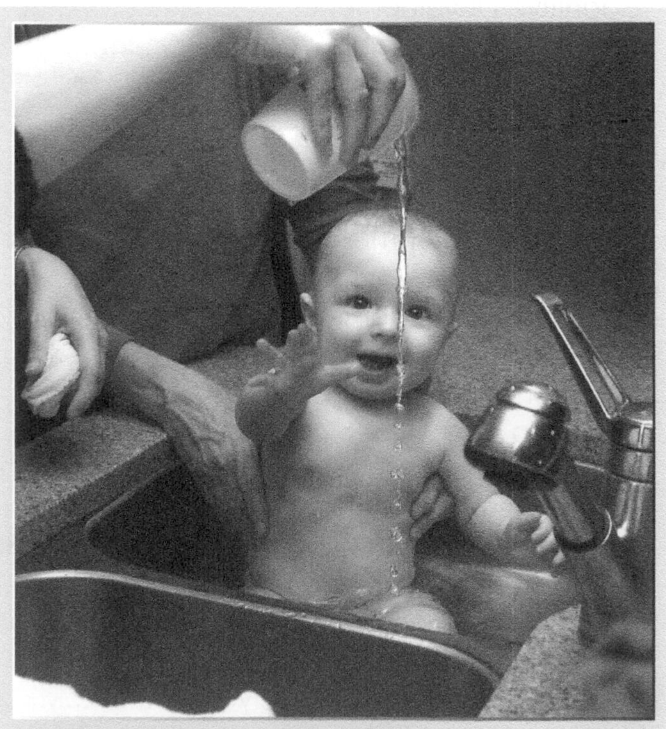

Figure 2.1 (Ursy Potter Photography)
Nolan explores the world through her senses during the sensorimotor stage from birth to 2 years.

Between the ages of 18 and 24 months, children develop *symbolic thinking*, which allows them to engage in pretend play. During pretend play, one thing or object can represent another and children begin to act out different roles such as mother, father, baby, teacher, doctor, etc.

They also begin to develop a sense of *object permanence* at this age. For example, 6-month-old children do not become upset when caregivers remove

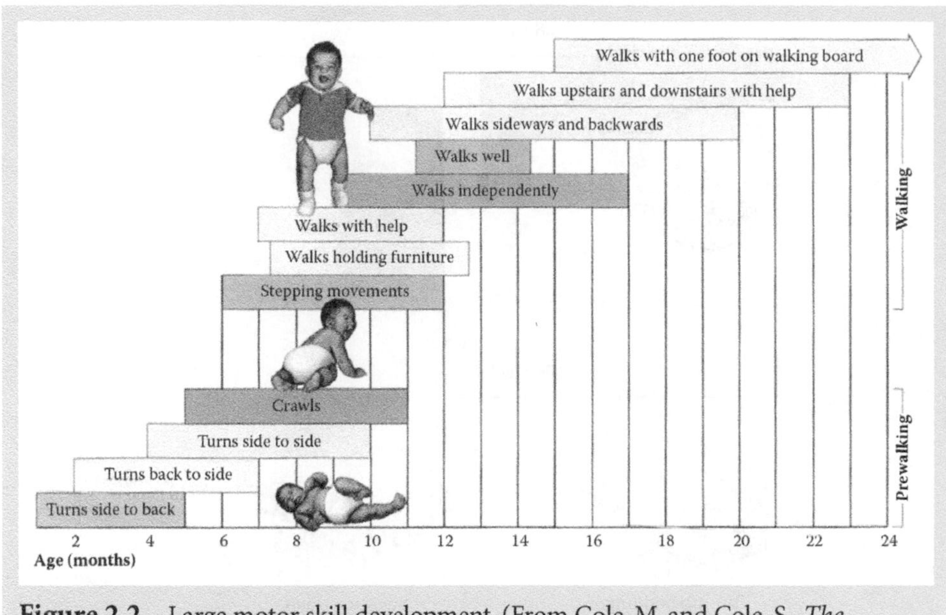

Figure 2.2 Large motor skill development. (From Cole, M. and Cole, S., *The Development of Children*, 4th edn., Worth Publishers, New York, 2001.)

toys because at this age, children do not remember objects they cannot see. By the time they reach 18 to 24 months, these children will remember and search for "hidden" toys and object permanence is fully achieved.

PHYSICAL DEVELOPMENT

Over their first 2 years, children's gross motor skills develop as they learn to roll over, sit up, crawl, stand, walk, run, and climb (Figure 2.2). Fine motor skills also progress rapidly. Children move from being unable to intentionally pick up objects, to picking up objects with a swipe of the hand and finally to using their fingers and thumb in a pincer grasp to manipulate small objects (Figure 2.3, Table 2.1).

SOCIAL AND EMOTIONAL DEVELOPMENT

Humans are instinctively social beings. They need to feel emotionally secure in order to develop healthy relationships (Figure 2.4). The need for emotional and physical security remains, even as children begin to explore their world, as noted in the following quote:

> *Perhaps more dramatically than any other age, toddlerhood represents the tension between two ever present yet opposing human impulses: the exhilarating thrust of carefree, unrestricted, uninhibited exploration, where one can soar free without looking back at those who are left behind and the longing to feel safe in the protective sphere of intimate relationships.*
>
> (Lieberman, 1991)

As children in this age group continue to struggle between their contradictory needs to be independent and feel secure, they alternately explore and return to their trusted parents or caregivers for physical comfort. Toddlers learn that they have the power to control some of their world by resisting it.

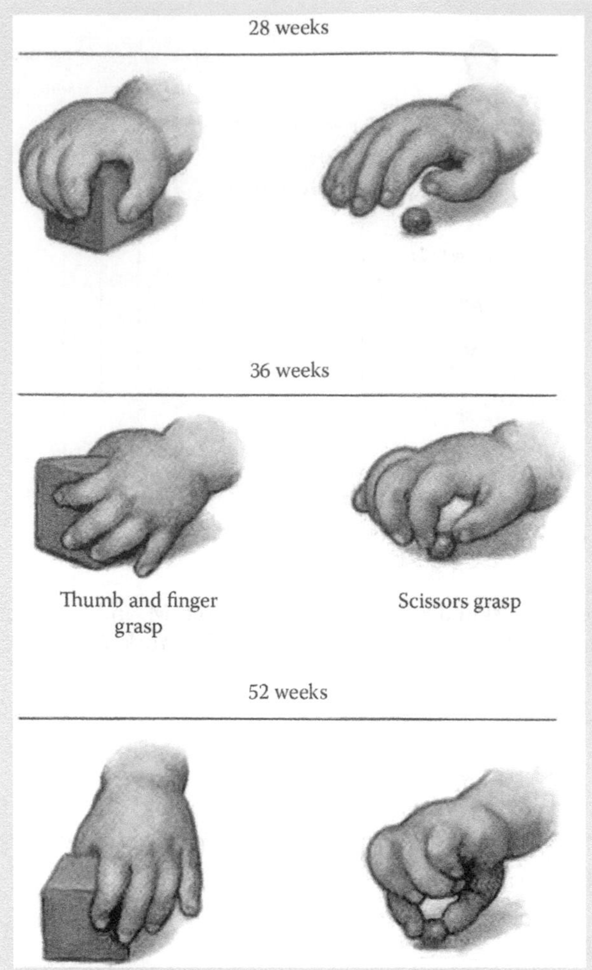

Figure 2.3 This figure shows thumb and finger grasping movements at 28, 36, and 52 weeks of life. In just 24 weeks, a child's coordination moves from a gross grasp to very fine, precise movements. (From Cole, M. and Cole, S., *The Development of Children*, 4th edn., Worth Publishers, New York, 2001.)

As they attempt to express their will and learn their limitations, they may say "no," scream, or have temper tantrums. Children resort to such "negative" behaviors because their language skills are too primitive to easily communicate their needs, desires, and fears (Lieberman, 1991) (Table 2.1).

Figure 2.4 (Ursy Potter Photography)
Infants and toddlers develop a sense of security when their parents and caregivers consistently meet their basic needs for food and comfort and by their communicating with touch, holding, seeing, and hearing.

LANGUAGE DEVELOPMENT

Children normally begin vocalizing around 2 months of age by cooing, babbling, and playing with different noises. Some children utter their first words before their first birthday.

By their second year, some children already make short 2- or 3-word sentences and may have vocabularies of several hundred words (including the ever famous, "mine" and "no") (Figure 2.5).

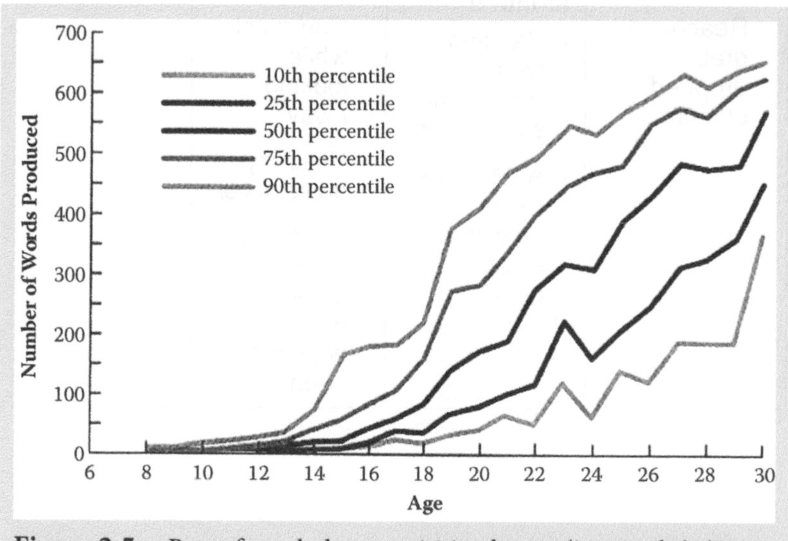

Figure 2.5 Rate of vocabulary acquisition by age (in months). (From Bates, E., in Bizzi, E., Calissamo, P., and Volterra, V., Eds., Frontiere della Biologia [Frontiers of Biology], Giovanni Trecanni, Rome, 1999.)

It is important to note that young children learn these incredible skills by interactively communicating with those around them, rather than solely listening to others, or listening to tapes, CDs, or television.

A summary of child development from birth through age two is shown in Table 2.1. A typical chart depicting milestones is shown in Figure 2.6.

Table 2.1 Summary of Child Development: Birth through 24 Months

	Birth to 3 Months	4–6 Months	7–9 Months	10–12 Months	13–18 Months	19–24 Months
Cognitive Development	**Month 1** • Prefers visual patterns over other visual stimuli • Alert one out of 10 hours **Month 2** • Visually prefers face over objects • Increased awareness of objects **Month 3** • Visually searches for sounds • Begins exploratory play of own body	**Month 4** • Senses strange people and places • Stares at place where an object is dropped **Month 5** • Begins to play with objects • Explores objects by mouthing and touching **Month 6** • Inspects objects • Reaches to grab dropped objects	**Month 7** • Visually searches for disappeared toy (at least briefly) • Imitates physical acts in their repertoire **Month 8** • Prefers new and complex toys • Explores weight, textures, properties of objects **Month 9** • Can find hidden objects they saw hidden • Anticipates return of objects and/or people	**Month 10** • Points to body parts • Uses trial-and-error approach to obtain a goal • Searches for a hidden object but normally in a familiar spot **Month 11** • Imitation increases • Associates properties with objects **Month 12** • Can reach while looking away • Uses toys appropriately • Searches for a hidden object where it was last seen	• Imitates movements • Uses toys inappropriately (e.g., wooden block like a real phone) • Recognizes self in mirror • Remembers where objects are located • Uses stick as a tool	• Recognizes shapes • Notices little objects and small sounds • Sits alone for short periods with book • More symbolic thinking allows simple "pretend" or make believe play • Object permanence is fully achieved (Valerie Rice)

Table 2.1 Summary of Child Development: Birth through 24 Months (Continued)

	Birth to 3 Months	4–6 Months	7–9 Months	10–12 Months	13–18 Months	19–24 Months
Physical Development	**Month 1** • Movements are mostly reflexive • Keeps hands in fist or slightly open **Month 2** • Gains head control • Grasping becomes voluntary **Month 3** • Swipes at dangling objects • Can turn over from back to stomach	**Month 4** • On stomach, can support head and chest on arms • Grasps small objects put into hand; will bring it to mouth **Month 5** • May roll from stomach to back • Passes objects from hand to hand • Sits supported for long periods • Reaches for toy and successfully grasps **Month 6** • May creep on belly • Fascinated with small items • Balances well while sitting • Looks at, reaches for, grasps and brings objects to mouth	**Month 7** • Holds objects in each hand, may bang together • Hands free while sitting • Pushes up on hands and knees; rocks **Month 8** • Uses thumb–finger opposition • Manipulates objects to explore • Crawls • Stands leaning against something **Month 9** • Walks while adult holds hands • Pulls self to standing; gets down again • Explores with index finger • Sits unsupported • Puts objects in containers	**Month 10** • Sits from a standing position • Momentary unsup-ported stand **Month 11** • Stands alone • Crawls and climbs upstairs • Feeds self with spoon **Month 12** • Cruises around furniture • Walks but may still prefer crawling • May climb out of crib or playpen • Has complete thumb opposition • Uses spoon, cup, and crayon	• Enjoys unceasing activity • Carries objects in both hands while walking • Points with index finger • Picks up small objects with index finger and thumb • Walks smoothly • Jumps with both feet off floor • Turns pages	• Walks up and downstairs with help of railing • Jumps, runs, throws, climbs • Begins to show hand preference • Walks smoothly; watches feet • Transitions smoothly from walk to run (Ursy Potter Photography)

(continued)

Table 2.1 Summary of Child Development: Birth through 24 Months (Continued)

	Birth to 3 Months	4–6 Months	7–9 Months	10–12 Months	13–18 Months	19–24 Months
Social and Emotional Development	**Month 1** • Eye contact with mother • Spontaneous smile **Months 2–3** • First selective “social smiles” • Strong interest in looking at human faces	**Month 4** • Returns a smile • Loves to be touched and play peek-a-boo • Smiles when notices another baby • Looks in direction of person leaving a room **Month 5** • Displays anger when objects taken away • Imitates some movements of others **Month 6** • Explores face of person holding them • Differentiates between social responses	**Month 7** • Raises arms to be picked up • Laughs at funny expressions **Month 8** • Responds very differently to strangers than family or caregiver(s) • May reject being alone **Month 9** • Explores other babies • Imitates play	**Month 10** • Displays clear moods • Becomes aware of social approval and disapproval **Month 11** • Seeks approval • May assert self **Month 12** • Prefers family members over strangers	• Imitates housework • Plays in solitary manner • Laughs when chased • Explores reactions of others • Responds to scolding and praise • Little or no sense of sharing • Emotional roller coaster-anger to laughter within moments • Children learn to say “no,” to scream and have temper tantrums	• Hugs spontaneously • Becomes clingy around strangers • Imagines toys have life qualities • Enjoys parallel play • Orders others around • Communicates feelings, desires, and interests (Ursy Potter Photography)

Table 2.1 Summary of Child Development: Birth through 24 Months (Continued)

	Birth to 3 Months	4–6 Months	7–9 Months	10–12 Months	13–18 Months	19–24 Months
Language Development	**Month 1** • Responds to human voice • Begins small throaty sounds **Month 2** • Distinguishes between speech sounds • Makes guttural "cooing" noises **Month 3** • Coos • Responds vocally to speech of others	**Month 4** • Begins babbling: strings of syllable-like vocalizing • Vocalizes moods • Smiles at person speaking to him/her **Month 5** • Utters vowel sounds • Watches people's mouths • Responds to name • Vocalizes to toys **Month 6** • Vowels interspersed with consonants • Vocalizes pleasure and displeasure • Responds to sounds of words, not meaning	**Month 7** • Tries to imitate sounds or sound sequences • May say mama or dada **Month 8** • May learn first words • Understands simple instructions • Uses 2-syllable utterances **Month 9** • Pays attention to conversation • May respond to name and "no" • Uses social gestures	**Month 10** • Obeys some commands • Learns words and appropriate gestures • May repeat words **Month 11** • Imitates inflection, facial expressions • Uses jargon (sentences of gibberish) • Recognizes words as symbols for objects **Month 12** • Practices words he or she knows in inflection • Speaks one or more words	• Points to named objects • Has 4- to 6-word vocabulary (13–15 months) • Uses 2-word utterances • Has a 20-word vocabulary (16–18 months) • Refers to self by name • Spontaneous humming and singing	• Likes rhyming games • Tries to "tell" experiences • Uses "I" and "mine" • Typical expressive vocabulary of 200 words • Repeats words or phrases of others • Uses some short incomplete sentences

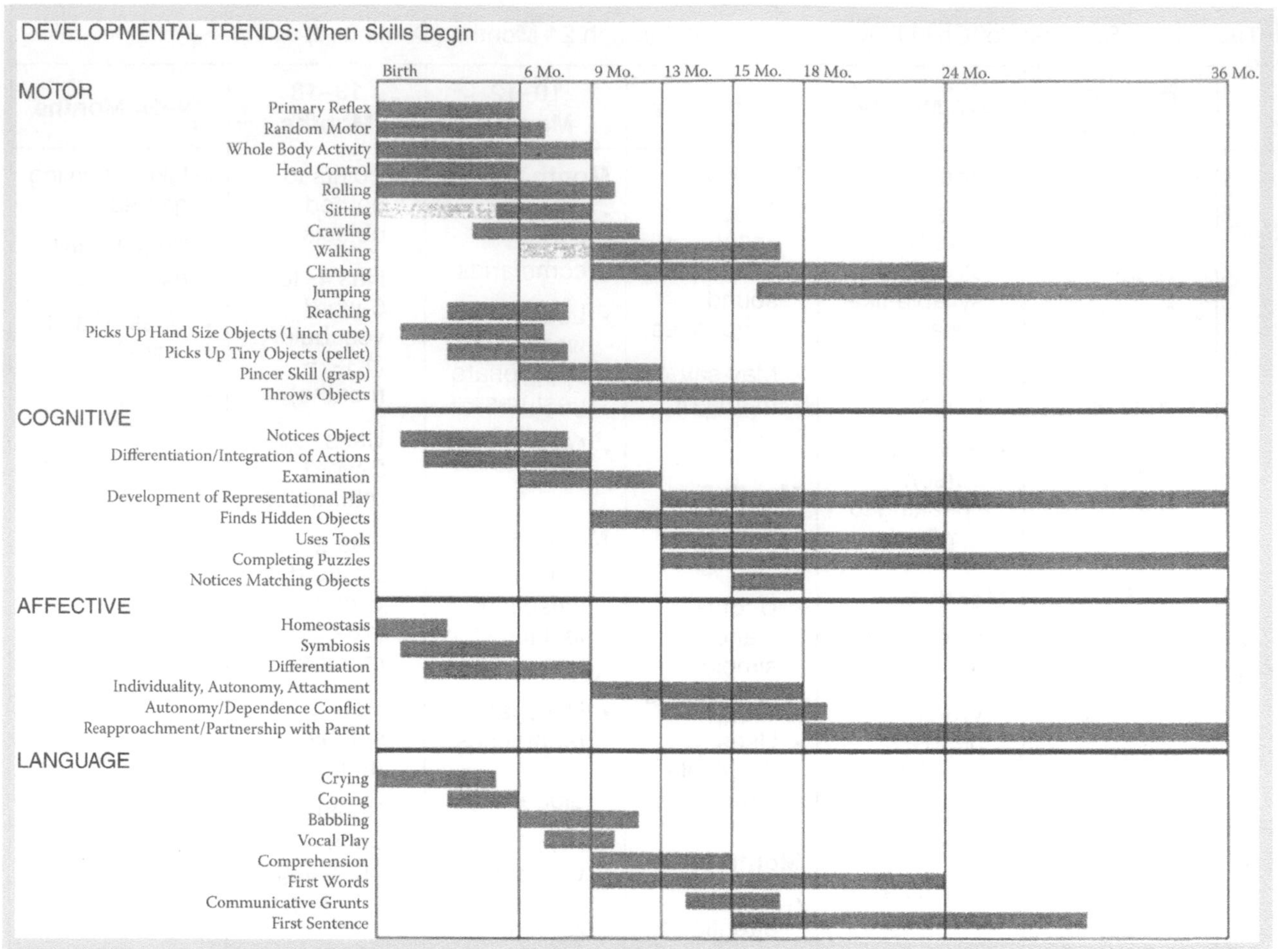

Figure 2.6 An overview of developmental milestones: Birth to 3 years. (From McCune, L., Developmental Trends: When Skills Begin. Presented at a Training Conference for Intertek, 1993.)

EARLY CHILDHOOD (2 TO 7 YEARS)

COGNITIVE DEVELOPMENT

Piaget described cognitive development for 2- to 7-year-olds as the *preoperational stage*. At this age, children learn representational skills and become more proficient in many areas, including language.

Children in this age group have difficulty imagining others' viewpoints. They are not selfish, but simply incapable of seeing the world through someone else's eyes. For example, children speaking on the telephone will discuss items in their immediate environment that the listener cannot see.

Children in the preoperational stage also cannot *reverse mental action*s. As a result, they cannot consistently understand the consequences of their own actions.

The American Academy of Pediatrics explains this type of thinking beautifully:

> *...Even if she's found out the hard way once, don't assume she's learned her lesson. Chances are she doesn't associate her pain with the chain of events that led up to it and she almost certainly won't remember this sequence the next time.*
>
> (Shelov and Hannemann, 1993, p. 266)

Figure 2.7 (Ursy Potter Photography)
Children between the ages of 2 and 7 love sliding down the slide, riding their tricycle/bicycle, jumping, and climbing. They improve their new abilities during their play, developing balance, eye–hand coordination, and interactive skills.

PHYSICAL DEVELOPMENT

Children between the ages of 2 and 7 like to practice their developing motor skills (Figure 2.7). Their equilibrium and gross and fine motor skills improve, and they become proficient at activities such as hopping on one foot, pumping a swing, and skipping. Refinement of fine motor skills occurs simultaneously as children demonstrate independent self-care skills as they button, snap, and zip the zippers on their clothes, as well as begin to hold a pencil, write, and draw.

SOCIAL AND EMOTIONAL DEVELOPMENT

Parallel play is when two children play alone, but in close proximity to each other. Toddlers begin to interact by following, imitating, and chasing one another, as well as exchanging toys.

Two-year-olds want to be independent, yet feel safe and secure. They are very possessive and have difficulty sharing with friends.

However, children become progressively more independent by the time they reach 3, 4, and 5 years of age. They are better at communicating, sharing, and taking turns. They enjoy dramatic play, move from *parallel play* to interactive play, and tend to develop close relationships with one or two "best friends" (Figure 2.8).

For the first time, children begin to exhibit more interest in other children than in adults. As children enter formal schooling, social acceptance from someone other than their parents or family takes priority.

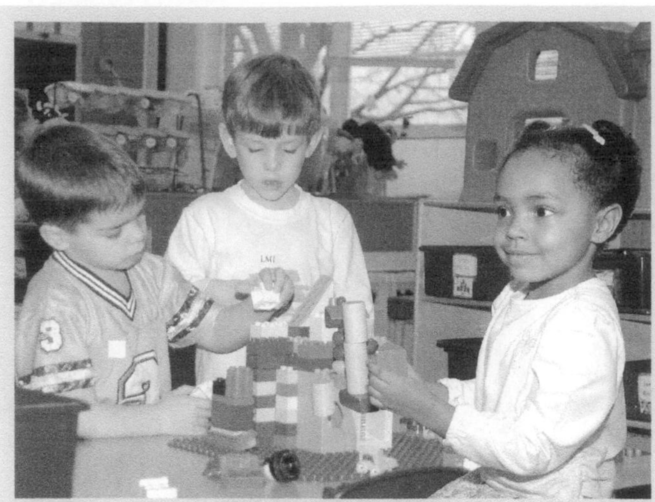

Figure 2.8 (Ursy Potter Photography)
Children work together to accomplish a task during cooperative play that develops between the ages of 3½ and 5.

LANGUAGE DEVELOPMENT

Three- and four-year-old children often know between 900 and 1600 words. By the time they are 6, they may be able to speak 2600 words. They answer simple questions, have conversations, and tell stories toward the end of early childhood (Owens, 2001). Children between the ages of 2 and 7 have fun learning language by playing word games using rhymes, songs, and chants.

Table 2.2 contains a summary of child development from two through four years of age.

Table 2.2 Summary of Development for Children 2 to 7 Years

	2–3 Years	4–5 Years	6–7 Years
Cognitive Development	• Match primary shapes and colors • Understand concept of "two" • Enjoy make-believe play • Point to body parts • Interested in learning to use common objects • Match objects to pictures	• Able to categorize • Know primary colors • Often believe in fantasy. Accept magic as an explanation • Develops a concept of time • Recognizes relationships between a whole and its parts • Focus on only one aspect of a situation • Base much of their knowledge on how the situation appears	• Longer attention span • Remember and repeat three digits • Have difficulty imagining other's point-of-view • Cannot consistently understand the consequences of their actions

Table 2.2 Summary of Development for Children 2 to 7 Years (Continued)

	2–3 Years	4–5 Years	6–7 Years
Physical Development	• Learn to use toilet • Walk backward, stoop, and squat • Toss or roll large balls • Dress self with help • Throw balls overhead and kick balls forward • Brush teeth, wash hands, and get drinks • Full set of baby teeth • Explore, dismantle, and dismember objects	• Uses a spoon, fork, and dinner knife • Walk straight lines, hop on one foot, run, climb trees and ladders, and turn somersaults • Skillfully pedal and steer tricycles • Buttons and unbuttons • Can cut on a line with scissors • Left- or right-hand dominance is established • Can skip, jump rope, and run on tiptoe • Can copy simple designs and shapes	• Develop permanent teeth • Tie shoelaces • Enjoy testing muscle strength and skills • Skilled at using scissors and small tools • Ride bicycles without training wheels (Valerie Rice)
Social and Emotional Development	• Like to imitate parents • Affectionate: hugs and kisses • Easily frustrated. Can be aggressive and destructive • Fears and nightmares • Spends much time watching and observing • Seeks approval and attention of adults • Likes to be the center of attention	• Understands and obeys simple rules, but often changes rules of a game as they go along • Persistently asks "Why?" • Can communicate, share, and take turns • More interested in children than adults • Enjoys doing things for themselves • Enjoys dramatic play with other children • Develops "best friend" • Organizes other children and toys in pretend play • Basic understanding of right and wrong • Good sense of humor. Enjoys sharing jokes and laughing with adults • Plays simple games • Interested in group activities	• Tends to play with same gender play mates • Strong desire to perform well and do things right • Interested in rules and rituals • Enjoys active games • Sensitive and emotionally vulnerable • Tries to solve problems through emotions

(continued)

Table 2.2 Summary of Development for Children 2 to 7 Years (Continued)

	2–3 Years	4–5 Years	6–7 Years
Language Development	• Early 2- to 3-word sentences; later 3- to 5-word sentences • Enjoys simple stories, rhymes, and songs • Hums, attempts to sing, play with words and sounds • Enjoys repeating words and sounds • 75% to 80% of their speech is understandable • Follows two-step commands • "Swears" • Talks about the present	• Recognizes some letters. May be able to print own name • Recognizes familiar words in simple books or signs • Speaks fairly complex sentences • Expressive vocabulary of 1500 to 2200 words • Can memorize own address and phone number • May understand 13,000 words • Uses 5 to 8 words in a sentence • Understands that stories have a beginning, middle, and end • Tells stories • By the end of this age range, children attain 90% of their adult grammar • Can follow three-step commands • Understands before and after	• Expressive vocabulary of 2600 words • Receptive vocabulary of 20 to 24,000 words • May reverse printed letters • Speaking and listening vocabularies double • Becomes interested in reading (Valerie Rice)

MIDDLE CHILDHOOD (7 TO 14 YEARS)

COGNITIVE DEVELOPMENT

Around the 5th to 7th year, children enter what Piaget referred to as the *concrete operations stage of development*. At this stage, children begin to understand others' viewpoints.

They also understand concrete, hands-on problems, and start to apply basic logic. However, they are still unable to think or perform in an abstract manner. In fact, they still have problems considering *all* of the logical, possible outcomes of their actions. For many children, the ability to use foresight and understand the consequences of their actions *before* engaging in them does not consistently appear until they reach 11 to 14 years of age (continuing into their early twenties).

After the age of 11 or 12, some children (but not all) enter Piaget's final stage of mental development, the *formal operations stage*.

This is when children become capable of true abstract thinking, including:

- Speculating about future events
- Understanding cause and effect
- Developing and testing hypotheses
- Reasoning scientifically

(Kagan and Gall, 1997)

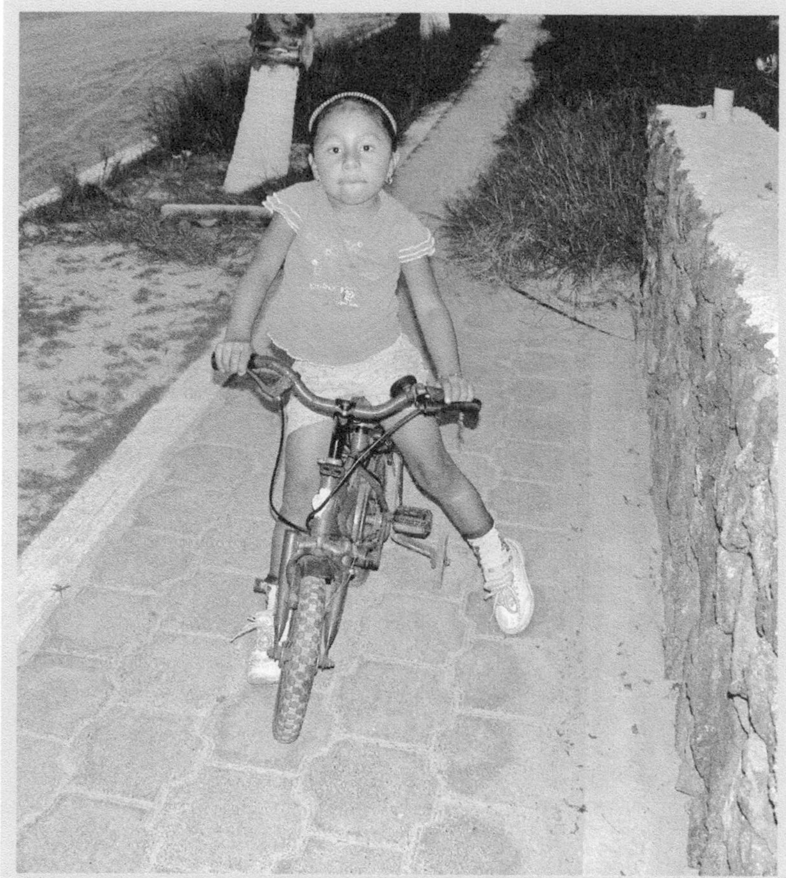

Figure 2.9 (Ursy Potter Photography)
Between the ages of 7 and 14, children are integrating their many skills into coordinated, fluid movements. Riding a bicycle means integrating balance, both large and small muscle groups, and visual-perceptual-motor skills.

PHYSICAL DEVELOPMENT

In middle childhood, children's fine or small and gross or large motor skills are well on their way to full development. They can usually tie their shoelaces, print their name, and balance on their bike (Figure 2.9).

During this stage of development, children enjoy organized sports. The ability to judge distances, eye–hand coordination, strength, and endurance all improve in adolescence, allowing children to begin to master the skills needed to play more adult-oriented sports (Kagan and Gall, 1997).

SOCIAL AND EMOTIONAL DEVELOPMENT

In these middle years, children value acceptance by friends (normally of the same sex) more than their own independence or obeying parental wishes and guidelines. They become more daring and adventurous (Figure 2.10).

Figure 2.10 (Valerie Rice)
Between the ages of 7 and 14, children are more daring and adventurous. Succeeding at daring acts boosts their self-esteem and helps them deal with fears. (Zuckerman and Duby, 1985.)

Rules and rituals are important during middle childhood, as they give children a sense of comfort and stability. Unlike younger children who engage in pretend play, children between the ages of 7 and 14 prefer real tasks and real activities, depending on the temperament or personality of the child. They spend much of their day interacting and socializing with peers; these may include indoor games such as video games as well as outdoor play such as sports and bike riding.

Friendships offer security and intimacy. Through friendships, children develop a sense of trust and attachment to others of their own age and usually, of their own gender (Santrock, 1986).

LANGUAGE DEVELOPMENT

Communication skills move beyond direct verbal and nonverbal interaction, as middle school children continue to prove their reading and writing skills. Their play includes secret codes, word meanings, and made-up languages. Through these activities, they learn more about language and bond with their friends.

> *Receptive vocabulary* refers to all the words a person understands.
>
> *Expressive vocabulary* refers to all the words an individual uses when speaking.
>
> By the age of 12, children typically have receptive vocabularies of 50,000 words (Owens, 2001).

By the completion of middle school, communication skills are almost equal to those of adults. By the age of 15, most youth are masters of language and competent communicators.

Table 2.3 Summary of Development for Children 7 to 14 Years

	7–8 Years	9–11 Years	12–14 Years
Cognitive Development	• Increased problem-solving ability • Longer attention span • May still have difficulty considering *all* the logical, possible outcomes of their actions • Understands concrete, hands-on problems or situations and applies basic logic	• Plans future actions • Solves problems with minimal physical output • May still have difficulty considering *all* of the logical, possible outcomes their actions	• Engages in abstract thought • Uses hypothetical reasoning • Speculates about future events
Physical Development	• Good sense of balance • Better manipulative skills (e.g., can easily tie shoelaces) • Catches small balls • Nearly mature brain size	• Girls are generally ahead of boys in physical maturity • Improved coordination and reaction time • Increased body strength and hand dexterity	• Enjoys playing organized sports • Better able to judge distances • Endurance improves • Body hair begins to emerge • Body shape changes for boys (broad shoulders) and girls (curves and breasts)
Social and Emotional Development	• Enjoys an audience • Sees things from other children's points of view, but still very self-centered • Finds criticism or failure difficult to handle • Views things as black and white, right or wrong, good or bad with little middle ground	• Begins to see parents and authority figures as fallible human beings • Better understands other people's perspectives instead of only their own • Enjoys being a member of a club; peers become very important	• Different interests than the opposite sex • Values peer acceptance highly • Becomes daring and/or adventurous • Interested in real tasks and real activities (Valerie Rice)
Language Development	• Talks a lot • Brags • Communicates thoughts and ideas	• Often has rituals, rules, secret codes, and made-up languages • Likes to read fictional stories, magazines, and how-to project books • May be interested in their eventual career • Uses cursive writing	• Has receptive vocabulary of 50,000 words • Communication skills almost equal to an adult • Competent communicators

OPTIMAL STIMULATION

Optimal stimulation is the force that drives children to explore their environments by using objects in new and different ways. The optimal level of stimulation varies between children and depends on the child's temperament, their family, and environmental situations.

Adults (parents, teachers, and caregivers) can help children discover their optimal level of stimulation. Through exploration and trial and error, children find their optimal level and put it into practice.

As children explore their world, they test their limits and the limits of objects they use. Confined only by their imaginations and developmental capabilities, they try activities that are not within their physical capabilities. They learn both by accomplishing tasks and through failing at tasks. As they invent new ways to use objects and try different methods of communicating, they begin to understand who they are, how they fit in, and how they can have an affect on their world.

EXPLORATION STRATEGIES

Exploration strategies are conscious or unconscious behaviors that children use to learn about their world. These vary in frequency, duration, and intensity depending on a child's stage of development and environment. However, all children use the same sequence of exploration strategies and typically use all exploration strategies available to them.

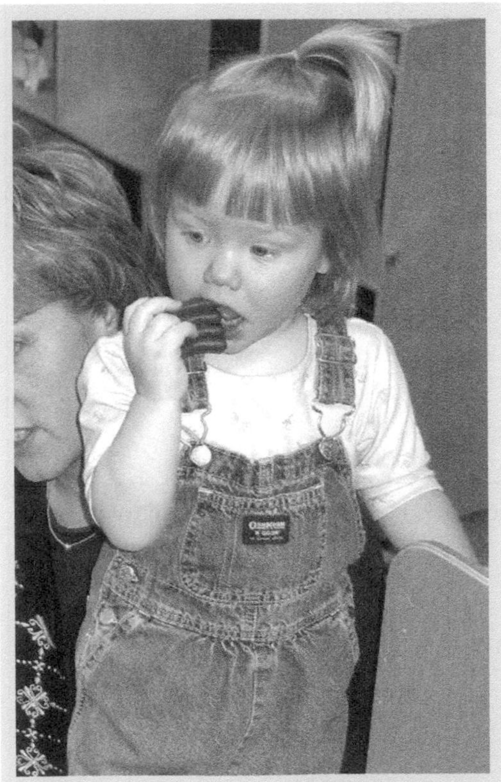

Figure 2.11 (Intertek)
Children use mouthing to learn about the texture, size, consistency, and shape of objects. Young children use this technique of exploring their world more than any other.

MOUTHING

Infants suck on and mouth objects out of frustration, a need for contact or to explore their world (see Figure 2.11). Children mouth objects for up to 1 to 1½ hours per day (DTI, 2002), but only mouth nonpacifier objects for about 36 minutes each day (Juberg et al., 2001). Mouthing gives infants more information about an object than any other exploration strategy.

When faced with a new object, children as old as 5 use mouthing as an exploration strategy. Children with developmental delays may do so well into the school-age years.

The mouth is a source of pleasure for children (and adults), which often leads to mouthing for extended periods of time. Young children may explore objects, while older children and adults chew gum. Furthermore, tooth development continues into the teenage years and one source of relief during "teething" is chewing.

MOUTHING AND LOOKING

Infants and toddlers often integrate vision into their exploration of objects by looking at an object and then mouthing it and vice versa. In this way, they actively connect the visual image of an object with the way it feels in their mouth (Figure 2.12 and Figure 2.13).

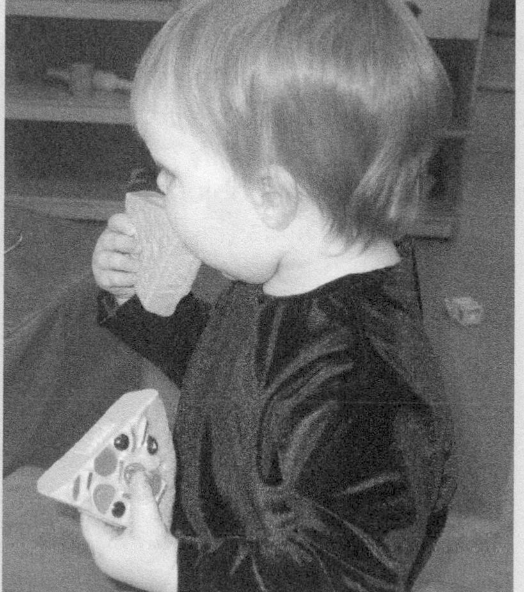

Figure 2.12 (Intertek) **Figure 2.13** (Intertek)
This 9-month-old child is alternately looking at a toy and then mouthing it. This helps the child connect visual and tactile information to understand the toys construction.

ROTATING AND TRANSFERRING HAND-TO-HAND

Children quickly learn about the three-dimensional qualities of objects during play. This newfound discovery leads them to rotate an object with one or both hands as they visually inspect it (Figure 2.14 and Figure 2.15).

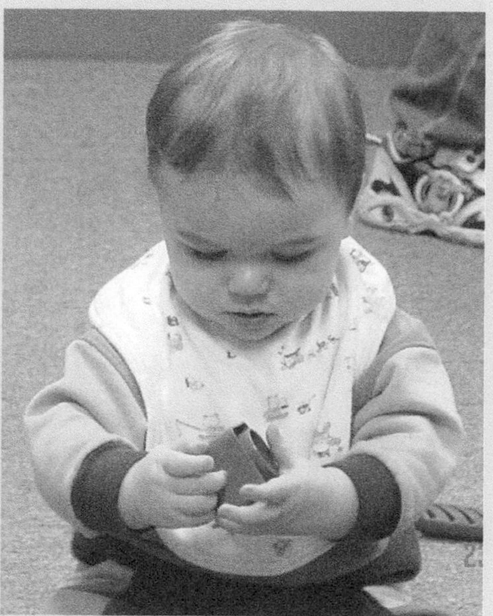

Figure 2.14 (Intertek) **Figure 2.15** (Intertek)
This 2 year old and 10 month old demonstrate learning through rotation and transfer of objects. Initially, children rotate objects with one hand, turning them back and forth while examining different sides.

As motor coordination improves, children use both hands to rotate objects and pass them from one hand to the other. Doing so helps children learn properties of the object (size, shape, weight, and consistency), while they practice new motor skills such as releasing, grasping, and using two hands together.

INSERTION (BODY INTO OBJECT AND OBJECT INTO BODY)

When infants and toddlers learn to isolate one finger without extending the others, they begin to explore by insertion. They explore objects by putting fingers in objects or running fingers along their outside edges.

As children explore they also insert other body parts (hands, feet, legs, head, etc.) and their entire bodies into objects (Figure 2.16 and Figure 2.17) or they may insert items into their own body cavities (Figure 2.18). The last insertion strategy is used most extensively by (but not limited to) 2- and 3-year-old children.

Figure 2.16 (Intertek)

Figure 2.17 (Rita Malone)

Figure 2.18 (Intertek)

Children explore by inserting their fingers, hands, feet, and even their head into various objects, such as this 2-year-old child inserting her head into a bowl.

They also love to insert their entire bodies into cabinets, baskets, and in this case, into a clay pot! Children also try inserting objects into their own body cavities. This 2-year-old is inserting drumsticks into his ears.

BANGING, THROWING, AND DROPPING

Banging is a developmental strategy that does not readily lend itself to explanation. Children appear to bang objects together in order to hear sounds and evaluate their weight and textures (Figure 2.19).

Children under the age of 5 like to throw objects. The act of throwing helps the child to evaluate the object's weight and weight distribution (Figure 2.20).

Infants and toddlers drop objects to learn more about object permanence (discussed earlier in cognitive development). When infants or toddlers notice that caring adults will retrieve a toy, dish, or cup they drop from their high chair, they begin to understand that objects still exist when they no longer see them.

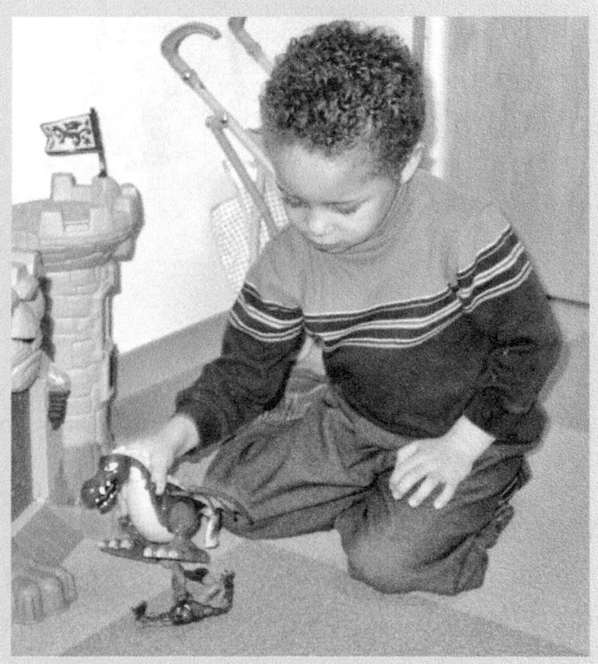

Figure 2.19 (Intertek)
Children love to bang objects together. They seem to love the sounds and seeing one object make another skid away or fall. This is one of their first lessons in cause-and-effect. They see, first hand, how their own actions affect the world around them.

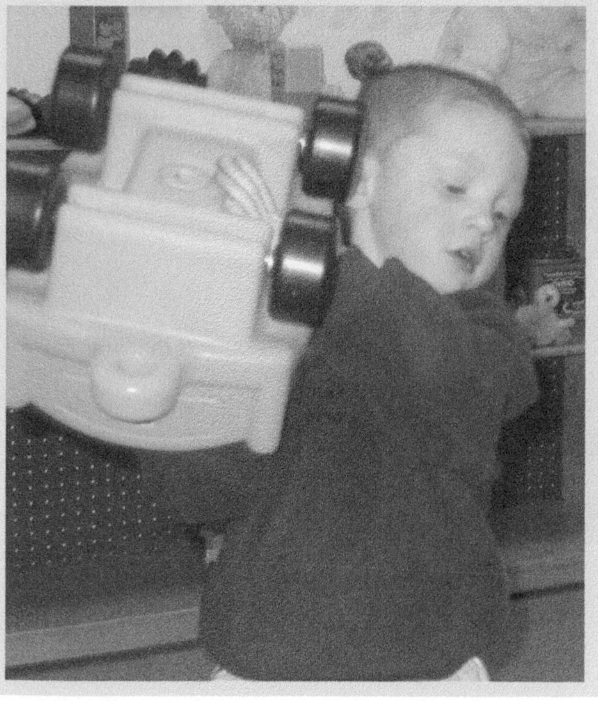

Figure 2.20 (Intertek)
This 2-year-old is exercising his motor skills and power by throwing. While this activity seems destructive to adults, children learn about object weight, lifting strategies, and their own strength by throwing.

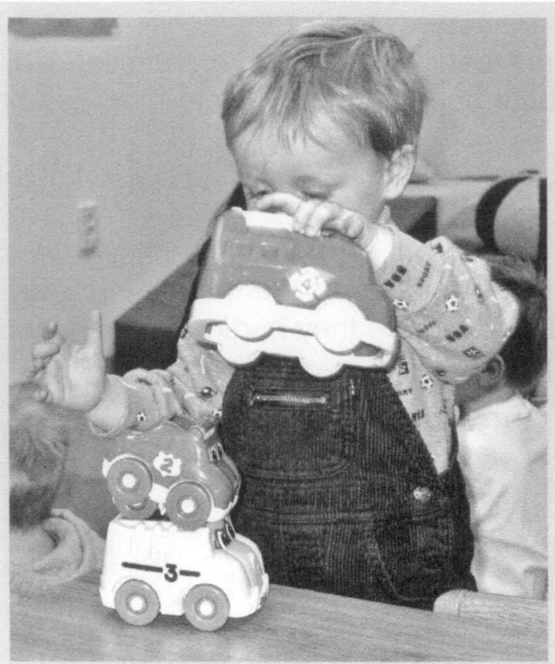

Figure 2.21 (Intertek)
This 1-year-old is experimenting with toys to see how they interact. In this way, he answers questions such as "Can this toy fit on top, inside, or over another object?"

COMBINING OBJECTS

Toddlers, in particular, enjoy exploring how objects fit together. They attempt to view and understand the different ways that objects interact (Figure 2.21).

MATCHING OBJECTS

By their 18th month, children begin to differentiate between the concepts of "same" and "different." Children try to group objects with similar characteristics or functions (Figure 2.22).

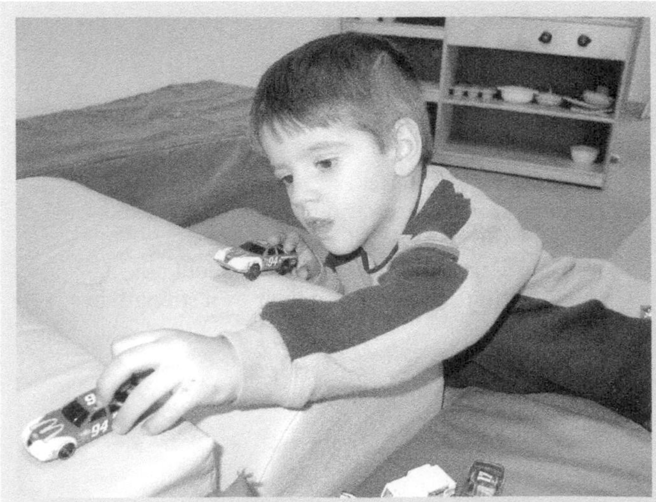

Figure 2.22 (Intertek)
Children learn by grouping objects that are the same.

This 2-year-old is matching similar cars that are the same size, shape, and color.

USING OBJECTS APPROPRIATELY

Children use objects as intended—but also experiment by using them in alternate ways. For example, children explore roles by talking into toy telephones or "feeding" dolls. When children begin using objects appropriately, they will not understand and sometimes even become upset, if another player uses the object in a way that does not make sense to the child (Figure 2.23 and Figure 2.24).

Figure 2.23 (Intertek)

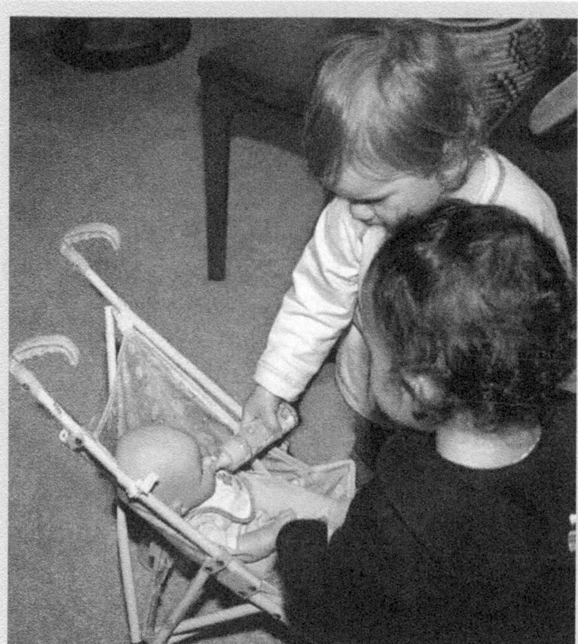

Figure 2.24 (Intertek)
Once children learn how to use objects "correctly," they are quite proud of their knowledge. They sometimes become upset or angry when someone uses them inappropriately.

These pictures show children playing appropriately, stacking and inserting blocks into a container and feeding their doll.

REPRESENTATIONAL PLAY

Representational play or "roleplay" recreates real life. The child puts together real-life dramas using objects as props (Figure 2.25 and Figure 2.26). Common examples include playing dress-up, pretending to be "mommy," "daddy," or "baby" and playing school with teachers and students. Another example is pretending to be a character witnessed on television—for many young children what they see on television is "real-life."

Figure 2.25 (Intertek) **Figure 2.26** (Valerie Rice)
During representational play, children recreate life as they see it. Here one little girl plays dress up, while another pretends to be a teacher. In this way, children explore social roles by "trying them on."

IMAGINATIVE PLAY

In imaginative play, the child is no longer content to use objects in their intended ways. Instead, they experiment and discover other ways to use objects. For example, they may use a block as a telephone or their living room furniture can become a "magical" castle (Figure 2.27).

Figure 2.27 (Intertek)
Imaginary play lets children explore relationships in unexpected ways. This 5-year-old is using his imagination to bring different objects into his game, creating a unique scenario.

TESTING THE LIMITS

Testing one's own limits and the limits of one's own environment is also a type of exploration strategy (Figure 2.28). School-age children often test limits by increasing risks. For example, 10-year-olds might drop an object from a high place, set something on fire, or run over something with a bicycle. This form of exploration helps children understand how to transform objects.

However, children have a poor grasp of cause and effect. Adult supervision is essential. The risks they take may be greater than they recognize, putting themselves or others in danger. Children often perceive themselves as invincible.

Figure 2.28 (Intertek)
This 5-year-old is testing his balancing abilities. This trial-and-error strategy increases along with a child's physical and cognitive development.

SUMMARY

Each child is unique, but passes through universal stages. Individuals who design for children must understand how children explore and learn about their surroundings as they progress through each developmental stage.

When we design to accommodate each stage, we protect children even while we challenge their abilities. In this type of environment, children can develop and test their skills one step at a time, building one success upon another, gathering momentum and confidence as they go (Figure 2.29).

ACKNOWLEDGMENTS

We would like to thank all of the children and their families who allowed us to use their pictures in this chapter.

Figure 2.29 (Valerie Rice)

REFERENCES

Bates, E. (1999). On the nature and nurture of language. In: E. Bizzi, P. Calissano, and V. Volterra (Eds.), *Frontiere della bilogia [Frontiers of Biology]*. Rome: Giovanni Trecanni.
Cole, M. and Cole, S. (2001). *The Development of Children*, 4th edn. New York: Worth Publishers.
DTI (Department of Trade and Industry) (2002). Research into the mouthing behavior of children up to five years old. www.dti.gov.uk/homesafetynetwork/ck_rmout.htm.
Juberg, D.R., Alfano, K., Coughlin, R.J., and Thompson, K.M. (2001). An observational study of object mouthing behavior by young children. *Pediatrics*, 107(1), 135–142.
Kagan, J. and Gall, S. (Eds.) (1997). *Gale Encyclopedia of Childhood and Adolescence*. Farmington Hills, MI: Gale Research.
Lieberman, A.F. (1991). Attachment and exploration: The toddler's dilemma. *Zero to Three*, 11(3), 6–11.
McCune, L. (1993). Developmental Trends: When Skills Begin. Presented at a Training Conference for Intertek.
Owens, R.E. (2001). *Language Development: An Introduction*, 5th edn. Boston, MA: Allyn and Bacon.
Santrock, J.W. (1986). *Life-Span Development*. Dubuque, IA: Brown Publishers.
Shelov, S.P. and Hannemann, R.E. (Eds.) (1993). *The Complete and Authoritative Guide: Caring for Your Baby and Young Child Birth to Age 5*. The American Academy of Pediatrics. New York: Bantam Books.
Zuckerman, B.S. and Duby, J.C. (1985). Developmental approach to injury prevention. *Pediatric Clinics of North America*, 32(1), 17–29.

OTHER REFERENCES

McCune, L. and Ruff, H. (1985). Infant special education: Interactions with objects. *TECSE* 5(3), 59–68.
Sewell, K.H. and Gaines, S.K. (1993). A developmental approach to childhood safety education. *Pediatric Nursing*, 19, 464–466.

OTHER WRITINGS ABOUT PIAGET'S THEORIES

Bringuier, J.-C. (1980). *Conversations with Jean Piaget*. Chicago, IL: University of Chicago Press (original work published 1977).
Chapman, M. (1988). *Constructive Evolution: Origins and Development of Piaget's Thought*. Cambridge: Cambridge University Press.
De Lisi, R. and Golbeck, S. (1999). Implications of Piagetian theory for peer learning. In: A. O'Donnell and A. King (Eds.), *Cognitive Perspectives on Peer Learning*. Mahwah, NJ: Lawrence Erlbaum Associates.
DeVries, R. and Zan, B. (1994). *Moral Classrooms, Moral Children: Creating a Constructivist Atmosphere in Early Education*. New York: Teachers College Press.
Gallagher, J.M. and Reid, D.K. (Foreword by Piaget and Inhelder) (1981/2002). The Learning Theory of Piaget and Inhelder. Order from online publisher: www.iuniverse.com
Ginsburg, H. (1997). *Entering the Child's Mind*. New York: Cambridge University Press.
Gruber, H.E. and Vonèche, J. (1995). *The Essential Piaget*. Northvale, NJ: Jason Aronson.
Lourenco, O. and Machado, A. (1996). In defense of Piaget's theory: A reply to 10 common criticisms. *Psychological Review*, 103(1), 143–164.
Montangero, J. and Maurice-Naville, D. (1997). *Piaget or the Advance of Knowledge*. Mahwah, NJ: Lawrence Erlbaum Associates.
Piaget, J. (1964). *Six Psychological Studies*. New York: Vintage. [The first 70 pages].
Piaget, J. (1973). *The Child and Reality: Problems of Genetic Psychology*. New York: Viking.
Piaget, J. (1983). "Piaget's theory". In: P. Mussen (Ed.), *Handbook of Child Psychology*. New York: Wiley.
Piaget, J. and Inhelder, B. (1969). *The Psychology of the Child*. New York: Basic Books. (Original work published 1966.)
Piaget, J. (1985). *Equilibration of Cognitive Structures*. Chicago: University of Chicago Press.
Piaget, J. (1995). *Sociological Studies*. London: Routledge.
Wadsworth, B.J. (1989). *Piaget's Theory of Cognitive and Affective Development*, 4th edn. New York: Longman.

CHAPTER 3

CHILD ANTHROPOMETRY

BEVERLEY NORRIS AND STUART A. SMITH

TABLE CONTENTS

INTRODUCTION

Anthropometry is the scientific measurement of sizes and shapes of the human body. Product designers use anthropometric data to ensure that

- products are easy to use (for example, that toys are small enough for children to grasp),
- controls are easy to reach (e.g., children can reach and operate the brake on a bicycle),
- hands or fingers will not reach around or through guards (such as fire guards or into a moving fan),
- adequate movement and reach zones (as in the design of playpens and cots) are available, and

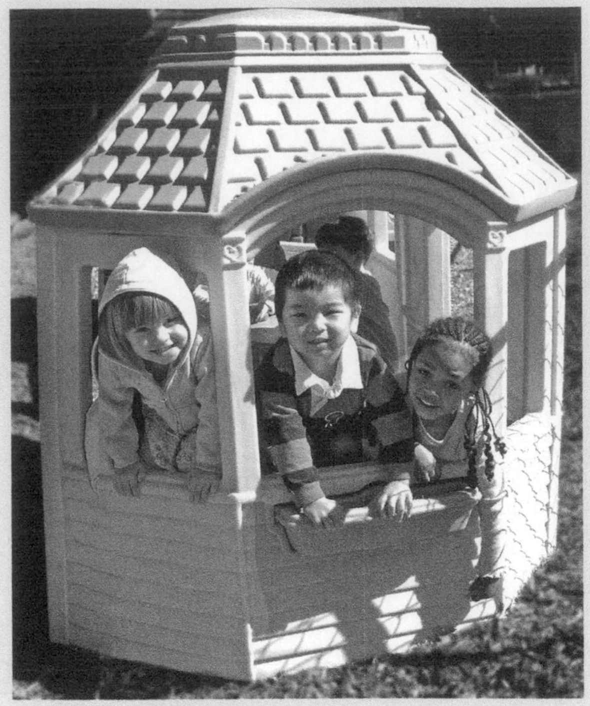

Figure 3.1 (Ursy Potter Photography)
Children's anthropometric dimensions vary considerably within age groups and designing for children requires adjusting the product dimensions to fit these ranges.
It also means rounded edging, avoiding openings that could entrap children or their limbs, and planning for children to use the items in unexpected ways.
In this playhouse, the windowsills and even the flower box must support the weight of playing children.

- gaps and openings (such as railings or banisters) do not trap fingers, hands, and heads (Figure 3.1).*

In collecting ***static anthropometric data***, researchers measure children's (or adult's) body dimensions in standardized postures, using equipment such as anthropometers and tape measures. Recently, several large-scale surveys have used three-dimensional scanning techniques,† which offer many advantages over more traditional methods.

Static data tell us little about how people actually use a product. This is because "real life" typically involves nonstandard postures and movement (Figure 3.2).

Figure 3.2 (Ursy Potter Photography)
Many anthropometric tables use static dimensions, yet children at play are dynamic! Designers must consider children's dynamic ranges of motion in products for children.
Also, anticipate "unexpected" uses of products. Many children such as Charlie try out their newly developing balance skills.

Functional or ***dynamic anthropometry*** describes the limits of movement, such as how far different sizes of people can reach overhead, in front of, or to the side of the body. Although the title "functional" or "dynamic" would seem to indicate that these measurements represent real life, this is not

* Other applications for data on human body dimensions and shapes include clothing design, medical assessments, tracking nutritional status, and growth.

† See http://store.sae.org/caesar/ as an example.

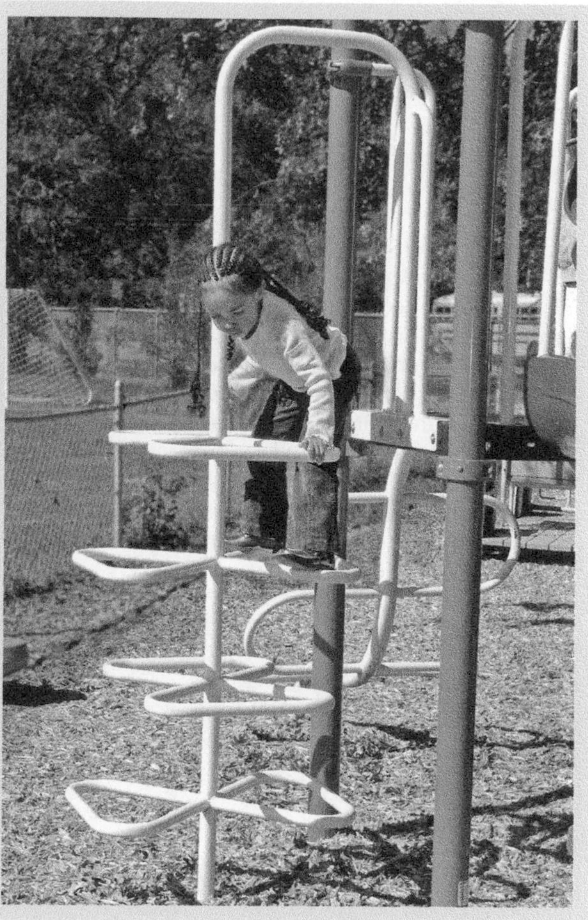

Figure 3.3 (Ursy Potter Photography) One reason that dynamic anthropometric data measured in standardized postures may not represent how children realistically interact with products or furniture is that most movements involve a number of body movements. For example, Jeda is holding with one hand, reaching with the other, and bending at the waist.

These issues are particularly important when designing for children. Anthropometric surveys have historically focused on adult working populations. Data on children were rare until recently and may not be disseminated as widely. One source of collated anthropometric data on children is Childata (Norris and Wilson, 1995), which brings together some of the larger and more comprehensive surveys on children into one resource.

necessarily the case. For instance, people increase their overhead reach by twisting, standing on their toes, or jumping.

Yet, data on maximum overhead reach are usually measured with the person's feet flat on the floor and not straining. Such standardization is important so that data can be compared across sources yet these sorts of dimensions, while intended to be applicable to design, can still be unrepresentative of real life (Figure 3.3).

Measurements are taken with the child or adult only lightly clothed, for example in their underwear, and barefoot. Clothing can alter a person's dynamic anthropometry, for example by restricting their posture (Figure 3.4). A correction of up to 25 mm (1 in.)* for shoes for children up to age 12 and boys over 12, and 45 mm (1¾ in.) for girls over 12 years and women is recommended (Norris and Wilson, 1995). Corrections for other clothing are likely to be negligible, with the exception of outdoor wear, nappies, gloves, and protective clothing such as cycle helmets.

DESIGNING FOR CHILDREN OF DIFFERENT AGES

Children are typically described by their age groups. For instance, when designing for children, we often differentiate between preschool and school-age children. However, childhood is a time of continual and rapid change in physical and psychological abilities (perhaps like in old age—the other end of life!).

* The original data sources cited in this chapter present their data in millimeters and so for the rest of this chapter data will be presented in millimeters. To convert millimeters to inches: 1 in. = 25.4 mm.

Figure 3.4 (Stuart Smith)
The thick snowsuit and layers of clothes that Sami is wearing to keep him warm also restrict his movement. This means that he cannot keep both feet on the footrests (which he needs to do for safe use of the car) and finds it difficult to play with the toy-bar in front of him. When he is dressed in normal clothing he can keep his feet on the footrests.

This highlights the importance of thinking of the use of the product, and the anthropometric data that are used to design it.

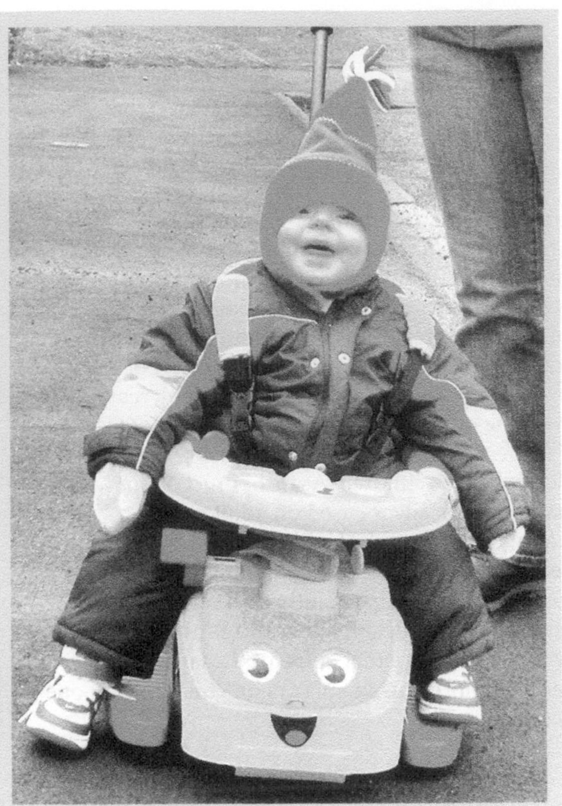

This means that using age categories for design purposes (in years or in months for very young children) will likely misrepresent the true picture (Figure 3.5). The large variations between children of the same age, particularly around adolescence, make it more difficult to design for children than for adults. Designers of products for children must account for such key stages in child development and important developmental factors.

The rate of growth varies during childhood. Growth is rapid during the first few years of life and then slows until the adolescent growth spurt when it increases again.

The timing of the growth spurt varies between individuals and is different in boys and girls. In general, this growth spurt occurs between the age of 9 and 14 years for girls and between 11 and 16 years for boys, although there is some evidence that the adolescent growth spurt is starting earlier in today's youth (Van Wieringen, 1986; Smith and Norris, 2001) (Figure 3.6).

Such differences in the age at which puberty and menarche are reached means that some children may have finished growing while others have barely begun. In statistical terms, this is demonstrated by an increase in the standard deviation (SD) of anthropometric data for adolescents.

Figure 3.5 (Ursy Potter Photography) The large variation in size among children of the same age, as well as their rapid growth, make designing for children challenging. Perhaps even more challenging is buying products for children. For example, clothing is often labeled by the year-of-age of the children it is expected to fit; yet, it rarely fits children of that age.

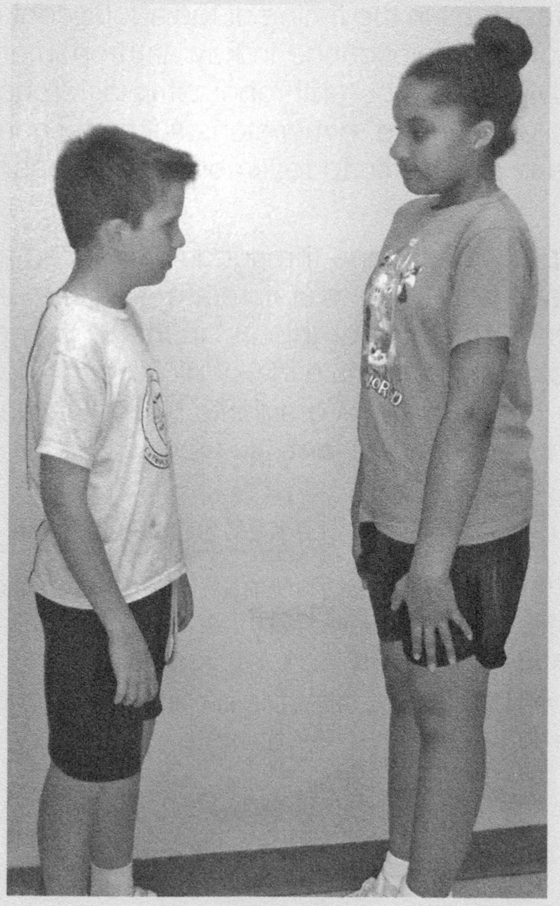

Figure 3.6 (Valerie Rice) Simply considering age is not sufficient for design, as gender and development are also important. For example, girl's growth spurts occur earlier than boys (much to the chagrin of both). Here, Elia at age 12 is shorter than Alyssa aged 11.

Children's body dimensions not only increase with age but also become more variable within age groups. In statistical terms, this means that the SD of anthropometric measurements increases with age.

Wide variation within age groups can lead to an overlap across age groups. For instance, a tall (95th percentile) 7-year-old boy may be taller than a short (5th percentile) 10-year-old (132 and 129 cm, respectively [Pheasant, 1986]). This variation within age groups sometimes makes it more appropriate to design according to variables other than age; weight for instance is sometimes used to specify the suitability of baby carriers, car seats, and nappies or diapers. Some anthropometric data sources, particularly those produced for growth studies are in fact classified by weight, rather than stature.

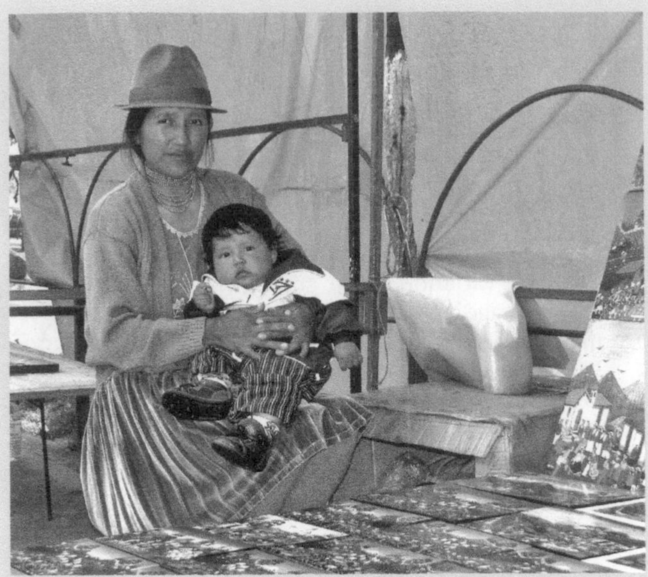

Figure 3.7 (Ursy Potter Photography)
Children's body proportions change as they grow. For example, a baby's head is very large in comparison to its body size.
A newborn's head length is about 25% of their stature, but decreases to about 10% in adults.

Anthropometric differences between the sexes appear slowly after birth and increase with age. Differences in the timing of the adolescent growth spurt mean that girls can overtake boys around adolescence in key anthropometric dimensions such as stature. After approximately age 13, the anthropometric differences between the sexes become more pronounced; for example, dimensions such as shoulder breadth in boys and hip breadth in girls increase and begin to resemble adult proportions.

Body proportions change throughout childhood and adolescence. At birth, the ratios of head length and leg length to stature are greater than in older children or adults. Children's bodies evidence "maturity gradients" in that in young children the upper body, particularly the head, is closer to adult size than their lower bodies (Cameron et al., 1982; Tanner, 1962) (Figure 3.7). These maturity gradients direct the developmental transition from children's large-headed, short-legged forms to typical adult proportions (Pheasant, 1996).

KEY SOURCES OF DATA

Table 3.1 summarizes the key sources of anthropometric data on children that are currently available internationally; however, the list is not exhaustive. New sources are published or become available regularly and, as is described in "Secular Trend," it is important to use the most up-to-date and applicable data available. Full references and a wider bibliography are provided at the end of the chapter.

Table 3.1 Key Sources of Anthropometric Data

Nation	Reference	What is it?
France	Coblentz et al. (1993)	• The most comprehensive data available on French children • Provides data on 18 anthropometric dimensions for children aged 4 to 18 years, with a nationally representative sample of 532 boys and 524 girls
Germany	DIN[a] (1981)	• This German Standards organization compiled data for children's body measurements between 1968 and 1977 • This data represents "the regional and social variation of children from the Federal Republic of Germany" at that time
Japan	HQL (1997)	• The Japanese Body Size Dataset houses the primary source of data for Japanese children • This dataset of 178 anthropometric dimensions for children aged 7 to 18 was collected from 1992 to 1994 on 178 children. It includes a nationally representative sample size of 6096 males and 5032 females
Latin America[b]	Chaurand et al. (2001)	• The Mexican data cover 50 dimensions for children aged from 2 to 18, measured on 351 males and 229 females • The Cuban data cover 24 dimensions measured on children aged 6 to 18, measured on 7028 males and 6971 females • The Chilean data cover eight dimensions of children aged 6 to 18, measured on a total of 4611 children
The Netherlands	Steenbekkers (1993)	• This primary source of Dutch data represents the most recent anthropometric data on European children available to the public at the time of writing • The sample was geographically representative of the Netherlands and included children born both in the Netherlands and abroad and those born to non-Dutch parents. The sample was not considered sufficiently representative of socioeconomic factors • Sampled 2245 children aged 2 to 12 years, measured between February 1990 and February 1991 • These researchers previously measured 633 children aged 0 to 5.5 years, including some dimensions not recorded in their 1993 study. Monthly age groups in infants were classified as 30 days (Steenbekkers, 1989) • A sample of 87 infants aged 0 to 14 months were also measured in one geographical area of the Netherlands a few years earlier (1987–1989)
Poland	Nowak (2000)	• Data on Polish children includes 43 anthropometric dimensions for children aged from 4 to 18 years and also on a disabled population aged 15 to 18 years. Also projects anthropometric data for 2010

(continued)

Table 3.1 Key Sources of Anthropometric Data (Continued)

Nation	Reference	What is it?
UK	Pheasant (1986); DES (1972); BSI (1990); Prescott-Clarke and Primatesta (1999)	• Data from the United Kingdom (UK) were produced using the ratio scaling method of estimating anthropometric measurements from stature (Pheasant, 1986). The stature data used were measured in 1970/1971 (DES, 1972) in a nationwide sample of 15,000 children, representative of the UK school population aged 3 to 18 years. The stature data were considered still valid in 1985 when the data were re-published (DES, 1985). The data were calculated using body part or stature ratios from North American sources (Martin 1960; Snyder et al., 1977). All data are calculated to the nearest 5 mm • The most recent, large-scale anthropometric surveys of children publicly available in Britain was conducted for the clothing industry and many of the measurements have limited application to design, for example, bent arm length (BSI, 1990). The survey included 2000 infants (0 to 5 years) in 1987, 4770 girls (5 to 16) in 1986, and 3428 boys (5 to 16) in 1978. Subjects were nationally representative (excluding Northern Ireland), from mainly urban areas and socioeconomic factors were disregarded. Most importantly, the results represent Caucasian subjects only (BSI, 1990) • The Health Survey for England measured males and females aged 2 to 24 years during 1995, 1996, and 1997. These are the most up-to-date data on UK stature and weight. Children from regions throughout England were measured to ensure nationally representative data (Prescott-Clarke and Primatesta, 1999)
United States (US)	NHANES[c] Snyder et al. (1975, 1977)	• NHANES provides nationally representative data of the US child population. The NHANES III survey measured children aged from 2 months to 18 years. These are the most recent, available data on US stature and weight (DHHS, 1996) • The most comprehensive collection of design-related child anthropometric data was collected between 1975 and 1977 on a sample of 4127 children aged between birth and 19 years. They were nationally representative of ethnic, demographic, and socioeconomic variables (Snyder et al., 1977) • An earlier report from the same US authors provides additional measurements, recorded on a sample of 4027 children aged between birth and 13 years, again nationally representative and measured between 1972 and 1975. Monthly age groups in infants were rounded to the nearest month (Snyder et al., 1975)

[a] The Deutsches Institut fur Normung (DIN) is the German Standards organization, which has compiled data from the German Standard for body measurements of children.

[b] These Latin American countries include Mexico, Cuba, and Chile.

[c] The US Department of Health and Human Services runs the National Health and Nutritional Examination Survey (NHANES).

INTERNATIONAL DIFFERENCES IN CHILD ANTHROPOMETRY

Appendix A presents the stature and weight data for two of the tallest and heaviest populations (United States and UK) and two of the shortest and lightest populations (Mexico and Japan).

At age 2, the largest difference in average stature for males or females between the UK, United States, and Mexico is only 12 mm, but by age 7, this increases to 32 mm. If Japanese children are included, the difference is 67 mm. These differences increase as children get older. The greater differential in sizes of children as they age means that an international product designer needs to make greater use of anthropometric data for both clothing design and marketing.

A similar comparison for body weight shows the largest difference for males or females between the UK, United States, and Mexico at age 2 is 0.5 kg (1.1 lb) and by age 7 is only 1.3 kg (2.9 lb). If Japan is considered, the difference is 4.2 kg (9.3 lb). Again, the difference between the heaviest and lightest increases as children get older.

Table 3.2 compares the most recent stature data on UK children (Prescott-Clarke and Primatesta, 1999) to the most comprehensive data set on children from the United States (Snyder et al., 1977). The current UK child population is approximately 2% taller than the same-age US children (data from Snyder are still considered representative of the current US population [Ochsman and Van Houten, 1999]). The percent difference in body weight is larger however, with the current UK child population being approximately 7% to 10.5% heavier than US children.

Differences in anthropometric measurements between countries are usually expressed as height, weight, and body mass index (BMI) or skinfold measurements, as these are considered the most useful indicators of children's growth. However, the designer needs additional information regarding other dimensions such as head breadth, finger diameter, and overhead reach.

One study comparing head circumference found a difference between UK and Japanese children (UK children's head circumference being greater than that of Japanese children) and suggested that this primarily reflected the differences in stature. In fact, the ratio of head circumference to stature was almost identical in UK, North American, and Japanese children (Tsuzaki et al., 1990).

It has been suggested that differences in stature can be used to estimate differences between populations in other body parts, by calculating the ratio of the body part to stature. This assumes that body proportions are similar in both populations and has been

Table 3.2 Comparison of the Latest Stature and Weight Data for UK Children (Prescott-Clarke and Primatesta, 1999) to Data on US Children (Snyder et al., 1977)

	Percent Difference between the US and UK Data			
	Stature		**Weight**	
Source	Male (2–18)	Female (2–18)	Male (3–18)	Female (3–18)
USA (Snyder et al., 1977)	–2.3%	–2.2%	–6.9%	–10.4%

Note: A negative value indicates that the US data are lower than the UK data.

demonstrated to be an acceptable method of estimation for body lengths, but less so for those measurements subject to body-fat deposits such as body breadths and circumferential measurements (Pheasant, 1982).

Where there are small differences in stature between countries (i.e., those considered to have comparable populations such as the UK and US), these differences might be considered negligible for other, smaller body dimensions such as finger measurements. This is especially true, given the confidence limits within which all anthropometric data are reported; for example, stature data is often only measured to within the nearest 5 mm (Lindsay et al, 1994; Chinn and Rona, 1984) and some data are rounded to the nearest 5 mm (Pheasant, 1986).

For larger body dimensions such as shoulder height, the adjustment is still small compared to the amount of variation found within each age group (shoulder height varies by 180 mm between 5th and 95th percentile 18-year-old boys). However, there can be larger differences between countries in weight and this will have a greater impact on difference in "fatty" dimensions. When designing for an international population, it is important to consider all the indicators available (i.e., stature and weight) and to be aware of how these translate across all parts of the body.

DESIGN IMPLICATIONS

Designing for children means thinking not only about the products that are intended for use by children (such as toys, nursery products, and playground equipment), but also about all of the products and environments with which children will come into contact (Figure 3.8 and Figure 3.9).

This means that the designers of homes, education, leisure, and transport systems and the products within them, should have children in mind, particularly when it comes to safety. Architectural features such as stairs, door and window handles, guards, railings, and balustrades as well as household consumer products such as kitchen and electrical goods must

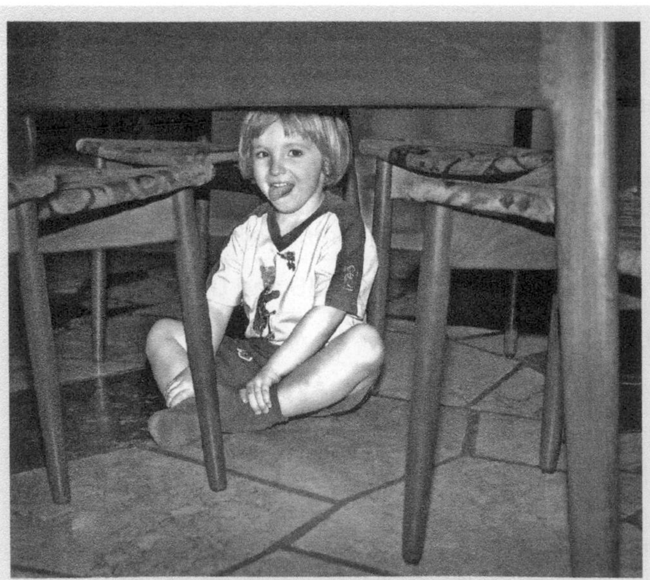

Figure 3.8 (Ursy Potter Photography) Most designers concentrate on the original purpose of the product and obviously, kitchen tables and chairs are for sitting and eating.

However, has there ever been a kitchen table and set of chairs in a home with children that has not also served as a place to hide or build a "tent" or "fort?" It might be worth advertising a table and chairs as "child friendly," with rounded edges and guards over hardware underneath to prevent children from receiving cuts or scratches. It might be a selling point for parents or caregivers.

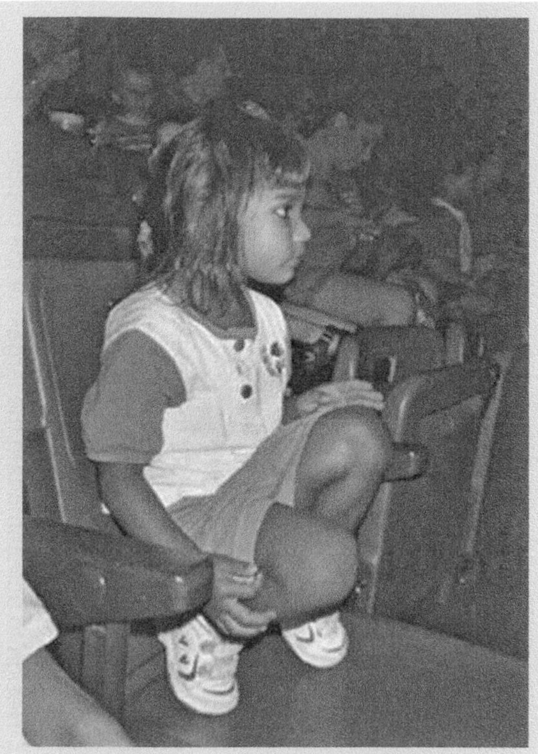

Figure 3.9 (Valerie Rice) Theaters for movies, plays, or musical presentations can be particularly difficult for young children. Certainly, young children attend presentations in these venues, yet there is little to no accommodation for them.

accommodate children (Figure 3.10). The designs could prevent use and operation by children, such as with guards for electrical outlets. Table 3.3 shows how anthropometric data can be applied to different design requirements.

Although many designs aim to include the majority of the population and exclude the extremes, there are times when this is unacceptable. There is a risk of injury in these critical situations.

Here, safety margins or a safety tolerance (i.e., an extra percentage of the dimension or a fixed dimension) should be added to the maximum percentile to ensure the entire population is included. For example, if a fireguard is to be effective, there must be no risk of any child being able to pass their fingers through, or around, the guard to reach the fire. By a combination of minimizing gaps in the guard and increasing the distance between the guard and the fire, the risk of injury can be considerably reduced, even for children with the smallest diameter fingers and with the longest arms.

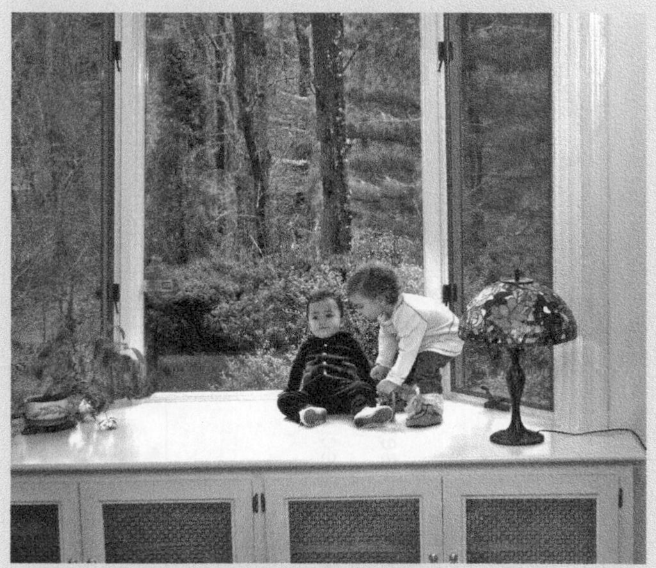

Figure 3.10 (Ursy Potter Photography) Designers must consider whether children will come into contact with their products.

For example, this bay window sits close enough to the floor that innovative small children can climb onto the window shelf. They cannot, however, unlock the windows as they have both lower and higher locking mechanisms. They could fall against the glass, burn themselves on the hot lamp or light bulb, or knock over and break the objects on the window shelf break (and possibly cut themselves).

Table 3.3 Ways to Apply Anthropometric Data to Designs for Children

Design Scenario	Aim	Examples	Design to Accommodate
Fit	Design to ensure user-product match and appropriate and effective use	• Childcare products (cots, beds, high-chairs, walkers, pushchairs, strollers, pacifiers, bottles, etc.) • Toys (playground equipment, bicycles, ride on toys, games, puzzles, rattles, developmental toys, etc.) • Safety equipment (cycle helmets, stair gates, child-resistant locks, car seats, car restraints, etc.) • School furniture, feeding equipment	Design to accommodate the maximum range of the population For example, use at least 5th to 95th percentile or greater extremes and add a safety tolerance if it is a safety-critical design scenario
Reach	Placement to ensure access and appropriate and effective use	• Ensure handrails, handles, controls, information or labels, and instructions are within reach • Step height	Design to accommodate the smallest of the target population, e.g., use 5th percentile values or smaller and add a safety tolerance if it is a safety-critical design scenario
Clearance	Placement to avoid undesirable or unintentional contact	• Access hatches • Gap between desk and seat • Moving parts and hazards are out of reach	Design to accommodate the largest of the population, e.g., use 95th percentile values or greater and add a safety tolerance if it is a safety-critical design scenario
Entrapment	Avoid unintentional retention of the whole body or body parts	• Whole body: railings, washing machines, etc. • Head: cot sides, bunk beds, ladders and open stairs, banisters or balustrades, playground equipment, etc. • Hand or finger: doors, cupboards, folding pushchairs or strollers, fireguards, etc.	Design to ensure the smallest of the population, e.g., 5th percentile values or smaller, cannot pass the body part into gaps or apertures or that the largest of the population, e.g., 95th percentile, can pass into and through the gap or aperture safely; use more extreme values and add a safety tolerance if it is a safety-critical design scenario
Exclusion	Ensure inaccessibility and inoperability so that children are excluded from using products not intended for them	• Design and placement of barriers, railings, guards, etc. • Controls on stoves etc., are inoperable • Child-resistant packaging	Design to exclude the smallest or weakest of the population e.g., use 5th percentile values or less and add a safety tolerance if it is a safety-critical design scenario

USE ANTHROPOMETRIC DATA WITH CAUTION

Published anthropometric data usually present mean, SD, and a selection of percentile values (typically 5th and 95th percentiles). Percentile values, although useful, may cause unintended design problems, such as reduced accommodation afforded by a design, if applied incorrectly. Table 3.4 presents some of the more common errors in using percentiles (more discussion of these issues may be found in Roebuck, 1995):

Table 3.4 How Can We Apply Anthropometric Data

Anthropometric Principles	Do Not (for example)	Why?
1. A percentile value is a point on a scale for a specified population	• Assume that a child with a 95th percentile arm length has longer arms than 95% of other populations	• Different ethnic groups may have different body proportions, so one group may naturally have longer arms. This means that using 95th percentile arm length from a population with shorter-arm proportions will not accommodate 95th percent of children's reaches • Use care when using data from another population to ensure that the two populations are similar
2. Percentile values indicate the rank order of the data,[a] but not the magnitude of change between percentiles	• Assume that the difference between the 5th and 10th percentile is the same as the difference between the 90th and 95th percentile[b]	• We cannot assume that the difference between percentile values for one dimension will be the same for other dimensions, nor is the magnitude of change the same between various percentiles
3. Body dimensions, and body-part dimensions, are not consistently related	• Add 95th percentile shoulder-to-wrist length to 95th percentile hand length to calculate 95th percentile arm length	• This will not give a correct 95th percentile shoulder-to-fingertip length, because people's body dimensions differ in relationship to each other
4. Percentiles describe a size on *one* dimension only	• For example, the 95th percentile person does not exist in reality, as no person is 95th percentile on all dimensions	• Two people of 95th percentile stature may be of the same height for different reasons. One may have long legs and a short torso, while the other may have short legs and a long torso • Even if one allows for a plus or minus 15% tolerance level, no individual person's complete set of anthropometric measurements could be included when designing at the 50th percentile level (Roebuck et al., 1975; Daniels, 1952; Vasu et al., 2000)

[a] That is, ordered in rank from lowest to highest, or the 1st to 99th percentile.

[b] For example, the difference between 5th and 10th percentiles may be 3 mm, while between 90th and 95th it may be 10 mm.

The selection of the most appropriate dimension is critical when using anthropometric data. However, users of the dimensions must be aware of how the measurements were taken and what they mean. Anthropometric surveys do not always follow standardized protocols for naming and defining variables. This may mean that two surveys use the same name for two different dimensions.

For example, the Childata handbook contains two entries for hip breadth, one measured at the trochanters (the bony prominences near the thigh bone), and the other being the maximal measurement while seated. These dimensions are both called hip breadth but provide different data. For this reason, designers should read the definition of a dimension when selecting it for use in design. This is especially important when comparing data across different populations, so comparisons are made on the same dimensions.

REASONS FOR HUMAN VARIABILITY

Anthropometric data are used to ensure that products are produced to fit and suit as many people as possible. As described above, it is readily apparent that children vary in size, shape, physical and psychological abilities, as well as in personal preferences.

Even so, it is important to determine exactly how people vary and how many people will be accommodated, or excluded, with a certain set of design parameters. It is important to understand the reasons for and the extent of human variability and how the group of children for whom we have data may differ from those we are trying to accommodate in our designs (Figure 3.11). Table 3.5 summarizes some of the reasons for human anthropometric variability.

Figure 3.11 (Ursy Potter Photography)
Anthropometric variability at any given age can be more extreme in children than in adults. Although this creates a challenge, there are certainly ways to accommodate children of all sizes and shapes.

In this example, the stool is a precarious long-term answer for reaching the water fountain. The lowered water fountain should have been positioned carefully to include the shortest children, as taller children would still have been able to use it.

Table 3.5 Major Sources of Anthropometric Variability		
Sources of Variability		**Implications**

Gender	Body proportions can vary between the sexes during childhood	• When designing to accommodate the range of a mixed population, examine data carefully before automatically using the 5th percentile female and 95th percentile male values to set design limits, as is often accepted practice when designing for adults.
		• There may be instances when these will not accommodate the range of children at a particular age, especially during adolescence.
		• For example, 12-year-old girls tend to be bigger than boys. They can be
		  • taller—95th percentile UK female stature is 16.3 cm compared to 16.2 cm for males (Pheasant, 1996),
		  • heavier—95th percentile UK female weight is 56kg (123.4 lb) compared to 54 kg (119 lb) for males (Pheasant, 1996), and
		  • have wider hips—97th percentile Dutch female hip breadth is 344 mm compared to 318 mm for males (Steenbekkers, 1993).

Age Groups	Children do not develop at a steady rate throughout childhood	• Trends in data in one age group may not apply to another. That is, the proportional change in a particular body dimension over one age range (say 12 to 24 months) will not be the same over another age range (such as 5 to 6 years). • This is particularly important with infants and adolescents, when anthropometric measurements can change rapidly. • Data on infants are sometimes presented in 6-monthly age bands, during which a rapid growth can occur, and so mean data for such groups should be used with caution. • It is also important to check the details of how data were collected and categorized. The specification of ages of children used to collect data can vary between sources. For example, some studies define age in calendar years so that age 5 includes children aged 60 to 72 months. Others use the 5th birthday as a midpoint and categorize children aged 5 as those from 54 to 66 months.
Disability	Anthropometric and performance measurements are nearly always carried out on children without any form of physical or psychological impairments	• This is because much behavioral research is carried out for medical applications, such as establishing growth standards or medical reference "norms" and not for design. • Most anthropometric data are intended to represent the "majority" of children, meaning it will not include those with impairments. • Designs using these data will not necessarily accommodate children with a disability. • This is a serious inadequacy, as design data on children with disabilities is quite limited.

(continued)

Table 3.5 Major Sources of Anthropometric Variability (Continued)

Sources of Variability		Implications
Cultural Differences	Children around the world differ in size and shape (Figure 3.12) (also see Table 3.1)	• Some researchers suggest that worldwide variation in stature is nearly 40 cm (15.75 in.), between the mean height of the tallest and shortest adults in the world (Pheasant, 1986). • There are also anthropometric differences between children of different ethnic groups, which have relevance for multiracial or ethnic communities such as in Europe and North America (Gatrad et al., 1994). • Anthropometric differences are not only caused by genetic differences, but can also vary due to differences in socioeconomic conditions, even within the same country and the same ethnic background. This is particularly so in developing countries, but even within the UK, small differences in the rate of growth have been observed between children in Scotland and England (Monteiro et al., 1994; Chinn and Rona, 1994).

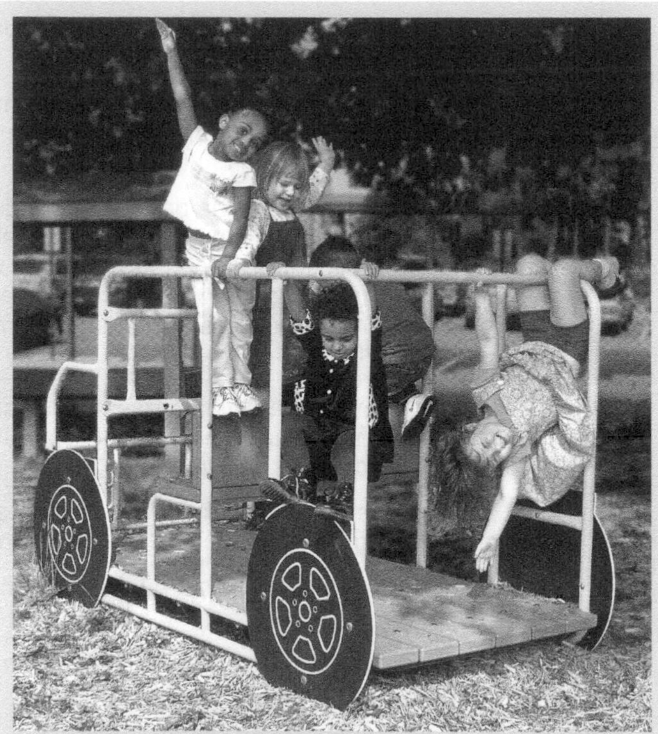

Figure 3.12 (Ursy Potter Photography)
For anthropometric data to be representative of a population, they should account for all racial or ethnic, demographic, and socioeconomic variables within that population. Very few studies are able to do this, and it is extremely difficult to keep up with the flux in society.

SECULAR TREND

The term "secular trend" refers to historical changes over time. In anthropometry, secular trends refer to changes in the body sizes of children over the past decades. It is generally accepted that a secular (or historical) growth trend exists for adults. The average secular increase in height is thought to be ~10 mm per decade in Europe and North America (Eveleth and Tanner, 1990). In the UK, mean adult-male stature increased by 17 mm and mean female stature by 12 mm during the period 1981 to 1995 (Peebles and Norris, 1998). However, some authorities believe that such growth is slowing or has virtually ceased for developed countries (Roebuck, 1995; Pheasant, 1996).

There is some suggestion of secular growth evidence for children also. The rate of changes in growth for multiple countries between 1962 and 1982 appear to indicate that the growth has slowed and the reasons behind the growth have reached their peak of influence (Tanner, 1962, 1978; Tanner et al., 1982). In the UK, it was concluded that a trend of increased stature in 5- to 11-year-olds had ceased during the period 1970–1986 (Chinn et al., 1989) and one study concluded that no evidence existed for secular growth trends in the UK after 1959 (Rona, 1981). Similarly, in 1985, Pheasant stated, "there is no evidence that the secular trend in growth is continuing at the present time or has operated in the recent past" (DES, 1985, p. 1). The same has been found for secular growth change at birth (Roche, 1979). Ochsman and Van Houten (1999) report that the overall changes found in body sizes of US children were not considered to be of a large enough magnitude to update the data of Snyder et al. (1977), which was one of the first comprehensive surveys on children.

Similar work by Smith and Norris (2001, 2004) examined the growth of children in the UK and United States over the past three decades to assess secular growth trends. Stature increases were generally less than body weight increases (as a percentage) at 5th percentile, mean, and 95th percentile levels for UK children. Also, UK children now were found to be closer in size to US children than they were 30 years ago. Table 3.6 shows the largest increases in stature for UK children.

In summary, the available evidence would appear to indicate that secular growth trends are slowing significantly across all ages for stature and other "bony" dimensions. However, weight and other circumference and breadth dimensions have increased dramatically over the past decades. For example, Table 3.7 shows the largest increases in body weight for UK children over the past 25 years.

Table 3.6 Largest Increases in the Stature of UK Children between the Early 1970s (DES, 1972) and the Mid-1990s (Prescott-Clarke and Primatesta, 1999)

Percentile	Males	Females
5th	56 mm, 4% (age 13)	51 mm, 4% (age 12)
Mean	47 mm, 3% (age 13)	41 mm, 3% (age 11)
95th	51 mm, 3% (age 11/13)	42 mm, 3% (age 11)

Table 3.7 Largest Increases in the Body Weight of UK Children between the Early 1970s (DES, 1972) and the Mid-1990s (Prescott-Clarke and Primatesta, 1999)

Percentile	Males	Females
5th	7 kg (15.4 lb) at age 15 16%[a] (age 15)	6 kg (13.2 lb) at age 13 24% (age 10)
Mean	6 kg (13.2 lb) at age 15 12% (age 13)	5.5 kg (12.1 lb) at age 12 13% (age 12)
95th	15 kg (33 lb) at age 17 20% (age 10)	16 kg (35.2 lb) at age 18 24% (age 7)

[a] The largest percentage increases in weight occur in younger children as the proportional increases are greater for younger children.

A pattern was identified for changes in absolute body weight at 5th, mean, and 95th percentile levels. Once past 6 years of age, 95th percentile weight increases the most (at each year of age), often by more than double the rate of increase of mean or 5th percentile at that age. This indicates that the heaviest 5% of the population are increasing in weight at a greater rate than the rest of the population. Such large increases in weight and other "fatty" dimensions, coupled with the leveling out of stature increases, means that the body shape of children will become more endomorphic in nature, reflected by the fact that levels of overweight and obese children are increasing in many countries.

The Center for Disease Control (CDC) in the United States reports that there are nearly twice as many overweight children and almost three times as many overweight adolescents as there were in 1980.* In 1999, 13% of children and adolescents were classified as overweight (DHHS, 1999). There appear to be similar increases in the UK (Prescott-Clarke and Primatesta, 1999).

Little is known about how a secular increase in stature affects other anthropometric measurements. Some suggest that a secular increase in stature in some countries can be almost fully attributed to an increase in leg length (Van Wieringen, 1986; Gerver et al., 1994). However, we need more precise information using comparable data over long periods of time on a complete range of body dimensions and from a large sample of countries. Such data are not yet available, so current interpretations of secular trend are based solely on stature and weight data.

SUMMARY

Childhood is a time of rapid and inconsistent growth, when the variance in a body dimension at a given age will be greater than at any time in later life. The timing of adolescent growth spurts can mean that younger children can temporarily "overtake" their seniors and designing for even one age group can be difficult, let alone using the generic concept of "designing for children."

* These charts can be seen at www.cdc.gov/nccdphp/dnpa/obesity/faq.htm#children.

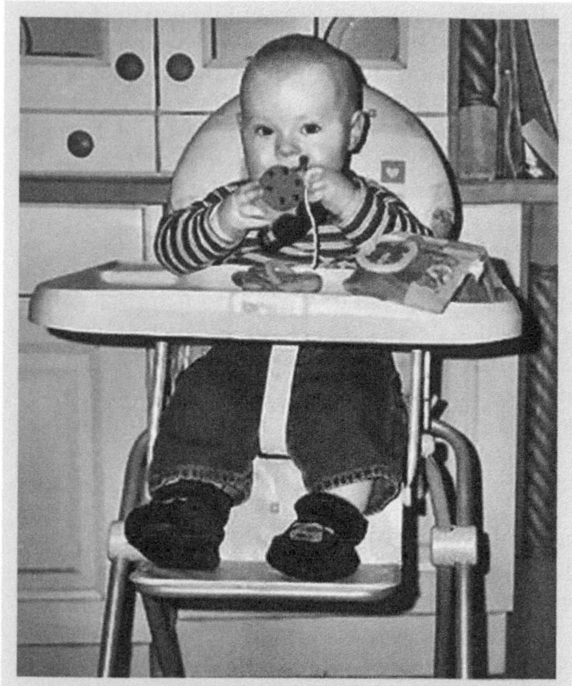

Figure 3.13 (Stuart Smith)

This chapter has also emphasized international differences in child anthropometry, which given the growing internationalization of markets, means that an even wider spectrum of anthropometric requirements need to be met.

Children are one of the vulnerable sections of society, in terms of the potential serious outcome of minor accidents, but this needs to be balanced by their need to explore and learn about their environment and the need for play. This means designing for children must focus on a balance between safety and stimulation, a solid foundation for design, and a careful application of reliable and valid anthropometric data, backed by thorough user testing.

REFERENCES AND INTERNET SOURCES

BSI (British Standards Institute). (1990). Body measurements of boys and girls from birth up to 16.0 years, BS7231: Part 1. British Standards Institute, 389 Chiswick High Road, London, UK.

Cameron, N., Tanner, J.M., and Whitehouse, R.A. (1982). A longitudinal analysis of the growth of limb segments in adolescence. *Annals of Human Biology*, *9*, 211–220.

Chaurand, R.A., Leon, L.R.P., and Munoz, E.L.V. (2001). Dimensiones Antropometricas de Pobliacion Latinoamericana, Universidad de Guadalajara, Mexico.

Chinn, S. and Rona, R.J. (1984). The secular trend in the height of primary school children in England and Scotland from 1972–1980. *Annals of Human Biology*, *11*(1), 1–16.

Chinn, S. and Rona, R.J. (1994). Trends in weight-for-height and triceps skinfold thickness for English and Scottish children, 1972–1982 and 1982–1990. *Paediatric and Perinatal Epidemiology*, *8*, 90–106.

Chinn, S., Rona, R.J., and Price, C.E. (1989). The secular trend in height of primary school children in England and Scotland 1972–79 and 1979–86. *Annals of Human Biology*, *16*(5), 387–395.

Coblentz, A., Martel, A., and Ignazi, G. (1993). Enquete biometrique sur une population d'enfants et d'adolescents scolarises en France, Rep. AA 292/93, Laboratoire d'Anthropologie Appliquee, University Rene Descartes, Paris.

Daniels, G.S. (1952). *The "Average Man?"* Technical Note WCRD 53-7. WPAFB, Ohio: Wright Air Developement Centre, USAF (AD-10203).

DES (Department of Education and Science). (1972). *British School Population Dimensional Survey*, Building Bulletin 46, Department of Education and Science, London: HMSO.

DES (Department of Education and Science). (1985). *Body Dimensions of the School Population*, Building Bulletin 62, Department of Education and Science, London: HMSO.

DHHS (Department of Health and Human Services). (1996). National Center for Health Statistics. Third National Health and Nutrition Examination Survey (NHANES), 1988–94. Hyattsville, MD: Centers for Disease Control and Prevention.

DHHS (Department of Health and Human Services). (1999). National Center for Health Statistics (NHANES). Prevalence of Overweight among Children and Adolescents: United States. Hyattsville, MD: Centers for Disease Control and Prevention.

DIN (Deutsches Institut fur Normun). (1981). DIN 33402: *Body Dimensions of People*, June 1981. Berlin, Germany: Deutsches Institut fur Normun e. V. (German Standards Institute).

Eveleth, P.B. and Tanner, J.M. (1990). *Worldwide Variation in Human Growth*, 2nd edn. Cambridge: Cambridge University Press.

Gatrad, A.R., Birch, N., and Hughes, M. (1994). Preschool weights and heights of Europeans and five subgroups of Asians in Britain. *Archives of Diseases in Childhood*, *71*, 207–210.

Gerver, W.J.M., De Bruin, R., and Drayer, N.M. (1994). A persisting trend for body measurements in Dutch children. The Oosterwolde II study. *Acta Paediatrica*, *83*, 812–814.

HQL. (1997). Japanese Body Size Data 1992–1994, Japan: HQL. HQL/Research Institute of Human Engineering for Quality Life, Dokita-Daibiru Bldg, 3F 1-2-5 Dojima, Kita Ku, Osaka 530-0003, Japan. www.hql.or.jap

Lindsay, R., Feldkamp M., Harris, D., Robertson, J., and Rallison M. (1994). Utah growth study: Growth standards and the prevalence of growth hormone deficiency. *Journal of Pediatrics*, *125*(1), 29–35.

Martin, W.E. 1960. *Children's Body Measurements for Planning and Equipping Schools.* Special Publication No. 4. Bethesda, MD: US Department of Health, Education, and Welfare.

Monteiro, C.A., D'Aquino Benicio, M.H., and Da Cruz Gouveia, N. (1994). Secular growth trends in Brazil over three decades. *Annals of Human Biology*, *21*(4), 381–390.

Norris, B. and Wilson, J.R. (1995). *Childata: The Handbook of Child Measurements and Capabilities—Data for Design Safety*. UK: Department of Trade and Industry.

Nowak, E. (2000). The anthropometric atlas of the Polish population—Data for design. Warsaw: Institute of Industrial Design.

Ochsman, R.B. and Van Houten, D.T. (1999). Physical dimensions of US children: Have they changed? *Proceedings of the 7th International Conference on Product Safety Research*, 30th September–1st October, Washington D.C.: European Consumer Safety Association/Consumer Product Safety Commission.

Peebles, L. and Norris, B. (1998). *Adult Data: The Handbook of Adult Anthropometric and Strength Measurements—Data for Design Safety*. UK: Department of Trade and Industry.

Pheasant, S.T. (1982). A technique for estimating anthropometric data from the parameters of the distribution of stature. *Ergonomics*, *25*, 981–992.

Pheasant, S.T. (1986). *Bodyspace: Anthropometry, Ergonomics, and Design*, 1st edn. London: Taylor & Francis.

Pheasant, S.T. (1996). *Bodyspace: Anthropometry, Ergonomics, and Design*, 2nd edn. London: Taylor & Francis.

Prescott-Clarke, P. and Primatesta, P. (Eds). (1999). *Health Survey for England: The Health of Young People '95–97*. London: The Stationary Office.

Roche, A.F. (1979). Secular trends in stature, weight, and maturation. *Monographs of the Society for Research in Child Development*, Serial No. 179; *44*(3–4), 3–27.

Roebuck, J.A., Kroemer, K.H.E., and Thomson, W.G. (1975). *Engineering Anthropometry Methods*. New York: John Wiley and Sons.

Roebuck, J.A. (1995). *Anthropometric Methods: Designing to Fit the Human Body*. Santa Monica, CA: Human Factors and Ergonomics Society.

Rona, R.J. (1981). Genetic and environmental factors in the control of growth in childhood. *British Medical Bulletin*, *37*(3), 265–272.

Smith, S.A. and Norris, B. (2001). Assessment of validity of data contained in Childata for continued use in product design. Institute for Occupational Ergonomics, University of Nottingham. Available from www.virart.nott.ac.uk/pstg/validity.pdf

Smith. S.A. and Norris, B. (2004). Changes in the body size of UK and US children over the past three decades. *Ergonomics*, *47*(11), 1195–1207.

Snyder, R.G., Spencer, M.L., Owings, C.L., and Schneider, L.W. (1975). Physical characteristics of children as related to death and injury for consumer product safety design. Report no. UM-HSRI-BI-75-5. Bethesda, MD: Consumer Product Safety Commission.

Snyder, R.G., Schneider, L.W., Owings, C.L., Reynolds, H.M., Golomb, D.H., and Schork, M.A. (1977). Anthropometry of infants, children, and youths to age 18 for product safety design. Report no. UM-HSRI-77-17. Bethesda, MD: Consumer Product Safety Commission.

Steenbekkers, L.P.A. (1989). KIMA—Ergonomische gegevens voor kinderveiligheid, Product Ergonomics Group. Delft, The Netherlands: Delft Technical University.

Steenbekkers, L.P.A.(1993). *Child Development, Design Implications and Accident Prevention,* Physical Ergonomics Series. Delft, The Netherlands: Delft University Press.

Tanner, J.M. (1962). *Growth at Adolescence*. Oxford: Blackwell.

Tanner, J.M. (1978). *Foetus into Man*. London: Open Books.

Tanner, J.M., Hayashi, T., Preece, M.A., and Cameron, N. (1982). Increase in leg length relative to trunk in Japanese children and adults from 1957–1977: A comparison with British and Japanese Americans. *Annals of Human Biology*, *9*, 411–414.

Tsuzaki, S., Matsuo, N., Saito, M., and Osano, M. (1990). The head circumference growth curve for Japanese children between 0–4 years of age: Comparison with Caucasian children and correlation with stature. *Annals of Human Biology*, *17*(4), 297–303.

Van Wieringen, J.C. (1986). Secular growth changes, in: *Human Growth,* vol. 3: *Methodology; Ecological, Genetic, and Nutritional Effects on Growth*, edited by Tanner J.M. and Faulkner F. London: Plenum Publications.

Vasu, M., Mital, A., and Pennathur, A. (2000). Evaluation of the validity of anthropometric design assumptions. *Proceedings of the XIVth Triennial Congress of the International Ergonomics Association and 44th Annual Meeting of the Human Factors and Ergonomics Society 2000*, pp. 6-304–6-306. Santa Monica, CA: Human Factors and Ergonomics Society.

KEY INTERNET RESOURCES

www.itl.nist.gov/iaui/ovrt/projects/anthrokids/ This site contains the data and additional information on the work of Snyder et al. (1975, 1977).
www.archive.official-documents.co.uk/document/doh/survey97/hse95.htm This site contains the full text of the Health Survey for England 1995–97 (Prescott-Clarke and Primatesta, 1999).
www.cdc.gov/nchs/nhanes.htm This is the gateway to the NHANES series of reports, featuring data tables and graphs, methodology details, and more.
www.cdc.gov/growthcharts/ This site is the Center for Disease Control (CDC) in the United States, specifically presenting the latest growth charts for children.
www.virart.nottingham.ac.uk/pstg Information and ordering information on the publication Childata (Norris and Wilson, 1995), a comprehensive collection of anthropometric data on children, is available on this site, along with information on other strength and safety critical data for children.
www.humanics-es.com/recc-children.htm Humanics ErgoSystems, Inc. Web site contains extensive research and resources of interest on ergonomics, safety, and health for children around the world.

APPENDIX 3A. ANTHROPOMETRIC DATA

Table 3.A1 Children's Stature (mm)[a] for UK, USA, Japan, and Mexico

Country	Sex	Age	Mean	5th	95th	Sex	Age	Mean	5th	95th
UK	Male	2	910	837	985	Female	2	901	831	975
		3	992	916	1066		3	983	912	1061
		4	1059	982	1136		4	1055	983	1130
		5	1122	1039	1212		5	1120	1029	1210
		6	1189	1109	1274		6	1182	1085	1264
		7	1250	1168	1339		7	1241	1154	1346
		8	1311	1222	1412		8	1297	1211	1392
		9	1359	1260	1454		9	1358	1252	1473
		10	1413	1304	1518		10	1420	1319	1532
		11	1473	1360	1586		11	1481	1358	1612
		12	1523	1402	1659		12	1539	1421	1645
		13	1597	1456	1751		13	1579	1464	1687
		14	1656	1506	1800		14	1611	1512	1714
		15	1720	1602	1848		15	1624	1510	1725
		16	1750	1627	1868		16	1638	1541	1742
		17	1760	1646	1870		17	1632	1531	1737
		18	1764	1662	1881		18	1634	1532	1739
United States	Male	2	909	846	978	Female	2	897	825	969
		3	988	919	1057		3	982	913	1060
		4	1052	981	1133		4	1051	972	1125
		5	1123	1042	1214		5	1122	1039	1208
		6	1189	1082	1294		6	1179	1091	1278
		7	1260	1150	1356		7	1243	1147	1383
		8	1313	1224	1407		8	1311	1193	1412
		9	1377	1275	1490		9	1366	1269	1486
		10	1420	1300	1530		10	1427	1329	1547
		11	1474	1369	1602		11	1502	1384	1617
		12	1555	1433	1683		12	1555	1433	1662
		13	1616	1465	1756		13	1599	1494	1730
		14	1690	1562	1812		14	1612	1501	1727

Table 3.A1 Children's Stature (mm)[a] for UK, USA, Japan, and Mexico (Continued)										
Country	Sex	Age	Mean	5th	95th	Sex	Age	Mean	5th	95th
USA		15	1728	1588	1844		15	1628	1520	1713
		16	1750	1619	1870		16	1631	1541	1734
		17	1765	1629	1822		17	1634	1535	1740
		18	1773	1664	1938		18	1632	1519	1749
Japan	Male	7	1193	1117	1280	Female	7	1186	1102	1274
		8	1246	1169	1332		8	1236	1148	1321
		9	1303	1215	1395		9	1300	1210	1403
		10	1358	1261	1453		10	1358	1266	1461
		11	1406	1317	1517		11	1427	1314	1531
		12	1473	1354	1590		12	1485	1371	1579
		13	1552	1424	1674		13	1531	1435	1625
		14	1610	1489	1722		14	1556	1475	1641
		15	1656	1558	1748		15	1572	1483	1661
		16	1687	1597	1776		16	1573	1490	1666
		17	1691	1595	1786		17	1578	1488	1669
		18	1693	1603	1786		18	1581	1497	1671
Mexico	Male	2	898	832	958	Female	2	897	831	963
		3	970	905	1043		3	970	892	1044
		4	1048	963	1120		4	1039	960	1112
		5	1118	1029	1191		5	1108	1016	1188
		6	1175	1086	1264		6	1167	1087	1256
		7	1228	1134	1322		7	1218	1129	1307
		8	1279	1185	1373		8	1269	1167	1371
		9	1334	1233	1435		9	1318	1194	1442
		10	1381	1270	1492		10	1399	1288	1510
		11	1437	1325	1549		11	1457	1340	1574
		12	1480	1358	1602		12	1500	1384	1616
		13	1542	1410	1674		13	1533	1442	1624
		14	1611	1482	1740		14	1555	1456	1654
		15	1685	1571	1799		15	1577	1486	1668
		16	1700	1594	1806		16	1588	1496	1680
		17	1705	1599	1811		17	1582	1486	1678
		18	1707	1608	1816		18	1572	1478	1666

[a] 1 in.= 25.4 mm.

Table 3.A2 Children's Weight (kg)[a] for UK, United States, Japan, and Mexico

Country	Sex	Age	Mean	5th	95th	Sex	Age	Mean	5th	95th
UK	Male	2	14.2	11.6	17.5	Female	2	13.7	11.2	16.7
		3	16.4	13.4	20.2		3	16	12.8	20.2
		4	18.4	15	22.5		4	18.3	14.7	23.4
		5	20.4	16.3	25.5		5	20.4	16.2	25.5
		6	22.9	18	29.1		6	22.8	17.7	30.2
		7	25.8	20.5	33.7		7	25.9	19.8	37.9
		8	29.1	22.7	39.4		8	28.8	21.7	41
		9	32	24	43.9		9	32.7	24.2	45.7
		10	35.6	26.7	51.8		10	37.1	27.8	52.9
		11	40.2	29.3	55.2		11	42.4	28.9	62.3
		12	44.8	31	63.1		12	47.5	34.4	66
		13	50.8	35	71.6		13	51.8	38.9	70.6
		14	56.4	39	79.7		14	56.7	41.3	80.3
		15	62.9	47.1	85.6		15	58.4	43.6	79.2
		16	66.7	49.5	89.7		16	60.3	46.4	79.4
		17	70.1	52.1	95.9		17	60.2	46.5	77.2
		18	70.5	55.5	92.1		18	61.6	46.7	83.9
United States	Male	2	13.6	11.3	16.3	Female	2	13.2	10.8	16.3
		3	15.8	12.8	18.7		3	15.4	12.4	19.1
		4	17.7	14.4	21.8		4	17.9	14.0	24.7
		5	20.1	15.8	25.6		5	20.2	15.7	27.5
		6	23.3	17.5	36.0		6	22.6	16.4	32.8
		7	26.3	19.5	36.5		7	26.4	19.1	43.3
		8	30.2	21.3	47.4		8	29.9	21.8	44.8
		9	34.4	23.9	49.6		9	34.4	23.3	50.9
		10	37.3	25.7	51.1		10	38	26.8	58.2
		11	42.5	29.2	65.5		11	44.2	28.5	62.6
		12	49.1	34.5	73.0		12	49	34.0	72.3

Table 3.A2 Children's Weight (kg)[a] for UK, United States, Japan, and Mexico (Continued)										
Country	**Sex**	**Age**	**Mean**	**5th**	**95th**	**Sex**	**Age**	**Mean**	**5th**	**95th**
USA		13	54	36.6	80.0		13	55.8	37.0	81.8
		14	60.6	43.6	81.6		14	58.5	44.4	81.5
		15	66.1	49.1	97.6		15	58.2	45.7	78.3
		16	68.8	51.8	98.1		16	61.7	46.2	85.6
		17	72.9	53.6	107.8		17	62.4	47.9	81.5
		18	71.3	55.5	100.9		18	61.5	44.1	92.5
Japan	Male	7	22.8	18.2	29.1	Female	7	22.2	17.2	29.0
		8	25.6	20.2	34.5		8	24.4	19.5	31.7
		9	28.8	22.4	39.5		9	28.3	21.6	39.2
		10	32.7	24.6	46.1		10	31.4	24.2	42.3
		11	36.2	26.9	51.1		11	36.1	27.3	48.0
		12	40.6	29.2	57.5		12	41.2	29.8	53.9
		13	46.2	33.0	63.9		13	44.7	33.9	56.8
		14	51.4	37.3	68.8		14	48.1	37.8	62.3
		15	56.8	43.2	73.6		15	50.8	41.1	63.2
		16	60.4	47.0	78.8		16	51.6	41.8	63.4
		17	61.7	49.2	79.1		17	52.7	42.2	64.6
		18	62.9	49.8	81.2		18	53.3	42.3	67.0
Mexico	Male	2	13.7	10.4	17.0	Female	2	13.2	10.7	17.5
		3	16.6	12.1	18.7		3	15.26	12.0	18.6
		4	17.5	14.4	21.0		4	17.3	13.7	20.3
		5	20.2	15.0	24.9		5	19.7	14.6	24.5
		6	22.8	16.2	29.4		6	22.4	15.8	27.9
		7	25.8	17.6	34.0		7	25.1	16.9	33.4
		8	29.3	19.4	39.2		8	28.4	18.5	38.3
		9	32.8	21.3	44.4		9	32.3	19.1	45.5
		10	36.3	21.5	51.2		10	36.3	23.1	49.5
		11	40.6	25.8	55.5		11	42.3	25.8	58.8
		12	42.7	26.2	59.2		12	45.6	29.1	62
		13	49.4	31.3	67.6		13	48.6	33.8	63

(continued)

Table 3.A2 Children's Weight (kg)[a] for UK, United States, Japan, and Mexico (Continued)										
Country	**Sex**	**Age**	**Mean**	**5th**	**95th**	**Sex**	**Age**	**Mean**	**5th**	**95th**
Mexico		14	55.5	39	72		14	53.1	38.3	67.9
		15	65.0	44.2	85.8		15	54.2	38.7	69.7
		16	65.6	45.3	85.9		16	56.4	42.5	70.3
		17	66.9	45.6	88.2		17	57.4	42.9	71.9
		18	68.1	48.9	87.2		18	54.9	43.8	65.6

Note: Children's weight (kg) for UK (Prescott-Clarke and Primatesta, 1999), United States (DHHS, 1996), Japan (HQL, 1997), and Mexico (Chaurand et al., 2001).

[a] 1 kg = 2.2046 lb.

CHAPTER 4

VISUAL ERGONOMICS FOR CHILDREN

KNUT INGE FOSTERVOLD AND DENNIS R. ANKRUM

TABLE OF CONTENTS

INTRODUCTION

Don't sit so close to the TV. You'll ruin your eyes!

(Mother, 1969)

Mothers often admonish their children in this way. In one sense mother was right; visual environments affect how children's vision develops. Intensive near work can contribute to myopia or nearsightedness. Yet in another sense mother was wrong; intensive near work does not usually harm visual receptors in the eye or the visual part of the brain.

Children not only watch television, they play games on their televisions and use computers for homework, schoolwork, games, and social interaction. Is this desirable? Should parents set guidelines or encourage their children to use technologies?

Caregivers (and scientists) want to know if intensive computer use will affect their children's health, safety, and ability to learn. There is particular concern for very young children, whose visual systems are still developing.

Although we recognize that stimulating visual environments promote healthy visual systems (Figure 4.1), little is known about what "stimulating environments" are and how to design them (Figures 4.2 and 4.3).

Visual ergonomists study children, their tasks, and their settings in order to evaluate and design healthy visual environments.

When is there too much visual stimulation? Can stimulating environments have adverse effects? How do we know when visual input is too little, too much, or just right? How do we know if a child sees correctly, and how do we recognize visual problems?

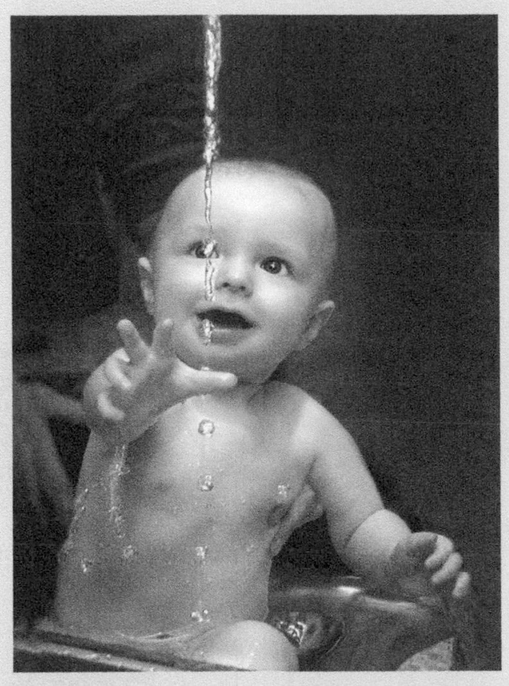

Figure 4.1 (Ursy Potter Photography)

I was 9 or 10 when I first realized I needed glasses. I could read. I hadn't had problems in school. My father was describing leaves on a distant tree. I replied, with all the indignity that a child feels when they know they are being told an untrue story, "you can't see leaves from here!" I saw exactly what a child draws: a trunk with a large green blob on top. He asked me about dew on the grass. "No one sees dew on the grass; it's just something they write about in books!" My father took me to an optometrist that same week.

(V. Rice, 2005, Personal communication.)

Teach children to communicate about symptoms and discomfort.

Children often have difficulty communicating what they see and feel. Many lack the vocabulary to adequately describe their visual problems. Instead, they avoid uncomfortable visual tasks or ignore their discomfort.

Children may not be able to distinguish between normal and abnormal sight. How can they realize their vision is blurry if they have never seen the world in sharp focus? Since children may not communicate their visual problems, adults must watch for the signs and teach children to recognize and describe their symptoms.

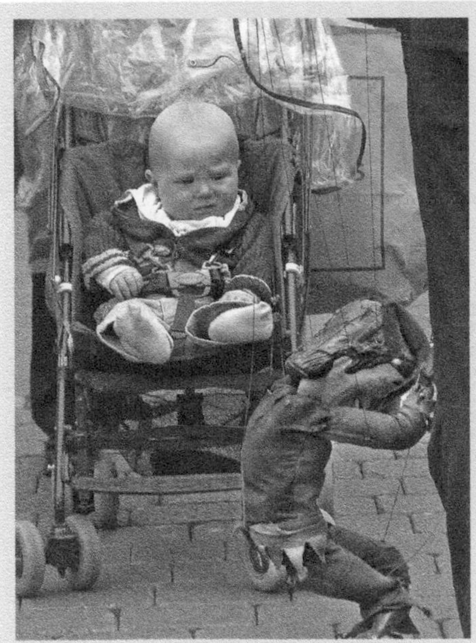

Figure 4.2 (Ursy Potter Photography) **Figure 4.3** (Ursy Potter Photography)
Children learn through their senses about the natural and social world around them. They need rich, yet not overwhelming, visual information.

THE ANATOMY OF THE EYE

Table 4.1 Anatomy of the Eye

The eye is a sphere, about 25 mm (1 in.) in diameter. The front part of the eye is a transparent, dome-shaped structure called the *cornea* (Figure 4.4). The color of the *iris* is our "eye color." The *pupil*, the "black part" of our eye, is in the center of the iris.

The pupil changes size according to the light level. When it is dark, the pupil enlarges to let more light in. When it is bright, the pupil becomes smaller to reduce the amount of light entering the eye. We refer to this process as ***adaptation*** (Table 4.2).

Light rays that fall on the *macula*[a] provide sharp images. The *fovea* is in the center of the macula. Packed densely with cones, the fovea provides the sharpest detail. When we "look" at something, we turn our eyes so light rays from the area of interest fall on the fovea.

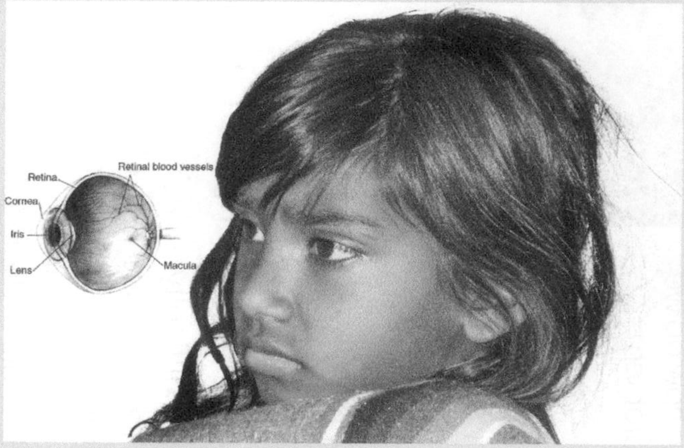

Figure 4.4 (Eye image by eyeSearch.com. Original photo of the child by Valerie Rice. Adapted composite by Ursy Potter Photography.)

In order to focus on the retina and provide sharp images, light rays must bend as they pass through the eye. This process is called *refraction*.[b] Most of this bending takes place as light rays pass through the cornea. Minor adjustments in the shape of the lens (*accommodation*) bend light to enable us to see sharp images (see Table 4.2 and 4.3).

Normal, clear vision occurs when parallel rays of light converge to focus at a single point on the surface of the retina. This focuses distant objects and allows us to see sharp images without the need for accommodation. This normal condition of the eye is *emmetropia*.

[a] The macula is a 3–5 mm (0.1–0.2 in.) diameter, oblong area at the center of the retina.

[b] In optometry, this term is commonly used when measuring spectacle prescriptions.

Table 4.2 How We "See"

1. Light rays pass though the pupil and lens.
   - These light rays bend both as they pass through the cornea and then the lens.
   - The lens is inside a lens capsule.
     Ciliary muscles circle and connect the lens capsule with zonular fibers. These fibers normally hold the lens capsule in a somewhat flattened shape.
   - The rays then pass though the aqueous humor, a clear, watery fluid in the center of the eye.
2. Bent light rays then focus on the retina.
   - The retina is a thin membrane at the back of the eyeball. It contains millions of photoreceptors called rods and cones.
   - The rods and cones connect to the optic nerve with nerve fibers. Rods are more sensitive to light and cones are more sensitive to detail and color.
3. Nerve fibers transmit information contained in the light to the brain.
   - The brain interprets this information and we "see."

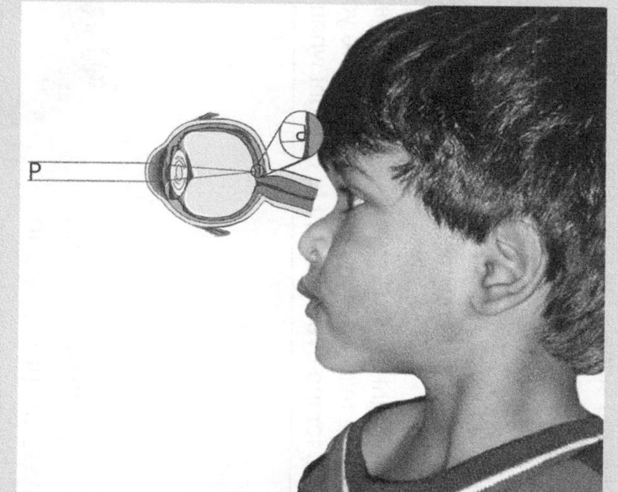

Figure 4.5 (Eye image by eyeSearch.com. Original photo of the child by Valerie Rice. Adapted composite by Ursy Potter Photography.)

Bending of light rays in the eye.

Most of the bending of the light rays (refraction) occurs at the cornea.

Light also bends as it passes through the lens, but to a lesser extent.

The lens fine tunes the bending to ensure that sharp images are projected on the retina. Notice that the image projected at the retina is turned up side down. The inversion takes place as the light rays travel through the cornea and the lens.

Table 4.3 Visual Accommodation and Adaptation	
Accommodation	*Accommodation* enables us to view close objects (such as when reading) and to shift our focus between near and far objects. a. Our eyes must visually *accommodate* to objects at different viewing distances. The ciliary muscle changes the shape of the lens in order to bend the incoming rays of light so that they strike a point on the retina and enable us to see clearly. b. Vision blurs when light rays converge either too far in front of or too far behind the retina. The brain reacts to this blur and signals the ciliary muscles to change the shape of the lens to bring the object into sharper focus.
Adaptation	During visual *adaptation,* the visual system adjusts to different light levels. a. In low light, our pupils widen to let more light enter our eyes. When light levels increase, the pupils constrict in order to reduce the light entering the eye. b. It takes longer to adapt to low light than to bright light. For example, when you enter a dark movie theater it takes a relatively long time to see where you are going. Yet it takes only a few seconds for the eyes to adapt to daylight when exiting.

NORMAL VISUAL DEVELOPMENT

The various components of the eye and the kinds of vision each mature at different rates. As the eyeball increases in size, it begins to lengthen. The most rapid increase takes place during the first year of life (Figure 4.6). The growth process then slows down, with a gradual increase in length until full growth—during the teen years (Brown, 1996; Pennie et al., 2001). Tables 4.1–4.3 describe the anatomy of the eye and normal vision. Table 4.4 describes the developmental stages of vision. Table 4.5 provides a list of definitions associated with vision.

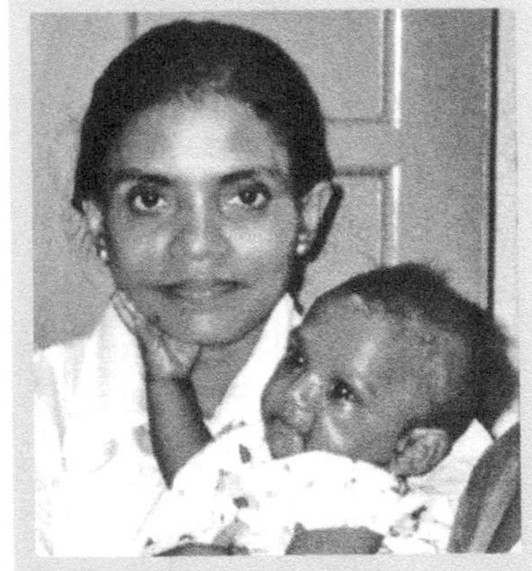

Figure 4.6 (Valerie Rice)

Table 4.4 Developmental Stages of Vision (PBA, 2003; TSB, 2003)

Age		
Newborns	**Newborns do not see clearly.** • Pupils cannot fully dilate. • Lens is nearly spherical. • Retina and macula not fully developed. • Somewhat farsighted with some astigmatism. • Generally between 6/60 (20/200) and 6/120 (20/400) acuity.	**Newborns:** • See all colors except blue. • Initially respond to brightness and high contrast and later shift to details and patterns. • Prefer black and white (high contrast). • Briefly focus on close objects such as faces.
1 Month	• Ability to focus on near objects (*accommodation*) improves.	**By 1 month, they:** • Can follow a slow moving object intermittently (not smoothly). • Move eyes and head together.
2 Months	• Accommodation approaches maturity.	**At 2 months, infants:** • Focus on a face. • Watch people at a distance. • Shift their focus between objects. • Attend to objects 6 ft away. • Follow vertical movements better than horizontal movements.
3 Months	**They visually focus and search.** • Eye movements coordinate most of the time. • Visual attention and searching begins. • Begins to associate visual stimuli with events (e.g., bottle and feeding).	**By 3 months, infants:** • Can follow a slowly moving object and stop moving their head or eyes when the object stops. • Prefer color (especially red and yellow), black and white, and faces. • Glance at objects as small as 2.5 cm (1 in.) diameter. • Begin to associate visual objects with events.
3 to 6 Months	**Full retinal development** • Eyes accommodate for near and far objects and back again as they begin to shift their gaze. • Depth perception (distance) begins to develop. • By about 4 months, acuity is typically between 6/60 (20/200) and 6/90 (20/300). • By about 5 months, their eyes can converge.	**Infants of this age:** • Discern sharp detail. • Develop an interest in using their hands. • Like to look at other babies and into mirrors. • Show an interest in novel sights. • Recognize and respond to familiar faces.
4 to 5 Months	**Distance and depth perception still developing**	**By 4 months, infants (Figures 4.7, 4.8, and 4.9) follow objects with their eyes:** • Past midline (horizontal), vertical, and in a circle. **By about 5 months, they:** • Begin to visually explore. • Study objects with near vision.

(continued)

Table 4.4 Developmental Stages of Vision (PBA, 2003; TSB, 2003) (Continued)

6 Months	**Distance and depth perception still developing** • Eyes are $^{2}/_{3}$ of adult size. • Acuity is $^{6}/_{60}$ ($^{20}/_{200}$) or better. • Eyes begin to work together for binocular vision. • Eye movements are well coordinated for near and distant viewing.	**These children can:** • Recognize faces at a 6 ft distance. **Between 6 and 9 months, the child:** • Begins hand-to-hand transfers. **Between 9 and 12 months, the child will:** • Search for objects after they have been hidden (object permanence).
1 Year	**Children focus and accommodate** • Visual acuity is about $^{6}/_{15}$ ($^{20}/_{50}$). • Depth perception is mature.	**At this age, children can:** • Pick out simple forms. • Visually track a 180° arc. • Practice eye–hand coordination and eye–body movements by grasping, reaching, and placing.
2 to 5 Years	**At about 2 years:** • Optic nerve myelinization is completed,[a] • Visual acuity is between $^{6}/_{6}$ ($^{20}/_{20}$) (normal) and $^{6}/_{9}$ ($^{20}/_{30}$). • Can orient upright (vertically). **By about 3 years, the retina matures.** **Between 2 and 4, corneal thickness is mature.**	**Preschoolers can:** • Practice eye–hand coordination by drawing and studying pictures. • Imitate. • Match colors and shapes. • Associate names with objects. **By about 3 years, they:** • Use visual memory for puzzles and drawing simple shapes.

[a] Myelinization is the forming of the fatty sheath surrounding some nerves.

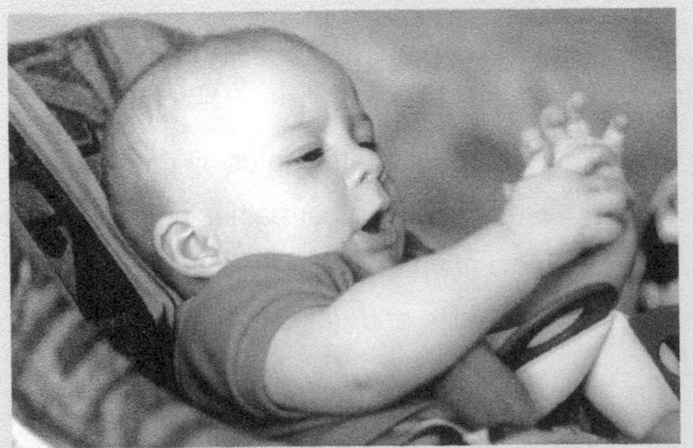

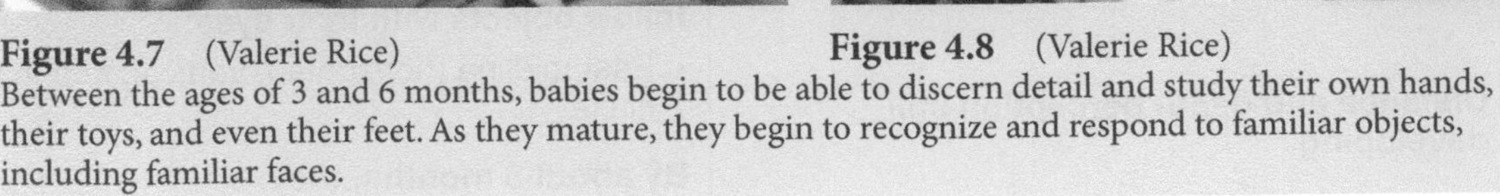

Figure 4.7 (Valerie Rice) **Figure 4.8** (Valerie Rice)
Between the ages of 3 and 6 months, babies begin to be able to discern detail and study their own hands, their toys, and even their feet. As they mature, they begin to recognize and respond to familiar objects, including familiar faces.

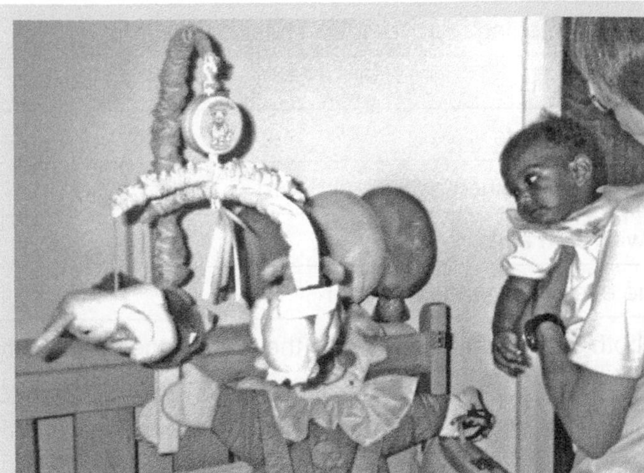

Figure 4.9 (Valerie Rice)
Newborns are more auditory than visual; their visual systems lag behind their auditory systems in development. Although visually stimulating environments are important for normal development, too much stimulation may prematurely shift an infant to a visual dominance. Children need balanced environments that stimulate each of their senses without bombarding them with visual stimuli.

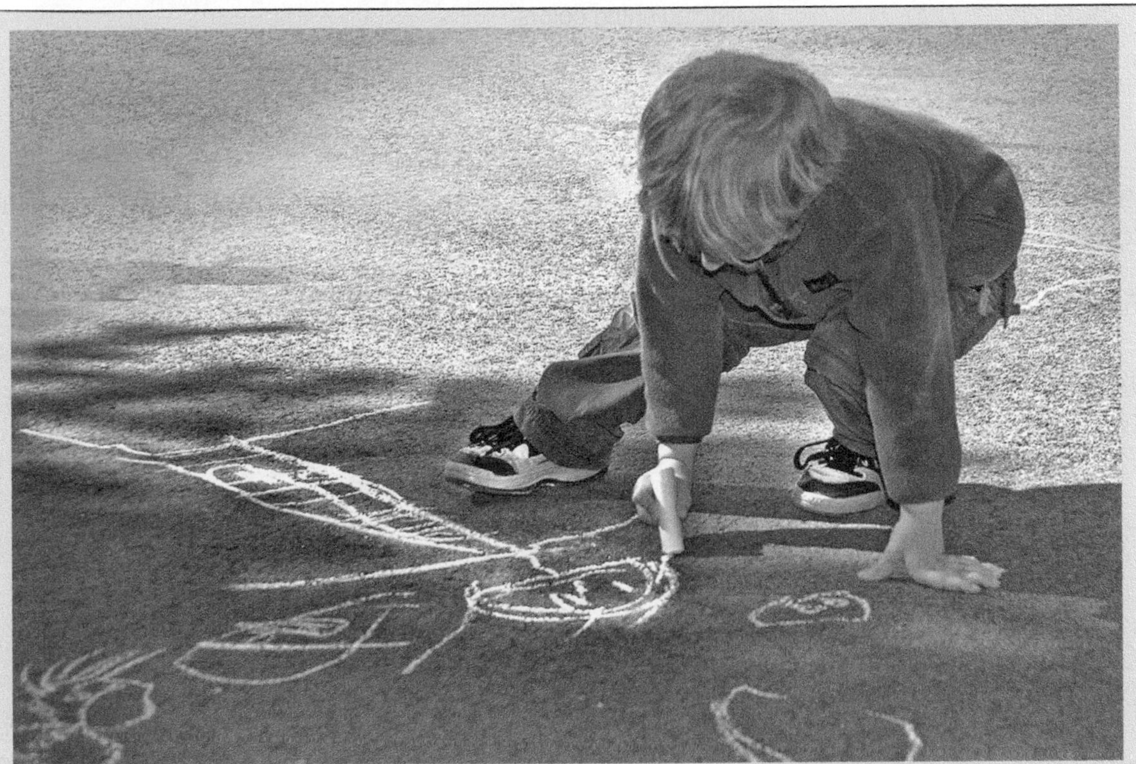

Figure 4.10 (Ursy Potter Photography)
Visually rich environments help children's visual–motor systems mature.

By about this age	***Children can...***
33 to 36 months	sort colors and identify a stick that is 5–8 cm (2–3 in.) longer than another stick.
28 and 35 months	cut with a single "snipping" movement at a time on a drawn line and hold a pencil with an adult-like grasp (Furuno, et. al., 1997).
48 months	draw a person with a head and four features.
54 months	draw a person with a head and six features along with simple pictures of things seen or imagined.
60 months	draw a person with a head and eight features, such as the one drawn above by Zander (Johnson-Martin, et. al., 2004).

Table 4.5 Normal Vision Involves More than Just "Seeing", as Noted in These Definitions	
Visual acuity	Seeing fine details
Near vision	Seeing close objects
Distance vision	Seeing far objects, such as a blackboard
Binocular coordination	Seeing with both eyes
Eye–hand coordination	Using the eyes and hands together
Eye movement coordination	Moving both eyes simultaneously so they work together quickly and accurately, such as when reading
Focusing or accommodation	Maintaining clear vision at different distances
Peripheral vision	"Seeing" objects outside of the central area of vision
Color vision	Distinguishing colors

DESCRIBING VISION AND VISUAL PROBLEMS

The following section describes vision and visual problems, as well as how they relate to function and disability (Figure 4.11).

VISUAL ACUITY

Visual acuity is the ability to see clearly

Parents and caretakers should...

- use health care professionals to evaluate their child's vision.
- monitor their newborn's vision in the hospital nursery.
- make visual development part of periodic "well-baby" checks and before the child begins to read.
- provide infants with balanced sensory environments.

Figure 4.11 (Ursy Potter Photography)
Vision affects posture, in that children (and adults) will often adopt a particular posture to be able to see clearly without even recognizing they are leaning or hunched. Their focus is on being able to see, and they only recognize the posture as fatiguing after they become sore.

Table 4.6 How We Measure Visual Acuity

Optometrists and ophthalmologists typically describe visual acuity as a fraction based on $^{6}/_{6}$ (international) and $^{20}/_{20}$ (in the United States) vision:[a] • The top number in the fraction represents the distance in meters (feet) at which the person can identify an object. • The lower number represents the distance from which a person with normal eyesight could identify the same object. • With the international 6/6 system, a person with 6/10 vision can identify at 6 m what a person with normal vision can see at 10 m. • Using the 20/20 system in the United States, a person with 20/400 vision can identify at 20 ft what a person with normal vision can see at 400 ft. Reduced visual acuity may result from diseases and organic malfunctions and refractive errors.	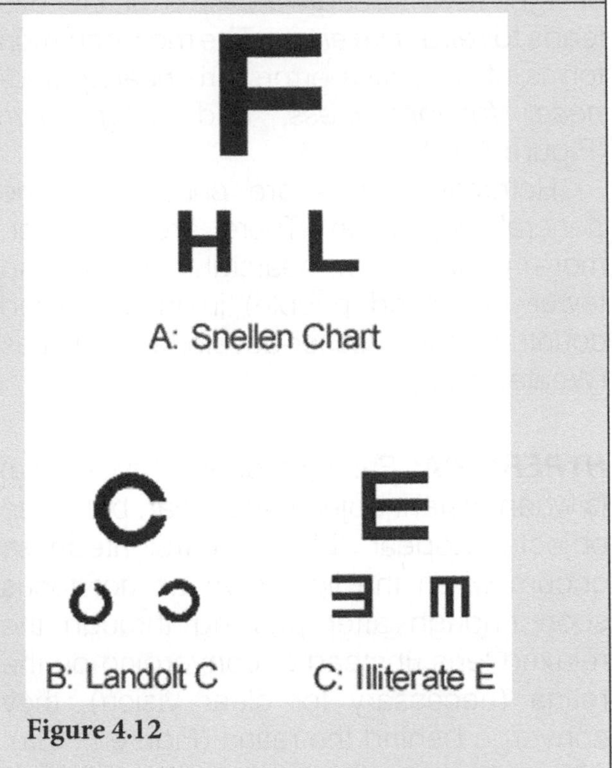 **Figure 4.12** We measure visual acuity with standardized tests. The Snellen, Landolt C, and Illiterate E charts are the most common. (Fostervold, 2003a.)

[a] The fraction is sometimes presented as the resulting decimal digit. An visual acuity of 6/6 (20/20) is then 1.00.

DISEASES AND ORGANIC ANOMALIES: Diseases and organic malfunctions may result from congenital disorders during pregnancy (such as Congenital Rubella Syndrome*), injuries at birth (such as birth-related trauma), or disease and organic anomalies. Blindness or severe visual impairments are usually caused by disease or organic malfunctions, rather than from other causes such as traumatic accidents.

* Congenital Rubella Syndrome is characterized by abnormalities of the heart and nervous system, the eyes, and the ears.

REFRACTIVE ERRORS: Incorrect bending of light rays through the lens of the eye leads to *refractive errors*. The most common forms of refractive errors are *nearsightedness*, *farsightedness*, and *astigmatism* (Figure 4.13).

Refractive errors are common in the general population. There are, however, more people with nearsightedness (and fewer farsighted people) in industrialized countries than in lesser-developed countries (Weale, 2003).

HYPEROPIA: Farsightedness (*hyperopia*) is when distant objects are clear, but close objects appear blurry. Farsightedness occurs when the light rays do not focus soon enough after passing through the relaxed lens. Instead of converging on the retina (necessary for clear vision), they converge behind the retina (Figure 4.13a). Although research is not consistent, it is likely that 20%–40% of preschool children and 7%–15% of juveniles are farsighted (Negrel et al., 2000; Weale, 2003).

MYOPIA: Nearsightedness (*myopia*) is when close objects are in focus, but objects at a distance appear blurry. The light rays focus too soon after passing through the relaxed lens and converge in front of the retina (Figure 4.13b).

About 1%–4% of preschool children and 15%–40% of young adults are nearsighted (Maul et al., 2000; Negrel et al., 2000; Weale, 2003). Concave lenses can usually correct nearsightedness.

Many disagree as to whether nearsightedness is caused by genetics or the environment (Mutti et al., 1996). Most eyecare specialists maintain that children inherit genes that predispose them to visual dysfunctions such as myopia. This suggests that we can only "fix" nearsightedness in children with appropriate glasses.

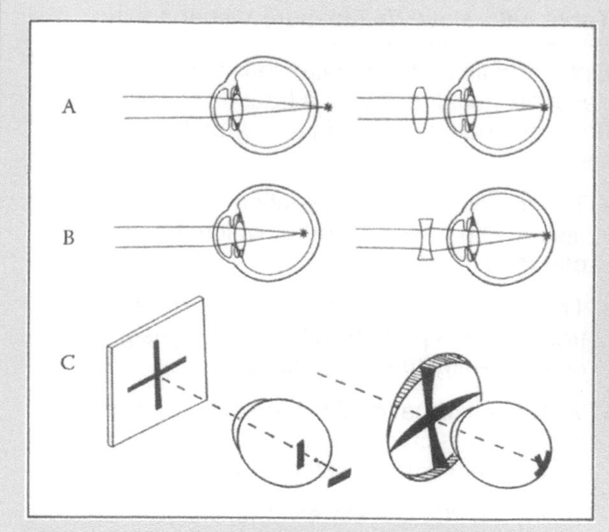

Figure 4.13 (With permission of Lie, 1986.)

Common refractive errors

(a) Farsightedness (*hyperopia*)	• Image falls behind the retina • Corrected with ***plus*** lenses
(b) Nearsightedness (*myopia*)	• Image falls in front of the retina • Corrected with ***minus*** lenses
(c) Astigmatism	• Image lacks a single point of focus • Corrected with ***cylindrical*** lenses

Table 4.7 Diet and Vision: Environment at Work?

Diet influences health. Some people believe nutrition can contribute to nearsightedness.

Foods high in sugar (glycemic foods) promote acute and chronic hyperinsulinaemia (excessively high insulin levels in the blood), which may accelerate scleral tissue growth and lead to a thickening of the lens.

This might contribute to myopia (Cordain et al., 2002; Furushima et al., 1999). However, definitive information linking the two is not yet available.

Figure 4.14 (Valerie Rice)
Children's vision results from genetics as well as their environment.

There is support for this view. For example, nearsightedness among juveniles is largely associated with heredity. Further, identical twins' refractive errors tend to be more similar than nonidentical twins (Teikari et al., 1988). A probable candidate gene coding for high myopia has been identified (Lam et al., 2003).

Others maintain that the environment plays a large role in nearsightedness (see Table 4.7 for example). Near work, higher school achievement, and less time spent playing sports have been found to contribute to nearsightedness (Mutti et al., 2002). The abrupt increase in nearsightedness in most Western countries after World War II seems to support this view (Richler and Bear, 1980; Young et al., 1969).

It is difficult to explain away such dramatic and sudden increases in nearsightedness with genetics alone, unless we assume there is an epidemic of genetic mutations; the gene pool simply does not change that abruptly. The environment, on the other hand, can change rapidly and the pace of change has been immense over the last century.

Both genetic and environmental factors seem to be at work. Most likely, we each inherit genes that specify the upper and lower limits of possible refraction. The environment then influences where (within the genetically determined boundaries) that person's actual refraction will fall (Figure 4.14).

The key is to determine which environmental factors contribute to the development of refractive errors and how we can manipulate them to improve eyesight.

ASTIGMATISM: *Astigmatism* occurs when the surface of the cornea (and in some conditions, the surface of the lens) curves more in one direction. Normally, the cornea curves equally in all directions, resembling a basketball. With *astigmatism*, the shape of the cornea resembles an American football. Instead of individual light rays focusing at one point at the back of the retina, they focus at two different points (Figure 4.13c). People with astigmatism usually experience blurred vision at all distances. For some, blurred vision occurs only when they *accommodate*.

Astigmatism is common. As many as 70% of children under 1 year of age have astigmatism (Weale, 2003). This prevalence decreases and stabilizes during the first 2 years of life. With most age groups, rates of astigmatism range between 18% and 32% (Dobson et al., 2003; Maul et al., 2000; Weale, 2003).

Most astigmatism is inherited, since differences in corneal shape appear to be genetic. Acquired astigmatism may be caused by injury, disease, or changes in the physiology of the eyes. Glasses or surgery can correct nearly all cases of astigmatism. Surgery should be the absolute last solution for children (if nothing else solves the problem).

BINOCULAR VISION

Binocular vision is the ability to see with both eyes. When looking at objects, both eyes must aim at the same place to avoid seeing double images (Table 4.8 and Figure 4.15).

Figure 4.15 (Ursy Potter Photography)
Eye image by K.I. Fostervold (Modified CorelDRAW® clipart.)

Six extraocular muscles control eye movements and direct the eyes so the object appears as a single image.

1. Obliquus Superior
2. Obliquus Inferior
3. Rectus Superior
4. Rectus Inferior
5. Rectus Lateralis
6. Rectus Medialis

Table 4.8 Binocular Vision

Binocular vision involves at least three interrelated factors:

1. *Simultaneous perception*: The brain processes information from both eyes.
   That is not the case for everyone; some people have sustained damage or loss of vision in one eye.
   They suppress information from one eye to avoid double images (see example of strabismus, below).
2. *Sensory fusion*: Images projected on the retina of the two eyes converge into one image.
   People with adequate sensory fusion have better acuity when they use both eyes instead of one eye.
3. *Depth perception*: Adequate depth perception creates a perception of depth or stereopsis. Depth perception is important to assist motor control, such as when reaching for an object or playing football.
   We may experience "false stereopsis" with certain color combinations (e.g., red background and blue text). Using them together on a display may confuse children by creating a perception of depth where there is none.

VERGENCE: Vergence refers to opposing movements of the eyes, as when they converge or diverge from each other to maintain single vision at different viewing distances.

Resting point of vergence When there is nothing to look at (as in total darkness), our eyes naturally converge on a distance called the **"resting point of vergence" (RPV)** (Figure 4.16, Table 4.9).

KEY POINTS:

1. Those with a near *resting point of vergence* can more easily sustain close viewing distances, yet do not experience difficulty viewing farther distances.
2. Our eyes work harder when viewing objects closer than our resting point of vergence (Owens and Wolf-Kelly, 1987).
3. Research on adults shows that the resting point of vergence moves closer as we lower our gaze (Heuer and Owens, 1989) (see Figure 4.16).
4. Children have much closer resting points than adults.

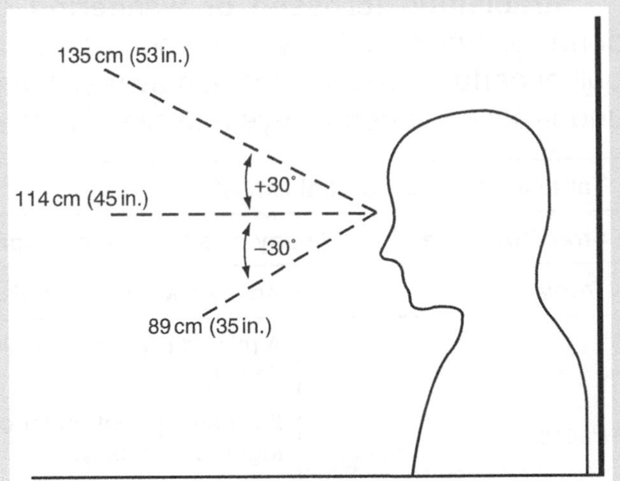

Figure 4.16 We see close images better at lower viewing angles because our visual resting point of vergence naturally moves closer as we lower our gaze.

Table 4.9 Adult and Children's Near Point and Resting Point

	Resting Point of Vergence (RPV)	Near Point of Convergence (NPC)[a]
Adult	• 114 cm (45 in.) from the eyes when looking straight ahead. • 135 cm (53 in.) when looking up 30°. • 89 cm (35 in.) when gazing down 30°.	• An NPC <10 cm (4 in.) is considered normal in young adults.
Young Children (under 8)	• RPVs for children have not been established.	• Less than 5 cm (2 in.). • NPCs greater than 9 cm (3.5 in.) are considered abnormal in children.

[a] The figures presented for near point of convergence, both for adults and especially for children, vary depending on the method used. Such differences in measurement techniques are particularly controversial regarding children's near points.

Near point of convergence (NPC) The near point of convergence (NPC) is the closest viewing distance that a person can focus on an object with both eyes while continuing to see it as one image. If a person views an object closer than this, the object will appear as a double image.

The near point of convergence becomes farther with age. At age 2, the NPC averages 1 cm (0.4 in.). At age 10, the NPC averages 2.5 cm (1 in.) (Chen et al., 2000).

The number of children with farther-than-normal NPCs increases with age. This is important, as a far NPC often causes eye stress and suggests there are underlying problems with binocular vision.

Strabismus (crossed or wandering eye) Strabismus is a common term for the condition where both eyes are not aimed at the same point due to an over-pull or under-pull of certain eye muscles. It may result from misalignments of the eyes or over-action or under-action of certain eye muscles (Table 4.10).

Table 4.10 What is Strabismus?	
Strabismus may include any or all of these conditions:	
Tropia	Another word for *strabismus.*
Phoria	A related condition in which the misalignment is present, but not apparent (latent). It is usually kept under control so that the eyes appear normal and work together normally. However, the misalignment may be unmasked by covering either one of the eyes or may become apparent under prolonged visual stress.
Esotropia/phoria	One eye turns inward and appears crossed (Figure 4.17a).
Exotropia/phoria	One eye turns outward (Figure 4.17b).
Hypertropia/phoria	One eye turns upward (Figure 4.17c).
Hypotropia/phoria	One eye turns downward.
Cyclophoria/phoria	The vertical poles of the cornea are rotated (Figure 4.17d).
Concomitant strabismus	The misdirection of the eye is consistent in all directions of gaze.
Noncommitent strabismus	The angle at which the eye turns away from the line of sight differs in different directions of gaze.

Children who have a latent form of strabismus (*phoria*) usually adjust their extraocular muscles in order to focus both eyes on the same spot. These extraocular muscles, however, fatigue with sustained visual load, leading to symptoms that may include double vision and impaired performance.*

There are two nonsurgical approaches for the remediation of strabismus:

1. Help the child adapt to the problem or
2. Optometric treatment
   Popular optometric treatment methods include the *Optometric Extension Program* (OEP) and the *Measuring and Correcting Methodology* (MKH) after H.-J. Haase.†

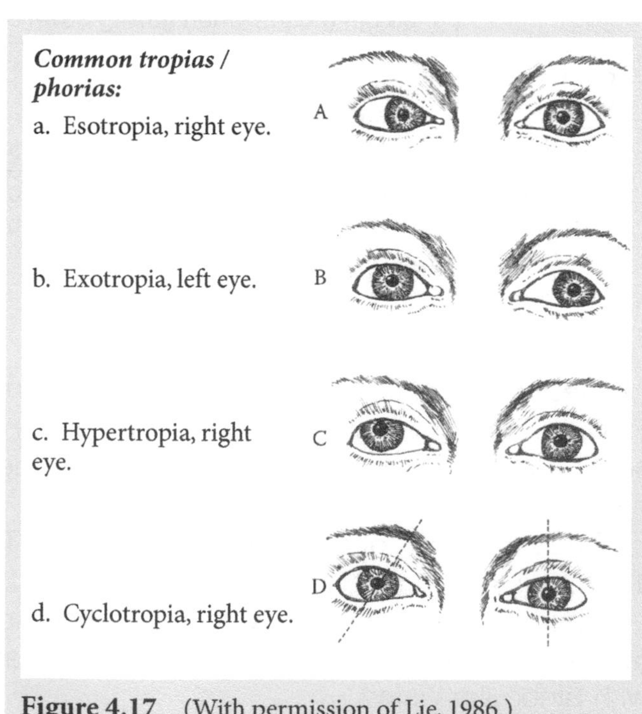

Figure 4.17 (With permission of Lie, 1986.)

* See subsequent section for more on its effect on development and learning.

† (see Tables 4.11 and 4.12).

Table 4.11 Optometric Extension Program[a] (OEP)	
Background	• Developed in the 1920s by optometrist Aleksander Sheffington. • The OEP is fairly widespread in the United States and is known and used in parts of Europe.
The concept	• Visual difficulties result from a misfit between the visual demands of the environment and the person's visual skills. • Children learn these visual skills through a series of motor developmental programs.
The problem	• Prolonged near point stress results in functional adaptations such as evasive behavior, phorias or tropias, and accommodation problems.
The solution	• Visual training, either alone or in addition to glasses. Since vision is a learned skill, the individual needs to relearn the system. • Surgery may be necessary for extreme cases and convergence therapy may be used postsurgery to train the eyes to maintain binocular fixation.
For more information	www.healthy.net/oep www.oep.org

[a] Also referred to as Behavioral Optometry.

Table 4.12 Measuring and Correcting Methodology after H.-J. Haase (MKH)[a]	
Background	• Developed in the 1950s by German professor Hans-Joachim Haase. • The MKH is used mostly in the German speaking parts of Europe. • It is not particularly well known in English speaking countries.
The concept	• The oculomotor system is at rest when the eyes look at distant objects. • The visual system adapts to primary problems such as phorias and farsightedness. However, the continual adaptation for primary problems may eventually lead to secondary visual problems such as visual suppression, reduced vergence capacity, and reduced accommodation.
The problem	• Continual adaptation to primary visual problems puts one at risk for developing secondary visual problems. • For example, intensive visual work at close distances requires sustained muscular effort that increases the demand on the visual system. • The adaptation required to cope with this increased demand increases the danger of developing visual problems.
The solution	• Prescribe glasses with prism corrections for constant wear. The glasses give the eye muscles an opportunity to "relax" and restore the ability to sustain unrestricted or effortless stereoscopic vision (often successive re-corrections are required). • In cases with large phorias, MKH may be used to uncover the actual extent of the misalignment prior to surgery.
For more information	www.ivbv.org (mostly in German) www.healthy.net/oep www.oep.org.

[a] Also referred to as the method of Optical Full Correction.

COLOR VISION

As mentioned previously, cones are more sensitive to color. There are three types of cones on the retina: short, medium, and long wavelength receptors. Each type is sensitive to a wide range of wavelengths, but most sensitive to a narrower region of the color spectrum.

COLOR BLINDNESS: Color blindness is an inaccurate term that refers to the inability to distinguish between certain colors (Figure 4.18, Table 4.13). About 8% of males and 0.5% of females are considered color deficient. Such color deficiencies are usually congenital.*

It is impossible to know what people with color deficiencies actually experience. However, some tell us that they see the world as shades of yellow or red (Sciffman, 2001).

Figure 4.18 (Ursy Potter Photography)
Signs of color deficiencies may include:

1. difficulty in sorting blocks of various colors,
2. consistently drawing familiar objects in unfamiliar colors (e.g., a light green sun), or
3. inability to identify ripe berries.

Children with such symptoms should see a visual specialist.

Table 4.13 Color Deficiencies	
Dichromats	• Dichromats lack one of the three types of cones in part of, or the entire eye. As a result, they are unable to distinguish between certain colors (wavelengths) in that area of the eye. • The most common deficiency is red–green.
Anomalous trichromats	• The person does not lack any cones, but one type of cone demonstrates abnormal absorption of light waves.
Monochromats	• Monochromats lack two or all three types of cones. As a result, they see the world as mostly shades of gray.[a]

[a] The Norwegian scientist Knut Nordby (1990) has published an interesting autobiographical account of his life as a monochromat.

* Present at birth due to heredity or environmental influences.

Color blindness can interfere with life functions. For example, colorblind children may combine clothing colors that subject them to ridicule by their classmates. It can affect performance at computers that display certain colors.

Colorfield Insight's website at www.colorfield.com simulates color blindness to help designers determine if their designs will be legible to people with color deficiencies.

COLOR AND VISUAL ERGONOMICS: Humans can distinguish between more than a thousand different colors when they are placed side-by-side (Sekuler and Blake, 1994). Color can be an effective way to convey information; for example, red often signals danger (Table 4.14).

Too many colors, however, can produce a "Christmas tree" effect, making it difficult to determine which information matters. This is especially important for children, as they lack the training and ability to handle complex information. Children may pay more attention to the attractiveness of a multicolored message than to the content.

Blindness Blindness is believed to affect 50 million people worldwide, including 1.4 million children 15 years of age and younger (SSI, 2003; see Table 4.15 on "blindness"). We cannot be certain of the actual prevalence of blindness among children worldwide because research studies have not been consistent in how they define blindness. Further, we lack an international consensus on how to identify children with visual difficulties.

We do know, however, that the prevalence of blindness varies. In particular, blindness is more common in impoverished areas. Even so, around 57% of cases of childhood blindness are unavoidable and the rates are similar between developed and undeveloped countries. About 28% of childhood blindness is preventable; further, preventable blindness is more prevalent in developing countries.

Table 4.14 Using Color to Convey Information	
Using colors to convey information to children	1. Avoid using too many colors on the same page. 2. If a color conveys a specific type of information, be consistent. Ensure the color always means the same thing throughout the application.
Using color to convey information to children with color deficiencies	1. Avoid relying only on color to provide contrast. 2. Some color contrasts, such as grey and yellow, do not transfer well. Consider changing the intensity of those colors instead. 3. Locate important information in the same place each time it is presented. 4. Use redundant forms of coding to convey important information. For example, the "Don't Walk" signal for pedestrians is on top, in red and often depicts a person standing still.

Low vision In the United States, vision problems affect 1 in 20 preschoolers and 1 in 4 school-aged children. Low vision is thought to affect about 4–5 million children worldwide (PBA, 2003). Table 4.16 lists definitions and some treatments for low vision.

Table 4.15 What Do We Mean by "Blindness?"

Visual acuity	Vision of 3/60 (20/400) or worse in the better eye while wearing corrective lenses (WHO, 2004).
Visual field	A visual field of 10° or less with corrective lenses (180° is considered a normal visual field) (WHO, 2004).
Legal blindness in the United States and most of Europe	Vision of 6/60 (20/200) or worse in the better eye while wearing corrective lenses or a visual field of 20° or less with corrective lenses.
"Functional blindness"	Relying on the sense of touch when writing and reading. Only about 10% of those who are considered blind have no sight whatsoever.
Causes of blindness	
Cataracts	Clouding of the lens of the eye or surrounding transparent membranes that obstruct the passage of light.
Glaucoma	Increased pressure in the eyeball that damages the optic disk and causes a gradual loss of vision.
Trachoma	Corneal infections, irritations, and inflammations that result in corneal scarring.
Vitamin A deficiency	Especially in children under 5.
Causes of avoidable blindness	
Corneal scarring	Scarring of the cornea.
Cataracts	See above.
Glaucoma	See above.

Table 4.16 Definition and Treatments for Low Vision

What is low vision?	• Optimum corrected vision of from $^{6}/_{18}$ ($^{20}/_{60}$) to $^{3}/_{60}$ ($^{20}/_{400}$) in the better eye. • "Best case": From 6 m (20 ft) away, a standard eye chart looks like what someone with normal $^{6}/_{6}$ ($^{20}/_{20}$) vision sees from 18 m (60 ft). • "Worst case": From 3 m (20 ft) away, a standard eye chart looks like what someone with normal $^{6}/_{6}$ ($^{20}/_{20}$) vision sees from 60 m (400 ft) away. A person with eyesight that is any worse than this is considered blind.
Causes	Can be caused by many conditions, including river blindness, corneal scarring, and albinism.
The problem	Many rehabilitation specialists and special education teachers are taught to work with blind children, but not with children with low vision.
Treatments	***Distance aids*** • Closed circuit television • Computer scanners • Image magnifiers ***Magnifying aids*** • Bar prisms • Magnifying glasses • Powerful glasses In poorer countries: plastic drainpipes fitted with an appropriate lens (With this simple aid, over a third of children with low vision can read.) ***Surgery***
Environmental modifications	• Tilted desks to help with close reading. • Provide highly contrasting colors for reading and daily living tasks. • Use larger font sizes for reading materials, including computer screens. • Provide appropriately high light levels. • Provide highly contrasting colors along pathways and steps for increased mobility.

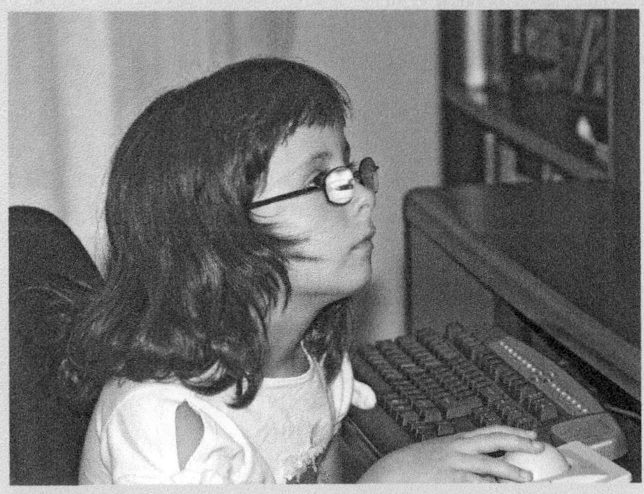

Figure 4.19 (Ursy Potter Photography)
Many children with low vision can benefit from using larger font sizes on the computer screen and higher light levels.

Equal rights Both the *Americans with Disabilities Act* in the United States and the *General Framework for Equal Treatment in Employment and Occupation* in the European Union countries mandate equal rights for visually impaired citizens in most areas of society. To fulfill this goal, society must provide adequate education and leisure activities for visually impaired children in ordinary social settings.

Providing appropriate individual aids and suitable environments is a substantial design challenge (Figure 4.20). The goals of the National Agenda for the Education of Children and Youths with Visual Impairments, Including Those with Multiple Disabilities of the American Association for the Blind can be seen in Table 4.17.

Figure 4.20 (Ursy Potter Photography)
It is difficult for most teachers, who have learned primarily through their own vision, to understand the intricacies of teaching children who are blind or have low vision.

Table 4.17 National Agenda Goals:
• Mechanisms of reading (e.g., Braille, large print, audiotapes, and electronic aids)
• Orientation and mobility aids and training
• Independent living skills (e.g., handling money and finances, working in the kitchen, and transportation services)
• Career education
• Assistive technology (see Chapter 9)
• Visual efficiency skills
• Social interaction
• Recreation and leisure

VISUAL WORK

NEAR WORK AND MYOPIA

The concern about "ruining your eyes" from excessive close work is older than television, and it is not only our mothers who have expressed this concern.

Sitting hurts the loins, staring, your eyes.
(Plautus, Roman Playwright*)

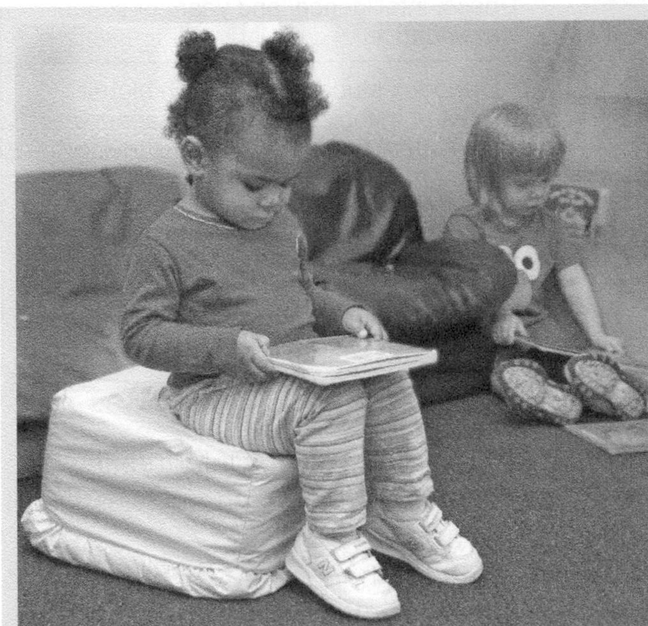

Figure 4.21 (Ursy Potter Photography)

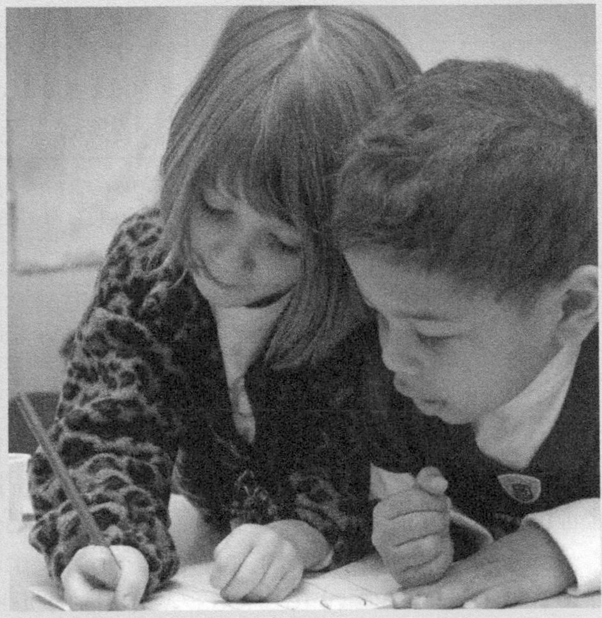

Figure 4.22 (Ursy Potter Photography)

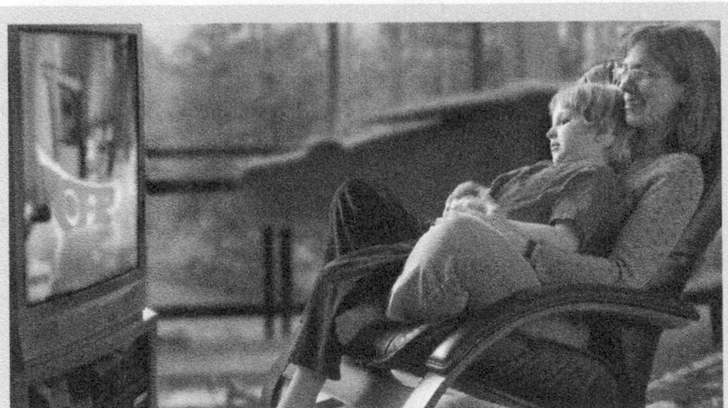

Figure 4.23 (Ursy Potter Photography)
We are still learning about how nearsightedness among children affects their ability to sustain close viewing such as when reading, writing, doing computer work, playing video games, and watching television.

* In 1713, Bernardino Ramazzine (1713/1983) thought that constant glaring at small objects had an impact on eyesight, with inevitable outcomes of myopia and eventual blindness.

Table 4.18 Near Work and Myopia

Author (date)	Subjects and Study	Findings
Young et al. (1969)	570 Alaskan Eskimos Studied 3 generations	Older Eskimos with less classroom education were less nearsighted (more farsighted). • 3.5% grandparents nearsighted • 10.4% parents nearsighted • 58.6% children nearsighted
Rosner and Belkin (1987)	157,748 male recruits in the Israeli Army	Strong association between nearsightedness, intelligence, and years of school attendance.
Zylbermann et al. (1993)	830 Orthodox and Secular Jewish students	Orthodox students had much higher rates of myopia, apparently because they spent much more time studying.
Kinge et al. (2000)	223 engineering students participated in a 3 year longitudinal refraction study	Dose–response relationship between time performing near work tasks and visual changes associated with myopia.
Research findings suggest that prolonged near work contributes to myopia. Further, animal studies repeatedly demonstrate that continuous near vision, especially when young, induces myopia.		

Effective prevention requires an understanding of cause and effect. Unfortunately, we do not understand all the factors that lead to myopia. Table 4.19 lists characteristics of children who seem to be prone to becoming nearsighted.

The early stages of nearsightedness may include certain changes in visual functions that predict nearsightedness. We do not know, however, whether these characteristics cause or are only correlated with nearsightedness.

Table 4.19 Predictors of Nearsightedness in Children

Author	Present (predictive)	Not Present (not predictive)
Mutti et al. (2002)		• Refraction in infancy • Refraction at school entry • Parental history of myopia
Drobe and de Saint-Andre (1995)	• Loss of physiological farsightedness • Tend to squint inward • Less able to shift visual distances (accommodation) • Near and far comfort zones overlap	

Table 4.20 Progressive Lenses to Prevent Myopia		
Author	**Subjects**	**Results**
Leung and Brown (1999)	Hong Kong schoolchildren • 36 with progressive lenses • 32 with single lenses	Progressive lenses reduced incidence of myopia.
Edwards et al., 2002	Chinese children (aged 7–10.5) • 138 with progressive lenses. • Control group of 160	Progressive lenses did not reduce incidence of myopia.

It is difficult to evaluate interventions aimed at preventing myopia. This is because myopia develops over a long period of time, and it is impossible to control for all the variables.

One approach postulates that progressive lenses can lessen the development of myopia. Inconclusive research results (Table 4.20) have led the National Eye Institute to conclude that the benefits are insufficient to warrant the use of progressive lenses by children.*

Other proposed preventive measures include (Choo, 2003; Saw et al., 2002):

- visual training,
- intermittently stopping to view distant objects while reading,
- removing distance eyewear during sustained near tasks,
- bifocals and contact lenses,†
- special purpose eyewear,‡
- eye drops that block accommodation,
- biofeedback and traditional Chinese treatments.

Although some report positive effects, most of the evidence is anecdotal and inconclusive.

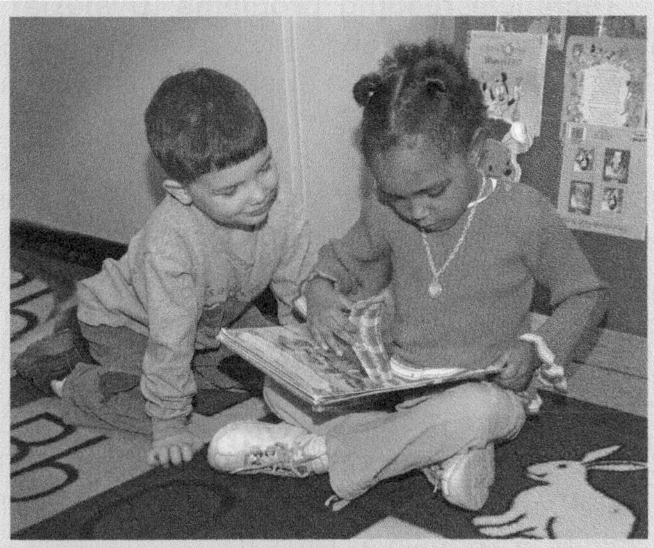

Figure 4.24 (Ursy Potter Photography)
Reading becomes increasingly important as a child nears adulthood. In order to obtain a driver's license, adults must be able to read road signs, instruction books, and driving tests. To safely take medicines, children must read medicine bottles, warning labels, and information sheets. In both developed and undeveloped countries, one's reading ability can be the difference between earning a living wage and poverty.§

* The National Eye Institute's position on progressive lenses for children can be seen at www.nei.nih.gov/news/statements/comet.htm

† Particularly rigid and oxygen-permeable contact lenses.

‡ For example, eyewear that alters the viewed image or that reduces the need for accommodation.

§ The United Nations uses literacy rates as a primary indicator of social and economic development.

READING

Reading is a critical milestone in a child's path toward independence. Reading is necessary to play most computer and video games. Even in first grade, students read to understand written instructions and to do their homework assignments.

READING PROBLEMS: Public education reduces illiteracy. Some professionals, however, are concerned that many children (although not illiterate) do not read as well as they should for their age and level of education. There is some evidence that visual and ocular deficits among children impair their ability to read (Simons and Grisham, 1987). This issue is controversial; some specialists maintain that ocular deficits do not play a role in reading difficulties as long as a child can see writing on the page.

Although individuals with low vision usually find it difficult to read, reading depends on a variety of interactions between our visual, cognitive, and motor systems. Impairments in acuity, contrast sensitivity, and visual field status (particularly loss of the central visual field) affect reading performance, but do not accurately predict reading ability. Visual factors can affect both technical and endurance reading problems. It is important to distinguish between technical reading problems (what we often term dyslexia) and endurance problems.

Technical reading problems (Dyslexia) Children who have a technical reading difficulty find it difficult to sight-read. They cannot translate letters into meaningful words. Developmental dyslexia exists when, despite normal training, these difficulties persist in a child of normal intelligence.

We do not know what causes dyslexia. Some, however, attribute dyslexia to brain dysfunctions, phonological deficits,* and visual perception disorders. Phonological deficits are the most likely contributor and, in comparison, some consider vision relatively unimportant (Hulme, 1988).

Visual aspects are receiving increasing interest.† For example, malfunctions in the part of the brain that processes visual information (the magnocellular layer of the lateral geniculate nucleus (LGN) may cause reading problems (Cestnick and Coltheart, 1999).

Endurance problems Some children are able to read, but experience visual fatigue quickly. They cannot read accurately for very long and experience a variety of symptoms (Table 4.21).

Although no one knows why some children have difficulty with endurance, possible explanations include the following:

- The historical shift from outdoor scanning activities to intensive visual work at close range has resulted in less dynamic and more static visual activity.
- Static visual activity for prolonged durations can lead to short-term pain and contribute to muscle fatigue of the oculomotor system.

Table 4.21 Signs of Endurance Problems
• Eye fatigue
• "Shivering" text
• "Jumping letters"
• Orbital aching
• Tearing
• Dry eyes
• Headache
• Neck aches
• Nausea
• Light-headedness
• Difficulty in concentration
• Intermittent blurring or double vision or both

* Phonological deficits include the inability to distinguish individual sounds or syllables and difficulties with sounds-symbol relations, as well as the storage and retrieval or sound-based information from memory. See Chapter 5 for more on this topic.

† The many websites dedicated to vision therapy underscore the new interest in vision problems and how they affect children's ability to learn.

- It is also possible that prolonged static "work" by the oculomotor system will lead to impaired visual function (Lie, 1989; Pestalozzi, 1993; Wulff, 1998).

The relationship between technical and endurance problems Reading remediation aims to help children master the technical aspects of reading. As children with endurance problems read well for short periods, teachers may not correctly identify their problem. Instead, they may label them as lacking motivation, concentration, or as having behavioral problems. Although phonological difficulties contribute to technical reading problems, visual or ocular factors may dominate in endurance difficulties. Because many children have both technical and endurance difficulties, helping them read requires evaluation and treatment of both vision and phonological awareness.

Figure 4.25 (Valerie Rice)
Children with visual endurance problems do not know what is wrong. They only know it is uncomfortable to read for long periods. They question their own intelligence and become discouraged by their inability to learn.

Unfortunately, many children become lost in the educational system when parents and teachers fail to recognize the visual component of their reading difficulties. The children themselves do not understand why they cannot read as well as their classmates.

Teachers and parents should look for these symptoms.

Teachers, caregivers, and healthcare professionals must look for symptoms of reading problems, such as lagging behind in reading, complaints of fatigue, and discomfort or boredom while reading. If a child has a problem, it must be determined whether it is technical or endurance (or both). In any case, children with symptoms should be examined by a pediatric optometrist or ophthalmologist who can identify the cause of their vision difficulties.

If the problem is technical, the child will need a reading specialist who treats dyslexia. If the problem is endurance, the child will need to see an optometrist or ophthalmologist who specializes in the treatment of near point visual stress, along with a therapist who specializes in visual therapy.*

The main challenge in reading endurance remediation programs is how to maintain stereoscopic vision without stressful symptoms. Both the OEP and MKH systems have optometric correcting procedures that promote visual near work while reducing stress.

* If no specialist is available, make the optometrist aware of the symptoms. They should take special notice of binocular problems and problems with near-vision. An examination by a developmental pediatrician can help rule out other diagnoses that could interfere with reading.

Table 4.22 Meares–Irlen Syndrome (Scotopic Sensitivity)

Vision specialists evaluate children for Meares–Irlen Syndrome (scotopic sensitivity). With Meares–Irlen Syndrome, printed text appears distorted. There is less distortion when the text or the background has a particular color.

Symptoms include:

1. Reading problems
2. Visual and ocular problems (particularly involving near-vision)
3. Visually related symptoms
4. Additional symptoms:
   a. Words appear to "move," "jumble," or "fall off the page"
   b. The page appears too bright or the words too close together
   c. They only read in dim lighting

Treatment may include:

a. Colored filters to alleviate difficulties. This remedy is controversial. It may help to try colored filters when children underachieve in reading and report visual problems.
b. The theory is that the retinal receptors may be over-stimulated by specific wavelengths of light, causing distortions and that a colored lens can block out the troublesome wavelengths.

Recommendations for children with reading endurance problems:

1. Large text
2. Short, well-spaced lines
3. Frequent breaks
4. Proper lighting

COMPUTERS

Media use among children varies from country to country, regarding both the type of media and the amount of use (Roe, 2000). However, large numbers of children in developed countries clearly use computers on a regular basis.

Children are a new and lucrative market for technology. New hardware, software, accessory technologies, and video games are arriving on the market at an ever-increasing speed. Marketing strategies for technology increasingly include parents of toddlers and infants. When the marketing is directed toward the parents, the message is that anyone who "deprives" their child of technological products retards their development. What parent would actively choose not to provide their child with a head start in life? Yet, the question of the influence of excessive near work and near point stimulation on child development in general and visual development in particular, remains. Parents (and teachers) continue to struggle with the question of what is best for their children (Table 4.23).

Table 4.23 Home Computers

Most popular child activities on the computer (Mumtaz, 2001)	• Playing games • Entertainment • Chatting
Parent's reasons for purchase (Sutherland et al., 2000)	• To help children learn in school • To compensate for perceived shortages of up-to-date school technologies • To help children prepare for technologically rich futures in the world of work and leisure • To keep up with "other families" • To explore technologies • To support parents' work or study • To encourage children to stay at home • For entertainment (computer games)

Although the concern over the impact of visual displays on children's development remains, a moratorium on children using computers simply would not work. Computers are already an inevitable part of people's lives, especially in developed countries. A better question is how ergonomics can help improve the "fit" between the child, the computer equipment, and the tasks (see Chapters 16 and 22).

VISION AND THE COMPUTER: Although both reading books and work on computers require close vision, computer work has a greater effect on eyestrain. One reason is that people work longer hours on computers and take fewer breaks.

Other factors that contribute to eyestrain include monitor location, screen and text legibility, screen flicker and lighting, and glare. Spending a long time deciphering hard-to-read screens causes visual fatigue (Collins et al., 1990) (Table 4.24). Table 4.25 provides suggestions to improve screen quality and Table 4.26 describes differences in computer monitors.

Figure 4.26 (Ursy Potter Photography)
Children of all ages are attracted to the interactive and immediate nature of working on computers.

Table 4.24 Children and Visual Fatigue	
Symptoms of visual fatigue	• Blurred vision • Dry eyes • Burning eyes • Watery eyes • Red eyes • Double vision • Headaches • Reduced effectiveness • Mistakes
How a child might show visual fatigue	• Rubbing eyes • Squinting • Excessive blinking • Tilting head back or to the side • Leaning close to the screen • Irritability while working on the computer

SCREEN QUALITY:

Table 4.25 Suggestions to Improve Screen Quality	
Improve legibility	1. Character size 2. Brightness 3. Contrast 4. Resolution 5. Stability 6. Color and polarity[a] 7. Viewing distance and angle 8. Cleanliness of the screen 9. Glare
Improve brightness contrast	1. Ensure children know how to and are able to easily adjust brightness and contrast controls on the computer screen. 2. Use picture-based tutorials that children and adults can use together. These should explain how and when to adjust brightness and contrast. 3. Dark characters with a light background (similar to the printed page) are much preferred to light characters with a dark background.

[a] ***Positive polarity*** (*negative contrast*) presents dark text against a light background (e.g., the printed page). ***Negative polarity*** (*positive contrast*) presents light characters on a dark background (e.g., the negative of a photograph).

Table 4.26 How Do Computer Monitor Screens Differ?	
CRTs (cathode ray tube)	• Used in standard monitor screens • Basically the same technology used in television sets[a]
Flicker	• The screen may appear to flicker if the refresh rate is too low. • Flicker can cause headaches and visual fatigue. • A refresh rate of at least 85 Hz is recommended to avoid flicker.
Misconvergence	• Shadows that appear around text and graphics due to improper convergence of the electronic beams.
Color variations	• Color may vary across the screen.
Contrast	• Generally, the contrast of VDU text is lower than that of text on paper.
Flat panel screens[b]	• These monitors eliminate flicker, increase contrast, and reduce misconvergence[c] and color variations across the screen.

[a] The main components are a cathode that produces the electronic beam and a glass tube. The image is created when electronic beams strike the fluorescent layer at the back of the screen. The phosphor reacts by transforming electronic energy to light. The image is made visible by light emitted by the phosphor. In raster scanning, the most common method for generating images on a CRT, an electronic beam is projected on the screen in horizontal lines, starting from the top of the monitor. The number of complete screens drawn per second is the refresh rate, expressed in hertz (Hz). A typical VDT displays between 500 and 1000 scan lines with refresh rates that vary from 50 to 100 Hz.

[b] The image on an LCD (liquid crystal display) flat panel is created by passing white light through a layer of liquid crystals sandwiched between two layers of polarized material. The crystals react to electric current in much the same way as a conventional shutter, either allowing light to pass through or blocking it. The red, green, and blue elements of a pixel are achieved through simple filtering of the white light. This technology eliminates flicker, increases contrast, and reduces misconvergence and color variations across the screen.

[c] Misconvergence is the appearance of individual red, green, and blue lines (horizontal or vertical) where there should be one single-colored line.

MONITOR LOCATION: Inappropriate monitor positions contribute to eyestrain. The placement of many monitors—high and close to the eyes—contradict the capabilities of the visual system.

Horizontal monitor placement—viewing distance The stress of converging causes your eyes to work harder when viewing close monitors than when looking off into the distance. You can verify this for yourself by holding your finger in front of you at arm's length; bring it slowly towards your nose, following it with your eyes. Notice that the closer your finger comes, the more eyestrain you feel.

The closer the viewing distance, the more the eyes must converge or turn inward. Viewing distances that are too close lead to eyestrain.

Naturally, if it is stressful to view the screen up close, one solution is to place it farther away. But, how far is far enough? As noted in the section "Resting Point of Vergence," the eyes have a *resting point of vergence* (*RPV*) that represents the distance the eyes most easily converge.

Viewing a monitor substantially closer than the RPV causes eyestrain. However, RPV distances are not absolute; we only need to keep in mind that it is better to avoid close distances and that farther viewing is better (at least up to the RPV). Viewing objects farther than the RPV has not been found to cause any problems.

Figure 4.27 (Ursy Potter Photography)
Video games and other popular electronic equipment are seldom scaled for children. High screen positions, which are not uncommon, contradict the capabilities of the visual system.

Although some guidelines specify a range (maximum and minimum) of acceptable viewing distances, for instance 61–76 cm (24–30 in.), there is no physiological basis to limit the maximum viewing distance. Screen and character sizes, however, often restrict the options—a far viewing distance does no good if one cannot read the screen.

As a rule, viewing distance should never be closer than 65 cm (25 in.). If the character height restricts the viewing distance to closer than that, increase the zoom level so the child can easily read the screen. Never discourage farther viewing distances when the child can easily read screen text.

Many people suggest placing the monitor at arm's length. The authors know of no studies that have found an association between a person's arm length and visual function, nor would one be remotely logical. This is an especially inappropriate recommendation to follow for smaller children.

The type of text also plays a role. Small, narrow characters are more difficult to see than large, thick text. Small text forces the user either to bring the screen closer or to lean forward. Children, caregivers, and manufacturers should use thicker, larger fonts and images to allow for greater viewing distances.

Vertical monitor placement Many guidelines recommend locating the top of the monitor at eye level or even above eye level. That location, however, does not match the capabilities of the visual system.

As we mentioned above, the effort of converging the eyes can contribute to eyestrain. Lowering the computer screen reduces the demand on convergence. As the eyes gaze down, the effort to converge becomes easier. See Table 4.27 for a way to demonstrate this for yourself.

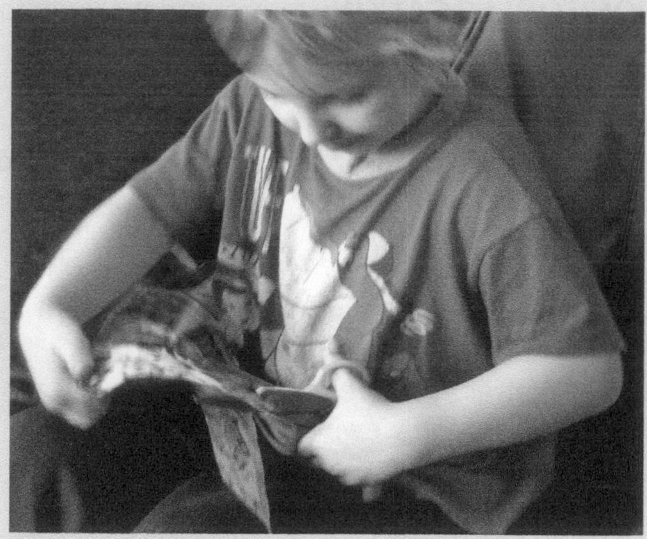

Figure 4.28 (Ursy Potter Photography)
Through evolution, the visual system has primarily been optimized for far-distance viewing at high gaze angles and near-distance viewing at low gaze angles.

The average RPV is 114 cm (45 in.) when looking straight ahead, but when looking down at about 35° (such as when reading a book) the RPV moves in to about 89 cm (35 in.). The closer in you view something from your RPV, the harder your eyes work (Owens and Wolf-Kelly, 1987). Therefore, a downward gaze angle makes the same physical viewing distance less stressful to the visual system. The ability to accommodate (focus sharply on near objects) also improves with a downward gaze. See Table 4.28 for a way to demonstrate this.

Low screen-locations avoid neck discomfort, improve eye–hand coordination, and benefit the visual system (Lie and Fostervold, 1995; Fostervold, 2003b). When using a screen at eye level, viewers bend their necks backward from neutral (Ankrum and Nemeth, 2000). Lower monitor positions, on the other hand, allow users to adopt a wide variety of nonstressful neck postures (Ankrum and Nemeth, 1995). When performing tasks that require eye–hand coordination with a mouse, accuracy improves with a lower screen (Leoni et al., 1994). Although, again, these results are from adults, they should apply equally to children.

Table 4.27 Try it: How lowering your gaze helps your eyes converge on an image
• With your head erect, hold a pencil at arm's length and well above your eye level.
• Raise your eyes and follow the pencil as you bring it closer to your eyes.
• Note how close you can bring it before it becomes a double image.
• Also, note how much discomfort you feel.
• Next, try the same exercise at eye level.
• Finally, bring the pencil in from below your chin.
• Which gaze angle allowed you to bring the pencil closest before it became a double image?
• Which gaze angle caused the least discomfort?
• Lowering your gaze has decreased the demand on convergence.

There does not appear to be any scientific evidence on the visual system that supports placing computers at eye level or higher (Straker and Mekhora, 2000). In fact, people who work with lower computer screen placements have better accommodative flexibility and convergence than do people who work with their monitors placed higher (top edge of the computer screen even with the eyes or higher) (Fostervold et al., 2006).

How low should we set the screen? The top of the screen should be at least 15° below horizontal eye level (Ankrum and Nemeth, 1995). However, it should not be so low that the child cannot easily see the bottom of the monitor in a variety of postures, including reclining.

MONITOR TILT: Notice how you are holding this book. Most likely, you have it set low and tilted away at the top.

Now, while you are reading, rotate the book so that the top comes closer to your eyes than the bottom. Keep rotating.

The more you rotate the top towards you, the more uncomfortable it becomes to read. Some people tilt a monitor down (so the top comes toward the user) to avoid glare, however this can cause that same type of discomfort.

Table 4.28 Try it: Lowering your gaze improves accommodation

- Hold a business card at arms' length in front of your eyes.
- Bring it toward you until the letters just begin to blur.
- Without moving your head, gradually lower the business card in an arc, at the same distance from your eyes.

The letters get sharper!

The eyes have improved their ability to accommodate by looking downward.

Tilting the monitor down (so that the top of the screen is closer to the eyes than the bottom) can increase visual and postural discomfort, especially for the neck (Ankrum and Nemeth, 1995). Rather than tilting the screen down to avoid glare, the source of glare must be eliminated.

It is difficult to place traditional monitor screens low and tipped back on standard desks. You can place flat panel displays low enough, and their thin profile allows them to tilt back sufficiently. However, the pedestals on many flat panels do not allow them to be located low enough to conform to ISO 9241 part 5. Flat panels with removable pedestals and 75/100 VESA-compatible mounts (VESA, 1999) should be specified to allow for appropriate screen positions (Ankrum, 2001).

Why it works: The horopter

Loosely translated, *horopter* means "horizon of vision."

The horopter is the locus, or grouping, of points in our field of view that appear to us as single images. These points appear as double images if seen anywhere else in space.

Why does this happen? When we look out at the world, objects in the upper part of our peripheral vision are generally farther away than the point we are looking at. Objects in the lower part of our peripheral vision are usually closer.

For example, if you focus on a spot on the ground, what is below the spot will usually be closer to you and what is above the spot will usually be farther away.

In response to the environment, the visual system has evolved to perform best when the visual plane tilts away from us at the top (Ankrum, et al., 1995).

People feel more comfortable viewing items that do not "contradict" the horopter. It is especially uncomfortable to look at low computer screens (below eye level) without tilting it back. It is easier and visually more comfortable to look at that same screen after tilting it back so the top edge is farther from the eye than the bottom edge (Ankrum, et al., 1995).

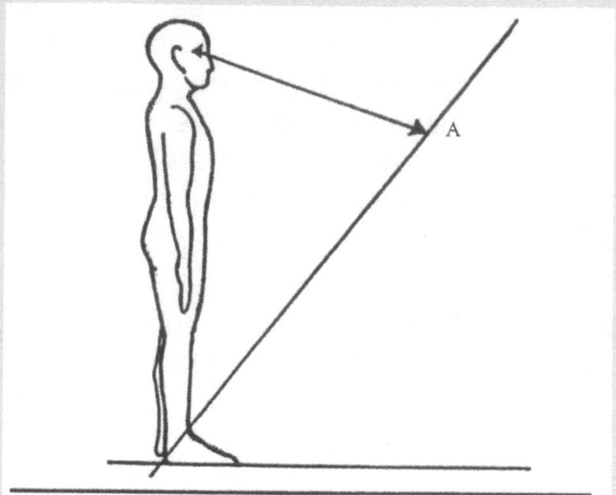

The vertical horopter for fixation at point A.

Figure 4.29

If reading material is flat on a horizontal surface, it is tipped back much farther than the horopter. However, when the material is on a slope, it more closely approximates the horopter. (Ankrum, et al., 1995.)

Table 4.29 Try it: Find your vertical horopter
• The vertical horopter is spatially located at the midline, in front of the observer and forms a straight line passing through the feet with the top inclined outwards.
• Hold a piece of rope vertically at arm's length. Look at the center of rope.
• Both ends of the line will appear as double images in your peripheral vision. When you tilt the rope backward (with its top farther away), the ends will appear as single images.

Note: The horizontal horopter is curved with the sides coming closer to the observer.

Table 4.30 Guidelines for Parents, Manufacturers, and Child Users	
Parent or teacher	1. For each degree that the center of the screen is below eye level, tip the screen back 1° from vertical. The top of the monitor should not be closer to the eyes than the bottom. (If you need a measuring tape, it is close enough.) 2. For written work, supply desk surfaces that slope upward (front to back) approximately 20° (use slope boards if sloping desks are not available). 3. Teach children about proper sitting postures and viewing. These include appropriate viewing distances, glare, fonts, and how to recognize possible causes of their discomfort. 4. Encourage children to take frequent breaks. Provide them with specific activities to do during those breaks, especially when they are intensely involved in playing games. 5. Place the top of the computer screen at least 15° below the child's eye level. If this is not possible, make sure that the child does not have to bend his or her neck back. 6. If more than one child uses the computer, ensure that the seat, computer screen, or both adjust in height. Teach the child how to make the appropriate adjustments.
Manufacturers	1. Ensure computer screens permit tilt up to 35°. 2. Make flat panels VESA compliant. Allow removable pedestals or pedestals that allow the bottom of the housing to touch the work surface and tilt back. 3. Provide guidance to avoid glare from ceiling lights when tilting the monitor. a. Indirect lighting b. Positive polarity screen (dark letters on a light background) c. Supplemental task lighting (to reduce the need for overhead lighting)
Children at computers	1. Place the computer screen at least 65 cm (25 in.) from the eyes. Farther is even better as long as the screen is legible. Increase zoom level if necessary. 2. Take frequent breaks. 3. Take short vision breaks by looking out in the distance. 4. Adjust the screen brightness so the eyes are comfortable; readjust throughout the day if the lighting changes.

SCHOOLWORK AND HOMEWORK: Designers should consider children's vision when designing physical environments. This is often not the case; for example, a proposed Dutch standard for school furniture did not consider eye height (Molenbroek et al., 2003).

VISION—A PREMISE FOR POSTURE

Vision is the main source of information for normally sighted children and adults. People of all ages adapt postures so they can see what they want.

Children move into positions that enable them to see, often staying in uncomfortable, hunched, and craned postures for long periods of time, despite their discomfort. Such postures may become habitual and affect growth patterns. Productivity and comfort improve when children can see easily, their bodies have support and their eyes are not stressed.

Inappropriate postures may also lead to other problems. For example, Table 4.31 describes what happens when a child adapts less desirable postures to accommodate visual demands.

Design the layout of furniture and equipment to support children's activities. For example, position equipment so that it is easy to see and use.

Some researchers maintain that changes in neck and trunk muscle tension (such as when children sit in postures that require close viewing) can contribute to myopia (near sightedness), hyperopia (farsightedness), esophoria (inward squint), and exophoria (outward squint) (Birnbaum, 1993).

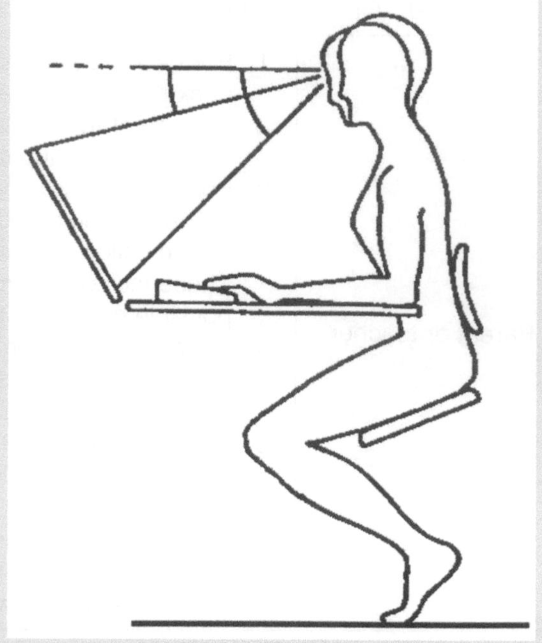

Figure 4.30 When lowering the screen, it is necessary to tilt the top of the monitor (away from the eyes) to ensure an angle of at least 90° between the line-of-sight and the screen. (Ergonomidesign AS. With permission.)

Table 4.31 How Our Tasks Affect How We Sit and How We "See"

Sitting posture	Adaptive posture	Postural problem	Vision problem
Leaning forward when writing	Extreme neck flexion with the face nearly parallel to the flat work surface	Neck and upper back tension and fatigue	Shorter viewing distances, possibly leading to reduced visual acuity and accommodative flexibility (Marumoto et al., 1998)
Sitting upright to view a high monitor	Increased neck extension and higher gaze	Neck, shoulder, and back tension	Maintaining static postures with a higher gaze
Reclining to view a high monitor	Buttocks slid forward toward front of seat. May tilt head back to view screen	No lumbar support. Possible back pain	Gazing upward creates greater accommodative demand

SLOPED DESKS: The visual system functions in a world that slants away (recall the section on the vertical horopter) and is more comfortable viewing computer screens that tilt away at the top. The same is true with desks.

Adults who use desks that slope about 20° sit straighter and flex their necks less (Bridger, 1988). We expect the findings and the principles of sloping desks also apply to children. Binocular vision is better as both eyes are approximately the same distance from the task. Fine-motor coordination improves as the task is brought closer to the eyes and within a comfortable "working envelope" for the child (Birnbaum, 1993). Finally, children prefer desks that slope (Agaard-Hansen and Storr-Paulson, 1995; Callan and Galer, 1984).

VISION AND LIGHTING

Most children accept inferior lighting conditions as long they can see what they are doing. They may not know how to improve their environment, or even that it can be improved.

Because they are less likely to complain than adults, many caregivers and teachers assume that lighting is not a problem for children. However, lighting conditions that strain adult vision also strain children's vision.

Since a child's visual development occurs, in part, in response to visual demands, proper lighting is perhaps even more critical for children. Lighting can positively or negatively influence their visual development. In fact, future designs may help children's visual systems "learn" functions we once considered part of "normal development." For example, varying proportions of direct and indirect lighting could create luminous environments that increase the sensation of depth in kindergarteners (Nersveen, 1999).

HOW MUCH LIGHT IS RIGHT? More is not always better; different tasks require different levels of light. Because we do not know children's preferences in lighting, most designs assume that what is good for adults is also good for children.* For example, School and University Lighting standard ANSI-IESNA RP-3 (IESNA, 2000) does not mention possible differences in children's visual systems.

LIGHTING AND CHILDREN: COMMONLY ASKED QUESTIONS:

1. Does lighting affect behavior?

For years, lighting designers assumed that increasing light levels would improve performance. However, as long as light levels are within normal limits, simply changing illumination will not affect reading performance (Veitch, 1990). In most cases, the primary objective of lighting is to make objects visible. The visual system can function in a variety of well and dimly lit environments.

* Current recommendations for adults range between 500 to 1000 lux (Kroemer & Grandjean, 1997), preferences can be lower (Veitch & Newsham, 2000) or higher (Kroemer & Grandjean, 1997). Preferences can even exceed levels found to be comfortable (above 1000 lx) (Kroemer & Grandjean, 1997).

2. Will a nightlight affect a child's vision?
Depriving a baby chicken (chick) of form vision (sharp contrasts in the visual field) causes the axial length of their eyes to lengthen and causes abnormal refraction (Papastergiou et al., 1998). Some researchers compare keeping a light on for humans with deprivation of form vision as in chicks. However, no association was found between the prevalence of myopia and sleeping with nightlights in children under 2 years of age (Saw et al., 2001).

3. Infants love to stare at bright lights. Will glare harm their vision?
When glare is irritating for adults and children, they can move away to get relief. Infants cannot move away as easily. On the contrary, young children seek sources of light.

Designers should avoid lighting with mirror reflectors in maternity wards and other areas where often there are infants. Side window shades can reduce glare in the backseats of cars. Indirect lighting can reduce glare in areas of high use by children. Since the design and installation of most lighting considers only adult requirements, designers may need to "think again" to avoid glare for children.

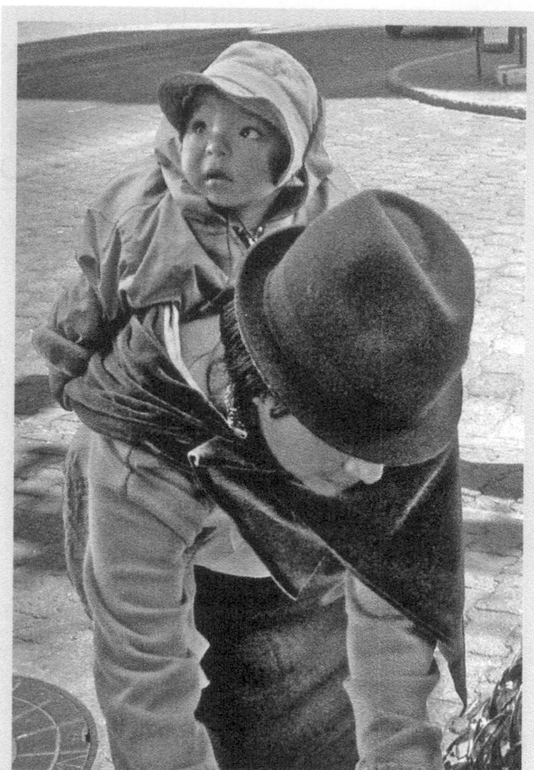

Figure 4.31 (Ursy Potter Photography) Infants seek sources of bright light as early as the first week of life, perhaps because they represent interesting contrasts in the environment. Although no studies with infants directly implicate short-term exposure to glare with long-term harmful effects, it would be wise to eliminate glare whenever possible.

4. How do lights that "flicker" affect infant's vision?
Infants are vulnerable to flicker. The larger and brighter the flickering light, the easier it is to perceive flicker. When using indirect lighting, the light reflected from the ceiling must be extremely bright in order to distribute sufficient light to the room.

Infants lying on their backs looking at the ceiling cannot avoid the resulting flicker. Flicker-free lighting systems, with electronic ballasts and a combination of indirect or direct lighting are preferable in maternity wards and other locations of high use by infants and small children.

5. When can children accurately judge distance and speed?
Many people consider poor judgment to be a sign of a lack of psychological maturity. Although that is true in some areas, physiological factors can play a role. The dorsal and ventral streams are two parallel visual pathways in the brain.*

* The dorsal stream plays a critical role in processing motion and the planning and execution of actions. The ventral stream processes information about objects, features, and relationships.

These two systems do not work separately until sometime between 5 and 12 years of age. The ability to perceive distance reaches maturity later than the perception of location (Rival et al., 2004). This lack of separation may explain some of the difficulties children experience in estimating speed, movement, and distance in three-dimensional space (Figure 4.32).

Training programs to improve the judging ability of children offer promising anecdotal reports, but the effects seem to be transient (Connelly et al., 1998). If the underlying reason for poor judgment is physiological, training may do nothing more than give parents and teachers a false sense of security.

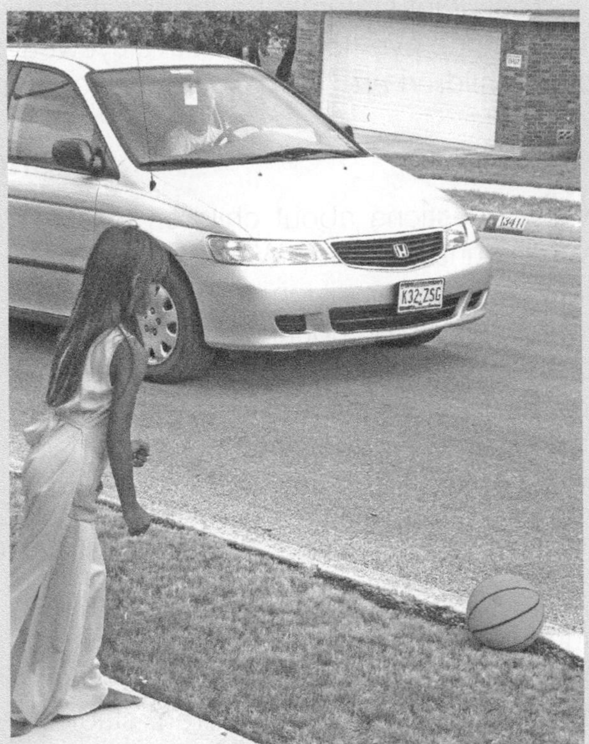

Figure 4.32 (Valerie Rice)
Five- to nine-year-olds have difficulty judging whether a gap in oncoming traffic is adequate to safely cross a street. Only some 11- to 12-year-olds can cross safely! Most children rely on distance to make their judgment and do not consider the speed of oncoming traffic.

6. Are vision and balance related?

Vision is important for movement and balance. Although there are many possible reasons for a child's "clumsiness" or poor balance, there is a link between physiological immaturity and dorsal and ventral stream processing.*

Children with "developmental clumsiness" have more difficulty detecting global visual motion (ventral) and global visual form (dorsal) (Sigmundsson et al., 2003). Thus, many seemingly clumsy children may not be clumsy because they fail to concentrate, but rather because they have difficulties processing visual information.

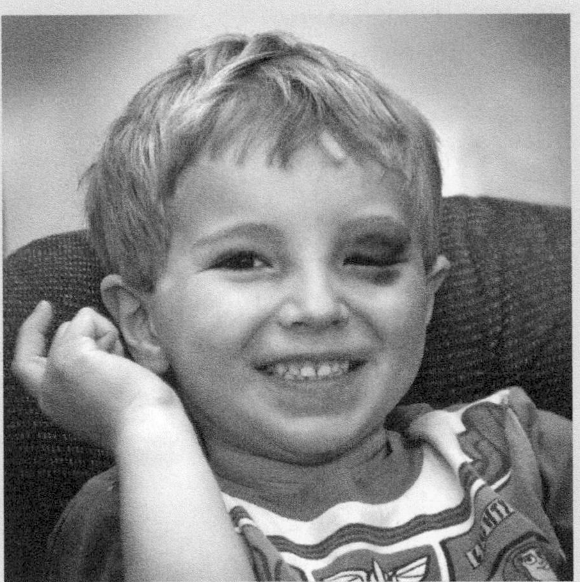

Figure 4.33 (Ursy Potter Photography)
Protective eyewear is easy to use and does not interfere with performance. It is a small price to pay to protect loss of vision or loss of an eye.

7. Is protective eyewear during sports really necessary?

The American Academy of Pediatrics and the American Academy of Ophthalmology have issued a joint policy statement on protective eyewear for young athletes. They strongly recommend protective eyewear for athletes who are functionally one-eyed and for those who have undergone eye surgery or trauma. The statement also lists the risks associated with different sports and indicates which eyewear is appropriate for what sport.

* Question 5 explains dorsal and ventral processing.

Their joint statement suggests that people with previous surgeries or who see primarily from one eye need protective eyewear the most. This is does not however suggest that other children are "safe" without it; for example, vigorous sports have safety risks.

CONCLUSIONS

Many questions about children's vision remain unanswered. We do not know how much near vision work will cause problems. We do not know what viewing distances are "safe." We can only attempt to improve the odds by changing our children's viewing environment so their near viewing tasks are "farther," "lower," "for shorter times," and so on.

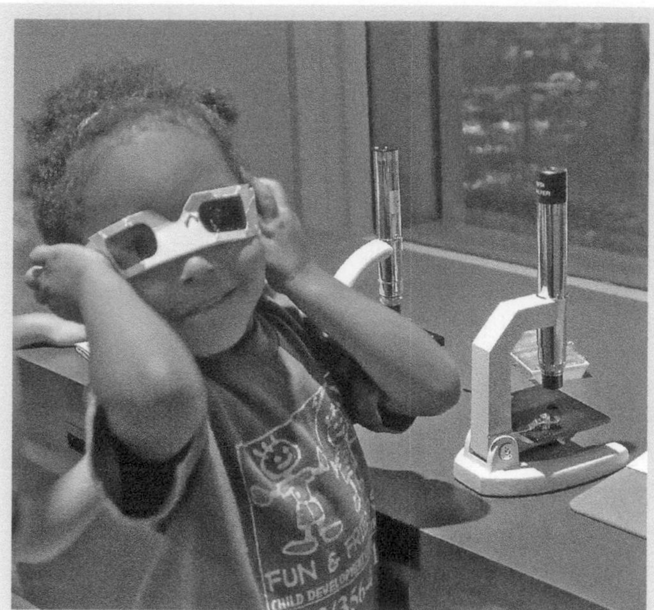

Figure 4.34 (Ursy Potter Photography)

Many of the studies cited in this chapter were conducted with adult subjects. Although nothing is known that would make it likely that the results would be different for children, studies with children as subjects are needed.

When one considers the time children spend reading, writing, and on the computer, it is clear that a good deal of this time involves "near vision tasks" in artificially lit environments. As parents and educators, we must do everything we can to lessen the potentially negative impact of the "computer" age on our children and encourage healthy vision.

SUGGESTIONS:

1. The best way to reduce the effects of “near vision” is to make it “less near.” Try to get the monitor at least 65 cm (25 in.) away from the child’s eyes, or the farthest distance possible without sacrificing legibility.
   a. Adjust the fonts to increase the size of the text.
   b. Use the zoom and “word wrap” features to facilitate reading at farther viewing distances.
   c. Use larger monitors to allow for farther viewing distances.
2. Lower the computer screen.
   Ensure the top of the screen is at least 15° below the child’s horizontal eye level.
3. Tip monitors back so that the top of the screen is a bit farther from the eyes than the bottom of the screen. This is especially important when the screen is lowered.
4. Place the monitor so that windows and bright lights are to the side, not behind or in front of the child.
5. Use sloped desks for reading and noncomputer work tasks.
6. Develop good work habits that involve postural and visual variation. Take regular breaks and look into the distance often (perhaps every 10 minutes).
7. Use good quality lighting equipment, preferentially flicker free systems with electronic ballasts. Place lights or lamps to prevent shadows and glare.
8. Get regular eye exams and look for signs of visual fatigue. If visual fatigue occurs, take action! Change the environment, take breaks, and rest the eyes. Remember that children are less likely than adults to complain about discomfort. Therefore, adults who work with children must look for warning signs that indicate a child is having problems.

The American Optometric Association (AOA) lists these child warning-signs.

- Lose their place while reading
- Avoids close work
- Holds reading material closer than normal
- Tend to rub their eyes
- Has headaches
- Turns or tilts head to use one eye only
- Makes frequent reversals when reading or writing
- Uses finger to maintain place when reading
- Omits or confuses small words when reading
- Consistently performs below potential

REFERENCES

Agaard-Hansen, J. and Storr-Paulson, A. (1995). A comparative study of three different kinds of school furniture. *Ergonomics*, 38(5), 1025–1035.

Ankrum, D. R. (2001). Flat panel potential for conformance to the display location requirements of ISO 9241 Parts 3 and 5. *Human–Computer Interaction: Proceedings of HCI International '01.*

Ankrum, D. R., Hansen, E. E., and Nemeth, K. J. (1995). The vertical horopter and the angle of view. In A. Grieco, G. Molteni, E. Occhipinti, and B. Piccoli (Eds.), *Work with Display Units 94, Selected Papers of the Fourth International Scientific Conference on Work with Display Units* (pp. 131–136). Amsterdam: Elsevier Science.

Ankrum, D. R. and Nemeth, K. J. (2000). Head and neck posture at computer workstations—what's neutral? Proceedings of the 14th Triennial Congress of the International Ergonomics Association, San Diego, (vol. 5, pp. 565–568).

Ankrum, D. R. and Nemeth, K. J. (1995). Posture, comfort and monitor placement. *Ergonomics in Design*, 3, 7–9.

Birnbaum, M. H. (1993). *Optometric Management of Nearpoint Vision Disorders.* Boston, MA: Butterworth-Heinmann.

Bridger, R. S. (1988). Postural adaptations to a sloping chair and work surface. *Human Factors*, 30(2), 237–247.

Brown, P. (1996). Amazing emmetropia. *The Optician*, 212(5563), 28–32.

Callan, M. J. and Galer, I. A. R. (1984). Ergonomics is Kid's Stuff: The ability of primary school children to design their own furniture. In E. D. Megaw (Ed.), *Contemporary Ergonomics 1984: Proceedings of the Ergonomics Society's Conference*, 2–5 April 1984, (pp. 167–172). London: Taylor & Francis.

Cestnick, L. and Coltheart, M. (1999). The relationship between language-processing and visual-processing deficits in developmental dyslexia. *Cognition*, 71(3), 231–255.

Chen, A. H., O'Leary, D. J., and Howell, E. R. (2000). Near visual function in young children. Part I: Near point of convergence. Part II: Amplitude of accommodation. Part III: Near heterophoria. *Ophthalmic and Physiological Optics*, 20(3), 185–198.

Choo, V. (2003). A look at slowing progression of myopia: Solutions for this common ocular disorder might come from cellular biology. *Lancet*, 361(9369), 1622–1623.

Collins, M., Brown, B., Bowman, K., and Carkeet, A. (1990). Workstation variables and visual discomfort associated with VDTs. *Applied Ergonomics*, 21, 157–161.

Connelly, M. L., Conaglen, H. M., Parsonson, B. S., and Isler, R. B. (1998). Child pedestrians' crossing gap thresholds. *Accident Analysis and Prevention*, 30(4), 443–453.

Cordain, L., Eaton, S. B., Brand Miller, J., Lindeberg, S., and Jensen, C. (2002). An evolutionary analysis of the aetiology and pathogenesis of juvenile-onset myopia. *Acta Ophthalmologica Scandinavica*, 80(2), 125–135.

Dobson, V., Miller, J. M., Harvey, E. M., and Mohan, K. M. (2003). Amblyopia in astigmatic preschool children. *Vision Research*, 43(9), 1081–1090.

Drobe, B. and de Saint-Andre, R. (1995). The pre-myopic syndrome. *Ophthalmic and Physiological Optics*, 15(5), 375–378.

Edwards, M. H., Li, R. W., Lam, C. S., Law, J. K., and Yu, B. S. (2002). The Hong Kong progressive lens myopia control study: Study design and main findings. *Investigative Ophthalmology and Visual Science*, 43(9), 2852–2858.

Fostervold, K. I. (2003a). Health consequences and behavioural changes associated with work with visual display units (VDU): An empirical investigation (Doctoral dissertation). Oslo, Norway: University of Oslo, Institute of Psychology.

Fostervold, K. I., Aarås, A., and Lie, I. (2006). Work with visual display units: Long-term health effects of high- and downward line-of-sight in ordinary office environments. *International Journal of Industrial Ergonomics*, 36(4), 331–343.

Fostervold, K. I., Aarås, A., and Lie, I. (2006). Work with visual display units: Long-term health effects of high- and downward line-of-sight in ordinary office environments. *International Journal of Industrial Ergonomics.*

Furuno, S., O'Reilly, K. A., Hosaka, C. M., Inatsuka, T. T., Allman, T. L., and Zeisloft, B. (1997). *Hawaii Early Learning Profile*. Palo Alto, CA: VORT Corporation.

Furushima, M., Imaizumi, M., and Nakatsuka, K. (1999). Changes in refraction caused by induction of acute hyperglycemia in healthy volunteers. *Japanese Journal of Ophthalmology*, 43(5), 398–403.

Heuer, H. and Owens, D. (1989). Vertical gaze direction and the resting posture of the eyes. *Perception*, 18, 363–377.

Hulme, C. (1988). The implausibility of low-level visual deficit as a cause of children's reading difficulties. *Cognitive Neuropsychology*, 5(3), 369–374.

IESNA. (2000). *Guide for Educational Facilities Lighting. Document Number: IESNA RP-3-00*: Illuminating Engineering Society of North America.

Johnson-Martin, N. M., Attermeier, S. M., and Hacker, B. (2004). *The Carolina Curriculum for Preschoolers with Special Needs*. Baltimore, MD: Brooks Publishing Co.

Kinge, B., Midelfart, A., Jacobsen, G., and Rystad, J. (2000). The influence of near-work on development of myopia among university students. A three-year longitudinal study among engineering students in Norway. *Acta Ophthalmologica Scandinavia*, 78(1), 26–29.

Kroemer, K. H. E., and Grandjean, E. (1997). *Fitting the Task to the Human: A Textbook of Occupational Ergonomics* (5th ed.). London: Taylor & Francis.

Lam, D. S. C., Lee, W. S., Leung, Y. F., Tam, P. O. S., Fan, D. S. P., Fan, B. J. et al. (2003). TGFß-induced factor: A candidate gene for high myopia. *Investigative Ophthalmology and Visual Science*, 44(3), 1012–1015.

Leoni, F. M. Q., Molle, F., Scavino, G., and Dickmann, A. (1994). Identification of the preferential gaze position through evaluation of visual fatigue in a selected group of VDU operators. *Documenta Ophthalmologica*, 87, 189–197.

Leung, J. T. and Brown, B. (1999). Progression of myopia in Hong Kong Chinese school children is slowed by wearing progressive lenses. *Optometry and Vision Science*, 76(6), 346–354.

Lie, I. (1986). *Syn og Synsproblember*. [Vision and Visual Problems]. Oslo, Norway: Universitetsforlaget (in Norwegian).

Lie, I. (1989). Visual anomalies, visually related problems and reading difficulties. *Optometrie*, 4, 15–20.

Lie, I. and Fostervold, K. I. (1995). VDT—Work with different gaze inclination. In A. Grieco, G. Molteni, B. Piccoli, and E. Occhipinti (Eds.), *Work with Display Units 94, Selected Papers of the Fourth International Scientific Conference on Work with Display Units* (pp. 137–142). Amsterdam, Netherlands: Elsevier Science.

Marumoto, T., Sotoyama, M., Villanueva, M. B. G., Jonai, H., Yamada, H., Kanai, A. et al. (1998). Significant correlation between school myopia and postural parameters of students while studying. *International Journal of Industrial Ergonomics*, 23(1–2), 33–39.

Maul, E., Barroso, S., Munoz, S. R., Sperduto, R. D., and Ellwein, L. B. (2000). Refractive error study in children: Results from La Florida, Chile. *American Journal of Ophthalmology*, 129, 445–454.

Molenbroek, J. F. M., Kroon-Ramaekers, Y. M. T., and Snijders, C. J. (2003). Revision of the design of a standard for the dimensions of school furniture. *Ergonomics*, 46(7), 681–694.

Mother. (1969). Personal communication.

Mumtaz, S. (2001). Children's enjoyment and perception of computer use in the home and the school. *Computers and Education*, 36(4), 347–362.

Mutti, D. O., Mitchell, G. L., Moeschberger, M. L., Jones, L. A., and Zadnik, K. (2002). Parental myopia, near work, school achievement and children's refractive error. *Investigative Ophthalmology and Visual Science*, 43(12), 3633–3640.

Mutti, D. O., Zadnik, K., and Adams, A. J. (1996). Myopia. The nature versus nurture debate goes on. *Investigative Ophthalmology and Visual Science*, 37(6), 952–957.

Negrel, A. D., Maul, E., Pokharel, G. P., Zhao, J., and Ellwein, L. B. (2000). Refractive error study in children: Sampling and measurement methods for a multi-country survey. *American Journal of Ophthalmology*, 129, 421–426.

Nersveen, J. (1999). Belysning og børns udvikling [Lighting and children's development]. *Lys*, 3 (in Danish).

Nordby, K. (1990). Vision in a complete achromat: A personal account. In R. F. Hess, L. T. Sharpe, and K. Nordby (Eds.), *Night Vision: Basic, Clinical and Applied Aspects* (pp. 290–315). Cambridge: Cambridge University Press.

Owens, D. A. and Wolf-Kelly, K. (1987). Near work, visual fatigue, and variations of the oculomotor tonus. *Investigative Ophthalmology and Visual Science*, 28, 743–749.

PBA (2003). Prevent Blindness America. www.preventblindness.org/children/baby_developing.html

Papastergiou, G. I., Schmid, G. F., Riva, C. E., Mendel, M. J., Stone, R. A., and Laties, A. M. (1998). Ocular axial length and choroidal thickness in newly hatched chicks and one-year-old chickens fluctuate in a diurnal pattern that is influenced by visual experience and intraocular pressure changes. *Experimental Eye Research*, 66(2), 195–205.

Pennie, F. C., Wood, I. C. J., Olsen, C., White, S., and Charman, W. N. (2001). A longitudinal study of the biometric and refractive changes in full-term infants during the first year of life. *Vision Research*, 41(21), 2799–2810.

Personal Communication. (2004). Conversation with ergonomist Valerie Rice, September 2004.

Pestalozzi, D. (1993). Ophthalmologic aspects of dyslexia: Binocular full correction of dyslexics with prismatic glasses. In S. F. Wright and R. Groner (Eds.), *Facets of Dyslexia and Its Remediation* (pp. 585–600). Amsterdam: Elsevier Science.

Ramazzini, B. (1983). Diseases of workers: Latin text of 1773 revised with translation and notes by Wilmer Cave Wright. New York: The Classics of Medicine Library, Division of Gryphon Editions.

Richler, A., and Bear, J. C. (1980). Refraction, near work and education. A population study in Newfoundland. *Acta Ophthalmologica (Copenhagen)*, 58(3), 468–478.

Rival, C., Olivier, I., Ceyte, H., and Bard, C. (2004). Age-related differences in the visual processes implied in perception and action: Distance and location parameters. *Journal of Experimental Child Psychology*, 87(2), 107–124.

Roe, K. (2000). Adolescents' media use: A European view. *Journal of Adolescent Health*, 27(2 Supplement 1), 15–21.

Rosner, M. and Belkin, M. (1987). Intelligence, education and myopia in males. *Archives of Ophthalmology*, 105(11), 1508–1511.

SSI (2003). Sight Savers International. www.sightsavers.org/html/eyeconditions/childhood.htm

Saw, S.-M., Shih-Yen, E. C., Koh, A., and Tan, D. (2002). Interventions to retard myopia progression in children: An evidence-based update. *Ophthalmology*, 109(3), 415–421.

Saw, S.-M., Wu, H.-M., Hong, C.-Y., Chua, W.-H., Chia, K.-S., and Tan, D. (2001). Myopia and night lighting in children in Singapore. *British Journal of Ophthalmology*, 85(5), 527–528.

Sciffman, H. R. (2001). *Sensation and Perception: An Integrated Approach* (5th ed.). New York: John Wiley & Sons.

Sekuler, R. and Blake, R. (1994). *Perception* (3rd ed.). New York: McGraw-Hill.

Sigmundsson, H., Hansen, P. C., and Talcott, J. B. (2003). Do 'clumsy' children have visual deficits? *Behavioural Brain Research*, 139(1–2), 123–129.

Simons, H. D. and Grisham, D. J. (1987). Binocular anomalies and reading problems. *Journal of the American Optometric Association*, 58(7), 578–586.

Straker, L. and Mekhora, K. (2000). An evaluation of visual display unit placement by electromyography, posture, discomfort and preference. *International Journal of Industrial Ergonomics*, 26, 389–398.

Sutherland, R., Facer, K., Furlong, R., and Furlong, J. (2000). A new environment for education? The computer in the home. *Computers and Education*, 34(3–4), 195–212.

Teikari, J. M., Kaprio, J., Koskenvuo, M. K., and Vannas, A. (1988). Heritability estimate for refractive errors—A population-based sample of adult twins. *Genetic Epidemiology*, 5(3), 171–181.

TSB, Texas School for the Blind. (2003). Monitoring visual development. Available at www.tsbvi.edu/index.htm

VESA (1999). *VESA Flat Panel Monitor Physical Mounting Interface Standard*. Milpitas, CA: Video Electronics Standards Association.

Veitch, J. A. (1990). Office noise and illumination effects on reading comprehension. *Journal of Environmental Psychology*, 10, 209–217.

Veitch, J. A., and Newsham, G. R. (2000). Preferred luminous conditions in open-plan offices: Research and practice recommendations. *Lighting Research and Technology*, 32, 199–212.

WHO (2004). World Health Organization, Fact sheet no. 282. http://www.who.int/mediacentre/factsheets/fs282/en/index.html

Weale, R. A. (2003). Epidemiology of refractive errors and presbyopia. *Survey of Ophthalmology*, 48(5), 515–543.

Wulff, U. (1998). Impaired binocular vision and learning difficulties. *Neues Optikerjournal*, 1–15. English translation at www.ivbv.org/Literatur.htm

Young, F. A., Leary, G. A., Baldwin, W. R., West, D. C., Box, R. A., Harris, E. et al. (1969). The transmission of refractive errors within Eskimo families. *American Journal of Optometry and Archives of American Academy of Optometry*, 46, 676–685.

Zylbermann, R., Landau, D., and Berson, D. (1993). The influence of study habits on myopia in Jewish teenagers. *Journal of Pediatric Ophthalmology and Strabismus*, 30, 319–322.

CHAPTER 5

HEARING ERGONOMICS FOR CHILDREN: SOUND ADVICE

NANCY L. VAUSE

TABLE OF CONTENTS

Section A: Hearing and Noise

INTRODUCTION

WHY IS NOISE A PROBLEM?

Most of us do not think of sound (noise) as an ergonomic issue. Yet loud and distracting "toxic" noise is dangerous and can damage our ears. Even "safe" classroom sounds (noise) can hinder children from doing *their jobs*: listening, thinking, talking, creating, and learning (Shield and Dockrell, 2003a,b).

According to the World Health Organization (WHO), about 580 million people (40 million in the United States) have some degree of hearing impairment. Further, WHO believes that this number underestimates the real prevalence of hearing loss. Noise-induced hearing loss (NIHL) caused by toxic noise accounts for one third of all permanent hearing loss in developed countries.

Figure 5.1 (Ursy Potter Photography)
Teachers, parents, and students need to recognize that noise exposure at school, home, work, travel, and in recreational activities can harm children's hearing.

Noise can adversely affect a child's ability to communicate and learn and their general health, behavior, performance, and safety.

Worldwide, hearing loss in children is increasing at an alarming rate (WHO, 1997). In the United States, 5.2 million children between 6 and 19 years old have a hearing loss directly linked to noise exposure (Niskar et al., 2000).

> Noise creates an ergonomic mismatch with educational, health, and hearing consequences.

Not only is ***toxic noise*** a public health threat, it is an invisible, insidious, and often forgotten ergonomic issue (Box 5.1). The good news is most noise-related health and learning problems are usually **preventable**!

Box 5.1 Toxic Noise

Exposure to unsafe decibel levels of sound can be toxic, causing temporary or permanent hearing damage and hearing loss. While all the structures of the inner ear that support hearing are delicate, the hair cells are especially fragile. These begin working while the fetus is still in the womb.

The ears do not make new hair cells after birth; once damaged, hair cells cannot regenerate. All you will ever have are those present at birth. A one-time exposure to loud noise or repeated exposures to toxic noise at various unsafe levels can damage and eventually "kill" hair cells.

Remember:

- You cannot "toughen" your ears by constant exposure.
- You cannot "tune out" loud noises that you hear over time.
- If a person feels they have "gotten used to" the loud sounds they hear over time, they have probably already suffered some amount of permanent hearing loss.

(*Source:* Deafness Research Foundation.)

Figure 5.2 (Nancy Vause)
Most people know someone with age-related hearing loss (presbycusis) or occupational hearing loss, but they often do not realize the risk noise poses in children. Toxic noise can cause "old ears" in young bodies.

"SAFE" NOISE LEVELS

Toxic noise (inside and outside the classroom) can damage hearing, but even "safe" noise levels can disrupt learning (Figure 5.3).

Figure 5.3 (Kurt Holter Photography, Frederick, MD)* **Figure 5.4** (Valerie Rice)*
Noise makes it difficult for children to hear important sounds such as speech and warning signals. Most children are still learning how to become listeners. It is difficult for them to develop listening skills that require them to hear speech, concentrate, read, and learn in noisy environments. Depending on the classroom environment, a child may (or may not) not feel comfortable continuously asking for clarification or may not even realize he has missed important information.

On the other hand, some school environments are exceptionally noisy. The noise levels during lunch in a cafeteria, while working in shop class, or while in band can reach toxic levels.

Noise does not affect everyone equally; children have differing levels of sensitivity to noise. Its impact is particularly deleterious for children with English as a second language (ESL) or who exhibit attention deficits, developmental delays, speech, language, or learning disabilities, as well as children with temporary or permanent hearing loss.

* Figure 5.3 reprinted with permission of Banner School parents. Figure 5.4 reprinted with permission from St. Thomas Moore Catholic School.

Temporary hearing impairments are more common than most people realize. At any given time, several children in a primary school classroom may have some form of temporary hearing impairment (Nelson, 1977) (Box 5.2).

Even hearing loss in one ear can create problems; having one "good" ear does not fully compensate or make "everything okay" for learning. There is a higher incidence of grade failure among children with unilateral hearing loss (Bess et al., 1998). A good listening environment and preferential seating will improve the child's ability to listen and learn.

Acoustics can enhance learning or hinder the educational process.

BOX 5.2 TEMPORARY HEARING LOSS

- Temporary hearing loss interferes with hearing, making it more difficult to learn may result from ear fluid (otitis media), colds, flu, or the buildup of earwax.
- Teachers may also notice a child pulling on their ears, talking louder or softer, turning up the television or radio to higher loudness levels or inattentiveness to soft sounds; fever, ear fluid draining, or balance problems; or children exhibiting unusually irritable, tired, or inattentive behaviors.
- Temporary hearing loss may delay speech and language development.

Negative effects of poor classroom acoustics (such as noise) include:

- teachers must talk louder, placing them at risk for vocal fatigue and voice disorders, and
- teachers and students are distracted from auditory and visual tasks such as teacher instruction, white or blackboards, books, or conversations (Anderson, 2001a).

School and classroom locations or designs can contribute to noise problems. *One size does not fit all.* Consider the shape, size, and surfaces. The acoustical requirements for small or large classrooms, music or shop rooms, auditoriums, cafeterias, and gymnasiums are different and must receive individual consideration. Effective (and cost-effective) listening environments require careful planning during the design process, rather than as an afterthought. Locating schools near noise sources such as public transportation systems (e.g., airports, railroads, etc.) also can expose children to unwanted noise.

Unfortunately, many classrooms are so noisy they interfere with communication required for teaching and learning (Knecht et al., 2002). Teachers report noise as one of the most unsatisfactory environmental classroom conditions (GAO, 2001).

Classroom acoustics standards have been in place only since 2002 in the United States. The American National Standards Institute (ANSI) approved ANSI S12.60 2002 as the first American standard and is available free: http://asastore.aip.org. A multidisciplinary group of educators, subject matter experts, and parents thoughtfully developed the standard. The Access Board also recognized noise and reverberation as significant barriers to listening and learning. They proposed this standard as a guideline to meet American with Disability Act (ADA) requirements.*

Although the Occupational Safety and Health Act of 1970 (OSHA) protects adults from noise, no equivalent protection exists for schoolchildren. In one survey of teachers, 75% reported their school did not have a noise policy or program (Dockrell and Shield, 2002). Surprisingly, many health textbooks do not include information about noise, noise hazards, NIHL, or strategies to prevent hearing injuries.

In a sense, a classroom with good acoustics is a critical teacher tool required to facilitate instruction much like books or desks. If no teacher expects a child to read without adequate light, no teacher should expect children to listen and learn in the equivalent of auditory darkness.

* Of course, this standard meets only one portion of ADA requirements, as other physical and mental disabilities need other types of accommodations.

TOXIC NOISE LEVELS

"Toxic" noise can damage ears by destroying delicate hair cells of the inner ear. While many people assume hearing loss only occurs with exposure to very loud levels of impulse noise (like firecrackers),* more frequently hearing damage occurs from long-term exposure to toxic noise.

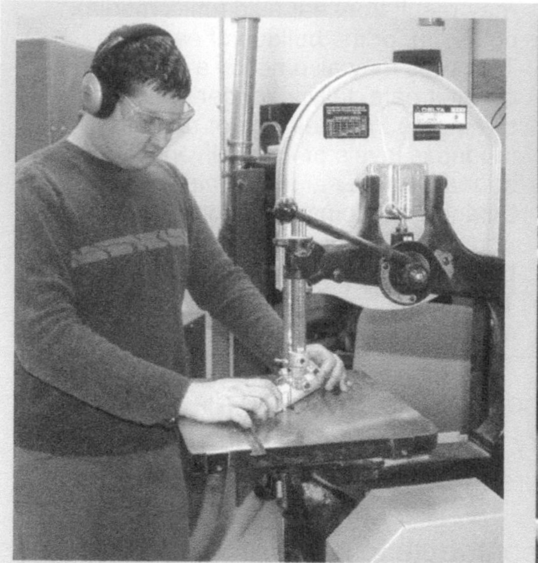

Figure 5.5 (Michael Younkins) **Figure 5.6** (Valerie Rice)
Few people realize noise exposure during routine school activities (e.g., bus, lunch, gym, music, shop, band, etc.) can exceed that of an 8 hour workday in a factory (WHO, 1997). Twenty-six percent (26%) of students in band or vocational classes (e.g., woodworking, body repair, shop) exhibit some degree of noise-induced hearing loss (NIHL) (Montgomery, 1990).

Ear injuries can also occur at home (e.g., motorcycles, CD players, toys), during recreational activities (e.g., band practice, video game arcades, sporting events, car and truck races, snowmobiling, target practice), or while working after school (e.g., fast-food headsets, lawn and farm equipment, power tools, chain saws).

Most people do not realize the danger of toxic noise (Figure 5.8 through Figure 5.10). No one may immediately notice these injuries because ears do not usually hurt or bleed. All the while repeated exposure to loud toxic noise continues to affect these delicate inner ear structures—disrupting their blood supply and may cause permanent hearing loss.

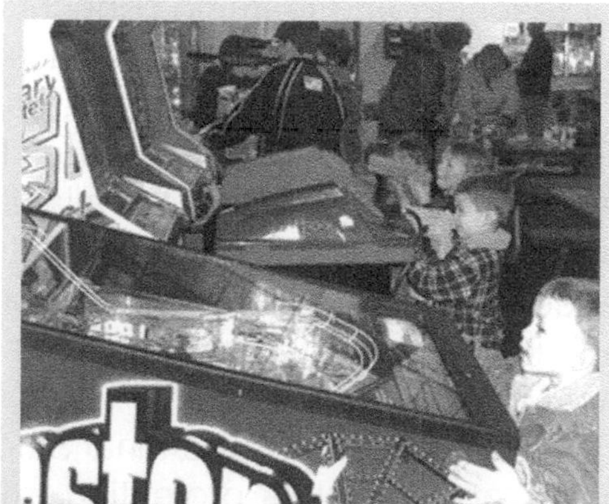

Figure 5.7 (Nancy Vause)
Game sounds are loud; children and parents yell above them to communicate in video game arcades. These arcades can be extremely noisy and may contribute to hearing damage.

* Permanent NIHL can result from loud impulse noise (≥140 dBP) and from lower levels (≥85 dBA) of continuous noise exposure over time in adults.

Since NIHL ear damage initially decreases hearing in the higher frequencies around 4000 Hz, many people (children and adults) do not realize they have even lost hearing. Often, listeners do not realize they are missing sounds and may accuse family or friends of "mumbling."

By the time teachers, students, or parents notice a problem it is too late. Damage is often already irreversible and permanent. The good news is—usually hearing loss injury from noise exposure is easily preventable.

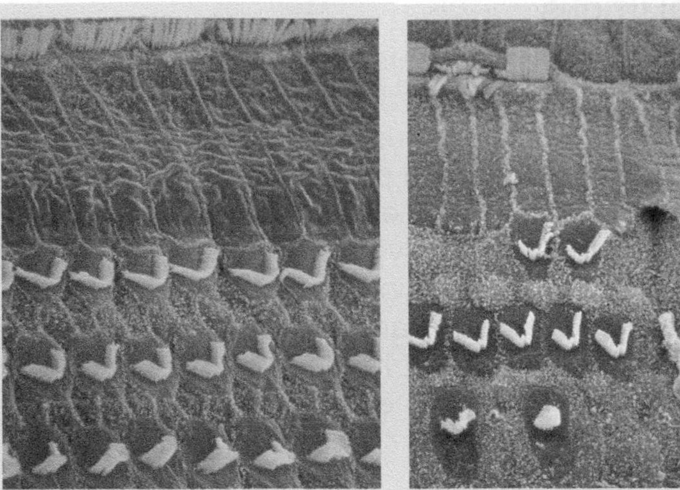

Figure 5.8 (NHCA) **Figure 5.9** (NHCA)
Figure 5.8 shows normal hair cells within the inner ear, while Figure 5.9 shows damaged hair cells.*

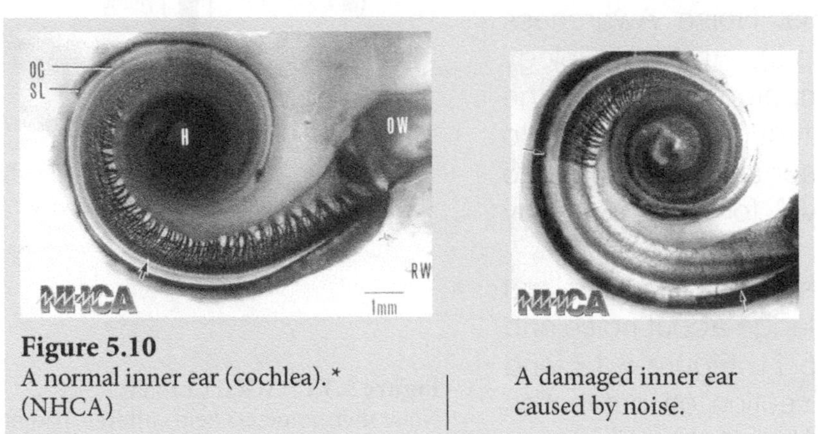

Figure 5.10
A normal inner ear (cochlea).* (NHCA)

A damaged inner ear caused by noise.

THE RESULT

Toxic noise places children, teachers, and parents at risk. Rates of children experiencing NIHL are increasing (Bess et al., 1998). In the United States, about one in eight 6- to 19-year-olds experience loss of hearing from noise exposure (Niskar et al., 2001). The problem exists in other countries as well, including Sweden, China, and France (Costa et al., 1998; Morioka et al, 1996; Meyer-Bisch, 1996).

Exposure to noise hazards affects 97% of third-graders (Blair et al., 1996). Although we do not know how much of children's NIHL is caused by classroom exposure, it definitely causes concern. This is particularly so since children do not intuitively know how to protect hearing and prevent hearing loss, and they may not believe damage can happen to them (Watkins, 2004).

* Poster graphics Figures 5.8, 5.9, and 5.10 reprinted with permission of the National Hearing Conservation Association (NHCA).

THE SOLUTIONS

Teachers, parents, students, and communities must first learn about noise and the risk of ear damage. Then, they must use this knowledge to address the noise pollution problem both in side and outside the classroom.

The European Agency for Safety and Health at work is educating the European public through a public awareness campaign entitled "Stop that Noise! It can cost more than your hearing" in an effort to educate workers.

Resources are also found in Appendix C with descriptions of the programs or products. In the US, several professional government and nonprofit groups are also tackling the noise pollution problem. Projects like Better Speech and Hearing Month in May, International Noise Awareness Day (every April), the Audiology Awareness Campaign's "Get on the right track—protect your hearing" to promote consumer awareness in auto racing, or "HEAR" for awareness by musicians.

There are many ways to gain and implement knowledge about noise and hearing (Figure 5.11, Figure 5.12, and Table 5.1). The section on prevention below provides additional details, especially on reducing classroom noise.

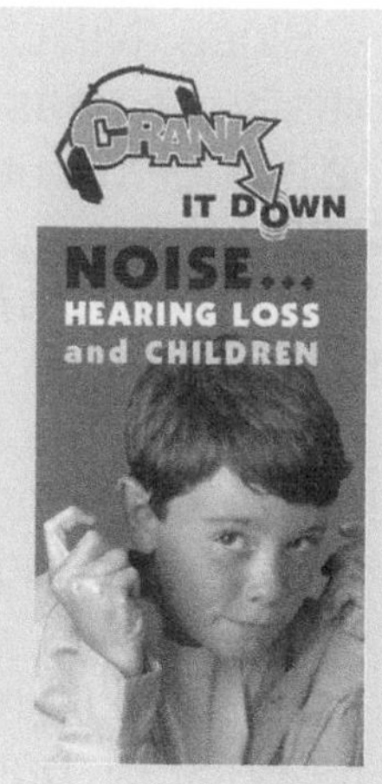

Figure 5.11 There are many ways to educate the public about hearing loss. However, to be effective with children, the programs need to target children, as well as, adults. Marketing techniques can include posters, classroom education, and public service announcements by popular individuals or characters. (From NHCA and AAA.)

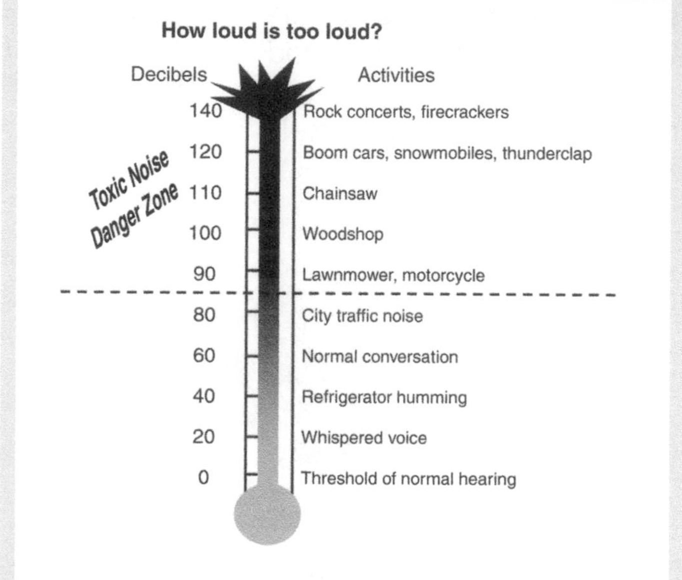

Figure 5.12 (Geri Corona)
Noise thermometers help children, parents, and teachers understand which sounds are in the toxic noise "DANGER ZONE," in order to avoid injury and hearing loss.

Table 5.1 Common Household Items can Cause Ear Injury after Relatively Short Exposures without Hearing Protection. These Risks are Based on Adult Allowable Exposure Levels, as there are not Comparable Guidelines for Children

Home Equipment	Sound Level	Risk of Hearing Loss
Chain saw	110	1 minute
Hair dryer	90	2 hours
Leaf blower	110	1 minute
Subway train	90	2 hours
Amplifier rock concert at 4–6 ft	120	Immediately

Source: From National Institute for Occupational and Safety (NIOSH), U.S. Department of Health and Human Services, Public Health Service, Centers for Disease Control and Prevention Publication 98.126. "Occupational Noise Exposure Revised Criteria 1998," June, Cincinnati, OH, 1998, pp. 24–25.

THE BOTTOM LINE

We often take hearing for granted. Yet, we enjoy our good hearing in daily communication, learning, safety, and quality of life (Figure 5.13 and Figure 5.14).

Helen Keller provides a thought-provoking perspective. She could neither see nor hear. When asked to compare her loss of vision with her loss of hearing she said (Walker, 1986):

Blindness cuts people off from things,

Deafness cuts people off from people.

Who can envision cutting our children off from friends and family? Can we even imagine the isolation and miscommunication that hearing loss creates?

Figure 5.13 (Kurt Holter Photography, Frederick, MD)* **Figure 5.14** (Kurt Holter Photography, Frederick, MD)*
Hearing ergonomics for children provides teachers, parents, and students "sound advice" on the basics of sound, hearing, toxic noise, classroom acoustics, hearing conservation prevention strategies, and protective measures to enable children to listen for a lifetime.

Sound Advice

Increase public awareness!

Spread the word, emphasize noise destroys hearing and can hinder learning!

Start a school hearing conservation program TODAY!

If children don't understand why they should protect their hearing—they probably won't.

Form a team to examine classroom acoustics!

* Photo reprinted with permission of Banner School parents and photographer Kurt Holter Photography, Frederick, MD.

SOUND SCIENCE

People should understand basic acoustics—the science of sound—(e.g., production, transmission, reception, and effect), before they can improve acoustics in their surroundings.*

OBTAINING INFORMATION

Many foundations and organizations (e.g., Audiology Awareness Campaign, Wise Ears, League for the Hard of Hearing, Hearing Education and Awareness for Rockers [HEAR], Self Help for Hard of Hearing People [SHHH]) provide useful consumer information. The National Hearing Conservation Association represents the noise and hearing conservation professionals (academic, clinical, industry, and government) and provides an abundance of information in various formats for use by individuals or groups (Figure 5.15 and Figure 5.16).

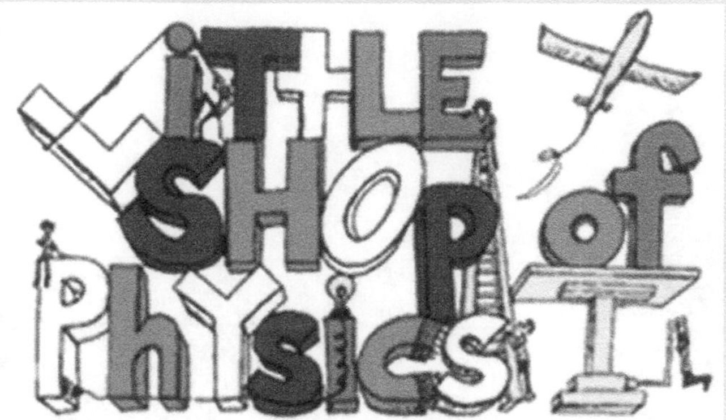

Figure 5.15 (Adam Beehler) **Figure 5.16** (Adam Beehler, Colorado State University)
The National Hearing Conservation Association provides educational advocacy, information, and useful teacher tools. The popular "Sounds like Fun" program (developed by "Dangerous Decibels" and "Little Shop of Physics") offers age-appropriate interactive activities and physical demonstrations (like Adam's Weird Ears/Big Ears) ready for inclusion in health or science curriculum ideas, lesson plans, videos, and audio demonstrations and detailed activities are available on the NHCA Web site.†

Involving children in activities helps them to learn the concepts. Since sound energy is invisible, the instruction and interactions must be creative to help children understand the underlying concepts (Figure 5.17 through Figure 5.20). Some of these demonstrations and activities include:

- *Good Vibrations*: understanding the connection between sound and vibration (Figure 5.17);
- *Sound Moves*: sound is energy that moves through the air to ears (Figure 5.18);
- *Holy Interfering Waves, Batman!*: hear the effects of sound produced next to a nearby object;
- *Memory Foam*: observe the different absorption effects of materials;

* Several textbooks (Harris, 1998; Berger, 2000; Yost, 2000) provide technical explanations of sound's properties beyond the scope and intent of this discussion.

† Photo and logo reprinted with permission of Little Shop of Physics (http://littleshop.physics.colostate.edu).

- *Bubble Trumpet*: see loud damaging noises do not always produce noticeable visual changes;
- *Shake it Break it*: illustrate the impact of noise on delicate inner ear hair cells (Figure 5.19 and Figure 5.20);
- *Weird Ears/Big Ears*: experience sound gathering from different directions (Figure 5.15); and
- *Ear Model/Photo Phone*: observe how an ear can channel sound to the eardrum.

Figure 5.17 (Nancy Vause) **Figure 5.18** (Nancy Vause)
Engage in activities that teach children about sound properties, the consequences of toxic noise, and what they can do to prevent injury and hearing loss. Children learn by doing, as well as seeing and hearing.

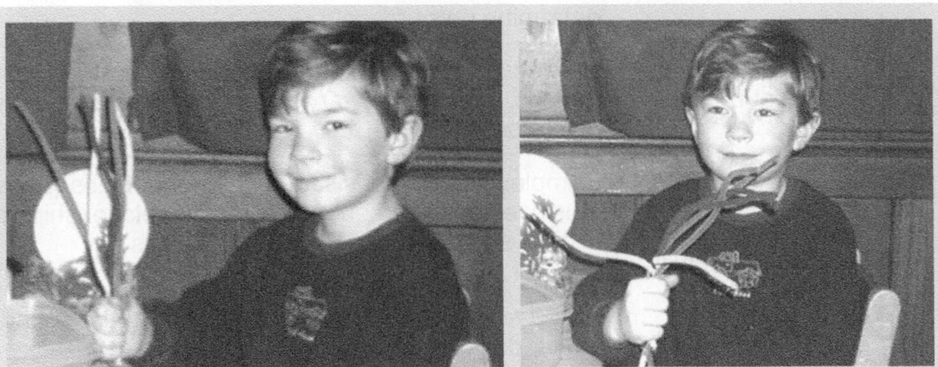

Figure 5.19 (Nancy Vause) **Figure 5.20** (Nancy Vause)
Age-appropriate fun activities should begin in kindergarten or earlier to increase awareness while promoting healthy hearing. This boy is learning that when the pipe cleaners: he holds, representing the tiny hairs in the inner ear are shaken (vibrated due to noise), they "break" and hearing is damaged. It is a powerful, hands-on, message a young child can understand.

SOUND PARAMETERS

Familiar parameters of sound include: (1) frequency, (2) intensity, and (3) duration.

Musicians often include timbre (Table 5.2). These parameters distinguish one sound from another. The timbre allows easy recognition of unique musical instruments or a friend's voice.

Table 5.2 Basic Acoustic Terminology, Concepts, and Definitions

Terminology	Perception (If Applicable)	Definition
Sound	Sound/noise	1. Pressure variations in any moveable (elastic) medium like air, water, and solids 2. Pressure changes in the form of vibration
Noise	Noise/sound	1. Random sound without clearly defined frequency components 2. Unwanted, disagreeable, or undesirable sound
Frequency	Pitch	Number of vibrations or complete cycles per second (the faster the cycles, the higher the frequency or pitch) Measured in hertz (Hz)
Intensity or sound pressure level (SPL)	Loudness	The sound wave's strength, amplitude, or pressure displacements. The larger the amplitude—the higher the intensity or amplitude Originally measured in bels after Alexander Graham Bell. Current use divides the bel into 10 more manageable units, decibels (dB)
Duration	Exposure time	The amount of time a sound is present or the amount of time a listener listens to the sound. The latter value is important to determine noise "dose"
Temporal properties of sound/noise	Beat/rhythm	How sound is distributed over time (short bursts, long steady periods, intermittent, varying, etc.). Important to determine noise "dose"
Timbre	Sound quality	Complex sounds consist of many frequencies (speech, noise, music, etc.). The frequency–intensity relationships create distinct sound characteristics or qualities Timbre is mainly determined by the harmonic content of a sound and the dynamic characteristics of the sound (vibrato and attack-decay envelope)

NOISE AND NOISE ABATEMENT

Unwanted sound is noise to most listeners. Noise can cover a broad distribution along the frequency spectrum and be a combination of several frequencies. Therefore, determining injury risk and noise reduction strategies requires evaluation of the sound parameters (intensity, frequency, and duration).

Most noise involves such a large frequency range it is impossible to manage all frequencies at once. Devices such as *sound level meters* can measure the overall intensity including all frequencies using an A-weighted sound level (dBA) intensity measurement. These devices can also narrow the frequency range (bandwidth) by limiting the range between two frequency limits.

Noise control engineers, industrial hygienists, and audiologists measure noise intensity broken down into even smaller more manageable frequency ranges called *octave bands* (OB). These groups of frequencies or OB help to identify effective reduction strategies. The noise OB characteristics help determine ear injury risk (higher frequencies are usually more dangerous) and classroom noise problems (lower frequency noise can vibrate through the walls).

In determining "How much noise is too much noise?" intensity and OB measurements are not enough. The noise–time relationship (temporal characteristic) and the duration of exposure are also important (Table 5.3). Noise levels vary in duration and intensity during an adult or child's "workday." Dosimeters measure individual noise exposure levels and calculate an average noise level or "noise dose" over a specific time limit.

One study placed dosimeters on 116 children (6 to 14 years) from the time they got up in the morning until bedtime. The children's noise exposures reached levels as high as 115 dBA during recess. Levels of toxic noise existed during bus rides, lunch, gym, shop, and hockey games. In the end, the average noise exposure for all children over a 24 hour day was 87.5 dBA. This level exceeds the permissible exposure level (PEL) for factory workers (WHO, 1997).

Recreational activities are a primary source of noise (Kullman, 1999). Sporting events (fireworks, fans, and football) reach peak levels of 135 and 143 dBA (Epstein, 2000). Even a nice movie can range from 72 to 86 dBA or higher. Monster truck rallies have the highest intensity levels, peaking at 146 dBA (Berger and Kieper, 1994).

Table 5.3 Description and Examples of Different Types of Noise Temporal Characteristics

Noise Temporal Characteristics	Description	Example
Continuous	Describes a constant presence of noise. Steady state and varying describe noise intensity parameters	See below
Steady state	Intensity remains constant	• Motorcycles • Lawn mowers • Tractors • Generators
Varying	Intensity levels fluctuate with no significant quiet periods in the environment (never totally quiet). This is just as damaging as continuous noise	• Woodworking equipment (louder when cutting wood) • Professional sporting events
Intermittent	Intensity varies significantly with periodic interruptions of quiet between noise episodes Some evidence suggests this type of noise is not as hazardous since the ear has some time to recuperate	• Dental drills • Highways • Fast-food headsets • Heating, ventilation, and air conditioning (HVAC) systems
Impulse or impact	Distinctive single pressure peak or burst (<1 s) which can repeat after a period of quiet (>1 s) Impulse and impact noise are not exactly the same, but most people use the terms interchangeably One intense exposure can cause permanent damage	• Thunder burst • Firecracker • Popgun • Shotgun • Hammers • Pneumatic tools

Most experts consider high levels of impulsive noise more hazardous than continuous noise because the sound wave can immediately damage tissues, rupture eardrums, disrupt middle ear bones, disturb blood supply, and cause permanent damage to the inner ear with only ***one exposure***. Furthermore, without prior warning the short duration can prohibit use of prevention strategies such as inserting earplugs or putting on earmuffs (Figure 5.21).

SIGNAL-TO-NOISE RATIO

In the best acoustical classroom designs, competing background noise does not significantly interfere with a students' ability to hear and understand speech. This ability depends upon the ***signal-to-noise ratio*** (SNR).

SNRs compare the intensity of speech such as a teacher's voice with competing background noise. For example, if a teacher's voice (at the student's ears) is 55 dBA and the background noise is 50 dBA, the SNR = +5 dBA. The teacher's voice (signal) is 5 dB louder than the background noise (Figure 5.22).

Classroom SNR can range from +5 to −7 dB (Finitzo-Hieber, 1988; Markides, 1986). Unfortunately, this is an educational ergonomic mismatch since students require SNRs at least +15 dB louder than background noise to hear teacher instruction (Nelson et al., 2002).

Simply talking louder does not solve the problem. Children have difficulty understanding speech when the overall background noise exceeds 69 dBA, even with an ideal SNR (Studebaker et al., 1999).

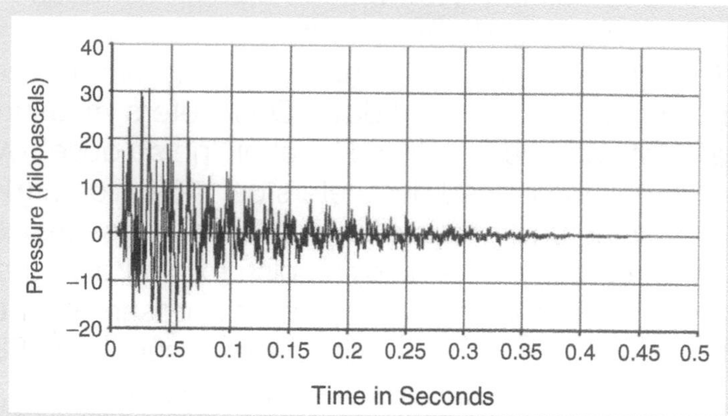

Figure 5.21 (Joel Kalb and Dick Price)
An impulse sound wave. Impulse noise measurements use a peak SPL measurement (dBP) to differentiate between continuous, varying, or intermittent noises that do not suddenly peak. The impulse noise sound wave reflects this very sharp rise and fall of pressure.

Figure 5.22 (Valerie Rice)
The SNR varies as a teacher walks around the room, turns her head or has objects between her and the listening children that block the movement of some of the voiced sound waves.

Although this would make it seem like teachers should stand in one place and project their voices in a single direction, some teachers use it to their advantage, addressing their students from different vantage points throughout the room. They can tailor the SNR to the individual student and learning activity.

STUDENT EDUCATIONAL ACTIVITY

SOUND MEASURES

Grade Level (5–9) Time required (55 minutes)
Effect of Distance on Intensity Calculate SNR
(Adapted from www.dangerousdecibels.org)

Obtain a sound level meter (SLM), blender or radio, tape measure.
(SLMs are available at electronics stores like Radio Shack)

Introduce the decibel chart:

- Set the SLM at 80 dB, select A weighting, Set Response to SLOW
- Have students measure the existing "background noise" in the classroom. (Reduce the SLMs decibel range as needed. Record the sound level in dBA and reset to 80 dBA)
- Ask one student to whisper softly. Have nine nearby students join in
- Ask what happens to sound if you move closer or farther away adjusting the decibel range as needed

Place a blender (radio) near the edge of a table facing the classroom:

- Adjust the speed of the blender to a steady 80 dBA—this will serve as the baseline intensity and zero distance. This is the level students will compare changes as students move around and remeasure.
- Without changing the blender speed, move the SLM at standard unit of measure (e.g., 12 in.) further away and record the sound level in dBA (reduce the SLM decibel range as needed).
- Continue moving the SLM further away in 12 in. increments and record the sound levels (in dBA) until the decibel levels fall below the 50 dBA SLM limit.
- Graph the values with distance as the independent variable (*x*-axis) and sound level as the dependent variable (*y*-axis).
- Repeat using the C-weighting scale if you have time.
- Advanced students can measure at each location with blender on and off to determine the SNR for various locations and conditions (teacher moving around, different activities occurring simultaneously).
- Have students develop recommendations for the best methods to help everyone hear and understand in the classroom.

REVERBERATION

Reverberation (or echo) is the persistence of sound in a room after the source of the sound has stopped. The direct sound continues to bounce off floors, ceilings, and walls, resulting in indirect sound waves. Reverberation time (RT) is the time required for a sound to decrease (decay) to 60 dB from the original onset level. RT depends on room volume (size), shape, and surface area of materials in the room and absorption capabilities of room materials.

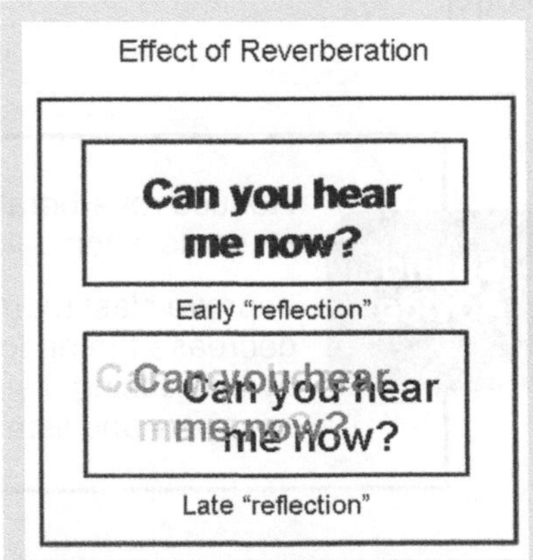

Figure 5.23 (Geri Corona)
Long RTs introduce more noise and blend successive speech syllables together into continuous sound. The effect of reverberation on speech intelligibility is illustrated above. While the top frame is clear and crisp, the bottom frame is "blurred," just as sound is "blurred" by long RTs.

You can hear sound reflections or reverberation by clapping your hands in a large, empty cafeteria, or gymnasium. Without sound-absorptive materials, sound continues to bounce off walls, windows, ceilings, furniture, and floors creating longer RTs or echoes (Figure 5.24). Room reverberation adds to the overall noise level of the room since the sound energy from reflections combines with original sounds.

Long RTs make one's voice sound better when singing in the shower. However, they also make speech harder to understand and degrade the quality of speech (Figure 5.23).

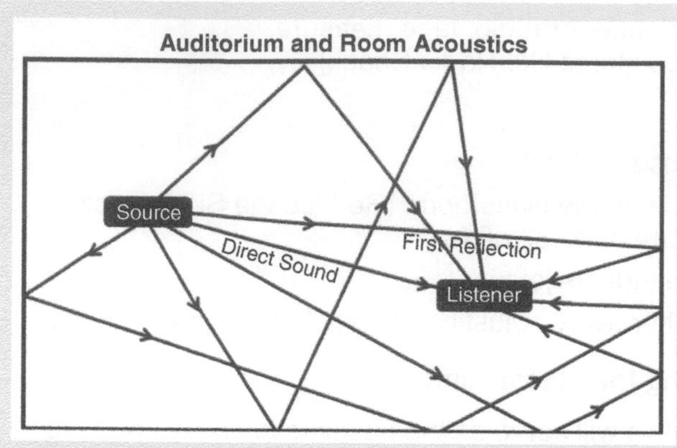

Figure 5.24 (Central Michigan University)
Sound pathways of direct sound and reflected sound bouncing off walls, ceilings, and floors before reaching the listener.

Most schoolchildren need listening environments with RTs under 0.6 s, yet classroom RTs range from 0.4 to 1.2 s (Knecht et al., 2002). The combination of long RTs and background noise (low SNR) further degrades communication, especially for children (Crandell and Bess, 1986).

Figure 5.25 (Kurt Holter Photography, Frederick, MD)
Reverberation is particularly difficult for children to overcome, as they are inexperienced and inefficient listeners. They are still learning to listen and discriminate small units of speech (phonemes). Many of these small units of speech are not very loud. In fact, they are often very soft and easily masked by classroom noise and reverberation.

Reduce reverberation by facing students when talking.

Improve classroom SNR and decrease the effects of reverberation by decreasing distances between speakers and listeners.

EARS AND HEARING

The process of hearing involves a remarkable series of events as the ear converts the frequency and intensity of sound energy into electrical signals that it transmits to the brain. This enables the brain to hear, recognize, discriminate, and interpret sounds. Our ears simultaneously listen to the sounds of life as music plays, voices laugh, alarms blast, computers hum, floors creak, babies cry, pots clang, birds chirp, automobiles race, wind blows, insects buzz, and dogs bark.

Ears never go "off duty." They operate continually, even during sleep. Furthermore, the sense of hearing has the advantage of spanning in all directions around a listener because sound (unlike light) is relatively free to travel around objects (obstructions do not totally block sound waves).

ANATOMY OF THE EAR AND "HOW IT WORKS"

See interactive ear anatomy of the outer, middle, and inner ear sections at:

"Let's Hear it for the EAR" www.kidshealth.org/kid/body/ear_noSW.html or

Virtual Tour of the Ear: www.augie.edu/perry/ear/ar.htm

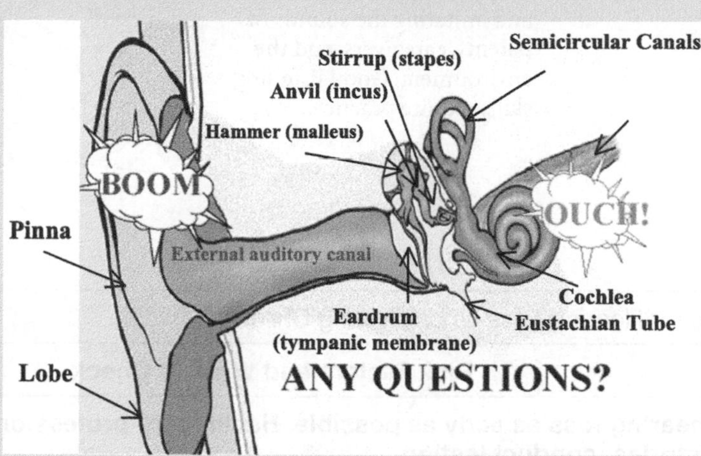

Figure 5.26 Anatomy of the ear. (Adapted by Nancy Vause from US Army Center for Health Promotion and Preventive Medicine graphic.)

The outer ear is more than the *pinna* on the outside of your head ready to keep your glasses from falling off! The pinna behaves like radar antennas collecting sound waves and directing them into the ear canals outer (external) *auditory canals*.

The *outer ear canal* is a short tube that directs and funnels sound pressure waves to the eardrum.* As the sound pressure waves reach the *tympanic membrane* (eardrum), they set it in motion by striking the surface of this very thin membrane. The tympanic membrane vibrates in proportion to the intensity and frequency of the sound. The sound vibrations move through the air-filled *middle ear* using the smallest bones in the body, the *ossicles*.

These middle ear bones form the *ossicular chain*. The bones themselves are the *malleus, incus*, and *stapes* (also called the *hammer*, *anvil*, and *stirrup*). The ossicles serve as a bridge and transmission device connecting the tympanic membrane to the fluid-filled *cochlea* (*inner ear*). The *Eustachian tube* keeps the *middle ear* at atmospheric air pressure by connecting the middle ear with the back of the throat.†

The inner ear houses anatomy responsible for balance (*semicircular canals*) and hearing (*cochlea*). The fluid-filled cochlea resembles a snail shape lined with about 16,500 microscopic, delicate outer (13,000) and inner (3,500) hair cells (*stereocilia*). The vibration from the stapes moving in and out sets the cochlear fluid in motion and stimulates hair cells generating electrical (neural) impulses at the *auditory nerve* (*VIII Cranial Nerve*), which, in turn, sends the neural signals through several processing stations along the central auditory pathway to the brain.‡

* Earwax (cerumen) is commonly found in this part of the ear. Cerumen is actually good as long as it does not block the sound transmission to the eardrum. Cerumen traps dirt before it can reach the eardrum and repels insects from making your ear canal their home.

† A feeling of stuffiness may occur if air pressure changes, such as when ascending or descending in an aircraft. Swallowing, yawning, or even chewing gum can open the Eustachian tube equalizing middle ear pressure with the atmospheric pressure.

‡ This simplified description of auditory anatomy and physiology should not minimize the complexity and sophistication of this elegant sensory system. Please see additional references at the end of this chapter for more information.

AUDITORY DEVELOPMENT

The following risk factors (Table 5.4) and developmental milestones (Table 5.5) will help caregivers and teachers distinguish between children who are developing hearing normally and those who may be at risk for hearing loss.

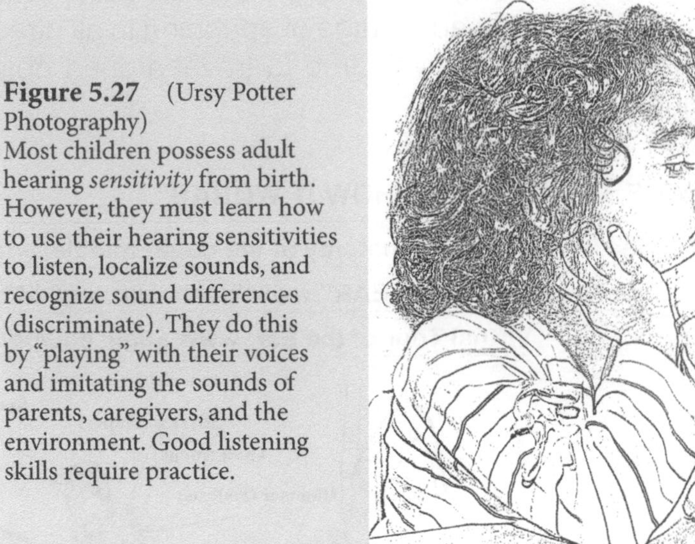

Figure 5.27 (Ursy Potter Photography)
Most children possess adult hearing *sensitivity* from birth. However, they must learn how to use their hearing sensitivities to listen, localize sounds, and recognize sound differences (discriminate). They do this by "playing" with their voices and imitating the sounds of parents, caregivers, and the environment. Good listening skills require practice.

Table 5.4 Risk Factors for Hearing Loss among Young Children[a]

Age	Risk Factors and What to Check
Newborns <28 days	**Identify hearing loss as early as possible. Health care professionals, assisted by family histories, conduct testing.** • Family history of hereditary childhood sensorineural hearing loss • Deformities of the head and face, including those affecting the pinna and ear canal • Birth weight less than 1500 grams (approximately 3.5 pounds) • Apgar (general health) scores of 0–4 at 1 minute or 0–6 at 5 minutes after birth • Jaundice at a serum level requiring transfusion (Hyperbilirubinemia) • In utero infections, such as cytomegalovirus, rubella, syphilis, herpes, and toxoplasmosis • Medicines that can damage hearing administered to the mother or infant (ototoxic medications), including aminoglycosides (some antibiotics, e.g., gentamicin) • Bacterial meningitis • Mechanical ventilation lasting 5 days or longer • Findings associated with syndromes known to include sensorineural hearing loss • Any occasion when the baby is unable to breathe (asphyxia) • Fetal alcohol syndrome • Maternal diabetes • Parental radiation
Infants <2 years	**At this age, caregivers may notice symptoms of hearing loss.** • Concern regarding communication or developmental delay • Fluid in the ear for 3 months or longer • Head trauma associated with loss of consciousness or skull fracture • Bacterial meningitis or other infections known to cause sensorineural hearing loss • Use of medications that may damage hearing (aminoglycoside antibiotics like gentamicin, neomycin, streptomycin, etc.) • Findings associated with a syndrome known to include sensorineural hearing loss • Measles, encephalitis, chicken pox, influenza, and mumps

[a] Adapted from the NICDC (http://www.nidcd.nih.gov/health/hearing/index.asp) retrieved 10 January 2005.

Table 5.5 Normal Auditory and Language Developmental Milestones

Age	Milestones	Image
Birth to 5 Months	• Reacts to loud sounds • Turns head toward a sound source • Watches your face when you speak • Makes noise in response to others talking to him • Vocalizes sounds of pleasure and displeasure (laughs, giggles, cries, or fusses)	 (Nancy Vause)
6 to 11 Months	• Understands "no" • Babbles (says "ba-ba-ba" or "ma-ma-ma") • Tries to communicate by actions or gestures • Tries to repeat your sounds	
12 to 17 Months	• Attends to a book or toy for about 2 minutes • Follows simple directions accompanied by gestures • Answers simple questions nonverbally • Points to objects, pictures, and family members • Says two to three words to label a person or object (pronunciation may not be clear) • Tries to imitate simple words	 (Ursy Potter Photograhy)
18 to 23 Months	Enjoys caregivers reading to them • Follows simple commands without gestures • Points to simple body parts such as mouth and nose • Understands simple verbs such as "eat" and "sleep" • Correctly pronounces most vowels and phonemes n, m, p, h, especially in the beginning of syllables and short words • Begins to use speech sounds • Says 8 to 10 words (pronounciation may still be unclear) • Asks for common foods by name • Makes animal sounds such as "moo" or "meow" • Starting to combine words such as "more milk" • Begins to use pronouns such as "mine"	

(continued)

Table 5.5 Normal Auditory and Language Developmental Milestones (Continued)

Age	Milestones
2 to 3 Years	• Knows about 50 words at 24 months • Says around 40 words at 24 months • Knows some spatial concepts such as "in" and "on" • Knows pronouns such as "you," "me," "her" • Knows descriptive words such as "big" and "happy" • Speech is more accurate but may still leave off ending sounds • Strangers may not be able to understand what the child says • Answers simple questions • Begins to use more pronouns such as "you," "I" • Speaks in two- to three-word phrases • Uses question inflection to ask for something (e.g., "My ball?") • Begins to use plurals such as "shoes" or "socks" and regular past tense verbs such as "jumped"
3 to 4 Years	• Groups objects such as foods, clothes, etc. • Identifies colors • Uses most speech sounds but may distort some of the more difficult sounds such as l, r, s, sh, ch, y, v, z, and th. These sounds may not be fully mastered until age 7 or 8 • Uses consonants in the beginning, middle, and ends of words. Some of the more difficult consonants may be difficult to understand, but attempts to say them • Strangers are able to understand much of what the child says • Able to describe the use of objects such as "fork," "car," etc. • Has fun with language. Enjoys poems and recognizes language absurdities such as, "Is that an elephant on your head?" • Expresses ideas and feelings rather than just talking about the world around him or her • Uses verbs that end in "ing," such as "walking," "talking" • Answers simple questions such as "What do you do when you are hungry?" • Repeats sentences
4 to 5 Years	• Understands spatial concepts such as "behind," "next to," and "under" • Understands complex questions • Speech is understandable but makes mistakes pronouncing long, difficult, or complex words such as "hippopotamus" • Says about 200 to 300 different words • Uses some irregular past tense verbs such as "ran" and "fell" • Describes how to do things such as painting a picture • Defines words  (Ursy Potter Photography)

Table 5.5 Normal Auditory and Language Developmental Milestones (Continued)

5 Years	• Understands more than 2000 words • Understands time sequences (what happened first, second, third, etc.) • Carries out a series of three directions • Understands and uses rhyming • Engages in conversation • Sentences can be eight or more words in length • Uses compound and complex sentences • Describes objects • Uses imagination to create stories

HEARING EVALUATIONS

Caregivers or teachers can request hearing evaluations on infants as well as older children (Box 5.3). Some hearing evaluation techniques do not require a response or reaction (electro-acoustic or electrophysiological). For example, infants typically lie still or sleep during hearing screenings in the newborn nursery or well baby clinic. As a child's motor and auditory skills mature, the testing may involve behavioral responses to calibrated sounds in a "sound-treated room."

Most schools employ traditional annual school screenings where children listen and raise their hands when they hear a tone. At any stage, if responses to sound are inconsistent, an audiologist should conduct a comprehensive evaluation.

Knowledge of hearing tests will help educators understand what is available and how to explain the evaluations and their benefits to caregivers or parents. Box 5.4 shows some methods to evaluate hearing.

BOX 5.3 WHY TEST A CHILD'S HEARING?

- Prevalence varies from one to six of every 1000 children in the United States are born with a hearing loss that can interfere with learning speech and responding to sounds. Newborn screening can identify most problems, but hearing testing may not identify some congenital hearing loss until later in childhood.
- Even though a child hears from birth, children can acquire a temporary or permanent hearing loss at any time during childhood. By identifying hearing impairment early, health care professionals can implement effective interventions to improve communication skills, language development, and behavioral adjustment.
- Is newborn hearing screening a requirement? Many states do require newborn hearing screening prior to hospital discharge. If not, parents should seek a newborn hearing screening within the first month. A list of state requirements is available at www.asha.org.
- The Healthy Hearing 2010 objectives of the National Institute on Deafness and other Communication Disorders (NIDCD) includes initiatives to detect hearing loss early, reduce the incidence, and treat or rehabilitate hearing loss as soon as possible (www.nidcd.nih.gov/health/hearing/index.asp).
  Research indicates that hearing aids can help 80%–90% of all hearing losses that cannot be treated with medicine or surgery.
- Parent concern is sufficient to prompt a visit to an audiologist as parents initially suspect a hearing problem with 70% of children with hearing loss.

Do not delay—get your annual hearing check up today!

Do not accept advice that a child is too young for a hearing test.

BOX 5.4 HEARING TESTS: WHAT THEY MEASURE, WHY, AND THE PROCESS

Test Measures	What is it?	Indications	Process
Otoacoustic emissions (OAE)	Measures the ear's "echo"	• An echo response is normal and can indicate hearing sensitivity for different parts of the cochlea • No echo may indicate hearing loss or damage to the cochlear hair cells	• The examiner inserts a tiny sponge-like earphone into the ear canal Some school systems and hearing conservation programs are beginning to employ OAE technology to detect NIHL and damage to the inner ear (cochlea) hair cells
Auditory brain stem response (ABR)	Checks the electrical impulses of the nerve	• Determines whether the auditory nerve and brain stem are transmitting sound to the brain • Very useful in testing infants, young children, and other difficult to test patients	• Children lie still or sleep wearing headphones and small electrodes taped to their head and ear Activity and noise can introduce extraneous neural activity, so doctors may choose to give a mild sedative to active children to help calm them during testing
Immittance	Evaluates how well the vibrating system of the ear works and the integrity of auditory and facial nerves	• Routine part of a comprehensive hearing evaluation consisting of tympanometry, static immittance, and acoustic reflex thresholds • If a child has fluid in his or her middle ears, examiners may monitor tympanograms to track progress • Can detect perforations (holes) in the tympanic membrane (eardrum)	The child sits quietly and still for the following tests: • ***Tympanometry*** The examiner places a small ear probe in the ear canals. The examination varies air pressure and presents sounds to each ear to produce a dynamic picture of the function of the middle ear • ***Static immittance*** Tympanogram produces equivalent volumes. Compares an isolated picture of the middle ear to the overall acoustic immittance of the auditory system. Detects small perforations in the tympanic membrane • ***Acoustic reflex testing*** Sound causes the middle ear muscle to flex This measurement can help confirm pure tone audiometry results, hearing sensitivity, and the location of an auditory pathway disorder
Pure tone audiometry	Tests the child's ability to hear to quietest possible sound (threshold)	• Listeners judge their ability to "barely" hear the lowest intensity (threshold) across a range of sound frequencies important for communication for each individual ear (e.g., 250, 500, 1000, 1500, 2000, 3000, 4000, 6000, and 8000 Hz)	• Listeners wear headphones or insert earphones, but may also listen through a "bone conduction" transducer to measure hearing "thresholds" for each ear Schools and hearing conservation programs use pure tone audiometry. The frequency by intensity thresholds produce a graph called the audiogram (Figure 5.28) Schools usually "screen" a child's ability to hear a limited number of frequencies (500, 1000, 2000, and 4000 Hz) at intensity levels louder than threshold but still within normal limits (20 dB HL) Screening usually does not identify initial hair cell damage unless damage has already decreased hearing sensitivity. Schools may not include frequencies first affected by noise injury (3000, 4000, or 6000 Hz) delaying identification of noise-induced hearing loss in children

Table 5.6 Terms that Describe the Degree of Hearing Loss and Its Impact on Communication and Safety[a]

Degree of Hearing Loss	Decibel Level in HL	Examples of Loudness	Impact on Communication and Safety
Normal hearing	–15 to 10 dB	Rustling leaves, "coo coo clock" ticking	• No hearing-related problem
Borderline normal	11 to 25 dB	Faint speech	• Difficulty hearing very quiet speech, especially in noise • Some guidelines consider this range of hearing sensitivity 'Minimal' hearing loss
Mild hearing loss	25 to 40 dB	Quiet or whispered speech, clicking fingers	• Difficulty hearing faint or distant speech, even in a quiet environment • Difficulty hearing in noisy classrooms the teacher or talkers • May not hear parent voice or warning signal alerting the child of danger
Moderate hearing loss	40 to 55 dB	Quiet or normal speech, radio at a normal level	• Only hears conversational speech when the speaker is close. These children find it difficult to hear in class with even low levels of background noise. They will need special assistance and amplification in a classroom • May not hear distant warning signals particularly in noise • Will not hear distant caregiver voice alerts
Moderately severe hearing loss	55 to 70 dB	Normal or loud speech, doorbell	• Hears only close loud conversational speech. Will have difficulty in a classroom, especially if the speaker does not face the student. This child will need special assistance and amplification in a classroom • Will not hear distant honking car horn, warning signals, or caregiver voice alerts particularly in noise
Severe hearing loss	70 to 90 dB	Telephone ringing, thunder	• Cannot hear conversational speech. May hear close loud voices. This child will need special assistance and amplification in a classroom • Cannot hear honking car horn, warning signals, or caregiver voice alerts
Profound hearing loss	90 dB or more	Power tools, chain saw	• May hear loud sounds; hearing is not the primary communication channel. This child will need special assistance and amplification in a classroom • Cannot hear honking car horn, warning signals, or caregiver voice alerts

[a] A comprehensive hearing examination includes additional diagnostic testing not described in this discussion. The examples only offer an approximate guide. Consult your family doctor, audiologist, or ENT doctor for more information.

Hearing care professionals do not usually use percentages when describing hearing impairments (e.g., 50% hearing loss). Instead, they describe audiometric results by the degree of hearing loss and by the shape of the audiogram (configuration).

Audiograms portray the shape and range of frequencies of a person's ability to hear the softest sounds. Considering the shape of an audiogram together with the person's degree of hearing loss helps us understand how the hearing impairment affects their functioning.

The zero point on the intensity scale shown in Figure 5.28 corresponds to normal hearing.*

The smaller the intensity decibel (dB) value, the more sensitive a persons' hearing. Audiologists measure the value for each individual frequency for each ear.

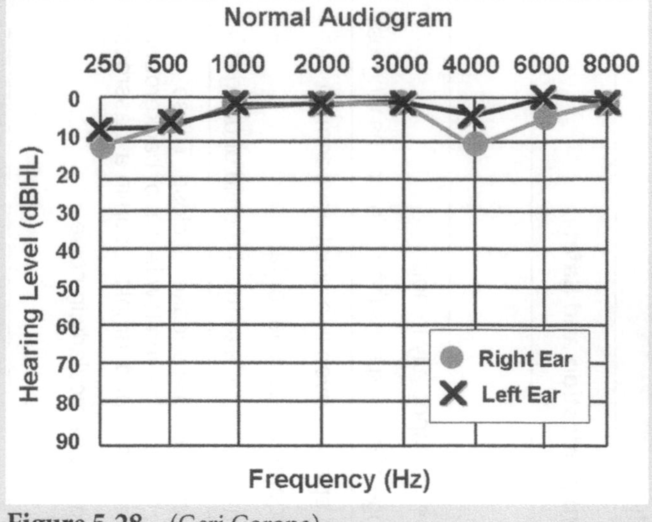

Figure 5.28 (Geri Corona)
A normal audiogram.

HEARING IMPAIRMENTS

A hearing impairment can occur from a number of conditions. These include obstruction, disease, or injury in any part of the auditory pathway disrupting sound transmission.

We characterize hearing disorders by

- the type of hearing loss (where the problem occurs in the auditory system), and
- the degree of the impairment (extent of problem compared to normal function).

Children may not tell us that they are having difficulty in hearing. Instead, they may initially complain of a feeling of "stuffiness," a hollow-sounding voice, ringing or buzzing (tinnitus) in the ear, having difficulty localizing sounds or state that others are "mumbling." Parents or teachers may notice hearing impairments through inattentive behavior or speech or language delays, before a child realizes s/he is not hearing well.

Medical or surgical intervention can usually resolve problems in the outer and middle ear. However, damage to the inner ear (cochlear) or auditory nerve usually will result in permanent sensorineural hearing loss.

* The audiogram decibel scale calibrates zero corresponding to the lowest level detected by normal listeners for each frequency. The graph distinguishes the sound level meter dB SPL values by referring to the audiogram intensity values as hearing level or dB HL. Some children hear intensity levels better than average. Negative dB HL values indicate hearing sensitivity more acute than the lowest detected sound by average normal listeners.

TYPES AND CAUSES OF HEARING LOSS (ETIOLOGY)

There are two basic types of hearing impairments:

1. Disorders that reduce sensitivity to sound (conductive, sensorineural, and mixed hearing loss)
   This is the more common of the two. Listeners require increased intensity levels (louder sounds) to perceive sounds.
2. Disorders of the auditory nervous system (retrocochlear, central auditory processing)
   This may or may not include a loss of sensitivity. In some cases, children may not hear or process sounds appropriately even though the sounds are loud enough to detect.

DISORDERS REDUCING SENSITIVITY TO SOUND

Conductive Hearing Loss A ***conductive*** hearing loss occurs when an outer or middle ear condition or obstruction keeps sound from reaching a normally functioning inner ear (Box 5.5). Most conductive disorders are mild-to-moderate in degree, temporary, and treatable, but may require preferential seating until hearing returns to normal sensitivities.

You can simulate a conductive hearing loss by turning a radio on to a soft but comfortable volume and inserting earplugs. The reduced and "muffled" sound similar to what someone with a conductive hearing loss would hear. If you turn up the volume, you can hear more clearly.

BOX 5.5 CAUSES OF CONDUCTIVE HEARING LOSS

Type	Etiology	Description
Conductive	**Otitis media (middle ear effusion or middle ear infection)**	• Otitis media (OM) describes a variety of conditions affecting the middle ear • It is the most common cause of conductive hearing loss in children and is often due to Eustachian tube dysfunction • Estimates indicate 76%–95% of all children will have at least one OM ear infection by 6 years • Ear infections are second only to well-baby checks as the reason for office visits to a physician • Some children are more at risk; these include children with cleft palate, craniofacial abnormalities or Down's syndrome, native populations, children who live in inner cities, attend day-care centers, and those passively exposed to cigarette smoke • When fluid remains in the middle ear and impedes tympanic membrane (eardrum) vibrations and middle ear bone movement, a hearing loss can occur and teachers should consider preferential classroom seating • In very young children, this hearing loss may hinder spoken language development • The duration characterizes OM acute episodes defined as a single bout lasting fewer than 21 days • Chronic OM persists beyond 8 weeks and can result in permanent damage to the middle ear mechanism. Subacute lasts from 22 days to 8 weeks • OM is considered recurrent if three or more episodes occur in a 6 month period. A child is "prone" if OM occurs before 1 year or the child has six bouts before they are 8 • Recurring or chronic OM that disrupt auditory inputs during the critical period for auditory development can place the child at risk of developing central auditory processing disorders, language, or psychoeducational delays (Stach, 1998) but generally children are very resilient (Paradise, 2007)

Conductive	**Foreign objects**	• Small objects placed in the ears (e.g., beads, beans, rocks, or food) can block sound or rupture the tympanic membrane (eardrum). Children do this more often than caregivers expect
	Wax (cerumen)	• Wax can pack in the ear canal and act like an earplug, blocking sound waves from striking the eardrum • Never use Q-tips to remove wax; this may push the wax deeper into the canal or (if inserted too deeply) even puncture the eardrum
	Otitis externa	• Bacteria, a virus, or a fungus can cause inflammation of the external ear. This may swell the ear canal shut, resulting in a temporary hearing loss • One common condition is *Swimmer's Ear*, a painful bacterial infection that occurs after bathing or swimming during hot, humid weather
	Physical trauma	• Physical trauma can separate the ossicular chain (middle ear bones) or rupture of the tympanic membrane • *Barotraumas*, unequal air pressure in the inner ear and the atmospheric pressure outside the body can also result in physical trauma • Barotrauma occurs with altitude changes, congestion, and other blockages of the Eustachian tube

* A more comprehensive listing of hearing loss etiologies is available in the references.

Sensorineural Hearing Loss A ***sensorineural*** hearing loss occurs when some condition, injury, or damage stops the intricate cochlea (inner ear) from "translating" middle ear sound energy into a representative neural impulse in the auditory nerve. The cochlea is a fluid-filled intricate sensory system. Neural impulses will not transmit correctly if there is damage or destruction of the tiny rows of hair cells. The loss is usually irreversible and permanent. About 30% of hearing impaired children also have an additional disability (Wolff and Harkins, 1986). Unfortunately, the Department of Education does not report data by disability for children birth to 5 years of age.

Sensorineural hearing loss can be caused by a variety of congenital or acquired conditions include: syndromes and inherited disorders (e.g., Alport, Pendred, Usher syndromes); infections (e.g., rubella, syphilis, HIV, meningitis, mumps); ototoxic medications; some antibiotics (e.g., gentamicin); autoimmune disease; aging (presbycusis); and noise exposure.

Other than aging, noise is the most common cause of ear injury resulting in sensorineural hearing loss.

Ear injury usually occurs after chronic long-term exposures to toxic noise, but damage can also occur immediately from a single exposure to high intensity impulse. ***Acoustic trauma*** can damage eardrums and middle ear ossicles in addition to delicate inner ear hair cells. Either way, the result is ear injury resulting in NIHL. Since the ear injury often occurs gradually and without blood or pain, children are often unaware of the damage.

Risk of inner ear injury or acoustic trauma is generally determined by:

- How loud: intensity of sound. The louder the more hazardous.
- How long: duration of sound exposure. The longer the more hazardous.

Frequency and temporal noise variables also influence risk, that is, higher frequency is more dangerous and impulse noise is more hazardous.

The complexity of sensorineural hearing losses is greater than conductive hearing losses because the cochlea influences the entire range of sounds. If the inner ear hair cells lose sensitivity, their ability to "fine tune" results in a frequency blurring or distortion.

A person with a conductive hearing loss can overcome the loss by increasing sound loudness. Unfortunately, the solution is not as simple for those with permanent sensorineural hearing loss.

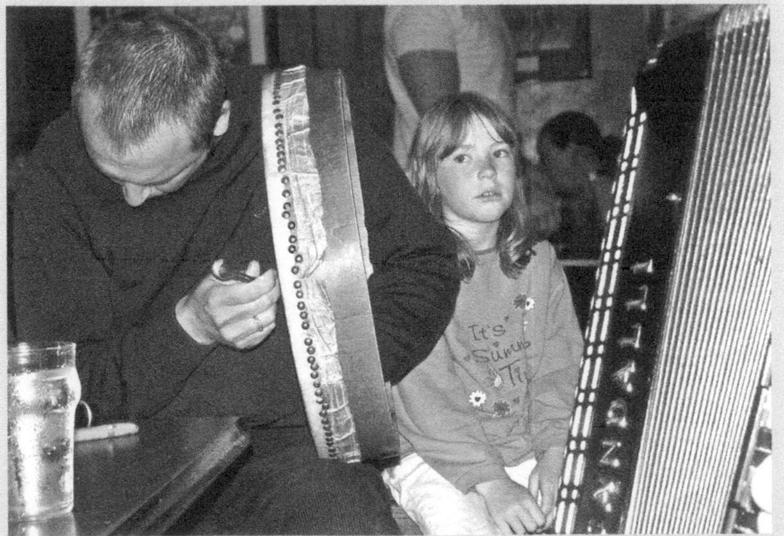

Figure 5.29 (Ursy Potter Photography)
Even sounds we enjoy can injure our ears and destroy hearing. Some suggestions to protect your hearing:

- Check how loud and long you can listen to sounds or music you like to hear
- Crank it down
- Give your ears a 10 minute break every hour
- Consider musician's earplugs

Sensorineural hearing loss damages the ear's ability to transmit an exact replica of sounds to the brain. This impairs a child's ability to detect the individual frequency components of a sound and distorts their ability to understand speech. A child may "hear" but not understand the speech, particularly in noisy environments. Depending on the degree of impairment, teachers and family members may accuse these children of "not paying attention or listening."

If a child has a sloping high-frequency hearing loss, they will be able to hear most vowel energy and rhythm of speech but may have difficulty in understanding or listening consistently. This is because many of the sounds that distinguish one word from the next are found in the mid to higher frequencies.

Children with sensorineural hearing loss benefit particularly from good classroom acoustics, preferential classroom seating, and regular audiometric evaluations. They may require auditory rehabilitation such as amplification, FM systems, etc. As technology continues to improve, children with severe or profound sensorineural hearing loss are often candidates for cochlear implants* if they do not benefit from conventional amplification. Optimized classroom acoustics is a critical ergonomic issue for all children but particularly for children using amplification (hearing aids).

You can simulate sensorineural hearing loss by turning the radio to a soft level, inserting earplugs, and tuning in slightly off station. This causes you to hear the music or speech, but it is muffled and unclear. Increasing the volume makes it louder, but not necessarily any easier to understand. Without fine-tuning, the result is louder static.

Children (and adults) with sensorineural hearing loss often do not recognize the extent of their loss of hearing; they may need encouragement to evaluate, treat, or rehabilitate their hearing. While, protecting precious hearing is the priority, if a child (or adult) experiences a sensorineural hearing loss, amplification technology (hearing aid) can usually help improve detection and communication abilities.†

Mixed hearing loss A child with **mixed hearing loss** possesses both conductive and sensorineural components. It is a combination of factors affecting both the middle ear and the inner ear (cochlea).

DISORDERS OF THE AUDITORY NERVOUS SYSTEM: AUDITORY PROCESSING DISORDERS: Children with Auditory Processing Disorders (APD) are particularly at "high risk" if they are younger than 13 year old. Auditory processing is what the brain does with the sounds the ears hear (Figure 5.30 and Figure 5.31).

Another (synonymous) term used is Central Auditory Processing Disorder (CAPD) (ASHA, 1996). CAPD is different from Attention Deficit Hyperactivity Disorder (ADHD). Some experts consider ADHD an output problem with a child unable to control behaviors whereas APD interferes in the input of auditory information (Chermak et al., 1999). Studies estimate APD affects between 2% and 3% of children with twice as many boys experiencing APD than girls (Chermak and Musiek, 1997).

* Cochlear implants are different from conventional hearing aids. Rather than amplifying the sound, this device is implanted and bypasses the outer and middle ear delivering signals directly to the auditory nerve.

† If a person experiences a sensorineural hearing loss, studies show 80%–90% can benefit from amplification (hearing aids). Technology continues to advance. Even though the intricate abilities, performance, and perception of the human ear are complex and difficult to replicate, consult your audiologist for specific hearing care options. Technology information and local audiologist referrals are available at www.audiologyawareness.com, by calling 888-833-EARS (3277), or by referring to Resources in Appendix C.

Figure 5.30 (Kurt Holter Photography, Frederick, MD)

Figure 5.31 (Kurt Holter Photography, Frederick, MD)

A child with an auditory processing disorder (APD) often will behave like a child with a hearing loss, but in reality the child possesses normal hearing sensitivity for sound.

One way to differentiate hearing loss from APD is that hearing is "ear-based," while listening is "brain-based." A child with an APD needs a tailored intervention strategy or individual education plan (IEP) that focuses on improving listening and learning skills, providing strategies to address environmental challenges and remediation. They will probably need preferential seating and individual attention to be certain they hear and understand what they are to do for their classwork and homework.

Florida Department of Education has an excellent list of intervention strategies to assist teachers and parents, available at www.firn.edu/doe/bin00014/pdf/y2001-9.pdf.

APD can result from head trauma, brain tumors, neurologic, or vascular changes. There also is some concern that APD is hereditary, although many professionals also believe a cause is inconsistent or poor auditory input during the critical period of auditory perceptual development.

Even though their intelligence and hearing sensitivity are normal, children with APD have difficulty attending to, discriminating, recognizing, or understanding auditory information. Although APD children are capable of hearing faint sounds, they exhibit poor listening, memory, spelling, reading or phonic skills, distractibility, short attention span, impulsiveness, and disorganization.

They also have difficulty understanding speech (particularly in groups and noisy environments), localizing sounds, separating dichotic stimuli, processing normal and altered temporal cues, and understanding speech with reduced redundancy. They may find it difficult to follow directions and exhibit a discrepancy between their intelligence scores (IQ) and verbal skills. Children can possess APD in isolation or along with attention deficit disorders, learning disabilities, and language disorders.

Each child is different. Therefore, a multidisciplinary team (e.g., audiologist, speech pathologist, psychologist, learning disability specialists, social worker, regular classroom teacher, and parent) will develop intervention strategies to meet the specific needs of the individual child.

The classroom-learning environment is a critical part of the management of APD since noise and reverberation exacerbate the disorder. In fact, some studies suggest these children require an SNR ranging from +12 to +20. Each child's tailored plan should include accommodation strategies (e.g., note takers, study guides) or modifications of the learning process (reduced language or repetition of concepts).

AREAS OF INTERVENTION

PREFERENTIAL SEATING
INSTRUCTIONS
PREVIEW, REVIEW AND SUMMARIZE
TIME
CLASSROOM ADAPTATIONS
SELF ADVOCACY
ORGANIZATION

Section B: Prevention

TOXIC NOISE AND ACOUSTIC TRAUMA

The incidence of NIHL in children is increasing at almost epidemic proportions. Approximately 97% of all third graders experience noise hazards (Blair et al., 1996). Even sound we may not consider "too loud" can have negative effects. Studies link noise levels (less than 75 dBA) to nonauditory effects such as changes in arousal, sleep patterns, blood chemistry, cardiovascular impacts, and effects on a fetus.*

CONTINUOUS NOISE

Again, the risk of NIHL depends upon the intensity (loudness), duration (time), frequency, and nature of the noise. Continuous toxic noise exposure can damage hearing gradually over time.

Since the loudness of noise frequently varies over time, professionals use noise-sampling equipment (dosimeters) to calculate the average exposure over a day. The National Institute of Occupational Safety and Health (NIOSH) recommends adults limit noise exposure to an *average* of 85 dBA for an 8 hour workday.† A child's routine day (bus, band, sports, and shop) is noisier than a routine day for most adults (Clark, 1994; WHO, 1997).

There are no comparable standards for children. Application of adult guidelines to children is worrisome, especially in light of the smaller sized anatomical features and developing physiology.

Without guidelines for children, it may be best to use NIOSH's most conservative average exposure level of 80 dBA for an 8 hour day. This is considered a "safe limit" for *most* adults.

We do not know the long-term cumulative impact of childhood noise exposure to a "child's lifetime noise dose" as they mature into adulthood.‡ However, we do know that the louder the noise, the less exposure time is safer for children. For example, exposure time is halved for every 3 dBSPL increase in intensity. This is because sound is energy; we calculate it by its logarithmic (exponential) progression. This means that a 10 dBSPL increase does not simply add 10 to the previous level. It multiplies the previous level by 10.

If a child listens to an MP3 player for 8 hours at 80 dBA (100% noise dose) for one day and increases to 83 dBA the next day, he should limit his listening time during the second day to 4 hours (decrease time by ½ for every 3 dB increase). A 4 hour exposure at 83 dBA delivers the same energy to the ear as listening to an 80 dBA noise for 8 hours (Table 5.7).

* Read more about nonauditory effects in Jones and Broadbent (1998); Jansen (1998); Bronzaft (1997); Evans et al. (1995); Cohen et al. (1980); Gerhardt et al. (1999).

† The Occupational Safety and Health Act (OSHA) 29 CFR 1910.95 monitors safety and health in industry. Some industries are not covered by OSHA (mining, farming, and construction); mining is monitored by the Mine Safety and Health Administration (MSHA). NIOSH (Department of Health and Human Services) conducts research and makes recommendations to OSHA. They recommend 85 dBA as the criteria for safe occupational exposures. Yet, some individuals are more susceptible than others and it is possible to experience hearing loss even after following the recommended guidelines.

‡ A noise "dose" combines time with loudness.

Unfortunately, many MP3 players deliver toxic noise to children's ears sometimes exceeding 130 dB peak pressures. Researchers at Boston Children's Hospital recommended "Cranking it Down" to 60% of the potential volume to achieve a relatively safe listening level for 1 hour following NIOSH's adult standards (Fligor and Fox, 2004).

Table 5.8 lists some examples of the sound energy range of our ears. A few added decibels dramatically increase the amount of sound energy (Figure 5.32).

Table 5.7 Intensity–Time Tradeoff: Examples of "Noise Dose" using NIOSH Guidelines

Intensity Level	Exposure Duration (Continuous Noise)	Noise Dose for Children
Louder (more dBA)	Longer (more time)	More noise dose
74 dBA	8 hours	25%
77 dBA	8 hours	50%
80 dBA	8 hours	100%
83 dBA (+3 dB)	8 hours (same time)	200% (dangerous noise dose)
83 dBA (+3 dB)	4 hours (reduced time ½)	100%
86 dBA (+3 more dB)	2 hours (reduced time ½)	100%
89 dBA (+3 more dB)	1 hours (reduced time ½)	100%
92 dBA (+3 more dB)	30 minutes (reduced time ½)	100%

Table 5.8 Common Sounds and Examples of Intensity–Pressure Relationships

Common Sources of Sound	Intensity	Change in Pressure	dB from Baseline
Threshold of hearing	0 dBA		
Whisper	35 dBA		Baseline
Quiet classroom	45 dBA	=35 dBA × 10	(+10 dB)
Conversation	55 dBA	=35 dBA × 100	(+20 dB)
City traffic noise	65 dBA	=35 dBA × 1000	(+30 dB)
Cafeteria	75 dBA	=35 dBA × 10,000	(+40 dB)
Hair dryer	85 dBA	=35 dBA × 100,000	(+50 dB)
Motorcycle or lawn mower	95 dBA	=35 dBA × 1,000,000	(+60 dB)
Band concert	105 dBA	=35 dBA × 10,000,000	(+70 dB)
Monster truck rally	115 dBA	=35 dBA × 100,000,000	(+80 dB)

Figure 5.32 (Valerie Rice)
Noise levels during aerobic classes, sporting events, monster truck rallies, or from auto factory-installed sound systems (such as in a vehicle) of 115 dBA have 100 million times more sound energy than 35 dBA. They are capable of causing 10 million times more damage to hearing. Experts agree any exposure to continuous noise above 129 dBA (even for less than a second) is hazardous (AAA, 2003).

Sound Advice

How to tell if you are in toxic noise (without a sound level meter)

- Place a friend at arm's length
- If you must yell so your friend can hear you talk,
- You are in the ***toxic noise danger zone***
- Turn it down... Walk ***away***...
- ***Protect your ears***

IMPULSE NOISE

Impulse noise can cause immediate and permanent damage to our hearing. The adult limit for impulse noise exposure is 140 dBP (NIOSH, 1998).

While most childhood noise hazards do not reach this intensity, there are some exceptions (Figure 5.33). Some toy noises measure over 110 dBA and cap guns are 138 dBP (Nadler, 1997a,b). A recent survey of eighth graders revealed 17% participate in recreational shooting (Watkins, 2004).

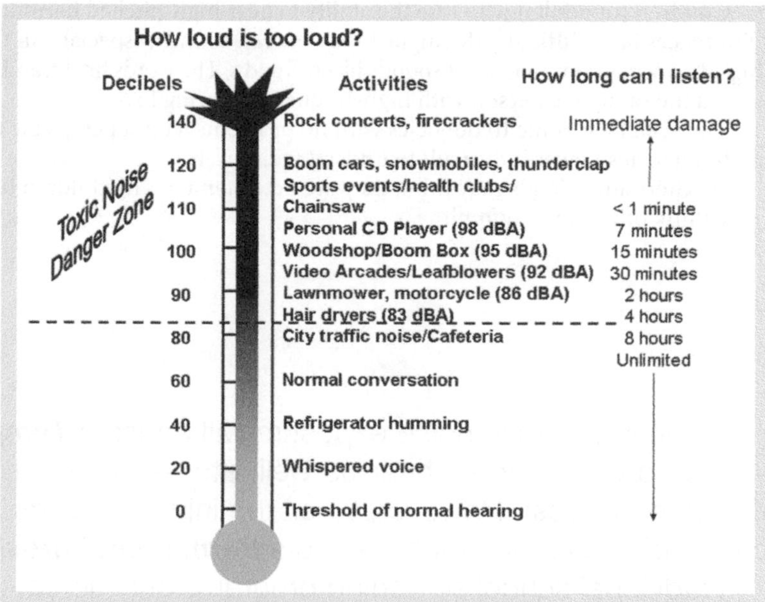

Figure 5.33 (Geri Corona)
Common sources of noise and their intensity levels, along with permitted exposures.

Sound Advice

Limit noise hazard exposure to:

- **80 dBA continuous noise**
- **140 dBP impulse noise**

RECOGNIZING NOISE-INDUCED HEARING LOSS (NIHL)

Typically, hearing loss due to noise exposure occurs first at the higher frequencies, particularly around the 4000 Hz level (3000–6000 Hz), where a child might not immediately notice the effects.

In fact, a child may have difficulty understanding speech, complain of noises or ringing in the ears (tinnitus), turn the volume up on the radio or TV, have difficulty locating sounds, and never realize the reason for these behaviors is hearing loss.

If you **think** you have a hearing loss—
you are probably ***right***

If you **think** no one has noticed
you are probably ***wrong***©.

Don't delay—Get your hearing tested by your audiologist TODAY!

Reprinted with permission of the Audiology Awareness Campaign Foundation www.audiologyawareness.com

Something caught the attention of the child and distant man, but the man on the right did not seem to notice.

Children (or adults) who lose the ability to hear high-pitched harmonic differences have difficulty distinguishing between sounds, especially in the high-frequency components of sounds like h, f, and s. The words hit, fit, and sit may sound the same to a person with high-frequency hearing loss.

They can hear some frequencies with no problem; yet not being able to hear all frequencies makes it difficult to understand speech.

Teachers and friends may think they are not listening. The children themselves may think others are "mumbling."

Figure 5.34 (Kurt Holter Photography, Frederick, MD)

Sometimes, toxic noise exposure will cause a ***Temporary Threshold Shift*** (TTS). This is when a child cannot hear as well after exposure to noise. Typically, hearing recovers after auditory rest. However, when the injury exceeds the ear's ability to recover, hearing thresholds no longer improve and a ***Permanent Threshold Shift*** (PTS) results.

Traditional school screening programs only detect when hearing is outside the "normal range" limits. Typically, the testing is not sensitive to small but significant changes in hearing sensitivity below 25 dB HL that could serve as early warning indicators of a hearing loss.

A progressive ***School Hearing Assessment Program*** (SHAP) should test children's hearing thresholds at targeted grades and institute hearing health education at all grade levels. School-based programs are shown later in this chapter.

Unfortunately, permanent damage to the inner ear hair cells can occur long before a pure tone audiometric test can detect any change in hearing, even using a threshold test procedures. Fortunately, the ***Otoacoustic Emissions*** (OAE) test can often detect inner ear damage immediately even before the pure tone hearing thresholds decline. Some schools and employers use OAE technology to detect NIHL.*

* Hearing evaluations section contains additional information on hearing tests.

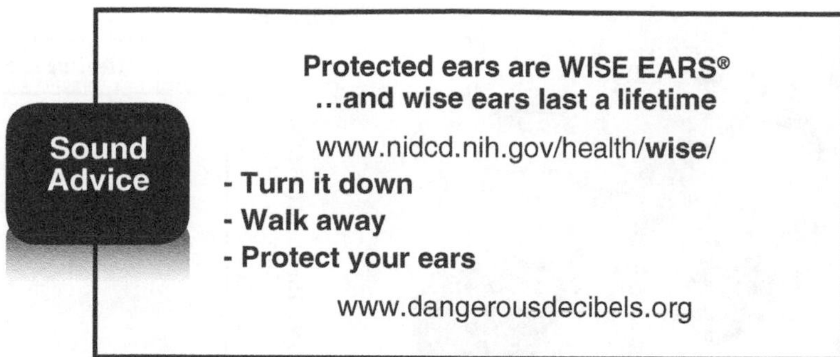

CLASSROOM NOISE

Toxic noise destroys hearing, however noise in the classroom even at "safe" levels creates an ergonomic mismatch for everyone.

Noise degrades performance

EFFECT ON TEACHERS

It is difficult to teach in noisy classrooms. Teachers often compensate for background noise by raising their voices risking strained vocal cords.* Teachers report voice strain and fatigue and missed work due to voice problems (Smith et al., 1998).

Noise affects adults' ability to perform these tasks (Kryter, 1994; Belojevic et al., 1992):

- Concentration, learning, or analytic processing
- Speaking and listening
- Fine muscular movements
- Simultaneous tasks
- Continuous performance
- Prolonged vigilance
- Performance of any task involving auditory signals
- Paying attention to multiple channels

It is reasonable to assume noise affects children in similar ways.

EFFECT ON CHILDREN

Even though a child is born with the ability to hear, children must learn how to listen. To learn basic phonemic awareness† skills requires children must learn to listen (Johnson, 2000; Flexer, 1999).‡ Children's auditory listening skills continue developing until adolescence (Figure 5.35 and Figure 5.36).

---

* This automatic raising of one's voice is known as the Lombard Effect (Lane and Tranel, 1971).

† Phonemes are small units of speech that correspond to letters of an alphabetic writing system. The awareness that language is composed of these small sounds is phonemic awareness.

‡ This may be due to their lack of verbal experience and the fact that children's brains do not fully develop until approximately age 15. A phoneme is the smallest contrastive unit in the sound system of a language or the smallest unit of speech.

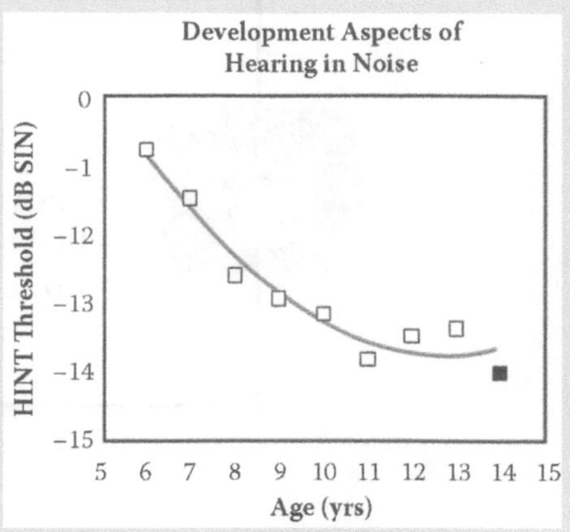

Figure 5.35 (Kurt Holter Photography, Frederick, MD) **Figure 5.36** (Reprinted with permission from ASA)

In order to learn to listen, children need optimized classroom acoustics. This means reducing the effects of noise and reverberation, so there is less background noise. As the child matures, their ability to listen more efficiently continues to improve though early childhood.

Children do not reach adult performance until they become teenagers. Figure 5.36 shows mean improvement in Hearing-in-Noise-Test thresholds as a function of age for groups of normally developing children. Average adult thresholds are shown as a filled square. (Adapted from Soli, S.D. and Sullivan, J.A., *J. Acoustical Soc. Am.*, 101, S3070, 1997.)

Children's auditory abilities differ from adults' (Werner and Boike, 2001; Johnson, 2000; Litovsky, 1997). Adults can distinguish speech when the noise is louder (SNR = −4), whereas young children in the same room would experience great difficulty understanding the same conversation. Children with English as a Second Language (ESL) and children with hearing loss would experience even greater difficulties. Because of these differences, adults may not recognize a poor listening environment for children when they hear it.

Phonemic awareness eludes about 25% of middle-class first graders and an even greater portion of children from lower socioeconomic backgrounds. Such differences are largely a result of children's exposure to speech patterns. The importance of an optimized classroom listening environment is critical since children without phonemic awareness find it difficult to learn to read and write (Adam et al., 1990).

Children who cannot hear well (***audibility***) have difficulty doing their "work" of listening accurately (Figure 5.37). Lacking adults' extensive "language library," children often cannot "fill in the blanks" when hearing new words and partial phrases.

Noise has a particularly negative effect for 10- to 11-year-old children with ESL (Figure 5.38). Children with English as their primary language correctly identified speech in the presence of higher noise levels than children with ESL.

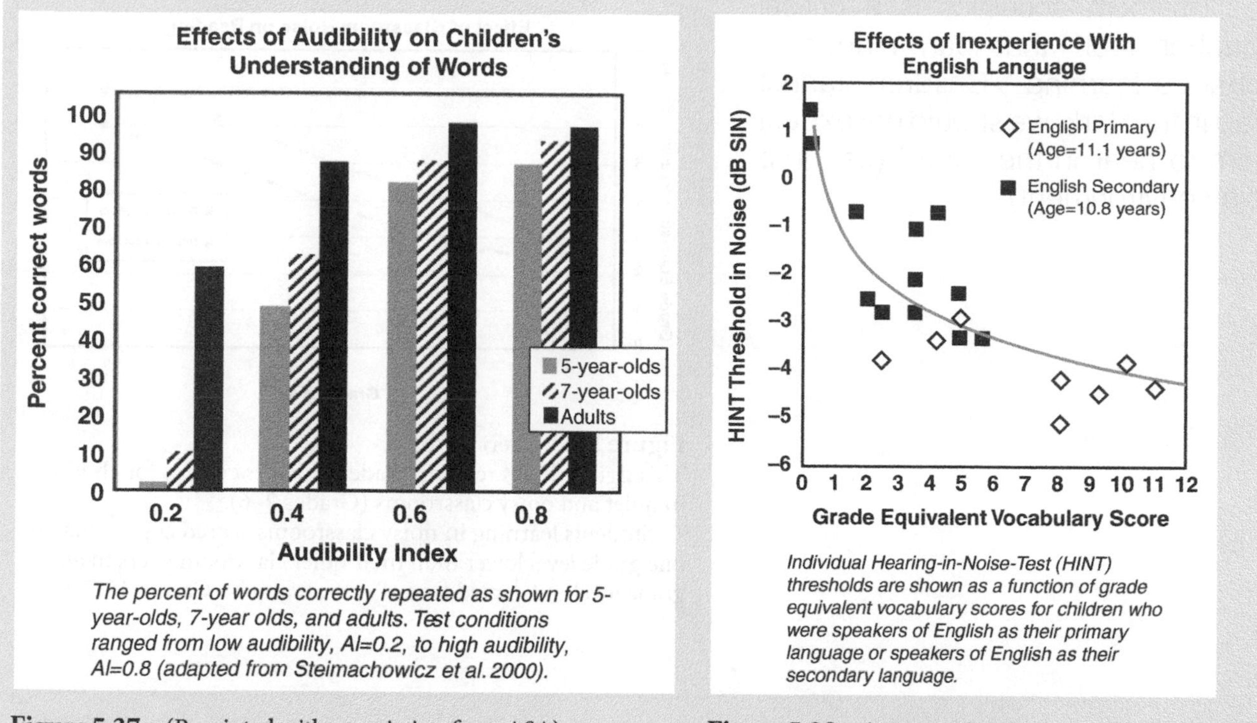

Figure 5.37 (Reprinted with permission from ASA)

Figure 5.38 (Reprinted with permission from ASA)

Given that children have difficulty hearing and understanding speech in noisy environments, it follows that their academic performance also suffers. Children's academic scores have dropped due to train or traffic noise (Lucas et al., 1981) (Figure 5.39). Reading scores also tend to be lower for children in noisy classrooms compared with quiet ones (Bronzaft and McCarthy, 1975) (Figure 5.40).*

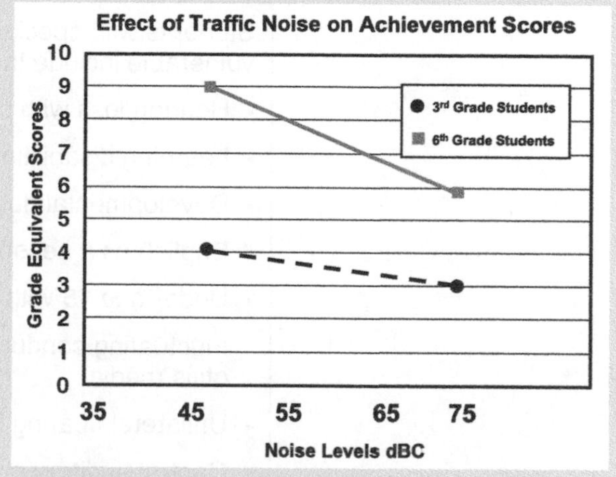

Figure 5.39 This graphic shows achievement scores as a function of increasing noise levels for third and sixth graders. Third-grade students in quiet classrooms scored one grade level higher than third graders in noisy classrooms.

The differences were even more dramatic for the sixth graders. These older children have more language experience and listening experience. However, quiet classroom sixth graders scored three grade equivalent levels higher than sixth graders in noisy classrooms. (Adapted from Lucas, J.S., Dupree, R.D., and Swing, J.W., *Learn. Mem. Cogn.*, 20(6), 1396, 1981.)

* For a review of the effects of noise on children at school, see Shield and Dockrell (2003a,b). For more information on hearing, noise-induced hearing loss, and classroom acoustics, see Heller (2003). For information on children with sensitivities to sound and other senses, see Kranowitz (1998) and Fischer et al. (1991).

Classroom acoustics is a critical "teacher tool" that can hinder or enhance learning.* Children cannot read in the dark, we should not expect them to hear in the "dark" (i.e., in a noisy environment).

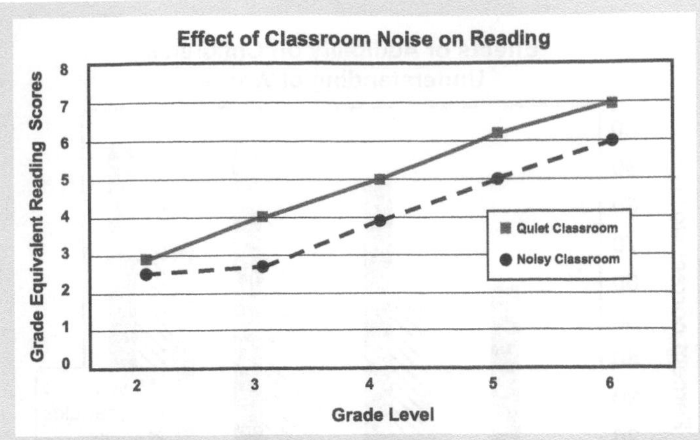

Figure 5.40 (Geri Corona)
This figure shows reading grade equivalent scores for children in quiet and noisy classrooms (Grades 2–6).

Students learning in noisy classrooms scored approximately one grade level lower than their quiet classroom peers in all grade levels. (Adapted from Bronzaft and McCarthy, 1975.)

Special needs students are especially susceptible to educational difficulties when acoustics are poor.†

Students with special needs that are particularly vulnerable include those with‡:

- Hearing loss with or without amplification
- Learning disabilities
- Developmental delay
- English as a second language
- Under age 15 with immature speech and language
- Fluctuating conductive hearing impairments (e.g., otitis media)
- Unilateral hearing impairments
- Central auditory processing disorders
- Chronic illness
- Emotional difficulties
- Behavior difficulties
- Cochlear implants
- Cognitive disorders
- Articulation disorders
- Sensory integrative disorders

* Other factors may interact with noise as well, and influence academic performance (Shield and Dockrell, 2003a,b).

† The need for good acoustics in all classrooms is important because the American's with Disabilities Act and the UK Disability Discrimination Act encourage mainstream education for special needs children as possible.

‡ See for example Heller (2003); Nelson et al. (2002); Cunningham et al. (2001); Erikis-Brophy and Ayukawa (2000); Crandell and Scalding (1995); Gelnett et al. (1994); Schappert (1992); Flexer et al. (1990); and Ross (1990).

Techniques for improving acoustics in a classroom include facing the children, moving closer to the children, and puting yourself on their level. This preschool teacher has her children sitting on the floor around her.

Notice how she faces them while pointing out and talking about photos, rather than looking at the book herself. In the traditional classroom on the right, the teacher moved her desk so close to her students that it touches their desks.

These techniques benefit all children but are especially effective for those children with difficulties in hearing, listening, or following directions. Also, give those children preferential seating by placing them directly in front of your desk.

Figure 5.41 (Ursy Potter Photography)

Figure 5.42 (Valerie Rice)

Sound Advice

Use preferential seating.

When speaking, face children with hearing, listening, or following directions difficulties.

Design classroom activities to reduce distances betweet talkers and listeners.

CLASSROOM ACOUSTICS STANDARDS

Children need to hear clearly in class so they can develop academically and socially (Evans and Maxwell, 1997). Two scientists recognized the need for good classroom acoustics over a half century ago.

> *The school was established to promote learning, which is acquired largely by word of mouth and listening. Therefore, acoustics is one of the most important physical properties that determine how well the school building can serve its primary function. Thus, the exclusion of noise and the reduction of reverberation are indispensable in adapting classrooms to the function of oral instruction.*
>
> (Knudsen and Harris, 1950)

Over 50 years later, many schools are still struggling to resolve the noisy classroom problem (Figure 5.43 and Figure 5.44). Yet, the cost of incorporating good acoustics from the beginning in new construction is miniscule and its benefits are priceless (Lubman, 2004).

Prior to approval of the first comprehensive classroom noise standard in 2002, noise levels in US schools often exceeded those permitted in European schools by more than 20 dBA (Lubman, 2004). Popular US architectural trends of the 1960s and 1970s did not help, as the "open-floor plans" produce long RTs and poor S/N ratios.

The standard (*ANSI S12.60-2002 Acoustical Performance Criteria, Design Requirements and Guidelines for Schools*) provides acoustical performance criteria, as well as design requirements and guidelines for new classrooms and for renovation of existing classrooms

Figure 5.43 (Valerie Rice)

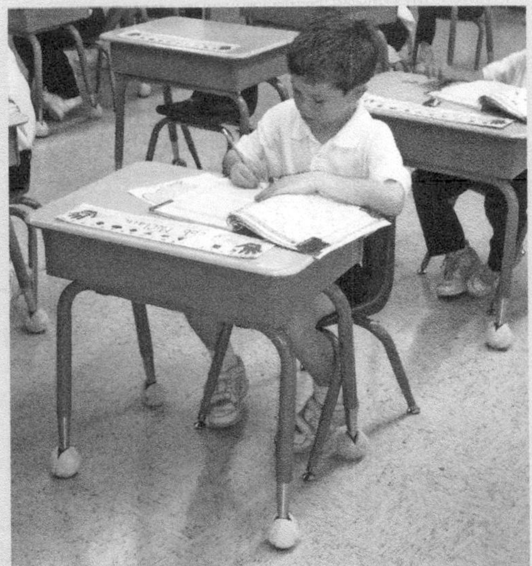

Figure 5.44 (Valerie Rice)

Classroom design can both create and improve acoustical problems.* Classrooms need low noise, low reverberation times, and high sound isolation to provide classroom acoustics conducive to communication and learning. Often schools resort to creative stopgap measures (such as placing tennis balls under table and chair legs) to reduce classroom noise.

* Classroom design particularly affects hearing (audibility) and understanding (speech intelligibility) (Cavanaugh and Wilkes, 1999).

to avoid classroom noise (Table 5.9 and Figure 5.45). Now educational facilities planners, builders, and architects must consider noise when selecting school site locations, designing classrooms or flow plans and purchasing construction materials, equipment, or air-handling systems. This standard attempts to provide universal design guidelines optimizing classroom acoustics for everyone (Lubman, 2004). The standard is available free of charge (http://asastore.aip.org) through a unique partnership between the Acoustical Society of America, Armstrong Ceiling Systems, Owens Corning, and Trane.

The annexes offer design and construction practices, installation methods, and optional procedures to demonstrate conformance to performance and design requirements. Although the recommendations are good, they focus on children and adults with normal hearing. Some special needs students (e.g., hearing loss) may still require additional classroom sound optimization, signal-enhancing technology like hearing aids, FM systems, or sound amplification/distribution systems.

Table 5.9 ANSI S12.60 (2002) Recommended Classroom Acoustics

Unoccupied Background Noise	Overall Sound Levels	Signal-to-Noise Ratio	Unoccupied Reverberation Times
35 dBA	≤70 dBA	+15 dB	<0.6 s

The World Health Organization (WHO) recommends ambient noise levels of 35 dB (30 dB for children with hearing impairments) in unoccupied classrooms. The new ANSI S 12.60 standard brings the United States into agreement with WHO and many other countries in the world. European standards are available through the National Hearing Conservation Association and teacher Web sites.

(http://www.hearingconservation.org/rs_regulatoryInfo.html#europeanStandards and (www.teachernet.gov.uk/acoustics).

Figure 5.45 (Valerie Rice)
Table 5.9 guidelines refer to core learning spaces <20,000 ft^3.

- The standard recommends a level of 40 dBA for ancillary and core learning areas with a larger volume.
- The recommended level in corridors not used for formal learning is 45 dBA.
- Additional standards are provided by the Department for Education and Employment (London).*

* Available at http://asa.aip.org/classroom.html, Two helpful booklets: "Classroom Acoustics, a resource for creating learning environments with desirable listening conditions" (free) and "Acoustical Barriers to Learning" (nominal fee) are available at: http://asa.aip.org/classroom/booklet.html. For some European standards on noise in educational buildings, see Vallet (2000) and British Associations for Teachers of the Deaf (2001).

PUTTING PREVENTION INTO PRACTICE

Prevention efforts need to touch individuals, families, schools, and communities. All persons should understand the sources and solutions to prevent hearing loss due to toxic noise. Then they need to implement the solutions!

Teacher, parent, and student awareness is crucial to prevention programs. Often the first efforts occur in children's schools.

SCHOOL HEARING LOSS PREVENTION (HEARING CONSERVATION) PROGRAMS

The insidious nature of NIHL, the unavoidable exposure to noise during daily activities, and the increasing number of children with NIHL provide sufficient evidence to justify a hearing loss prevention (hearing conservation) program. Studies indicate this type of effort is worthwhile, as children who participate in school hearing loss prevention (hearing conservation) programs increase prevention practices and knowledge (Chermak et al., 1996).

A school or school district hearing conservation team (audiologist, industrial hygienist, safety engineer, noise control engineer, teachers, parents, and nurses) can develop, plan, and manage a program to educate children.* The program can resemble existing smoking, pregnancy, and drug abuse prevention programs.

COMPREHENSIVE SCHOOL HEARING LOSS PREVENTION PROGRAMS *SHOULD* INCLUDE:

1. Noise hazard identification (How to identify sounds that are "too loud")
2. Engineering controls to abate noise (How to reduce the noise)
3. Monitoring audiometry. Table 5.10 provides a template for a progressive hearing testing program designed to detect noise damage before the hearing loss occurs, while also identifying the common middle ear problems potentially resulting in "temporary" hearing loss (Langford, 2002)
4. Preventive measures (Behavior strategies such as Turn it down, Walk away, Wear earplugs or muffs, Crank it Down and Hearing Protection Device [HPD] Solutions such as appropriate selection, fit, use, and care of HPDs)
5. Hearing health education should include:
    - The anatomy of the ear
    - Toxic noise. The impact of exposure to noise on hearing, and causes of hearing loss
    - The audiogram
    - Examples of noise hazards
    - How to recognize when sound is too loud
    - Warning signs of hearing damage
    - Age-appropriate activities and information
    - Preventive measures, including the use of HPDs (earplugs or muffs)

    Teachers can easily incorporate these topics in existing health, science, music, math, or physics curriculums. Audio demonstrations of simulated hearing loss are particularly effective (find several in the teacher resource appendix)
6. Evaluate hearing conservation program effectiveness
(Outcome measures—Does it work?)

* Schools may send a nurse for certification as an occupational hearing conservationist by the Council for Accreditation in Occupational Hearing Conservation (CAOHC). OSHA recognizes CAOHC (www.caohc.org) as the training experts offering short courses across the United States to train occupational hearing conservationists and professional supervisors.

Innovative Programming

- A local parent flew a helicopter to the school playground to kick off International Noise Awareness Day.
- They called the event BANG—Be Aware of Noise Generation.
- Students measured the sound, participated in educational activities and programs ending with awards for the best noise awareness poster (with donated prizes).
- The local supermarket displayed the posters during May (Better Speech and Hearing Month) reminding all adult shoppers "noise destroys" and to protect their ears.

http://www.militaryaudiology.org/bang/index.html

Table 5.10 Guidelines for a Progressive School Hearing Assessment Program (SHAP)

Grades	Testing	Education
K-5	Traditional suprathreshold screening and immittance testing to identify hearing loss and middle ear dysfunction	Age-appropriate information in a problem-based learning approach on physical structure and function of the ear and the physics of sound noise hazards Lesson plans and educational materials like "Crank it Down" or "Sounds like Fun" are available from resources shown in the reference section
6–8	Hearing threshold assessment to include 2000, 3000, 4000, 6000, and 8000 Hz to detect the presence or absence of noise notches. OAE testing would confirm damage to inner ear hair cells If testing identifies a noise notch or inner ear hair cell damage, specific counseling regarding toxic noise hazards, exposure limits, preventive actions, and hearing protection is appropriate	Age-appropriate educational programs with special emphasis on noise hazards commonly encountered by the student population (e.g., farm hazards in rural communities or urban noise hazards in metropolitan areas)
9–12	Threshold and OAE testing, as risk of NIHL is greater for adolescents	Educational components should include individual hearing protection fitting and identification of noise hazard risk Special attention to school noise (band, vocational, sports), recreational noise hazards (MP3 players, personal stereos, cars/motorcycles, hunting), as well as after school employment (lawn care, fast-food communication headsets, construction)

Source: From Langford, J., *NHCA Spectrum*, 19(3), 3, 2002.

Creative scheduling, insert earphones or mobile test facilities are solutions for schools with logistical limitations in implementing a threshold SHAP testing protocol.

INVOLVE THE STUDENTS

If children don't know or understand, why they should walk away, turn it down, or protect their ears
They probably won't.

Hearing instruction will be more effective if teachers' communications with children reflect their needs, interests, and everyday experiences. Avoid "preaching" to children about hearing; it will not work (Figure 5.46).

Eighth graders report:

- Their most common source of noise exposure is stereo headsets, video arcades, concerts, musical instruments, and lawn mowers.
- 67% say they never or rarely wear hearing protection during exposures.
- 73% do not understand that NIHL results in permanent damage, untreatable by medical or surgical intervention.
- 78% incorrectly believe cotton is a good hearing protector (earplug).
- Many do not think they need to use preventive measures (Watkins, 2004).

Figure 5.46 (Nancy Vause)
Children's education must be age appropriate. This photo shows young children going through a "cochlea" as they learn about the physiology of the ear.
Attempting to scare or force children into healthy hearing habits usually does not work. Discussions and interactive experiences are more effective. See the teacher resource appendix for information on activities, lesson plans, and experiences to involve the children in the learning process (Haller and Montgomery, 2004).

Problem-based learning (PBL) lets children use the information they learn to develop solutions to problems (Bennett and English, 1999). It also makes them feel they are contributing to the "greater good" of their school.

HEARING PROTECTION

When children cannot "walk away" or "turn it down," they can easily prevent virtually 99% of NIHL by using hearing protection such as earplugs or noise muffs. Many people, adults, and children alike believe using hearing protection will reduce their ability to hear and communicate. In noisy environments, the use of hearing protection actually improves the ability to understand conversations.

This is because the HPD reduces (attenuates) the overall SNR, allowing the cochlea to respond normally. The inner ear can transmit sounds without "overdriving" the hair cells, resulting in sound distortion. An equivalent example is using sunglasses to protect eyes on a very bright day.

Figure 5.47 (Ursy Potter Photography)

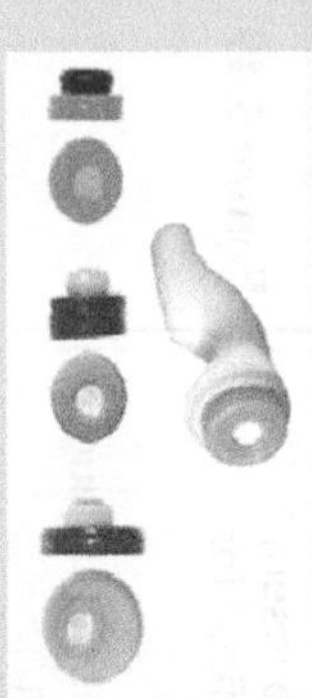

Figure 5.48 (Courtesy of USA Human Research and Engineering Directorate)

By the time a professional musician reaches 30, he often has a noise "worklife" exposure of over 25 years. It does not matter if the music is classical, country, rock, or rap. The intensity and duration of the music is what matters.

Musician's earplugs are available in premolded sizes or as a custom fit. Custom earplugs require an audiologist to take ear impressions. This type of earplug reduces toxic sound hazard levels while simultaneously preserving sound quality and frequency signatures.

Musicians consider the expense a good investment to protect their precious hearing. The custom earplug is gaining popularity among professional musicians (Landau-Goodman, 2001).

Since sunglasses decrease the overall illumination, the eyes function more efficiently and effectively. In a similar manner, hearing protection reduces the acoustic "glare" of high or toxic noise levels. Children and adults can still hear and enjoy the music or conversations more clearly because the ear is not *distorting* the sounds.

What is the best hearing protection? The hearing protection a child wears properly!

Appropriate HPD selection depends upon on the user's requirements for noise reduction, comfort, cost, durability, ease-of-use, communication requirements, and fit.

Involve children in the selection process and find the noise reduction rating (NRR) of the HPD. Products with this rating have undergone design and testing for noise reduction. However, do not expect too much of this rating.

The NRR value does not describe the actual attenuation of the HPD (as one would expect). Many people do not realize that an NRR of 25–30 dB may only really provide 10 dB of protection. If the person does not fit the HPD properly, the protection is reduced even more.

There are two basic types of HPDs: earplugs and noise (ear) muffs. Both are effective in reducing continuous and impulse noise.*

* The US Center for Health Promotion and Preventive Medicine (CHPPM) provides a thorough list and explanation of HPDs available at: http://chppm-www.apgea.army.mil/hcp/devices.aspx

	What is it?	What about it?	Advantages	Disadvantages	Example
Earplugs	Earplugs are available in many sizes, materials, and colors. Some styles offer connecting cords to minimize lost earplugs. Options include pre-formed, hand-formed, or custom molds. All require a good fit to provide adequate protection from noise hazards.	• The "Ready to Fit" or "Preformed" earplugs are suitable for adolescents and young adults (Figure 5.49). They are not the best option for young children. They require individual sizing and fitting by medically trained personnel. The ability to wash and reuse make them less expensive in the long run, and they provide excellent noise reduction capabilities. • The hand-formed (foam) earplug is a common type of earplug that is appropriate for children. Hand-formed plugs are tightly hand rolled and inserted into ear canal. Children can usually wear foam-type earplugs since they conform to the size and shape of their ears (Figure 5.55 through Figure 5.57). Generally, these earplugs are only available in one size fits most, but manufactures are recognizing the need for smaller "child" size earplugs. • Hand-formed earplugs are not practical if children must frequently take their plugs in and out several times during the exposure. Also, a child's hands should be clean, as contaminates could transfer to the ear from the child's hand when they insert the earplugs. Children can wash and reuse this type of HPD until they are too dirty or until the material loses its ability to roll down or expand. • The most important aspect of any hearing protective device is proper fit. Before inserting any earplugs, it is helpful for a CAOHC certified hearing care professional to examine the ear canal for obstructions that might interfere in the use or wear of hearing protection. They can also help select the type of earplug most appropriate for the size of the ear canal (Figure 5.58 through Figure 5.60). • Tips for properly fitting hearing protection are available at www.e-a-r.com/pdf/hearingcons/earlog19.pdf or http://chppm-www.apgea.army.mil/hcp/devices.aspx	• Inexpensive (short-term use) • Relatively comfortable • Excellent overall attenuation when inserted properly • Good at blocking low-frequency sounds • Medical personnel fitting not required to fit hand formed ear plugs • Can wear under hats or protective helmets (motor-cycle, safety, etc.) to provide double hearing protection	• Useless if inserted incorrectly • Should not insert with soiled hands (dirt or chemicals) • Takes the most expertise and effort to insert, possibly difficult for children • Most expensive earplug in the long run	Figure 5.49

Noise (ear) muffs	Noise (ear) muffs consist of plastic ear cups, ear cup seals, and a headband. The ear cup seals should completely cover the outside of the ear and be flush with the face and head. Hair, glasses, or earrings can break the seal and decrease the protection up to 3–7 dB. The headband should be comfortable and hold the ear cups in place. Noise earmuffs offer passive or active protection options.	• Noise (ear) muffs are easy for children to use (Figure 5.50) and maintain. The ear cup does need to be checked periodically, as normal skin oil and perspiration will cause the seals to harden and reduce protection. • Noise muffs block the noise in a passive manner, no battery required. • However, active noise reduction (ANR) technology is now available to the public. Also called active noise cancellation (ANC) is possible through advanced computer technology by creating and generating a mirror image of the surrounding steady-state noise. Playing both noises at the same time "cancels" the environmental noise! • ANR works best in low-frequency continuous noise and does ***not*** work at all for impulse noise. HPDs with ANR are more expensive, but the price is dropping as more people buy the technology advances.	• Good for intermittent noise exposures • Good at blocking high-frequency sounds • Fit not required by medical personnel (users should be instructed on their proper wear) • Easy to monitor use • Easy to fit correctly • One size fits most adult users—check headband sizing for children	• Expensive when compared to earplugs • Not suited for extended periods of use • Difficult to wear when working in tight spaces • Not compatible with all types of head gear	**Figure 5.50** (Nancy Vause)
Ear canal caps	One size fits most and may fit into the ear canal or just "cap" the outside of the ear canal. Those that fit into the ear canal provide the best protection. Depending on the design, the headband fits over the head, behind the head, or under the chin. Eyeglass temples do not interfere with proper fit.	• Ear caps eliminate the need to handle and roll an earplug and reduce the bulk of a noise muff. • Ear canal caps are another suitable choice for children if fit correctly (Figure 5.51). • Ear canal caps are good for short intermittent exposures, but are only effective for noise levels up to 95 dB (A).	• Relatively easy to fit • Medical fit is not required (individuals should be instructed on the proper fit and wear of ear canal caps) • One size fits most • Can rest around neck when not worn	• Cannot use when noise levels exceed of 95 dB(A) • More expensive than earplugs • Can become uncomfortable when worn for long periods of time	**Figure 5.51** (Courtesy of USA CHPPM)

(continued)

	What is it?	What about it?	Advantages	Disadvantages	Example
Musician's earplugs	The custom earplug offers several attenuation filter options. It is also available as a preformed ready-to-fit earplug.	• Musician's earplugs are appropriate for children as they participate in bands and music classes (Figures 5.47, 5.48, and 5.52). They offer a way to turn down the volume and still enjoy the music. • They decrease sound equally at all frequencies while maintaining the natural sound quality. • The equal attenuation allows musicians to hear all of the instruments, but at a reduced intensity level. The flat attenuation reduces the artificial low frequency, brassy sound of speech, and music experienced with traditional HPDs. • Musician earplugs are a favorite of professional musicians, conductors, singers, and sound engineers. They are available and becoming more popular with the general public, amateur musicians, and school band members, conductors, and sound engineers.	• Comfortable for long periods of time • Preserves sound quality • Protects hearing • Several filter options allows listeners to tailor the same custom mold to various attenuation levels (9, 15, or 25 dB) (Figure 5.52) • Economical pre-formed option (prices range from $9 to $25 per pair)	• Requires medical fit and sizing • Expensive (prices range from $150 to $225 per pair custom) • Growing children will grow out of a good fit and require a new set • High maintenance. Care must be taken to keep filter free of wax (may require more than one filter)	**Figure 5.52** (Courtesy of USA HRED.)
Hunter's earmuff	These special use earmuffs incorporate electronics controlling a microphone and amplifier into the earmuff design. They attempt to instantaneously attenuate high-level impulse sound and in quiet environments amplify soft sounds.	• Children who accompany their parents on hunting expeditions often resist wearing HPDs since they must hear very soft sounds in the quiet forests or fields. Yet, a single exposure to the impulse noise of a shotgun (without hearing protection) can damage a child's hearing.	• Provides a hearing protection option for those children and adults who participate in hunting activities • Decreases the effects of hearing protection on situational awareness	• Expensive and requires batteries • The protective aspects of the earmuff electronics are not simultaneous. As "attack time" increases (even 1 s) the risk of noise hazard exposure and hearing loss also increases	**Figure 5.53** (Nancy Vause)
Headphones	Earbuds or headphones that accompanying tape, CD, or digital music players.	• Recreational headphones offer almost no protection from toxic noise. In fact, they are often a source of noise, as children often set the levels too loud (e.g., 100 dBA). Turning up the volume of a personal stereo or MP3 player to "drown" out the noise will only exacerbate the problem and increase the risk of hearing loss.		Some styles increase intensity levels.	

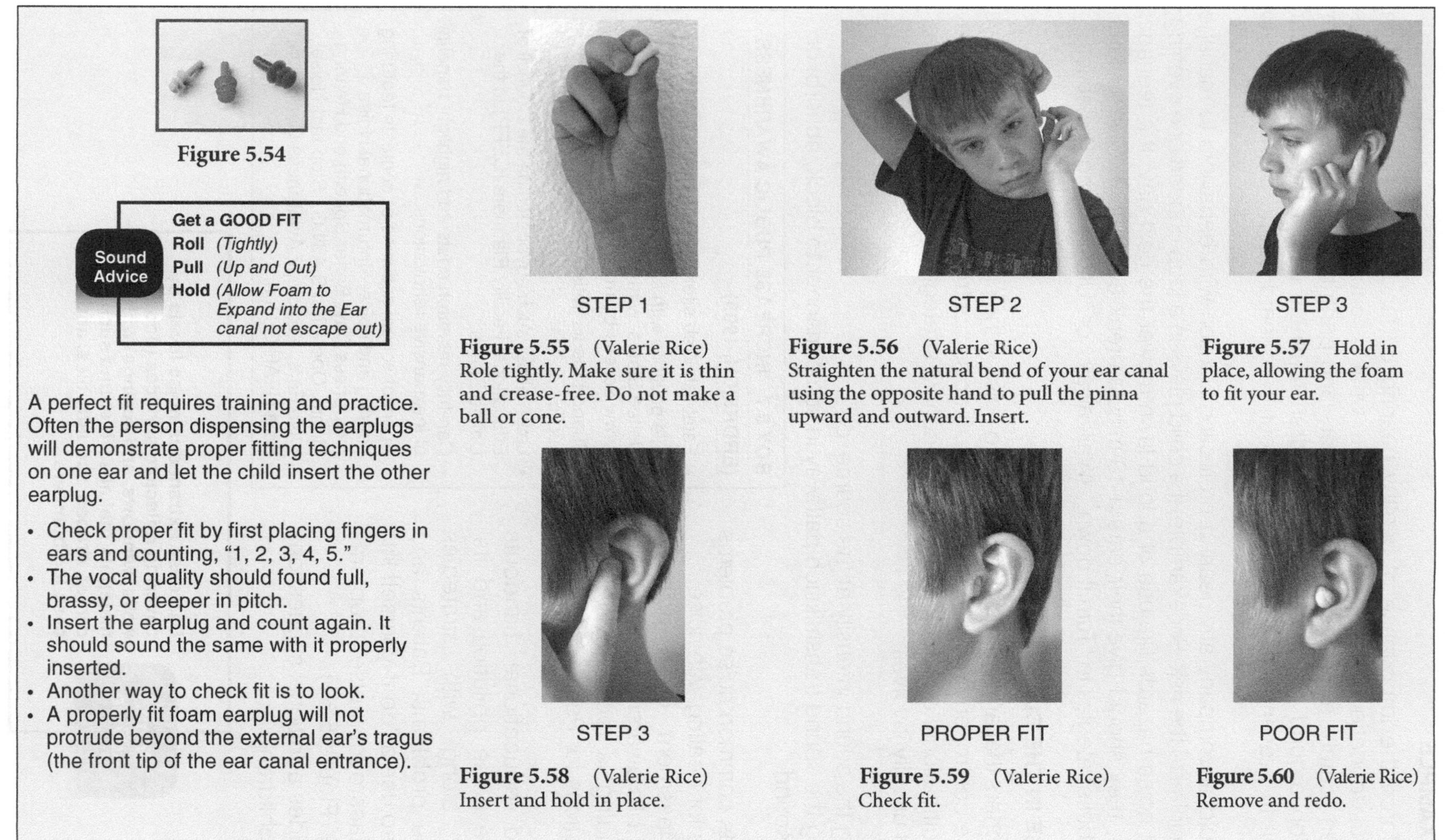

Figure 5.54

Sound Advice

Get a GOOD FIT
Roll *(Tightly)*
Pull *(Up and Out)*
Hold *(Allow Foam to Expand into the Ear canal not escape out)*

A perfect fit requires training and practice. Often the person dispensing the earplugs will demonstrate proper fitting techniques on one ear and let the child insert the other earplug.

- Check proper fit by first placing fingers in ears and counting, "1, 2, 3, 4, 5."
- The vocal quality should found full, brassy, or deeper in pitch.
- Insert the earplug and count again. It should sound the same with it properly inserted.
- Another way to check fit is to look.
- A properly fit foam earplug will not protrude beyond the external ear's tragus (the front tip of the ear canal entrance).

Figure 5.55 (Valerie Rice) Role tightly. Make sure it is thin and crease-free. Do not make a ball or cone.

Figure 5.56 (Valerie Rice) Straighten the natural bend of your ear canal using the opposite hand to pull the pinna upward and outward. Insert.

Figure 5.57 Hold in place, allowing the foam to fit your ear.

Figure 5.58 (Valerie Rice) Insert and hold in place.

Figure 5.59 (Valerie Rice) Check fit.

Figure 5.60 (Valerie Rice) Remove and redo.

TEACHING BY EXAMPLE

Adults can reinforce the importance of healthy hearing by leading by example. Often the most powerful educational message comes from what is "caught" rather than what is "taught." Children "catch" a message when teachers and parents routinely wear proper hearing protection around hearing hazards. Children routinely use a bike helmet, car seat, or seat belt to prevent potential injuries without question. Hearing protection is no different.

Children should accompany any hearing protection solution with behaviors to manage toxic noise and minimize the risk. For example, if a child plays in a loud concert one evening, they might avoid noise hazards like use of a loud lawn mower the next day. If a teenager attends a dance, they should give their ears a 10-minute break every hour. Give children Dangerous Decibel strategies to "Turn it down," "Walk away," or "Use Hearing Protection."

REDUCING NOISE IN THE CLASSROOM

Schools and teachers that are serious about reducing noise in the classroom find it is always best to hire noise control engineers who specialize in noise control and abatement. They consider all specifications, develop effective noise control strategies, and cost estimates.

Noise control usually involves:

1. decreasing the sound intensity at the source, or
2. interrupting the sound transmission pathway in some manner to block, absorb, or divert the sound.

Section C lists common noise problems and suggestions for dealing with noise.

Reducing classroom noise pollution often requires public awareness, especially if additional funding is required. Box 5.7 contains suggestions for Public Awareness Programs.

Notices and educational handouts should describe noise pollution and its' consequences, along with strategies to resolve noise problems. Parents and teachers can also use a "do-it-yourself kit" to diagnose classroom noise problems (Lubman, 2004) (Figure 5.61).

Box 5.7 provides additional suggestions for parents and children.

BOX 5.7 INCREASE PUBLIC AWARENESS (ERDREICH, 1999)

- Establish Web sites. Write articles.
- File petitions with school districts and in the United States, with the US Access Board (www.access-board.gov).
- Support media initiatives that educate the public.
- Learn about such advocates as the Council of Educational Facility Planners (CEFP) in the United States.
- Carefully research costs and highlight benefits of inexpensive renovations.
- Sponsor school activities or contests featuring hearing and noise—International Noise Awareness Day, Better Speech and Hearing Month, Operation B.A.N.G., Stop that Noise!, Wise Ears, Audiology Awareness Campaign, Be a HEAR O Run.

Decrease extraneous noise levels by covering reflective surfaces (floor, walls, windows, and ceilings) with materials that absorb sound such as carpet, drapes, acoustic tiles, and desk leg bumpers.

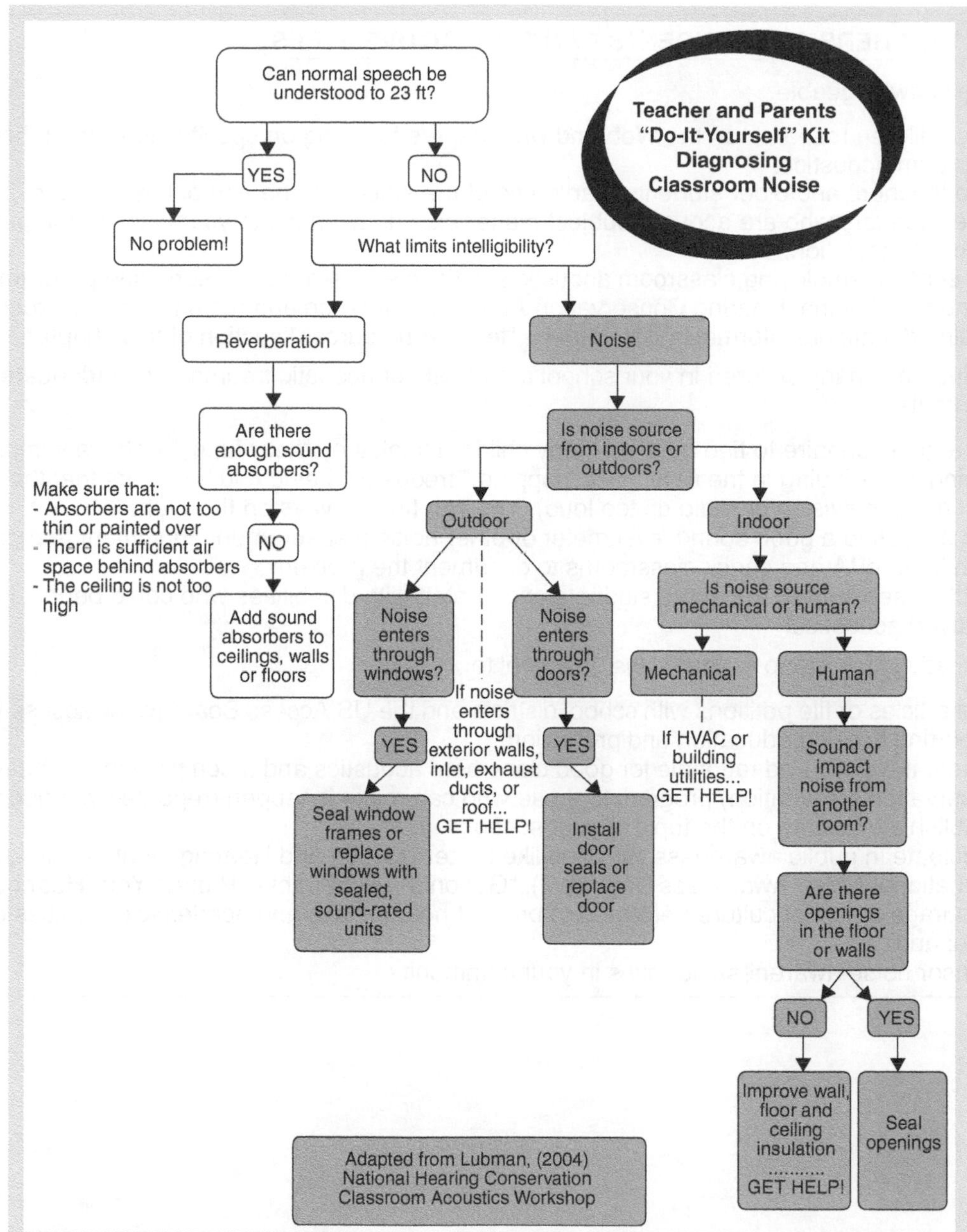

Figure 5.61 "Do-it-yourself Kit" for diagnosing classroom noise. (Reprinted with permission from Lubman, D., *Spectrum*, 19(Suppl. 1), 20, 2004.)

BOX 5.8 TEACHERS AND STUDENTS CAN TAKE ACTIVE STEPS

1. Become knowledgeable.
   a. Have children research on the Web and write papers focusing on specific aspects on hearing and classroom acoustics.
   b. Have teachers and older students read some of the articles at the end of this chapter.
   c. Invite speakers who are acoustic subject matter experts to speak to your class or the parent teacher association.
   d. Visit schools employing classroom acoustic solutions and hearing conservation programs.
   e. Attend the National Hearing Conservation Annual Conference and meet the noise experts.
   f. Obtain educational information listed in the "teacher resources" section of this chapter.
2. Determine how many children in your school are at risk for acoustic trauma and inadequate classroom environments.
   a. Use a questionnaire to find out how many children think they "hear funny" or have symptoms of hearing loss (ringing in the ears, ears "popping," frequently being told by others that they are turning the television or radio up too loud). You can find surveys on the Web.
   b. Gain access to a good sound level meter or other noise measurement equipment that can measure down to 35 dBA and check classrooms to document the problem.
   c. Check to see if there are some students and/or staff with disabilities who could benefit from improved acoustics.
3. Form an advocacy group in your class or school to:
   a. Write articles or file petitions with school districts and the US Access Board (www.access-board.gov) supporting hearing education and protection;
   b. Present the need and rationale for good classroom acoustics and a comprehensive hearing conservation (prevention) program to those who can make it happen (repeatedly, if necessary);
   c. Establish a Web site on the topic at your school;
   d. Participate in public awareness activities like Better Speech and Hearing Month (May) and International Noise Awareness Day (April); "Get on the right track—Protect Your Hearing".
   e. Encourage a school culture dedicated to prevent hearing loss and academic difficulties caused by noise; and
   f. Sponsor noise awareness activities in your community.

Section C: Design Applications

Noise Problems and Solutions*

Noise originating from inside or outside the classroom can disrupt learning. Acoustic treatments can reduce the amount of sound in a classroom (intensity) and the length of time the sound waves bounce around off ceilings, walls, floors, etc. (RT). Sound waves, unlike a beam of light, can pass through, up, over, under, and around objects (walls, doors, windows, and floors) simultaneously. Noise control can prove difficult as implementing one solution may create another noise problem. When in doubt, consult a noise control expert.

NEW SCHOOL AND CLASSROOM CONSTRUCTION

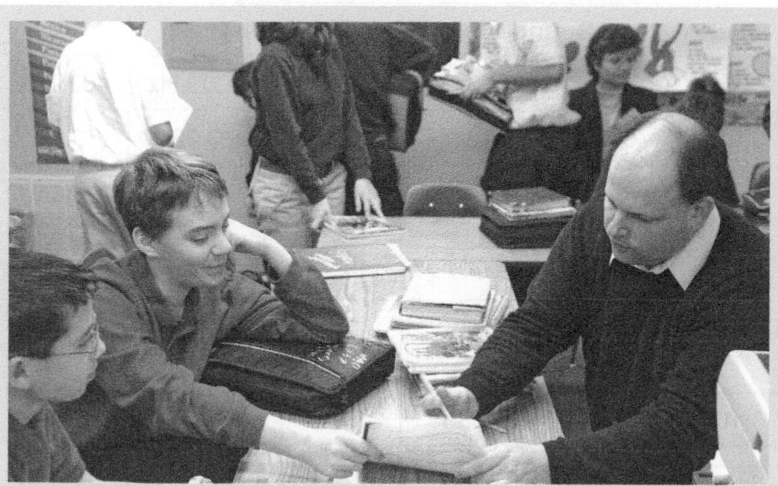

Figure 5.62 (Kurt Holter Photography, Frederick, MD)
Solutions: Noise control professionals (noise control engineers, industrial audiologists, and industrial hygienists) can assist school facility planners to determine size, shape, and surfaces of classrooms. They can also help select the building location and orientation, conduct acoustical surveys (background noise and reverberation levels), and recommend engineering controls and product/equipment specifications. Consultants can also meet student representatives.)

- Give a free copy of ANSI S12.60 to school officials, architects, facility planners, and builders (http://asastore.aip.org). Use acoustic architectural design.
- Select the school site and design with hearing in mind (seek quiet areas, avoid major transportation, and industrial noise sources).
- Locate high-traffic hallways away from classrooms (cafeterias, break areas, and auditoriums).
- Position noisy special purpose rooms away from other classrooms (music, technical/vocational areas, and gymnasium).
- Incorporate acoustic treatments in classrooms to reduce noise. Include noise control treatments in special purpose areas such as cafeterias, gyms, theaters, shops, and music rooms.

* This compilation is not comprehensive. It is intended as a general guideline only (Seep et al., 2000; Driscoll and Royster, 2000; Siebein et al., 2000).

MECHANICAL EQUIPMENT: HVAC (HEATING, VENTILATION, AND AIR CONDITIONING)*

Solutions:

- Incorporate noise control design specifications during planning. Noise control engineers will consider the size, shape, and surface of a classroom.
- Locate rooftop equipment away from classrooms (e.g., possibly position over hallways).
- Install air handlers with low air speed and sound level ratings.
- Install duct silencer or ducts with sound-absorptive duct liner.
- Plan large and long ducts for low air velocities.
- Use a slower fan speed.
- Avoid unit ventilators, fan coil units, and ductless split systems.
- Service, upgrade, or replace noisy individual heating units with forced air central heating systems.
- Consult an acoustical expert for most recent information and technology (e.g., silencers, adequate duct length, vibration isolators, and adequate duct and diffuser sizes).
- Use sound-absorptive materials (see Table 5.11). Appropriate ceiling and floor treatments can absorb approximately 70% and 20% of the ambient sound, respectively.

  Larger coefficients indicate more effective absorption properties. These materials will absorb more sounds, reduce reverberation, and improve the signal-to-noise ratios in the classroom.

Table 5.11 Sound Absorption Coefficients for Classroom Materials

Material	Sound Absorption Coefficient[a]					
	125 Hz	250 Hz	500 Hz	1000 Hz	2000 Hz	4000 Hz
Concrete	0.01	0.01	0.015	0.02	0.02	0.02
Ordinary window glass	0.35	0.25	0.18	0.12	0.07	0.04
Carpet-woven wool loop 1.2 kg/m^2 (35 oz/yd^2), 2.4 mm (3/32 in.) (no pad)	0.10	0.16	0.11	0.30	0.50	0.47
Loop pile tufted, 0.07 kg/m^2 (20 oz/yd^2) (no pad)	0.04	0.08	0.17	0.33	0.59	0.75
With 1.4 kg/m^2 (40 oz/yd^2) hair pad	0.10	0.19	0.35	0.79	0.69	0.79
Linoleum floor	0.02	0.03	0.03	0.03	0.03	0.02
Drapes (medium velour (475 g/m^2)	0.07	0.31	0.49	0.75	0.70	0.60
Mineral spray on 2.6 cm (1 in.)	0.16	0.45	0.70	0.90	0.90	0.85
Plaster	0.13	0.015	0.02	0.03	0.04	0.05
Plywood paneling	0.28	0.22	0.17	0.09	0.10	0.11
Gypsum board	0.29	0.10	0.05	0.04	0.07	0.09

[a] Part of the sound wave striking a surface is absorbed by any material. Some materials absorb better than others. The sound absorption coefficient is the measure used to compare sound absorption qualities and efficiency of a material. A 70 value indicates that 70% of the sound energy is absorbed and NOT reflected. These materials reduce noise and reverberation. This list is a representative sample (Harris, 1998; Berger, 2000).

* This is the most commonly cited noise problem (Siebien et al., 2000).

EXTERIOR NOISE: PLAYGROUNDS, STREET TRAFFIC, AIRPLANES, AND TRAINS

Solutions:

- Consult noise control experts (noise control engineers, industrial audiologists, and industrial hygienists).
- Install a suspended acoustical tile ceiling in the classroom if indicated.
- Install or replace door and window seals.
- Ensure the exterior noise source is within allowable levels noted in the local noise ordinances.
- Check with your local government to determine if a noise ordinance exists, the noise limits, and the responsible department.
- Provide noise isolation materials for the building shell with appropriate sound transmission class (STC) ratings.
- Install noise reduction windows.
- Relocate a classroom to another area within the school away from the exterior noise.

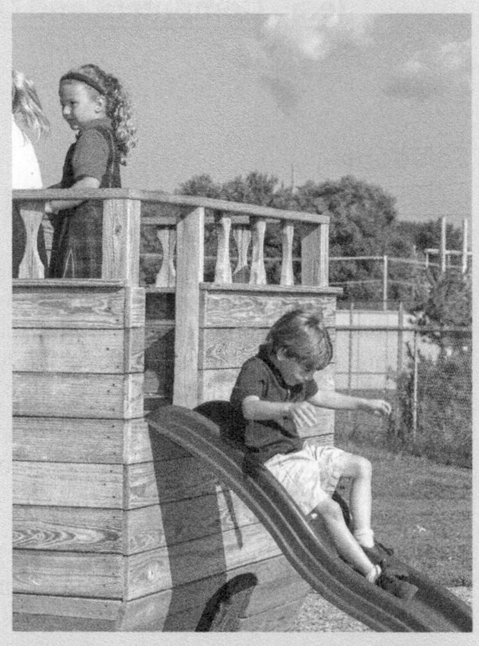

Figure 5.63 (Kurt Holter Photography, Frederick, MD)
Noise from outdoor playgrounds can interfere with classroom work, especially if they are located directly outside the class.
Yet, this location is convenient and often preferred for day care and preschool age children. Using noise reduction strategies within the classroom may be the best noise reduction techniques in these situations.

HALLWAY OR ADJACENT SPACE NOISE

Solutions:

- Ensure proper installation of existing door seals.
- Properly install doors and check adequacy of door and wall sound transmission class (STC) materials.
- Keep doors and windows closed.
- Avoid placing classrooms adjacent to high traffic or special-purpose classrooms (music, gym, or computer).

Figure 5.64 (Valerie Rice)
While monitoring and directing children's behaviors will certainly help keep things quiet, a better solution is to design the building using materials that resist sound transmission.

EXCESSIVE INTERIOR CLASSROOM NOISE
(e.g., Computers and Other Equipment, Fans, Fish Tanks, People)

Solutions:

- Consider size, shape, and surfaces of the classroom.
- Install acoustic-absorptive materials on ceilings and walls (e.g., suspended sound-absorptive ceiling tiles, curtains) equal in area to the floor area of the room. A properly acoustically designed ceiling can absorb 70% of the noise.
- Install carpeting. A floor with properly installed noise reduction treatments can absorb 20% of the noise.
- Support teacher control.
- Select, purchase, and use quiet equipment.

Figure 5.65 (Ursy Potter Photography)
Consider classroom furniture placement and teaching techniques to decrease distances between teacher and students. These include small group instruction, story time at the front of the room, and facing students during verbal instruction.

Classroom techniques that also reduce noise include:

1. separate aisles to facilitate moving closer to children;
2. teacher movement closer to the children during class;
3. nontraditional seating; and
4. placing rubber fittings (such as tennis balls) on student desk legs.

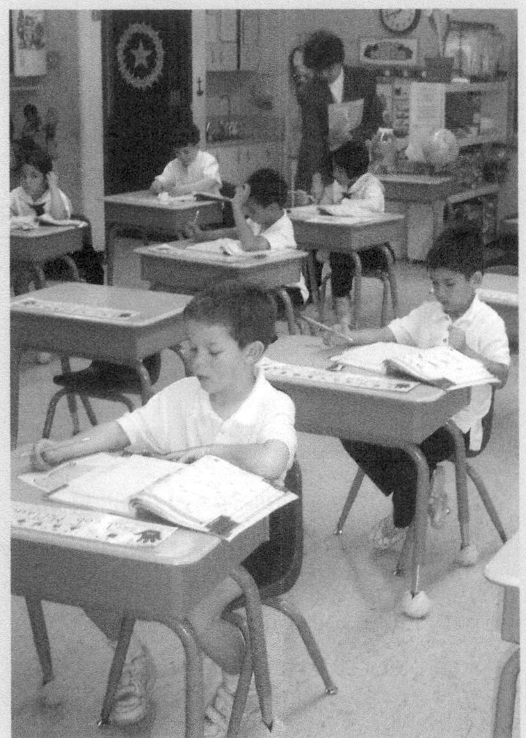

Figure 5.66 (Valerie Rice)

LONG REVERBERATION TIMES (ECHOES)

Figure 5.67 (Kurt Holter Photography, Frederick, MD)

Solutions:

- Consider size, shape, and surfaces of the classroom.
- Install durable acoustic-absorptive materials on the classroom ceiling, floor, and walls (use sound-absorbing products with high STC ratings).
- Reduce room volume with suspended ceiling of sound-absorbing tile.

UNACCEPTABLE CLASSROOM ACOUSTICS (NO AVAILABLE ENGINEERING OR MATERIAL SOLUTIONS)

- Install a classroom Sound Field Amplification and Distribution System.

  These systems are similar to a small, wireless, high-fidelity public address system that is self-contained to an individual classroom.

  These systems uniformly amplify the speaker's voice but not other classroom sounds through distributed loudspeakers strategically placed; this strategy increases SNR for all children and reduces teacher vocal strain. It is a cost-effective educational option in many classrooms (Flexor et al., 2002; Nelson and Rosenberg, 2001). Good acoustics will benefit all students and teachers.

SUMMARY

The sound of an ocean wave against the shore, music's rhythmic beat, a friendly voice on the other end of a phone, or a songbird chirp are wonderful sounds. On the other hand, continuous or high-impact noise makes it difficult for teachers and students to hear and interferes with learning.

Figure 5.68 (Ursy Potter Photography)

Noise interrupts teachers' ability to teach and students' ability to hear and learn. All students and teachers benefit from appropriate classroom acoustics.

Schools should immediately incorporate measures to reverse the increasing incidence of NIHL in children. Parents and teachers can initiate Hearing Health Education Programs that empower students with knowledge about noise hazards, the risks from various types of noise, the consequences of toxic noise exposure, and the action options to prevent NIHL.

We must *change knowledge; change attitudes* and most importantly *change behavior*, so everyone can

LISTEN FOR A LIFETIME.

Appendix A: Frequently Asked Questions (FAQs)

Q What is noise?

A Any unwanted sound.

It is the most common health hazard in the workplace. The National Institute of Occupational Safety and Health (NIOSH) reports about 30 million Americans are exposed to daily noise hazards in the workplace.

Q Does noise cause hearing loss?

A YES! Hearing loss is caused when your ear is damaged from a noise injury or acoustic trauma. The loss is usually gradual and painless, but permanent. Damage can also occur suddenly with impulse noise.

According to the National Institute of Health, more than 10 million Americans already have some permanent NIHL. One in eight children between ages 6 and 19 already have some degree of hearing loss.

Q What type of noise is hazardous to hearing?

A *Impulse noise* (>140 dBP) can cause permanent hearing loss with a one-time exposure.

Continuous noise, repeated exposures or long exposure to loud sound (>85 dBA) can cause temporary or permanent hearing loss.

Extended exposures (>8 h) to moderate levels (>75–80 dBA) of *continuous noise* can also damage hearing.

Q I do not listen to noise, I play in a band. Can music hurt my ears?

A Yes, particularly if you practice or listen to loud music over 20 h/week.

The loudness of the noise and exposure time are important variables. Musicians playing different types of music (Classical, Rock, Blues, Hip Hop, Country, and Jazz) can experience NIHL. In fact, Hearing Education and Awareness for Rockers (H.E.A.R.) promotes healthy hearing for musicians (www.hearnet.com). Artists with hearing loss: Ludwig von Beethoven, Pete Townshend, Neil Young, Brian Wilson, Sting, Eric Clapton, Bono, Keanu Reeves. If can hearing loss can happen to them, hearing loss can happen to you.

Q How does noise cause hearing loss?

A Sound travels through the outer ear (pinna and external auditory ear canal) to the middle ear (eardrum, ossicular chain) and is transmitted to the inner ear.

The inner ear has fluid-filled chambers lined with thousands of delicate hair cells. The hair cells bend, move or contract, signaling the auditory nerve to send electrical impulses to the brain, which interprets the meaning of the sound. Loud or long noise exposures can damage the inner ear hair cells and can permanently destroy the sound transmission pathway to the auditory nerve. Impulse noise can also rip inner ear chamber membranes, as well as destroy the hair cells.

Q Do these hair cells grow back with Rogaine or minoxidil?

A NO.

Although scientists are working on methods to repair or encourage regrowth—it is not an option for humans at this time.

Q Can a doctor prescribe medications or do surgery to repair the hearing damage caused by noise?

A No. Noise injures your ears causing damage to the structures and results in permanent hearing loss. However, scientists are working to develop pharmaceutical preventions and interventions. Sudden hearing loss, however, is a medical emergency as interventions are time sensitive.

Q Can I get used to noise or will it toughen my ears?

A NO!

You cannot make your ears tough by repeated exposure. If you are already "accustomed" to the noise—you may already have a hearing loss.

Figure 5.69 (Kurt Holter Photography, Frederick, MD)

Q So what if I lose a little hearing? How does the loss of hearing really affect me?

A You will probably not lose your hearing all at one time. Most people with noise induced hearing loss do not even realize it until their family or friends notice the change.

Speech consists of several frequencies (low- medium-, and high-frequency sounds). Unfortunately, noise destroys higher frequencies first, which are critical for understanding speech.

You may have trouble understanding conversations particularly in noisy situations, especially hearing the difference between words like hit, fit, and sit. Most people with hearing loss report they hear just fine—they just have trouble understanding conversation.

If you lose hearing and do not seek help, after a period, your speech will also deteriorate. Hearing yourself talk provides valuable feedback. Since hearing loss distorts the sounds you hear, your voice may start to change. Eventually, you may sound like you need to start speech therapy unless you seek rehabilitation options (e.g., amplification).

Figure 5.70 (Nancy Vause)

Figure 5.71 (Kurt Holter Photography, Frederick, MD)

Q Isn't hearing loss just an old person's problem?

A No, noise-induced hearing loss can happen to anyone, any time, and at any age.

In fact, some researchers believe young children may be more vulnerable to noise-induced hearing loss.

Q Am I exposed to noise hearing hazards?

A If you operate power tools, lawn mowers, stereo headsets, MP3 players, musical instruments, household appliances, and firearms, you may be exposed to noise hazards. Some toys even introduce hearing hazards to children.

Q How loud is too loud?

A Think how long, how much, and how high.

This answer depends upon how long you are exposed to the noise and the loudness levels. The longer the exposure, the more risk of hearing loss.

Higher decibels indicate a louder noise. Continuous noise >80 dBA is considered potentially hazardous for longer than 8 hours for most adults. Impulse noise levels >140 dBP can cause damage with just one exposure.

Q How can I tell noise is too loud without sound measuring equipment?

A Sound is too loud if:

- You must raise your voice to be heard
- You have difficulty hearing someone less than an arm's length away (approximately 2 ft)
- You must turn the radio volume louder after attending a concert or party
- Speech sounds dull, muffled, or as if people are mumbling after leaving a noisy environment
- You experience pain or ringing in your ears after noise exposure.
- You adjust the volume control of an MP3 player to more than 60% of the potential max volume for more than 1 hour in a day

Q How can I protect myself?

A Solutions:

- Crank it down—turn down the volume
- Walk away, avoid loud noise
- Leave a noisy area and reduce the time your ears are exposed to noise. Give your ears a rest or break
- If you cannot avoid noise:
    - Protect your ears and wear adequate hearing protection (e.g., earplugs or earmuffs). Cotton, cigarette butts, modeling clay are NOT adequate hearing protection.
    - Notice the noise in your life at work and play. Take control of the toxic noise in your life. Check out the local noise ordinance in your town or school. Some schools prohibit loud noise from cars, motorcycles, dances, and concerts to protect students.
- See an audiologist for a hearing test every year

Q Can I hear people talk or warning signals while wearing hearing protection?

A Yes!

In fact, hearing protection (earplugs or muffs) does not block all sound, but instead lowers frequency noise allowing you to hear other people and warning sounds more easily. Hearing protection is the sound equivalent of sunglasses to light.

Q What should I do if I think I have a hearing problem?

A Do not delay, visit a board certified and state licensed audiologist for a hearing evaluation. If hearing loss is a result of disease or a medical condition, the audiologist will refer you to your physician or an ENT specialist. If you think you have a hearing problem, you are probably right…If you think no one has noticed, you are probably wrong.

Q What is this ringing in my ears? What can I do about it?

A Another name for the ringing is "tinnitus."

Millions of people experience some form of tinnitus. First—avoid noise and practice prevention. Today, numerous treatments and services are widely available. If you experience tinnitus, contact your audiologist or otolaryngologist! (ENT doctor)

Q Can you prevent noise induced hearing loss?

A Absolutely.

Everyone can help prevent noise-induced hearing loss by understanding the toxic noise hazards and by practicing good hearing health, whether in the home, at work, on travel, or at play.

Q HOW?

A Dangerous Decibels will advise

- Know which noises cause damage (80 dB and above)
- Turn it Down
- Walk Away
- Protect Your Ears
  - Wear earplugs, earmuffs, or other hearing protection devices when involved in a loud activity such as snowmobiling, shooting, auto racing, or using power tools, lawn or farm equipment.
  - Be alert to hazardous noises in the environment even when having fun "relaxing."
- Make family, friends, and teachers aware of noise hazards. If you're having trouble hearing or if sounds you hear are muffled and distorted or there is a ringing or roaring sound in your ears, see a doctor at once!
- Your doctor may refer you to an otolaryngologist, a doctor who specializes in the ear, nose, and throat (ENT).
- Have your hearing tested by an audiologist.

Q What is a hearing conservation program?

A A program designed to identify hearing hazards, pursue engineering noise controls, monitor hearing, provide hearing protection, raise awareness of hearing noise hazards and educate listeners about noise hazards, and preventive measures required to protect hearing for life!

Q What can I do?

A

- Make sure early education includes the dangers, types, and countermeasures against hearing noise hazards that can contribute to NIHL.
- Promote healthy hearing in your school, community, state, and nation.
- Investigate local noise ordinances.
- Know potential noise hazards and intervention strategies.
- Monitor your hearing annually.
- Protect your hearing and properly use hearing protection when avoiding noise hazards is impossible or impractical.

Q How do I know if I have a have a hearing loss and should be tested?

A Take this quick test adapted from the Wise ears Campaign of NIDCD.

Figure 5.72
(Wise ears Campaign of NIDCD)

If you answer yes on any question, you need to make an appointment with an audiologist.	
Yes ☐ No ☐	Do you have a problem hearing when you listen over the telephone?
Yes ☐ No ☐	Do you have trouble listening to conversations when two or more friends talking at the same time?
Yes ☐ No ☐	Do your parents complain that you turn the TV volume up too high?
Yes ☐ No ☐	Do you have to listen very hard or strain to understand the teacher or your friends?
Yes ☐ No ☐	Do you have trouble hearing when it is noisy in the classroom?
Yes ☐ No ☐	Do you find yourself asking parents, teachers, or friends to repeat themselves?
Yes ☐ No ☐	Do your parents, teachers, or friends seem to mumble (or not speak clearly)?
Yes ☐ No ☐	Do you misunderstand what the teacher, your parents, or friends say to you? Do you respond in the wrong manner?
Yes ☐ No ☐	Do you have trouble understanding the speech of other children and female talkers?
Yes ☐ No ☐	Do teachers, your parents, or friends get annoyed because you frequently misunderstand what they say?

WISE EARS!®

Hearing Matters—Protect It

Appendix B: Properties of Sound

WHAT IS SOUND?*

Sound† is "vibrating air molecules" that stimulates our sense of hearing. Leaving aside the effect of temperature, the speed of sound varies depending upon the density of the medium through which it passes. Sound travels approximately at 344 m/s (1130 ft/s) at sea level.

Sound travels faster through liquids, gases, and solids with greater densities. For example, sound travels approximately 1494 m/s (4900 ft/s) in water and much faster in steel 6096 m/s (20,000 ft/s) (Berger, 2000).

SOUND AND DISTANCE

Increasing the distance between the ear and a sound source causes a fairly predictable decrease in the intensity of the sound.‡ Each doubling of the distance decreases the sound level by 6 dB SPL.

This is particularly important in classrooms—notably, for children who sit on the other side of the room as they listen to a teacher's voice. A child close to the teacher may hear the voice at a comfortably loud intensity level; however, the teacher's voice arrives much softer to children sitting farther away.

> Sound waves are characterized by their ***frequency***, ***sound pressure levels*** (***intensity***), and ***timbre***.
>
> 1. A sound wave's ***Sound Pressure Level* (SPL)** determines its loudness (***intensity***).
>   SPLs represent the height (***amplitude***) of the sound wave. We express ***Sound Pressure Levels*** (SPL) in ***decibels***.
> 2. The ***frequency*** (speed of vibration as cycles per second) determines its ***pitch***.
> 3. ***Timbre*** helps us distinguish between sounds.
>   Most sounds are complex and consist of a variety frequencies of different intensities (***overtones***).
>   We enjoy listening to sounds that are harmonic; these include simple higher frequencies that are multiples of the fundamental tone (2×, 3×, 4×, etc.).
>   In contrast, noise is usually a random combination of frequencies.

* Sound is formally defined as the fluctuations in pressure above and below the ambient pressure of a medium that has elasticity and viscosity. We also define sound as the auditory sensation evoked by the oscillations in pressure described above (ANSI S1.1-1994).

† Several textbooks (Harris, 1998; Berger, 2000; Yost, 2000) provide technical explanations of sound's properties beyond the scope and intent of this discussion. Additionally, numerous fun acoustic demonstrations are available in the context of the Web.

‡ This is called the *inverse square law*. See basic texts for further information (Harris, 1998; Berger, 2000; Yost, 2000).

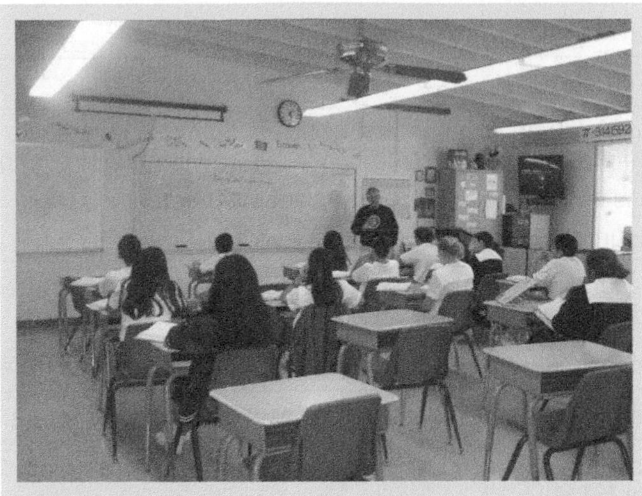

Figure 5.73 (Valerie Rice)

For example, in a perfect classroom environment, if a child hears the teacher's voice at a 55 dB SPL when sitting 3 ft away, the intensity of the voice for children sitting 6 ft away is 49 dB SPL (55 − 6 = 49). Children listening 12 ft away hear a voice that is 43 dB SPL while those children listening in the back of the classroom (24 ft away) hear a voice with an intensity level merely louder than a whisper (37 dB SPL).

The teachers' voice is significantly softer to children listening on the opposite side of the room from the teacher. It is no surprise that these children (and adults) are easier to distract and fatigue as they strain to hear. Listening is particularly difficult (and tiring) for children with hearing impairments (Hicks and Tharpe, 2002) especially since background noise levels are often much louder.

The primary consideration is the location of the speaker relative to the listener. There is no point to placing a child next to the teacher's desk if instruction takes place at the other side of the classroom. Noise and seat placement can also interfere with a child's ability to hear other children's questions and answers.

Table 5.12 Definitions of Terms Used in Acoustics

Acoustic Term	Definition
Weighting	• Sound intensity is measured using a filter to exclude some low frequency and very high-frequency sounds—to mimic the way the human ear perceived sound • dBA is used in occupational noise measurements • Measured in dBA, dBB, dBC
Pure tone	• Variations of pressure occur at only one frequency—few real-world pure tones, but used in hearing testing
Spectrum	• Describes sound as a function of frequencies and intensities
Sound level	• Used to describe intensity or loudness of sound in general
Sound pressure level (SPL)	• Used to describe sound levels measured by a measuring device (sound level meter [SLM])
Hearing level (HL)	• Used when describing decibel levels produced by an audiometer. Calibrated to human hearing frequency sensitivities
Time-weighted average	• Average exposure level for the day adding up the amount of time spent at each noise level if the noise exposure levels fluctuate • Usually measured by a dosimeter, usually normalized to 8 hours • OSHA and MSHA use 5 dB exchange rate—NIOSH and most European countries recommend 3 dB exchange rate
Permissible exposure level (PEL)	• A weighted sound level for a particular time (usually 8 hours) • Accumulates to 100% noise dose
Recommended exposure level (REL)	• The amount of noise exposure recommended by NIOSH for particular time (usually 8 hours) is 85 dBA for continuous noise and 140 dBP for impulse noise. These recommendations acknowledge adults with over a 40-year lifetime exposure 8% may experience hearing loss whereas exposure at 80 dBA only placed 3% of the population exposed at risk for material hearing loss. This author adopted the more conservative REL for children although more research is needed to understand the physiological effects of continuous and impulse noise on the ears and hearing of children
Time-weighted average	• Average exposure level for the day adding up the amount of time spent at each noise level if the noise exposure fluctuates
Exchange rate	• Relationship between permitted noise exposure levels and the exposure durations • Sometimes called the time–intensity tradeoff, trading ratio, or doubling rate
Noise dose	• Allowable noise exposure at any particular point in time • Considers PEL for an 8 h day AND the 3 dB exchange rate For example: • For Adults TWA 85 dBA for 8 h = 100%; TWA 88 dBA for 4 h = 100%; and TWA 91 dBA for 2 h = 100%. For children using the REL of 80 dBA. TWA 80 dBA for 8 h = 100%; TWA of 83 dBA for 4 h = 100%; TWA of 86 for 2 h = 100%; and TWA of 89 for 1 h = 100% * These values are more conservative than those recommended for adults by the Occupational Safety and Health Act where 25% of the population is predicted to experience some degree of hearing loss over a 40 year worklife using a 5 dB exchange rate

Appendix C: Resources

WEB SITES, ORGANIZATIONS, PRODUCT SOURCES

The following list of helpful Web sites and organizations that offer resources, products, standards, classroom lesson plans, ideas, and activities. It is not comprehensive but intended to serve as a starting point for teachers, parents, children, teenagers, and anyone interested in healthy hearing.

Organization	Web Address	Phone	Resource Description
Aero Technologies	www.aearo.com	317-692-6666	Hearing conservation video descriptions and reviews; hearing protection devices, hearing loss, and tinnitus simulation CD.
Academy of Dispensing Audiologists	www.audiologist.org/consumers/resources.cfm	866-493-5544	Information on hearing loss: awareness, prevention, research & disease, find an audiologist who dispenses state-of-the-art hearing aids.
Acoustical Society of America	http://asa.aip.org http://asa.aip.org/classroom.html http://asa.aip.org/classroom/booklet.html http://asastore.aip.org	631-390-0215	Classroom Acoustics I, A resource for creating learning environments with desirable listening conditions; Classroom Acoustics II, Acoustical Barriers to Learning; New Classroom Acoustics Standard.
American Academy of Audiology	www.audiology.org	800-AAA-2336	Information, posters, brochures, audiocassette (Say What?).
American Academy of Otolaryngology—Head and Neck Surgery	www.entnet.org/healthinfo/hearing/noise-hearing.cfm http://www.entnet.org/kidsent/content/accessinfo/consumer/consumerbros.htm	703-836-4444	Hearing, noise, and hearing protection information (Web and brochures); videos (sensorineural hearing loss and tinnitus); classroom aids for children with hearing loss.
American Industrial Hygiene Association	www.aiha.org/GovernmentAffairs-PR/html/publicrelations.htm#oor	703-849-8888	Consumer brochures, noise awareness, and hearing loss information.
American Speech Language and Hearing Association	www.asha.org/public/hearing	800-498-2071	Hearing information and noise brochures, sponsor of Better Speech and Hearing Month (MAY) Web site for kids.
American Tinnitus Association	www.ata.org	503-248-9985	Hear for a Lifetime; teacher training video, I love what I hear video.

(continued)

Appendix C: Resources (Continued)			
Organization	**Web Address**	**Phone**	**Resource Description**
Audiology Awareness Campaign (AAC)	www.audiologyawareness.com AUDIOLOGISTS Caring for America's Hearing	888-833-EARS (3277)	Not for profit multiorganizational foundation dedicated to educate public about value of hearing. Free brochures, free earplugs, awareness posters, public service announcement videos, online hearing test, hearing related articles, find an audiologists.
Better Hearing Institute	www.betterhearing.org	703-684-3391	Noise awareness information, posters, brochures, video (people vs. noise), better speech and hearing month materials.
Caldwell Publishing Company	www.caldwellpublishing.com	800-284-7043	videos (Human Hearing, Vols. 1–4). very detailed auditory anatomy.
Classroom Acoustics	www.classroomacoustics.com	Not applicable	Home page for classroom acoustics with links to Federal Register. Resources, issues, news, suggestions, noise busters information.
Classroom Acoustics in New Construction (Guidelines)	www.quietclassrooms.org/library/guide.htm	Not applicable	Guidelines for Classroom Acoustics in New Construction based on a 1997 Acoustics and Learning Workshop (http://www.quietclassrooms.org/library/dec97shp.htm).
Colorado State University Fort Collins, CO 80523-1875	http://littleshop.physics.colostate.edu	970-491-6206 fax:970-491-7947	CSUs Little Shop of Physics developed a number of FUN physical demonstrations to communicate a variety of fundamental concepts.
Council for Occupational Hearing Conservation	www.caohc.org	414-276-5338	Provides adult education, information and guidance to industry and those serving industry regarding the successful implementation of an occupational hearing conservation program. Occupational Hearing Technician Accreditation, Hearing Conservation Manual.
CRANK IT DOWN	www.hearingconservation.org	303-224-9022	GREAT elementary, middle, and high school adaptable curriculum; activities, descriptions.
Deafness Research Foundation: Hear US (National Campaign for Hearing Health)	http://drf.org	866-454-3924 703-610-9025	DRFs National Campaign for Hearing Health, a multi-year public education, government relations, and advocacy initiative to ensure that all Americans—especially children—will benefit from these breakthroughs. By advocating for detection, prevention, intervention, and research. The campaign promotes a lifetime of hearing health for all America: baby, children, teen, adult, and seniors.

Appendix C: Resources (Continued)

Organization	Web Address	Phone	Resource Description
The Ear Foundation	www.earfoundation.org	800-545-HEAR	Hearing related activities and information. Interactive Ear Anatomy on Web.
Educational Audiology Association:	www.edaud.org	(800) 460-7322	Hearing, noise and classroom information & videos (classroom acoustics: Listening vs. Learning); Find an educational audiologist feature if you need help to coordinate hearing screening programs and provide community awareness about hearing.
Gallaudet University National Information Center on Deafness	www.gallaudet.edu/	202-651-5050	Hearing related activities and information.
Healthy Hearing	www.healthyhearing.com/ healthyhearing/newroot/default.asp	Not applicable	General consumer information, testimonials, articles.
Hearing Education and Awareness for Rockers H.E.A.R.	www.hearnet.com	415-431-EARS (3277)	Noise awareness information, resources for musicians, video: Can't Hear you Knocking.
Hearing is Priceless (HIP) HIP Talk House Ear Institute	www.hei.org/htm/hipstart.htm	213-483-4431	Prevention curricula (elementary, middle, and high school) quizzes, diagrams, evaluations, video (hip talk with demo of simulated hearing loss).
Howard Leight Industries	www.howardleight.com	800-327-1100	Hearing related activities and information, posters, hearing protection devices, ear model, brochures, video (Maxman, Defender of Hearing).
It's About Time	www.its-About-Time.com	888-689-8463	Active physics; medical text with information on hearing and hearing loss, classroom activities.
International Institute of Noise Control Engineering—US INCE	www.i-ince.org www.inceusa.org	Fax 515-294-3528	Noise and acoustic information. A worldwide consortium of organizations concerned with noise control, acoustics, and vibration. The primary focus of the institute is on unwanted sounds and vibrations.
JeopEARdy	Acoustical Testing Laboratory Web site at www.grc.nasa.gov/www/acousticaltest/ hearingconservation/resources/ jeopeardy.htm and look in the section on Hearing Conservation for the *Auditory Demonstrations II* on-line request form.	Not applicable	Interactive multimedia hearing conservation training resource via a Microsoft PowerPoint file (accompanied by additional linked files containing sounds, videos, and other resources) that can be used as a unique interactive "game" by hearing conservationists. This resource may be used either in its basic ("ready-to-use") form, or it can be customized to meet specific needs of the audience and instructor.

(continued)

Appendix C: Resources (Continued)			
Organization	**Web Address**	**Phone**	**Resource Description**
League for the Hard of Hearing STOP THAT NOISE!	www.lhh.org/noise/index.htm	888-NOISE-88 917-305-7700	Information, posters, brochures. Educational anti noise program, hands on multimedia program with lesson plans including activities, VHS video (Stop that Noise), weekly reader quiz, audiocassettes (Unfair Hearing Test), handouts for teachers and students. Developed by Lea.
Military Audiology Association Operation BANG (Be Aware of Noise Generation)	www.MilitaryAudiology.org/bang/index.html	Not applicable	Curriculum; activities description; material lists
Musician's Earplugs	www.grc.nasa.gov/acousticaltest/ www.audiologyawareness.org www. audiology.org www.etymotic.com/pro/emlas.aspx(authorized earmold labs) www.grc.nasa.gov/www/acostics/education	Not applicable	Find a licensed audiologist familiar with custom or ready fit musician's earplugs on the AAC or AAA Web site or in the phone directory. Ensure the acoustic integrity by ordering from an authorized earmold lab listed on the Web site.
NASA Center Operation of Resources for Educators & NASA Glenn Research Center Acoustical Testing Laboratory	www.core.nasa.gov (Videos) http://acousticaltest.grc.nasa.gov (CDs) www.osat.grc.nasa.gov (info, activities, Web site dB chart)	866-776-2673	NASA CORE Connect Video Series Program 5: "Quieting the Skies" (#099.20-05) NASA Glenn—Auditory Demonstrations CDs (Vols. I and II) Great auditory demonstrations and simulated hearing loss in spacecraft interiors, automobile passenger compartments, aircraft.
National Clearinghouse for Educational Facilities	www.edfacilities.org/rl/acoustics.cfm	Not applicable	Annotated list of links, books, and journal articles on classroom acoustics.
National Hearing Conservation Association (GREAT RESOURCE!)	www.hearingconservation.org/ NHCA	303-224-9022	Links to: Crank it Down; Operation BANG (Be Aware of Noise Generation); Sounds like Fun Program; Dangerous Decibels; CSU Little Shop of Physics; Noise Destroys Educational Programs for various age groups.
National Institute on Deafness and Other Communication Disorders (NIDCD) NIDCD Information Clearinghouse WISE Ears Campaign	www.nidcd.nih.gov/health/education/index.asp www.nidcd.nih.gov/health/wise (Wise Ears)	800-241-1044	Hearing related information, educational resource guide, classroom activities, video (I love what I hear), online videos, online games, surveys, and activities. visit the kids and teachers page; play with the interactive sound ruler; hear 30 s radio spots.
National Institute for Occupational Safety and Health (NIOSH)	www.cdc.gov/niosh/topics/noise	800-35-NIOSH	Publications and other information on work-related hearing loss. General issues, such as practical guides to preventing hearing loss, and specific issues, such as noise levels in teen-related work environments (fast food, lawn care, etc.).
Noise Pollution Clearinghouse	www.nonoise.org/	888-200-8332	Noise awareness information, noise news, community and classroom information, teaching students sound hygiene.

Appendix C: Resources (Continued)

Organization	Web Address	Phone	Resource Description
Occupational Safety & Health Administration (OSHA)	www.osha.gov/dts/osta/otm/noise/index.html	800-321-OSHA (6742)	This new tool offers practical information to help eliminate hearing loss for millions of workers who are exposed to high noise levels on the job. Offers tips for teen workers exposed to hearing hazards.
Oregon Hearing Research Center Dangerous Decibels	www.dangerousdecibels.org	888-674-6674 503-494-4649	Museum, traveling exhibits, elementary, middle and high school curricula, teacher training.
Perry Hanavan/ Augustane College	www.augie.edu/perry/	605-361-5251	Virtual ear tour, comprehensive anatomical illustration Web sites.
PBS	www.pbs.org/teachers	Not applicable	Lesson plans on noise, hearing, and classrooms. Video (Future of Our Schools: Inside and Out—1 h). Public awareness of classroom acoustics.
PBS	http://pbskids.org	Not applicable	Kids site. Information on noise, sound, and hearing. Great site!
Hearing Loss Association of America, Formerly Self Help for Hard of Hearing People (SHHH): Operation SHHH	www.hearingloss.org	301-657-2248	Posters, brochures, standing noise thermometer, videos (Operation SHHH with SHHHerman and The Lion That Does Not Roar).
Sertoma International Quiet Please	www.sertoma.org	816-333-8300	Posters, brochures, videos (For Your Ears Only and Listen Up!) information.
Sight and Sound Association: Know Noise	www.sightandhearing.org	800-992-0424 651-645-2546	Teacher resource guide; posters, pamphlets, virtual ear model on Web; video (Know Noise), audiocassette (Unfair Hearing Test); lesson plans, activities, illustrations, and transparencies for third to sixth grade. Know Noise video (1993-14 min), supplemental.
Sound Distributions Systems	www.audioenhancement.com www.comtek.com www.lightspeed-tek.com www.phonicear.com www.sennheiser.com www.telex.com/hearing www.williamssound.com	see individual vendors	Vendors offering sound distributions systems.

(continued)

Appendix C: Resources (Continued)			
Organization	**Web Address**	**Phone**	**Resource Description**
The Soundry (Think Quest)	http://library.thinkquest.org/19537	866-600-4357	Interactive and educational Web site about sound. Covering everything from the most basic concepts of what sound actually is to the specifics of how humans perceive sound (developed by students).
Teachers.net	www.teachers.net/lessons/posts/1726.html	858-272-3274	Elementary student lesson plans.
Think Quest—Oracle Education Foundation	www.thinkquest.org/	866-600-4357	Think Quest, an international Web site-building competition, sponsored by the Oracle Education Foundation. Teams of students and teachers build Web sites on educational topics including hearing and noise.
US Architectural and Transportation Barriers Compliance Board (or Access Board)	www.edfacilities.org/rl/acoustics.cfm www.edfacilities.org/acoustic/index.htm www.access-board.gov/adaaq/about/bulletins/als-a.htm	202-272-0023 604-279-7408 800-872-2253	Acoustics online coursework, presentations by architects, interior designers, and engineers. Manufacturers of acoustical materials develop and sponsor seminars on acoustical issues and publish guides and manuals for design professionals.
Workers' Compensation Board of British Columbia	www.worksafe.bc.ca	604-276-3100 604-276-3068 888-621-SAFE	Excellent video (The Hearing Video) demonstrating how toxic noise can damage inner ear hair cells.

REFERENCES

Adam, A.J., Fortier, P., Schiel, G., Smith, M., Soland, C., and Stone, P. (1990). *Listening to Learn: A Handbook for Parents with Hearing Impaired Children.* Washington, D.C.: Alexander Graham Bell Association for the Deaf.

American Academy of Audiology (2003). Best Practices for Preventing Noise-Induced Occupational Hearing Loss. A Position Statement, October.

American Speech-Language and Hearing Association (1996). Central auditory processing. Status of research and implications of clinical practice. *American Journal of Audiology*, 5(2), 41–54.

ANSI S12.60-2002 (2002). American National Standard, Acoustical Performance Criteria, Design Requirements and Guidelines for Schools. Melville, NY:ANSI.

Anderson, K.L. (2001a). Voicing concerns about noisy classrooms. *Educational Leadership,* 58(7), 77–79.

Belojevic, G., Öhrström, E., and Rylander, R. (1992). Effects of noise on mental performance with regard to subjective noise sensitivity. *International Archives of Occupational and Environmental Health*, 64, 293–301.

Bennett, J.A. and English, K. (1999). Teaching hearing conservation to schoolchildren: Comparing the outcomes and efficacy of two pedagogical approaches. *Journal of Educational Audiology*, 7, 29–33.

Berger, E.H. (2000). Hearing protection devices. In: E.H. Berger, L.H. Royster, J.D. Royster, D.P. Driscoll, and M. Layne (Eds.), *The Noise Manual.* Fairfax, VA: American Industrial Hygiene Association.

Berger, E.H. and Kieper, R.W. (1994). Representative 24-hour Leqs arising from a combination of occupational and non-occupational noise exposures. *Journal of the Acoustical Society of America*, 95(5) (pt. 2), 2890.

Bess F.H., Dodd-Murphy, J., and Parker R.A. (1998). Children with minimal sensorineural hearing loss: Prevalence, educational performance and functional status. *Ear and Hearing,* 19(5), 339–354.

Blair, J.C., Hardegree, D., and Benson, P.V. (1996). Necessity and effectiveness of a hearing conservation program for elementary students. *Journal of Educational Audiology*, 4, 12–16.

British Association of Teachers of the Deaf (2001). Classroom acoustics—Recommended standards. *BATOD Magazine*, January, 2001.

Bronzaft, A.L. (1997). Beware: Noise is hazardous to our children's development. *Hearing Rehabilitation Quarterly*, 22, 4–13.

Bronzaft, A.L. and McCarthy, D.P. (1975). The effect of elevated train noise on reading ability. *Environment and Behavior*, 7(4), 517–528.

Cavanaugh, W.J. and Wilkes, J.A. (1999). *Architectural Acoustics: Principles and Practice.* New York, NY: John Wiley & Sons.

Chermak, G.D., Curtis, L., and Seikel, J. (1996). The effectiveness of an interactive hearing conservation program for elementary school children. *Language, Speech and Hearing Services in Schools*, 27, 29–39.

Chermak, G., Hall, J., and Musiek, F. (1999). Differential diagnosis and management of central auditory processing disorder and attention deficit and hyperactivity disorder. *American Journal of Audiology*, 10, 289–303.

Chermak, G. and Musiek, F. (1997). *Central Processing Disorders: New Perspectives*. San Diego, CA: Singular Publishing Group.

Clark, W.W. (1994). School related noise exposure in children. *Spectrum* 11(Suppl. 1), 28–29.

Cohen, S., Krantz, D.S., Evans, G.W., and Stokols, D. (1980). Community noise and children: Cognitive motivational and physiological effects. In: J. Tobias, G. Jansen, and D. Ward (Eds.), Proceedings of the Third International Congress on Noise as a Public Health Problem, ASHA Report 10.

Costa, O.A., Axelsson, A., and Aniansson, G. (1998). Hearing loss at age 7, 10 and 13—An audiometric follow-up study. *Scandinavian Audiology*, (Suppl. 30), 25–32.

Crandell, C. and Bess, F. (1986). Speech recognition of children in a "typical" classroom setting. *American Speech-Language and Hearing Association*, 29, 82–87.

Crandell, C. and Scalding, J. (1995). The importance of room acoustics. In: R.S. Tyler and D.J. Schum (Eds.), *Assistive Devices for Persons with Hearing Impairment*, pp. 142–164. Needham Heights, MA: Allyn & Bacon.

Cunningham, J., Nicol, T., Zecker, S.G., Bradlow, A., and Kraus, N. (2001). Neurobiological responses to speech in noise in children with learning problems. *Clinical Neurophysiology*, 112, 758–767.

Dockrell, J.E. and Shield, B.M. (2002). Children's and teachers' perceptions of environmental noise in classrooms. *Proceedings of the Institute of Acoustics*, 24(2).

Driscoll, D. and Royster, L.H. (2000). Noise control engineering. In: *The Noise Manual*, pp. 279–378.

Epstein, K.I. (2000). Sound offense, no defense, *Washington Post*, December 19, 2000, *Health* 9–15.

Erdreich, J. (1999). Classroom acoustics. *CEFPI IssueTrak*, June 1999. Available at: www.cefpi.org/pdf/issue9.pdf Retrieved 15 January 2003.

Eriks-Brophy, A. and Ayukawa, H. (2000). The benefits of sound field amplification in classrooms of Inuit students of Nunavik: A pilot project. *Language, Speech and Hearing Services in Schools*, 31(4), 324–335.

Evans, G.W., Hyge, S., and Bullinger, M. (1995). Chronic noise and psychological stress. *Psychological Science*, 6, 333–337.

Evans, G.W. and Maxwell, I. (1997). Chronic noise exposure and reading deficits: The mediating effects of language acquisition. *Environment and Behavior*, 29(5), 638–656.

Finitzo-Hieber, T. (1988). Classroom acoustics. In: R. Roser (Ed.), *Auditory Disorders in School Children*, 2nd ed., pp. 221–223. New York, NY: Thieme-Stratton.

Fischer, A.G., Murray, E.A., and Bundy, A.C. (1991). *Sensory Integration Theory and Practice*. Philadelphia, PA: F.A. Davis Company.

Flexer, C. (1999). *Facilitating Hearing and Listening in Young Children*, 2nd ed. San Diego, CA: Singular Publishing Group.

Flexer, C., Biley, K.K., and Hinkley, A. et al. (2002). Using sound field systems to teach phonemic awareness to preschoolers. *The Hearing Journal*, 55(3), 38–44.

Flexer, C., Millen, J., and Brown, L. (1990). Children with developmental disabilities: The effects of sound field amplification in word identification. *Language, Speech and Hearing Services in Schools*, 21, 177–182.

Fligor, B. and Fox, C. (2004). Output levels of commercially available compact disc players and potential risk to hearing. *Ear and Hearing*, 25(6), 513–527.

Gelnett, D., Sumida, A., and Soli, S.D. (1994). The development of the Hearing in Noise Test for Children (HINT-C). Paper Presented at Annual Convention of the American Academy of Audiology, Richmond, VA.

Gerhardt, K.J., Pierson, L.L., Huang, X., Abrams, R.M., and Rarey, K.E. (1999). Effects of intense noise exposure on fetal sheep auditory brain stem response and inner ear histology. *Ear & Hearing*, 20(1), 21 32.

Haller, K.A. and Montgomery, J.K. (2004). Noise induced hearing loss in children. What educators need to know. *TEACHING Exceptional Children*, 36(4), 22–27. www.speechandlanguage.com/article/june2004.asp

Harris, C. (1998). *Handbook of Acoustical Measurement and Noise Control*. New York, NY: McGraw-Hill.

Heller, S. (2003). *Too Loud, Too Bright, Too Fast, Too Tight: What to Do If You Are Sensory Defensive in an Overstimulating World*. New York, NY: Harper Collins.

Hicks, C.B. and Tharpe, A.M. (2002). Listening effort and fatigue in school-age children with and without hearing loss. *Journal of Speech, Language and Hearing Research*, 45(3), 573–584.

Jansen, G. (1998). Physiological effects of noise. In: C. Harris (Ed.), *Handbook of Acoustical Measurement and Noise Control*, 3rd ed., pp. 25.1–25.25. New York, NY: McGraw-Hill.

Johnson, C.E. (2000). Children's phoneme identification in reverberation and noise. *Journal of Speech, Language and Hearing Research*, 43, 144–157.

Jones, D. and Broadbent, D. (1998). Human performance and noise. In: C. Harris (Ed.), *Handbook of Acoustical Measurement and Noise Control*, 3rd ed., pp. 24.1–24.24. New York, NY: McGraw-Hill.

Kranowitz, C.S. (1998). *The Out-of-Sync Child*. New York, NY: Skylight Press.

Knecht, H.A., Nelson, P.B., Whitelaw, G.M., and Feth, L.L. (2002). Background noise levels and reverberation times in unoccupied classrooms: Predictions and measurements. *American Journal of Audiology*, 11, 65–71.

Knudsen, V. and Harris, C. (1950). *Acoustical Designing in Architecture*. New York, NY: American Institute of Physics.

Kryter, K.D. (1994). *The Handbook of Hearing and the Effects of Noise, Physiology, Psychology and Public Health*. New York, NY: Academic Press.

Kullman, L. (1999). What'd you say? U.S. News Online, 1–7. Retrieved, July 5, 2003, from http://usnews.com/usnews/health/articles/990426/nycu/26hear.htm

Landau-Goodman, K. (2001). Attitudes and Use of Musician's Earplugs. Unpublished Dissertation, Central Michigan University.

Lane, H. and Tranel, B. (1971). The Lombard sign and the role of hearing in speech. *The Journal of Speech and Hearing Sciences*, 14, 677–709.

Langford, J. (2002). Ensuring "prevention" in school hearing screening/threshold programs. *NHCA Spectrum*, 19(3), 3–4.

Litovsky, R.Y. (1997). Developmental changes in the precedence effects: Estimates of minimal audible angle. *Journal of the Acoustical Society of America*, 102, 1739–1745.

Lubman, D. (2004). Classroom acoustics: Removing acoustical barriers to learning. *NHCA Spectrum*, 19(Suppl. 1), 20.

Lucas, J.S., Dupree, R.D., and Swing, J.W. (1981). Report of a study on the effects of noise on academic achievement of elementary school children and a recommendation for a criterion level for a school noise abatement program. *Learning, Memory and Cognition*, 20(6), 1396–1408.

Markides, A. (1986). Speech levels and speech-to-noise ratios. *British Journal of Audiology*, 20, 115–120.

Meyer-Bisch, C. (1996). Epidemiological evaluation of hearing damage related to strongly amplified music (personal cassette players, discotheques, rock concerts) high-definition audiometric survey on 1364 subjects. *Audiology*, 35, 121–142.

Morioka, I., Luo, W.Z., and Miyashita, K. et al. (1996). Hearing impairment among young Chinese in a rural area. *Public Health*, 110, 293–297.

Nadler, N.B. (1997a). Noisy toys: Hidden hazards. *Hearing Health*, 13, 18–21.

Nadler, N.B. (1997b). Noisy toys: Some toys are not as much fun as they look. *Hearing Rehabilitation Quarterly*, 22, 8–10.

National Institute for Occupational and Safety (NIOSH) (1998). U.S. Department of Health and Human Services, Public Health Service, Centers for Disease Control and Prevention Publication 98.126. "Occupational Noise Exposure Revised Criteria 1998," June, pp. 24–25, Cincinnati, OH.

Nelson, P. (1997). Impact of hearing loss in children in typical school environments. Paper presented at 133rd ASA Meeting, State College, PA.

Nelson, P.B., Soli, S.D., and Seltz, A. (2002). *Classroom Acoustics II: Acoustical Barriers to Learning*. Melville, NY: Acoustical Society of America.

Niskar, A.S., Kieszak, S.M., Holmes, A., Esteban, E., Rubin, C. and Brody, D. Estimated prevalence of noise-induced hearing threshold shifts among children 6 to 19 years of age: The third National Health and Nutrition Examination Survey, 1998–1994, U.S. NHCA Spectrum, 2000; 17 Suppl 1: 23.

Niskar, A.S., Kieszak, S.M., Holmes, A.E., Esteban, E., Rubin, C., and Body, D.J. (2001). Estimated prevalence of noise-induced hearing threshold shifts among children 6 to 19 years of age: The Third National Health and Nutrition Examination Survey, 1988–1994, United States. *Pediatrics*, 109(1), 40–43.

Ross, M. (1990). Definitions and descriptions. In: J. Davis (Ed.), *Our Forgotten Children: Hard-of-Hearing Pupils in the Schools*, pp. 3–17. Washington, D.C.: U.S. Department of Education.

Schappert, S. (1992). Office visits for otitis media: United States, 1975–1990. Advance Data from Vital and Health Statistics, National Center for Health Statistics, No. 214, pp. 1–15.

Seep, B., Glosemeyer, R., Hulce, E., Linn, M., Aytar, P., and Coffeen, R. (2000). Classroom acoustics: A resource for creating learning environments with desirable listening conditions. Melville, NY: Acoustical Society of America. Available in print from ASA or online at: http://asa.aip.org/classroom/booklet.html

Shield, B.M. and Dockrell, J.E. (2003a). The effects of noise on children at school: A review. *Journal of Building Acoustics*, 10(2), 97–116.

Shield, B.M. and Dockrell, J.E. (2003b). External and internal noise surveys of London primary school. *Journal of the Acoustical Society of America*, 115, 730–738.

Siebein, G.W., Gold, M.A., Siebein, G.W., and Ermann, M.G. (2000). Ten ways to provide high-quality acoustical environments in schools. *Language, Speech and Hearing Services in Schools*, 31, 376–384.

Smith, E., Lemke, J., Taylor, M., Kirchner, H.L., and Hoffman, H. (1998). Frequency of voice problems among teachers and other occupations. *Journal of Voice*, 12, 480–488.

Soli, S.D. and Sullivan, J.A. (1997). Factors affecting children's speech communication in classrooms. *Journal of the Acoustical Society of America*, 101, S3070.

Stach, B.A. (1998). *Clinical Audiology: An Introduction*, p. 128. San Diego, CA: Singular Publishing.

Studebaker, G.A., Sherbecoe, R.L., McDaniel, D.M., and Gwaltney, C.A. (1999). Monosyllabic word recognition at higher-than-normal speech and noise levels. *Journal of the Acoustical Society of America*, 105, 2431–2444.

U.S. General Accounting Office (2001). BIA and DOD Schools: Student Achievement and Other Characteristics Often Differ from Public Schools', GAO-01-934, p. 25.

Vallet, M. (2000). Some European standards on noise in educational buildings. Proceedings of the International symposium on Noise Control and Acoustics for Educational Buildings, Proceedings of the Turkish Acoustical Society, Istanbul, May 2000, pp. 13–20.

Walker L. (1986). *A Loss for Words*, p. 20. New York, NY: Harper and Row.

Watkins, D. (2004). The prevalence of high frequency hearing loss in eighth grade children and their attitudes and knowledge regarding noisy leisure time activities. *NHCA Spectrum*, 21(Suppl. 1), 27.

Werner, L. and Boike, K. (2001). Infants' sensitivity to broadband noise. *Journal of the Acoustical Society of America*, 109, 2103–2111.

Wolff, A.B. and Harkins, J.E. (1986). Multihandicapped students. In: A.N. Schildroth and M.A. Karchmer (Eds.), *Deaf Children in America*, pp. 55–82. San Diego, CA: College-Hill Press.
World Health Organization (1997). Prevention of Noise Induced Hearing Loss. WHO: Prevention of Deafness and Hearing Loss 98.5, October, Geneva, Switzerland.
Yost, W.A. (2000). *Fundamentals of Hearing: An Introduction*. San Diego, CA: Academic Press.

ADDITIONAL REFERENCES

Acoustical Society of America (2002). Technical Committee on Speech Communication. *Classroom Acoustics II, Acoustical Barriers in Learning*. Melville, NY: Acoustical Society of America.
Airey, S.L., MacKenzie, D.J., and Craik, R.J.M. (1998). Can you hear me at the back? Effective communication in classrooms. In: Seventh International Congress of Noise as a Public Health Problem, Sydney, Vol. 1, pp. 195–198.
American National Standards Institute (1994). *Acoustic Terminology*. New York, NY: American National Standards Institute, S1.1-1994.
American Speech-Language and Hearing Association (2000). *Healthy People 2010—Health Objectives for the Nation and Roles for Speech-Language Pathologists and Speech Language Hearing Scientists*. Rockville, MD: American Speech-Language and Hearing Association.
American Speech-Language-Hearing Association (2004). *Acoustics in Educational Settings: Technical Report*. Rockville, MD: American Speech-Language-Hearing Association.
American Speech-Language-Hearing Association (1995). Position statement and guidelines for acoustics in educational settings. *American Speech-Language and Hearing Association*, 37(Suppl. 14), 15–19.
American Speech-Language-Hearing Association (2002) Guidelines for fitting and monitoring FM systems. In: *ASHA Desk Reference*, pp. 151–171. Rockville, MD: ASHA.
Anderson, K.L. (2001b). Hearing conservation in the public schools revisited. *Seminars in Hearing*, 12(4), 340–364.
Axelsson, A. and Jerson, T. (1985). Noise toys: A possible source of sensorineural hearing loss. *Pediatrics*, 76(4), 574–578.
Bess, F. and Tharpe, A. (1986). An introduction to unilateral sensorineural hearing loss in children. *Ear and Hearing*, 7, 3–13.
Bess, F.H. and Poynor, R.E. (1972). Snowmobile engine noise and hearing. *Archives of Otolaryngology*, 19(5), 164–168.
Boothroyd, A. (1997). Auditory development of the hearing child. *Scandinavian Audiology*, 26(Suppl. 46), 9–16.
Boothroyd, A. (2003). Room acoustics and speech reception: A model and some implications. In: Proceedings from the 1st International FM Conference. ACCESS: Achieving Clear Communication Employing Sound Solutions, p. 207. Warrenville, IL: Phonak AG.
Bradley, J. (1986). Speech intelligibility studies in classrooms. *Journal of the Acoustical Society of America*, 80, 846–854.
Bradlow, A.R., Kraus, N., Nicol, T., McGee, T., Cunningham, J., Zecker, S.G., and Carrell, T. (1999). Effect of lengthened formant transition duration on discrimination and representation of cv syllables by normal and learning disabled children. *Journal of the Acoustical Society of America*, 104(4), 2086–2096.
Broadbent, D.E. (1981). The effects of moderate levels of noise on human performance. In: J. Tobias and E. Schubert (Eds.), *Hearing: Research and Theory*, Vol. 1. New York, NY: Academic Press.
Bronzaft, A.L. (1981). The effect of a noise abatement program on reading ability. *Journal of Education Research*, 1, 215–222.
Brookhouser, P.E., Worthington, D.W., and Kelly, W.J. (1992). Noise-induced hearing loss in children. *Laryngoscope*, 102, 645–655.
Chermak, G.D., Curtis, L., and Seikel, J.A. (1996). The effectiveness of an interactive hearing conservation program for elementary school children. *Language Speech Hearing Services in Schools*, 27, 29–39.
Cohen, S., Evans, G.W., Stokols, D., and Krantz, D.S. (1986). *Behavior, Health and Environmental Stress*. New York, NY: Plenum Press.
Cohen, S., Krantz, D.S., Evans, G.W., and Stokols, D. (1980). Physiological, motivational and cognitive effects of aircraft noise on children: Moving from the laboratory to the field. *American Psychologist*, 35, 231–243.
Cohen, S., Krantz, D.S., Evans, G.W., Stokols, D., and Kelly, S. (1981). Aircraft noise and children: Longitudinal and cross-sectional evidence on adaptation to noise and the effectiveness of noise abatement. *Journal of Personality and Social Psychology*, 40, 331–345.
Crandell, C. (1991). Classroom acoustics for normal-hearing children: Implications for rehabilitation. *Educational Audiology Monograph*, 2(1), 18–38.
Crandell, C. (1992). Classroom acoustics for hearing-impaired children. *Journal of the Acoustical Society of America*, 92, 2470.
Crandell, C. (1993). Noise effects on the speech recognition of children with minimal hearing loss. *Ear and Hearing*, 14, 210–216.
Crandell, C. and Smaldino, J. (1994). Room acoustics. In: R. Tyler and D. Schum (Eds.), *Assistive Devices for the Hearing Impaired*. Needham Heights, MA: Allyn & Bacon.
Crandell, C. and Smaldino, J. (1996). Speech perception in noise by children for whom English is a second language. *American Journal of Audiology*, 5(3), 47–51.
Crandell, C. and Smaldino, J. (2000). Classroom acoustics for children with normal hearing and with hearing impairment. *Language Speech and Hearing Services in Schools*, 31, 362–370.

Department for Education and Employment (1997). *Guidelines for Environmental Design of Schools* (Building Bulletin 87). London: The Stationary Office.

Department for Education and Skills (2003). Building Bulletin 93 Acoustic Design of Schools. www.teachernet.gov/acoustics

Dockrell, J.E. and Shield, B.M. (2003). Acoustic guidelines and teacher strategies for optimizing learning conditions in classrooms for children with hearing problems. In: Proceedings from the 1st International FM Conference: Achieving Clear Communication Employing Sound Solutions, pp 217–228. Warrenville, IL: Phonak AG.

EAR, EarLogs. Cabot Safety Corporation: Southbridge, MA. www.e-a-r.com/hearingconservation/ www.e-a-r.com/hearingconservation/earlog_main.cfm, www.aearo.com/thml/industrial/tech01.asp#noise

Evans, G., Bullinger, M., and Hygge, S. (1998). Chronic noise exposure and physiological response: A prospective study of children living under environmental stress. *Psychological Science*, 9(1), 75–77.

Evans, G. and Tafalla, R. (1997). Noise, physiology and human performance: The potential role of effort. *Journal of Occupational Health Psychology*, 2(2), 144–155.

Evans, G.W. and Lepore, S.J. (1993). Nonauditory effects of noise on children: A critical review. *Children's Environments* 10(1), 31–51.

Flexer, C. (2002). Rationale and use of sound field systems: An update. *The Hearing Journal*, 55(8), 10–32.

Flexer, C. (2003). Integrating sound distribution systems and personal FM technology. In: Proceedings from the 1st International FM Conference. ACCESS: Achieving Clear Communication Employing Sound Solutions, pp. 121–131. Warrenville, IL: Phonak AG.

Folmer, R.L., Griest, S.E., and Martin, W.H. (2002). Hearing conservation education programs for children a review. *Journal of School Health*, 72(2), 51–57.

Gupta, D. and Vishwakarma, S.K. (1989). Toy weapons and firecrackers: A source of hearing loss. *Laryngoscope*, 99, 330–334.

Haines, M.M., Stansfeld, S.A., Berglund, B., and Job, R.F.S. (1998). Chronic aircraft noise exposure and child cognitive performance and street. In: N. Carter and R.F.S. Job (Eds.), Proceedings of the 7th International Conference on Noise as a Public Health Problem, Vol. 1, pp. 329–335. Sydney: Noise Effects '98 Pty Ltd.

Hellstrom, P.A., Dengerink, H.A., and Axelsson, A. (1992). Noise levels from toys and recreational articles for children and teenagers. *British Journal of Audiology*, 26, 267–270.

Hetu, R., Getty, L., and Quoc, H.T. (1995). Impact of occupational hearing loss on the lives of workers. In: T.C. Morata and D.E. Dunn (Eds.), *Occupational Medicine: State of the Art Reviews*, Vol. 10(3), pp. 495–512. Philadelphia, PA: Hanley & Belfus Inc.

Hygge, S. (2000). Effects of aircraft noise on children's cognition and long-term memory. Presentation at the FICAN Symposium, February, San Diego, CA.

Institute for Environment and Health (1997). The non-auditory effects of noise. Report R10.www.le.ac.uk/ieh/pdg/exsumR10.pdf

Jager Adams, M. et al. The Importance of Phonemic Awareness. Excerpted from Phonemic Awareness in Young Children. http://www.readingrockets.org/article.php?ID=415

Kodaras, M. (1960). Reverberation times of typical elementary school settings. *Noise Control*, 6, 17–19.

Lass, N.J., Woodford, C.M., and Lundeen, C. et al. (1987a). A survey of high school students' knowledge and awareness of hearing, hearing loss and hearing health. *The Hearing Journal*, 15–19.

Lass, N.J., Woodford, C.M., and Lundeen C. et al. (1987b). A hearing conservation program for a junior high school. *The Hearing Journal*, 32–40.

Lubman, D. (1997). Satisfactory classroom acoustics. Paper presented at Acoustical Society of America, State College, PA.

Maxwell, L. and Gary, E. (1997). Chronic noise exposure and reading deficits: The mediating effects of language acquisition. *Environment and Behavior*, 29(5).

Meinke, D. (2001). Crank it down! Update. *Spectrum*, 18(3), 1–4.

Montgomery, J.K. and Fujikawa, S. (1992). Hearing thresholds of students in the second, eighth and twelfth grades. *Language, Speech and Hearing Services in the Schools*, 23, 61–63.

Nabelek, A. and Nabelek, I. (1994). Room acoustics and speech perception. In: J. Katz (Ed.), *Handbook of Clinical Audiology*, 3rd ed., pp. 834–846. Baltimore, MD: Williams & Wilkins.

Nabelek, A. and Pickett, J. (1974). Reception of consonants in a classroom as affected by monaural and binaural listening, noise, reverberation and hearing aids. *Journal of the Acoustical Society of America*, 56, 628–639.

National Center for Education Statistics. U.S. Department of Education. Condition of American's Public School Facilities, 1999. NCES 2000-032, Washington, D.C., June 2000.

National Center for Education Statistics (2001). National Assessment of Educational Progress (NAEP): 1992–2000 Reading Assessments. Washington, D.C.: U.S. Department of Education Office of Educational Research Improvement.

National Institute for Occupational and Safety (NIOSH) (1996). U.S. Department of Health and Human Services, Public Health Service, Centers for Disease Control and Prevention, Publication 96-115. "National Occupational Research Agenda", Cincinnati, OH.

Nelson, P.B. (2003). Sound in the classroom—Why children need quiet. *ASHRAE Journal*, February 2003, 22–25.

Nelson, P.B. and Rosenberg, G.G. (2001). Classroom Acoustics: Standards & Technology Update. Presentation at American Academy of Audiology, San Diego.

Nelson, P.B. and Soli, S. (2000). Acoustical barriers to learning: Children at risk in every classroom. *Language, Speech and Hearing Services in Schools*, 31, 356–361.
NIH (National Institutes of Health) (1990). Consensus conference: Noise and hearing loss. *Journal of American Medical Association*, 263(23), 3185–3190.
Niskar, A.S., Kieszak, S.M., and Holems, A.E. et al. (1998). Prevalence of hearing loss among children 6–19 years of age. *Journal of the American Medical Association*, 279(14), 1071–1075.
Nober, L.W. (1973). Auditory discrimination and classroom noise. *Reading Teacher*, 27(3).
Nober, L.W. and Nober, E.H. (1975). Auditory discrimination of learning disabled children in quiet and classroom noise. *Journal of Learning Disabilities*, 8(10), 656–773.
Occupational Safety and Health Administration (1983). Occupational Noise Exposure; Hearing Conservation Amendment; final rule. 29CFR 1910.95 Federal Register.48(46):9738–9785.
Palmer, C.V. (1997). Hearing and listening in a typical classroom. *Language, Speech, and Hearing Services in Schools*, 28(3), 213–218.
Plakke, B.L. (1985). Hearing conservation in secondary industrial arts classes: A challenge for school audiologists. *Language, Speech and Hearing Services in Schools*, 16, 75–79.
Rosenberg, G. and Balke-Rahter, P.E. (1995). In-service training for the classroom teacher. In: C. Crandell, J. Smaldino, and C. Flexer (Eds.), *Sound-field FM Amplification*, pp. 149–190. San Diego, CA: Singular Publishing Group.
Shield, B.M. and Dockrell, J.E. (2002). The effects of environmental noise on child academic attainments. *Proceedings of Institute of Acoustics*, 24(6).
Siebert, M. (1999). Educators often struck by voice ailments. *The DesMoines Register*, p. 4.
Smith, A.W. (1989). A review of the effects of noise on human performance. *Scandinavian Journal of Psychology*, 30, 185–206.
Stansfeld, S. and Haines, M. (2000). Chronic aircraft noise exposure and children's cognitive performance and health: The Heathrow Studies, Presentation at FICAN Symposium, February, San Diego, CA.
Stelmachowitz, P.G. et al. (2000). The relation between stimulus context, speech audibility, and perception for normal-hearing and hearing-impaired children. *Journal of Speech, Language, and Hearing Research*, 43, 902–914.
Suter, A. (1992). ASHA monographs on communication and job performance in noise: A review. Rockville, MD, American Speech-Language-Hearing Association, Monograph, 28, 53–78.
United States Access Board Request for Information on Classroom Acoustics (1998). www.access-board.gov/pulications/acoustic-factsheet.htm
Universal Design (1999). Ensuring access to the general education curriculum. *Research Connections in Special Education*, 5, 1–2.
U.S. Bureau of the Census (1990). Population estimates and projections. Washington, D.C.: National Center for Education Statistics. www.census.gov/population/www/socdemo/lang_use.html
U.S. Department of Health and Human Services (2000). *Healthy People 2010: National Health Promotion and Disease Prevention Objectives*. DHHS Publication No. 91-50121. Washington, D.C.: US Government Printing Office, Superintendent of Documents.
U.S. General Accounting Office, Health, Education and Human Services Division (1995). School Facilities: Condition of America's Schools, Document #: GAO/HEHS-95-61, Report # B-259307, February 1.
U.S. Public Health Service (1990). *Healthy People 2000*. Washington, D.C.: Government Printing Office.
U.S. Public Health Service (2000). *Healthy People 2010*. Washington, D.C.: Government Printing Office.
Weinstein, C.S. and Weinstein, N.D. (1979). Noise and reading performance in an open space school. *The Journal of Educational Research*, 72, 210–213.
White, S. (2003). Providing an educational hearing conservation program for kids. *The Hearing Review*, 10(10), 24–26, 63.
WHO (1995). Community Noise, Archives of the Center for Sensory Research 2(1), Gerglund, B. and Lindvall, T. (Eds.), Prepared for the World Health Organization by Stockholm University, Sweden. www.who.int/peh
Woodford, C.M. and O'Farrell, M.L. (1983). High-frequency loss of hearing in secondary school students: An investigation of possible etiological factors. *Language, Speech and Hearing Services in Schools*, 14, 22–28.

Figure 5.74
(Nancy Vause)

The author is grateful to "Mema" and Greg for providing unwavering support to educate the public about the value of hearing health care. They encourage curiosity, honesty, excellence, mercy, generosity, unselfish love, and communication by example. I wish to thank an anonymous reviewer and Geri, both generous and talented friends. Thanks to numerous individuals regularly offering thoughtful critiques, creative suggestions, innovative ideas, peaceful solace, artistic talent, and escarole soup. I appreciate the assistance and permissions from all contributors including but not limited to: Audiology Awareness Campaign, Banner School (Frederick, MD), Suzanne Roose (Director of Admissions), Kurt Holter Photography, National Hearing Conservation Association, Mike and Patrick Younkins, Dr. David Lubman, Dianne and Greg Huyck, Adam Beehler, Deanna Meinke, Valerie Rice, Rani Lueder, Ursy Potter Photography. Each graciously provided numerous photos. Don Driggers, Karen Ruch Mohney, and Tracy Hargus happily supplied the no notice technical expertise. Thank you.

CHAPTER 6

PHYSICAL DEVELOPMENT IN CHILDREN AND ADOLESCENTS AND AGE-RELATED RISKS

RANI LUEDER AND VALERIE RICE

TABLE OF CONTENTS

INTRODUCTION

The world is changing, and these changes cause us to revisit many of our long-held assumptions. Children are subject to quite different sensory, cognitive, and physical demands than they have faced before.

Today's children are also different from previous generations in many countries. They are often taller, heavier, and less fit. Many children (particularly girls) are experiencing puberty earlier than previous generations, which can impact their potential for musculoskeletal pain and disorders.

New technologies and new methods can introduce ergonomic risks. We do not know what ergonomic risks we are exposing children to and what the impact of exposure may mean.

Although it is obvious that cognitive changes through childhood are quite dramatic, many assume the physical risk factors bear some correspondence to those adults face. Many also assume children risk incurring the same sorts of musculoskeletal disorders that adults experience in today's workplaces—despite strong evidence to the contrary. In addition, certain ergonomic risk factors may have consequences that are more serious for children than for adults.

Our language may also lead to confusion. Despite having one name, childhood is a dynamic process of moving through a number of stages, each with its own characteristics and risk factors that affect physical development and well-being. While other chapters in this text have addressed many issues on this topic,* the aim of this chapter is to revisit physical development and risk of musculoskeletal disorders.

Figure 6.1 (Ursy Potter Photography)

GROWTH AND DEVELOPMENT

BODY SIZES

American and European children are taller (Juul et al., 2006) more overweight (Table 6.1), and less fit than their predecessors. Traditional assumptions about children's developmental stages of growth do not always reflect differences between nations and cultures (Bass et al., 1999).

Children's developmental patterns of growth are changing as well. Such changes in the onset of puberty have design implications. For example, children who are getting taller require higher seats. Those who are wider (heavier) require deeper and wider seat pans. Nations where children experience puberty at an earlier age have a greater need for postural support than previous generations.†

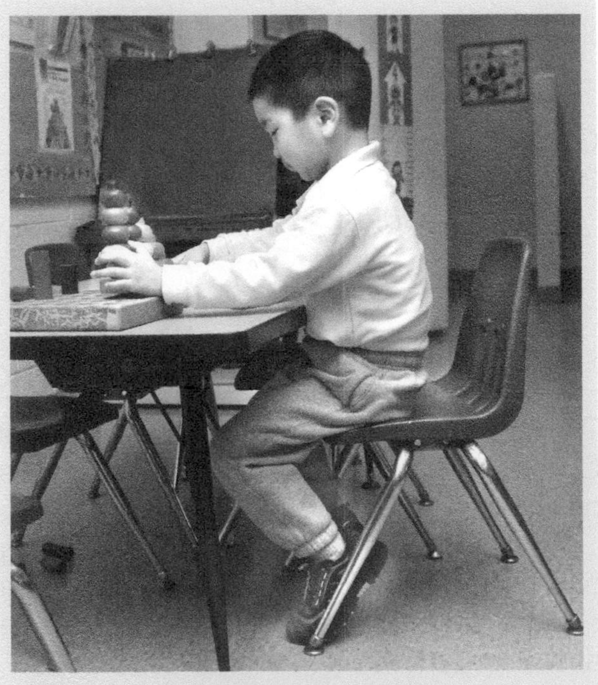

Figure 6.2 (Ursy Potter Photography)

* While all of the chapters in this book address children's development, specialized examples of development can be found in the following chapters: Chapter 2 (Child development); Chapter 3 (Anthropometrics); Chapter 4 (Vision); Chapter 7 (Physical fitness); Chapter 8 (Injuries); Chapter 14 (Warnings); Chapter 21 (School furniture); and Chapter 26 (Wayfinding).

† Please see Table 6.2 (Developmental changes of the spine), Table 6.3 (Musculoskeletal injures), and Table 6.6 (Back pain among children).

ONSET OF PUBERTY: Although boys and girls exhibit a similar rate of growth through most of childhood, that changes during upon and during puberty. Girls in Europe and the United States are attaining puberty at an earlier age than previous generations (Biro et al., 2006; Juul et al., 2006). Given that the onset of puberty has a dramatic effect on the rate and pattern of growth and development (Table 6.3), understanding these developmental patterns is critical when considering ergonomic risk factors that children experience.

Girls who are heavier* seem to undergo puberty earlier than their thinner counterparts (Biro et al., 2006; Juul et al., 2006; Trentham-Dietz et al., 2005). US children (who tend to be heavier) experience puberty a full year before schoolchildren in Denmark, who tend to be thinner (Juul et al., 2006).

Left-handed girls have been found to experience menarche a year before right-handed girls (Orbak, 2005). Early onset of puberty is associated with greater risk of musculoskeletal disorders (see Tables 6.2, 6.3, and 6.6). Given that left-handed girls are already at a disadvantage in our right-handed world, these children may need particular postural support in their early adolescence.

OBESITY: The number of overweight and obese children aged 6–19 years has increased dramatically over the last two decades, but appears to be leveling off (Hedley et al., 2004) (Table 6.1).

The increase in obesity is not exclusive to the United States, although children in the United States are among the heaviest. While children are heavier in many countries, the national rate of increase in weight varies.† For example, while children in the United States are heavier than British children, British children seem to be gaining weight faster than children in the United States (Smith and Norris, 2004; also see Reilly et al., 1999; Reilly and Dorosty, 1999).

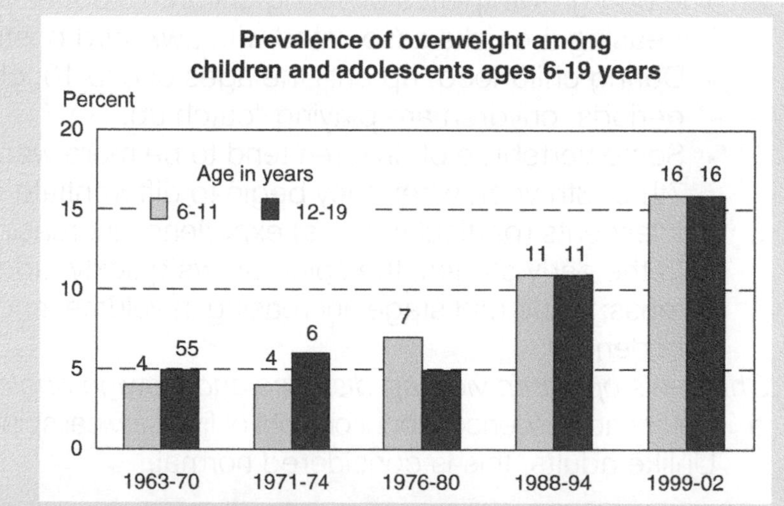

Table 6.1 Prevalence of Overweight US Children and Adolescents Aged 6 to 19 years, from 1963–1965 through 1999–2002[a]

Age (Years)	NHANES 1963–65 1966–70	NHANES 1971–74	NHANES 1976–80	NHANES 1988–94	NHANES 1999–2002
6–11	4%	4%	7%	11%	16%
12–19	5%	6%	5%	11%	16%

Note: There continue to be disparities in rates between boys and girls, and between racial and ethnic groups (Hedley et al., 2004)[b]

[a] For more detailed estimates, see Hedley et al. (2004) and Ogden et al. (2002). Data for 1963–65 are for children 6 to 11 years of age; data for 1966–70 are for adolescents 12 to 17 years of age, not 12 to 19 years

[b] www.cdc.gov/nchs/products/pubs/pube/hestats/ovrwght99.htm

* That is, they have higher body mass index (BMI) scores.

† NHANES (2002) provides nationally representative data of the US child population. The NHANES III survey measured children aged from 2 months to 18 years. These are the most recent, available data on US stature and weight (DHHS, 2000). See also www.cdc.gov/growthcharts.

DEVELOPMENTAL CHANGES

It is helpful to understand the implications of how children develop and mature (Table 6.2), when designing for children. These considerations include:

1. *Allow for children's different patterns of growth at different ages* (Bass et al., 1999).
   a. Younger children tend to grow more in their extremities.
   b. Adolescent growth (following puberty) largely affects the spine.
2. *Accommodate movement through a range of neutral postures for all children.*
   a. Allow children to move freely through a range of good postures.
   b. Encourage neutral postures.
      - Although the ideal posture may change somewhat as children mature,* they are also developing postural habits that should help protect them when they are adults.
      - It is very difficult (nigh impossible) to teach adults to "unlearn" bad postural habits.
3. *Some children need particular postural support:*
   a. The end growth plates of young children's bones are soft and vulnerable increasing the risks associated with awkward postures.†
      - During childhood, up until the ages of 8 to 10, children's spine is in growth periods, children are playing "catch up."
      - Some vertebrae of children tend to be more wedge shaped until around their 7th or 8th year, when they begin to differentiate and square off.
   b. Adolescents (particularly girls) experience increasing risk of back and neck pain.
      - In the early stages, the spine grows quickly, adding length without adding mass; in the mid stage, increasing in volume, and in the later stages, increasing in density.
4. *Children's optimum viewing distance and viewing angles change through childhood*‡
   a. Before adolescence, about one out of five cervical spines is kyphotic (Dormans, 2002). Unlike adults, this is considered normal.

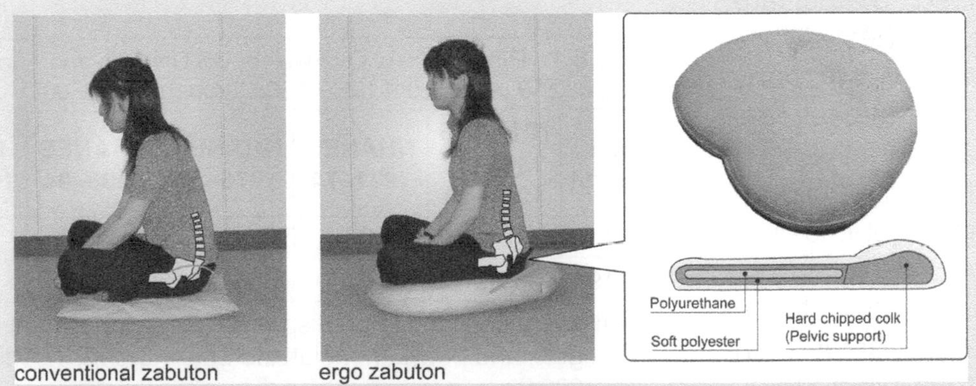

Figure 6.3 (Photos and drawing by Noro Seating Research Labs)
At birth, the spine is C-shaped. The spine does not develop the S-shape, which we associate with the spine, until puberty (see Table 6.2). These age-related changes affect design solutions. Above, an adolescent in Japan sits on a contoured Zabuton mat designed to reinstate the natural curve of the low back.

* Please see Table 6.4 about children's developmental stages.

† Bass and Bruce (2000) note "in children, most ligaments are attached to the epiphyses (growing ends of bones) above and below. This anatomical arrangement concentrates forces on the physis (horizontal growth plate). While low-velocity trauma can result in ligamentous injury, high-velocity trauma is more likely to result in physeal disruption, because forces applied rapidly do not allow time for ligaments to stretch."

‡ Please see Chapter 4 of this book on children and vision.

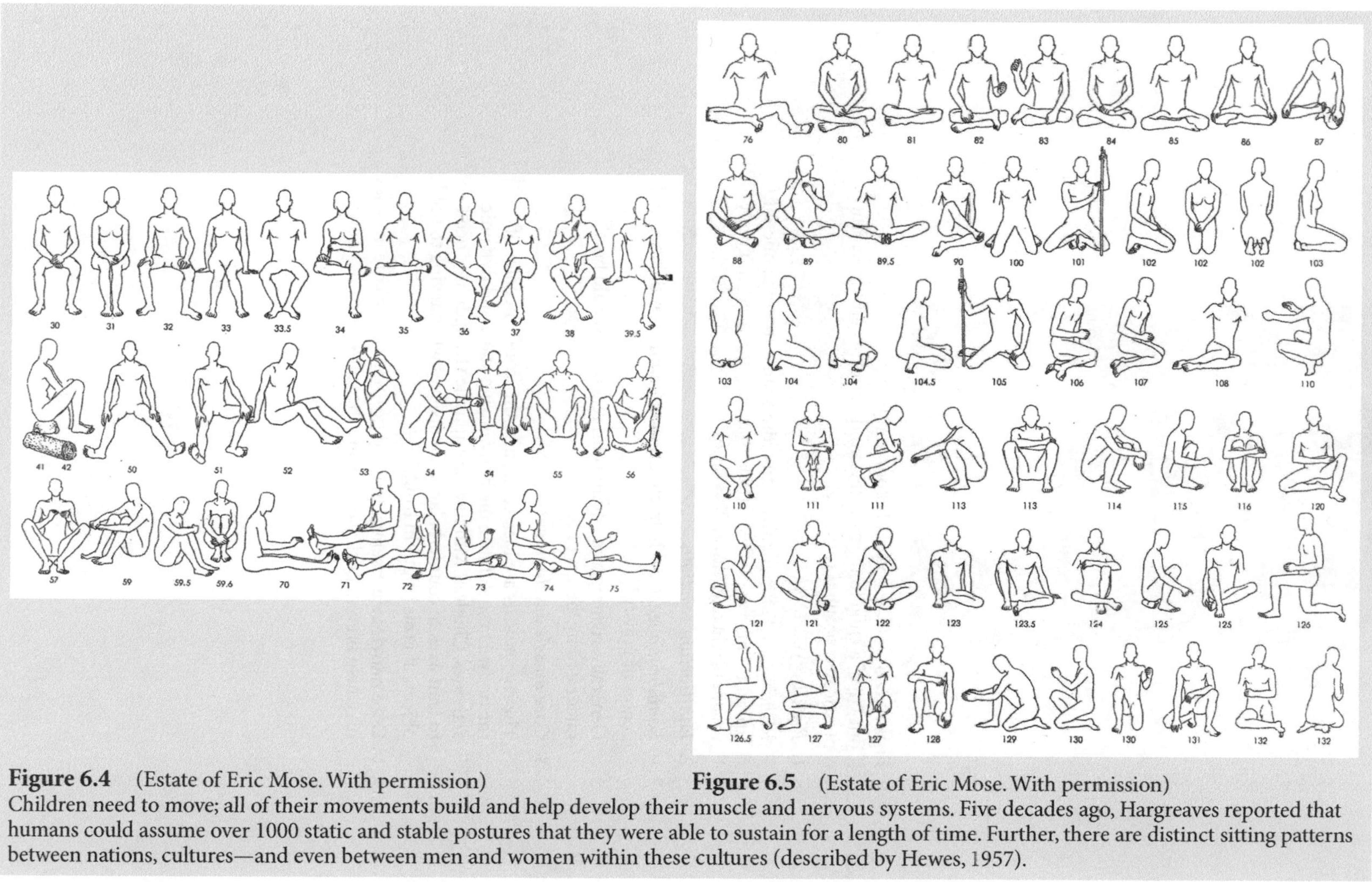

Figure 6.4 (Estate of Eric Mose. With permission)

Figure 6.5 (Estate of Eric Mose. With permission)

Children need to move; all of their movements build and help develop their muscle and nervous systems. Five decades ago, Hargreaves reported that humans could assume over 1000 static and stable postures that they were able to sustain for a length of time. Further, there are distinct sitting patterns between nations, cultures—and even between men and women within these cultures (described by Hewes, 1957).

Figure 6.6 (Ursy Potter Photography)
If one "tips" a child to the side, the child performs an innate, but sophisticated postural response:

1. ***Righting reactions***

   *Righting reactions** involve lifting the head to re-align oneself upright. These are reflex actions that keep the head and body in an upright and "normal" position.

2. ***Equilibrium reactions***

   Equilibrium reactions help children maintain their balance and center of gravity.†

   Generally, righting and equilibrium reactions should be integrated by the age of 3.

3. ***Core muscle development***

   The *core muscles* are located around the "core" or center of the body. These stabilize the body during static and dynamic activities. Children with low core-muscle strength may need to lean back to counterbalance themselves when trying to hold objects at arms length in front.

   Core muscles are essential to good posture; strong core muscles may help prevent back pains where the causes are not known.

* The five types of *righting reactions* include labyrinth righting, body-righting (acting on the head), neck-righting, body-righting (acting on the body), and optical righting.

† ***Equilibrium reactions*** include protective reactions of the arms and legs, tilting reactions, and postural fixating reactions.

Table 6.2 Child and Adolescent Changes in the Spine

<table>
<tr><th></th><th></th><th>Design Implications</th></tr>
<tr><td>Rates of Growth</td><td>1. Children's rate of growth begins to stabilize after their second year of infancy, then continues at about the same pace until the growth spurt that follows the onset of puberty (Kiczmarski et al., 2000).
2. As puberty approaches, growth velocity slows (preadolescent dip) and then accelerates during mid-puberty, where it compares with the rate of growth during infancy (Spear, 2002).
3. Puberty is associated with an increase in weight.
Children gain 50% of their adult weight in adolescence. Girls often experience weight gains before the growth spurt, whereas boys generally gain weight during this time.
4. There is considerable variability in rates of growth between children of the same age.
A better predictor is the child's maturity relative to their age[a] (Spear, 2002).</td><td>• High growth periods following the onset of puberty deserve particular concern.
• These years are characterized by rapid changes in body length and weight.
• Age is a poor predictor of a child's size.</td></tr>
<tr><td>Pattern of Growth</td><td>1. Growth is a dynamic process.
A child's pattern of growth and maturation reflect an interplay of dynamic processes between body components and at the molecular, cellular, and somatic levels.
2. Age
a. Childhood
• Boys undergo longer periods of growth before reaching adolescence (Spear, 2002).
• Before puberty, the legs grow more rapidly than the trunk (Bass et al., 1999).
b. Adolescence
Adolescents undergoing their peak increase in height are at greater risk of low back pain.[b]
• During puberty, teenagers gain about 15% of their ultimate stature and 45% of their final skeletal mass (Spear, 2002).
• The growth spurt among adolescents is largely in the torso (Bass et al., 1999).
• Boys experience their peak rate increase in height and weight (velocity) at the same time.
• Boys attain a greater peak velocity growth in height.
• In girls, their peak weight velocity is six to months before the rapid increase in height.</td><td>• Both the rate of growth and the relative growth of body parts differs between children and adolescents following puberty.
• High-growth periods such as in late adolescence deserve particular concern.
• Boys undergo longer periods of growth before reaching adolescence.</td></tr>
</table>

[a] Based on the Tanner's (1978) stages of sexual maturity. This example from Spear (2002) may be instructive: "In boys, a faint but noticeable mustache at the corners of the lip corresponds to sexual maturity rating. This is the beginning of the peak height velocity, the time when he will need the peak in energy and nutrient intake."

[b] In their review of the research, Duggleby and Kumar (1997) concluded that two types of adolescents are at risk for low back pain. The first group represented adolescents at the peak of the growth spurt who compete in sports and train for more than 15 hours a week. The second group represents adolescents just past the growth spurt who tended to be inactive and might smoke.

(continued)

Table 6.2 Child and Adolescent Changes in the Spine (Continued)

		Design Implications
Bone Development	**Interplay between components** • Bones are soft and the relationship between child bones and muscles changes over time. • Unlike adults, children's spines (intervertebral discs) can actively "feed" on nutrients and eliminate waste. • Each bone surface has a unique development pattern. • Developmental differences appear to vary by regions, affecting bone development in ways that may predispose children living in some regions to specific deficits in bone size, content, and density.[c] • Injuries of the spinal cord in children tend to occur at different sites on the spine than similar injuries in adults. **Age** Most bone mass is accrued during childhood and adolescence—particularly during puberty (Arikoski et al., 2002). Yet, age is a poor predictor of growth. There is a complex interaction between the child's age, height, weight, rate of growth, and age of onset of puberty (Warner et al., 1998).[d] 1. **Childhood** Young children's bones grow in volume and mass, but to a lesser extent than during adolescence (Arikoski et al., 2002). • The end growth plates of young children's bones are soft and vulnerable, increasing the risk of awkward postures. • Some vertebrae of children tend to be more wedge shaped until around their 7th or 8th year, when they begin to differentiate and square off. 2. **Early adolescence** • Bones and spine are weak and vulnerable; tendons, ligaments, and muscles are still developing. • Growth plates are often in overdrive; bones and spine grow much longer without substantially adding mass. 3. **Mid-adolescence** • The spine increases in volume without a corresponding increase in mass. 4. **Late adolescence** Child spines "catch-up," increasing mass as growth slows.	• Young children are at greater risk because their bones are still developing. • Adolescents (after puberty) are at risk because they are playing "catch up."

[c] Bass et al. (1999) report that each bone surface has a unique developmental pattern that appears to be specific to the region where the child resides. These may affect the differing tempo and rate of growth in bone size and mass and direction of growth of the periosteal and endocortical surfaces, and the differing tempo of growth of the axial and appendicular bones.

[d] Complex interaction also includes, hormones (Alexy et al., 2005), obesity, gender and nutrition—notably protein (Alexy et al., 2005), calcium, and vitamin D. Consumption of sodas has been found to adversely affect bone mineral content accrual in girls, but not boys. Some, but not all, studies found differences between children of different ethnic origins and race, which largely attributed to differences in bone size.

Table 6.2 Child and Adolescent Changes in the Spine (Continued)

		Design Implications
Lumbar Curve	1. **Contours of children's low back** The spine is "C-shaped" at birth. Young children lack low back (lumbar) curves, which largely develop during adolescence following puberty. 2. **Flexibility of the spine** a. **Development** The lumbar spine in boys and girls begins to stiffen from ages 15 to 18 (Burton et al., 1996; Haley et al., 1986). The range of motion and overall flexibility of the low back increases with age, for both children and adolescents, stabilizing around the 15th year (for boys), and a bit earlier for girls (Burton et al., 1996). b. **Girls and boys differ** Girls' lumbar spines are more flexible than those of boys (5 to 9 year olds, Haley et al., 1986). Girls exhibit a greater range of lumbar extension than boys and adults.	• Children and adults differ in their need for lumbar support.
Cervical Spine (Neck)	• Children under 8 years have more neck motion. • Ligaments are more lax, muscles are weaker, the orientations of the shallow bony protrusions of vertebra (facet joints) and cartilage have not ossified, etc. (Dormans, 2002). • Young children's vertebrae are more wedge shaped. • By their 8th year, the cervical spine approaches the size, shape, and vertical orientation of adults as their vertebra gradually lose their oval or wedge shape and become more rectangular (Dormans, 2002). • Before adolescence, about one out of five cervical spines is kyphotic. Unlike adults, this is considered normal.	• Children of different ages have differing requirements for viewing angles and distances.
Gender	• Girls reach puberty before boys. • Before puberty, boys and girls tends to have similar amounts of lean body mass and fat to lean body mass ratios (about 15% for boys and 19% for girls) (Spear, 2002). • Following puberty, boys gain more lean body mass and proportionately more muscle than fat. Among boys, lean muscle mass doubles between the ages of 10 and 17 years.	• Until the onset of puberty, boys and girls have similar rates of growth. • After puberty, the differences between genders are greater meaning a greater range of design sizes is needed.

(continued)

Table 6.2 Child and Adolescent Changes in the Spine (Continued)

		Design Implications
Movements	• Postural support and freedom of movement often conflict. • These tradeoffs vary at different ages.	• Adaptable products and environments that encourage a range of age-appropriate postures and movements.

MUSCULOSKELETAL SYMPTOMS

Children report high rates of pain and discomfort, particularly at the neck and lower back.* Adults may ignore their children's reports of symptoms for a variety of reasons. Adults may assume that:

- children are adaptable
- children will "grow out of it"
- everyone has an ache or pain
- children exaggerate
- if left alone, it will resolve itself

Figure 6.7 (Valerie Rice) Children—particularly adolescents and girls—report high rates of musculoskeletal discomfort, particularly for the neck and back.†

Children who report pain are more likely to experience pain or report symptoms as young adults. Such symptoms may impair children's short-term and long-term well-being.

* For a summary of much of the research literature on children and back pain, please refer to Table 6.3.

† Please see also Table 21.2 about schoolchildren and school furniture.

Never take children's pain for granted:

1. *Discomfort and pain adversely affects children's ability to function.*
   - Children who report discomfort often have more difficulty in performing life functions, are more tired, and less healthy and happy.
   - Medical evaluations have determined that many children and adolescents who report pain also experience medical disorders.
2. *Pain symptoms may also signify problems that affect the child's long-term well-being.*
   - Many researchers suggest the strongest predictor of future back pain is having experienced symptoms in the past. Back pain in children appears to increase the risk of back pain during adulthood.
   - As children enter and progress through adolescence, their symptoms often become more frequent and intense. Discomforts in specific body-parts often become more generalized as these progressively affect adjacent body parts.
3. *Children who report pain are more likely to experience pain or report symptoms as young adults.*
4. *Discomfort hinders learning (Evans et al., 1992).*

   Static postures may introduce new sources of distraction and discomfort can interfere with a child's ability to learn.

It is helpful to understand the implications of discomfort symptoms in light of how children develop and mature.

Even if child symptoms seem similar:

1. Children and adults may exhibit the same symptoms despite having entirely unrelated disorders that reflect differences in their ages (Burton et al., 1996).
   For example, similar complaints may suggest a tendon disorder in an adult but a disorder affecting the growth plates of a child's bones.
2. Children and adults may experience the same disorder, but the implications of these disorders may differ considerably due to age-related differences.
   That is, the same disorder may be more hazardous for growing children. For example, back pain in children may deserve particular concern for children, whose bones are still developing.
3. Exposure to the same ergonomic factors may be differentially affected by the person's age.
   That is, the same exposure may affect children and adults differently.

Table 6.3 Children and Health Disorders

		What Is It?	Issues	Secondary Effects	Symptoms and Diagnosis
Skeleton[a]	Bones	The skeleton • supports body structures • protects internal organs • together with muscles, facilitates movement	Children's skeletal systems differ from those of adults: 1. Infants and children have different (Bass et al., 1999) and more variable types of bone damage than adults. These are often subtle and difficult to identify (Swischuk and Hernandez, 2004). • The relationship between child bones and muscles changes over time. • There are regional differences in the rate and pattern of bone development.[b] 2. Bone fractures in children require particular attention: • Children's bones are still growing, particularly at the growth plates. • There appears to be a direct relationship between the bone mineral content and the compressive strength of the vertebrae (Gallagher et al., 2005). 3. Bone degeneration in children differs from adults. • For example, cervical disc calcification is common in adults, but unusual among children. • Further, calcification in children differs from that in adults (Dai et al., 2004). • Nonsurgical treatments have better prognoses for children than for adults, particularly with young children. • Because girls mature before boys, their growth plates change into hard bone at an earlier age.	Pain in young children is often referred (Bass and Bruce 2000). • Hip pain may signify back problems. • Knee pain often signifies hip problems.[c]	• Do not ignore child reports of pain. • High growth periods such as in late adolescence deserve particular attention. • Bone fractures are often subtle and difficult to identify (Swischuk and Hernandez, 2004). • Pediatric patients who report pain related to the spine can have serious spinal abnormalities (Lipton et al., 2001). • Damage to the growth plates at the ends of bones are particularly important and require immediate attention. • Children often have difficulty communicating the nature of their symptoms. • A physician is needed to pinpoint the nature of the disorder (Payne and Ogilvie, 1996).

Soft Tissues[d]	Ligament/joint Capsule	• Ligaments connect bones to other bones.
		They are comprised of short bands of tough fibrous connective tissue mostly made of long, stringy collagen fibers.
		• In children, ligaments often appear as condensations of joint capsules (Bass and Bruce, 2000).
			• Joint pain in children is not normal or to be expected (Bass and Bruce, 2000; Bass et al., 1999).
			• Young children's muscle tendons and joints tend to be more elastic than those of adults (Cornu and Goubel, 2001).
			  • For example, children under 8 years have more neck motion because ligaments are more lax, muscles are weaker, and the orientations of the shallow facet joints and cartilage have not ossified (Dormans, 2002).
			• Some ligaments limit the mobility of articulations, or prevent certain movements altogether.
			• Child ligaments typically attach to the growing ends of the bones (epiphyses) (Bass and Bruce, 2000).
			  • Low force-traumas often disrupt these areas; there is insufficient time for ligaments to respond.
			  • Traumatic forces on the joint tend to injure the weakest link, i.e., the growth plate of the bone, rather than the ligaments—that may sprain but rarely rupture.
			• Repeated small stresses applied to the joint capsule or ligaments can lead to chronic pain around a joint (periarticular pain) (Bass and Bruce, 2000).
			• Tears to menisci,[e] which absorb shocks at the knees, may cause the knees to buckle or "lock-up."
				Joint dislocation:
				• Can lead to secondary damage of splinter bone fragments.
				• Immediate response is critical to avoid further damage to joints.
				Mechanical symptoms may include (Bass and Bruce, 2000):
				• Destabilized joints
				• Abnormal joint movements
				• Feeling unstable and a sense of "giving away"
				• Locking of joints
					• Do not take pain in any joint for granted.
					• Many children may only experience joint irritability.
					• It is often difficult to diagnose young children.
					• Cartilage may not be visible in x-rays (Eastwood, 2001).
					  • This has led to disagreement about the appropriateness of minimally invasive surgeries.
					• MRIs may be used to view joints of older children whose bone has ossified (Eastwood, 2001).
					• A clinical exam may be critical. Prior to diagnosing and stabilizing joints.

[a] Bone disorders may include fractures, instability, dislocation, and abnormal or damaged cells (necrosis).

[b] Such regional differences affect susceptibilities of the populations. Bass et al. (1999) notes "the differing tempo and direction of growth of the periosteal and endocortical surfaces, and the differing tempo of growth of the axial and appendicular skeleton may predispose to regional deficits in bone size, bone mineral content (BMC), and volumetric bone mineral density (vBMD)."

[c] Further, damage to the SCFE (slipped capital femoral epiphysis) of the hip often presents as knee pain (Bass and Bruce, 2000).

[d] Disorders may affect the ligaments, joint capsule, tendons, menisci that serve as shock absorbers, and synovial fluid that lubricates joints.

[e] The menisci are layers of shock-absorbing cartilage at the middle of the knee.

(continued)

Table 6.3 Children and Health Disorders (Continued)

		What Is It?	Issues	Secondary Effects	Symptoms and Diagnosis
Soft Tissues[d]	Tendons	• Tendons are tough bands of fibrous connective tissue that connect muscle to bone, direct muscle power, and help stabilize joints dynamically. • Most tendon strength is due to its arrangement of densely packed fibers.	• As children grow, their increased ability to generate force appears to take place in parallel with reductions in the elasticity of their muscle, tendons, and joints (Cornu and Goubel, 2001). • Tendon injuries often result from adolescent sports activities. ◦ Violent muscle-contractions during forceful sports activities can cause tendons to pull bone fragments at the point of origin or insertion at the bone (Bass and Bruce, 2000). ◦ Injuries can also result from abnormal gait patterns. • Chinese children have more joint laxity than Caucasian children of all ages[f] (Tan and Ong, 2000). • There are regional and ethnic differences in the prevalence of joint disease in children of Asiatic origins (Tan and Ong, 2000).		Damage to tendons may cause: • Instability, lack of coordination, or inability to function. • Unexplained joint irritability.

Soft Tissues	Muscles and Blood Vessels	• Muscles are elastic tissues that support the joint and allow the body to move by contracting	• Muscle growth increases dramatically following puberty, when muscle mass increases ~1.5-fold for boys and girls (Alexy et al., 2005). • The anatomy of infants differs from older children (McCarthy et al., 2004). • As in adults, Raynaud's Phenomenon principally affects girls and often is unrelated to connective tissue disease (Nigrovic et al., 2003).	• While childhood connective tissue diseases are rare, they are high risk and complex.[g] • Many connective tissue diseases involve inflammation of blood vessels (Sills, 2000).	
Whole Body			• Musculoskeletal infections affect bone, joint, and soft tissues simultaneously are common (McCarthy et al., 2004).	• Effects of infection in children may continue long after the acute episode.	• Treatment and close follow-up are essential.

[f] Tan and Ong (2000) reported, "Chinese children were found to have 100% laxity at age 3 years, 67% laxity at age 6, and 28% laxity at 12 years." They compare these findings to a different study of Caucasian children, by Cheng et al. (1991), who had 50% at 3 years, 5% at 6 years, and 1% at 12 years

[g] Diseases affecting the connective tissues include Raynaud's phenomenon, juvenile dermatomyositis (a rheumatic disease); scleroderma (chronic disorder marked by skin hardening and thickening); and systemic lupus erythematosus (SLE)—a chronic (long-lasting) rheumatic disease that affects joints, muscles, and other body parts.

Ergonomics and Safety as Related to Physical Pain and Injury	
Considerations	**Ergonomic and Safety Implications**
1. Pain in children is often referred (Bass et al., 1999). a. Hip pain may signify back problems. b. Knee pain often signifies hip problems.[a] 2. High-growth periods such as in late adolescence deserve particular concern. 3. Bone fractures are often subtle and difficult to identify (Swischuk and Hernandez, 2004). 4. Pediatric patients who report pain related to the spine can have serious spinal abnormalities (Lipton et al., 2001). 5. Damage to the growth plates at the ends of bones are particularly important and require immediate attention. 6. See Table 6.4, classes of growth bone fractures (Bass and Bruce, 2000; Salter and Harris, 1963).	***Ergonomics*** 1. Design exercise and sports programs that a. build a child's bones and muscles progressively and b. permit rest periods between exercise sessions to permit bone remodeling and muscles to repair. 2. Design of chairs used at computers, in classrooms, and for computer games a. Include backrests to support children's spines, particularly during late adolescence b. Provide adjustable furniture, supplement back supports, or change furniture as children physically mature ***Safety*** 1. Do not ignore a child's report of pain. 2. Do not assume based on experience with adults; the child's pain may reflect something different from that experienced by an adult. 3. "When in doubt, check it out." 4. Pay particular attention to injuries that may involve growing ends of bones (growth plates). 5. Recognize that children often have difficulty communicating the nature of their symptoms and that they experience referred pain. • A trained professional or physician is often needed to pinpoint the problem (Payne and Ogilvie, 1996).
Injuries may be more severe than they appear. *Injuries may be quite different than they seem.* *Injuries may have greater consequences than may immediately be apparent.* *Pain may indicate abnormalities rather than injuries.*	

[a] Further, damage to the slipped capital femoral epiphysis (SCFE) of the hip often presents as knee pain (Bass and Bruce, 2000).

OVERUSE SYNDROMES

Overuse syndrome* is a name given to a number of conditions that affect soft tissues, particularly the muscles, tendons, ligaments, and nerves. The condition is characterized by discomfort and persistent pain that becomes more regular and stronger as the disorder progresses. Risk factors for this disorder include constrained and awkward postures, repetitive or forceful movements, vibration, and cold.

Overuse syndromes have become quite prevalent among adults at many workplaces. Perhaps the most commonly recognized of these disorders is carpal tunnel syndrome (CTS). This has caused many people to assume that children who use computers are at similar risk of the same disorders.

This does not appear to be the case at this time. Although many children are at risk of acquiring an overuse disorder while (for example) playing competitive sports on a regular basis, there is scant evidence that children experience the same sorts of disorders that adults experience.

This is not to suggest that children are not at risk of these disorders. Many of our common assumptions about children's degree of risk and the risk factors associated with these disorders appear to be unfounded.†

This also does not suggest that children are free of discomfort. On the contrary, most studies have found that children report high rates of back and neck discomfort and pain (see "Back Pain" and Table 6.3). However, such symptoms appear to be quite different from those of adults and may result from different risk factors.

CARPAL TUNNEL SYNDROME‡: Carpal tunnel syndrome (CTS) is rare among children. In their review, Van Meir and De Smet (2003) note that only 215 cases of CTS (total) have ever been reported in the research literature. Research on pediatric CTS typically describes individual and unusual cases (Alfonso et al., 2001; al-Qattan et al., 1996; Bona et al., 1994; Coessens et al., 1991).

Children and adolescents also present different characteristics than do adults. Most published cases of CTS have been attributed to genetic disorders, such as relating to the shape of the carpal tunnel (Van Meir and De Smet, 2003 & 2005; Goldfarb et al., 2003).

- Ensure that children who report even modest complaints such as clumsiness and weakness, wrist or hand pain receive prompt medical attention.
- Provide a thorough review of the family history.

To date, research is lacking that suggests that children develop CTS from typing on a computer, in the same way that adults do. While there have been anecdotal reports from Britain about children at risk from performing text messaging,§ the research literature has not yet substantiated their findings. CTS is more commonly caused by genetic and metabolic disorders¶ and less commonly by impact forces such as sports injuries when seen in children.**

* Also referred to as Cumulative Trauma (CTDs) and Repetitive Strain Injuries (RSIs).

† Bright and Bringhurst's (1992) warning about "Nintendo elbow" based on one case of a 12-year-old boy did not herald a wave of such injuries in the 15 years that followed it.

‡ Detailed reviews of the research on Carpal Tunnel Syndrome among children are available from Goldfarb et al., (2003); Haddad et al. (1996); Lamberti and Light (2002); Van Meir and De Smet (2005); and Van Meir and De Smet (2003).

§ See for example, BBC (2002); Narain (2005); and Potter (2006).

¶ See, for example, Danta (1975), De Smet and Fabry (1999), Kayali et al. (2003), Musharbash (2002), Sanchez-Andrada et al. (1998), Unal et al. (2003), Yensel and Karalezli (2006) and Van Meir and De Smet (2003, 2005).

** See for example, Lamberti and Light (2002), Martinez and Arpa (1998) and the review by Van Meir and De Smet (2003).

RAYNAUD'S PHENOMENON: Although Raynaud's is rare among children, it more commonly affects girls (Nigrovic et al., 2003). In Raynaud's, impaired circulation is characterized by changes in finger color (Duffy et al., 1989). Its causes are not well understood.*

Cleary et al. (2002) described the case of a 15-year-old boy who apparently developed Raynaud's from a vibrating video game. However, these writers have not found similar cases where the disorder links Raynaud's to the design of a product.†

> Raynaud's phenomenon is a circulatory condition associated with spasms in the blood vessels of the fingers and toes that cause them to change color. It occurs both in children and adults.

BLISTERS: Hand blisters, cuts, and abrasions have been reported in children and associated with playing certain electronic games with joystick controls (Casanova, 1991; Osterman et al., 1987; Wood, 2001). One manufacturer (Nintendo) made free protective gloves available for child users‡ (CR, 2002).

Blisters and abrasions on the external surface of the hand may be considered an overuse injury. However, it is not the same sort of disorder as those we commonly think of such as CTS (affecting the median nerve of the hand) or tendonitis (affecting the tendons).

Blisters (unless infected) are more benign and certainly differ from other sorts of overuse injuries. Even so, products that produce blisters in children require redesign.

> ***Examples of applications for injury and design***
>
> 1. Design of products
>   a. Sports equipment
>   Protection from breaking a bone, especially involving the growth plates are critical.
>   - Design and test protective gear.
>   - Adjustable equipment may enable children to use equipment as they grow and develop.
>
>   b. Playgrounds
>   c. Emergency phone booths
>   d. Age-related furniture that provides support while promoting proper movement
> 2. Design of physical education programs (see Chapter 7)
>   These should progressively build and limit overuse of individual muscle groups.
> 3. Design of injury prevention programs

BACK PAIN§

Children report high rates of discomfort, particularly at the neck and lower back. Table 6.4 shows individual risk factors for back pain among children. Table 6.5 shows other risk factors for back pain, and Table 6.6 lists studies about back pain in school children.

* The findings of a large review of 123 pediatric cases from the Children's Hospital in Boston (Nigrovic et al., 2003) conflict with prior studies that attributed Raynaud's in pediatrics to connective or related diseases (Nigrovic et al., 2003).

† This is not to suggest there is no risk, only that to our knowledge it appears to be rare.

‡ Nintendo's Web site provides health and safety guidance for using their products www.nintendo.com/consumer/manuals/healthsafety.jsp.

§ Gardner and Kelly (2005) provides an excellent review of the research on back pain among children. Other good reviews include Balague et al., (2003), Balague and Nordin (1992), Balague et al. (1999), Barrero and Hedge (2002), Jones and Macfarlane (2005), King (1999), Murphy et al. (2004), Steele et al. (2006), and Trevelyan and Legg (2006).

Table 6.4 Individual Child Risk-Factors for Back Pain[a]

Age	• Existing discomfort increases as children mature. • Puberty can be a time when pain is first reported. • Discomfort increases during peak growth period in adolescence.[b] • Risk is greater from early-to-mid adolescence, particularly for girls. • Symptoms among children tend to progress as children become older. • Discomfort and pain often becomes more frequent and intense. • Further, discomfort in specific body parts may affect adjacent body parts. Discomfort becomes progressively more generalized as children enter adolescence.
Gender	• Girls experience higher rates of discomfort than boys do, and earlier. • Risk among boys is greater if they regularly participate in competitive sports.
Family	• Parents or siblings experiencing back pain increases risk. • A family history of other soft tissue disorders may also suggest that the child is at great risk.
Other	• Stress, low self-esteem, depression[c] • Smoke • Have outside jobs • Ride (rather than walk) to school • Time spent watching television

[a] Please consider this review as general guideline. While much of the research literature is consistent, it often is not unanimous in its conclusions (see Table 6.6 on child and adolescent back pain).

[b] Feldman et al. (2001) defined the peak growing stage as child growth exceeding 5 cm (2 in.) over 6 months.

[c] This is not to suggest the child's pain is necessarily "all in their head." The experience of pain is inherently stressful; children's feelings of stress may result from the pain signals that have medical foundations. Further, the experience of stress manifests itself in ways that may exacerbate physical disorders, such as by muscle tension and releasing stress hormones. One must also keep in mind that the two may sometimes be indirectly related.

Table 6.5 Back Pain Risk Factors among Schoolchildren

Classroom	• Extended sitting in static postures • Prolonged work on the computers • Finding their chair uncomfortable • Hard, inflexible work areas that limit movement
Other	• Inadequate storage • Overly heavy book bags • Intensive competitive sports activities

SCHOOL CHILDREN AND BACK PAIN FINDINGS:

Table 6.6 Back Pain in Schoolchildren: Selected Research Findings[a]

	Citation	Subjects	Findings	Conclusions
Australia	Grimmer and Williams (2000)	• 1269 adolescents in the 8th through the 12th grades in 12 volunteer high schools	***Findings include:*** • 15% of boys reported LBP in the 8th grade, compared to 25% of girls • 20% of boys reported LBP in the 12th grade, compared to 44% of girls	***Low back pain (LBP) risk factors:*** • Age (older children → more LBP) • Gender Girls reported more LBP than boys • Sitting postures 8th and 10th grade females who sat for long periods[b] • Long amounts of time carrying loaded backpacks[c] • Hours playing sports
Australia	Harris and Straker (2000a,b)	• Surveyed 314 5th–12th graders (aged 10 to 17) about their use of laptops • 251 children responded online, 63 on printed surveys • 13% male and 87% female • Also interviewed and watched 20 children using laptops	***Findings include:*** • On average, children worked 3.2 h/day on their laptop computer • 60% child users reported discomfort when using their laptop • 61% reported discomfort from carrying their laptop	Children reporting discomfort: • were often older (9th, 10th, and 12th graders report more discomfort) and • varied between schools (may have resulted from child's exposure to computers and length of time in class in different schools)
Belgium	Gunzburg et al. (1999)	• Surveyed 392 9-year-old schoolchildren • Compared these responses with lumbar spine medical examinations	***LBP findings:*** • 36% schoolchildren reported at least one case of LBP, 14% reported LBP the previous day ◦ Of these, 23% sought medical help • 64% children with back pain had parents who experienced back pain ***Factors that did not increase risk:*** • Carrying book bags by hand versus back • Watching television >2 h/day • Playing sports[d] The only clinical sign associated with LBP was pressure on the iliac crest of the iliolumbar ligament.[e]	***LBP risk factors:*** • Overly heavy book bags • Playing video games (>2 h/day) • Riding to school (rather than walking) Children who reported LBP were more tired, less happy, and worse sleepers.

Belgium	Szpalski et al. (2002)	• Surveyed 287 children aged 9 to 12 years and again after 2 years • Compared the results with medical findings	***Findings include:*** • 36% of 9-year-old children reported prior LBP • 2 years later, LBP rates remain unchanged	***LBP risk factors:*** • Children who walk to school were at less risk of new LBP
Canada	Feldman et al. (2001)	• 502 7th to 9th graders • Recorded rate of growth, muscle strength, activity, subjective responses, and medical findings • Recorded three times in 12 months (initial, at 6 months, and at 12 months)	Primary measure was having LBP at least once a week in last 6 months ***Factors that did not increase risk:*** • Poor abdominal strength • Physical activity • Lumbar or sit and reach flexibility	***LBP risk factors:*** • Following a high growth spurt (>5 cm in last 6 months) • Smoking • Outside work in last 6 months • Poor mental health • Tight hamstrings and quadriceps
	Feldman et al. (2002)	• Feldman (2001) study • Analyzed neck or upper-limb pain at least once a week in previous 6 months	***Findings include:*** • 28% incidence neck or upper limb pain over year ***Factors that did not increase risk:*** • Participating in sports	***Neck or upper limb risk factors:*** • Outside work • Lower mental-health • Participating in childcare

[a] Here and elsewhere, the acronym "***LBP***" signifies the term "***low back pain***."
[b] These researchers noted there were similar but nonsignificant trends among other age groups. They also noted their low back pain findings were specific to girls in the study because the boys did not sit for long periods after school.
[c] Please read Chapter 13 for more detailed information.
[d] Unlike others, these researchers did not appear to separate children who participated in moderate versus competitive sports.
[e] More details about this dimension are available from Kennedy et al. (2004).

(continued)

Table 6.6 Back Pain in Schoolchildren: Selected Research Findings (Continued)

	Citation	Subjects	Findings	Conclusions
China	Szeto and Lau (2002)	• Surveyed 463 students aged 11 to 17	***Findings include:*** • An association between symptoms in adjacent body part • Strongest association between elbow and finger discomfort	***Risk factors:*** Longer computer use
Denmark	Harreby et al. (1999)	• Surveyed 671 boys and 718 girls aged 13 to 16 in 8th and 9th grades • Also underwent medical examination	These researchers emphasized child reporting of moderate-to-severe, recurrent, or continuous LBP (***SLBP***). • 59% ever experienced LBP • 51% LBP in the last year • 6.4% increase LBP from 14 to 15th year ***Medical findings:*** • 14% “hyper-mobile” • 12% tight hamstring muscles (>40°) • 19% SLBP	***Risk factors severe or recurrent LBP:*** • Being female • Daily smoking • Heavy job work in leisure time • Intensive sports activities ***Positive medical LBP findings:*** • More often female • Have BMI scores >25 kg/m^2 • Competitive sport (boys) • Poor physical fitness • Daily smoking • Heavy jobs in leisure time

Denmark	Kjaer (2004)	• Surveyed 439 13-year-old children and 412 40-year-old adults • Subjects underwent MRI scans	***LBP findings:*** • 22% children and 42% adults experienced LBP last month • 41% children and 69% adults experienced LBP the previous year ***Medical findings:*** • 20% children and 50% adults evidenced degeneration of disks • 3% of children and 25% of adults showed disk herniation • 6% children and 30% adults had endplate changes Close to two-thirds of children had "abnormal" discs, but the changes were milder than in the adults ***Girls and boys:*** • In boys, degenerative disc changes in the upper lumbar are more strongly associated with LBP than in girls. • In girls, degenerative changes were stronger in the lower lumbar-spine than boys	***Positive medical LBP findings:*** • In children, degenerative changes in the discs and endplates were more strongly associated with child reports of LBP • Most evidenced changes in disc signal and irregularity of the nucleus' complex • Changes in disc height found in almost half of the children • When ignoring the lowest ratings, about one-fifth of the children had signs of disc degeneration • Disk herniation in less than 3% of the children • Nerve root compromise in 9%
	Leboeuf-Yde et al. (2002)	• Surveyed school children: 481 children aged 8 to 10 and 325 aged 14 to 16	• Mild negative association between parental education and back pain in children but not in adolescents	• Parental education was a poor predictor of children's risk of back pain
	Leboeuf-Yde and Kyvik (1998)	• Surveyed 34,076 twins aged 12 to 41 (86% response rate)	• Rates of low back pain increased greatly in the early teen years (earlier for girls than for boys) • Most girls aged 18 years and boys aged 20 experienced at least one LBP episode • 7% of 12-year-olds reported low back pain	***Conclusions:*** • High rates of children report LBP • Research on back pain should start with children and adolescents
	Wedderkopp et al. (2001)	• Interviewed 481 school-children (aged 8 to 10) and adolescents (aged 14 to 16)	***Findings:*** • 39% reported back pain in previous month • Thoracic pain most common in children • Adolescents reported thoracic and lumbar pain equally • Children rarely reported neck pain and pain in less than one area of spine • No gender differences found	***Conclusions:*** • Children of different ages report different kinds of back symptoms

(continued)

Table 6.6 Back Pain in Schoolchildren: Selected Research Findings (Continued)

	Citation	Subjects	Findings	Conclusions
Denmark	Wedderkopp et al. (2005)	• Interviewed 254 girls aged 8 to10 and 165 girls aged 14 to 16	***Findings:*** • Investigated rates of pain relative to puberty • Low back pain, pain in the mid back, and neck pain were not associated with stage of puberty	• Girls' LBP rates increased after puberty until they reach maturity • The biggest increase takes place between the beginning and mid puberty
Denmark	Mikkelsson et al. (1999)	• Selected children from 1756 3rd and 5th graders • Followed groups of children with widespread pain and neck pain, and children who were pain free to evaluate changes in symptoms over a year	***Findings:*** • Children in the initial pain groups continued to experience fluctuating pain symptoms • More than half the children in the two pain groups experienced pain on a weekly basis • In contrast, only one-tenth of children from the original pain-free group reported pain symptoms at end of year • Depressive, behavioral, and psychosomatic symptoms were poor predictors of pain outcomes • The experience of expanding pain (from neck pain to widespread neck pain) was associated with psychosocial symptoms	• Children with widespread pain or neck pain continued experiencing regular intermittent pain • In contrast, 9 out of 10 in the pain-free child population remained pain free • In contrast to other studies, social and psychosocial factors were poor predictors of the persistence of pain
Finland	El-Metwally et al. (2004)	• Surveyed 1756 3rd and 5th grade school children • Performed hypermobility and endurance tests at start	***Musculoskeletal pain findings:*** • 54% reported pain at least once a week after first year (64% after fourth year) • 64% reported musculoskeletal pain • As children became older, localized symptoms of pain tended to become more generalized	***Musculoskeletal risk factors:*** • Older children reported higher rates • Girls reported more than boys • More persistent preadolescent pain • Greater joint laxity in preadolescence • More than one kind of pain
Finland	Stahl et al. (2004)	• Surveyed 366 preadolescents aged 9 to 12, who were originally free of pain, after 1 and 4 years	***Findings include:*** • 21% students developed neck pain by first year • Pain became more intense as it became more frequent	***Neck pain risk factors:*** • Older children reported higher rates • Girls reported more than boys

Finland	Salminen et al. (1999)	• Surveyed 1503 8th graders • Performed MRIs on a matched-group with and without back pain • Followed to track differences	***Findings:*** • 3-year follow-up indicates that children with LBP experienced an earlier and more frequent onset of degenerative process in lower lumbar disks than those who did not initially report LBP • A risk of persistency of recurrent LBP is highest among those showing the first signs associated with degenerative disks • Children experiencing LBP are also more likely to experience protruding disks[f]—yet this relationship becomes more ambiguous as they enter adulthood	***LBP risk factors:*** • Those who report back pain earlier are more subject to degenerative disks
	Tertti et al. (1991)	• Performed MRIs on two groups of 39 15-year-olds with and without LBP out of 1503 children	• 38% children with LBP and 26% of those without displayed disk degeneration • Concluded that disk degeneration is common among children with LBP at this age. Some children without back pain appear to be asymptomatic	• Disk degeneration in young people is not necessarily pathological. However, it may predispose them to spinal symptoms.
Iceland	Kristjansdottir and Rhee (2002)	• Surveyed 2173 school children aged 11 to 12 and 15 to 16		***LBP risk factors in schoolchildren:*** • Older children reported more LBP • Physical condition (e.g., chronic health disorders, morning tiredness, fitness) • Behavior scores (sports participation, television viewing, eating habits, smoking) • Low social support
Israel	Limon et al. (2004)	• Surveyed 10,000 children in elementary schools (filled out by nurses)	***Ergonomic risk factors in schools included:*** • Shortcomings in all areas investigated • 30%–54% students carried >15% of body weight • Improper chairs in15% 1st grade and 20% 6th grade • In three in four classes, students sat with their side facing teacher; 35% of students sat with their backs facing teacher • 6% schools offered no physical activity in recess	***LBP risk factors:*** • Weight of student and their backpack • Insufficient storage facilities • Improper student chair and desk height • Classroom seating arrangements • Physical activity at recess

[f] This should not suggest that protruding discs in children and adults have the same implications. Animal studies suggest that unlike adults, forward slippage in immature spines are not clearly related to degeneration of the disks (Sairyo et al., 2001, 2004).

(continued)

Table 6.6 Back Pain in Schoolchildren: Selected Research Findings (Continued)

	Citation	Subjects	Findings	Conclusions
The Netherlands	Diepenmaat et al. (2006)	• 3485 teens aged 12 to 16	***Findings include:*** • Reported 12% neck or shoulder pain, 8% LBP, and 4% arm pain	***Musculoskeletal risk factors:*** • Girls reported more neck and low back pain • Not living with both parents • Stress (neck or shoulder pain, back pain)
Spain	Kovacs et al. (2003)	• Surveyed 16,394 school children aged 13 to 15 and parents in Mallorca	***Findings include:*** • 51% boys and 69% girls reported having LBP • LBP in adolescents in south Europe resembles child rates in north Europe and LBP in adults ***Factors that did not increase risk:*** • Body mass index, book-carrying posture, hours sitting, smoking, alcohol intake, and family LBP • Scoliosis, but not LBP related to heredity	***LBP risk factors:*** • Being female • Pain in bed when rising • Scoliosis, different leg-lengths • Participating in sports more than twice a week
Sweden	Petersen et al. (2006)	• Surveyed 1155 children in grades 0–6 aged 6 to 13 (helped by parent)	***Findings include:*** • Children frequently experienced recurrent pain in adjacent body-parts • Two-thirds of the children reported pain at least once a month • One-third reported pain at least once a week • 6% reported daily-pain symptoms • Half the children with recurring symptoms reported pain in several body-locations	***Back pain risk factors:*** • Age (older children report more multiple pain symptoms) • Being female • Higher prevalence of multiple weekly pain symptoms in girls

Sweden	Brattberg (2004)	• Surveyed 335 children aged 8, 11, and 14 in 1989 • Followed the children in 1991 and 2002 (2 and 11 years later) • The first two times, the surveys were in school; the third time when they were adults, they responded by mail.	***Findings include:*** • 20% reported back pain in all three studies • In the 2002 study, 59% of the women and 39% of the men reported pain at 21, 24, and 27 years of age	***Back pain risk factors:*** • Reported back pain in 8- to 14-year-olds • Headaches once a week or more • Children aged 10 to 16 report "feeling nervous"
Switzerland	Balague et al. (1994)	• Surveyed 1755 children aged 8 to 16 (98% response rate)	***Findings include:*** • 34% of the children reported at least one episode of spinal pain • 67% of these children experienced LBP • 33% reported cervical or thoracic pain • 11% of the children had seen a doctor for back pain • Children reported the most painful positions were sitting and leaning forward	***LBP risk factors for children:*** • Age • LBP increased with age, particularly once they reached the age of 13 • Parents had been treated for LBP • Competitive sports (24% versus 16% ≤ twice a week) • Time watching television
	Jones et al. (2003)	• 1046 schoolchildren aged 11 to 14 years who were free of LBP (89% responded) • Measured onset of new LBP in following year	***LBP risk factors:*** • Psychosocial factors were better predictors than physical risk factors • High psychosocial difficulties • Conduct problems • Those reporting somatic symptoms were more prone to developing LBP, abdominal pain, headaches, and sore throats ***Factors that did not increase risk of LBP:*** • Baseline height, weight, and body mass index • Heavy book bag weights[g]	• Children who are free of LBP but experience psychosocial or somatic symptoms are at higher risk of developing LBP.

[g] This study may have been confounded, since they did not control weights after baseline.

(continued)

Table 6.6 Back Pain in Schoolchildren: Selected Research Findings (Continued)

	Citation	Subjects	Findings	Conclusions
Switzerland	Bejia et al. (2005)	• Surveyed 622 school children aged 11 to 19 • The first 201 children also underwent a spinal exam	***LBP findings:*** • 28% reported having experienced LBP • 8% experienced chronic LBP • 23% of school absenteeism and 29% for sports absenteeism due to LBP ***Medical findings:*** • 31% children had objective findings of LBP	***LBP risk factors:*** • Parent or sibling history of LBP • Extended sitting durations • Held back for 1 year in school ***Chronic LBP risk factors:*** • Not satisfied with height and comfort of class chair • Playing football
Tunisia	Burton et al. (1996)	• Surveyed 216 11-year-old children • Followed up with these children annually for 4 years • Measured lumbar mobility in first and last year	***LBP findings:*** • 11.8% at age 12+ to 21.5% at 15+ years • Lifetime prevalence rose from 11.6% at age 11+ to 50.4% at age 15+ years ***Lumbar flexibility:*** • Increased lumbar extension with age over the 4 years (boys and girls) • Reduced ability to flex in girls with age • This change in flexibility did not predict LBP risk	***LBP risk factors:*** • Increasing age • Boys more at risk than girls • Sports (boys)
United Kingdom	Jones et al. (2004)	• Surveyed 249 boys and 251 girls aged 10 to 16	***LBP findings:*** • 40% reported having experienced LBP • 13% reported recurrent and disabling LBP • Rates increased with age, about 20% of teens, 14 years or older, reported recurrent LBP • 23% visited a medical practitioner • 30% had loss of physical activity or sports • 26% were absent from school due to LBP	

United Kingdom	Murphy et al. (2004)	• Surveyed 66 school children aged 11 to 14, 12 schools
		• Recorded classroom-sitting postures
		  ◦ 30-minute trials in class
		  ◦ Measured forward trunk flexion (> 20°, >45°, >60°), trunk rotation, neck flexion, etc.
		• Recorded height, weight
			Findings include:
			• 44% reported LBP in last month
			• 26% in the last week
			• 51% reported neck pain in the last month
			• 21% in the last week
				LBP risk factors:
				• Longer class-lessons (>1 h)
				• Extended time trunk >20° forward flex
				Upper back pain risk factors:
				• Static sitting postures increased risk of upper back pain and neck pain over the previous week.
				• >20° trunk flexion increased risk of LBP over the previous week
				Neck pain risk factors:
				• Taller children reported more pain

	Olsen et al. (1992)	• Surveyed 1242 adolescents aged 11 to 17 over 4 years (92% response rate)	***Findings include:*** • 30% of adolescents experienced LBP • 22% reported discomfort in last year ◦ One in three of these reported restricted activity (missed school or could not participate in sports) • 7% sought medical attention • By age 15, rates of LBP increased to 36%	

	Watson et al. (2002)	• Surveyed 1446 children aged 11 to 14 in North West England (97% response rate)
		• Compared responses with those of their parents
			• 24% of children reported LBP previous month
			  ◦ 42% of these indicated it only lasted 1–2 days
			  ◦ 15% reported it lasted over 7 days
			• Girls reported higher rates than boys (28% versus 19%)
			• Moderate agreement between child and parent's reports of the child's back pain
				LBP risk factors:
				• Discomfort increased with age for both girls and boys
				• 94% of children reporting LBP report some disability, especially when carrying school bags
				• Few sought medical attention

	Watson et al. (2003)	• Surveyed 1446 school children aged 11 to 14 (97% response rate) • Children also filled out a 5-day bag-weight diary	***Factors that increased risk:*** • Having a part time job—but not lifting at work ***Factors that did not increase risk:*** • Body mass index • Percent body weight • Average load or method of carrying book bag • On average, carried 8 kg weight	***Psychosocial more important than physical risk factors:*** • Children reporting conduct or emotional problems were three times more likely to report LBP • High hyperactivity and peer problem scores related to 50% increase in LBP • Headaches for more than 3 days in previous month related to a threefold increase in LBP

(continued)

Table 6.6 Back Pain in Schoolchildren: Selected Research Findings (Continued)

	Citation	Subjects	Findings	Conclusions
United States	Sheir-Neiss et al. (2003)	• Surveyed 1125 school children aged 12 to 18 • Also recorded children's weight, height, and backpack weight	***Findings include:*** • 74% reported neck or back pain previous month[h] • School children in the back-pain group reported poorer general health, more limited physical functioning, and more body pain	***Risk factors:*** • More backpack use and weight • Watching television for more than 2 h/day • Girls more than boys • Older more than younger • Higher body mass index (BMI)
	Jacobs and Baker (2002)	• Surveyed 152 6th graders	***Findings include:*** • Most children reported some musculoskeletal discomfort in at least one body part within the last year	***Musculoskeletal discomfort risk factors:*** • More hours on the computer moderately increased the overall risk of musculoskeletal discomfort • No appropriate computer furniture at home • Children who do not use computer furniture at home report 1.89 times more overall discomfort • More time on the computer (even touch-typing is a risk factor) • Children who could touch-type were 54% less likely to report discomfort than those who "hunt and peck"

[h] They defined back pain as neck and back pain (a) rated 2 or more on a scale 0 to 10 that (b) interfered with school or leisure, (c) required medical treatment, or (d) exemption from physical education or sports. These researchers noted the low participation rate might have biased the sample toward more back pain respondents.

GROWTH PLATE INJURIES*

Growth plate injuries occur in children and adolescents. These fractures warrant particular concern because of the potential harm to growing children. Yet, adults may not recognize the seriousness of such injuries in children because they seem to resemble the less hazardous "sprains" of adults.

> The growth plate (also called the *epiphysis*) is the area of growing tissue near the end of the long bones in children and adolescents. Each long bone has at least two growth plates: one at each end. The growth plate determines the future length and shape of the mature bone. Upon maturity, growth is complete and growth plates close and are replaced by solid bone.
>
> NIAMS (2001b)

Growth plate injuries

- 15% of childhood fractures
- Twice as common among boys as girls
- Most common among boys aged 14 to 16 and girls aged 11 to 13
- Less common among older girls because their bodies mature and bones stop growing at an earlier age

In a growing child, a serious injury to a joint is more likely to damage a growth plate than the ligaments that stabilize this same joint. The growth plate is the weakest area of the growing skeleton, weaker than the nearby ligaments and tendons that connect bones to other bones and muscles.

Approximately half of all growth plate injuries occur at the lower end of the outer bone of the forearm (radius) at the wrist. Growth plate injuries are also common in the lower bones of the leg (tibia and fibula) and can occur in the upper-leg bone (femur) or in the ankle, foot, or pelvic bone.

Most growth plate injuries result from acute events such as falls or a blow to a limb. However, chronic injuries can also result from overuse. In one large study of growth plate injuries in children, the majority resulted from a fall, usually while running or playing on furniture or playground equipment.

Warning signs in children (NIAMS, 2001a)

- Cannot keep playing due to pain after acute or sudden injury
- Unable to play for long because of persistent pain following a previous injury
- Visible deformity of child's arms or legs
- Severe pain from acute injuries that prevent using the arm or leg

Other causes of growth plate injuries

- Child abuse
- Extreme cold (e.g., frostbite)
- Radiation or chemotherapy
- Sensory deficit or muscular imbalance
- Genetics

Competitive sports, such as football, basketball, softball, track and field, and gymnastics, accounted for one-third of the injuries. Recreational activities, such as biking, sledding, skiing, and skateboarding, accounted for one-fifth of all growth plate fractures, while car, motorcycle, and all-terrain vehicle accidents accounted for only a small percentage of fractures involving the growth plate.

Children who participate in athletic activities often experience some discomfort as they practice new movements. Muscles get sore as due to small tears during use, followed by repairs during consequent rest periods. Muscle soreness after exercise is generally greater one to two days after exercise and then subsides (known as "delayed muscle soreness").

Yet, a child's complaint always deserves careful attention. Some injuries, if left untreated, can cause permanent damage and interfere with proper growth of the involved limb. Growth plates are also susceptible to other disorders, such as bone infection that can alter their normal growth and development. Persistent pain in children that affects athletic performance or impairs their ability to move normally should be examined by a doctor.

* Much of this content is available in more detail in NIAMS (2001a,b) Growth Plate Injuries. NIH Publication 02-5028. National Institute of Arthritis and Musculoskeletal and Skin Diseases. US Dept. Health and Human Services. www.niams.nih.gov/hi/topics/growth_plate/growth.htm

LONG-TERM EFFECTS: About 85% of growth plate fractures heal without any lasting effect. The long-term impact on growth depends on these factors, in order of importance:

1. Severity of the injury
   Growth may be stunted if the injury cuts off the blood supply to the epiphysis. If the growth plate is shifted, shattered, or crushed, a bony bridge is more likely to form and the risk of growth retardation is higher. An open injury in which the skin is broken carries the risk of infection, which could destroy the growth plate.
2. Age of the child
   Bones of young children have more growing to do and are therefore more serious. However, younger bones also have a greater ability to remodel.
3. The affected growth plate
   Some growth plates, such as at the knee, have a greater impact on bone growth.
4. Type of growth plate fracture

This section lists six fracture types; Types IV and V are the most serious.

How are growth plate injuries treated?

Begin treatments as soon as possible after injury. These usually involves a combination of

1. **Immobilization**

   The affected limb is often put in a cast or splint.
2. **Manipulation or surgery**

   The need for manipulation or surgery depends on the location and extent of the injury, its effect on nearby nerves, blood vessels, and the child's age.
3. **Strengthening and range of motion exercises**

   Treatments may continue after the fracture heals.
4. **Long-term follow-up**

   Long-term follow-up is often necessary to track the child's recuperation and growth.

The most frequent complication of a growth plate fracture is premature arrest of bone growth. The affected bone grows less than it would have without the injury, and the resulting limb could be shorter than the opposite, uninjured limb. If only a part of the growth plate is injured, growth may be lopsided and the limb may become crooked.

Growth plate injuries at the knee are at greatest risk of complications. Nerve and blood vessel damage occurs most frequently there. Injuries to the knee have a much higher incidence of premature growth arrest and crooked growth.

Table 6.7 Additional Resources
NIAMS National Institutes of Health One AMS Circle Bethesda, MD 20892–3675 Ph: (301) 495–4484 (877) 266–4267 www.niams.nih.gov
American Academy of Pediatrics 141 Northwest Point Boulevard Elk Grove Village, IL 60007-1098 Ph: (847) 434–4000 www.aap.org

Do not allow or expect the child to "work through the pain."

Further Reading on Growth Plates:

1. Bright, R. W., Burstein, A. H., Elmore, S. M. (1974) Epiphyseal-plate cartilage: A biomechanical and histological analysis of failure modes. *J Bone Joint Surg (Am)*. 56, 688–703.
2. Broughton, N. S., Dickens, D. R., Cole, W. G., Menelaus, M. B. (1989) Epiphyseolysis for partial growth plate arrest. Results after four years or at maturity. *J Bone Joint Surg (Br)*. 71(1), 13–6.
3. Iannotti, J. P., Goldstein, S., Kuhn, J., Lipiello, L., Kaplan, F. S., Zaleske, D. S. (2000) The formation and growth of skeletal tissues. *Orthopaedic Basic Science: Biology and Biomechanics of the Musculoskeletal System*. Edited by J. A. Buckwalter, T. A. Einhirn, S. R. Simon American Academy of Orthopaedic Surgeons, Rosemont, IL, pp. 78–109.
4. Moen, C. T., Pelker, R. R. (1984) Biomechanical and histological correlations in growth plate failure. *J Pediatr Orthop*. 4, 180–84.
5. Ogden, J. A. (1981) Injury to the growth mechanisms of the immature skeleton. *Skeletal Radiol.* 6, 237–53.

Table 6.8 The Salter–Harris System for Classifying Growth Plate Injuries[a]

Type	What Is It?	Considerations	
SH Type I	Separation of the growth plate (*epiphysis*) without fracture to the bone.	• Type I fractures typically need to be put back into place and immobilized. • Unless there is damage to the blood supply to the growth plate, there is a chance of recovery.	
SH Type II	Partial separation through growth plate The fracture line is the same as SH Type I, except that it extends into the shaft of the bone (*metaphysis*).	• The most common growth plate fracture. • As with Type I, Type II fractures typically must be put back in place and immobilized.	
Type III	Complete fracture through growth plate. where the line of the fracture extends into the joint.	• This type of fracture is rare. • Usually occurs at one of the long bones of the lower leg. • Surgery is sometimes necessary to restore the joint surface to normal. • Good prognosis, if blood supply is intact and fracture is not displaced.	
Type IV	Fracture line extends through the growth plate (epiphysis), and into the shaft of the bone and the joint.	• Most common at the end of the upper-arm bone (humerus) near the elbow. • Surgery is needed to restore the joint surface and align the growth plate. • Prognosis for growth is poor (likely to arrest growth) unless perfect alignment is achieved and maintained during healing.	
Type V	The end of the bone is crushed and the growth plate is compressed.	• This injury is uncommon. • Most likely to occur at the knee or ankle. • Prognosis is poor. Premature stunting of growth is to be expected.	
Type VI Peterson Classification	A portion of the epiphysis (growth plate) and metaphysis is absent.	• This injury is rare. • Bone growth is almost always stunted. • May lead to considerable deformity. • Usually occurs with open wounds or compound fractures. • Often involves lawnmowers, farm machinery, snowmobiles, or gunshot wounds. All type VI fractures require surgery, and most will require later reconstructive or corrective surgery.	

[a] Many classification systems have been used throughout the world, but the Salter and Harris (1963) (SH) classification has become the most accepted standard among health care professionals for more than 40 years.

CONCLUSIONS

This chapter described some dimensions of children's physical growth and development.

Designers must take children's physical development into account when creating products and places for children to encourage appropriate growth and to help keep them safe. The severity of an injury may be greater if it occurs during the developmental process. This means there may not be a direct comparison of a child's injury compared with similar injury in an adult.

Further, injuries and pain may have different implications in children:

1. Children often do not report pain, as they do not know it is abnormal to have pain during an activity.
2. Pain, as experienced by a child, may suggest a quite different disorder than similar pain reported by an adult.
3. Pain indicates that something may be wrong; adults should pay attention and "check it out" when a child reports pain.

Caregivers may need to solicit information from children, rather than waiting for the child to step forward with the information concerning their discomfort. Designers, when testing their products with age-appropriate children, may also need to directly ask about comfort and ease-of-use.

More information is needed on childhood injuries, soreness, and discomfort. Additional studies are needed regarding the long-term effects of overuse, repetitive patterns of pain, and even of acute injuries. It is well known that the best predictor of injury among adults is a previous injury, yet it is not known if the previous injury extends into childhood.

REFERENCES

Alexy, U., Remer, T., Manz, F., Neu, C.M., and Schoenau, E. (2005) Long-term protein intake and dietary potential renal acid load are associated with bone modeling and remodeling at the proximal radius in healthy children. *Am. J. Clin. Nutr.*, 82, 1107–1114.

Alfonso, D.T., Sotrel, A., and Grossman, J.A. (2001) Carpal tunnel syndrome due to an intraneural perineurioma in a 2-year-old child. *J. Hand Surg. [Br].*, 26(2), 168–170.

al-Qattan, M.M., Thomson, H.G., and Clarke, H.M. (1996) Carpal tunnel syndrome in children and adolescents with no history of trauma. *J. Hand Surg. [Br].*, 21(1), 108–111.

Arikoski, P., Komulainen, J., Voutilainen, R., Kroger, L., and Kroger, H. (2002) Lumbar bone mineral density in normal subjects aged 3–6 years: A prospective study. *Acta Paediatr.*, 91(3), 287–291.

Balague, F. and Nordin, M. (1992) Back pain in children and teenagers. *Baillieres Clin. Rheumatol.*, 6(3), 575–593.

Balague, F., Nordin, M., Skovron, M.L., Dutoit, G., Yee, A., and Waldburger, M. (1994) Non-specific low-back pain among schoolchildren: A field survey with analysis of some associated factors. *J. Spinal Disord.*, 7(5), 374–379.

Balague, F., Troussier, B., and Salminen, J.J. (1999) Nonspecific low back pain in children and adolescents: risk factors. *Eur. Spine J.*, 8(6), 429–438.

Balague, F., Dudler, J., and Nordin, M. (2003) Low-back pain in children. *Lancet,* 361(9367), 1403–1404.

Barrero, M. and Hedge, A. (2002) Computer environments for children: A review of design issues. *Work,* 18, 227–237.

Bass, A.A. and Bruce, C.E. (2000) The persistently irritable joint in childhood: An orthopaedic perspective. *Eur. J. Radiol.*, 33(2), 135–148.

Bass, S., Delmas, P.D., Pearce, G., Hendrich, E., Tabensky, A., and Seeman, E. (1999) The differing tempo of growth in bone size, mass and density in girls is region-specific. *J. Clin. Invest.*, 104(6), 795–804.

BBC (2002) British Broadcasting Corporation. Thumbs take over for text generation. Health section. *BBC News*, March 25, 2002.

Bejia, I., Abid, N., Ben Salem, K., Letaief, M., Younes, M., Touzi, M., and Bergaoui, N. (2005) Low back pain in a cohort of 622 Tunisian schoolchildren and adolescents: An epidemiological study. *Eur. Spine J.*, 14(4), 331–336.

Biro, F.M., Khoury, P., and Morrison, J.A. (2006) Influence of obesity on timing of puberty. *Int. J. Androl.*, 29(1), 272–277.

Bona, I., Vial, C., Brunet, P., Couturier, J.C., Girard-Madoux, M., Bady, B., and Guibaud, P. (1994) Carpal tunnel syndrome in Mucopolysaccharidoses. A report of four cases in child. *Electromyogr. Clin. Neurophysiol.*, 34(8), 471–475.

Brattberg, G. (2004) Do pain problems in young school children persist into early adulthood? A 13-year follow-up. *Eur. J. Pain*, 8(3), 187–199.

Bright, D.A., Bringhurst, D.C. (1992) Nintendo elbow. *West J. Med.*, 156(6), 667–668.

Burton, A.K., Clarke, R.D., McClune, T.D., and Tillotson, K.M. (1996) The natural history of low back pain in adolescents. *Spine*, 21(20), 2323–2328.

Casanova J. (1991) Nintendinitis [Letter]. *J. Hand Surg. (Am).*, 16, 181.

Cheng, J.C., Chan, P.S., and Hui, P.W. (1991) Joint laxity in children. *J Pediatr Orthop.*, 11(6), 752–756.

Cleary, A.G., McKendrick, H., and Sills, J.A. (2002) Hand-arm vibration syndrome may be associated with prolonged use of vibrating computer games. *BMJ*, 324(7332), 301.

Coessens, B., De Mey, A., Lacotte, B., and Vandenbroeck, D. (1991) Carpal tunnel syndrome due to an hamangioma of the median nerve in a 12-year-old child. *Ann. Chir. Main Memb. Super.*, 10(3), 255–257.

Cornu, C. and Goubel, F. (2001) Musculo-tendinous and joint elastic characteristics during elbow flexion in children. *Clin. Biomech.*, 16, 758–764.

CR (2002) Consumer Reports Online. Product Recalls Nintendo Mario Party video game cartridge for N64 play system.

Cruz Martinez, A. and Arpa, J. (1998) Carpal tunnel syndrome in childhood: Study of 6 cases. *Electroencephalogr. Clin. Neurophysiol.*, 109(4), 304–308.

Dai, L., Ye, H., and Qian, Q. (2004) The natural history of cervical disc calcification in children. *J. Bone Joint Surg. Am.*, 86(7), 1467–1472.

Danta, G. (1975) Familial carpal tunnel syndrome with onset in childhood. *J. Neurol. Neurosurg. Psychiatry*, 38(4), 350–355.

De Smet, L. and Fabry, G. (1999) Carpal tunnel syndrome: Familial occurrence presenting in childhood. *J. Pediatr. Orthop. B*, 8(2), 127–128.

DHHS (2000) *CDC Growth Charts: United States.* U.S. Department of Health and Human Services. Centers for Disease Control and Prevention. National Center for Health Statistics. DHHS Publication no. (PHS) 2000-1250 0-0431. Number 314. December 4, 2000 (revised). 28 ps.

Diepenmaat, A.C., van der Wal, M.F., de Vet, H.C., and Hirasing, R.A. (2006) Neck/shoulder, low back, and arm pain in relation to computer use, physical activity, stress, and depression among Dutch adolescents. *Pediatrics*, 117(2), 412–416.

Dormans, J.P. (2002) Evaluation of children with suspected cervical spine injury. *J. Bone Joint Surg.*, 84-A1, 124–132.

Duffy, C.M., Laxer, R.M., Lee, P., Ramsay, C., Fritzler, M., and Silverman, E.D. (1989) Raynaud's syndrome in childhood. *J. Pediatr.*, 114(1), 73–78.

Duggleby, T. and Kumar, S. (1997) Epidemiology of juvenile low back pain: A review. *Disabil Rehabil.* 19(12), 505–512.

Eastwood, D.M. (2001) Elbow injuries in childhood—A source of disquiet. *J. Bone Joint Surg.*, 83-B, 469–470.

El-Metwally, A., Salminen, J.J., Auvinen, A., Kautiainen, H., and Mikkelsson, M. (2004) Prognosis of nonspecific musculoskeletal pain in preadolescents: A prospective 4-year follow-up study till adolescence. *Pain*, 110(3), 550–559.

Evans, O., Collins, B., and Stewart, A. (1992) Is school furniture responsible for student sitting discomfort? Unlocking potential for the future productivity and quality of life. In: Hoffmann, E. and Evans, O. (Eds.), Proceedings of the 28th Annual Conference of the Ergonomics Society of Australia, Melbourne, Australia, 2–4 December 1992. Ergonomics Society of Australia Inc., Downer, ACT, Australia, pp. 31–37.

Feldman, D.E., Shrier, I., Rossignol, M., and Abenhaim, L. (2001) Risk factors for the development of low back pain in adolescence. *Am. J. Epidemiol.*, 154(1), 30–36.

Feldman, D.E., Shrier, I., Rossignol, M., and Abenhaim, L. (2002) Risk factors for the development of neck and upper limb pain in adolescents. *Spine*, 27(5), 523–528.

Gardner, A. and Kelly, L. (2005) Back pain in children and young people. BackCare, London. Available at www.backcare.org.uk/catalog/detail.php?BID=20&ID=2938

Goldfarb, C.A., Leversedge, F.J., and Manske, P.R. (2003) Bilateral carpal tunnel syndrome after abductor digiti minimi opposition transfer: A case report. *J. Hand Surg. [Am].*, 28(4), 681–684.

Grimmer, K. and Williams, M. (2000) Gender-age environmental associates of adolescent low back pain. *Appl. Ergon.*, 31(4), 343–360.

Gunzburg, R., Balague, F., Nordin, M., Szpalski, M., Duyck, D., Bull, D., and Melot, C. (1999) Low back pain in a population of school children. *Eur. Spine J.*, 8(6), 439–443.

Haddad, F.S., Jones, D.H., Vellodi, A., Kane, N., and Pitt, M. (1996) Review of carpal tunnel syndrome in children. *J. Hand Surg. [Br].*, 21(4), 565–566.

Harreby, M., Nygaard, B., Jessen, T., Larsen, E., Storr-Paulsen, A., Lindahl, A., Fisker, I., and Laegaard, E. (1999) Risk factors for low back pain in a cohort of 1389 Danish school children: An epidemiologic study. *Eur. Spine J.*, 8(6), 444–450.

Harris, C. and Straker, L. (2000a) Survey of physical ergonomics issues associated with school children's use of laptop computers. *Int. J. Ind. Ergon.*, 26(3), 337–346.

Harris, C. and Straker, L. (2000b) Survey of physical ergonomics issues associated with school children's use of laptop computers. *Int. J. Ind. Ergon.*, 26, 389–398.

Hedley, A.A., Ogden, C.L., Johnson, C.L., Carroll, M.D., Curtin, L.R., and Flegal, K.M. (2004) Overweight and obesity among US children, adolescents, and adults, 1999–2002. *JAMA*, 291(23), 2847–2850.

Jacobs, K. and Baker, N.A. (2002) The association between children's computer use and musculoskeletal discomfort. *Work*, 18(3), 221–226.

Jones, G.T. and Macfarlane, G.J. (2005) Epidemiology of low back pain in children and adolescents. *Arch. Dis. Child.*, 90(3), 312–316.

Jones, G.T., Watson, K.D., Silman, A.J., Symmons, D.P., and Macfarlane, G.J. (2003) Predictors of low back pain in British schoolchildren: A population-based prospective cohort study. *Pediatrics*, 111(4) Pt 1, 822–828.

Jones, M.A., Stratton, G., Reilly, T., and Unnithan, V.B. (2004) A school-based survey of recurrent non-specific low-back pain prevalence and consequences in children. *Health Educ Res.*, 19(3), 284–289.

Juul, A., Teilmann, G., Scheike, T., Hertel, N.T., Holm, K., Laursen, E.M., Main, K.M., and Skakkebaek, N.E. (2006) Pubertal development in Danish children: Comparison of recent European and US data. *Int. J. Androl.*, 29(1), 247–255.

Kayali, H., Kahraman, S., Sirin, S., Beduk, A., and Timurkaynak, E. (2003) Bilateral carpal tunnel syndrome with type 1 diabetes mellitus in childhood. *Pediatr. Neurosurg.*, 38(5), 262–264.

Kennedy, E., Cullen, B., Abbott, J.H., Woodley, S., and Mercer, W. (2004) Palpation of the iliolumbar ligament. *NZ J. Physiother.*, 32(2) 76–80. http://nzsp.org.nz/index02/index_Welcome.htm

King, H.A. (1999) Back pain in children. *Orthop. Clin. North Am.*, 30(3), 467–474.

Kjaer, P. (2004) Low back pain in relation to lumbar spine abnormalities as identified by magnetic resonance imaging (Report 1). PhD Thesis. Back Research Center, University of Southern Denmark.

Kovacs, F.M., Gestoso, M., Gil del Real, M.T., Lopez, J., Mufraggi, N., and Mendez, J.I. (2003) Risk factors for non-specific low back pain in schoolchildren and their parents: A population based study. *Pain*, 103(3), 259–268.

Kristjansdottir, G. and Rhee, H. (2002) Risk factors of back pain frequency in schoolchildren: A search for explanations to a public health problem. *Acta Paediatr.*, 91(7), 849–854.

Lamberti, P.M. and Light, T.R. (2002) Carpal tunnel syndrome in children. *Hand Clin.*, 18(2), 331–337.

Leboeuf-Yde, C. and Kyvik, K.O. (1998) At what age does low back pain become a common problem? A study of 29,424 individuals aged 12–41 years. *Spine*, 23(2), 228–234.

Leboeuf-Yde, C., Wedderkopp, N., Andersen, L.B., Froberg, K., and Hansen, H.S. (2002) Back pain reporting in children and adolescents: The impact of parents' educational level. *J. Manipulative Physiol. Ther.*, 25(4), 216–220.

Legg, S.J., Pajo, K., Sullman, M., and Marfell-Jones, M. (2003) Mismatch between classroom furniture dimensions and student anthropometric characteristics in three New Zealand secondary schools. In: Proceedings of the 15th Congress of the International Ergonomics Association, Ergonomics for Children in Educational Environments Symposium, 6, 395-7, Seoul, Korea, 24–29 August 2003.

Limon, S., Valinsky, L.J., and Ben-Shalom, Y. (2004) Children at risk: Risk factors for low back pain in the elementary school environment. *Spine*, 29(6), 697–702.

Lipton, G., Riddle, E., Grissom, L., Fitru, T., Marks, H., and Kumar, S.J. (2001) An unusual cause of low-back pain in children: A report of two cases. *J. Bone Joint Surg. Am.*, 83-A, 10, 1552–1554.

McCarthy, J.J., Dormans, J.P., Kozin, S.H., and Pizzutillo, P.D. (2004) Musculoskeletal infections in children: Basic treatment principles and recent advancements. *J. Bone Joint Surg.*, 86(4), 850–860.

Mikkelsson, M., Sourander, A., Salminen, J.J., Kautiainen, H., and Piha, J. (1999) Widespread pain and neck pain in schoolchildren. A prospective one-year follow-up study. *Acta Paediatr.*, 88(10), 1119–1124.

Murphy, S., Buckle, P., and Stubbs, D. (2004) Classroom posture and self-reported back and neck pain in schoolchildren. *Appl. Ergon.*, 35(2), 113–120.

Musharbash, A. (2002) Carpal tunnel syndrome in a 28-month-old child. *Pediatr. Neurosurg.*, 37(1), 32–34.

Narain, J. (2005) Testing health warning. This is London, London Cuts Section. *Daily Mail*, August 23, 2005.

NIAMS (2001a) National Institute of Arthritis and Musculoskeletal and Skin Diseases. Questions and answers about growth plate injuries. NIH Publication 02-5028. US Department of Health and Human Services. Adapted from *Play It Safe, a Guide to Safety for Young Athletes*. American Academy of Orthopaedic Surgeons.

NIAMS (2001b) National Institute of Arthritis and Musculoskeletal and Skin Diseases. Questions and answers about growth plate injuries. NIH Publication 02-5028. US Department Of Health and Human Services. Adapted from Play It Safe, a Guide to Safety for Young Athletes. American Academy of Orthopaedic Surgeons. Adapted from *Disorders and Injuries of the Musculoskeletal System*, 3rd Edition. Robert B. Salter, Baltimore, Williams and Wilkins, 1999.

Nigrovic, P.A., Fuhlbrigge, R.C., and Sundel, R.P. (2003) Raynaud's phenomenon in children: A retrospective review of 123 patients. *Pediatrics*, 111(4), 715–721.

Ogden, C.L., Flegal, K.M., Carroll, M.D., and Johnson, C.L. (2002) Prevalence and trends in overweight among US children and adolescents, 1999–2000. *JAMA*, 288, 1728–1732.

Olsen, T.L., Anderson, R.L., Dearwater, S.R., Kriska, A.M., Cauley, J.A., Aaron, D.J., and LaPorte, R.E. (1992) The epidemiology of low back pain in an adolescent population. *Am. J. Public Health*, 82(4), 606–608.

Orbak, Z. (2005) Does handedness and altitude affect age at menarche? *J. Trop. Pediatr.*, 51(4), 216–218.

Osterman, A.L., Weinberg, P., and Miller, G. (1987) Joystick digit. *JAMA*, 257, 782.

Payne, W.K. III and Ogilvie, J.W. (1996) Back pain in children and adolescents. *Pediatr. Clin. North Am.*, 43(4), 899–917.

Petersen, S., Brulin, C., and Bergstrom, E. (2006) Recurrent pain symptoms in young schoolchildren are often multiple. *Pain*, 121(1–2), 145–150.

Reilly, J.J. and Dorosty, A.R. (1999) Epidemic of obesity in UK children. *Lancet*, 354(9193), 1874–1875.

Reilly, J.J., Dorosty, A.R., and Emmett, P.M. (1999) Prevalence of overweight and obesity in British children: Cohort study. *BMJ*, 319(7216), 1039.

Sairyo, K., Katoh, S., Ikata, T., Fujii, K., Kajiura, K., and Goel, V.K. (2001) Development of spondylolytic olisthesis in adolescents. *Spine J.*, 1(3), 171–175.

Sairyo, K., Katoh, S., Sakamaki, T., Inoue, M., Komatsubara, S., Ogawa, T., Sano, T., Goel, V.K., and Yasui, N. (2004) Vertebral forward slippage in immature lumbar spine occurs following epiphyseal separation and its occurrence is unrelated to disc degeneration: Is the pediatric spondylolisthesis a physis stress fracture of vertebral body? *Spine*, 29(5), 524–527.

Salminen, J.J., Erkintalo, M.O., Pentti, J., Oksanen, A., and Kormano, M.J. (1999) Recurrent low back pain and early disc degeneration in the young. *Spine*, 24(13), 1316–1321.

Salter, R.B. and Harris, R.W. (1963) Injuries involving the growth plate. *J. Bone Joint Surg.* (A), 45, 587–621.

Sanchez-Andrada, R.M., Martinez-Salcedo, E., de Mingo-Casado, P., Domingo-Jimenez, R., Puche-Mira, A., and Casas-Fernandez, C. (1998) Carpal tunnel syndrome in childhood. A case of early onset. *Rev. Neurol.*, 27(160), 988–991.

Sheir-Neiss, G.I., Kruse, R.W., Rahman, T., Jacobson, L.P., and Pelli, J.A. (2003) The association of backpack use and back pain in adolescents. *Spine*, 28(9), 922–930.

Sills, J.A. (2000) Overview of the clinical presentation of connective tissue diseases in children. *Eur. J. Radiol.*, 33, 112–117.

Smith, S.A. and Norris, B.J. (2004) Changes in the body size of UK and US children over the past three decades. *Ergonomics*, 47(11), 1195–1207.

Spear, B.A. (2002) Adolescent growth and development. *J. Am. Diet. Assoc.*(JADA), March 2002, S23–S29.

Stahl, M., Mikkelsson, M., Kautiainen, H., Hakkinen, A., Ylinen, J., and Salminen, J.J. (2004) Neck pain in adolescence. A 4-year follow-up of pain-free preadolescents. *Pain*, 110(1–2), 427–431.

Steele, E.J., Dawson, A.P., and Hiller, J.E. (2006) School-based interventions for spinal pain: A systematic review. *Spine*, 31(2), 226–233.

Swischuk, L.E. and Hernandez, J.A. (2004) Frequently missed fractures in children (value of comparative views). *Emerg. Radiol.*, 11(1), 22–28.

Szeto, G.P.Y., Lau, J.C.C., Siu, A.Y.K., Tang, A.M.Y., Tang, T.W.Y., and Yiu, A.O.Y. (2002) A study of physical discomfort associated with computer use in secondary school students. In Thatcher, A., Fisher, J., Miller, K. (eds.) *Proceedings of CybErg'2002. The Third International Cyberspace Conference on Ergonomics*. International Ergonomics Association Press.

Szpalski, M., Gunzburg, R., Balague, F., Nordin, M., and Melot, C. (2002) A 2-year prospective longitudinal study on low back pain in primary school children. *Eur. Spine J.*, 11(5), 459–464.

Tan, T.H. and Ong, K.L. (2000) Joint disease in children of Asiatic origin. *Eur. J. Radiol.*, 33, 102–111.

Tanner, J.M. (1978) Foetus into man: Physical growth from conception to maturity. Cambridge: Harvard University Press, pp. 60–77.

Tertti, M.O., Salminen, J.J., Paajanen, H.E., Terho, P.H., and Kormano, M.J. (1991) Low-back pain and disk degeneration in children: A case-control MR imaging study. *Radiology*, 180(2), 503–507.

Trentham-Dietz, A., Nichols, H.B., Remington, P.L., Yanke, L., Hampton, J.M., Newcomb, P.A., and Love, R. R. (2005) Correlates of age at menarche among sixth grade students in Wisconsin. *WMJ*, 104(7), 65–69.

Trevelyan, F.C. and Legg, S.J. (2006) Back pain in school children—Where to from here? *Appl. Ergon.*, 37(1), 45–54.

Unal, O., Ozcakar, L., Cetin, A., and Kaymak, B. (2003) Severe bilateral carpal tunnel syndrome in juvenile chronic arthritis. *Pediatr. Neurol.*, 29(4), 345–348.

Van Meir, N. and De Smet, L. (2003) Carpal tunnel syndrome in children. *Acta Orthop. Belg.*, 69(5), 387–395.

Van Meir, N. and De Smet, L. (2005) Carpal tunnel syndrome in children. *J. Pediatr. Orthop. B.*, 14(1), 42–45.

Warner, J.T., Cowan, F.J., Dunstan, F.D., Evans, W.D., Webb, D.K., and Gregory, J. W. (1998) Measured and predicted bone mineral content in healthy boys and girls aged 6–8 years: Adjustment for body size and puberty. *Acta Paediatr.* 87(3), 244–249.

Watson, K.D., Papageorgiou, A.C., Jones, G.T., Taylor, S., Symmons, D.P., Silman, A.J., and Macfarlane, G.J. (2002) Low back pain in schoolchildren: Occurrence and characteristics. *Pain*, 97(1–2), 87–92.

Watson, K.D., Papageorgiou, A.C., Jones, G.T., Taylor, S., Symmons, D.P., Silman, A.J., and Macfarlane, G.J. (2003) Low back pain in schoolchildren: The role of mechanical and psychosocial factors. *Arch. Dis. Child.*, 88(1), 12–17.

Wedderkopp, N., Leboeuf-Yde, C., Andersen, L.B., Froberg, K., and Hansen, H.S. (2001) Back pain reporting pattern in a Danish population-based sample of children and adolescents. *Spine*, 26(17), 1879–1883.

Wedderkopp, N., Andersen, L.B., Froberg, K., and Leboeuf-Yde, C. (2005) Back pain reporting in young girls appears to be puberty-related. *BMC Musculoskelet. Disord.*, 6, 52.

Wood, J. (2001) The How sign—A central palmar blister induced by overplaying on a Nintendo console. *J. Arch. Dis. Child.*, 84, 288.

Yensel, U. and Karalezli, N. (2006) Carpal tunnel syndrome and flexion contracture of the digits in a child with familial hypercholesterolaemia. *J. Hand Surg. [Br].*, 31(2), 154–155.

FURTHER READING

Aagaard-Hensen, J. and Storr-Paulsen, A. (1995) A comparative study of three different kinds of school furniture. *Ergonomics*, 38, 1025–1035.

Bruynel, L. and McEwan, G.M. (1985) Anthropometric data of students in relation to their school furniture. *NZ J. Physiother.*, December, 7–11.

BSI (1980) British Standards Institution BS 5873—Part1: Educational Furniture. Specification for Functional Dimensions, Identification and Finish of Chairs and Tables for Educational Institutions.

Cakmak, A., Yucel, B., Ozyalcn, S.N., Bayraktar, B., Ural, H.I., Duruoz, M.T., and Genc, A. (2004) The frequency and associated factors of low back pain among a younger population in Turkey. *Spine*, 29(14), 1567–1572.

Carson, S., Woolridge, D.P., Colletti, J., and Kilgore, K. (2006) Pediatric upper extremity injuries. *Pediatr. Clin. North Am.*, 53, 41–67.

CDC (2000) Centers for Disease Control and Prevention, National Center for Health Statistics. CDC Growth Charts: United States. www.cdc.gov/growthcharts. May 30, 2000.

CDC (2005) Centers for Disease Control and Prevention. *Health, United States, 2005 with Chartbook on Trends in the Health of Americans.* National Center for Health Statistics. Health, United States, 2005.

CEN/TC 207-Standards—prEN 1729-1 (2006) Furniture—Chairs and Tables for Educational Institutions—Part 1: Functional dimensions, Under Approval.

Evans, W.A., Courtney, A.J., and Fok, K.F. (1988) The design of school furniture for Hong Kong schoolchildren. An anthropometric case study. *Appl. Ergon.*, 19(2), 122–134.

Fallon, E.F. and Jameson, C.M. (1996) An ergonomic assessment of the appropriateness of primary school furniture in Ireland. In: Ozok, A.F. and Salvendy, G. (Eds.), *Advances in Applied Ergonomics*. USA Publishing, West Lafayette, IN, pp. 770–773.

Gibson, C.T. and Manske, P.R. (1987) Carpal tunnel syndrome in the adolescent. *J. Hand Surg. [Am].*, 12(2), 279–281.

Gschwind, C. and Tonkin, M.A. (1992) Carpal tunnel syndrome in children with mucopolysaccharidosis and related disorders. *J. Hand Surg. [Am].*, 17(1), 44–47.

Habibi, E. and Hajsalahi, A. (2003) An ergonomic and anthropometrics study of Iranian primary school children and classrooms. In: Luczak, H. and Zink, K.J. (Eds.), *Human Factors in Organizational Design and Management—VII. Re-designing Work and Macroergonomics—Future Perspectives and Challenges.* IEA Press, Santa Monica, CA, pp. 823–830.

Habibi, E. and Mirzaei, M. (2004) An ergonomic and anthropometric study to facilitate design of classroom furniture for Iranian primary school children. *Ergonomics SA*, 16(1), 15–20.

Hewes, G.W. (1957) The anthropology of posture. *Sci. Am.*, February, 123–132.

Hodgkinson, P.D. and Evans, D.M. (1993) Median nerve compression following trauma in children. A report of two cases. *J. Hand Surg. [Br].*, 18(4), 475–477.

Jeong, B.Y. and Park, K.S. (1990) Sex differences in anthropometry for school furniture design. *Ergonomics*, 33(12), 1511–1521.

Kayis, B. (1991) Why ergonomic designs in schools? Ergonomics and human environments. In: Popovic, V. and Walker, M. (Eds.), Proceedings of the 27th Annual Conference of the Ergonomics Society of Australia, Coolum, Australia, 1–4 December 1991. Ergonomics Society of Australia Inc., Downer, ACT, Australia, pp. 95–103.

Kilroy, A.W., Schaffner, W., Fleet, W.F. Jr., Lefkowitz, L.B. Jr., Karzon, D.T., and Fenichel, G.M. (1970) Two syndromes following rubella immunization. Clinical observations and epidemiological studies. *JAMA*, 214(13), 2287–2292.

Knight, G. and Noyes, J. (1999) Children's behavior and the design of school furniture. *Ergonomics*, 42, 747–760.

Knusel, O. and Jelk, W. (1994) Pezzi-balls and ergonomic furniture in the classroom. Results of a prospective longitudinal study [Sitzballe und ergonomisches Mobiliar im Schumzimmer. Resultate einer prospektiven kontrollierten Langzeitstudie] *Schweizerische Rundschau fur Medizin Praxis*, 83(14), 407–413.

Koskelo, R. (2001) Function responses upper secondary school students considering the saddle chair and the adjustable table. Department of Physiology, University of Kuopio, Finland, pp. 1–24.

Kostera-Pruszczyk, A., Rowinska-Marcinska, K., Ryniewicz, B., Olszewicz-Dukaczewska, M., and Gola, M. (2005) Carpal tunnel syndrome or congenital hand anomaly: A clinical and electromyographic study. *J. Peripher. Nerv. Syst.*, 10(3), 338–339.

Krum-Moller, D., Jensen, M.K., and Hansen, T.B. (1999) Carpal tunnel syndrome after epiphysiolysis of the distal radius in a 5-year-old child. Case report. *Scand. J. Plast. Reconstr. Surg. Hand Surg.*, 33(1), 123–124.

Laeser, K.L., Maxwell, L.E., and Hedge, A. (1998) The effects of computer workstation design on student posture. *J. Res. Comput. Educ.*, 31, 173–188.

Lin, R. and Kang, Y.Y. (2000) Ergonomic design for senior high school furniture in Taiwan, ergonomics for the new millennium. In: Proceedings of the XIVth Triennial Congress of the International Ergonomics Association and the 44th Annual Meeting of the Human Factors and Ergonomics Society, San Diego, CA, July 29–August 4, 2000. Human Factors and Ergonomics Society, Santa Monica, CA, Vol. 6, pp. 39–42.

Linton, S.J., Hellsing, A.L., Halme, T., and Akerstedt, K. (1994) The effects of ergonomically designed school furniture on pupils' attitudes, symptoms and behaviour. *Appl. Ergon.*, 25, 299–304.

Mandal, A.C. (1982) The correct height of school furniture. *Hum. Factors*, 24, 257–269.

Mandal, A.C. (1984) The correct height of school furniture. *Physiotherapy*, 70(2), 48–53.

Mandal, A.C. (1997) Changing standards for school furniture. *Ergon. Des.*, 5, 28–31.

Marschall, M., Harrington, A.C., and Steele, J.R. (1995) Effect of workstation design on sitting posture in young children. *Ergonomics*, 38, 1932–1940.

McAtamney, L. and Corlett, E.N. (1993) RULA: A survey method for the investigation of work-related upper limb disorders. *Appl. Ergon.*, 24, 91–99.

McMillan, N. (1996) Ergonomics in schools. Contemporary ergonomics in New Zealand—1996. In: Darby, F. and Turner, P. (Eds.), Proceedings of the Seventh Conference of the New Zealand Ergonomics Society, Wellington, August 2–3, 1996. New Zealand Ergonomics Society, Palmerston North, New Zealand, pp. 87–93.

Mikkelsson, L.O. Nupponen, H., Kaprio, J., Kautiainen, H., Mikkelsson, M., and Kujala, U.M. (2006) Adolescent flexibility, endurance strength, and physical activity as predictors of adult tension neck, low back pain, and knee injury: A 25 year follow up study. *Br. J. Sports Med.*, 40(2), 107–113.

Milanese, S. and Grimmer, K. (2004) School furniture and the user population: An anthropometric perspective. *Ergonomics*, 47(4), 416–426.

Molenbroek, J. and Ramaekers, Y. (1996) Anthropometric design of a size system for school furniture. In: Robertson, S.A. (Ed.), *Contemporary Ergonomics 1996*. Taylor & Francis, London, pp. 130–135.

Molenbroek, J.F.M., Kroon-Ramaekers, Y.M.T., and Snijders, C.J. (2003) Revision of the design of a standard for the dimensions of school furniture. *Ergonomics*, 46(7), 681–694.

Mosher, E.M. (1899) Hygienic desks for school children. *Educ. Rev.*, 18, 9–14.

Mououdi, M.A. and Choobineh, A.R. (1997) Static anthropometric characteristics of students age range 6–11 in Mazandaran Province/Iran and school furniture design based on ergonomics principles. *Appl Ergon.*, 28(2), 145–147.

NCHS (2003) US Department of Health and Human Services. Prevalence of Overweight among Children and Adolescents: United States, 1999–2002. www.cdc.gov/nchs/products/pubs/pubd/hestats/overwght99.htm

Oates, S., Evans, G.W., and Hedge, A. (1998) An anthropometric and postural risk assessment of children's school work environments. *Comput. Schools*, 14(3/4), 55–63.

Okamoto, H., Kawai, K., Hattori, S., Ogawa, T., Kubota, Y., Moriya, H., and Matsui, N. (2005) Thoracic outlet syndrome combined with carpal tunnel syndrome and Guyon canal syndrome in a child. *J. Orthop. Sci.*, 10(6), 634–640.

Ong, B.C., Klugman, J.A., Jazrawi, L.M., and Stutchin, S. (2001) Bilateral carpal tunnel syndrome in a child on growth hormone replacement therapy: A case report. *Bull. Hosp. Jt. Dis.*, 60(2), 94–95.

Panagiotopoulou, G., Christoulas, K., Papanckolaou, A., and Mandroukas, K. (2004) Classroom furniture dimensions and anthropometric measures in primary school. *Appl. Ergon.*, 35(2), 121–128.

Parcells, C., Stommel, M., and Hubbard, R.P. (1999) Mismatch of classroom furniture and student body dimensions. *J. Adolesc. Health*, 24, 265–273.

Paschoarelli, L.C. and Da Silva, J.C.P. (2000) Ergonomic research applied in the design of pre-school furniture: The Mobipresc 3.6. In: Proceedings of the XIVth Triennial Congress of the International Ergonomics Association and 44th Annual Meeting of the Human Factors and Ergonomics Society, Vol. 2, San Diego, pp. 24–27.

Potter, N. (2006) Digital-age hazard: Sore thumbs. Survey warns of repetitive stress from cell phones. *ABC News*, February 21, 2006.

Prado-Leon, L.R., Avila-Chaurand, R., and Gonzalez-Munoz, E.L. (2001) Anthropometric study of Mexican primary school children. *Appl. Ergon.*, 32(4), 339–345.

Pronicka, E., Tylki-Szymanska, A., Kwast, O., Chmielik, J., Maciejko, D., and Cedro, A. (1988) Carpal tunnel syndrome in children with mucopolysaccharidoses: Needs for surgical tendons and median nerve release. *J. Ment. Defic. Res.*, 32(1), 79–82.

Schindler, C.C.A. (1892) United States Patent US0483266 A1. Issued September 27, 1892. United States Patent Office.

Shinn, J., Romaine, K.A., Casimano, T., and Jacobs, K. (2002) The effectiveness of ergonomic intervention in the classroom. *Work*, 18(1), 67–73.

Smith-Zuzovsky, N. and Exner, C.E. (2004) Effect of seated positioning quality of typical 6- and 7-year-old children's object manipulation skills. *Am. J. Occup. Ther.*, 58(4), 380–388.

Storr-Paulsen, A. and Aagaard-Hensen, J. (1994) The working positions of schoolchildren. *Appl. Ergon.*, 25, 63–64.

Taylour, J.A. and Crawford, J. (1996) The potential use and measurement of alternative work stations in UK schools. In: Robertson, S.A. (Ed.), *Contemporary Ergonomics*. Taylor & Francis, London, pp. 464–469.

Tolo, V.T. (2000) Orthopaedic treatment of fractures of the long bones and pelvis in children who have multiple injuries. *Instr. Course Lect.*, 49, 415–423.

Troussier, B., Tesniere, C., Fauconnier, J., Grison, J., Juvin, R., and Phelip, X. (1999) Comparative study of two different kinds of school furniture among children. *Ergonomics*, 42, 516–526.

Van Heest, A.E., House, J., Krivit, W., and Walker, K. (1998) Surgical treatment of carpal tunnel syndrome and trigger digits in children with mucopolysaccharide storage disorders. *J. Hand Surg. [Am].*, 23(2), 236–243.

Wilson, K.M. and Buehler, M.J. (1994) Bilateral carpal tunnel syndrome in a normal child. *J. Hand Surg. [Am].*, 19(6), 913–914.

CHAPTER 7

PHYSICAL EDUCATION AND EXERCISE FOR CHILDREN

BARBARA H. BOUCHER

TABLE OF CONTENTS

INTRODUCTION

CHILDREN'S PHYSICAL ACTIVITY

Inactivity is a worldwide health problem (WHO, 2004). Without exercise and proper nutrition, bones can be weak and susceptible to fractures, even as children pass adolescence and enter their adult lives.

Children benefit mentally, physically, and socially from physical activity. Physical exercise is essential for proper development of children's bones, muscles, and joints. Regular physical activity can help control body weight, keep heart and lungs working efficiently, improve alertness, improve self-esteem, and enrich a child's outlook of life.

A child who participates in regular activity is more likely to remain active as an adult. Unfortunately, activity tends to decline with age (Armstrong, 1998). There are more barriers to physical activities for children of parents with lower incomes and education levels (Duke et al., 2003). These facts make it even more important to teach children that regular physical activity is part of a healthy, desirable lifestyle. Hopefully, children will carry this knowledge and these behaviors with them to create a healthier, more active society.

PROGRAM DESIGN

Physical activity improves muscle strength, coordination, and fitness. It also relieves stress, reduces anxiety and depression, and improves self-esteem and mental clarity (Ekeland et al., 2004; Tomporowski, 2003). Well-designed Physical Education (PE) programs assure the benefits of physical activity because they focus on the child and his or her environment.

Well-designed PE programs are "user-centered" and support children's developmental levels. This developmental focus aims instructions and activities at or slightly beyond a child's current capabilities. In this way, children take part in activities they can do, while pushing to attain greater achievement. This helps children learn and enables them to better respond to challenges, recognize their own achievements, and develop self-confidence. When PE programs reflect their development, children are less likely to become discouraged.

Well-designed PE programs complement children's environments. For example, prolonged sitting in class leads to poor postures and unbalanced strength and weakness; PE programs can help reverse the effects of prolonged flexion postures by stretching and strengthening children's hips, back, and shoulder muscles. Targeted PE activities strengthen muscles for such daily activities as carrying a backpack.

User-centered PE programs are holistic. A holistic orientation considers the child's health now and in the future. For example, PE helps avoid health risks such as obesity through physical activity, and provides lifestyle education (healthy eating and wellness). User-centered programs focus on emotional (e.g., teamwork) and physical development.

WHAT PHYSICAL EDUCATION *IS* AND *IS NOT*

Physical education is physical, and it is education. Although PE programs may include sports and play, these terms are not synonymous. The benefits and outcomes for sports, play, and physical education differ, and substituting one for the other reduces the benefit of each (CPEC, 2001). Understanding how sports, play, and PE interact, overlap, and enhance children's physical development is critical to designing effective PE programs.

SPORTS: Most children and their parents prize sports activities. Outside of school, sports abound in little league and community programs. Almost 40% of US children aged 9 to 13 years participate in organized physical activity outside of school (CDC, 2002). Sports are organized (rule-based) and competitive (performance score-based). Sports can help children learn teamwork, build physical skills, and develop self-esteem (Figure 7.1).

Although sports activities generally promote child health, these activities alone are not enough. This is because adult leaders of sports seldom have sufficient education to guide children's development, the design of sports activities seldom accounts for children's development, and sports do not typically include information about lifestyle education.

Physical education teachers' training includes child development. Adults who manage sports programs are often parents or other volunteers or with little knowledge of child developmental maturation processes. Volunteers are necessary for most sports programs, but they may not know how to nurture the physical or emotional development of the children they coach.

The design of most sports focuses on adult skills and abilities. Children of ever-younger ages now participate in activities historically reserved for adults. Yet sports that benefit adults are not necessarily good for children. Children are developmentally different from adults; some sports are more appropriate for adults, who have fully developed bones, muscles, and coordination (see Table 7.1).

Figure 7.1 (Valerie Rice)
The purpose of sports is to excel—usually by winning a competition. We often assume that sports participants benefit from exercise. While taking part in sports does not necessarily improve overall fitness or promote a healthy lifestyle, it can help if it is part of a regular regimen of physical exercise. It is even better if regular exercise is coupled with an educational program on healthy lifestyles. Children who participate in regular activity are likely to remain active as adults.

Table 7.1 Cautions Regarding Sports-Oriented Physical Education Programs

Potential Problem	Rationale
Introducing a sport to children who are too young to play well	• Children who are not developmentally ready for the activity find it difficult to perform successfully. • If children's strength and coordination have not sufficiently developed, their risk of injury increases relative to performing developmentally appropriate activities.
Playing the same sport year after year, striving to develop ever-increasing skills	• Playing the same sport often means playing the same position on the team. Children learn a limited set of skills that will not necessarily prepare them for the many physical activities in adulthood. • Playing one sport also increases the child's risk of developing repetitive motion injuries. Although we do not understand the long-term effects, it is a concern since child bones are still developing.
Competitive games produce both winners and losers. The children who are not competitive may suffer social stigmas	• Children who have negative experiences in sports may choose to refrain from physical activity. These children may suffer life-long consequences from being the "last kid chosen for a team."

Playing sports is not the same as participating in PE programs that prepare children for life-long fitness. PE lifestyle education helps children learn to maintain "ideal" weights, reduce their susceptibility to illness, and optimize physical performance (see Table 7.1).

In addition, sports programs do not complement the child's environment. That is, they do not target the child's muscle development or coordination needed for everyday activities.

Clearly, there are physical and emotional benefits from competitive team sports. This chapter focuses on the larger issues of how to design well-balanced PE programs that help children achieve a range of goals.

Figure 7.2 (Ursy Potter Photography)
Play is the "work" of infants, toddlers, and preschoolers. It is how they learn, as well as how they develop their musculoskeletal systems. In play, they exercise their imagination, learn to interact with others, solve problems, and practice the roles they will eventually assume as adults.

PLAY: Children (and adults) play because it is fun. Play is unstructured and loosely organized. Rules are flexible, and children make them up as they go along. Rewards are intrinsic; they do not need to achieve a goal (Figure 7.2; see Chapter 10 for more about play).

Play involves creativity. Through play, children exercise imagination. Social skills develop as children create their own roles, responsibilities, and interactions. Self-discovery occurs as children identify their own abilities and gradually improve them. They learn how high they can jump, how long they can balance, and how tense they need to be to hang from a bar by their knees!

Play is critical for early child development. As children enter school, play is systematically limited. Once children enter kindergarten, they begin to live by structured programming and rules. They follow a schedule and walk in a line. Yet, elementary school children continue to benefit from unstructured play (Table 7.2). Unfortunately, almost 40% of schools have reduced daily recess (CPEC, 2001).

Table 7.2 Benefits of Unstructured Play

Recess (unstructured play):

- Motivates children to come to school to play with their friends.
- Enables children of different cultures to interact.
- Relieves stress.
- Provides time alone to think and meditate.
- Enables teachers to observe children interacting with each other.
- Allows children to select physical activities that promote circulation and attentiveness.
- Provides an opportunity to be creative and to create new games or play.
- Helps children to learn cooperative play and taking turns.

Source: From CPEC, Council on Physical Education for Children and the National Association for Sport and Physical Education position paper, Recess in Elementary Schools, July 2001. Available at: http://naecs.crc.uiuc.edu/position/recessplay.html

PHYSICAL EDUCATION: The "physical" portion of PE may look like a sport or child's play. Yet physical education programs have learning outcomes. A resurgence of interest in children's physical education is evident, primarily in response to health concerns. China's PE programs illustrate this new emphasis on PE in schools (CHERN, 2004) (Table 7.3), as do the programs for some other countries.

New mandates for specific physical performance outcomes are in effect in the United States, due to public demand for more PE time in the public schools (CSBE, 1999). Sixty percent of states in the United States require schools to follow national or state PE guidelines (Burgeson et al., 2001). Most require that education programs reflect the children's developmental level and physical performance outcomes.

The Texas Essential Knowledge and Skills for Physical Education (TEKS PE) is one example (Texas Administrative Code, 1999; see Table 7.4). The TEKS PE contains annual requirements for each child for movement, physical activity and health, and social development.

Table 7.3 PE Requirements in Elementary School in Selected Countries

Country	PE Requirements
China	• 1 hour per day • Includes morning exercises, preparatory exercises, and exercise during class breaks • Some extracurricular physical training
Brazil	• 1 hour per week • Some programs help poor, rural communities teach children games with exercise
Japan	• 90 periods in each school year
India	• No national requirements for PE in elementary education
United States	• Requirements regulated at the state level, and vary state to state
Saudi Arabia	• Two periods per week for boys only • Girls have classes in home economics instead of PE
France	• 3 hours per week

Source: From WES (2004).

Table 7.4 Learning Outcomes (Texas Administrative Code, 1999)
Texas Essential Knowledge and Skills for Physical Education
In Grades K-5, children learn fundamental movement skills and begin to understand how the muscles, bones, heart, and lungs function during physical activity. Students begin to develop a vocabulary for movement and apply concepts dealing with space and body awareness.
Students are engaged in activities that develop basic levels of strength, endurance, and flexibility. They also learn to work safely in both group and individual movement settings. A major objective is to present activities that complement their natural inclination to view physical activity as challenging and enjoyable.

Source: From Texas Administrative Code, Title 19, Part II, Chapter 116. Texas Essential Knowledge and Skills for Physical Education, 1999. Available at: www.tea.state.tx.us/rules/tac/chapter116/ch116a.pdf

Ensuring the balance of academics and support for physical health in the daily lives of children is a challenge for teachers and parents. Unfortunately, in the mix of regulations, funding issues and academic requirements, the implementation of PE varies widely.

Despite the excellent intent of PE regulation and curricula, its implementation is fraught with confusion regarding sports, play, and physical education. Most PE programs need more emphasis on children's physical and physiological development. Children need to learn how to achieve the benefits of practicing a healthy lifestyle. Children who remain active are more likely to continue that lifestyle as adults.

DEVELOPMENTAL CHANGES IN BOYS AND GIRLS

Designing effective PE programs requires an understanding of how children develop physiologically, cognitively, and socially between the ages of 5 and 15 years.

SKELETAL DEVELOPMENT

Immature skeletal systems include ***epiphyseal plates*** (also known as ***growth plates***). ***Growth plates*** are responsible for the longitudinal growth of bone. They are located between the shaft and the larger end of the bone (called the ***epiphyses***).

The longitudinal growth of long bones occurs primarily at one growth plate.

For example, the thighbone (*femur*) grows primarily at its distal (furthest from the body) growth plate. The bone of the upper arm (*humerus*) grows primarily at its proximal (closest to the body) growth plate and the forearm bones grow distally.

The growth plate's vascular (blood) supply nourishes growing cells. Injury to this area can impair vascular supply and thereby stunt growth. On the other hand, an increased vascular supply can lead to abnormal increases in cell production (Rarick, 1973).

An untreated or improperly treated *epiphyseal* fracture can cause one leg to grow longer than the other or uneven growth within a limb. The potential for growth impairments among children deserves concern.

Most long bones in immature skeletons have epiphyses at both ends. Pressure epiphyses respond to forces caused by muscle tension and weight bearing. On the other hand, traction epiphyses are at muscle attachments, where bones respond to muscle pulling. For example, there is a traction *epiphysis* where tendons connect the kneecap (*patella*) to the lower leg bone (*tibia*).

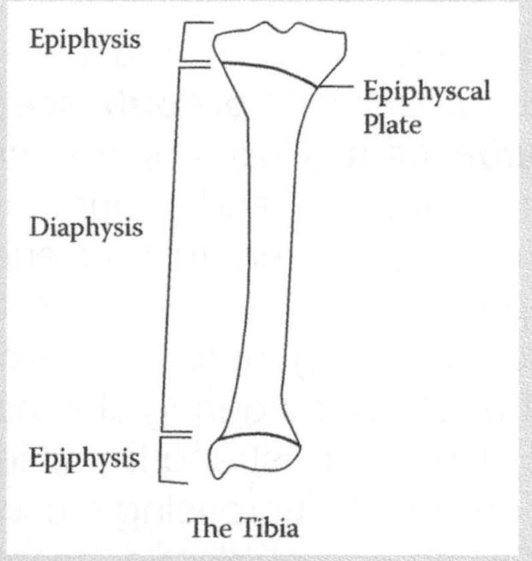

Figure 7.3 (Barbara H. Boucher)

Both pressure and traction epiphyses shape bones as they respond to forces (Rarick, 1973). The pull of a muscle (*traction*) and compression (*pressure*) are outside forces that stress bone and cause new bone to form. The shapes of bones reflect forces applied to them. For example, pulling or contracting the anterior thigh muscles (*quadriceps*) causes the tendon (via the patella) to pull on the *epiphysis;* this forms a prominence just below the knee.

Bones need a certain amount of stress (force) in order to develop or grow normally. As muscles develop, they generate mechanical forces of traction and pressure; these play a central role in stimulating the formation of the skeletal system (Schonau, et al., 1996).

A child's skeleton is most vulnerable during periods of rapid growth.

Children are at particular risk of injuring a growth plate (compared to adults). Growth plate injuries can occur at any age, but occur most frequently during the period of a growth spurt (England and Sundberg, 1996). A growth plate injury can mean a temporary period of pain, swelling, and decreased range of motion about the joint or it can mean many problems including growth arrest.

One example is excessive throwing, which can affect the growth plate in the proximal humerus, a condition known as *Little Leaguer's Shoulder*. The mechanism of injury may be due to the amount of rotational torque placed on the proximal humeral growth plate during throwing (Bernhardt-Bainbridge, 1995). Complete growth arrest is rare and its severity depends on the skeletal age of the child. A complete growth arrest has the greatest impact on younger children and may result in a limb length discrepancy and functional impairment (Larson et al., 1976).

Children need active muscle movement for their bones to develop fully. They also need to activate all their muscles, regularly and somewhat evenly, for their bones to develop correctly. That is, if a child repeatedly activates only a limited range of movement patterns, the bones will develop alignments that reflect the muscles used. Level of activity largely determines bone mass increases in childhood (MacKelvie et al., 2003). Accumulated bone mass has lasting effects into adulthood, when it is much more difficult to build or rebuild bone mass (Heaney et al., 2000).

MUSCLE DEVELOPMENT

The body has 650 muscles that make up almost half of our body weight. The muscle *composition* of ***prepubescent*** (***Prp***) boys and girls is essentially *equal* to that of ***post-pubescent*** (***Pop***) children and adults. However, the *ability* of Prp muscle to increase in strength, size, and endurance is significantly *different* from Pop muscle.

Muscle strength and endurance increase as children age from early childhood to approximately 14 years. Between the ages of 7 and 12 years, muscle strength nearly doubles. In general, a muscle's strength reflects its cross-sectional area or size. Some strength gains during childhood are a result of typical muscle growth in size.

However, strength-training enables Prp children to develop force output without significantly increasing muscle mass (Stout, 1995; Rowland, 2000). *Even with appropriate training and guidance, coaches and parents should not expect child athletes to become muscular.*

A child's muscles appear to relax more slowly than those of an adult (Lin et al., 1994). Slower relaxation times affect the speed of performance during coordinated movements. Understanding normal muscle development is key to setting appropriate performance expectations for children.

Flexibility refers to the ability to lengthen muscles at the joints during movement. Children differ considerably in their flexibility. Girls are more flexible than boys between the ages of 5 through 14 years and girls' flexibility changes little from the ages of 5 to 11 years. Boys lose flexibility between the ages of 5 and 15 years (Stout, 1995). Compare photos Figure 7.4 and Figure 7.5; the boy's spine exhibits greater curvature and his feet are less upright. His position depicts less flexibility in the muscles on the back of his legs, compared to the position of the girl.

Bone growth typically takes place before muscles develop. Muscles also become less flexible during periods of rapid growth. Boys and girls may avoid growing pains by managing flexibility. For example, children who grow rapidly may experience pain where the patellar tendon attaches to the tibia. Stretching to lengthen the quadriceps can help alleviate the strong pull this tendon exerts on the epiphysis. Stretching also provides sensory feedback that helps children experience the extremes of joint motions. This 'inner knowledge' helps shape the child's self-image, helping them to learn purposeful movements.

Figure 7.4 (Barbara H. Boucher)
The ability of this 11-year-old girl to reach her toes with her feet up depicts greater flexibility in the muscles on the back of her legs, hamstrings, and gastrocnemius, compared to her lower spine. Her lower spine is straighter than her upper spine, depicting flexibility in her thigh muscles over her back muscles.

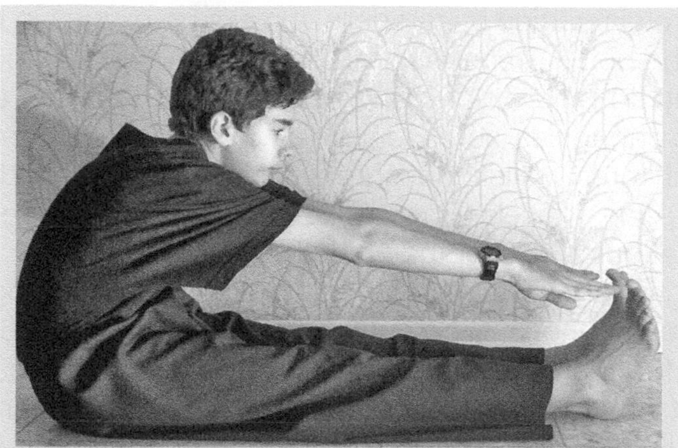

Figure 7.5 (Barbara H. Boucher)
The inability of this 14-year-old boy to reach his toes with his feet in this position depicts inflexibility in the muscles on the back of his legs. He has developed more flexibility in his low back muscles, as evidenced by the smooth, continuous curve in his spine.

CARDIOPULMONARY SYSTEM DEVELOPMENT

As mentioned earlier, regular physical activity keeps all muscles fit, including ***voluntary*** (*skeletal*) muscles and ***involuntary*** cardiac (heart) and smooth (stomach, intestines, etc.) muscles. Part of "fitness" is the ability to function over time (***endurance***). We can measure endurance by ***maximal aerobic power*** or VO_{2max}. This measure represents the greatest amount of oxygen used to create energy for sustained muscle activity (by aerobic metabolism). Prp and adult muscles are similar in their ability to sustain muscle activity (aerobic capacity, based on VO_{2max} measures). ***Aerobic capacity*** improves with training (Stout, 1995) and increases from ages 8 to 16 years in boys and from ages 8 to 13 years in girls (Roemmich and Rogol, 1995).

Prepubescent children who have more muscle than fat (larger proportions of ***fat-free mass***) have greater aerobic capacities than Prp children with proportionally more fat (Roemmich and Rogol, 1995). Exercise programs for groups of children must consider children who are overweight or who have low fat-free mass.

Body mass index (BMI) can help determine whether an individual's weight is appropriate for their height. Calculate a child's BMI by measuring their height and weight ($BMI = wt/ht^2$). By plotting the correct age and gender on the chart available at the CDC Web site, one can quickly identify underweight, at-risk, or overweight children, according to percentile cutoff points (CDC, 2000) (Table 7.5).

Table 7.5 Body Mass Index (BMI) Cutoffs

Underweight	BMI-for-age <5th percentile
At risk of overweight	BMI-for-age 85th percentile to <95th percentile[a]
Overweight	BMI-for-age ≥95th percentile

[a] A child who has an 85th percentile BMI (body mass index) is larger in that dimension than about 84% of the children of that gender and age group.

Hemoglobin is a protein carried by red cells. It picks up oxygen in the lungs and delivers it to cells throughout the body. Prepubescent children have less hemoglobin in their blood. This affects their ability to bind oxygen and causes them to reach a peak VO_{2max} earlier. Prp children's aerobic capacities increase as they develop more hemoglobin as they grow. This means younger children will fatigue more quickly than older children, who can perform aerobic work at lower VO_{2max} levels (Rowland, 2000).* Exercise programs should account for younger children fatiguing sooner than older children do.

Endurance training is an important component of an exercise program. However, aerobic capacity may not be a functional component of a sport (Table 7.6). Most physical skills required in baseball, for example (batting, throwing, and running bases) consist of short bursts of energy that rely on a child's anaerobic capacity. Anaerobic capacity in Prp children is significantly lower than that of young adults. Anaerobic power increases with body size as children mature, leveling off at age 15 for girls and 17 for boys (Roemmich and Rogol, 1995; Clippinger-Robertson, 1987).

Muscle fiber type may also limit children's anaerobic capacity.† Developing children are physiologically primed for aerobic performance vs. anaerobic performance. Children can improve both their aerobic capacity and anaerobic metabolism (Rowland, 2000). We describe endurance training subsequently in more detail.

Table 7.6 Aerobic and Anaerobic Activities
Aerobic (endurance) activities are long lasting activities of such intensity that the body is able to meet the demands for oxygen and fuel. Aerobic exercise causes the heart and lungs to work harder than when they are at rest. Body fat supplies energy during aerobic activities. Examples of aerobic activities include jumping rope, skating, walking, running, cycling, and swimming.
Anaerobic activities are high-intensity, short-term activities where the demand for oxygen exceeds the supply. Glycogen (glucose stored in muscles) supplies energy during anaerobic activities. Examples of anaerobic activities are short-term sports plays (kickball, baseball, and volleyball), weight training, relays, and sprints.

Table 7.7 Physical Changes During Development

1. Muscle activity influences bone growth and alignment.
2. Both pressure and traction epiphyses form the shape of bones in response to forces.
3. Levels of activity largely determine increases in bone mass during childhood.
4. Muscle strength and endurance increase throughout childhood.
5. Increases in strength do not usually affect muscle mass among Prp children.
6. Aerobic capacity of Prp muscle is very similar to adult muscle when measured by VO_{2max} and can improve with training.
7. Anaerobic power increases as children grow and mature and improves with training.
8. Younger children fatigue quicker than older children do.
9. Young children are better able to handle aerobic activity than anaerobic activity.
10. Overweight (over fat, higher BMI) children fatigue more quickly than children with appropriate weight for their height (lower BMI).

This table summarizes physical changes covered in this section.
Each point is important for designing physical education programs.

* The pulmonary system or the size and function of the lungs, has little influence on the changes in endurance during development.

† Slow-twitch (aerobic) fibers seem to be recruited (activated to contract) more than fast-twitch (anaerobic) fibers in children. Fast-twitch fibers activate more quickly and assume a greater role during physical activity as a child matures.

COGNITIVE AND SOCIAL DEVELOPMENT

Most children perform a variety of physical activities. They play, participate in structured exercises, engage in sports, or take recreational classes such as gymnastics, dance, or martial arts. Such activities can be fun and build self-esteem. Yet, they can also provoke anxiety if design of the program or the goals of the parents and teachers do not reflect the child's developmental level.

An understanding of children's cognitive, emotional, and social abilities at different ages helps in designing effective programs and setting realistic goals for participating children.

The Province of British Columbia, Ministry of Education lists social goals among their objectives for quality physical education programs (PBC, 1995):

- Foster the development of positive attitudes
- Encourage active participation
- Reinforce problem-solving skills
- Recognize differences in students' interests, potential, and cultures
- Develop personal and career-planning skills

FIVE TO 7 YEARS: Five-year-old children are easily distracted, forget instructions, lose interest quickly, and may not care to compete or win. This can frustrate performance-oriented adults!

Five-year-old children must think about motor tasks before executing complex movements. Games with simple rules and game strategies are best for this age. The typical 6-year-old is similar to the 5-year-old, except at 6 they tend to be more restless and develop a need for independence.

Decision-making ability changes at about age 7. Seven-year-old children can integrate physical and cognitive skills while playing games. They can pay attention to more than one facet of a game and make instantaneous play decisions. At 7 years, children may begin to manage ball play with rules, goals, and play for offense and defense (Figure 7.6).

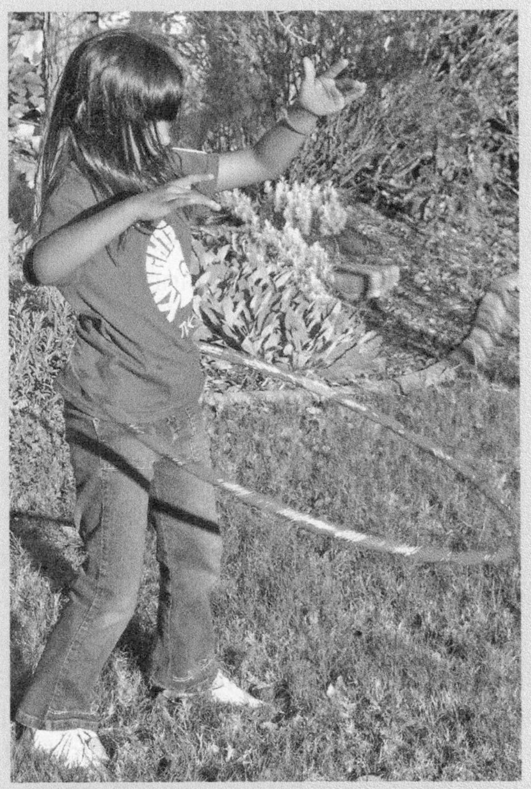

Figure 7.6 (Valerie Rice)
Physically, young children between the ages of 4 and 7 years need regular activity of all their muscle groups for their bones and muscle strength to develop. They can follow simple rules, but are still easily distracted and lose interest quickly. Young children benefit most from noncompetitive games where they are reinforced for participating and by the enjoyment of the activity.

SEVEN TO 11 YEARS: Seven- to 11-year-old children understand concrete but not symbolic ideas. They tend to focus on the present rather than the future. Thus, at these ages, knowledge of the long-term benefits of exercise has little influence over their behavior. Seven- to 11-year-olds can accept responsibility for their actions, are susceptible to "hero worship," and believe in teamwork and the need for rules (Cooper, 1992). A child in this age range should not blame others for his performance. At the same time, the child may place a high value on the opinion of the teacher (hero) and teachers should give performance critiques carefully.

By their 8th year, children may begin to compare themselves with their peers. Pre-teens may prefer the company of their friends to family. Children often associate athletic ability with social popularity. They believe athletic ability is determined by a person's level of effort and the amount they practice (Cooper, 1992).

Pre-teen children may judge their less-competent peers as making less effort or "lazy." The designer of a PE program should carefully consider how to include competitive activities for this age group. Peer judgments can be harsh and children are just learning cooperative team building.

In the 8- to 10-year-old range, athletic or physically competent children begin to move out of relaxed sports activities such as tee-ball. More physically competent 8- to 10-year-old children begin to join competitive activities like little leagues. Parents and children make comparisons between children. Comments from adults and children on athletic performance influence how a child's self-esteem develops (Cooper, 1992).

Physical education should provide an atmosphere of acceptance that encourages children to develop and accomplish their aims without comparison to others. Training can improve fitness (as measured by running speed and agility) and children's social status among their peers (Gross et al., 1985).

The impact of incremental success for each child cannot be overemphasized. Accomplishment will influence their motivation for activities as children, their involvement in activities as an adult, and their sense of self.

EFFECTS OF PUBERTY*: Before puberty, boys and girls have relatively similar body shapes and compositions. Physical maturity has little impact on a child's social life before puberty.

However, as puberty approaches, physical maturity becomes important for self-perception, self-esteem, and interaction with peers (Stout, 1995). Adolescents want to be like their peers. They want to look and act alike. This may cause a problem for those who mature early or late.

The social implications for maturing "early" or "late" is different for boys and girls. Elementary school girls who mature "early" are likely to have negative feelings about their bodies, when they stand out from their peers.

Boys who mature early tend to be taller than their peers, often developing positive feelings about their bodies and a sense of social dominance. In contrast, boys who mature "late" and lag in physical stature may develop negative feelings about their bodies. Such differences magnify as male peer relationships tend to center around athletic activities (Bancroft and Reinisch, 1990).

Early-maturing girls and late-maturing boys may be more reluctant to participate in PE. They will need supportive encouragement and programming that helps them feel like they fit in with their peers. Adult leaders should openly identify children's strengths, as children may not see their own or their peers' strengths.

* Puberty is the stage of physical and sexual maturation, marking the advance from childhood to adolescence. It results from hormonal changes.

DESIGNING A PHYSICAL EDUCATION PROGRAM

Effective PE programs "fit the child-user." This is no small task, given the range of differences among children. A program based on understanding the developmental and physiological changes in children's bodies can help children avoid injuries and develop whole-body skills and a long-term commitment to an active lifestyle. To do so, they must reflect several developmental dimensions. These developmental dimensions or "themes" include core strength, endurance, weight bearing, coordination, complexity, and social impact.

While many PE programs address more than one theme, the most important issue is to balance all themes. The matrix table "Managing Child Variables with PE Themes" may help plan or evaluate PE programs (see Table 7.10).

THEMES

CORE STRENGTH: Core strength refers to trunk muscle strength. The trunk muscles include abdominal and spinal muscles. Muscles that control the hips and shoulders are also important for core strength. Core muscles are critical for coordinated movement. Strong core muscles provide a stable base for moving arms and legs.

> *Core strength is critical for coordinated movement.*

Child development occurs proximal to distal. That is, an infant develops control of the trunk muscles before they control their legs and arms. Control of the shoulder precedes hand and finger control.

Coordinated and fluid movements are impossible when core muscles are weak. Unfortunately, many sports instructors forget this principle and focus on developing movement patterns required in sport before they develop children's core strength. While this approach may develop a strong "splinter skill," it fails to develop a stable base for movement.

For example, according to one text:

> *The trunk goes through three stages of development when a child is learning how to throw a baseball correctly. In the first stage, the trunk is completely passive and all motion occurs at the arm. The second stage occurs when the trunk stabilizers increase in activity to support the arm during the throw. In the final stage, the trunk actively provides force to project the ball. However, when the demands of a task (i.e., velocity, accuracy, and distance) change or when fatigue sets in, one body segment may regress while others remain at the advanced level.*
>
> (Campbell, 1995)

This description of developmental stages of throwing a ball makes it sound like trunk stability will occur naturally, developmentally as the child learns to throw. The implied focus is a concentration on teaching the child to throw.

However, conversion from the first to the second stage (above) will occur only when the core muscles are sufficiently strong for the task. This means that specific exercises and activities to strengthen the core must be part of the training. Strong core muscles are necessary to develop distal skills such as throwing and batting balls (see Table 7.8).

Table 7.8 Characteristics of Children with Low Core Strength
Children with low core strength: 1. Tire easily 2. Have poor posture; appear to 'slump' and bend forward at waist and shoulders 3. Their head rests forward of their shoulders 4. May often rest their head on their desks 5. May rest their head on their desk or arm while writing 6. Try to support their feet above floor level, making their knees higher than their hips

Figure 7.7 (Valerie Rice)
This child may lack the core strength needed to swing the bat with force and accuracy. Her trunk bends back (hyperextends) and hips flex as she prepares to swing.

Visualize her arms moving forward to swing the bat while her back flexes, she bends forward, and her hips extend.

All of this trunk movement indicates low core muscle strength. In the child with weak core musculature, trunk movements are extreme and poorly controlled.

A basic principle of physics is also useful: For every action, there is an equal and opposite reaction. Children with strong core muscles are able to contract and relax muscles in rhythm with movements of their arms. During movement such as throwing, the trunk is stable as trunk muscles tighten and react to forward forces from the arms. Thus, the equal and opposite reaction occurs while maintaining an erect posture, fluidly twisting their body and matching their arm movements while throwing.

Figure 7.8 (Barbara H. Boucher)
Core muscles are inactive when sitting unsupported on the floor. Inactive core muscles allow the body to relax into gravity, resulting in a forward curved spinal posture. Other consequences are a hyperextended neck and compression of the chest and abdominal organs.

Core strength is important in light of the fact that bone growth responds to both traction and pressure forces acting on them. The lower or lumbar vertebrae respond to their alignment as the weight of the head and upper trunk rests on the spine (pressure force).

Children sit for long periods, in class or at play, hunched forward with their postures constrained, back bent forward (flexed) and breathing restricted. If the back is slumped and the head leans forward, there is less pull at the back of the spinal column and greater pull at the front (anterior) of the spinal vertebrae. Thus, the vertebrae can become misaligned. Slumped postures also misalign internal organs and restrict lung expansion. Children require core strength to encourage anterior pelvic tilt. Designing furniture to support the back helps but does not eliminate risks associated with low core strength.

Supporting the pelvis at or slightly above the lower legs encourages the pelvis to tilt forward (see Figure 7.7 and Figure 7.8). This anterior pelvic tilt enables the sacrum to support the spine more effectively and activates abdominal and spinal muscles. The development of an anterior pelvic tilt also promotes the natural development of the lumbar curve. Children need to sit with an anterior pelvic tilt to promote the natural development of a lumbar curve.

The abdominal muscles attach to the lower rib cage. When the abdominal muscles contract, the rib cage is pulled down (traction force), allowing the lungs to expand. The spinal muscles attach to the vertebrae along the spine; when the spinal muscles contract, they pull the trunk upright and align the head vertically. Sitting unsupported on the floor has the opposite effect and spending prolonged periods in this position can be harmful to children (Figure 7.8). On the other hand, sitting on a ball during instruction can help strengthen core muscles (Figure 7.9).

Figure 7.9 (Barbara H. Boucher) Sitting at or slightly above the lower legs encourages the pelvis to tilt forward (anterior tilt) and activates core muscles. This position promotes natural and good spinal posture, including in the neck area. Additionally, the expanded chest and abdomen cavities correctly allow optimum function of the internal organs.

ENDURANCE: Well-designed PE programs for children gradually and intentionally increase muscle endurance requirements. Endurance activities are the same as aerobic activities. Endurance or sustained muscle activity indicates the body's ability to move oxygen to muscles. The heart, lungs, and skeletal muscles all benefit from endurance activities.

Higher endurance = Less fatigue

Children with greater endurance can run, bike, dance, ride horses, and play active sports longer than children with less endurance do. Aerobic exercise strengthens the cardiopulmonary (heart and lungs) system, improves oxygen delivery, muscle use of oxygen, and the body's ability to remove waste products.

Aerobic exercise also burns fat and calories and reduces the risk of heart disease, high blood pressure, and diabetes. Aerobic exercises include running, power walking, ice and roller-skating, dancing, bicycle riding, jumping on trampolines, jumping rope, swimming, choreographed calisthenics, and soccer.

Well-planned PE programs incorporate endurance activities into each school day.* With good planning, children can participate in graded† activities that increase endurance appropriate for their age and general condition.

Measure time in activity vs. intensity of activity when planning endurance activities.

* Children who are active for 60 minutes, five days a week for 6 weeks earn the U.S. Presidential Active Lifestyle Award! (See *Get Fit: A Handbook for Youth Ages 6–15*. http://fitness.gov/getfit.pdf).

† "Graded" activities incorporate sequential gradations. For example, the first graded physical activity may be easy and become slightly more challenging as a child succeeds at each level. Graded cognitive tasks may begin as simple activities that require little instruction and sequentially increase in degree of difficulty. The challenge is to determine the correct increment of difficulty between one task and the next that will help the child achieve success.

Teachers often base PE schedules on factors other than optimizing aerobic activities. They need to be creative in order to provide daily aerobic activity. For example, children might take longer routes from classrooms to the cafeteria while skipping or jumping during parts of the route.

Incorporating music into PE is fun, especially for younger children. It also provides variety and can promote endurance. Simple movements designed to increase core strength become "dance." Having children chant or sing during exercise increases the aerobic demand of the activity.

Talking or singing during exercise can indicate how hard the child is working. Children should be able to sing or talk throughout low-level endurance exercise. Both the instructor and the child can use this as a gauge and reduce the level of activity if the children cannot talk or sing while exercising.

For part of the curriculum, PE should focus on walking and running to challenge each group of children based on their development. The children's age and general condition can gauge the gradual increases of sustained muscle work.

The time spent actively moving, resting, and waiting is important. While games such as baseball and kickball can build perceptual motor skills and promote teamwork, these games require substantial standing and waiting during the game and fail to serve as aerobic/endurance activities. This is also true for relay races and bowling where there are long times in between turns. PE programs fail to develop endurance if children spend more time waiting between turns than participating.

WEIGHT BEARING: PE programs must include weight-bearing activities to promote bone growth and create bone mass. Preventing the loss of bone mass (osteoporosis) in old age begins in childhood.

Weight-bearing builds bone.

Generally, weight-bearing activities refer to movements done in an upright and standing position. However, if one body part is weight-loaded over another body part, the supporting body part is "weight-bearing" and the skeleton receives bone-forming pressure.

The arms are weight bearing in crawling, but the most efficient position to benefit from weight bearing is upright—sitting or standing. Gravity pulls the weight of the body when sitting and standing. Multiple muscle groups work to keep the body upright and to move the body in the upright position. Jump rope games, trampolines and stairs, step exercises, and walking over foam surfaces build bone mass and develop balance skills. Graded gymnastic activities are an excellent means of incorporating weight bearing into PE programs. Teachers can simultaneously incorporate core strength and endurance training.

Coordination is described through three developmental sequences.

COORDINATION: Several developmental sequences govern the maturation of coordinated movement. These sequences involve gross motor movements, movements through the planes of the body, and using each side of the body.

Table 7.9	Developmental Sequence: Flexion/Extension ⇒ Abduction/Adduction ⇒ Rotation
Flexion	Bringing two connected body parts closer together (e.g., bringing the hand to the shoulder flexes the elbow)
Extension	Straightening a joint (e.g., straightening the leg is extending the knee)
Abduction	Moving body parts away from the center of the body or body part (e.g., moving the arm up and away from the side of the trunk or spreading the fingers)
Adduction	Moving body parts toward the center of the body or body part (e.g., moving the arm down and toward the side of the trunk or moving the fingers together)
Rotation	Moving a body part around a central axis (e.g., rotating the forearm by turning the palm up (*supination*) or down (*pronation*))

Developmental sequence: Gross motor skills precede fine motor skills.

Gross motor (large muscle) skills develop before fine motor skills (precise, small muscle movements). Children can reach with their arms before they can grasp objects with a fine pinch grip; they can walk before they can "dribble" a soccer ball.

Physical Education programs for young children should begin with activities that exercise the large muscles and gradually increase coordination requirements. Since children develop at slightly different rates from one another, instructors must assess a child's readiness to progress.

You can plan coordination of movements with another developmental sequence: flexion/extension develops first, then abduction/adduction and finally, rotation (Table 7.9).

Younger children are able to rotate their trunk and extremities through isolated movements, but they cannot complete a smooth sequence of rotation movements. For example, they find it difficult to perform a smooth windup and pitch of a baseball.

Physical education programs for younger children, 5- and 6-year-olds, should include movements that strengthen flexion and extension of their trunks and extremities, gradually combining flexion and extension of more than one body part. Emphasis on abduction/adduction or lateral movements can begin as children approach age 7 years. Practice in combining rotation with other movements can also begin at about age 7.

PE programs should promote skill coordination by exercising each side of the body alone, simultaneously and alternatively at different times. This is particularly valuable for children at very young ages, which tend to use one side of the body at a time. They need practice to begin to use both sides of the body together in a coordinated manner.

Developmental sequence: Unilateral activities precede bilateral activities.

Well-planned calisthenics are an example of coordination-specific exercises. A simple sequence of movements, organized within an obstacle course combines coordination skills with specific flexion/extension and abduction/adduction movements.

Near the age of 7 years, most children have developed all the movement components and have the cognitive potential to begin practicing increasingly complex coordination movements (Figure 7.10). Components of martial arts, Tai Chi, or dance offer appealing and beautiful movement exercise.

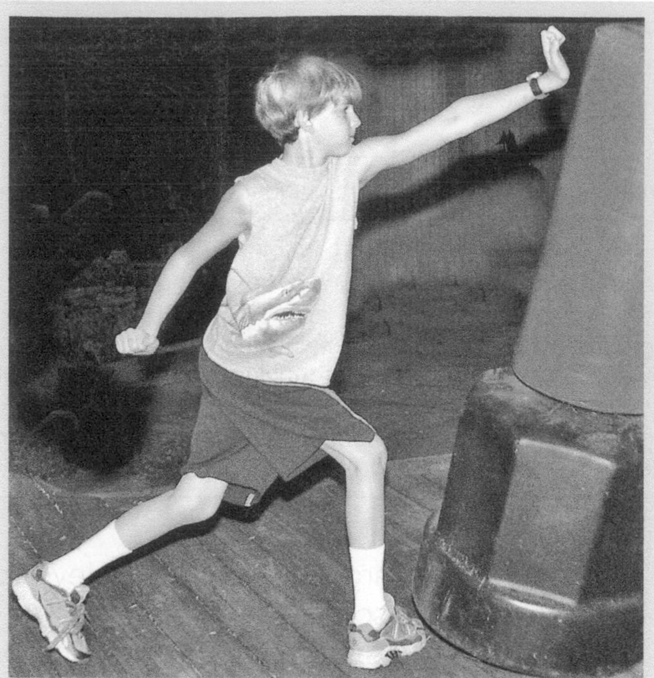

Figure 7.10 (Ursy Potter Photography)
There is movement or muscle activation at each joint in Shea's limbs while his trunk (using core muscles) must provide a stable base for limb movement. Rotating each limb and the trunk into smooth movement requires maturation of coordination.

COMPLEXITY: Activities that involve multiple ideas or movements at one time are more complex. The more rules, movements, and response options a child has to manage during the activity, the more complex the activity. Activities that require a quick decision between several responses—including time or speed of performance—are especially complex.

Younger schoolchildren perform best with simple, repetitive movement sequences. Obstacle courses are a good means for providing a variety of movements and keeping a large group active for specific times. Individual ball play allows each child to work individually on gross skills.

Children can progress from playing with a ball individually, to ball play with a partner, and then to play within a group of four. This introduces a gradual progression to more complex physical and mental skills. Participation in team sports such as basketball and soccer requires a large set of skills and an understanding of play options and complex sequences of movements.

Young children are not as ready for complex team sports as older children or adults. Five and 6-year-old children have difficulty remembering rules, paying attention, and performing the complex movements team sports require. Adults who watch children play organized sports often comment that they are inattentive. For example, young children may wander off the playing field or toss their glove in the air and catch it while playing baseball.

SOCIAL IMPACT: Well-designed PE programs engage children cognitively and promote positive social behaviors. Essential social behaviors include self-control, ability to self assess, and respect for differences among peers. The ability to think through behavior and make choices (self-control) improves with maturity and through multiple influences.

Physical education is not the sole or primary influence on developing a child's self-control. Even so, well-designed programs set developmentally appropriate expectations for self-control. Adults should not expect younger children to wait for long periods between activities. They need more support on behavioral control. They benefit from small group interactions and low adult-to-child ratios. The adult or teacher provides guidance, gives expectations, and offers feedback to the children on their behavior as acceptable or not acceptable.

Seven-year-old children want immediate rewards for their efforts. Some children will always compare unfavorably if rewards (praise or peer appraisal) are earned by scores. Educators and parents must evaluate positive and negative effects on individual children. If a game is too complex for a child, they are set up for failure. Despite best efforts, a child may be unable to improve his or her performance, to succeed or to be a valued player. A child's developmental stage may limit what they can achieve if inappropriate standards are set.

Assess children's strengths and limitations and teach them to assess themselves. Self-assessment is a critical skill for promoting life-long exercise and health. Children who learn basic competencies and/or how to modify skill levels are better able to make decisions about physical activity throughout their lives.

Skill assessments early and at the end of the school year demonstrate how children improve over time. A first-grader may learn that he cannot run faster than others can but is skilled at jumping. At the end of the school year, he will be happy to see improvements in both areas. Skill assessment early and at the end of the school year also helps to "red flag" children who need special attention. More support, targeted activities, or possibly adapted PE can take place during the school year or over the summer.

Respecting differences among peers is one of the first steps to learn cooperation, teamwork, and using each person's strengths to achieve a goal. Although most children in an age group have similar physical competencies, children themselves usually realize their differences. A few will be less competent and a few will be especially competent. If the PE program rewards only the more competent children, other children feel incompetent or strive for a level of performance that may exceed their ability. PE has the best social impact when children realize their own potential through gains in physical activity.

The PE teacher is the children's role model. He or she must show respect for children and for their abilities. Rewards or praise from the PE teacher should recognize effort, tenacity, and the gains of each individual child.

SAFETY

No protective equipment or turf can substitute for the presence of an adult during children's PE. Younger children need to work in smaller groups, with a smaller child to adult ratio than older children do. Personal equipment should fit, provide protection, and absorb impact, while other equipment, such as baseball bases should breakaway upon impact (see Chapter 7 and Chapter 12 on injuries and consumer products for additional information).

PROGRAM EVALUATION

Each activity or lesson plan in a PE curriculum should address one or more of the above themes. Table 7.10 can help educators plan their PE programs. Educators can fill in the boxes with the activities that meet the requirement or use it as a "check list." If an activity in the kindergarten curriculum develops core strength, then the teacher lists the activity in the box next to Core Strength and below Five Years Old. For example, rolling the length of an exercise mat with arms at the sides is an activity that might fit in this block.

After a curriculum review, professional PE teachers may find some activities that do not fit into any of the blocks. If the activity does not meet a theme or is inappropriate for the

Table 7.10 Managing Child Variables with Physical Education Themes

Themes	Five Years Old	Six Years Old	Seven Years Old	Eight to 10 Years Old
Core Strength				
Endurance				
Weight-bearing Activity				
Coordination				
Complexity				
Social Impact				

PE teachers may evaluate their curriculum by fitting each lesson plan or activity into the appropriate blocks in this table.

age groups in the table, eliminate the activity from the program. Using this table can reveal missing themes for particular age groups and direct the PE teacher to include activities to fill in the blocks.

SUMMARY

Figure 7.11 (Valerie Rice)

User-centered PE programs focus on children in the program. This type of design incorporates principles of child development, counteracts or builds on the environmental influences, and takes the entire child (physical as well as psychosocial, now and in the future) into account.

The concept of designing PE in accordance with the principles of child development is not new. However, PE programs have favored other goals (emulation of sports) in recent years. A good PE program intentionally incorporates developmentally appropriate activities that enable children to achieve health-related goals at school, home, and during play. At the same time, teachers of good programs consider the changing abilities of children.

Viewing a child's school-based activities and postures to guide PE may be a new concept for some. Good PE programs counteract the negative impact of awkward or constrained postures common among children in elementary schools. They also target the strengths needed for a child's daily activities, such as carrying their backpacks or books.

Physical education and physical activity are essential components of a child's learning environment. They promote physical and mental health, improve concentration, and foster social growth. Physical education sets the stage for adult health and wellness.

REFERENCES

Armstrong, N. (1998). Young people's physical activity patterns as assessed by heart rate monitoring. *Journal of Sports Sciences*, 16, S9–S16.

Bancroft, J. and Reinisch, J.M. (Eds.) (1990). *Adolescence and Puberty*. New York, NY: Oxford University Press.

Bernhardt-Bainbridge, D. (1995). Sports injury. In: S.K. Cambell (Ed.), *Physical Therapy for Children* (pp. 383–422). Philadelphia, PA: WB Saunders.

Burgeson, C.R., Wechsler, H., Brener, N.D., Young, J.C., and Spain, C.G. (2001). Physical education and activity: Results from the school health policies and programs study 2000. *Journal of School Health*, 71(7), 279–293.

Campbell, S.K. (1995). The child's development of functional movement. In: S.K. Campbell (Ed.), *Physical Therapy for Children* (pp. 4–37). Philadelphia, PA: WB Saunders.

CDC (2000). Centers for Disease Control (CDC) Growth Charts, National Center for Health Statistics. Available at: www.cdc.gov/growthcharts

CDC (2000). School Health Policies and Program Study 2000. Fact Sheet: Physical Education and Activity. US Centers for Disease Control and Prevention. Available at: www.cdc.gov/nccdphp/dash/shpps/factsheets/fs00_pe.htm

CDC (2002). Physical Activity and Health: Report to the Surgeon General. US Centers for Disease Control and Prevention. Available at: www.cdd.gov/nccdphp/sgr/adoles.htm

CERN (2004). China Education & Research Network. Available at: www.edu.cn/HomePage/english/education/paedu/physical/index.shtml

Clippinger-Robertson, K. (1987). The child's response to aerobic exercise: Physiologic differences and vulnerability to injury. *Top Acute Care Trauma Rehabilitation*, 2, 83–92.

Cooper, K.H. (1992). *Kid Fitness: A Complete Shape-up Program from Birth through High School*. New York, NY: Bantam Books.

CPEC (July 2001). Council on Physical Education for Children and the National Association for Sport and Physical Education position paper, *Recess in Elementary Schools*. Available at: http://naecs.crc.uiuc.edu/position/recessplay.html

CSBE (1999). California State Board of Education Policy Physical Education Requirements. Policy # 99-03, June.

Duke, J., Huhman, M., and Heitzler, D. (2003). Physical activity levels among children aged 9–13 years—United States, 2002. *Morbidity and Mortality Weekly Report*, 52(33), 785–788.

Ekeland, E., Heian, F., Hagen, K.B., Abbott, J., and Nordheim, L. (2004). Exercise to improve self-esteem in children and young people. *Cochrane Database System Review*, 1, CD003683. Available at: www.update-software.com/cochrane/abstract.htm

England, S.P. and Sundberg, S. (1996). Management of common pediatric fractures. *Pediatric Clinics of North America*, 43(5), 991–1011.

Gross, A.M., Johnson, T.C., and Wojnilower, D.A. (1985). The relationship between sports fitness training and social status in children. *Behavioral Engineering*, 9, 58–65.

Heaney, R.P., Abrams, S., Dawson-Hughes, B., Looker, A., Marcus, R., Matkovic, V., and Weaver, C. (2000). Peak bone mass. *Osteoporosis International*, 11, 985–1009.

Larson, R.L., Singer, K.M., Bergstrom, R., and Thomas, S. (1976). Little League survey: The Eugene study. *American Journal of Sports Medicine*, 4(5), 201–209.

Lin, J-P., Brown, J.K., and Walsh, E.G. (1994). Physiological maturation of muscles in childhood. *Lancet*, 343, 1386–1389.

MacKelvie, K.J., Khan, K.M., Petit, M.A., Janssen, P.A., and McKay, H.A. (2003). A school-based exercise intervention elicits substantial bone health benefits: A 2-year randomized controlled trial in girls. *Pediatrics*, 112, e447–e452.

PBC (1995). Province of British Columbia, Ministry of Education, Curriculum Branch. Available at: www.bced.gov.bc.ca/irp/pek7/pre2k-7.htm

Rarick, G.L. (1973). *Physical Activity: Human Growth and Development*. New York, NY: Academic Press.

Roemmich, J.N. and Rogol, A.D. (1995). Physiology of growth and development: Its relationship to performance in the young athlete. *Clinics in Sports Medicine*, 14(3), 483–502.

Rowland, T.W. (2000). Exercise science and the child athlete. In: W.E. Garrett and D.T. Kirkendall (Eds.), *Exercise and Sports Science* (pp. 339–349). Philadelphia, PA: Lippincott Williams and Wilkins.

Schonau, E., Werhahn, E., Schiedermaier, U., Mokow, E., Schiessl, H., Scheidnauer, K., and Michalk, D. (1996). Influence of muscle strength on bone strength during childhood and adolescence. *Hormone Research*, 45 (suppl 1), 63–66.

Stout, J.L. (1995). Physical fitness during childhood. In: S.K. Campbell (Ed.), *Physical Therapy for Children* (pp. 127–156). Philadelphia, PA: WB Saunders.

Texas Administrative Code (1999), Title 19, Part II, Chapter 116. *Texas Essential Knowledge and Skills for Physical Education*. Available at: www.tea.state.tx.us/rules/tac/chapter116/ch116a.pdf

Tomporowski, P.D. (2003). Effects of acute bouts of exercise on cognition. *Acta Psychologica* (Amst), 112(3), 297–324.

WES (2004). World Education Sevices-Canada. Available at: www.wes.org/ewenr

WHO (2004). World Health Organization. Physical Activity. Available at: www.who.int/hpr/physactiv/country.strategy.shtml

WHO (2004). World Health Organization. Physical Activity for Children. Available at: www.who.int/hpr/physactiv/children.youth.shtml

SECTION C

Injuries, Health Disorders, and Disabilities

CHAPTER 8

CHILDREN AND INJURIES

VALERIE RICE AND RANI LUEDER

TABLE OF CONTENTS

OVERVIEW

A child's "work" is to develop physically, emotionally, socially, and cognitively. In order to develop, a child must continually face new challenges, hopefully in small enough steps so he or she can master the task and move on to new trials.

Simply presenting a challenge will not make a child want to participate. Most children (and most adults) learn more and learn more quickly when learning situations are "fun."

Designing tasks and equipment for children can be complex because the designs must achieve several goals simultaneously. They must:

- be self-reinforcing ("fun")
- build on previous skills
- challenge children to develop new skills
- be usable by children of various ages and abilities
- be "safe"

Figure 8.1 (Ursy Potter Photography)
Children need challenges to develop their physical and mental skills. Each challenge should occur incrementally, that is, it should build on previously mastered skills and not be so difficult that it overwhelms the child.

While the focus of this chapter is safety, eliminating all risks would also eliminate challenges that are essential to children's growth and development (Figures 8.1 through 8.5). It is therefore prudent to carefully evaluate products, places, and tasks and to educate children and their caregivers about safety issues.

Child safety practices help prevent injuries. They are also important because injury and illness can disrupt a child's learning and development; a child with a broken leg cannot run or climb, thereby delaying development of their gross motor skills and sense of balance.

A child who cannot fully participate in activities may also experience a setback in social skills, as other children and adults often treat them differently. They may be excluded from activities they can handle or may be seen as fragile, uncoordinated, or even less competent.

Injuries occur for many reasons. A poorly designed product or task design can be at "fault," as can a caretaker, or a child. Often, injuries occur due to a combination of factors.

The National Center for Injury Prevention and Control (CDC, 2005a) lists the 10 leading causes of ***unintentional injuries***, as well as ***unintentional injuries that result in death*** (Table 8.1). Guided by these findings, this chapter will cover falls, burns, poisoning, choking or suffocation, and drowning.

Figure 8.2 (Ursy Potter Photography)
Eliminating all risks would also eliminate challenges that are essential to children's growth and development.

Table 8.1 Ten Leading Causes of Unintentional Injuries (2005) and Unintentional Injury Deaths (2004), United States, All Races, Both Genders[a]

	Age Groups				
Rank	**Under 1 Year**	**1–4**	**5–9**	**10–14**	**15–24**
1	*Falls* **Suffocation**	*Falls* **Motor Vehicle Accident, traffic**	*Falls* **Motor Vehicle Accident, traffic**	*Falls* **Motor Vehicle Accident, traffic**	*Struck by/Against* **Motor Vehicle Accident, traffic**
2	*Struck by/Against* **Motor Vehicle Accident, traffic**	*Struck by/Against* **Drowning**	*Struck by/Against* **Fire/Burn**	*Struck by/Against* **Drowning**	*Fall* **Poison**
3	*Bite/Sting* **Drowning**	*Bite/Sting* **Fire/Burn**	*Cut/Pierce* **Drowning**	*Overexertion* **Fire/Burn**	*Motor Vehicle Accident, Occupant* **Drowning**
4	*Fire/Burn* **Fire/Burn**	*Foreign Body* **Suffocation**	*Pedal/Cyclist* **Suffocation**	*Cut/Pierce* **Other Land Transport**	*Overexertion* **Other Land Transport**
5	*Foreign Body* **Fall**	*Cut/Pierce* **Pedestrian, other**	*Bite/Sting* **Other Land Transport**	*Pedal Cyclist* **Suffocation**	*Cut/Pierce* **Fall**
6	*Other Specified* **Unspecified**	*Overexertion* **Fall**	*Overexertion* **Pedestrian, other**	*Unknown/Unspecified* **Poison**	*Bite/Sting* **Firearm**
7	*Cut/Pierce* **Natural/Environment**	*Fire/Burn* **Natural/Environment**	*Motor Vehicle Accident* **Struck by/Against**	*Motor Vehicle Accident, Occupant* **Firearm**	*Unknown/Unspecified* **Suffocation**
8	*Suffocation* **Poison**	*Other Specified* **Unspecified**	*Foreign Body*[b]	*Bite/Sting* **Fall**	*Unknown/Unspecified* **Fire/Burn**
9	*Overexertion* **Other Specified**	*Unknown/Unspecified* **Other Specified**	*Dog Bite*[b]	*Other Transport* **Pedestrian, other**	*Other Transport* **Unspecified**
10	*Motor Vehicle Accident, Occupant* **Struck by/Against**	*Poison* **Poison**	*Unknown/Unspecified*[b]	*Dog Bite* **Other Transport**	*Poison* **Pedestrian, other**

Italic, Unintentional injuries; **Bold**, Unintentional injuries resulting in death.

[a] Modified from CDC, National Center for Injury Prevention and Control. Original tables available from http://cdc.gov

[b] Deaths for ages 5–9; #s 8, 9, and 10 are tied: Firearm, Natural/Environment, Other Transport, and Unspecified.

Injuries may occur when a child takes on challenges that are beyond his or her capabilities. Occasionally children will do this on their own, as they do not realize their limitations. At other times, they may try to copy the behaviors of adults or older children, as this little girl tries a task that is easy for her older sister, but a bit beyond her own abilities.

Figure 8.3 (Valerie Rice)

Figure 8.4 (Valerie Rice)

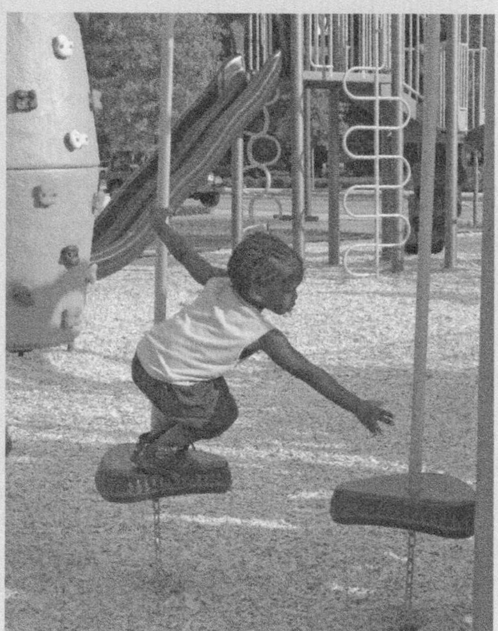

Figure 8.5 (Valerie Rice)

ERGONOMICS AND SAFETY

Ergonomics is the design of places, things, and materials to fit people. In order to decrease injuries among children, one must design the environment, place appropriate items in the environment, and provide education and educational experiences that adults and children can understand, remember, and use in times of emergency.

> Many publicly available documents cited in this chapter are also available for download at http://www.childergo.com/recc-children.htm

WHAT DO I DO FIRST?

There is a priority for efforts directed at preventing injuries in children:

1. Design the area, product, or task to remove or reduce the potential for injury.
2. Properly supervise children by having a responsible adult present. The adult must recognize the appropriate safety issues.
3. Educate and warn the child.

This sequence is especially important in injury prevention for children because children will not be aware of the dangers and consequences of their actions. Even children who are old enough to understand consequences often have implicit trust in their caregivers and believe that anything within their reach is safe. They believe that their caregivers' carefully selected the items in their environment.

Designing to eliminate or reduce the probability of injury removes the possibility of human error. The less dependent on the individual caretaker, the better, as there can be many levels of knowledge among caretakers. Yet, caretakers are infinitely better than depending on the decision making and knowledge of a child (Figure 8.6).

Figure 8.6 (Ursy Potter Photography)
Children act first and think later. They may not associate a warning about a place, item, or condition with another.

As children grow older, they can assume more responsibility. In fact, because they become very aware of rules and set ways to do things, children can become excellent sources of information. If they know and understand "safety rules," they eagerly share that knowledge and find great satisfaction in reminding each other and reminding adults of those rules! In this way, they can be of great assistance in creating a safe environment.

FALLS

INTRODUCTION

Children love to run, jump, and climb. They need to be physically active in these ways. The activity helps develop their bodies and minds, and the coordination between the two.

Figure 8.7 (Ursy Potter Photography) Children suffer contusions and bruises mainly to the head and face, fractures (mainly to the upper extremities), and concussions. The most serious consequences are head injuries. Usually, the gravity of the injury depends on how far the child falls and the type of surface he lands on.

BOX 8.1

1. Falls can be:
   a. from the same level, due to slips or trips
   b. from elevated heights such as a roof, ladder, stairs, or even from furniture
2. Slips generally occur when a child falls on a slippery surface.
3. Trips occur when a child's foot strikes an object and he looses his balance. Uneven surfaces or items left on the floor are common culprits.

As children push the boundaries of their physical limitations, they are going to fall. Most of the time, the falls are minor, perhaps resulting in a scratch or bruise (Figures 8.7 and 8.8). However, the consequences of falls can be serious, even when they are not from great heights. They may even be on level ground.

Most people think the elderly are at far greater risk of fall-related injuries. This is not true. Children under the age of 12 are nearly as likely to sustain a fall injury as seniors aged 64 and older. In the United States, nearly three million children visit emergency rooms annually for fall related injuries, with most of these injuries occurring among children under four years of age (NSC Injury Facts; Safe Kids).

While rates of injuries and hospitalizations vary among countries, falls remain the most common cause of injuries in children of all ages in much of Europe* and the United States (AIHW, 2002; CDC, 2005a; Nordin, 2004). Injuries are most common among children aged 5 and under.

* These European countries are Austria, Denmark, Greece, the Netherlands, Sweden, and Australia.

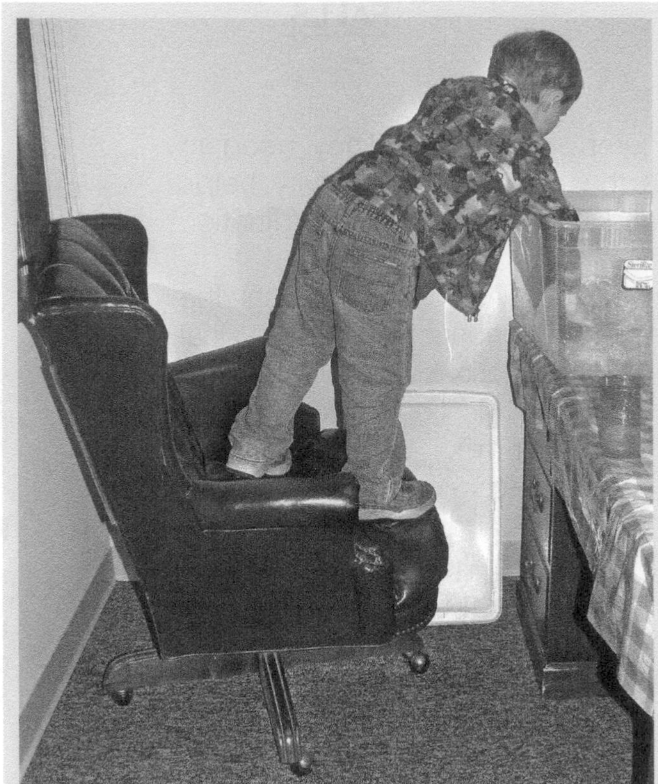

Figure 8.8 (Ursy Potter Photography)
Children can identify ingenious methods to achieve their goals, but they may not have the knowledge to understand the potential for mishaps or accidents.

Most falls take place at home during children's regular playtime, and between noon and early evening. In the home, children fall from furniture, such as high chairs, beds, and adult-size chairs and couches. They also fall from windows. Deaths can occur from hangings on ropes or cords, such as those on window blinds or curtains.

Outside and at home, children fall from family play equipment, often injuring themselves when they intentionally jump from swings or other equipment. Outside and away from home, children fall from playground equipment and from bleachers. In both places, at home and at school, children fall from stairs. While many children fall from a height, slips and trips are also considered "falls" (Box 8.1 and Tables 8.2 and 8.3).

Table 8.2 Where Do Falls Happen?

Group	Figure	Where falls happen	Group	Figure	Where falls happen
Babies	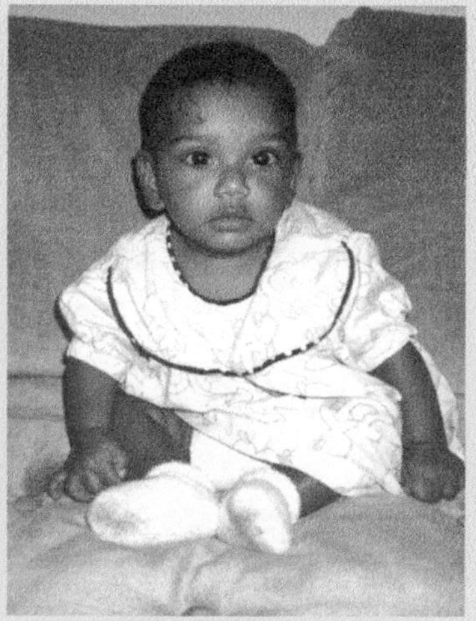 **Figure 8.9** (Valerie Rice)	• Roll off furniture	**Younger Children**	 **Figure 8.10** (Valerie Rice)	Fall during sports and recreation activities and while climbing or running on: • Playground equipment • Furniture and counters • Stairs • Ladders • Fences • Sports and recreation • Items that children can climb, with foot and handholds
Toddlers	 **Figure 8.11** (Valerie Rice)	Fall while climbing: • Nursery furniture and other furniture • Stairs[a] • Balconies • Windows • Playground equipment • Items not designed for climbing, but on which they can gain a hand and foothold	**Older Children**	 **Figure 8.12** (Ursy Potter Photography)	Slip, trip, or fall: • During sports or recreation • On playground equipment

[a] Please see Chapter 15 for more detailed information about children and stairs. Also see Figures 8.13 and 8.14.

CHILDREN AT RISK OF FALLING

Table 8.3 All Children Are at Greater Risk of Falling than the General Population, but Some Are at Particular Risk

1. Preschoolers	• Very young children are at greatest risk
2. Boys	• Boys fall more often than girls • Boys sustain more severe injuries
3. Children with balance disorders and who are less mobile	• Reduced coordination make it more difficult to recover balance
4. Wheelchair users	• Children who use wheelchairs are likely to sustain fall-related injuries
5. Low-income children	• Possibly due to: Less supervision Environmental problems such as aging or deteriorating housing
6. African-American and Hispanic children	• More likely to fall from heights • Possibly due to: Less supervision Greater exposure to urban, multiple-story, low-income housing

Figure 8.13 (Ursy Potter Photography)

Figure 8.14 (Ursy Potter Photography)
Stairs can present particular problems for falling (see Chapter 15 for more information about stairs).

PRODUCTS

Many falls are associated with particular products. Table 8.4 lists products most closely associated with falls.

Table 8.4 Products Associated with Children's Falls

	Background	Risk Factors	Prevention Strategies
Windows[a]	• Windows are a source of entertainment for children. • Children climb, stand on their toes, and ask a caregiver to lift them to peek through a window. • In the United States, each year nearly 5,000 children (mostly toddlers) sustain injuries falling from windows. Most of these are head injuries. • Most falls occur in the spring and summer.	Living in • apartments • urban, low-income neighborhoods • overcrowded, deteriorated housing Unsupervised boys under age 3 are at greater risk.	• Do not rely on screens to protect children (not strong enough). • Move items children can stack and climb away from windows. • Never let children play around windows or patio doors. • Ensure children cannot reach open windows. • Install window stops that block openings >10 cm (4 in.). • Set rules with children about playing near windows. • Close windows or install child-safety guards.
Playgrounds[b]	• Most playground injuries involve children falling from playground equipment. • Almost half of playground-related injuries are serious.[c]	• Most injuries occur on public play-grounds (especially schools and daycares). • Most deaths occur on home play-grounds.[d]	• Recommended height varies from 1.5–2.5 m[e] (4.9–8.2 ft). • One way to reduce height and injuries is to replace jungle gyms with rope climbing frames.[f] • Effective injury prevention steps include reducing fall height, providing absorptive ground surfaces, and maintaining grounds and equipment.
Stairs[g]	• Stair falls can be minor (sprains, strains, and bruises) or serious. • Among the most serious are head and neck injuries.	• Younger children: under 4 years of age are at greatest risk. • Children up to the age of 9 are also at increased risk.	• Install safety gates at the top and bottom of stairs, with a secure latch. • Mount safety gates so children cannot push them open. • Keep stairs free of clutter. • Supervise children and do not let them play on stairs. • If you allow a child to use a walker, never allow its use near stairs. • Design stairs, handrails, and side rails to fit children.

(continued)

Table 8.4 Products Associated with Children's Falls (Continued)

	Background	Risk Factors	Prevention Strategies
Walkers	• Baby walkers make children mobile before they are developmentally ready. • Walkers with wheels make a child more mobile, increasing access to hazards. • In emergencies, caregivers may not be able to respond quickly enough to a child's greater mobility and speed (see Figure 8.15).	• Tipping over • Access to stairs, porches, or floor-level changes • Stoves	• Use stationary bouncers and playpens in place of mobile baby-walkers. • When using a mobile walker, ensure its design is too wide to fit through a standard doorway and stops the walker at the edge of steps (ASTM).

[a] Spiegel and Lindaman (1977).
[b] See Chapter 28 for a detailed review of playgrounds and playground injuries. Also Mowat et al. (1998).
[c] These include fractures, internal injuries, concussions, dislocations, strangulation, and amputations.
[d] Phelan et al. (2001), Tinsworth and McDonald (2001).
[e] See Chalmers et al. (1996), Macarthur et al. (2000), and Laforest et al. (2001).
[f] See Mott et al. (1997) and Siebert et al. (2000).
[g] See Chapter 15 for a detailed review of stairs and stair design for children.

- In 2002, nearly 4600 children aged 4 and under were treated in hospital emergency rooms for baby walker–related injuries.*
- Two children each year die from injuries associated with walkers.
- Baby walkers cause more injuries than any other nursery product.
- About 80% of the time, the caregiver was actively supervising the child at the time of the injury.
- Neither the American Academy of Pediatrics (AAP) or the National Association for Children's Hospitals and Related Institutions (NACHRI) recommend walkers.

National SAFE KIDS Campaign

Figure 8.15 (Valerie Rice)

* See Chapter 11 for further information on walkers.

BLEACHERS

Children climb and sit on bleachers while attending sports events with parents or at school. Bleachers are tiered (stepped) and frequently do not have backs. Most often, they are outdoors or in gymnasiums.

See Figures 8.16 and 8.17, as well as Boxes on this page and the next for additional information about bleachers.

When do bleacher falls occur?

When the bleacher:

1. Is missing guardrails (sides or back)
2. Has openings between guardrails, seats, and footboards are large enough for a child to fit through
3. Collapses after being opened, closed, or maintained improperly
4. Has missing or broken assisting components (handrails, aisles, nonskid surfaces) are missing or broken
5. Is older*

Figure 8.16 (Valerie Rice)
Children fall from or through bleachers onto lower surfaces.

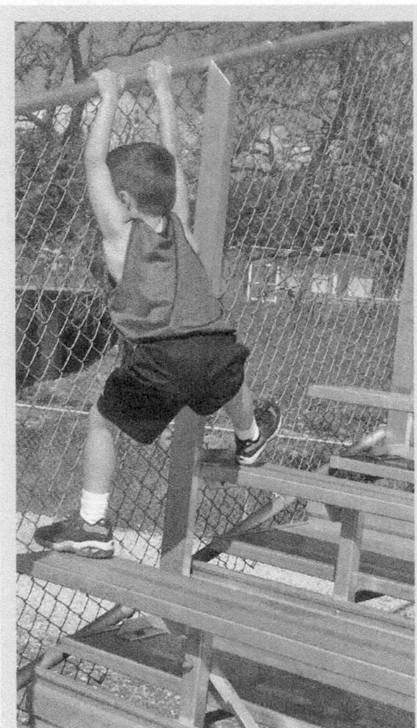

Figure 8.17 (Valerie Rice)
In 2002, about 3350 children under 15 years of age were seen in US emergency rooms for falls from bleachers.

Common types of bleachers

- Permanent and stationary (common in football fields)
- Portable and movable (common at little league games)
- Telescopic and folding (common in gymnasiums)
- Temporary bleachers

* Old bleachers were not required to provide guardrails or small openings to prevent a child falling through. Most codes today do not require retrofitting.

PREVENTION

In the United States, the Consumer Product Safety Commission (CPSC) provides *Guidelines for Retrofitting of Bleachers.*[*] Although voluntary, these have become de facto standards.

- Guardrails should be at the back and open ends of bleachers.[†]
- The top surface of the guardrail should be at least 106.6 cm (42 in.) above the leading edge of the footboard, seat board, or aisle beside it.
- Guard rails are not needed when bleachers are next to a wall at least as high as the recommended guardrail height.[‡]
- Children should not be able to pass under or through components of guardrails or seating.[§]
- Guardrails should not encourage climbing:
    - Avoid a ladder effect.
    - Use in-fill between vertical top and bottom rails.
    - Avoid openings that may provide a foothold for climbing.[¶]
    - Chain link mesh may not exceed 3.2 cm² (1.3 in.²).
- Incorporate aisles, handrails, and nonskid surfaces into retrofits and replacements.
- When retrofitting, avoid new hazards that cause bleachers to tip or collapse, tripping hazards, and protruding or sharp edges.
- Inspect with trained personnel quarterly.
- Only allow personnel with special training to operate bleachers.

* The CPSC complete guideline can be downloaded at http://cpsc.gov and also http://www.childergo.com/recc-children.htm. CPSC Publication #330.

† Where the footboard, seat board or aisle is at least 76.2 cm (30 in.) above the lower surface. The bleacher may be exempt if the top row is only at 76.2 m (30 in.).

‡ At this height, a 10.2 cm (4 in.) diameter sphere will not pass between bleachers and wall.

§ All openings between elements of the guardrail or seating should prevent a 10.2 cm (4 in.) sphere from passing through, anywhere the opening would permit. a fall of 76.2 cm (30 in.) or more. The 10.2 cm (4 in.) opening will prevent 95% of all children 4 months or older from completely passing through the opening. The 4.5 cm (1.75 in.) opening in guardrail in-fill (to prevent a ladder effect) is based on the foot width of a young child.

¶ The width should not exceed 4.5 cm (1.8 in.) where foot can rest.

SHOPPING CARTS

Having a child strapped in a cart while shopping is a great benefit for caregivers. They can move quicker to select their items and do not have to supervise their child quite as closely as when he or she is on his or her own feet!

Sometimes children like riding in carts and sometimes they do not. Sometimes they kick at the cart to keep from being put inside. They do not want limitations on their free-range movements! They may try to unbuckle their belts out of curiosity or in the desire to escape!

In 2003, just under 20,000 US children under the age of 5 were injured from falls associated with shopping carts (CPSC). Most sustained head injuries; some were serious injuries such as concussions and fractures.*

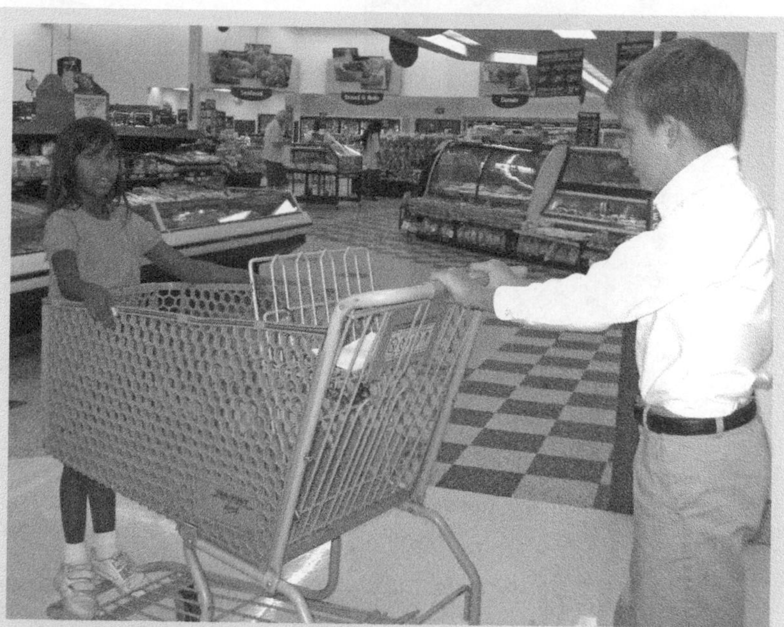

Figure 8.18 (Valerie Rice)
Many children still want to ride on carts when they get older. They want to stand on the front, back, or side of the cart or get inside the basket.
It is tempting for parents to do this because it seems easier to let a child ride on the cart than to supervise a curious, busy child. However, a child can easily fall when standing on a cart and sustain a serious injury.

Shopping-cart falls involve:

- falling from carts
- striking objects while riding
- being caught between objects in the cart (Smith et al., 1996; Friedman, 1986; Campbell, 1990)

Table 8.5 How to Prevent Falls from Shopping Carts

Caregivers	• Always use the restraint systems in shopping carts. • Never let a child stand in the cart, even during placement into or removal from the cart. • Never let a child ride in the main cart basket. • Never let children ride or climb on the sides, front, or back of the cart. • Never let one child push another child in a cart, without supervision.
Manufacturers	• Include a picture of proper seating and wear of the restraint system in the cart. • Include warnings on carts where parents can see them. • Include a lap belt and a strap between the legs, as children cannot as easily wiggle out of seat belts having both. • Wide belts exert less pressure over a small area. • Buckles should not easily entrap or pinch small fingers.
Store supplying cart	• Inspect the carts regularly for broken pieces and frayed or broken seat belts. • Post warnings for parents regarding the proper use of cart seats.
Communities	• Provide public education on how to avoid shopping-cart injuries.

* (See Figures 8.18 through 8.22 and Table 8.5.).

New shopping carts on the market

Figure 8.19 (Valerie Rice)
Cart seats resemble infant car seats. The seat and seat belts fray fairly quickly from near constant use.

Figure 8.20 (Valerie Rice)
This cart has a bench seat behind the basket for older children. Bench seats give older children a more appropriate restraint system for riding.

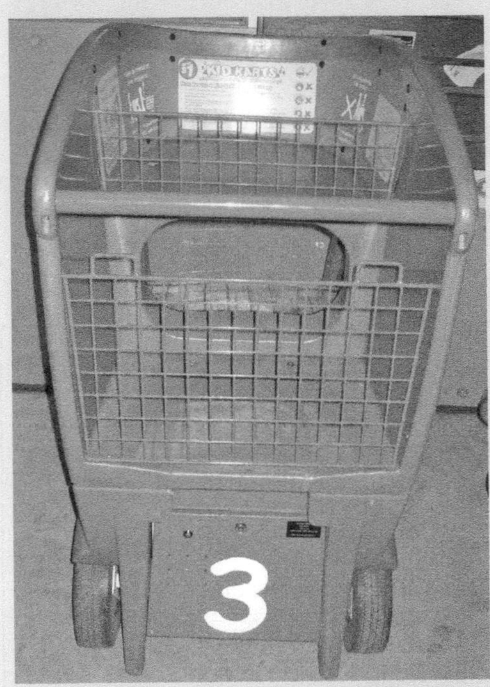

Figure 8.21 (Valerie Rice)
Miniature shopping carts for children may help; many children are happy to "help" shop for specific items for their own cart (not pictured).

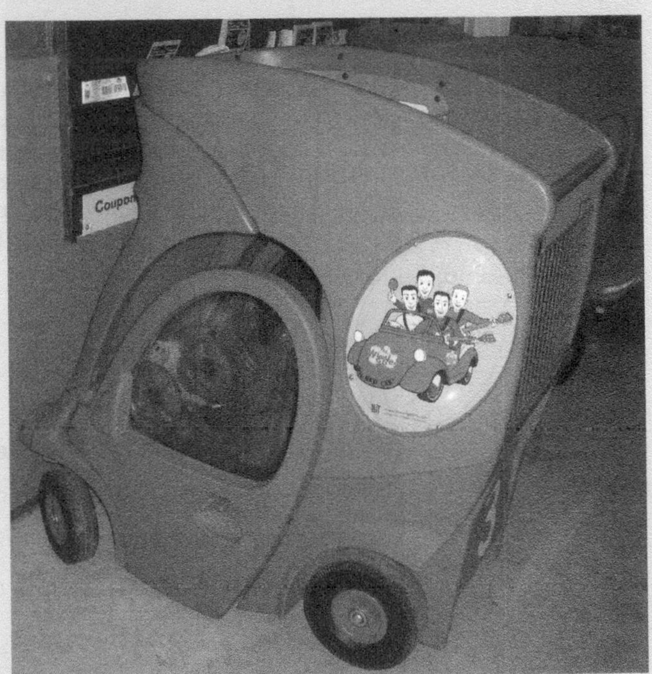

Figure 8.22 (Valerie Rice)
These carts have seats that look like cars (with restraints). The child feels more in control in the "car" and has something to do as they "drive."

PREVENTING FALLS

Table 8.6 provides general recommendations for preventing children's falls.

Table 8.6 Safety Recommendations to Prevent Falls[a]

Parents/Caregivers	Inside	• Infants
		  ◦ Never leave babies alone (unsupervised) on furniture even at a low height, including a bed, changing table, counter, sofa, or chair.
		  ◦ Keep crib sides up, even for babies who cannot yet push up on both hands or roll over.
		  ◦ Lower the crib mattress when a child can sit alone or stand.
		  ◦ Never use baby walkers on wheels.
		  ◦ Put baby bouncers on the floor, not on a table or high surface.
		• Do not let children climb or jump on furniture, play on stairs, or run and jump while holding sharp objects.
		• Always use safety straps on highchairs, preferably a five-point strap system.
		• "Child proof" your home.
		  ◦ Do not leave toys or other items on the floor or stairs.
		  ◦ Have clear paths for movement in the house, free of debris and furniture.
		  ◦ Pad sharp edges of furniture, especially coffee tables and fireplace edges.
		• Barricade stairs at the top and bottom, using secure wall attachments.
		• Prevent slipping
		  ◦ Secure area rugs with foam backing, double-sided tape, or other measures.
		  ◦ Clean spills immediately, especially on slick floors.
		  ◦ Consider installing nonskid stair runners.
		  ◦ Wear shoes or slippers that fit and will grip the floor or carpet surface. Socks do not work well on slippery floors.
		  ◦ Place slip-resistant mats or stickers in children's bathtubs.
		  ◦ Keep soap and other objects off the floor of the tub and shower. Remove soap buildup from the tub and shower floor.
		  ◦ Place wall-mounted handles next to toilets and in bathtubs or showers.
		• Do not use tables or chairs as makeshift ladders, be a good role model for your children.
		• Keep wiring away from walking areas.

	Outside[b]	• Always supervise a child on stairs, decks, and high porches. • Remove anything a child can climb that is near a balcony rail. • Do not let your child play on fire escapes, in halls with windows that do not have window guards, near elevators, and near or on stairs. • Always strap children into the stroller or grocery cart. • Do not let children play on roofs or ladders. • Remove damp leaves, moss, ice, and snow from walkways. • Cut back overhanging trees or bushes along walkways. • Make sure walking surfaces are clear and level (driveway, walkway), especially when they are tiles, bricks, or stepping stones. Fill in holes.

(continued)

Table 8.6 Safety Recommendations to Prevent Falls[a] (Continued)

Parents/Caregivers	Outside[b]	• Keep the home and walkways well lit, especially when outdoors at night.
		• Put tools and play equipment away after use, including ladders.
		• Keep your deck safe. ◦ Check local building codes. ◦ Build a 1 m (3.2 ft) handrail and guard or barrier on decks 1.2 m (3.94 ft) high. Use vertical bars or slats that children cannot climb. ◦ Inspect your deck for breakage, loose boards, and rotting wood.[c]
		• Always supervise children when using playground equipment.
		• Teach children to use equipment properly.
		• Separate older and younger children during play.
		• Teach children to walk inside and run outside.
		• Use gates to block access to stairs.
		• Be sure you can see where you are going when carrying large loads (or carrying children). Ask for help if you cannot.
		• Entrance mats ◦ Use water absorptive mats to reduce tracking in water (makes floors slippery). ◦ Use mats with beveled edges that do not curl or slide. ◦ Recessed mats that are level with the ground surface are best.
Manufacturers		• Adhere to child product regulations (e.g., slats on baby gates should be less than 3 cm [1.25 in.] apart; crib slats 3.5 cm [1.375 in.] apart).
		• Build entrance and room thresholds no more than 1.9 cm (0.75 in.) high to minimize tripping (check local building codes).
		• Design for how children actually use products, rather than how children should use them.
Communities		• Establish community education programs.
		• Target the most frequent and serious injuries that occur locally.
		• Encourage close supervision by caregivers.
		• Reward appropriate behaviors (e.g., "attaboys" with photos in the local paper).
		• Provide home checklists for parents.[d]
		• Provide information on safe or recalled products (available from http://cpsc.gov).
		• Develop a community safety program, much like a "community watch." Have them watch for safe practices and equipment (cracks in sidewalks, adequate lighting), assist in modifying playgrounds and check to be certain that playground equipment meets CPSC guidelines.
		• Have classes for parents and other caregivers (including babysitters).
		• Offer free classes for local schools (children often carry their learning home and teach their parents).
		(Sibert et al., 1999)

[a] See Figures 8.23 and 8.24 also.
[b] For further information, please refer to Chapter 12 (Toys); Chapter 15 (Stairs); Chapter 17 (Vehicle Safety), and Chapter 28 (Playgrounds).
[c] Especially where it attaches to the house and under the stairs.
[d] See for example, checklists at http://cpsc.gov

Figure 8.23 (Valerie Rice)
While it is not always possible to "design out" all risk factors, it is important for designers of products and places to consider how a child will view them. Above, designers have placed a drain next to a set of stairs. There is no barrier between them. In the photo, the adults are using the stairs while the children are climbing up (and playing on) the drain and the concrete area between the drain and the stairs. The area is steep and hazardous for children, yet attractive as a "play area."

Figure 8.24 (Valerie Rice)
Above right, a boy climbs atop a statue of a lion. While the area is matted, if the child fell and hit his head on the concrete statue or statue base, he could sustain a serious head injury.

BURNS

INTRODUCTION

Burns are one of the most horrifying and painful injuries a person can experience. Rehabilitation from burns can take months or even years of medical treatment, therapy, and counseling.

Resulting scar tissue can be fragile, influence joint movement, and interfere with normal processes such as sweating. This means the scar tissue will influence a child's ability to exercise or spend time outdoors.

More people die in fires than all natural disasters. Most could have been avoided. *Most burns to children are accidental.* A child reaches for her mother, pulls on her arm, and spills hot coffee. A youngster dashes through the kitchen and hits the handle of the pan on the stove and hot spaghetti sauce spatters. A 2-year-old pulls on a cord and a curling iron or clothing iron falls onto his skin. An adolescent "hangs out" with his friends in his bedroom while waiting for his parents to come home from work. They take turns seeing who can hold a lit match the longest, playing a deadly game of "chicken"—but no one wins in this game.

> ***Burns in the United States***
>
> - Burns are the third leading cause of unintentional injuries among children under the age of 1; eighth among children under the age of 4 (NCIPC, 2003).
> - Emergency-room practitioners treat nearly 100,000 children a year with burns. Most of these involve thermal burns (>½), scalds (~¼), and chemical and electrical burns.
> - Over 1000 children die from burn injuries (KKH, 2005).

We can prevent most accidental burns and scalds by changing the design of environments and the way we do things (Safekids, 2005). Table 8.7 contains more information about burn injuries among children.

Table 8.7 Burn Injuries among Children

Where and Why Do Children Sustain Burns?	• House fires cause most deaths among children, while most burn-related hospitalizations and outpatient treatments for burns result from scalds by hot liquids. ◦ Products that cause burns include hair curlers and curling irons, room heaters, ovens and ranges, irons, gasoline, and fireworks. ◦ Each cause has different risk factors, but human error—judgment, behavior, or both—is the primary culprit. • Children's errors occur in innocence. • Children are curious about everything. They reach; they run—if they can see it, they want to try it! They do not have the same fears and understandings that older children and adults have. • Children are injured more often and more severely because their skin is thinner than adults. • Children's skin burns at lower temperatures and the burns are deeper than the skin of adults who are exposed to the same temperature for the same time period.
What Children Are at Greatest Risk?	• Children under the age of 2 are most likely to be admitted to hospitals for burn injuries (Stockhausen and Katcher, 2001). ◦ These children are burned by scalds from spills or hot baths, contact with hot items (e.g., clothing or curling irons), and direct exposure to flame. ◦ Less commonly, they are burned by corrosive chemicals, high voltage, and other conditions. • Preschoolers sustain three out of four severe burns and scald injuries. ◦ These occur most frequently in kitchens and (secondly) in bathrooms. • Older children tend to have more access exposure to fire. • All children are at risk for electrical and chemical burns, particularly if homes lack smoke alarms. • Boys are at greater risk of burn injuries than girls.

FIRE

Most fire-related deaths occur in homes. In typical home fires, residents have only 2–3 minutes to get outside safely.

Most home fires start in the kitchen or from heating equipment. Kitchen fires can start from overheating cooking oil or accidentally placing something flammable near the cooking surface.

Most home fires occur in winter months, when occupants turn up the heat, build fires, and use portable heating devices (CDC, 2005). Heating appliances, such as portable space heaters, can inflame items that are too close to the source of heat.

Most victims of fires die from smoke or toxic gas inhalation, rather than fire. The effects of the inhalation can make it more difficult to think, move, and escape (Hall, 2001).

In the United States:

- Four out of five deaths from fire occur in homes (Karter, 2003 using 2002 data).*
- Twenty percent of deaths from fire occur to children (CPSC, 2005).
- Smoking is the leading cause of death from home fires (NFPA, 2005).
- Alcohol use contributes to about 40% of fire deaths in homes (Smith, 1999).

* In the United States, 4,000 people die in a home fire annually (USCPSC, 2005a); someone dies in a fire nearly every 3 hours and an injury occurs every 37 minutes (Karter, 2003 using 2002 data). Among industrialized countries, Canada and the United States have two of the highest death rates from fires. However, over the last 20 years rates have decreased in the United States more than in any other Western nation, now nearly equaling those in some European and Pacific Rim countries (USFA, 2000).

Figure 8.25 (Ursy Potter Photography)
Fire fascinates many children. While it is dangerous, it is also a source of joy as Zan blows out his candles or cooks over a campfire. Such mixed messages are difficult for young children to reconcile.

Only children of 18 months and over can strike a match.

Young children do not escape fires as easily as older children and adults. They frequently hide under a bed, in a closet, or in a wardrobe to escape the fire. They are often afraid of bright lights, loud noises, and uniforms of the firefighters, who try to rescue them.

CAUSES: Causes of fires include home heaters, kitchen cooking equipment, cigarettes and matches, candles, and electrical elements (see Figures 8.25 and 8.26 and Table 8.8) (USCPSC, 2005a).

Wood- or coal-burning stoves, and kerosene, gas, or electrical space heaters cause about a fifth of all home fires.* In all cases, read and follow the instructions that come with the heater. All of these heating systems are readily accessible to children. Many sit on the floor, within children's reach. Children can touch them or lay things on them, not realizing the danger of burning themselves.

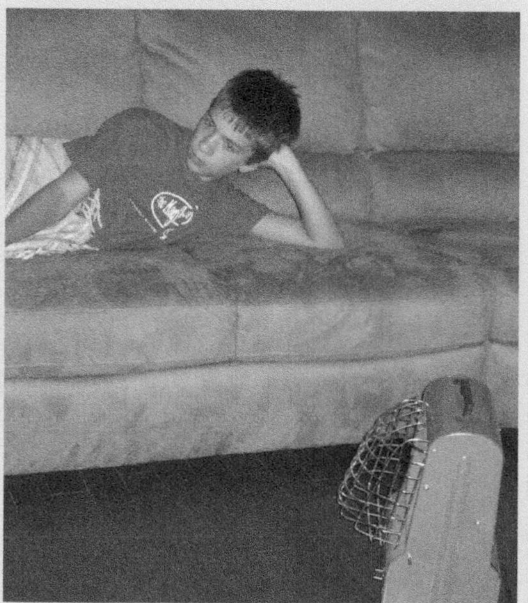

Figure 8.26 (Valerie Rice)
Space heaters are usually within children's reach. Children can touch them, lay things on them, or pull them too close to their seating, not realizing the risk.

Design a "kid-free" zone

- Never let children play close to heaters.
- Keep heaters out of children's reach.

Table 8.9 displays values for 95th percentile forward functional forward reach for boys and girls. This measure does not include stretching or moving the body into specific postures to increase the reach distance. Therefore, to be safe, add several inches to the 95th percentile–reach data.†

* Room heaters are potential sources of carbon monoxide poisoning. Every home that uses any of the following should have a U.L. approved carbon monoxide detector: unvented kerosene and gas space heaters, chimneys, furnaces, water heaters, wood stoves, or fireplaces. Carbon monoxide is colorless, odorless, and the symptoms are similar to the flu, so most people would not recognize exactly what is happening, unless they have an alarm. See CPSC's Web site fire safety checklist at http://cpsc.gov

† For example, a 7-year-old child can add 10 cm (4 in.) by stretching and 19 cm (7.5 in.) by both stretching and leaning to reach an object, while a 12-year-old can add 14 and 25.4 cm (5.5 and 10 in.), respectively. The added reach depends on the child's size and balance.

Table 8.8 Fire-Related Problems and Solutions

	Problems	Solutions
Gas-Fired Space Heaters	• If there is a problem with the heater or the pilot light is lit incorrectly, gas can escape, accumulate, and ignite explosively. • Many sit on the floor, within children's reach. Children can touch them or lay things on them, not realizing the danger of burning themselves.	• Read and follow manufacturer's instructions on using gas space heaters. • Do not use unvented heaters in small, enclosed areas. This way, if there are problems with the heater, you avoid carbon monoxide poisoning. • Propane heaters (LP) that have gas cylinders in the body of the heater are prohibited in most states and localities in the United States. Do not use them. • Light the pilot according to the manufacturer's instructions. • If matches are necessary for lighting the pilot, light them before turning on the gas so you do not let gas buildup. Keep matches out of reach of children. • Establish "kid-free" zone around heater (Table 8.9).
Kerosene Heaters	• Human error can result in an accidental fire, such as: ∘ refueling while the heater is operating, possibly starting a fire ∘ spilling flammable liquids, which consequently catch fire ∘ storing flammable liquids near the heater or near where people smoke • Heaters often sit on the floor, within the reach of children. They can be knocked over and children can touch them or lay things on them, not realizing they may burn themselves or start a fire.	• Know your local and state codes and regulations for using a kerosene heater. • Do not use gasoline in a kerosene heater. • Use containers with labels and check the labels before use to avoid mistaking gasoline for kerosene. • Place heater securely, so it cannot be knocked over. • Place heater where, if it falls over or catches fire, it will not trap you. • Only use grade I–K kerosene to avoid sulfur dioxide emissions. • Check and recheck the pump if you buy kerosene from a gasoline station. Use the kerosene pump, not the gasoline pump. • Refuel the heater when it is NOT operating. • Refuel the heater outside so spills of flammable material do not occur inside. • Ventilate the room where the heater operates (e.g., door open or the window ajar). • Do not store flammable liquids and fabrics near an open flame. • If a flare-up occurs, do not try to move the heater or smother the flames with a rug or a blanket. Use the manual shutoff switch and call the fire department. • Establish a "kid-free" zone around the heater (see Table 8.9). • Do not permit children to refuel or assist with refueling. Lock flammable liquids where children cannot access them.

Table 8.8 Fire-Related Problems and Solutions (Continued)

	Problems	Solutions
Portable Electric Heaters	• Most fire-related deaths and injuries due to portable electric heaters occur at night while owners are sleeping. • One problem may be that portable electric heaters require extension cords that are bigger than the owners choose to use (to handle the current). • Combustible materials stored too close may cause a fire. • Heaters sit on the floor, within reach of children who can touch them, knock them over, or lay things on them that may be flammable.	• Do not keep combustible materials near the heater, place them a minimum of 0.91 m (3 ft) away. This includes clothing, towels, and papers. Do not place heater too close to a bed. Making this part of the "kid-free" zone will encourage the same behaviors from children. • Do not use extension cords, if possible. If you use an extension cord, read the power rating on the cord to make certain it is as high as the heater requires. Do not place anything on top of the cord, including rugs and carpets. Use cord protectors to prevent children playing with or tripping over cords. • Do not place heaters on top of furniture, including tables, chairs, book cases, and cabinets. • Do not use heaters to dry wet clothing or shoes. • Place the heater where a person or pet cannot knock it over. • Establish a "kid-free" zone around the heater (Table 8.9).

Wood Stoves	• Combustible materials may either be part of the home structure or be mistakenly placed near the stove, causing a fire.
	• Children may play in or near the stove, try to burn items in the wood stove, or touch the hot stove.
	• Human error may contribute to a fire:
	  ∘ Burning trash in the stove
	  ∘ Using flammable materials to start the fire
	  ∘ Failing to clean the chimney properly
		• Install wood-burning stoves in accord with building codes.
		• Inspect stoves according to the instructions that accompany the stove, generally twice a month during the months of use.
		• Have professional chimney-sweeps clean the chimney.
		• Avoid smoldering fires to decrease creosote buildup.
		• Protect floors at least 45.7 cm (18 in.) beyond the stove using a code-specified or listed floor protector.
		• Locate the stove away from combustible walls (check stove instructions).
		• Do not burn household trash in a stove, as this can cause the stove to overheat and children may try to "help" burn trash.
		• Do not use gasoline or other flammable liquids to start wood-stove fires, as it can explode.
		• Do not use coal unless the manufacturer labels it as an appropriate use.
		• Teach children about dangers of combustible materials.
		• Do not let children play around or use stoves without supervision.
		• Establish a "kid-free" zone around the stove (Table 8.9).

(continued)

Table 8.8 Fire-Related Problems and Solutions (Continued)

	Problems	Solutions
Fireplace	• Children are intrigued by the moving, colorful flames and may play with the fire, try and burn things in the fire, or sit too close to the fire (Figure 8.27). • Human error can contribute to fires: ○ Burning trash creates sparks that may escape and cause a fire ○ Failing to clean the chimney properly	• Keep fireplaces clean. • Use a fireplace screen. • Burn only wood (or other approved material). • Always have a responsible adult supervising the fire. • Use professionals to clean the chimney annually. • Be certain the fire is completely out before stopping the supervision (leaving the house, going to bed, or simply ceasing to supervise). • Teach children about fires and fireplace safety. Store ignition sources in locked areas or where children cannot reach them (Table 8.9). • Establish a "kid-free" zone around the stove (Table 8.9).
Kitchens and Cooking	• More than 100,000 fires a year begin when people are cooking. It is the main cause of home fires (Ahrens, 2001). • Electric appliances account for nearly twice the number of fires as gas appliances.	• Do not place flammable items on or near the range, including pot holders, towels, plastic utensils, and cookbooks (paper). • Do not wear loose clothing while cooking. Fasten loose or long sleeves up with pins or elastic bands. They catch on items and catch fire more easily. • Do not reach across the range top while cooking. • Do not place items children might want over the stove, such as candy or cookies. • A responsible adult should supervise cooking (Figure 8.28). • Teach fire safety to all family members, including things like putting a lid on grease fires. • Turn pan handles inward, so they do not protrude into the travel area in front of the stove and children cannot accidentally bump them as they pass.
Candles	• Children are attracted by the flickering light produced by a candle. They may reach for the flame not realizing the danger. • As noted above, candles are associated with birthdays and younger children may attempt to blow them out, possibly either knocking them over or spreading a spark. • Young children do not realize the consequences of their actions and may try to reach over the candle or knock it over inadvertently. • Human error sometimes results in forgetting the lit candle and leaving it for hours or going to bed while it remains lit.	• Keep candles and ignition sources out of reach of children (Table 8.9). • Supervise the use of candles and put them out before leaving the area or going to bed. • Use sturdy holders that do not tip and are not flammable. Place them where people and pets cannot knock them over. • Never use candles near flammable material, such as clothing, bedding, and papers. • Be especially careful during holidays. Fires due to candles nearly double during Christmas (NFPA, 2005).

Table 8.8 Fire-Related Problems and Solutions (Continued)

	Problems	Solutions
Electrical	• Electrical cords, extension cords, and wall outlets can cause electrical burns in children. • Almost two-thirds of electrical burn injuries happen to children 12 years old and under. • Improper installation or use of electric wires and devices can also cause fires. • Children are curious and go through a stage when they like to insert things (including their fingers, keys, spoons, and other items) into openings. • Some older children even wonder how "much" of a "shock" they might get and try it on purpose.	• Childproof outlets. • Place plastic plugs in electrical outlets. • Replace cords or plugs that are loose or fraying. Replace appliances that spark, smell unusual, or overheat. • Do not overload outlets. • Use bulbs of the correct wattage for fixtures. • Be careful about home repairs as fires can start from improper installation of electrical wiring or devices. • Never run electrical wires under rugs. • Never place lamps or night-lights so they touch flammable materials such as bedspreads, curtains, or clothes. • Establish clear "no-touch" rules for younger children.
Materials	Some materials are more flammable than others.	• Their careful use may prevent fires. • See section below on clothing.
Clothing	• Almost all fibers in clothing can burn, but some burn more quickly than others. • Clothing that ignites most often includes pajamas, nightgowns, robes, shirts or blouses, pants or slacks, and dresses. • Clothing ignition starts most often from matches, cigarette lighters, and candles; however, it can also start from close exposure ranges, open fires, and space heaters. • Deaths and injuries of children due to fire were reduced by 95% following the Flammable Fabrics Act in the 1970s, which requires that children's sleepwear be flame retardant.	• Fabrics that are difficult to ignite and tend to self-extinguish include 100% polyester, nylon, wool, and silk. • Cotton, cotton–polyester blends, rayon, and acrylic ignite easily and burn rapidly. • Tight weaves or knits and fabrics without a fuzzy or napped surface ignite and burn slower than open knits or weaves or fabrics with brushed or piled surfaces. • Clothes that are easily removed can help prevent serious burns. • Follow manufacturer's care-and-cleaning instructions on products labeled "flame resistant" to ensure that their flame-resistant properties are maintained. • *Do not let children sleep in daytime clothing, always ask children to sleep in flame-retardant sleepwear*.
Flammable Liquids	• Flammable liquids often cause fires. Gasoline is the most frequent cause, but others include acetone, benzene, lacquer thinner, alcohol, turpentine, contact cements, paint thinner, kerosene, and charcoal-lighter fluid. • Their vapors are invisible and can ignite by a simple spark at a fair distance from the liquid.	• Store flammable liquids outside the home or in specially designed cabinets for flammable materials. • Make certain flammable liquids are out of reach of children.

(continued)

Table 8.8 Fire-Related Problems and Solutions (Continued)

	Problems	Solutions
Upholstered Furniture and Mattresses	• Many home fires start by cigarettes or lit ashes falling onto upholstered furniture. • No standards currently exist, although the CPSC has long considered them. • Standard methods of testing and classifying the ignition resistance of upholstered furniture can be purchased from NFPA.[a]	• Furniture made in the last 10 to 15 years and furniture that meets the requirements of the Upholstered Furniture Action Council's (UFAC) Voluntary Action Program have greater resistance to fire of cigarettes. Look for a gold tag with red lettering that states "Important Consumer Safety Information from UFAC." • Check furniture where smokers have been sitting for smoking materials. • Do not place ashtrays on arms of chairs where they can fall. • Purchase furniture made primarily with thermoplastic fibers (nylon, polyester, acrylic, and olefin). They resist ignition better than cellulose-type fabrics (rayon or cotton). • High thermoplastic contents have greater resistance to cigarette ignition. • Never smoke in bed. It is a frequent cause of fire deaths in homes. • Place heaters or other fire sources at a minimum of 0.91 m (3 ft) from the bed. • Use only mattresses manufactured since 1973. • Do not glorify smoking or use of fire to children, so they feel more "grown up" if they try either. • Try and limit smoking in front of and around children (this can help with their respiratory health also). Never have children responsible for empting ashtrays.

[a] NFPA Standards 260, 261, and 272 address flammability of upholstered furniture. NFPA Standard 284 addresses mattress flammability. See www.nfpa.org

Table 8.9 Forward Functional Reach Standing to Grip Recommendation for "Kid–Free" Zone around In Home Heaters (360°)[a]

(Snyder, 1977; all measurements rounded up in cm)

Age	*n*	Boys and Girls [cm (in.)]	5th–95th Percentile [cm (in.)]
2–3.5	62	39.70 ± 3.1 (14.84 ± 1.22)	35.50–45.40 (13.98–17.87)
3.5–4.5	77	43.10 ± 3.6 (16.97 ± 1.42)	37.80–49.70 (14.88–19.57)
4.5–5.5	74	45.40 ± 3.4 (17.87 ± 1.34)	40.4–51.5 (15.91–20.28)
5.5–6.5	76	48.50 ± 2.9 (19.09 ± 1.14)	44.2–53.3 (17.40–20.98)
6.5–7.5	74	50.70 ± 3.2 (19.96 ± 1.26)	46.2–56.1 (18.19–22.09)
7.5–8.5	64	52.90 ± 2.7 (20.83 ± 1.06)	47.8–56.8 (18.82–22.36)
8.5–9.5	80	55.50 ± 4.1 (21.85 ± 1.61)	49.2–64.1 (19.37–25.24)
9.5–10.5	74	57.70 ± 3.3 (22.72 ± 1.30)	52.8–64.0 (20.79–25.20)
10.5–11.5	94	59.90 ± 3.0 (23.58 ± 1.18)	53.4–66.6 (21.02–26.22)
11.5–12.5	91	62.10 ± 3.8 (24.45 ± 1.50)	56.3–67.7 (22.17–26.65)
12.5–13.5	97	64.90 ± 4.5 (25.55 ± 1.77)	57.6–71.6 (22.68–28.19)
13.5–14.5	80	67.00 ± 4.3 (26.38 ± 1.69)	59.4–74.1 (22.60–29.17)
14.5–15.5	87	67.70 ± 3.8 (26.65 ± 1.50)	61.3–73.4 (24.13–28.90)
15.5–16.5	63	69.60 ± 5.1 (27.40 ± 2.00)	60.9–76.4 (23.98–30.08)
16.5–17.5	70.6	70.60 ± 5.0 (27.80 ± 1.97)	62.0–78.7 (24.41–30.98)
17.5–19.0	44	70.70 ± 5.4 (27.83 ± 2.13)	63.0–80.90 (24.80–31.85)

[a] This does not account for stretching or leaning.

Table 8.10 Overhead Reach to Fingertip on Tip Toes (Jurgens, 1974) and Jumping (Branta, 1984). Recommend Items Placed "Out of Reach" for Children Exceed These Overhead Reaches. Note: jumping data needs to be added to overhead reach to fingertip

Age	Girls [cm ± SD (in.)]				Boys [cm ± SD (in.)]			
	Tip Toes	5th–95th Percentile	Jumping	5th–95th Percentile	Tip Toes	5th–95th Percentile	Jumping	5th–95th Percentile
3	135.5 ± 6.7 (53.39 ± 2.64)	124.48–146.52 (49.05–57.73)			136.9 ± 7.7 (53.94 ± 3.03)	124.23–149.57 (48.96–58.92)		
4	147.1 ± 7.8 (57.96 ± 3.07)	134.27–159.93 (52.91–63.01)			146.8 ± 6.0 (57.84 ± 2.36)	136.93–156.67 (53.96–61.72)		
5	155.6 ± 7.1 (61.31 ± 2.8)	143.92–167.28 (56.70–65.92)	15.24 ± 4.06 (6.03 ± 1.6)	8.56–21.92 (3.4–8.66)	155.0 ± 7.9 (61.07 ± 3.11)	142.01–167.99 (55.95–66.19)	14.83 ± 4.32 (5.84 ± 1.7)	7.72–21.94 (3.04–8.64)
6	161.8 ± 7.1 (63.75 ± 2.8)	150.12–173.48 (59.14–68.36)	17.88 ± 3.56 (7.04 ± 1.4)	12.02–23.74 (4.74–9.34)	163.0 ± 9.1 (64.22 ± 3.59)	148.03–177.97 (58.32–70.13)	18.75 ± 4.06 (7.38 ± 1.6)	12.07–25.43 (4.75–10.01)
7	171.3 ± 8.9 (67.49 ± 3.51)	156.66–185.94 (61.72–73.26)	21.39 ± 4.06 (8.42 ± 1.6)	14.71–28.07 (5.79–11.05)	173.4 ± 8.1 (68.32 ± 3.19)	160.08–186.72 (63.07–73.57)	21.67 ± 4.06 (8.53 ± 1.6)	14.99–28.35 (5.90–11.16)
8	178.0 ± 8.1 (70.13 ± 3.19)	164.68–191.32 (64.88–75.38)	23.62 ± 4.57 (9.30 ± 1.8)	16.10–31.14 (6.34–12.26)	177.9 ± 9.3 (70.09 ± 3.66)	162.60–193.20 (64.07–76.11)	24.87 ± 4.57 (9.79 ± 1.8)	17.35–32.39 (6.83–12.75)
9	187.2 ± 10.5 (73.76 ± 4.14)	169.93–204.47 (66.95–80.57)	26.39 ± 4.06 (10.39 ± 1.6)	19.71–33.07 (7.76–13.02)	186.5 ± 10.4 (73.48 ± 4.1)	169.39–203.61 (66.74–80.22)	26.59 ± 4.06 (10.47 ± 1.6)	19.91–33.27 (7.84–13.10)
10	199.1 ± 9.4 (78.45 ± 3.7)	183.64–214.56 (72.36–84.54)	29.16 ± 4.32 (11.48 ± 1.7)	22.05–36.27 (8.68–14.28)	194.8 ± 8.2 (76.75 ± 3.23)	181.31–208.29 (71.44–82.06)	29.41 ± 5.08 (11.58 ± 2.0)	21.05–37.77 (8.29–14.87)
11	200.6 ± 10.5 (79.04 ± 4.14)	183.33–217.87 (72.23–85.85)	28.58 ± 5.33 (11.25 ± 2.1)	19.81–37.36 (7.80–14.70)	201.2 ± 12.4 (79.27 ± 4.89)	180.80–221.60 (71.23–87.31)	29.69 ± 4.57 (11.69 ± 1.8)	22.17–37.21 (8.73–14.65)
12	211.1 ± 11.7 (83.17 ± 4.61)	191.86–230.34 (75.59–90.75)	31.34 ± 5.08 (12.34 ± 2.0)	22.98–39.70 (9.05–15.63)	208.2 ± 10.4 (82.03 ± 4.1)	191.09–225.31 (75.29–88.77)	32.13 ± 4.83 (12.65 ± 1.9)	24.19–40.07 (9.52–15.78)
13	218.5 ± 10.0 (86.09 ± 3.94)	202.05–234.95 (79.61–92.57)	33.96 ± 4.83 (13.37 ± 1.9)	26.02–41.90 (10.24–16.50)	219.6 ± 13.4 (86.52 ± 5.28)	197.56–241.64 (77.84–95.20)	35.23 ± 6.35 (13.87 ± 2.5)	24.79–45.67 (9.76–17.98)
14	221.3 ± 9.6 (87.19 ± 3.78)	205.51–237.09 (80.97–93.41)			228.2 ± 15.0 (89.91 ± 5.91)			

Figure 8.27 (Valerie Rice)
Daniel works in front of the family room fireplace. Although the fire is out, he sits much closer than a "kid-free" zone would permit.

This fireplace has a retractable "screen" pulled back on both sides. It helps prevent sparks from flying out of the fireplace but can still get very hot.

Figure 8.28 (Valerie Rice)
Trying to be helpful, Kiran climbs onto the kitchen shelf and reaches to turn off the alarm above the stove. While avoiding the range top, she could easily slip, or her hair or clothing could catch fire. Adult supervision requires constant vigilance around an active "helpful" child.

SOLUTIONS:

Early warnings: smoke detectors (alarms)

Time is critical when there is a fire.

Depending on people to notice there is a fire can cost precious minutes as they may not smell the smoke or they may be overcome by fumes. Time is lost when people try to put the fire out themselves, investigate to see if it is "really" a fire, or assume it is a false alarm (see Box 8.2 and Figure 8.29).

Automatic systems save time:

- Most automatic alarms change two steps (*detection and alarm*) into a single step.
- Automatic alarms that also alert fire departments change three steps (*detection, alarm for those in the home, alerting fire department*) into one.*

Although almost 92% of American homes have smoke alarms, about one-third do not work because batteries are worn or missing (FEMA).

BOX 8.2

What type of smoke detector?

- These basic types are fine:
    - Ion: Reacts faster to open flaming fires, usually the least expensive
    - Photoelectric: Reacts faster to smoldering fires, reacts less to cooking

Does the power source matter?

- A long life battery (lithium, 10 years) means you do not have to replace it as often.
- Regular batteries (typically 9 V) require annual replacement or when the alarm begins to "chirp" (often starts in the middle of the night when the temperature drops).
- Household electrical power. A "redundant system" helps, by having a battery backup in case the main power fails.
- "Hush" buttons let you turn sound off easily in false alarms.

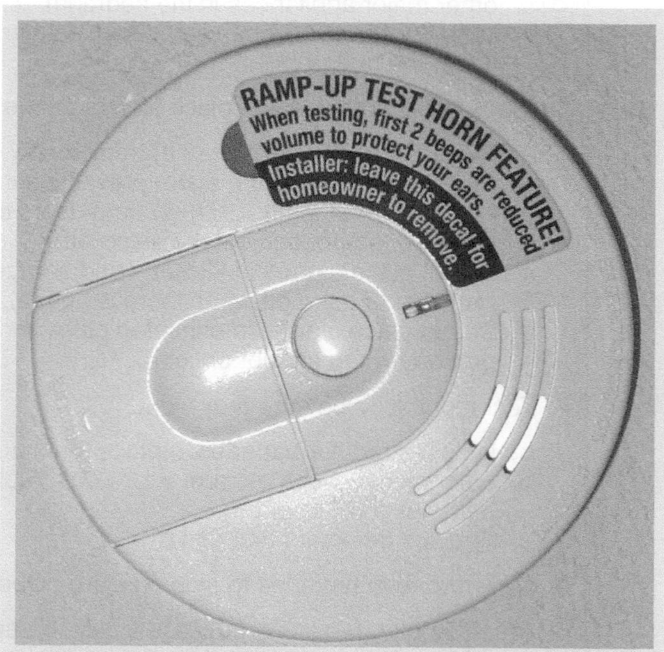

Figure 8.29 (Valerie Rice)
Two-thirds of home fires that kill children under the age of 5 happen in homes that do not have smoke detectors (USFA, 2005a).

- Smoke detectors with alarms reduce the risk of dying in a home fire by 50%.
- Installing both a smoke detector and sprinkler system cuts the chance of death by 82%.†

* Although occupants must cancel the alarm when the fire is insignificant or false, it is an effective method of prevention. Automatic sprinkler systems begin to fight fire even before the fire department arrives.

† A home fire-alarm system alarms doors, windows, and spaces within the home for break-ins; monitors for fire or intrusion by reporting to a security office (and checking with the owner by calling their phone), where they report the information to your local police or fire department. They can be expensive and are most often found in larger, more expensive homes. One benefit is that information is available for special circumstances, such as a disabled family member who may need special assistance.

Recommendations:

1. *Install* smoke detectors at home (Box 8.2).
   Check local codes and regulations; some require specific detectors.
2. *Use enough smoke detectors.*
   a. This may represent one on each floor, one near the bedrooms, and one near the kitchen.
   b. If you close the door to your bedroom when you sleep, put an alarm in the bedroom, particularly if someone smokes in the bedroom or there is a TV, an air conditioner, or other major appliances in the bedroom.
   c. If the door is open when you sleep, an alarm close outside the bedroom will detect a fire in the bedroom.
3. *Placement:* Smoke rises; place alarms on the ceiling.*
   a. Place alarm at least 1 m (3 ft) from supply registers of forced air heating systems (that might blow on the alarm preventing it from detecting smoke).
   b. Place alarm at least 1 m (3 ft) from doors to kitchens or bathrooms that contain showers (steam can set the alarm off).
   c. Do not place alarms in kitchens and garages (cooking fumes and car exhaust are likely to set them off when there is no fire).
   d. Do not place alarms in unheated attics and crawl spaces—electronics do not work properly when too hot or cold.
4. *Test* monthly to be sure they work.
   a. If it uses regular batteries, replace them annually.
   b. Select a day, such as "turn your clock, replace your batteries" or "spring clean your batteries too."
   c. Replace detectors that "chirp."
5. *Never remove* batteries to stop an alarm. Open windows and doors instead.
6. *Clean* smoke detectors. Excessive dust, grease, or other material may impair its function (vacuum grill).
7. *Replace* alarms every 10 years or as the manufacturer recommends.
   If you have persons with hearing disorders, get an alarm that also has a visual strobe.

* Alternatively, place it 15.2–30.5 cm (6–12 in.) below the ceiling on the wall, with the top of the alarm no closer than 10.16 cm (4 in.) or further than 30.5 cm (12 in.) from the ceiling.

Fire extinguishers (PASS):

1. Keep fire extinguishers where you can get them easily, but keep them out of children's reach (see Table 8.9 and Table 8.10).
2. Use them on small fires only, after calling the fire department.
3. Have an all-purpose extinguisher on each floor (Box 8.3), in the kitchen, basement, garage, and workshop. All-purpose means you can use them on grease and electrical fires. Use them properly (Box 8.4).
4. Know how to check them for function. Know when to replace them (Figure 8.30).
5. Do not fight a fire if you also have to fight to breathe.
6. Many fire departments recommend keeping a fire extinguisher in your car and boat.

Figure 8.30 (Ursy Potter Photography)
Young children should not operate fire extinguishers. Place them conveniently. Know how to use them. Have them inspected regularly.

BOX 8.3 SELECTING A FIRE EXTINGUISHER

Extinguisher types:

Class A: Ordinary combustibles: wood, cloth, paper, rubber, and many plastics
Class B: Flammable liquids: gasoline, oil, grease, tar, oil-based paint, lacquer, and flammable gas
Class C: Energized electrical equipment: wiring, fuse boxes, circuit breakers, machinery, and appliances

1. Look for the independent laboratory testing seal.
2. Size product to your needs. Portable extinguisher may empty in 8 seconds.
3. Match the type of fire extinguisher to its use. Labels on extinguishers show list of which of the three classes of fire it can extinguish:
   a. A red slash through the fire signifies it cannot put out.
   b. A missing symbol signifies the extinguisher was not tested for that class of fire.
   c. A number tells the size fire the extinguisher can put out.
   d. Larger numbers signify it can extinguish larger fires.
   e. Multipurpose labels indicate you can use it for all three.
4. Do not use extinguishers for Class A fires on grease or electrical fires!

BOX 8.4 MEMORY AID FOR USING FIRE EXTINGUISHERS

1. *Pull the pin*

   Release lock with nozzle pointing away from you.
2. *Aim low*

   Point extinguisher to the base of fire.
3. *Squeeze lever slowly and evenly*
4. *Sweep nozzle from side-to-side*

Sprinklers Home fire sprinkler systems can activate while the fire is in its early stages. These slowly discharge water from the home water system.

Homes do not flood as the result of discharging fire sprinkler systems. They are less expensive when being put into a new home than when retrofitting an older home.* Even with sprinkler systems, smoke alarms are necessary.

Escape plan Fires are emergencies. When it happens, there is no time to hesitate and think about what to do and where to go. Therefore, design the plan and practice in advance, so the process is automatic.

Table 8.11 Design a Plan for Escape

Establish a Family Escape Plan	• Have two ways out for every room. • Make certain the way out (egress) is clear of furniture or other objects. Have one escape ladder for each upper-floor bedroom—for those who can use it. • Screens and windows must open easily. • Make certain it is easy to remove the window safety bar with a "quick release." • Establish a place to meet outside. • Explain the plan to all family members; everyone must react automatically in a real fire. • Write, diagram, and post the plan. Ensure babysitters see and understand it. • Practice the plan. • Kids will enjoy the role-play, and it may save their lives. See if everyone can be out and at the meeting place in 3 minutes, the time it generally takes to get out safely. Hearing the alarm during practice will keep them from being afraid of the sound during a real fire.
Teach Children	• The plan (route and meeting place). • To alter their route, if the first route is blocked by smoke or fire. • To test the heat of a closed door by touching the back of their hand (lightly) to the top of the door, the doorknob, and the crack between the door and doorframe. • If the door feels cool, have them brace their shoulder against the door and open it slowly, if heat and smoke come in, they should slam it closed. • To low-crawl through smoke. • To stay in the room with the door closed if all routes of escape are blocked. • Not to stop to try to get anything: no toys, no special games, just get out! Not to hide inside. • To call 911 and give the home location. ◦ Young children can learn the important facts about their address and phone number if they learn it in a song. It does not need to rhyme and you can use any easy, familiar tune. • To stay out once they are out. • Not to be afraid of firefighters. Let them see pictures or photos of firefighters in full dress. • Whom to ask for if parents do not come out (relatives, neighbors).

* In the United States, current costs range per area of the country, but in general the cost is about $1.50 per sq ft in a new home and from $2.50 to $5.00 per sq ft for an older home.

WHAT CHILDREN NEED TO KNOW:

BOX 8.5 POINTERS FOR CHILDREN ABOUT FIRE

- Fire is a tool, not a toy.
- Fire is fast, hot, dark (smoke), loud, and deadly.
- Never play with matches, cigarettes, or fire.
- Never play with candles or hot candle wax.
- If children find matches or lighters, tell an adult rather than picking them up.
- How to stop, drop, and roll if their clothes catch fire. How to smother flames, such as with a blanket or with dirt, if outside.
- How to handle smoke: cover their mouths and noses with a damp towel or piece of clothing while leaving the building (but do not go look for something if it will slow the process of getting out).
- How to look for fire exits in buildings and planes as they enter.
- To avoid elevators if there is a fire.
- Not to go too close to heaters, stoves, fireplaces, grills, and ovens and not to leave items that could burn on, near or in them.
- Cook only with supervision.
- Not to play with electrical cords or sockets and to get help plugging or unplugging things into wall sockets.
- Not to place blankets, toys, towels, clothes, or anything else on top of a lamp, as the bulb can cause things to ignite.

BOX 8.6 WEBSITES TO HELP TEACH CHILDREN ABOUT FIRE

The United States Fire Administration's Kids' Web site at http://usfa.fema.gov shows a boy (Marty) and his turtle (Jett). They guide children through an interactive fire safety program. Children find fire hazards in the house, receive cheers for correct answers, and can even earn (and print) a junior fire marshal official certificate for themselves.

http://safetypub.com provides a Children's Fire Prevention handbook, free for individuals. It includes quizzes on fire safety, pages to color, and ideas for developing an escape plan.

http://redhotdots.net This site features Bud the Dog, Dan the Fire Man, and The Red Hot Dot.

http://firesafety.buffnet.net This site lists fire prevention children's activities that can include hands-on activities. Any educational program can use them, such as communities, schools, or parents.

http://firepreventionweek.org This site uses Sparky, the dog, to educate children about fire safety. It has a section on sleepover safety and many games featuring *Sparkyville*, *Danger Dog,* and *Tales from the Great Escape*.

ADULT CAREGIVERS:

1. Keep flashlights where you can find them: next to your bed or the children's bedrooms and in other strategic locations such as the living or family room or basement. They are useful to help getting out, but also for signaling for help.
2. Make certain emergency personnel can read the address of your house easily and quickly, day or night. Every second counts.
3. Put emergency numbers and your own address by your phone for children or babysitters to read. They may panic and forget.
4. Get everyone out and call emergency services before trying to fight a fire.
5. Never fight a fire if:
    - you have to fight to breathe
    - the fire may block your exit
    - you do not have the right equipment

6. Stock a first aid kit. Include ipecac. Teach family members where it is and how to use it.
7. Know basic first aid for burns.
8. "Fire-proof" your home. Build according to Code and:

   - use flame-resistant materials on walls and ceilings, use flammable materials sparingly
   - purchase flame-resistant furnishings
   - compartmentalize rooms (it limits the spread of fire and smoke) and have doors that close off various areas of the home
   - keep exit routes clear, without clutter

FIRE SAFETY AND PEOPLE WITH DISABILITIES: While it is not known how many people with disabilities are injured in fires, some estimate that 7% of fire victims have a physical disability, 3% have physical and mental impairments of aging, and 2% have a mental disability (NFPA CHO and NACFLSE, 2000).

Caregivers of children with disabilities must consider their child's abilities when planning fire prevention:

1. For children who have hearing difficulties or are deaf, use smoke alarms with flashing or strobing lights. The intensity must be high enough to wake a sleeping child.*
2. Purchase alarms with buttons that are large and easy to push. Those you can test with a flashlight or television remote assist people with low vision and mobility problems. Explain and demonstrate their use to children.
3. Make provisions for children with disabilities in the escape plan.
4. If quick mobility is a problem, seriously consider a residential sprinkler system.

FIRECRACKERS

Fireworks are a traditional way of celebrating certain events, such as the 4th July (American Independence Day) and New Year's Day. Some theme parks set off fireworks every evening to signal the end of the day and give visitors a spectacular view to end their outing.

There are individuals that study the art of explosives and become professionals at creating stunning displays. Yet, many people like setting off explosives themselves, adults and children alike.

The US Consumer Products Safety Commission (CPSC) regulates the sale of certain types of fireworks and after 7 years of data collection, their conclusion states:

US Fireworks Facts (USCPSC, 2005b):

- About 8800 people were treated in emergency rooms for injury in 2002 (half of them were under age 15).
- 60% of injuries occur around July 4th.
- Most injuries are burns to the hands, eyes, and head.
- Children under age 4 receive the most injuries from sparklers.
- Boys aged 10 to 14 receive the most injuries from fireworks.

* This follows U.L. standard 1971 and a number of companies have these alarms available including Ace Hardware Corporation, BRK Electronics, Gentex Corporation, Kidde Fire Safety, and Menards, Inc.

"In instances where legal types of fireworks were involved in accidents, either from misuse or malfunction, the resulting injuries were relatively minor and did not require hospitalization."

Illegal fireworks cause about one-third of the injuries, which can be more serious. All fireworks are potentially dangerous, even sparklers burn at over 1000° F. Children see only the fun and excitement; they do not understand the danger. It is up to the adults to supervise and protect children. Box 8.7 lists US Federal Regulations for Fireworks; Box 8.8 contains some internet sources of information.

Recommendations (USCPSC, 2005b):

- Leave the fireworks to professionals.
- If you must use them, do so with extreme caution.
  Read and follow all instructions that come with fireworks.
- Check local and state laws, as many US states prohibit consumer fireworks of a certain class.
- Light fireworks in areas that are open and clear. Keep away from houses and items that may "catch" a spark and ignite, such as dry grass or leaves.
- Light one at a time.
- Never hold any part of your body over the fireworks (in the line-of-fire).
- Wear eye protection when lighting.
- Keep water nearby for emergencies.
- If a firework malfunctions, soak it in water and throw it away. Do not continue to try to light it.
- Keep track of people who are around while lighting fireworks.
- Do not use containers for fireworks, especially glass or metal.
- Keep unused fireworks away from firing areas.
- Read and follow all instructions for storage, typically store in a dry, cool place.
- Never use homemade fireworks.
- Never carry fireworks in your pocket.
- Soak fireworks in water before throwing them away.

For Children:

- Closely supervise children during fireworks.
- Never let them play with or light the fireworks. Even sparklers can ignite clothing.
- Make certain they are out of range before lighting fireworks.
- Do not let children run around when using fireworks; confine their activity to a specific and safe area.
- Store fireworks out of reach of children.

BOX 8.7 US FEDERAL REGULATIONS FOR FIREWORKS	
Federal Hazardous Substances Act	Prohibits the sale of dangerous types of fireworks to consumers • Large reloadable mortar shells • Cherry bombs • Aerial bombs • M-80 salutes • Larger firecrackers containing more than two grains of powder • Mail-order kits designed to build these fireworks
USPSC Regulation, December 6, 1976	• Lowered the permissible charge in firecrackers to no more than 50 mg of powder. • Set specifications for fireworks other than firecrackers. • Fuses must burn at least 3 s but no longer than 9 s. • All fireworks must carry a warning label describing necessary safety precautions and instructions for safe use.
USPSC Regulation, March 26, 1997	• Set requirements to reduce tipover of large multiple tube mine and shell devices.

BOX 8.8 FIREWORKS SAFETY RESOURCES ON THE WEB

1. The National Council on Fireworks Safety Web site contains injury statistics and US state laws.
   http://fireworksafety.com
2. The Prevent Blindness America Fireworks Safety Web site provides guidance on injury prevention, specifically blindness.
   http://preventblindness.org
3. The Fireworks Safety Fact Sheet contains facts about safety, including a lesson plan for educating children.
   http://nfpa.org
4. This site by the American Academy of Ophthalmology concentrates on preventing eye injuries.
   http://medem.com

SCALDS

> *As a new Occupational Therapist, my treatment focus was children. A precocious black boy was brought to me one day for evaluation and treatment. He was three or four years old, his eyes sparkled, his laugh was infectious. The entire pediatric staff was in love with him. I was no exception. I helped him with hand skills. Each of his fingers had been partially amputated from being held under hot bath water. Yes, it was a punishment, but no, his caregiver had not anticipated the result. They did not understand the skin and tissue vulnerability of a small child.*
>
> (Rice, 2006)

> *A severe scald can kill a small child because their skin is thinner and more vulnerable than the skin of an adult. Exposure to hot tap water (140° Fahrenheit) for three seconds will result in a third degree burn, meaning the child will require skin grafts. About 80% of caregivers with young children in their homes do not know the temperature setting of their hot water heater.*
>
> (Safekids, 2005)

Table 8.12 shows scald injury information and potential solutions.

Table 8.12 Child Scald Injuries

	Problems	Solutions (See Also the Section on Kitchens in Table 8.8)
Kitchen	• Most scald injuries occur in the kitchen or other areas where adults are preparing and serving food. • Hot drinks are the most frequent cause of scalding burns among children. • Burns that occur due to microwave use are usually because of spilling liquids or foods. The injuries are most often to the child's face or trunk.	• Supervise children when cooking and serving food (especially outside grills, where it is easy for distracting conversation to catch caregiver's attention). • Consider having a "no child zone" around the stove. See Table 8.9 and Table 8.10 for guidance. • Test food temperature before serving to children, especially with a microwave since they heat food unevenly. For example, the container may be warm, while the chocolate is too hot to drink or the cheese inside can be scalding, while the breading is only warm. • Turn pot handles inward, so no one knocks them over when they walk past the stove. • Use back burners, instead of front burners when children are around. • Do not use tablecloths with young children; they easily pull them off the table. • Never hold a child in your lap while eating or drinking hot food or liquid. • Never walk around with a hot drink when children are playing underfoot. • Do not store children's "treats" such as cookies near the stove.
Bath	Hot baths are the second most frequent cause of scalding burns among children (Figure 8.31). Hot tap water: • Causes more deaths and hospitalizations than any other hot liquid. • Accounts for 25% of scalding burns among children. Burns often occur in the bathroom. • Burns are more severe and cover more of a child's body than other types of scald burns.	• Stay with young children during their bath. Do not leave, even for something brief, like answering the phone. • Check local ordinances and building codes. Many require antiscald plumbing in new construction. • Lower hot-water heater settings to 120°F. • Install antiscald devices in water faucets and shower heads, even if local codes do not require it. • Fill the tub with cold water first and then add hot water until you reach a comfortable temperature for your child. • Become familiar with how warm water feels. • Normally, 100°F is good for a child's bath. Use an outdoor or candy thermometer and take the temperature of the water to learn how it feels. • Respect your child, if they tell you the water is "too hot." • Do not let children turn on their own faucets until they are old enough to do so safely. • Use a cool-mist humidifier instead of a steam vaporizer.
Outside	Metal and other materials become hot in the sun and can burn children: • On playgrounds • In cars	• Take care around playground equipment, especially slides. • Check car seats and seat belt buckles before putting children inside. Consider using sunblinds for cars near children's seats. If parking in direct sunlight, cover children's seats with a blanket or towel. • Check metal and vinyl seats, before placing children in stroller.

Figure 8.31 (Valerie Rice)
Hot baths are the second most frequent cause of scalding burns among children. More deaths and hospitalizations are due to hot tap-water than any other hot liquid. Children's skin is thinner than adult skin, so children can burn at temperatures that are acceptable for adult bathing.

A SPECIAL CASE: CHILDREN WHO START FIRES

Children start fires. Most of them do it accidentally by playing with matches or lighters. These "games" mostly kill children under age 5. They are the primary cause of fire-related deaths among preschoolers. Preschoolers themselves are the ones who start these fires most frequently. In the United States, children set over 30% of the fires that kill children. The problem of children playing with and accidentally setting fires is smaller in Canada and much smaller in Japan than in the United States (relative to population) (NFPA, 2005).

By the age of 12, half of all children would have played with fire. Many fires start in bedrooms or closets, when children are unsupervised. Children who start fires accidentally are usually curious, under age 10, comprise 60% of all children's firesetters, can be boys or girls, and demonstrate little risk for future fire-setting (USFA, 2005b).

Some children start fires deliberately. In Sweden, people light 10,000 fires a year on purpose, between 50% and 80% are set by children (SRSA, 1999). In the United States, children under the age of 18 make up 55% of all arson arrests, almost half are 15 years old or less and 5% are 10 years or younger (USFA, 2005b). In England and Wales between 1995 and 1999, approximately half of the individuals found guilty or issued a warning for arson each year were boys under age 18 (BHO, 2005).

Children who set fires on purpose tend to be boys. In the United States they have a history of setting fires and may show other acts of vandalism. Their typical targets are schools, open fields, dumpsters, and buildings that are empty (USFA, 2005b).

Fire setting is also due to curiosity or boredom (SRSA, 1999; Fineman, 1995). Nearly 80% of schoolchildren in Sweden state they have played with fire, mostly 12- to 14-year-olds (SRSA, 1999). The largest group that starts fires in the United States is 5- to 10-year-old boys (Fineman, 1980) and some start before the age of 3 (Kafry, 1980).

For children who seem to deliberately set fires, consult a health care practitioner and seek information from specific programs such as:

- The National Juvenile Firesetter Arson Control and Prevention Program (http://usfa.fema.gov)
- The Arson Prevention Bureau (http://arsonpreventionbureau.org.uk)
- Project Aquarius, a service working with young firesetters (http://plusonefostercare.co.uk)

These children often need counseling and education, not punishment or scare tactics.

POISON

PROBLEM

Poisonings can occur from both naturally occurring environmental toxins (e.g., radon and arsenic) as well as from human-made toxins (e.g., cleaning agents and medicines). Children's risk of poisoning increases because of their behavior, knowledge, physiology, and exposures (see Table 8.13). As the Chairman of National Poison Prevention Week Council, Soloway (2004) wrote:

> *Every day, parents, grandparents, and caregivers are faced with children who develop new skills, seemingly overnight. We celebrate when children begin to crawl, walk, and climb. But those same developmental milestones coincide with an insatiable curiosity to explore the world—and one way that children explore the world is by picking things up and putting them into their mouths. Pills and berries look like candy, liquid cleaning products and automotive fluids look like soft drinks, and wild mushrooms just look like dinner.*
>
> (Soloway, 2004)

Other poisonings happen over time through passive environmental exposures. While these are less understood or recognized among the general population, they are no less damaging.

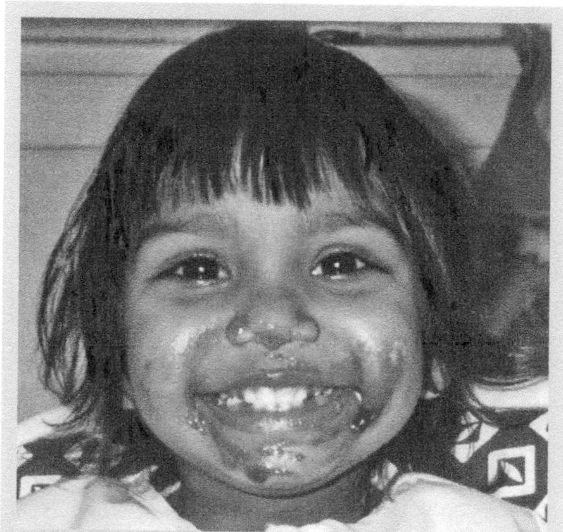

Figure 8.32 (Valerie Rice)
Children act quickly and so do poisons. Young children explore by putting things into their mouths. They cannot tell the difference between food and poisonous materials.

Table 8.13 Risk Factors for Poison-Related Injuries among Children

Behavior	• Very young children explore their world by putting things in their mouths (even dirt and bugs). • Young children love to mimic their parents and other adults. ° They may eat medicines they see adults eat or ingest chemicals or cleaning agents their parents use, because they do not know the difference.
Knowledge	• Most poison-related injuries occur because children cannot tell the difference between toxic materials and acceptable food and drink (Figure 8.33).
Physiology	• Children's bodies are less able to handle poisons because they are physically smaller and have faster metabolisms and respiration. • Children absorb nutrients (and poisons) more quickly than adults, given the same exposure level. ° For example, since a child's respiratory rate is faster, they inhale proportionally twice as much as an adult who is exposed to the same level of an airborne toxin (EPA, 2005). • As organs, the nervous and musculoskeletal systems are developing, they are more vulnerable to damage. Acceptable exposure guidelines, developed for adults, may not be acceptable for children.
Exposure	• Many contaminants are either in the soil or at higher concentrations at ground level—where many children like to play, including radon, pesticide vapors, and mercury. • Since small children crawl and play on the ground, they may come into greater contact with environmental toxins. • It is unknown exactly how early exposure to toxins will affect function in later life, yet scientists are setting "lifetime exposure rates." • This recognizes the fact that exposure to toxins, such as pesticides, may result in cancer many years later.

Table 8.14 Poisoning Statistics for Children

Deaths	• Annually, about 50,000 children up to 14 years of age die from unintentional poisonings worldwide (WHO, 2002);[a] In the United States, about two children under the age of 5 die each month from accidental poisonings (Stratton, 2004). • Low-income countries have more cases of unintentional poisonings of children because they lack the product-safety regulations and more effective education and emergency response programs found in more affluent nations.[b]
Injuries	• Most accidental poisonings occur in children. • Every 7 minutes a US child under the age of 5 arrives at an emergency room for accidental poisoning; this equates to 78,000 visits annually. • 90% of suspected exposures are from household products (CPSC, 2005). • Children under age 6 are at greatest risk, as they account for over half of the poison exposures (Litovitz, 2001). Even so, older children are also at risk; children over age 5 accounted for about 20% of emergency-room treatments during 2002 (NPPWC, 2005a,b).
Phone Enquiries to Poison Centers	• About half of all calls to poison centers in New Zealand are about children (NPC, 2001). • In the United States, emergency call staff receive 6500 emergency calls each day; half of these involve children under the age of 6 (Soloway, 2004). Over 1 million calls a year are for children under age 5 (CPSC, 2005).

[a] Children's deaths by poisoning account for about 5% of deaths by injury in less developed countries and about 2% in developed countries (Bates et al., 1997).

[b] For example, the death rate is 20 times higher in Vietnam than in Austria, Australia, or Germany.

WHEN AND HOW ACCIDENTAL POISONINGS OCCUR

Ninety percent of accidental poisonings occur in the home (Litovitz, 2001, Box 8.9). In the United States, poison control call centers receive the most phone inquiries regarding poison exposure of children between 4 and 10 p.m. and during the warmer months. Both of these times are when more children are at home and parents may be busier, such as preparing meals and getting ready for bed in the evening (NPPWC, 2005a,b).

Agents: Most deaths, injuries, and phone inquiries regarding poisons are for children's exposure to household agents, that is, plants and medicines found in and around the home (CPSC 2005; Box 8.9). This is followed by cosmetics and agricultural agents (CPSC, 2005; NPC, 2001; NPPWC, 2005a,b). In the United States, 60% of exposures are of nonpharmaceutical agents; 40% are pharmaceuticals.

BOX 8.9 COMMON HOUSEHOLD POISONS FOR CHILDREN

- Personal care products (cosmetics)
- Cleaners
- Over-the-counter pain relievers
- Hydrocarbons (lamp oil, furniture polish)
- Plants
- Pesticides
- Art supplies
- Alcohol
- Supplements (vitamins, iron)

Figure 8.33 (Ursy Potter Photography)
Many poisonous products look like and come in similar looking containers, to drinks or food. Cleaning agents and juices often look alike, as do some candies and pills. Store food, household, and chemical products in separate areas.

The risk of poisoning depends on:

- how easily children can find and access the product
- its toxicity
- how quickly the poison acts
- how long it takes to get emergency assistance

GENERAL SOLUTIONS

Preventing accidental poisonings requires design changes at micro (product packaging) and macro (national and community) levels.
The design changes include:

- Child-resistant packaging
- Labeling
- Product reformulation
- Caregiver education
- Nationally accessible poison control centers
- Appropriate interventions

Poison Centers

The World Health Organization lists worldwide poison control centers at their Web site: http://who.int

In all areas of the United States, a person can call a single number—(800) 222–1222—and talk with local experts including nurses, pharmacists, and physicians who will answer questions or talk a caregiver through helping a child who has been exposed to a potential poison.

There are 62 poison control centers in the United States. These also provide community education through handouts, Web sites, and on-site talks.

CHILD-RESISTANT PACKAGING: Child-resistant packaging will not keep all children out of containers holding potentially harmful substances; some children will find a way to open them. Even so, these designs provide an additional barrier that helps slow children down, enabling a caretaker to intervene and take the substance from the child before they have a chance to ingest it.

For children aged 4 and under who have ingested oral prescription drugs (NPPW, 2005b):

- 23% of the medications belong to someone who does not live with the child
- 17% belong to a parent or great-grandparent*

The lack of child-resistant packaging products and the misuse of packaging increase risk. For example, many daily and weekly pillboxes are not child resistant. Adults may not believe children can reach or ingest medications.

Packaging misuse occurs when adults take medications (or other household products) home and repackage them in nonresistant containers. Sometimes child-resistant tops are difficult to open, so adults leave the top slightly ajar. The convenience is not worth it when small children are in the house.

Poisonings also occur when children visit homes that do not normally have children living there, or when visitors who do not live with children come to visit a home with children.

* In the US, two million children live in a households headed by a grandparent (Cosby et al., 2004), so there could obviously be other reasons, in addition to children not being around as frequently, that contribute to this statistic.

Child-Resistant Packaging
The Poison Prevention Packaging Act*

Goal: Reduce accidental poisonings by providing a barrier to access for children under the age of 5.

Regulation: Packaging must prevent at least 80% of those children tested from opening the container during a 10 min test, according to the US Consumer Product Safety Commission (USCPSC).

The Environmental Protection Agency requires that most pesticides be in child-resistant packaging.

Has child-resistant packaging helped? Absolutely. The USCPSC estimates that since the requirements went into effect (1972), about 900 lives have been saved. The estimated fatality reduction is 45%.

Does this solve the problem? No! No packaging or cap can totally prevent children's access. Parents and other caregivers need to be careful to lock products where children cannot reach them and carefully supervise children when the products are in use.

National Poison Prevention Week Council FAQs
http://poisonprevention.org

LABELING: Labels provide information to users on how to use a product, as well as information on ingredients and potentially harmful effects.

In the United States, labels for hazardous materials must include:

- the words *Warning* or *Caution*
- a list of the ingredients that are at a hazardous level
- a description of possible harm, if the product is used improperly†
- instructions on how to use the product properly and safely
- a phone number for information (usually the manufacturer)
- if the product is not appropriate for children, the label must say so

Overall Prevention Tips

Physical designs are the best, because they eliminate human error. However, no label or packaging design will be 100% effective in preventing accidental poisonings. Education can help prevent mistakes due to lack of knowledge. Box 8.10 shows some common misconceptions about preventing poisonings among children.

* Products covered include aspirin and aspirin substitutes (acetaminophen, Tylenol), oral prescription drugs, drugs and dietary supplements containing iron, ibuprofen, antidiarrhea medicines (loperamide), preparations containing lidocaine and dibucaine (anesthetic medicines), mouthwash containing 3 g or more of ethanol (alcohol), naproxen, ketoprofen, certain types of liquid furniture polish, oil of wintergreen, drain cleaners, oven cleaners, lighter fluids, turpentine, paint solvents, windshield washer solutions, automobile antifreeze, fluoride-based rust removers, minoxidil, methacrylic acid, and hydrocarbons (hydrocarbons are in many automotive additives and degreasers and cosmetics such as baby oils and nail polish remover).

† In an attempt to help children who are too young to read and understand labels, the Children's Hospital of Pittsburgh invented Mr. Yuk. Parents and caregivers can explain the meaning of the symbol, obtain them from poison control centers, and place them on objects within the home. These stickers are free. *See Chapter 14 on Warnings for more details about "Mr. Yuk."*

Additional tips to prevent accidental poisonings:

- Keep poisonous products out of reach and in locked areas (Table 8.9 and Table 8.10 provide reach information for children).
- Keep toll-free nationwide poison control center number (in the United States: 800-222-1222) and other emergency medical service numbers near every telephone.
    - Information to include: the child's age, height, weight, and health history; what, when, and how much was ingested (inhaled or splashed on the skin); whether the child vomited and the child's current behavior. If told to go to the hospital emergency room, take the poison with you.
- Keep activated charcoal at home—use on advice of a poison control center or a physician only.
- Keep ipecac (ip-eh-kak) at home—use on advice of poison control center or a physician only.
- Stay informed.
    - Know which household products are poisonous. For example, the alcohol content in mouthwash can make it poisonous, if ingested in large amounts.
- Be cautious; supervise children esp. when adults are using poisonous products.
- Use extra caution during mealtimes or during disruptions of the family routine, many poisonings occur during these times.
- Know symptoms of poisoning (Box 8.11) (NSKC, 2005).

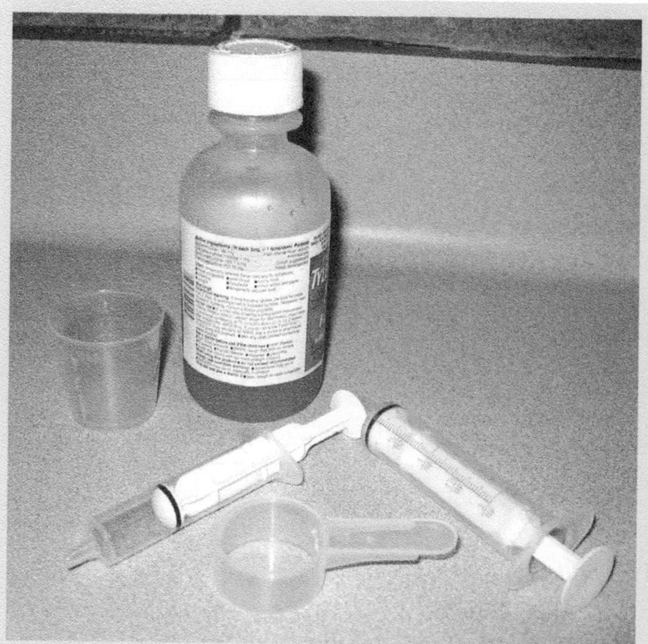

Figure 8.34 (Valerie Rice)

Figure 8.35 (Valerie Rice)

Often the instruments for measuring out dosages are difficult to read or use. One report revealed that 70% of parents or caregivers were not able to measure the correct dose when giving over-the-counter medicines to their children (NPPWC, 2005a,b).

BOX 8.10 COMMON MYTHS

Myth	Reality
As long as I keep medications in child-resistant containers, my children are sufficiently protected.	There is no such thing as a childproof container. Child-resistant containers are designed to slow a child's opening the container, so a caregiver can intervene before the child can ingest the contents.
Children will not drink or eat items that smell or taste bad.	Children eat and drink items that have strong, unpleasant odors and tastes, including pine oil cleaners, bleach, gasoline, and diapers.
As long as I am there, my children will not pick up things they should not to eat or drink. It is only when they are alone that I have to worry.	Most poisonings happen while the product is in use. Keep an eye on your child when cleaning or using cosmetics.
If I place cleaning products and other poisonous substances up high, my children will be safe.	Parents often believe they are placing items high enough, however: • Children often climb in order to access cabinets, even those above the stove and refrigerator. • Parents underestimate children's reach, climbing abilities, and curiosity.

COMMUNITY APPROACH

Steps that can affect poisonings on a local level:

1. Identify poisoning trends in your community or school.
2. Raise awareness among communities about local and national trends.
3. Teach parents, caregivers, and children how to identify causal agents and prevent accidental poisonings, and teach their children about poisons.
4. Participate in a National Event:
   - National Poison Prevention Week annually in March. Community- and school-based activities are available at http://aap.org
   - National Lead Poisoning Week in October. Information is available at http://epa.gov

BOX 8.11 SYMPTOMS OF POISONING

The child:

- is sleepy though it is not bedtime or naptime
- cannot follow you (or another object) with their eyes
- experiences nausea, dizziness, vomiting, diarrhea, rapid heart rate, or low blood pressure
- may have a rash or burn

The child's:

- eyes may go back and forth or in circles
- vision may be double or they may have no vision
- mouth appears burned or stained
- breath smells strange
- behavior may be different from normal

SPECIFIC CAUSAL AGENTS AND SOLUTIONS

Appendix A contains a poison supplement on causal agents and solutions. These include lead, arsenic, carbon monoxide, art supplies, and household items (Figure 8.36).

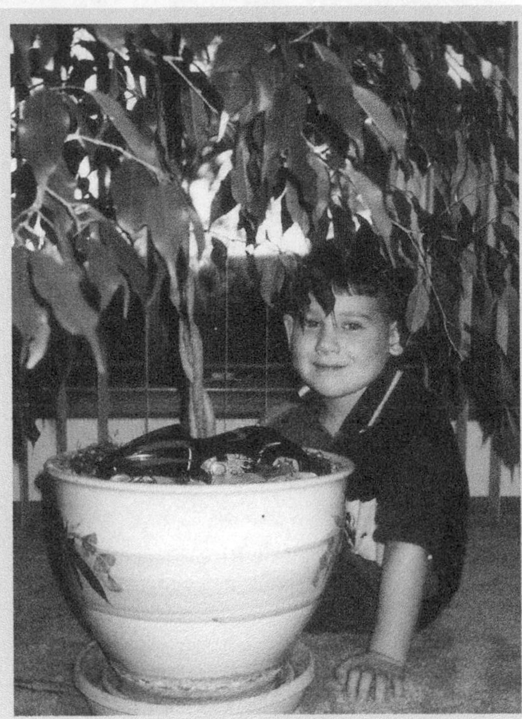

Figure 8.36 (Valerie Rice)
Children at greatest risk of accidental poisoning from eating plants are toddlers and young children who are still experimenting with putting items in their mouths. Ingestion of plants or plant parts can cause minor illness or risk of death, depending on the plant and the size and susceptibility of the child. Some cause highly allergic reactions. The best solutions are to know your plants and not keep poisonous plants in or around the home. If you must have them, fence them so that children cannot reach them (see Appendix A for more information).

CHOKING

INTRODUCTION

In the United States, mechanical airway obstruction from choking, suffocation, and strangulation is the leading cause of unintentional injury, resulting in death of children under a year old (Box 8.12). It is the fourth leading cause of death in children aged 1 to 9 years, surpassed only by motor vehicle injuries, drowning or submersion, and fire or burns (NPIPC, 2000).*

> My son had choked on a hot dog and I just couldn't believe it. He was gone within 5 or 6 minutes.
>
> Joan Stavros Adler
>
> Her son Eric was only 4 years old.
>
> Eric Stavros Adler Web site
> http://ericadler.net

Children less than 3 years of age are at highest risk of death and injury from mechanical airway obstruction. This is due to their immature anatomy and developmental stage. Young children:

- do not have a fully developed set of teeth for thorough chewing
- have airways of a smaller diameter and their swallowing mechanism is immature and less effective than an adult's
- lack experience and cognitive skills to avoid choking (for example, they forget to eat slowly and chew well)
- have a natural tendency to put everything in their mouths

* Some progress in the prevention of deaths from unintentional suffocation has been made through federal regulation and public education. In the United States, the death rate from unintentional suffocation for children aged 0 to 4 years has fallen from 3.46 in 1981 to 2.78 per 100,000 in 1998 (NCIPC, 2000).

Figure 8.37 (Valerie Rice)

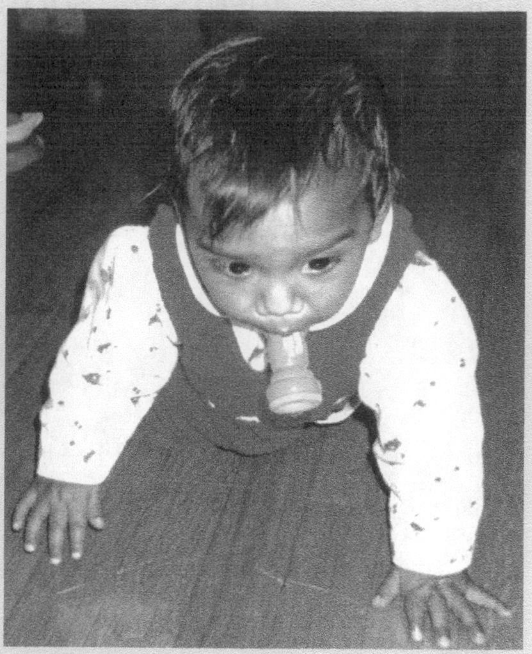

Figure 8.38 (Valerie Rice)

Figure 8.39 (Valerie Rice)

Children under age 1 are most at risk of choking. Children choke on everyday items, including common foods. It happens in a matter of minutes and more often than most people think. The implications are serious; choking can cause death or permanent brain damage.

The child above puts a noisemaker into her mouth and then crawls with it in her mouth. Although cute, it is small enough for her to choke on. Toys small enough to choke on will generally fit through a 1.25 in. (3.2 cm) circle or are shorter than 2.25 in. (5.7 cm) long for children under 4.

A better idea is to give children items that they can easily grasp, but are too big for them to choke on, such as the teething ring on the left.

BOX 8.12 CHILDREN AND CHOKING

Definitions

- *Choking*—the interruption of respiration by internal obstruction of the airway, usually by food or small toys in young children.
- *Suffocation*—obstruction of the airway from an external object that blocks the nose and mouth, such as a plastic bag.
- *Strangulation* also results from external compression of the airway from an object, such as a string caught around the neck.

Facts

- There were over 17,000 visits to US emergency rooms for nonfatal choking in children aged 14 or younger (60% on food and 31% on nonfood items).*
- Sixty-two percent of toy-related fatalities are from choking and suffocation or asphyxia.
- The greatest risk is among children under age 1; risk decreases as children grow older.
- Injury from unintentional suffocation is more common in males than in females.
- About every 5 days, a child in the United States dies from choking on an object.

WHY DO CHILDREN CHOKE?

MOUTHING: One reason why young children choke is because they put so many things into their mouths. Young children mouth items to explore their environment. It is a natural and normal part of their development (see Chapter 2 on developmental stages in children).

Children under age 3 and especially those under the age of 1, spend the most time mouthing objects. They also put the greatest variety and number of items in their mouths.

Children will put almost anything into their mouths, without discrimination, including their own diapers, restraints, bathing and cooking items, even furniture and safety gates. The risk for choking increases when they begin to crawl and walk, as they have greater access to household items (DTI, 2002).

What happens when a child chokes?

- Air enters the lungs via the trachea (windpipe). Choking occurs when an item goes into the trachea instead of the esophagus (food pipe).
- Coughing may remove the item if the object is not too far into the trachea (epiglottis).
- It is a medical emergency when an item totally blocks the airway.

EATING: Many parents do not realize that some foods are hazardous for their children, even some of the more common foods that children like, such as hot dogs and lollipops. Yet, these foods, as well as others, pose a danger to children (Table 8.15).

Young children's mouth muscles are not as strong as adults. Their molars (for grinding foods) come in at about 12 to 18 months of age, but it can take another 2 years or more until they are fully in. If they cannot chew their food, it means they will try to gulp their food and swallow it as a whole, putting them at risk of choking.

* Based on 2001 data. In this year there were 169 choking deaths among children ages 14 and younger (30% on food and 70% on nonfood items).

Some teething medications numb mouth and throat muscles. This could make it more difficult for children to swallow. Children may also eat too quickly, walk or talk while they eat, or gulp their food due to hurrying or distractions.

Some foods fit exactly into a child's windpipe, effectively blocking their breathing. Hot dogs are an example. They are soft and mold somewhat to fit the windpipe, especially when cut horizontally.

PREVENTING CHOKING

Prevention is the first line of defense for any hazard.

FOODS: Most childhood choking injuries occur while eating. Children are most likely to choke on small, round foods (Rimell et al., 1995). Table 8.15 lists food types that may pose choking hazards, Table 8.16 lists common foods children choke on and suggests choking prevention measures. Table 8.17 lists general prevention strategies (also see Figures 8.40 and 8.41).

Some children will try to pack many small items into their mouths at once, such as raisins. Cooking them until they are soft and mushy will prevent them from being a choking hazard.

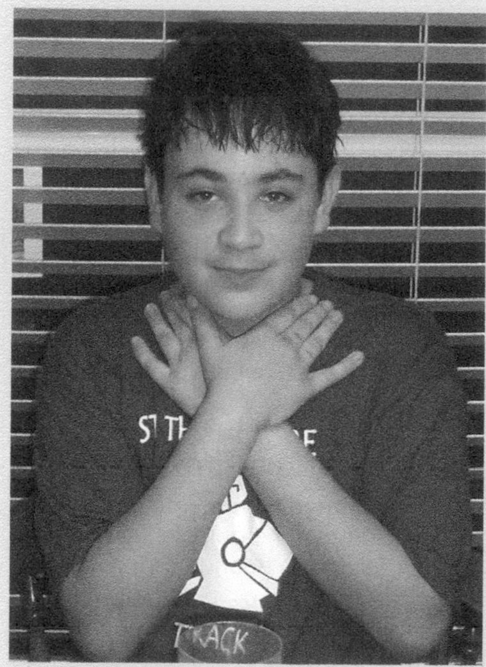

Figure 8.40 (Valerie Rice)
Teach children the international symbol for choking (above).

While prevention is the first line of defense for any hazard, preparation for emergencies can save a life. Know basic emergency first aid and cardiopulmonary resuscitation (CPR) techniques.*

Emergency medical treatments for infants and young children differ from those for adults.

Figure 8.41 (Ursy Potter Photography)
Each of the foods pictured above is a favorite of children: popcorn, grapes, hard candy, carrots, and hot dogs. Yet, each can also be a choking hazard. Table 8.16 and Table 8.17 give suggestions on choking prevention with these (and other) foods.

* The American Academy of Pediatrics recommends the American Heart Association's Pediatric Basic Life Support course (800-242-8721) or the American Red Cross's Infant and Child CPR Course.

Table 8.15 Characteristics of Foods that Pose Choking Hazards		
Food Characteristic	**Danger**	**Examples**
Firm, smooth, and slippery foods	Slides down a child's throat before they have a chance to chew it	Nuts and hard candies
Small, dry, and hard foods	Difficult for children to chew and they may attempt to swallow them whole	Popcorn, raw vegetables, hard pretzels
Sticky or tough foods	Get stuck in airway and do not break apart easily for removal from airway	Peanut butter, sticky candies (taffy, caramel), gum, meat, dried fruits, marshmallows
Fibrous, tough foods	Hard to chew so children may try to swallow them whole	Meats (steak, roast, and other fibrous meats, including meat jerky), bagels

Table 8.16 Serving Suggestions to Prevent Children from Choking on Foods	
Foods	**Suggestions**
Hot dogs and sausage links	Cut hot dogs lengthwise and then horizontally, not in round slices
Hard candies including jelly beans	Do not serve to children under age 5[a]
Grapes	Remove seeds, slice into 0.6 cm (0.25 in.) pieces and crush
Nuts	Chop into small pieces or avoid for children under age 5
Raw vegetables or fruits	Cook until soft and mash[b]
Popcorn	Do not serve to children under age 5
Raisins	Cook until soft and mushy
Fruit with seeds or pits	Remove seeds, chop into 0.6 cm (0.25 in.) pieces or cook until mushy
Meat	Remove bones, cut into 0.6 cm (0.25 in.) pieces. Serve only very tender meats
Frozen banana pieces	Do not serve to children under age 5
Cheese chunks	Cut into 0.6 cm (0.25 in.) pieces
Peanut butter	Never serve on a spoon, spread thinly over bread or crackers

Note: In general, caretakers should cut food into pieces that are ≤0.6 cm (0.25 in.) for infants and ≤1.3 cm (0.5 in.) for toddlers and children aged 5 and under.

[a] Some health care professionals suggest breaking the candies into smaller pieces, which a child can suck on, but this is not recommended.

[b] These include carrots, broccoli, cauliflower, and apples.

Table 8.17 General Suggestions to Prevent Choking on Food

Caregivers	
Supervision	• Always supervise children while they are eating. This includes babies using bottles, as a propped-up bottle can block a child's airway or a baby can choke on their own "spit-up" milk. • Be especially careful during gatherings when adult-type snacks are out and more easily available to children. • Do not allow siblings to feed younger children without adult supervision; they often do not cut food properly or mistakenly feed younger children inappropriate foods.
Serving	• Follow the serving recommendations mentioned above. • Foods that dissolve easily or are in tiny pieces are best: 0.6 cm (0.25 in.) for infants, 1.3 cm (0.5 in.) for toddlers up to age 5. • Cook, mash, grate, or avoid hard foods for young children. • Cut meat and cheese into small pieces. • Remove tough skins from sausages and frankfurters. • Do not feed infants while they are laughing or crying. • Have children sit while eating. • Teach children to take small bites and chew thoroughly.
Distractions	• Minimize distractions while eating. • Help children slow down and pay attention to their eating. • Do not allow children to rush, walk, run, talk, or laugh while chewing and swallowing. • Do not allow young children to eat in cars. It is difficult for a driver to notice a child choking and to pull over to assist a choking child in time.
Food Producers	
Warnings	• Put warnings on foods that pose a significant choking hazard to young children. • Put warnings in appropriate language for caregivers and in pictures for children. • Put suggestions for caregivers on foods that pose a choking hazard.
Design	• Redesign foods (especially those children prefer) into shapes, sizes, or consistencies that are not likely to block an airway.
Other Suggestions	
	Develop regulations for foods that pose choking hazards for children to include: • Food and Drug Administration authority for recall • Establish a national database of food choking incidents • Labeling of foods that pose choking hazards

TOYS AND OTHER NONFOOD CHOKING HAZARDS: Children also choke on items other than food. Most often, these include small items left around the house or small toys or toy parts (Table 8.18). Although most choking injuries occur in children under age 3, older children also put things in their mouths and choke on them (Figure 8.42).*

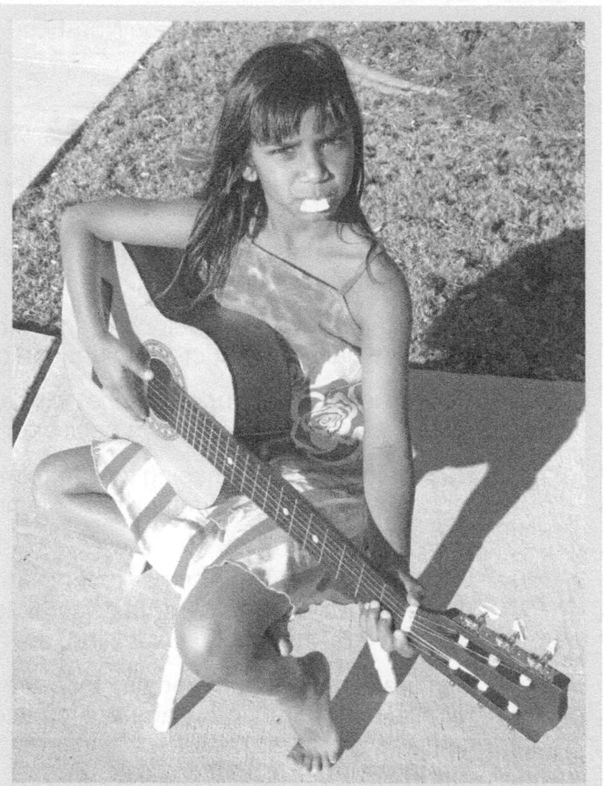

Figure 8.42 (Valerie Rice)
Choking injuries occur most often among children under age 3 and particularly among those under age 1 who are still "mouthing." Yet older children also put objects in their mouths, such as this little girl holding a guitar pick in her mouth.

Table 8.18 Children Choke on These Nonfood Items

Household Items	
Arts and crafts materials	Ballpoint pen caps
Ball bearings	Buttons
Coins	Jewelry
Paper clips	Pieces of plastic packaging or bags
Pieces of paper, foil, plastic, or metal	Safety pins
Small pieces of hardware (screws)	Watch batteries
Small Toys or Parts of Toys	
Toy cars with removable rubber wheels	Plastic "bricks" such as Legos
Plastic beads	Small balls (including marbles)
Deflated balloons (balloon strings can also be a strangulation hazard)	

In the United States, the *Child Safety Protection Act* (CSPA) bans toys that children under age 3 might choke on, breathe in, or swallow.† For similar toys for children from ages 3 to 6, the CSPA requires warning labels on the packaging. Toys small enough to fit through a 3.2 cm (1.25 in.) circle or toys that are smaller than 5.7 cm (2.25 in.) long are unsafe for children under 4.

* The risk of choking is higher in homes where there are older brothers or sisters. This may be because toys with small parts are more available.

† Other countries have similar standards, such as Australian Standard AS1647. See Chapter 12 on toys.

Table 8.19 Prevention of Choking on Non-Food Items	
Parents and Caregivers	

Purchasing	• Adhere to recommended age ranges when purchasing toys.
	• Some parents purchase toys that are not appropriate for their children, in spite of labeling (see Chapter 14). Reasons include:
	  ◦ Some parents do not understand why a toy is hazardous and view the warnings as informational only (Langlois et al., 1991).
	  ◦ Some parents think their children are more mature than the “average” child and therefore are able to handle products for older age children.
	• Do not use pacifiers or rattles that will easily break (be especially aware of using items not specifically made for use as a pacifier or rattle).
	• Be careful about purchasing toys from manufacturers that do not adhere to specific standards (such as from other countries or homemade toys). Ask the maker about the safety, breakage, and small parts.
	• Consider purchasing a “small parts tester” to evaluate whether small toys or other items may present a choking hazard. In general, toys that fit through a toilet paper tube (smaller than 3.17 cm x 5.71 cm (1.2 in. x 2.2 in.)) are hazardous for young children.
	• Be careful of toys from vending machines, they may not adhere to the same safety regulations for size nor do they have age-appropriate warnings.

Supervision	***“Childproof” the house:*** • Inspect children’s toys for lost parts. • Keep small items off the floor and make frequent checks for small items (including behind cushions and under furniture). • Regularly inspect old and new toys for damage. • Teach older children to keep their toys off the floor. • Never let a child under the age of 8 play with balloons without adult supervision. • Never let children chew on balloons or balloon pieces. • Keep small items out of reach.[a] • Do not let young children play with toys designed for older children. • Adhere to the safety labels on children’s products and toys. • Be careful of beanbag chairs or toys stuffed with polystyrene beads, children can easily inhale the beads should the covering rip.
Storage	• Keep plastic disposable diapers out of reach of children; cover the diapers they wear with clothing. • Hang mobiles and toys out of reach.

Table 8.19 Prevention of Choking on Non-Food Items (Continued)

Manufacturers	
Design	• Design products in the safest manner possible to prevent choking hazards for young children, that is: ◦ Adhere to standards regarding size of toys and toy parts. ◦ Anticipate the use and potential misuse of toys and toy parts by children. ◦ Consider the physical and cognitive development of the target (child) audience. ◦ Conduct user tests with toys. • Although children will put nearly anything into their mouths, especially when they are younger, there are some design principles that manufacturers can consider. These include: ◦ *Color:* Design items that do not target young children, that is, use less desirable colors (not in the bright, primary colors young children prefer). ◦ *Taste and smell:* Although taste and smell may not deter very young children, unpleasant tastes and odors may keep slightly older children from putting items in their mouths. ◦ *Texture:* There is no information available on appeal (or lack of appeal) of different textures for children during mouthing. Research might reveal some useful information for design.
Labeling	• Label products with safety warnings when it is not possible to design toys or toy parts to prevent choking or the product is for an older child that needs the smaller parts to match their developmental level (see Chapter 14 on Warnings). • Warnings that specifically explain why a recommendation is made are more effective. • Consider using an additional warning with a picture to designate that the product is for older children. Make the warning recognizable by children.
Community	
Parent Education	• Provide information to prevent choking through community wellness programs or health care professionals. • Well-baby clinic staff can give parents information on: ◦ Mouthing at the 4th to 6th month visit. ◦ Safe foods at the 6th month visit. ◦ Suggestions for childproofing the home as children gain mobility at the 9th to 12th month visit.
Daycare Providers & Baby sitters	• Parents are not the only caregivers who need this information. Daycare providers and babysitters also need to understand choking hazards, prevention strategies, and first aid.

[a] These include jewelry, buttons, safety pins, and hardware such as nuts, bolts, and screws.

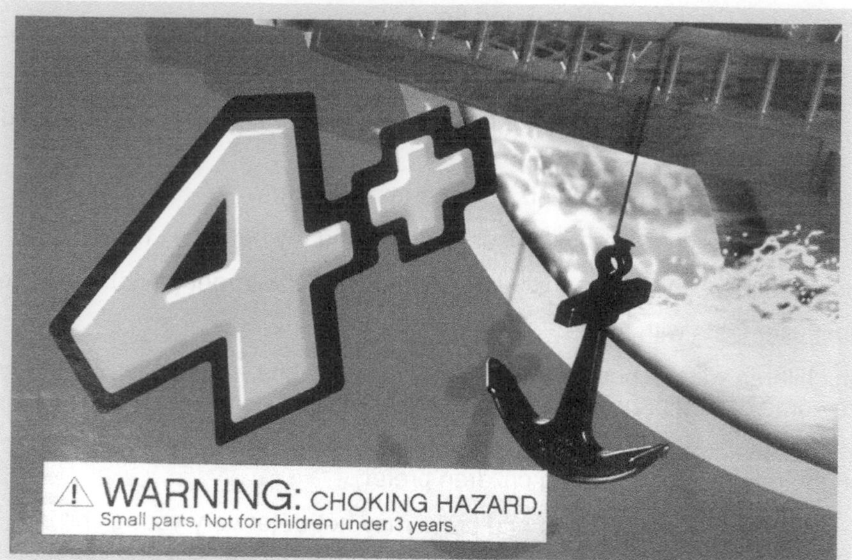
4+
WARNING: CHOKING HAZARD.
Small parts. Not for children under 3 years.

Figure 8.43 (Ursy Potter Photography)
This warning label appears on a box of Legos, which contains small "bricks" for building. The large "4+" makes it clear the toys are for children aged 4 and over. It is also clearly offering a choking hazard warning. Yet, the smaller print states "small parts, not for children under 3 years." Some parents may become confused about whether their 3-year-olds can use them safely.

DROWNING

INTRODUCTION

The story is familiar:

> *"I went into the kitchen to get some iced tea. It just took a few seconds."*
> *"I looked away for a minute; I turned the burgers on the grill, turned around and...."*
>
> *"In children, a lapse in adult supervision is the single most important contributory cause for drowning"*
>
> (WHO, 2005)

Most caregivers do not realize that young children do not tend to splash or cry out when they get into trouble in the water. Instead, they slip away quietly and without warning (CPSC, 2003; Brenner, 2003) (Figures 8.44 through 8.47).

While there are many contributing factors, alcohol consumption and falling into water are common contributory factors in childhood and adolescent drownings in many countries (WHO, 2005; Howland et al., 1995; Howland and Hingson, 1988). Water clarity can be a factor, as lifeguards and caregivers may not be able to see children who need help in muddy waters. Large crowds of people also make supervision difficult.

Figure 8.44 (Valerie Rice) **Figure 8.45** (Valerie Rice)
It takes only seconds for a child to drown and childhood drownings (or near-drownings) often happen while a parent or caregiver is present. There is simply a momentary gap in supervision.

Figure 8.46 (Valerie Rice)

Definitions

Drowning: Death by suffocation within 24 hours of immersion in water.

- *Wet Drowning*: The drowning victim inhales water, interfering with respiration and causing the collapse of the circulatory system.
- *Dry Drowning*: The victims' airway closes due to spasms caused by water. While less common than wet drowning, about 10% of submersed people become asphyxiated without ever inhaling water (WHO, 2005) (Fiore and Heidemann, 2004).

Figure 8.47 (Ursy Potter Photography)

Facts about Drownings

- *It takes 2 minutes after submersion for a child to lose consciousness.*
- *It takes 4–6 minutes for irreversible brain damage to occur.*
- *Children who survive are typically rescued within 2 minutes after submersion (92%).*
- *Most children who die are found after 10 minutes (86%).*
- *Nearly all who require CPR die or suffer from severe brain injury, if they live.* (NSKC, 2005)

PROBLEM

While information on childhood drownings are not collected in the same way in all countries, it is the fourth leading cause of death among 5- to 14-year-olds and the 11th cause of death among children under 5 years (Krug, 1999). Boys drown more often than girls, perhaps due to greater exposure and greater risk taking.

> About 9 out of 10 children who drowned were being supervised, usually by a family member (SafeKids, 2004).

Although drowning deaths have decreased (Branche, 1999), drowning remains the second leading cause of unintentional injury-related deaths among US children from ages 1 to 14 and the third leading cause for children under the age of 1. Most drownings occur in residential swimming pools (single-family homes or apartment complexes), yet children can also drown in any available water (CDC, 2002; Figures 8.48 through 8.50).

Figure 8.48 (Valerie Rice) **Figure 8.49** (Ursy Potter Photography)
Children can drown in as little as 1 in. of water. Infants often drown in bathtubs, toilets, and buckets. Younger children (1 to 4 years old) drown in swimming pools, wading pools, spas, and hot tubs. Older children drown more often in swimming pools and open water, such as oceans, lakes, and rivers (Brenner, et al., 1995).

Other US facts (NKSC, 2005):

- Four children are hospitalized for near-drowning for every child who drowns.
  - 15% of hospitalized children die
  - ~20% suffer severe, permanent disabilities
- Four children receive emergency-room treatment for every hospital admission.

> What is the cost?
>
> - Medical costs for a single near-drowning incident can range from $75K for emergency room treatment to $180K for long-term care in the United States.
> - Estimated costs for drownings and near-drownings of US children, 14 years old and younger, is $6.2 billion annually.

Who is at risk?*

- Children aged 4 and under account for about 60% of child-drowning deaths.
- The drowning rate for boys is two to four times that of girls.
- Girls drown in bathtubs twice as often as boys.
- Black children aged 14 and younger drown twice as often as white children.
- White children aged between 1 and 4 drown twice as often as black children, mostly from residential-pool drownings.
- Low-income children drown more often, from nonswimming pool incidents.
- Drowning fatalities are higher in rural areas, perhaps due to less access to emergency medical care.

(CPSC, 2003)

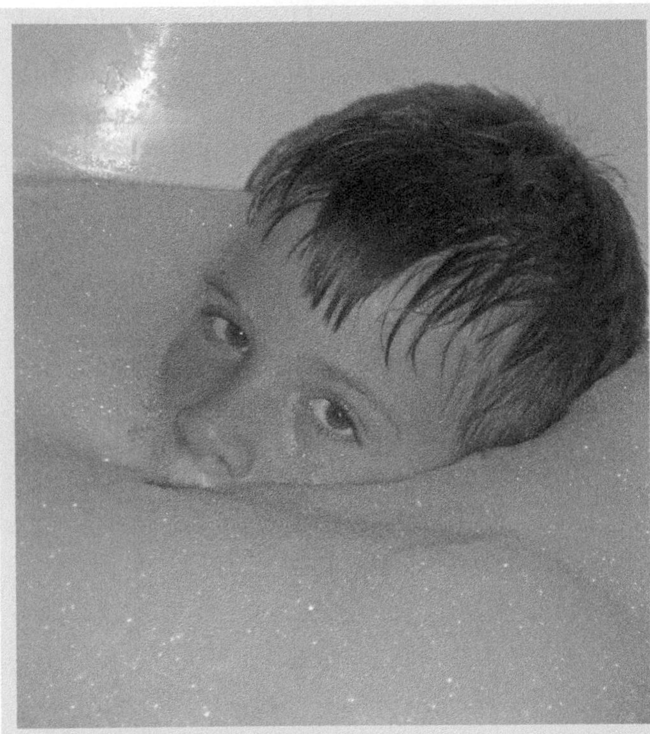

Figure 8.50 (Valerie Rice)

Facts about specific causes within homes (NSKC, 2005):

- About 10% of child drownings take place in *bathtubs*, most when there is temporarily no adult supervision.
- Since 1984, approximately 300 US children have drowned in *buckets* used for household cleaning. Most of these children were between 7- and 15-months-old and from 61 to 78.7 cm (24 to 31 in.) tall.

* See CDC (2005); Rodgers (1989); and NKSC (2005).

GENERAL SOLUTIONS FOR PARENTS:

- Know how to swim. Teach children how to swim (Figure 8.51).
- Never leave children without adult supervision when near water, even for a few seconds.
- Do not drink alcohol while supervising children's swimming or boating activities.
- Know how to give CPR.
- Swim where there are lifeguards.
- Have children wear life jackets if they have low swimming-skills.
- Teach children to swim with a friend. Have them learn water safety, including buddy rescue, as they mature.
- Be aware of water safety, including attempting activities that are beyond their capabilities.
- Know the water depth, especially before diving. Children (and adults) have sustained skull and neck fractures from diving into shallow water.
- Drowning occurs in the winter, too. Avoid walking, skating, or riding on weak or thawing ice on any body of water.

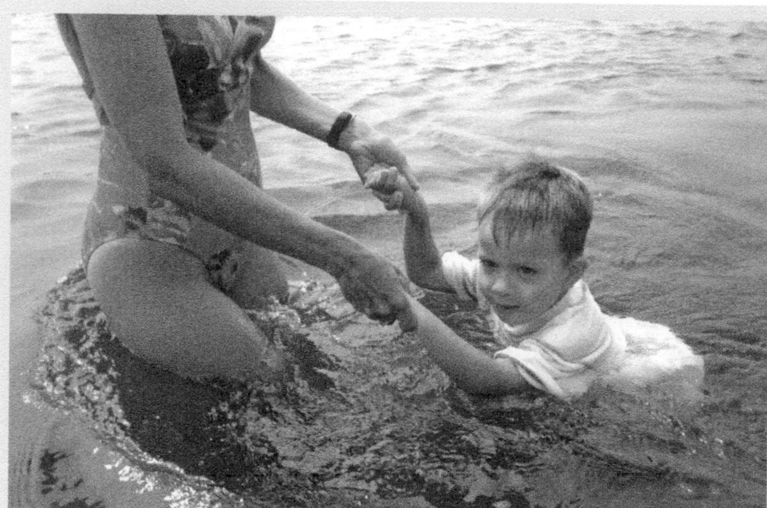

Figure 8.51 (Valerie Rice)

Of US children who drowned, 74% did not know how to swim (Safekids, 2004). Infant swimming or water adjustment programs do not guarantee an infant can swim, make appropriate decisions, or fail to panic in the water. The age to teach children to swim is the same as to ride a bike, age 5. Children under the age of 4 usually are not developmentally ready for formal instruction in swimming.

SOLUTIONS

***HOMES*:** Table 8.20 lists potential problems and solutions in homes for bathrooms, buckets, coolers, fish tanks, wading pools, spas, hot tubs, and whirlpool baths. Swimming pool suggestions follow.

Table 8.20 Suggestions for Preventing Home-Based Drowning among Children

	Problem	Solutions	
		Parents/Caregivers	Manufacturers
Bathrooms, Buckets, Coolers, Fish Tanks, and Wading Pools	Toddlers can easily lose their balance while exploring a water-filled container and fall in. Their arms are not strong enough to push and lift themselves out.	• Young children should not have free access to bathrooms, including toilets, sinks, and bathtubs. Place latches high on the bathroom door, above the reach of children. • Install childproof locks on toilet lids (similar to cabinet door locks). • Supervise children's bathing and toileting. Never leave a child under age 3 alone in a bathtub, even while in a "bathing ring." • Supervise children closely and limit their access while using five-gallon buckets. Empty the containers filled with liquids immediately after use (Figures 8.52 and 8.53). • Place wire mesh over fishponds and aquariums or place barriers around them (Figure 8.54).	• Place warning labels and instructions on infant "bathing rings." • Place warnings on products for children (wading pools) as well as other products (coolers, fish tanks). • Adhere to USCPSC voluntary guidelines for five-gallon buckets, for education, and for labeling.
Spas, Hot Tubs, and Whirlpool Bathtubs[b]	The *USCPSC* reports of the leading causes of deaths in children aged 4 or under in spas and hot tubs include: • Children's hair being pulled into the suction fitting drain of a spa, hot tub, or whirlpool bathtub, causing their heads to be held under water. • Strong drain suction causing body part entrapment.[a]	• Provide each spa or hot tub with a locked safety cover for when not in use. ∘ If missing or broken, drain and do not use the spa until the cover repair is complete. ∘ Covers designed to retain heat are not sufficient protection (children can slip into the water in spite of the cover). • Teach children to wear long hair up, to wear a bathing cap, and to keep their hair away from drains. • Each user (including children) should know where the cut-off switch is located (for the pump). • Keep water below 105°F. • Supervise children, always.	• Adhere to the US voluntary standards for drain covers *(ASME/ANSI A112.19.8M-1987* and Safety Vacuum Release Systems *(ASME/ANSI A112.19.17)* to avoid risk of hair entrapment. ∘ Safety vacuum release systems automatically shut off a pump upon detection of a blocked drain. ∘ Dome-shaped drain outlets and dual drains decrease the intensity of suction found with a single drain. • Locate cut-off switch in close proximity to users, that is, easy to reach and turn-off in an emergency.

[a] The ***USCPSC 2005 publication #363: Guidelines for Entrapment Hazards: Making pools and spas safer*** contains guidelines as well as a list of applicable standards.

[b] CPSC Documents #5067 and 5112.

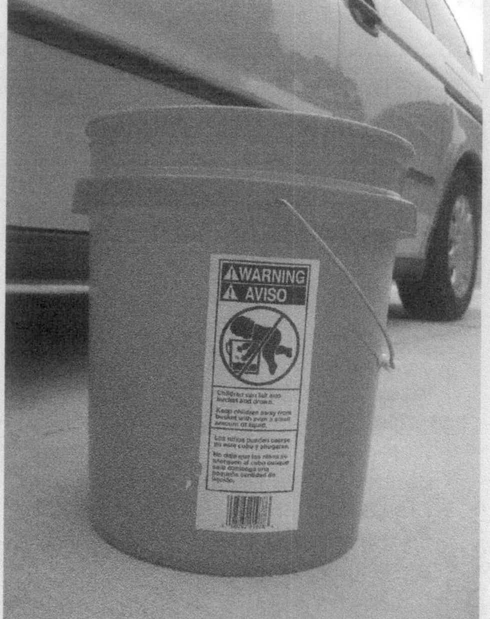

Figure 8.52 (Valerie Rice) **Figure 8.53** (Valerie Rice)
A child can drown in a few seconds. Caregivers must closely supervise children at all times when playing in water. Empty wading pools and buckets when not in use.

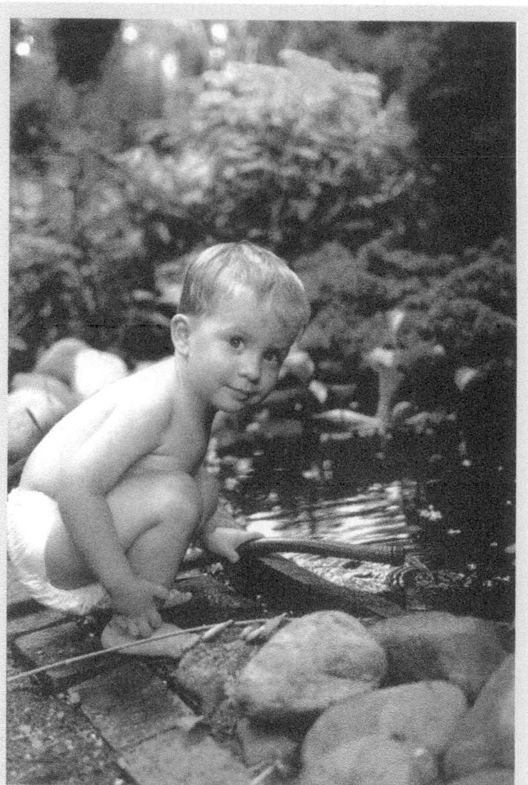

Figure 8.54 (Copy Berg)
Backyard fishponds are a source of delight to children. They can also be a drowning hazard. Consider placing wire mesh over them or barriers around them.

Swimming pools Most US childhood drownings of 1- to 4-year-olds occur in home pools (Brenner, et al., 2001) owned by parents (65%) and friends or extended family members (33%) (CPSC, #362) (Figure 8.55).

Most parents think that their homes are safer than they are:

A US study by the National Safekids Campaign found that of parents who own pools or spas (Safekids, 2004):

- 98% felt they had taken adequate measures to ensure children's safety, yet most had not made essential environmental changes.
- About 61% did not have isolation fencing around their pools or spas.
- 43% had no self-closing and self-latching gate.
- Installation and proper use of four-sided isolation fencing could prevent 50% to 90% of residential pool drownings.

Figure 8.55 (Valerie Rice)
In the United States, at the time of drowning, almost half of all children were last seen in the home. Most were missing for less than 5 minutes (77%) and in the care of one or both parents (CDC, 2005; SafeKids, 2005; Present, 1987). In 69% of cases, caretakers did not expect the children to be in or around the pool (CPSC, #359). Trespassing into pools accounted for less than 2% of drownings (CPSC, #362).

Above ground pools may be somewhat safer, as their construction serves as an additional entry barrier (Rodgers, 1989). Yet, there are unique concerns such as their shallow depth (eliminating diving activities) and side construction that does not support walking or sitting. One way to decrease pool access is to remove steps and ladders when the pool is not in use.*

Barriers While all swimming pools in Australia are required by law to be fenced, each state in the United States can have (or not have) its own regulations on fencing barriers. Barriers do not guarantee children's safety, but they do give parents additional time to find their child and possibly prevent a drowning. The CPSC has guidelines for barriers surrounding pools and some US localities have incorporated these into their building codes (CPSC, #362).

Children should not be able to get to a pool by going over, under, or through a barrier.

* Parents and caregivers should always keep a phone with emergency numbers, as well as rescue equipment, near any pool. The equipment should include a shepherd's hook, safety ring, and rope.

Alarms and Covers Door alarms, pool alarms, and automatic pool covers (power safety covers) add an extra level of protection (Fior and Heidemann, 2004).* Underwater pool alarms sound when something hits the water.

Table 8.21 provides guidelines for barriers, covers, and alarms.

Figure 8.56 (Valerie Rice)
Establish layers of protection for pools and water-play areas. For example, in this water-play area, there is a double barrier of thick bushes as well as a chain link fence.

There is a small ratio of life guards/supervisors to children and only a certain number of children can be in the area at any given time.

* It is best if alarms and pool meet standards such as those by the American Society for Testing and Materials (ASTM). For example, ASTM f-13-1491 applies to power pool covers.

Table 8.21 Guidelines for Barriers (Fences), Alarms, and Covers for Pools

Barriers (Fences)	***Barriers should include isolation fencing on all four sides:*** • A wall can provide a side of the barrier, as long as the doors and windows have an alarm that sounds when they open. It is preferable to keep the doors or windows shut and locked, similar to a locked gate. • The fence height should be at least 1.2 m (4 ft) high (121.9 cm [48 in.]) above grade (measured on the side of the barrier away from the pool) and should have self-closing and self-latching gates (Figure 8.57). • Fences should not have easy footholds or handholds to dissuade children from climbing over them. • Any openings in the fence should not be larger than 10.2 cm (4 in.) wide to prevent children from squeezing through (based on the head breadth and chest depth of young children) (Figure 8.57). • If fence construction consists of horizontal and vertical members, the distance between the horizontal surfaces should be greater than 114.3 cm (45 in.). If it is less than 114.3 cm (45 in.), then the horizontal surfaces should be inside the fence (same side as the pool). This helps prevent children climbing. • Self-latching devices should be located on the top of the gate, facing the pool, at least 137.2 cm (54 in.) above the bottom of the gate. If it is placed lower than 137.2 cm (54 in.), the release mechanism should be at least 7.6 cm (3 in.) below the top of the gate to prevent a child from reaching over the top of the gate and opening the latch. • Barriers should be closed and locked when the pool is not in use.
Alarms and Covers	***US recommendations state alarms should:*** • be heard in the house and in areas away from the pool • be in the control of adults and not located where children can disarm them • sound for 30 seconds or more, within 7 seconds after a door is opened or something hits the water[a] • sound different from other sounds in and around the home • be at least 85 dB, 3.1 m (10 ft) away from the alarm mechanism • have an automatic reset • have a temporary deactivation switch that children cannot access ***Pool Covers:*** • All pools should have covers. In the best circumstances, covers should be used when the pool is not in use. • Pool covers should withstand the weight of two adults and a child. This is for rescue purposes, as well as allowing for quick removal of standing water from the cover.

[a] UL 2017 General-Purpose signaling Devices and Systems, Section 77 provides guidelines for alarms.

Figure 8.57 (Ursy Potter Photography and Valerie Rice)

- The diamond shaped openings in chain link fences should be less than 3.8 cm (1.5 in.) (some suggest 1.75 in.) in diameter.
- A strong gauge mesh can cover the openings or slats to reduce their size, fastened at the top and bottom of the fence.
- If a strong mesh covers the chain link, it must be fastened securely and tight enough so that children cannot put their feet or hands into the openings.

Figure 8.58 (Valerie Rice)

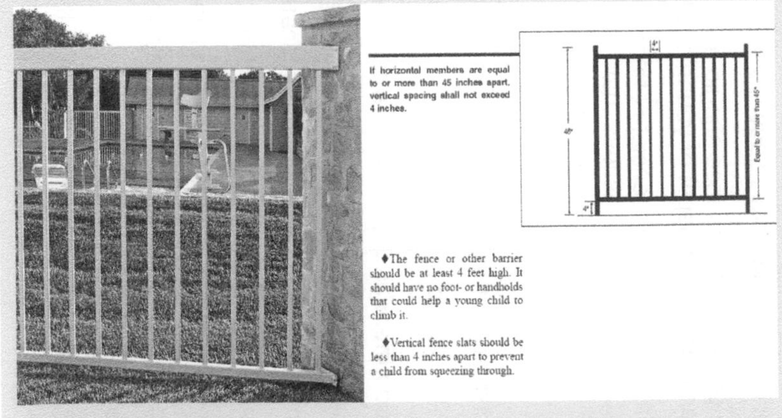

If horizontal members are equal to or more than 45 inches apart, vertical spacing shall not exceed 4 inches.
4"
48"
4"
Equal to or more than 45"
♦The fence or other barrier should be at least 4 feet high. It should have no foot- or handholds that could help a young child to climb it.
♦Vertical fence slats should be less than 4 inches apart to prevent a child from squeezing through.

Figure 8.59 (Valerie Rice; compilation with CPSC data by Ursy Potter Photography)
Space between vertical members should be less than 4.45 cm (1.75 in.), to reduce the possibility of a child getting a good foothold and climbing the fence (decorative cutouts should also be less than 4.45 cm [1.75 in.]).

BOATING: US data on injuries and deaths among children involved in boating-related incidents are seen in Table 8.22 (including yachts, canoes, Jet Skis, and other boats). Also see Figures 8.62 through 8.68.

Table 8.22 Suggestions for Preventing Home-Based Drowning among Children

Important Facts
• Most (85%) of children in drowning deaths were not wearing a personal flotation device (PFD) (ARC, 2005). • Twenty-eight states have boating safety laws requiring children to wear PFDs at all times when on boats or near open bodies of water (varying in age requirements, exemptions, and enforcement) (ARC, 2005). • Air-filled swimming aids, such as water wings, cannot substitute for a PFD. • Properly fitting life vests could prevent ~85% of boat-related drownings (Safekids, 2004).

Figure 8.60 (Valerie Rice)

Figure 8.61 (Valerie Rice)

Figure 8.62 (Ursy Potter Photography)

Figure 8.63 (Valerie Rice)

Figure 8.64 (Ursy Potter Photography)

Most US states and territories (except Hawaii, Idaho, and Guam) have mandates for wearing life jackets (PDFs). However, they vary in specifics and requirements for age. Some other countries do not have such mandates.

Over half of US boating incidents involve problems involving operator error, often through lack of attention, inexperience, unsafe speeds, or lack of caution. Boat operators had not attended boating safety courses in approximately three-fourths of boating-related deaths (ARC, 2005).*

Guidelines:

- Have children wear coast guard–approved PFDs at all times while boating.
- Have rescue aids available while boating.
- Train crews in rescue procedures and provide clear information to passengers.

OPEN WATERS AND FARM PONDS: Figure 8.65 through Figure 8.68 depict issues and solutions for open-water and farm ponds.

* The American Red Cross' Web site at http://redcross.org includes many water-safety guidelines. These include specific suggestions for locations (beaches, boating, pools, lakes, rivers, oceans, and waterparks), types of watercrafts (boats, Jet Skis, Windsurfers, and rafts/tubes), and activities (snorkeling, scuba diving, swimming, tubing, rafting, and water skiing).

Figure 8.65 (Ursy Potter Photography)
Dangers in open waters, such as the ocean, include swimming in conditions that are beyond a child's ability, such as large waves, strong tides (riptides), offshore winds, and going out too far in the water.

Figure 8.66 (Ursy Potter Photography) **Figure 8.67** (Ursy Potter Photography)
A child can still drown wearing a life jacket if it does not fit properly. In the photos above, a mother rescues her child when the jacket held him afloat, but did not keep his head out of the water. Float rings (that surround a child) can also be a problem as a child can slip out of the float. In some cases, children have turned upside down while in a float ring and have been unable to right themselves. Close supervision is always necessary.

Figure 8.68 (Ursy Potter Photography)
Most accidental drownings in farm ponds happen to children aged 4 and under. Children lose their balance and fall, step into deep holes while wading, or fall through thin ice (Murphy and Steel, 2005).

- Place barriers around the ponds.
- Post NO TRESPASSING signs, with pictures of the hazard.
- Eliminate hazards such as steep banks and submerged objects.
- Mark drop-offs.
- Keep a rescue post by the pond with an attached rope, a buoy, and a sign with emergency numbers.
- Fill abandoned wells, if possible. If not, solidly cover abandoned wells and check them regularly.

CARETAKERS: Although this chapter focuses on injuries to children, caretakers also experience injuries sometimes due to their care-giving duties. For example, mothers sometimes experience musculoskeletal injuries associated with lifting and carrying their children.

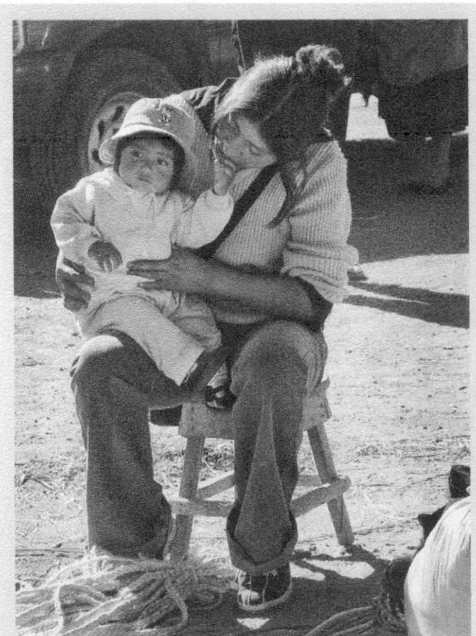

Figure 8.69 (Ursy Potter Photography)

Table 8.23 Childcare Practices that may Contribute to Musculoskeletal Symptoms and Injuries[a]

Practice
• Carrying a child in car seat
• Carrying a child on one hip
• Carrying a child while bending down
• Lifting a child up to or off a changing table
• Lifting a child into or out of a crib with high sides
• Lifting a child from the floor
• Standing, bent over to wash the child
• Changing a child on the floor, crib, or playpen
• Opening baby-food jars and cans
• Pushing a child on a seated toy
• Breast feeding in awkward positions
• Bottle feeding in awkward positions

Note: Most high-risk practices entail the use of the body, particularly the back or neck, in an awkward posture, often repetitively. An awkward posture is one that deviates from a neutral position (i.e., head upright, maintenance of vertebral curves, shoulders at one's side, elbows flexed to 90°, forearms not rotated, and wrists straight).

[a] Based on literature and expert opinion

Kroemer and Grandjean (2001); Sanders, (2004).

CONCLUSIONS

This chapter included a few of the more prevalent causes of unintentional injuries among children and adolescents. Certainly, the inclusion of additional sources and descriptions of injuries would be beneficial, such as overexertion and motor vehicle accidents or even unintentional injuries. Not including them is neither a comment on their gravity or a statement on the design implications for prevention. Instead, it is an acquiescence to time and space limitations.

The reasons children are at greater risk for injury are many and this chapter mentions each at least once. Knowledge of these reasons and risks are necessary to prevent children's injuries.

Children are curious—it is their "job" to explore, try new things, and learn through investigation. Exploration necessarily incurs surprises and discoveries—and some of them hurt.

For example, children of all ages test their physical prowess through jumping. At age 2, they jump from a step, at 8 they may leap from a swing, at 14 they take a running bound over a picnic table, and at age 16 they leave the ground and defy gravity on their skateboard!

Their actions seem to say, as Dr. Suess might: *I can jump. I can leap. I can jump with both my feet. I can jump with one or two. I can jump from high, can you? I can jump from a stair. I can jump from a chair. Over a bench, over a table. My jumps are great, my leaps are stable!...*until they fall, of course, and they will fall as it is part of the learning process.

Children are also more at risk because they are developing their muscles, tendons, bones, and nervous systems. This means their coordination is improving as they strive to harmonize their bodies and their minds into a fluid response. They may not react as quickly or accurately and if they are hurt, the injury may impact their development.

They are also developing mentally through trial and error experiences, as well as learning to trust the experiences of others as relevant to themselves. For a time, they will lack the ability to express themselves adequately. Later, they may fail to express themselves because of fear of loss, fear of punishment, or from a belief that no one will listen. Also, children do not understand the full consequences of their actions until they become young adults. Each of these puts them at additional risk of injury, intentional and unintentional.

Even their size increases their risk, as children are not easily visible from moving vehicles, and the design of much of their environment focuses on adult needs and use. Many of the tasks they do and the sports children play were developed for adults.

Adults easily misinterpret children's motivations and many discount their youthful opinions. Still others fail to recognize their physical, cognitive, and emotional strengths—and limitations. Whether children are seen as less or more competent than they are, both are failures to see them realistically and both put a child at greater risk.

Children depend on adults. If the adults (designers, caregivers, teachers) are unaware of dangers (and resolutions) associated with childhood injuries, how will the child avoid injury? For example, in the United States approximately 6% of children between ages 4 and 8 ride in booster seats, yet this is the recommended safety seat for this age group. This need cannot be answered by the children themselves. The answers must come from the adults who surround them.

Figure 8.70 (Valerie Rice)
Injuries also occur during play around the home. Children do not understand the consequences of hiding under, in, or on top of objects in and around the home.

Perhaps as children age, they will learn about and attend to injury information. However, they will need guidance for many years, as neurological brain development continues into their early 20s. Perhaps this is why:

- only 35% of high school students report they always wear their seat belt and
- alcohol is a major factor in 50% of drownings among young males

(US data, NCIPC, 2001)

How do we protect infants, children, adolescents, and teens from inevitable injury? By design.

- By designing products and places that fit their abilities, while challenging them to improve those same abilities
- By designing to eliminate human error when we can
- By designing safety and education programs so the right audience can take the information in, learn, remember, and act on it
- By designing the tasks to fit the child

REFERENCES

AAP (American Academy of Pediatrics) (1997) Fires and burns. In M. Widome (Ed.), *Injury Prevention and Control for Children and Youth*, 3rd edn. Committee on Injury and Poison Prevention, pp. 233–267.

Ahrens, M. (2001) The US fire problem overview report: Leading causes and other patterns and trends. Quincy, MA: National Fire Protection Association.

AIHW (Australian Institute of Health and Welfare). Available at: http://aihw.gov.au. Retrieved 2006.

Alliance to End Childhood Lead Poisoning. Available at: http://aeclp.org. Retrieved 2006.

American Academy of Pediatrics, American Public Health Association, and National Resource Center for Health & Safety in Child Care (2002) *Caring for Our Children National Health and Safety Performance Standards: Guidelines for Out-of-Home Child Care*, 2nd edn. Elk Grove Village, IL: American Academy of Pediatrics.

American Medical Association. Preventing Common Household Accidents. Available at: http://medem.com

Anderson, H. (1995) Anlagda branders omfattning, motiv och paverkande faktorer. Stockholms Universitet: Stockholm, as reported in SRSA, 1999.

ARC (2005) American Red Cross. Available at: http://redcross.org

ASTM (American Society for Testing and Materials). F977-03 Standard Consumer Safety Specification for Infant Walkers. Available for purchase at http://astm.org

Baker, M.D. and Bell, R.E. (1991) The role of footwear in childhood injuries. *Pediatric Emergency Care*, 7, 353–355.

Ball, D.J. and King, K.L. (1991) Playground injuries, a scientific appraisal of popular concerns. *Journal of the Royal Society of Health*, 111, 134–137.

Barlow, B., Niemirska, M., Gandhi, R.P., and Leblanc W. (1983) Ten years of experience with falls from a height in children. *Journal of Pediatric Surgery*, 18(4), 509–511.

Bates, N., Edwards, N., Roper, J., and Volans, G. (Eds.). (1997) *Handbook of Poisoning in Children*. London: Macmillan.

Bearer, C.F. (1995 Summer/Fall) Environmental health hazards: How children are different from adults. *The Future of Children*, 5(2), 11–26. Available at: http://futureofchildren.org

Beyond Pesticides. Available at: http://beyondpesticides.org

BHO (British Home Office) (2005, retrieved) Crime reduction toolkits: Offender profile: Age and gender. Available at: http://crimereduction.gov.uk

Blaxall, M.C.D. (1995) Child falls, morbidity, and mortality in the North Health Region. Auckland, NZ: Safekids, Starship Children's Hospital .

Branche, C.M. (1999) What is happening with drowning rates in the United States? In J.R. Fletemeyer and S.J. Freas (Eds.), *Drowning: New Perspectives on Intervention and Prevention*. Boca Raton, FL: CRC Press.

Brenner, R.A. (2003) Prevention of drowning in infants, children, and adolescents. *The American Academy of Pediatrics*, 112(2), 437–440.

Brenner, R.A., Trumble, A.C., Smith, G.S., Kessler, E.P., and Overpeck, M.D. (2001) Where children drown. *Pediatrics*, 108(1), 85–89.

Briss, P.A., Sacks, J.J., Addiss, D.G., Kresnow, M., and O'Neill, J. (1994) A nationwide study of the risk of injury associated with day care center attendance. *Pediatrics*, 93, 364–368.

Butts, D. and Peterson, J. (2004) Grandparents raising children. In A.G. Cosby, R.E. Greenberg, L.H. Southward, and M. Weitzman (Eds.), *About Children*. Elk Grove Village, IL: American Academy of Pediatrics.

Campbell, M., Ferguson, J., and Beattie, T.F. (1990) Are falls from supermarket trolleys preventable? *BMJ*, 301, 1370.

CCA Wood, United States Environmental Protection Agency. Available at: http://epa.gov
New York State Department of Environmental Conservation. Available at: http://dec.state.ny.us
Consumer Product Safety Commission. Available at: http://cpsc.gov

CDC. Centers for Disease Control and Prevention. Available at: http://cdc.gov

CDC (1992) Centers for Disease Control and Prevention. Unintentional deaths from carbon monoxide poisoning—Michigan, 1987–1989. *MMWR* 41(47), 881–883, 889, 1992.

CDC (1995) Centers for Disease Control and Prevention. Unintentional carbon monoxide poisonings in residential settings—Connecticut, November 1993–March 1994. *MMWR* 44(41), 765–767, 1995.

CDC (1996) Centers for Disease Control and Prevention. National center for injury prevention and control. Toy-Related Injuries among Children and Teenagers—United States, 1996.

CDC (1997) Centers for Disease Control and Prevention. Update: Blood lead levels—United States, 1991–1994. *MMWR* 46(07), 141–146.

CDC (1999) Centers for Disease Control and Prevention. Carbon monoxide poisoning deaths associated with camping—Georgia, March 1999. *MMWR* 48(32), 705–706, 1999.

CDC (2000) Centers for Disease Control and Prevention. Recommendations for blood lead screening of young children enrolled in Medicaid: Targeting a group at high risk. *MMWR* 49(RR14), 1–13.

CDC (2001) Centers for Disease Control and Prevention. Notice to reader: National lead poisoning prevention week—October 21-27, 2001. *MMWR* 50(42), 927–928.

CDC (2002) Centers for Disease Control and Prevention. National Center for Environmental Health (NCEH) Fact sheet what every parent should know about lead. Available at: http://cdc.gov and http://childergo.com/recc-children.com

CDC (2005a) Centers for Disease Control and Prevention. Web-based Injury Statistics Query and Reporting System (WISQARS) [Online]. National Center for Injury Prevention and Control, Centers for Disease Control and Prevention (producer). Available at: http://cdc.gov/ncipc/wisqars Ten leading causes of unintentional injury deaths, United States, all races, both sexes, 2005.

CDC (2005b) Centers for Disease Control and Prevention. (2005, retrieved). Available at: http://cdc.gov

Chalmers, D.J., Marshall, S.W., Langley, J.D., Evans, M.J., Brunton, C.R., Kelly, A.M., and Rickering, A.F. (1996) Height and surfacing as risk factors for injury in falls from playground equipment: A case-control study. *Injury Prevention*, 2(2), 98–104.

Chiaviello, C.T., Christoph, R.A., and Bond, G.R. (1994) Infant walker-related injuries: A prospective study of severity and incidence. *Pediatrics*, 93, 974–976.

Chicago Children's Memorial Hospital. Available at: http://chicagolead.org

Children's Environmental Health Network. Available at: http://cehn.org

Choking Episodes among Children. Available at: http://cdc.gov/ncipc

City of Milwaukee Childhood Lead Poisoning Prevention Program. Available at: http://ci.mil.wi.us

Coats, T.J. and Allen, M. (1991) Baby walker related injuries—A continuing problem. *Archives of Emergency Medicine*, 8, 52–55.

CFA (1998) Consumer Federation of America. *Playing It Safe, A Fourth Nationwide Safety Survey of Public Playgrounds.* United States Public Interest Research Group, 1998, pp. 1–15.

Cosby, A.G., Greenberg, R.E., Southward, L.H., and Weitzman, M. (2004) About children. Elk Grove Village, IL: American Academy of Pediatrics.

Cox, C. (Winter 2000) EPA takes action on diazinon: Too little, too late. *Journal of Pesticide Reform.* Winter 2000. Available at: http://pesticide.org

Cox, C. (Winter 2001) Ten reasons not to use pesticides. *Journal of Pesticide Reform.* Available at: http://pesticide.org

CPSC (2003) Consumer Product Safety Commission. News from *CPSC: CPSC Warns Backyard Pool Drownings Happen "Quickly and Silently"*. Available at: http://cpsc.gov

CPSC (2005) Consumer Product Safety Commission. National Poison Prevention Week Warns: Most child poisonings result from common household products. News Release 05-136. Washington, D.C.: Office of Information and Public Affairs, Consumer Product Safety Commission.

CPSC (2006) Consumer Product Safety Commission. Carbon Monoxide Questions and Answers—Document 466. Available at: http://cpsc.gov

CPSC (Consumer Product Safety Commission). These reports are available at http://cpsc.gov and also http://childergo.com/recc-children.htm

- CPSC Playground Surfacing. Washington, D.C.: US Government Printing Office Document 1990-261-040/12254, 1990.
- CPSC Guidelines for retrofitting bleachers. Publication #330.
- CPSC Document #359: How to plan for the unexpected: Preventing child drownings
- CPSC Document #362: Safety barrier guidelines for home pools
- CPSC Document #5067: Children drown and more are injured from hair entrapment in drain covers for spas, hot tubs, and whirlpool bathtubs: Safety alert
- CPSC Document #5112: Spas, hot tubs, and whirlpools

Davidson, L.L., Durkin, M.S., Kuhn, L., O'Connor, P., Barlow, B., and Heagarty, M.C. (1994) The impact of the safe kids/health neighborhoods injury prevention program in Harlem, 1988 through 1991. *American Journal of Public Health*, 84, 580–586.

DTI (2002) (Department of Trade and Industry) Research into the mouthing behavior of children up to five years old. Available at: http://dti.gov.uk

Environmental Protection Agency, Pesticides Program. Available at: http://epa.gov

EPA (2005, retrieved) (Environmental Protection Agency). Available at: http://epa.gov

Fazen, L.E. and Felizberto, P.I. (1982) Baby walker injuries. *Pediatrics*, 70, 106–109.

Ferrari. J.R. and Baldwin, C.H. (1989) From cars to carts: Increasing safety belt usage in shopping carts. *Behavior Modification*, 13, 51–64.

Fineman, K.R. (1980) Firesetting in childhood and adolescence. *Psychiatric Clinics of North America*, 3(3), 483–500.

Fineman, K.R. (1995) A model for the qualitative analysis of child and adult fire deviant behavior. *American Journal of Forensic Psychology*, 13(1), 31–60.

Fiore, M. and Heidemann, S. (2004) *Near Drowning*. Available at: http://emedicine.com

Friedman, J. (1986) Reported shopping cart accidents: Data from 1979–1985. Washington, D.C.: US Consumer Product Safety Commission.

Goldman, L.R. (Summer/Fall 1995) Case studies of environmental risks to children. *Critical Issues for Children and Youths*, 5(2). Available at: http://futureofchildren.org

Gumm, B. (2004) Integrated pest management in the home. *Home Energy*, 21(6), 36–39.

Hall, J.R. (2001) *Burns, Toxic Gases, and Other Hazards Associated with Fires: Deaths and Injuries in Fire and Nonfire_Situations*. Quincy, MA: National Fire Protection Association, Fire Analysis and Research Division.

Health Canada. Healthy Development of Children and Youth: The Role of Determinants of Health, December, 1999. Available at: http://hc-sc.gc.ca

HomeSafe (2005, retrieved). Available at: http://homesafe.com

Howland, J. and Hingson, R. (1988) Alcohol as a risk factor for drowning: A review of the literature (1950–1985). *Accident Analysis and Prevention*, 20, 19–25.

Howland, J., Mangione, T., Hingson, R., Smith, G., and Bell, N. (1995) Alcohol as a risk factor for drowning and other aquatic injuries. In R.R. Watson (Ed.), *Alcohol and Accidents: Drug and Alcohol Abuse Reviews*, Vol. 7, Totowa, NJ: Humana Press.

International Program on Chemical Safety, Commission of the European Communities. Principles for Evaluating Risks from Chemicals during Infancy and Early Childhood: The Need for a Special Approach. Available at: http://inchem.org

Istre, G.R., McCoy, M.A., Osborn, L., Barnard, J.J., and Bolton, A. (2001) Deaths and injuries from house fires. *New England Journal of Medicine*, 344, 1911–1916.

Kafry, D. (1980) Playing with matches: Children and fire. In D. Canter (Ed.), Fires and Human Behavior (pp. 47–60). Chinchester, U.K: Wiley.

Karter, M.J. (2003) *Fire Loss in the United States during 2002*. Quincy, MA: National Fire Protection Association, Fire Analysis and Research Division.

KKH (2005, retrieved) (Keep Kids Healthy). Available at: http://keepkidshealthy.com

Kolko, D.J. (1985) Juvenile firesetting: A review and methodological critique. *Clinical Psychology Review*, 5, 345–376.

Kotch, J.B., Chalmers, D.J., Langley, J.D., and Marshall, S.W. (1993) Child day care and home injuries involving playground equipment. *Journal of Pediatrics and Child Health*, 29, 222–227.

Krug, E. (Ed.). (1999) Injury: A leading cause of the global burden of disease. Geneva: WHO.

La Forest, S., Robitaille, Y., Lesage, D., and Dorval, D. (2001) Surface characteristics, equipment height and the occurrence and severity of playground injuries. *Injury Prevention*, 7, 35–40.

Landrigan, P.J. and Carlson, J.E. (Summer/Fall 1995) Environmental policy and children's health. *The Future of Children: Critical Issues for Children and Youths*, 5(2). Available at: http://futureofchildren.org

Langlois, J.A., Wallen, B.A., Teret, S.P., Bailey, L.A., Hershey, J.H., and Peeler, M.O. (1991) The impact of specific toy warning labels. *JAMA*, 265, 2848–2850.

Lesage, D., Robitaille, Y., Dorval, D., and Beaulne, G. (1995) Does play equipment conform to the Canadian standard? *Canadian Journal of Public Health*, 86, 279–283.

Lewis, L.M., Naunheim, R., Standeven, J., and Naunheim, K.S. (1993) Quantitation of impact attenuation of different playground surfaces under various environmental conditions using a tri-axial accelerometer. *The Journal of Trauma*, 35(6), 932–935.

Litovitz, T.L., Klein-Schwartz, W., White, S., Cobaugh, D., Youniss, J., Omslaer, J., Drab, A., and Benson, B. (2001) Annual report of the American association of poison control centers toxic exposures surveillance system. *American Journal of Emergency Medicine*, 19(5), 337–396.

Macarthur, C., Hu, X., Wesson, D.E., and Parkin, P.C. (2000) Risk factors for severe injuries associated with falls from playground equipment. *Accident: Analysis and Prevention*, 32 (3), 377–382.

MMWR (1999) Playground safety—United States, 1998–1999. *Morbidity and Mortality Weekly Report*, 48(16), 329–332.

MMWR (2002) Nonfatal choking-related episodes for children 0 to 14 years of age—United States, 2001. Available at: http://cdc.gov

Morrison, M.L. and Fise, M.E. (1992) Report and model law on public play equipment and areas. Washington DC: Consumer Federation of America.

Mott, A., Rolfe, K., James, R., Evans, R., Kemp, A., Dunstan, F., Kemp, K., and Sibert, J. (1997) Safety surfaces and equipment for children in playgrounds. *Lancet*, 349(9069), 1874–1876.

Mowat, D.L., Wang, F., Pickett, W., and Brison, R.J. (1998) A case-control study of risk factors for playground injuries among children in Kingston and area. *Injury Prevention*, 4, 39–43.

Murphy, D.J. and Steel, S. (2005) Farm pond safety. Available from Agricultural Engineering Department, the University of Pennsylvania, 246 Agricultural Engineering Building, University Park, PA, 16802.

Musemeche, C.A., Barthel, M., Cosentino, C., and Reynolds M. (1991) Pediatric falls from heights. *The Journal of Trauma*, 31, 1347–1349.

National Alliance to End Childhood Lead Poisoning. Available at: http://aeclp.org

National SAFE KIDS Campaign—Airway Obstruction. Available at: http://safekids.org

National Safety Council. Available at: http://nsc.org

Natural Resources Defense Council. Available at: http://nrdc.org

NCIPC (2000) National Center for Injury Prevention and Control. Centers for Disease Control and Prevention Web site. Available at: http://cdc.gov/ncipc/wisqars

NCIPC (2001) National Center for Injury Prevention and Control. *Injury Fact Book 2001–2002*. Atlanta, GA: Center for Disease Control and Prevention. Available at: http://cdc.gov or by calling 770-448-1506.

NCIPC (2003) National Center for Injury Prevention and Control. Ten leading causes of nonfatal unintentional injury, United States. Available at: http://cdcd.gov

NFPA (2005, retrieved) National Fire Prevention Association. Available at: http://nfpa.org

NFPA CHO and NACFLSE (2000) National Fire Prevention Association Center for High-Risk Outreach and the North American Coalition for Fire and Life Safety Education. Solutions 2000: A symposium. Available at: http://nfpa.org

Nordin, H. (2004) Falls among young children in five European countries: Injuries requiring medical attention following falls to a lower level among 0–4 year olds. A study of EHLASS data from Austria, Denmark, Greece, the Netherlands, and Sweden. Stockholm, Swedish Consumer Agency.

Northwest Coalition for Alternatives to Pesticides. Available at: http://pesticide.org

Northwest Coalition for Alternatives to Pesticides (Summer 1999) Does government registration mean pesticides are safe? *Journal of Pesticide Reform* 19(2). Available at: http://pesticide.org

NPC (2001) (National Poison Center). Thirty-sixth annual report. Available from the Department of Preventive and Social Medicine, University of Otago, P.O. Box 913, Dunedin, New Zealand. Available at: http://healthsci.otago.ac.nz

NPPWC (2005a, retrieved) (National Poison Prevention Week Council). Available at: http://poisonprevention.org

NPPWC (National Poison Prevention Week Council). (2005b). Fact Sheet: FAQs. Available at: http://poisonprevention.org

NSKC (2004) National SAFE KIDS Campaign. Washington, DC.

NSKC (2005, retrieved) National SAFE KIDS Campaign. Available at: http://safekids.org

NYC DHMH (New York City Department of Health and Mental Hygiene). Window falls prevention program. Available at: http://nyc.gov

Parker, D.J., Sklar, D.P., Tandberg, D., Hauswald, M., and Zumwalt, R.E. (1993) Fire fatalities among New Mexico children. *Annals of Emergency Medicine*, 22(3), 517–522.

Pesticide Action Network. Available at: http://pesticideinfo.org

Phelan, K.J., Khoury, J., Kalkwarf, H.J., and Lanphear, B.P. (2001) Trends and patterns of playground injuries in United States children and adolescents. *Ambulatory Pediatrics*, 1(4), 227–233.

Present, P. (1987) Child drowning study. A report on the epidemiology of drowning in residential pools to children under age five. Washington, DC: Consumer Product Safety Commission (US).

Qualley, C. (1986) *Safety in the Artroom*, Worcester, MA: Davis Publications.

Reider, M.J., Schwartz, C., and Newman J. (1986) Patterns of walker use and walker injury. *Pediatrics*, 78, 488–493.

Reilly, J.S., Walter, M.A., Beste, D. et al. (1995) Size/shape analysis of aerodigestive foreign bodies in children: A multi-institutional study. *American Journal of Otolaryngology*, 16, 190–193.

Rice, V.J. (2006) Personal experience and report.

Rimell, F.L., Thome, A. Jr., Stool, S., Reilly, J.S., Rider, G., Stool, D., and Wilson, C.L. (1995) Characteristics of objects that cause choking in children. *JAMA*, 274(22), 1763–1766.

Rivara, F.P., Alexander, B., Johnston, B., and Soderberg, R. (1993) Population-based study of fall injuries in children and adolescents resulting in hospitalization or death. *Pediatrics*, 2, 61–63.

Rodgers, G.B. (1989) Factors contributing to child drownings and near-drownings in residential swimming pools. *Human Factors*, 31(2), 123–132.

Roseveare, C.A., Brown, J.M., Barclay-McIntosh, J.M., and Chalmers, D.J. (1999) An intervention to reduce playground equipment hazards. *Injury Prevention*, 5(2), 124–148.

Runyan, C.W., Bangdiwala, S.I., Linzer, M.A., Sacks, J.J., and Butts, J. (1992) Risk factors for fatal residential fires. *New England Journal of Medicine*, 327(12), 859–863.

Sacks, J.J., Brantley, M.D., Holmgreen, P., and Rochat, R.W. (1992) Evaluation of an intervention to reduce playground hazards in Atlanta child-care centers. *American Journal of Public Health*, 82, 429–431.

Safekids (2004) Clear danger: A national study of childhood drowning and related attitudes and behaviors. Available at: http://usa.safekids.org

Safekids (2005, retrieved) National SAFE KIDS Campaign. Available at: http://safekids.org

Safer Pest Control Project. Available at: http://spcpweb.org

Sharp, R. (2005, retrieved) Cited in EPA, 2005.

Sheridan, R.L., Ryan, C.M., Petras, L.M., Lydon, M.K., Weber, J.M., and Tompkins, R.G. (1997) Burns in children younger than two years of age: An experience with 200 consecutive admissions. *Pediatrics*, 100, 721–723.

Sibert, J.R., Mott, A., Rolfe, K., James, R., Evans, R., Kemp, A., and Dunstan F.D.J. (1999) Preventing injuries in public playgrounds through partnership between health services and local authority: Community intervention study. *BMJ*, 318, 1595.

SJNAHS and TGF (2001) St. John's National Academy of Health Sciences and The George Foundation, Bangalore, India. Available at: http://leadpoison.net

Smith, G.A., Dietrich, A.M., Garcia, C.T., and Shields, B.J. (1996) Injuries to children related to shopping carts. *Pediatrics*, 97, 161–165.

Soloway, R. (2004) Statements made at the 2004 US National Poison Prevention Week news conference by the conference chair. Available at: http://poisonprevention.org

Sosin, D.M., Keller, P., Sacks, J.J., Kresnow, M., and van Dyck, P.C. (1993) Surface specific fall injury rates on Utah school playgrounds. *American Journal of Public Health*, 83, 733–735.

Spiegel, C.N. and Lindaman, F.C. (1977) Children can't fly: A program to prevent childhood morbidity from window falls. *American Journal of Public Health*, 67(12), 1143–1147.

SRSA (1999) Swedish Rescue Services Agency. Children and fire. Available at: http://srv.se

Stockhausen, A.L. and Katcher, M.L. (2001) Burn injury from products in the home: Prevention and counseling. *WMJ*, 100(6), 39–43.

Stoffman, J.M., Bass, M.J., and Fox, A.M. (1984) Head injuries related to the use of baby walkers. *Canadian Medical Association Journal*, 131, 573–575.

Stopford, W. (1988) Safety of lead-containing hobby glazes. *North Carolina Medical Journal*, 49(1), 31–34.

Stratton, H. (2004) *Statements Made at the 2004 US National Poison Prevention Week News Conference by the Chair of the Consumer Product Safety Commission*. Available at: http://poisonprevention.org

Tarrago, S.B. (2000) Prevention of choking, strangulation and suffocation in childhood. *WMJ*, 99(9), 42, 43–46. Available at: http://medem.com

Tinsworth, D. and McDonald, J. (2001) *Special Study: Injuries and Deaths Associated with Children's Playground Equipment*. Washington, DC: US Consumer Product Safety Commission.

United Nations Report on Arsenic in Drinking Water. Available at: http://who.int

US Centers for Disease Control and Prevention (2005) Second National Report on Human Exposure to Environmental Chemicals. Available at: http://cdc.gov

US Consumer Product Safety Commission (1979) Method for identifying toys and other articles intended for use by children under 3 years of age which pose choking, aspiration or ingestion hazards because of small parts. Washington, DC: General Services Administration; 16 CFR 1501.

US Department of Agriculture Food and Nutrition Service (2003) A Guide for Use in Child Nutrition Programs.

US Department of Housing and Urban Development Office of Healthy Homes and Lead Hazard Control. Available at: http://hud.gov

US Environmental Protection Agency. Available at: http://epa.gov/lead

USCPSC (2005a) United States Consumer Product Safety Commission (2005, retrieved). Available at: http://cpsc.gov

USCPSC (2005b) United States Consumer Product Safety Commission. Fireworks, CPSC Document #012 (2005, retrieved). Available at: http://cpsc.gov

USFA (2000) United States Fire Association. Federal Emergency Management Agency. Solutions 2000: A symposium to examine fire safety challenges of those who cannot take lifesaving action in a timely manner, in the event of a fire, specifically: Young children (under 5), older adults (over 65) and people with disabilities. Available at: http://nfpa.org

USFA (2005a) United States Fire Association. Federal Emergency Management Agency (2005a, retrieved). Available at: http://usfaparents.gov

USFA (2005b) United States Fire Association. Federal Emergency Management Agency. (2005b, retrieved). Arson Awareness Week. Available at: http://usfa.fema.gov

WHO (1993) World Health Organization. Summary of an Updated International Program on Chemical Safety Environmental Health Criteria Document on Arsenic. Available at: http://who.int.

WHO (2002) World Health Organization. World Health Report. Available at: http://who.int

WHO (2005) World Health Organization (2005, retrieved). Available at: http://who.int/en

Witheaneachi, D. and Meehan, T. (1997) Council playgrounds in New South Wales, compliance with safety guidelines. *Australian and New Zealand Journal of Public Health*, 21(6), 577–580.

Zahm, S.H. and Ward, M.H. (1998) Pesticides and childhood cancer. *Environmental Health Perspectives*, 106(3), 893–908 (June 1998). http://ehp.niehs.nih.gov

Also:

http://aapcc.org/teaching.htm Teaching aids and activity sheets

http://fda.gov Suggestions for home poison protection

http://cpse.gov

http://childergo.com/recc-child.htm

APPENDIX A: POISON SUPPLEMENT

CAUSAL AGENTS AND SOLUTIONS

Lead. Programs in developed countries, such as the United States, UK, and Germany have greatly reduced the incidence of lead poisoning. However, the problem does still exist. For example, in the United States, over 3 million children aged 6 or under have toxic blood lead levels (Homesafe, 2005). In the United States, children who are at greatest risk are financially poor and live in large metropolitan areas or older homes. In developed countries, most exposures are due to inhaling dust from lead-based paints (Box 8.A1).

Other exposures are more transitory. In 2004, four importers voluntarily recalled 150 million pieces of toy jewelry available from vending machines across the United States (in cooperation with the US Consumer Product Safety Commission). Half of these items contained dangerous levels of lead.

Lead poisoning is the top childhood environmental disease in developing countries. The World Health Organization estimates there are 15–18 million children in developing countries with permanent brain damage due to lead poisoning (SJNAHS and TGF, 2001). The occupational and environmental exposures include inadequate emission controls from lead smelters, battery recycling plants, and automobile exhaust.*

PHYSICAL RESULTS AND SYMPTOMS:

Consider this scenario:

> *Paul lives in the projects. He is 10 years old and already a failure in everyone's eyes. He is prone to "acting out" and becomes angry and aggressive when frustrated. He is small for his age. He still cannot read and has difficulty grasping concepts and remembering things. Neither his parents nor his teachers can figure out if he deliberately ignores their instructions or if he really cannot remember them from day to day. He does not feel good physically. He feels tired almost all the time.*

While Paul's problems could be the result of a number of causes, one possible cause is exposure to lead. It causes brain, kidney, and liver damage, resulting in lower intelligence, reading and learning problems, behavioral difficulties, and slow physical growth.

This is a child, who was born normal and is now damaged because of the products and environment around him. Treatment can lower blood lead levels, but any damage is permanent. Who knows how many assistive-care facilities or even detention centers house children from these circumstances?

Some say there is no safe level of lead in the body. Long-term effects occur with exposure to extremely small amounts, while having no definite, distinct symptoms. Table 8.A1 contains recommendations to reduce or eliminate lead exposure.

* Elimination of leaded gasoline resulted in: 77% reduction in blood lead levels in the United States in 10 years. A 50% drop in gasoline lead levels in the UK was associated with a 20% drop in children's blood lead levels. Yet, in India (as other developing countries) the permissible limits (0.56 g/L) are much higher than limits in developed countries. Information available at: http://leadpoison.net

BOX 8.A1 LEAD

What is it? Lead is a heavy metal. It is used in many materials and products. Unfortunately, it is not biodegradable and persists in the soil, air, and drinking water unless properly removed.

What does it do when absorbed? When absorbed into the body, it is distributed to the liver, brain, kidneys, and bones. It is toxic to organs and systems, especially to neurological development. Lead causes anemia by impairing oxygen-carrying molecules, beginning at exposures of around 40 μg/dL. Even at low levels it can reduce a child's IQ score (intelligence quotient) and attention span. It is associated with reading and learning disabilities, hyperactivity, slowed growth, hearing loss, sleep disorders, and other health, intellectual, and behavioral problems. High exposures can cause mental retardation, convulsions, coma, and death.

Who is most at risk? Children under six because they absorb up to 50% of the lead they ingest. It interferes with the development of their organs, systems, and their brain. By the time symptoms show (headache, lethargy or hyperactivity, nausea, stomach pain, vomiting, and constipation, but no diarrhea), the brain has suffered damage. Women of child-bearing age and pregnant women are also at risk because lead ingested by a mother can cross the placenta and affect the unborn fetus.

How does exposure occur? Lead can be ingested, inhaled, or absorbed through skin. In countries where lead-based paint is no longer used, primary exposure is through lead-contaminated dust from deteriorating lead-based paint in older homes, from children putting paint chips in their mouths (lead compounds taste sweet and a few chips the size of a thumbnail can raise levels to 1,000 times the acceptable limit), or from soil contaminated by flaking exterior lead-based paint (or sand/blasting). Soil contamination occurs from leaded gasoline disposal. Playing in the dirt can result in accidental ingestion and vegetables grown in the soil can absorb the lead. Older lead pipes or pipes soldered with lead leach lead as they corrode (faucets or fittings made of brass also contain lead). Those who work in construction, demolition, painting, with batteries, in radiator repair shops, lead factories, or with a hobby that uses lead can bring lead home on their clothing (such as painting and remodeling; working with auto radiators or batteries; auto repair; soldering; making sinkers; bullets; stained glass; pottery; going to shooting ranges; hunting; or fishing[a]). Rare exposures include food and drink storage in leaded crystal, lead soldered cans, or lead glazed ceramics.

Worldwide, exposures include gasoline additives, food can soldering, lead-based paints, ceramic glazes, drinking water systems, cosmetics, and folk remedies.

How do I know if my child has been exposed? A blood test shows recent exposure. Lead only stays in the blood about three weeks. Most is excreted in urine and the remainder goes into the bones.

Can an exposed child be treated? Yes and no. Damage that has been done can not be reversed. Medications can reduce high levels of lead in the body.

Table 8.A1 What to Do About Lead Exposure

Lead-Based Paint[a]	• Check to see if paint is chipping or peeling.
	• Check for places where surfaces rub together and potentially make lead dust: windows, doors, steps, and porches.
	• Dirt and dust:
	  ◦ Keep the places where children play clean and dust free. Regularly wet-wipe floors, windowsills, and other surfaces that may contain lead dust.
	  ◦ Wash children's hands, pacifiers, and toys often to remove dust.
	  ◦ Have children play on grass instead of bare dirt. Take off shoes when entering a home to avoid tracking in soil that may contain lead.
	• Design your work carefully when doing renovations and repair. Seek professional guidance on sealing off the work area, covering ducts and furniture, and keeping children away from the work site and the work itself, including scraping, sanding, and using heat guns on paint before repainting. For example, suggested processes include using 6 mL plastic for sealing off the area, misting surfaces before working on them, cleaning dust and chips with wet rags soaked in trisodium phosphate or phosphate-containing powdered dishwashing detergent, wearing rubber gloves, and properly disposing of cleaning materials.

Cooking, Eating, and Drinking	• Know what kinds of pipes supply your water[b]
	• Have the water tested for lead content.
	• If home pipes have lead in them:
	  ◦ Do not use water that has been sitting over 6 hours in the pipes. For example, let the water run for at least 2 minutes. After letting it run, put some in a container in the refrigerator for later use.
	  ◦ Hot water dissolves lead more easily. Use only cold water for drinking, cooking, or making baby formula. Run the water until it feels colder. Do not use folk remedies or imported, old, or handmade pottery to store food or drinks.
	• Provide meals high in iron, vitamin C, and calcium that help prevent young bodies from absorbing lead.

Work and Hobbies	If you work with lead in your job or hobby, change clothes and shower before you go home.

Note: If your children are at risk, have their blood lead level tested (Box 8.A2). Have a certified inspector assess your home

[a] Lead paint was banned in US residential paint in 1978. France started banning lead paint in the 1870s. Germany, Australia, Japan, and some other countries began banning in the early 1920s. According to US Housing and Urban Development: 90% of pre-1940, 80% of pre-1960, and 62% of pre-1978 buildings have lead in them.

[b] The Safe Drinking Water Act was enacted in the United States in June 1986, with states having until 1988 to comply. It requires the use of lead-free pipe, solder, and flux for plumbing systems connecting with public water systems. Before that, most solder contained about 50% lead. There is suspicion that some plumbers continued to use lead solder beyond that date. Faucets or fittings made of brass, lead pipes, or copper pipes with lead solder all contain lead.

BOX 8.A2 HOW MUCH LEAD DOES IT TAKE TO GET LEAD POISONING?

The amount is incredibly small: 10 µg/dL of blood.

If you are metrically challenged, here is a way to visualize what that means.

- A deciliter is about half a cup
- A packet of sweetener (either the pink or blue) is 1 gram
- There are 1 million micrograms in a gram
- So, divide the stuff from one packet into one million piles (pretend!)
- Now, discard 999,990 of those "piles"
- Take the remaining 10 piles and mix them into half a cup of liquid.

This is the amount of lead that can cause poisoning.

Blood lead levels of only 10 µg/dL have been associated with negative effects on cognitive development, growth, and behavior among children aged 1 to 5 years (CDC, 2000).

http://leadpro.com

Do home testing kits work? Yes, but they may not be reliable for detecting lead paint buried under layers of newer paint. They are useful for testing toys, pottery, and household items.

Do high efficiency particulate air (HEPA) filters work? HEPA filters trap small particles of dust, often too small for a regular vacuum. They will also collect lead dust. Thus, they may help, but are not a "cure" for lead dust in the air.

ARSENIC

Arsenic occurs naturally in the environment. It is in rocks, soil, water, and air and is released during events such as forest fires and volcanoes. It is carcinogenic to humans. Most industrial arsenic in the United States is found in pressure-treated wood, as a preservative. It is sometimes present in dyes and metals.

Pesticide-treated wood contains arsenic, in the form of chromated copper arsenate (CCA), to deter insect damage and decay. Some children are at greater risk of lung or bladder cancer as a result of playing on arsenic pressure-treated playground equipment (EPA, 2005; Figure 8.A1). A child playing on an arsenic-treated playset would exceed the lifetime cancer risk in only 10 days (according to US Federal Pesticide Laws) (Sharp, 2005).

Wood recently treated with CCA has a greenish tint that fades over time. In all likelihood, CCA-treated wood was used if the playground was not made of cedar or redwood.

Figure 8.A1 (Valerie Rice)
In the United States, pressure-treated lumber for playground equipment is being phased out. Yet, it continues to be sold and used for decks, backyard planters, and occasionally for playground equipment. The pesticide (arsenic) can leach into the soil or children touch and accidentally ingest it by putting their hands in their mouths or eating after touching it.

Drinking water is the greatest worldwide public health threat of arsenic. Absorption through the skin is minimal, therefore using the same water for bathing and laundry should be safe.

Children are more susceptible to arsenic in drinking water because their bodies are developing (higher heart rate and respiration), but also because they drink approximately 2½ times as much water as adults (proportionally). Unfortunately, most exposure standards are based on adults (EPA, 2005).

Figure 8.A2 (Valerie Rice) Following are the methods to reduce the risk of arsenic poisoning in children (EPA, 2005):

- Build playgrounds and decks with cedar, redwood, locust, or composite wood or use plastic or "plastic lumber."
- Apply an oil-based, semi-transparent stain to CCA-treated wood once a year.
- Wash children's hands thoroughly after being in contact with the wood, especially before eating or drinking.
- Benches and picnic tables can also be made from this wood, so never lay food directly on the wood and do not inadvertently touch wood while eating.

Long-term exposure from contaminated drinking water can cause cancer of the skin, lungs, urinary bladder, and kidney, as well as peripheral vascular diseases (black foot disease in China). First though are skin pigmentation changes, followed by hyperkeratosis.*

The World Health Organization's (WHO) Guidelines for Drinking-Water Quality establishes 0.01 mg/L as a provisional guideline value for arsenic (WHO, 1993).†

CARBON MONOXIDE (CO)

Carbon monoxide is an odorless, colorless gas. In the United States, approximately 3500 children receive emergency room treatment and 24 die annually from CO exposure (CDC, 2005). Carbon monoxide poisoning results in more fatal unintentional poisonings in the United States than any other agent (CDC, 1999).

The problem is particularly evident in colder climates during the winter, since some space heaters or malfunctioning heaters emit CO. If sufficient ventilation is not available, CO poisoning can result. Other sources can include water heaters, gas-operated dryers, stoves, and fireplaces; clogged chimneys or corroded flue pipes; wood-burning stoves and cars left running in a closed garage. The difficulty is not having sufficient ventilation to let the CO escape. For example, snow can cover a vehicle and block the tailpipe causing exhaust fumes to put a family inside at risk.

* Hyperkeratosis—a thickening of the outer layer of the skin. The thickening can occur as part of the skin's defense system.

† Arsenic levels in drinking water can be measured. Although the technology for removal from piped water exists, it requires technical expertise, can be relatively expensive, and may not be easily applied in some urban and many rural areas. Chemical household water treatments are being tested. Other suggestions include rainwater harvesting, pond-sand-filtration, and piped water supply from safe or treated sources (WHO, 2004). Countries with concentrations greater than the guideline value include Argentina, Australia, Bangladesh, Chile, China, Hungary, India, Mexico, Peru, Thailand, and the United States. Negative health-effects have been found in Bangladesh, China, India (West Bengal), and the United States.

Another source of danger occurs while camping. Carbon monoxide poisoning kills approximately 30 campers a year. Improper use of portable heaters, lanterns, and vehicles can result in CO poisonings (Safekids, 2005).

CO poisoning symptoms are similar to those of flu (headache, fatigue, nausea, and dizziness), but without fever. Exposure can cause both neurological damage and death. Once again, children are particularly at risk due to their higher respiratory and metabolic rates, resulting in the gas accumulating in their bodies faster than in adults. Unborn children have an even greater risk of birth defects if exposure occurs to their mothers when they are pregnant (see Figure 8.A3).

Figure 8.A3 (Valerie Rice)
CO detectors may prevent up to half of the deaths due to CO exposure (CDC, 2005; Safekids, 2005).*

Every home should have CO detectors in each sleeping area of the home 457 cm (15 ft) from fuel-burning appliances.†

- Never stay in a home after a CO alarm goes off, instead get out and seek medical attention.
- Have fuel-burning appliances inspected and chimneys cleaned each year.
- Do not let car engines run while parked in a garage or covered in snow.
- Do not use charcoal grills or hibachis inside a closed space (home or garage).
- Do not use an oven to heat a home.

ART SUPPLIES

Another source of accidental poisoning is art supplies. Poisoning can occur when children put the items in their mouths directly, put their fingers in their mouths, or rub their faces while working with projects, or eat without washing their hands.

The Labeling of Hazardous Art Materials Act (LHAMA) requires toxicological evaluation and labeling of art materials sold in the United States. The labels must conform to the labeling standard, ASTM D 4236 and include information on safe use, and potential acute and chronic health effects (Box 8.A3).

* Several states and some cities in the United States require CO detectors in homes.

† The Recreation Vehicle Industry Association requires CO detectors/alarms in motor homes and in recreational vehicles that can be towed and have a generator or are prepared to use a generator.

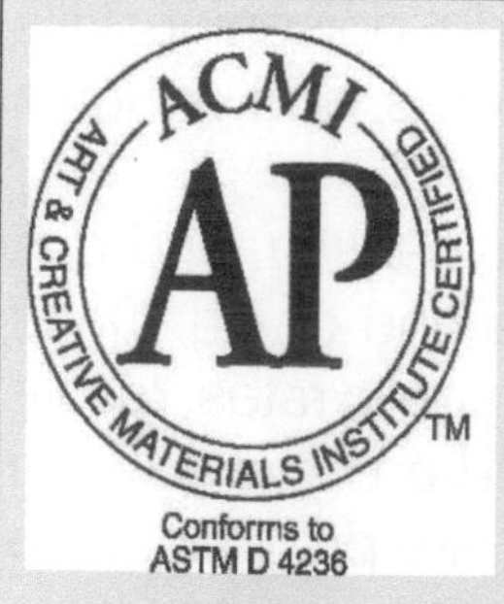

BOX 8.A3

The Art & Creative Materials Institute, Inc. (ACMI) is a nonprofit association of 210 manufacturers of art, craft, and other creative materials. ACMI was organized to assist its members in providing products that are nontoxic. ACMI sponsors a certification program for children's art materials. Certifications ensure that the products are nontoxic and meet voluntary standards of quality and performance.

Certifications that include performance also meet material, workmanship, working qualities, and color requirements, and labeling is recognized by the American National Standards Institute (ANSI) and the American Society for Testing and Materials (ASTM).

The CL Seal certifies products are properly labeled according to a toxicological evaluation by a medical expert for health risks and that the safe and proper use of these materials is included. These two seals appear on only 15% of the adult art materials in ACMI's certification program and on none of the children's materials.

Additional information on working with arts and crafts materials can be found at http://acminet.org

Labels used with permission from ACMI.

General recommendations for arts and crafts work with children:

- Buy only art supplies labeled as safe (nontoxic) for children.
- Never eat, drink, or smoke while using art and craft materials.
- Wash hands after using art supplies.
- Use products according to directions. Do not use products for skin painting or food preparation unless labeled for that use.
- Never put art materials in other containers.

HOUSEHOLD ITEMS

Table 8.A2 and Table 8.A3 describe household items that can be poisonous to children. Table 8.A4 and Table 8.A5 list outdoor items (pesticides and plants). Adhere to general recommendations to avoid accidental poisonings that are also listed in this chapter, such as "Mr. Yuk" labels and reach tables.*

* See Chapter 14 on Warnings for a detailed description of "Mr Yuk."

Table 8.A2 Household Items that Can Poison Children

Item	Examples	Solutions
Medicines and Vitamins	• Iron—found in vitamins[a] • Acetaminophen[b] • Antacids • Allergy medicines • Weight loss pills • Prescription drugs (especially antidepressants and medications for the heart, blood pressure, and diabetes) • Eye drops, nasal sprays, and inhalers	**Design—Storage and Placement:** • Keep out of reach of children. Use reach guidelines mentioned in this chapter and Chapter 3—Anthropometry. • Store in separate, locked areas. Never store medicines on bedside tables or in bedside drawers. • Manufacturers should use child-resistant containers and include warnings, even on products for adults. **Safety:** • Keep purses or diaper bags containing medicines, vitamins, or cosmetics out of children's reach. • Put medicines stored in the refrigerator in a special plastic container in the back of the refrigerator (out of reach of children). • Supervise children when taking vitamins or medicines. Never encourage a child to take vitamins or medicines by referring to it as candy. • Refrain from taking medications in front of children. • Throw away old medicines by flushing them down the toilet. • Read labels, follow instructions, administer medicines by weight and age, and use the dispenser that comes with the package. • Turn the lights on when administering medications. If some medicine spills or a child vomits, be careful not to give additional medicine that could cause an overdose. • Purchase only the amount of medicine needed and keep track of the amount of medicine in the bottle. • Be especially careful when normal household routines change, such as vacations and holidays.
Nicotine	Cigarettes, cigars, snuff, or gum containing nicotine	**Design—Advertising and Storage:** • Advertising should appeal to adults, not children (such as talking animals and cartoon characters). • Keep out of reach of children. Use reach guidelines in this chapter and Chapter 3—Anthropometry. **Safety:** • Clean ashtrays immediately after use. • Never let children clean ashtrays.
Alcohol	• Beer, wine, and liquor, as well as rubbing alcohol, mouthwash, aftershave, and some colognes • Baking ingredients such as vanilla and almond extracts may also contain high levels of alcohol	**Design—Advertising and Storage:** • Target advertising to adults, not children (such as talking animals and cartoon characters). • Keep out of reach of children. Use reach guidelines mentioned in this chapter and Chapter 3—Anthropometry. • Store in separate, locked areas and original containers. **Safety:** • Take special care during holidays and parties when adults may leave drinks around the house.

(continued)

Table 8.A2 Household Items that Can Poison Children (Continued)		
Cosmetics	• Check ingredients in cologne, perfume, hair spray, hair dye, shampoo, artificial fingernail remover, fingernail polish remover, deodorants, creams, lotions and baby oil, herbal products, and baby powder. • Toothpaste in large amounts can be toxic. • Muscle rubs containing oil or wintergreen or methyl salicylate and products containing camphor.	**Design—Storage:** • Keep out of reach of children. Store in locked cabinets and original containers. • Use Mr. Yuk stickers. **Safety:** • Supervise children's use of cosmetics, including brushing teeth. • Explain possible poisoning from ingesting these products when children are old enough to comprehend.
Cleaners and Detergents	• Mineral oil—found in furniture polishes. • Hydrocarbons—found in paint thinners, lamp oils, gasoline, and lighter fluid. • Lye or sodium hydroxide—found in oven cleaners. • Cellusolve (ethylene glycol butyl ether)—in concentrated glass cleaners and solvents. • Bleach—in cleaning agents. Also: dishwashing detergents, window cleaners, and toilet bowl deodorizers. • Swimming pool chemicals.	**Design—Storage:** • Keep out of reach of children. Use reach guidelines mentioned in this chapter and Chapter 3—Anthropometry. • Store in locked cabinets and original containers. • Use Mr. Yuk labels. **Safety:** • Supervise children's use of cleaning products. • Take special care when using these products, know where small children are at all times.
Pesticides	• Ant and roach sprays, roach "houses," flea and tick powders, sprays, and "bombs." • Materials designed for yard work: weed killers, fertilizers. • Mothballs. • See additional information below.	**Design—Advertising and Storage:** • Keep out of reach of children. Use reach guidelines mentioned in this chapter and Chapter 3—Anthropometry. • Store in areas that are inaccessible to children, in original containers. **Safety:** • Keep children and toys from the area where pesticides are used, according to package label recommendations. Keep children away while working with pesticides, even if you ventilate the area by opening windows and turning on the fan. • Do not let children in recently sprayed areas. • Never leave pesticides unattended when you are using them. • Set a good example: wear protective clothing and never mix products.
Car Products	Antifreeze, oil, de-icers, windshield washer fluid, leather and upholstery cleaner, wax, and gasoline.	Keep out of reach of children, store in locked areas and original containers.
Paints	Paint, paint thinner (even a small amount of paint thinner can be deadly to a child).	Keep out of reach of children. Store in locked cabinets and original containers.
Plants	Household plants and berries from plants can be toxic (see below).	Know which plants are poisonous and do not keep them in your home or outdoor garden (see Table 8.A3).

[a] A child can die swallowing as few as five vitamins containing iron.

[b] In asprin-free pain medicines, decongestants, aspirins, and nonsteroidal anti-inflammatory drugs.

Table 8.A3 Miscellaneous Household Items and Issues Regarding Poisons

Item	What Happens	Solution
Miniature "Button" Batteries	Tiny batteries, such as those in watches, calculators, cameras, and hearing aids can pass through a child without problem, but they can cause internal burns if they get lodged in the esophagus or intestinal tract.	Keep out of reach of children.
Holidays	• Fire salts, used in fireplaces to produce colored flames, are poisonous. • Artificial snow sprays to decorate can cause lung irritation if inhaled.	

Table 8.A4 Pesticides and Children in the United States (EPA, 2005)

Information	Risk	Symptoms	Solutions

• ~79,000 children were involved in common household pesticide poisonings or exposures in 1990.
• 47% of homes with children under 5 years old, store at least one pesticide product within reach of children.
• 75% of homes without children under age 5 store at least one pesticide product within reach of children.
• 13% of pesticide poisonings occur in homes other than the injured child's own home.
	• The main pesticide exposure for children occurs during home pesticide use, including lawn and garden care.
	• Some pesticides, including some used in homes, are likely carcinogens.
	• As with other poisons, children are at greater risks than adults.
		• Dizziness
		• Vomiting
		• Headaches
		• Sweating
		• Fatigue
		• Skin, eye, or respiratory tract irritation
			• Use pesticides when children are not home or restrict their access to garden and lawn areas free of pesticides.
			• Store pesticides out of reach of children:
			  ◦ In locked cabinets, over 4 ft off the ground.
			  ◦ See Table 8.9 and Table 8.10 for reach guidance.
			• Wash fruits and vegetables that may have been exposed to pesticides thoroughly.

Note: A product stored less than 4 ft (122 cm) off the ground in an unlocked area is typically "within a child's reach."

Table 8.A5 Poisonous Plants and Children

Typically Outdoors	Typically Indoors	Solutions
• Azalea/Rhododendron • Black locust • Boxwood • Buttercups • Caladium • Castor bean • Cherry trees (wild or cultivated) • Chinaberry • Daffodil • Elderberry • English ivy • Foxglove—Digitalis • Giant Elephant Ear • Gladiola • Holly • Hyacinth • Hydrangea • Iris • Jimson weed • Juniper • Larkspur • Lantana • Lily of the valley • Mayapple • Mistletoe • Monkshood • Moonseed • Morning glory • Nightshade • Oak • Oleander • Poison ivy • Pokeweed • Water hemlock	• Amaryllis • Arrowhead • Caladium • Candelabra cactus • Cyclamen • Devil's ivy • Dumbcane • Eucalyptus • Jerusalem cherry • Lily of peace • Pencil cactus • Philodendron • Rosary pea	• Know and label plants by name. • Put poison stickers on the labels of plants that are toxic (such as Mr. Yuk, mentioned above). • Move poisonous plants out of children's reach. • Teach children not to eat plants (leaves and berries) or mushrooms outside. • Know symptoms of poisoning, symptoms may be immediate or delayed by several hours. ◦ Nausea, stomach cramping, vomiting, and diarrhea. ◦ Some cause burning of the mouth and tongue, along with swelling of the tongue, irregular heartbeat, nervous excitement, and mental confusion. ◦ Some can cause death, such as the rosary pea or castor bean, foxglove, larkspur, daphne, rhubarb leaf blade, golden chain, lantana, laurels, rhododendrons, azaleas, jasmine, oak foliage, and acorns, foliage of wild cherries. • If you suspect poisoning: ◦ Call the poison control center. ◦ Take a leaf or stalk of the plant the child has ingested to the doctor's office with you.

http://ansci.cornell.edu/plants
http://aggie-horticulture.tamu.edu/
http://chw.edu.au/
http://chppm-www.apgea.army.mil/

CHAPTER 9

ASSISTIVE TECHNOLOGIES FOR CHILDREN

ROBIN SPRINGER

TABLE OF CONTENTS

DEFINING THE ISSUES

Many people associate *adaptive equipment* with a mysterious territory of wheelchairs, hospitals, inability, and dependence. They may believe adaptive equipment is cumbersome, complex, expensive, and used only by people with disabilities. Such perceptions reinforce differences among people rather than similarities.

The goals of adaptive equipment are to equalize, coalesce, and include. Many of these devices can be inexpensive, easy to use, and can seamlessly interface with one's environment, blurring lines between disabled and able-bodied, the *cans* and the *can nots*. Able-bodied people use adaptive equipment every day (Box 9.1).

BOX 9.1 PRODUCTS INVENTED AS ASSISTIVE TECHNOLOGY

The ***typewriter*** was created in 1808 to help a blind woman write legibly.

The ***tape recorder*** was created in 1948 as a "low-cost talking book machine for the blind."

Both leveled the playing field for persons *with* disabilities.

Both are used worldwide by persons *with**out*** disabilities.

WHAT IS A DISABILITY?

The word *disability* often conjures up images of people who are unable to "do things." Often "disability" has a negative connotation associated with it. However, disabilities are common and a natural part of life. In the United States, one out of five people has a disability. In India, this figure is approximately one in 30. Approximately 20,000 people with disabilities in Afghanistan receive assistance in rehabilitation and education (Turmusani, 2004). In Africa, one in 16 has a disability and only two percent have access to rehabilitation (Onasanya, 2004).

We move in and out of dependent and independent states throughout our lives. Infants are totally dependent on others to satisfy their needs. As adults, we may be relatively independent or we may experience injuries that render us temporarily or permanently unable to function without assistance. As we age, we may become dependent due to age-related illnesses. These times of dependence are times of *dis-ability.*

BOX 9.2 DEFINITION OF DISABILITY	
Americans with Disabilities Act (2004)	1. A physical or mental impairment that substantially limits one or more of the major life activities of the individual; 2. A record of such impairment; or 3. Being regarded as having such an impairment.
United Nations (1993)	Any restriction or lack (resulting from an impairment) of ability to perform an activity in the manner or within the range considered normal for a human being.

Dis-ability also occurs when external environments interfere with our ability to function. For example, children who are unable to hear due to loud construction outside the classroom are as disabled *at that particular time* as children who are physiologically incapable of hearing. Each instance leads to functional deafness.

We minimize our differences when we see people as having a spectrum of abilities (Box 9.3). Doing so allows us to focus on the most effective way to enable each child to become independent.

Implementing an appropriate *adaptive plan* for a child with a disability can enable the child to thrive, foster independence, and help cultivate skills that contribute to society. The absence of a well thought-out plan can result in the child abandoning new technology, resulting in frustration, discouragement, and fear.

No single technology configuration is right for all people with a specific disability. Usually, a range of options is available. Important considerations in choosing among alternatives for a child include the child's needs and abilities, objectives, the environments where the technology is used, and available alternatives.

BOX 9.3 THE ONUS OF DISABILITY

Shift the Focus of Disability from...

...an individual with a deficit, to

...an individual with a performance deficit, to

...an environment that contributes to or removes performance barriers.

(NIDRR, 2004)

WHAT IS ASSISTIVE TECHNOLOGY?

Assistive technologies are items that enable people with disabilities to participate in activities of daily living, helping to ensure equal opportunities (Krug, 2004). These range from highly technological to everyday products and include wheelchairs, adapted vans, communication devices, and modified computers.

Assistive Technology assists individuals in participating more fully in society. It levels the playing field between disabled children and their able-bodied counterparts (Box 9.4). Proper tools enable children who have disabilities to compete in a broader range of environments. They can excel at school and eventually at work as capable peers.

BOX 9.4 DEFINITIONS FROM THE TECHNOLOGY-RELATED ASSISTANCE ACT OF 1988*

Assistive Technology	Technology designed to be utilized in an assistive technology device or assistive technology service.
Assistive Technology Service	A service that helps disabled individuals choose, obtain, or use the device(s).
Services	Evaluating, providing, and maintaining the product(s) and coordinating other services with assistive technology and training.

* US Public Law 100-407, the Technology-Related Assistance for Individuals with Disabilities Act, was signed into law in 1988 and amended in 1994. The basic purpose of the law is to provide financial assistance to the states for needs assessments, identifying technology resources, providing assistive technology services and conducting public awareness programs.

AVAILABILITY OF SERVICES

The availability of services for the provision of assistive technologies varies from country to country. For example, the United States guarantees access to technology for the disabled (Box 9.4), yet many countries do not offer equal civil rights. Table 9.1 lists some governmental regulations from several countries that influence access to these services.

Approximately 98 percent of disabled children in developing countries do not receive an education (CRIN, 2004). Organizations including the United Nations Children's Fund (UNICEF) and the Child's Rights Information Network (CRIN)* work to protect all children, advocating to ensure children, including those with disabilities, receive an education.

The *United Nations' Convention on the Rights of the Child*[†] requires countries to "provide free, compulsory basic schooling that is aimed at developing each child's ability to the fullest." Doing so requires that children have access to schools and high-quality education (UNICEF, 1999).

Table 9.1 Provisions to Ensure Education and Services

Country	Reference	What it Says
Australia	The Commonwealth Disability Discrimination Act 1992 (DDA)	This law to combat discrimination includes discrimination in education and training. Draft Disability Standards are being created to clarify the goals of the DDA by describing how to comply with the Act.
Canada	Charter of Rights and Freedoms and the Canadian Human Rights Act	These are general antidiscrimination laws. There are no laws to mandate accommodating children with disabilities in schools. While there are "programs" and "policies" to support inclusion, participation is voluntary and varies among provinces.
European Union	A project is underway, with the title "Accessibility for All"	This project entails examining the need for standards to increase access to information and communication technologies.
France	Law 238 Law 89-486	Law 238 provides compulsory education for children with disabilities. Law 89-486 reaffirms Law 238 and further states that priority must be given to schooling in a mainstream environment.
United Kingdom	Education Act 1996	This Act states that children with special education needs are to be educated in mainstream schools.
United States	Public Law 100-407	Children who qualify for special education are entitled to receive adaptive technology services. In order to obtain services, a series of steps is required.[a]

[a] These steps include (1) a written request from the parent/guardian to the school stating the child's assistive technology needs, (2) the school's response including a proposed assessment plan, and (3) written consent from the parent or guardian agreeing to the school's plan. The school and the parent each have 15 days for their respective responses. Once consent is given, the school has 50 days to complete the assessment and hold an Individualized Educational Plan or IEP.

* CRIN is a global network that disseminates information about the Convention on the Rights of the Child and child rights among nongovernmental organizations (NGOs), United Nations agencies, intergovernmental organizations (IGOs), educational institutions and other child rights experts.

† This has been ratified by every country except the United States and Somalia.

BOX 9.5 U.S. LAWS THAT ADDRESS THE NEEDS OF INDIVIDUALS WITH DISABILITIES

The following is a list of notable laws enacted in the United States to address the needs of people with disabilities:

- Rehabilitation Act of 1973
- Education for All Handicapped Children Act of 1975 (PL 94-142)
- Rehabilitation Act Amendments of 1978 (PL 95-602)
- Education of the Handicapped Act Amendments of 1986 (PL 99-457)
- Rehabilitation Act Amendments of 1986 (PL 99-506)
- Developmental Disabilities Assistance and Bill of Rights Act Amendments of 1987 (PL 100-146)
- Technology-Related Assistance for Individuals with Disabilities Act of 1988—*"The Tech Act"*—(PL 100-407)
- Americans with Disabilities Act of 1990—*"ADA"*—(PL 101-336)
- Individuals with Disabilities Education Act of 1990—*"IDEA"*—(PL 101-476)
- Rehabilitation Act Amendments of 1992 (PL 102-569)
- Technology-Related Assistance for Individuals with Disabilities Act of 1994 (PL 103-218)
- Individuals with Disabilities Education Act Amendments of 1997 (PL 105-17)

EXAMPLES OF ASSISTIVE TECHNOLOGY

The proliferation of computers has led to a corresponding rise in assistive technology device (ATD) software and hardware designed for educational and therapeutic purposes. Most assistive technologies address functional needs such as communication, mobility, and accessibility. Such *ATD*s include alternative keyboards, speech-recognition software (voice input), word prediction, hands-free mice, and text-to-speech software (software that "reads" computer screen text and information).

Augment
To make or become greater in size, amount or intensity.

Augmentative Communication products allow children who are unable to speak to communicate with others. An example of a *low-tech augmentative communication product* is a picture board that enables users to communicate by pointing to objects or processes. A high-tech alternative provides real or synthesized electronic speech that articulates *dynamic* (developed at that time) or predefined sentences (Figure 9.1a through Figure 9.1c).

Figure 9.1a (Mayer-Johnson LLC)
A Symbol Overlay for Augmentative Communication. Symbol Overlays can be purchased or even made at home, to identify a child's commonly used words and phrases.

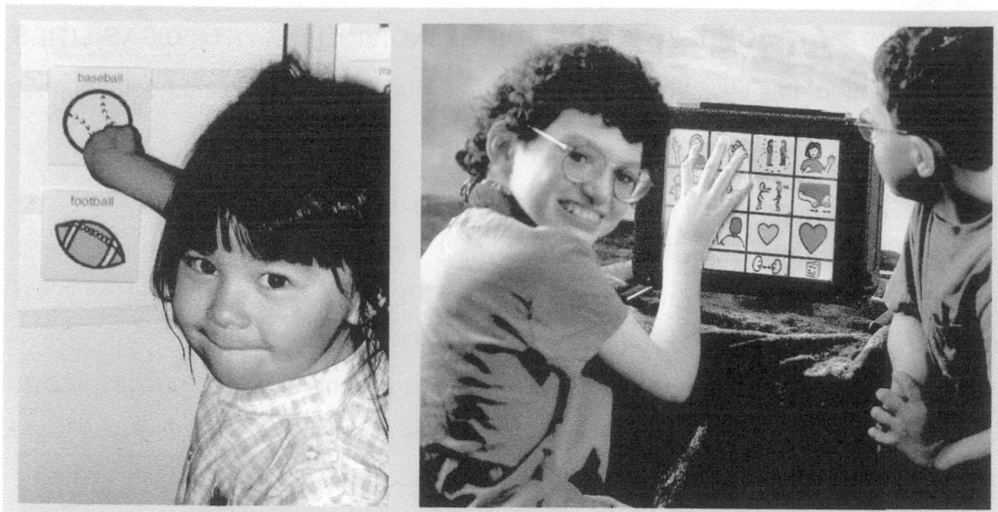

Figure 9.1b and Figure 9.1c (Mayer-Johnson LLC)
Children, caregivers, and therapists can select the pictures to include on a picture board. Matching what a child generally needs or wants with their method of communication gives the child a sense of control over his environments.

SELECTING AN ASSISTIVE TECHNOLOGY DEVICE

Selecting the appropriate assistive technology device is crucial. The wrong product may be more harmful to a child's well-being than not providing any assistive technology. The proper choice requires an understanding of the child's current and potential future needs including the child's prognosis, the ease and cost of maintaining the device, and the product life cycle.

Criteria to evaluate the ease of use of assistive devices

- Easy to understand
- Easy to transport
- Easy to operate

Understanding the child's specific situation and its implications is necessary to evaluate a plan's costs and benefits. For example, computers may cost from a few hundred dollars to a few thousand dollars, with a broad range in system capabilities, component qualities, warranties, and upgradeability.

DEVELOPING A PLAN

A good technology plan considers four aspects of design. These include the consumer's desired outcomes, preferences, abilities, and the environment in which the technology will be used. Results are matched with available devices.

THE CHILD USER'S DESIRED OUTCOMES: *Client-centered approaches* in occupational therapy intervention that consider patient objectives are more effective than therapies that exclude patients from the decision-making process. Similarly, integrating the child's input into the technology plan can help empower the child (Figure 9.2).

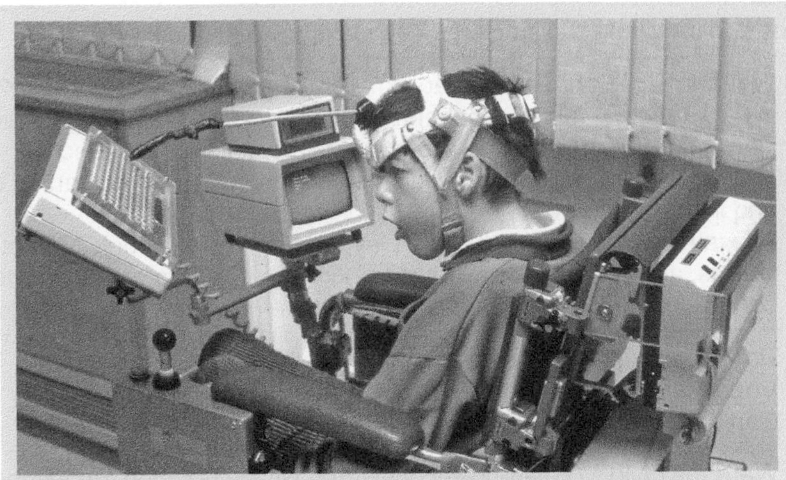

Figure 9.2 (Karl Dean and Hull & East Yorkshire Hospitals) Children have their own wants and desires, and they should play a primary role in selecting assistive devices. Children may want to read and write, but they may also want access to the games their friends play.

THE CHILD USER'S ABILITIES: Children are not their disabilities. A child's chance for success cannot be presumed by his diagnosis. Assessments by caregivers, physical, occupational, or speech therapists tell us what a child can do on an everyday basis, *at that time*. These assessments can facilitate the development of a strong team approach with common goals.

THE CHILD USER'S PREFERENCES: We all have likes and dislikes that influence our choices. A child's ideas about how products "fit" into his life will affect whether he uses them. He may resist products for several reasons, including not being involved in choosing the technology.

Realistic expectations are also important. Children need to understand in real terms what the device will and will not do. They need to know how long it will take to become proficient to avoid becoming discouraged and quitting.

Although the child will be the primary user of the technology, family members may also be involved in choosing, assisting the child in using, and maintaining the technology.

THE CHILD USER'S ENVIRONMENT: The environment should influence the design. For example, if the child intends to use a wheelchair outdoors, identify the specifications that will best meet the environment. Lighter wheelchairs' special design features let a child turn and move quickly, so he can race or participate in sports, while other wheelchairs will let him ride in the sand at the beach (Box 9.6, Figure 9.3a and Figure 9.3b).

Table 9.2 Questions to Ask When Developing a Technology Plan

Child User's Desired Outcomes	1. What does the child wish to accomplish in utilizing the technology? 2. What would the family, teachers, or therapists like the child to achieve? 3. Can the technology help in areas where the child cannot function independently? 4. Will the technology address major life issues that are important to the child's physical and intellectual development? These may include the ability to play, participate in school, interact with family and friends, and control his environment. Older teenagers or young adults may also consider securing and maintaining gainful employment.
Child User's Abilities	1. Does the child possess fine and/or gross motor skills? 2. Does the child have independent locomotion? 3. Can the child hear, see, and speak? If so, how clearly?
Child User's Preferences	1. Were the child's preferences considered when choosing the technology? 2. Does the child want the specific assistive technology product? 3. Does the child perceive the technology to have negative connotations associated with it?
Child User's Environment	1. Will there be exposure to variations in weather, such as rain, snow, or sand? 2. Will the environment need to remain quiet (e.g., a classroom)? 3. Will there be background noise? 4. Will there be electrical or other interference?

BOX 9.6 THE CHALLENGE OF SAND

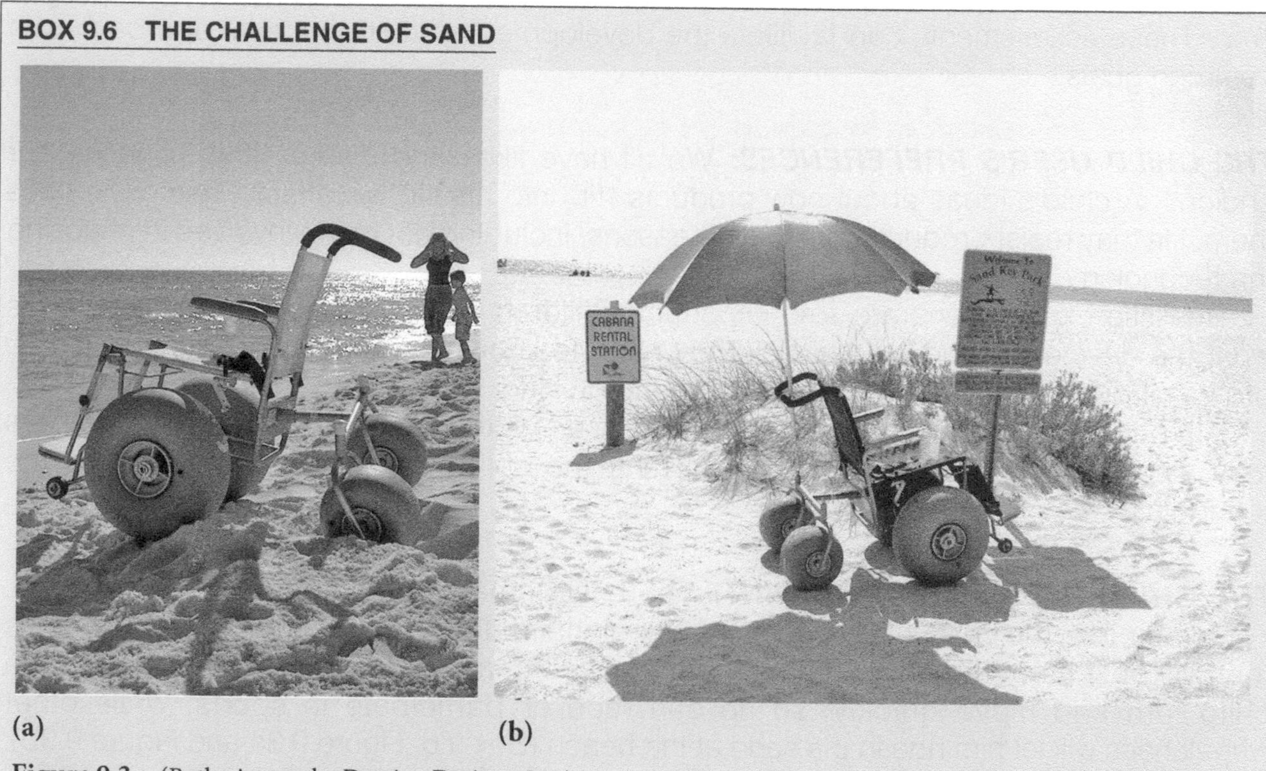

Figure 9.3 (Both pictures by Deming Designs, Inc.)
In the mid-1990s, Hawaii lagged behind other states in the United States in availability of ATDs. A conference was designed to introduce Assistive Technology Devices while addressing needs unique to their region. Much of Hawaii is comprised of beaches where family and community recreation takes place, but traditional wheelchairs are not conducive to use on the sand. The beach wheelchair was introduced (a and b), which had thick tires for riding over sand and canvas seating to withstand water. This example demonstrates how the environment made it impossible to use an established ATD but, with modifications dictated solely by the environment, an ATD could be successful.

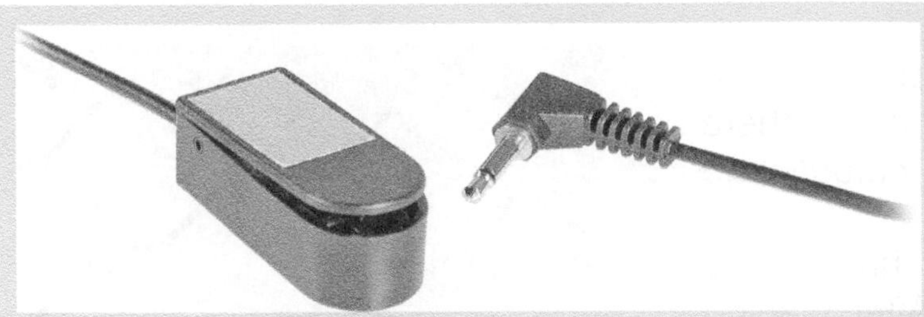

Figure 9.4a (Tash, Inc.)
An Accessibility Switch that requires very light pressure.

Figure 9.4b (AbleNet)
A General Purpose Accessibility Switch.

CONSIDERING THE FUTURE

Children's capabilities change over time, as they grow and mature or as their disabilities progress or lessen. These criteria will influence the selection of the technology.

Understanding the prognosis and disease process is essential. For example, a child with a degenerative illness may initially access his or her environmental control unit using a *general purpose accessibility switch* (requires 100 to 200 g pressure to activate (0.22 to 0.44 lb)), but in six months or a year may require a *switch* that requires a very light touch (5 g pressure (0.011 lb)) to compensate for reductions in upper extremity strength (Figure 9.4a and Figure 9.4b).

The future of the technology also influences the selection. Each child's plan should distinguish between a *minimum acceptability* and a *recommendation*. When selecting devices, minimum requirements represent the bare minimum capabilities the hardware must possess in order to use the products on the computer. However, these specifications may not be sufficiently powerful to achieve optimal or even acceptable results.

For example, historically, each new generation of speech recognition software needs more computer speed and power. A child may initially use speech recognition software on a minimally acceptable computer, but the computer may not accommodate future software upgrades. Further, a minimally acceptable system that enables software installation may not meet broader objectives, such as system response time or stability during hours of continual use.

COMPUTER PRODUCTS

Computers streamline workflow for adults and children with and without disabilities. However, while an able-bodied child may be comfortable using standard computers, it is often necessary to adapt computers with additional hardware and software to accommodate a child with a disability. While the following is not a complete list of types of adaptive computer products, it underscores the broad range of technologies that assist individuals with disabilities.

KEYGUARDS

A *Keyguard* fits over traditional or adaptive keyboards (Figure 9.5). Usually plastic or metal, there are two types of *keyguards,* one with *cutouts* and another offering a *protective layering.*

Cutouts. Keyguards with cutouts have holes that fit over keys on the keyboard. After placing the keyguard on top of the keyboard, users can only strike unprotected keys.

This product works well for individuals who lack fine motor control. It enables them to use a standard keyboard with greater accuracy because (1) some keys may be blocked from use and (2) users must do more than touch the keys to implement the keystrokes. They have to put their fingers inside the hole on the *keyguard* and exert pressure on the key to implement the choice.

Protective layering. The second type of *keyguard* provides a protective layer between the keyboard and the environment. It may have multiple colors to distinguish among types of keys on the keyboard or it may blank out certain keys that can distract users. These *keyguards* protect the keyboard from damage due to spills or excess moisture, including saliva.

Figure 9.5 (Viziflex Seels, Inc.)
Example of a Cutout Keyguard. This device offers a child more control over his keystrokes by requiring the child insert his finger into the opening to strike a key.

HEAD POINTERS

Head Pointers are sometimes referred to as *Hands-Free Mice.* These are not *pointing sticks* or *mouth sticks.* Hands-free mice are products that use infrared technology to simulate mouse movement.

To use *head pointers,* users put a reflective device on one's forehead, glasses, or headband. Moving one's head moves the cursor on the screen. Additional technologies such as *dwell selection software* and *accessibility switches* implement mouse clicks.

Dwell Selection Software allows users to implement mouse clicks using software instead of a mouse. For example, a child can move the mouse cursor over an object using a *head-pointing device,* and by leaving the pointer over the object for a specified amount of time (*dwelling*), the object is automatically selected.

Figure 9.6 (AbleNet, Inc.)
An accessibility switch provides a means for children to control devices. Pressing the accessibility switch may replace the need to turn a knob, flip a switch, or click a button.

Children use *Accessibility Switches* to implement choices on devices including wheelchairs, computers, or communication devices. For example, an accessibility switch may be a push plate, a radio transmitter, or *pneumatic switch* (also referred to as a *sip/puff switch*), by which children activate the switch by controlled breathing into the product.

JOYSTICKS AND TRACKBALLS

Joysticks and *Trackballs* provide alternatives to computer mice. Most people who play computer games or visit video arcades are familiar with these controls.

Joysticks are vertical stick-like controls that direct the mouse cursor up, down, sideways, and diagonally. *Trackballs* are essentially upside-down mice. Rather than holding one's hand on a mouse while rolling it over a surface, *trackballs* remain stationary while the user rolls the ball. *Trackballs* help some users who experience neck, shoulder, or arm discomfort caused by mousing. *Trackballs* can also benefit individuals who have limited neck, shoulder, or arm movement because users can place the *trackball* in their laps, eliminating the need to hold their arms up.

Trackballs enable the user's wrist to be held in neutral. It is important to be aware of finger fatigue when using a trackball and not simply exchange one form of fatigue for another.

ON-SCREEN KEYBOARDS

On-screen keyboards simulate standard keyboards by providing a *graphical user interface* (GUI) that portrays the keyboard on the computer screen (Figure 9.7). The user clicks the pictures of the keys. This enables students to key without having to use their hands or *mouthsticks* to key on a standard keyboard. Traditional *mouse*, *joystick*, *hands-free mouse*, *touch screen*, and *dwell selection software* work with on-screen keyboards.

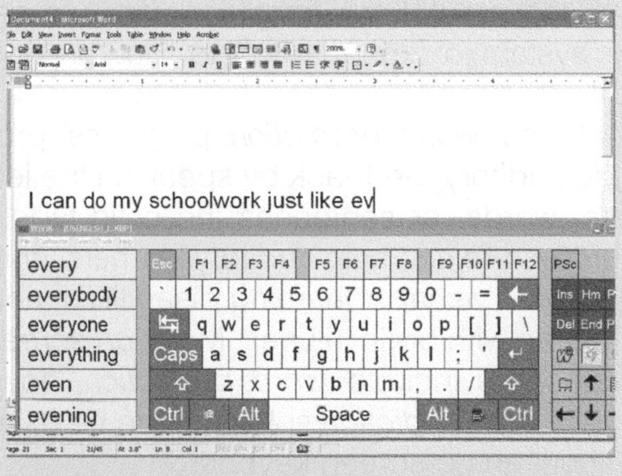

Figure 9.7 (Computer Talk)
On-Screen Keyboards enable children to type using a Graphical User Interface instead of a traditional keyboard.

On-screen keyboards benefit children with limited dexterity or fine motor strength. Many of these devices offer *dwell* selection. With *dwell selection*, users set the hover time from a fraction of a second to several seconds. A child who has good head control but finds it difficult to hold his head steady needs a *short dwell time* when using a *hands-free mouse*. Children with poor head control need *longer dwell times*. Child users will need to experiment to find the delicate balance between clicking the wrong key (when the dwell time is too short) and fatigue from sitting in fixed postures for longer than necessary (when the dwell time is too long).

Some *on-screen keyboard* programs allow users to customize keyboard colors, which is helpful for children with visual impairments and learning disabilities. This feature also encourages children to explore the software. For example, a child may customize the keyboard so it has a purple background, lime green keys, and red letters.

Allowing children to customize the look of the on-screen keyboard does not change the colors of the printed text. Additionally, children with disabilities do not typically use the keyboard for extended periods of time, and they often have little opportunity for creative expression. Enabling children to change the colors on the keyboard provides an opportunity for creative expression and facilitates interest in using the technology. There may be a trade-off between user preferences and professional recommendations on visual fatigue and readability of multicolored screens.

WORD PREDICTION

Word Prediction Software helps children increase typing speed. As the child types, *word prediction software* anticipates what the user wishes to key next and suggests a list of words on the computer screen. As each additional letter is typed, on-screen word choices change to reflect the new cues. When the word the user is typing appears in the word list, the user inserts the selection in the document. *Word prediction software* begins to predict the user's next word, based on the child's word patterns (Figure 9.8).

Another common feature in *word prediction* software is *Abbreviation Expansion*. This feature allows children to program the system, assigning a series of letters, words, or sentences to one or a few keystrokes. When the child types the abbreviation, the software inserts the preprogrammed information into the document. For example, a user can set his system to expand HAYT to "How are you today?"

Some *word prediction* programs provide auditory feedback by speaking the letters, words, or sentences the child types. This feature can help auditory learners, children with visual or learning disabilities, and children with brain injuries. Many *on-screen keyboards* include *word prediction*, however *word prediction* software can be used without an on-screen keyboard.

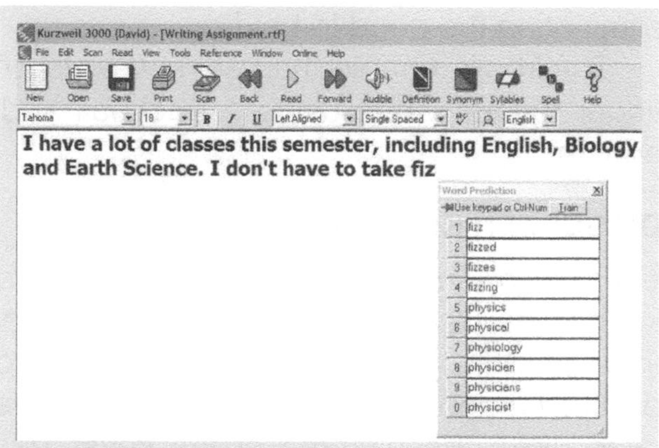

Figure 9.8 (Kurzweil Educational Systems) Word Prediction Software helps children find words and phrases more quickly by anticipating what they are going to "say" (type). In some programs, clicking individual words enables children to hear choices or a definition of the word before making their word selections.

SCREEN ENLARGEMENT

Screen Enlargement software magnifies the images on the computer screen, enabling individuals with low vision to use computers. Children can adjust magnification levels and choose text and background colors that help them distinguish images.

TEXT-TO-SPEECH

Children with visual impairment and blindness often use *text-to-speech software* with *screen enlargement software. Text-to-speech software* can read menus, dialog boxes, document names, and text. It can even announce the time, using the time and date settings in the computer.

Text-to-speech software helps children with dyslexia and other learning disabilities by reading letters, words, and sentences aloud as the child types. Some *text-to-speech software* will also highlight the words as they are read. This feature allows the student to follow along with the computer, reinforcing reading and spelling skills.

Table 9.3 Speech-Recognition Timeline	
Mid-1980s	Speech recognition required users to speak one letter at a time.
1995	Discrete speech programs enabled users to dictate complete words but users were required to pause between words.
1997	Continuous speech programs enabled users to dictate without pausing between each word.
2005	Speech recognition capabilities, including increased *recognition accuracy* and *intuitiveness* of commands, continue to improve.
In the future, the software may be easier to use, include commands that are more intuitive, and we may be given the ability to dictate using *conversational speech*.	

SPEECH-RECOGNITION

Speech-recognition software is an input device, just as keyboards, mice, and joysticks are input devices. *Speech-recognition* software enables children to use the computer by speaking instead of typing on a keyboard or using a mouse. Child users with upper extremity limitations, visual impairment, learning disabilities including dyslexia, and brain injuries may benefit from *speech-recognition software*.

There are two types of speech-recognition software: *discrete speech* and *continuous speech*.

DISCRETE SPEECH VS. CONTINUOUS SPEECH: *Discrete speech programs* require users to pause between words (usually for between 1/10th and 2/10th second). *Continuous speech programs* understand speech without distinct pausing between words. Individuals with profoundly affected speech often have less difficulty using *discrete speech* programs but *discrete speech* software is no longer available. It is therefore important to retain the services of a private consultant who has experience with *discrete speech software and continuous speech software*. These experts can teach children with affected speech methods to increase recognition accuracy, and they can modify continuous speech software to be more conducive for use by children who lack speech clarity.

AUGMENTATIVE AND ALTERNATIVE COMMUNICATION

Augmentative and Alternative Communication Devices (AAC) provide language for those unable to speak. The two most important factors for people who rely on *augmentative communication* are to be able to say what they want to say and to say it as fast as they can.

AAC uses three language representation methods: Minspeak, spelling, and single-meaning pictures. *Minspeak* or *Semantic Compaction* is the fastest and most common method. It consists of symbols strung together in short sequences. Fluency with its use can exceed that of spelling and word prediction within a

Figure 9.9 (Prentke Romich Company) Augmentative and alternative communication devices enable children to communicate.

few hours of instruction (Gardner-Boneau and Schwartz, 1989; Hill and Romich, 1999). Children with language capabilities of a two-year old can use it successfully (Romich, 1999). This method allows users to say exactly what they want because they create their sentences spontaneously, rather than using preprogrammed sentences.

Minspeak-Based Semantic Compaction is a language in itself. It may help to compare learning *Semantic Compaction* to learning Shakespeare or the shorthand court reporters use. *Semantic Compaction* uses fewer than 100 icons in combinations of no more than three pictures, making it possible to create more than 6000 words. For example, child users who want to eat might press the Apple icon and Verb icon, where the Apple icon represents food and the Verb icon represents action. A user could specify kiwi as the food by pressing the Apple icon two times and the Dog icon (because it is brown and furry).

EDUCATIONAL SOFTWARE

Educational software assists children in learning a wide variety of skills. *Educational software* can teach children how to use switches. This might represent teaching children cause and effect relationships (pressing the switch to achieve a result) or it can help them practice motor and cognitive skills (pressing the switch at the appropriate time).

Literacy software is a type of educational software that benefits children with learning disabilities and attention or hyperactivity disorders. These children may have difficulty with memory, writing, or comprehension. *Literacy software* incorporates word prediction, speech recognition, and text-to-speech technologies, providing a multisensory approach to help children process information more effectively.

Children who find it difficult to write may benefit from software programs that develop keyboard skills through picture mapping, letter recognition, word building, etc. Additional software helps students develop and organize ideas.

Math programs help children learn basic mathematical skills. Such programs teach skills such as number sequencing, telling time, and adding and subtracting.

SWITCHES AND MOUNTS

Switches come in almost every size and style imaginable to accommodate nearly every disability. Children use *switches* to control computers, wheelchairs, and augmentative communication devices. There are several types of *switches.* Their classifications are based on the amount or type of force required to produce an effect (see Box 9.7 for information on selecting switches for a particular child or need).

Switches that require more force are often larger and allow individuals with gross motor skills to hit the switch using heavy pressure. Children with good fine

Figure 9.10 (AbleNet, Inc.)
Switches can be used to implement choices and control devices.

motor skills or less movement often benefit from smaller, lighter switches that require a delicate touch. For example, an eyebrow switch can be mounted on a lightweight, adjustable visor. Activation occurs with a slight upward movement of the user's eyebrow. Children with movement limitations can also use a *tongue switch*, which is placed in the child's mouth and pressed with the tongue.

Box 9.7 FACTORS TO CONSIDER WHEN IDENTIFYING APPROPRIATE ACCESSIBILITY SWITCH(ES)

1. What is the goal in using the switch?
2. What type of movement does the user have to be able to engage the switch?
3. How much force can the user exert to execute the action?
4. What areas of the individual's body can be used to implement the switch?
5. Is it convenient and easy to use without interference?
6. Can more than one area be used to implement a switch?

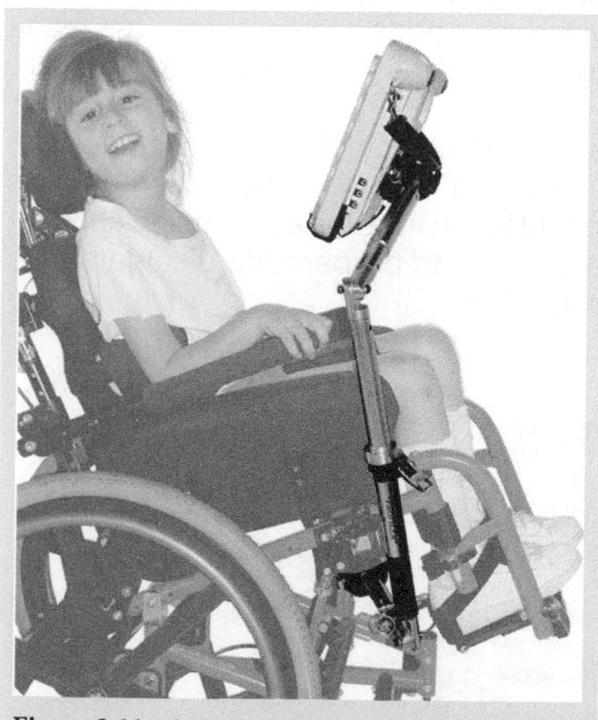

Figure 9.11 (CJT Enterprises, Inc.)
This type of mount has an angle attachment and positions easily.

Mounts (Figure 9.11) enable users to attach switches, keyboards, touch screens, and other devices in places that are easy to reach. For example, some mounts can position devices or switches without losing space. *Switch mounts* can allow for easy and precise adjusting of switch positions and mounts with goosenecks.

TRAINING

A child must modify his existing processes and learn new skill sets when he receives assistive technology. To be successful, the child *must* have appropriate training to learn how to make these modifications. Just as we would never purchase an automobile for our child before he learned how to drive, children should not be given assistive technology without training and ongoing technical support. The training should be provided by an expert in the specific technology the child is using. This will increase the probability of successful implementation.

For example, properly implemented, *speech-recognition software* enables children with disabilities to match or even exceed the typing speed and accuracy of their able-bodied counterparts. However, speech-recognition, like many other technologies, is not an out-of-the-box solution. Studies have shown that nearly two-thirds of *adult users* abandon speech-recognition software because it does not perform to their expectations (Box 9.8). Children may have even greater difficulties and be quicker to abandon the technology than adults.

Technologies, including speech-recognition software, must be taught in settings that allow children to work with the product properly. Experts in the nuances of the technology help the child user develop realistic expectations, minimize frustration, and use the product correctly.

BOX 9.8 DO USERS USE THEIR PRODUCTS?

A survey by Emkay Innovative Products* found: (STM, 2001)

- Fewer than three percent of their respondents used the software for one hour or more, three to five times per week.
- At least half of these respondents abandoned their software shortly after purchasing it.
- Most (62%) were not using the software.

SIMILARITY AWARENESS

When children with disabilities are integrated into "regular" classrooms, their needs are identified, adaptive technology devices are provided, and additional resources are made available for the classroom. However, often little or no preparation is provided for the able-bodied children.

Mainstreaming allows children with disabilities to interact in a regular classroom setting in which they may be included for all or part of the school day.

The first time able-bodied children have contact with a person who is disabled may be in the classroom. It is therefore as important to sensitize the able-bodied children about disabilities, as it is to prepare the disabled child with assistive technology.

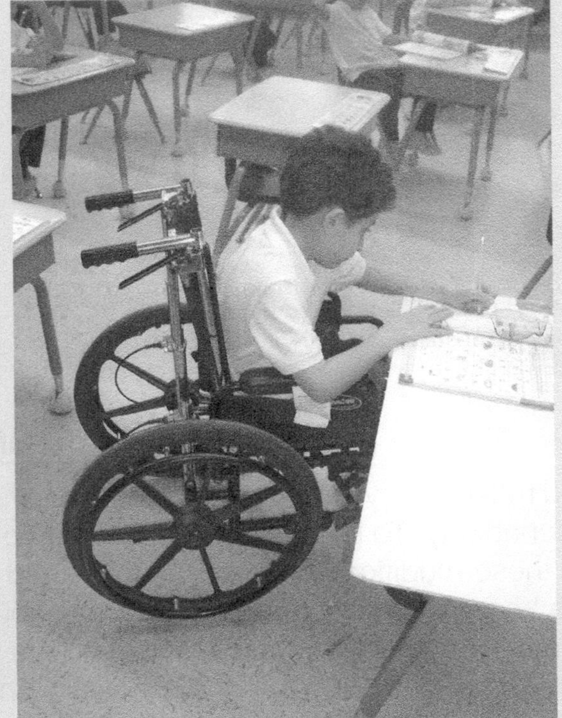

Figure 9.12a and Figure 9.12b (Valerie Rice)
Temporary disabilities, like permanent disabilities, require adjustments be made to the child's environment.

* Emkay Innovative Products is a microphone manufacturer. Their parent company is Knowles Electronics. Their findings were based on a survey of more than 13,000 people who purchased speech recognition software.

Mainstreaming (integrating students with disabilities into regular classrooms) is an opportunity to introduce children to concepts of ability, disability, separation, and inclusion. While exploring these terms, we can empower children with an understanding that, depending on how we define each of the above-mentioned terms, each of us can be considered disabled in one manner or another. Children can become aware that people can choose between terminology that separates and alienates or includes and supports.*

TECHNOLOGICAL ADVANCES

Technological advances continue to offer opportunities for people with disabilities but, because people with disabilities represent a relatively small market share (and children with disabilities comprise an even smaller market), research and development costs often make assistive technology products cost-prohibitive. As technology advances and the able-bodied community incorporates assistive technology and *Universal Design* into everyday life, the research and development costs will decrease, making the products more financially affordable.

SIMPLIFYING USE

People with disabilities who incorporate assistive technology into their lives may still need assistance assembling and maintaining the technology. Some students need aides to accompany them to class on an ongoing basis.

As products evolve, the responsibilities of the aides may decrease. For example, *plug and play* products make setup easier. Modulated products, especially those that self-diagnose problems, permit removal and replacement of a malfunctioning module. Wireless products, including *Bluetooth* technology, can simplify installation of *environmental control units*. They require a less invasive installation, without extensive wiring and involvement in household construction.

Multimodal devices give users more than one way to access information, which can be helpful for people with disabilities. For example, *Tablet PC*s offer flexibility by integrating a speech-ready system into a unit with a touch screen. Users can input data by traditional means (mouse and keyboard), by touching the screen with a stylus or one's fingers, or by voice.

The term *multimodal* can be misleading. A product can be *multimodal* without including *redundancy* in the design. If manufacturers incorporate *redundancy*, products will have all of the access options embedded in the device and users will simply activate the mode they wish to utilize. If a product is *multimodal* but does not offer redundant input, the device may require users to type, speak, *and* use a stylus, making the product more difficult or impossible for a disabled individual to access.

* Note from the editors. Chapter 9 reviews related issues in the context of developing countries.

UNIVERSAL DESIGN

Universal design is the design of products and environments so as many people can use them as possible. While it is primarily about *function,* it also includes consideration of attractiveness, affordability, and accessibility. Incorporating *universal design* makes products easier to use (Table 9.4).

Table 9.4 Principles of Universal Design[a]

Principle	Description	Examples of Universal Design[b]
1. Equitable Use	The design is useful and marketable to people with diverse abilities.	• A computer that incorporates speech recognition (speech input), text-to-speech (speech output), and graphical user interface (GUI).
2. Flexibility in Use	The design accommodates a wide range of individual preferences and abilities.	• Software that provides for alternative ways to execute tasks (e.g., a keyboard *or* a mouse *or* accessibility switch).
3. Simple and Intuitive	Use of the design is easy to understand, regardless of the user's experience, knowledge, language skills, or current concentration level.	• Instructions that include text and images to direct the user.
4. Perceptible Information	The design communicates necessary information effectively to the user, regardless of ambient conditions or the user's sensory abilities.	• An LED display on a telephone answering machine that indicates the number of messages, *and* a light that blinks when there is a message *and* a tone that announces when there is a message.
5. Tolerance for Error	The design minimizes hazards and the adverse consequences of accidental or unintended actions.	*Computer programs:* • With an "undo" function • That requires users to confirm an action before performing it (e.g., "do you really want to delete the selected file?").
6. Low Physical Effort	The design can be used efficiently and comfortably with a minimum of fatigue.	• Keyboard enhancements such as StickyKeys, RepeatKeys, and SlowKeys. • Word prediction software that reduces the amount of keying required.
7. Size and Space for Approach and Use	Appropriate size and space is provided for approach, reach, manipulation, and use regardless of user's body size, posture, or mobility.	• Smaller keyboards, designed for children. • Larger keyboards, designed for individuals who lack fine motor skills.

[a] Copyright ©1997 NC State University, The Center for Universal Design. www.design.ncsu.edu/cud

[b] ©2005 Robin Springer.

CHANNELS OF DISTRIBUTION

Assistive technology is a specialized field, just like physical therapy, occupational therapy, and medical disciplines. Doctors, counselors, educators, and therapists spend countless hours remaining current in their fields of expertise. So does the expert in assistive technology. Because assistive technology is a field in itself, doctors, educators, and others cannot be expected to be experts in assistive technology.

The increase in the amount of special product knowledge required to integrate assistive technology into a child's life has made *distribution networks* increasingly important. Although some health care professionals and teachers may have knowledge of assistive technology, when incorporating assistive technology into a child's education plan, an assistive technology expert needs to be included to ensure the child is receiving appropriate technology and services.

CONCLUSION

It is an exciting time in the world of assistive technology. *Automatic speech recognizer* systems enable child users to control applications by voice. Children can navigate phone systems by voice, touch-tone, or live operators. *Biometrics*, devices that use unique biological properties to identify individuals, including voiceprint, thermography, fingerprint, and retina scans, are more widely available. Such technologies allow individuals to access information and locations without the use of one's hands or eyes to swipe an identification card or type a password.

Voice over Internet Protocol, which converges multiple communications channels, enables people to use their computers to communicate telephonically. A blind individual can listen to e-mails using text-to-speech. A person who is deaf or hard of hearing can read the contents of a voicemail after converting it to text.

Understanding that all individuals encompass a spectrum of abilities, the only reason to label children as *disabled* is for access to the government programs and products they need to lead more productive lives. As technology becomes more widely used and accepted in mainstream environments and as we improve awareness and inclusion, the need for this label will diminish. We will live in a "What I Choose To Use" world in which all of the access options are built into the technology: speech-recognition, text-to-speech, word prediction, on-screen keyboard, touch screen, stylus, wireless, traditional mouse, and keyboard. It will be up to each individual to use the access that works best at any given moment in time. It will be a world that is accessible to all.

RESOURCES

Table 9.5 Resources on the Web

Center	Web site
Australia	
Australian Legislation	www.legislation.act.gov.au
Canada	
The Canadian Human Rights Commission	www.chrc-ccdp.ca
International	
World Health Organization (WHO)	www.who.int/en
Center for International Rehabilitation	www.cirnetwork.org
United Kingdom	
Her Majesty's Stationery Office (HMSO)	www.hmso.gov.uk
Special Education Needs and Disability Act 2001, Chapter 10	www.hmso.gov.uk
United States	
Centers for Disease Control	www.cdc.gov/ncbddd
Individuals With Disabilities Education Act Website	www.ed.gov/about/offices/list/osers
Americans With Disabilities Act Website	www.ada.gov
United States Website for Disability Information	www.disabilityinfo.gov
The Center for Universal Design (TCUD)	www.design.ncsu.edu/cud
Disability World: Bimonthly Web-zine of International Disability News and Views	www.disabilityworld.org
Mainstream products originally created as assistive technology for people with disabilities	www.accessiblesociety.org

REFERENCES

CRIN (Child's Rights Information Network) (2004). The child rights information network. www.crin.org

Cross, R.T., Baker, B.R., Klotz, L.V., and Badman, A.L. (1997). Static and dynamic keyboards: Semantic compaction in both worlds. Proceedings of the 17th Annual Southeast Augmentative Communication Conference, pp. 25–30. Birmingham: SEAC Publications. www.prentrom.com

Gardner-Bonneau, D.J. and Schwartz, P.J. (1989). A comparison of words strategy and traditional orthography. Proceedings of the 12th Annual RESNA Conference, pp. 286–287.

Hill, K. and Romich, B. (1999). Identifying AAC language representation methods used by persons with ALS. American Speech-Language-Hearing Association (ASHA) Convention, San Francisco.

Krug, E. (2004). Address at the International Paralympic Symposium on Disability Rights, Athens, Greece. www.who.int/en

NIDDR (2004) National Institute of Disability and Rehabilitation Research. www.ed.gov/about/offices/list/osers/nidrr

Onasanya, A. (2004). Disability in Africa. Pearls of Africa. www.pearlsofafrica.org

Romich, B. (Fall, 1999) Minspeak. Current expressions online: A newsletter for friends of Prentke Romich Company. www.prentrom.com

STM (2001) "Headsets make a difference in speech recognition accuracy". *Speech Technology Magazine.*

Turmusani, M. (2004) Afghanistan: Community based approach to parents with disabled children: Reality or ambition? *Disability World*, 23, 1. www.disabilityworld.org

UNICEF (1999) United Nations Children's Fund. The State of the World's Children 1999 Report. www.unicef.org/sowc99

CHAPTER 10

MEETING THE NEEDS OF DISABLED CHILDREN IN DEVELOPING COUNTRIES

DAVID WERNER

TABLE OF CONTENTS

GOODNESS OF FIT

In the villages of third world countries, no one talks about "ergonomics" and few know what it means. Even the term "appropriate technology" must be used carefully. In countries branded as "less developed," villagers wonder if "appropriate" means they are expected to settle for second-rate solutions, because what is appropriate in one setting is not necessarily appropriate in another.

Assistive technologies for disabled persons in developed countries are increasingly specialized, elaborate, and exorbitant. These advanced technologies are often ineffective or irrelevant in villages and low socioeconomic environments. Unfortunately, low-cost low-tech solutions developed from simple local resources are often overlooked and undervalued.

Helping users in these environments requires working with them to develop solutions that provide a "goodness of fit," i.e., matching solutions with their life circumstances, culture, and environment. Although they may be unfamiliar with the discipline of "ergonomics," this matching is the essence of designing to fit the person (i.e., ergonomics).

SEEING THE BIG PICTURE

What makes a solution work (or not) for a given circumstance depends on many factors including the local cultural, environmental, and economic conditions. Macroergonomics involves studying the "big picture" together with its parts to understand how each part might contribute to problems and influence the effectiveness of a solution.

This does not mean the microergonomic perspective is lost; it is merely one part of the whole. The individual user's needs, abilities, and limitations remain central to the macroergonomic perspective, but the conditions in which the individual lives and functions are also part of the solution.

A child's "job" is to learn emotionally and intellectually, to hone skills and capabilities, and to grow and develop into a healthy, productive adult. Children accomplish this through play as well as chores at home, in the community, and at school. Children's ability to make this transition is affected by their family, community, and government policies as well as the resources available. Disabled children are challenged in each of these arenas.

RESOURCES

UNDERSTANDING THE IMPOVERISHED COMMUNITY

Poverty often limits children with disabilities more than their disabling condition. Their needs are clear only within the context of their family and larger community. This means individuals who work with them must gain an understanding of their environment, culture, and the demands of their day-to-day lives. Framing potential solutions within their resource limitations yields a more successful resolution.

Involving the disabled individual (and their family) harnesses their ingenuity and empowers them within their home environment. It also helps ensure that local conditions are considered in the therapeutic process. For example, in many poor countries and communities, severely disabled children—and even many moderately disabled children (especially girls)—do not survive past early childhood. Rehabilitation professionals and social workers may reprimand destitute parents for neglecting disabled children or giving preferential treatment to healthier children. Rather than helping the situation, this can result in alienation (Figure 10.1). Instead, workers must collaborate with the family to find solutions to problems, within the context of their environment.

Figure 10.1 Rehabilitation professionals and extension workers often ignore the larger picture when conducting "outreach" services in poor communities. While focusing on the physical, mental, visual, or auditory impairments of the disabled child, they overlook the parents' difficulties in meeting the child's and family's basic needs.

In their typical environment, rehabilitation workers may not have to think about basic needs of their clients such as housing, food, and water. However, these issues may be of vital concern to families and disabled children in developing countries. For example, worldwide, malnutrition is a major cause of disability, especially for mothers, fetuses, and young children.*

* Chronic undernutrition (long-term hunger) before and during pregnancy increases the risk of premature birth, complications of labor, and postpartum crises. Each of these can cause brain damage that may delay development and lead to cerebral palsy, deafness, blindness, epilepsy, or multiple disabilities. Nutritional deficits during pregnancy can cause congenital syndromes such as cretinism from iodine deficiency as well as blindness and lowered resistance to infections from a shortage of vitamin A. Poverty contributes to under nutrition, poor nutrition, infectious disease, and lack of adequate health care—all of which impact the frequency, severity, and complications of disability, as well as the availability of rehabilitation.

ANALYZING AND HARNESSING THE SITUATION

WHEN IS A PERSON "DISABLED"?

As seen above, poverty is both a contributing cause of disability and an obstacle to meeting disabled people's needs. As the song goes "every cloud must have a silver lining,"* we must find such "silver linings" by looking for specialized and unique opportunities for including and rehabilitating disabled children.

For example, a person with mild to moderate mental retardation may not be seen as disabled in a poor farming community. This person can play an important role such as performing simple and repetitive jobs (Figure 10.2).

Figure 10.2 Edgar, a teenage boy with Down's syndrome, helped support his family by selling drums of water he hauled from the river. This gave Edgar great pride and let him contribute to his family and community. In time, a new public water system was built that benefited the community as a whole, but put Edgar out of work.

Children with impairments are more limited in some environments than others (Figure 10.3). Mentally slow children may fit well in poor farming communities, where livelihood depends largely on physical activities. On the other hand, slow children may find it difficult to adapt in middle-class city neighborhoods, where schooling is needed for a meaningful role in society.

A child who is mentally slow but physically strong,
in a village may not be very handicapped,
but in a city or in school may be very handicapped.
A child who is physically disabled but intelligent,
in a village may be very handicapped,
but in a city or in school may not be especially handicapped.

Figure 10.3 Depending on their disability, some children find it easier to adapt to one setting (rural or urban) than another.

* From Melancholy Baby by George A. Norton. www.bartleby.com/66/55/42655.html

WHEN IS A DEVICE "FUNCTIONAL"?

HIGH TECHNOLOGY IS NOT ALWAYS THE BEST SOLUTION

Low-income neighborhoods and rural areas provide unique opportunities to develop low-cost yet highly functional assistive devices using locally available materials. These devices are designed to adapt to the possibilities as well as limitations of the local situation.

Although costly rehabilitation equipment may look more impressive, simple locally made equipment, thoughtfully adapted to the specific child and situation, often works as well or better. Once again, the key is to take a larger viewpoint for evaluation, assessment, and intervention.

A CASE STUDY EXAMPLE—MOBILITY: Most large metropolitan areas in developed countries attempt to provide some accessibility of public buildings for people with disabilities. While the equipment to promote accessibility works well in those situations, it may be less functional in a rural environment.

Mike Miles, a writer experienced in Community Based Rehabilitation, tells the following story about a town in Bangladesh.*

Two people set out one day to mail a letter at the post office. First to arrive was a disabled young woman, obviously well-to-do. She pulled up near the post office in her big adapted van, lowered her electric wheelchair on the built-in lift, and buzzed up to the post office. There she encountered four steps. No ramp. Cursing the inaccessibility, she wheeled back to her van and drove home, her letter unmailed.

Next came a boy, both legs paralyzed by polio. Wanting to mail a letter, he climbed onto his homemade skateboard and hitched a ride by hanging to the back of a truck. On reaching the post office, he saw the four steps. No problem! He scuttled on his backside from one step to the next, dragging his skateboard behind him. At the top, he hopped back on his board and rolled into the building. Bypassing a line of people, he wheeled under the counter and purchased stamps from the clerk, who obligingly mailed his letter.

Mike then asks:

"Which of these two disabled persons is more independent?"

"Which mobility aid is more appropriate?"

"What barriers were involved: Physical? Cultural? Psychological?"

"Would the answers be the same in other countries or circumstances?"

* Referenced from a lecture by Miles, M. (September 2001). Community based rehabilitation, Amsterdam.

Figure 10.4 depicts a similar scenario.

COMPLEX
AND
COSTLY
This motorcycle with two back wheels has been adapted with a ramp at the back so that the wheelchair rider can wheel up into it and drive while seated in his wheelchair.
Developed by the Spastics Society of India, Madras.
SIMPLE
AND
CHEAP
Like this child from India who cannot walk, children (and adults) all over the world use skate-boards, trollies, or make home-made carts to move and play.

Figure 10.4 (Photos from *Nothing About Us Without Us*, by David Werner. Available online in several languages at www.healthwrights.org)
Two mobility aids used by children in India. Which is more appropriate? What might happen if the children swapped their vehicles?

A CASE STUDY EXAMPLE—"TECHNOLOGY" FROM THE GROUND UP: Uneven dirt floors impede some disabled children. Yet earth floors and courtyards are useful when constructing rehabilitation aids such as parallel bars, standing frames, and special seating.

For example, children with scissoring of the legs caused by spasticity may benefit from special seats that enable them to sit with their legs widely spread. In the United States or Europe, such children might be fitted with expensive and elaborate molded plastic seats. Equally functional seats can be made in village homes or yards by pounding wooden stakes into the ground (Figure 10.5 and Figure 10.6).

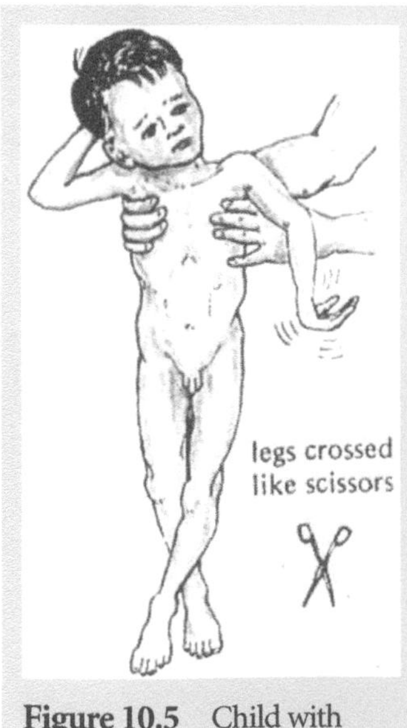

Figure 10.5 Child with cerebral palsy "scissoring legs."

Figure 10.6 Seating for a boy with cerebral palsy and "scissoring legs."

KEEP THE DISABLED CHILD IN THE FAMILY AND COMMUNITY "LOOP"

The psychological and emotional health of disabled children and their relationships to family and community are important, even when considering a rehabilitation device or activity. It is important to ask: "Does this therapeutic device or exercise help integrate or exclude the child from the day-to-day life of their family?"

It is best to avoid isolating children from their community and even rehabilitation exercises can ultimately isolate a child and emphasize how they are "different" from their peers. Figures 10.7 through Figure 10.9 show examples of keeping the child involved with their community. By creating work-based therapeutic activities for the child, you keep them involved, accomplish therapeutic goals, and build on strengths rather than limitations (Figure 10.10).

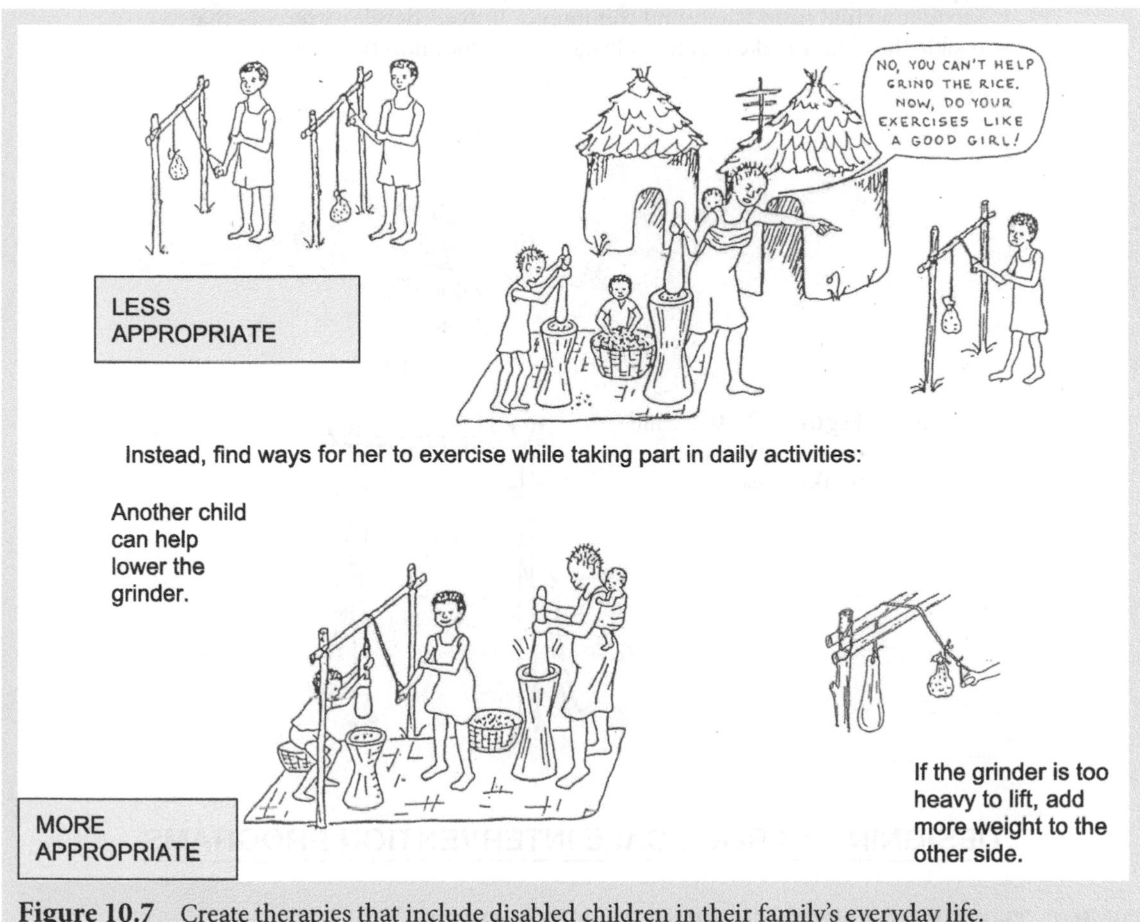

Figure 10.7 Create therapies that include disabled children in their family's everyday life.

Figure 10.8 In places where people grind grain with a hand mill, it can be used for exercises. So can grinding grain on a stone dish. The resistance of a mill can be adjusted from easy to hard.

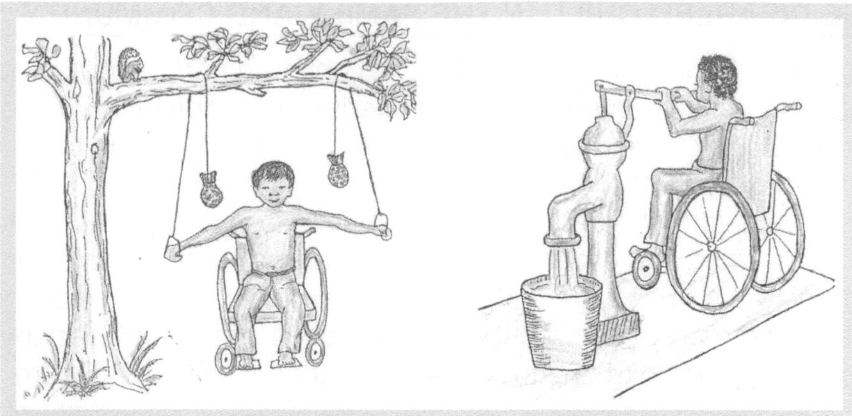

Figure 10.9 Therapeutic exercises to strengthen arms and shoulders can separate a child from family and community. Instead, develop exercises that enable the child to take part in daily family and community activities.

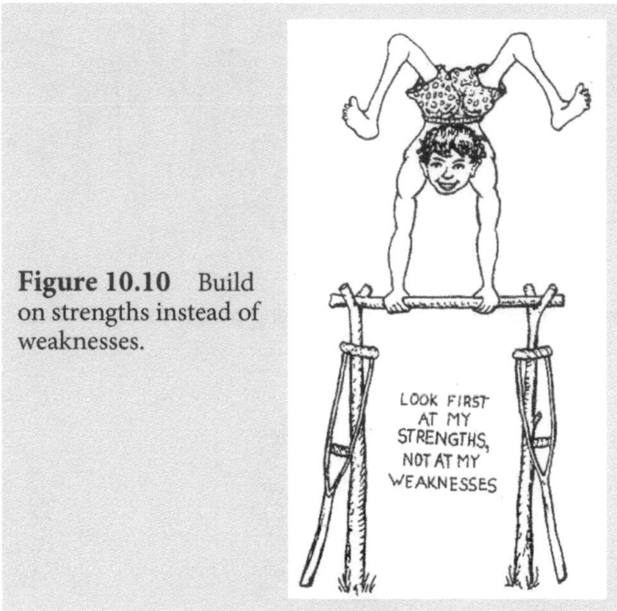

LOOK FIRST
AT MY
STRENGTHS,
NOT AT MY
WEAKNESSES

Figure 10.10 Build on strengths instead of weaknesses.

DESIGNING LARGE-SCALE INTERVENTION PROGRAMS

COMBINING THE BEST AND THROWING OUT THE REST

Large-scale intervention programs are most successful when they follow a macroergonomics approach using similar premises to those listed above:

1. Integrating interventions within each system that affects the child, such as the economic, health care, social, and educational systems.
2. Taking the local culture, resources, and environment into account.
   a. Define disability according to the culture, resources, and environment.
   b. Define function according to the culture, resources, and environment.
3. Keep the disabled child involved in the activities of their family and community.
4. Involve the child with the disability in the rehabilitation process.

For large-scale interventions, two additional principles are added:

1. Involve the individuals with disabilities, their families, and their communities throughout the process.
2. Educate the individuals with disabilities and their communities so they can independently run their own programs, calling upon outside experts as subject matter experts only.

Two international initiatives were recently introduced in third world countries. These are the **Independent Living Movement** (IL) and **Community Based Rehabilitation** (CBR). Each has strengths and weaknesses, given in Table 10.1.

Table 10.1 Two Large-Scale Intervention Programs

	How it began	Strengths	Weaknesses
Community Based Rehabilitation	Launched by rehabilitation experts working with the **World Health Organization** (WHO) in the early 1980s	• Tries to reach all disabled people, particularly the poorest and most needy • Often seeks home-based alternatives to institutions • Plan is inclusive, involving both government and private initiatives	• Disabled individuals are often not involved in decision making, organizing, and leading efforts that affect their lives. • Local supervisors, who are the nondisabled facilitators of the process, often receive minimal training. Therefore, they do not always know enough to help disabled individuals meet their greatest needs effectively.
Independent Living Movement (IL)	Started by disabled people in industrialized countries It has gradually made headway in poorer countries through organizations such as **Disabled People International (DPI).**	• Led by disabled activists. Disabled people determine the direction of its organizations and services. • Provides social action for equal opportunities Its rallying slogan has become "Nothing About Us Without Us" (Figure 10.11).	• Largely a middle-class movement that often neglects the poor. • Tends to leave out vulnerable groups such as the mentally retarded, mentally ill, and children. • Outreach often assumes the needs of the third world poor resemble those of disabled people in Western countries. • "Living independently" on one's own is a Western concept; living "interdependently" may be more appropriate for societies rooted in extended families with a strong sense of community.

Figure 10.11 The Independent Living Movement's slogan.

The challenge for rehabilitation workers and disabled individuals is to find ways to combine and adapt the empowering self-determination of the ILM with the broad outreach of CBR. The Program of Rehabilitation Organized by Disabled Youth of Western Mexico, known as PROJIMO, is a program that combines the best of the two programs listed in Table 10.1. Projimo is the Spanish word for neighbor.

The goal of PROJIMO is to promote self-determination while working within the person's cultural and socioeconomic environments. It does so by training individuals in the community, who have disabilities to be "crafts persons." When persons with disabilities learn to design and make assistive equipment and to include the user in the process, it helps both the creator and the recipient of the equipment.

PROJIMO AS AN EXAMPLE OF A SUCCESSFUL INTERVENTION PROGRAM

Birth of a Program Growing out of Need

Sometimes, the best programs develop from a "grass roots" need. PROJIMO grew out of a villager-run health program in the Sierra Madre in 1981. A number of mountain villages selected a local person with a disability to be trained as their village health worker. As years passed, others with disabilities became health workers.

Individuals with disabilities are some of the best health workers. Their health work is not just a job; it is a chance to gain importance and respect in their communities. Through their own experiences, they understand and relate to others in need. By becoming health workers, their weaknesses became their strengths.

Over time, some health workers with disabilities became leaders in the village health program. However, they wanted to know more about helping other persons with disabilities. This led to the formation of PROJIMO, a sister organization to the village health program. PROJIMO is primarily run and staffed with village health workers with disabilities and other village youth with disabilities who were invited to participate.

Today, nearly all the leaders, rehabilitation workers, and crafts persons in PROJIMO are disabled persons. Most came to PROJIMO for their own rehabilitation. They liked the program and stayed to learn skills and help others, eventually becoming capable rehabilitation workers themselves. Some returned home and started similar (but always uniquely different) community-based programs in their own towns or cities.

Having a disability provides team members with unique insight, creativity, and compassion when working with other disabled people—especially those with disabilities similar to their own. (Box 10.1 contains additional comments on the benefits of having individuals with disabilities as staff. Other types of rehabilitation efforts use this concept, such as Alcoholics Anonymous.)

BOX 10.1 BENEFITS OF DISABLED STAFF MEMBERS AND TECHNICIANS

Rehabilitation workers with disabilities of their own often have a deep understanding of the direct effect of disability on a person's life as well as its social, environmental, and emotional considerations. They see the person and not just the "impairment." Rather than looking at weaknesses, they encourage and build on strengths.

In the same way, technicians who are disabled (brace, limb, and wheelchair makers, etc.) appear more likely to make certain that assistive devices and therapeutic measures really fit the needs of the disabled child user. Because of their own experience, they think of a whole range of factors and possibilities that a nondisabled technician may be less likely to consider.

Finally, the success of the workers encourages the success of those they assist (Figure 10.12).

Training

PROJIMO workers receive hands-on training as apprentices. A wide range of rehabilitation professionals visit, teach, and learn from the program participants. These include physical and occupational therapists, brace and limb makers, rehabilitation engineers, special educators, wheelchair designers, etc.

The programmatic distinction between visiting professionals teaching rather than merely providing services is important. Every consultation is a learning opportunity for the village rehabilitation workers, the family members, the child, and the visiting professional. That way, when visiting professionals leave, they leave a team with upgraded skills, rather than creating dependency.

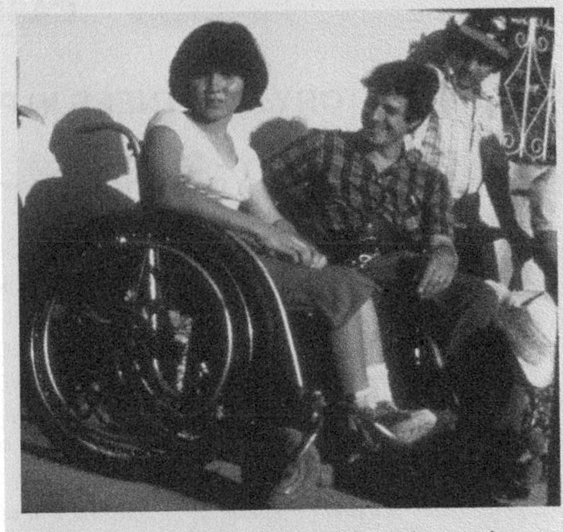

Figure 10.12 Ralf Hotchkiss (center), a paraplegic wheelchair designer, helped start PROJIMO's wheelchair shop. He has since helped disabled groups start shops in over 30 developing countries.

Mission

The PROJIMO team evaluates the needs of children and adults with a wide range of disabilities. They teach families about assistance, provide physical and occupational therapy, and custom design a wide range of low-cost assistive devices such as orthopedic appliances, prosthetics, special seating, and wheel chairs. Government rehabilitation centers in cities also contract with PROJIMO to make prosthetics, orthotics, and other assistive devices.

As the team members are also members of the local community, they are familiar with the needs, personalities, and culture within the community. The team provides a wide range of skills training and conducts community awareness events. These include child-to-child activities to help children without disabilities accept and welcome the child with a disability. Schoolchildren sometimes help make educational toys or assistive devices for disabled children whom they befriend. Again, the programs are unique to the needs of that community. Box 10.2. shows two books that describe PROJIMO.

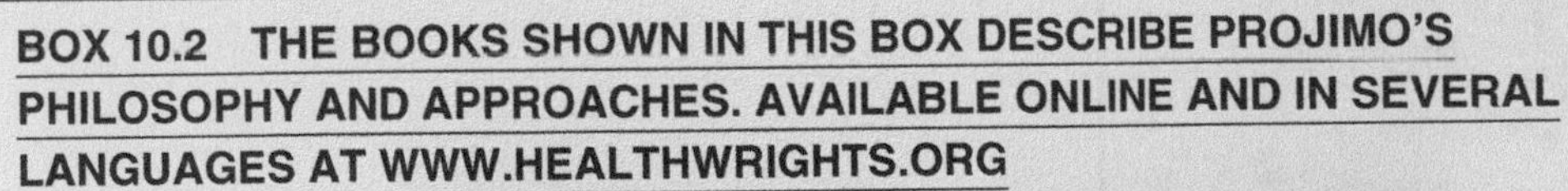

BOX 10.2 THE BOOKS SHOWN IN THIS BOX DESCRIBE PROJIMO'S PHILOSOPHY AND APPROACHES. AVAILABLE ONLINE AND IN SEVERAL LANGUAGES AT WWW.HEALTHWRIGHTS.ORG

CASE STUDY EXAMPLES

#1 MARIO'S STORY: PARTNERS IN PROBLEM SOLVING

BOX 10.3 PARTNERS IN PROBLEM SOLVING

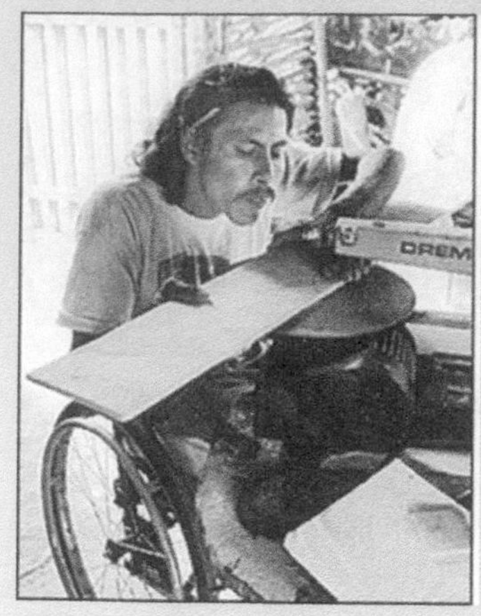

Mario Carasco, a former street boy who became paraplegic from a bullet wound, became a PROJIMO carpenter (left).

He includes children in the problem solving process as friends and equals (right).

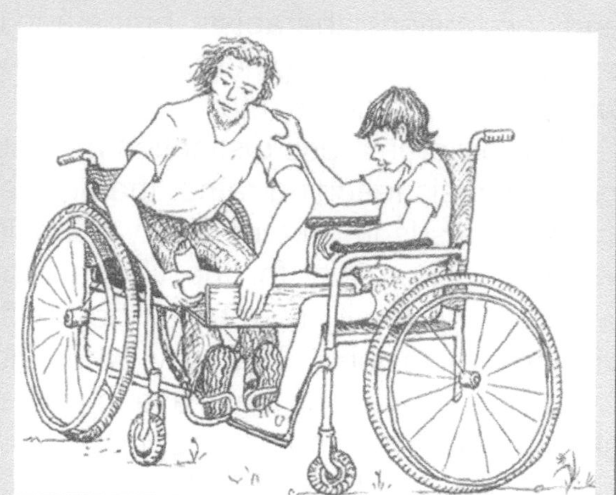

#2 MARCELO'S STORY: THE TOTAL PICTURE, FROM DISABILITY TO ABILITY

Marcelo's story demonstrates the success of the intervention program on an individual on a personal level. The story begins with his injury and follows with his selection as a health worker, training with PROJIMO, and finally to his employment with PROJIMO. Through this program, Marcelo became one of the local experts, treating others, and creating assistive devices for them.

I first met Marcelo years ago when he was 4 years old, in a remote village. He was sitting in the dust outside his family's hut, unable to walk because polio had paralyzed his legs as an infant (Figure 10.13).

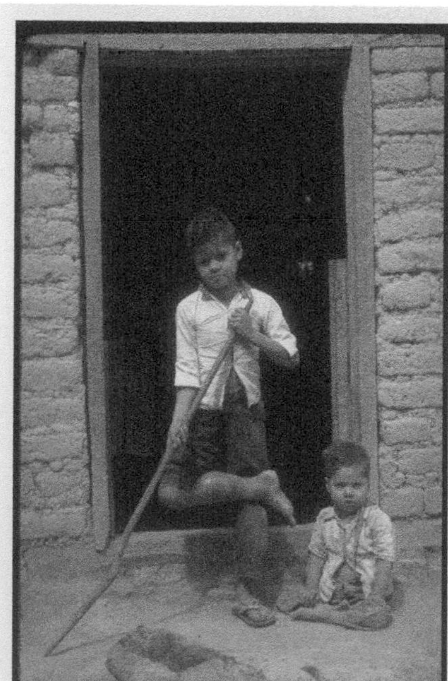

Figure 10.13 Marcelo, at 4, sits next to his brother. Marcelo had polio. Marcelo's brother (standing in photo) was injured when a tree fell on his leg.

Figure 10.14 As a health worker at 14, Marcelo taught his fellow villagers about dehydration and rehydration using a "gourd baby."

The village health workers did what they could for Marcelo. They helped him learn to walk with braces and crutches. They arranged for him to go to primary school. When he was 14, they trained him as a village health worker. Marcelo returned to his village and served his community (Figure 10.14).

In 1981, when PROJIMO was just beginning, the team invited him to join the program and arranged for him to apprentice in an orthotics (brace making) shop in Mexico City. He further extended his learning from the visiting professionals and in time became a first rate orthotist and prosthetist (Figure 10.15).

Figure 10.15 Marcelo and Conchita make limbs for a girl who had both her legs amputated.

Marcelo's secret to finding working solutions is his caring, humble approach. He relates to anyone, including children, in a friendly and cooperative way, always ready to listen and explore suggestions. Marcelo considers the people he assists to be his partners in the problem-solving process.

#3 BENO'S STORY: DEVELOPMENT OF A "LOW-TECH" ASSISTIVE DEVICE

Beno's story demonstrates why solutions may fail when physical side effects are not taken into account. It is important to observe children with disabilities in their own environments and to reobserve over time. Solutions that work in a rehabilitation setting may not work in home environments or may not continue to work successfully over time. In such situations, "low" technology solutions may be more functional for the children and their circumstances than "high" technology solutions.

Beno was born with spina bifida, a spinal defect that partially paralyzes the lower body. When he was 4 years old, he visited a rehabilitation center in a large city. There, a specialist prescribed a metal walker for him (Figure 10.16). Beno would halfway sit in the walker, while he pushed himself around. This walker cost Beno's family one-third of their yearly wages. Unfortunately, the walker was counterproductive for his physical condition.

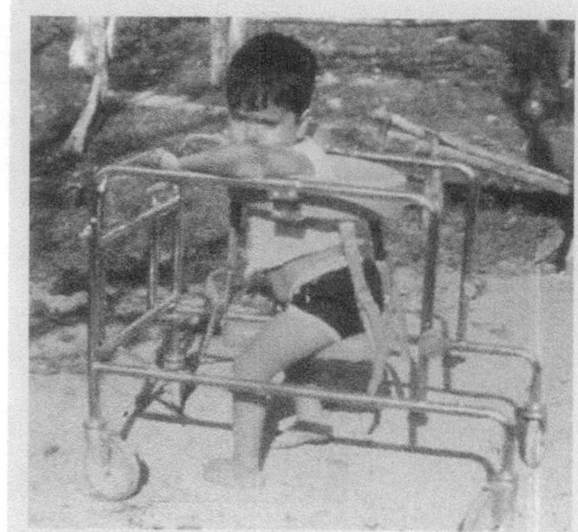

Figure 10.16 Beno with his metal walker.

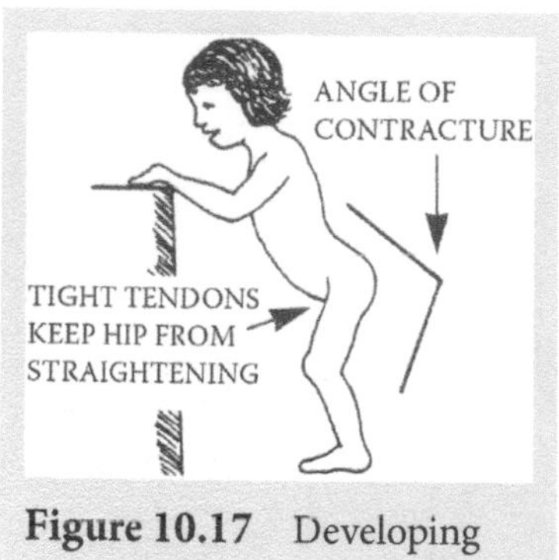

Figure 10.17 Developing contractures.

As happens with many children who have spina bifida, Beno was developing hip flexion contractures that kept him from walking upright (Figure 10.17). One of his PROJIMO workers, Mari, noted that the walker, instead of straightening his hip contractures, was making them worse. His thighs were nearly at a right angle to his body! Spending prolonged time in this position would strengthen the contractures, rather than stretching them out to permit him to straighten his legs and stand.

To help Beno begin to stand and walk, another of Beno's PROJIMO workers, Marcelo, developed a set of parallel bars for him. The purpose of parallel bars is to offer an area for a disabled person to walk in, developing lower body strength while they support themselves with their arms (Figure 10.18). Marcelo adjusted the parallel bars to straighten Beno's hips as he walked.

Figure 10.18 Beno in the parallel bars.

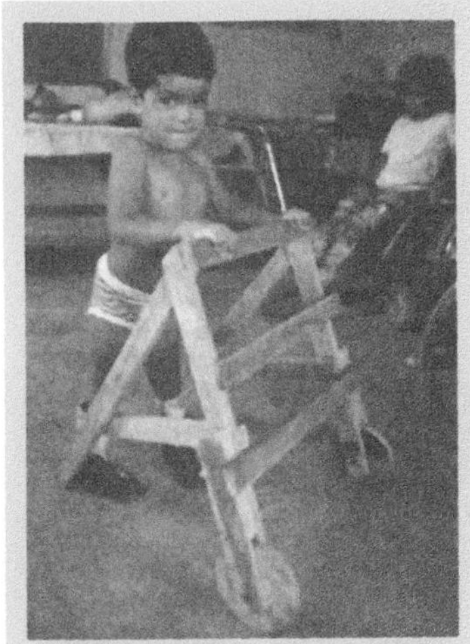

Figure 10.19 Beno in his wooden walker.

Once Beno could stand more upright, Marcelo helped Beno's father make him a wooden walker (Figure 10.19). Over a period of months, he began to stand straighter and his walk improved.

Although the metal walker, provided as an assistive device, was more sophisticated than the crude parallel bars and handmade wooden walker, the rustic aids better met Beno's needs.

#4 LINO'S STORY: CHILDREN WITH DISABILITIES DEVELOP ASSISTIVE DEVICES

This example demonstrates how having individuals with disabilities as workers can improve the development of assistive devices, along with treatment outcomes.

Lino was born with spina bifida. He was fortunate because his family stretched his hips regularly since he was a young child, thus helping him avoid contractures (Figure 10.20). He had enough strength above his knees to bear most of the weight on his legs. This eventually enabled him to learn to walk.

Several local PROJIMO leaders have spinal cord injuries, which helped them understand the needs and possibilities of children with spina bifida. When Lino was 3, PROJIMO workers taught him to use a walker, hoping this would later enable him to graduate to crutches.

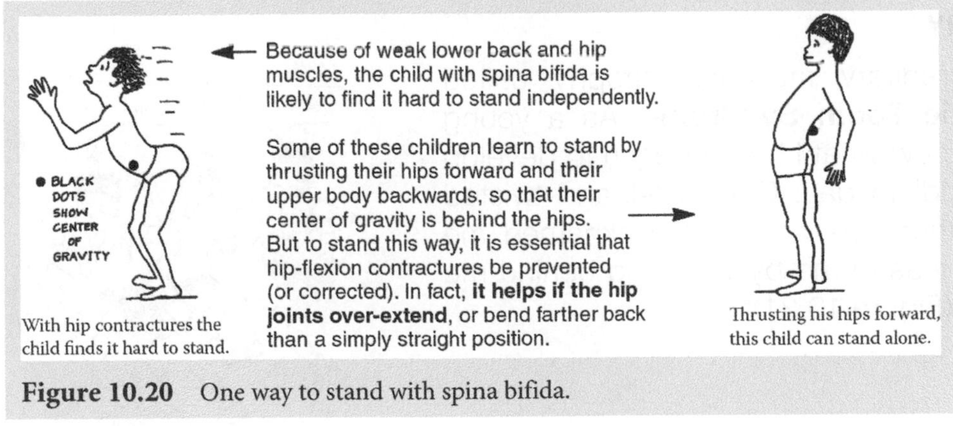

Figure 10.20 One way to stand with spina bifida.

Marcelo was one of the PROJIMO workers who helped Lino (see first example for an introduction to Marcelo). Having a disability himself, Marcelo understood how terrifying it can be for a child with weak legs to give up a walker and start using crutches. A walker provides stability, while crutches feel unstable. Crutches wobble and children may take some nasty falls as they learn to use them.

To spare Lino from such trauma, Marcelo designed a walker that gradually converts into crutches. This consisted of two forearm crutches, joined with cross struts held in place with butterfly nuts. Its bolts were gradually loosened, making the walker more wobbly. Over time, the cross struts and side supports are removed one by one, until only the crutches are left.

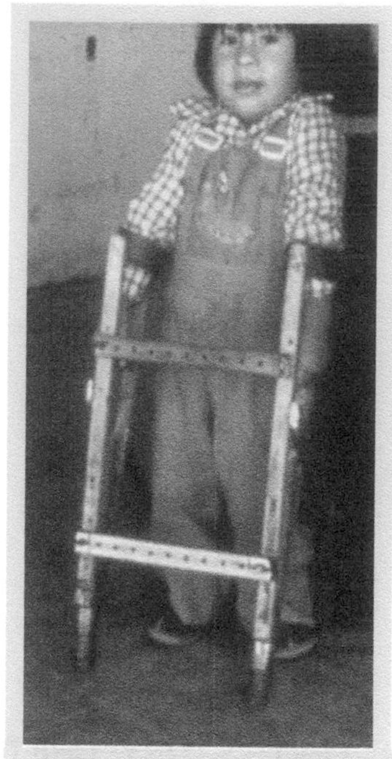

Lino's walker can gradually be converted into crutches.

#5 DAVID AND RICK WERNER'S STORIES: PARTNERS IN PROBLEM SOLVING

These two examples demonstrate how results tend to be better when service providers work together with disabled clients as equals to solve problems. While these examples are of adults, the process improves outcomes with children also.

Third world programs run by disabled villagers are sometimes more effective than those run by highly trained specialists. Several reasons for this are: they listen, work with the individual who is disabled as a partner, and keep adapting the product until it works.

DAVID'S STORY

I have a hereditary muscular atrophy called **Charcott Marie Tooth Syndrome.** As a young boy, nobody knew what I had. I began to develop a strange waddling gait. Often I fell and twisted my ankles. Other kids at school nicknamed me "Rickets" and teased me by waddling behind me in a duck gait (Figure 10.21).

Figure 10.21 Duck gait of Charcott Marie Tooth syndrome.

When I was 10 years old, my parents took me to a specialist. Finding my feet weak and floppy, he prescribed arch supports (Figure 10.22). I hated the things! It was harder to walk with them than without. By pushing up the arch, they tilted my weak ankles outward. I fell and sprained my ankles more than ever.

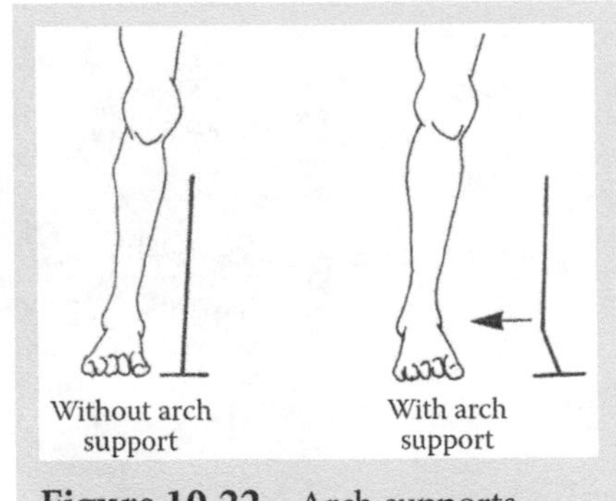

Figure 10.22 Arch supports.

Figure 10.23 The author as a child with metal braces.

My walking steadily worsened, so the doctor prescribed metal braces (calipers) (Figure 10.23). These held my ankles firmly but made me more awkward than ever. They were heavy and uncomfortable, once causing deep, painful sores that took months to heal. I stopped wearing them.

As the years went by, my feet became deformed (Figure 10.24). Decades later, while working at PROJIMO, Marcelo* suggested that I might walk better with plastic leg braces.

"Leg braces!" I exclaimed, "I know they help some kids walk better, but they made me walk worse!"

"Maybe the braces you had as a child didn't work because you weren't included in developing them," said Marcelo. "What do you say if you and I design your braces together? We can keep experimenting until we create something that works. And if they don't work, at least we'll have the adventure of trying."

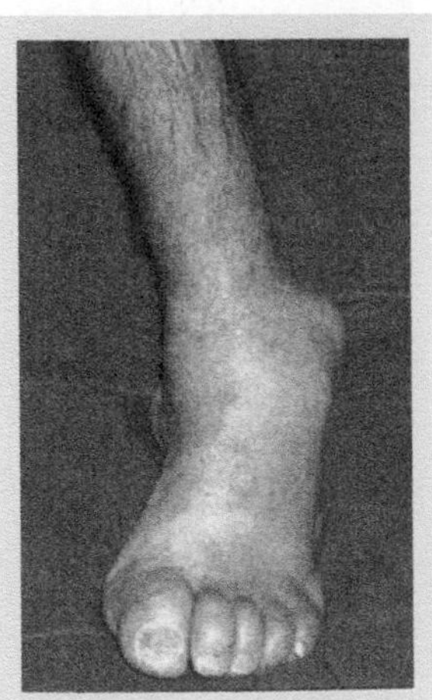

Figure 10.24 My left foot.

* Marcelo is PROJIMO's brace and limb maker. See Case Study #1.

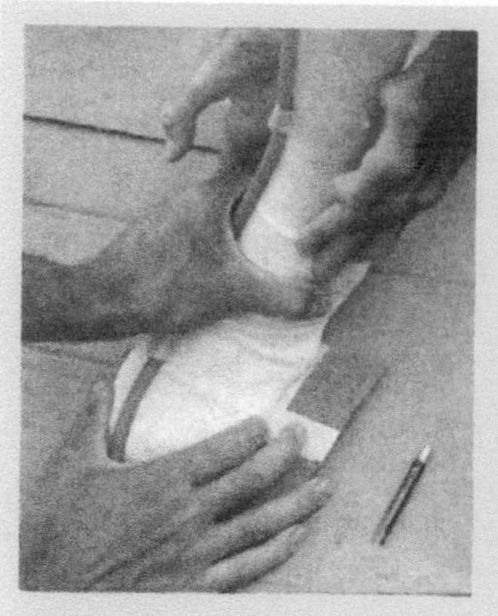

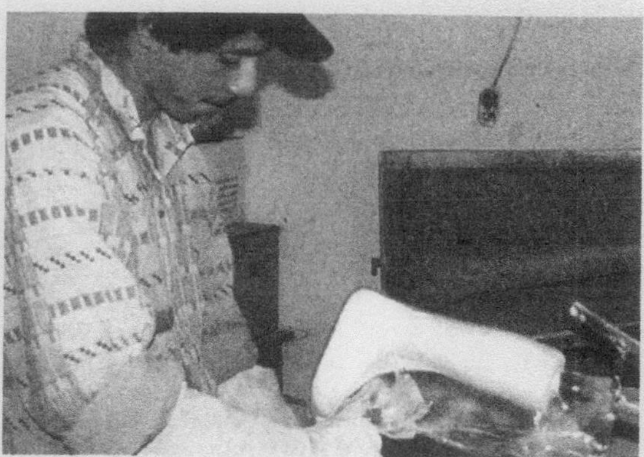

Figure 10.25 and Figure 10.26 Developing ankle-foot orthoses as a team. Four hands are better than two when making the cast (Figure 10.25).

So we worked together making plastic ankle-foot orthoses (AFOs), by trial and error, learning from our mistakes (Figure 10.25 and Figure 10.26). Because of my lateral deformity and contracted foot drop, it was challenging to develop a functional support that was also comfortable.

Through trial and error, we succeeded. We found that rigid braces with no ankle flexibility provided the least painful and firmest support. However, the rigidity caused an awkward, choppy gait, and tended to push my ankles backward at the end of each step (Figure 10.27).

We experimented with footwear until we came up with rocker bottom shoes with inward-sloping soles. I now walk better than I did 30 years ago; I can even climb mountains (Figures 10.28 and Figure 10.29b).

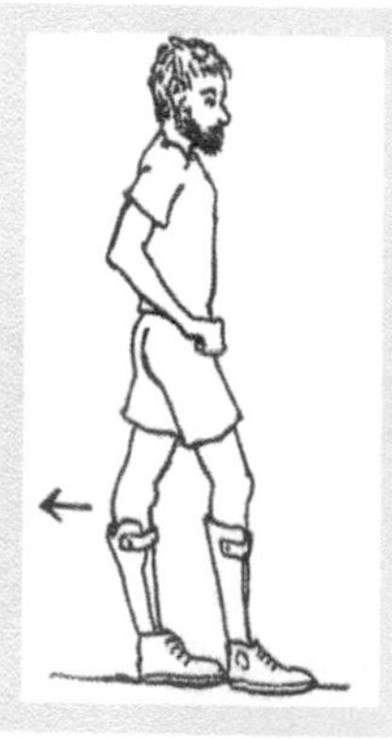

Figure 10.27 The final product, a functional ankle-foot orthosis! In an early trial, the rigid braces caused an awkward, choppy gait.

Figure 10.28 A provisional wooden wedge was used to experiment with a rocker-bottom sole for a smoother gait.

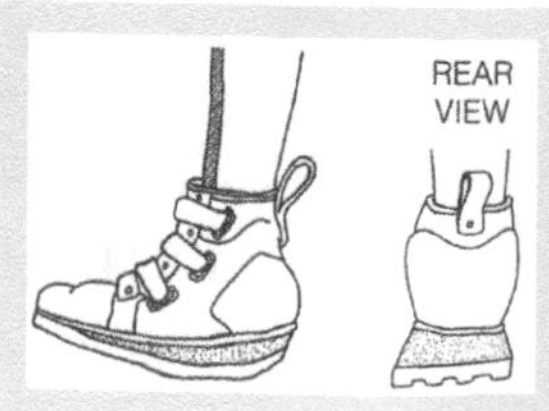

Figure 10.29a Finally, the most comfortable rocker-bottom design was fixed into the sole.

Figure 10.29b David Werner climbs a mountain with Efraín Zamora.

RICK'S STORY

My brother Rick had this same hereditary muscular atrophy. One night, while crossing the street on his crutches, a car ran over him. One leg had many fractures. The other was amputated. A prosthetist had an expensive artificial leg made for him. A year later, his prosthesis still did not function.*

PROJIMO, the community rehabilitation program in rural Mexico, agreed to help. On his arrival, they fitted him with a new prosthesis. This prosthesis needed a series of adjustments (as did his prior device). But this time Marcelo was readily available as needed.

Rick was soon walking on his new prosthesis: first on parallel bars, then on crutches (Figure 10.30a, Figure 10.30b and Figure 10.31). The team made a set of steps for him to practice climbing on.

Figure 10.30a

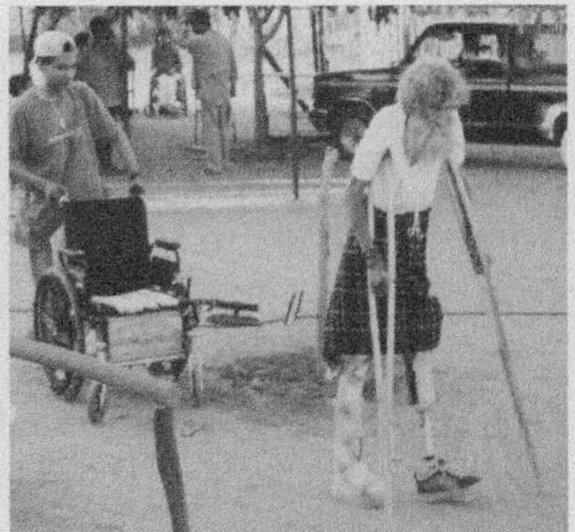

Figure 10.30b

Figure 10.31

Rick begins walking, first in between parallel bars, then with crutches, and then on stairs.

In keeping with the PROJIMO philosophy of improving self-determination of the disabled individual, they involved Rick in the community, by encouraging him to teach English to village schoolchildren. Thus, he partnered in his own rehabilitation and the well-being of the entire community (Figure 10.32).

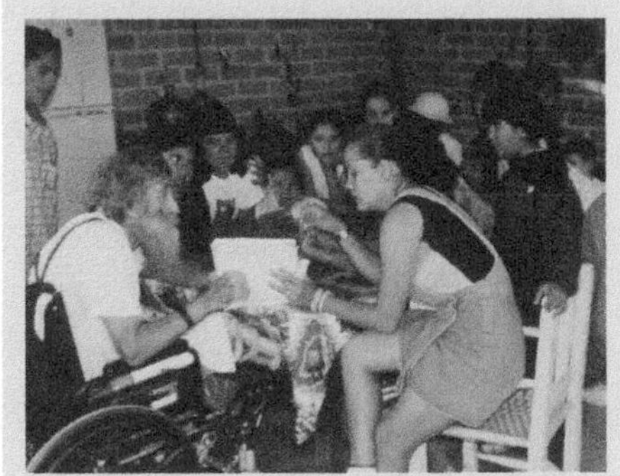

Figure 10.32 Rick teaching English to village children.

* Medicare paid the costs for this prosthetist, who made him an expensive artificial leg. A year later, his prosthesis still did not function. Because of all the red tape, adjustments on the leg that should have been completed in a couple of days were delayed for weeks and sometimes months. As a result, more than a year passed and Rick still did not have a functional prosthesis.

#6 CARLOS' STORY: FROM HELPLESS TO HELPER

This story demonstrates how community-based programs promote flexible, innovative, and holistic approaches that are often difficult to achieve in large rehabilitation institutions.

In small, community-based programs where disabled persons and mostly disabled staff live together and relate as friends, problem solving often becomes a creative adventure. Breakthroughs come in unexpected ways, because doors are left open and the world is exciting. Even the child with multiple disabilities can become a source of encouragement and assistance for another child.

Carlos is the son of poor migrant farm workers in Oaxaca, Mexico. When he was 10 years old, he was hit by a truck. Brain damage left him physically disabled, mentally impaired, and almost blind. His family abandoned him. Social workers took him to PROJIMO, which became his new home.

A wheelchair was built and adapted to Carlos' needs because his muscles were spastic and he could not walk. At first, Carlos had trouble pushing it himself. The shop workers experimented with a design where the big wheels are in front (Figure 10.33), so he could push it more easily, given his spasticity.

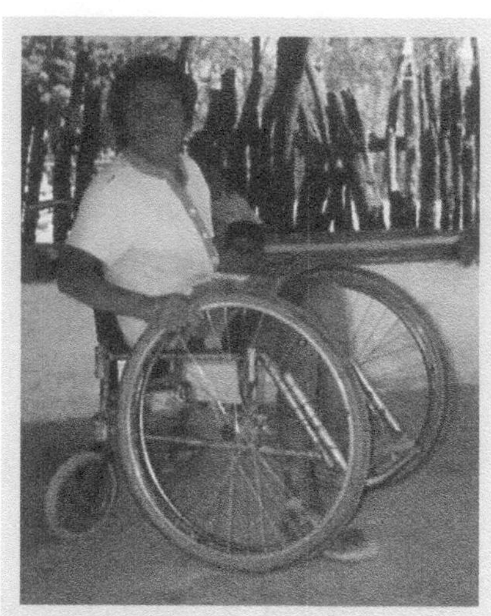

Figure 10.33 Wheelchair with large wheels in front.

Carlos discovered how to navigate his wheelchair by using one foot like a blind man's cane, to feel his way along the pathways and ramps (Figure 10.34).

At first, when the team encouraged Carlos to try to stand, he would not cooperate. However, Carlos had made friends with Teresa, a girl with multiple sclerosis.

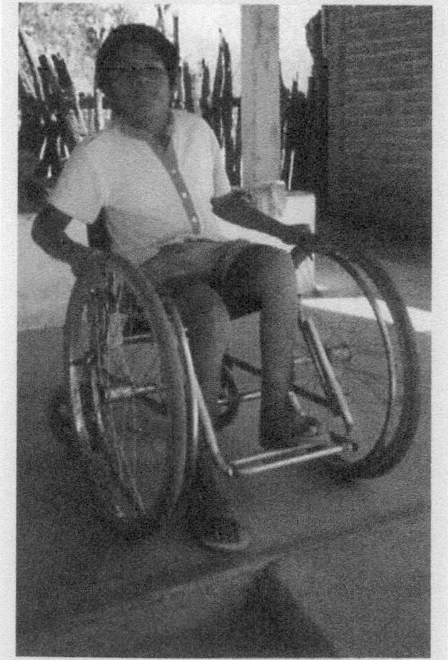

Figure 10.34 Carlos learns to propel himself.

One day when a PROJIMO worker and a schoolgirl were helping Teresa stand with the parallel bars, Carlos suddenly said, "Me want to stand, too" (Figure 10.35 and Figure 10.36). Soon Carlos was standing and learning to walk between the bars (Figure 10.37 and Figure 10.38).

Figure 10.35 Carlos wants to stand.

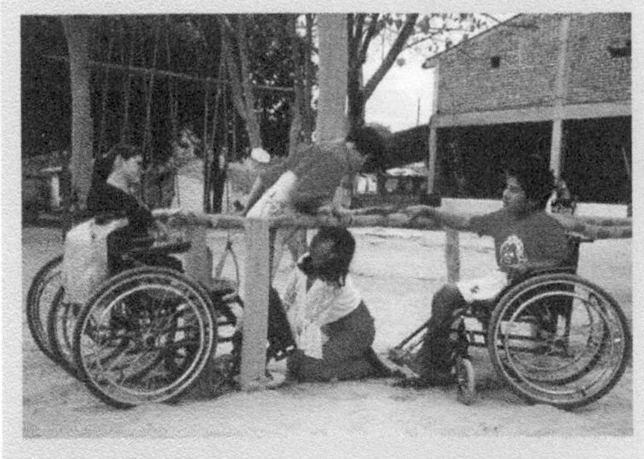
Figure 10.36 Carlos wants to stand.

Figure 10.37 Carlos stands!

Figure 10.38
Walking in parallel bars.

Next, the shop workers fitted Carlos with a metal walker.

First, they experimented to find the height and position to help him stand straight with the greatest stability (Figure 10.39).

Carlos loved his walker and was constantly using it. But he would soon tire and because he was virtually blind, he could not find his wheelchair when he wanted to sit.

Figure 10.39
Carlos and his metal walker.

Then, the team made a simple wooden walker with a seat, so Carlos could sit when he got tired (Figure 10.40 and Figure 10.41). Carlos learned to transfer from his walker to his wheelchair and back again (Figure 10.42).

Figure 10.40 and Figure 10.41 Carlos and his wooden walker with a seat.

Figure 10.42 Carlos transfers from his walker to his wheelchair.

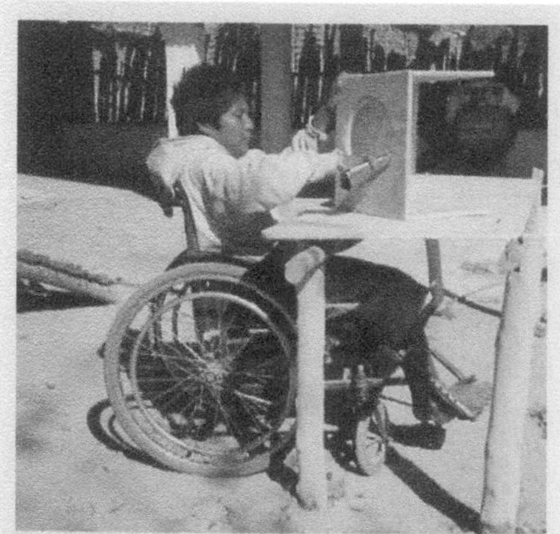

Figure 10.43 Carlos took great pride in sanding his toilet seat.

Carlos was not toilet trained and no effort to train him succeeded. So, the team tried a new approach. Carlos loved to "work" in the carpentry and toy making shops. He would patiently sand a wooden puzzle or block until it was smooth. So they invited Carlos to help make his own portable toilet (Figure 10.43).

Then he wanted to use it. Soon Carlos was looking for every opportunity to use his toilet and would even get up at night to use it (Figure 10.44 and Figure 10.45).

Carlos constantly wanted to use the toilet he had helped to make, but again, he often had difficulty finding it in time. So Martín, a schoolboy with asthma, made him a new walker with a built-in toilet (Figure 10.46 through Figure 10.48).

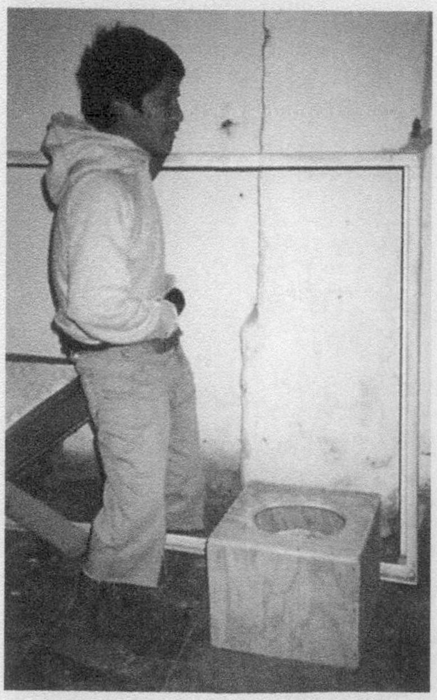

Figure 10.44 Standing momentarily to lower his pants helped him to improve his balance.

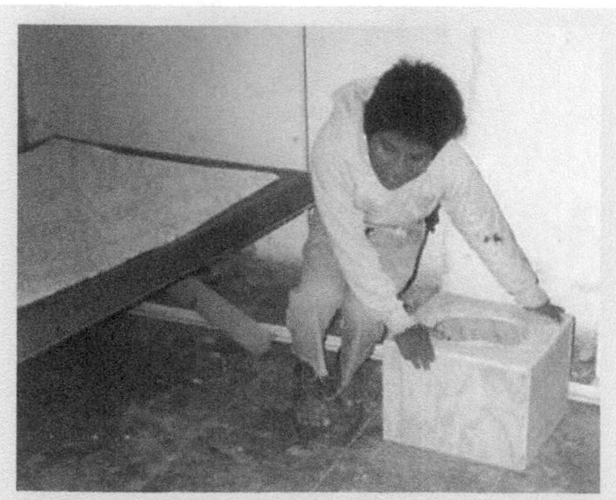

Figure 10.45 Carlos even got up at night to use his toilet.

Figure 10.46 Martín builds a walker with a built-in toilet seat.

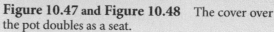

Figure 10.47 and Figure 10.48 The cover over the pot doubles as a seat.

To keep the seat dry, the whole toilet swings up out of the way when necessary (Figure 10.49). In this way, Carlos can urinate on the ground (not on the seat) or turn and sit on the pot (Figure 10.50). Carlos thrived on the praise he got for using his toilet correctly. Figure 10.51 shows Carlos using his walker.

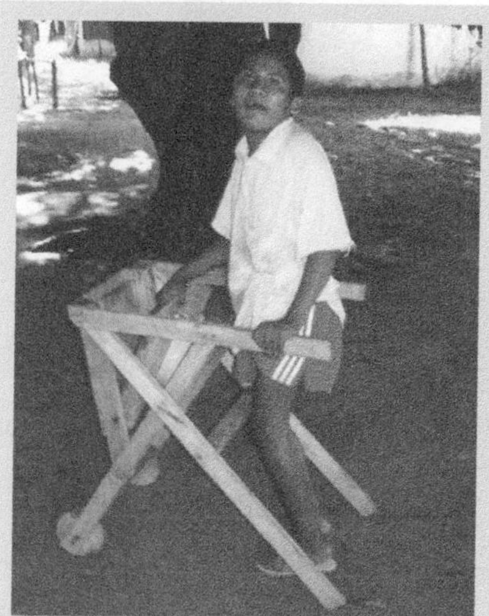

Figure 10.49 The seat swings up.

Figure 10.50 Carlos demonstrates use of the pot.

One of the major breakthroughs in the rehabilitation of a disabled person or in the psychosocial growth of any child (or adult) is when the disabled child begins to look beyond satisfying his or her personal needs and finds deeper satisfaction in helping others.

This breakthrough is especially rewarding when it happens with a child who is mentally disabled or has multiple disabilities. Such breakthroughs cannot be planned. Rather, rehabilitation workers need to be perceptive when a seminal situation presents itself, and build on its liberating features.

One day, while playing, Carlos began pushing Teresa's wheelchair. Seeing potential in this act of mischievous independence, workers Rosa and Inéz encouraged Carlos to continue (Figure 10.52).

Figure 10.51 Carlos using the walker.

When her chair nearly ran into a tree, Teresa (who usually is stubbornly silent) cried out to Carlos to turn to one side. Laughing gleefully, Carlos followed her instructions.

In this way, Teresa and Carlos discovered that combining their different skills enabled them to get around better than either could alone. Teresa could not push her own chair, but she could see where she was going. Carlos could not see, but he could push Teresa around, and her wheelchair provided the stability of a walker (Figure 10.53).

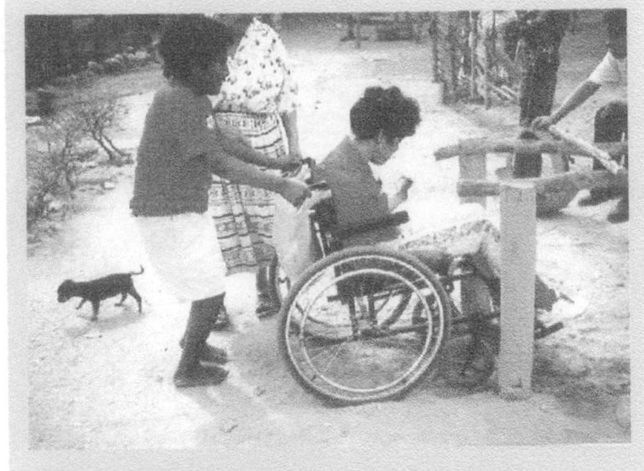

Figure 10.52 Carlos pushes Teresa.

So together, they had both mobility and vision, and could go where they chose. They discovered this possibility through friendship and play. By helping Teresa push, Carlos' balance and self-confidence improved until at last he could stand and take small steps by himself (Figure 10.54).

Carlos discovered the benefits of mutual self-help, where people's different abilities complement one another in ways that compensate for each other's contrasting impairments. After learning that he and Teresa could get around better by combining their respective abilities, Carlos began to assist others similarly.

Here he pushes the wheelchair of Rosita, a multiply disabled girl (Figure 10.55).

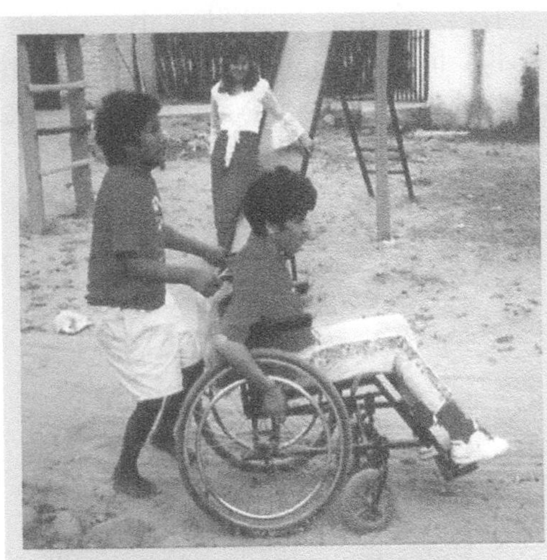

Figure 10.53 Together they help each other.

Carlos and Alonzo provide a third example of children helping others.

Alonzo was born with hypothyroidism (cretinism) because of lack of sufficient iodine in his mother's diet. He is developmentally delayed and was unable to walk when his mother took him to PROJIMO. He spoke little and frowned a lot.

When his mother tried to make him stand up, he would sink to his knees. He stubbornly refused to try the parallel bars.

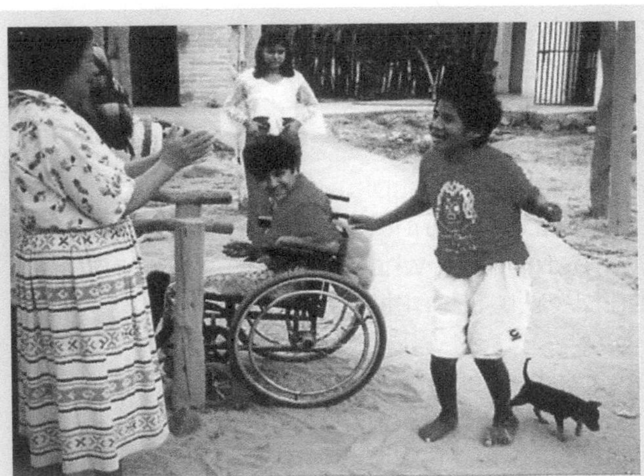

Figure 10.54 Carlos takes a few steps.

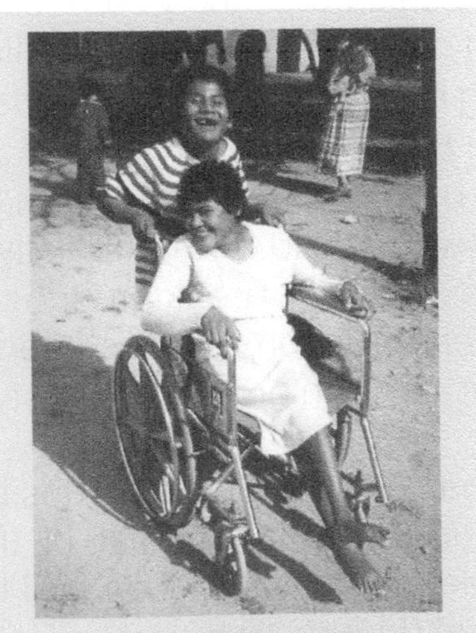

Figure 10.55 Carlos pushes Rosita.

Inéz, who helps with physical therapy, asked Carlos to show Alonzo how he walked in the bars. Carlos cheerfully walked back and forth, calling out to Alonzo, "Look at me!" (Figure 10.56).

Perhaps Carlos' awkwardness and bent-kneed, spastic gait gave Alonzo courage. With his mother's help, he stood at the bars (Figure 10.57).

Carlos reached out and guided Alonzo's hands.

With his mother stabilizing him from behind, Alonzo began to take steps. Walking backward between the bars (a good unplanned exercise for Carlos as well) Carlos gradually led Alonzo forward (Figure 10.58).

Figure 10.56 Carlos shouts, "Look at me!"

To improve Alonzo's balance, Inéz suggested that his mother encourage him to walk holding a rope stretched between two trees. As with the parallel bars, again Alonzo was afraid to try and refused.

Figure 10.57 Alonzo's mother helps him stand.

Figure 10.58 Carlos and Alonzo's mother help him walk.

Once again, Carlos came to the rescue. Carlos walked back and forth holding the rope while Alonzo sat watching (Figure 10.59).

Now, Alonzo was willing to try. At first he hung fearfully to the rope without moving. But Carlos coaxed him into a "follow the leader" game, back and forth along the rope. In the end they both had a good time (Figure 10.60).

Figure 10.59 and Figure 10.60 Carlos and Alonzo walking along a rope.

Carlos became a good role model for Alonzo, who worshiped his new, equally awkward therapy assistant. Watching Carlos walk with his walker, Alonzo wanted to try a walker, too. Inéz chose a suitable wooden walker that Alonzo eagerly tried.

Figure 10.61 Alonzo tilted forward.

But problems arose. Alonzo's feet didn't keep up with his body while he pushed his walker. He tilted forward until he fell (Figure 10.61). Alonzo grew discouraged and frightened. Meanwhile, Carlos paraded back and forth in his own walker-with-a-seat.

Figure 10.62 Alonzo's feet have to keep up!

Alonzo, scowling, pointed to Carlos' walker-with-a-seat. Clearly, he wanted to try it. To everyone's delight, Alonzo was able to use it to walk without falling forward.

As with the first walker, he would begin to lean forward precariously. But before he fell, his legs bumped into the seat behind him, acting like a brake (Figure 10.62).

With new confidence Alonzo barreled around the playground. He was so excited he paid no attention to his new bruises from repeated bumps against the seat edge.

Based on Alonzo's discovery, Polo (another worker) built a walker for Alonzo with a padded "backstop" that functioned like the seat on Carlos' walker.

The backstop worked fairly well on the first trial, but Alonzo still tilted forward (Figure 10.63). So Polo adjusted the backstop farther forward (Figure 10.64). This time Alonzo walked more upright (Figure 10.65).

Figure 10.63 Alonzo walks, but leans forward.

Figure 10.64 Polo adjusts the walker.

Figure 10.65 Alonzo walks more upright.

COMMUNITIES AS PARTNERS

Not only do the disabled village workers view their clients as their partners, they also consider all of the individuals within the community as their partners. For example, when they considered starting a community-based program for disabled children, they invited the whole village to a meeting.

Nearly everyone was enthusiastic. Men agreed to help turn an overgrown lot at the end of town into a community center. Mothers offered to open their homes to families with disabled children from neighboring settlements.

Schoolchildren offered to help build a rehabilitation playground—provided they could play there, too. This was a particularly good idea because it meant shared play between disabled and nondisabled children (Figure 10.66 and Figure 10.67).

Figure 10.66 To build the playground the children bring poles from the forests and old tires from the trash.

Although no one in the village was aware of the concept of "universal design," i.e., designs that can be used by both able-bodied and disabled individuals, in effect, that is what they created in their **Playground for All Children**.

In many rehabilitation programs and on most playgrounds, a boy like the one pictured below (on the ramp) would be separated from his playmates so that a therapist could work with him to stretch his tight heel cords—a sort of therapeutic torture! Here the team encourages the boy to play with other children in a way that provides equivalent stretching (Figure 10.68). This makes therapy functional and fun.

Figure 10.67 Some of the schoolchildren build a ramp while others clean the grounds.

Figure 10.68 A boy with spastic cerebral palsy stretches his tight heel-cords playing on the ramp made by the children.

Figure 10.69 PROJIMO's "Playground for All Children" was built using low-cost materials such as trees and old tires.

Old tires are especially helpful in creating playgrounds and playground equipment as seen in both the photograph of the finished playground and in the drawings (Figure 10.69 and Figure 10.70).

One advantage of community-based programs (and treating the community as a potential partner) is that it provides a flexible, innovative, and inclusively holistic approach. This is often difficult to achieve in large, formal rehabilitation institutions.

OLD TIRES *have a hundred uses in*
A PLAYGROUND FOR ALL CHILDREN

Figure 10.70 Old tires serve as swings, seats, tunnels, levers, and things to climb!

CHILD-TO-CHILD: A NEW TYPE OF PROGRAM

"Child-to-Child" is an educational approach where school-age children collaborate to help disabled children, particularly children who are younger or have special needs. It was launched during the International Year of the Child and is now used in more than 60 countries.

Many community-based rehabilitation programs use child-to-child activities to help nondisabled children relate to disabled children in a friendly, helpful way. This program raises awareness and teaches compassion for children who are "different." One of Child-to-Child's goals is to help nondisabled children understand disabled children, be their friends, include them in games, and help them become more self-reliant.

RAISING AWARENESS

One technique to raise awareness is to "give" school-age children temporary handicaps to help them experience what it is like to have disabilities. To simulate a lame leg, a stick is tied to the leg of the fastest runner, stiffening his knee (Figure 10.71 and Figure 10.72).

After the children run a race and come in last, the facilitator asks the temporarily handicapped child what it feels like to be left behind. The children try to think of games they can play with children with a lame leg: for example marbles or checkers.

Skits and socio-dramas can also help youngsters appreciate the abilities of children who are different. In one short but effective skit, a girl approaches a boy on crutches. "Pedro," she says, "I can't open this bottle. Will you open it for me? You have such strong hands because you use crutches." Pedro easily unscrews the lid and hands the bottle back to the girl, smiling (Figure 10.73). This short skit teaches children to look at these children's strengths, rather than weaknesses.

Figure 10.71 and Figure 10.72 Simulating a right leg disability.

Sometimes disability-related Child-to-Child activities are conducted in ways that exclude disabled children from central roles. These activities tend to be about disabled children, not with them. It is far better to involve disabled children in central roles.

Figure 10.73 "Pedrito, your hands are so strong! Can you open this bottle?"

CASE STUDY: RAMONA FROM NICARAGUA

Teachers and health workers from all over Nicaragua each brought one or two children from their area to a Child-to-Child training program. A few of these children were disabled.

In a game that helps children learn about disability, all children race on only one leg, with the other tied behind them. They could use crutches, poles, or whatever walking aids they could find.

Ramona, who had one leg paralyzed by polio, reluctantly agreed to enter the race. Used to walking with crutches on one leg, she flew ahead of the others, winning easily (Figure 10.74). The other children admired Ramona's agility.

Figure 10.74 Ramona flew ahead.

While the other children learned a lesson about when a "disability" actually makes a person "disabled," the workshop was a turning point in Ramona's life. She participated in activities with a new sense of confidence. She went on to become a dynamic community organizer and gifted facilitator of Child-to-Child workshops.

Ramona used Child-to-Child techniques to facilitate another child's schooling as follows:

Jesús was born with spina bifida and lost one leg due to pressure sores on his foot. He lost most of his vision due to meningitis.

At age 11, he had never attended school. Jesús came to PROJIMO to heal the pressure sores on his buttocks and to attend the village school.

Until his pressure sores healed, he went to school on a gurney (narrow wheeled cot) and later in a wheelchair (Figure 10.75). He had been eager to learn, but his visual impairments made it difficult to manage in school.

Figure 10.75 Jesús attending school on a gurney.

Ramona suggested using the Child-to-Child approach to help Jesús' classmates and teacher understand his problem and find ways to assist his studies. She described the Child-to-Child process to the children and said she wanted to explore with them what it was like to be blind or partially blind.

Two student volunteers played the role of students with visual impairments and had their eyes blindfolded. Another student played the role of teacher. The "blind" students tried to find their way around the classroom and to follow the "teacher's" instructions.

To simulate partial blindness, Ramona covered a girl's head with a cotton shirt (Figure 10.76). (She experimented with cloths until she found one that limited vision similar to Jesús' visual loss.)

Figure 10.76 Simulating partial blindness.

The "teacher" asked the girl to read from her book. Only by holding it very close to her face could she read the largest letters.

Then the "teacher" wrote on the blackboard and said, "Read it!" To read, the blindfolded pupil had to go very close to the blackboard.

Then the pretend teacher asked Jesús to read a large letter on the blackboard. Jesús rolled forward. To read the letter he had to lift himself high in his wheelchair until his nose almost touched the letter. He read it proudly (Figure 10.77 and Figure 10.78).

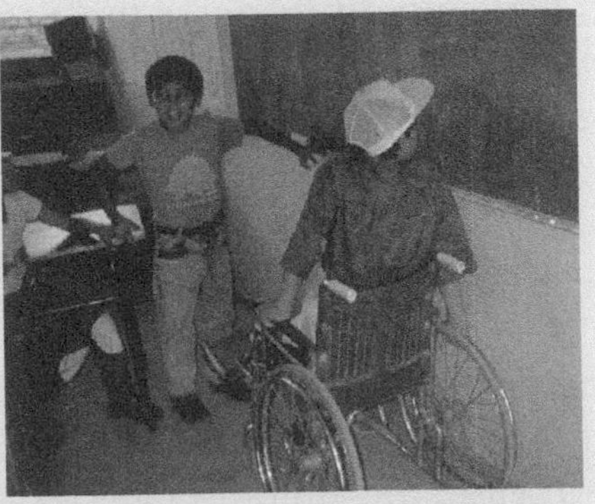

Figure 10.77 and Figure 10.78 Jesús rolls forward and reads from the blackboard.

The children saw the difficulty Jesús had reading, both from the blackboard and his books. Ramona asked, "Can you think of ways that you, Jesús' classmates, can help him understand his lessons and get the most out of school, in spite of his disability?"

The students provided a wide range of suggestions to help Jesús.

Figure 10.79 Jesús sits and hears the words from a friend.

- Allow him sit up front, near the blackboard.
- Write large and clear on the blackboard.
- Allow him to sit next to a friend who whispers to him what is on the blackboard (Figure 10.79).
- The teacher could read aloud while she writes on the blackboard.
- A nearby child could copy the information the teacher writes on the blackboard into Jesús' notebook—in big, dark, clear letters.

Figure 10.80 A magnifying glass might help him read.

- Supply Jesús with a magnifying glass (Figure 10.80).
- Children could take turns after school, helping Jesús with his homework and reading from his books.
- Children could take turns helping bring him to school.

Figure 10.81 Jesús could use a tape recorder.

With prompting, the children came up with more ideas:

- Jesús' lessons could be recorded on a tape recorder so he can study whenever he wants (Figure 10.81).
- Allow him to take exams orally.

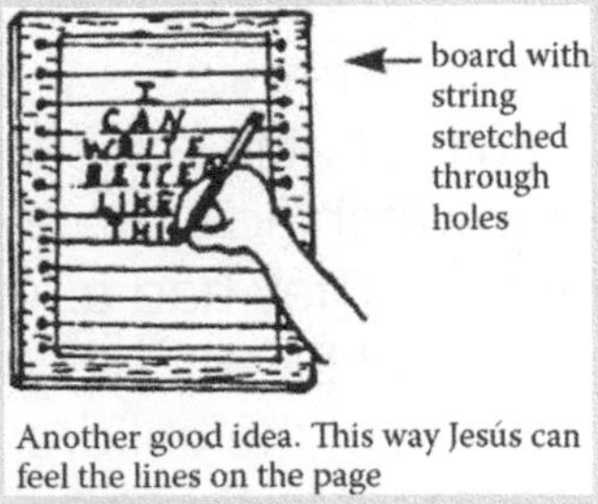

After this discussion, Ramona asked a visually impaired visitor for ideas. She suggested writing paper with extra dark lines—or darkening the pale lines on ordinary paper might help Jesús to write. She showed them how to stretch string through holes on a board to make lines Jesús could feel (Figure 10.82).

board with string stretched through holes

Another good idea. This way Jesús can feel the lines on the page

Figure 10.82 Lines Jesús can feel.

The blind visitor told the class Jesús could learn to read with his fingers. She showed them a book with Braille script and read to them with her fingertips. She let the pupils feel its tiny bumps and had Jesús guide his fingers over the page.

She gave Jesús a sheet with the Braille alphabet. Next to each Braille letter, she printed a large letter, so Jesús could learn them. The children were fascinated. Jesús was so excited he trembled. The visitor explained the Braille system was named after a blind boy, Louis Braille, who invented it.

The Child-to-Child activity was a big success. The teacher asked him to sit up front next to a boy who provided many suggestions to help Jesús learn.

A few of his classmates began to accompany Jesús to and from school. This became more fun after the PROJIMO shop made him a hand-powered tricycle, on which his friends could hitch a ride (Figure 10.83).

Figure 10.83 Jesús gives a friend a ride on his hand-powered tricycle.

Child-to-Child activities do not solve every problem right away. For example, rather than helping Jesús do his own homework, his helpers at first tended to do it for him! However, both Jesús and his classmates learned about the joy that comes from bridging barriers, creative problem solving, and helping one another.

In accordance with PROJIMO and Child-to-Child philosophies, Jesús was encouraged to help others. His first opportunity was to help facilitate a Child-to-Child interaction in a nearby village.

Two brothers, Chiro and Ricardo, aged 9 and 11, had muscular dystrophy. They walked awkwardly and were shy around other children. Their parents tried to send them to school, but the boys were terrified. Jesús understood how they felt.

For the first Child-to-Child activity, children from the neighborhood were invited to help build a playground and some therapy equipment in the family's backyard. This helped the brothers gradually lose their fear of being around other kids (Figure 10.84).

Figure 10.84 Neighborhood children helped build rehabilitation equipment in the backyard.

Next, Child-to-Child activities were held at the village school while Ricardo and Chiro watched from a distance. Jesús led the simulation games. The team brought several wheelchairs and Jesús organized and took part in a wheelchair race. An experienced wheelchair rider, Jesús easily won the race. The other kids were impressed and applauded.

Ecstatic, Jesús began to dance and do stunts in his wheelchair. Tipping it back in a wheelie, he spun in circles, tipping precariously forward and backward with incredible balance (Figure 10.85 and Figure 10.86). The more daring students tried to do the same—with calamitous results.

With all this excitement, Chiro and Ricardo gradually overcame their fear. They had been watching from a distance but slowly came forward until they stood with the other children. They had a hard time believing that a boy their age who could not walk at all and who could barely see could be such a star of the show.

Some of the children introduced themselves to the two brothers and invited them to join their class. In this way, Jesús found the joy of assisting other "children who are different" through innovatively sharing the Child-to-Child process. It had made a big difference to him. Now he could help it make a difference to others.

Figure 10.85 and Figure 10.86 Jesús racing and performing.

DISABILITIES, HUMAN RIGHTS, AND POLITICS

Helping poor and disabled persons in developing countries often intertwines with questions of human rights and social justice. This can take the form of helping to secure the rights of an individual with a disability or tackling government regulations.

CASE STUDY: ALEJANDRO

Twelve-year-old Alejandro was shot and paralyzed by a policeman for asking a casual question.

PROJIMO helped young Alejandro with medical care, rehabilitation, mobility aids, schooling, and skills training. They also tried to get a semblance of justice within existing power structures by getting the city government to assume—for a while—some of the boy's medical care and disability-related costs (Figure 10.87). PROJIMO also helped his impoverished family to find ways to generate income (Figure 10.88).

Alejandro eventually became part of the PROJIMO team, learning to weld and custom-design wheelchairs for children (Figure 10.89).

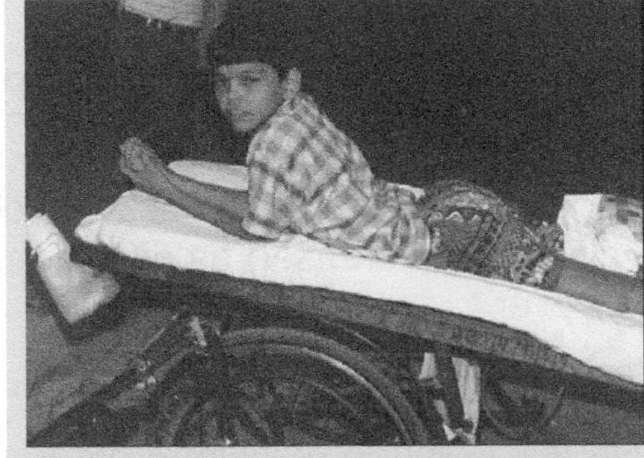

Figure 10.87 Alejandro uses a gurney to help pressure sores heal.

Figure 10.88 Alejandro's brothers collect and sell cardboard and scrap metal using a tricycle luggage cart.

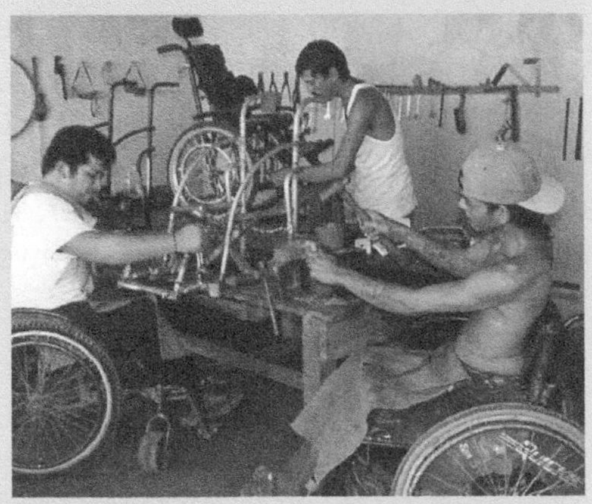

Figure 10.89 Alejandro (on right) 5 years later, building a child's wheelchair.

CASE STUDY: DISABLED WOMEN AND CHILDREN PROTECT THE VILLAGE DOCTOR

Taking a stand for others can give individuals and groups a feeling of self-determination. Persons with disabilities often feel helpless, especially in controlling their environment. Yet, individually and collectively, they can show great courage.

One time a group of soldiers burst into a village health center and arrested the village doctor. He was accused of providing emergency care to a man they were pursuing. As the soldiers marched the doctor to the back of their pickup truck, villagers watched silently through closed doors and windows. They were worried for the doctor, but afraid.

Figure 10.90 Disabled women and children surround the police truck.

The disabled workers, who were residents of the village, acted (Figure 10.90). In wheelchairs and on crutches, women and children alike surrounded the soldiers' truck. Even when ordered away from the truck, they refused. Finally, reluctant to attack women and children in wheelchairs, on crutches, and on gurneys, they released the doctor.

These actions won the participants great respect and appreciation in the village. One of the village elders commented proudly, "What they lack in muscle, they make up for in guts!"

THE POLITICS OF POVERTY

Poverty can lead to disability, through malnutrition and violence. PROJIMO began as a peaceful program to help disabled children in rural Mexico. However, the need for services in cities has increased as disabilities from poverty, malnutrition, and violence are greater than ever.*

CASE STUDY: JULIO

With the violence resulting from poverty and unmet needs, many families keep guns for protection. Julio became a quadriplegic person when his sister accidentally shot him with their father's gun. By the time Julio arrived at PROJIMO, a pressure sore had destroyed the base of his spine; this deep sore healed by packing it daily with a paste made from bee's honey and granulated sugar.†

As a part of the rehabilitation, disabled persons learn new skills that enable them to earn a living. Spinal cord–injured persons who have full use of their hands can develop a range of work skills. However, a quadriplegic person with little formal education and minimal use of their hands has fewer options.

* To enter the North American Free Trade Agreement (NAFTA), Mexico changed its constitution and annulled the agrarian reform laws that had protected the holdings of small farmers. As land has concentrated into giant plantations, over 2 million peasants have moved to cities. The competition for jobs and cuts in services and subsidies has left many people in poverty. Since NAFTA began in 1994, PROJIMO has attended to the needs of over 400 spinal cord-injured persons, mostly from gunshot wounds.

† This ancient Egyptian cure, rediscovered by Western medicine, is used in some modern hospitals.

Because Julio had little physical power, he was asked to supervise others' work. He kept careful records of the hours each person worked and calculated their weekly wages. To write these records, the team made him a plastic penholder he could attach with Velcro to his hand. A small desk was attached to his gurney (Figure 10.91).

Over time, Julio developed skills that made him more self-reliant. He learned to dress and bathe himself, transfer to and from his wheelchair, and propel his wheelchair. He teaches these skills to others (Figure 10.92).

He learned how to grip things by bending his wrist back to close his partially contracted fingers (tenodesis). He also learned to trigger spastic stiffening of his legs in order to prevent pressure sores by periodically lifting his buttocks off his wheel-chair cushion (Figure 10.93).

Julio is now the head teacher of what has turned into a successful enterprise that generates income for both himself and PROJIMO* (Figure 10.94).

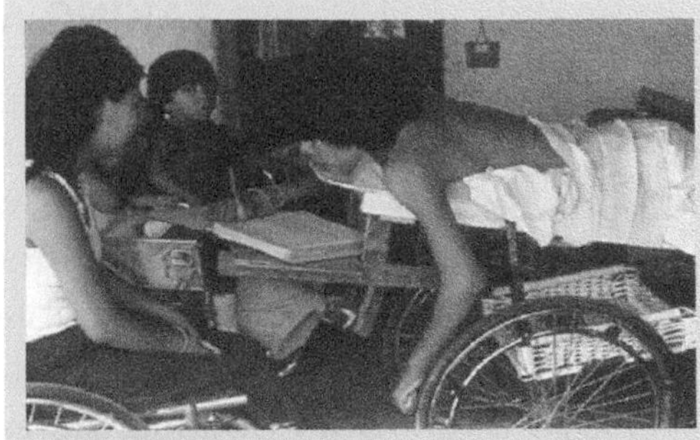

Figure 10.91 Julio records workers' hours.

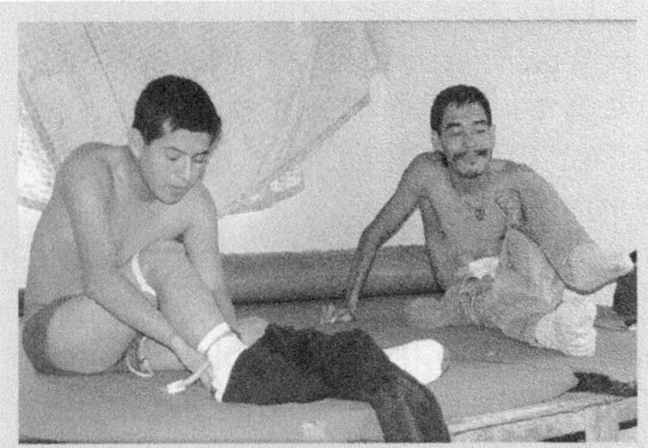

Figure 10.92 Julio (right) teaches Romeo to dress himself.

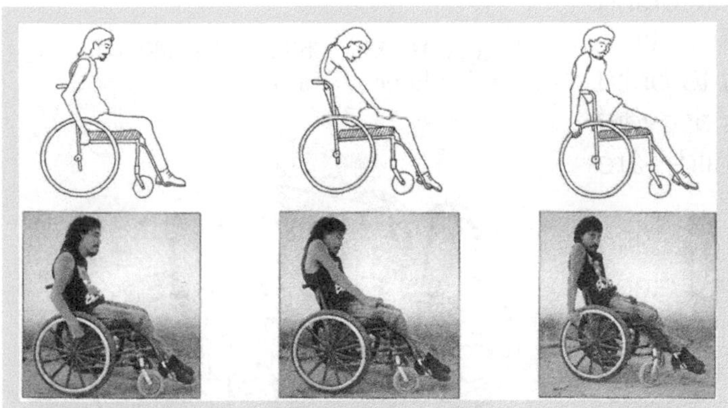

Figure 10.93 Julio taps his thigh to trigger spasticity that lifts his backside off the seat.

Figure 10.94 Julio teaches Spanish to a visiting physiotherapist.

* This is an intensive, one-to-one conversational Spanish training program. The course caters especially to persons from other countries who want to learn Spanish and at the same time to volunteer in the Community-Based Rehabilitation Program. Students in the course live with a family at PROJIMO or in the community. Participants have included physical and occupational therapists, doctors, medical students, and social change activists.

TAKING A ROLE IN SOCIAL CHANGE

The thoughts and feelings of the persons who are disabled must be foremost in the planning process when designing a rehabilitation or intervention program. The programs are for them, and they should have as much control of the process and program as possible.

The most urgent problem for the majority of persons with disabilities living in developing countries, and especially for disabled children, is getting enough to eat. Although it can be difficult to believe, some poor families will let their disabled child starve, not because they don't love them, but because there is not enough food to go around. Sacrifice is a way of life...and death.

Too often, the goal of rehabilitation is to "normalize" persons with disabilities so they will fit better into the existing social order. However, as persons with disabilities become self-determined, many of them question the concept of normalization.

They see the existing social order as a system that enables the strong to dominate the weak, where greed overshadows the spirit of compassion, and where some people live with tremendous surplus while others struggle to survive. These disabled persons, rather than being "normalized" to fit into such a society, would rather work to change it (Figure 10.95).

Figure 10.95 Children spreading the word of hope and pride.

Disabled activists increasingly see the quest for equal rights and equal opportunities as inseparable from the struggle of other categories of disadvantaged and marginalized people (Figure 10.96). They are committed to building solidarity with all oppressed groups in an effort to build a social order based less on self-seeking acquisition and more on caring and sharing. They want to help build a society that puts less emphasis on "fitting in" or "being normal" and more on being equal, one that is inclusive, yet at the same time celebrates diversity. Above all, they want to build a society that makes sure the basic needs of every human being—and certainly every child—are met.

Figure 10.96 Disabled and non-disabled work together to build a fair and equitable society.

Disabled activists in many countries are joining international grassroots movements to advance a model of socioeconomic development that is sustainable, equitable, and which provides opportunity for all people to meet their basic needs, participate fully in community life, and realize their full potential.

While these goals may be beyond the mission of a rehabilitation or intervention program, they will surely become part of the lives of the individuals with disabilities and their families. For this reason, rehabilitation planners and workers must be aware, ready, and have their own policies for the level of official involvement disabled workers will have.

CASE STUDY: ROBERTO'S INVOLVEMENT

Roberto Fajardo is one PROJIMO's founders. Incapacitated by severe juvenile arthritis as a child, he has become a defender of the underdog.[*] As a disabled village health worker and leader in Community-Based Rehabilitation, he organized landless farmers to demand their constitutional land rights.[†]

He and his followers succeeded in equitably redistributing over 50% of illegally large landholdings in the region surrounding Ajoya. As a result, rates of child malnutrition and mortality declined substantially, as did the incidence of maternal malnutrition, premature birth, and children born with cerebral palsy and other disabilities.[‡]

Figure 10.97 Teaching children about their strengths.

Teaching individuals to be self-sufficient, to work together, and to respect each other's strengths does not stop with their being able to do their daily life activities. Most will want to fully participate in the community and society in which they live. They want to make life better for all people (Figure 10.97).

CONCLUSIONS

This chapter has described both macro- and microergonomic methods to develop and implement rehabilitation, training, and intervention programs in rural, low-income settings. Rather than using the term ergonomics, the term "goodness-of-fit" was used to describe both the devices and the processes, but the messages are the same regardless of the terminology.

Moving from the big picture to the individual:

- Use a systems approach.
- Understand the community in which you work, including the people, the culture, the socioeconomic circumstances, and the politics.
- Tailor your interventions to that community.
- Look at and use the strengths available, at all levels—individual, collective group, and the community.
- Empower the people most likely to benefit from the program. Give them independence, not dependence.

[*] Roberto recalls a cold night in winter when his grandmother took the blanket off him and put it on the other children, "because he was just going to die anyway."

[†] Without land, families go hungry. Malnutrition is not only a leading cause of disability, it is the most urgent and life-threatening problem of millions of disabled persons, especially children.

[‡] As mentioned earlier, the changes in the Mexican Constitution required for Mexico to join the North American Free Trade Agreement (NAFTA) has resulted in land throughout Mexico concentrating into fewer hands. Millions of new landless peasants have moved to the slums of the cities. Without sufficient work, the result is an increase in disability, related both to malnutrition and to growing social unrest and violence.

- Use train-the-trainer techniques, train individuals with disabilities to become trainers/workers themselves.
- Fit the tasks and tools to the person. Keep working on it (iteratively), until it fits and is functional.
- Use a holistic approach with the individual (physical, psychosocial, emotional, and cognitive).
- Treat individuals and communities as partners.

More important than all of the suggestions above is to make decisions with a caring attitude; when you have the welfare of others at heart, the right decisions seem to result.

Note: Nearly all of the drawings and photos in the chapter are by the author and most have been taken from the handbook *Nothing About Us Without Us: Developing Innovative Technologies For, By and With Disabled Persons* by David Werner, published by HealthWrights, Palo Alto, California, 1998. (e-mail: healthwrights@igc.org) www.healthwrights.org.

Further ideas and designs of innovative, low-cost equipment for disabled children can be found in David Werner's two books: *Disabled Village Children* and *Nothing About Us Without Us,* Both books are freely accessible online (and can be purchased in print) at www.healthwrights.org.

SECTION D

Children and Product Design

CHAPTER 11

DESIGNING PRODUCTS FOR CHILDREN

VALERIE RICE AND RANI LUEDER

TABLE OF CONTENTS

FIRST PRINCIPLES IN DESIGNING FOR CHILDREN

The focus of this chapter is product design and fitting products to the children who will use them. When designing products for children, there are additional considerations to those used when designing for adults. The basic ideas that follow demonstrate why designing for children is somewhat different from designing for adults.

CHILDREN ARE *NOT* LITTLE ADULTS

1. They differ physically*

Childhood is a dynamic process as children move through a number of stages, each with its own characteristics and risk factors. Not only do their physical proportions differ from adults, their patterns of growth change as they mature. Younger children tend to grow more in their extremities, whereas adolescent growth largely affects the spine.

Their bones need stimulation to grow normally, yet too much or improper use may impair their growth. In young children (up to about 8–10 years), the growing ends of the bones (growth plates) are soft, vulnerable and subject to damage.

* See Section B, Child Abilities and Health and Chapter 8 on children's injuries in this text.

Adolescents face different challenges; in the early stages, the spine is growing quickly adding length without adding mass; in the mid-stage, increasing volume and in later stages increasing in density.

Adolescents, particularly females, are prone to developing back and neck pain. This has implications, for example, in the design of children's sports activities and equipment.*

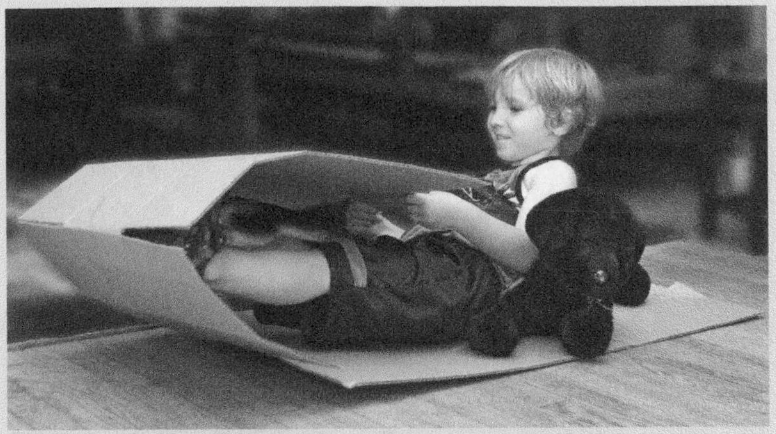

Figure 11.1 (Ursy Potter Photography)
Children often use one object for another; here, Zander transforms a box into a racecar. Such differing perspectives of children present unique design challenges.

Children move from being less coordinated than adults to seeming to have great coordination—all within a few short years. Yet, their movements are often unplanned and rarely conserve energy as they throw themselves into their activities.

Children sustain different types of injuries. In addition, their injuries heal differently than adults.

2. Children differ cognitively

Children take risks and often lack judgment. They test their skills, their adult supervisors, and their environment in their quest for independence.

This process reflects their brain development. Children's brains are more plastic‡ than adult brains. This is why, as an example, children can learn certain new skills such as novel languages more readily than adults.

> Babies experience a "synaptic surge" between birth and 12 months and continue through their fifth year.† A "pruning" process that discards unused connections (synapses) follows this. The child's life experiences shape this neural process. Designers and caregivers can influence how the brain develops at this young age.
>
> (Eliot, 2000)

> *Myelin* is the fatty material that forms a sheath around brain cells that send messages to other neurons, to muscles, and other parts of the body. It helps "contain" electric impulses and contributes to their speed of transmission.
>
> Children have less myelin, which contributes to their brain plasticity. The myelination process continues until their early twenties.
>
> This lag provides children with longer opportunities to learn totally new and unique skills. It may also hinder the development of judgment and decision-making.
>
> (Straugh, 2004)

There are critical times during which particular environmental input is necessary for a child to move into his or her next stage of development. In fact, not getting the appropriate experiences can result in the child not developing normally.

This means designers and caregivers can affect a child's development through the experiences they and their products provide. It also means children will use products and places in ways that may make absolutely no sense to the adults who created those same products and places.

* See Chapter 6 on children's musculoskeletal development and age-related risk in this text.

† That is, the greatest growth of synapses and dendrites (neuron branches that exchange signals with other brain cells) occurs between birth and age 5.

‡ *Plasticity* refers to the brain's ability to reorganize neural pathways according to new experiences.

Figure 11.2 (Ursy Potter Photography)
Children's perceptions are very different than those of an adult. This little girl is hiding under a table. She believes that if she can't see you, you can't see her!

Figure 11.3 (Valerie Rice)
Children depend on adults to provide safe environments.

3. Children differ emotionally

Children's motivations, interests, and fears are quite different from those of adults. Young children need distinct kinds of challenges. Their environments must afford them opportunities to succeed and sometimes fail; and with practice and perseverance, the ability to turn their failures into successes. They must be given safe environments, where failure is "ok" and seen as part of the learning process.

Games and learning environments can help children develop the courage and tenacity necessary for other life events in adulthood. Those who design for children can provide such opportunities in the things and experiences they design for children: games and products for academics, recreation, and travel.

4. Children's perspectives differ

Children see things in a "different light," both physically and metaphorically. Physically, for a crawling baby, the floor, legs of the couch and other furniture, and adult legs and feet take up a good portion of their visual perspective.

Children learn through their senses. They want to see, touch, taste, experience, and experiment. The world seems exciting, interesting, and worth exploring.

Children have little fear; they lack the experience that comes with encountering painful events such as falling, burning a hand, or pinching a finger. Places and things that might seem scary from an adult

Figure 11.4 (Kiran Rice)
Six-year-old Kiran Rice took this photo of her friend and neighbor. It demonstrates her own perspective. What children see and attend to, how they make decisions and chose to use products continue to change as they mature.

viewpoint are interesting to children; the inside of a kitchen cabinet or a clothes dryer may seem attractive; it is new, dark, and it holds many new things to explore.

Yet, even as they explore, children watch the adults around them interact with things far above them physically (Figure 11.4). Children mimic these giant creatures (adults), climbing and reaching to try to achieve the same heights as those who seem to command their environment. Children enjoy impersonating adults with child-size "adult" products.

DESIGNING FOR CHILDREN POSES SPECIAL CHALLENGES

Children depend on adults to provide a safe environment, especially when they are very young. Designers must attempt to understand and anticipate how children will use products and what dangers may present themselves to an unknowing child.

1. Children are "moving targets"

The rapid pace of change in children presents special challenges. Effective design requires an understanding of its intended users. Designing effective products and environments for children requires an understanding of their physical, cognitive, and emotional stages of development in order to accommodate their growth.

Products that grow with children are more attractive to the adults who purchase them, as children use them longer. Products that adjust, such as in size or design, can be appropriate for children of different ages, as seen with many playground designs.

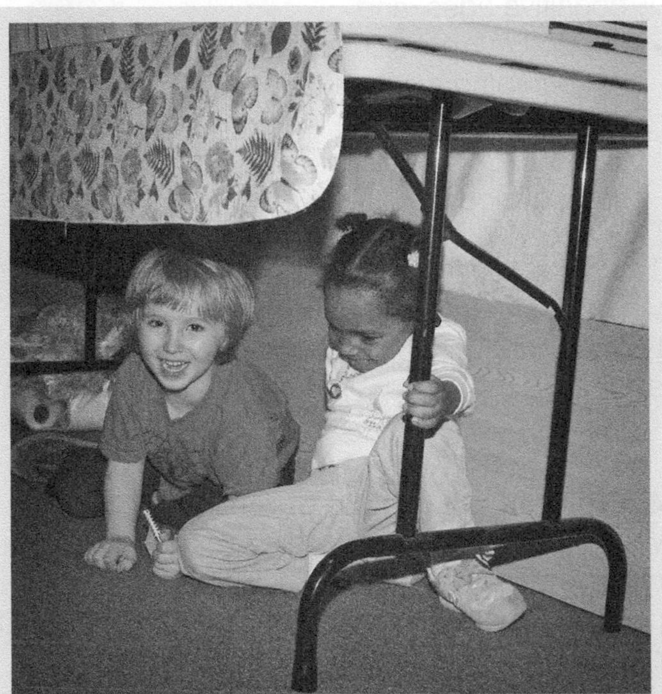

Figure 11.5 (Ursy Potter Photography)
Children use products differently than intended and in doing so, they may perform unsafe behaviors. Hopefully, the locking mechanism on the folding table shown above is in place and will not slip. Otherwise, it (and whatever is on it) could fall on and injure these happy children.

2. Standard usability approaches do not always apply to children

Human factors engineers and ergonomists* commonly test and interview to evaluate the needs and preferences of intended user populations. Children have difficulty articulating their needs and preferences.

Very young children cannot yet talk and young children lack the necessary language and experience to sufficiently describe what they require. Older children may confuse their needs with their wants, especially as they face peer pressure to have the most up-to-date ("cool") products.

Observations, without verbal feedback, may sometimes be useful. Yet, even this information may not be sufficient to adequately design for children. Knowledge of developmental growth patterns is essential.

"Productivity" for children is often their play, which enables them to explore and learn about their world.

* Human Factors Engineering and Ergonomics are considered synonymous terms. Hereafter, only the term ergonomics will be used.

3. Typical design tenets are not enough when designing for children

Ergonomists and designers fit the task or product to the person in order to promote comfort, productivity, and safety. When designing for children, designers need to fit the task to both the child as he is now, as well as "pulling" the child into his next stage of development. For example, playgrounds must:

- meet user preferences and comfort, while ensuring "productivity" and safety,
- accommodate children of different chronological, physical, and cognitive ages, and
- support current skill levels while encouraging the acquisition of new skills that enable the child to continue to move forward as he or she develops.

This final consideration is essential when designing for children, but is typically absent when designing for adults.

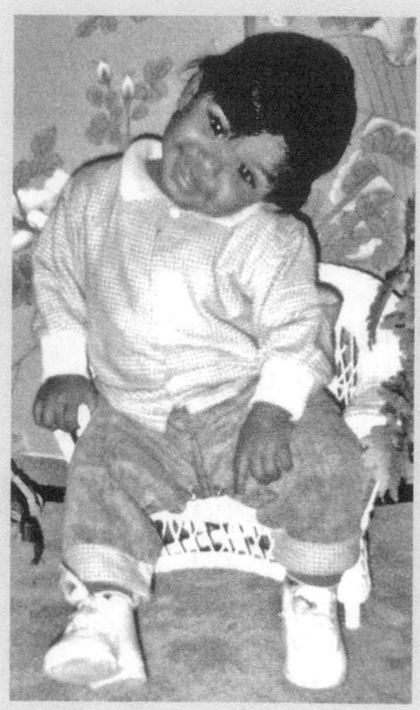

Figure 11.6 (Valerie Rice) When the objects we use and the environments in which we operate fit us, we are more comfortable, make fewer mistakes, and are more likely to enjoy what we are doing (Bennett, in press). The same is true for children.

4. Children are not the purchasers

When designing for most adults, the designer often knows the end user will likely purchase the product. Therefore, marketing can target the adult population.

When designing products and places for children, their parents, teachers, and caregivers will do the purchasing. Therefore, the products and places must appeal equally to the child and to adults. The adult's knowledge, their own childhood, their concern for the child's growth and development, as well as for their happiness—all must be considered. The product must appeal to both the adult who pays for it and the child who will use it.

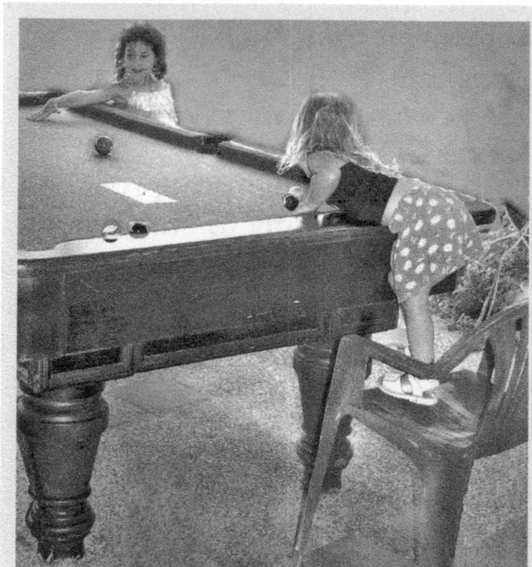

Figure 11.7 (Ursy Potter Photography) Children want to play with adult products. If there is not a "child-adapted" product to assist them, they will discover an access on their own. Unfortunately, it may not be safe.

> Many publicly available documents cited in this chapter can be downloaded at http://childergo.com/recc-children.htm

BABIES AND TODDLERS

Examples of product features that may harm infants and toddlers include attachments such as buckles that break while in use (particularly if the child is being carried in a front or backpack) and strings and cords that tie around infants' necks or become caught on cribs, playpens, or other products. Products also become hazardous when adults use the child products differently than intended.

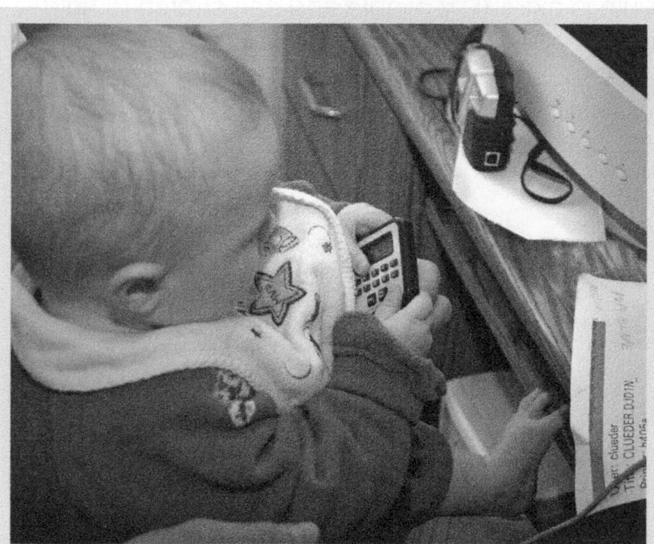

Figure 11.8 (Ursy Potter Photography)
Infants under the age of 1 are at the greatest risk of choking. Give children items they can easily grasp, but are too big for them to choke on. Toys small enough for children under 4 to choke on will generally fit through a 3.2 cm (1.25 in.) circle or are shorter than 5.7 cm (2.25 in.).

Figure 11.9 (Consumer Product Safety Commission)
Never tie strings around an infant's neck. Do not leave cords of any kind near an infant.

> The ***Consumer Product Safety Commission* (CPSC)** is a United States Federal regulatory agency responsible for protecting the public from unreasonable risks of serious injury or death from consumer products.
>
> Its activities include performing research, educating the public, responding to consumer inquiries, developing voluntary standards with industry, issuing and enforcing mandatory standards, and banning unsafe products.
> CPSC's website is http://cpsc.gov
> The US hotline is (800) 638-CPSC
> Hearing impaired: (800) 638–8270

RATTLES

Figure 11.10 (Figure from Consumer Product Safety Commission)
CPSC tests whether rattles exceed this minimum permissible size with a device with an oval opening that measures approximately 35 mm (1 3/8 in.) × 50 mm (2 in.) × 30 mm (1/16 in.) deep. The rattle is considered hazardous if any portion of the rattle or its handle can pass through this opening. CPSC recommends that parents select rattles that are larger than these dimensions.

Rattles

Although CPSC established safety requirements for the design of rattles in 1978, some rattles sold as party favors or decorations might be given to children. Rattles must be:

1. large enough so they cannot enter an infant's mouth and lodge in the back of the throat. An infant's mouth is extremely pliable and can stretch to hold larger shapes than one might expect
2. made such that they will not separate into small pieces that infants can swallowed or inhale.

Figure 11.11 (Figure from Consumer Product Safety Commission)
Remove all crib toys that stretch across the crib with strings and cords by the time the infant reaches 5 months or begins to push up with his hands or knees (whichever is first) to avoid entanglements that may lead to strangulation. These should be out of reach of the infant at all times. Avoid catch points.

INFANT CUSHIONS

CPSC banned infant cushions in 1992 that had these cushion characteristics that led to infant suffocations.
Infant cushions that are:

1. covered with soft fabric
2. filled with loose granular material such as plastic foam beads or pellets
3. flattens easily to create a "nest"
4. conforms to the infant's face or body
5. intended or marketed for infants under 1 year of age

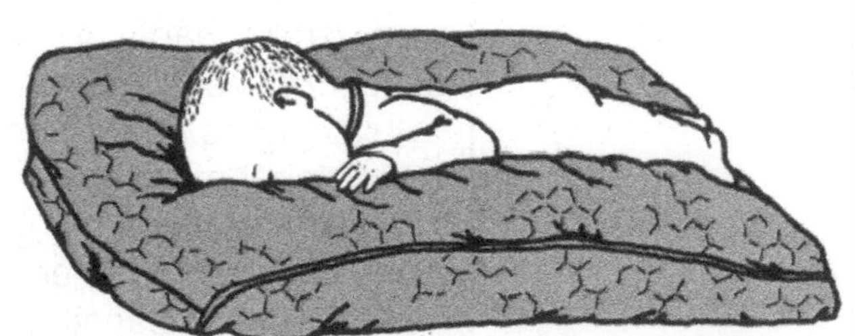

Figure 11.12 (Figure from Consumer Product Safety Commission)
According to CPSC, the key feature of these cushions in contributing to infant suffocation deaths is its ability to conform to an infant's face and body. Almost all reported deaths involved children lying in a prone, stomach-down position. In all but two of the incidents, the infant was less than 4 months old.

Figure 11.13 (Consumer Product Safety Commission) The "Crib Cuddle" was recalled after reports of infant injuries and death. It is similar to a hammock that attaches to the side rails of a crib by six adjustable straps. A plastic heart-shaped battery-operated box pulsates and vibrates to simulate a heartbeat and is inserted into a soft pad in the hammock.

Infant sleeping guidelines

The CPSC, AAP, and NICHD* recommendations:

- Place baby on his or her back on a firm tight-fitting mattress in a crib that meets current safety standards.
- Remove pillows, quilts, comforters, sheep-skins, pillow-like stuffed toys, and other soft products from the crib.
- Consider using a sleeper or other sleep clothing as an alternative to blankets, with no other covering.
- If using a blanket, put baby with feet at the foot of the crib. Tuck a thin blanket around the crib mattress, reaching only as far as the baby's chest.
- Make sure your baby's head remains uncovered during sleep.
- Do not place baby on a waterbed, sofa, soft mattress, pillow, or other soft surface to sleep.

Subsequent research associates *sudden infant death syndrome* (SIDS) with infants sleeping prone (on their stomach) on soft cushioning or comforters.

BABY CARRIERS

The CPSC† reports that millions of baby carriers have been recalled for hazardous features that have in some products caused infant deaths and injuries.

These deaths occur when infants become tangled in restraining straps, when carrier seats topple on beds or other soft surfaces, or when unrestrained children fall from the carrier. Infants can be quite active; their movements can cause the carrier seats to move or tip. Infants may also move the seat when they kick adjacent objects. *Typically, infants are left unattended when these events occur.*

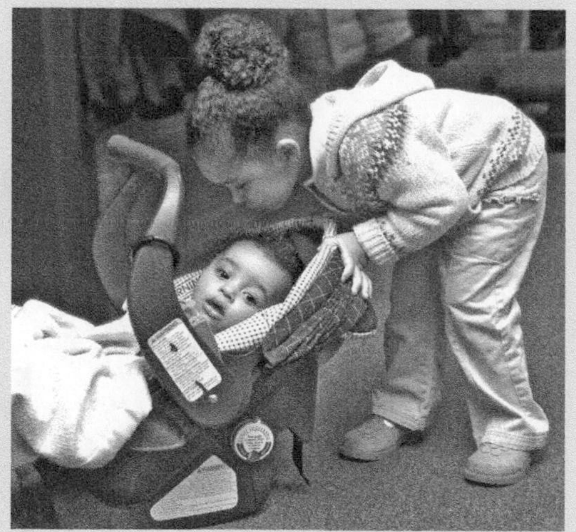

Figure 11.14 (Ursy Potter Photography) Caregivers need to know the rules about what infant products to use and how to use them.

* These reports include the *Consumer Product Safety Commission* (CPSC), the *American Academy of Pediatrics* (AAP), and the *National Institute of Child Health & Human Development* (NICHD).

† The CPSC reports there were at least 5 deaths and 13,000 estimated injuries in a recent year attributed to baby carriers. Their Web site is http://cpsc.gov For further information, consumers may call the CPSC's toll-free hotline on 800-638-CPSC.

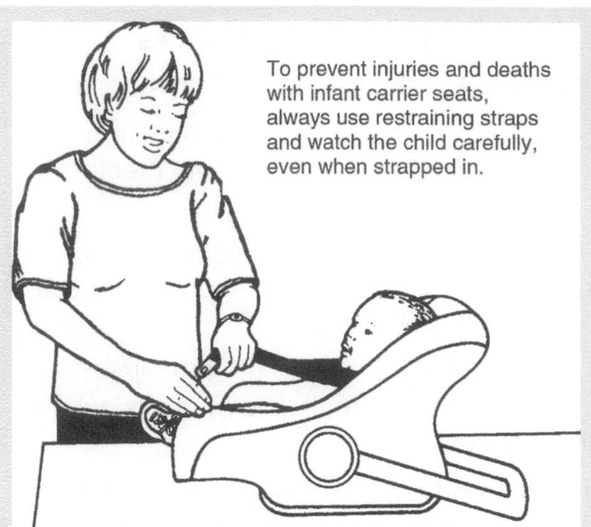

To prevent injuries and deaths
with infant carrier seats,
always use restraining straps
and watch the child carefully,
even when strapped in.

Figure 11.15 (Figure courtesy of CPSC)
CPSC recommends:

- Carriers have a wide, sturdy, and stable base.
- Parents stay within arm's reach of the baby when the carrier is elevated.
- Never place carriers on soft surfaces that make it unstable, as a baby's movements can move the infant seat.
- Always use the products' safety belts.
- Do not use infant carriers as infant car seats (unless the product is specifically designed for both purposes).

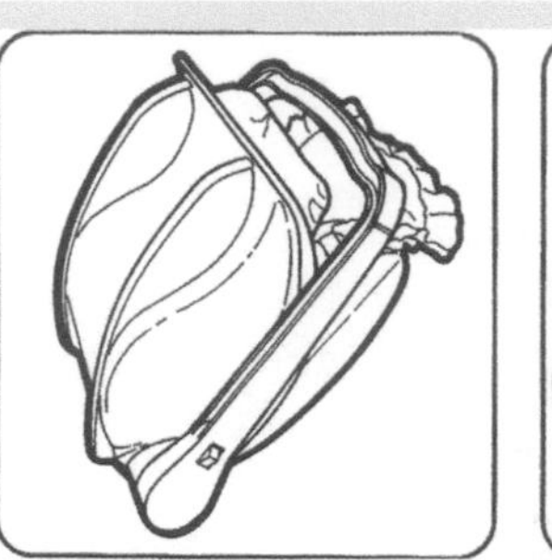

Figure 11.16 (Figure courtesy of CPSC)
This baby carrier was recalled because some infants sustained serious head injuries after falling from the carrier.

Voluntary manufacturer recall in cooperation with CPSC

Figure 11.17 (Figure courtesy of CPSC)
The back support buckle of this baby carrier can detach from the shoulders, potentially causing the infant to fall.

Voluntary manufacturer recall in cooperation with CPSC

BABY CARRIERS (ON THE CAREGIVER): Infant carriers made of fabric may benefit the infant and caregiver. They hold the baby close, while allowing the caregiver to move about and perform daily activities. In addition to encouraging emotional ties, they may reduce physical demands by placing the load close to the caregiver's center of gravity.

Evaluate carrier designs carefully. Design features to watch for include buckles that break or slip, clasps that slip or tear, and openings that may cause the infant to slip.

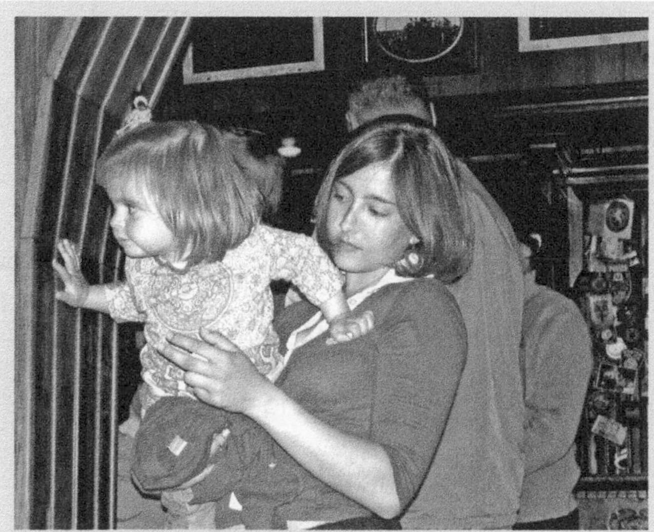

Figure 11.18 (Ursy Potter Photography)
Carrying infants is physically difficult. There is some risk because of the physical load and the demands on the mother or caregiver's attention.

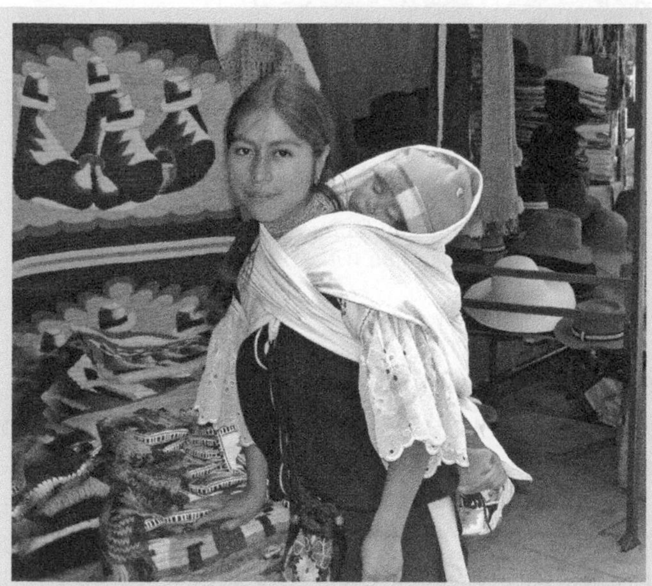

Figure 11.19 (Ursy Potter Photography)
Baby carriers may reduce some of the risks associated with carrying an infant in one's arms.

Figure 11.20 (Consumer Product Safety Commission)
Baby carriers are available that either have the infants face toward or away from the caregiver. All carriers should support the infants head.

INFANT BATH SEATS

Infant bathtub seats that the infant sits in while bathing are plastic chairs with three or four attached "legs" or suction cups. Caregivers may use these products to assist the baby's sitting in the bathtub while the caregiver washes the child.

These are used with infants who are physically capable of sitting without the product. They give the child additional support, while restraining his movements.

When used properly, these products are safe. However, these bathtub seats are not safety devices. They may increase the likelihood that caregivers will leave them unattended and the child may potentially drown (Rauchschwalbe et al., 1997).

Infants should never be left unattended in bathtubs. Small children can drown quickly and silently in a small amount of water. Checking bath water temperature with a wrist or elbow before putting your baby in to bathe will prevent burn injuries.*

Children should also not be left alone near any standing water. This includes toilets, 5 gallon (19 liter) buckets, and pools.

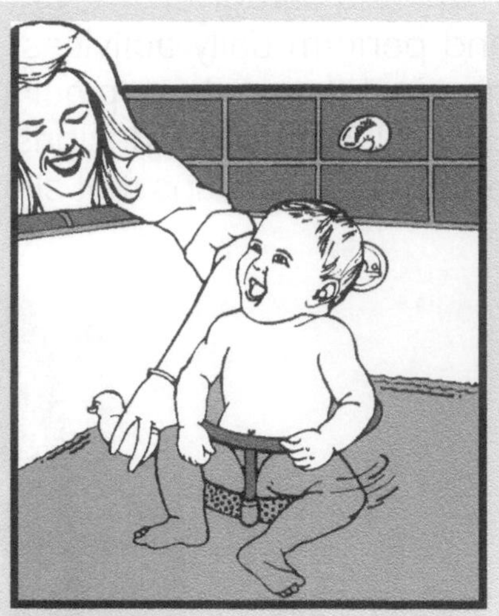

Figure 11.21 (Figure courtesy of Consumer Product Safety Commission)
Never leave a baby unattended or with a sibling in a tub of water. Do not rely on a bath ring to keep your baby safe.

Figure 11.22 (Figure courtesy of the Consumer Product Safety Commission)
This infant bath seat was recalled because some infants sustained serious head injuries after tipping and falling.
Voluntary manufacturer recall in cooperation with CPSC

HIGH CHAIRS

High chairs are used when feeding toddlers or letting them play nearby while mothers are working in the kitchen. Both a waist and "crotch" strap are necessary to prevent toddlers from climbing or slipping out of the seats. Using only one of the two straps is dangerous as a child can slip and strangle either in the waist strap (if a crotch strap is not used) or between the tray and the chair seat. The tray is not a restraining device. High chairs can topple if a child pushes away from a counter top, wall, or table; rocks the chair back and forth; or kneels or stands in the chair.

---

* See Chapter 8 on children's injuries for additional information on recommended water temperatures, as well as on scalding and near-drowning injury prevention.

Although it may seem easy to allow a child to climb into or out of a chair on his or her own, it is dangerous. The chair can easily tip and fall.

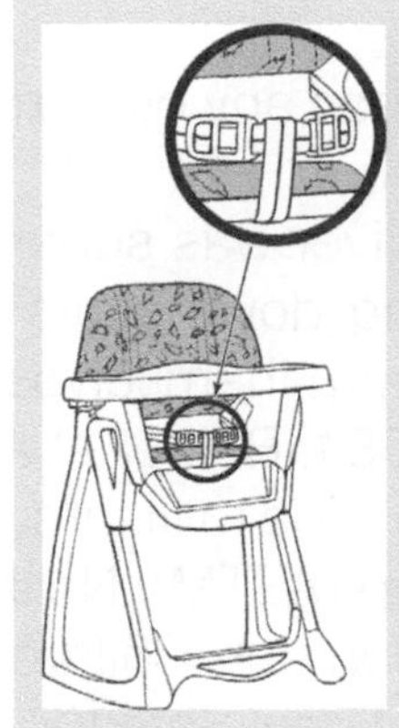

Figure 11.23 (Figure courtesy of Consumer Product Safety Commission) High chairs should have both waist and crotch straps.

Figure 11.24 (Photos courtesy of Stokke LLC.) Child seating with a footrest decreases their "fidgeting," improves body stabilization, and increases reach distance (Hedge et al., 2003). Having an adjustable chair that "grows" with the child eliminates the times when a child is "in between" chair sizes.

Figure 11.25 (Stuart Smith) Tips for purchasing high chairs (CPSC):

1. Get a wide base for stability.
2. Select a strap system that can be fastened only if both the waist and crotch straps are included.
3. Look for wide straps (comfort) that are easy to use and fastenings that are unlikely to pinch small fingers or pieces of skin.
4. Look for a chair with a post between the legs to prevent slipping.
5. Examine straps and attachments for security and resistance to fraying.
6. Check the locking device on folding chairs to be certain it locks firmly.
7. Examine the footrest to see if the chair will adjust as the child grows.

Remember that most manufacturers will provide replacement straps, if they become frayed, lost, or destroyed.

BABY WALKERS

Baby walkers cause more injuries than any other nursery product. Injuries to children using baby walkers are frequent and serious.

About 80% of the time, a caregiver was supervising the child at the time of the injury. Most injuries involve children falling down stairs or off porches. Some children are hurt when the walker tips over; others burn themselves pulling objects off a hot stove (Fazen & Felizberto, 1982; Stoffman et al., 1984; Reider et al., 1986; Chiaviello et al. 2005).

Baby walkers must either be too wide to fit through a standard door or have features that stop the walker at the edge of a step (ASTM). Neither the American Academy of Pediatrics (AAP) nor the National Association for Children's Hospitals and Related Institutions (NACHRI) recommend walkers (National Safe Kids Campaign).

Children lean forward and "walk" on their toes in walkers, both of which are unlike a normal walking gait. Consider using stationary bouncers or activity centers in place of baby walkers. They exercise the child's muscles and encourage bone growth without promoting an abnormal gait.

Figure 11.26 (Ursy Potter Photography)
Baby walkers make children mobile before they are developmentally ready. This hinders their learning to develop a sense of balance and to use postural muscles to walk. These children develop forward-leaning postures rather than upright and centered walking.

Walkers with wheels may also make a child more mobile, potentially exposing them to new risks. In emergencies, caregivers may not be able to respond quickly enough to account for the child's greater mobility and speed.

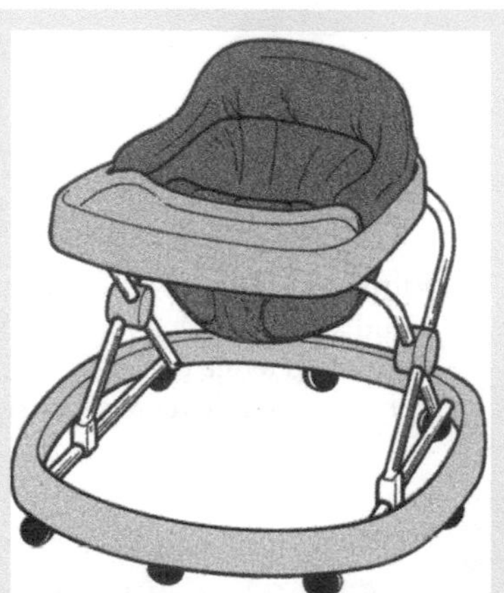

Figure 11.27 (Figure courtesy of Consumer Product Safety Commission)
This baby walker was recalled because some infants sustained serious head injuries after falling from the walker.

Voluntary manufacturer recall in cooperation with CPSC

BABY STROLLERS

Baby strollers are invaluable to the caregiver. Yet stroller injuries are common, particularly during an infants' first year. The greatest risk is that children fall from strollers, sometimes sustaining head injuries.* Older toddlers (probably unrestrained) may fall as they try to stand in the strollers (Powell et al., 2002).

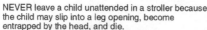

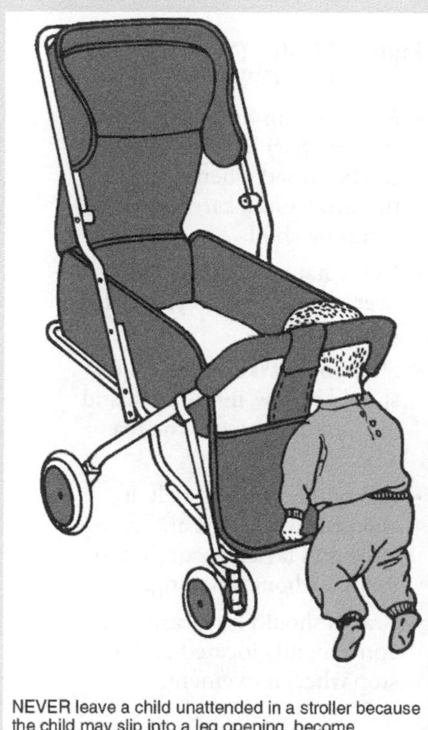

NEVER leave a child unattended in a stroller because the child may slip into a leg opening, become entrapped by the head, and die.

Figure 11.28 (Figure courtesy of Consumer Product Safety Commission)
This type of accident occurs when caregivers inadvertently leave a child sleeping in a stroller, unsupervised. The baby somehow wriggles toward the front of the stroller and slips through the handrest and the bottom of the seat and strangles.

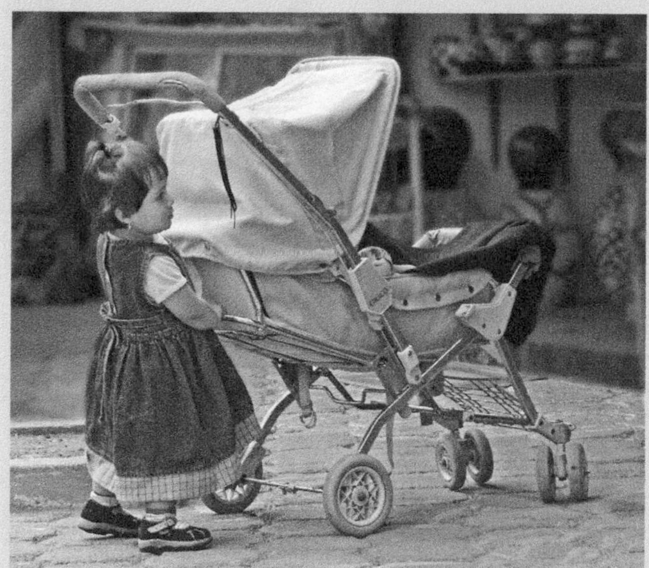

Figure 11.29 (Ursy Potter Photography)
This little girl imitates her mother pushing a stroller. Stroller-tips when pushing a stroller often involved young children who were unable to physically balance the stroller (Powell et al., 2002).

* Almost 13,000 children 3 years old and younger are treated each year in the United States for injuries caused by baby strollers. Most of these involve infants in their first year. Most injuries involve the head or face and usually are minor. About 2% of children with stroller injuries are admitted to the hospital (Powell et al., 2002).

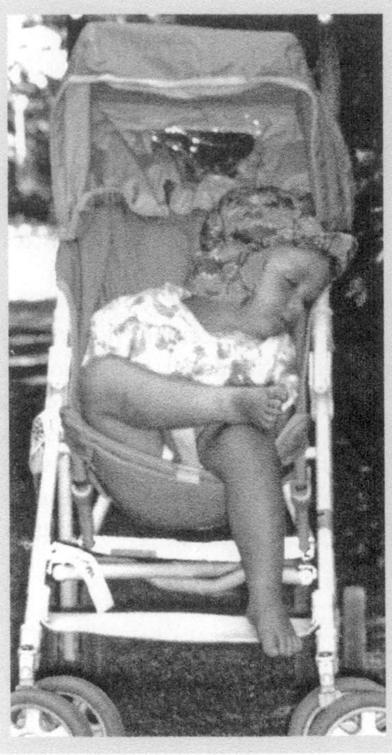

Figure 11.30 (Valerie Rice)
Stroller Tips (CPSC):

- Make certain that the handrest (or grab bar) in front of a seat can be closed when using the stroller as a carriage for a reclining child.
- Never hang bags or purses over stroller handles, as it may cause tipping.
- Shopping baskets on strollers should be low, in the back and in front of (or over) the rear wheels.
- Always use the seat belt. It should have a waist and crotch strap and fasten securely and easily without pinching.
- Brakes should lock easily, be conveniently located and fully stop wheel movement.

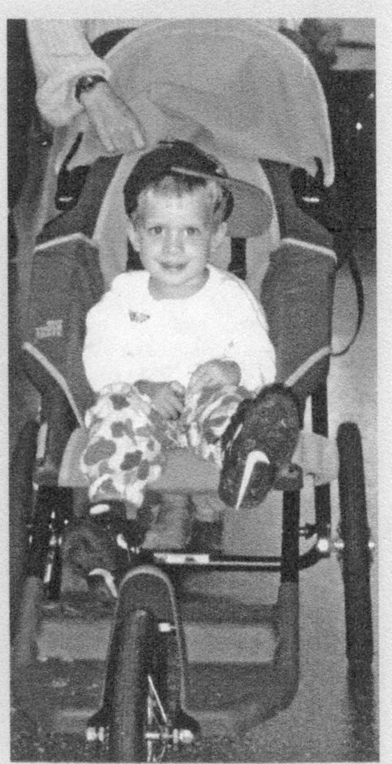

Figure 11.31 (Valerie Rice)
Specialty strollers are available, such as this one for jogging. As with other strollers, these need good restraint systems, braking devices, and locking devices to prevent collapsing (folding) during use. Some also come with rain hoods. These need to be both light and sturdy.

PLAYPENS AND PLAY YARDS

Play yards provide enclosed spaces for young children to play in safety.
To prevent playpen injuries and deaths:

- always put babies younger than 12 months to sleep on their back
- carefully set up the playpen before putting a child in it and be certain it locks in place
- use only the manufacturer's mattresses', do not add additional padding
- if using mesh-sided playpens, make sure the mesh is less than ¼ in., and ensure it is securely attached
- make sure the playpen is in good shape
- do not use a playpen with catch points
- place playpens away from windows with blinds and cords for adjusting blinds, as they can entangle around the child and become a choking hazard

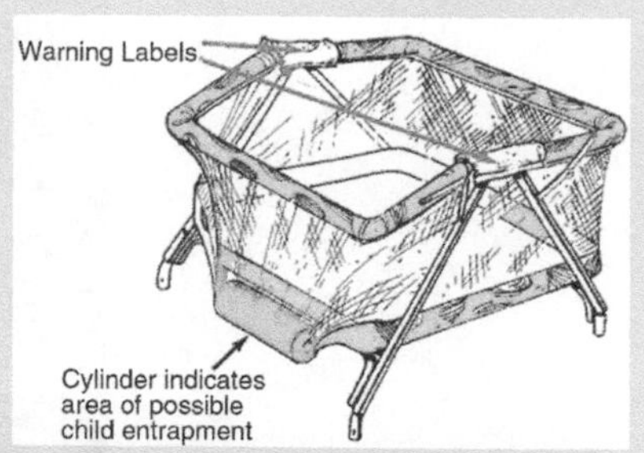

Figure 11.32 (Photo courtesy of Consumer Product Safety Commission)
Playpens used with sides down can present a suffocation hazard to infants. When the drop side is down, the mesh net material forms a loose pocket, leaving a gap between the edge of the playpen floor and the mesh side. Infants can easily fall or roll into this pocket. Once an infant's head is entrapped in this pocket, the infant can suffocate.
Consumer Product Safety Commission

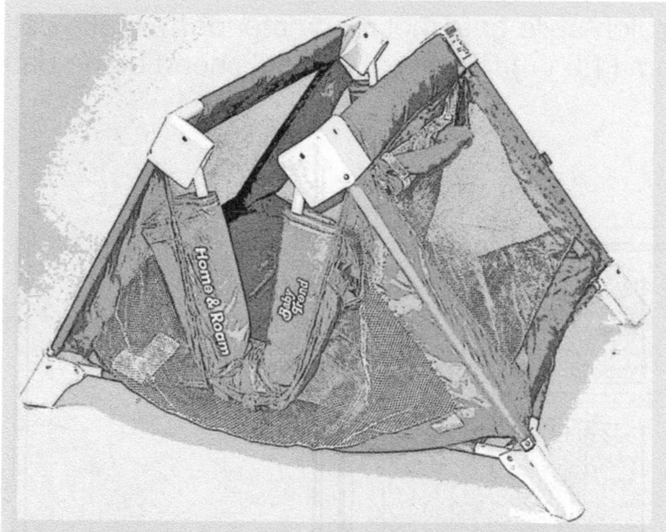

Figure 11.33 (Photo courtesy of Consumer Product Safety Commission)
Many playpens such as this product were made in the 1990s. These collapsed at the hinges, putting the child at risk of entrapment or strangulation. CPSC voluntary standard in 1999 requires that hinges lock when rails are pulled up.
Consumer Products Safety Commission

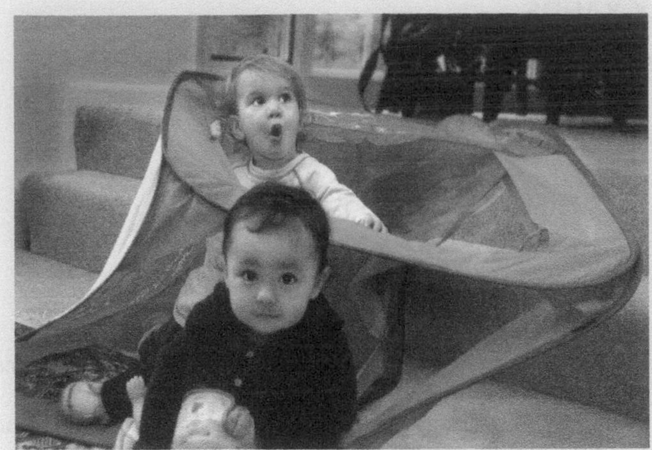

Figure 11.34 (Ursy Potter Photography)
Children love tunnels, tents, and things they can crawl into. However, caregivers must be careful with them, as many can be unstable.

BABY GATES

Safety gates can help keep toddlers and children out of hazardous areas, particularly stairs. Ensure that children are not able to dislodge the gates. For the top of stairs, gates that screw to the wall are more secure than pressure gates.

Safety Gate Guidance (CPSC):

- Select a gate with a top edge that is straight with a mesh screen or rigid bars.
  If using an accordion-style gate (not recommended), make certain the "V" shape on top is less than 38 mm (1.5 in.) to prevent a child's head being caught.

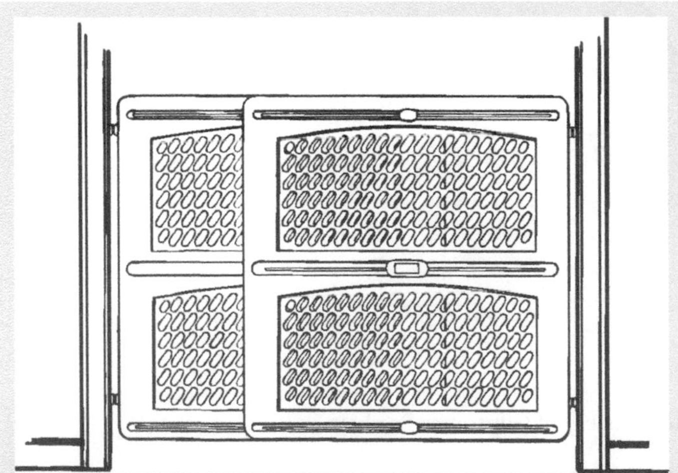

Figure 11.35 (Figure courtesy of Consumer Product Safety Commission)
New safety gates that meet safety standards display a certification seal from the Juvenile Products Manufacturers Association (JPMA).

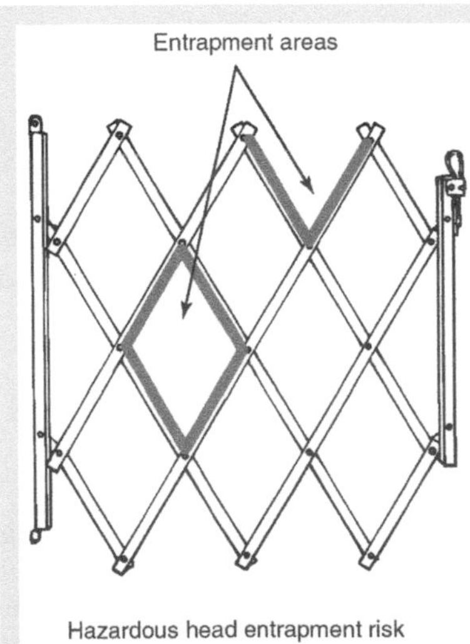

Figure 11.36 (Figure courtesy of Consumer Product Safety Commission)
Deaths have occurred with older, accordion style safety gates when children's heads are trapped in the "V" shape on top or the diamond shaped openings.

- Make certain the gate is secure.
- If a pressure bar is included, place it on the side away from the child, so the child cannot stand on it to try to climb over. Pressure bar fixtures are not recommended for stair tops.

BABY SWINGS

Baby swings can help keep a baby occupied while adults are busy. Most are relatively light and can be moved to the adult activity. Most infants love the rocking motion and many fall asleep after a short while!

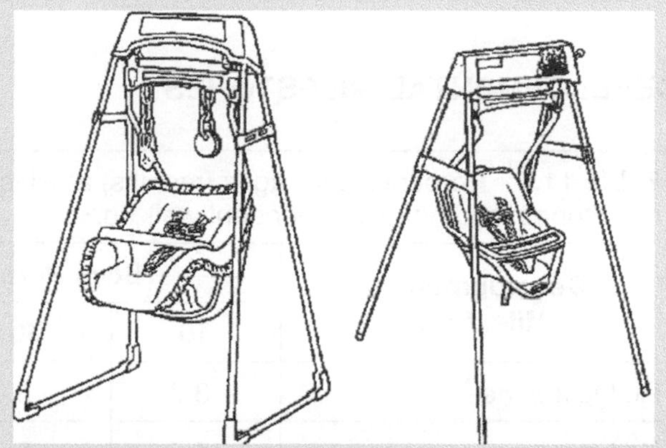

Figure 11.38 (Figures courtesy of Consumer Product Safety Commission)
These swings have shoulder harness straps that are placed over each shoulder and buckled between the child's legs. If the straps on these swings loosen or are unbuckled, a child can become tangled in the straps and strangle.

Figure 11.37 (Photo courtesy of Consumer Product Safety Commission)
This product was recalled after reports of infant deaths caused by the infants becoming entangled in the swing's shoulder harness straps.
Voluntary manufacturer recall in cooperation with CPSC

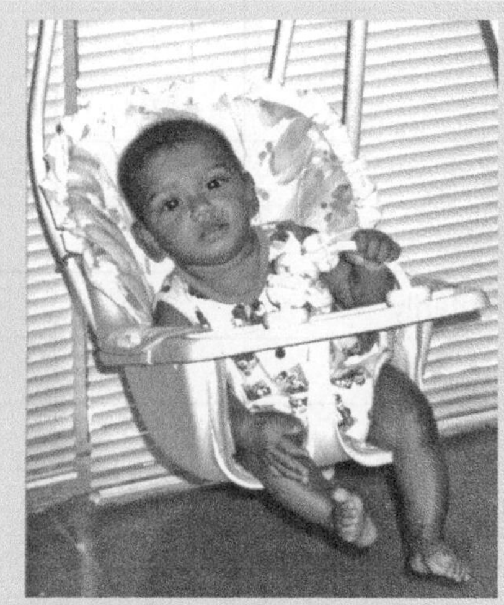

Figure 11.39 (Valerie Rice)
Never put a child in an upright swing if they do not have the muscle tone and development to sit without their head flopping to the side or slouching. Instead, use a swing that can recline for newborns. For older babies, a swing that a caregiver can recline after the baby has fallen asleep is a desirable feature.

Be certain the front tray can be easily swung to the side or removed for quickly taking the baby out of the swing. Alternately, an open top swing allows for removing the child without scrapping their legs on the tray edge or bumping their head on the top bar.

- Ensure the product is well constructed and stable.
- Ensure the design will prevent a baby from sliding off the seat (waist and crotch bar or strap).
- Five-point systems with shoulder, waist, and crotch straps should reduce the chances of babies sliding or climbing out.
- Comply with the manufacturer's age and weight recommendations.
- Swings can be battery operated or wind-up. Battery swings may have additional features such as alternate settings for speeds or music selections.

DEVELOPMENTAL MILESTONES

Table 11.1 Percentiles for ages (months) at which healthy, full term infants attain developmental milestones leading to independent walking.[a]

Developmental Milestone	Boys Percentiles			Girls Percentiles		
	10	50	90	10	50	90
Roll to supine	3.7	5.0	6.2	3.8	5.2	6.7
Roll to prone	4.8	6.0	8.8	4.6	6.0	8.7
Pivot[b]	5.0	6.4	8.6	4.7	6.8	8.8
Crawl on stomach[c]	6.1	7.0	8.7	5.8	6.9	8.9
Creep on hands and knees	6.8	8.3	10.4	6.5	8.8	11.8
Creep on hands and feet	8.1	9.0	11.8	7.7	9.6	11.9
Sit up	7.3	9.1	11.8	6.8	8.9	11.5
Stand at rail	7.4	8.4	9.8	7.1	8.7	10.5
Pull to stand	7.5	8.5	11.3	7.3	8.9	11.5
Cruise at rail[d]	8.0	9.5	11.7	8.0	9.8	11.8
Stand momentarily	9.9	12.4	15.6	9.8	12.5	14.9
Walk, one hand held	9.2	11.5	13.0	9.5	11.8	13.5
Walk alone	10.8	13.0	15.9	11.0	13.1	15.7
Walk up and down stairs	16.2	18.9	23.7	16.0	19.7	23.8

Source: As emphasized throughout this text, products and places should be designed to fit the children who will use them. Design for children will only be effective if the designer understands children and the ages at which they are likely to achieve new skills, physically, cognitively, emotionally, and socially. Therefore, those who design for children must remain aware of developmental milestones, such as those listed above. Individuals who care for children, use and purchase children's products also need to be aware of child development, especially when contemplating issues such as whether a child is physically ready to use a particular product or when to progress from one product to another. Also, see Chapter 2 of this book.

[a] Adapted from Largo et al. (1985). Data are based on 56 boys and 55 girls from the second Zurich Longitudinal Study.

[b] Pivoting – child on stomach moves in circular manner by actions of arms and legs.

[c] Crawling on stomach – child pulls forward by coordinated action of arms and legs.

[d] Cruise at rail – sideways walking while holding on to a table or similar stable object.

NURSERY EQUIPMENT SAFETY CHECKLIST

From the beginning of a child's life, products, such as cribs, high chairs, and other equipment intended for a child must be selected with safety in mind. Parents and caretakers of babies and young children need to be aware of the many potential hazards in their environment—hazards occurring through misuse of products or those involved with products that are not well designed for use by children.

This checklist from the *Consumer Product Safety Commission* (CPSC, 2005) is a safety guide for buying new or secondhand nursery equipment. It also can be used when checking over nursery equipment now in use in your home or in other facilities that care for infants and young children.

Assess whether the equipment has the safety features in this checklist. If not, can missing or unsafe parts be easily replaced with the proper parts? Can breaks or cracks be repaired to give more safety? Can older equipment be fixed without creating a "new" hazard? If most answers are "NO," the equipment is beyond help and should be discarded. If the equipment can be repaired, do the repairs before allowing any child to use it.

Table 11.2 Nursery Equipment Safety Checklist

Back Carriers	**Yes**	**No**
1. Carrier has restraining strap to secure child.	___	___
2. Leg openings are small enough to prevent child from slipping out.	___	___
3. Leg openings are large enough to prevent chafing.	___	___
4. Frames have no pinch points in the folding mechanism.	___	___
5. Carrier has padded covering over metal frame near baby's face.	___	___
CPSC RECOMMENDS: Do not use until baby is 4 or 5 months old. By then, the baby's neck is able to withstand jolts and not sustain an injury.		
Bassinets and Cradles	**Yes**	**No**
1. Bassinet or cradle has a sturdy bottom and a wide base for stability.	___	___
2. Bassinet or cradle has smooth surfaces—no protruding staples or other hardware that could injure the baby.	___	___
3. Legs have strong, effective locks to prevent folding while in use.	___	___
4. Mattress is firm and fits snugly.	___	___
Follow manufacturer guidelines on the weights and sizes of babies who can safely use the product.		
Baby Bath Rings or Seats	**Yes**	**No**
1. Suction cups securely fastened to product.	___	___
2. Suction cups securely attach to SMOOTH SURFACE of tub.	___	___
3. Tub filled only with enough water to cover baby's legs.	___	___
4. Baby never left alone or with a sibling while in bath ring, even for a moment!	___	___
Never leave a baby unattended or with a sibling in a tub of water. Do not rely on a bath ring to keep your baby safe.		

(continued)

Table 11.2 Nursery Equipment Safety Checklist (Continued)

Carrier Seats	**Yes**	**No**
1. Carrier seat has a wide sturdy base for stability.	___	___
2. Carrier has nonskid feet to prevent slipping.	___	___
3. Supporting devices lock securely.	___	___
4. Carrier seat has crotch and waist strap.	___	___
5. Buckle or strap is easy to use.	___	___
Never use the carrier as a car seat.		
Changing Tables	**Yes**	**No**
1. Table has safety straps to prevent falls.	___	___
2. Table has drawer or shelves that are easily accessible without leaving the baby unattended.	___	___
Do not leave baby unattended on the table. Always use the straps to prevent the baby from falling.		
Cribs	**Yes**	**No**
1. Slats are spaced no more than 2 3/8 in. (60 mm) apart.	___	___
2. No slats are missing or cracked.	___	___
3. Mattress fits snugly—less than two finger width between edge of mattress and crib side.	___	___
4. Mattress support is securely attached to the head and footboards.	___	___
5. Corner posts are no higher than 1/16 in. (1.5 mm) to prevent entanglement of clothing or other objects worn by child.	___	___
6. No cutouts in the head and footboards, which allow head entrapment.	___	___
7. Drop-side latches cannot be easily released by baby.	___	___
8. Drop-side latches securely hold sides in raised position.	___	___
9. All screws or bolts that secure components of crib are present and tight.	___	___
Do not place crib near draperies or blinds where child may become entangled and strangle on the cords. Replace the crib with a bed when the child reaches 89 cm (35 in.) in height or can climb and/or fall over the sides.		
Crib Toys	**Yes**	**No**
1. No strings with loops or openings having perimeters greater than 14 in. (356 mm).	___	___
2. No strings or cords longer than 7 in. (178 mm) should dangle into the crib.	___	___
3. Crib gym has label warning to remove from crib when child can push up on hands and knees or reaches 5 months of age, whichever comes first.	___	___
4. Components of toys are not small enough to be a choking hazard.	___	___
Avoid hanging toys across the crib or crib corner posts with strings long enough to strangle the child. Remove crib gyms when the child is able to pull or push up on hands and knees.		

Table 11.2 Nursery Equipment Safety Checklist (Continued)		
Gates and Enclosures	**Yes**	**No**
1. Openings in gate are too small to entrap a child's head.	___	___
2. Gate has a pressure bar or other fastener that will resist forces exerted by a child.	___	___
Avoid head entrapment; do not use accordion-style gates or expandable enclosures with large v-shaped openings along top edge or diamond-shaped openings within.		
High Chairs	**Yes**	**No**
1. High chair has waist and crotch restraining straps that are independent of the tray.	___	___
2. Tray locks securely.	___	___
3. Buckle on waist strap is easy to use.	___	___
4. High chair has a wide stable base.	___	___
5. Caps or plugs on tubing are firmly attached and cannot be pulled off and choke a child.	___	___
6. If it is a folding high chair, it has an effective locking device to keep the chair from collapsing.	___	___
Always use restraining straps; otherwise, child can slide under tray and strangle.		
Hook-On Chairs	**Yes**	**No**
1. Chair has a restraining strap to secure the child.	___	___
2. Chair has a clamp that locks onto the table for added security.	___	___
3. Caps or plugs on tubing are firmly attached and cannot be pulled off and choke a child.	___	___
4. Hook-on chair has a warning never to place the chair where the child can push off with feet.	___	___
Do not leave a child unattended in a hook-on chair.		
Pacifiers	**Yes**	**No**
1. No ribbon, string, cord, or yarn attached to pacifier.	___	___
2. Shield is large enough and firm enough to not fit in child's mouth.	___	___
3. Guard or shield has ventilation holes so the baby can breathe if the shield does get into the mouth.	___	___
4. Pacifier nipple has no holes or tears that might cause it to break off in baby's mouth.	___	___
To prevent strangulation, never hang pacifier or other items on a string around a baby's neck.		

(continued)

Table 11.2 Nursery Equipment Safety Checklist (Continued)		
Playpens	**Yes**	**No**
1. Drop-side mesh playpen or crib has label warning never to leave side in the down position.	___	___
2. Mesh has small weave (less than ¼ in. (7.5 mm) openings).	___	___
3. Mesh has no tears, holes, or loose threads.	___	___
4. Mesh is securely attached to top rail and floorplate.	___	___
5. Top rail cover has no tears or holes.	___	___
6. Wooden playpen has slats spaced no more than 2 in. (60 mm) apart.	___	___
7. If staples are used in construction, they are firmly installed and none are missing or loose.	___	___
Never leave an infant in a mesh playpen or crib with the drop-side down. Even a very young infant can roll into the space between the mattress and loose mesh side and suffocate.		
Rattles, Squeeze Toys, Teethers	**Yes**	**No**
1. Rattles, squeeze toys, and teethers are too large to lodge in a baby's throat.	___	___
2. Rattles are of sturdy construction that will not break apart in use.	___	___
3. Squeeze toys do not contain a squeaker that might detach and choke a baby.	___	___
Remove rattles, squeeze toys, teethers, and other toys from the crib or playpen when the baby sleeps to prevent suffocation.		
Strollers and Carriages	**Yes**	**No**
1. Wide base to prevent tipping.	___	___
2. Seat belt and crotch strap securely attached to frame.	___	___
3. Seat belt buckle is easy to use.	___	___
4. Brakes securely lock the wheels.	___	___
5. Shopping basket is low on the back and directly over or in front of rear wheels for stability.	___	___
6. When used in carriage position, leg-hold openings can be closed.	___	___
Always secure the seat belts. Never leave a child unattended in a stroller. Keep children's hands away from pinching areas when stroller is being folded or unfolded or the seat back is being reclined.		
Toy Chests	**Yes**	**No**
1. No lid latch that could entrap child within the chest.	___	___
2. Hinged lid has a spring-loaded lid support that will support the lid in any position and will not require periodic adjustment.	___	___
3. Chest has ventilation holes or spaces in front or sides or under the lid should a child get inside.	___	___
If you already own a toy chest or trunk with a freely falling lid, remove the lid to avoid a head injury to a small child or install a spring-loaded lid support.		

Table 11.2 Nursery Equipment Safety Checklist (Continued)		
Walkers	**Yes**	**No**
1. Wide wheelbase for stability.	___	___
2. Covers over coil springs to avoid finger pinching.	___	___
3. Seat is securely attached to frame or walker.	___	___
4. No x-frames that could pinch or amputate fingers.	___	___
Place gates or guards at top of all stairways or keep stairway doors closed to prevent falls. Do not use walker as baby sitter.		

Source: From CPSC, 2005.

TRANSPORTATION TOYS FOR CHILDREN

Children love vehicles they can ride on, whether self-propelled or motorized. They are empowered by their ability to move on their own and their development of motor and balancing skills. Younger children enjoy imitating adults in toys that resemble an adult car or truck.

These "toy vehicles" require varying levels of skills, especially balance, equilibrium, and visual-perceptual-motor skills. The design of the vehicle can influence the difficulty level. For example, more wheels means greater stability (tricycle, bicycle, and unicycle).

Lower level vehicles* with wider spaced wheels are more stable (and the ground is closer in a fall). Thicker tires are more stable when riding in sand. Other design features that influence difficulty level are wheel size, ease of turning, speed potential, and brakes.

Figure 11.40 (Ursy Potter Photography) Children tend to be less cautious, learn quickly, and rebound swiftly. They take risks before sufficiently achieving a skill, learning the rules, or developing judgment. The "car" is stable and the correct size for this toddler.

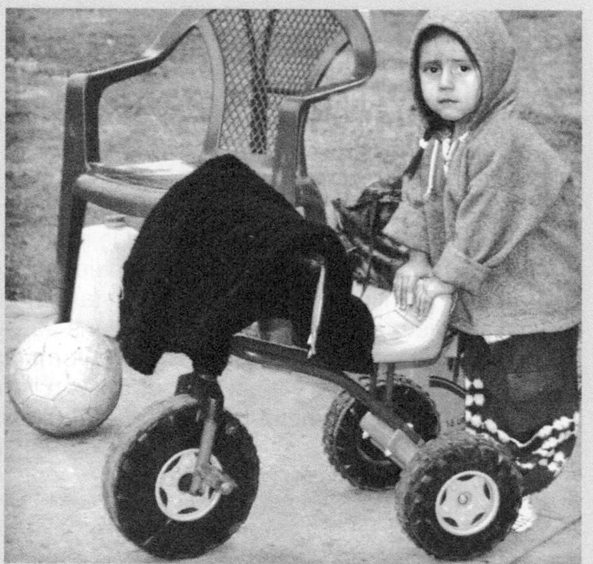

Figure 11.41 (Ursy Potter Photography) This little girl is ready for her first ride-on toy. However, the tricycle is probably a little too big for her. Younger toddlers like to walk behind their vehicles or even behind strollers and push them.

* Lower to the ground.

Table 11.3 Skill Development in Children and Riding Products that Move[a]

Age	Physical and Cognitive Skills in Children
12–18 Months	Toddlers who are proficient at walking should also be ready for their first ride-on toy. This might be a rocking horse or bouncing-type ride (on protected springs). This initial toy will enable them to develop balance as they put their weight on their bottoms, rather than on their feet. The next ride-on toy is likely to be a three- or four-wheeled vehicle without pedals. Children straddle these vehicles and their feet should reach and be flat on the floor, while seated. Initially, they may push a self-propelled vehicle with both feet and many children start by pushing themselves backwards! Casters allow the vehicle to move in any direction. Most of these vehicles will have a device (probably a wheel) for steering, so the child can begin to coordinate feet and hand movements.
19–23 Months	Maneuvering vehicles that children sit inside and push with their feet is more difficult than the ones they straddle. An older toddler may be able to use such a vehicle, as most are capable of walking and climbing easily. As children at this age are capable of symbolic thinking, they also enjoy pretend play. Realistic toys are fun for them, including buttons to push such as horns or lights, and places to put other toys into and take them out of such as pretend gas tanks.
2–3 Years	By age two, most children can walk, run, stoop and rise, maneuver around objects, and walk backwards. In other words, their balance is better. They may be ready for three- or four-wheeled vehicles, such as tricycles, tractors, or fire engines with pedals. They will not immediately be good at using pedals, but proficiency will develop over this age range. They are still very interested in imitating adult behavior with realistic toys. They understand and know to avoid common dangers, such as busy streets. They can follow simple rules or instructions, but forget them relatively easily. By age three, most children will be able to pedal and steer fairly well. The vehicle selected for them should fit the child—their feet should easily reach the floor while sitting and their hands reach the handlebars with a relaxed reach (without full arm extension). They can also steer a slow-moving battery-operated vehicle, however they are likely to need assistance with "changing gears" (many go forward and backward), driving backwards, and steering on surfaces that provide resistance to the wide wheels.
4–5 Years	Four- to five-year-olds have fairly good coordination. Most can skip and jump rope. They are good at pedaling and steering tricycles. Many are ready to ride a two-wheeled bicycle with training wheels and some are ready to ride without training wheels. It is important to work with a child at his own speed and not push him to accomplish a task, if he is not ready. Parents and caregivers can judge, with the child's input, his readiness to attempt riding without training wheels. If uncertain, caregivers can consult with the child's physical education teacher. They should be able to ride appropriate sized, three-wheeled scooters, but they will not be able to maintain their standing balance for long. Children of this age understand and obey simple rules. However, they are easily distracted and forget instructions easily. They still cannot perform complex motor tasks without thinking about their movements. They will also not understand the risks involved in bicycling. For these reasons, even if they are physically able to balance, steer, and maneuver their bicycles they should not do so without close supervision. They will not be capable of making accurate judgments about distance and speed either of themselves or of other vehicles (cars, other bikes, or scooters).
6–7 Years	Children between six and seven like to test their skills and physical strength, as they seek greater independence. They should be capable of riding a bicycle without training wheels. Although they may have the coordination to use hand brakes, they may not have sufficient hand strength. Have them try the bicycle (and the hand brakes) before purchasing. They should be capable of "driving" slower four-wheeled motorized vehicles. Some may be capable of riding non-motorized scooters, with supervision. Seven-year-olds can integrate their cognitive and physical skills during activities and can follow rules. Their problem solving ability is improved, but they still are not able to think through to all the possible consequences of their actions. Nor are they able to judge distances or speeds (both their own and those of other vehicles).

Table 11.3 Skill Development in Children and Riding Products that Move[a] (Continued)	
Age	**Physical and Cognitive Skills in Children**
8–9 Years	Eight- to nine-year-olds have good balance and fine motor skills. They are more daring and willing to engage in more risky behaviors such as stunts, tricks, and activities they previously thought were dangerous. They are better able to handle nonmotorized scooters by age eight, which requires simultaneous balancing, steering and use of a rear-located foot brake. They may be more interested in and skilled with beginning skateboarding. They can pay attention longer, but are still not capable of abstract thinking and being aware of all future consequences of their actions. They are very concerned about the opinions of their friends and may adopt behaviors they think will lead to acceptance. This could include not wearing protective gear such as helmets, engaging in risky behavior, and purchasing brand name gear that is popular with their friends.
10–11 Years	Ten- to eleven-year-olds have better gross and fine motor coordination, are stronger and have faster reaction times. This means they will respond quicker to situations, such as obstacles or cars. They should be able to use hand gears and brakes. They should be able to operate a slow-moving wheeled vehicle (10 mph). They may be more interested in skateboarding and skateboard parks and their balance and coordination is such that they should be more skilled in this regard. They are still testing their strengths and abilities and are willing to engage in behaviors that are more adventurous. They can think through future actions, but still will not consider all of the possible consequences of those actions.
12–14 Years	Twelve- to 14-year-olds have good balance and coordination and more endurance. This means they can ride for longer periods of time. Reaction time is quickest between the ages of 13 and 14. They are able to judge distances better, as well as speed. This means they should be able to anticipate where they (or another vehicle) will be, if they continue at a given speed. Riding a fast-moving motorized vehicle (scooter or cycle) will still be difficult for a 12-year-old because of the need to balance and steer while moving quickly on a heavier vehicle. Their ability to think abstractly and visualize future actions and potential consequences of their actions develops between the ages of 11 and 14. Formal thinking and logical problem solving also occurs between 12 and 14. They enjoy organized activities and may be interested in bicycling or skating clubs or competitions. They remain bold in their behavior and still highly value their friends' opinions. Children should not operate ATVs until age 14.

Source: From CPSC. *Age Determination Guidelines: Relating Children's Ages to Toy Characteristics and Play Behavior.* Smith, T.P. (Ed.), 2002. www.cpsc.gov

[a] See also Chapter 2 of this text on developmental stages in children.

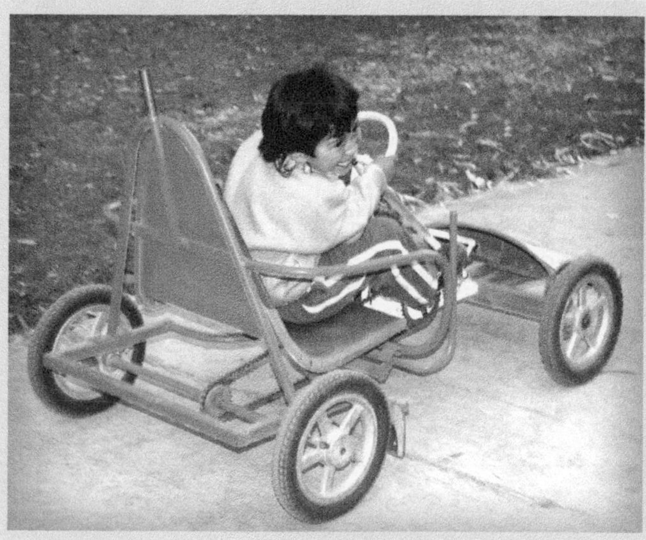

Figure 11.42 (Ursy Potter Photography)
This older toddler is more advanced physically and well able to pedal and steer his car simultaneously. Since toddlers enjoy imitating adults and realistic activities, he may especially enjoy this vehicle that looks like a car, rather than a tricycle.

SCOOTERS

Table 11.4 Scooter Injuries, Design Considerations, and Safety[a]

	Problem	Injury Types and Causes

Injuries	**Nonpowered**
	• Approximately 84,000 US emergency room cases over 9 months in 2001, 85% of children <15 years old.
	• Injuries most often among 5–12 year-olds.
	• 2/3 are boys, except in the United Kingdom the gender distribution is equal (Adebove & Armstrong, 2002).
	• Younger children suffer the most severe injuries (Abbot, et. al, 2001).
	• Few children wear protective gear, even if they own it—less than with skates or skateboards.
	• Helmets are worn incorrectly more often among scooter riders than skating, skateboarding, or bicycle riding (Forjuoh et al., 2002).
	• Wear of protective gear tends to decrease with age (2nd–6th grade measured only; Anderson-Suddarth & Chande, 2005).
	Powered
	• Approximately 4,000 US emergency room cases over 9 months in 2001, 39% of children <15 years old.
		Types
		• Fractures or dislocations of the arm or hand
		• Head, face, and extremity injuries (upper and lower)
		  ○ Abrasions, lacerations, contusions
		  ○ Sprains, strains
		• Deaths also result from motor vehicle collisions, falls, and head injuries
		Similar injury patterns in Finland and Austria (Schalamon, 2003), Australia (Chapman, et al., 2001), Puerto Rico (Melendez, et al., 2004), and Switzerland (Mankovsky, et al., 2002)
		Causes
		• Falls
		  ○ Striking stationary hazard
		  ○ Losing balance
		  ○ Wheels of the scooter getting caught by uneven ground
		  ○ Difficulty braking
		• Collisions with a car or truck cause the most serious accidents

Design Recommendations	• Lightweight aluminum scooters typically weigh 4.55 kg (10.03 lb) or less. They should be tested and approved for the rider's weight. • Younger children may need a wider base. • Most scooters have small wheels (10 cm or 4 in.). Braking is not always quick, as the brakes must grip the surface of the small wheel. • Low clearance may mean the scooter will catch on large cracks in the sidewalk or on sticks or rocks. • Avoid sharp edges and check for loose parts. • Check strength of folding and locking mechanism, so it does not give way during use.
Safety Recommendations	• Wear a helmet, kneepads, elbow pads, and sturdy shoes. • Wrist guards (with brace) may interfere with handgrip and steering (Cassell, et al., 2005). • Make certain folding and locking mechanism is fully locked before riding. • Provide close supervision of riders of nonmotorized scooters under the age of 8. • Children should not use motorized scooters until approximately age 12. • Teach children to avoid gravel, uneven pavement, and traffic. • Never ride at night. • Check local laws regarding scooter use, especially for motorized scooters. In some US states, they are considered motor vehicles and are not allowed on sidewalks or roadways, but only on private property.

[a] Based on Chapman et al. (2001); Fong and Hood (2004); Gaines et al. (2004); Levine et al. (2001); Montagna (2004); Parker et al. (2004); Powell and Tanz (2004); and Rutherford et al. (2000).

Figure 11.43 (Valerie Rice) **Figure 11.44** (Valerie Rice)
Kiran's Scooter. A nonmotorized scooter is powered by pushing with one foot, while balancing on the scooter with the other. Here, Kiran is balancing by resting her kicking foot far forward on the front, angled portion of the attachment bar, while simultaneously steering.

Should she need to brake, she would have to move her kicking foot to the back and press down on the metal bar above the wheel. This movement takes considerable coordination and the small wheel and wheel cover can make the stop slower than a child might expect. Although she is wearing a helmet, she is not wearing other protective gear that could help, such as knee and elbow pads and hand gloves, nor is she wearing closed, sturdy shoes. Notice the low clearance of the scooter.

SKATE BOARDS

Table 11.5 Skate Board Injuries, Design Considerations, and Safety[a]

	Problem	Injury Types and Causes
Injures	• There were about 50,000 US emergency room treatments to 5–14-year-olds in 2003. • 50% to 60% of skateboard injuries are among children <15 years old. • Most injuries are to those who have been riding less than a year. • Injuries occur most often to boys and to those on roads. Injuries are typically more serious than those of skaters (Osberg, et al., 1998). Many younger children do not have adequate balancing skills, core strength, and gross motor control for skateboarding. • Younger children have higher center of gravity. • The *American Academy of Orthopedic Surgeons* does not recommend skateboarding for children age 5 and younger. • Many children do not have the strength and skills to move beyond beginning stages of skateboarding until ages 10–14. • Wear of protective gear tends to decrease with age (2nd–6th grade measured only, Anderson-Suddarth and Chande, 2005).	**Types (in order of frequency)** • Sprains, fractures, contusions, and abrasions of the extremities • Wrist and ankle fractures • Head injuries, most serious injuries are head injuries. • Open fractures occur more frequently in skateboarding than skating or scooter riding (Zalavras et al., 2005). • Deaths also occur from collisions with motor vehicles, falls, and head injuries. **Causes** • Falls ◦ Striking debris, such as holes, bumps, rocks ◦ Irregular surfaces cause half of fall-related injuries ◦ Losing balance ◦ Stunts and trick riding • Collisions with car or truck • Falling on outstretched arm or hand can result in upper extremity fractures.

Table 11.5 Skate Board Injuries, Design Considerations, and Safety[a] (Continued)

Design Recommendations	• Skateboards have three parts, the board (deck), the mechanism to which the board is attached (truck), and the wheels. A shorter board (deck) is suggested for beginners, as it is easier to use, balance, and handle. • Select the appropriate skateboard for the task (slalom, free style, speed). • Select the appropriate skateboard for the child's weight. • Padded clothing is available, including special skateboarding gloves. Clothing must be loose enough to permit free movement and tight enough to prevent slippage. Protective padding can safeguard hips, knees, elbows, wrists, hands, and head. • Avoid sharp edges and check for loose parts—maintain the board. • Current skateboards are not designed for two people to use simultaneously—one person should ride the skateboard (no tandem riding).
Safety Recommendations	• Wear a helmet, kneepads, elbow pads, wrist guards, and closed toe, slip-resistant shoes. • Be certain the helmet fits and is adjusted properly. • Stunt riding should be done in specified areas. • Learn to skate before trying stunts (learn first: balancing, slowing, stopping, turning, falling). • Teach children how to fall. • Teach children to check for and avoid gravel, uneven pavement, traffic. • Never ride at night. • Never "hitch" a ride by grabbing onto a passing car or truck. • Skateboarders should not wear headphones while skateboarding—they need to hear what is going on around them.

[a] *Source:* CPSC (2006); AAOS (2004); Anderson-Suddarth and Chande (2005); Cody et al. (2005); and references listed in table.

Figure 11.45 (Ursy Potter Photography)

Figure 11.46 (Ursy Potter Photography)

Younger children may not have adequate strength (including core strength), balance, and reaction times for skateboarding. Requirements for protective gear are not as highly regulated as they are for bicycles.

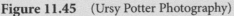

Figure 11.47 (Ursy Potter Photography)

Figure 11.48 (Ursy Potter Photography)

Figure 11.49 (Ursy Potter Photography)

Multitasking can also be a problem; children try to use their cell phones, text message, or take photographs as they ride! As children will fall, enforce the use of protective gear. Additionally, children should be taught "how to fall":

- Roll rather than absorb the force with your body (especially arms)
- Land on fleshy part of body
- Crouch if you begin to loose your balance, so the fall distance is lessened
- Relax as you fall, rather than stiffen
- Have children "practice" their falls

Figure 11.50 (Ursy Potter Photography)
Children are injured most often when they first begin riding and after they have ridden for a year or more. The more experienced riders tend to be injured during stunt riding. Children should always ride with protective gear, including helmets, kneepads, elbow pads, wrist guards, and closed toe and slip-resistant shoes.

IN-LINE SKATING

Children and adults enjoy in-line skating for recreation, sport, exercise, and transportation. Approximately 17.3 Americans skated at least once in 2004 and 3.9 million skated about once every 2 weeks (SGMA, 2006).

In-line skating provides children (and adults) with an excellent aerobic work out. It also helps develop muscles in the hip, buttock, upper leg, and lower back regions. Many adults prefer this low-impact activity, and it is an excellent form of exercise for children and adults to do together.

Injuries occur most often, not among the novices or the most experienced, but among intermediate level skaters (CPSC, 2006). The idea that injuries occur only among beginners is a myth.

Many US in-line skaters do not take lessons, even though some of the motions are not easy or "natural." This applies to stopping and to falling properly.

The greatest number of skating injuries (90%) for both children and adults involve forward falls where they attempt to break their fall with outstretched arms and hands (Knox and Comstock, 2006). This reaction is an automatic, protective reflex. In fact, several reflexes that persist throughout life may be involved (Table 11.6). The significance is that training a child to react differently (such as roll during a fall) can be very difficult, as he will have to override these reflexes.

Figure 11.51 (Ursy Potter Photography)
Many children can begin in-line skating by the age of 6, if they have the coordination, strength, and motor ability and can follow instructions.

Figure 11.52 (Ursy Potter Photography)
Although wrist and lower arm injuries are common among skateboarding, scootering, and in-line skating, they occur most often among in-line skaters. Injuries occur most often on sidewalks, driveways, streets and parks, or bike paths (adults and children).

As with other situations, designing the product to prevent an injury is best followed or reinforced by preventive training. In this instance, the design solution is to have a child wear a well-designed wrist brace. Approximately a third of serious skating injuries can be prevented by using a wrist brace (CPSC, 2006). Wrist braces allow the wrist area to strike the surface and slide along the surface rather than absorbing the energy directly.

Many in-line skaters wear no protective gear at all. Younger children wear helmets and other protective gear more often than teens, and females (any age) wear protective gear more often than males (Osberg, 2001; Warda et al., 1998). The reason most often mentioned for not wearing a helmet while in-line skating is that the helmet is not comfortable (Osberg, 2001).

Table 11.6 Reflexes a Child Must Overcome in Learning to Roll during a Fall[a]

Reflex	Evaluation (typically done with an infant or young toddler)	Response
Visual Placing	A child is held vertically (under the arms and around the chest) and advanced toward a surface.	Child lifts arm and hands, extending and placing them on the support with fingers extended and abducted.[b]
Protective Extension (forward, backward, sideways)	A child is held vertically (for forward protective extension evaluation, for others child can be sitting) and then moved quickly downward toward a flat surface.	Child lifts arms and hands, extending and abducting arms and extending fingers, as if to break a fall.
Asymmetrical Tonic Neck	The young child is often positioned on hands and knees, with elbows slightly bent. Turn the child's head slowly to one side and hold it in that position with the jaw directly above the shoulder.	Arm and leg on the jaw side will extend, while the arm and leg on the skull side will flex.

[a] Visual Placing and Protective Extension continue throughout life. Asymmetrical Tonic Neck Reflex (ATNR) should be integrated in young children, some do retain an ATNR influence.
[b] Abduction means movement away from the center of the body.

Body Sizes and Capabilities Such as Reach

Figure 25.25 (Valerie Rice) **Figure 25.26** (Valerie Rice)

- See Chapter 3 on *anthropometry* (children's body dimensions and capabilities such as reach).
- More on *child reach* in Chapters 8 (injuries) and 19 (preschool).
- More about child-environment fit in Chapters 10 (disabled village children), 15 (stairs), 16 (home computing), 21 (school furniture), and 27 (museums).

Built Environments

Figure 1.24 and 25.1
(Ursy Potter Photography)

- See Chapter 25 on *designing cities for children.*
- See also Chapters 15 (stairs), 18 (family farms), 19 (preschool), 24 (new schools), 26 (wayfinding), 27 (museums), and 28 (playgrounds).

Cars and Other Vehicles

Figure 17.6 (NHTSA)

- See Chapter 17 on *children in vehicles.*
- More on *child pedestrian risk* in Chapters 8 (injuries), 25 (cities for children), and 26 (wayfinding).

Computers

Figure 1.3 (Ursy Potter Photography)

- See Chapters 6 (physical development), 16 (home computers), 21 (school furniture), 22 (usability), and 23 (school technologies) on children and computers.
- See also Chapters 9 (assistive technologies for children), 13 (bookbags), and 24 (learning).

Development in Children

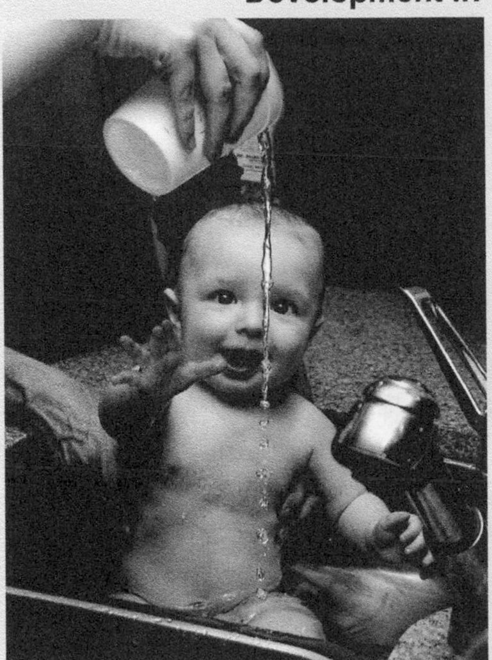

Figure 2.1 (Ursy Potter Photography)

- See Chapter 2 on *child development.*
- See also Chapters 3 (anthropometrics), 4 (vision), 5 (hearing), and 6 (physical development and risk).
- You will find more on this topic in Chapters 8 (injuries), 11 (products), 12 (toys), and 21 (handwriting).

Disabilities

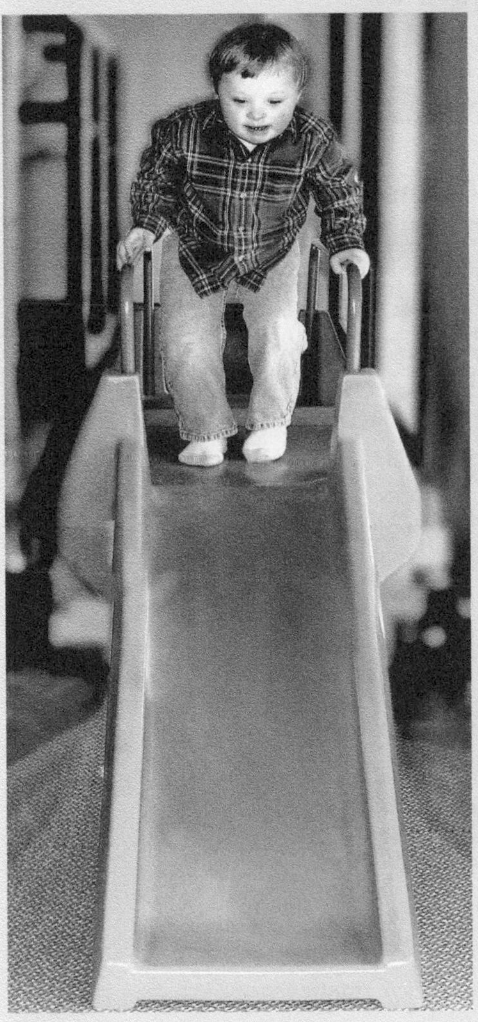

Figure 19.8 (Ursy Potter Photography)

- See Chapters 9 (assistive technologies for children), 10 (disabled village children), and 25 (cities) on *children with disabilities.*

Exercise and Physical Education

Figure 11.58
(Ursy Potter Photography)

- See Chapter 7 on *exercise and physical education.*
- See also Chapters 6 (physical development), 11 (products), 25 (neighborhoods), and 28 (playgrounds).

Farms

Figure 1.17 and 18.8 (Ursy Potter Photography)

- See Chapter 18 on *children and adolescents in farms.*
- See also Chapter 10 on disabled village children.

Handwriting

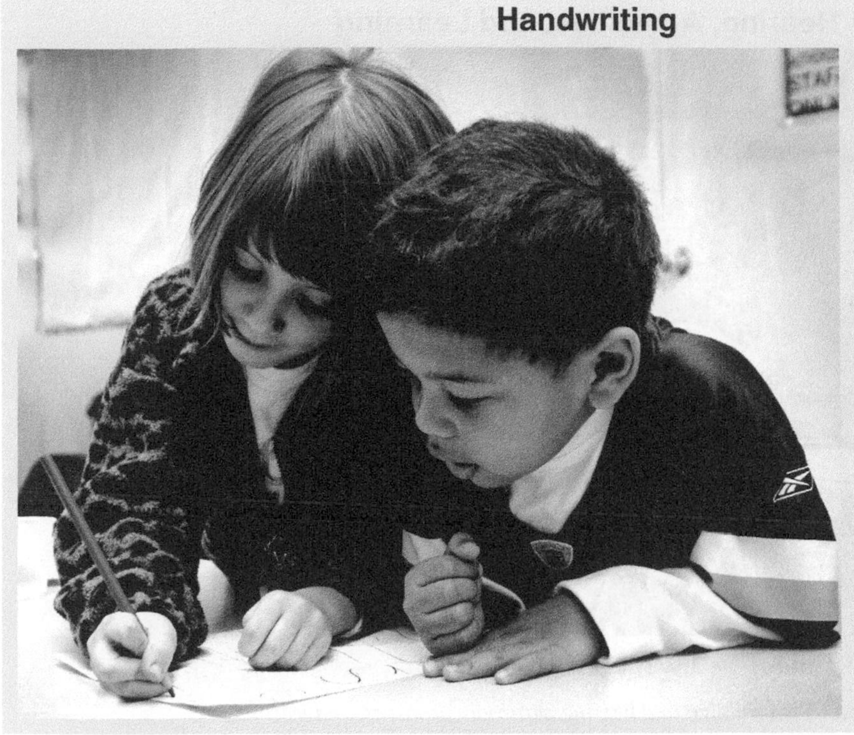

Figure 1.19 and 20.1 (Ursy Potter Photography)

- See Chapter 20 on *children and handwriting.*
- More on the *development of motor coordination* in Chapters 2 (child development), 7 (physical education), 11 (products), 22 (child-friendly interfaces), and 28 (playgrounds).

Health, Growth, and Physical Development

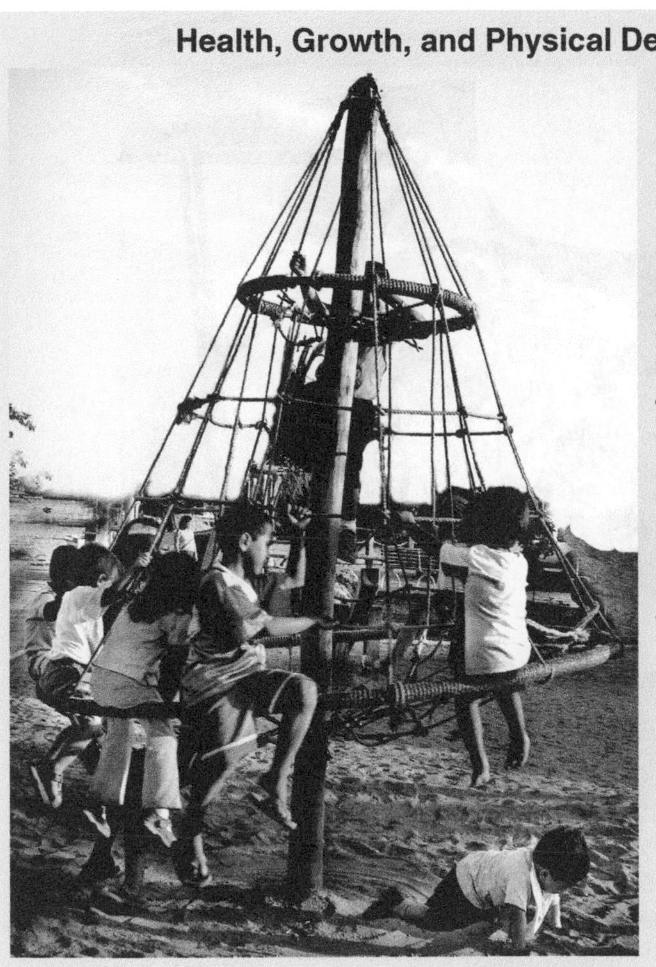

Figure 6.1 (Ursy Potter Photography)

- Read about *child development and health* in Chapters 2 (development), 3 (anthropometrics), 6 (physical development and risk), 7 (physical education), and 8 (injuries).
- See also Chapters 11 (products), 12 (toys), 18 (farms), 23 (technologies), and 25 (cities).

Hearing, Acoustics, and Learning

Figure 1.4 and 5.1 (Ursy Potter Photography)

- See Chapter 5 on *children and hearing, classroom acoustics, and learning.*

Home

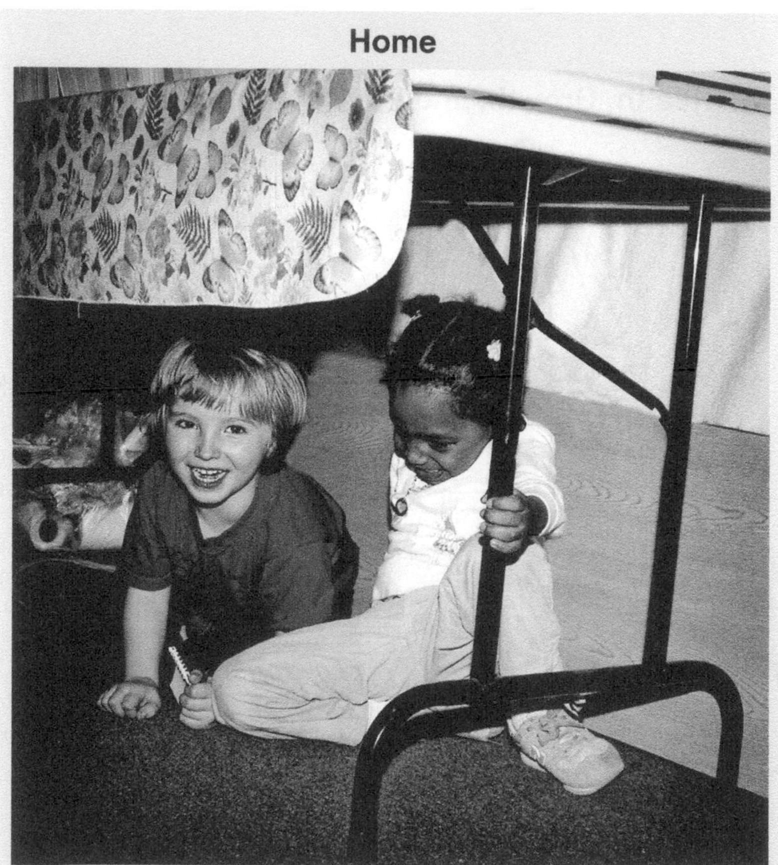

Figure 11.5 (Ursy Potter Photography)

- For more about *children at home* see Chapters 8 (injuries), 11 (products), 14 (warnings), 15 (stairs), and 16 (home computing).
- See also Chapter 25 (cities for children) on *residential environments.*

Developing Countries

Figure 25.49 (Ursy Potter Photography)

- For more about ergonomics, children and communities in developing countries see Chapters 10 (disabled village children) and 25 (cities for children).

Injuries

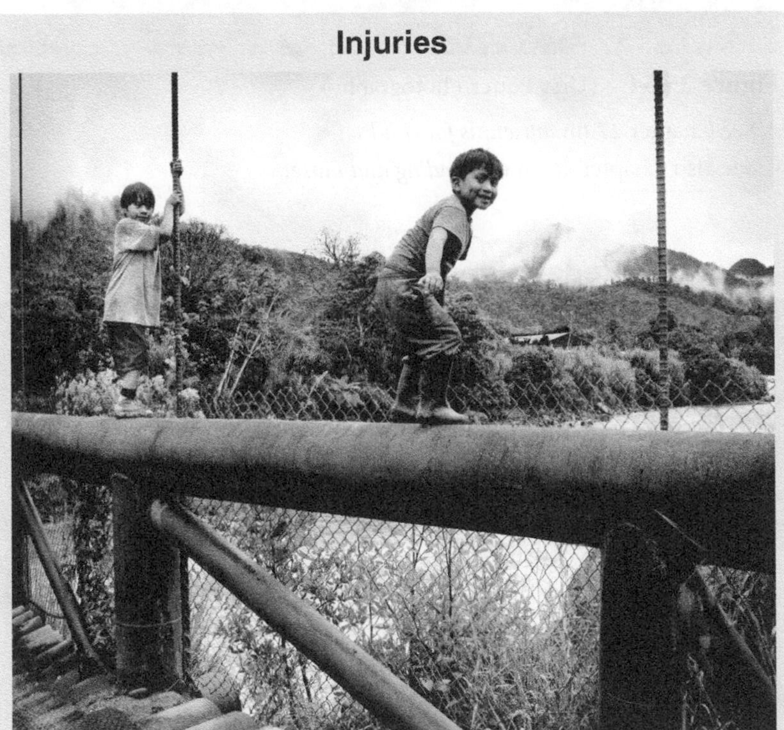

Figure 25.21 (Ursy Potter Photography)

- For information on *injuries in children*, see Chapters 6 (physical development and risk), 8 (injuries), and 11 (handgun inset in products).
- See also Chapters 2 (development), 7 (physical education), 13 (bookbags), 14 (warnings), 15 (stairs), 17 (vehicles), 18 (farms), and 25 (cities for children).

Museums for Children

Figure 25.54 (Ursy Potter Photography)

- See Chapter 27 on *museums for children*.
- See also Chapter 26 on *wayfinding and museums*.

Pedestrian Safety and Children

Figure 17.1 (Ursy Potter Photography)

- See Chapters 8 (injuries), 11 (products), and 25 (cities for children) on *child pedestrian safety*.
- See also Chapters 4 (vision), 11 (products), and 26 (wayfinding).

Play

Figure 25.3 (Ursy Potter Photography)

- Read about *children at play* in Chapters 11 (products), 12 (toys), and 28 (playgrounds).
- More on this topic is in Chapters 2 (development), 7 (physical education), 8 (injuries), 22 (usability), and 25 (cities).

Playgrounds

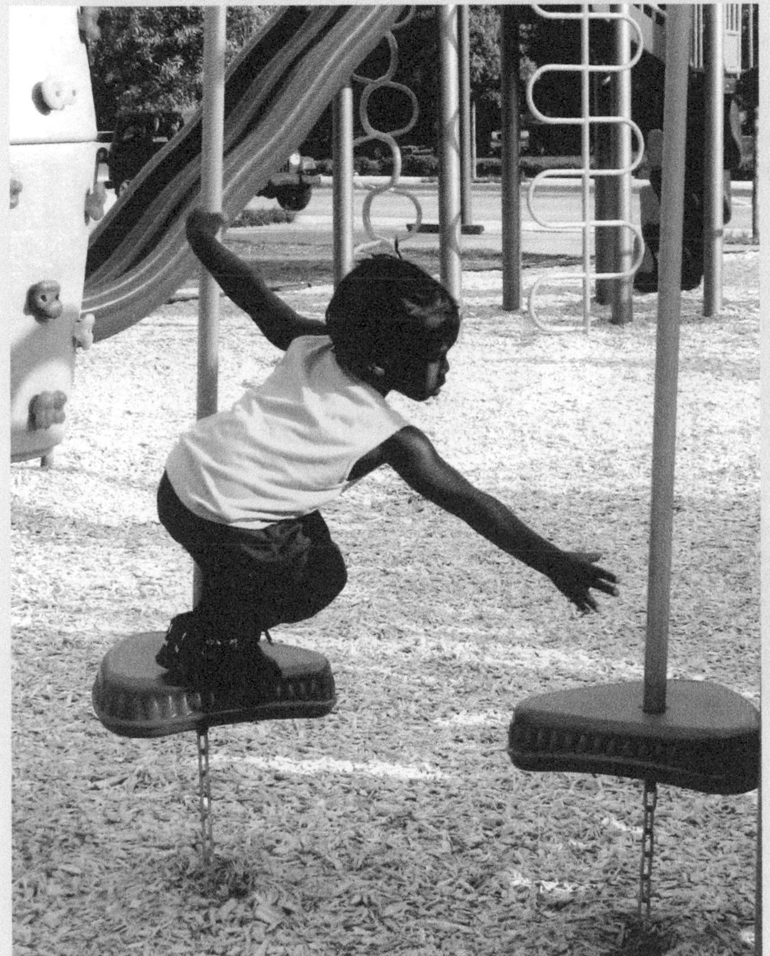

Figure 1.7 (Valerie Rice)

- See Chapter 28 on *children and playgrounds.*
- See also Chapters 2 (development), 3 (anthropometrics), 7 (physical education), 8 (injuries), 11 (products), 14 (warnings), 25 (cities), and 26 (wayfinding).

Posture

Figure 11.7 (Ursy Potter Photography)

- *Children and posture* are described in Chapters 6 (physical development), 7 (physical education), 13 (bookbags), 16 (home computing), 20 (handwriting), 21 (school furniture), and 23 (school technologies).
- See also Chapters 3 (anthropometrics), 4 (vision), and 18 (farms).

Products

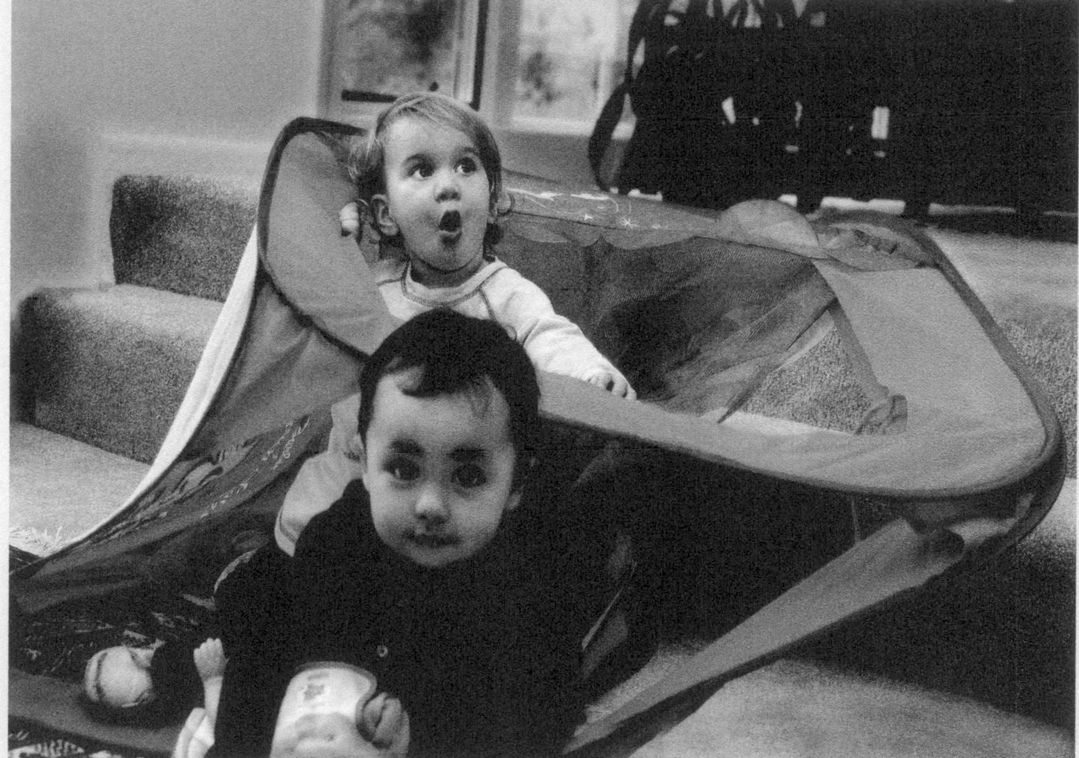

Figure 1.10 and 11.34 (Ursy Potter Photography)

- Reach about *products for children* in Chapters 11 (products), 12 (toys), 13 (bookbags), 14 (warnings), and 22 (child-friendly interfaces).
- See also Chapters 8 (injuries), 16 (home computers), and 23 (school technologies).

Schools

Figure 1.20 and 21.2 (Ursy Potter Photography)

- Read about *ergonomics for schoolchildren* in Chapters 5 (hearing), 19 (preschool), 21 (school furniture), 23 (school technologies), and 24 (new schools).
- See also Chapters 6 (physical development), 7 (physical education), 13 (bookbags), 15 (stairs), 20 (handwriting), 22 (usability), and 28 (playgrounds).

Vision

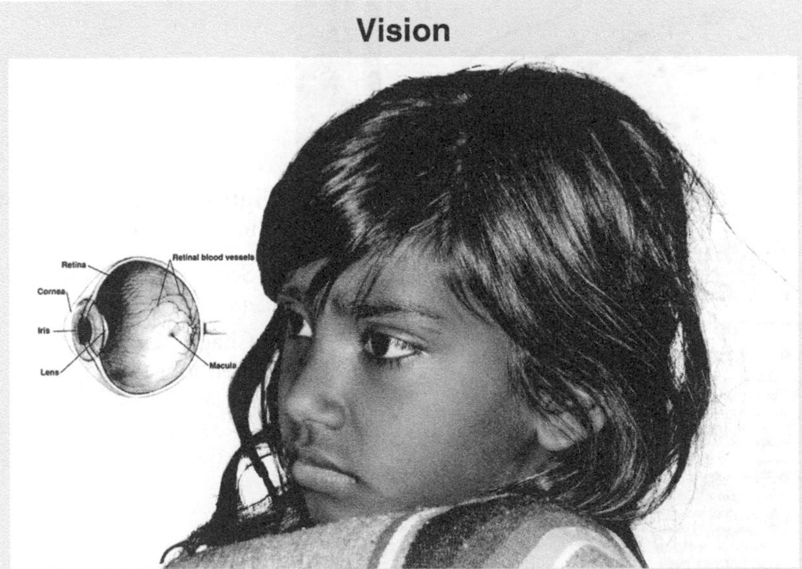

Figure 4.4 (Eye image by eyeSearch.com. Original photo of the child by Valerie Rice. Adapted composite by Ursy Potter Photography)

- See Chapter 4 on *children, vision, and ergonomics.*
- See also Chapters 2 (development), 11 (products), 20 (handwriting), and 26 (wayfinding).

Warnings

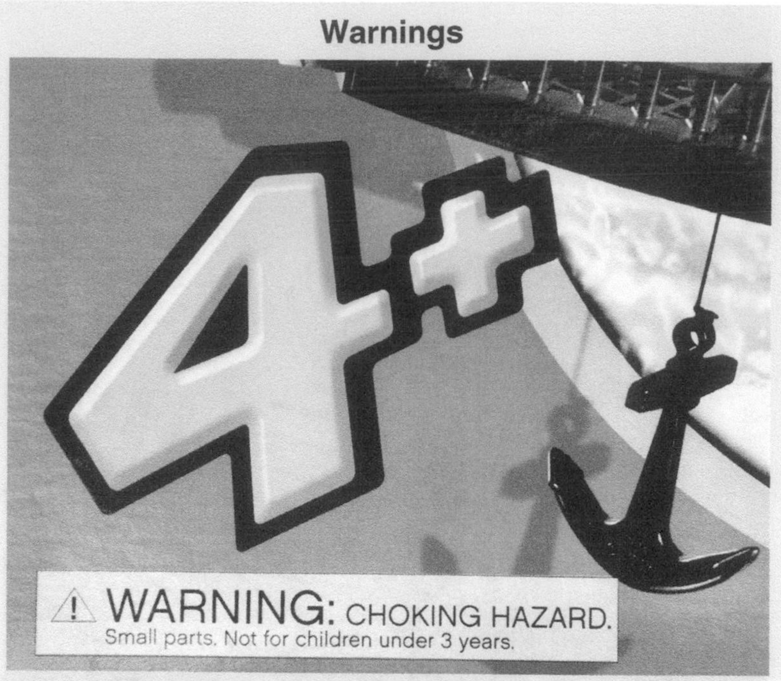

4+
WARNING: CHOKING HAZARD.
Small parts. Not for children under 3 years.

Figure 8.43 (Ursy Potter Photography)

- See Chapter 14 on *warnings for children.*
- Also Chapters 8 (injuries), 11 (products), 12 (toys), and 25 (cities).

Wayfinding

Figure 26.15 (Ursy Potter Photography)

- See Chapter 26 on *children and wayfinding.*
- Chapter 25 (cities) also mentions this topic.

Table 11.7 In-line and Roller Skating Injuries: Design and Safety Considerations

	Problem	Injury Types and Causes
Injuries[a]	**Injury rates** • In 2003, in the US, there were about 27,000 emergency room treatments to 5–14-year-olds for in-line skating accidents.[b] • Boys are four times as likely to be injured as girls (Ellis et al., 1995). • Children 11–12 years old are at greatest risk. **Observational studies (adults and children)** • Skaters are more likely to wear wrist guards than helmets, elbow, or knee pads (Beirness et al., 2001). • Beginners and advanced skaters wear more gear than others. • Children wear wrist guards more often than helmets. **Concerns** • Younger children have less ankle strength for in-line skating. • Stabilizing and supporting the ankle is an important part of skate design. Younger children can start with a broader base, such as roller-skating with traditional double-row skates while they develop balancing skills, core and lower extremity strength, and gross motor control. • Younger children have higher center of gravity.	**Types** • Fractures, primarily of the distal forearm • Sprains, fractures, contusions, abrasions • Head injuries are the most serious • Deaths also result from motor vehicle collisions, falls, and head injuries **Location of Injuries** • Wrist, forearm, elbow • Knee, ankle, foot • Face **Causes** • Falls ◦ Loss of balance ◦ Striking something, such as debris, holes, bumps, rocks ◦ Inability to stop ◦ Out-of-control speeding ◦ Tricks or stunts • Striking stationary or moving objects

(continued)

Table 11.7 In-line and Roller Skating Injuries: Design and Safety Considerations (Continued)

Design Recommendations	• The parts of a skate include the shell, cuff, liner, and frame. The boot shell surrounds the lower leg; the cuff surrounds the ankle. Attached to the frame are the wheels, brake, axle, and bearings. **A child's boots should be:** • comfortable and fit well • snug in heels and support ankles • moderately stiff for support • soft or hard shell, soft shell must offer good ankle support • well ventilated • supportive of the arch (may want removable, replaceable arch support) **Parts** • ***Liner and foot bed:*** Crucial to fit and are often removable and washable. During fitting, the foot can be placed in the liner first and then, if it is comfortable, placed back into the boot and tried again for comfort and fit. They should "breathe" or permit air circulation. • ***Tongue:*** Should be well padded and stay in place. Moderates pressure of laces, buckles or lace and buckle combination. Latter offers most support. • ***Frame:*** Generally rigid plastic or aluminum. User should check alignment and secure attachment. For quick turns and pivots, consider a frame that will allow adjusting the middle two wheels slightly lower ("rockering"). Most racing and hockey skates are metal. Some frames can be loosened from the boot and aligned in a slightly different direction. Some frames allow the skater to change wheel size slightly. Since smaller wheels are easier for beginners (see below), this allows progression of skater according to improved abilities, with the same skates. • ***Wheels:*** Smaller, thicker wheels roll slower, and are more stable and controllable, thus they may be better for beginners. Most are made of urethane. Hardness is measured from 0 to 100, soft to hard, respectively. Softer wheels have better shock absorption (better for outdoor use) but do not last as long. Some skates mix wheels (hard and soft) to promote both durability and shock absorption. • ***Brakes:*** Generally a rubber pad on the heel of one boot that drags on the ground, when the toe of the boot is lifted. The brake should be on the heel of a child's dominant foot. About 10% of the population is "left-footed." Eye, hand, and foot dominance are not always the same. To check a child's foot dominance, roll a ball to them so that it approaches them from the middle of their body, the foot they choose to kick the ball with should be their dominant foot. Brakes that look like rollers may be difficult for children, as they may not have the strength and coordination to apply sufficient pressure to stop easily. • ***Bearings:*** Wheels roll on ball bearings inside the wheel hub. Bearing quality is rated on manufacturing precision. Ratings between ABEC 1 and 5 are not likely to make a significant difference for a beginning skater.

Safety Recommendations	**Purchasing** • Select skates to match a child's style and abilities. • Types include hockey, freestyle, racing, fitness (also used for cross training), recreation, aggressive (designed for stunts), and cross training. For example, racing skates have a longer wheelbase, low-cut leather boot and typically do not have a brake! • Most new skaters begin with recreational boots. **Skating Behavior** • It is not enough to learn to skate; learn to stop and fall. • Walk skating paths first to look for hazards. Avoid gravel, uneven pavement, and traffic. Take sharp curves slowly, some are too sharp for in-line skates. Stunt riding is safest in specified areas. • Learn proper skating etiquette: pass on the left, skate on the right, voice intent of where you will skate, and point to where you intend to go, if facing others. **Equipment** • Maintain equipment, check the wear of the wheels, rotate as necessary. • Wear a helmet, kneepads, elbow pads, wrist guards, and closed toe, slip-resistant shoes. • Wear a well-fitting, well-adjusted helmet. • Avoid wearing headphones while skateboarding—skaters need to hear what is going on around them.

[a] Based on Bierness et al. (2001); CPSC (2006); ISRC (2006); Knox and Comstock (2006); Powell and Tanz (1996); Schieber et al. (1996); Schieber and Branche-Dorsey (1995); Zalavras et al. (2005).

[b] This represents about 41,000 injuries involving children and teenagers younger than 20. This rate represents 31 out of 100,000. There were about 27,600 US emergency room treatments to 5–14 year olds for roller skating accidents in 2003.

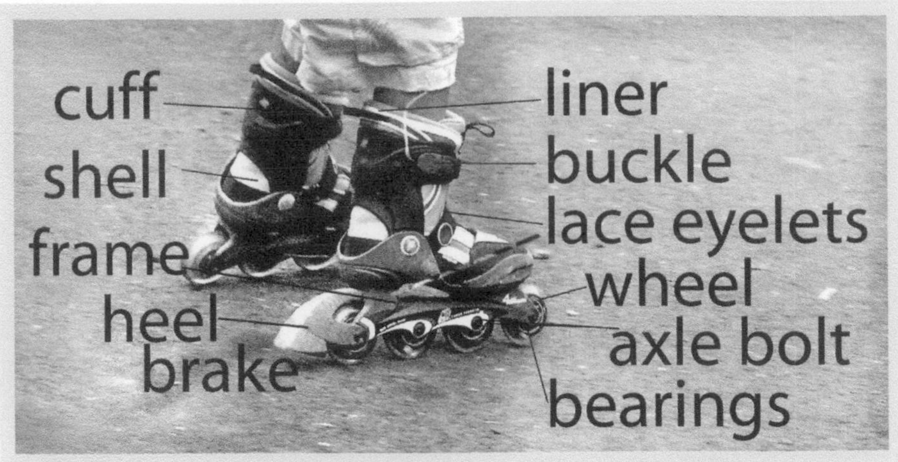

Figure 11.53 (Ursy Potter Photography)
The parts of a skate are shown here and described in Table 11.7 for in-line skates.

There is an European standard (DIN EN 13843) for in-line skates. It specifies safety requirements; testing methods; and markings, information, and warnings required by the manufacturer. It is for in-line skates used by skaters of over 20 kg (44.1 lbs) (those for skaters weighing less than 20 kg (44.1 lbs) are considered toys). This standard does not apply to traditional roller skates (EN 13899).

In-line skates are no longer "play vehicles" in some US States; they are a means of transportation and skaters can use them on the street. There may be restrictions as to locations allowed for use, such as crowded areas.

Newer designs make braking easier.

The Activated Braking Technology (ABT) allows skaters to brake without lifting the toe of their skate. Instead, they bend their knees and slide the braking foot forward. The resulting pressure against the boot cuff pushes a rod (attached to the cuff), which, in turn, pushes the brake. Other designs use springs to activate a brake or a rotating rubber disk.

Figure 11.54 (Ursy Potter Photography)
Never purchase skates a child can "grow into" as loose skates do not offer enough ankle support and may contribute to a child's developing blisters.

Alternatives include purchasing skates with liners of two different sizes (meaning the child can use the skates for two full sizes) or to purchase an extendable skate. Skates do not stretch, but some liners are made to conform to a child's foot. When in doubt, try it out—and rent several types to see what works.

Resources on In-line Skating	
http://iisa.org	In-line skating resource center. Provides information on gear, instruction, places to skate, events, and skating club information, touring vacations, research results, and US laws.
http://inlinenow.com	Information on nontraditional in-line skating.
http://sgma100.com	Sporting Goods Manufacturers Association (SGMA)

BIKES

Bicycles are a wonderful source of physical exercise for children's play. They also provide transportation to and from neighbor's homes, schools, parks, and shopping. An estimated 33 million US children ride bicycles.

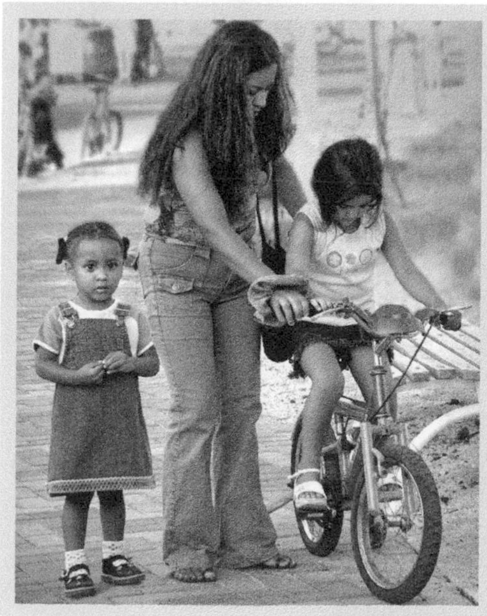

Figure 11.55 (Ursy Potter Photography)
Bicycle riding requires use of a complex set of skills. Children must simultaneously balance; visually scan for obstacles and direction; react to visual, kinesthetic, and auditory information; pedal, steer, and react quickly to unexpected events by turning, braking, or both! This can be a challenge, especially when riding among other children or when cars are present.

A child needs to be both physically and cognitively ready for such a challenge. Working up to the full challenge, a little at a time, will help children develop the skills and judgment necessary to keep them safe.

THE PROBLEM: Unfortunately, this great source of exercise, play, transportation, and sport can also result in injury or death. Over 500,000 individuals seek emergency room treatment for bicycle-related injuries in the United States annually. In 2003, about 286,000 of these injuries were to children aged 14 and younger (Safekids, 2006).

Head injuries account for nearly half of the bicycle-related injuries and 60% of the approximately 100 deaths (Safekids, 2006). Head injuries that do not result in death can leave the injured child with a permanent disability. Wearing a helmet can greatly reduce the risk of head injury (see the section below on helmets).

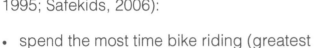

Figure 11.56 (Ursy Potter Photography) Not all countries have regulations for bicycles, bicycle helmets, or even rules of the road for transportation.

Those at greatest risk (Hu et al., 1995; Carlin et al., 1995; Safekids, 2006):

- spend the most time bike riding (greatest exposure time)
- tend to be boys more than girls
- are children aged 14 or younger (five times more likely to be injured than older riders)
- ride on sidewalks frequently
- children between the ages of 10 and 14 (incur injuries that most often result in death)

How and where children are injured:

- Riding at dawn, dusk, or night*
- Within a mile from home
- Children under age 4 are injured riding in the garage, driveway, or yard

Fatal accidents occur:

- when children engage in risky behavior such as running a stop sign, swerving into traffic, or turning left across a lane of traffic (about 80% of fatal accidents)
- during bicycle collisions with motor vehicles
- when riders are not in intersections (66%)
- between May and September and between 2 and 8 pm
- on minor roads within a mile from home

Loss of control occurs from:

- riding a bike that is too big for the rider
- braking difficulties
- riding too fast for the conditions or for the skills of the rider
- stunt riding

* Children under the age of 14 are four times more likely to be injured at night than during the day.

- riding with more than one person on a bike at one time (this changes the balancing required and the person in control of the bike often cannot adjust their balance quickly enough to prevent a fall)
- the tire wheel striking something, such as a rock, a rut, or sand
- riding on or through slippery surfaces
- hands slipping from the handle bar or foot slipping from the pedal

Mechanical difficulties occur from:

- brake failure
- slipped chain
- disengagement of the wheel or steering mechanism
- gear shifting difficulties
- loose pedals
- breaks in a wheel spokes
- getting something caught in the bike, especially in the spokes of the wheel, such as a persons clothing or even their limbs (fingers, hands, or feet)

REGULATION AND STANDARDS CPSCs regulations for all new bikes include (CPSC, 2006):

- construction (and performance) of the brakes, wheels, steering system, and frame
- reflectors on the front, back, sides, and pedals to increase visibility at night
- elimination of uncovered sharp edges and jutting parts
- require brakes on all bikes with seat height of 22 in. (55.88 cm) or more

DESIGN RECOMMENDATIONS:

Overall

- Men and boys generally have a longer torso, therefore their bikes and handlebar placement may differ slightly from girls.
- Beginning riders will probably benefit from an upright posture with a "high-rise" handlebar.

Brakes

- For small children (5–8) coaster (back-pedaling) brakes offer a mechanism for braking when their hands are relatively weak and the brakes are sturdy (unlike hand brakes).
- For small children just learning to ride, it may be beneficial to teach them to balance first on a flat surface, with a low bike without pedals (because children want to put their feet out when learning to fall). This means children will learn to balance at low speeds when pushed by a parent or on a gradual incline, adding pedals when they can balance and steer.
- Add a handbrake to the coaster brake as children's hand strength increases.

Seating

Seating is a matter of personal comfort. However, pressure should be taken on the ischial tuberosities*, but with sufficient cushioning to allow a distribution of weight to the soft tissues of buttocks. Excessive padding can cause chafing and pressure. For short rides, children may be able to use nearly any seat; however, a child that rides frequently will want a seat that fits well. Note that seat designs are different for men and women, although designers do not always include these differences on children's bikes.

- **Height:** A seat is too high when a child must rock his or her hips to pedal. The knee should be slightly bent at the bottom of the rotation.
- **Angle:** The seat should stay fairly horizontal in order to keep pressure on ischial tuberosities. Even a slight forward tilt can result in pressure on the upper extremities as weight is carried on the arms and handle bar.

Spokes: Covering the spokes with a 10 mm (0.39 in.) hexagonal mesh with a plastic shield between the fork and horizontal bar will prevent children's feet from being caught between the spokes or between the wheel and fork (D'Souza et al., 1996).

Purchase to fit the child's needs, based on the following:

- Most frequent terrain they ride on:
  - Roads and sidewalks—comfort and hybrid bikes, for fast or long rides on pavement a road bike is lighter than others.
  - Off-road paths—mountain bikes with good shock absorption.
- Time (short versus long rides)
- Fit
  - Have the child straddle the frame; there should be 1–3 in. (25.4–76.2 mm) between the crotch and the top tube for hybrid and comfort bikes and 3–6 in. (76.2–152.4 mm) for mountain bikes.
  - Handlebars should be easy and comfortable to reach.
  - The knee should be slightly bent at the lowest rotation of the pedal and the child should be able to place the ball of the foot on the ground while seated.

Table 11.8 New Retractable Handlebar

Problem	In cycling accidents that do not involve a car, up to 80% of internal injuries to abdominal and pelvic organs are associated with hitting the handlebars.
Mechanism of Injury	The child loses control of the bike and falls. During the fall, the front wheel rotates into a right angle with the child's body and the child lands on the end of the handlebar. Common injuries are to the spleen, kidney, liver, and pancreas.
Potential Solution	Reduce injury severity during impact with a handlebar that retracts upon impact and "dampens" the impact by absorbing energy. The prototype reduced forces at impact by about 50%.
More Testing is Necessary For more information: traumalink.chop.edu	

Sources: Winston, et al., 1998; Aborgast, et al., 2001.

* Ischial tuberosities are the bony protrusions within the buttocks that people typically "sit upon."

MAINTENANCE: Read the owners manual regarding maintenance. Periodically check for loose, cracked, or rusted parts.

- *Brakes:* If they make noise (squeal) or do not stop the bike quickly, have them checked and repaired by a bicycle mechanic.
- *Handlebars:* Check the alignment of the handlebars with the front tire. Check for tightness and security of the fastenings.
- *Chain:* Clean by removing dried grease and rust; oil as directed in the owners' manual.
- *Tires and Wheels:* Keep tires inflated properly. Replace tires that are worn. Check tire security and alignment, the wheel should not rub against the frame as they rotate.
- *Seat:* Check the alignment and security of the seat (does not twist or turn). Adjust seat height so the child's leg bends slightly at the knee when the pedal is at the bottom of the rotation.
- *Bike height:* Be certain the child can stand comfortably, with his or her feet flat on the ground while straddling the bike. He or she should have 1–2 in. (25.4–76.2 mm) between his or her crotch and the center bar of the bike.
- Get periodic adjustments by bike shop or knowledgeable person.

SAFETY RECOMMENDATIONS

Use protective gear:

- Helmets—always
- Gloves—especially when learning or riding for competition
- Knee pads—especially when learning

Supervise children, especially those who are younger:

- Be aware that children up to age 8 do not judge speed and distance well. They will not be able to adequately judge complex traffic situations.

Educate children on:

- how to ride
- rules of the road
- cautions, such as not being easily seen by cars
- protective gear, including reflectors, reflective clothing and lights for night use (however, it is best not to let children ride at night)
- how to carry gear
- how to adjust their helmets

Figure 11.57 (Valerie Rice)
Safe bike-riding is a pleasure and builds coordination and balance.

Additional resources

Table 11.9 Additional Resources

http://www.bikewalk.org	***Bicycle Federation of America.*** Provides information on public health, planning for bicycle friendly communities, land use and schools, and wellness and training.
http://www.bhsi.org	***Bicycle Helmet Safety Institute.*** Provides detailed information on helmet standards, fitting, and research. Also provides handouts and educational materials for teachers, parents, and children.
http://www.canadian-cycling.com	***Canadian Cycling Association.*** Provides information on bicycling, as well as handouts, texts, videos, and CDs. Has a section called "sprockids," which offers free instructor training.
http://www.cdc.gov/ncipc/bike	***Center for Disease Control,*** National Center for Injury Prevention and Control. Provides information on injuries, safety, helmets, injury prevention, and community programs.
http://www.cich.ca	***Canadian Institute of Child Health.*** Provides safety information, fact sheets, and posters.
http://www.cpsc.gov	***Consumer Product Safety Commission.*** Provides information on bicycle safety and helmet standards. Information on product recall is also available. The second site listed has a comic book entitled "Sprocketman" for children.
http://www.bikeleague.org	***League of American Bicyclists.*** Provides information and links to safety, travel, and other cycling information. Includes advocacy, community, current events, and education related to cycling.
http://www.nhtsa.dot.gov	***National Highway Traffic Safety Administration,*** Department of Transportation. Provides pedestrian–bicycle safety information by region and an interactive "safety city" for children. The children "travel" through the city, stopping to check their helmet, etc. along the way.
http://www.safekids.org	***National Safe Kids Campaign.*** Provides a variety of safety information, fact sheets, and abbreviated statistics on injuries. Although not their own statistics, they keep them up-to-date.
http://www.pedbikeinfo.org http://www.bicyclinginfo.org	***Pedestrian and Bicycle Information Center.*** Provides information on both bicycle and pedestrian issues including education, safety, tips for teaching children to ride, and community programs.
http://sickkids.ca/safekidscanada	***Safe Kids Canada.*** Provides safety tips, information on bicycles, helmets, and child carriers.

RECREATION

Recreation and sports give children a chance to try their skills and hone their abilities (Figure 11.58). They also give alternate opportunities (from school lessons) to learn necessary social and work-related skills such as following rules, self-discipline, and teamwork.

Such activities can also result in injury. Injuries can occur for many reasons, but this section will address a few products that children use and their design and safety implications.

Recreation (physical activity)

- Develops muscles and bones
- Increases gross motor coordination
- Improves eye–hand and perceptual–motor coordination
- Encourages social skills
- Can stimulate imagination and creativity
- Promotes a healthy lifestyle
- Fights obesity
- Has a positive effect on the brain's executive processing skills such as paying attention, decision making, and planning

Regular physical activity can even improve academic performance.

Figure 11.58 (Ursy Potter Photography) Younger children lack the strength and coordination to play adult sports with adult rules, regulations, and equipment. Scaled down versions of lesser heights, weights, and distances can allow them to learn the game at levels that also give them a chance at success.

PLAYGROUNDS

Many children are injured on playgrounds. In the United States, approximately 200,000 children are injured on playgrounds annually (NCIPC, 2001).

A few playground facts:

- Girls (55%) are injured more often on playgrounds than boys (45%) (Tinsworth, 2001).
- Most playground injuries occur to 5–9 year-olds and at school (Phelan 2001).
- Playgrounds in low-income areas have more maintenance-related hazards than playgrounds in high-income areas, such as trash, broken and rusted equipment, and damaged surfacing material (Suecoff, 1999).

Injuries and design features for playgrounds are covered in Chapter 28 of this book. (See also Figures 11.58 through 11.61)

Figure 11.59 (Ursy Potter Photography)
Young children love playgrounds that incorporate things they can climb into and out of, such as tunnels, doors, and slides.

Figure 11.60 (Ursy Potter Photography)
Injuries on public playgrounds occur most often on climbing equipment (Tinsworth, 2001).

Figure 11.61 (Valerie Rice)
Injuries at home occur more often on swings (Tinsworth, 2001). These two children are pretending, as they slowly fell after getting extremely dizzy!

TRAMPOLINE

Children younger than 6 are at the highest risk for injuries on trampolines (see Table 11.9). When children attempt stunts such as a flip they can incur the most serious injuries, such as a spinal cord injury from landing on their head and neck.

Although injuries are frequent on trampolines, deaths are rare (CPSC, 2006). Injuries from mini-trampolines follow similar injury patterns as full-sized, but with fewer hospital admissions and greater numbers of head lacerations among children younger than 6 (Shields et al., 2005).

Problems especially occur with multiple, simultaneous users. When two children are on a trampoline, the lighter child is five times more likely to be injured than the heavier child (CPSC, 2006).

Supervision does not solve the problem. Over half of the injuries occur while the children are being supervised by adults. The American Academy of Pediatrics does not recommend using home trampolines.

RECOMMENDATIONS: In the United States, the CPSC has worked with the industry to develop standards for trampolines.

Four new requirements were added in 1999:

- Padding must completely cover the frame, hooks, and springs.
- The box containing the trampoline must have a label stating trampolines over 20 in. (508 mm) tall are not recommended for children under 6 years of age.
- Ladders cannot be sold with trampolines to prevent access by young children.
- Warning labels must be placed on the trampoline bed to warn consumers not to allow more than one person on the trampoline at a time and to warn that somersaults can cause paralysis and death.

According to a literature review and study of Manitoba trampoline-related injuries, none of the recommendations (warning labels, padding, spotters, limiting use to a single jumper, avoiding stunts, or recessing the trampoline to ground level) has sufficiently reduced injury risk (Warda and Biggs, 2004). The authors point out the support for a ban, iterating the stand taken by the American Board of Pediatrics.

> A *spotter* supports another person during a particular physical exercise or when learning a stunt. The emphasis is on assisting the person to do more, safely, than he or she normally could alone. A spotter should know when and how to help with a motion or lift and when it is safe to begin to reduce (or remove) their input, as the individual improves his or her skills.

Table 11.10 Trampolines: Problems, Injuries, Standards, and Recommendations[a]

Problem	Types of Injuries	Location of Injuries	Causes of Injuries (CPSC, 2006)	Standards
• Injuries to the head, neck, face, legs, and arms. • In 2003, 98,000 treated in US emergency rooms were primarily children. • In 2002, 11,500 injuries were reported in the United Kingdom—a 50% increase over a 5 year period.	• Broken bones (most frequently to the upper extremity) • Sprains and strains • Bruises, scrapes, and cuts • Concussions and other head injuries	Most injuries occur: • at private homes • on full-size trampolines • when multiple children are on the trampoline at one time (75%)	• Collisions with another person • Unexpected, improper landings on the trampoline • Falling off or jumping off the trampoline • Falling onto the springs or frame	**US:** F 381, 1997. Safety Specification for Components, Assembly, Use, and Labeling of Consumer Trampolines, strengthened to include warnings about supervision, multiple jumpers, and padding. ASTM F 2225, 2003. Safety Specification for Consumer Trampoline Enclosures. **UK:** Before 2001: BS 1892-2-8:1986 Trampolines. After 2001: BS EN 13219:2001 Trampolines
Recommendations	**Design**	• Use energy absorption ground cover (see Chapter 28 on playgrounds in this text) • Use shock-absorbing pads over springs, hooks, and frame • Place trampoline into a pit to lower height • Use trampoline enclosures to prevent falls off trampolines • Place trampoline about 1.5 m (59 in.) from objects such as playgrounds, fences, poles, etc. • Safety harnesses and spotting belts, when appropriately used, offer added protection for young athletes		

	Safety	• Use spotters
		• Set the rules. Explain why they are important. Discuss them with all children who use the trampoline. Post them for children to read. Include the following:
		  • Have children remove jewelry (especially necklaces) while jumping
		  • Have children learn complex stunts (such as flips) from trained, competent individuals under supervised conditions with spotters (preferably in a gymnastics class)
		  • Do not allow multiple children on the trampoline at the same time
		  • Do not allow children to bounce off the trampoline onto the ground
		• Do not let children under age 6 on a full size trampoline
		• Remove ladders so young children cannot access trampolines
		• Always supervise children on trampolines		

[a] See also Furnival et al. (1999) and Smith (1998).

CHILD PROTECTIVE DEVICES

Wearing protective gear can and does prevent injuries. Many children are willing to engage in risky behavior. If they are wearing protective equipment or if they are more experienced, their parents are more willing to allow their risky behavior (Morrongiello and Major, 2002).

Yet, though injuries do occur frequently to beginners, the rates often follow a U-shaped curve. That is, injuries also occur among athletes (and children) who have more experience. This could be because they are more lackadaisical in their use of safety precautions, but is likely due to their trying more difficult stunts.

Although wearing protective equipment does not prevent all injuries, it certainly helps. It helps most when buyers understand what it does and how it works. Additionally, not all products are equally beneficial, nor are there regulations or standards in all cases. Finally, if a child does not wear protective equipment correctly or it does not fit, it may not protect any better than having no equipment at all.

> Just wearing safety gear does not ensure safety!
>
> Gear must fit the activity and the child, and the child must wear it correctly.

HELMETS

BICYCLE HELMETS:

Problem

The most severe injuries sustained while cycling are head injuries, which can result in permanent damage or death. A child who rides a bicycle without a helmet is three times more likely to suffer a head injury (Henderson, 1995) and 14 times more likely to be involved in an accident that is fatal (NHTSA, 2004). Depending on the type of impact, the risk of sustaining a head injury is reduced between 45% and 88% simply by wearing a properly fitted bike helmet (Kedjidjian, 1994; Henderson, 1995; NHTSA, 2004).

Other facts (NHTSA, 2004, Safekids, 2006):

- Helmet use by children would prevent between 57,000 and 100,000 head, scalp, and face injuries each year and 75% of bike-related fatalities.
- About 70% of fatal bicycle crashes involve head injuries, but only one out of five cyclists aged 5–14 usually wears a helmet.
- The best way to increase helmet use is through legislation and noticeable enforcement, coupled with education.

Figure 11.62 (Ursy Potter Photography)
Wearing a helmet is the "single most effective way to reduce head injuries and fatalities resulting from bicycle crashes" (NHTSA, 2004). Notice that both the little girl on the scooter and the little boy are wearing their helmets too far back on their heads.

Most brain injuries to cyclists are diffuse injuries, with bleeding under the skin (subdural hematoma) with axonal injury being associated with fatalities (Henderson, 1995). Although the probability of these injuries is greater should the head hit an object (as the change in momentum is greater), they can occur without direct impact due to the rapid acceleration or deceleration of the head.

Children's heads are considered flexible and deform upon impact (Corner, et al., 1987 cited by Henderson, 1995). This is especially true for younger children whose bones are still developing. It is debatable whether this deformation can be a help or a problem, depending on the impact mechanics. Some of the thicker helmet designs (compared to adult helmets) were created to address this issue (Halstead, 2006).

What is known is that brain damage may not be easily observed upon initial evaluation, but may be seen later. This is why a child with a seemingly mild head impact may be kept at the hospital for observation (Henderson, 1995).

Some believe that it is not solely the helmet that protects a child, but that the child (or adult) who wears a helmet is more cautious and therefore suffers fewer accidents and injuries. This is not true, as the types of accidents and injuries experienced by those who wear helmets are similar to those who do not wear helmets, with the exception of head injuries (Maimaris et al., 1994).

- Children should not wear helmets on a playground as the straps can catch on equipment and are a strangulation hazard.
- Bicycle helmets should not be used for activities requiring specialized helmets (such as football or skiing).

One study examined the location of bicycle-related head injuries by examining the damaged helmets. Injuries were primarily located in the frontal area (40%), the rear of the head (25%), and the sides (Fisher and Stern, 1994). The impacts were against hard flat surfaces. The data were not examined to see if the locations were different for adults versus children. However, children are more likely to sit upright, travel more slowly (the average speed in the above study was estimated at 33 km/h), and be less skillful in balancing. This may mean their injuries are at different sites, such as side impacts.

Helmets should be worn during any activity during which a fall could occur. For example, while using scooters, inline or other roller skates, skateboards, and riding horses. They are even suggested while snowboarding and skiing.*

STANDARDS: Standards for helmets exist in a number of countries (for example, Australia, Canada, Finland, Iceland, New Zealand, and the United States). The Australian standard requires a test for "localized loading." This test ensures helmet integrity during impact over a small, localized area. This permits use of a lighter outer shell, but also restricts use of helmets with hard foam padding and large ventilation openings (Henderson, 1995).

The Consumer Product Safety Commission published 16 CFR Part 1203, Safety Standard for Bicycle Helmets that applies to helmets manufactured after March 1999. It includes specific performance tests for impact protection; greater head coverage area for children aged 1–5; performance tests for chin strap strength, helmet, and strap stability;† peripheral vision of 105°—right and left of center; and a label indicating the helmet meets CPSC's standard (CPSC, 2006). The CPSC standards cover adult, youth, and infant helmets (age one and younger).

* Recent evidence shows helmets may reduce head injuries during skiing and snowboarding by 29%–60%. Additional research is needed to determine the relationship between helmets and neck injuries (Hagel et al., 2005; Sulheim et al., 2006).

† Likelihood to remain on the head during a collision.

During testing, impacts thought to be similar to those the helmet wearer might encounter are put on the helmet. The helmet is on a headform, modeled on a human head. Energy levels, environmental factors, and impact characteristics are considered and linear acceleration is monitored.

Figure 11.63 (Ursy Potter Photography)
Helmet standards and laws do not exist in all countries. All except 13 states within the United States have laws requiring the use of helmets for children. Most apply to those under age 16 (BSHI, 2006).

Most use a vertical drop test, during which the headform and helmet are raised and released. The impact is monitored. Although there has been criticism that the headform does not recreate the deformable characteristics of a human head, especially a child's, no sufficient replacement has been identified (Henderson, 1995).

Consumer reports tested adult and toddler buckle-strap systems using a test force that was a little less than the federal standard (Consumer Reports, 2004). Although the helmets passed the federal safety standard for impact and buckle-strap strength, the buckles broke in 4 of 12 samples of toddler buckle-strap systems.

The U.S. Department of Transportation standard is a minimum requirement. The Snell Standard is more stringent.

> Some US States require all children under age 18 to wear a helmet while skateboarding, skating, or riding scooters or bicycles.
> Those violating the law are subject to monetary fines.

LEGISLATION: As stated above, the best way to increase helmet use and decrease head injuries, is through legislation, noticeable enforcement, community education, and discount purchase programs.* In a study in Maryland, with no formal program, helmet use increased from 7% to 11%, with education alone it increased from 8% to 11%, but when coupled with legislation, the increase went from 11% to 37% (Dannenberg et al., 1993). Henderson (1995) postulates legislation is necessary to increase helmet wearing up to 80%.

UNDERSTANDING HELMET DESIGN: The goal of a helmet is to reduce injury. It does this by absorbing energy and distributing impact loading. According to the laws of physics, energy is not destroyed; rather, it is transferred or absorbed. The bending of helmet padding is a demonstrable absorbing of energy in order to prevent the energy from reaching and harming the child's head.

Additionally, the force in a deceleration is inversely proportional to the time it takes for the deceleration to occur. This means that if the velocity reduction occurs over a longer

* Similar results have been published in a number of studies (Bergman et al., 1990; Cameron et al., 1994; Graitcer and Kellermann, 1994; Kedjidjian, 1994; Rivara et al., 1994; Wood and Milne, 1988).

time period, the deceleration also changes and the force is reduced. If wearing a helmet can increase the time period by a factor of 2, the force is reduced by 50%. The goal is to reduce the force to the skull and human brain. Forces greater than 400 G (400–700 G) can result in concussion and unconsciousness, while 700 G can result in permanent brain damage (Straub, 2006).*

Most helmet padding is composed of plastic and elastic materials. Plastic materials will not recover from a change of shape when they absorb energy and must be replaced after a single impact. Elastic materials will crush, thus bringing a child's head to a safer (slower) stop. Although the elastic material will recover from a change in shape and move the child's head in the opposite direction (bounce), it should not recover too quickly or the energy will simply transfer to the child's head. The material also must have sufficient thickness and stiffness to not collapse to its full capacity too quickly, as this will also result in a transfer of energy to the child's head. Most liners are made of expanded plastic foam.

Thicker pads that cover a larger area absorb more energy. Maximizing the contact area between padding and the head will, in turn, maximize the absorption of energy and reduce the energy transferred to the head. Therefore, caregivers placing "stick on" pads in helmets need to maximize their use and aim for consistent contact between the padding and the child's head.

The outer shell of a helmet is typically constructed of a harder, less elastic material. This helps to spread the impact force over a wider area, so the force of a blow to the head is not concentrated in a single small area. Other methods to help spread the force of impact are increasing the stiffness or thickness of the padding.

The outer shell can also help protect a child's head from penetrating objects, protect the padding during everyday use, and offer less resistance should the child slide on a surface. A softer material may "catch" during a slide, thus creating a potential rotational (twisting) injury to the neck (Andersson, 1993; Hodgson, 1990a, b).

Most shells are rigid, hard, and smooth. They are often made of fiberglass and resin composites or thermoplastics. The task (football, cycling, snow skiing, etc.), environment (hot, cold), and suspected impact forces and mechanisms help determine the most appropriate materials. Thermoplastics are typically used for bicycle helmet outer shells or microshells.

A helmet can only protect if it covers the head. A child must wear a helmet that fits and stays in the proper position on the head during an accident. This is a particular difficulty for children, who often cannot figure out how or where to stick their pads or how thick to make them. They also have difficulty adjusting their straps and may not have the patience or reasoning ability to follow the instructions. When strap and fastener systems come apart or break, a child may use the helmet without proper adjustment.

Caregivers and parents must teach the child how to adjust the straps and how the helmet should fit. It may help to have a mirror handy, so the child can check his or her fit before venturing out to play.

> Liners (padding) compress to absorb energy during an impact.
>
> After a serious crash, a bicycle helmet must not be used again even if it looks fine!

* During testing, headform acceleration cannot exceed a peak of a predetermined amount. For example, the Australian standard prohibits exceeding a peak of 400 g, 200 g for a cumulative duration of three milliseconds and 150 g for a cumulative duration of six milliseconds (Henderson, 1995).

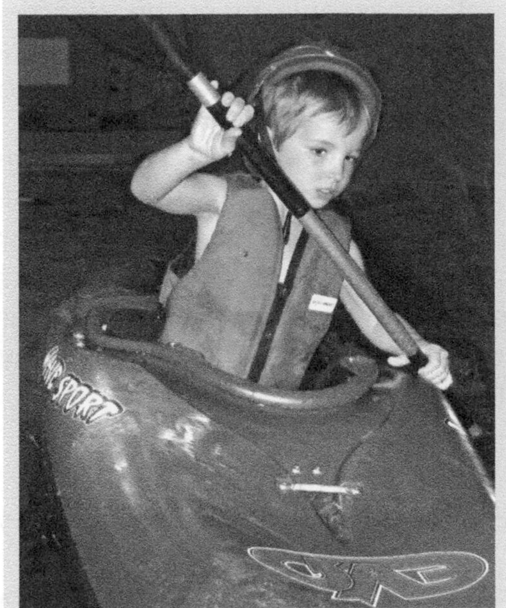

Figure 11.64 (Ursy Potter Photography)
Even the best helmet cannot protect a child if it is not worn properly. Children often wear helmets too loosely, positioned on the back of the head (as shown), or without attaching the chinstraps. Many children are satisfied simply to have a helmet on their heads, to satisfy their parents' request! Strap systems need to be easy to adjust and hold the helmet snug and in place.

Whitewater sports have their own helmets, although there are no standards. The impacts are more frequent than bicycling, they must drain water and the rock hazard requires greater coverage.

Retention failure occurs when a helmet fails to stay on a child's head during an impact. This can occur due to poor fit, inadequate adjustment or fastening, or poor design. Poor design can include a weak strap or snap.

Obviously, there are tradeoffs in weight (outer shells are heavier), comfort, and protection. Although it is impossible for a helmet to protect a child's head in all circumstances, new materials bring new potential. Research regarding the materials, mechanisms of injury, and prevention continue.

PURCHASING: Helmets for children are designed for infants, toddlers, and youth. Most are similar to adult helmets but fit a smaller head using a circumference measurement. Helmets for very young children tend to have added thickness, especially in the rear of the helmet. Some come with buckle guards to prevent pinching the child's skin when fastening.

> Never buy a helmet your child can "grow into."
>
> Always buy a helmet that fits!

- Check the label to make certain it is certified by a standards and testing organization, such as ANSI, Snell, or ASTM International. (In the United States, they must also have a CPSC sticker.)
- Let your child pick the helmet, if he or she likes it, he or she is more likely to wear it.
- Check for fit, comfort, and ease of use (see below).
- If the rider rides a lot, look for good ventilation. Many children's models do not have good ventilation and a small-sized adult helmet might fit.

The foam padding in a helmet will lose some absorption properties over time. The manufacturers' recommendation is that replacement is necessary every 3 years, even if it has never been in a fall or crash.

OTHER HELMETS: Skate-style helmets do not necessarily meet the standards for a skateboard helmet. The CPSC standard for bicycles and ASTM F1492 standard for skateboard helmets are different (see Standards below). Skateboard helmet recommendations also differ from inline skating helmets.

Skating helmets (for roller skates and inline skates) are sometimes hotter as they often have fewer vents. Their liners (padding) also tend to be thin. Most are bicycle helmets are in the skate shape. The bottom line is that the inside labels must be checked and the CPSC is the minimum standard for bicycle helmets in the United States. Only those with an ASTM F-1492 or the Snell N-94 sticker have been tested for multiple impacts, which is important during stunt skating or skateboarding with frequent falls. If a helmet does not have one of these stickers, it should be replaced after a single impact.

STANDARDS: In the United States, each bicycle helmet must pass the CPSC standard and have a label inside the helmet stating that it meets the standard. There are also voluntary standards, i.e., ASTM, Snell, or ANSI.

There are no US helmet laws for skateboarding or roller-skating. This means a manufacturer can sell any helmet for skating and skateboarding—it does not have to meet any standard.

There is, however, a voluntary standard for skateboarding helmets, ASTM 1492 and Snell N-94. Helmets that meet these standards will also have a sticker inside stating the helmet met the standard. A helmet that has both an ASTM 1492 or Snell N-94 sticker and a CPSC sticker is dual certified for bicycling and skateboarding.

The difference in the two standards (CPSC and ASTM 1492) is a difference in the testing. The drop height for skateboard helmets is 1 m (39.37 in.), while for bicycle helmets it is 2 m (78.74 in.). In addition, skateboard helmets are tested for multiple impacts, that is, more than one impact in the same area of the helmet.

Some helmets do not meet either the CPSC standard or the ASTM 1492 standard, but will have a sticker indicating they have met a European whitewater helmet standard. The difficulty is, when only a number is listed, it is difficult for the consumer to understand what type of standard the helmet has met and make an informed, knowledgeable decision about its purchase and use for their child (BHSI, 2005).

Although there are no required standards in the US for skiing or snowboarding, there is a CEN 1077 European standard. There are also voluntary standards that are more stringent than the European standard: ASTM F2040 and Snell Memorial Foundation standards RS-98 for recreational skiing and S-98 for snow sports (BHSI, 2006).

Check the stickers inside the helmet! Sometimes even purchasing a different size will mean the helmet no longer meets one of the standards.

GETTING CHILDREN TO WEAR A HELMET: One of five children, 5–14 years old, wears a helmet while bicycling. Just because children believe that wearing a helmet is protective and could save their life, does not mean they will wear it! The choice to wear a helmet, for children between ages 9 and 14, is associated with their concerns about its social consequences and whether their friends and family members wear helmets (Gielen et al., 1994*).

A few of many reasons why a child might not want to wear a helmet:

- It itches
- It feels bad
- It is ugly
- It does not fit

* Children are least likely to wear a helmet (in areas that do not require a helmet) while riding a scooter, compared with riding a bicycle, in-line skating, and skateboarding (Forjuoh et al., 2002).

Who is more likely to wear a helmet (Cody et al., 2004)?

- Girls (45% versus 33% for boys)
- Younger children (51% for 5–9 year olds versus 35% for 10–14 year olds)
- Children riding bikes (42%), skateboards (40%), and skates (40%) rather than scooters (30%)
- Children riding in sections designed for wheeled recreation (42%–50%) rather than other areas, such as schools (38%) or residential streets (33%)
- Children in sites requiring helmet use (52% versus 38%)
- Children accompanied by adult supervisors (47% versus 37%)
- Children accompanied by adult supervisors who are wearing helmets (67% versus 50%)
- Children riding in states with helmet laws (45% versus 39%)

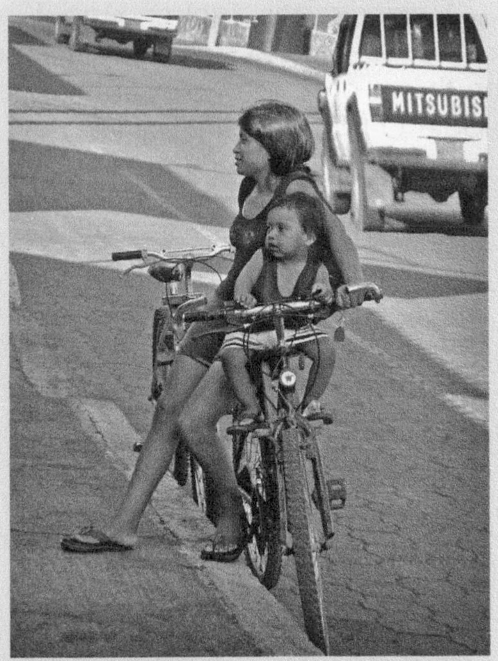

Figure 11.65 (Ursy Potter Photography) It is difficult to protect children if their caregivers do not understand the risks themselves. Community education, helmet laws, and incentives can improve the chances that protective gear will be used.

Adults need to role model the behavior, wearing a helmet for each activity that warrants it (biking, riding motorcycles, etc.). There should be no question, as it is simply part of the activity. Just as wearing a seat belt is not an option; it is part of the ride.

FIT: When a group of physicians evaluated bicycle helmets, helmet fit, and parents and children's ability to fit the helmets, the results were not good. Only 4% passed the standards for condition and fit (Parkinson and Hike, 2003).

The results revealed:

- 55% passed for being in acceptable condition
- 52% passed the stability test of moving a maximum of 1 in. (25.4 mm) front to back
- 73% passed the stability test of moving a maximum of 1 in. (25.4 mm)
- 33% passed for proper strap position
- 80% (collectively) failed the basic stability portion of the assessment

In essence 96% of the 479 participating children and adolescents wore helmets in unacceptable condition or fit or both, in spite of high acceptance of helmet use. A second, observational study showed that 35% of those wearing helmets, wore them incorrectly (tilted or straps not secured) (Cody et al., 2004).

Some helmets use a "dial fit." A dial at the back of the helmet can be twisted clockwise to tighten and counterclockwise to loosen. This may be easier for a child to fit, but should still be checked by an adult.

A chinstrap is tight enough if the children can open their mouth wide enough to eat without feeling the strap is choking or pinching them. The chinstrap should not be against the throat or at the point of the chin and the loose ends should be held in place by the rubber O-ring near the buckle.

A helmet fits when the straps are adjusted and:

- the child's forehead skin moves as the helmet is gently twisted to the left and right (if it does not, it is too loose)
- the child rolls the helmet forward with their hands and it does not come off the head or block their vision
- the child rolls the helmet backward with their hands and it does not expose their forehead

It is not easy or quick to get a proper fit. Do not try to fit the helmet quickly; it can take as much as a half an hour to fit it properly!

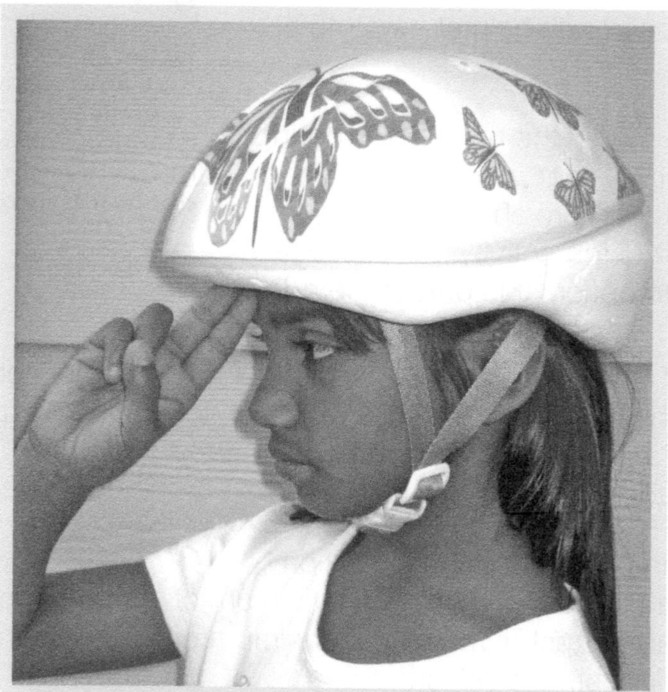

Figure 11.66 (Valerie Rice)
Safe Kids recommends the "Eyes, Ears, and Mouth" test:

The rim of the helmet should be one to two finger-widths above the eyebrows, the straps should form a "V" just below the ear lobe, the buckle should be flat against the skin and the strap should feel snug when the rider's mouth is open.

Notice that the "V" should be slightly higher and that the forward straps tighten slightly when Kiran opens her mouth.

Table 11.11 Five-Step Helmet Fit Test (BHSI[a])

Step	Problem	Solution
1. With one hand, gently lift the front of the helmet up and back.	Helmet moves back to uncover the forehead.	Tighten front strap to junction. Also, adjust padding thickness, position, or both, especially in back. Make sure chinstrap is snug. If this does not work, the helmet may be too big.
2. With one hand, gently lift the back of the helmet up and forward.	Helmet moves forward to cover the eyes.	Tighten back strap. Make sure chinstrap is snug. Also, adjust padding thickness, position, or both, especially in front.
3. Put a hand on each side of the helmet and rock from side to side. Shake the head "no" as hard as possible.	Helmet slips from side to side.	Check padding on sides and make sure straps are evenly adjusted.
4. Open the mouth (lower jaw) as wide as possible, without moving the head. The top of the helmet should pull down.	Helmet does not pull down when opening your mouth.	Tighten chinstrap. Make sure junction is under each ear.
5. Check to see if the front edge of helmet covers the forehead. The front edge of the helmet should not be more than 1–2 finger widths from the eyebrows.	Helmet does not cover the forehead.	Position helmet no more than 1–2 finger widths above eyebrows. Tighten any loose straps. Make adjustments so the helmet stays over the forehead.
Have someone else test the helmet fit by doing the 5-Step Test outlined above. Be certain the head is held still during the test. The helmet should pass each of the 5 steps.		

[a] Available at http://helmets.org

For further information	
http://helmets.org	Bicycle Helmet Safety Institute (BHSI): Information on helmets for numerous activities, including skydiving, sledding, soccer, water skiing, etc. In addition, safety tips; research results; guidance for parents, teachers, and children.
http://kidshealth.org	Safety tips in language and with pictures that children can understand.
teamoregon.orst.edu	Suggestions for proper fit of helmet etc., but real claim to fame is their offering of bicycling classes to all levels of riders.
CPSC.gov	Terrific handouts and pamphlets for children on helmet fitting and safety.
http://lidsonkids.org	This site is dedicated to the provision of information on helmet safety, put together by the National Ski Areas Association (NSAA), with others in the ski industry.
http://waba.org	Helmet advocacy program of the Washington Area Bicyclist Association. This site provides technical and resource information on helmet safety and promotes better helmets through improved standards.

WRIST GUARDS

Wrist guards or braces are recommended for recreational activities such as inline skating, skateboarding, and snowboarding. They absorb some of the impact (attenuate the impact), rather than the wrist taking the full impact. They also provide a protective cover on which a child can "slide" (which also dissipates impact) should they fall and strike their wrist or forearm.

A review of the literature in Table 11.12 suggests wrist protectors may be more effective during low-impact conditions. However, they degrade grip strength and steering maneuvering and should not be worn while playing on playgrounds or riding bicycles. Although they appear to decrease hand, wrist, and forearm injuries, they may increase more proximal upper extremity injuries for snowboarders.

Table 11.12 Findings Regarding the Use of Wrist Guards or Braces[a]

Reference and Purpose	Process	Findings	Conclusions
Moore, et al. (1997) Purpose: compare injury patterns between wrists with and without commercial wrist guards.	20 arms from 10 cadavers Loads of 16 kg (35.27 lb) from a height of 78 cm (30.7 in.) were applied to wrists (gravity driven drop). Radiographs and dissections were analyzed for fracture patterns and ligamentous integrity.	***Fracture Types*** • Distal radius: 7 guarded, 8 unguarded • Intraarticular: 4 guarded, 7 unguarded • Carpal: 1 guarded, 7 unguarded ***Ligament Injuries*** • Carpal intrinsic ligament injuries: 3 guarded, 8 unguarded • Extrinsic: 1 guarded, 4 unguarded • Capsular tears: 1 guarded, 8 unguarded	Injury patterns differ with and without wrist guards.
Giacobetti, et al. (1997) Purpose: Determine effectiveness of commercial wrist guards in preventing wrist fractures. Interest: inline skating	40 arms from 20 cadavers Experimental = 20 with wrist guard. Control Group = 20 without guard Tested to fracture using an Instron Servohydraulic Material Testing System.	Without wrist guards fractured at 2245N, with wrist guards fractured at 2285N ($p > .05$). Fracture patterns not different.	These wrist guards are not helpful in these experimental conditions.
Greenwald, et al. (1998) Purpose: compare the dynamic impact response of commercial wrist guards.	12 arms from 6 cadavers Experimental = 6 with wrist guard. Control Group = 6 without guard Tested to fracture using a modified guillotine-type drop fixture placed over a force platform.	Identified four dynamic loading phases: initial linear loading, nonlinear phase, rapid linear loading, final nonlinear loading to failure. Three transition points between loading phases identified. Vertical force and impulse higher during transition points and at failure with wrist guard ($p < .01$). Time to transition and to failure did not differ between wrists with and without guards.	Wrist guards may have preventive effect during low-energy dynamic impacts.

Table 11.12 Findings Regarding the use of Wrist Guards or Braces[a] (Continued)

Reference and Purpose	Process	Findings	Conclusions
Staebler et al. (1999) Purpose: examine subfailure load sharing of wrist guards. Interest: Inline skating and snowboarding.	Cadaver arms Experimental 1 = commercial wrist guard 1; Experimental 2 = commercial wrist guard 2; Control = no wrist guard Measured: (1) bone strain in distal radius, distal ulna, and radial midshaft and (2) guard stiffness and energy absorption.	Wrist guards increased energy absorption and reduced dorsal and volar distal radius bone strain. One guard, with an elevated volar place (off heel of hand) reduced dorsal distal ulnar bone strain.	It appears that wrist guards both load-share and absorb energy during low-impact falls.
Grant-Ford et al. (2003) Purpose: examine the biomechanical effects of a wrist guard on the wrist and ulnocarpal joints during mechanical loading. Interest: pommel horse use by athletes.	6 male and 6 female arms Experimental 1 = Ezy probrace; Experimental 2 = Ezy probrace with palmar pad; Control = no brace 32.13 kg (70.83 lb) compressive load applied along long axis of pronated forearm with dorsiflexed wrist in contact with support surface.	Both experimental conditions reduced wrist joint dorsiflexion angles. Probrace with pad ulnocarpal joint intraarticular peak pressure.	Ezy ProBrace with palmar pad reduced pressure attenuation, and may decrease cumulative effects of pommel horse exercise training.
Cassell et al. (2005) Purpose: Examine effectiveness of wrist guards during recreation needing hand strength and dexterity (bikes, scooters, monkey bars).	Tested the potential of wrist guards with 48 children, 5–8 year-olds performed functional recreational activities while wearing wrist guards.	Wrist guards decreased grip strength, bicycle steering, and skilled monkey bar use, but did not decrease scooter steering.	Wrist guards for skating are not recommended for use during bicycle riding or on playgrounds. More research is required to determine usefulness while riding scooters.
Hagel et al. (2005) Purpose: examine the effect of wrist guards on injuries. Interest: snowboarding injuries.	Cases: 1066 upper extremity (UE), snowboarding injuries occurring in the 2001–2002 season in Canada. Controls: 970 nonupper extremity injuries during the same season. Matched by date, age, and gender. Cases were compared with controls with regard to wrist guard use.	Those who used wrist guards: • 1.6% of those with distal UE injuries • 6.3% of those with proximal UE injuries • 3.9% of controls Risk reductions due to wrist guard use (odds ratios): • 0.15 for distal UE injuries (reduced risk by 85%) • 2.35 for proximal UE injuries	Wrist guards reduce the risk of distal UE injuries (hand, wrist, and forearm), but may increase the risk of proximal UE injuries (elbow, upper arm, and shoulder).
Kim et al. (2006) Purpose: Examine shock attenuation of wrist protectors.	Impact force from free-fall device was assessed using four falling heights and five hand conditions (bare hand, commercial wrist guard, Sorbothane glove, air cell, air bladder).	Fall height and hand conditions had significant effects. The padded conditions had a smaller peak impact compared to bare hand. The wrist guard became ineffective with fall height above 51 cm (20.1 in.). The air bladder was the only protective device to remain below the critical value in all conditions (< 45% of bare hand impact force).	Commercial wrist guards can provide limited impact attenuation. Damped pneumatic springs could substantially improve attenuation.

[a] These studies are listed by date of publication.

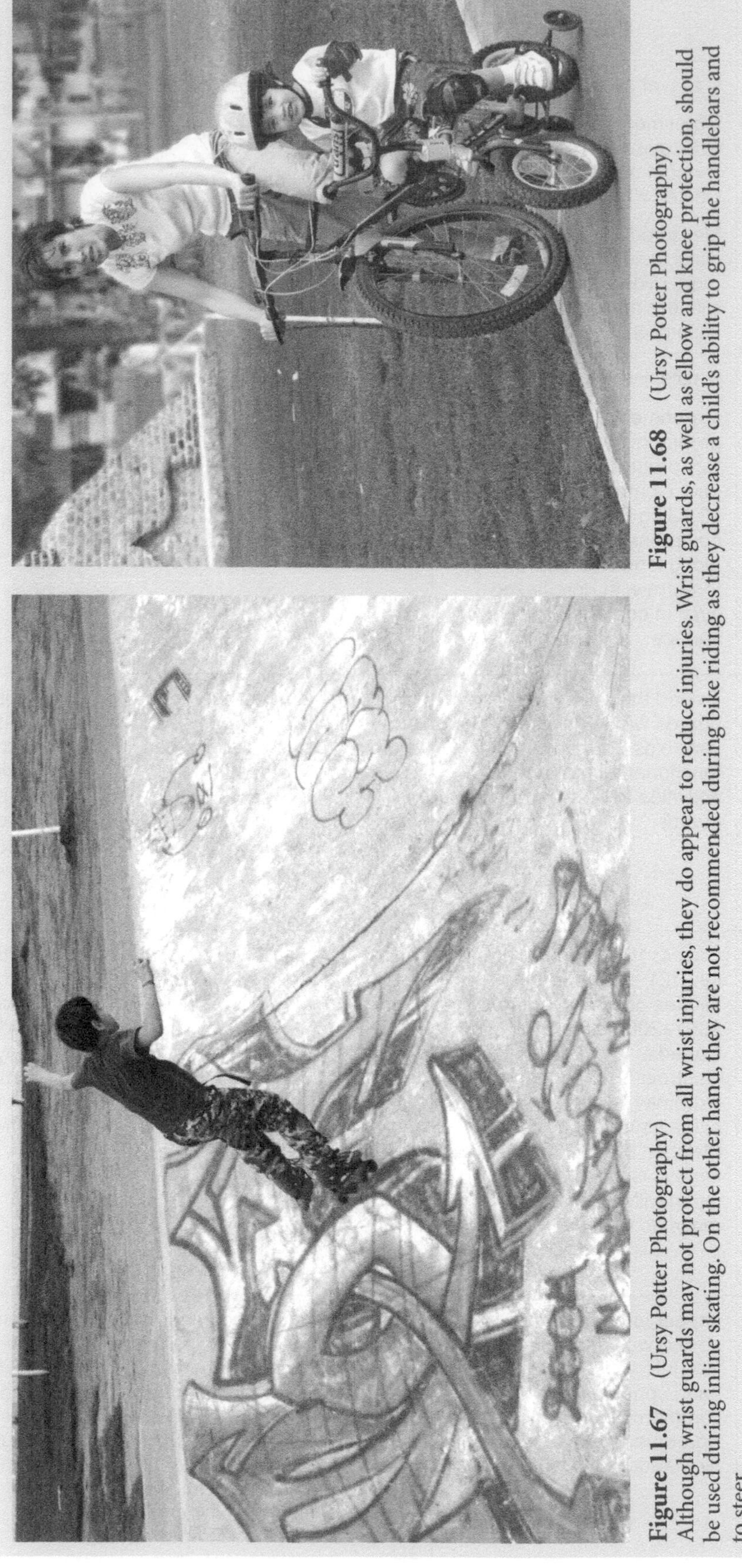

Figure 11.67 (Ursy Potter Photography)

Figure 11.68 (Ursy Potter Photography)

Although wrist guards may not protect from all wrist injuries, they do appear to reduce injuries. Wrist guards, as well as elbow and knee protection, should be used during inline skating. On the other hand, they are not recommended during bike riding as they decrease a child's ability to grip the handlebars and to steer.

FIREARMS

INTRODUCTION

Child deaths due to gunfire are higher in the United States than in other industrialized countries.* Yet, US deaths to children and teens from firearms are declining, dropping 42% since 1994.

Firearm accidents, suicides, and intentional shootings injure children. Unintentional shootings account for 20% of the firearm-related fatalities to those 14 and younger; however, these may be under reported, due to medial examiner coding mechanisms (Shaechter et al., 2003).

Figure 11.69 (Valerie Rice)
Many children enjoy playing with water guns and with games in video arcades and home videogames that involve shooting.

> In 2001, US emergency room treatment was provided to children ≤14:
>
> - 1400 for firearm-related injuries; 21% required hospitalization
> - 8000 for nonpowder gun injuries (BB guns, pellet guns)
>
> (NSK, 2003)

Most gun-related accidents occur:

- in the home or in the home of a friend (Li et al., 1996)
- when children are home and unsupervised, that is, after school, on weekends and during the summer and holiday seasons (Li et al., 1996)
- with a handgun (70%)

Intentional assault injuries tend to happen between 8 and 9 pm on roads or public places (Li et al., 1996). More gun-related incidents occur in rural areas than urban or suburban areas. Urban gun-related incidents tend to occur indoors with a handgun, and rural injuries occur outdoors with a shotgun or rifle.

Exposure and access to loaded firearms increase the risk of accidental firearm injury and death to children (Safekids, 2005), including suicides (Miller and Hemenway, 1999). Additional risk factors, for death, injury, and gun use are listed in the table below.

* CDC reports the rate of firearm deaths for children 15 and younger is 12 times higher in the US than 25 other industrialized countries combined. US children are 16 times more likely to be murdered, 11 times more likely to kill themselves, and 9 times more likely to be killed in an accident by gunfire (CDC, as reported by CDF, 2001).

Table 11.13 Who Owns Handguns? How Do They Store Them?
Slightly over one-third of US families with children have a gun at home: • 43% store at least one unlocked (61% found by Senturia et al., 1996) • 9% store at least one loaded and unlocked (7% found by Senturia et al., 1996) • 4% store them unlocked, unloaded, but with ammunition stored in the same place • 39% store them locked, unloaded and with guns separate from ammunition
Why do people own handguns? • Protection and crime prevention: 40% • Hunting and recreation: 60%
• Those that keep a gun loaded are more likely to own it for protective purposes, own a gun as part of their work requirements, have a handgun, and have no men living in their home (Senturia et al., 1996).

Source: From (Schuster et al., 2000).

Table 11.14 Risk Factors for Death, Injury, and Use of Guns among Youth

At Risk for:	Risk Factors

Death
61% of children killed by gunfire between 1979 and 1999 were white, 35% were black
	• Deaths for young black males (ages 15 to 19) are four times that of white males of the same age group.
	• Children aged 10–14
	• Homicide: Black children are more at risk
	• Suicide:
	  ∘ White children are more at risk
	  ∘ Boys are seven times more likely to commit suicide with a gun than girls

Unintentional Shooting Injury	• Boys • Black children • Children living in the south
Gun use[a]	• 10–12 year-olds • Children whose friends use guns are more likely to use guns • Those who drink early are more likely to use guns (Stevens et al., 2001)

Sources: NSK, Report to the Nation: Trends in unintentional childhood injury mortality (1987–2000) http://usa.safekids.org, 2003; Safekids, *Safety tips.* http://usa.safekids.org, 2005.

[a] Risk factors for unsupervised gun handling among adolescents in California are being male, black, living in a rural area, having a gun in the home, having no parental supervision in the afternoons (parents don't know where they are) and risk behaviors of smoking, drinking and being the victim of a gun-related threat (Hemenway, 2004).

BEHAVIORAL IMPLICATIONS

As stated in the beginning of this chapter, children's judgment is not as astute as that of an adult. Many parents and caretakers have unrealistic expectations in this regard.

Children are curious and interested. They are "hands-on" learners. In addition, many children play with toy guns and pistols. They shoot during video games and in the water with "water guns" or "squirt pistols."

Young children may not be able to distinguish what is real and what is not, both in terms of the activity (is what happens on television real or pretend?) and when evaluating a gun. They may also not think it is loaded.

In one study, toy water-guns and a real handgun were placed in a drawer in a room in which two- and three-person pairings of boys waited. Researchers observed the boys, aged 8–12. Of the 72% that found the real gun, 76% handled the gun and 48% pulled the trigger, about 50% were unsure if it was real or a toy. This is in spite of the fact that 90% of the boys reported receiving prior instruction regarding guns and gun use (Jackman et al., 2001). This finding indicates that most boys will handle a gun, if they find it.

BB or Pellet guns

- Injuries from use of BB or pellet guns are highest at age 13 & decline after 13.
- Injuries from use of guns are highest at age 13 and increase sharply until age 18.
- Both show similar declines since the 1990's.

(Nguyen et al., 2002)

Unsupervised handling of guns by adolescents in California usually involves shooting the weapon with friends, away from the home (Hemenway, 2004).

Adults' unrealistic expectations of children's behavior can include:

- not believing a child will gain access to a gun
- not believing a child will handle a gun, if one is found
- not believing a child will try to fire a gun
- not believing a young child is capable of pulling a trigger (some children can pull a trigger at the age of 3)
- assuming children can tell the difference between real and toy guns
- believing children can and will exercise sound judgment in regard to guns
- believing children will do as they are told regarding guns

Even with education about guns and how to behave should a gun be found, children still may not behave according to adult expectations (Wintemute, 1999, citing an ABC news program).

We cannot accurately predict which child will bring a weapon to school or be violent.
(Furlong and Bates, 2001).

SOLUTIONS

The solutions are to remove the danger (eliminate access), design to take away the danger (improve gun technology), provide guarding (improve storage and gun locks), and to warn or educate adults and children.

Gunlocks and load indicators could prevent 30% of unintentional injuries. Safety locks could prevent essentially all accidental firearm deaths committed by children age 5 and under (Safekids, 2005). The following section, "Children's Safety and Handguns" provides information on design issues related to prevention.

Legislation that makes gun owners criminally liable if a child gains access and unintentionally injures someone, appears to decrease unintentional shooting deaths among children under age 15. Yet, the evidence is not conclusive (Cummings et al., 1997).

Education of children needs to include active learning, including modeling, rehearsal, and feedback (6–7-year-olds). This type of education is more effective than purely education-based teaching regarding safety skills and guns (Gatheridge et al., 2004).

However, it does not help younger children (4–5) transfer their learning to situations different from those presented during the teaching session (Himle et al., 2004). Situational training based on real life and reinforced over time might help, but has not been adequately tested.

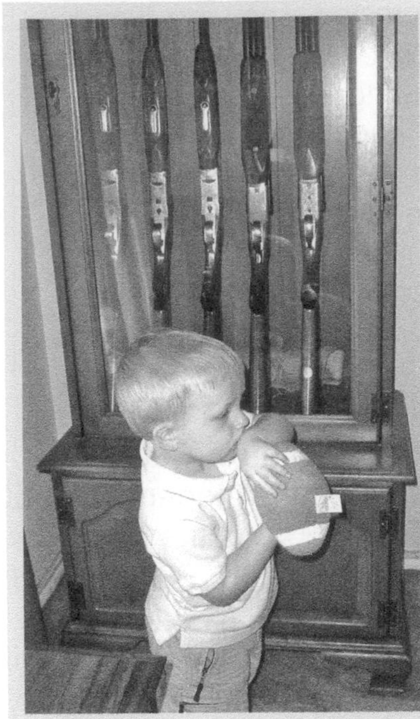

Figure 11.70 (Hersh Moody) Guns should be stored in locked, sturdy gun cases and stored unloaded. Ammunition should be stored separately from the guns, also in a locked cabinet or case (Grossman et al., 2005; Sidman, 2005).

CHILDREN'S SAFETY AND HANDGUNS

Hal W. Hendrick, Ph.D., CPE, DABFE

I saw the parents in the courtroom. Their 18-year-old son was charged with 1st degree murder of his 15-year-old friend. I looked at the boy. He looked so young and so scared. I looked back at the parents, their faces worn with worry. The shooting was an accident. It happened because the boy was "fooling around" with a loaded weapon. This did not have to happen.

(Hendrick, 2005)

THE PROBLEM

Fatal and nonfatal suicide and unintentional firearm shootings are a major problem in the United States. It is especially traumatic when the shooting involves children:

In the U.S., an injury from firearms is the second leading cause of death among children 10 to 19 years of age (Fingerhut, 1993).

An average of one child per day dies because of an accidental shooting. Among 10- to 14-year-old children, there are three times more serious injuries from firearms than there are deaths (Annest et al., 1995). Many of the injuries result from someone using another person's firearm without their permission.

Figure 11.71 (Ursy Potter Photography)

CAUSAL FACTORS

At least five major factors contribute to these statistics, as described in Table 11.15.

Table 11.15 Five Major Factors that Contribute to Unintentional Firearm Shootings

Risk Factors	Explanation
Not securing handguns in the home	Research shows that 32% to 35% of children in the United States live in households with guns (Flanagan, 1997; Stevens et al., 2001) and approximately 54% have access to unlocked guns (Stevens et al., 2001). Sixty-one percent of households with guns report keeping them unlocked (Senturia et al., 1996) and 1.4 million homes with 2.6 million children had firearms accessible to children (Schuster et al., 2000).
Unrealistic Parent Perceptions about Children and Gun Safety	Often, parents underestimate the risk of injury from handguns to their children (Pediatrics, 1999).[a] Some parents' self-reports about bringing their children to ambulatory centers reveal that 28% believe they can trust their child (under 12) with a loaded gun and that 58% of those owning guns store them either loaded or unlocked.
Lack of Gun Safety Training	**Parents who do not secure their guns and underestimate the dangers to their children show a lack of gun safety training. Hundreds of accidental shooting investigations tell us that shooters either lacked any handgun safety training or their training was very old and of questionable content (Paradis and Hendrick, 2003).** Nearly all persons, when first taking a class on handgun safety, demonstrate unsafe actions that could result in injury. It is far better for these mistakes to occur during a class where a professional can immediately make corrections than for the mistakes to occur at home. Presently, except for some localities where firearms safety training is a requirement for carrying a concealed weapon, there is no requirement for firearms safety training in order to either own or use a handgun.
Handgun Design–Induced Human Error	A number of handgun characteristics can induce human error—especially for those who have not had handgun safety training, including periodic refresher training (see Table 11.16).
Stress	People are more likely to unintentionally fire handguns in high-stress situations. Three of the most common high-stress situations are domestic arguments, suspecting there is an intruder in one's home, and being threatened. When the situation involves drugs or alcohol, the likelihood that someone will fire a handgun is even higher. Under conditions of excessive stress, individuals experience: 1. a narrowing of attention, attending only to the most obvious and seemingly most relevant cues in the threat situation (Organ and Bateman, 1991; Szilagyi and Wallace, 1990; Selye, 1976) and 2. as part of the automatic physiological alert response to threat; the sympathetic-adrenal medullary neuroendocrine system activation increases one's strength and causes the individual to tense his or her muscles (Selye, 1976). Anything that startles the handgun holder will likely cause the holder to automatically clench his or her fist. With the holder's finger on the trigger and their narrowed attention focused elsewhere, they fire the gun. This is a very common type of occurrence in unintentional shootings (Hendrick, 2005; Paradis, 2005).

[a] Paul Paradis, a gun expert and crime scene investigator, has taught handgun safety to several thousand persons and investigated hundreds of shooting cases for the Colorado Public Defender's Office. He states that parents frequently underestimate the risk of injury from handguns to their children (personal communication, 2001).

Table 11.16 Handgun Design Induced Human Errors

		Solution
Trigger pull	• Firmly gripping a typical revolver or semiautomatic handgun requires the holder to apply about 0.9 kg (2 lb) pressure. Some target pistols require as little as 1.13 kg (2.5 lb) pressure to engage the trigger and fire. • Merely swinging and stopping a handgun requires about 1.35 kg (3 lb) pressure. If a person's attention focuses elsewhere while his or her finger is on the trigger, she or he will likely apply the 1.35 kg (3 lb) pressure on the trigger, thus unintentionally firing the weapon. • Under high stress, even highly trained police officers will unintentionally fire a handgun in single action mode and sometimes even when in double action mode (see below). • Some triggers also require only 0.16 cm (1/16 in.) or less movement to fire. This is particularly true of some guns used for precision target shooting. This is too little to sense the movement in time to stop and avoid firing the gun. • **Double-action** versus **single-action** mode firearm triggers: ◦ **Shooting sequence:** Double action semiautomatics first shoot in double action mode. Subsequent shots are in single action mode. Pistols typically fire in double action unless one first cocks the hammer. ◦ **Trigger pressure**: Handguns typically require significantly more pressure to fire when in double action than when in single action. Double-action mode in self-defense guns usually requires 4.5–5.4 kg (10–12 lb) pressure to fire. Single action mode in self-defense guns most often requires 2.3–2.7 kg (5–7 lb) pressure to fire. Target pistols often require less than 1.3 kg (3 lb) to fire.	Mandate that all triggers of handguns used for home protection require 4.5 kg (10 lb) and ¼ in. movement to fire (author's recommendation in the absence of more precise research data).

Target handguns for home protection	• Target guns used for precision shooting are designed to be fired by trained shooters on supervised ranges.[a] Consequently, these gun triggers typically require very light pull pressure and very small travel to discharge the weapon. Although such guns are ideal for target shooting, they are extremely dangerous if kept in the home for home protection.	Never use a precision handgun for home protection purposes.
Lack of a loaded chamber indicator	The greatest single cause of accidental shootings is likely when people handle what they believe are unloaded guns. In almost all cases, the only way to determine there is a cartridge in the handgun's chamber is to open the weapon and directly view the chamber. Some pistols even give the impression that the cylinder is empty: • When casually viewing the front of the cylinder, gun handlers appear to see only black holes because the tips of bullets are recessed. • This is particularly true when the pistol design prevents viewing the rear of the cylinder chambers when the cylinder is in place for firing. • If some cylinder chambers are loaded and others are not, the pistol handler may also experience confusion as to which cartridge will fire next and believe that the next firing will be an empty chamber. • With semiautomatic handguns, an untrained gun handler may believe that removing the clip will result in the gun being empty. In reality, there still is a bullet in the chamber.	Provide chamber indicators that can be easily seen and felt on all handguns. Users need to be able to feel, as well as see, the loaded chamber indicator because home self-defense situations often occur at night.
Lack of standardization of handgun safeties	There is a wide variety of safety devices on handguns. Contrary to the belief of untrained laypersons, the primary purpose of many of these safety devices is to prevent the gun from accidentally discharging if someone drops it—not to prevent the gun from firing when one pulls the trigger. However, a safety device often can prevent a gun from firing when someone pulls the trigger. Even when the external portions of the handgun safety mechanism may appear similar, internally the mechanisms may work quite differently. With some semiautomatics, removing the clip containing the cartridges prevents the weapon from firing; with most, it does not. Thus, a person familiar with a semiautomatic where removing the clip does prevent it from being fired may mistakenly believe the same is true on another semiautomatic and thus, pull on the trigger, and have the gun unintentionally discharge.	Standardize safety devices. Because of this lack of standardization, a person somewhat familiar with the safety mechanism on one handgun may incorrectly assume it is the same on another handgun and believe the safety is "on" when it is actually "off." This means the person can fire the handgun, when they believe they cannot fire it.

[a] It was a fatal shooting case. The weapon was a 22-caliber target pistol that was unsecured in the home. The pistol trigger had a 2.5 lb pressure pull in single action and approximately a 1/16th in. travel in order to discharge the weapon. The teenager had been playing with the gun, cocking the hammer, while watching TV. While his attention still was on the TV set, he swung the cocked pistol around to hand it to a friend. As he stopped his movement in front of his friend, the gun discharged!

(continued)

Table 11.16 Handgun Design Induced Human Errors (Continued)

		Solution
Lack of a comfortable position for one's trigger finger when outside the trigger guard	Unfortunately, the only comfortable position for one's trigger finger is when it is on the trigger. Although handgun safety courses consistently teach the basic principle of keeping one's trigger finger outside of the trigger guard until the moment one decides to actually shoot the weapon, the reality is that doing so is uncomfortable and seems "unnatural." Thus, even for someone with training who does not consistently practice, there is a natural tendency to place one's trigger finger on the trigger. Someone without training will almost invariably place their finger on the trigger, even though they did not intend to fire the gun.	Have professional ergonomists team with gun manufacturers to develop a design that provides a reasonably comfortable resting place for one's trigger finger outside of the trigger guard.
Gun grips that are not properly sized for the individual	When it comes to handguns, one size does not fit all. Thus, a child playing with their parents' handgun is likely to have hands too small for the weapon's grip. This can cause the child to inadvertently fire the weapon or to point in one direction but actually fire it in another.	Lock guns where children cannot access them. As part of gun safety training, teach those with an interest in purchasing a handgun how to select one that fits their hand size and strength. Have ergonomists and handgun manufacturers' team to design handguns to fit specific adult anthropometric ranges, considering hand size and strength.

PREVENTING UNAUTHORIZED ACCESS

One of the most effective ways to prevent handgun child suicides and unintentional shootings is to keep children (and others) from access to handguns without permission and supervision. Two different types of measures can accomplish this in the home: (1) external restraints and (2) handgun internal personalized access measures.

Figure 11.72 (Valerie Rice)
One way to limit children's access to guns is to store them in a locked, secure case.

Table 11.17 Preventing Unauthorized Access to Firearms

	Action	Advantages	Disadvantages	Some Manufacturers
External measures prevent useful access to the handgun				
Trigger Locks	• Encase the trigger portion of the handgun with a guard that must be unlocked by a key, combination lock, or keypad in order to fire the weapon.	• Low tech • Relatively inexpensive	• Users may lose or misplace the key or forget the combination lock in a crisis. • Children are notoriously skillful at eventually discovering the location of keys or gaining access to combination codes. • It takes time to set the combination or find and use the key—a particular problem at night if one does not want to turn on a light, such as when there is an intruder in the house.	
Lock Boxes Gun Vaults	• Lock boxes and gun vaults for storing handguns and ammunition are readily available commercially. • Like trigger locks, they require a key, combination lock, or keypad to open them. In some cases, homeowners can use a padlock. • Other variations, such as lockable gun cabinets or gun racks, are also available commercially.	• Similar to trigger locks	• Similar to trigger locks	

Internal measures may more effectively prevent unauthorized handgun use with designs that only authorized users can fire.				
Magnetic	• Requires handgun users to wear a ring with a tiny magnet embedded in the bottom. • Operates by a flag-shaped steel block, which blocks the internal firing mechanism until close proximity of the magnetic ring rotates the block, thus enabling the trigger to depress. • This happens when the person grips the gun with the hand having the ring.	• May be the most advantageous of available internal devices. • No one else can fire the handgun except the one wearing the ring. • As it is a passive device, it requires no additional action on the part of the gun handler. • Ordinary magnets cannot operate the device.	• Users must wear the ring (although the magnet can be embedded in any ring the person normally wears). • One cannot use metal grips on the gun handle that block the magnet's pull (nonmetallic grips can readily replace most metal grips).	
Magnetic or Electric	• It is also called "solenoid use limitation device." • It involves a magnetically activated solenoid, which requires a power source. • As with the magnetic type, users wear a magnetic ring. When users grip the gun, the ring magnet activates the solenoid via electronic encoding and unblocks the firing mechanism. • A custom grip is a part of the system.	• Similar to magnetic type	• Similar to magnetic type • Also, more complex and the power source will eventually go dead (which might happen at a critical time)	
Grip Cocking	• Firing a handgun with this system requires users to exert considerable pressure on the grip to enable the gun to fire.	• The grip pressure level is set high enough that young children cannot activate the handgun (approximately 4.5 kg /10 lb), its major advantage. • Persons unfamiliar with the system will not likely grip the gun tight enough to enable it to fire.	• Small or otherwise weak adult users may not be able either to fire the gun or to fire it accurately. • Many older adolescents and teenagers have sufficient strength to enable the firing mechanism.	

(continued)

Table 11.17 Preventing Unauthorized Access to Firearms (Continued)

		Advantages	Disadvantages	Some Manufacturers
Keypad Entry	• A three-digit keypad is embedded on the side of the handgun. • Users install their own personal code. • To enable firing of the gun, users enter preset three-digit codes by pressing the appropriate keys. • Users press each key until the desired digit shows (e.g., 3 presses for a "3," etc.). For quick response, two keys can be preset, leaving only one key to press to enable the gun to fire. • Some conclude the system is suitable for home defense handguns.	• People who do not know the code cannot readily activate the system. • As it is fully mechanical, it does not require a power source (which might go dead at a critical time). • It is quicker to access the gun than with lock boxes, as users do not have to open the box and take the handgun out. • If someone forcefully takes the gun away from the owner or user before he or she enters the full code, that person cannot fire it.	• As with keypad lock boxes, users may forget the code. • Others, such as their children may eventually discover the code.	Florida-based Saf-T-Lock systems
Key-locking Systems	• Key-locking firing mechanisms on some models of semiautomatic pistols and revolvers	• Adds little to the handgun cost. • Only persons with the key can enable the handgun to fire.	• User may misplace key. • Key may not be readily available when user needs it. • Children may find and use the key.	Steyre of Austria Taurus of Brazil
Radio Frequency Identification (RFID)	• Handguns with these devices are also called "smart guns." • Users wear a watch-like wristband or other emitter with encoded close-range radio wave and microchip inside the gun's grip that receives signal causing a tiny aerospace stepper motor to activate the firing mechanism. • Some manufacturers consider it more practical to incorporate RFID technology into true electronic guns; some gun manufacturers are considering (and developing) this advancement.	• Sandia laboratories gave RFIDs its highest rating in its evaluation of this and the above systems.	• RF links add weight and cost. • Limited battery maintenance, status indicators and reliability. • Difficult to modify current guns.	SmartLink, Inc. of Oakland, California

CONCLUSIONS

Designing consumer products for children uses the same design principles used for adult products—plus some! Designers must be aware of child development from infancy through teen years to produce adequate designs for children.

Designing products for children is important not only for sales revenue and injury prevention, but also because product design can influence a child's development. Designs must appeal to both the children who use them and the adults who purchase them.

Although there are numerous products not included in this chapter, incorporating all that are pertinent would require a text in itself. Instead, this chapter gives the reader a taste of the complexity involved in designing for children, the issues to consider, and descriptions of some consequences of poor design.

Figure 11.73 (Valerie Rice) Adult-size products do not fit children. Yet, many times children must use them because there is not a readily available alternative. Here a young boy is "laying" on a folding chair.

REFERENCES

AAOS (2004) American Academy of Orthopedic Surgeons. *Skateboarding Safety*. Available at: http://orthoinfo.aaos.org

Abbott, M.B., Hoffinger, S.A., Nguyen, D.M., and Weintraub, D.L. (2001) Scooter injuries: A new pediatric morbidity. *Pediatrics*, 108(1).

Aborgast, K.B., Cohen, J., Otoya, L., and Winston, F.K. (2001) Protecting the child's abdomen: A retractable bicycle handlebar. *Accident Analysis and Prevention*, 33, 753–757.

Adebove, K. and Armstrong, L. (2002) Pattern and severity of injuries in micro-scooter related accidents. *Emergency Medicine Journal*, 19(6), 571–572.

Anderson-Suddarth, J.L. and Chande, V.T. (2005) Scooter injuries in children in a midwestern metropolitan area. *Pediatr. Emerg. Care*, 21(10), 650–652.

Andersson, T., Larsson, P., and Sandberg, U. (1993) (as reported by Henderson, 1995) Chin strap forces in bicycle helmets, Swedish National Testing and Research Institute, SP Report, 1993 (BHSIDOC # 487, Bicycle Helmet Safety Institute, Arlington VA, 1990).

Annest, J.L., Mercy, J.A., Gibson, D.R., and Ryan, G.W. (1995) National estimates of non-fatal firearm related injuries. *Journal of the American Medical Association*, 273, 1749–1754.

Bennett, C. (personal communication).

Bennett, C., Woodcock, A., and Tien, D. (2006) Ergonomics for students and staff. In R. Geller, W. G. Teague, H. Frumkin and J. Nodvin (Eds.), *Safe and healthy school environments*. New York, New York USA: Oxford University Press.

Bergman, A.B., Rivara, F.P., Richards, D.D., and Rogers, L.W. (1990) The Seattle children's bicycle helmet campaign, *American Journal of Diseases of Children*, 144, 727–731.

Bierness, D.J., Foss, R.D., and Desmond, K.J. (2001) Use of protective equipment by in-line skaters: An observational study. *Injury Prevention*, 7, 51–55.

BSHI (2006) Bicycle Safety Helmet Institute. Available at: http://helmets.org

Cameron, M.H., Vulcan, A.P., Finch, C.F., and Newstead, S.V. (1994) Mandatory bicycle helmet use following a decade of helmet promotion in Victoria, Australia: An evaluation. *Accident: Analysis and Prevention*, 26, 325–337.

Carlin, J.B., Taylor, P., and Nolan, T. (1995) A case-control study of child bicycle injuries: Relationship of risk to exposure. *Accident: Analysis and Prevention*, 27(6), 839–844.

Cassell, E., Ashby, K., Gunatilaka, A., and Clapperton, A. (2005) Do wrist guards have the potential to protect against wrist injuries in bicycling, micro scooter riding and monkey bar play? *Injury Prevention*, 11, 200–203.
CDF (2001) Children's Defense Fund. Protect children instead of guns. Available at: http://childrensdefense.org.
Chapman, S., Webber, C., and O'Meara, M. (2001) Scooter injuries in children. *Journal of Paediatric and Child Health*, 37(6), 567–550.
Chiaviello, C.T., Cristoph, R.A., and Bond, G. R. (2005) Infant walker-related injuries: A prospective study of severity and incidence. *Pediatrics*, 93(6), 974–976.
Cody, B.E., Quraishi, A.Y., and Mickalide, A.D. (2004) Headed for injury: An observational survey of helmet use among children ages 5 to 14 participating in wheeled sports. Washington (DC): National SAFEKIDS Campaign. Available at: http://usa.safekids
Consumer Product Safety Commission (CPSC). Available at: http://cpsc.gov
Consumer Reports (2004, July) Bike helmets for safety's sake.
Corner, J.P., Whitney, C.W., O'Rourke, N., and Morgan, D.E. (1987) Motor cycle and bicycle protective helmets: Requirements resulting from a post crash study and experimental research. School of Civil Engineering, Queensland Institute of Technology, for the Federal Office of Road Safety, Department of Transport and Communications, Report CR 55.
CPSC (2002) United States Consumer Product Safety Commission. Age determination guidelines: Relating children's ages to toy characteristics and play behavior. Smith, T.P. (Ed.). Available at: www.cpsc.gov
CPSC (2005) United States Consumer Products Safety Commission. Tips for your baby's safety: Nursery equipment safety checklist. CPSC Document #200 Washington D.C.: CPSC.5 p. Available at: www.cpsc.gov
CPSC (2006) *Skateboards*. Publication #93. Available at: http://cpsc.gov
CPSC (Consumer Product Safety Commission) (2005) Tips for your baby's safety, CPSC Document #200. Available from http://www.cpsc.gov
CPSC (Consumer Product Safety Commission) (2006, retrieved) CPSC to in-line skaters: Skate but skate safely—Always wear safety gear. Document #5014. Available at: http://cpsc.gov
Cummings, P., Grossman, D.C., Rivara, F.P., and Koepsell, T.D. (1997) State gun safe storage laws and child mortality due to firearms. *Journal of the American Medical Association*, 278(13), 1084–1086.
D'Souza, L.G., Hynes, D.E., McManus, F., O'Brien, T.M., Stephens, M.M., and Waide, V. (1996) The bicycle spoke injury: An avoidable accident? *Foot and Ankle International*, 17(3), 170–173.
Dannenberg, A.L., Gielen, A.C., Beilenson, P.L., Wilson, M.E.H., and DeBoer. (1993) Bicycle helmet laws and educational campaigns: An evaluation of strategies to increase children's helmet use, *American Journal of Public Health*, 83(5).
Eliot, L.S. (2000) *What's Going on in There? How the Brain and Mind Develop in the First Five Years of Life*. New York: Bantam.
Ellis, J.A., Kierulf, J.C., and Klassen, T.P. (1995) Injuries associated with in-line skating from the Canadian hospitals injury reporting and prevention program database. *Canadian Journal of Public Health*, 86, 133–136.
Fazen, L.E. and Felizberto, P.I. (1982) Baby walker injuries. *Pediatrics*, 70, 106–109.
Fingerhut, L.A. (1993) Firearm mortality among children, youth and young adults 1–34 years of age, trends and current status: United States, 1985–1990. Advance data: Vital and health statistics of the CDC. Hyattsville, MD: National Center for Health Statistics, 231, 1–20.
Fisher, D. and Stern, T. (1994) Results of a study on 1100 bicycle helmets involved in accidents. In: *Proceedings, 38th Annual Conference, Association for the Advancement of Automotive Medicine.*
Flanagan, T.J. (Ed.). (1997) *Sourcebook of Criminal Justice Statistics*. Washington D.C.: U.S. Department of Justice, Bureau of Justice Statistics, 1997.
Fong, C.P. and Hood, N. (2004) A paediatric trauma study of scooter injuries. *Emergency Medicine Australasia*, 16(2), 139–144.
Forjuoh, S.N., Fiesinger, T., Schuchmann, J.A., and Mason, S. (2002) Helmet use: A survey of 4 common childhood leisure activities. *Archives of Pediatric Adolescent Medicine*, 156(7), 656–661.
Furlong, M.J. and Bates, M.P. (2001) Predicting school weapon possession: A secondary analysis of the youth risk behavior surveillance survey. *Psychology in the Schools*, 38(2), 127–139.
Furnival, R.A., Kelle, A.S., and Schunk, J.E. (1999) Too many pediatric trampoline injuries. *Pediatrics*, 103(5), 5.
Gatheridge, B.J., Miltenberger, R.G., Huneke, D.F., Satterlund, M.J., Mattern, A.R., Johnson, B.M., and Flessner, C.A. (2004) Comparison of two programs to teach firearm injury prevention skills to 6 and 7 year old children. *Pediatrics*, 114, 294–299.
Gaines, B.A., Shultz, B.L., and Ford, H.R. (2004) Nonmotorized scooters: A source of significant morbidity in children. *J. Trauma*, 57(1), 111–113.
Giacobetti, F.B, Sharkey, P.F., Bos-Giacobetti, M.A., Hume, E.L., and Taras, J.S. (1997) Biomechanical analysis of the effectiveness of in-line skating wrist guards for preventing wrist fractures. *American Journal of Sports Medicine*, 25(2), 223–225.
Graitcer, P.L. and Kellermann, A.L. (1994) Is legislation an effective way to promote bicycle helmet use? In: *Proceedings, 38th Annual Conference, Association for the Advancement of Automotive Medicine.*

Grant-Ford, M., Sitler, M.R., Kozin, S.H., Barbe, M.F., and Barr, A.E. (2003) Effect of a prophylactic brace on wrist and ulnocarpal joint biomechanics in a cadaveric model. *The American Journal of Sports Medicine*, 31, 736–743.

Greenwald, R.M., Janes, P.C., Swanson, S.C., and McDonald, T.R. (1998) Dynamic impact response of human cadaveric forearms using a wrist brace. *The American Journal of Sports Medicine*, 26, 825–830.

Grossman, D.C., Mueller, B.A., Riedy, C., Dowd, M.D., Villaveces, A., Prodzinski, J., Nakagawara, J., Howard, J., Thiersch, N., and Harruff, R. (2005) Gun storage practices and risk of youth suicide and unintentional firearm injuries. *Journal of the American Medical Association*, 293, 707–714.

Hagel, B., Pless, B., and Goulet, C. (2005) The effect of wrist guard use on upper-extremity injuries in snowboarders. *American Journal of Epidemiology*, 162(2), 149–156.

Hagel, B.E., Pless, I.B., Goulet, C., Platt, R.W., and Orbital, Y. (2005) Effectiveness of helmets in skiers and snowboarders: Case-control and case crossover study. *British Medical Journal*, 330, 281.

Halstead, P.D. (2006) Personal communication. Additional information is available from P.D. Halstead, Director Sports Biomechanics Impact Research Laboratory, University of Tennessee, Knoxville, TN.

Hedge, A., Ruchika, J., Jagdeo, J., Ruder, M., and Akaogi, M. (2003) Effects of chair design on toddler behavior. *Proceedings of the Human Factors and Ergonomics Society 47th Annual Meeting*, October 13–17, Denver, CO, pp. 937–941.

Hemenway, M.M. (2004) Unsupervised firearm handling by California adolescents. *Injury Prevention*, 10, 163–168.

Henderson, M. (1995) *The Effectiveness of Bicycle Helmets: A Review*. Motor Accidents Authority of New South Wales, Australia. ISBN 0 T310 6435 6.

Hendrick, H. (2005) Recounting of personal experiences as an expert witness.

Himle, M.B., Miltenberger, R.G., Gatheridge, B.J., and Flessner, C.A. (2004) An evaluation of two procedures for training skills to prevent gunplay in children. *Pediatrics*, 113, 70–77.

Hodgson, V.R. (1990a) Impact, skid and retention tests on a representative group of bicycle helmets to determine their head-neck protective characteristics, Department of Neurosurgery, Wayne State University, Michigan, (BHSIDOC # 310, Bicycle Helmet Safety Institute, Arlington VA).

Hodgson, V.R. (1990b) Skid tests on a select group of bicycle helmets to determine their head-neck protective characteristics, Department of Neurosurgery, Wayne State University, Michigan (BHSIDOC # 357, Bicycle Helmet Safety Institute, Arlington VA).

Hu, X., Wesson, D.E., Chipman, M.L., and Parkin, P.C. (1995) Bicycling exposure and severe injuries in school-age children: A population-based study. *Archives of Pediatrics & Adolescent Medicine*, 149(4), 437–442.

ISRC. In-line Skating Resource Center. (2006, retrieved). Available at: http://iisa.org

Jackman, G.A., Farah, M.M., Kellermann, A.L., and Simon, H.K. (2001) Seeing is believing. What do boys do when they find a real gun? *Pediatrics*, 107, 1247–1250.

Kedjidjian, C.B. (1994) Bicycle helmets save lives. *Traffic Safety*, 94(3), 16–19.

Kim, K., Alian, A.M., Morris, W.S., and Lee, Y. (2006) Shock attenuation of various protective devices for prevention of fall-related injuries of the forearm/hand complex. *The American Journal of Sports Medicine*, 34, 637–643.

Knox, C.L. and Comstock, R.D. (2006) Video analysis of falls experienced by paediatric ice-skaters and roller/in-line skaters. *British Journal of Sports Medicine*, 40, 268–271.

Levine, D.A., Platt, S.L., and Foltin, G.L. (2001) Scooter injuries in children. *Pediatrics*, 107(5), E64.

Li, G., Baker, S.P., DiScala, C., Fowler, C., Ling, J., and Kelen, G.D. (1996) Factors associated with the intent of firearm-related injuries in pediatric trauma patients. *Archives of Pediatrics & Adolescent Medicine*, 150(11), 1160–1166.

Maimaris, C., Sommers, C.L., Browning, C., and Palmer, C.R. (1994) Injury patterns in cyclists attending an accident and emergency department: A comparison of helmet wearers and non-wearers. *British Medical Journal*, 308, 1537–1540.

Malina, R.M. (2004) Motor development during infancy and early childhood: Overview and suggested directions for research. *International Journal of Sport and Health Science*, 2, 50–66.

Mankovsky, A.B., Mendoza-Sagaon, M., Cardinaux, C., Hohlfeld, J., and Reinberg, O. (2002) Evaluation of scooter-related injuries in children. *J. Pediatr. Surg.* 37(5), 755–759.

Melendez, J.A., Mercado-Alvardo, J., Mercado-Alvarado, J., and Garcia-Gubern, C. (2004) Most common injuries associated to scooters seen in the emergency department. *Boletin de la Asociacion Medica de Puerto Rico*, 96(3), 173–177.

Miller, M. and Henenway, D. (1999) The relationship between firearms and suicide: A review of the literature. *Aggression and Violent Behavior*, 4(1), 59–75.

Montagna, L.A., Cunningham, S.J., and Crain, E.F. (2004) Pediatric scooter-related injuries. *Pediatr. Emerg. Care*, 20(9), 588–592.

Moore, M.S., Popovic, N.A., Daniel, J.N., Boyea, S.R., and Polly, D.W. (1997) The effect of a wrist brace on injury patterns in experimentally produced distal radial fractures in a cadaveric model. *American Journal of Sports Medicine*, 25(3), 394–401.

Morrongiello, B.A. and Major, K. (2002) Influence of safety gear on parental perceptions of injury risk and tolerance for children's risk taking. *Injury Prevention*, 8, 27–31.

NCIPC (National Center for Injury Prevention and Control). (2001) *Injury Fact Book 2001–2002*. Atlanta, GA: Center for Disease Control and Prevention. Available at: http://cdc.gov or by calling 770-448-1506.

Nguyen, M.H., Annest, J.L., Mercy, J.A., Ryan, G.W., and Fingerhut, L.A. (2002) Trends in BB/pellet gun injuries in children and teenagers in the United States, 1985–1999. *Injury Prevention*, 8, 185–199.

NHTSA (2004) National Highway and Traffic Safety Association. *Traffic Safety Facts*. Available at: http://nhtsa.dot.gov

NSK (2003) National Safekids Campaign. Report to the nation: Trends in unintentional childhood injury mortality (1987–2000). Available at: http://usa.safekids.org

Organ, D.W. and Bateman, T.S. (1991) *Organizational Behavior*. Homewood, IL: Irwin.

Osberg, J.S. (2001) Safety behavior of in-line skaters. *The Western Journal of Medicine*, 174, 99–102.

Osberg, J.S., Schneps, S.E., DiScala, C., and Li, G. (1998) Skateboarding: More dangerous than roller-skating or in-line skating. *Archives of Pediatric and Adolescent Medicine*, 152(10), 985–991.

Paradis, P. (2005) Personal Communication, September 24.

Paradis, P. and Hendrick, H.W. (2003) Human factors deficiencies in handgun safety training. In: *Proceedings of the Human Factors and Ergonomics Society 47th Annual Meeting* (pp. 1782–1786). Santa Monica, CA: Human Factors and Ergonomics Society.

Parker, J.F., O'Shea, J.S., and Simon, H.K. (2004) Unpowered scooter injuries reported to the consumer product safety commission: 1995–2001. *Am. J. Emerg. Med.* 22(4), 273–275.

Parkinson, G.W. and Hike, K.E. (2003). Bicycle helmet assessment during well visits reveals severe shortcomings in condition and fit. *The Journal of Pediatrics*, 112(2), 320–323.

Pediatrics (1999) Firearms in the home, parental perceptions. *Pediatrics*, 10(5), 1059–1063.

Phelan, K.J., Khoury, J., Kalkwarf, H.J., and Lanphear, B.P. (2001) Trends and patterns of playground injuries in United States children and adolescents. *Ambulatory Pediatrics*, 1(4), 227–233.

Powell, E.C. and Tanz, R.R. (2004) Incidence and description of scooter-related injuries among children. *Ambul. Pediatr.*, 4(6), 495–499.

Powell, E.C. and Tanz, R.R. (1996) In-line skate and rollerskate injuries in childhood. *Pediatric Emergency Care*, 12(4), 259–262.

Powell, E.C., Jovtis, E., and Tanz, R.R. (2002) Incidence and description of stroller-related injuries to children. *Pediatrics*, 110(5), pe62.

Rauchschwalbe, R., Brenner, R.A., and Smith, G.S. (1997) The role of bathtub seats and rings in infant drowning deaths. *Pediatrics*, 100(4), E1.

Reider, M.J., Schwartz, C., and Newman, J. (1986) Patterns of walker use and walker injury. *Pediatrics*, 78, 488–493.

Rivara, F.P., Thompson, D.C., and Thompson, R.S. (1994) The Seattle children's bicycle helmet campaign: Changes in helmet use and head injury admissions. *Pediatrics*, 93, 567–569.

Rutherford, G.G. (2000) Unpowered scooter-related injuries – United States – 1998–2000. MMWR, 49(49), 1108–1110. Available at http://www.cdc.gov/mmwr

Safekids (2005). *Safety Tips*. Available at: http://usa.safekids.org

Safekids (2006). Available at: http://safekids.org

Schalamon, J., Sarkola, T., and Nietosvaara, Y. (2003) Injuries in children associated with the use of nonmotorized scooters. *J. Pediatr. Surg.*, 38(22), 1612–1615.

Schieber, R., Branche-Dorsey, C.M., Ryan, G.W., Rutherford, G.W., Stevens, J.A., and O'Neil, J. (1996) Risk factors for injuries from in-line skating and the effectiveness of safety gear. *New England Journal of Medicine*, 335, 1630–1635.

Schieber, R.A. and Branche-Dorsey, C.M. (1995) In-line skating injuries: Epidemiology and recommendations for prevention. *Sports Medicine*, 19, 427–432.

Schuster, M.A., Franke, T.M., Bastian, A.M., Sor, S., and Halfon, N. (2000) Firearm storage patterns in homes with children in the United States. *American Journal of Public Health*, 90, 588–594.

Selye, H. (1976) *The Stress of Life*. New York: McGraw-Hill.

Senturia, Y.D., Christoffel, K.K., and Donovan, M. (1996) Gun storage patterns in U.S. homes with children. *Archives of Pediatrics and Adolescent Medicine*, 150, 265–269.

Senturia, Y.D., Kaufer, K., and Donovan, M. (1996) Gun storage patterns in US homes with children: A pediatric practice-based survey. *Archives of Pediatrics & Adolescent Medicine*, 150 (3), 265–270.

SGMA (Sporting Goods Manufacturers Association) (2006) Available at http://www.sgma.com/

Shields, B.J., Soledad, A.F., and Smith, G.A. (2005) Comparison of minitrampoline and full-sized trampoline related injuries in the United States, 1990–2002. *Pediatrics*, 116(1), 96–103.

Sidman, E.A., Grossman, D.C., Koepsell, T.D., D'Ambrosio, L., Britt, J., Simpson, E.S., Rivara, F.P., and Bergman, A.B. (2005) Evaluation of a community-based handgun safe-storage campaign. *Pediatrics*, 115, 654–661.

Smith, G.A. (1998) Injuries to children in United States related to trampolines, 1990–1995: A national epidemic. *Pediatrics*, 101(3), 406–412.

Stoffman, J.M., Bass, M.J., and Fox, A.M. (1984) Head injuries related to the use of baby walkers. *Can. Med. Assoc. J.*, 131, 573–575.

Staebler, M.P., Moore, D.C., Akelman, E, Weiss, A.C, Fadale, P.D., and Crisco, J.J. (1999) The effect of wrist guards on bone strain in the distal forearm. *The American Journal of Sports Medicine*, 27, 500–506.

Stevens, M.M., Gaffney, C., Tosteson, T.D., Mott. L.A., Olson, A., Ahrens, M.B., and Konings, E.K. (2001) Children and guns in a well child cohort. *Preventive Medicine*, 32, 201–206.

Straugh, B. (2004) *The Primal Teen: What the New Discoveries about the Teenage Brain Tell Us about Our Kids*. New York: Anchor.

Straub, D. (2006) Personal communication on helmets. Available at: http://davisstraub.com/Glide/helmets.htm

Suecoff, S.A., Avner, J.R., Chou K.J., and Crain E.F. (1999) A comparison of New York City playground hazards in high- and low-income areas. *Archives of Pediatrics & Adolescent Medicine*, 153, 363–366.

Sulheim, S., Holme, I., Ekeland, A., and Bahr, R. (2006) Helmet use and risk of head injuries in alpine skiers and snowboarders. *Journal of the American Medical Association*, 295(8), 919–924.

Szilagyi, A.D. and Wallace, M.J. (1990) *Organizational Behavior and Performance*, 5th edn. Glenview, IL: Scott, Foresman.

Tinsworth, D. and McDonald, J. (2001) *Special Study: Injuries and Deaths Associated with Children's Playground Equipment*. Washington, D.C.: U.S. Consumer Product Safety Commission.

Warda, L. and Briggs, G. (2004) Trampoline injuries in Manitoba: Investigating trends. Report submitted by The Injury Prevention Center of Children's Hospital. Available from Children's Hospital NA335-700 McDermot Ave., Winnipeg, Manitoba, R3E OT2.

Warda, L., Harlos, S., Klassen, T.P., Moffatt, M.E.K., Buchan, N., and Koop, V.L. (1998) An observational study of protective equipment use among in-line skaters. *Injury Prevention*, 4, 198–202.

Winston, F.K., Shaw, K., Kreshak, A.A., Schwarz, D.F., Gallagher, P.R., and Cnaan, A. (1998) Hidden spears: Handlebars as injury hazards to children. *Pediatrics*, 102, 696–601.

Wintemute, G.J. (1999) The future of firearm violence prevention: Building on success. *Journal of the American Medical Association*, 281(5), 281–285.

Wood, T. and Milne, P. (1988) Head injuries to pedal cyclists and the promotion of helmet use in Victoria, Australia. *Accident: Analysis and Prevention*, 20, 177–185.

Zalavras, C., Nikolopoulou, G., Essin, D., Manjra, N., and Zionts, L.E. (2005) Pediatric fractures during skateboarding, roller-skating and scooter riding. *American Journal of Sports Medicine*, 33, 568–573.

CHAPTER 12

CHILDREN'S PLAY WITH TOYS

BRENDA A. TORRES

TABLE OF CONTENTS

INTRODUCTION

Children cannot distinguish between play, learning, and work. Yet most theorists see play as distinct and separate from other activities in that it (Garvey, 1990a; Rogers and Sawyers, 1988a):

- is pleasurable
- is internally motivated
- focuses on process
- requires active engagement
- is not real—but carried out as if it were
- is voluntary, spontaneous, and without externally imposed rules

Play also supports nonplay activities by promoting creativity, problem solving, and the development of language and social roles. Pleasant and familiar toys attract children and stimulate their senses. Children play with toys when they are intrigued and remain occupied as long as they are having fun.

Toys are generally designed for specific ages or they may aim to grow with the child. This second type of toy becomes more challenging as children mature.

Playing with toys can contribute to children's development. An awareness of child development can help in selecting toys for children.

In the past, the primary focus of human factors/ergonomics in toy design centered on safety and comfort. The concept has broadened to designing toys to fit the child user. "User friendliness" is no longer a novel idea; consumers expect products to be user friendly. They also expect toys to be functional, easy to use, pleasurable (Jordan, 2000), and to contribute to growth and development!

Young children have insatiable curiosities. As they play with all kinds of toys, they try to figure out what an object is, how it works, and what to do with it (Garvey, 1990b). When they encounter a new object, preschool children use a predictable sequence of stages known as "the three Ms": *manipulation, mastery,* and *meaning* (Beaty, 2002).

Manipulation is an exploratory stage

In their quest for stimulation, children explore objects in new and different ways. Child observation studies reveal 15 exploration strategies that children use, described in Chapter 2. For example, infants and toddlers put blocks in their mouths, rotate them in their hands, and bang them. They sit and transfer blocks from one container to another. Children use whatever skills they have available. They often seek all of the possibilities that play objects offer before moving to the next stage.

Discovering new things and skills excites children. Once familiarized with an object, children re-enact activities. In this stage, children ask to have the same stories reread. Practice and repetition lead to mastery (Figure 12.1).

Mastery follows familiarization

Figure 12.1 (Ursy Potter Photography)
Children learn to master objects and skills through repetition. They stack and restack blocks after they topple. Children may create seemingly endless lines of single blocks across the floor.

Final Stage. The add meaning

Once children master a challenge, they begin to add their own meaning to activities and find pride in their accomplishments. Stacks of blocks become towers. Block lines become walls (Figure 12.2).

Preschool children engage in representational play or use objects imaginatively. Older children create new meanings by testing limits. For example, older children may try stacking blocks in different orientations to evaluate the impact on height. They introduce new characters or conditions to favorite stories to change the endings. First hand experience is about struggling, exploring, discovering, practicing, problem solving, and control (Bruce, 1993).

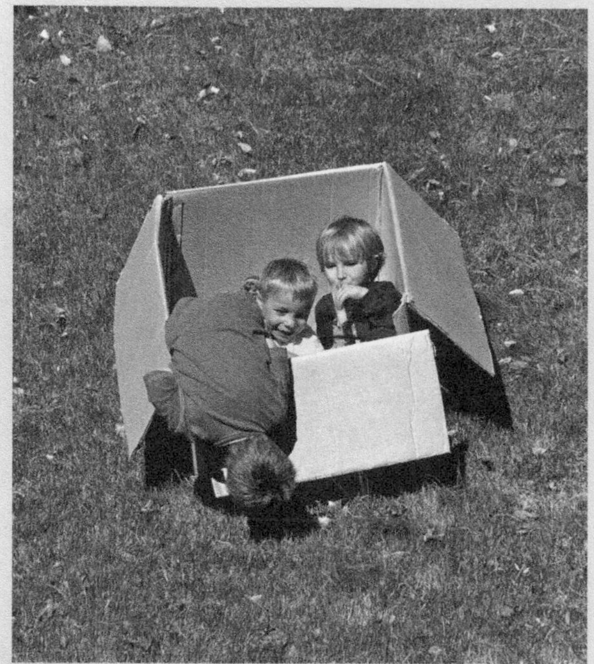

Figure 12.2 (Ursy Potter Photography)
Playing is more involved than simple interaction between the child and their play object; playing provides many links to a child's environment. It helps children discover how objects relate to them and their surroundings. Playing with objects can serve as an invitation for others to play, such as in a 'tea party'. Objects can provide channels for children to express feelings, concerns, or interests. Toys can be manufactured or children can make their own toys from other objects, such as this packing box.

THE UNIVERSE OF PLAY

While there are many developmental theories and theorists, most agree with these guiding principles for how children learn through play.

Developmental stages are sequential

As children develop necessary skills for each stage, they move on to the next stage.* This does not mean the child no longer resorts back to or enjoys the previous stage (Garvey, 1990a; Rogers and Sawyer, 1988b). Practice play and symbolic play become resources that they can return to repeatedly. Further, as children progress through each stage of play, stages overlap.

Practice makes "perfect"

The practice play stage, generally spanning 0 to 24 months, is a time for repetition. Young children engage in mouthing, banging, rotating, throwing, and matching. They ask repetitive questions. Parents may hear "Why?" more often than they would like!

* The section "Universe of Play" describes general developmental progressions. The section "Toys by Age and Development" provides additional detail by age.

For young children, almost every object is new; practicing helps them learn about the objects and their own capabilities and limitations. During this time, babies learn muscle control. Children of all ages need to practice their skills to improve and gain competence (Figure 12.3).

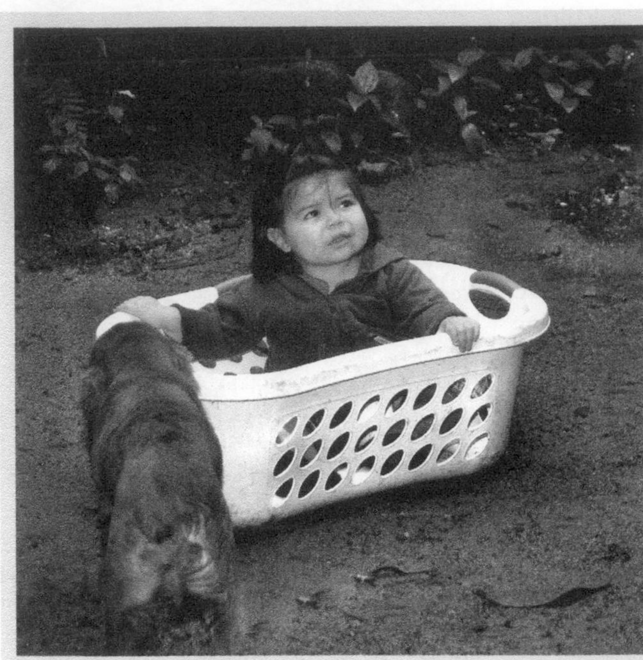

Figure 12.3 (Ursy Potter Photography)
A toddler's "practice" may include putting her body into objects, transferring objects between hands, and rotating objects. Motion activities include banging, dropping, and throwing. While adolescents and adults use more complex skills and motor actions, they must also practice to excel. This stage prepares them for the future.

Piaget theorized that cognitive development depends on direct sensory experiences and motor actions. However, the considerable intellectual capabilities seen in children with physical disabilities such as cerebral palsy suggest this theory does not apply in all cases. Children also learn from observation and social interaction.

Young children between 2 and 4 years enjoy games that encourage cooperation and repetitive, specific actions (e.g., "*Ring around the Rosies*," the "*Hokey Pokey,*" or "*Hide and Seek*"). They do not like to compete.

Fantasy fuels symbolic play

During symbolic or pretend play, children imitate others' actions (see Figure 12.4). As children develop language skills, they learn object names and begin to understand symbols. Labels allow for symbolic coding. Such representation enables children to use one object to represent another or to substitute objects.

As children mature, representational play (e.g., using a toy cup as a real cup) progresses to pretend play (e.g., crayons become nails to build a house; the TV remote becomes a cell phone). This progression moves from single simple transformations (e.g., using a teapot as an iron) to multiple and more complex actions (e.g., becoming a space ranger and their bedroom into the planet Zorg).

Initially, pretend play requires realistic replicas, such as a toy cup for a real cup. As children gain experience in substitutions and transformations, they transform objects to suit their needs. A large box becomes a palace, car, or fort.

Older children can draw from more experiences during pretend play. However, as children grow older (4 to 7 years), their needs for "appropriateness" increase; things need to look real and have more detail (Garvey, 1990b).

Between the ages of 6 and 10, their interest in games with rules peaks, while symbolic play declines. During these ages, children plan, coordinate, and base symbolic play on reality.

For example, children may organize plays and puppet shows with predetermined themes or plots (Rogers and Sawyer, 1988b). Even playing fantasy games (e.g., acting out battles as Yu-Gi-Oh™ characters) requires them to carefully follow rules of engagement and only enact behaviors they expect of their character.

Figure 12.4 (Valerie Rice)
During symbolic play, children's first imitated actions generally involve personal interactions with others, such as when feeding or changing clothes. Later, they imitate actions such as washing dishes or mowing lawns. Above, Kiran imitates her mother by pushing a stroller and wearing her shoes.

Play often has rules. A game is play when players agree to the competition and rules. Players are free to change these rules and create new games at any time. However, games are not "play" where nonplayers enforce rules (as with sports) or if players engage in activities solely to win or to achieve specific goals.

"Playing by the rules" requires complex cognitive skills

Games with rules require sophistication in a child's cognitive and social adaptation. By 6 or 7, children develop mental representations of their actions, understand, and group items in categories (e.g., smaller or lighter), and recognize that mass remains the same regardless of its orientation. At this age, children use logical reasoning in concrete situations.* They begin to understand that others may have a different viewpoint from their own.

* Concrete in that the situation involves objects or events in the child's immediate environment.

TOYS BY AGE AND DEVELOPMENT

Children's developmental levels influence what they need and how they like to play. The reciprocal is also true; play can influence a child's development. Children's play can contribute to their sense of identity and social expectations. Children become bored or frustrated when they play with toys that do not offer the appropriate type or level of stimulation. For these reasons, toys are designed towards the child's level of development (Figure 12.5).

Toys must:

- allow for manipulation, exploration, and practice
- encourage children to pretend
- support solitary play as well as play with others
- provide opportunities for expressing ideas, feelings, and relationships
- help children develop their own sense of meaning or understanding

Figure 12.5 (Valerie Rice)

Children's caregivers and their culture also affect their preferences. Parents control what they make available to a young child. Cultures define acceptable behaviors and practices. In Western or urban middle class communities, parents value play as an integral part of development and education (Goncu et al., 2000). Typically, these parents take time to play with their child. In non-Western or low-income communities, parents attribute little value to playing. Adults do not play with children. Children play with children.

Age ranges based on child development will break down the following sections. Each section will provide highlights of developmental milestones and safety tips based on uses expected at that stage of development and legal requirements (Christensen and Stockdale, 1991; Fisher-Thompson et al., 1995; Goodson and Bronson, 2001; Therrell et al., 2002).

FIRST 3 MONTHS

Up to their third month, babies lack the skills to engage in active play. Rather, they explore with their eyes and ears (Figure 12.6). A newborn's visual focus is best at a close range, between 20.3 and 30.4 cm (8 to 12 in.) from their face. They prefer bright colors such as red, blue, and green and like high contrasts such as black and white patterns. They prefer faces and basic shapes such as circles, squares, and triangles (Easterbrook et al., 1999; Morton and Johnson, 1991).

For these infants, raise toys above them so they can look at and begin to bat them. An infant's hearing is acute and as neck muscles mature, they turn their heads in the direction of sounds, or to gaze at objects. As their caregivers talk to them, babies watch their mouths and eyes. Toys for this age provide sensory stimulation and may include pictures, mirrors, and mobiles.

Babies enjoy repetitive flexing and extending of their arms and legs. They do not begin to initiate purposeful movements until 3 or 4 months of age. Toys that the child activates must do so at random locations by low forces.

At 3 months, babies begin to direct their reach and mouth objects. Toys should be easy to grasp and easy to wash. Personal interactions involving touch and sound are very important to infants. Holding the child while talking, singing, and placing objects with various tactile sensations on their skin or in their hands (soft cloth, blocks, cotton, etc.) helps them learn, as well as bond with their caregiver. Varying your own vocalizations, such as whispers, tongue clicking, and humming also interests infants (see Table 12.1).

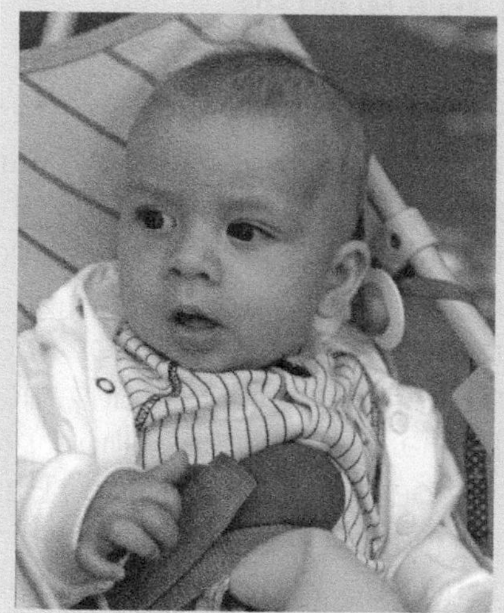

Figure 12.6 (Valerie Rice)
As infants begin to mature, they turn their heads to visual (and auditory) stimuli. Try this:

- Move visual items slowly to either side of their head (still above them).
- They will follow the item by turning their entire head, rather than just their eyes.

Table 12.1 Toys for Infants up to 3 Months

	Sensory Stimuli	Suggestions
Visual	• High contrast • Bright primary colors • Moving objects • Faces	• Mobiles in bright colors • Large, unbreakable mirror • Happy face drawn on white paper plate • Colorful objects: rattles, teething rings with thin handles for grasping • Soft dolls • Cloth toys
Auditory	• Simple patterns • Soft voices • Soft music	• Musical mobiles • Wind chimes • Tapes and music • Introduce sounds (one at a time): ticking clock, crinkling of paper, etc.
Sensory stimulation that does not require them to make physical contact. Toys for infants up to 3 months old should primarily provide visual and auditory stimulation.		

3 TO 6 MONTHS

Children between 3 and 6 months of age can reach, grasp, pull, and shake. These infants continue to put objects into their mouths. They can sit without support, roll over, and visually track objects. They enjoy social interactions; distinguish among familiar people, and show preferences among people. Basic language starts with cooing, gurgling, and laughter. Table 12.2 provides guidelines for toys for 3- to 6-month-old children.

Table 12.2 Toys for 3 to 6 Months

Goals: Touching, Holding, Batting, Turning, Shaking, Kicking, Mouthing, and Tasting	
• Easy to grasp • Self-activated • Safe for mouthing • Visually and auditorily stimulating	• Rattles or teething rings with thin handles to fit in their grasp • Suction cup rattles • Ankle rattles • Bright, shiny, colorful objects: stuffed toys, large balls • Measuring spoons • Rolling large balls with internal chimes • Crib and floor gyms • Activity quilts • Soft, squeaky toys

6 TO 12 MONTHS

Between 6 and 12 months, children become much more mobile. Many children move from creeping, crawling, pulling up to stand, and cruising (walk holding furniture), to independent walking—all within this period!

Fine motor skills improve and many children develop a pincer grip (thumb and forefinger) by the end of their first year. In addition to mouthing, children bang, rotate, insert, drop and throw objects, and transfer objects between hands.

Table 12.3 Guidelines for 6 to 12 Months

Goals: Action and Manipulation	
• Visually complex patterns and primary bright colors • Physical action (gross and fine motor) • Self-effect	• Wind-up toys • Stacking toys • Blocks and boxes (inserting and taking out) • Activity boards • Bath toys

By 11 months, infants can remember people, objects, games, and actions with toys. They show persistence and interest in novelty. Socially, children from 6 to 12 months like a familiar person nearby. They watch, imitate, and recognize their effect on others. They enjoy social games such as "peek-a-boo" and "bye-bye." They babble, imitate sounds, and may even have a few words in their vocabulary. See Table 12.3 guidelines for toys for 6- to 12-month-old children (Fox and Riddoch, 2000).

Figure 12.7 (Valerie Rice)
Between the ages of 6 and 12 months, many children progress through creeping, crawling, pulling up to stand, and cruising (walking holding onto furniture) to independent walking. Their fine motor skills develop rapidly also, moving from a gross full-hand grasp to fine motor pincer grasp with the thumb and forefinger.

Therefore, the range of toys that are appropriate varies along with the individual child's skills and abilities. In general, they enjoy toys that allow for grasp and release exploration, banging, inserting/removing, and manipulating. They are just grasping the fact that they can "cause" an effect, so they particularly enjoy visual and auditory feedback that result from their actions with their toys.

Figure 12.8 (Valerie Rice)

12 TO 24 MONTHS (1-YEAR-OLDS)

Toddlers are busy! They walk, run, climb, push, pull, and carry. By the time they are 2, they can kick and sometimes catch large balls.

Their small muscle skills are also improving. Toddlers stack, mark, scribble, poke, and insert objects into holes. At first, they can only manage round cylinders into circular holes. Later, they can match angles.

Toddlers love to explore and enjoy seeing their effect on the world. They like to combine, match, sort, and arrange objects. Different materials are fun, such as sand, water, and pretend 'snow.' This is the time when they even enjoy processes such as taking objects out to play and putting them back. They still put everything in their mouths, so avoid small parts and small objects.

Figure 12.9 (Ursy Potter Photography) Children between 12 and 24 months are perfecting their physical skills. They enjoy jumping, bumping, climbing, sliding, and swinging. Although some children are tentative with these activities, the activities themselves influence the child's development, such as balance, equilibrium, and coordinated movements.

Communication for 12- to 24-month-old children includes strings of sounds, gestures, pointing, and reaching. Cars that beep, bells that ring, and blocks that crash when they fall attract them now. Toddlers play next to or "parallel" to another child, rather than interactively (Fox and Riddoch, 2000) (see Table 12.4).

Table 12.4 Guidelines for Toys Intended for 12 to 24 Months

Goals: Skill Improvement	
• Gross Motor ◦ Walk, jump, run, push, pull, carry • Balance ◦ Ride tricycle, run, jump • Fine Motor ◦ Stacking, inserting, grasping, holding, eating with utensils • Perceptual Motor ◦ Drawing, painting, kicking/hitting a ball, catching • Imitation play	• Swings, stairs, backyard gym • Obstacle courses (tubes, tunnels, boxes) (Figure 12.9) • Large balls 1.75 in. (44.5 mm) in diameter or greater • Activity tables (water, sand) • Bubbles • Scooters and later tricycles • Blocks, nesting cups • Pop up boxes • Activity boards • Ring stackers • Picture books of heavy cardboard or plastic • Games ◦ Follow the leader ◦ Music and dance

24 TO 36 MONTHS (2-YEAR-OLDS)

Two-year-olds are refining their large and small muscle skills. They want to jump from heights, hang from their arms, run, tumble, and play "rough-house." They push, pull, and steer well. Their fine motor skills enable them to enjoy manipulating small objects and drawing.

Two-year-olds understand counting and numbers. They are interested in pattern, sequence, order and size, as well as textures, shapes, and colors.

Two-year-olds have a large vocabulary. Nursery rhymes and finger games such as "the itsy bitsy spider" are fun and challenge their little fingers. They recognize characters on television. They desire independence and feel strongly about their accomplishments and failures, yet they also need quiet time (see Table 12.5 and Figures 12.10 and 12.11).

Figure 12.10 (Ursy Potter Photography) Imagination, fantasy, and imitation are all part of a 2-year-old's play. They like to pretend to cook, vacuum, drive the car, mow the lawn, and change and feed the baby. They like their toys to look similar to adult objects, such as pots, pans, vacuums, cars, etc.

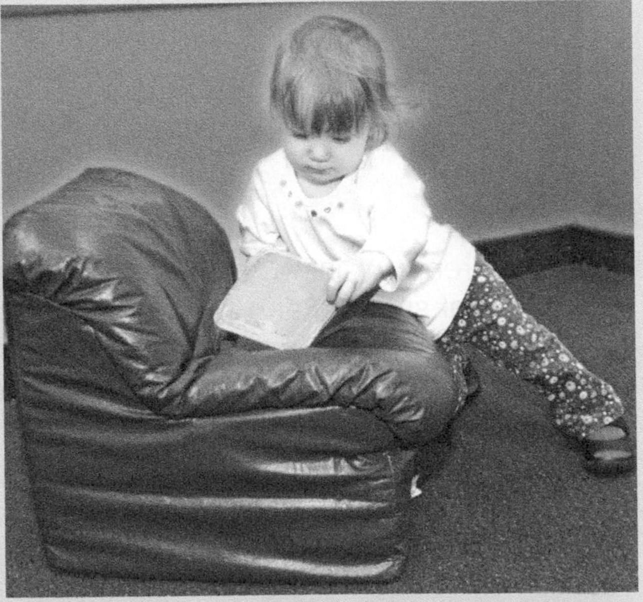

Figure 12.11 (Ursy Potter Photography) Quiet time, such as with books, is important and some children need this time to settle down before a nap. Reading aloud to children helps them develop prereading skills by building language concepts and vocabulary. Toddlers enjoy books with simple, colorful pictures, pop-ups, and sensory pages, such as tactile additions, i.e., items they can feel that are slick, furry, or fuzzy.

Table 12.5 Guidelines for Toys Intended for 2- to 3-Year-Old Children (Figures 12.12 and 12.13)	
Goals: Practice for Skill Refinement	
• Gross Motor • Fine Motor • Balance • Perceptual-Motor • Imitation Play: "Realistic" representations	• As above for 2 years • Playground equipment • Wagons, scooters, tricycles • Dough, finger paints, water colors • Simple shape sorters • Large connecting blocks • Large plastic balls and bats • Wading pool and water toys • Toy kitchens, lawn mowers, tools, grocery cart, phone • Dolls that can be bathed, fed, changed • Stuffed animals • Simple story books and 3 to 5 piece puzzles • Play sets of people, cars, animals

3 TO 5 YEARS

Preschool children run, jump, and balance with assurance. They like to take risks and test their physical strength and skills. Fine motor control improves, enabling them to cut with scissors, grip pencils, manipulate computer mouses, and string beads. Creating is important as they paint, draw, and build. They draw representational pictures and want to know how things work.

Figure 12.12 (Ursy Potter Photography) **Figure 12.13** (Ursy Potter Photography)
Goal-directed activities appear at around 4 years of age. Between the ages of 3 and 5, children can plan and carry out their plan to achieve their goals. Their coordination is good, and they enjoy both realistic and fantasy play.

Dramatic play peaks; they love to use costumes and props to create fantasy scenes. Gender differences appear in both their play and their assumption of various roles.

By 5, play has become cooperative and practical. They may not be ready for competitive play, but preschoolers enjoy playing games of chance with others (see Table 12.6).

Table 12.6 Guidelines for Toys Intended for 3 to 5 Years

Goals: Improve Skills, Encourage Imagination	
• Creativity and dramatic play • Toys for fantasy or representational play • Realistic detail and working parts • Gross, fine, perceptual motor, and balance skills • Games of chance not strategy	• (As above for 24 to 36 months) • Dress-up clothing and costumes • Life-size play houses and forts • Toy cash registers, money, food, cars, tools, carriages, dolls, etc. • Push, pull, and riding toys • More complex shape sorters • Pegs, pegboards, and hammers • Crayons, paper, blunt-tipped scissors • Connecting blocks or logs • Toy musical instruments • Recorded songs to sing and dance along • Games of chance • Books, puzzles with more pieces • Computer games or activities

6- TO 8-YEAR-OLDS

Six- to eight-year-old children engage in a variety of physical activities (Figure 12.14 and Figure 12.15). Some engage in competitive sports such as gymnastics, baseball, soccer, hockey, etc. (Huston et al., 1999).

At age 6, they may grip their pencil tightly and close to the tip, but with too much or too light a pressure. However, by their eighth year they are less tense when holding pencils, utensils, or tools.

Communication skills now go beyond listening and talking. Children between 6 and 8 can read, spell, and print. They are interested in games and sports, because they now understand rules and strategies, and they can read the directions.

Magic tricks fascinate 6- to 8-year-olds because they border between fantasy and reality. They still enjoy dramatic play, but prefer it planned and organized rather than creating situations and fantasies as they go along.

Children between ages 6 and 8 have an acute sense of mastery. They want to do well, and they enjoy seeing (and showing off) their creations. By 8, children find performing for an audience more fulfilling, such as during sports or recitals. They sustain their interests enough to begin hobbies or collections.

Team sports interest them, because they are learning to interact, trust, and discover that group effort can achieve more than singular efforts. Most develop a "best friend" and their special friends and cliques satisfy the new felt need to "belong." They play cooperatively with others and enjoy group activities. They expect their peers to play fair and live up to particular standards (their own and their group's). Their play is generally with same sex peers.

Unfortunately, they allow media such as TV, movies, and magazines to influence them (Harrell et al., 1997; Lee, 2002). At this age, they begin to favor "real people" over cartoon characters (see Table 12.7).

Table 12.7 Guidelines for Toys Intended for 6 to 8 Years

Goals: Final Product, Results, and Goals	
• Realistic detail and working parts • Gross, fine, perceptual motor, and balance skills • Creativity and construction • Promotion of friendships or social interactions • Games of chance, but also strategy	• Sports equipment: roller skates, bikes, bat, ball, and glove • Button, sticker, and jewelry making kits • Building systems • Sports teams • Magic kits • Video and computer games

Figure 12.14 (Valerie Rice) **Figure 12.15** (Valerie Rice)
Six- to eight-year-old children can ride bikes, jump rope, hopscotch, climb (trees, trapeze), and hop on one foot. Many are proficient at roller skating, ice skating, and swimming. By the time children are 6, their movements are deliberate but may still be a bit clumsy. However, by 8, their body movements are more rhythmic and their eye–hand coordination is quick and smooth. They cut, paste, hammer, and use tools. Their combined proficiency with fine motor skills, eye–hand coordination, and ability to reason and read leads many of them into beginning computer games or hand-held games. Many 8-year-olds can out-perform their parents on video and computer games!

9- TO 12-YEAR-OLDS

Nine- to 12-year-olds are interested in strength, skills, and teams (Figure 12.16 and Figure 12.17). Their physical abilities are honed, although they may still be somewhat awkward due to acceleration and asymmetries of anatomical growth. Depending on the child, they may be active or passive, depending on both their experiences and availability of activities.

At this age, children can think critically, evaluate ideas and people, and form their own opinions. However, acceptance of their peers is critical to most preteens and teens and they often bend their own opinions to those of the group around them. They allow athletic, musical, and dramatic celebrities to influence them.

Social groups become more structured. Boys tend to be more involved with peer group activities while girls tend to have more intimate and individual relationships. Significant adults, such as teachers and group leaders can exert a strong influence on group acceptance and self-confidence of adolescents (see Table 12.8).

Table 12.8 Guidelines for Toys for 9- to 12-Year-Olds

Goals: Complexity; Final Product, Results and Goals; Social Interaction	
• Sports and recreation	• Sports equipment
• Creativity and construction	• Kits: models, jewelry
• Thinking, problem solving, and strategy	• Games of strategy
• Social interactions	• Sports teams
• Collectables	• Cards, stamps, toys, dolls
• Electronics devices	• Videogames, computer games, Web-based games

Figure 12.16 (Valerie Rice) **Figure 12.17** (Valerie Rice)
Many 9- to 12-year-olds are computer savvy and can take apart, use, and repair electronic devices and gadgets. Video and computer games let them have a mechanism to play alone, in teams, or in competition with one another. They are also interested in team sports and many take lessons in dance, gymnastics, sports, music, or art.

EDUCATIONAL TOYS

Children engage in activities that are fun and somewhat challenging. Toys can promote learning of (a) general skills, traits, and values and/or (b) intellectual skills. General skills, traits, and values include broad skills such as eye–hand coordination, concentration, creativity, social cooperation, problem solving, language development, and negotiation (Rubin, et al., 1983). Intellectual skills include reading, writing, arithmetic, history, and science.

> *"We like to buy things that are learning-based, more than just toys", said Dianne Chodaczek of Riverside, Conn., as she made her way out of the crowd with a $49.99 LeapPad tucked under her arm. She said she had come on an errand for her grown daughter, who had already bought a LeapPad for her 4-year-old son and had decided to get one for her 6-year-old son, too.*
> ("Toy Story: Looking for Lessons," *New York Times,* January 3, 2002.)

Parents often prefer educational toys for their children. Parents of preschool and kindergarten children often prefer broad-based educational toys, but they are uncertain if the toys really have an effect on their children (Nelson-Rowe, 1994).

Company marketing firms recognize parental preferences and use them to attract customers. Testimonials of "satisfied" parents comprise a portion of their advertising. While some toy companies employ child development experts to help design educational toys (Ben-Ezer, 2001), they conduct limited research to demonstrate the effectiveness of those designs.

Educational toys make intuitive sense. While we would like to believe they help our children move through each developmental stage, there is little research to substantiate those claims.

TOY SAFETY

Consumers expect children's products, especially toys, to be safe. The following sections present an overview of toy safety.

INJURY AND FATALITY

A review of injury and fatality incidents with toys reveals that children under 5 years are at the greatest risk for injuries (CPSC NEISS data, 1996–2003) and fatalities (CPSC NEISS data, 1996–2002). During their first year of life, infants are most at risk when they become mobile: turning, rolling, crawling, walking, etc. Also, their use of mouthing as a primary exploration strategy increases their risk.

Choking is the leading cause of toy-related deaths (1993–2003, CPSC Web site) (Rider and Wilson, 1996) (Rimell et al., 1995). Intense, active mouthing moves objects to the back of the oral cavity, triggering the swallow reflex. Although this first action is voluntary, subsequent phases are vigorous, involuntary, and cannot be stopped (Torres and Stool, 1999) (see Chapter 7).

Every child will experience his share of cuts, bumps, and bruises while growing up. However, these types of experiences can be minimized. People can design toys to minimize the potential for injuries and avoid all potential for accidental fatalities. Some hazards that are likely to result in a fatality are choking (airway obstruction), suffocation, strangulation, and drowning.

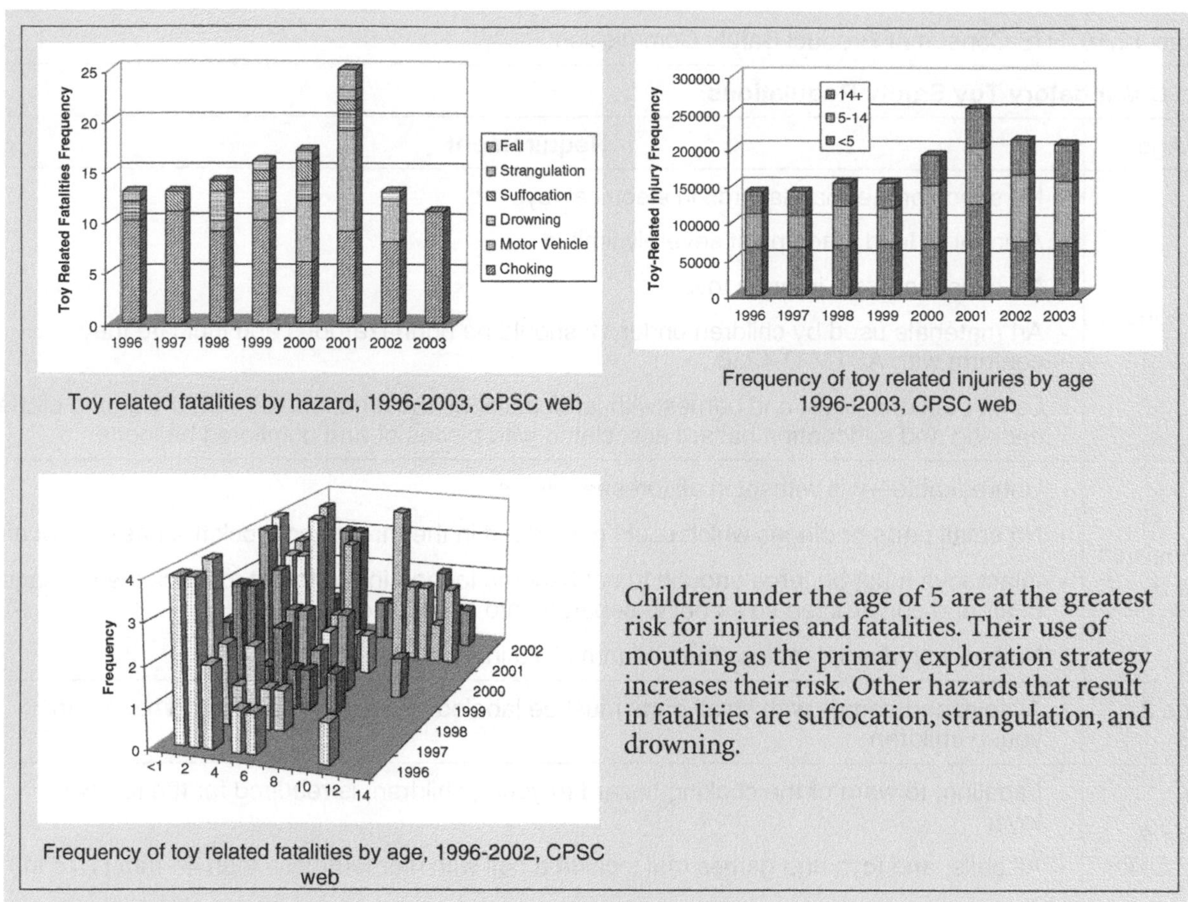

Toy related fatalities by hazard, 1996-2003, CPSC web

Frequency of toy related injuries by age 1996-2003, CPSC web

Frequency of toy related fatalities by age, 1996-2002, CPSC web

Children under the age of 5 are at the greatest risk for injuries and fatalities. Their use of mouthing as the primary exploration strategy increases their risk. Other hazards that result in fatalities are suffocation, strangulation, and drowning.

REGULATIONS

Most developed countries regulate toys via voluntary and mandatory standards. Table 12.9 lists governing bodies in the United States and Europe and the most common standards applied to all toys. Latin American countries, Australia, Hong Kong, to name a few and even the International ISO toy standard have adopted and/or referenced the European or United States toy standards. Table 12.10 summarizes United States CPSC general mandatory toy safety regulations. Please refer to the specific laws for complete details on these requirements and others.

Table 12.9 Toy-Regulating Bodies in the United States and Europe

United States	Europe
• Consumer Product Safety Commission (CPSC) Toy Distribution: Title 16, Code of Federal Regulations (16 CFR) for Commercial Practices, Subchapter C—Federal Hazardous Substances Act Regulations, Part 1500 to 1512. • American Society of Testing and Materials (ASTM) Voluntary Toy Safety Standard: F963-96a, Standard Consumer Safety Specification on Toy Safety.	• European Committee for Standardization COUNCIL DIRECTIVE of 3 May 1988 on the approximation of the laws of the Member States concerning the safety of toys (88/378/EEC) • EN-71 Safety of Toys

Table 12.10 US Consumer Product Safety Commission	
CPSC Mandatory Toy Safety Regulations	
Age	**Requirement**
All Ages	• No shock or thermal hazards in electrical toys • Amount of lead in toy paint severely limited • No toxic materials in or on toys • Art materials used by children under 12 should be nonhazardous and indicate they conform with ASTM D-4236 • Latex balloons, toys, and games with latex balloons must be labeled with a warning of the choking and suffocation hazard associated with pieces of and uninflated balloons
Under Age 3	• Unbreakable—will withstand all foreseeable uses • No small parts or pieces which could get lodged in the throat and block the passage of air • Infant toys must be large enough to not become lodged in the throat to block the passage of air and constructed so as not to separate into small pieces • No ball with diameters less than 44 mm (1.75 in.)
Ages 3 to 6	• All toys and games with small parts must be labeled to warn of the choking hazard to young children
Ages 3 and Older	Labeling, to warn of the choking hazard to young children, is required for the following toys: • All balls, and toys and games that include a ball with diameter less than 44 mm (1.75 in.) • All marbles, and toys and games with marbles
Under Age 8	• No electrically operated toys with heating elements • No sharp points on toys • No sharp edges on toys

Governments review injury and fatality trends and will try to specifically regulate the most frequent types of hazards with the most fatal consequences. Regulations have identified some performance tests to help evaluate the integrity of the toy assembly and materials. Because of this, some companies think they can design, produce, and then attempt to "test safety" into the toy. For all other potential hazards, in the United States, the CPSC will cite Section 15 of the Federal Hazardous Substance Act and in Europe, the European Union will uphold the 92/59/EEC of 29 June 1992 General Product Safety Directive. These laws simply state that a consumer product, not just toys, must be essentially safe. Therefore, not only must toy designers comply with all specific laws that apply to their product, but they must also take measures to prevent injuries that are not specifically regulated. Laws that specifically recognize every possible hazard with every possible toy cannot exist. Essentially, safety means that toy companies build safety in as a necessary and important characteristic of the toy.

It takes a proactive approach that starts at the conceptual stages of product development. A toy may meet or even exceed industry regulations, but still be unsafe due to risks inherent in some toy characteristics. Toy designers and manufacturers can achieve essential safety by understanding how a child will interact with the toy, including intentional use, foreseeable use, and child ergonomics. Then they can design the toy to be safe for those interactions. They can either prevent the interaction or make it safe to do so. The same type of information designers use to figure out what will make a toy fun can be used to make the toy safe.

EXPECT AND PREDICT INTERACTIONS: Child development milestones help identify how a child will interact with a toy. Cognitive and motor skills develop in a sequential manner. Usability studies help confirm these behaviors and identify other possible uses.

Children, with their innate sense of curiosity, seek the most stimulation possible from their environment. If the toy does not provide the needed stimulation, they will use innovative strategies to provide it themselves. It is unrealistic to think children can stop or curtail this innate curiosity and process of discovery. While toy companies cannot control children's behaviors, designers can predict those behaviors based on the child's development stage and environment. They can then design products that are safe as possible, in each circumstance.

CHILD HUMAN FACTORS: Once the designer knows what the interaction will be, he or she must understand what the consequence of the interaction is. For example, we know that toddlers and preschoolers will insert their body parts into objects.

What is the consequence of this child inserting his finger into a small round hole? Child human factors such as anatomy, physiology, anthropometry, strength data, etc. help designers determine what the outcome of the predicted interaction can be.

Child anthropometrics include the sizes of children's fingers. This information will allow the designers to make the openings in a toy large enough so that all children can insert and remove their fingers easily. Alternatively, the designer can make the openings smaller and not allow any of the fingers to enter.

LEARN FROM HISTORY: Review of historical injuries or government recalls of similar toys or products with similar characteristics or function can confirm or identify interactions and their consequences. In the United States, the CPSC can provide some of this information through their Web site.

LABELING AND WARNINGS

Regulatory agencies mandate that toys include warning labels and manufacturer recommendations for appropriate child user ages. Several studies show that compliance to warning labels is poor at best (Table 12.11) (Dejoy, 1994) and warnings should be secondary to well-thought out, safe design (see Chapter 14).

Here is an excerpt from The Toy Industry Foundation, Inc. guide, Fun Play, Safe Play.

> *..Toys are labeled based on four criteria: the safety aspects of the toy; the physical capabilities of the child (ability to manipulate the toy); the cognitive abilities of the child (understand how to use the toy); and the child's interest.*
>
> *The most common safety label warns against choking hazards. Any toy or game manufactured for children ages three to six is required to carry such a warning if the toy contains small parts, small balls, marbles, or a balloon. Such toys are not for children under three or any child who is still mouthing objects. Other common labels to look for include "flame retardant/flame resistant" on fabric products, "surface or machine washable" on stuffed toys and dolls and "UL Listed" (for Underwriters Laboratories) on electrically operated toys.*

Some manufacturers add other warnings to the package and/or instructions advising parents that special care should be taken. Toys that would have cautionary labels might include: science toy sets with toxic chemicals, craft kits with sharp or breakable items, and crib gyms and mobiles, which should be removed when a baby reaches five months of age or begins to push up on hands and knees…

Table 12.11 Warning Label Studies

Study	Noticed	Read	Complied
Friedmann (1988)	88%	46%	27%
Otubso (1988)	64.3%	38.8%	25.5%
Strawbridge (1986)	91%	77%	37%

Unfortunately, many parents and relatives are unaware that age labeling recommendations are designed to promote child development and safety. Most parents describe their children as above average (Brown et al., 2002) and many parents and relatives buy their child toys that are designed for an older child.

Children may become frustrated if their toys are too advanced for them. They may give up trying to use the toys (discouraging their use and the gift-giver) or resort to exploration strategies of younger children, which may introduce safety concerns. Toy designers must consider parental perceptions as a foreseeable use.

CONCLUSIONS

Designing toys for children is not easy, but it is exciting and fulfilling. Balancing safety issues with a slight "pushing of the envelope" to encourage a child's development, while simultaneously making a product fun for a child takes thought, problem solving, and creativity (Figures 12.18 through 12.20). The trick is not to lose sight of the user: the child. That is why product testing, with the appropriate age and gender of children for whom the product is intended, is essential.

Figure 12.18 (Valerie Rice)

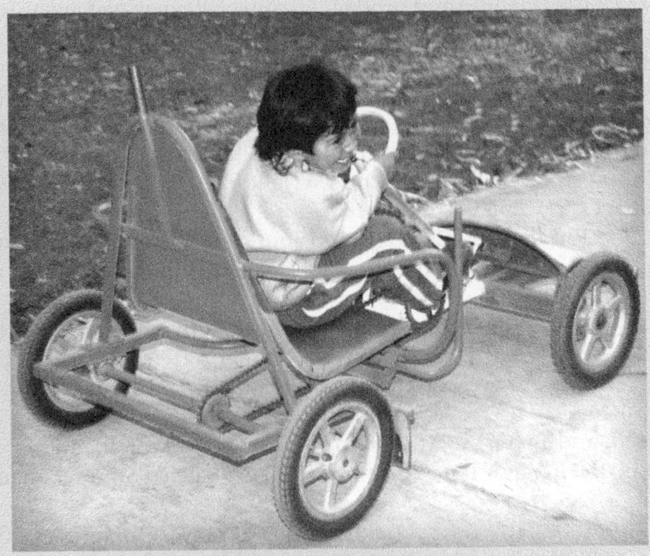

Figure 12.19 (Ursy Potter Photography)

Children of all ages enjoy play and toys. The key is finding toys that are not only fun, but also encourage continued physical, emotional, and cognitive development. Toys that offer the chance to show competence, while still providing challenge are what every parent and teacher seeks for their children.

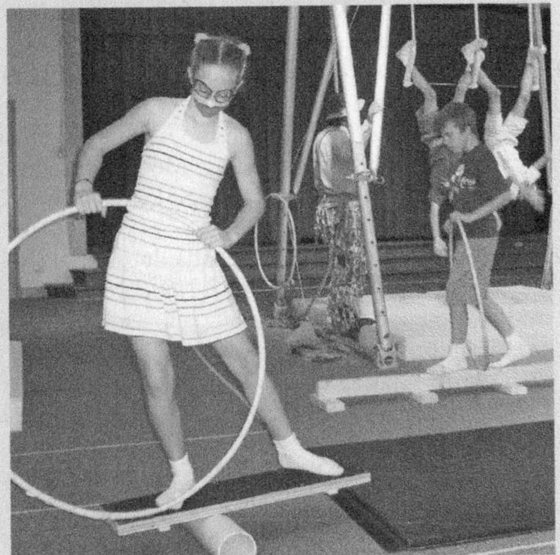

Figure 12.20 (Valerie Rice)

REFERENCES

Adams, R.J. (1987) An evaluation of colour preferences in early infancy. *Infant Behaviour and Development*, 10, 143–150.

American Standard for Testing and Materials. (1996) Standard Consumer Safety Specification on Toy Safety. ASTM Designation: F 963-96a.

Beaty, J. (2002) Cognitive development: Classification, number, time and space. *Observing Development of the Young Child*, 5th Edn. Upper Saddle River, NJ, Columbus, OH: Merrill Prentice Hall, pp. 254–294.

Ben-Ezer, T. (2001) Development is child's play: Developmental toys for children. *Playthings*, 99, 8, 2001-08-00, 42.

Bredekamp, S. and Copple, C., (1997) Developmentally appropriate practice in early childhood programs (revised edition). Washington, D.C.: National Association for the Education of Young Children.

Brown, T., Lepsis, M., Chen, X., and Harris, D. (2002) Caregiver vigilance national survey. Unpublished. Oak Brook, IL: RAM Consulting.

Bruce, T. (1993) The role of play in children's lives. *Childhood Education*, 69(4), 237–239.

Christensen, K.E. and Stockdale, D.F. (1991) Predictors of toy selection criteria of preschool children's parents. *Children's Environments Quarterly*, 8(1), 25–36.

DeJoy, D.M. (1994) Consumer product warnings: Review and analysis of effectiveness research. In Laughery, K.R. Sr., Wogalter, M.S., and Young, S.L. (Eds.), *Human Factors Perspectives on Warnings*, Santa Monica, CA: Human Factors Society, pp. 16–20.

Easterbrook, M., Kisilevsky, B.S., Hains, S.M.J., and Muir, D.W. (1999) Faceness or complexity: evidence from newborn visual tracking of facelike stimuli. *Infant Behaviour & Development,* 22(1), 17–35.

European Committee for Standardization (1998) Safety of Toys. EN-71 parts 1–5.

Fisher-Thompson, D., Sausa, A.D., and Wright, T.F., (1995) Toy selection for children: Personality and toy request influences. *Sex Roles*, 33(3–4), 239–255.

Fox, K.R. and Riddoch, C. (2000) Symposium on Growing Up with Good Nutrition: A focus on the first two decades. Charting the physical activity patterns of contemporary children and adolescents. *Proceedings of the Nutrition Society*, 59, 497–504.

Garvey, C. (1990a) What is play? *Play.* Cambridge, MA: Harvard University Press, pp. 1–15.

Garvey, C. (1990b) Play with objects. *Play.* Cambridge, MA: Harvard University Press, pp. 41–57.

Goncu, A., Mistry, J., and Mosier, C. (2000) Cultural variations in the play of toddlers. *International Journal of Behavioral Development,* 24(3) 321–329.

Goodson, B. and Bronson, M. (2001) Which toy for which child, A consumer's guide for selecting suitable toys ages birth through five. Washington, D.C.: United States Consumer Product Safety Commission, Publication number 285.

Guernsey, L. (2002) Toy story: Looking for lessons. *New York Times*, 2002-01-03, pp D1 (N) pG1 (L).

Harrell, J.S., Gansky, S.A., Bradley, C.B., and McMurray, R.G. (1997) Leisure time activities of elementary school children. *Nursing Research*, 46(5), Sept–Oct, 246–253.

Huston, A.C., Wright, J.C., Marquis, J., and Green, S.B. (1999). How young children spend their time: television and other activities. *Developmental Psychology*, 35(4), 912–925.

Jordan, P.W. (2000) Pleasure with products: Beyond usability. *Designing Pleasurable Products*. London and New York: Taylor & Francis, pp. 1–10.

Lee, J. (2002) Making toys for children too mature for most toys. (New toys for 8–12-year-olds feature influence of pop stars 'N Sync, Britney Spears, Mandy Moore, and Destiny's Child). *New York Times*, 2002-02-12, pp C1 (N) pC1 (L).

Morton, J. and Johnson, M.H. (1991) CONSPEC and CONLERN: A two-process theory of infant face recognition. *Psychology Review*, 98, 164–181.

Nelson-Rowe, S. (1994) Ritual, magic, and educational toys: Symbolic aspects of toy selection. In Best, J. (Ed.), *Troubling Children: Studies on Children and Social Problems*, New York: Aldine de Gruyter, Vol. VI, pp. 117–131.

Rider, G. and Wilson, C.L. (1996) Small parts aspiration, ingestion and choking in small children: Findings of the small parts research project. *Risk Analysis,* 16(3), 321–330.

Rimell, F.L., Thome, Jr., A., Stool, S., Reilly, J.S., Rider, G., Stool, D., and Wilson, C.L. (1995) Characteristics of objects that cause choking in children. *Journal of the American Medical Association*, 274(22), 1763–1766.

Rogers, C.S. and Sawyers, J.K. (1988a) Play is life for young Children. In *Play in the Lives of Children.* Washington DC: National Association for the Education of Young Children, pp. 1–11.

Rogers, C.S. and Sawyers, J.K. (1988b) Different ways to look at play. In *Play in the Lives of Children.* Washington DC: National Association for the Education of Young Children, pp. 12–22.

Rubin, K.H., Fein, G., and Vandenberg, B. (1983) Play. In E.M. Hetherington (Ed.). *The Handbook of Child Psychology: Social Development.* New York: Wiley, pp. 693–774.

Therrell, J.A., Brown, P., Sutterby, J.A., and Thornton, C.D. (2002). In T.P. Smith (Ed.), *Age Determination Guidelines: Relating Children's Age to Toy Characteristics and Play Behavior.* Washington, D.C.: United States Consumer Product Safety Commission.

Torres, B.A. and Stool, D.K. (1999) The application of human factors to the design of a toy: Little People® then and now. Proceedings of the Human Factors and Ergonomics Society 43rd Annual Meeting—1999, pp. 511–514.

Toy Industry Foundation, Inc. (2005) *Fun Play, Safe Play: A guide from the Toy Industry Foundation.* New York: www.toy-tia.org/Content/NavigationMenu/Toy_Industy_Foundation/Toy_Industy_Foundation.htm

United States Consumer Product Safety Commission. Federal Hazardous Substances Act Regulations. Title 16 of the Code of Federal Regulations (16 CFR) for Commercial Practices under Subchapter C, Part 1500 to 1512.

CPSC Web site: www.cpsc.gov/library/data.html

CHAPTER 13

BOOKBAGS FOR CHILDREN

KAREN JACOBS, RENEE LOCKHART, HSIN-YU (ARIEL) CHIANG,
AND MARY O'HARA

TABLE OF CONTENTS

INTRODUCTION

Schoolchildren around the world carry personal and school materials in book bags. In the United States alone, more than 40 million youths use book bags to carry textbooks, binders, laptop computers, bottles, clothes, athletic equipment, makeup, toys, and more.

Backpacks are a practical way to transport schoolwork. Even so, research demonstrates a strong correlation between musculoskeletal injuries and improper backpack use.

For example, a 1999 study found that the use of book bags resulted in more than 6000 injuries in the United States (Pain, 2001). The US Consumer Product Safety Commission reports that 7860 emergency room visits were related to backpacks and book bags (NEISS, 2002).

Symptoms

Physical symptoms from backpacks that were packed or worn improperly include:

1. shoulder, back, hands, neck, and leg pain (Iyer, 2000);
2. muscle soreness and numbness (Pascoe et al., 1997); and
3. physiological strain related to heart rate, blood pressure, and metabolic cost (Pascoe et al., 1997).

When left untreated, these can lead to serious musculoskeletal injuries.

CONTRIBUTING FACTORS

Early detection can prevent injuries and identify risky habits such as

Improper Use

1. overloading backpacks (packing too many things),
2. unevenly distributing backpack contents,
3. wearing a two-strap backpack on only one shoulder,
4. carrying backpacks for long time periods or too frequently,
5. using poorly designed book bags or backpacks, and
6. lifting backpacks with one hand or from an elevated position.

It is often possible to determine whether a backpack is too heavy by observing the child's postures and behaviors (Figure 13.1). Behaviors that indicate a backpack is overloaded, loaded unevenly, or worn improperly include the following:

Behaviors

1. Forward lean and tilt of the trunk and neck (Pascoe et al., 1997)
2. Raising one shoulder higher than the other while carrying a pack on one side or with one strap (Pascoe et al., 1997; Whittfield et al., 2001)
3. Hyperextension of the back or leaning backwards (Goodgold et al., 2002)
4. Using hands to pull shoulder straps (Goodgold et al., 2002)

These behaviors are risky because

1. awkward postures impair the spine's natural shock absorption abilities, increase muscle work, and may lead to fatigue and musculoskeletal injuries (Goodgold et al., 2002) and
2. prolonged loading of the spine increases the risk of low back pain in youths (Negrini et al., 1999).

Figure 13.1 (Valerie Rice)
Children change their postures to adapt to backpacks that are too heavy or improperly worn. Kiran compensates for carrying a backpack over one shoulder (left) by leaning toward the opposite side (right) and widening her stance. Daniel compensates by leaning slightly backward.

PREVENTING DISORDERS FROM IMPROPER BACKPACK USE

SELECTION

Popular fashions often influence youths' purchasing decisions. Fashion is important in getting children to use their backpacks, which come in many shapes, sizes, colors, and materials (see Figure 13.2).

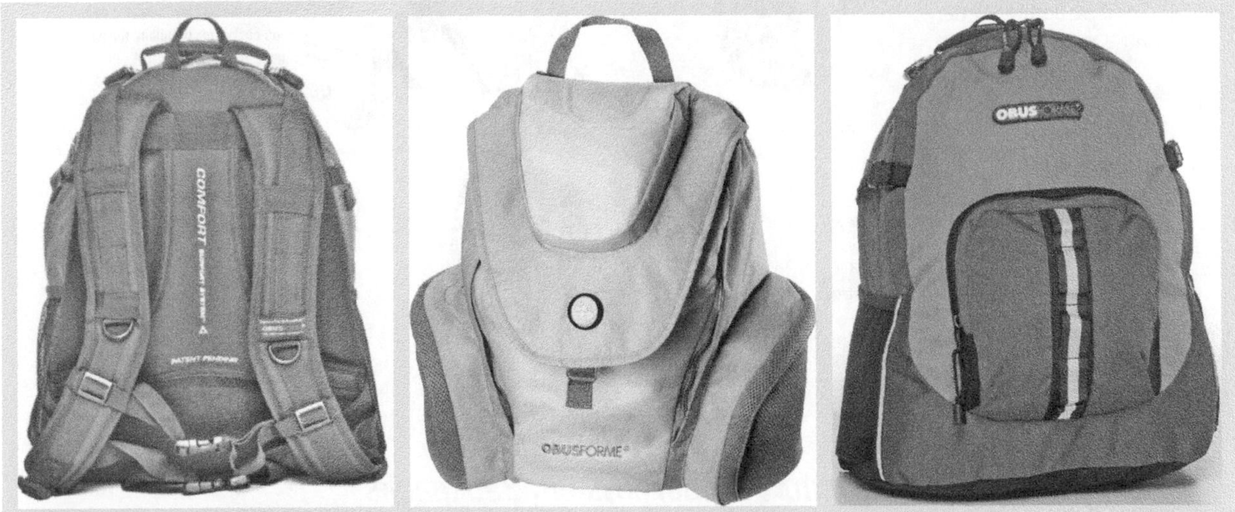

Figure 13.2 (Obus Forme Ltd.)
Fashion often means more to a child than whether the padding makes it easier to carry, fits their body size, or has room and pockets for their supplies. Caregivers and teachers will have to help children match their fashion desires with a pack that it is appropriate for their size, age, and the materials they intend to carry.

Selecting a backpack that is appropriate for the youth's size and age and "fit." Ideally, backpacks should

What youths, parents, and practitioners can do

1. rest on the child's back, between her neck and the curve of her low back;
2. include two wide contoured (curved), padded, and adjustable shoulder straps;
3. have a padded back for increased comfort;
4. hip and chest belts provide stability and transfer weight from shoulders to torso and hips; and
5. provide multiple compartments (especially on the sides) for more even weight distribution (Jacobs et al., 2002).

Some manufacturers include guidance to help users adjust their packs to fit them. Figure 13.3 shows a step-by-step guide.

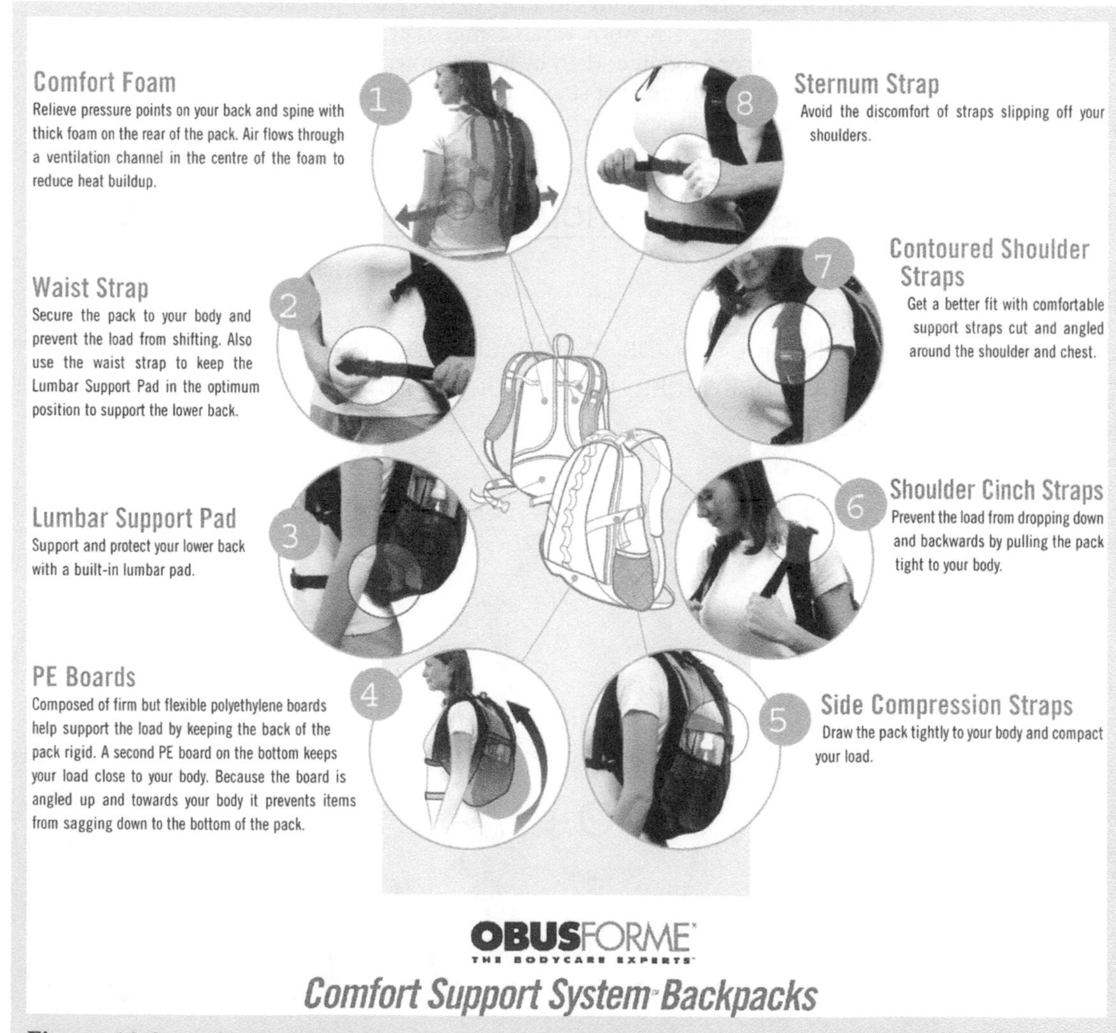

Comfort Foam
Relieve pressure points on your back and spine with thick foam on the rear of the pack. Air flows through a ventilation channel in the centre of the foam to reduce heat buildup.
1
Waist Strap
Secure the pack to your body and prevent the load from shifting. Also use the waist strap to keep the Lumbar Support Pad in the optimum position to support the lower back.
2
Lumbar Support Pad
Support and protect your lower back with a built-in lumbar pad.
3
PE Boards
Composed of firm but flexible polyethylene boards help support the load by keeping the back of the pack rigid. A second PE board on the bottom keeps your load close to your body. Because the board is angled up and towards your body it prevents items from sagging down to the bottom of the pack.
4
Sternum Strap
Avoid the discomfort of straps slipping off your shoulders.
8
Contoured Shoulder Straps
Get a better fit with comfortable support straps cut and angled around the shoulder and chest.
7
Shoulder Cinch Straps
Prevent the load from dropping down and backwards by pulling the pack tight to your body.
6
Side Compression Straps
Draw the pack tightly to your body and compact your load.
5
OBUSFORME
THE BODYCARE EXPERTS
Comfort Support System Backpacks

Figure 13.3 (Obus Forme Ltd.)
This manufacturer's guide takes older children and caretakers through a step-by-step process of adjusting the book bag. It emphasizes features that will help users be safe and comfortable while carrying loads.

LOAD

Some children carry book-bag loads of 20 pounds (9.07 kg), approximately 22% of their body weight (Negrini et al., 1999). Yet, experts recommend that book bags weigh no more than 10%–20% of a youth's body weight. For example, a child's ability to fully expand his lungs is restricted with book bags that are 9.4% or greater than his body weight (Legg and Cruz, 2004).

Obviously, this effect amplifies with weights greater than 10% of body weight (Lai and Jones, 2001). Research also demonstrated that a schoolbag load heavier than 10% of a child's body weight was associated with a restrictive effect on lung volume (Lai and Jones, 2001). Besides, tightening shoulder and hip straps even 3 cm (1.18 in.) tighter than comfort level was roughly equivalent to the effect on lung function of wearing a backpack loaded to a 15 kg load (33 lbs) (Bygrave et. al., 2004).

These recommendations may require further analysis because the percentage of body weight does not account for a child's muscle strength, height, and percentage of body fat. They also do not account for special cases, such as the prevalence of child obesity in the United States (Goodgold et al., 2002).

To help minimize risk:

1. load heaviest items closest to the youth's back (the back of the book bag),
2. do not place pointed or bulky objects against the back,
3. arrange books and materials so they will not slide around,
4. replace heavier items with lighter alternatives (e.g., a plastic lunch box instead of a metal one),
5. avoid packing heavy or bulky items like soda, water, shoes, and sports equipment,
6. ensure the youth carries only the items required for that particular day,
7. if the book bag is too heavy, consider using a book bag on wheels (Figure 13.4), and

Figure 13.4 (Ursy Potter Photography)
Perhaps one of the best methods to deal with the heavy loads of backpacks is not to have children carry them at all, but to have them pull them instead.

This works except for getting into or out of vehicles (especially public transportation), walking over non-paved or uneven terrain, and manipulating the busy halls of a junior or senior high school. A final concern for the children themselves is whether pulling a bag "looks cool" and is acceptable in their social circles.

8. be sure the backpack fits by tightening the shoulder and hip straps to where they feel comfortable for the child. Avoid wearing straps that are too loose or too tight.

WEAR

Most students use two-strap backpacks. In one study, about 93% of 11 to 13-year-old students used standard, two-strap backpacks; 7% used one-strap athletic bags or tote bags. However, 75% of students using two-strap backpacks report carrying the backpack with only one strap (Pascoe et al., 1997).

Wearing and fastening both straps increases stability (but not too tightly). Position book bags below the youth's shoulders while resting on hips and pelvis. Hip and chest belts help distribute the weight.

Figure 13.5 (Valerie Rice)
Even carrying a relatively light pack can change a child's posture, especially on stairs or hills. The weight of the pack and the strength of the child both influence posture. Here Daniel leans forward and slightly flexes his neck.

LIGHTEN THE LOAD

What schools and teachers can do

Individual teachers and schools can help lighten the load that children transport to and from school. For example, schools can encourage children to use textbook alternatives that decrease the weight of books. They can use books on CD ROM, Web-based books, or books bound in sections or paperbacks that can lighten the load. Although costly, schools can also issue a second set of books for home use. Some schools sell textbooks, enabling parents to buy a second set of books for their child's use at home.

Schools can limit materials and books that need to go home on a daily basis. Teachers can coordinate with each other when homework requires heavier textbooks. Teachers could also encourage lightweight alternatives to school supplies, such as by using spiral bound or composition books and pocket folders in place of heavy three- ring binders (Figure 13.5). Finally, schools with lockers can assign lockers or cubicles close to the classroom or other convenient locations (Jacobs et al., 2002).

CHANGE THE DESIGN

What book-bag manufacturers and legislators can do

Manufacturers can sponsor ergonomic research that evaluates backpack designs and design requirements. This is especially important because most studies have focused on adults and on the military.

Although some studies comparing the benefits of backpacks with internal and external frames seem to favor internal frames for perceptions of comfort, balance, and ease of gait (Legg et al., 1997), the results will depend on the pack and the users.

The demands for adult soldiers, such as climbing or walking up steep grades, running with a fully loaded pack, and performing soldiering tasks (such as low crawls and weapon firing), are quite different from the demands on a child's school day. Pivoting hip-belts that allow heavy-load carriage with minimal lower-body interference on steep hills and ultraloc straps to minimize pack movement (Legg et al., 1997) may simply not be necessary for school children.

Another study that focused on the effects of carrying different kinds of backpacks and load weight of children during stair-walking found that loads of 10% of body weight or greater induce trunk forward inclination during stair descent. Therefore, two-strap backpacks were safer than one-strap athletic bags for children stair-walking (Hong et al, 2003). When backpack loads are the same, one-strap backpacks also restrict ventilatory lung function more than two-strap backpacks (Legg and Cruz, 2004).

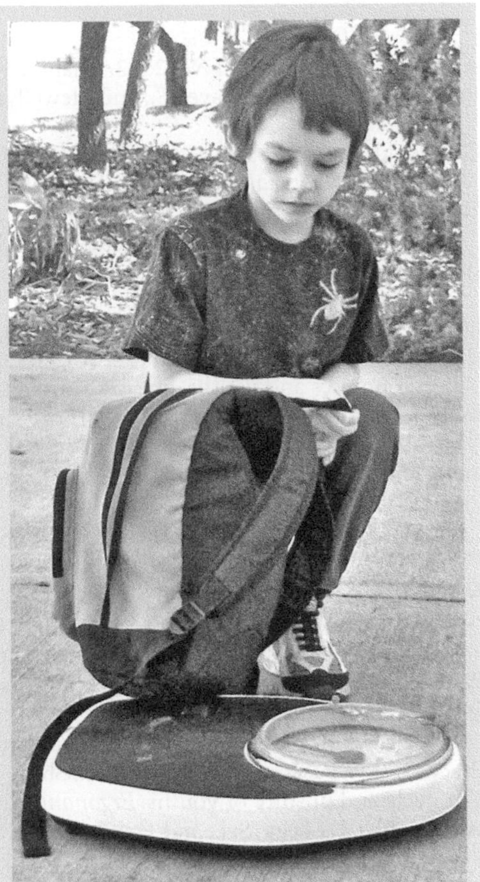

Figure 13.6 (Valerie Rice)
Students from the Occupational Therapy Department at Boston University visited local elementary schools during National Backpack Awareness Day. They taught students and teachers how to fit and wear backpacks, as well as weighing students' packs as part of the educational process.

As mentioned, book-bag manufacturers can also educate users in selecting, loading, and carrying book bags properly. LL Bean, a well-known out-door-equipment company, partnered with the American Occupational Therapy Association (AOTA) by providing brochures and hang tags on how to select, load, and carry book bags correctly. They also supported the Association's National Backpack Awareness Day in the United States in 2002 (Figure 13.6). Lands End has also partnered and helped produce public health information with the American Physical Therapy Association (APTA). These are just two examples of manufacturers' collaboration and health promotion with (and for) the public.

Health concerns associated with backpacks have provided the catalyst for legislative action in one US state. On September 29, 2002, California passed an Act that adopts maximum weight standards for elementary and secondary school textbooks (AB 2532, 2002; Kelly, 2002).

Section 49415 that was added to the California Education Code states:

> *"On or before July 1, 2004, the State Board of Education shall adopt maximum weight standards for textbooks used by pupils in elementary and secondary schools. The weight standards shall take into consideration the health risks to pupils who transport textbooks to and from school each day" (AB 2532, 2002).**

It is uncertain whether California legislation will become a trend in the United States.

* Resolution Number 10, introduced into the California legislature in 2003–2004, aimed to increase public awareness of the relative public health issues of heavy book bags.

REFERENCES

AB (2002). Assembly Bill No. 2532, Chapter 1096. Available at: www.childergo.com/recc-child.htm#schoolchildlaw

Bygrave, S., Legg, S. J., Mayer, S., and Llewellyn, M. (2004). Effect of backpack fit on lung function. *Ergonomics, 47,* 324–329.

Goodgold, S., Corocran, M., Gamache, D., Gillis, J., Guerin, J., and Coyle, J. S. (2002). Backpack use in children. *Pediatric Physical Therapy, 14,* 122–139.

Hong, Y., Lau, T. C., and Li, J. X. (2003). Effects of loads and carrying methods of school bags on movement kinetics of children during stair walking, *Research in Sports Medicine, 11,* 33–49.

Iyer, S. R. (2000). Musculoskeletal pain in school children. Proceedings of the IEA 2000/HFES 2000 Congress, 5-419–5-422.

Jacobs, K., Bhasin, G., Bustamante, L., Buxton, J., Chiang, H. -Y, Greene, D., Ford, S., Frank, R., Moczydlowski, E., Naidu, J., Nielsen, E., Savini, A., Singer, B., and Wieck, A. (May 27, 2002) Everything you should know about ergonomics and youths, but were afraid to ask, *Occupational Therapy Practice*, pp 11–14, 16–19.

Kelly, J. (October 22, 2002). Shaking off the strain: Your school backpack and home computer shouldn't be a pain. *KidsPost, Washington Post.* p. C14.

Legg, S. J. and Cruz, C. O. (2004). Effect of single and double strap backpacks on lung function. *Ergonomics, 47,* 318–324.

Legg, S. J., Perko, L., and Campbell, P. (1997) Subjective perceptual methods for comparing backpacks. *Ergonomics, 40,* 809–817.

Negrini, S., Carabalona, R., and Sibilla, P. (1999). Backpack as daily load for school children. *Lancet, 354,* 1974.

NEISS (2002) U.S. Consumer Product Safety Commission National Electronic Injury Surveillance System (NEISS) database, 2002. Numbers quoted are the national estimated figures.

Pain & Central Nervous System Week Editors (2001, September 22). Backpack misuse causes multiple problems, chiropractors say. *Pain & Central Nervous System Week.* Tufts University, Health Reference Center Academic Website: http://web6.infotrac.galegroup.com/

Pascoe, D. D., Pascoe. D. E., Wang, Y. T., Shim. D. M., and Chang, K. K. (1997). Influence of carrying book bags on gait cycle and posture of youths. *Ergonomics, 40,* 631–641.

Whittfield, J. K., Legg, S. J., and Hedderley, D. I. (2001). The weight and use of schoolbags in New Zealand secondary schools. *Ergonomics, 44,* 819–824.

OTHER RESOURCES

Backpacks: Do they cause aches in schoolchildren? (Feb 2000). *Child Health Alert, 1.* http://web6.infotrac.galegroup.com/

Baclawski, K. (1995). *The Guide to Historic Costume.* New York, NY: Drama Book Publishers.

Balague, F., Dutoit, G., and Waldburger, M. (1988). Low backpain in school children. *Scandinavian Journal of Rehabilitation Medicine, 20,* 175–179.

Burton, K. A., Clarke, R. D., McClune, T. D., and Tillotson, K. M. (1996). The natural history of low back pain in adolescents. *Spine, 21,* 2323–2328.

Grimmer, K. A., Williams, M. T., and Gill, T. K. (1999). The association between adolescent head-on-neck posture, backpack weight, and anthropometric features. *Spine, 24,* 2262–2267.

Grimmer, K. and Williams, M. (2000). Gender-age environmental associates of adolescent low back pain. *Applied Ergonomics, 31,* 343–360.

Guyer, R. L. (2001). Backpack=Backpain. *American Journal of Public Health, 91,* 16–19.

Hamilton, A. (2001). Prevention of injuries from improper backpack use in children. *Work, 16,* 177–179.

Kinoshita, H. (1985). Effects of different loads and carrying systems on selected biomechanical parameters describing walking gait. *Ergonomics, 28,* 1347–1362.

Lai, J. P. and Jones, A. Y. (2001). The effect of shoulder girdle loading by a school bag on lung volumes in Chinese primary school children. *Early Human Development, 62,* 79–86.

Leboeuf-Yde, C. and Kyvik, K. O. (1998). At what age does lower back pain become a common problem? A study of 29,424 individuals aged 12–41 years. *Spine, 23,* 228–234.

Legg, S., Cruz, C., Chaikumarn, M., Kumar, R. Efficacy of subjective perceptual methods in comparing between single and double strap student backpacks. (2002) In Thatcher, A., Fisher, J., Miller, K. (eds.) Proceedings of CybErg'2002. The third International Cyberspace Conference on Ergonomics. International Ergonomics Association press.

Legg, S., Criz, C. Influence of backpack straps on pulmonary function. (2002) In Thatcher, A., Fisher, J., Miller, K. (eds.) Proceedings of CybErg'2002. The third International Cyberspace Conference on Ergonomics. International Ergonomics Association press.

Lohman, E. B. and Wong, M. (2002, Mar/April). Packs & backs: Treating and preventing backpack-related pain is vital to avoiding long-term spinal deformities. *Physical Therapy Products*, 24–46.

Merati, G., Negrini, S., Sarchi, P., Mauro, F., and Veicsteinas. A. (2001). Cardio-respiratory adjustments and cost of locomotion in school children during backpack walking (the Italy backpack study). *European Journal of Applied Physiology, 85,* 41–48.

Negrini, S. and Carabalona, R. (2002). Backpacks on! Schoolchildren's perceptions of load, associations with backpain and factors determining the load. *Spine, 27,* 187–195.

Roth, C. (2001). Parents & teachers need a lesson in ergonomics. *Industrial Safety & Hygiene News, 35,* 62.

Troussier, B., Davoine, P., de Gaudemaris, R., Fauconnier, J., and Phelip, X. (1994). Backpain in school children: A study among 1178 pupils. *Scandinavian Journal of Rehabilitation Medicine, 26,* 143–146.

Vacheron, J. J., Poumarat, G., Candezon. R., and Vanneuville, G. (1999). The effect of loads carried on the shoulders. *Military Medicine, 164,* 597–560.

Wang, Y. T., Pascoe, D. D., and Weimar, W. (2001). Evaluation of book backpack load during walking. *Ergonomics, 44,* 858–869.

Wiersema, B., Wall, E., and Foad, S. (2003). Acute backpack injuries in children. *Pediatrics, 111,* 163–166.

CHAPTER 14

WARNINGS: HAZARD CONTROL METHODS FOR CAREGIVERS AND CHILDREN

MICHAEL J. KALSHER AND MICHAEL S. WOGALTER

TABLE OF CONTENTS

INTRODUCTION

Injury prevention is especially important for young children. They have not developed cognitive abilities, which may be necessary both to appreciate the magnitude of the hazards they encounter and to know how to avoid them (Figure 14.1). Parents and other caregivers should take appropriate action to protect young children from hazards. This may not occur for various reasons:

- People differ. Caregivers and the extent of care they give will differ, too.
- People rely on their own experiences. By seeing through the looking glasses of their experiences, caregivers may not be attuned to how a child sees the same thing.
- People may be overconfident. As a child matures and extends his or her abilities, the parents may believe he or she is capable of appropriately dealing with situations or common consumer products.
- Children behave in unexpected ways. Caregivers may falsely believe their children are better equipped to deal effectively with hazardous situations than is actually warranted (Hiebert and Adams, 1987).

Examples of hazards inside the home
Cleaning supplies
Medicines
Kitchen appliances
Electrical outlets
Toys
Weapons

Figure 14.1a (Alison Vredenburgh) **Figure 14.1b** (Kalsher & Assoc. LLC)
Examples of hazards present inside and outside the home. As shown here, products and situations both inside and outside the home can be dangerous. The problem, of course, is that many children do not view these products or situations as dangerous. Manufacturers and caregivers can both play important roles in helping to reduce childhood injuries.

HIDDEN HAZARDS

As consumer products become progressively more complex, it becomes increasingly difficult to identify product hazards. When products have "hidden" hazards (see Box 14.1), manufacturers bear a responsibility for warning caregivers (Figure 14.2).

Unintentional injuries are a significant cause of death and injury for children. Of these, motor-vehicle accidents and drowning are the leading causes of death among children (CPSC, 2002a; 2000a). Many children are also severely injured or killed following exposure to household chemicals, nursery products, toys, and playground equipment as shown in Table 14.1.

Reported injuries and deaths do not always list a cause or the exact circumstances surrounding these events. The injury statistics cited in Table 14.1 may actually underestimate their occurrence. Threats to children's safety clearly pose a significant problem.

BOX 14.1 EXAMPLES OF "HIDDEN HAZARDS"

Airbags	During the 1990s, the passenger-side airbags in many vehicles presented a hidden hazard to small children. At that time, it was not possible to discern the hazards of passenger-side airbags just by looking at vehicle's dashboard. Many consumers did not know, for example, that the air bags in many cars at that time deployed at rates exceeding 100 mph, which can cause serious injury or death if a person sits too close to the vehicle's dashboard at the time of a crash. As this danger became evident, manufacturers redesigned the airbag systems and placed warnings on vehicle sun-visors and on child-restraint systems.
Child Safety Seats	Many caregivers do not understand they must anchor these devices properly to the vehicle's seat belt system for them to be effective in a crash.
Window Blinds	Caregivers may not realize the cord that raises and lowers the blinds can wrap around a child's neck and choke them. Some kinds of window blinds can even cause lead poisoning.

Table 14.1 Injuries and Deaths among Children under the Age of 5 in the United States from Commonly Used Products

Products	Child Injuries
Poisoning	>1 million
Toy-Related Products	212,400
Playground Equipment	200,000
Nursery Products	69,500

⚠ WARNING

DEATH or SERIOUS INJURY can occur
- Children 12 and under can be killed by the air bag.
- The BACK SEAT is the SAFEST place for children.
- NEVER put a rear-facing child seat in the front.
- Sit as far back as possible from the air bag.
- ALWAYS use SEAT BELTS and CHILD RESTRAINTS.

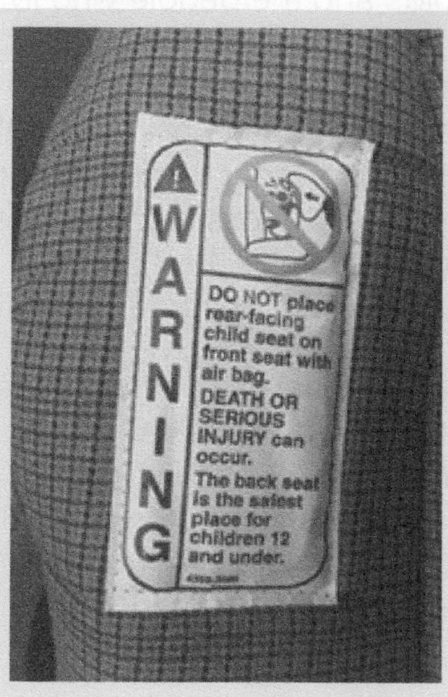

Figure 14.2 Warning about the potential hazards of airbags. It is helpful to place warnings where they can be seen with each use of a product. The warning depicted on the left is present on the sun visor of many late model U.S. vehicles with passenger side airbags. On the right side is a similar warning sewn onto a child car seat (U.S. Dept. Transportation NHTSA).

CONTROLLING HAZARDS

MANUFACTURERS

To control hazards, caregivers must first identify possible sources of risk and then take active steps to control or eliminate them. *Hazard analysis* is a step-by-step approach to identify hazards.

The term *hazard* refers to any part of a product or activity that might lead to injury.

Manufacturers use hazard analysis to evaluate foreseeable uses and misuses of their product in order to identify and eliminate possible hazards. This analysis could include data from government, trade association and company databases, news reports, and legal documents from cases involving injuries resulting from the use or exposure to consumer products. Analysis can also draw on expert predictions from health professionals, chemists, engineers, and human factors professionals about the risks these products pose to users.

Unfortunately, some manufacturers may perform this analysis in a cursory fashion without considering the human factors involved. Moreover, they may fail to recognize dangers associated with products and their uses as they relates to children. Without insight about children's capabilities and limitations, hazard analyses performed by manufacturers may fail to uncover some important potential dangers to children.

CAREGIVERS

Parents and caregivers can conduct their own hazard analysis of products and environments.

Therefore, parents and other caregivers should not rely solely on manufacturers to identify and resolve all of the potential hazards associated with the products they produce. Instead, they need to establish consistent procedures for evaluating products and environments to identify hazards and take active steps to eliminate them. Caregivers can use a less-structured approach to systematically identify and eliminate hazards.

For example, caregivers could observe children using the product in their own environment. Observant adult caregivers ordinarily carry out this process informally or "on the fly," to detect potential hazards that could harm their children. Because households and the people who live in them differ in many ways, these environments may also contain unique combinations of hazards.

Caregivers need, therefore, to carry out more formalized hazard analyses of their homes and environments in which their children live and play (including childcare facilities and schools) in order to detect potential hazards. By regularly identifying and correcting hazards in their homes and elsewhere, caregivers can take a major step toward protecting their children (Figure 14.3).

To protect their children, caregivers should:

- think about past "close calls" and potential future situations that might be hazardous. They should then think about the factors that were, or could have been, responsible for each "near-miss."
- try to predict future scenarios where children might get hurt.
  Although there are probably an infinite number of ways that a child could be injured, the caregiver should try to make as many "educated guesses" as they can.

- seek information from outside sources:
  - Professionals such as teachers, physicians, and other health-care providers can serve as a source of information about hazards and they themselves should conduct the hazard audits.
  - Government agencies such as the U.S. Consumer Product Safety Commission (CPSC) systematically collect data on products that have harmed children in past incidents. Other similar organizations are also knowledgeable about additional aspects of safety and much of this information is increasingly available on the Internet.
  - Childcare books may contain sections about household hazards or be devoted primarily to this topic.

Figure 14.3 (Kalsher & Assoc. LLC)
Playgrounds may be a potential threat to child safety. Parents and other caregivers need to establish consistent procedures for evaluating products and environments to identify potential hazards. Even equipment designed for children, such as the playground equipment depicted in the photo above, may pose a danger to very young children or children allowed to play on the equipment without proper adult supervision.

HAZARD CONTROL HIERARCHY

Once hazards are identified, how should caregivers deal with them? The *hazard-control hierarchy* offers a useful framework to guide hazard-related decisions that manufacturers, government agencies, and caregivers can use (The three basic steps are shown in Box 14.2).

When none of the steps comprising the hazard control hierarchy is likely to work effectively, a fourth step might be to remove the product from the marketplace entirely; for example, to issue a recall or to ban its sale. In instances in which a product has multiple hazards, all of the first three strategies may be used, but they are ordinarily applied in the same ordered sequence (e.g., Sanders and McCormick, 1993; Laughery and Hammond, 1999).

BOX 14.2 STRATEGY FOR ELIMINATING HAZARDS

Basic Hazard Control Hierarchy

1. Design out hazards.
2. Guard against hazards that cannot be prevented with design.
3. Warn about hazards that cannot be eliminated with design or guarding or both.

This hierarchy is ordinarily most useful to manufacturers who create and sell products and to government agencies that issue regulations to protect the public. However, it also offers strategies that caregivers may find relevant in their role as "protectors" of their children.

ELIMINATE HAZARDS WITH DESIGN

The primary and most fundamental method of hazard control is to eliminate the hazards through design. This is particularly important for products that children will use or otherwise contact. A good product design is one of the most effective ways to reduce injury potential. Here hazards are designed out before the product is offered for sale (refer to Box 14.3).

Unfortunately, it is not always possible to eliminate the hazards of certain products without affecting its intended function and performance (refer to Box 14.4). For example, it is not possible to eliminate all of the hazards associated with certain chemical solvents, knives, or power tools without reducing their usefulness. When one cannot eliminate or design out the hazard, there are other strategies. We will discuss these strategies in more detail later in this chapter.

BOX 14.3 ALTERNATIVE DESIGNS THAT ACCOMPLISH A GOAL WHILE ELIMINATING THE HAZARD	
Steam Vaporizers	
Product Use	Caretakers use steam vaporizers to relieve children's breathing difficulties when they have a cold or similar illness. Placing vaporizers near children brings the treatment closer to the child.
Potential Problem	The heating elements produce hot vapor. It is possible that a young child might touch a hot component of the vaporizer or the steam coming from it.
Alternative Designs	• Menthol–eucalyptus pads to ease breathing. • "Cool-mist" humidifiers produce a fine spray and disperse vapor by moving air over or through a wick immersed in water.

Most households contain a wide variety of cleaning agents, many of which can be harmful to children. For these products, manufacturers might consider replacing a hazardous ingredient in a cleaning mixture with a safer alternative that is similarly effective. Obviously, this requires that there are good alternatives available and a market of caregivers willing to purchase them if they are made available.

Clearly, the primary role of a caregiver is not to redesign products. However, caregivers can still take an active role in preventing injury to their children by exercising caution before allowing them to participate in certain activities (e.g., paintball), to play with certain toys (e.g., those with choking hazards), or to visit the homes of playmates that may contain unacceptable sources of risk (e.g., homes containing unsecured firearms or other dangerous products).

BOX 14.4 IT CAN BE DIFFICULT TO ELIMINATE A HAZARD COMPLETELY	
Lead Paint	
Problem	Young children who eat lead-based paint or inhale lead dust can experience a variety of severe, permanent health-related problems. Elevated levels of lead can result in learning disabilities, behavioral problems, mental retardation, and in extreme cases, death.
Goal	Eliminate lead-based paint in residential settings.
Proponent	U.S. Environmental Protection Agency (EPA) and other organizations.
Description of Results	Prior to the EPA's 1978 ban on lead-based paints, approximately 3–4 million U.S. children had elevated levels of lead in their blood. After the wide-scale introduction of safer paints, the number of children with elevated levels of lead dropped dramatically (EPA, 2004).
Complications	Total elimination of lead paint in residences did not occur because many older buildings contain layers and layers of old lead paint. Removal of lead paint is a hazardous process and can cause injury if workers or others ingest paint chips or breathe lead dust in the air. Professionals must remove the lead-based paint and the home must be evacuated during the removal process.

ELIMINATE HAZARDS THROUGH GUARDING

For hazards that cannot be designed out, the next best hazard control strategy is to guard against contact with the hazard. There are several forms of guarding. The enclosures around most electronic components are examples of hardware guarding. The "dead-man" switch that shuts off the power when a lawn mower handle is released is an example of procedural guarding. Requiring a prescription for drugs and limiting who can use them is an example of expert-referent guarding. There are many examples of guarding to prevent children from coming into contact with potential hazards.

CHILD-RESISTANT CAPS: One example of guarding is the use of child-resistant caps to prevent children from consuming certain medicines or coming into contact with other types of toxic materials (e.g., cleaning products). Several examples of child-resistant caps are shown in Figure 14.4.

Using child-resistant caps reduces the risk of a child inappropriately consuming the product. According to the CPSC, packaging can be called "child resistant" only if testing confirms that the packaging is sufficiently difficult for children to open, yet reasonably easy for adults to open (CPSC, 1995). According to the CPSC, *sufficiently difficult* is based on criteria in which testing of the packaging shows that no more than 15% of children under the age of 5 are able to open the package in fewer than 5 minutes and only 20% are able to open the package in under 10 minutes.

The CPSC first started requiring child-resistant caps in 1974. Since that time, child-resistant packaging of potentially dangerous household chemicals has been an effective guard against accidental poisoning and overdoses and has saved many children's lives (PSC, 2002b; Rodgers, 2002).

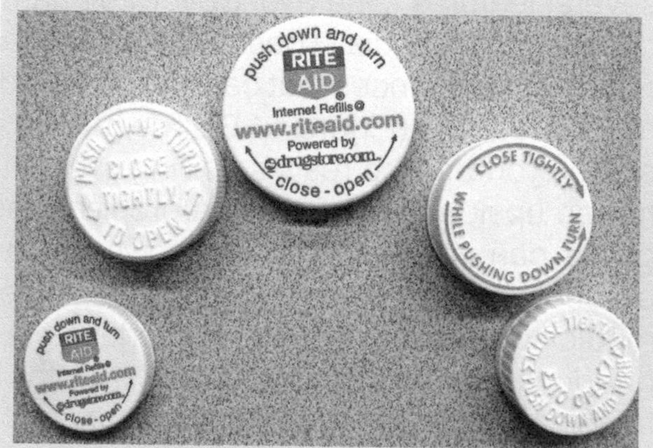

Figure 14.4 (Kalsher & Assoc. LLC)
Child-resistant caps: An example of guarding. Child-resistant caps decrease the chances that children will contact or consume products that could be hazardous, such as medicines or house-cleaning solutions.

However, though child-resistant caps can be an effective deterrent, products so equipped are not "childproof." Children can open child-resistant caps if the caps are not replaced properly; the safety mechanism only works when it is properly secured. Since these caps tend to be more difficult to open by older adults and disabled persons, they may also pose a hazard to children when such a person fails to close the cap securely or puts the contents (e.g., pills or other potentially harmful materials) into another, easier-to-open, container.

Child-resistant packaging can lead to unintended consequences—for example, well-meaning adult caregivers may assume that secured caps make them safe and leave them within reach of children. In other words, while child-resistant packaging can substantially reduce the likelihood of injury, the potential for harm still exists. Child-resistant caps thus show that guards may be only partially effective in controlling hazards.

MEDICATION DELIVERY AIDS: Devices that assist caregivers in serving a proper dosage of medication to their children also illustrate guarding. For example, pouring aids slow the flow of liquid products and measured dose delivery systems deliver a set volume of liquid each time a set procedure is completed.

One children's mouthwash (*ACT*, a Johnson & Johnson product) comes in a plastic bottle with a metered dosing dispenser on top (please refer to Figure 14.5). When the bottle is squeezed, the liquid rises through a tube in the center of the bottle and fills the dosing cup on top, which is marked with a fill line. The liquid can then be poured directly into the mouth or into a separate cup. Each time, 10 ml of liquid is delivered, which reduces the risk of overdose should a child inadvertently swallow the mouthwash.

Figure 14.5 (Kalsher & Assoc. LLC)
ACT Mouthwash: is an example of a metered dosing dispenser. Metered dosing dispensers, such as the one illustrated above, help to prevent children from consuming too much of a product.

A measured-dose drug delivery system may have several features. One (*NonSpil*, Taro Pharmaceuticals) comes with a specially designed dosage teaspoon that reduces the chances that caregivers will administer too much medicine (Figure 14.6). The gel-like consistency of the product also resists spilling even if the spoon is turned upside down. This is a problem because caregivers can easily spill an unknown portion of a dose of medication and then end up giving the child more medicine to compensate. Unfortunately, when this happens, caregivers frequently give the child more medicine than was lost. By making sure the medicine cannot be spilled in the first place, the product design reduces the risk of overdose.

It is important to note that neither pouring aids nor measured-dose delivery systems completely eliminate the potential for overdose. However, they do reduce the severity of harm by limiting exposure.

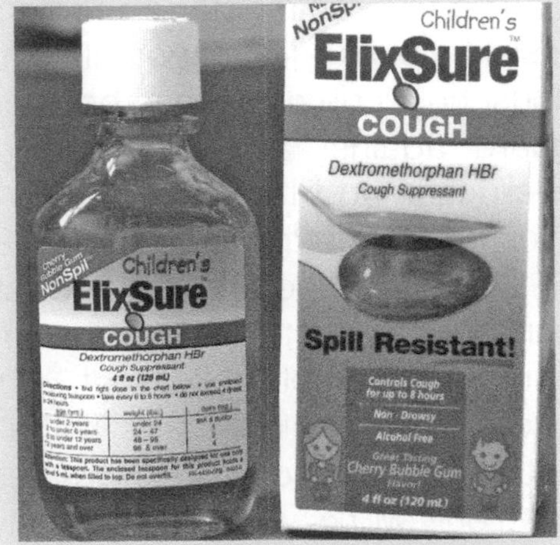

Figure 14.6 (Kalsher & Assoc. LLC)
Spill-resistant medicines are a useful tool for preventing drug overdose. Drug delivery systems that reduce spillage can help parents and other caregivers avoid giving too much medicine to their children.

ENGINEERING DESIGN: A different approach to hazard control is to engineer a way to keep children away from the hazardous materials. As an example, consider a product discussed in the preceding section: steam vaporizers. To keep children away from the burning steam, vaporizers possibly could include an elongated steam vent and a screen over the area where the steam is emitted. This added guard could prevent contact with the steam without much reduction of the vaporizer's effectiveness in humidifying the surrounding air. Again guarding does not eliminate the hazard entirely because there is still the possibility that the humidifier could be knocked over and burn a child if not placed in an out-of-reach location. Indeed the elongated vent could actually increase the likelihood of this kind of accident.

HOUSEHOLD CHEMICAL CONTAINERS: A final example of guarding involves a very different method of prevention. Colorful containers and pleasant odors may attract children to household chemical containers. Adding an extremely bitter taste, such as Bitrex (Macfarlan Smith), to household chemicals by manufacturers may help to avoid some severe poisonings. Children tend to spit it out before they consume much of it because it tastes bad (Jacob, 2003).

GUARDING AND CAREGIVERS: Although manufacturers can use the guarding methods above, caregivers have other forms of guarding to limit exposure to an existing hazard. Caregivers can conduct hazard analyses or safety audits in their home and other places their children might visit. In so doing, caregivers should:

- remove hazardous items. Place known hazardous items—such as household cleaners, other toxic materials, tools, equipment, and power cords—in areas such as locked cabinets away from a child's reach.
- avoid creating new hazards. Take care not to create hazards when using or storing items, such as placing hazardous chemicals in cabinets above a stove or oven upon which a child might attempt to climb.
- create and maintain safety zones. Use fencing, guards, or other barriers to prevent children from coming into contact with known hazards, such as stairways and pools.

Caregivers can help to reduce the chances of injuries at home by taking active steps to keep dangerous products out of reach of young children.

In this and the previous sections, we have considered two of the three approaches of the hazard control hierarchy: *design* and *guarding*. If hazards cannot be eliminated through these two methods, the next step in the hierarchy is to warn.

If design, guarding, and warning do not resolve the problem, the only remaining alternative may be to terminate sales and recall the product. For example, in January 1994, the CPSC issued a safety alert cautioning parents not to use soft bedding products in their infants' cribs. Several infants had been found dead, having suffocated by being placed on their stomach with their face, nose, and mouth covered by the bedding. The CPSC issued the safety alert after an investigation revealed that none of the traditional approaches to injury control—design, guarding, and warning—would ensure injury prevention (CPSC, 1994). Fortunately, all manufacturers of infant cushions voluntarily agreed to recall these products and cease future production of them.

Likewise, with certain products, such as those with small parts that pose a choking hazard, caregivers may refrain from purchasing them; place them in out-of-reach locations; or remove them from their home entirely. These possible actions are akin to the first two steps of the hazard control hierarchy; they remove the hazard or guard against the hazard by preventing children from having any contact with the hazard.

WARNINGS AS AN INJURY PREVENTION TOOL

Warnings have three main purposes (e.g., Wogalter and Laughery, 2005).

1. To communicate safety-related information so that people can make better, more informed decisions regarding safety issues
2. To persuade persons to perform safe behaviors that help to reduce or prevent injury and health problems
3. To remind individuals who may already know the hazards but may not be consciously aware of them at the time that knowledge is necessary. In other words, they can serve to cue pre-existing knowledge

Warnings come in a variety of forms, including signs, labels, product inserts and manuals, tags, audio and videotapes, and face-to-face verbal statements. In fact, safety professionals tend to think of warnings as a broad area of communications. They use the concept of *warning systems*, a term used to convey the idea that safety information is conveyed most effectively through multiple, overlapping components using a variety of media, formats, and messages.

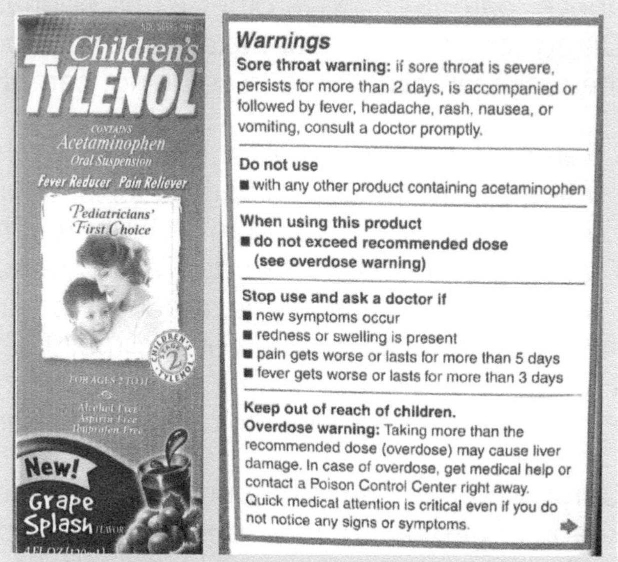

As shown in Figure 14.7, the warning system for children's pain relievers (e.g., Children's Tylenol™, McNeil) typically includes printed warning statements on the product packaging and container that address the most serious hazards posed by the product. More detailed information is presented in a printed package insert. The warning system for these products may also include warnings in print advertisements, in television and radio commercials, and on the Internet.

Figure 14.7 (Kalsher & Assoc. LLC) Warnings presented on the product packaging of Children's Tylenol™. Manufacturers should place the most important warning information directly on the product container and its packaging. More detailed information can be provided through other means, including package inserts and through the mass media.

As illustrated by this example, the individual components of a warning system may differ in content or purpose. Some components capture attention and direct the person to another component containing detailed hazard information. Outside packaging may give general information to serve primarily in assisting purchase decisions. Similarly, different components may be intended for different target audiences. For example, some components of a warning system for prescription drugs may be directed to the prescribing pediatrician (e.g., information contained in the Physicians Desk Reference or PDR); other components may target the pharmacist who fills the prescription and others may target parents who will ultimately administer the medicine to the child. The systems approach to warnings helps to ensure that different end-users in different situations receive the safety information most relevant to them.

It is important to note that while product warnings—or warning systems—can be used as injury prevention tools, they should never be used to replace other hazard control efforts such as hazard elimination and effective guarding.

HOW TO MAKE EFFECTIVE WARNINGS

This section reviews some basic principles for designing effective warnings. It considers the context in which warnings are used and then presents guidance on their format and content. When applying these principles, consider the following:

a. When to warn
b. Where and how to warn
c. How to prioritize warnings
d. Whom to warn

WHEN TO WARN: In general, a warning is necessary if one or more of the following exist:

When is a warning necessary?

- A significant hazard is present
- The hazard is not open and obvious
- The hazard, its potential consequence, and the actions needed to avoid it are unknown to the people who are likely to be exposed
- A reminder would help to ensure awareness of the hazard at the proper time

According to case law in the U.S., manufacturers have a duty to warn about hazards associated with both the foreseeable uses—and foreseeable misuses—of their products. Manufacturers must analyze their products for hazards and keep-up on the state of knowledge that relates to their products. In other words, manufacturers are expected to be in a superior position with respect to knowledge about their products (Madden, 1999; 2006).

Consumers (and certainly children) are not expected to have the extent of knowledge that manufacturers have about their own products and if hazards associated with these products are not completely designed out or guarded against, manufacturers have an obligation to adequately warn consumers about those hazards. According to the doctrine of strict liability in the U.S. (American Law Institute, 1998), if a product needs instructions or warnings to operate safely, but the product lacks adequate instructions and warnings, then the product can be considered defective.

WHERE AND HOW TO WARN: The best place for product warnings is usually directly on the product where users will notice them (Figure 14.8). People generally expect warnings to be on or near hazards they describe (Wogalter et al., 1991). Where there is not enough space to include all relevant safety related information, some hazard information may appear in accompanying materials—such as packaging and inserts.

Lack of space is not the only consideration that may lead a manufacturer to decide against placing warnings directly on a product. It may not be practical or may interfere with the use of the product as illustrated the baby-bottle tooth decay example presented in Box 14.5.

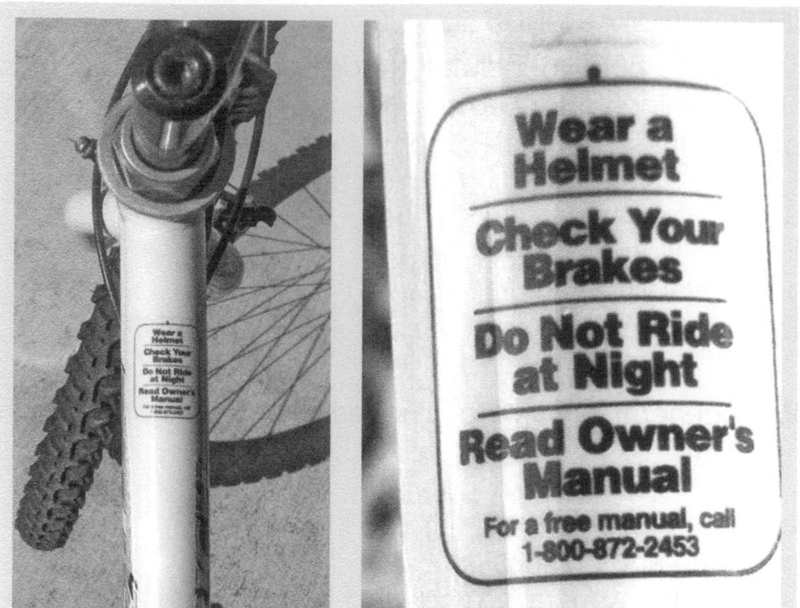

Figures 14.8 (Valerie Rice)
Warning placement is an important consideration. Although there is little room on a bicycle for warnings, here the manufacturer manages to include a few. Systematic testing could determine whether the messages are effective and whether they are the correct selection of message. The owner's manual probably includes other warnings.

BOX 14.5 BABY BOTTLES AND TOOTH DECAY	
The Problem	Baby-bottle tooth decay is the only severe dental disease common in children under 3 years of age (CPSC, 2000). It occurs when liquids containing carbohydrates such as milk, juice, or formula interact with normal mouth bacteria, which ferment the sugars and convert them to acids. The acid then etches the tooth enamel with prolonged contact. Children should not drink sweetened drinks too frequently or leave them in their mouths for too long. It is especially harmful for children to fall asleep while drinking sweet drinks.
Who Gives the Warning?	Baby-bottle manufacturers place warnings on the packaging and in instructional materials that accompany the product.
Why Not Place the Warning on the Bottle?	Caregivers sterilize baby bottles exposing them to heat and moisture. Warning labels would be likely to come off the bottle over time.
Other Considerations	
Sources of Information	Caregivers can receive information about baby-bottle tooth decay from a variety of sources (e.g., pediatrician, nurses, and baby books).
Severity of Injury	The potential injury (tooth decay of primary teeth) is not life threatening and the likelihood of long-term physical injury is small since permanent teeth do not erupt until later. However, expensive dental care and social stigma associated with unattractive teeth during early formative years are potential outcomes.

HOW TO PRIORITIZE WARNINGS: All relevant warnings may not fit directly on a product due to lack of space or number of potential hazards. Trying to fit information about too many hazards onto a product could make the most important warnings difficult to notice and read.
In this case, manufacturers must decide:

> *Prioritization:* A process that rank orders warnings by relative importance to safety.

a. which information to include directly on the product or container and which information to place elsewhere (e.g., in accompanying materials)
b. how to sequence the information
c. how much prominence to give a particular warning component relative to others

A hazard analysis or safety audit creates a list of hazards that manufacturers use for a process called *prioritization*. A warning about a hazard that is known and understood or one that is open and obvious should have relatively low priority. In contrast, a warning about a hazard that is relatively unknown (to persons at risk) that has the potential for severe injury should have relatively high priority.

Manufacturers cannot assume the types of hazards apparent and known by adults will be obvious to children—or even that adult caregivers will realize children may not understand a particular hazard. To illustrate this last point, consider the hazards associated with window screens (Box 14.6 and Figure 14.9). The window screens used for most homes are relatively inexpensive and have a mechanism that holds it onto window frames quite

BOX 14.6 WINDOW SCREENS IN US HOMES	
Problem	The mechanism that holds window screens into the window frame (in most newer U.S. homes) is quite weak.
Adult Understanding	Most adults know why it is important to avoid leaning on window screens (i.e., because of the risk of falling out of the window).
Children's Understanding	Young children do not understand the weakness of the screen attachment and may not even understand the risk of falling in any circumstance.
Result	Every year, over 4000 U.S. children aged 10 and younger are seen in emergency rooms for injuries related to window falls. Many of these children fell from a window with a screen in place. Each year approximately 12 children die, as a result (OHSU, 2004).
Potential Solutions	
Design	Use stronger window screen attachments.
Guard	Use window guards to limit how far the window will open. In the case of double-hung windows, allow only the upper portion to slide fully open.
Warnings	Window-screen manufacturers place warnings onto window screen frames. These warnings tell caregivers that children can be seriously injured.
Problem	Warnings are present but are not easily seen by caregivers. This is because the warnings are usually very small, inconspicuous, difficult to read, and their placement does not promote noticeability.
Bottom Line	Poor implementation of a warning means caregivers do not get the message that could help to prevent a serious injury.

Figure 14.9 (Kalsher & Assoc. LLC) Window screen warnings. Every year, thousands of children aged 10 and younger in the U.S. are seriously injured or killed following injuries related to window falls. Many of these children fell from a window with a screen in place. Window screen manufacturers place warnings onto window screen frames, but they are not easily seen because they are usually very small, inconspicuous, difficult to read, and their placement does not promote noticeability. Poor implementation of a warning means caregivers do not get the message that could help to prevent a serious injury.

weakly. Although most adults know why it is important to avoid leaning on window screens, young children may not. Every year, there are numerous reports of young children being seriously injured or killed after falling out of upper story windows in which the screen pops out. Solutions to this problem include using stronger window screen attachments, window guards that limit the extent to which the window can be opened and in the case of double-hung windows, allowing only the upper portion to slide fully open. Warnings can also be used, but only if they meet specific effectiveness criteria; they should be noticeable, understandable to the target audience, produce accurate beliefs about the nature of the hazard and the magnitude of the potential consequences, and motivate consumers to exercise appropriate precautionary behavior.

Many manufacturers do not consider the special characteristics and limitations of children when they design products or the warnings that accompany them. They apparently also fail to consider that caregivers may not fully appreciate the level of risk their product poses for children who may use or otherwise come into contact with it. Indeed, many manufacturers merely include the innocuous statement "Keep out of reach of children."

Several factors should be considered when prioritizing hazard warnings (Vigilante and Wogalter, 1997a,b), including:

- likelihood of injury
- severity of injury
- knowledge of the user
- overall importance

Apart from hazards that are unknown and have a high likelihood of producing an injury or that the injury can be particularly severe, manufacturers should also consider practical considerations. Sometimes, space limitations (e.g., a small label) or time limitations (e.g., a 30 second television advertisement) do not permit them to address all hazards. Warnings for hazards with higher priority should be placed in the most conspicuous locations—usually on the product itself—and possess features and characteristics that make it more likely users will notice it. Lower priority warnings might be placed in secondary components, such as in package inserts or manuals.

Manufacturers should be wary of placing warnings only in a product manual. The reason is that, as a method of hazard control, the effectiveness of warnings in product manuals can be quite limited. Low priority warnings may end up being one of a long list of items in a product manual that many people do not read thoroughly (Mehlenbacher et al., 2002). For these reasons (and others), it is usually preferable to place relevant warnings directly on the product whenever it is feasible to do so. However, accomplishing this goal sometimes requires the use of alternate label designs (e.g., peel-off labels, tags) to increase the available surface space (e.g., Wogalter and Young, 1994). Formatting also becomes more critical in this context as better formatting can help make the labeled information easier to grasp. We will discuss formatting in detail later in this chapter.

Since manufacturers may not always include a complete listing of the hazards associated with their products directly on the labeling, caregivers should take care to read any materials that accompany products very carefully and seek out additional sources (e.g., CPSC alerts).

WHOM TO WARN: In general, warnings should either address people exposed to the hazard or those in the best position to do something about it. People in the best position to provide assistance—such as caretakers—may not themselves be directly at risk. Some examples include warnings about the choking hazards posed by certain foods and toys (see Figure 14.10) and the side effects and contraindications of children's medicines.

Since 1994, the Child Safety Protection Act (CSPA) has required that toys intended for children between the ages of 3 and 6 years that contain small parts include an explicit choke hazard warning, such as the ones depicted in Figure 14.11.

Warnings for many of these types of hazards are directed to adult caregivers. Therefore, it is critical that characteristics of the target audience or receivers of the warnings be taken into account in their design (Laughery and Hammond, 1999). Sometimes older children are able to assist in their own protection.

The warning pictogram depicted at right started to appear on toy package labels in 1995 in European Union countries. It warns that the toy is unsuitable for children under 3 years. The pictogram was designed with the intent of avoiding the need to use many languages on the packaging (Aakerboom et al., 1995; Van Duijne et al., 2006). Similar warning labels are required for other toys that present a choking hazard to small children, including small balls, marbles, and balloons.

Figure 14.10 Toy safety pictogram. The prohibition symbol (circle slash) is red.

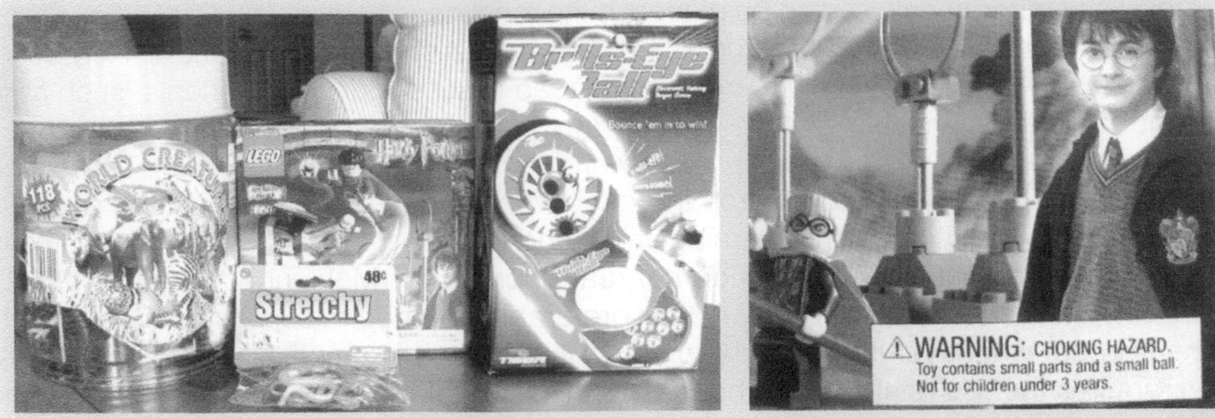

Figure 14.11 (Kalsher & Assoc. LLC)
Choking hazards. Certain toys pose a significant choking hazard to young children because they contain small parts. Warnings can help inform or remind parents of the risk and encourage them to take appropriate precautions.

ANSI Z535.4 WARNING DESIGN GUIDELINES

The American National Standards Institute's Z535.4 standard is a set of minimum guidelines for the design of warning labels in the United States (ANSI, 2002).
According to these guidelines, product safety labels should communicate:

- the nature or type of hazard (e.g., shock, cut, burn)
- the seriousness of the hazard
- the likely consequence(s) of coming into contact with the hazard
- how to avoid the hazard

ANSI Z535.4
The guidelines for warnings change when new information becomes available. They were first published in 1991, but were based on earlier guidelines (FMC, 1978; Westinghouse, 1981). The standard was revised in 1998 and again in 2002.

BOX 14.7 ANSI Z535.4 GUIDELINES FOR SIGNAL WORDS ON WARNING LABELS

Signal Word	Warning Indication	Colors
DANGER	An imminently hazardous situation, which *will* result in death or serious injury, if not avoided. For the most extreme situations.	White letters against a red background
WARNING	A potentially hazardous situation that *could* result in death or serious injury, if not avoided.	Black letters on an orange background
CAUTION	A potentially hazardous situation that may result in minor or moderate injury, if not avoided. Can also alert against unsafe practices.	Black letters on a yellow background

A sample ANSI-style warning is presented in Figure 14.12. The warning conveys the seriousness of the hazard (i.e., drowning) with a signal word printed in a separate panel. All letters of the signal word should be capitalized and the size of the font should be 50% larger than all other information presented on the warning label. The ANSI guidelines recommend the use of the following three signal words—DANGER, WARNING, or CAUTION—to convey high to low levels of hazard, respectively. Although many people do not know the intended difference between WARNING and CAUTION, most adult, English-speaking users appreciate that both denote a somewhat lower degree of hazard as compared to DANGER (e.g., Wogalter and Silver, 1990, 1995). ANSI Z535 also recommends the use of comprehensible pictorial symbols.

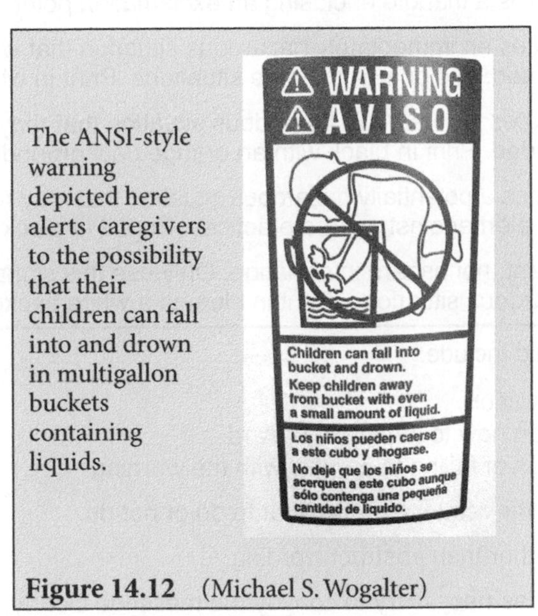

The ANSI-style warning depicted here alerts caregivers to the possibility that their children can fall into and drown in multigallon buckets containing liquids.

Figure 14.12 (Michael S. Wogalter)

ADDITIONAL WARNING DESIGN GUIDELINES

Although ANSI standards provide a good starting point for designing effective warnings, there are other sources of guidelines available, albeit in a more diffuse form. Helpful tips are available in the extensive body of research on warnings or in specific regulations.

Table 14.2 presents a set of guidelines regarding the format and content of components that comprise product warnings. These recommendations may be useful in increasing warning effectiveness in terms of conspicuity, comprehension, and compliance (ANSI, 2002; Laughery et al., 1994; Wogalter, 2006; Wogalter et al., 1999).

Warning Component	Table 14.2 Warning Design Guidelines
Format	• Warning must be large enough to be seen and read by the intended audience. • Left-justify text. • Consistently position component elements. • Orient messages to read from left to right. • List the most important warning statements first. • Use bullet lists, indenting, and white space to organize and communicate points or steps.
Signal Words	• Place alert symbols in front of the signal words. The alert symbol is a triangle enclosing an exclamation point. • *Danger*—Indicates an immediately hazardous situation that will result in death or serious injury if not avoided; use only in extreme situations. Print in white with a red background. • *Warning*—Indicates a potentially hazardous situation that may lead to death or serious injury if not avoided. Print in black with an orange background. • *Caution*—Indicates a potentially hazardous situation that may result in minor or moderate injury. Can also alert against unsafe practices. Print it in black on a yellow background. • *Notice*—Important, nonhazard information. Only use this signal word to inform users. It is not for hazardous situations. Print in blue on a white background.
Message Wording	• Messages should include: 1. hazard information 2. instructions on how to avoid the hazard 3. consequences of failing to comply with the warning • Be explicit—tell the reader exactly what to do or not do. • Use concrete rather than abstract wording. • Use as little text as necessary to convey the message clearly. • Use short statements rather than long complicated ones. • Use short and familiar words. • Avoid using abbreviations that the intended population may not understand. • Use active rather than passive voice. • Avoid using words or statements that might have multiple interpretations. • Evaluate or test the message with a target-audience sample.
Pictorial Symbols	• Use symbols that have been evaluated via comprehension testing. • Don't use symbols that are likely to generate confusion. • Use bold figures that have sufficient detail to communicate the message effectively. • Components of the symbol should be legible by persons in the target audience and at expected reading distances.
Font	• Use mixed case letters. Avoid all caps except for signal words or to emphasize specific words. • Use simple and familiar fonts. • Use san serif fonts (Arial, Helvetica, etc.) for signal words and large size print. • Use serif (Times, Times New Roman, etc.) fonts for smaller type sizes. • Characters should be thick and bold and should not be crowded or spaced awkwardly.

INTERNATIONAL ORGANIZATION FOR STANDARDIZATION AND AMERICAN NATIONAL STANDARDS INSTITUTE: The leading standard organization for the design of safety labels, signs, colors, and symbols outside of the U.S. is the International Organization for Standardization (ISO). The ISO standards relevant to the design of warnings are ISO 3864 and ISO 9186. ISO 3864 specifies international standards for safety labels. ISO 9186 describes test methods for assessing the comprehensibility of graphical symbols.

One of the most substantial differences between American National Standards Institute (ANSI)-style and ISO-style warnings concerns the use of text. Current ISO standards recommend graphical symbols—*not* text—in safety warnings. According to ISO, using text to convey hazard information can pose problems for people who are unable read the language in which the text is written. ANSI allows a symbol to be substituted for all or a portion of a text-based safety message, as long as the symbol has been tested and meets specific criteria: 85% comprehension with fewer than 5% critical confusions (misunderstandings of the intended meaning). If a symbol is substituted for all or a portion of a text message, ANSI still requires the use of a signal word (DANGER, WARNING, CAUTION) to communicate the hazard's level of seriousness. If a text message is used, ANSI recommends that the message describes the hazard, consequences, and instructions for how to avoid injury (e.g., use personal protective equipment).

The existence of more than one set of standards and guidelines can lead to conflicting recommendations. Although there are ongoing efforts to harmonize the ANSI and ISO standards, manufacturers must make the ultimate decisions regarding the design of warnings for their product.

WARNINGS THAT TARGET CHILDREN

As children get older, they develop a larger repertoire of cognitive abilities and can understand hazardous situations and product warnings. However, warnings intended for young children must be designed differently than those aimed at older children or adults.

Most guidelines tend to focus on warnings that target adults because:

a. adults bear primary responsibility for their children's safety
b. very young children lack cognitive abilities to understand the ramifications of a warning and even older children have significant variability in cognitive abilities across ages and within age groups (Nixon et al., 1980)
c. adults purchase most products for children and use warnings to guide their purchasing decisions

Nevertheless, well-designed, understandable warnings may assist caregivers to keep children safe. For example, Figure 14.13 shows example signs indicating to drivers of motor vehicles that children may be playing in the area. The danger is that younger children may enter the street without looking for oncoming vehicles. Usually signs like these are placed in residential areas because children commonly play in yards adjacent to streets.

Such signs would not be expected in business districts and major freeways. Signs like these serve as a reminder to drivers to be mindful that children may be in the area, but certainly drivers should realize the potential of children playing in other areas where no signs are posted. Similarly, airbag warning stickers in the U.S. state that children 12 and under are at risk of death and serious injury when sitting in a passenger position fitted with an airbag. Obviously, this sticker would not communicate to an infant, but most 8–12-year-olds

Figure 14.13 (Michael S. Wogalter)
Children-At-Play signs. Signs like these are posted to indicate that children may be playing in the area and may enter a street without looking.

are capable of reading it. The mere presence of the warning sticker may help to bolster the caregiver's authority when they ask children to sit in the back seat.

Developing several different versions of a warning—or a warning with multiple components—can help to accommodate different age groups and ability differences present within an age group.

DESIGNING WARNINGS FOR CHILDREN

Designers of warnings for children need to consider their unique strengths and limitations. They also need to recognize that children vary considerably across a wide range of dimensions, including their level of maturity. This is true with children even within a relatively narrow age range. As a result, it is critical that designers take steps to systematically evaluate warnings developed for children with a representative target group to ensure these materials exert the intended effects.

The mere presence of warnings, as opposed to their absence, could have unintended effects on children (Schneider, 1977). That is, some research suggests that some kinds of warnings in some kinds of situations may actually have a "boomerang" effect on children engaging in risky behaviors (Bushman and Stack, 1996). Similarly, some children may be especially attracted to items or activities they view as "forbidden fruit," particularly if they see an older sibling or adult using the item—for example, an inappropriate movie or video game (Bushman and Stack, 1996; Handelman and Parent, 1995). In these situations, some older children may view product warnings as messages that restrict their personal freedom as opposed to sources of information that can help them make better decisions. As a result, they may be inclined to respond in a manner inconsistent with the warning (e.g., Brehm, 1966). As Resnick (2006) describes, adolescents are frequently attracted to the warnings associated with certain music labels, movies, and TV shows (e.g., Christenson, 1992).

Figure 14.14 (Kalsher & Assoc. LLC)
Product packaging similarity and safety. Children may have difficulty distinguishing certain household chemicals from consumable foods and beverages because of similarities in the product's appearance and packaging. As shown in this photo, the degree of similarity in the appearance of edible and nonedible products varies considerably.

Warnings may produce unintended effects if they are misunderstood. For example, children may not avoid poisonous substances labeled with the skull and crossbones symbol if they do not know it is intended to mean "poison." Children may also be mistakenly attracted to dangerous products due to characteristics of the product's packaging including color, the container's shape (see, for example, Wogalter et al., 1997), and the presence of symbols or other characteristics. Pediatricians and poison control specialists note that young children are often poisoned when they mistake dangerous chemicals and their packaging for consumable foods and beverages (Ustinova, 2004). For instance, citrus-scented chlorine bleach may be mistaken for orange juice; blue cleaning fluid for a similarly colored sports drink; and powdered cleanser for parmesan cheese (see Figure 14.14). Similar types of errors might also occur with adults who cannot read the warning text due to insufficient language skill.

Suggestions for those who design warnings for children

Since relatively little is known about how to design warnings for younger children, any warning that targets children should be evaluated thoroughly. Yet, some guidelines for adult populations (e.g., Table 14.2) —with the appropriate cautions—may apply:

- *Use unambiguous language.* For older children and adolescents, warnings probably should include explicit language (e.g., "*Cigarettes Kill: One in every 3 smokers will die from smoking!*"). Warnings that do not have explicit language (e.g., "*Surgeon General's Warning: Quitting Smoking Now Greatly Reduces Risk to Your Health*") are less effective (Krugman et al., 1994).
- *Make warnings "stand out."* Warnings should generally incorporate formatting characteristics such as bright colors and the use of contrast to capture children's attention. However, it is important to avoid labeling characteristics that may draw younger children to hazardous products or lead them to believe that the product is safer than it really is (e.g., color pictures; cartoon characters).

- *Evaluate the warnings.* Manufacturers should systematically examine the warnings associated with their product to determine if they are achieving the intended goal of hazard control and not exerting any harmful effects (e.g., attracting children to potential hazards rather than deterring them from the hazards). Since evaluation is such a critical step in hazard control, we provide more information about evaluating warnings later in this chapter.

Some of the most effective warnings come from a child's caregivers. Sharp rebukes (e.g., "Don't touch that!") issued by parents to their children can be effective vehicles for preventing injury. Although very young children may not understand the reasons for a scolding, they quickly learn what they can and cannot touch especially when an admonition is paired with other safe, parent-imposed consequences (e.g., timeout; praise for safe behavior when it occurs).

Fortunately, as children mature, they increasingly learn and begin to understand why some "things" may be unsafe. In general, warnings should (cue or) convey information about the hazard, likely consequences (and their severity), and instructions on how to avoid injury. We believe there is some merit to using this format as a framework for teaching children about safety. In other words, caregivers can take the basic information from well-designed warning labels and "translate" it to their children.

As children get older and gain more experience with potentially hazardous products and situations, caregivers might also enlist their participation in developing and implementing procedures for conducting home safety audits and reporting emergencies.

PICTORIAL SYMBOLS

The best approach for conveying hazard information to very young children may be pictorial symbols or *pictograms* (Department of Trade and Industry, 1999; see Box 14.8).

Since some pictograms and the concepts they represent can be quite abstract, caregivers should explain the meaning of relevant pictograms to children so they will be able to avoid the respective hazards on their own. That is, children who are familiar with a pictorial are more likely to understand and comply than children who are not (DeLoache, 1991).

> **BOX 14.8 PICTOGRAM WARNINGS (PICTORIAL SYMBOLS)**
>
> Pictograms are pictures designed to convey hazard information normally presented via a text warning. Among adults, pictograms reduce search time when reading warning information on product labels (Bzostek and Wogalter, 1999).

> Mr. Yuk is a pictograph designed to prevent accidental poisoning of young children.*

Research settings have shown some pictorials are effective in conveying information to children as young as 30 months of age (DeLoache and Burns, 1994). However, the true effectiveness of pictorials with children is unknown because laboratory settings lack the types of distractions ordinarily present in everyday life.

An example of a pictograph developed for use directly with small children is Mr. Yuk. The Pittsburgh Poison Center at Children's Hospital of Pittsburgh created Mr. Yuk in 1971 to educate children and adults about poison prevention and to promote poison awareness. His image appears on bright green stickers that are to be placed on containers of poisonous substances. The idea is that with training, children learn to avoid products labeled with the Mr. Yuk stickers. However, tests of the effectiveness of Mr. Yuk stickers have produced mixed results (Ferguson et al., 1982; Oderda and Klein-Schwartz, 1985).

* Obtain stickers by sending a self-addressed stamped business size envelope to Mr. Yuk c/o Pittsburgh Poison Center, Children's Hospital of Pittsburgh, 3705 Fifth Avenue, Pittsburgh, PA 15213-2583, U.S.A.

Designers must carefully test pictorials to determine whether children notice, comprehend and comply with the warning's directive (ANSI, 2002; ISO, 1988). For example, as we noted previously, children who are not told otherwise may interpret the ubiquitous skull and crossbones pictograph (Figure 14.15) to mean "pirate food" (Schneider, 1977).

This simple example highlights the importance of careful tests of symbols or warnings for children. By ensuring that symbols communicate their meanings effectively, testing reduces the chances that children will misunderstand their meaning and as a result be injured. Parents can also play an important role by teaching their children that the skull and crossbones symbol is sometimes used to indicate poison.

Figure 14.15 The Skull and Crossbones Pictorial. Children may misinterpret some pictographs, such as the skull and crossbones symbol depicted here. This highlights the need for careful testing of this and other pictographs before use on hazardous products.

ALTERNATIVE WARNING METHODS

Warnings should focus on more than a child's visual and language comprehension capabilities. Olfactory and auditory warnings can also convey hazard information to children who might not otherwise be reached by written or pictorial warnings. Children who have difficulty reading or interpreting pictorial symbols may benefit from warnings they can hear, taste, or smell.

- *Sound.* Some warnings work by emitting a loud sound, such as a fire alarm.
- *Taste.* Adding a bitter taste to household chemicals not only guards against but also warns about the hazard.
- *Smell.* Utility companies use a similar approach, adding an odor to otherwise odorless natural gas to warn people of the presence of a gas leak.

Odors can also alert children to the presence of other types of hazards, such as by adding undesirable odors to potentially poisonous products. Children are fairly good at determining whether something is edible based on its odor (Wijk and Cain, 1994). For this reason, manufacturers should avoid pleasant odors, such as "fresh" or "lemon scented." If used, manufacturers should also add other methods of guarding or warning, such as child resistant caps, bitter taste, and unattractive container labels.

Advances in technology have made it possible to develop voice warnings to alert children to potentially hazardous situations. Inexpensive voice chips and digitized sound processor technology could be used in novel ways to create effective warnings for children. Speech-based warnings can sometimes be more effective than print-based warnings (Wogalter et al., 1993; Wogalter and Young, 1991)—particularly with very young children who comprehend spoken language before they can produce or read it.

Figure 14.16 (Michael S. Wogalter) New technologies have given rise to the development of alternative approaches to warnings, such as the recordable smoke detector depicted here. Careful testing can highlight the strengths—and limitations—of these products.

Manufacturers must make sure the target group understands not only the warning words but also the entire intended message. Manufacturers also need to determine the conditions in which speech-based warnings will work. For example, *recordable smoke alarms* (Figure 14.16) allow a caregiver to record a message (e.g., "Get up! There's a fire!") that plays when the device's alarm sounds.

EDUCATING CHILDREN ABOUT HAZARDS

Individual

Education is an important injury prevention tool that can help caregivers and children appreciate hazards. We have already mentioned using the ANSI components to tell older children more fully about hazards, consequences, and instructions on how to avoid hazards. Commonly, verbal warnings from parents or teachers educate children—especially very young children—by conveying important safety rules (Peterson and Saldana, 1996; Zeece and Crase, 1982). Parents' and teachers' views of safety are therefore especially strong determinants of a child's ability to recognize and respond appropriately to hazards they encounter (Uchiyama and Ito, 1999).

Group

Public education also informs children and their parents about safety. Individuals and groups can receive specific safety-related instruction via presentations or handouts. Teachers often use multiple methods of conveying information to children in their classes, including safety videos, visits to or from subject-matter experts (firefighters, police, emergency medical technicians) and safety-related activities (e.g., safety-oriented coloring books, development of home-safety checklists for their parents, and school-based emergency drills).

Large Scale Programs

Large-scale hazard implementation involves more than just education. It can include safety prevention programs at community, state, or national levels. Box 14.9 shows a list of potential components in a large-scale program focusing on child car seats.

Additional examples are provided in the sections that follow.

Mr. Yuk. The Pittsburgh Poison Center devised a large-scale educational program with their symbol, Mr. Yuk. Caregivers teach their children about hazards at home by placing Mr. Yuk stickers on various products (e.g., poisonous chemicals) and other potentially dangerous objects (e.g., electrical outlets). Then, each time the child attempts to touch anything with Mr. Yuk on it, they receive a mild scolding (e.g., "Don't touch that!"). Children soon learn not to touch dangerous objects.

BOX 14.9 CHILD CAR SEATS

A safety program on children's car seats could contain a variety of educational products and demonstrations, for adults and for children, such as:

- banners across busy streets announcing *Car Seat Safety Week*
- news briefs in the local media
- fliers from daycares and schools sent home to parents
- demonstrations on how to install child safety seats in vehicles for parents and teaching children about bicycle safety by having them bring their bicycles and helmets in for inspection during a "bike parade"
- distribution of coupons for financial savings on purchases of car seats
- a school visit by a local policeman who talks about safety
- instruction to children about how to fasten their own seat belts. Have older children measure themselves to see if they should be in a youth car seat or not.

Problems with the Mr. Yuk program:

- Caregivers must remember to attach Mr. Yuk stickers. Mr. Yuk would be a more effective poison prevention tool if all manufacturers of potentially dangerous household chemicals placed cautionary pictorials on their products prior to sale.
- Once children learn to avoid products that contain the sticker, they may falsely assume that other products are safe.
- If children contact an object labeled with Mr. Yuk and do not experience a negative consequence, they may falsely conclude that doing so in the future—perhaps with a more dangerous product—is okay. Nevertheless, it is usually better to use warnings than not to if the product is poisonous to children.

Safe Kids. The National Safe Kids Campaign illustrates a media-based safety program designed to educate caregivers about potential dangers to children (Tamburro et al., 2002). Using multimedia such as television and other advertising to convey hazard information has the benefit of reaching a wider audience than simply using a warning label or sign. The National SAFE KIDS coalition develops injury prevention strategies and conducts public outreach and awareness campaigns throughout the United States.

> The SAFE KIDS Campaign is a national nonprofit organization dedicated to the prevention of unintentional childhood injury. *(www. safekids.org).*

The spokesperson who delivers safety-related information can have a significant effect on adult consumers and their children. It is important to select a spokesperson who the audience considers relevant when conveying messages via the mass media (Wogalter et al., 1997). A physician, school principal, well-known celebrity, or other authority figure may facilitate credibility and believability of hazard-related messages.

Energy Kids. Mass media programs such as the National Safe Kids campaign and *Energy Kids* employ the Internet to disseminate safety-related information. Alliant Energy, an energy holding company in Madison, Wisconsin, created *Energy Kids* to teach children about energy safety and conservation (www.powerhousekids.com).

Energy Kids is an interactive Web-based program that targets children in third and fourth grades. Currently, the *Energy Kids* program has three modules that address electrical safety, natural gas safety, and storm safety respectively. As students use a computer mouse to move the cursor over different rooms of a house, a relevant safety-related message pops up when the cursor moves over a potential hazard. Each module is designed to complement a classroom presentation by a guest speaker.

Self-help materials. Self-help and related education materials written for new parents can also provide a great deal of information on a wide variety of topics, including childcare, health, and safety. Many childcare books offer extensive reviews of common hazards present in homes and other settings (e.g., playgrounds). Unfortunately, this information may only reach parents who are more highly educated and motivated to search for safety-related information concerning their children.

> Alliant Energy Kids: An interactive Web-based program designed to teach elementary children about energy safety *(www.powerhousekids.com).*

Fortunately, many parents receive free safety information packets from hospitals and birthing class instructors, as well as through product promotions. Some hospitals have in-room televisions with programs relevant to child safety for mothers who have just given birth.

In summary, there are numerous child safety resources available to children and their caregivers. The best way to ensure child safety is probably through a combination of warnings, prevention programs, and educational materials.

TESTING WARNING EFFECTIVENESS

Although guidelines for warning design are a useful first step, developing effective warnings does not stop there. Testing with a target population can help ensure effectiveness.

Many designers think they know when particular warning designs work and then find the target users interpret the entire design differently during testing. For example, warnings that are effective with adults might have very different results with children.

Comprehension and Behavioral Testing

Two major types of testing are *comprehension* and *behavioral* testing. *Comprehension* testing helps determine if members of the target audience understand the textual and pictorial warning components. ANSI and ISO have developed guidelines for testing the comprehension of pictorial symbols.

It is important to evaluate warnings and warning components (e.g., symbols) before use (Kalsher et al., 2000). Well-designed pictorial symbols can communicate hazards effectively to consumers with varied educational and cultural backgrounds. Pictorial symbols are especially valuable when combined with verbal warnings and instructions (e.g., Soujourner and Wogalter, 1997).

Just as well-designed pictorials can contribute significantly to the effectiveness of a warning, poorly designed symbols can result in dangerous comprehension errors. Thus, in order to produce effective warnings, pictorial symbols must be designed and tested repeatedly for an acceptable level of comprehension among the target group.

Testing Marshmallow Warnings. People do not usually consider it hazardous to eat marshmallows, despite the fact that some children have died through suffocation or are permanently injured after choking on them or related foods (Kalsher et al., 1999). Why are marshmallows hazardous? Because marshmallows possess characteristics that make them especially dangerous to children: (1) they are sweet and therefore attract children; (2) young children do not chew food completely before swallowing; (3) marshmallows appear soft and therefore, innocuous to parents; (4) marshmallows become stickier and swell when they contact the moisture present in the mouth; (5) an aspirated piece of marshmallow can be very difficult to dislodge because it continues to expand after entering the airway, thereby efficiently obstructing airways, including the trachea; and (6) marshmallows are light and can therefore be easily inhaled into the respiratory system (e.g., Kalsher et al., 1999; Rothman and Boeckman, 1980).

One study compared the effectiveness of warnings that depict choking hazards from eating marshmallows (Kalsher et al., 2000). Researchers asked participants to rate how well

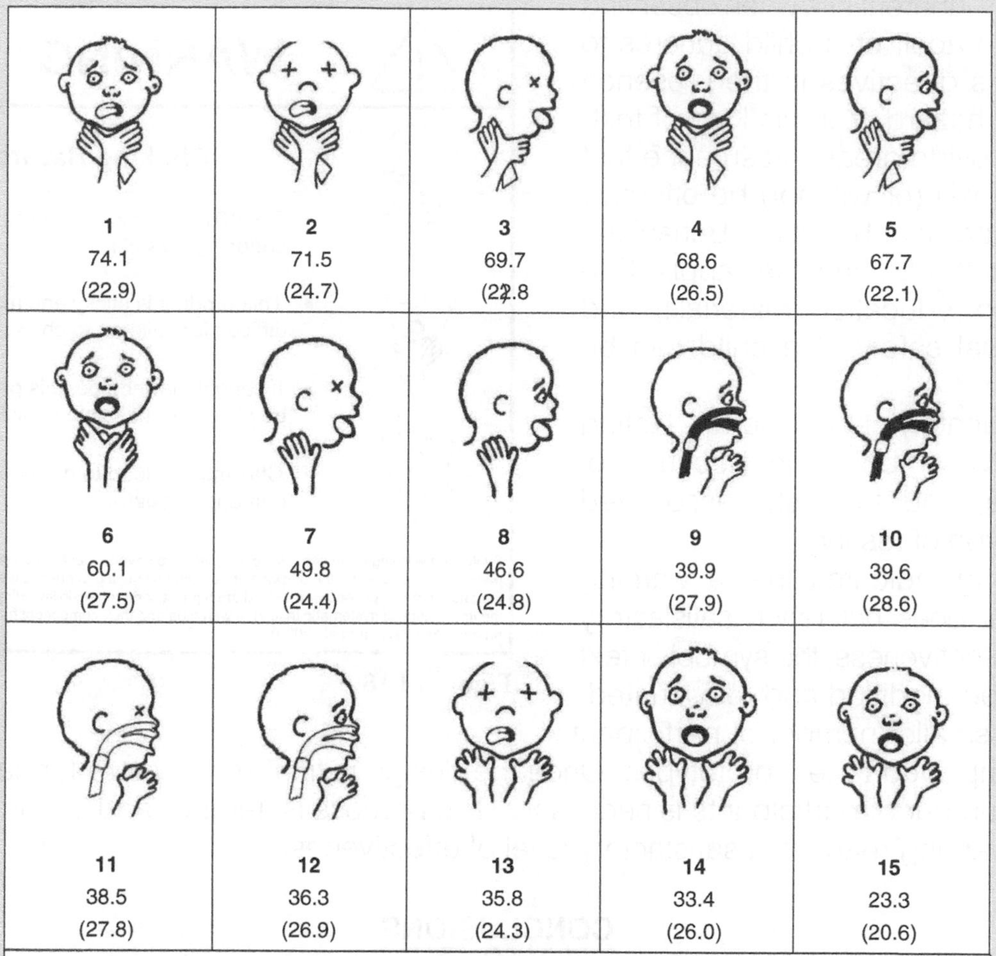

1	2	3	4	5
74.1	71.5	69.7	68.6	67.7
(22.9)	(24.7)	(22.8	(26.5)	(22.1)
6	**7**	**8**	**9**	**10**
60.1	49.8	46.6	39.9	39.6
(27.5)	(24.4)	(24.8)	(27.9)	(28.6)
11	**12**	**13**	**14**	**15**
38.5	36.3	35.8	33.4	23.3
(27.8)	(26.9)	(24.3)	(26.0)	(20.6)

Marshmallows and Choking:Testing Warning Effectiveness. Fifteen choking symbols were tested. The symbol number, mean comprehension estimates and standard deviations are below each symbol (Based on Kalsher et al., 2000).

Figure 14.17

pictorial symbols communicated the idea of a choking hazard. As shown in Figure 14.17, the comprehensibility estimates covered a fairly large range (from 74.1% to 23.3%), suggesting that the symbols differed dramatically in their ability to communicate.

Participants then assembled a combination of any three of the candidate symbols to depict a choking "sequence." The symbols they selected for this purpose were not always the ones that had previously received the highest comprehensibility estimates (Figure 14.18). Interestingly, participants reported that they constructed the multiple-panel pictorials according to how they perceived a choking event would progress over time. Thus, the last symbol in the sequence frequently conveyed loss of consciousness, such as limp or uncrossed hands and the crossed representation of eyes.

As illustrated by these examples, conducting careful evaluation of warnings and warning components is critical to ensure that warnings achieve their intended goals and among those in the intended target audience.

Behavioral testing enables researchers to observe how a participant responds to a particular warning. It is sometimes considered the "gold standard" method of determining warning effectiveness.

There is inherent risk when observing whether an adult or a child adheres to a warning's directives in the presence of a (real) hazard. Yet this kind of testing can provide greater assurance that a warning will (or will not) be effective in promoting safe behavior. Behavioral testing methods can be applied to safety and education materials, and when actual safety of a child can be assured.

Unfortunately, behavioral testing cannot always be done because of cost, time, and the risks associated with this type of testing.

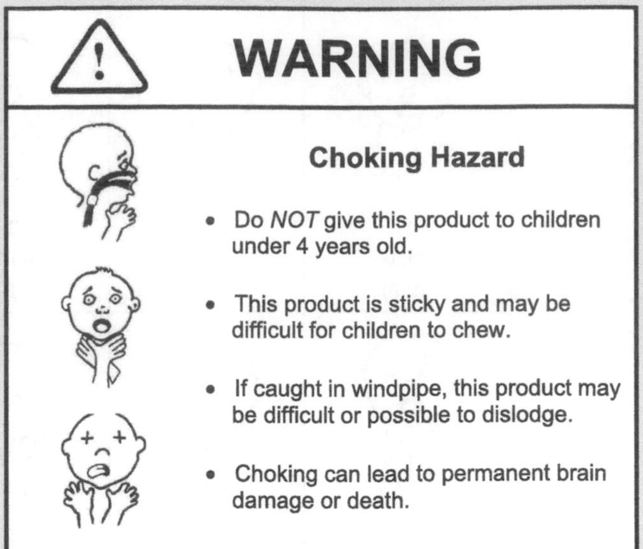

If testing indicates that a warning component does not reach satisfactory levels of effectiveness, the symbol or text needs to be modified and reevaluated. Initially, a smaller number of participants can be employed to test prototypes. Once the design is finalized, a more formal test with a larger number of participants is necessary. The process is iterative and should continue until the warning reaches a satisfactory level of effectiveness.

WARNING

Choking Hazard

- Do *NOT* give this product to children under 4 years old.
- This product is sticky and may be difficult for children to chew.
- If caught in windpipe, this product may be difficult or possible to dislodge.
- Choking can lead to permanent brain damage or death.

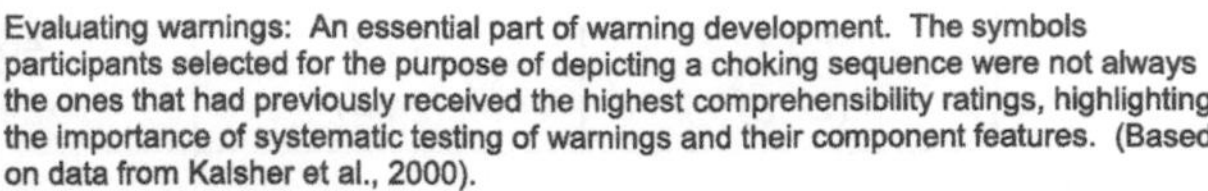

Evaluating warnings: An essential part of warning development. The symbols participants selected for the purpose of depicting a choking sequence were not always the ones that had previously received the highest comprehensibility ratings, highlighting the importance of systematic testing of warnings and their component features. (Based on data from Kalsher et al., 2000).

Figure 14.18

CONCLUSIONS

Manufacturers should eliminate hazards and dangers from children's products whenever possible. Since eliminating all risk is not possible, however, caregivers should also learn about effective injury prevention methods.

This chapter has presented a broad overview of ways to inform children and their caregivers of hazards. It has considered the roles of both manufacturers and caregivers in protecting children through designing out the hazard, guarding against the hazard, and warning about the hazard.

Most warnings for child-related hazards are directed to caregivers, who are then responsible for using that information to keep children safe. They must often "translate" warnings so that their children can comprehend their intended meanings. Manufacturers can assist by designing warnings that have clear and unambiguous meanings to both parents and (whenever feasible) their children. Unfortunately, relatively little is known with regard to developing warnings specifically for children, particularly younger children.

Most children above the age of 8 can read and therefore, may be able to take part in assisting in their own safety. As children get older, they tend to take on increasingly greater responsibility for their safety (e.g., Laughery et al., 1996; Lovvoll et al., 1996). Well-designed symbols may also help if they have attributes that make them understandable to children.

* Note that in Figures 14.17 and 14.18, symbol creators Jennifer Snow Wolff and Michael S. Wogalter grant free "copyleft" access to these figures. That is, anyone is free to copy, modify, and redistribute the figure so long as the new versions grant the same freedoms to others.

When possible, it is best to test the effectiveness of warnings with caretakers and if possible, children. Testing can help to determine whether a warning accurately communicates its intended message. Testing may also be useful for assessing whether a warning or other aspects of the product's labeling gives rise to misunderstandings or has attractive value to very young children. One benefit of designing warnings directed to older children is that it can bolster a caregiver's admonitions. A caregiver is assisted in telling their child to sit in the back seat of the vehicle because is safer will have some support because they can point to the strong relevant manufacturer's warning on the vehicle's sun visor.

In summary, although warnings and injury prevention programs will not eliminate childhood injuries, they can provide caregivers and their children with the opportunity to make informed decisions about the products they will use and the knowledge of how to use them safely.

ACKNOWLEDGMENT

We gratefully acknowledge the assistance of Deane Cheatham, Lorna Ronald, Shanna Ward, Carolyn Atwell, and Kevin Williams for their assistance with an earlier version of this chapter.

REFERENCES

Akerboom, S.P., Mijksenaar, P., Trommelen H., Visser, J., and Zwaga, H.J.G. (1995) *Products for Children: Development and Evaluation of Symbols for Warnings*. Amsterdam: Consumer Safety Institute.

American Law Institute (1998) *Restatement of the Law Third: Torts—Products Liability*. St. Paul, AK: American Law Institute.

ANSI (American National Standards Institute) (2002) American National Standard for Product Safety Signs and Labels, ANSI Z535.4-2002. Rosslyn, VA: National Electrical Manufacturers Association.

Brehm, J.W. (1966) *A Theory of Psychological Reactance*. New York: Academic Press.

Bushman, B.J. and Stack, A.D. (1996) Forbidden fruit versus tainted fruit: Effects of warning labels on attraction to television violence. *Journal of Experimental Psychology: Applied, 2*, 207–226.

Bzostek, J.A. and Wogalter, M. S. (1999). Measuring visual search time for a product warning label as a function of icon, color, column, and vertical placement. *Proceedings of the Human Factors and Ergonomics Society, 43*, 888–892.

Christenson, P. (1992) The effects of parental advisory labels on adolescent music preferences. *Journal of Communication, 42*, 106–113.

CPSC (U.S. Consumer Product Safety Commission) (2004) CPSC warns about pool hazards, reports 250 deaths of young children annually: Federal agency launches drowning prevention initiative, holding public hearings. www.cpsc.gov

CPSC (U.S. Consumer Product Safety Commission) (2003) National poison prevention week 16–22 March 2003, CPSC Document 386. www.cpsc.gov

CPSC (U.S. Consumer Product Safety Commission) (2002a) Consumer Product-Related Statistics. www.cpsc.gov/library/data.html

CPSC (U.S. Consumer Product Safety Commission) (2002b) More Products to Get Child-Resistant Packages. www.consumerreports.org

CPSC (U.S. Consumer Product Safety Commission) (2000a) Hazardous substances and articles, administration and enforcement regulations (Title 16, Chapter II, Part 1500). Washington, D.C.: U.S. Government Printing Office.

CPSC (U.S. Consumer Product Safety Commission) (2000b) Nursing Products Report. Retrieved on July 26, 2004 from http://www.cpsc.gov.

CPSC (U.S. Consumer Product Safety Commission) (1999) Annual report on pediatric exposures to products related to the poison prevention packaging act (PPPA) reported to the American Association of Poison Control Centers Toxic Exposure Surveillance System (TESS) during calendar year 1998. www.cpsc.gov.

CPSC (U.S. Consumer Product Safety Commission) (1995) Requirements for the Special Packaging of Household Substances; Final Rule. Originally published in the Federal Register, July 21, 1995 60, 37709–37744. Retrieved on July 26, 2004 from http://www.cpsc.gov.

CPSC (U.S. Consumer Product Safety Commission) (1994) CPSC safety alert relates soft bedding products and sleep position to infant suffocation deaths. Release #94-030. Retrieved on July 26, 2004 from http://www.cpsc.gov.

DeLoache, J.S. (1991) Symbolic functioning in very young children: Understanding of pictures and models. *Child Development, 62*, 736–752.

DeLoache, J.S. and Burns, N.M. (1994) Early understanding of the representational function of pictures. *Cognition, 52*, 83–110.
DTI (Department of Trade and Industry) (1999) *Development of Further Warning Symbols for Packaging.* London.
EPA (U.S. Environmental Protection Agency) (2004) Protecting Your Child from Lead Poisoning. www.epa.gov/lead
Ferguson, D.G., Cullari, S., Davidson, N.A., and Breuning, S.E. (1982) Effects of data-based interdisciplinary medication reviews on the prevalence and pattern of neuroleptic drug use with institutionalized mentally retarded persons. *Education and Training of the Mentally Retarded, 17*, 103–108.
FMC Corporation (1978) *Product Safety Signs and Labels*, 2nd edn. Santa Clara, CA: FMC Corporation.
Handelman, J. and Parent, M. (May 1995) Protecting or stimulating? The effects of warnings on television viewing by adolescents. Presentation at the Public Policy & Marketing Conference, Atlanta, GA.
Hiebert, E.H. and Adams, C.S. (1987) Fathers' and mothers' perception of their preschool children's emergent literacy. *The Journal of Experimental Child Psychology, 44*, 25–37.
ISO (International Organization for Standardization) (1988) *General Principles for the Creation of Graphical Symbols*, ISO 3461-1. Geneva: International Organization for Standardization.
Jacob, T. (2003) *Taste—A Tutorial.* Retrieved on August 3, 2004 from http://www.cf.ac.uk/biosi/staff/jacob/teaching/sensory/taste.html.
Kalsher, M.J., Brantley, K.A., Wogalter, M.S., and Snow-Wolff, J. (2000) Evaluating choking child pictorial symbols. Proceedings of the IEA 2000/HFES 2000 Congress, 4.790–4.793.
Kalsher, M.J., Wogalter, M.S., and Williams, K.J. (1999) Allocation of responsibility for injuries from a "hidden" hazard. *Proceedings of the Human Factors and Ergonomics Society, 43*, 938–942.
Krugman, D.M., Fox, R.J., Fletcher, J.E., Fischer, P.M., and Rojas, T.H. (1994) Do adolescents attend to warnings in cigarette advertising? An eye-tracking approach. *Journal of Advertising Research, 34*, 39–52.
Laughery, K.R. and Hammond, A. (1999) Overview of warnings and risk communication. In M.S. Wogalter, D.M. DeJoy, and K.R. Laughery (Eds.), *Warnings and Risk Communication* (pp. 3–13). Philadelphia, PA: Taylor & Francis.
Laughery, K.R., Lovvoll, D.R., and McQuilkin, M.L. (1996) Allocation of responsibility for child safety. *Proceedings of the Human Factors and Ergonomics Society, 40*, 810–813.
Laughery, K.R., Wogalter, M.S., and Young, S.L. (Eds.) (1994) *Human Factors Perspectives on Warnings, Volume 1: Selections from Human Factors and Ergonomics Society Annual Meetings 1980–1993.* Santa Monica, CA: Human Factors and Ergonomics Society.
Lovvoll, D.R., Laughery, K.R., McQuilkin, M.L., and Wogalter, M.S. (1996) Responsibility for product safety in the work environment. *Proceedings of the Human Factors and Ergonomics Society, 40*, 814–817.
Madden, M.S. (1999). The law relating to warnings. In M.S. Wogalter, D.M. Deloy, & K.R. Laughery (Eds.). *Warnings and Risk Communication.* (pp. 315–329). London: Taylor and Francis.
Madden, M.S. (2006). Warning duties for risks to children (Chap. 47). In M.S. Wogalter (Ed.) *Handbook of Warnings* (pp. 597–603). Mahwah, NJ: Lawrence Erlbaum Associates.
Mehlenbacher, B., Wogalter, M.S., and Laughery, K.R. (2002) On the reading of product owner's manuals: Perceptions and product complexity. *Proceedings of the Human Factors and Ergonomics Society, 46*, 730–734.
Nixon, J., Bryce, M., and Pearn, J. (1980) Do children understand warning symbols? Proceedings of the Ergonomics Society of Australia and New Zealand 17th Annual Conference Ergonomics in Practice, Sydney, pp. 43–50.
Oderda, G.M. and Klein-Schwartz, W. (1985) Public awareness survey: The Maryland Poison Center and Mr. Yuk, 1981 and 1975. *Public Health Reports, 100*(3), 278–282.
OHSU (Oregon Health and Safety University) (2004) Window Safety. Available at www.ohsu.edu
Peterson, L. and Saldana, L. (1996) Accelerating children's risk for injury: Mothers' decisions regarding common safety rules. *Journal of Behavioral Medicine, 19*, 317–331.
Resnick, M. (2006). Risk communication for legal, financial, and privacy agreements and mass media (Chap. 62). In M.S. Wogalter (Ed.) *Handbook of Warnings* (pp. 771–782). Mahwah, NJ: Lawrence Erlbaum Associates.
Rodgers, G. (2002) The effectiveness of child-resistant packaging for aspirin. *Archives of Pediatrics Adolescent Medicine, 156*(9), 929–933.
Rothman, B.F. and Boeckman, C.R. (1980) Foreign bodies in the larynx and tracheobronchial tree in children. *Ann. Otol. Rhino. Larynx. 5*(1), 434–436.
Sanders, M.S. and McCormick, E.J. (1993) *Human Factors and Engineering Design*, 7th edn. New York: McGraw-Hill.
Schneider, K.C. (1977) Prevention of accidental poisoning through package and label design. *Journal of Consumer Research, 4*, 67–74.
Sojourner, R.J. and Wogalter, M.S. (1997) The influence of pictorials on evaluations of prescription medication instructions. *Drug Information Journal, 31*, 963–972.
Tamburro, R.F., Shorr, R.I., Bush, A.J., Kritchevsky, S.B., Stidham, G.L., and Helms, S.A. (2002) Association between the inception of a SAFE KIDS coalition and changes in pediatric unintentional injury rates. *Injury Prevention, 8*, 242–245.
Uchiyama, I. and Ito, A. (1999) Children's view of safety and estimated behavior and their mothers' view of safety and cautions. *Perceptual and Motor Skills, 88*, 615–620.
Ustinova, A. (2004) Look-alike labels create lethal choice for children. www.jsonline.com

Van Duijne, F.H., de Koning, A.J., Heijne, G.M., van Wilgenburg, E., and van Buuren, R. (2006) Warning texts and pictograms on consumer products in the Netherlands. In R.N. Pikkar, E.A.P. Koningsveld, and P.J.M. Settels (Eds.), *Proceedings of the International Ergonomics Association*. Amsterdam: Elsevier Ltd. (ISSN 0003-6870).

Vigilante, W.J. Jr. and Wogalter, M.S. (1997a) On the prioritization of safety warnings in product manuals. *International Journal of Industrial Ergonomics, 20*, 277–285.

Vigilante, W.J. Jr. and Wogalter, M.S. (1997b) The preferred order of over-the-counter (OTC) pharmaceutical label components. *Drug Information Journal, 31*, 973–988.

Westinghouse Electric Corporation (1981) *Product Safety Label Handbook*. Trafford, PA: Westinghouse Printing Division.

Wijk, R.A. and Cain, W.S. (1994) Odor identification by name and by edibility: Life-span development and safety. *Human Factors, 36,* 182–187.

Wogalter, M.S. (2006). Scope of warnings (Chap. 1). In M.S. Wogalter (Ed.) *Handbook of Warnings* (pp. 3–9). Mahwah, NJ: Lawrence Erlbaum Associates.

Wogalter, M.S. and Laughery, K.R. (2005). Warnings. In G. Salvendy (Ed.), *Handbook of Human Factors/Ergonomics*, 3rd edn. New York: Wiley.

Wogalter, M.S. and Silver, N.C. (1990) Arousal strength of signal words. *Forensic Reports, 3*, 407–420.

Wogalter, M.S. and Silver, N.C. (1995) Warning signal words: Connoted strength and understandability by children, elders and non-native English speakers. *Ergonomics, 38*, 2188–2206.

Wogalter, M.S. and Young, S.L. (1991) Behavioural compliance to voice and print warnings. *Ergonomics, 34*, 79–89.

Wogalter, M.S. and Young, S.L. (1994) Enhancing warning compliance through alternative product label designs. *Applied Ergonomics, 25*, 53–57.

Wogalter, M.S., Brelsford, J.W., Desaulniers, D.R., and Laughery, K.R. (1991) Consumer product warnings: The role of hazard perception. *Journal of Safety Research, 22*, 71–82.

Wogalter, M.S., DeJoy, D.M., and Laughery, K.R. (Eds.) (1998) *Warnings and Risk Communication*. Philadelphia, PA: Taylor & Francis.

Wogalter, M.S., Kalsher, M.J., and Racicot, B.M. (1993) Behavioral compliance with warnings: Effects of voice, context and location. *Safety Science, 16*, 637–654.

Wogalter, M.S., Kalsher, M.J., and Rashid, R. (1997) Effect of warning signal word and source on perceived credibility and compliance likelihood. Proceedings of 13th Triennial Congress of the International Ergonomics Association, Tampere, Finland. pp. 478–480.

Wogalter, M.S., Young, S.L., and Laughery, K.R. (Eds.) (2001) *Human Factors Perspectives on Warnings, Volume 2: Selections from Human Factors and Ergonomics Society Annual Meetings 1994–2000*. Santa Monica, CA: Human Factors and Ergonomics Society.

Zeece, P.D. and Crase, S.J. (1982) Effects of verbal warning on compliant and transition behavior of preschool children. *Journal of Psychology, 112,* 269–274.

SECTION E

Children at Home

CHAPTER 15

STAIRWAYS FOR CHILDREN

JAKE PAULS

TABLE OF CONTENTS

INTRODUCTION

Stairways are one of the first and last physical challenges we face in life. Climbing and descending stairs requires a combination of strength, balance, timing, and equilibrium. It takes coordinated effort to avoid missteps, falls, and consequent injury.

Children learn to use stairs as a distinct experience from learning to walk. The gait for stairs requires simultaneous vertical and horizontal steps, in contrast to the primarily horizontal steps used while walking on floors or even terrain. In addition, when using stairs, the geometry of the stairs greatly affects gait. As children grow in size and coordination, they change their use of stairs significantly (see Figure 15.1 through Figure 15.3). Some of these changes are mirrored in later life as stair use becomes an even more dangerous challenge.

This chapter focuses on design factors that contribute to ease-of-use and safety for children and adults on stairs. Particularly important considerations include stairway visibility, step dimensions, and the provision of handholds. Such issues are relevant to designers, builders, parents, teachers plus school officials, and especially to the children themselves.

Figure 15.1 (Ursy Potter Photography)

Figure 15.2 (Jake Pauls)

Figure 15.3 (Ursy Potter Photography)

Children first climb stairs using crawling motions (all four extremities). The two toddlers pause to examine something as they move up the stairs to the landing, while the young girl in Figure 15.2 ascends easily up stairs with open risers. When descending, young children will often "bump" down the stairs on their backsides, however, as they begin to walk, they want to try coming down the stairs standing. Their first efforts need supervision and they need to be taught to use handrails. It is best to have easily used (or usable) handrails on both sides of the stairway for the adult and child.

In Figure 15.3, Sofia's grandmother helps her descend the stairs. Notice that the stairs are not uniform; the second stair has a significantly lower riser. Fortunately, Sofia's grandmother has already twisted to the side to get better footing in a "crablike" gait.

STAIR-RELATED INJURIES

Falls are a leading cause of injuries in most developed countries. For fall injuries associated with stairs, the trend has been for hospital-treated injuries in the United States to increase at about twice the rate of population growth. There is about one hospital-treated, stair-related injury for every 3 million uses of stairs (Pauls, 2001). These rates are even higher for high-risk populations such as very young children (Table 15.1).

Table 15.1 Fall-Related Injuries in the United States[a]

Ages	Falls (Quinlan et al., 1999)	Stair-Related Injuries (CPSC/NEISS, 1998)
0 to 4 Years	850,000	103,000
All Age Groups	7,000,000	915,000

[a] The data in Table 15.1 is based on data before and after 1998.

Stair-related injuries are particularly common among children aged 0–9 and adults 35 or older. These falls are serious and involve more than simple ankle twists and sprains. Head and neck injuries are common, especially for children under 4 years of age (Joffe and Ludwig, 1988).

Risk increases when adults carry children on stairs. In fact, this is the second most common scenario for stair-related injuries, representing about 11% of the total (Alessi et al., 1978) (see Figure 15.4 and Figure 15.5).

> *Adult women who carry children, groceries, or other objects down flights of stairs may have their view of the treads obstructed or attempt to descend stairs in the dark. Victims overstep the nosing (leading edge) and fall, often fracturing the lower arm, back, skull, or ankle. Their attempt to save what they are carrying (especially children) often increases the severity of injuries.*
>
> (Alessi et al., 1978)

Many fall injuries result from infants using certain hazardous mobile walkers as the child—unless prevented by a gate or door—rolls onto stairs.* Recognizing that the vast majority of children injured due to walkers involved tumbles down stairs caused the American Academy of Pediatrics (AAP) to recommend banning these infant walkers.

Alternatively, ensure walkers adhere to ASTM F977-96 standard (AAP, 2001). Walkers that comply with this standard stop moving when they encounter stairs.

Some reports of fall injuries on stairs are inaccurate. Some caretakers attribute child injuries to stair falls to avoid admitting to child abuse. Biomechanical modeling can help show whether the number of steps, slope, surface friction, and path of the alleged fall matches the child's injuries (Bertocci et al., 2001). To assist medical practitioners, the Cincinnati Children's Hospital Medical Center has a "professional's tool kit" that includes a review of literature and guidance for assessing children's fall injuries.†

* US hospital emergency staff treated nearly 8800 children younger than 15 months for injuries associated with infant walkers in 1999. The majority of those injuries, often head injuries, occurred from falls down stairs. Before regulatory and industry changes in 1994 the annual toll averaged 25,000 treatments.

† Their Web site is www.cincinnatichildrens.org/svc/prog/child-abuse/tools/falls.htm

Figure 15.4 (Jake Pauls) **Figure 15.5** (Jake Pauls)
Adults carrying children on stairs experience difficulty seeing the steps and risk overstepping a stair due to the steepness of short steps with high risers. The stairs in these photos are not too different from those encountered in many homes, that is, stairs in many homes are equally steep and pose the same dangers of overstepping, especially when carrying children.

PHYSICAL CHALLENGE

As mentioned above, the gait required on stairs differs from normal walking on a level surface. The gait for ascent also differs significantly from that for descent. People's speed, gait, posture, pacing length, and stepping rhythm are more restricted on stairs than on inclines or level ground (Templer, 1992) (Figure 15.6).

> *Young children, whose legs are too small to manage a rhythmic gait on stairs designed for adults and elderly or disabled people with stiff joints struggle to maintain their balance on stairs. Ascent requires both physical strength and agility, compounding difficulties for the old, young, and disabled.*
>
> (Templer, 1992)

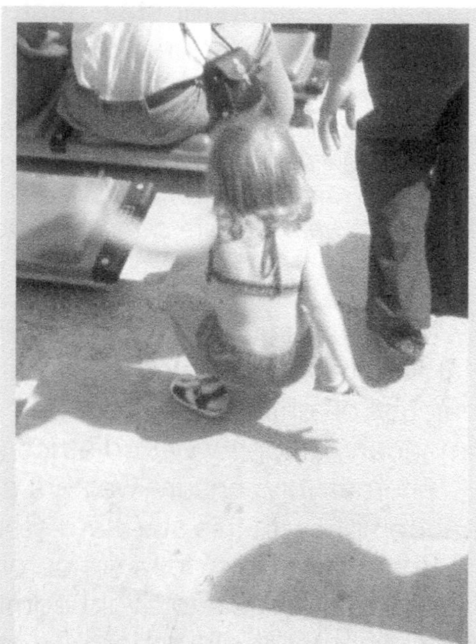

Figure 15.6 (Jake Pauls)
While the speed, gait, posture, pacing length, and stepping rhythm for all people are more restricted on stairs and make it more difficult for all ages to use stairs, they are especially serious for young children and older adults. Stepping down a high step can cause a child to lose his or her balance when legs are short or lack strength and agility.

Stairs may be difficult to climb, but young children relish the challenge (Figure 15.7 and Figure 15.8). Among elderly users, however such challenge can bring discomfort, displeasure, and fear. To deal with limitations of experience, agility, strength, and balance, children and elderly individuals become more cautious (Figure 15.9 and Figure 15.10). Very young children may crawl or bump their backsides, while somewhat older children and older people may

Figure 15.7 (Ursy Potter Photography)

Figure 15.8 (Ursy Potter Photography)

Most children love stairs. It takes great mastery for them to be able to simultaneously balance themselves and move forward as they ascend or descend. They can begin to feel their own sense of physical competence and freedom to explore their world when they "defy their own limitations" and conquer the challenge of stairs.

plant each foot on every stair tread. This contrasts with the rest of us, who typically lift or lower each foot over two-step heights.

It is especially important to use stairs safely and effectively when evacuating buildings. Although we have much to learn about how young children manage in such evacuations, data collected during apartment building evacuation drills suggest that young (preschool age) children's descent speed on stairs compares with descent speeds of older people and is about half that of younger adults (Proulx, 1995).

We need not be overly concerned about mixing evacuation flow of adults and children because when evacuations are efficient—and hence most desirable—they entail relatively slow speeds that should not be difficult for children. Inefficient evacuation can occur if movement speed slows even more due to crowding (Pauls, 1995). More information is needed on descent speeds and how to evacuate kindergartens and schools using stairs.

School-age children must often cope with physical challenges such as heavy backpacks or athletic bags that weigh a significant portion of their body weight. This affects their posture and general safety on stairs, unless they compensate by slowing their rate of descent. Heavy loads are best carried with backpacks that can help keep loads symmetrical (Hong et al., 2003; see Chapter 11 of this text for additional information on children and backpacks).

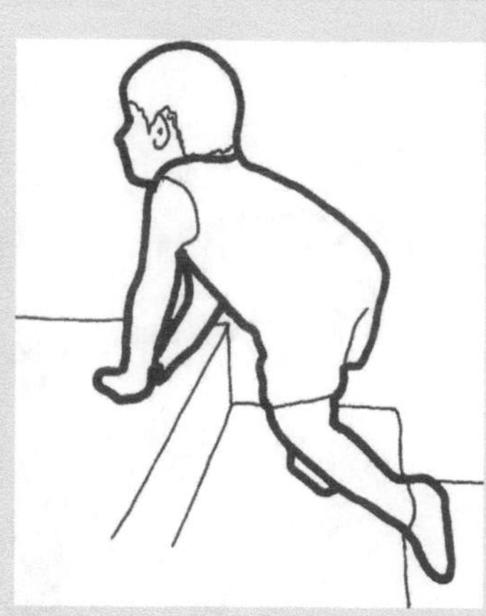

Figure 15.9 (Jake Pauls)

Figure 15.10 (Jake Pauls)

Children's early ascent of stairs uses a movement pattern similar to early crawling (on stomach) and creeping (on hands and knees).

8½ to 9 Months	Children crawl using coordinated arm and leg movements: right arm, left knee, left arm, right knee.
10 to 12 Months	Children move to creeping on their hands and feet, with their arms and legs extended (also known as *plantigrade creeping* or *bear-walking*).
13½ to 15 Months	Children begin creeping or hitching (moving while in a sitting position) up stairs (Figure 15.9). At this point, they still need to back down the stairs on their hands and knees.
17 to 19 Months	Most children can walk up stairs while holding one hand of an adult. At this point, they place one foot up and then move the other foot to the same step. Learning this skill on child-size stairs, rather than standard stairs, helps the learning process (see "Step Geometry Appropriate for Children" in this chapter).
22 to 24 Months	Children begin to walk up stairs while holding a rail rather than an adult's hand, still taking one step up and moving the other foot to the same step.
24 to 25½ Months	Children can now walk up stairs without holding onto a railing or an adults hand; however, the steps still do not alternate.[a]
24 to 26 Months	Children can walk downstairs holding onto a railing, still moving both feet to one step.
25½ to 27 Months	Children can walk downstairs alone without holding a railing or an adult's hand, still moving both feet to the same step.
30 to 34 Months	A child of this age can walk up standard size stairs without using a railing or holding an adults hand.[a] They can alternate steps, placing one foot on each step. This activity requires considerable skill and moving from one stair to another is a feat, as the riser height can be nearly to a child's knee (Figure 15.10).
34 Months and Older	A child of this age can typically walk down a standard stair without using a railing or holding an adults hand, and using alternating feet.[a]

[a] Although children can navigate without holding a hand rail or an adult's hand at these stages, using a handrail is always recommended.

(Furuno et al., 1997)

KEY STAIRWAY TERMS

Table 15.2 Stairway Terms (Figure 15.11)	
Stairway	• One or more stairs, usually arranged in flights between landings, along with handrails, guardrails, and other railing elements. • An example of a railing element is the construction below a railing that structurally supports the railing and may prevent passage under or through the rail.
Stair	• A series of one or more risers (the vertical element of a step). • Even a single-riser step is a stair. • There are safety problems even with single steps.
Step	• A riser combined with adjacent horizontal surfaces, either treads or landings. • Step dimensions are correctly measured as the vertical distance between nosings of adjacent steps (the "step rise") and the horizontal distance between nosings of adjacent steps.
Risers[a]	• The vertical element of a step that connects treads or ***landings***.
Landings	• The larger horizontal elements at the top and bottom of a flight of stairs usually—except in the case of a single riser—using intermediate treads (the horizontal element of a step).
Nosing	• The front, leading edge of a tread.
Tread Depth	• The horizontal distance between nosings of adjacent steps. • Also called the "***run***" or "***going***" in some countries.
Pitch	• The angle that the line of the nosings makes with the horizontal. • Stairs in traditional homes typically have a pitch of 42° to comply with US or British building code requirements. Many public stairs in newer buildings have a pitch of less than 33° because building code changes taking effect in the United States during the 1980s called for less-steep stairs (this did not generally apply to home stairs). • The fact that home stairs everywhere are still built to a less-safe and less-usable, steeper, pitch has been debated extensively in code-change deliberations since the 1980s in the United States and to a lesser extent in Canada.
Handrails	• Reachable, graspable rails located at the sides of stairs and, sometimes, at intermediate points on wider stairs. • Handrails provide several kinds of guidance and support, notably by helping prevent stair falls.
Guards or Guardrails	• Barriers located at the open side of stairs or walkways. • Guardrails provide a barrier to help prevent falls from stairs or walkways to lower surfaces.

(continued)

Table 15.2 Stairway Terms (Figure 15.11) (Continued)	
Misstep	• A departure from normal gait. • This does not mean the person experiencing the misstep made a mistake. Common stair missteps include overstepping, understepping, airsteps, heel scuffing, tripping, and slipping. • Even when people say, "I slipped," it is unlikely they actually have experienced a true stair slip, due to the biomechanics of the action. • Even so, slipping may occur secondarily, following an overstep. While descending stairs, the ball of the user's foot may land ahead of a nosing (overstep), causing—secondarily—a rotation and/or slipping of the foot at the nosing followed by complete loss of footing and balance.
Air-Steps	• The misstep people have when they fail to detect single-riser stair descents. • Caused by stepping forward onto what they expect to be a continuation of the level they have been on. Instead, the user's foot unexpectedly drops with a fairly hard impact. • This type of drop can injure the person's ankles, knees, hips, or back, even when they do not completely fall to the walking surface. • Injuries from air-steps often occur with unexpected drops of less than 18 cm (7 in.) because people cannot react fast enough to cushion the impact.

[a] Risers can be closed or open. Where risers are open (as shown in Figure 15.2), pay attention to the opening size as discussed in the section "Guards" of this chapter.

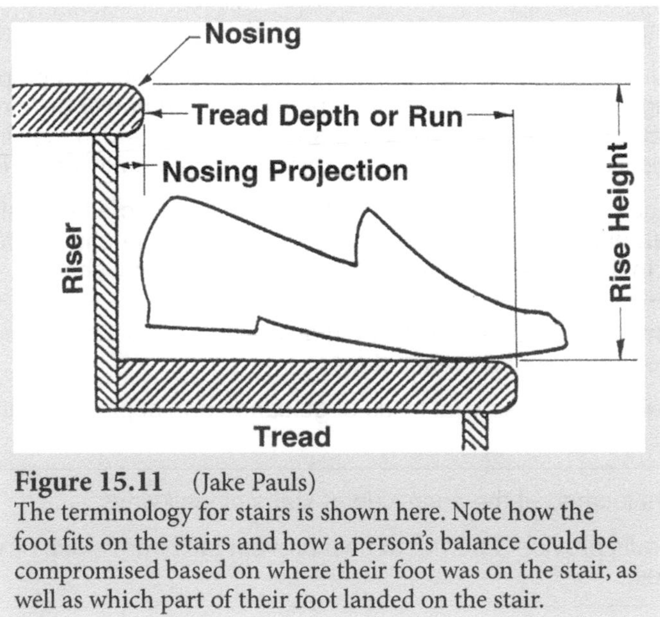

Figure 15.11 (Jake Pauls)
The terminology for stairs is shown here. Note how the foot fits on the stairs and how a person's balance could be compromised based on where their foot was on the stair, as well as which part of their foot landed on the stair.

THREE PRINCIPLES FOR STAIRWAY USABILITY AND SAFETY

> *The key to stair safety lies not so much in the hazard itself as it does in the users' awareness of their vulnerability to it. . . . So long as the users know they are coming to a stair, where to place their feet on the treads, and where to grab if they should momentarily lose their balance, then they are most likely to make it to the top or bottom of the flight without an accident.*
>
> (Asher, 1977)

The above statement points to three critical features of a stairway: visibility, appropriate step geometry, and functional handrails. These three features relate to stairway use, safety, and design, including universal design (Pauls, 2001).

Universal design is a design that "universally" applies to all users. This philosophy is at the heart of ongoing efforts to improve building code requirements for safer stairways. Although much of this effort has focused on the elderly, design features such as lower step risers and easy-to-grasp handrails benefit both children and adults.

VISIBILITY

Generally, because of their critical role in safety, stairways should be the most visually prominent features in the environment.

Users should be able to see stairways and the nosing of individual steps. Of course, being able to see the stair depends on the stairs' surface materials and illumination as well as the user's vision, expectations, and activity at the time (Figure 15.3 and Figure 15.4). Avoid tread materials or coverings that camouflage the step nosings (Box 15.1).

Oftentimes, people do not notice small differences between two surfaces, such as a single step. It helps to make stair handrails visually noticeable, as they are more prominently located—higher than the steps—within an approaching person's visual field.

Generally, visual features of stairways have been of greatest concern for older adults (Archea, 1985). Less is known about how young children see their environments as they approach and use stairs. One aspect of visibility highlighted in US stairway safety research is "orientation gradient" or "orientation edge" (Figure 5.12.) (Archea et al., 1979; Asher, 1977; Alessi et al., 1978).

A stair user's line of sight passes the edge of an adjacent wall or ceiling and abruptly reveals a distracting object, view, or activity distracting

Figure 15.12 (Jake Pauls)
In this drawing, an *orientation gradient* occurs as the user descends the stairs and begins to see a larger vista, including another person, to the side of the stairs.

Suddenly seeing something distracting after visual field has been blocked by a wall or other object, may cause a person to 'startle' or may disrupt the rhythm of their feet as they descend the stairs. While little information of this effect is available for children, it is likely that their descent (should they be the one on the stairs) might be influenced even more than that of adults because they are more easily distracted, in general, compared with adults.

BOX 15.1 TREAD MATERIALS THAT CAMOUFLAGE STAIRS

- Patterned carpets
- Monolithic stone and terrazzo
- New concrete
- Prominent wood grain

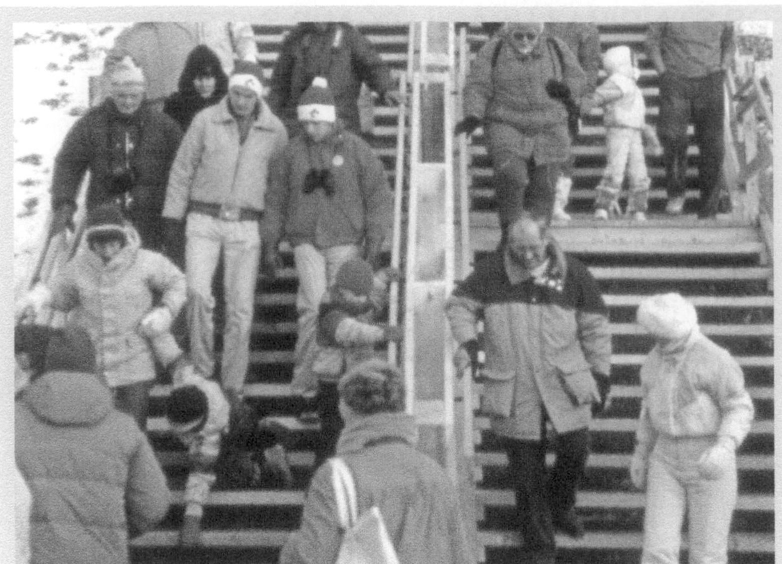

Figure 15.13 (Jake Pauls)
On a crowded stair, a mother arrests the fall of her son (left side of photo), while holding securely to the handrail with her right hand.
The boy on the right side of the stairway on the left descends while holding the handrail with both hands.
The difficulty of coming down stairs with a child is further shown on the stairway on the right of this photo. The child on the landing is distracted and has turned to look at something in the distance. While an adult would understand they needed to keep moving with the crowd and would curtail their reactions, a child may not have this understanding or level of self control.

the person and disrupting the rhythm of his or her foot movements. This causes the person to overstep or understep the nosing and fall (Alessi et al., 1978).

Information is lacking on how such orientation changes influence children. However, the presence of children could complicate the stair use task of others. Attention of people could be distracted even more as they experience a changing view while using a stairway.

Traveling up or down stairs with a child can impair visibility (see Figure 15.2 and Figure 15.6). It is generally prudent for adults to walk beside or in front of the child so they can help arrest or block the child's fall. This also reduces the likelihood of falling onto the child if the (possibly distracted) adult should fall. Crowded stairs pose their own step visibility and safety challenges (Figure 15.13).

STEP GEOMETRY

RISER AND TREAD DIMENSIONS: Stair step dimensions are the most-debated of all stair design features. Step dimensions should be appropriate and consistent with each other* (see Figure 15.3).

In general, the tread depth is the most critical dimension. One's foot needs an adequate area for support after stepping down the height of two risers during normal stair descent. A leading cause of fall-related stair injuries is overstepping of treads that are undersized (Alessi et al., 1978). This is true whether the undersizing occurs consistently within a stair flight or from one step to another.

However, for children and older adults, the greater challenge is the height of risers. Many children and older adults cannot easily climb or descend steps with risers as high as those commonly provided in homes or in certain public buildings such as stadiums with their aisle stairs.

In some US homes, risers are as high as 21 cm (8.25 in.)! In the UK, even higher risers of 22 cm (8.66 in.) height are permitted for "private stairs" within dwellings. However, the UK building regulations' pitch limit of 42° usually means that slightly lower risers are provided.

* Small differences in stair heights and tread depths are acceptable, such as 0.5 cm (3/16 in.).

Table 15.3 A Riser-Tread History Lesson			
1 BC, Vitruvius recommended riser-tread ratios of 3 to 4			
	Minimum	**Maximum**	**Preferred**
1570, Palladio recommended:			
Riser Height	11.4 cm (4.5 in.)	17.3 cm (6.8 in.)	
Tread Depth[a] (general range)	34.3 cm (13.6 in.)	52 cm (20.4 in.)	
We now recommend:			
Riser Height	10–12.5 cm (4–5 in.)	18 cm (7 in.)	15–16.5 cm (6–6.5 in.)
Tread Depth	28 cm (11 in.)	41 cm (16 in.)	33 cm (13 in.)
Concern about step dimensions has a very long history. This concern includes sensitivity to usability issues, which led several centuries ago to recommended dimensions very similar to those preferred today.			

[a] Run or going dimension.

STEP GEOMETRY APPROPRIATE FOR CHILDREN: *Step geometry should support universal design* (design for all). Current recommendations for step risers facilitate use by young children through older adults.

The current recommendation for even maximum riser height represents a decrease of about 3.2 cm (1.25 in.) from that permitted by some traditional building codes for home stairs. Table 15.4 shows preferred dimensions considering space limitations, younger and older stair users, plus the need for safety and usability improvements in homes.

As of 2003, one of the two model building codes in the United States (the NFPA Building Construction and Safety Code) adopted the "preferred" dimensions in Table 15.4, in part because its scope includes usability in addition to the traditional building code scope of safety. Codes permitting home stairs with higher risers and shorter tread depths typically require a projecting nosing. This is based on the assumption that such nosings somehow compensate for a shorter tread depth, however, this does not make such stairs as usable or safe as those with the "preferred" dimensions shown in Table 15.4.

Scaled-down stair geometries for children are **not** recommended, even in preschool-age facilities. New construction in the United States must comply with a "7–11" standard (7 in./18 cm riser heights with tread depths of at least 11 in./28 cm). This works reasonably well for children.

Table 15.4 Recommendations of Step Dimensions for Homes		
	Preferred	**Traditional Building Code Rules**
Riser Height	18 cm (7 in.)	Up to 21 cm (8.25 in.)
Tread Depth	28 cm (11 in.)	As little as 23 cm (9 in.)
Tread Including Projecting Nosing	28 cm (11 in.)[a]	25 cm (10 in.)

[a] No nosing projection needed or even desirable.

This should have resolved the debate about scaled-down stairs for educational settings. However, at least one building code, in use for many years in New York City, allowed special small-step stairs for some educational facilities.

Such school guidelines are problematic because both adults and children use school stairs; stair design must compromise to suit both. Although lower riser heights are better for all users, smaller stair treads for children may mean more missteps and injuries for adults. Thus, the move should be to "universal design" standards that so far have correctly focused on the "7–11" (maximum–minimum, respectively) dimensions as the best minimum standard for the widest range of users.

The exception is when building "learning" stair-steps for children's classrooms or therapy. Smaller stair dimensions are useful when stairs are specifically designed for very young children's play, exercise, and rehabilitation. Such stairs—not constructed as part of the building—sometimes lead up to a play area, an inside slide or simply permit the children to climb up one side and down the other, thus learning to go up and down the stairs while using handrails (Figure 15.14 through Figure 15.16).

Figure 15.14 (Valerie Rice)

Figure 15.15 (Valerie Rice)

Matching stair geometries to children's anthropometric size is not recommended, as older children and adults need to use the same stairs. The best minimum standard for the widest range of users is the "7–11" standard.

The only exceptions are stairs especially constructed for young children's learning and therapy. These stairs are not part of building construction. They offer an environment for children as they learn to move ahead one step at a time, keeping their balance, alternating their feet and developing gait proficiency. Therapists use them to teach, as well as to measure children's progress in functional mobility and balance (Zaino et al., 2004).

Figure 15.16 (Ursy Potter Photography)

EFFECTIVE STEP GEOMETRY WITH CARPETED STAIRS: Many people consider thick stair carpeting and padding plush and attractive. Many also believe that this helps to cushion falls. In practice, such carpets actually aggravate risks as they create insufficiently deep treads and excessively high risers. A common 2.5 cm (1 in.) thickness of carpet and padding reduces the effective tread depth by that same inch because the carpet in front of the step nosing does not support weight. In Canada, where building codes permit even shorter treads than for US home stairs, the effective tread depth with such thick carpeting is as little as 18 cm (7 in.).

With time and usage of the stairs, this carpet projection may increase by as much as 5 cm (2 in.) as the carpet becomes looser, causing the steps to have a large curvature and slope at the nosing in addition to not supporting ones foot.* This nonsupportive projection and slope increase the risk of overstep, followed by a slip or rotation of the foot around the nosing that may lead to a loss of balance, forward-pitching falls, and serious injuries.

Thick resilient step coverings also increase problems associated with riser heights. Effective riser heights increase by the difference between the uncompressed and compressed carpet thickness. This difference can be about 1.5 cm (0.6 in.) for typical carpet installations in homes; thus effective stair riser heights can range from 20 cm (8 in.) to as much as 23 cm (8.9 in.) in US homes.

This difference in riser height increases everyone's risk of tripping during stair ascent, although older adults are at greater risk of injury. Children who trip going up steps fall forward over a relatively small distance to the cushioned steps (e.g., toddler's crawling stance in Figure 15.1 and Figure 15.2). In contrast, loss of balance and falls among older adults can be catastrophic, particularly if they have osteoporosis or osteopenia.†

Ultimately, stair-carpeting decisions depend on the (1) increased risk of missteps and (2) the potential for cushioned surfaces to limit stair injuries. One compromise is to remove thickly padded carpets and carefully install and maintain a thinner, denser carpet on stairs (if necessary to cover otherwise unsuitable treads). The thickness and projection beyond the structural step nosing, as well as the radius of curvature of the carpet-walking surface, should be less than 1.3 cm (0.5 in.) (Pauls, 2001).

UNIFORMITY OF STEP GEOMETRY: We mentioned in "Riser and Tread Dimensions" that uniform step dimensions are crucial for safety. The recommended tolerance factor is 0.5 cm (0.2 in.). Irregular step dimensions easily contribute to missteps and falls (Templer, 1992).

RAILINGS

The function of a handrail is to help arrest a fall while on a stair or walkway. If high enough, handrails can also serve as "guards" to prevent falls from elevated stairs or walkways.

Other guards include full-height walls or partial-height walls (as with a banister). A guard can also be a railing system with vertical members—balusters or pickets—that provide structural support as well as forming a semi-open barrier below the handrail. For both railings, height above the adjacent walking surface is an important characteristic. Figure 15.4 illustrates a proper height guard, complete with vertical pickets or balusters onto which a (slightly low) handrail is attached.

* This problem is particularly common in the absence of regular and proper stair maintenance.

† A condition of bone in which decreased calcification, decreased density, or reduced mass occurs.

HANDRAIL HEIGHT: Children are clearly not "scaled down adults," as their proportions differ as well. Handrail heights for children should be based on how they use railings. Values for children's stature and traditional rules in the United States for handrail heights might be interpreted as suggesting that handrails for children could be as low as 43 cm (17 in.), relative to adult handrails of 76 cm (30 in.). Yet such heights would be far too low to protect children and adults from falls.

Canada has performed extensive handrail research on functional stairway handrail heights and the biomechanics of how people move. Their field studies included the full size range of children and adults. In addition, they performed laboratory studies of a full range of adults.

Canada's research demonstrates that traditional handrail height requirements in older building codes are unrealistically low and, in testing, they fared poorly (Maki et al., 1984, 1985; Pauls, 1985, 1989, 2001).

Testing with adults showed the functional benefits of and preferences for heights as high as 107 cm (42 in.)—the minimum permitted guardrail height in many building codes and safety standards. The most preferred and functional height for older adults was about 94 cm (37 in.), measured from the step nosing to the top of the handrail. When choosing from variable-height handrails, children often decide to use the higher one, even if this means using a handrail at head height.

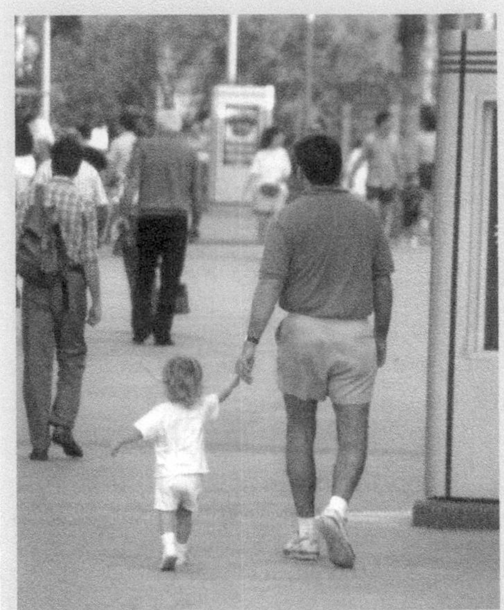

Figure 15.17 (Jake Pauls)
Children are accustomed from early walking to holding onto adult hands at head height and this works well for children's handrail usage also. Holding onto something that is higher also offers a biomechanical advantage.

Other research seems to confirm children's preference for higher handrails (Table 15.5). These preferences are close to the average (mean) stature of 3- to 6-year-old children (about 100 cm/39.4 in.) (Norris and Wilson, 1995). Overall, younger children appear to feel comfortable using handrails at shoulder height (about 77 cm or 30 in.) to head height.

This is not surprising, considering how small children hold onto adult hands (Figure 15.3 and Figure 15.17). Even during children's prewalking stage, "cruising," they occasionally hold furniture items at their shoulder to head height.

Table 15.5 Preferred Handrail Heights of Children in Laboratory Studies

Children's Ages	Preferred Heights	1988 Australian Standard for Adult-Height Handrails
3 to 6½	87 cm (34 in.)	87–90 cm (34–35.4 in.)
6½ to 10½	89 cm (35 in.)	
10½ to 14½	92 cm (36 in.)	

Source: From Seeger B.R. and Bails J.H., Ergonomic building design for physically disabled young people. *Assist. Technol.*, 2, 79, 1990.

However, it is not merely the socialization (handholding and cruising along existing household furniture) that influences their preferred handrail height. Holding onto something higher is biomechanically effective; it works to prevent falls. Adults who hold bus and transit train vertical rails at chest to shoulder height can attest to such effectiveness.

As part of rule making for the Americans with Disabilities Act Architectural Accessibility Guidelines (ADAAG), the US Architectural and Transportation Barriers Compliance Board (Access Board) reviewed literature and public comments on handrails. For children, they concluded:

> *ADAAG requires a mounting height of 34 to 38 in. (86.3 to 96.5 cm) for handrails based on adult dimensions. They did not include a requirement for secondary handrails in their final rule.*
>
> *However, advisory information on lower handrails has been added to the appendix at A4.8 and A4.9. This information recommends a secondary set of handrails at ramps or stairs in facilities that primarily serve children, such as elementary schools. They recommended a maximum handrail height of 28 in. (71 cm) and that the vertical clearance between handrails is at least 9 in. (23 cm) in order to reduce the risk of entrapment.*
>
> (ATBCB, 1998)

Obviously, there is conflicting evidence and opinion. However, the information provided in this section suggests that it is not necessary to provide additional, supplementary handrails, even if there are many children in the user group. This is because the earliest independent, unrestricted stair use should begin at about age 3. Handrails at the lower end of the permitted range (86–97 cm or 34–38 in.) should be adequate for 3-year-olds. Even handrails at the upper end of the range are still within the "vertical reach range to grip" of about 115 cm (45 in.) (Norris and Wilson, 1995).

Installing a second, lower-height handrail at 71 cm (28 in.) is still an option, although many children prefer higher heights of 94–97 cm (37–38 in.), which could, incidentally in combination, also satisfy the 23 cm (9 in.) handrail spacing criterion suggested in the ADAAG appendix.

Regardless, take measurements from stair nosings to the top of the handrail and carefully consider grasp characteristics for each handrail.

HANDRAIL GRASPABILITY: Unfortunately, as with the relatively steep stairs permitted in homes, handrails installed in homes are often not the best size or shape for grasping by children or even adults (Pauls, 1989). US stairway and handrail manufacturers, dealers, and homebuilders have effectively pushed for building code requirements for handrail size and shape that render them almost meaningless in terms of grasp effectiveness.

Handrails with smaller and circular shapes are dramatically more effective. Children can grasp them from below (Figure 15.18). These also enable children to grasp them from any direction of the hand. Unfortunately, the US building industry has not supported or encouraged the use of these excellent handrails.

Figure 15.18 (Jake Pauls)
Small-diameter, round handrail section facilitates grasp from any direction and permits children's effective use of handrails at head height.

Biomechanical research results on handrail shape and size are available for adults but not for children (Maki, 1985). Nor is there systematic research on children's grab response for handrails not already in hand at the beginning of the fall as has been investigated for adults (Maki et al., 1998).

However, there is anecdotal evidence such as that represented in Figure 15.19 that children can effectively grab for and grasp—even with both hands—a handrail not grasped before a fall.

Most adults can grasp round handrails of 4 cm (1.6 in.) diameter or less with a full encircling power grip. Most children above age 3 can grasp them with a 75%, encircling power grip.

Table 15.6 lists US adult handgrip diameters, measured between the thumb and middle fingers. Table 15.7 provides handgrip diameters for children; these suggest the extent to which children's hands can wrap around 4.7 cm (1.85 in.) diameter handrails, the size that permits a full power grip for half of US adults.

Table 15.6 US Adult Hand Grip Diameter

Body Size	Diameter
5th Percentile Female	4.0 cm (1.6 in.)
50th Percentile Adult	4.7 cm (1.85 in.)
95th Percentile Male	5.6 cm (2.2 in.)

Source: From Peebles, L. and Norris, B. in *Adultdata: The Handbook of Adult Anthropometric and Strength Measurements*, Department of Trade and Industry, UK, 1998.

Figure 15.19 (Jake Pauls)
Although there is no systematic research on the reaction times for children to grab a handrail during a fall, when they were not holding it prior to the fall, there is anecdotal evidence that this is possible.

This boy successfully arrested his fall on stairs with a double hand grab. He had been running and chasing a younger boy down a busy stairway in a shopping center. The stairway had two flights of about 16 risers each. He was not using either of the stairway's two handrails. Near the bottom he misstepped but caught himself with a double handhold on the handrail which was located to his left at a height of 81 cm (32 in.) and a circular diameter of 5 cm (2 in.). This photograph was taken at the instant he arrested his fall with the successful double grab for the handrail.

Strength testing for various grasp diameters is not available for children. In the interim, the recommendation is to use smaller-section handrails (about 3.2 cm or 1.25 in. diameter) for young children.

Another alternative is to use, for graspable support, smaller stair rail components such as vertical pickets or balusters with diameter smaller than 2.5 cm (1 in.) (as in Figure 15.2).

Table 15.7 Grasp Diameters for Children

Age (years)	cm	in.	Ratio of Adult Mean
1–1.5	2.38	0.94	0.50
1.5–2	2.58	1.02	0.55
2–2.5	2.74	1.08	0.58
2.5–3	2.89	1.14	0.62
3–3.5	2.98	1.17	0.63
3.5–4	3.03	1.19	0.64
4–4.5	3.16	1.24	0.67
4.5–5	3.25	1.28	0.69
5–5.5	a	a	a
5.5–6	3.32	1.31	0.71
6–6.5	3.43	1.35	0.73
6.5–7	3.55	1.40	0.75
7	3.62	1.42	0.77
8	3.79	1.49	0.81
9	3.97	1.56	0.84
10	4.09	1.61	0.87
11	4.33	1.70	0.92
12	4.47	1.76	0.95
13	4.71	1.85	1.00

Source: Morris and Wilson, 1995, Table 80 combining U.S. male and female data.

[a] Data not provided in reference.

Ordinary dimensioned lumber, often used for residential stair railings, is not functional for handrails for children or adults. It does not match the shape or size of human hands, and the edges are uncomfortable for grasping.

Open risers should also not be used where children use the stairs. It is too easy for children's limbs to slip in between the stairs. This can occur as a result of many factors, including their being less proficient in their balance and locomotion skills.

Figure 15.20 (Jake Pauls)

Children (and adults) cannot effectively use large, sometimes decorative, railing sections as functional handrails. These include relatively crude railings that are created with dimensioned lumber (Figure 15.20).

Although the circular-shaped handrail is excellent, oval-shaped handrails that are oriented vertically with horizontal dimensions of 3.8 cm (1.5 in.) and vertical dimensions of 5 cm (2 in.) are another possibility. Biomechanical testing with younger and older adults showed this shape to be effective to develop the forces and moments needed to help arrest a fall (Maki, 1985). Moreover, while not yet tested with children, it might be an appropriate shape for a child's hand extended from below and hooked over the top of the railing.

GUARDS: Two aspects of guards have been vigorously debated in model building code development hearings and in the associated literature. The first characteristic relates to the maximum permitted spacing of balusters, pickets, or other construction elements below a guardrail. The concern here is the need to prevent children from entrapment or passage through the guard openings (see Figure 15.21 and Figure 15.22). The second characteristic relates to whether children can easily climb the elements below the guardrail, often as part of play activity that stairs and railings may encourage.

Guard opening size Prior to 1988, building codes in the United States permitted guard openings of a size large enough for a 15 cm (6 in.) diameter sphere to pass through. Even earlier, some codes permitted passage of a 23 cm (9 in.) diameter sphere.

However, these codes failed to reasonably protect children, as a child could fit through the opening. Guard openings should not permit the passage of a 10 cm (4 in.) sphere (Stephenson, 1988, 1991, 1993) (Figure 15.21 and Figure 15.22).

These facts support the recommendations (Stephenson, 1988, 1991, 1993):

1. Virtually all children under 6 can pass completely through 15 cm (6 in.) wide openings.
2. Virtually no child 1 year or older can pass completely through 10 cm (4 in.) wide openings.
3. About half of all 10-year-olds can pass completely through 15 cm (6 in.) wide openings.
4. About half of children between 13 and 18 months old can pass through 12.7 cm (5 in.) wide openings.
5. About 5% of 1-year-olds can pass completely through 11.4 cm (4.5 in.) wide openings.

Although the code-change effort was successful at the national model code and standards level in the United States (just as there had been successes in other countries), this does not mean children and parents in the United States are totally protected. First, the building code requirements apply only to new construction. Second, homebuilders and others in some state and local jurisdictions launched effective counter campaigns.

Initially, in some US locations such as the Commonwealth of Virginia, the new rule applied everywhere but homes, the very place where young children were most likely to need the protection. In essence, the counter arguments emphasized controlling child behavior and advocating parental responsibility, rather than ergonomically justified changes in building construction.

However, children will explore, roam, and get into predicaments that can cause them harm (e.g., crawling beneath open stair risers). Parents will not be able to observe and protect their children during every second of their lives. Quite often, when a child has been hurt, the parents say, "I only turned away for a minute." They turned to respond to a question, to do a task, or to attend to another child. The absolutely best way to prevent harm is to design environments that are safe for all. While it is often necessary to rely on human

Figure 15.21 (Jake Pauls) **Figure 15.22** (Jake Pauls)
The design and construction of guards should prevent passage by children wherever children are expected to be present—whether or not there is constant parental supervision. Openings in guards, such as the spacing between vertical baluster should not permit the passage of a 10 cm (4 in.) sphere, a measure based on children's anthropometrics.

The same 10 cm (4 in.) limitation applies to openings between treads (with open risers as shown in Figure 15.2 and Figure 15.20) and in bleacher construction (with openings between footboards and seatboards).

intervention, humans must attend to a multitude of concurrent tasks. Railings and guards should provide protection independent of human intervention.

Some residential building codes* permit guard openings on the sides of stairs that do not allow passage of a 10.7 cm (4.38 in.) sphere, rather than the 10 cm (4 in.) sphere.

The rationale is not ergonomic, nor even economic (in terms of injury costs). Rather, some homebuilders were concerned about having to add a third picket to each tread to meet the 4 in. (10 cm) rule.

This rationale was linked to keeping a minimum tread depth requirement of 23 cm (9 in.) in homes, while 28 cm (11 in.) is the now-common model code requirement for all except home stairs. Had a larger tread depth been adopted, there would be no question that three pickets per tread would be the best solution—and, indeed, many traditional home stairways had three pickets per tread based on aesthetic factors among others.

Other railing designs avoid this issue, such as with pickets extending down to an element parallel to the pitch of the stair flight. In this case, codes permit the resulting triangular opening at the intersection of the tread and riser to prevent the passage of a 15 cm (6 in.) sphere, based on the untested assumption that a child cannot wriggle through such a triangular space.

Guard climbability The second guard issue, climbing of guards by children was more challenging (Stephenson, 2001; Riley et al., 1998). National injury estimates from the CPSC/NEISS system indicated a problem without providing much detail: "In 1988 and 1989, the estimated number of emergency room-treated injuries to children less than five years of age, who were reported to have fallen or jumped on or from a porch or balcony was approximately 15,000 to 16,000 per year" (Stephenson, 2001).

* The 2003 edition of the *International Residential Code* (the 2nd edition of a relatively new model code in the United States).

Of many different guard designs tested by Stephenson and colleagues in other countries, only one appeared to prevent climbing and be appropriate for building applications. It consisted of vertical pickets (with the 10 cm or 4 in. opening size) for the entire guard height except for the lowest and highest 10 cm (4 in.) portions of the height.

Figure 15.23 (Jake Pauls)
Guards should not be climbable by children. When a guard is shaped like a ladder, some children will attempt to climb it. For intermediate railings, vertical pickets or balusters are preferred over horizontal ones.

US model codes and safety standards have not yet adopted anything beyond relatively simple performance language for a requirement or even prudent guidance to control climbing. For example, the Life Safety Code, NFPA 101 (an ANSI American National Standard) notes in the nonenforceable appendix, "Vertical intermediate rails are preferred to reduce climbability" (NFPA, 2003a). Some other codes reference avoidance of a "ladder effect." Figure 15.23 provides an example of the "ladder effect."

No other single individual has done more than Elliott "Steve" Stephenson (pictured in Figure 15.24) to work with and educate building code officials on the dangers of and countermeasures for guardrails to which children have access. In one of his last articles, directed specifically to building code authorities, he described the safety of buildings as follows.

> *For many decades during the last century the needs of young children in buildings have been given inadequate consideration by our building codes with the result that there are literally millions of unsafe guards in our existing homes, apartments, motels, hotels and schools, as well as in other existing buildings at locations where young children can be expected to be present. This is an unfortunate legacy that has been left by building code authorities for the 400 million or more children who will be born in the United States during the 21st century.*
> (Stephenson, 2001)

Figure 15.24 (Jake Pauls)

Mr. Stephenson, however, did give up in despair. He went on to speak optimistically about the importance of facts to help guide building code authorities. For example, the injury facts and anthropometric data he published extensively in building industry magazines led to major building code changes (see "Guard Opening Size") regulating guard openings sizes for new construction.

> *During my 55 years of participation in the code development and enforcement activities, I have learned one very important and significant lesson: once building officials have been given needed information and facts concerning a particular problem warranting their attention and a positive action on their part, they will act to solve and correct that problem to the extent possible. Without pertinent and significant facts, however, no action will be taken. . . . From my personal experiences with the resolution of such problems, it has been clearly demonstrated that those who oppose effective action on the part of building officials solely on the basis of their commercial interests will eventually find that their sometimes selfish and untenable objectives are unacceptable and that those with the needed facts will obtain theirs.*
> (Stephenson, 2001)

CONTROLLING STAIRWAY USAGE

People can and will fall on stairs. Stairway design, construction, and maintenance can only help prevent falls and injuries; it will not totally eliminate them. While stairway design can help prevent injuries, prohibition of stairway use is the best strategy for very young children.

The key question is at what age or developmental stage should independent, unsupervised stairway use be permitted? Examinations of the literature and discussions with pediatricians suggest that an age of 3 years is a good starting point for independent stairway use* (Figure 15.25).

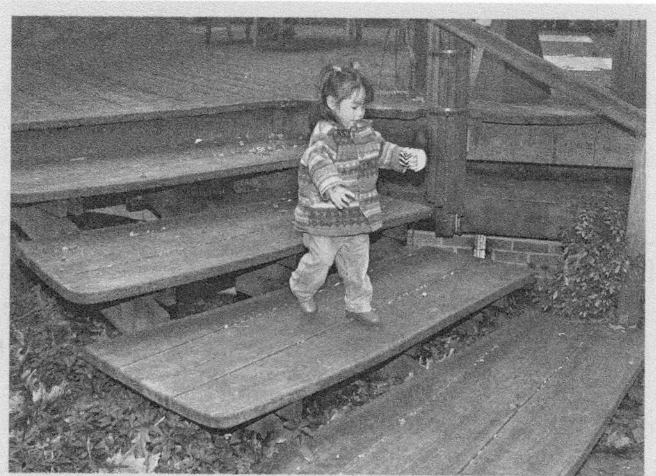

Figure 15.25 (Ursy Potter Photography)
Many children can begin independent use of stairs at about age 3. However, many stairs in and around homes will still tax their abilities and offer little or nothing to mitigate dangers, such as child-friendly handrails, needed even in the case of the shallow-pitch stair shown above.

Before age 3 and perhaps even after in the case of unintended nighttime use, caretakers should block stair access with proper gates or doors. Obviously, children should not be capable of opening these gates or doors. Gates should not be the folding, accordion type, which hold entrapment dangers for children.

One solution is home design that facilitates blocking undesirable access to stairways. Overall, home planning could also benefit by arranging spaces more effectively to minimize the need for children and parents to move between floor levels. These considerations will not only benefit children and their parents; much later in life it may be necessary to control intentional and unintentional movement by family members with dementia or who are debilitated by medications.

* While much younger children can crawl up stairs, they will not be able to descend—with a crawl—until about 18 months of age. The ability to use one foot per tread—walking upright—is achieved by about age 3.

As mentioned at the beginning of this chapter, mobile infant walkers are a special hazard in combination with stairs, even those with gates.

The American Academy of Pediatrics cautions (AAP, 2001):

Adult supervision also cannot be relied on to prevent infant walker-related injuries. Moving at more than 3 ft/sec, an infant can be across the room before an adult has time to react. In one study, 78% of children were being supervised at the time of the injury, including supervision by an adult in 69% of cases.

Other studies have also shown that many of these events occur with one or both parents in the room. Stair gates are not uniformly effective even when present; one study found that more than one third of falls down stairs occurred with stair gates in place, but the gates were either left open or improperly attached.

DESIGN REQUIREMENTS IN BUILDING CODES AND STANDARDS

Professional, consumer, and governmental organizations that focus on children's safety include the American Academy of Pediatrics, the Child Accident Prevention Trust, the Safe Kids Campaign, and the Centers for Disease Control and the US Consumer Product Safety Commission. However, applying information on children's safety to building codes and standards has generally received little attention. Notable exceptions are with guards, swimming pool enclosures, playground equipment, and bleachers.

Ergonomists are seldom involved with the development of model building codes and related safety. This is not healthy for the codes/standards development process or for the profession of ergonomics. Although ergonomists engaged in forensics activities have expertise in injury causation, little of this information makes its way into building codes and regulations.

The challenge for making stairs and stairways safer is in the application of principles of universal design. Universal design applies standards and requirements developed for special populations (or physical conditions) to the general environment so that everyone can benefit from user-friendly designs. In the United States, the growing concern about safety for older adults drives the universal design of stairs and railings. However, the design improvements for the aging population will also meet the needs of children. Lower stair risers and easily grasped handrails are two prominent examples.

There is a great need for cooperation among homebuilders, public health organizations (such as the American Public Health Association), and the model code development organizations (such as NFPA and ICC in the United States), to ensure standards are set using valid research results and that the standards ensure safety for everyone. Also, once standards are set, they need to be enforced (Figure 15.26 and Figure 15.27).

Figure 15.26 (Ursy Potter Photography)

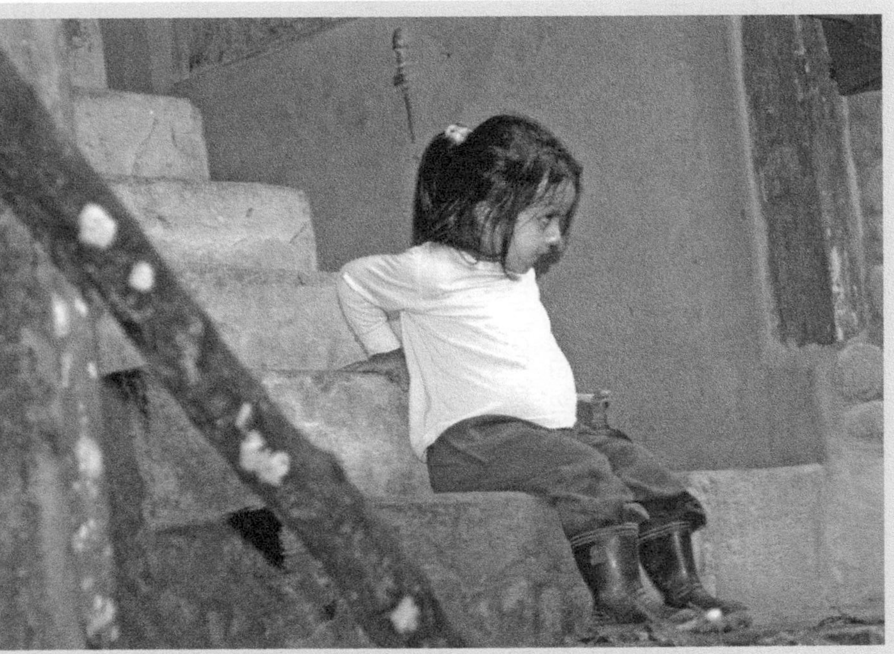

Figure 15.27 (Ursy Potter Photography)

Some stairs in traditional housing evolved from, or replaced, even more dangerous and difficult-to-use ladders. Such early stairs are sometimes relatively steep with high risers and they lack handrails and guards to mitigate or prevent falls. Appropriate design provides protections without constant reliance on humans to behave safely, as these behaviors may be inconsistent among children and parents are not always near by to supervise.

RETROFIT GUIDELINES

There have been numerous attempts to provide advice on retrofitting stairways to improve usability and safety. Unfortunately, many contain advice that can make stairways more dangerous, if adopted literally.

One example is the suggestion to retrofit with "sturdy handrails." To an average person, this might imply using a "two by three," a "two by four"—or worse—piece of dimensioned lumber. These are unsuitable handrails for adults and for children (this is especially clear from the data summarized in Table 15.5 for children and from depictions of common domestic railings such as in Figure 15.20).

A second example is the suggestion to add special friction tape to stair treads to mitigate a presumed, but likely misdiagnosed problem of "slipping." If the problem is actually due to other missteps or excessively narrow treads, the change could make stairs more dangerous, while preventing proper diagnosis and mitigation. The first rule of retrofitting stairways is "Do no harm"; the solution should not introduce new and potentially worse problems!

Six principles for retrofitting include (Archea et al., 1979):

1. Upgrade the stairs most frequently used by the most vulnerable people first. Children under the age of 5 have twice as many stair accidents as their proportion of the population suggests.
2. Avoid piecemeal repairs and temporary patches on stairs.
3. Do not try to learn new skills while fixing stairs.
4. There are upper, as well as lower limits to safe conditions on stairs.
5. Compensate for all defects that cannot be corrected.
6. Avoid repairs or renovations near the stairs that could create new hazards.

Other retrofit recommendations for stairways mentioned earlier relate to stair carpeting ("Effective Step Geometry with Carpeted Stairs"), handrails for children ("Handrail Height" and "Handrail Graspability"), and guards ("Guards").

Commercially available products can help fix problems with guards that have too large a spacing between balusters, which include clear plastic panels and tight-mesh netting. References for retrofitting stairways are in Box 15.2.

BOX 15.2 REFERENCES FOR STAIRWAY RETROFITTING

Checklist: *Universal Design Handbook* (Pauls, 2001) and at www.homemods.org

"Avoiding the causes of stair falls: Designing for injury reduction" (Templer, 1992, Chapter 7)

SUMMARY

Stairways are one of our earliest and one of our last physical challenges in life. Key considerations for safe and usable stairways include stairway visibility, appropriate and consistent dimensions of risers and treads, and provision of functional handrails and guards.

Building codes and standards have not adequately addressed these key factors, especially for homes. This is due, in part, to inadequate technical information. It is also due to political realities of the construction industry and the associated, highly interdependent private sector and governmental systems of building code development, adoption, and enforcement, especially in the United States.

While this chapter has emphasized the operation of these processes in the United States, there are comparable processes that short-change the consumer in other countries. There will not be easy solutions to the major problems of safety and functionality faced by stairway users so long as ergonomics is not adequately considered in the original provision of environments for children.

WHAT CAN DESIGNERS AND BUILDERS DO?

First, they must recognize that codes and most design guides (even some that are relatively up-to-date) do not adequately address ergonomic issues of building design for children. Codes, for example, have focused on certain traditional hazards such as fire that result in less frequent, but catastrophic results. The focus needs to broaden to include hazards that more commonly lead to injury; falls injure a hundred times more people than do fires.

This means that design and construction of built facilities for human movement—especially stairways—should not merely comply with minimum standards and code rules. Compliance with such standards and codes is necessary but not sufficient. The courts dealing with premises liability cases as well as with product liability recognize this criterion.

There are excellent guides available on Universal Design. The philosophy of inclusive design takes into account usability as well as safety.* Applying principles of Universal Design does not mean a building or facility will be less attractive or even more expensive. It means, among other things, that a design will work relatively well for the largest number of users.

Clearly, from the many technical details provided in this chapter, design for children's capabilities and needs is not incompatible with design for other vulnerable populations, especially elderly persons, nor is it incompatible with the needs or aesthetic desires of other adults. Moreover, regarding economics, "*Good Ergonomics is Good Economics.*"

WHAT CAN PUBLIC HEALTH AND BUILDING SAFETY AUTHORITIES DO?

In the early years of the twenty-first century, the community of public health professionals is rediscovering its roots in environmental health, especially in the health impacts of the built environment.

The design and retrofit of existing stairways in homes and public buildings pose special challenges for the public health community which sees—first among public authorities—the results of environmental failures. Public health officials must move beyond merely recording the injury toll, a toll that disproportionately affects children in relation to stairways. This will mean new (or rediscovered) collaborations with building safety authorities.

Organizations such as the American Academy of Pediatrics (AAP) and National Safe Kids Campaign have participated as advocates and technical resources to building safety authorities on issues of children's safety with railings.

* Pauls (1991) provides but one of many extremely comprehensive and useful chapters in the *Handbook on Universal Design.*

The American Public Health Association (APHA) has advocated, through its public policy process, sweeping improvements in the system of building code development, adoption, and enforcement. Stairways were a focus of at least two of its public policy initiatives in 1999 and 2000. These and APHA's many other roles such as advising the public directly on public health and safety issues are important for home stairway safety.

Children are a natural focus for such advocacy and public education because the public health messages are relatively well received by the public as well as by political bodies. Building safety authorities also need to rediscover their roots; they are, after all, doing public health work. Like most successful safety advocacy, it is most effective at the local level—something that children's safety organizations, like Safe Kids, understand well. The bottom line is that new and renewed collaboration is a key to improved safety for children, including their safety on stairs.

WHAT CAN SCHOOLS DO?

Stairways are a special challenge in older schools, which are more likely to involve multiple floor levels and many stairways. Handrail provision is a particular concern as many of the older schools were constructed when codes had antiquated rules for provision of intermediate handrails on wide stairs that permitted spacings of up to 224 cm (88 in.) between handrails.

Based on ergonomics, even for adults, the maximum spacing of such handrails should not exceed 150 cm (60 in.). For children, it would be prudent to keep handrails no more than 120 cm (48 in.) apart. Additionally, all handrails should be easy to reach and grasp by all users. This chapter provides detailed guidance on choosing such functional handrails.

Also, as many older stairs have deteriorated treads, it is important to recognize that any fixes to the tread surfaces should not introduce new hazards while trying to mitigate or eliminate old ones. For example, attaching "slip-proof" finishes—especially tapes—is not a good idea. It does not solve a significant problem as slipping is, on careful investigation, typically not the leading cause of stair-related falls.

Dimensions should be uniform—measured very meticulously nosing-to-nosing on each-and-every step. The school stair maintenance program should aggressively tackle the issue of structural settlement, as well as erosion from years of use, because nonuniformity is the leading cause of injurious falls on stairs. Merely providing warning of such hazards is not sufficient. The stairs must be adjusted or rebuilt in case of dimensional nonuniformities exceeding 0.5 cm (0.2 in.). Stairs with especially high risers should undergo rebuilding. Risers should not be higher than 18 cm (7 in.), even for adults and 15 cm (6 in.) is better for children.

Finally, stairway lighting often needs retrofitting, even with the limitations of reduced operating budgets that many schools face. Modern lighting systems provide high-quality lighting with greatly reduced energy demands. Automatic, occupancy-sensing switching systems can make sure the stair lighting is at best levels only during periods of use. It would be good economics—and ergonomics—for school administrators to employ a skilled illumination engineer, especially one who is familiar with the latest Illuminating Engineering Society (IES) *Lighting Handbook*, which has much useful information on these topics.

Regarding bleachers, which are common in schools, see also the next section on public buildings.

OPERATORS OF PUBLIC BUILDINGS

Some public buildings have stairs that must be used under the most difficult of circumstances, including mass egress in case of emergency. Stadiums, arenas, and theaters are examples. In such buildings, there is extensive use of stairs, including aisle stairs. For some seating systems, there may also be extensive use of bleachers or tiered seating without the degree of architectural finish and seating provision found in newer buildings.

Bleachers are especially a problem in schools and some other facilities for public assembly. Bleachers are essentially large assemblies of steps. Older bleachers lack proper aisles with better-proportioned steps and handrails. Guards at the elevated side of bleachers may also be deficient. Openings between footboards and seatboards are sometimes oversized relative to modern rules based on the 10 cm (4 in.) sphere rule discussed at length in the section on guards.

Crowds pose special hazards and a whole book could be devoted to the problems of crowd safety at concert events attracting older children. Stairs and crowds are not a good mix. Therefore, operators of public facilities where crowds congregate will need to pay extra attention to stairways. A mishap on a stair, even a relatively minor one, such as a misstep by one individual, can quickly become catastrophic if the highest standards of design and operation are not employed. Having crowds of children, as well as mixed crowds of adults and children (in family groups or not), pose particular challenges.

WHAT CAN PARENTS DO?

Parents know that children learn through "the school of hard knocks." They also know, first-hand, that injuries upset everyone and that some "hard knocks" are permanently disabling. Head injuries are a prime example. Falls from heights often result in head injuries, especially among children.

Avoiding well-known hazards—such as mobile walkers—is an important first step for the child's early mobility. It is also important to correctly select, install, and use stairway gates or barriers. Appreciate that children are challenged and intrigued by stairs. Learning to use stairs is part of a child's development. Parents need to consider carefully the timing and degree of independence along with the assistance needed by young children to use stairways. These are matters to discuss with pediatricians, who are versed in childhood development.

Home stair improvements can help all users, even though the improvements are intended to address the needs of young children. For example, careful fitting and maintenance of carpeting on stairs helps everyone. Excessive stair cushioning might ease the impact of a fall but it also increases the likelihood of falls for all users. Handrails that are more effective for small hands, than are commonly used decorative railing systems, help everyone.

While small children may be able to climb and descend stairs holding onto the railing pickets, this option is not available for adults who must have crucial support at a higher height. Eventually young children will want to mimic adult's use of handrails, therefore, making sure the railings are graspable is important for children and everyone else.

STAIRWAYS: A CHALLENGE FOR EVERYONE (INCLUDING PARENTS OF PARENTS)

Stairs are one of the oldest architectural elements. They are also the one causing the largest number of injuries in developed countries today. We must address the challenges posed by stairs better than we have been doing historically.

Children are disproportionately at risk for injury on stairs. Making stairways work better for children will help us all. Ergonomically appropriate stairways will be especially appreciated by some of children's best friends, their grandparents! It will help both of them be together and play together safely.

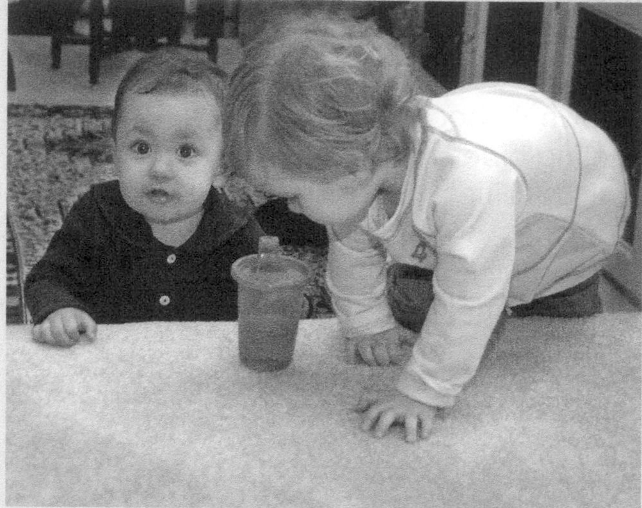

Figure 15.28 (Ursy Potter Photography)

Figure 15.29 (Ursy Potter Photography)

REFERENCES

AAP (2001). American Academy of Pediatrics, Committee on Injury and Poison Prevention. Injuries associated with infant walkers. *Pediatrics*, 108(3), 790–792. www.aap.org/policy/0102.html

Archea, J.C. (1985). Environmental factors associated with falls by the elderly. *Clinics in Geriatric Medicine* (Symposium on Falls in the Elderly: Biologic and Behavioral Aspects), 1(3), 555–569.

Archea, J.C., Collins, B.L., and Stahl, F.I. (1979). Guidelines for stair safety. NBS-BSS 120, Gaithersburg, MD: National Bureau of Standards.

Asher, J.K. (1977). Toward a safer design for stairs. *Job Safety and Health*, 5(9), 29.

ATBCB (1998). Americans with Disabilities Act (ADA) Accessibility Guidelines for Buildings and Facilities; Building Elements Designed for Children's Use. Architectural and Transportation Barriers Compliance Board, *Federal Register*, January 13, 1998 (http://www.access-board.gov/adaag/kids/child.htm).

Bertocci, G.E., Pierce, M.C., Deemer, E., and Aguel, F. (2001). Computer simulation of stair falls to investigate scenarios in child abuse. *Archives of Pediatrics and Adolescent Medicine*, 155(9), 1008–1014.

CPSC (2001). *Guidelines for Retrofitting Bleachers*. Pub. No. 330. Bethesda, MD: US Consumer Product Safety Commission. www.cpsc.gov/CPSCPUB/PUBS/330.pdf

CPSC/NEISS (1998). National Electronic Injury Surveillance System of the Consumer Product Safety Commission 1997 estimates report for stairs or steps, excluding pull-down and folding stairs. US Consumer Product Safety Commission, National Electronic Injury Surveillance System, National Injury Information Clearinghouse, Bethesda, MD.

Furuno, S., O'Reilly, K.A., Hosaka, C.M., Inatsuka, T.T., Allman, T.L., and Zeisloft, B. (1997). *Hawaii Early Learning Profile—Activity Guide*. PaloAlto, CA: VORT Corporation.

Hong, Y., Lau, T.C., and Li, J.X. (2003). Effects of loads and carrying methods of school bags on movement kinematics of children during stair walking. *Research in Sports Medicine*, 11, 33–49.

ICC (2002). *ICC Standard on Bleachers, Folding and Telescoping Seating, and Grandstands*, ICC 300-02. Falls Church, VA: International Code Council.

Joffe, M. and Ludwig, S. (1988). Stairway injuries in children. *Pediatrics*, 82(2), 457–461.
Maki, B., Perry, S.D., and McIlroy, W.E. (1998). Efficacy of handrails in preventing stairway falls: A new experimental approach. *Safety Science*, 28(3), 189–206.
Maki, B.E. (1988). Influence of handrail shape, size and surface texture on the ability of young and elderly users to generate stabilizing forces and moments. Report for Biomedical Engineering Research Program, DEE, National Research Council of Canada, Ottawa, NRCC 29401.
Maki, B.E., Bartlett, S.A., and Fernie, G.R. (1984). Influence of stairway handrail height on the ability to generate stabilizing forces and moments. *Human Factors*, 26(6), 705–714.
Maki, B.E., Bartlett, S.A., and Fernie, G.R. (1985). Effect of stairway pitch on optimal handrail height. *Human Factors*, 27(3), 355–359.
NFPA (2003a). *Life Safety Code*, NFPA 101. Quincy, MA: National Fire Protection Association International.
NFPA (2003b). *Building Construction and Safety Code*. Quincy, MA: National Fire Protection Association International.
Norris, B. and Wilson, J.R. (1995). *Childata: The Handbook of Child Measurements and Capabilities*. Department of Trade and Industry, UK.
NYC (2004). *Building Code of the City of New York* (1968 edition updated through 2004). City of New York. Binghamton, NY: Gould Publications, Table 6-4.
Pauls, J. (1985). Review of stair-safety research with an emphasis on Canadian studies. *Ergonomics*, 28(7), 999–1010.
Pauls, J. (1989). Are functional handrails within our reach and our grasp? *Southern Building*, September/October, 20–30.
Pauls, J. (1995). Movement of people. *SFPE Handbook of Fire Protection Engineering*. Quincy, MA: National Fire Protection Association International, pp. 3.263–3.285.
Pauls, J. (2001). Life safety standards and guidelines focused on stairways. In Preiser, W.F.E. and Ostroff, E. (Eds.), *Universal Design Handbook*. New York: McGraw-Hill, pp. 23.1–23.20.
Peebles, L. and Norris, B. (1998). *Adultdata: The Handbook of Adult Anthropometric and Strength Measurements*. Department of Trade and Industry, UK.
Proulx, G. (1995). Evacuation times and movement times in apartment buildings. *Fire Safety Journal*, 24(3), 229–246.
Quinlan, K.P., Thompson, M.P., Annest, L.J., Peddicord, J., Ryan, G., Kessler, E.P., and McDonald, A.K. (1999). Expanding the National Electronic Injury Surveillance System to monitor all nonfatal injuries treated in United States hospital emergency departments. *Annals of Emergency Medicine*, 34(5), 637–645.
Riley, J.E., Roys, M.S., and Cayless, S.M. (1998). Initial assessment of children's ability to climb stair guarding. *Journal of the Royal Society of Health*, 118(6), 331–337.
Seeger, B.R. and Bails, J.H. (1990). Ergonomic building design for physically disabled young people. *Assistive Technology*, 2(3), 79–92.
Stephenson, E.O. (1988). The silent and inviting trap. *The Building Official and Code Administrator*, November/December 1988, 28–33; March/April 1989, 2. *Southern Building*, November/December 1988, 7–15; *Building Standards*, January/February 1989; March/April 1989, 27.
Stephenson, E.O. (1991). Update on the silent and inviting trap. *Building Standards*, January/February, 4–5.
Stephenson, E.O. (1993). The elimination of unsafe guardrails: A progress report. *The Building Official and Code Administrator*, January/February, 22–25. *Building Standards*, March/April, 12–15. *Southern Building*, May/June, 6–10.
Stephenson, E.O. (2001). Climbable guards: Special enemy of the world's children. *Southern Building*, September/October, 32–35.
Templer, J.A. (1992). *The Staircase: Studies of Hazards, Falls and Safer Design*. Cambridge, MA: MIT Press.
Wilson, M.H., Baker, S.P., Teret, S.P., Shock, S., and Garbarino, J. (1991). *Saving Children: A Guide to Injury Prevention*. New York: Oxford University Press, 1991.
Zaino, C.A., Marchese, V.G., and Wescott, S.L. (2004). Timed up and down stairs test: Preliminary reliability and validity of a new measure of functional mobility. *Pediatric Physical Therapy*. 16(2), 90–98.

CHAPTER 16

CHILD USE OF TECHNOLOGY AT HOME

CHERYL L. BENNETT

TABLE OF CONTENTS

INTRODUCTION

In many parts of the world today, technology surrounds our children. Infants watch television from their parents' arms, children play electronic games, adolescents gossip by email, and students use their laptops to submit homework over the Internet (Figures 16.1 and 16.2).

Little is known about the long- and short-term effects of children's exposure to ergonomic risk factors associated with using technologies such as computers, video games, cell telephones, and MP3 players on their physical and cognitive development. Research on how children use technologies in the home (Cummings and Kraut, 2002; Subrahmanyam et al., 2001; Sutherland et al., 2000) has not assessed the physical impact.

Figure 16.1 (Ursy Potter Photography) **Figure 16.2** (Ursy Potter Photography)
Although some believe children benefit from using technologies in the home,* others fear that too much exposure at an early age may cause harm. Some fear technology use may impair children's communication and interactive skills, while proponents maintain the skills simply differ.

The truth is probably somewhere in-between; the key may be to limit children's exposure time, the programs they see, and to monitor (and encourage) their social interactions. Although reduced exposure is not the answer to eliminating the effects of the physical stressors, it will help decrease negative effects of postures such as holding the arm out from the side (Figure 16.1) or twisting to view the monitor and use the keyboard (Figure 16.2).

* Technology in the home includes television, computers, hand-held devices, and electronic games.

One reason we know so little about the impact of children using technology at home is that most studies have involved adults. As technologies change, so do the surrounding challenges. In addition, epidemiological studies to date have not allowed us to evaluate "cause and effect" relationships of musculoskeletal disorders for children.

We still do not know how many children experience musculoskeletal disorders. Research surveys suggest children experience high rates of back pain, but causes and the meaning of this pain is not clear (Balagué et al., 2003). Research has not shown a "statistically significant association between computer use and physical symptoms or clinical syndromes" (Gillespie, 2002).

We do not know how using technology affects our children now or will in their future (Table 16.1). We also do not understand its impact on the health of developed societies and future workforces. We need more research to identify trends and stave off adverse consequences.

This chapter includes guidance to help parents, policymakers, and children benefit from technologies in the home. It also refers to some family activities outside the home, such as using a cell phone or watching television in a vehicle.

Table 16.1 An Analysis of the Impact of Technology on Children Would Require Us to Consider Physical, Social, Psychological, and Environmental Factors

Physical Factors	Preexisting musculoskeletal disorders (MSDs), age, gender, stature, and physical activity levels
Psychological Factors	Sustained attention, computer anxiety, internalization, and control
Social Factors	Socioeconomic status, as this may relate to a child's exposure and access to technology, the school they attend—along with accessible technologies, and available leisure options
Environmental Factors	Physical aspects (workstation set-up), task aspects (frequency, duration, types of technology and purpose), and social aspects (working alone or with others)

Some researchers use scientific models that try to predict the complex relationship between musculoskeletal risk factors and symptoms in adults. Such models are not perfect for adults and may not apply in the same way to children, since children's cognitive and physical abilities differ so markedly from adults.

Models that study the effects on children and their environments must obviously be quite complex. Attempts to identify precise causal relationships regarding the risks associated with children using technologies become elusive as technologies change and children continue to grow.

Source: From Harris, C., Straker, L., Pollock, C., and Trinidad, S. (2002).

TECHNOLOGY IN CHILDREN'S HOMES

Children embrace and explore new technologies. They enjoy things that are new (Figure 16.3)! They interact with electronics in much the same way they approach other new activities or objects; as a learning experience, a challenge, and a new way to have fun!

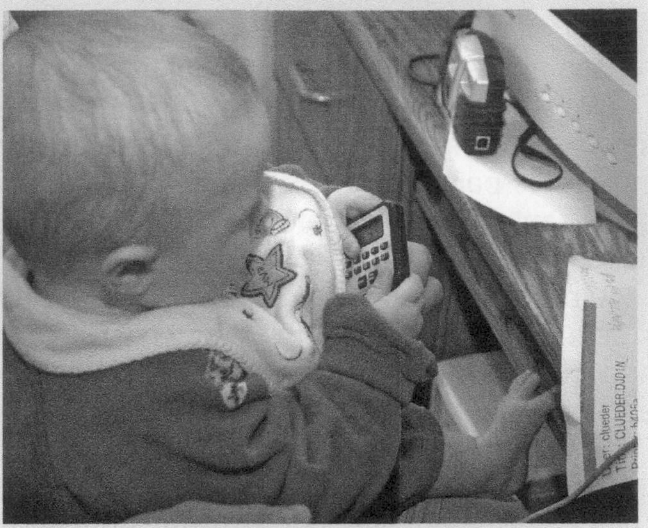

Figure 16.3 (Ursy Potter Photography)
Children are often attracted to technologies at a very young age. Buttons and "gadgets" enable them to experience the "cause and effect" of their actions (see Chapter 2).

During the early part of the twentieth century, there was a surge in communication, entertainment, and information devices for the home. Telephones, phonoraphs, radios, and televisions fascinated children and adults alike. These new technologies also created fears and concerns over their ultimate impact.

In 1936, Azriel L. Eisenberg warned of the dangers of using the radio in the home:

> *The popularity of this new pastime among children has increased rapidly...This new invader of the privacy of the home has brought many a disturbing influence in its wake. Parents have become aware of a puzzling change in the behavior patterns of their children. They are bewildered by a host of new problems, and find themselves unprepared, frightened, resentful, and helpless. They cannot lock out this intruder because it has gained an invincible hold of their children.*
>
> (As cited in Wartella and Jennings, 2000)

Today some parents feel similarly about computers, electronic games, and the Internet. Today's parent might say:

> *The good news is, he isn't watching as much television. The bad news is, I can't see what he's watching and the phone bill has gone through the roof. Even more worrying is that he's now got a group of friends that I don't know. One of his net-friends lives in Russia!*
>
> (Fuller, 2002)

This controversy over the social, psychological, and physical impact of children's use of technology continues.

MUSIC

In most cultures, music weaves into the everyday fabric of people's lives. New technologies such as MP3 players, digital radio, music videos, and the computer afford us with a broad range of opportunities to bring music into the home.

When parents express concern about their children and media, music (75%) is second only to television (84%) and parents of teenagers are more concerned than parents with younger children. Such concerns often relate to the message or the associated lifestyle (Woodard and Gridina, 2000). Yet, listening to music or watching television with teens can provide an opportunity for a discussion of values.

Children often do not recognize the risk of losing hearing from loud music. They may know the volume at a concert or on their radio hurts their ears, but not its long-term impact. They also may not identify headsets as a potential source of injury. Parents should check to ensure the volume of headphones remains at a moderate level; if others can hear the sound from a headset, it is likely too loud for the child. Even lower level, constant sounds can damage hearing (see Chapter 5 on hearing). Parents can educate their children directly. Pediatricians or teachers may serve as additional resources on this subject.

Most people enjoy music. Many of us attend music performances and listen to music while driving and performing housework and homework. Background music can be enjoyable, annoying, or distracting, depending on the situation and the individual. It can also affect performance.

Although we do not fully understand how background music affects us at home or in leisure activities, we do know that auditory distractions (e.g., background music and television) impair introverts' ability to perform complex cognitive tasks such as reading comprehension (Furnham and Strbac, 2002). Parents can help their children identify environments that are conducive to learning and studying. Children who are introverts should avoid auditory distractions.

TELEVISION

More than 2.5 billion radios and 1.5 billion television sets bring news, music, and entertainment to homes around the world. Three quarters of the world's population access television and radio (Annan, 2000). Television can monopolize a child's day. Close to 98% of American households have at least one television.

Children who have televisions in their bedroom tend to watch more than those who do not (Woodard and Gridina, 2000). They are also less likely to attend family dinners or have parental limits (Wiecha et al., 2001).

On average, American children watch four hours of television a day (AAP, 2004). Children from low-income homes are more likely to have television sets in their bedrooms than those with higher incomes (Woodard and Gridina, 2000).

> Children with few restrictions may use more than one media intensively.
>
> - Consider all the media they use.
> - Limit your child's "screen time" to one or two hours per day of nonviolent, educational programs (AAP, 2004).
> - Encourage a variety of activities.
>   - Emphasize tasks that require physical activity.
>   - Two hours of television followed by playing electronic games or using the computer is not much of a change in activity.

> Avoid television for children under the age of two (AAP, 2004).
>
> Attention Deficit Disorder (ADD) has been linked to the amount of exposure to television in young children (Christakis, 2004).
>
> This research raises questions about how the fast-paced visual images of television affect the development of the brain.

Intensive television viewing exacts a cost. Viewing four or more hours per day relates to poor school achievement (Williams et al., 1982) and childhood obesity (Crespo et al., 2001).

Video games sometimes displace television (Suzuki et al., 1997). Yet children who play video games also seem to watch more television (Woodard and Gridina, 2000).

Research clearly connects exposure to violence in media with aggressive and violent behaviors (Strassburger and Donnerstein, 1999). Frequent

PHYSICAL FACTORS WHILE VIEWING TV

Figure 16.4 (Cheryl Bennett) Children watch television from positions and locations that contribute to awkward postures and poor sightlines.

Figure 16.5 (Valerie Rice) Floor level and closer viewing positions may cause awkward neck and back postures, as children tilt their heads back and arch their backs. Many children watch television from the floor, leading to high viewing angles (Nathan et al., 1985).

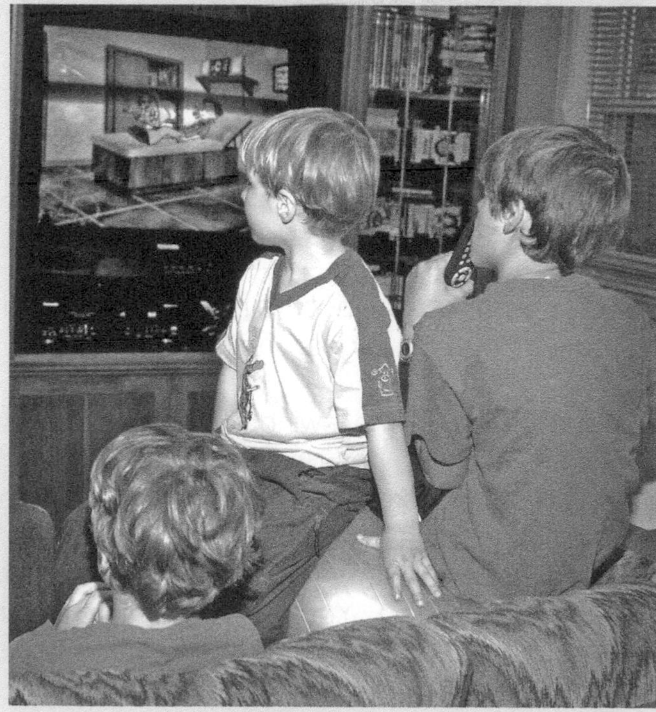

Figure 16.6 (Ursy Potter Photography) They also watch from beds, couches, and chairs and while sitting, laying prone, or while playing, resulting in any number of awkward positions.

exposure to violent content at a young age can foster lasting traits such as aggressive responses, a sense that the world is a mean and scary place and desensitization to real violence (Robinson et al., 2001).

Parental limits on content and exposure time are the most effective measures to reduce the effects of violence in the media (Robinson et al., 2001). Ratings of movies and television programs can help to identify programs with violence and sexual content. Watching together provides an opportunity for "teachable moments" when content is not in alignment with family values.

The effect of television viewing on physical health also warrants consideration. Look at children while they watch television. They often view from awkward viewing postures and viewing angles (Figures 16.4 through 16.6). These positions may be maintained for quite some time when a program is engrossing. The amount of time spent watching television should be considered in the overall health of a child.

SUGGESTIONS

1. Avoid providing children with a television in their bedroom, regardless of their age.
2. Limit their watching television, videos, and DVDs to one or two hours a day.
3. Encourage children to sit at least a few feet from the television.
4. Use size-appropriate furniture.
5. Encourage children to stretch, change positions, and vary their activities.
6. Look for alternative chairs that might help their muscles develop.

For example, intermittent use of sports or therapy balls* may help promote balance and stability.

PERSONAL COMMUNICATION

Each new technology brings a new set of challenges. For example, although cell phones enable us to communicate almost anywhere, their use (even with hands-free headsets) is associated with accidents and impaired responsiveness while driving.

It is likely that children are similarly distracted when they talk on cell phones while walking, running, riding bikes, and skateboarding. Encourage children to monitor and evaluate risks in their surroundings (such as traffic) before and while they use a cell phone. Since it may be unrealistic to expect many children to have such "situational awareness," encourage safe behaviors with clear rules about when and where to use cell phones.

E-mail and online exchanges provide additional mechanisms for communication. Children aged 12 and over are more likely to communicate with friends online (Figure 16.7), than younger users. The older the youth, the more likely they are to have a "buddy list" (Table 16.2) that lets them know when others are online. The greater ability to write and interest in communication during the teenage years often means more time spent chatting online.

Table 16.2 Use of Online "Buddy Lists," by Age

Child Age (Years)	Percent Use (%)
15–17	74
12–14	66
9–11	31

Source: From Roper Starch Worldwide. *America Online/Roper Starch Youth Cyberstudy* (Electronic report): AOL (America Online), 1999.

* Children who sit on sports or therapy balls should be able to support their feet on the floor. These balls are sold in different sizes and should match the size of the user.

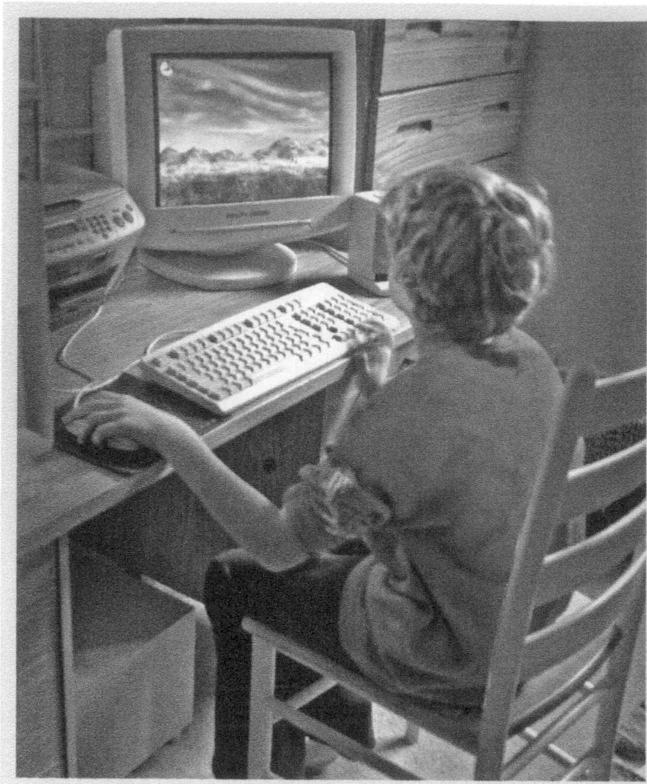

Figure 16.7 (Ursy Potter Photography)
Children aged 12 and older like to communicate on computers—sometimes in chat rooms with people they do not know. Caution is warranted, but avoid prohibiting them from these activities.

- Learn about your child's virtual friendships.
- Establish an understanding with your child about privacy expectations.
- If concerned, review your child's "Buddy List" (easier to do if the computer is not in the child's bedroom).
- Limit how much time your child spends chatting online.
- Teach children to identify discomforts and recognize them as warning signals that they should take a break.

The posture shown may lead to discomfort in the back (from hunching), neck (from looking up at the computer screen), shoulders, or arms (from working at a desk that is too high).

ACTIVITY BREAKS

Students who spend more time on the computer and less time in activity breaks (i.e., breaks that involve physical activity) express more discomfort (Szeto et al., 2002). Secondary students report that when not using the computer they are more likely to sleep or rest rather than participate in sports or other physical activities (Szeto et al., 2002).

Regular short changes of activity can help children avoid discomfort and reduce the impact of risk factors (Figure 16.8). Students who exercise during breaks report less discomfort (Szeto et al., 2002).

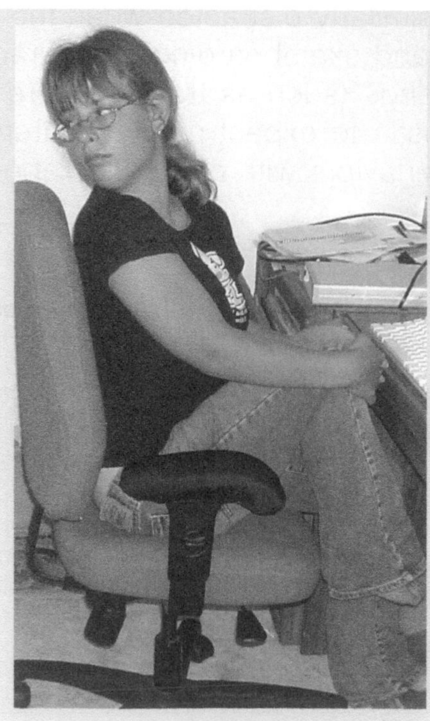

Figure 16.8 (Cheryl Bennett)
Children need to take regular breaks from playing or working on computers.
During activity breaks teach children to:

1. move physically through full range of motion
2. change visual focus and look at objects that are far away (20-20-20, see vision section below)
3. move vigorously to promote circulation
4. stretch
5. give your brain a break so you return refreshed to the topic
6. breathe! Sitting posture tends to be leaning forward and tends to make breathing more shallow. Sit up and stretch or stand up and take full breaths occasionally.

Children often ignore discomfort or keep pain to themselves when using video games or computers (Figure 16.9). Children are more likely to continue while in pain during sporting events than while using computers and video games.* Yet, children experiencing pain while playing a sport are more likely to tell a parent or adult about their pain than are children who experience pain during video or computer use.

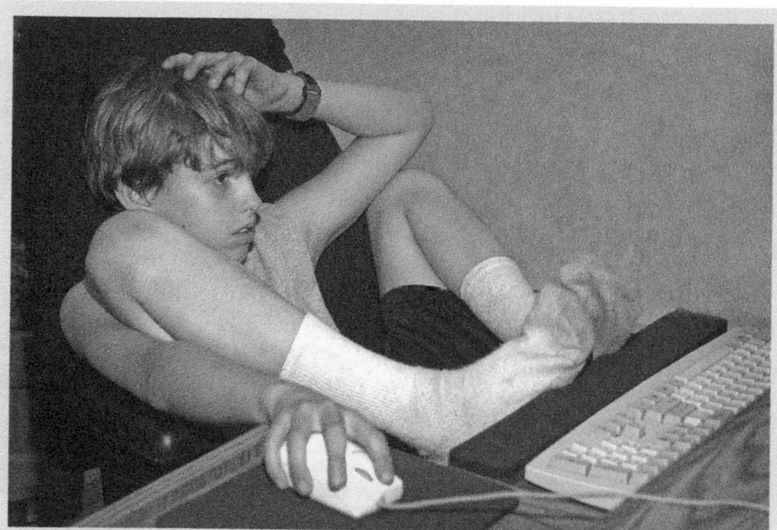

Figure 16.9 (Ursy Potter Photography)
Children who are engrossed at the computer are often unaware of how they are sitting. It is difficult to teach them to take regular breaks. They may become irritated when interrupted and find it difficult to establish self-imposed limits. The earlier children learn to take breaks and change their activities, the more readily they develop good habits that will continue into adulthood. Set a timer or use break software.

Most games and DVDs enable them to "pause" the action while they take a break. Game manufacturers should provide opportunities to build activity breaks into the structure of games intended for children (Table 16.3).

Table 16.3 Guidelines for Manufacturers and Parents

Recommendations for Manufacturers	1. Integrate breaks into programs. 2. Structure the breaks to involve physical activities, such as acting out parts of the programs.
Recommendations for Parents	1. Ask children about possible discomfort or pain (and be specific). 2. Educate children about possible consequences of ignoring their symptoms, such as longer lasting aches and pains, falling from stepping on a foot that is "asleep," etc. Let them know repeatedly performing the same action can lead to injury, especially if in an awkward posture. 3. Set time limits. 4. Set specific times for activity breaks (use timers, if necessary).

* Royster and Yearout (1999) found that 82% of children in pain were likely to continue their sports activity, compared to 67% of computer users and 53% of video game users. However, many more of those playing sports would report the pain to a parent or guardian than those using computers or video games.

VISION

Most technology uses visual displays at short focal distances. Constant near viewing may adversely affect vision (Marumoto et al., 1999). When using a cell phone, playing electronic games, or watching television, the eyes are working to focus on the display (Figure 16.10).

When focusing at a distance, the eye muscles are more relaxed. A good visual break mnemonic is the "20-20-20 rule"; for every 20 minutes of visual display focus, look at least 20 feet away for 20 seconds (Sen and Richardson, 2002). See Chapter 4 of this text for more information on vision and visual development of children.

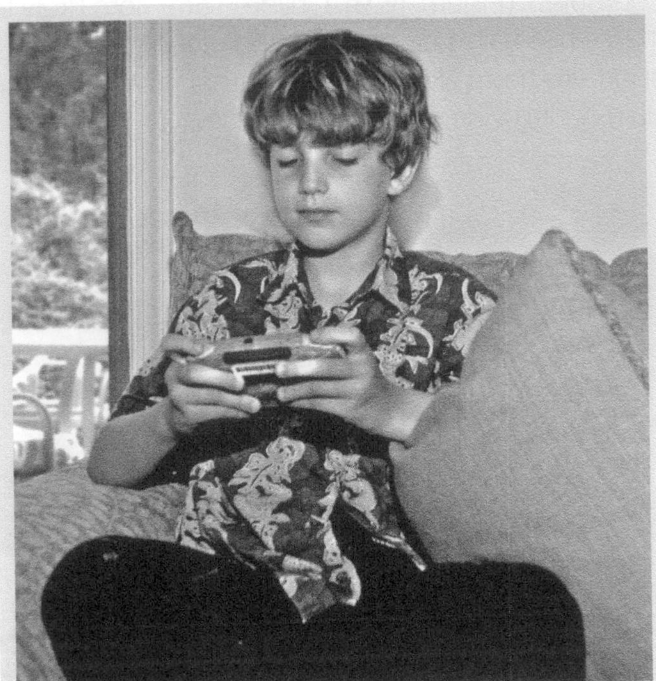

Figure 16.10 (Ursy Pottery Photography)
Most children take their vision for granted. Viewing portable electronic games and computers requires focusing at closer distances.

Teach children to take regular breaks and periodically use their distance vision.

PHYSICAL ASPECTS OF HOME COMPUTING

Furniture, equipment, and even the placement of computers can affect the child and the dynamics of the family.

LOCATION OF HOME COMPUTERS

Computers are relatively new within family households. Especially when families share a single computer, questions arise about where to place it (living room, bedroom, family room) (Cummings and Kraut, 2002; Jonsson, 1997; Venkatesh, 1996). Families typically place them wherever space is available. However, when there is only one computer, it is likely to be placed in a common room.

Families with several computers tend to keep them in individually controlled spaces. Many children would rather have a computer in their own room (Frohlich et al., 2001), yet this does not encourage social interaction or facilitate supervision by parents (Figure 16.11). Although it is common to have computers, televisions, or other electronic devices in children's bedrooms,* it is not recommended. Children who have televisions and similar media in their bedroom are more likely to be overweight and spend more time watching television (Dennison et al., 2002).

* Among children aged 6 and under, 77% watch television daily, and 43% have televisions in their bedrooms. Children in heavy television-use households read less (Rideout et al., 2003).

Figure 16.11 (Cheryl Bennett) **Figure 16.12** (Valerie Rice)
Some children are more technically informed than their parents. The process of teaching their parents may foster a sense of self-confidence and otherwise change the dynamics of the family (Sutherland et al., 2000). Such interactions will more likely occur if computers are in common spaces, rather than in the child's bedroom. Obviously, the child on the right is using a computer and workstation designed for an adult!

CHAIRS

Children spend a lot of time sitting at school, and school furniture has been associated with back pain (Linton et al., 1994; Mandal, 1994; Troussier et al., 1999). In part, this is because school chairs often do not fit the child users (Aagaard-Hansen, 1995; Jeong and Park, 1990; Parcells et al., 1999). Although we don't fully understand the causal relationships, we do know that children report more discomfort when there is a mismatch between the children and the school furniture (Milanese and Grimmer, 2004; see also Chapter 21).

Furniture mismatches are commonplace in the home (Figure 16.12), yet there has been almost no research on their impact on children. Home furniture design typically uses adult body sizes (anthropometric dimensions). The effect of adult-sized furniture on children is unknown.

A chair is "ergonomic" when it is comfortable for the user, fits their body size, and supports their activity. The proportions of body segments as well as stature affect this fit (Molenbroek, 2003); for instance, some individuals have longer torsos and some have longer upper legs. Finding a chair that fits several people of different sizes equally well is not easy. Modifications are typically necessary to make an available chair fit a child. A chair that adjusts for seat height and depth is best.

We have come to expect our chairs at work to adjust, but may be reluctant to purchase adjustable furniture for our homes. Even adjustable chairs may not adjust sufficiently to fit children (Figure 16.13).

Figure 16.13 (Valerie Rice) Most adult chair seats do not allow small children to sit high enough to use the keyboard or see the monitor properly—yet they are too high for their feet to reach the floor. When children's legs dangle over the edge of a chair with feet unsupported, there is pressure on the back of the knees or thigh or both. This pressure, when prolonged, can affect circulation and comfort. A footrest can provide support for the feet and help keep pressure away from the back of the knee where blood vessels and nerves are near the surface.

Appropriate seat height is determined by the popliteal height (distance from the back of the knee to the bottom of the foot in a sitting position, see Figure 16.14), and being certain that the distance from the seat pan to the floor matches this measurement. This matching will ensure a child's feet can rest comfortably on the floor.

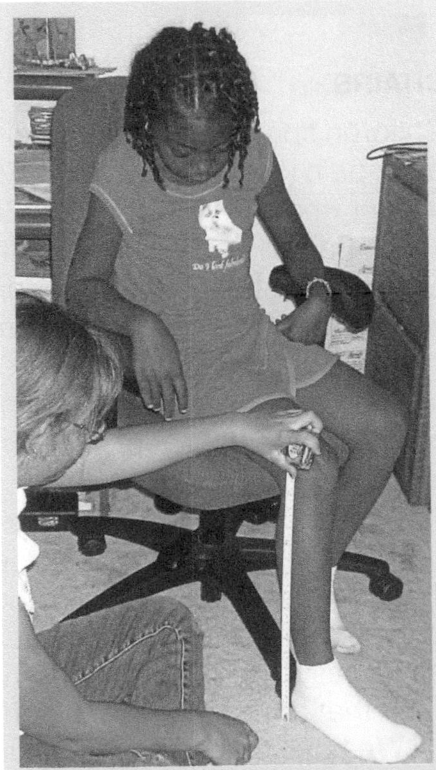

Figure 16.14 (Cheryl Bennett) The popliteal height (from the heel to the back of the knee) is the best gauge of appropriate seat height.

Seat pan depth in an adult chair is often too deep for a child, so their back does not contact the back of the chair while their knees bend over the front edge of the chair. Children with lower muscle tone, or low core strength, have even more difficulty and begin to assume a rounded back posture. A good rule of thumb is to have three fingers or so of space between the edge of the chair and the back of the knee when sitting against the backrest. Placing a cushion between the child's hips and the backrest can provide back support while shortening the seat depth. Medical supply stores (among others) sell cushions of various sizes and shapes, or upholstery foam cut into appropriate shapes and sizes can help align the child.

Ultimately, children should sit in comfortable, stable postures with back support and their feet on the floor or a footrest.

People in general, including both children and adults, do not automatically sit "properly" in ergonomic furniture, nor do they automatically know how to adjust their furniture (Linton et al., 1994). Displaying posters near their desks can help remind them to take breaks and to sit

Figure 16.15 Even child-size furniture may not work perfectly for computer use. There are few well-designed ergonomic workstations available for children!

Yet knowing how children's posture should look will help caregivers, parents, and children to select furniture and set up their workstation in the best way possible.

A few rules of thumb:

- Knees and hips do not need to be exactly at right angles.
- Encourage your child to sit in comfortable positions that allow him or her to move through a range of postures.
- Seat and armrest padding should distribute the child's weight to avoid direct pressure.
- Children's elbows should be at or slightly greater than 90°, with wrists fairly close to straight.
- A cushion may help raise the child's sitting height to a level where the arms can be used without raising the upper arms (this may also help them when viewing the computer screen).
- Well-designed and well-positioned chairs with arm- and wrist-rests may help encourage neutral (straight) wrists while relaxing shoulder and arm muscles.

in and move through a range of good postures (Figure 16.15). Having screensavers with pictures of seating postures is another way to provide reminders.

Although it may feel normal to slouch, children (and adults) should recognize that doing so indicates fatigue and serves as a reminder to change positions or take a break. Educating children about the results of slouching (shallow breathing, displacement of internal organs) and twisting (pressure on spine, reduced strength from the biomechanical disadvantages of these postures) can help them feel informed and respected. Arming children with information on "why" provides them with the opportunity to "do it right" on their own.

If chairs adjust, parents and teachers should demonstrate and enable the child to practice proper adjustment. Keep product instructions near the chair. If different people use the same chairs, make sure they understand they are not "messing up" chairs by adjusting them (Tien, 2002).

TABLE OR DESK HEIGHT

Adults who work with keyboards above elbow level experience discomfort (e.g., Sauter et al., 1991). We lack such information for children, but they are likely to be similarly uncomfortable when elevating arms to reach the keyboard and mouse (Figure 16.16). There is less muscle tension in the low back and shoulders when desk and table heights fit the child user (Hänninen and Koskelo, 2003). Keyboards close to elbow height provide children with an opportunity to rest their forearms.

Tables for children are more likely to be too high than too low. Even so, they may still have knee clearance problems. Children need to be able to fit their knees underneath and to sit close to the edge of the table. If a desk is too low, try placing wooden blocks (or devices designed for raising desks) under the legs. It is usually easier to raise the seat and provide foot support than to lower a desk that does not adjust.

An adjustable, extended keyboard tray is a platform that attaches to a desk or table on an arm that allows the keyboard and pointing device to be positioned independent of the level of the monitor. Such trays should have room for both the keyboard and pointing device on the same level, so the child will not have to use an extended reach for the mouse. Raising, lowering, and tilting an adjustable keyboard tray accommodates different positions and users.

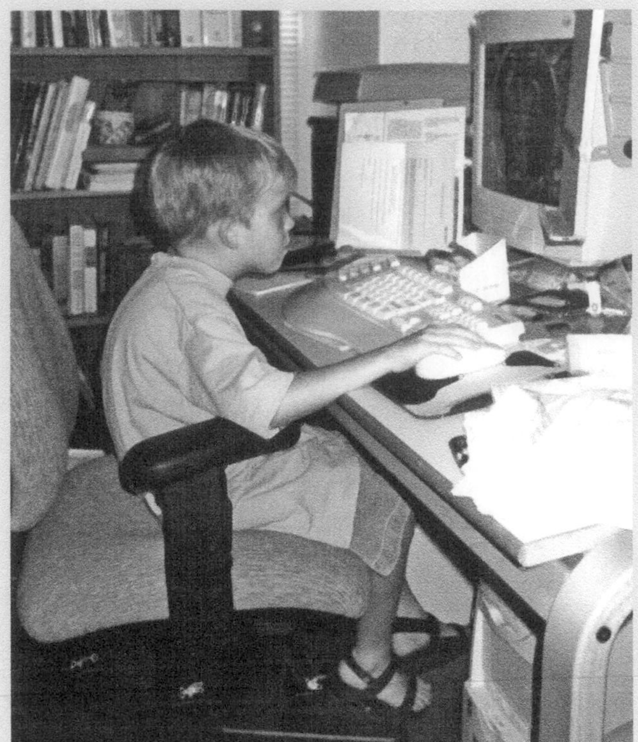

Figure 16.16 (Valerie Rice)
This adult chair is too high for the child's feet to touch the floor, yet not high enough for proper viewing of the computer screen. The armrest is high thus forcing him to raise his upper arm to the side (abduction). The table is high, so he elevates his forearm to reach the mouse. The seat depth is deep, so he has no back support, resulting in a "slumped" posture and neck extension. In short, he will not be comfortable for long!

KEYBOARD

Prolonged keyboard use is known to expose adults to a range of risk factors for musculoskeletal disorders such as static postures and repetitive finger movements while keying. We are only beginning to study the corresponding effects on children.

QWERTY keyboards continue to predominate.* Most keyboards are designed for adults and not adjustable. However, smaller keyboards are available and should be more comfortable for the smaller hands of children. A keyboard without the number keypad extension on the right side is a good option to reduce reaching for the mouse (discussed below). A toggle switch or "Y" port can allow alternate use of two keyboards at the same computer.

Learning to touch type can reduce head movements that accompany visual search for keys. Young children start typing in a "hunt and peck" style. As children get older, their typing education should focus on the type of keyboard they will use.

Even with standard QWERTY keyboards, children can learn alternatives to the standard keying technique taught for decades. Allowing the fingers to fall more naturally, for example, with the middle and ring fingers on the row above the home row can put the hands into a more neutral posture with less tension (Kahan, 2002).

Learning keyboard commands can also reduce the frequency of mouse or trackball use. Posting a list of keyboard commands near the computer facilitates learning to use functions that are easily forgotten (Shaw et al., 1985).

* QWERTY keyboards were named after the order of the six letter keys on the left top row of keys.

Computer and touch-typing classes are a good idea (Box 16.1).

BOX 16.1 TOPICS COVERED IN A GOOD TOUCH-TYPING CLASS

1. Keyboard techniques, keyboard commands and touch-typing
2. Risk factors for musculoskeletal discomfort
   a. What they are
   b. Methods to avoid them
3. Workstation setup
4. Correct postures
5. The importance of taking breaks
6. Keyboard principles: Maintain relaxed hands and fingers with a straight wrist and use a light touch on the keys (no pounding!)
7. Keyboard shortcuts to reduce frequency of using mouse or trackball

THE MOUSE AND OTHER INPUT DEVICES

Nonkeyboard input devices (such as mice) help children perform a range of tasks on the computer and navigate the Internet. Keyboards allow users to input alphanumeric information into the computer, whereas nonkeyboard devices such as a mouse, joystick, or trackball allow one to move or manipulate a cursor or pointer on a computer screen.

With an input device, a child can pull down a menu, move a cursor, "point and click," highlight, switch between open programs, and perform a variety of word processing operations. A child's hand is almost continually on a pointing or input device while "surfing" the Web or navigating Internet sites. Fine motor control is correlated with age and experience (Lehnung et al., 2003).

The numeric keypad found on the right of most keyboards requires other input devices to be placed farther to the right. Considering the smaller size of children, the distance to reach a right side input device is a greater stretch than for adults. It also requires lifting the upper arm to the side (abduction) and slightly to the front (flexion), along with a constant (static) hold, as the child keeps his or her hand on the input device. Over time, this can result in muscle fatigue and pain in the neck, shoulder, and elbow (Figures 16.17 and 16.18).

Figure 16.17 (Cheryl Bennett)

Figure 16.18 (Cheryl Bennett)

When the pointing device is on the right (16.17) of an extended keyboard (with a number pad and other keys on the right) the arm must also extend to the right. When the device is on the left side (16.18) the arm can be closer to the body.

Despite the advantages of using an input device on the left, most adults use input devices with their right hand, even those who are left-handed (Woods et al., 2002). Children can easily learn to use either hand to input and select according to their comfort and performance. When the keyboard is not being used (as is the case for some games and child software), the mouse or trackball may be located midline in front of the keyboard so either hand can use it. This puts the keyboard farther away but allows the arm to be more relaxed for frequent use of the pointing device.

Mouse and trackball pointing devices come in a range of shapes, sizes, and configurations. Although most have two or three buttons, some may have as many as five or six. For children, simple designs and smaller sizes are desirable (Figure 16.19a and Figure 16.19b).

Children may appreciate being involved in selecting their mouse, trackball, or other pointing devices. Shop where they can put their hands on the product. Having more than one device permits the child to change occasionally. Input devices shaped like a pencil can help children learn and practice proper grips for writing. Together, parents and children should select a device that is comfortable and keeps the wrist fairly straight.

Voice recognition software provides children with another non-keyboard input alternative (see Chapter 8 of this text, on Assistive Technologies).

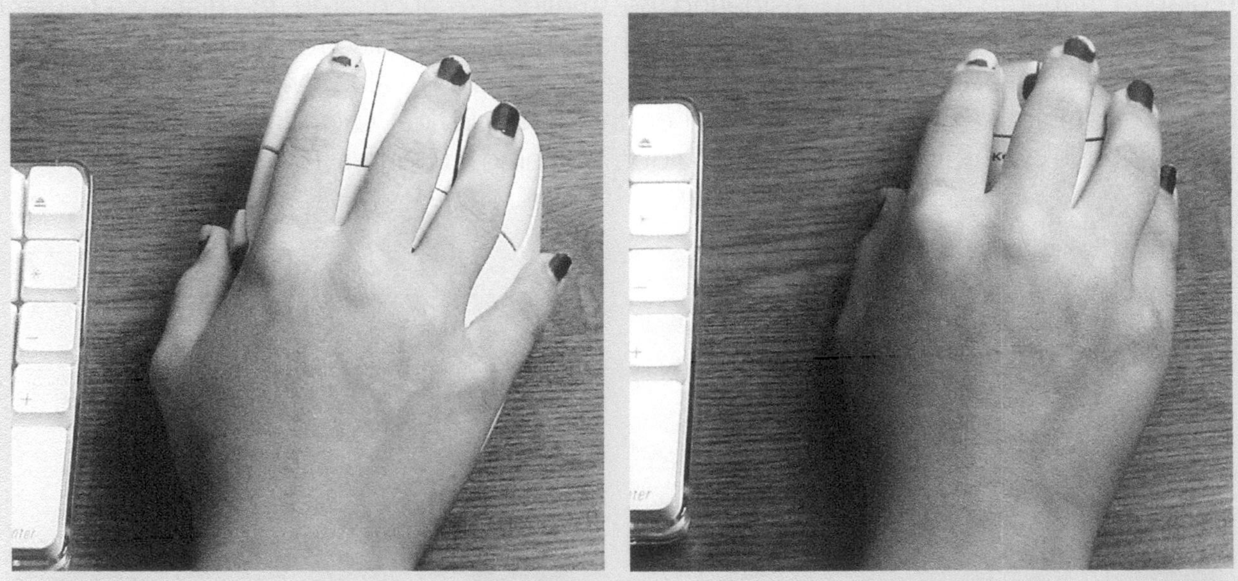

Figure 16.19a and Figure 16.19b (Cheryl Bennett)
A child's hand at age five is only 60% the size of an adult's hand and 86% by age 14. The smaller size can result in awkward postures (Figure 16.19a) if using adult size input devices (Blackstone and Johnson, 2004).

Children need input devices that fit their hand sizes (Figure 16.19b) and allow them to keep their wrist straight, without having to stretch or strain their fingers.

MONITOR

Position the monitor to give children a good, clear view from a comfortable posture. Avoid awkward postures, such as tilting the head backwards to see the screen (Figure 16.20). The eyes can scan down easily without affecting the position of the head but looking even slightly above horizontal tends to make the head tilt back.

Low screens are mainly a concern when they promote poor postures such as hunching forwards. It is easier to raise monitors than to lower them.

Placing a ream of paper, books, or a shelf beneath a monitor will raise it. If it is not possible to lower the monitor to the correct level for the child, raising the seat or providing a cushion should take care of the problem.

Adjustable height tables can help to place the monitor at the correct height—if it adjusts low enough to position the monitor properly for the child.

Adjustable computer tables are available and are relatively inexpensive. If using an existing, nonadjustable table, shortening the legs will accomplish the task.

Check for glare from windows or indoor lighting by turning off the computer to see the reflections on the screen. Position the computer to minimize glare and direct light from shining into the child's eyes.

Figure 16.20 (Cheryl Bennett)
Avoid placing the top of the monitor viewing screen higher than eye level (preferably set it lower).

- Establish eye level from a comfortable, balanced head position. Facing the monitor, imagine a horizontal line extending from the corner of the eye to the monitor.
- If eye level is lower than the top third of the screen, then the monitor is too high. When the head tilts back, there can be compression of the spinal cord as it exits the skull.
- If the monitor is too high, lower the monitor or raise the seat.
- If the monitor cannot be lowered, raise the seat and provide a footrest under the feet. If neither adjusts, consider raising the child with a pillow, stack of books or cut foam to the appropriate height.

SUMMARY OF HOME COMPUTING PHYSICAL PARAMETERS

The physical set up for using desktop computers is a matter of individual comfort (Psihogios et al., 2001). When more than one person uses the same computer, adjustability is even more important. Parents or educators should work with the child, helping them to understand the recommendations for placement, as well as helping them get into a habit of adjusting the workstation for their comfort.

Summary of Home Computing Guidelines
1. Fit the workspace to the child. 2. Adjustable furniture or child size furniture is best. 3. Provide adaptations to adult size furniture.
Chairs
a. Pillows or cushions will raise the child to be able to see the monitor. b. Pillows or cushions will provide back support. c. Have wide armrests on the chair for arm support. d. Have a footrest so legs do not dangle, if the chair is high.
Table
a. Use adjustable tables or keyboard trays with room for the mouse. b. Raise the child to the appropriate height for the table using pillows or cushions on their chair if needed. c. Provide wrist rests.
Keyboards and other input devices
a. Use small "child" size devices that are functional and appropriate. b. Have children try devices before purchase. c. Encourage use of left hand for mouse or trackball. d. Have children take a computer or keying class.
Computer screen
a. Position so resting eye level is no lower than the upper third of the monitor screen. b. Minimize glare. c. Adjust font size and screen contrast to maximize viewing area.
Education
a. Teach children about proper posture, and why it is important. b. Provide visual reminders (posters, screen savers). c. Demonstrate how to adjust furniture and let children practice. d. Teach children to recognize signs of fatigue. e. Take before and after photos to show your children how they sit (many do not have the proprioceptive feedback (information from their own muscles and nervous systems) to fully understand where their body and body parts are in space. They believe they are doing it "right." f. Encourage ergonomics education for children, parents, and teachers at schools.

FITTING A WORKSTATION TO A CHILD

Fitting the furniture at a workstation to a child can be a challenge. It is not as simple as purchasing child size furniture and you cannot simply adjust one item. Typically, one adjustment calls for another. Also, children keep growing so what fits this month may not in another month. Adjustability can allow furniture to fit for a longer time or for more than one person using the same workstation.

Figures 16.21 through 16.29 demonstrate the process of fitting a workstation to a child. The first series (Figures 16.21 through 16.23) demonstrates using readily available furniture for children, whereas the second (Figures 16.24 through 16.29) shows the use of an adult-size chair and work area.

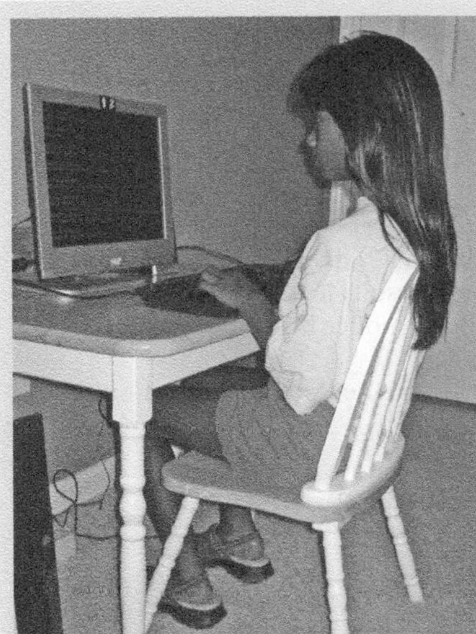

Figure 16.21 (Valerie Rice)
This child-sized table and chair were purchased as a set. The monitor is at a good height. The child's feet are flat on the floor. She has some back support. However, notice the angle of her elbow and wrist—she has to reach up to the keyboard and her wrists are resting on the hard edge of the wooden table.

Perhaps a computer tray would help to make the keyboard and mouse a little lower, but most keyboard trays will not fit on her small table. Another solution is to raise her seat.

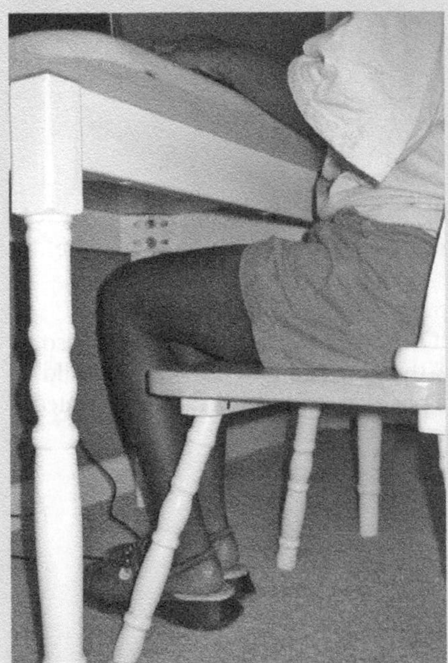

Figure 16.22 (Valerie Rice)
Does she have room underneath or will her thighs hit the table edge if we raise her?

She has plenty of room. In fact, notice the amount of space between the front edge of her seat and the back of her knee. The chair may even be slightly low for her.

Figure 16.23 (Valerie Rice)
Here, she sits on a pillow. It helps, seen by the decrease in her elbow and wrist angles, but she still may need to be a little higher. The next steps to try could be a phone book, or a phone book coupled with a pillow or a thicker pillow. Fortunately, her monitor adjusts easily up and down, as well as tilting. Once she is a little higher, she will need a small wrist rest at the front edge of her desk to protect her wrists from direct contact with a hard surface.

Figure 16.24 (Valerie Rice)
An adult-size computer table and an adult-size folding chair, even with a pillow, results in a keyboard and monitor that are too high (notice his elbow and wrist angles).

Since the monitor cannot be lowered, his seat must be raised.

Figure 16.25 (Valerie Rice)
With an adjustable adult-size office chair, he is now the right height for his eyes to gaze at the monitor, but his feet dangle, putting pressure on the back of his thighs. His arm is raised, but now may be slightly too high. Also, when the feet dangle, the pelvis can tilt causing the child's back to extend into a position of lordosis (an exaggerated lumbar curve or "swayback").

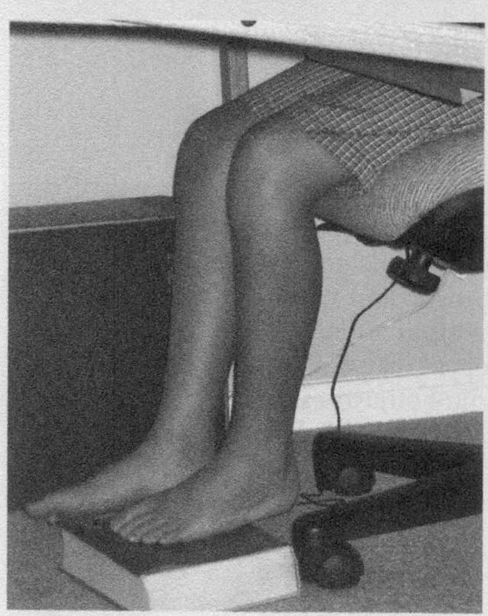

Figure 16.26 (Valerie Rice)
Here, he puts a book beneath his feet for a footrest, but his feet still do not rest solidly on the book. Pressure remains on the backs of his thighs.

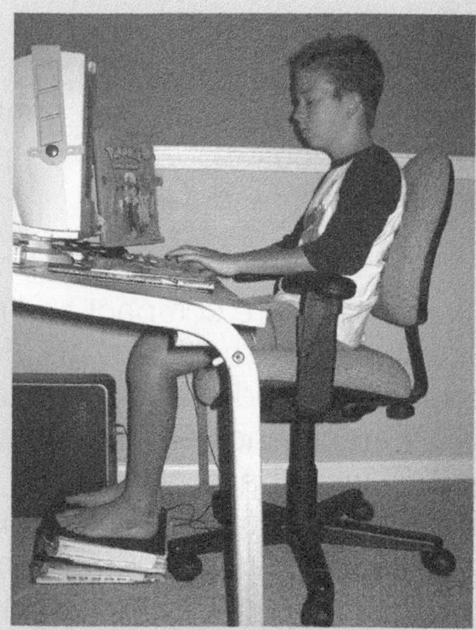

Figure 16.27 (Valerie Rice)
Although the best idea is to purchase an adjustable footrest, it is still a good idea to support his feet in the meantime. Here, he has put two notebooks under his feet, to get the right height and slant.

Now his monitor is at the right height, His arms and wrists look straighter, and his back and feet are supported.

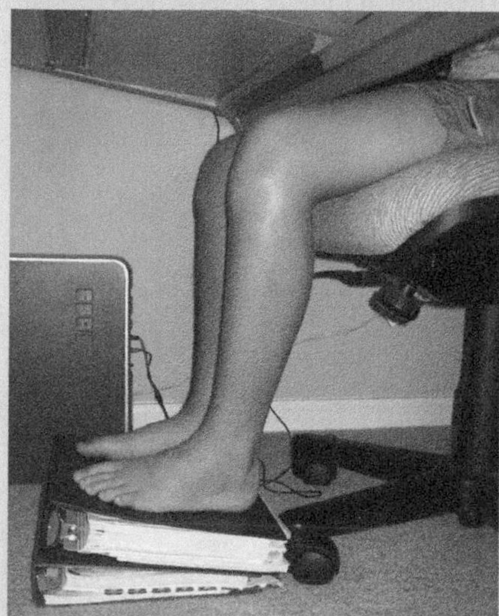

Figure 16.28 (Valerie Rice)
His feet rest flat on the notebooks and there is no pressure on the back of his thighs. There is also very little room between the top of his thighs and the desk to allow either the chair to move higher or his feet to be raised higher.

Figure 16.29 (Valerie Rice)
The arms on an adult-size chair are often spaced too wide for a child, causing them to abduct (raise to the side) their arms at the shoulder joint. Here, he has turned the armrest in so it can support his lower arm while held closer to his body.

"Quick fixes" such as phone books, cushions, and chair-top booster seats may enable a child to use computer tables or chairs designed for adults. However, such band-aid solutions are less than ideal and may even introduce new risk factors that affect children's vision, discomfort, and short- or long-term risk of musculoskeletal disorders. Yet, not all parents and caregivers can afford (or even find) appropriately sized, adjustable computer furniture for children.

PORTABLE DEVICES

Portable devices include laptops along with a growing variety of devices. Computers are merging with telephones, organizers, cameras, and other keying devices with internet wireless capabilities. Children may use them at school, home, bus stops, and on public transportation. Children can use them while sitting, standing, or lying down!

LAPTOPS

Laptop computers (also called notebooks or portable computers) are compact and portable. Although not pervasive, some private schools require students to have laptops (Fraser, 2002; Harris and Straker, 2000) and some public schools have funds specifically to provide laptops for students (Corcoran, 2002) (Figure 16.30). Parents and caregivers also find them helpful for long car trips, as children can watch a DVD while traveling.

Although convenient, typical laptops with attached monitors can compromise users' postures (Figure 16.30). When the keyboard is at a comfortable typing height, the monitor may be too low. If the child chooses a comfortable head and neck posture for viewing the monitor, the keying height is often uncomfortably high (Straker, et. al., 1997). In order to view a low screen, the child must flex neck forward (Straker et al., 1997), as do adults using laptops (Szeto and Lee, 2002).

Children can place their keyboard and monitor according to their own dimensions by connecting an external keyboard and pointing device (mouse, etc.) to the laptop. This is an advantage for adolescents who assume numerous postures such as sitting on the floor, sitting on a stool, lying on their bed or on the floor, or sitting with their laptop on their lap (Harris and Straker, 2000).

Another alternative is to connect the laptop to an external monitor, so the child can position the keyboard separately from the monitor. When possible, children should help select the laptop and attachments they will use. They may find the smaller keyboards on laptops fit their hands more comfortably than a typical keyboard.

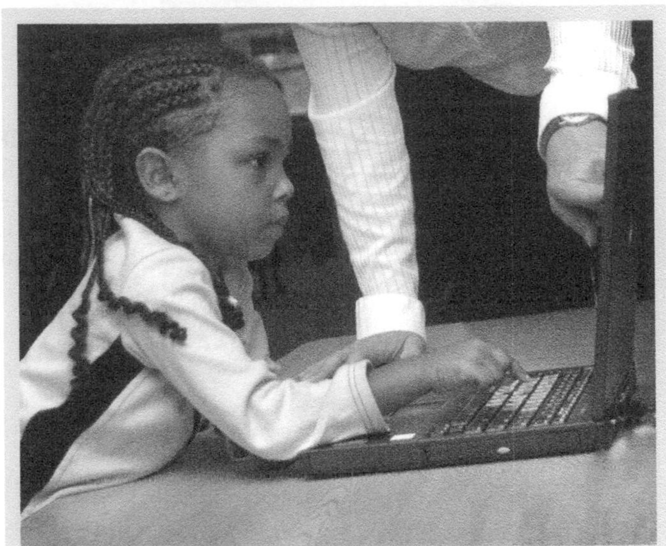

Figure 16.30 (Ursy Potter Photography)
This little girl is standing and leaning on the table as she uses the laptop. The monitor height is fine but she needs the support of the table for her arms.

The original flat screen technologies for laptop displays were simple liquid crystal; but increasingly, the flat screen technology is more rugged and reflects less (Straker, 2000). However, the wide range of angles at which the screen can be set (as well as the conditions in which children may use a laptop) make contrast and glare issues likely (Jonai, et al., 2002). Children can learn the simple technique of viewing a blank screen to determine the amount of glare. They will need to set the screen angle to best accommodate comfortable postures and to reduce glare.

PORTABLE COMPUTING DEVICES

Portable computing devices (PCDs) are generally smaller than laptops, such as subnotebooks, personal digital assistants (PDA), organizers, palmtops, and pocket computers.

They are currently about the size of an adult hand and typically have liquid crystal display screens, small keys, and stylus input (Figure 16.31 and Figure 16.32).

Technology continues to evolve but it is likely that more products of this size will develop. PCDs have risk factors for musculoskeletal discomfort similar to those for laptops and portable electronic games.

Once again, the numerous locations and lighting conditions during use will introduce visual contrast and glare difficulties that can cause visual fatigue. Also the small monitors can contribute to both visual fatigue and poor posture (Jonai et al., 2002).

Adults using notebook and subnotebook computers tend to lean forward and bend their necks forward (Szeto and Lee, 2002). Children may also lean forward to see the display screen and smaller fonts more clearly or, according to their visual range, may hold it at arm's length.

Perhaps even more than laptops, PCDs are easy to use in a variety of locations. Often the children's arms are unsupported, especially when holding the PCD closer to the eyes to aid viewing. If a child understands the risk associated with awkward positions (those that require muscular effort to maintain) and static (holding still) postures, they are more likely to identify and use simple solutions such as propping pillows to align their posture.

The PCD stylus commonly supplied with the device is often smaller in diameter than a pencil. This makes it difficult to grip and control, especially for younger children. A larger stylus is easier for most people (Figure 16.32).

Recommendations for Laptops and PCDs

1. Use external attachments to help laptops fit the child (keyboards, input devices, monitor).
2. Teach children about glare and contrast.
3. Teach children about correct postures and risk factors for discomfort.
4. Select a larger diameter stylus for the PCD.

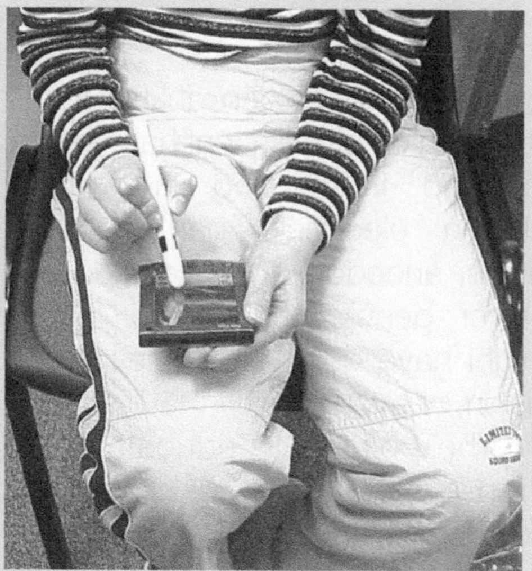

Figure 16.31 (Ursy Potter Photography) **Figure 16.32** (Diane Tien)
PCDs have less power than laptops, but enough to run word processing, spreadsheet, and database software (Straker, 2000). Through use of cradles or infrared wireless synchronizing devices, PCDs can become an extension of more powerful computers. Some schools already provide PCDs to students as alternatives to laptops (Corcoran, 2002). Their size and weight makes them attractive for children to carry to and from school.

Using a small stylus can create tension in the fingers and hands, especially during long sessions. The purchase and use of a larger grip stylus (with diameter larger than most pencils) or the use of an external input device or keyboard may prevent discomfort that can develop from a prolonged fine pinch grip.

ELECTRONIC GAMES

Today's youth sometimes seems obsessive about electronic games. They appear on computers, televisions, self-contained hand-held computers, and in arcades. Children and adults too spend hours playing various types of electronic games. Revenues generated by the electronic entertainment industry surpassed theater box office revenues* (Gentile et al, 2004).

Despite the worries of adults, some game-playing skills might be transferable to the workplace.† Research in this area has not yet been conducted with child subjects and with development-related cognitive and physical issues.

Increasingly, we perceive our world indirectly through our computer screens. Children begin using computers before they can read or write and would often rather watch a screen (whether DVD or a game) than look out the window in a car or interact with others. This may help them learn to remotely operate vehicles or weapon systems, but how will it affect their relationships with people they meet face to face or their ability to feel compassion?

The sedentary nature of most game playing is also a concern (see Chapter 7 on Physical Exercise).

As with computer use, musculoskeletal risk factors involve awkward postures, sustained postures, and repetition (Figure 16.33). In one assessment, individuals who played computer games had shoulder muscle stiffness and elevated scapula on the nondominant side (left if they were right-handed or vice versa) (Tazawa et al., 1997). According to their physicians, stopping their computer play relieved the symptoms. Once again, there is little information available on the physical impact of electronic games on children and adolescents (Gillespie, 2002).

Figure 16.33 (Ursy Potter Photography)
The natural tendency of young children to move from one position or activity to another usually provides some protection from static and awkward postures. However, electronic games require high focus; and interaction quickly overcomes the impulse to move and change postures, often for long sessions.

Anecdotal reports include sprained and broken ankles from stepping onto a foot that has "gone to sleep" from sitting on it while playing electronic games. Yet, similar anecdotes can be cited by earlier generations whose focus might have been on movies, television, or even radio.

The point is to educate children regarding the risks of stationary behavior during gaming and encourage movement, taking breaks, and good supported posture. Parents must also be

* This analysis was conducted on data collected in 1999.

† Dexterity and eye–hand coordination are positive attributes of electronic game playing (Yuji, 1996) and the skills may be similar to those used when driving automobiles or flying airplanes (Fery and Ponserre, 2001). Enhancement of information processing, essential to many work places, may even occur (Yuji, 1996).

alert for developments in the technology of games. Manufacturers may not have identified all the risks of new features.

Parents of children and adolescents frequently describe their children as engrossed in games, resisting interruptions, and churlish when asked to stop playing. Moderation and breaks are the best preventative measures.

Thankfully, most games incorporate a pause feature; so not all is lost when a break is taken. Definitive, active intervention is often necessary to enforce breaks and set limits on playing time, and parents need to supervise game playing, just as they supervise television viewing. Even so, parents report supervising television-viewing more often than Internet or video games (88% of parents compared with 50% and 48%, respectively) (Woodard and Gridina, 2000).

Children who play electronic games tend to be more sedentary, especially boys (DiNubile, 1993; Ho and Lee, 2001). Some prefer playing electronic games to sports and physical activities. Although playing video or arcade games increases the heart rate, blood pressure, and oxygen consumption, these do not provide sufficient cardiorespiratory stress to improve physical fitness (Dorman, 1997).*

As with television, there is concern that "playing violent computer games may increase aggressiveness and desensitize a child to suffering, and that the use of computers may blur a child's ability to distinguish real life from simulation" (Subrahmanyam, et al., 2000). Computer and video games can portray realistic and brutal animated scenes. Less aggressive behavior was seen in elementary students (mean age, 8.9 years) who decreased their exposure to violent material in video games and television during a 6-month randomized control study (Robinson et al., 2001).

Recommendations for Electronic Games

1. Limit daily game playing (screen) time.
2. Enforce breaks (have the child set a timer for 20–30 minutes when they begin to play an electronic game.
3. Educate children about risk factors for musculoskeletal pain or symptoms.
4. Supervise children's choices.
5. Encourage participation in other physical and cognitive activities and games.
6. Pay attention to the social and interactive skills of children.

Parents should screen games for violent or sexual content before purchase (Funk, 1993) and both parents and teachers should be aware of their role in influencing a child's choices with respect to selecting games and limiting playing time (Jirasatmathakul and Poovorawan, 2000). The Entertainment Software Rating Board (ESRB) rating symbols give age appropriate recommendations and characterize content that may be of concern. Use of these ratings can help determine what games are complementary to the values a parent wants to promote in his family.

We do not know whether playing electronic games adversely affects children's ability to interact and form close relationships. What is clear is that this issue raises considerable concern among parents (Shimauchi, 1999) and both positive (Dorman, 1997) and negative (Jirasatmathakul and Poovorawan, 2000) findings have been cited. By paying attention to the emotional and social skills of children in the classroom and at home, teachers and parents can determine when and how to intervene. They must watch children's interactions, observe the give-and-take of verbal and nonverbal exchanges and evaluate the child's ability to form and maintain relationships among family and friends.

* Limitations on game playing are included in the time limits for "screen time" mentioned in the section on television above.

VIDEO OR TELEVISION GAMES

Children play video games through platforms or consoles (such as those made by Sega, Nintendo, or Play Station) that connect to a television. The controller that connects to the television through cables is typically held with two hands and operated by the thumbs and one or more fingers on each hand. Thus, the player is "tethered" by the controller cables. That is, they can only sit as far away from the television as the cables will permit (Figure 16.34).

Figure 16.34 (Cheryl Bennett)
Video-game controls and wiring require children to sit in close proximity to the television. This can result in neck extension and neck and shoulder fatigue as they tilt their heads back to view the screen. Sitting on a therapy or sport ball can provide some active engagement of muscles and reduce the tendency for a fixed static posture.

This constraint tends to result in children sitting closer to the television when playing video games than when watching television. Living room furniture (couches and chairs) may not be positioned close enough for children to sit on while playing videogames. Therefore, children sit on the floor or at the edge of a couch or chair, unless they have other options.

"Nintendinitis" refers to palm lesions (Brasington, 1990; Macgregor, 2000) from intensive video-game playing. Tendonitis is an inflammation of the tendons. Overuse injuries are typically treated with anti-inflammatories such as ibuprofen and abstinence from activity that caused the problem. Often, only several days away from video games will clear the condition! In order to avoid overuse injuries, children (and adults) must be aware of the messages their bodies convey to them.

Pain signals communicate a need for rest. Weight lifters and other athletes know that intensive exercise results in normal muscle tenderness signifying microscopic tears in the muscle. The muscle needs a day or two of rest to repair before rebuilding through another bout of exercise. Muscles and other tissues also need rest and a break between intensive musculoskeletal "stress" during game playing.

Parents must actively question game players about discomfort, then listen to complaints of soreness or other physical problems and determine the source of the problem. If there is an indication of repetitive strain (from overuse), they must insist the child reduce the time spent in the activity causing the pain. If different activities of a child use similar motions, such as using a computer at school for schoolwork, playing videogames at home, or playing a musical instrument, the tendons and muscles are especially at risk of being overworked.

Children experience other musculoskeletal aches and pains associated with video-game playing. For example, primary school children who play video games for more than 2 hours a day report more back pain than children who watch television for a similar amount of time (Gunzburg et al., 1999). Compared with children who play less than 1 hour a day, children who play video games more (than 1 hour a day) have dark rings under their eyes and shoulder muscle stiffness (Tazawa and Okada, 2001). Playing video games should not hurt or make you look like a zombie.

A small number of children experience epileptic seizures when playing video games, possibly due to photosensitive epilepsy. Photosensitive epilepsy is a relatively rare form of epilepsy in which the individual is sensitive to flicker or repetitive patterns.

Children who are prone to such seizures should avoid sitting close to the screen, playing in a dark room, games with bright or flashing geometric figures, and screen patterns with a slow, 50 Hz frame rate (Furusho et al., 2002; Fylan, et al., 1999; Trenite et al., 2002). The best "treatment" is always prevention. Although it gradually declines as they get older, children who have experienced seizures triggered by video games simply should not play video games. If they still choose to play (and parents and health care providers agree), then anticonvulsant medications may be considered* (Graf et al., 1994).

Video movies and games can be useful for occupying children's attention. Some pediatricians and orthodontists use video movies or games to keep children still during procedures or while waiting. In one case, computer games were used after surgical procedures. A monitor was inverted and placed behind the head of the bed, achieving three goals; proper positioning of the neck for healing, improving compliance, and decreasing boredom (Foley, 2001). Some hair salons allow children to watch or play while wailing or while having their hair cut.

Video games can be a positive or negative influence on a child. They can be negative if children play too often, with too little supervision, or without learning about proper use and posture. Video games and viewing should not interfere with a child's social, physical, or emotional growth. The content of the games should not promote a view of the world as a mean or scary place (Robinson et al., 2001). They should enhance their life. They should be fun.

COMPUTER GAMES

Children can install software on a personal computer or download it from the internet. Interactive computer games allow users to play online simultaneously with other users anywhere in the world. Some computer games include "chat rooms," so players can communicate while playing.

In the United States, younger children (9–11 years) play online games more than older children do (15–17 years) and boys play more than girls (RSW, 1999). Yet, in Hong Kong, the 14-year-old age group plays the most (Ho and Lee, 2001). This may reflect cultural differences.

Children can use a mouse, trackball, or keyboard as their controller. They can also use joysticks or steering wheels for more realistic representations of cars and planes. The physical and social issues associated with computer games are similar to those for videogames and the physical workstation design should follow the recommendations for computer workstations.

PORTABLE GAMES

Portable games can go anywhere: on parent's errands to the bank or grocery store, to the ball game, and on long trips in a car or on a plane. Thanks to headphones, they can be relatively unobtrusive. Gameboy-type games are designed specifically for that use, but cellular telephones can also contain games.

Parents may have to squint to see the small screens, but that does not deter children. Backlighting provided in displays makes direct viewing easier. Most children hold portable games on their laps, resulting in pronounced neck flexion and rolling forward of the shoulders (see Figure 16.35a and Figure 16.35b).

As with other games, ask about physical discomfort and encourage children to take breaks. Breaks are good for concentration, focus, and obtaining a child's "best" performance. Knowing that taking frequent breaks will actually improve their game will go a long way to gaining a child's cooperation!

Cognitive effects of these frequent companions need more study. For instance, what are the implications of the preference for concentrating on short focal distance activities,

* The use of medication requires consultation with a physician well-acquainted with the particular illness.

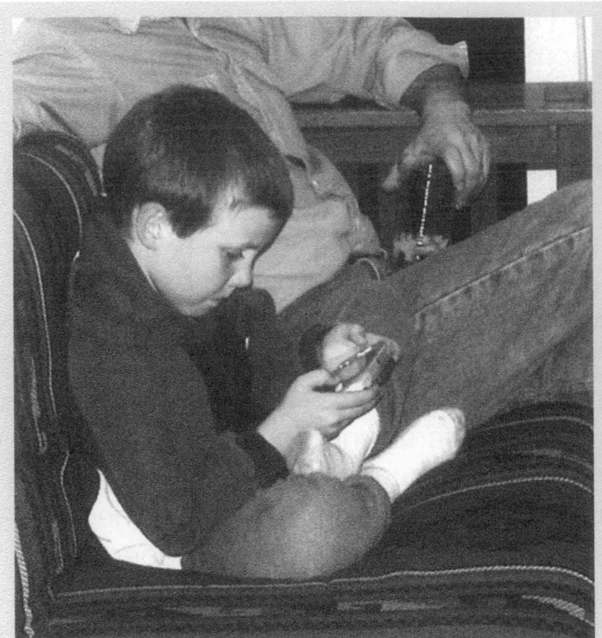

Figure 16.35a and Figure 16.35b (Cheryl Bennett)
Children adopt postures that can result in pain, discomfort, and perhaps even long-term problems when playing hand-held games. Supporting their arms on a pillow will permit them to hold their head erect and let them rest their forearms. When a pillow is not available, encourage children to be creative in finding a comfortable spot to play.

When at home, they can sit on the floor and rest their arms on the seat of a low chair. A child who understands correct postures and viewing angles may easily assume these postures, with a few gentle reminders.

rather than looking out at the surroundings? Does the ability to entertain oneself atrophy? And what happens when the games are not available?

ELECTRONIC GAMES SUMMARY

Suggestions for avoiding overuse symptoms from playing electronic games are located in Table 16.4.

Table 16.4 Guidelines for Avoiding Overuse SymptPoms from Game Playing

- Have your child take a break every 30 minutes, even if they think they do not need it.
- If a child's fingers, hands, wrists, or arms become tired, sore, or numb while playing, have them stop playing and rest their arms and hands for several hours.
- If symptoms of pain, discomfort, or numbness persist for several hours after playing, have the child stop playing for a day or more (24 to 48 hours).
- If the symptoms return, the child should stop playing and consult a health care practitioner.
- If a child is experiencing pain from one activity, do not let them engage in a similar activity, using similar motions with the same body parts. It can aggravate the symptoms.
- Teach children to notice the posture of their bodies, including their hands and arms.
- While playing, have the child relax their hands, hold the controller lightly, and use minimal force to operate the controls. Reward the child who holds the controller so lightly that you can pull it from their hands when you reach for it unexpectedly.
- Give children the tools they need. Purchase furniture that is designed for the child and the task at hand or that adapts to their environment.
    - For example, teach children to support their forearms on a cushion to reduce the load on the shoulders and neck when using portable devices; or provide a chair with a backrest or a sports or therapy ball for sitting while playing video or television games.

CONCLUSION

As technologies continue to develop, our challenge is to make certain the new emerging technologies are easy for its child users to operate, while accommodating their physical and mental abilities. It is important for children to understand basic principles of posture, body mechanics, and the risks and benefits associated with using electronic devices. Further, imparting an awareness of the importance of position, posture, and comfort to children at an early age can establish habits that will serve to empower them throughout their lives.

In addition, knowing how children learn and grow (see Chapter 2 on child development) can help parents and teachers to determine when and if a child's involvement in technology is interrupting other areas of their lives. A balanced lifestyle, incorporating social interactions with adults and children of various ages, physical activities, and plenty of opportunities for creativity, along with the use of technology should support the growth and health of our children.

Figure 16.36 (Cheryl Bennett)
Taking breaks can be combined with social values such as taking turns.

Figure 16.37 (Ursy Potter Photography)
It is the surrounding social environment and typical patterns of behavior that determine whether this is a healthy, happy social exchange or not.

ACKNOWLEDGMENTS

Much gratitude to Rani Lueder and Valerie Rice for the creation of this book and their inspirational generosity of spirit. Appreciation to the many who have contributed to the creation and development of the IEA Technical Committee Ergonomics for Children and Educational Environments. Thanks to Diane Tien for pioneering integrated ergonomics for children in schools and for her photos. And with much appreciation for the support and encouragement of my husband Charles and the originating ideas and patience of my daughter Caitlin.

REFERENCES

Aagaard-Hansen, J. (1995). A comparative study of three different kinds of school furniture. *Ergonomics*, 38(5), 1025–1035.

AAP (2004). American Academy of Pediatrics. http://www.aap.org/family/tv1.htm [March 2004]

Annan, K. (2000). United Nations World Television Forum 2000 (Final Report). United Nations Headquarters, New York City: United Nations.

Balagué, F., Dudler, J., and Nordin, M. (2003). Low-back pain in children. *Lancet*, 361(9367), 1403–1404.

Blackstone, J. M. and Johnson, P. W. (2004). Comparison of child and adult anthropometry: considerations for input device design. Human Factors and Ergonomics Society's 48th Annual Meeting, September 2004, New Orleans.

Brasington, R. (1990). Nintendinitis. *New England Journal of Medicine*, 332, 1473.

Christakis, D. A., Ebel, B. E., Rivara, F. P., and Zimmerman, F. J. (2004). Television, video, and computer game usage in children under 11 years of age. *Journal of Pediatrics*, 145(5), 652–656.

Corcoran, K. (April 4, 2002). New school tools: Schools plan for mobile technology. *San Jose Mercury News*, p. 1A and 12A.

Crespo, C., Smit, E., Troiano, R., Bartlett, S., Macera, C., and Andersen, R. (2001). Television watching, energy intake, and obesity in US children: results from the third National Health and Nutrition Examination Survey, 1988–1994. *Arch. Pediatr. Adolesc. Med.*, 155, 360–365.

Cummings, J. N. and Kraut, R. (2002). Domesticating computers and the internet. *Information Society*, 18(3), 221–232.

Dennison, B. A., Erb, T. A., and Jenkins, P. L. (2002). Television viewing and television in bedroom associated with overweight risk among low-income preschool children. *Pediatrics*, 109(6), 1028–1035.

DiNubile, N. A. (1993). Youth fitness—Problems and solutions. *Preventive Medicine*, 22(4), 589–594.

Dorman, S. M. (1997). Video and computer games: Effect on children and implications for health education. *Journal of School Health*, 67(4), 133–138.

Fery, Y.-A. and Ponserre, S. (2001). Enhancing the control of force in putting by video game training. *Ergonomics*, 44(12), 1025–1037.

Foley, K. H., Kaulkin, C., Palmieri, T. L., and Greenhalgh, G. D. (2001). Inverted television and video games to maintain neck extension. *Journal of Burn Care & Rehabilitation*, 22(5), 366–368.

Fraser, M. (2002). Ergonomics for grade school students using laptop computers. Paper presented at the Proceedings of the XVI International Occupational Ergonomics and Safety Conference, Toronto, Canada, June 10–12, 2002.

Frohlich, D. M., Dray, S., and Silverman, A. (2001). Breaking up is hard to do: Family perspectives on the future of the home PC. *International Journal of Human-Computer Studies*, 54(5), 701–724.

Fuller, A. (2002). *Raising Real People—Creating a Resilient Family* (2nd edn.). Melbourne, Australia: Australian Council for Educational Research Press.

Funk, J. B. (1993). Reevaluating the impact of video games. *Clin. Pediatr.*, 32(2), 86–90.

Furnham, A. and Strbac, L. (2002). Music is as distracting as noise: The differential distraction of background music and noise on the cognitive test performance of introverts and extroverts. *Ergonomics*, 45(3), 203–217.

Furusho, J., Suzuki, M., Tazaki, I., Satoh, H., Yamaguchi, K., Iikura, Y. et al. (2002). A comparison survey of seizures and other symptoms of Pokemon phenomenon. *Pediatric Neurology*, 27(5), 350–355.

Fylan, F., Harding, G. F., Edson, A. S., and Webb, R. M. (1999). Mechanisms of video-game epilepsy. *Epilepsia*, 40(Suppl. 4), 28–30.

Gentile, D. A., Lynch, P. J., Linder, J. R., and Walsh, D. A. (2004). The effects of violent video game habits on adolescent hostility, aggressive behaviors, and school performance. *Journal of Adolescence*, 27, 5–22.

Gillespie, R. M. (2002). The physical impact of computers and electronic game use on children and adolescents, a review of current literature. *Work: A Journal of Prevention, Assessment, and Rehabilitation*, 18(3), 249–259.

Graf, W. D., Chatrian, G. E., Glass, S. T., and Knauss, T. A. (1994). Video game-related seizures: A report on 10 patients and a review of the literature. *Pediatrics*, 93(4), 551–556.

Gunzburg, R., Balague, F., Nordin, M., Szpalski, M., Duyck, D., Bull, D. et al. (1999). Low back pain in a population of school children. *European Spine Journal*, 8(6), 439–443.

Hänninen, O. and Koskelo, R. (2003). Adjustable tables and chairs correct posture and lower muscle tension and pain in high school students. Paper presented at the Triennial International Ergonomics Association Congress, Seoul, South Korea.

Harris, C. and Straker, L. (2000). Survey of physical ergonomics issues associated with school children's use of laptop computers. *International Journal of Industrial Ergonomics*, 26(3), 337–346.

Harris, C., Straker, L., Pollock, C., and Trinidad, S. (2002). Musculoskeletal outcomes in children using information technology—The need for a specific etiological model. Paper presented at the CybErg 2002. Johannesburg: International Ergonomics Association Press.

Ho, S. M. Y. and Lee, T. M. C. (2001). Computer usage and its relationship with adolescent lifestyle in Hong Kong. *Journal of Adolescent Health*, 29, 258–266.

Jeong, B. Y. and Park, K. S. (1990). Sex differences in anthropometry for school furniture design. *Ergonomics*, 33(2), 1511–1521.

Jirasatmathakul, P. and Poovorawan, Y. (2000). Prevalence of video games among Thai children: Impact evaluation. *Journal of the Medical Association of Thailand*, 83(12), 1509–1513.

Jonai, H., Villanueva, M. B. G., Takata, A., Sotoyama, M., and Saito, S. (2002). Effects of the liquid crystal display tilt angle of a notebook computer on posture, muscle activities and somatic complaints. *Industrial Ergonomics*, 29, 219–229.

Jonsson, C. (1997). Aspects on the Swedish provisions on work with VDUs in telework and at school. Paper presented at the Work with Display Units 1997, Japan.

Kahan, N. J. (2002). Technique addresses computer-related RSI: Part II. *Occup. Med. Clin. Care Update*, 8(18), 1–3.

Lehnung, M., Leplow, B., Haaland, V. O., Mehdorn, M., and Ferstl, R. (2003). Pointing accuracy in children is dependent on age, sex and experience. *J. Environ. Psychol.*, 23, 419–425.

Linton, S. J., Hellsing, A. -L., Halme, T., and Akerstedt, K. (1994). The effects of ergonomically designed school furniture on pupils' attitudes, symptoms and behavior. *Applied Ergonomics*, 25(5), 299–304.

Macgregor, D. M. (2000). Nintendonitis? A case report of repetitive strain injury in a child as a result of playing computer games. *Scottish Medical Journal*, 45(5), 150.

Mandal, A. C. (1994). The prevention of back pain in school children. In: R. Lueder and K. Noro (Eds.), *Hard Facts About Soft Machines* (pp. 269–278). Bristol, PA: Taylor & Francis.

Marumoto, T., Sotoyama, M., Villanueva, M. B. G., Jonai, H., Yamada, H., Kanai, A. et al. (1999). Significant correlation between school myopia and postural parameters of students while studying. *International Journal of Industrial Ergonomics*, 23, 33–39.

Milanese, S. and Grimmer, K. (2004). School furniture and the user population: An anthropometric perspective. *Ergonomics*, 47(4), 416–426.

Molenbroek, J. F. M., Kroon-Ramaekers, Y. M. T., and Snijders, C. J. (2003). Revision of the design of a standard for the dimensions of school furniture. *Ergonomics*, 46(7), 681–694.

Nathan, J. G., Anderson, D. R., Field, D. E., and Collins, P. (1985). Television viewing at home: Distances and visual angles of children and adults. *Human Factors*, 27(4), 467–476.

Parcells, C., Stommel, M., and Hubbard, R. P. (1999). Mismatch of classroom furniture and student body dimensions empirical findings and health implications. *Journal of Adolescent Health*, 24, 265–273.

Psihogios, J. P., Sommerich, C. M., Mirka, G. A., and Moon, S. D. (2001). A field evaluation of monitor placement effects in VDT users. *Applied Ergonomics*, 32, 313–325.

Rideout, V. J., Vandewater, E. A., and Wartella, E. A. (2003). *Zero to Six: Electronic Media in the Lives of Infants, Toddlers and Preschoolers* (No. 3378). Menlo Park, CA: The Henry J. Kaiser Family Foundation.

Robinson, T. N., Wilde, M. L., Navracruz, L. C., Haydel, K. F., and Varady, A. (2001). Effects of reducing children's television and video game use on aggressive behavior: A randomized controlled trial. *Archives of Pediatrics & Adolescent Medicine*, 155(1), 17–23.

RSW (1999). Roper Starch Worldwide. The America Online/Roper Starch Youth Cyberstudy (Electronic report): AOL (America Online).

Royster, L. and Yearout, R. (1999). A computer in every classroom – are school children at risk for repetitive stress injuries (RSIs)? In: Lee, G. (Ed.), *Advances in Occupational Ergonomics and Safety.* The Netherlands: IOS Press, pp. 407–412.

Sauter, S. L., Schleifer, M., and Knutson, S. (1991). Work posture, workstation design, and musculoskeletal discomfort in a VDT data entry task. *Human Factors*, 33, 151–167.

Sen, A. and Richardson, S. (2002). Some controversies relating to the causes and preventive management of computer vision syndrome. Paper presented at the Cyberg 2002. Johannesburg: International Ergonomics Association Press.

Shaw, D. G., Swigge, K. M., and Herndon, J. (1985). Children's questions: A study of questions children ask while learning to use a computer. *Computers & Education*, 9(1), 15–19.

Shimauchi, Y. (1999). Children in Japan and multimedia. *Turkish Journal of Pediatrics*, 41(Suppl), 7–12.

Straker, L. (2000). Laptop computer use. In: W. Karwowski (Ed.), *International Encyclopedia of Ergonomics and Human Factors* (1st edn, p. 1953). London: Taylor & Francis.

Straker, L., Jones, K., and Miller, J. (1997). A comparison of the postures assumed when using laptop computers and desktop computers. *Applied Ergonomics*, 28(4), 261–268.

Strasburger, V. C. and Donnerstein, E. (1999). Children, adolescents and the media: issues and solutions. *Pediatrics*, 103(1), 123–139.

Subrahmanyam, K., Greenfield, P. M., Kraut, R. E., and Gross, E. F. (2000). The impact of home computer use on children's activities and development. *The Future of Children: Children and Computer Technology*, 10(2), 123–144.

Subrahmanyam, K., Greenfield, P., Kraut, R., and Gross, E. (2001). The impact of computer use on children's and adolescent's development. *Journal of Applied Developmental Psychology*, 22(1), 7–30.

Sutherland, R., Facer, K., Furlong, R., and Furlong, J. (2000). A new environment for education? The computer in the home. *Computers & Education*, 34(3–4), 195–212.

Suzuki, H., Hashimoto, Y., and Ishii, K. (1997). Measuring information behavior: A time budget survey in Japan. *Social Indicators Research*, 42, 151–169.

Szeto, G. P. Y., Lau, J. C. C., Siu, A. Y. K., Tang, A. M. Y., Tang, T. W. Y., and Yiu, A. O. Y. (2002). A study of physical discomfort associated with computer use in secondary school students. Paper presented at the CybErg 2002. Johannesburg: International Ergonomics Association Press.

Szeto, G. P. and Lee, R. (2002). An ergonomic evaluation comparing desktop, notebook, and subnotebook computers. *Archives of Physical Medicine and Rehabilitation*, 83(4), 527–532.

Tazawa, Y. et al. (1997). Excessive playing of home computer games by children presenting unexplained symptoms. *Journal of Pediatrics*, 130, 1010–1011.

Tazawa, Y. and Okada, K. (2001). Physical signs associated with excessive television-game playing and sleep deprivation. *Pediatrics International*, 43(6), 647–650.

Tien, D. (2002). Personal communication. Redmond, WA: Blackwell Elementary School.

Trenite, D. G., Silva, A. M., Ricci, S., Rubboli, G., Tassinari, C. A., Lopes, J. et al. (2002). Video games are exciting: A European study of video game-induced seizures and epilepsy. *Epileptic Disorders*, 4(2), 121–128.

Troussier, B., Tesniere, C., Fauconnier, J., Grison, J., Juvin, R., Phelip, X. et al. (1999). Comparative study of two different kinds of school furniture among children. *Ergonomics*, 42(3), 516–526.

USNTIA (2002) United States National Telecommunications and Information Administration. A Nation Online: How Americans are Expanding Their Use of the Internet. www.ntia.doc.gov/ntiahome/dn/anationonline2.pdf

Venkatesh, A. (1996). Computers and other interactive technologies for the home. *Communications of the ACM*, 39(12), 47–54.

Wartella, E. A. and Jennings, N. (2000). Children and computers: New technology—Old concerns. *The Future of Children: Children and Computer Technology*, 10(2), 31–43.

Wiecha, J. L., Sobol, A. M., Peterson, K. E., and Gortmaker, S. L. (2001). Household television access: Associations with screen time, reading, and homework among youth. *Ambulatory Pediatrics*, 1(5), 244–251.

Williams, P. A., Haertel, E. H., Haertel, G. D., and Walberg, J. (1982). The impact of leisure-time television on school learning: A research synthesis. *American Educational Research Journal*, 19, 19–50.

Woodard, E. H. I. and Gridina, N. (2000). *Media in the Home—The Fifth Annual Survey of Parents and Children* (No. Survey Series No. 7). Washington, D.C.: The Annenberg Public Policy Center of the University of Pennsylvania.

Woods, V., Hastings, S., Buckle, P., and Haslam, R. (2002). Ergonomics of Using a Mouse or Other Non-keyboard Input Device (Research Report No. 045). Prepared by the University of Surrey & Loughborough University for the Health and Safety Executive.

Yuji, H. (1996). Computer games and information-processing skills. *Perceptual and Motor Skills*, 83, 643–647.

CHAPTER 17

CHILDREN IN VEHICLES

RANI LUEDER

TABLE OF CONTENTS

OVERVIEW

In the United States, motor vehicle injuries are the leading cause of death in children at every age after their first birthday.* Children's injuries are also more severe than those of adults; their small size and developing bones and muscles make them more susceptible to injury in car crashes if not properly restrained (s.f. CDC, 2006). Many of these deaths could have been avoided, such as by proper use of child restraints in vehicles.

* Motor Vehicle accidents are the leading cause of death in toddlers aged 1 to 4 and in children aged 5 to 14. In 2004, 1638 children died and 214,000 children younger than 15 years were injured in motor vehicle crashes in the United States. That averages 5 deaths and 586 injuries each day (NHTSA, 2005). http://cdc.gov

> Every 90 s, a child is killed or injured in a motor vehicle in the United States.
> Centers for Disease Control

Figure 17.1 (Ursy Potter Photography)
The child riding in the back (above) is unsafe. Children are often impulsive and unable to follow common safety rules or judge the safety of a situation.

Children are also injured by other vehicles, such as when running in front of a vehicle to retrieve a ball. As they are small, they are hard to see when walking in traffic (Agran et al., 1990).

CAR RESTRAINTS

CHILD SAFETY SEATING

One-third of children younger than five years old who die in motor vehicle accidents were riding unrestrained (NHTSA, 2005). Many children who ride in child safety seats are also improperly secured; one large-scale survey found that only 15% of children were correctly restrained in their child safety seats (Taft et al., 1999).

Figure 17.2 (CDC/AAP)
This infant is positioned properly and buckled in an *infant-only car seat.*

Important features include

1. harnesses are snug;
2. plastic harness clip is positioned at the armpits to hold shoulder straps in place;
3. straps lie flat;
4. the baby's clothes allow positioning straps between legs; and
5. the infant sits upright and centered while facing the rear of the car.

The American Academy of Pediatrics (AAP) recommends that infants ride facing the rear until they reach their first year and weigh at least 9 kg (20 lb). Never place rear-facing seats in the front seat of a vehicle with an air bag.

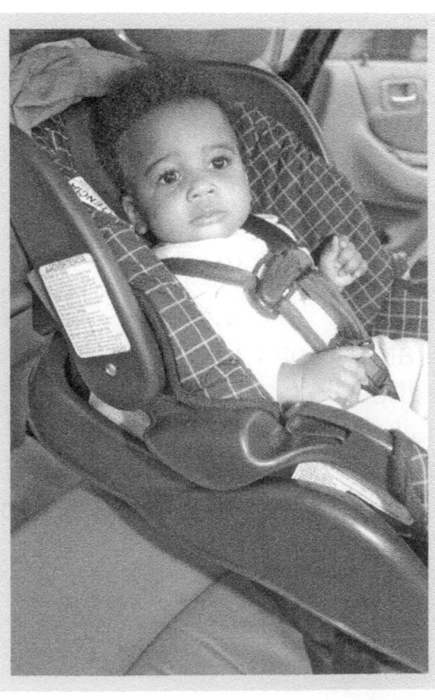

Nearly half of children under 5 years killed in motor vehicle crashes had not been restrained. Many of these children would have been saved by properly restrained child safety seats. These products reduce the risk of death by about 70% for infants and by about 55% for toddlers aged 1 to 4.

Children in night and daytime crashes were more likely to sit in the front than children in morning crashes. Children in nighttime crashes were also less likely to be properly restrained than those in daytime crashes.
(Chen et al., 2005)

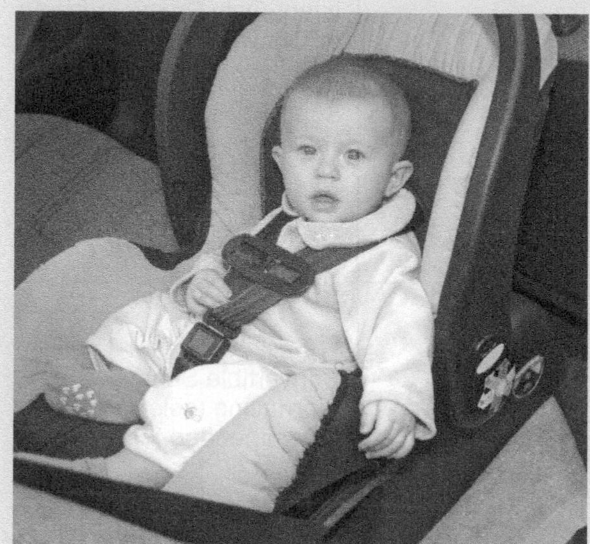

Figure 17.3 (CDC/AAP)
Both *infant-only seats* (above) and *convertible seats* are acceptable. However, convertible seats last longer as these adapt to the growing child; these seats are larger and include extra-harness slots to continue to secure growing children. Older children can also sit in these products while facing forward.
American Academy of Pediatrics

Children riding with at least two additional passengers were less likely to be properly restrained.
(Chen et al., 2005)

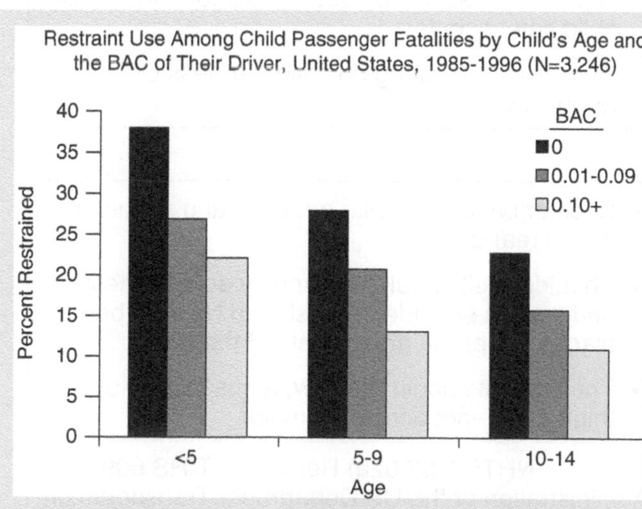

Figure 17.4 Depending on age, 15%–40% of children regularly ride unrestrained in motor vehicles and this rate increases when drivers have high blood alcohol contents (CDC, 2000).

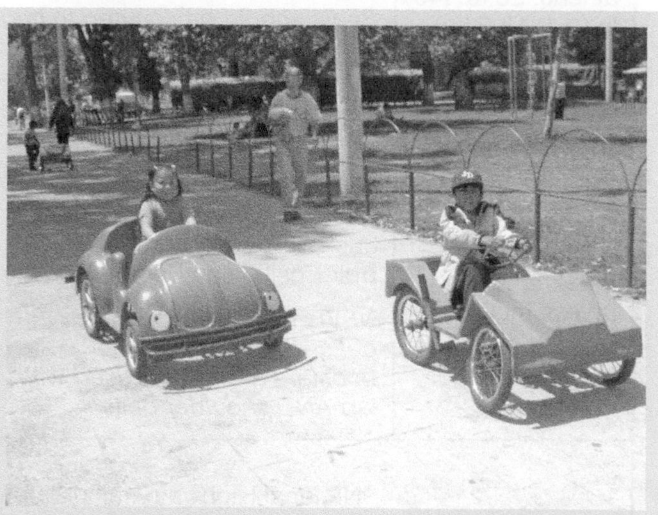

Figure 17.5 (Ursy Potter Photography)
Children are eager to drive at very young ages. Prepare them by instilling safe riding and driving practices early.

Table 17.1 General Child Seat Use Information

Always refer to the child seat and vehicle manufacturers' instructions for proper use and installation.

Age/Weight	Seat Type/Seat Position	Usage Tips
Infants		
Birth to 1 year and at least 20 lb (9.1 kg)	Infant-Only Seat/rear-facing **or** Convertible Seat/used rear-facing *Seats should be secured to the vehicle by the seat belts or by the LATCH system*[a]	• Never use in a front seat where an air bag is present • Tightly install child seat in rear seat, facing the rear • Child seat should recline at about 45° • Harness straps/slots at or below shoulder level (lower set of slots for most convertible child safety seats) • Harness straps snug on child; harness clip at armpit level
Less than 1 year and 20–35 lb (9.1–15.9 kg)	Convertible Seat/used rear-facing (select one recommended for heavier infants) *Secure seats to the vehicle with seat belts or the LATCH system*	• Never use in a front seat where an air bag is present • Tightly install child seat in rear seat, facing the rear • Child seat should recline about 45° • Harness straps/slots at or below shoulder level (lower set of slots for most convertible child safety seats) • Harness straps snug on child; harness clip at armpit level
Preschoolers/Toddlers		
1 to 4 years, and at least 20 to ~40 lb (9.1–18.2 kg)	Convertible Seat/forward-facing **or** Forward-Facing Only **or** High Back Booster/Harness. *Seats should be secured to the vehicle by the seat belts or by the LATCH system*	• Tightly install child seat in rear seat, facing forward • Harness straps/slots at or above child's shoulders (usually top set of slots for convertible child safety seats) • Harness straps snug on child; harness clip at armpit level
Young Children		
4 to at least 8 years, unless they are 4 ft 9 in. tall (144.8 cm)	Belt-Positioning Booster (no back, base only) **or** High Back Belt-Positioning Booster *NEVER use with lap-only belts—belt-positioning boosters are always used with lap AND shoulder belts*	• Booster base used with adult lap and shoulder belt in rear seat • Shoulder belt should rest snugly across chest, and rest on shoulder; and should NEVER be placed under the arm or behind the back • The lap belt should rest low, across lap or upper thigh area—not across stomach

NHTSA (2002a) Report DOT HS 809 245.
National Highway Traffic Safety Administration of the US Department Transportation

[a] See subsequent section on the LATCH system to facilitate installing safety seats in the car.

Figure 17.6 (NHTSA)
When to replace infant seats...

The National Highway Transportation Safety Administration modified its recommendation to replace child safety seats after each crash.

NHTSA now maintains that these may not need to be replaced following minor crashes when meeting **ALL** of the following criteria:

1. A visual inspection of the child safety seat and under easily movable seat padding does not reveal any cracks or deformation that might have been caused by the crash.
2. The vehicle in which the child safety seat was installed could be driven from the scene of the crash.
3. The vehicle door nearest the child safety seat was undamaged.
4. There were no injuries to any of the vehicle occupants.
5. Air bags (if available) did not inflate.

LOWER ANCHORS AND TETHERS FOR CHILDREN (LATCH) SYSTEM

Eighty percent of child safety seats are used incorrectly, often because of the poor fit between child safety seats and the design of the vehicle. The Lower Anchors and Tethers for Children (LATCH) system is designed to make it easier to install child safety seats without using the vehicle's seat belt system (see Table 17.2).

Since September 2002, new vehicles and new child safety seats in the United States must include new features. These vehicle and safety seat features together make up the LATCH system.

- Cars, minivans, and light trucks must include anchor points between the vehicle's seat cushion and seat back in at least two rear seating positions and (excepting convertibles) a top tether anchor.
- Child safety seats must include a lower set of attachments that fasten to vehicle anchors. Most forward-facing child safety seats have a top tether strap that attaches to the top anchor in the vehicle.

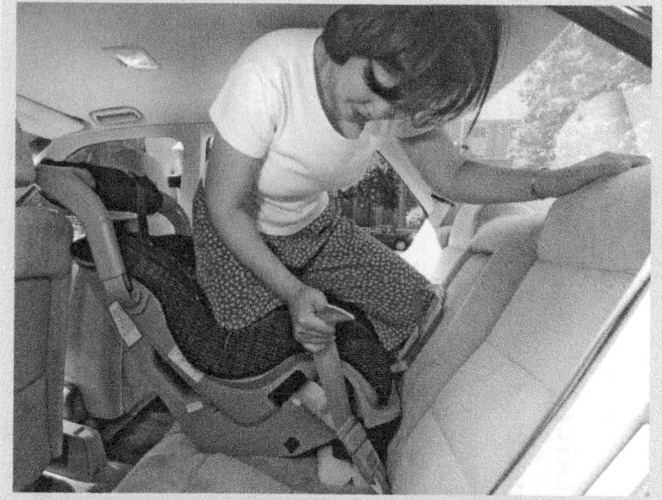

Figure 17.7 (NHTSA)
The *Lower Anchors and Tethers in Children* (LATCH) system requires manufacturers of new vehicles and child safety seats to incorporate features that make it easier to install child safety seats without the vehicle's seat belts.

<table>
<tr><th colspan="4">Table 17.2 Installing the LATCH System (Lower Anchors and Tethers for Children Standard for Anchoring Vehicles)
The LATCH system enables child safety seats to be installed without using seat belts</th></tr>
<tr><th colspan="4">Installing a Rear-Facing Convertible LATCH Seat</th></tr>
<tr>
<td>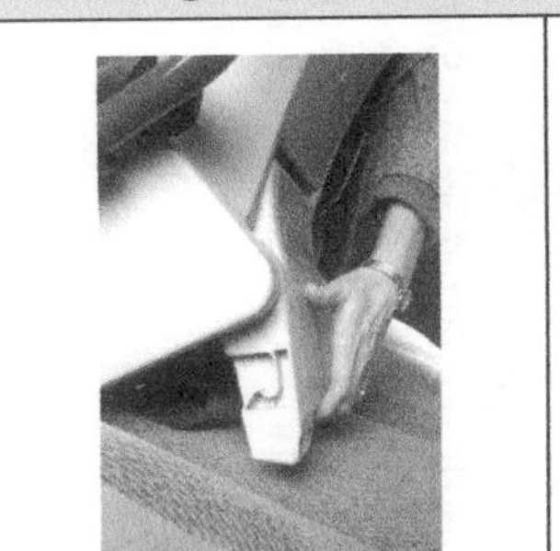
Figure 17.8 (NHTSA)</td>
<td>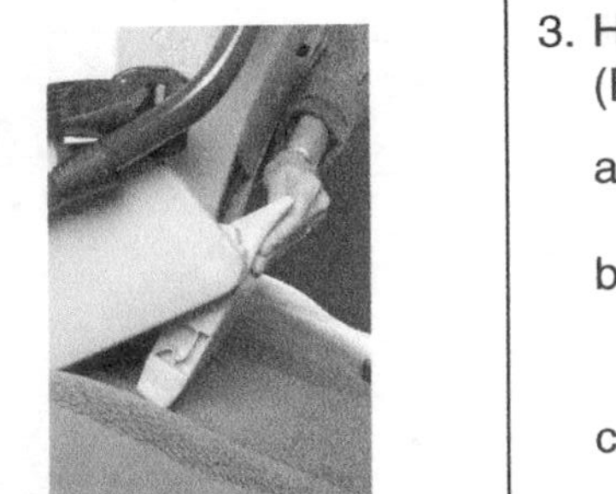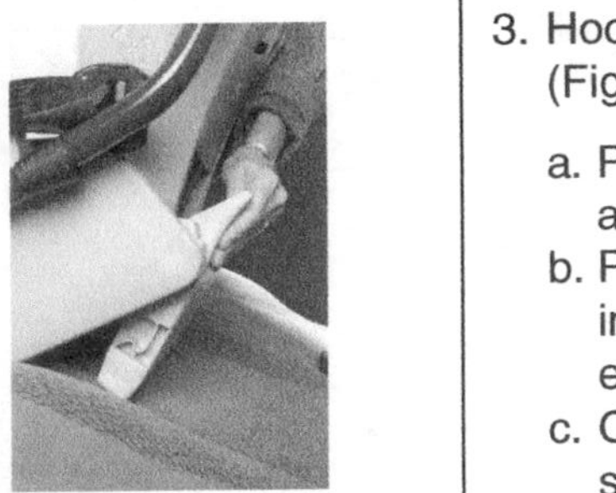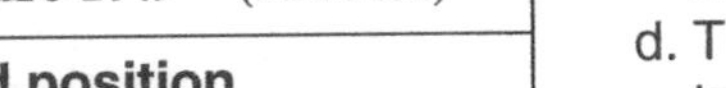
Figure 17.9 (NHTSA)</td>
<td rowspan="2">3. Hook LATCH attachments to the vehicle anchors (Figure 17.12)
a. Push with knees to assure that child seat is pressed against the vehicle seat back.
b. Put your body weight into the child safety seat (stand in front of seat and put one knee in seat), then pull the excess webbing.
c. Check for secure fit by placing your hands on the child safety seat, near the base of the child safety seat (near the latch attachments), and pull (Figure 17.13).
d. The child safety seat should not move more than an inch forward or sideways.
e. Child safety seat should be secured at 45° angle.</td>
<td>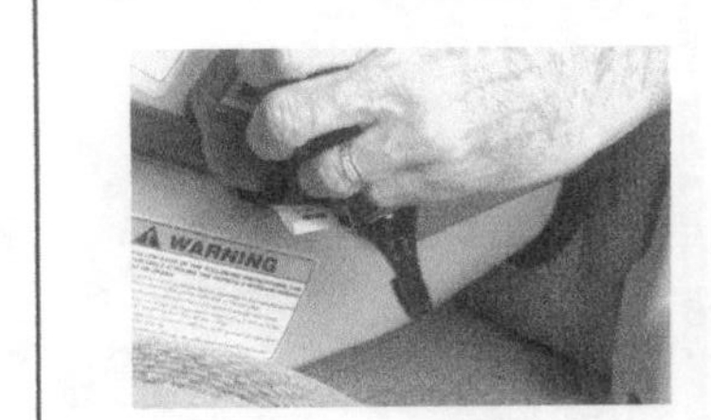
Figure 17.12 (NHTSA)</td>
</tr>
<tr>
<td colspan="2">1. Adjust seat to most reclined position
(Figure 17.8 and Figure 17.9)</td>
<td rowspan="2">
Figure 17.13 (NHTSA)</td>
</tr>
<tr>
<td rowspan="2">
Figure 17.10 (NHTSA)</td>
<td rowspan="2">Figure 17.11 (NHTSA)</td>
<td>
Figure 17.14 (NHTSA)</td>
</tr>
<tr>
<td rowspan="2">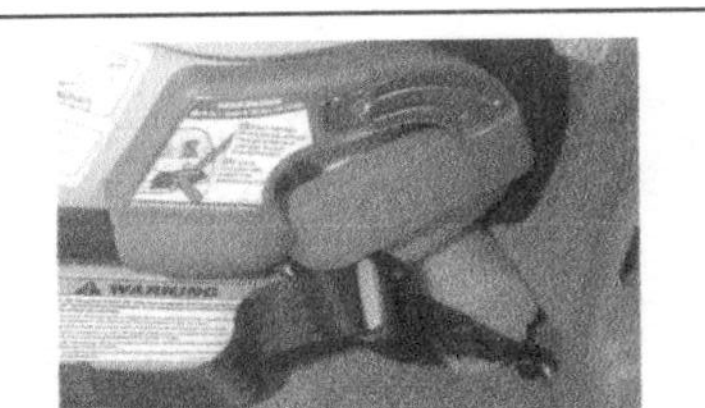
Figure 17.15 (NHTSA)</td>
<td rowspan="2">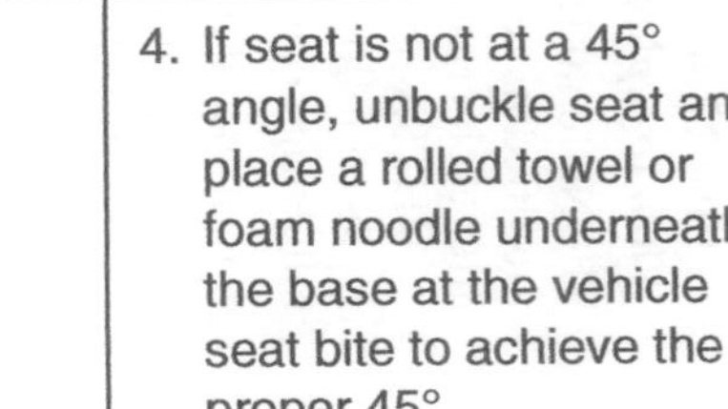4. If seat is not at a 45° angle, unbuckle seat and place a rolled towel or foam noodle underneath the base at the vehicle seat bite to achieve the proper 45°.</td>
</tr>
<tr>
<td colspan="2">2. Locate the lower LATCH attachments
(Figure 17.10 and Figure 17.11)</td>
</tr>
</table>

Installing a Forward-Facing Convertible LATCH Seat

1. Check to ensure that harness straps are at or above child's shoulders (in most seats harness straps should be in upper-most slots—read manufacturer instructions).
2. Adjust seat to upright or semi-reclined position (according to manufacturer's instructions).
3. Locate the lower LATCH attachments (Figure 17.16).
4. Hook the LATCH attachments to the vehicle anchors (Figure 17.17).
5. Put your body weight into the child safety seat (stand in front of seat and put one knee in seat), then pull the excess webbing.
6. Attach tether straps to tether anchor in vehicle and pull excess webbing to tighten (Figure 17.11).
   Check for secure fit by placing your hands on the child safety seat, near the base of the child safety seat (near the LATCH attachments), and pull —the child safety seat should not move more than an inch forward or sideways.

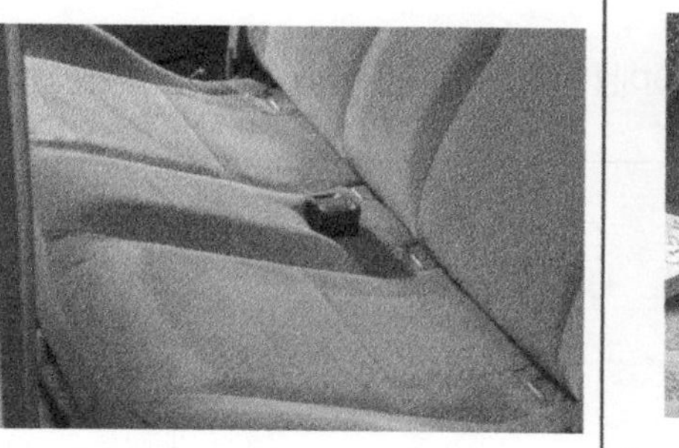

Figure 17.16 (NHTSA)

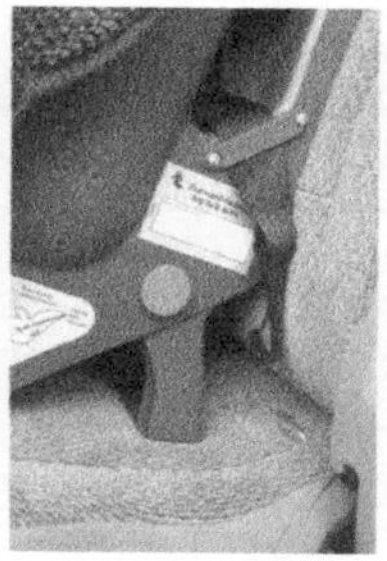

Figure 17.17 (NHTSA)

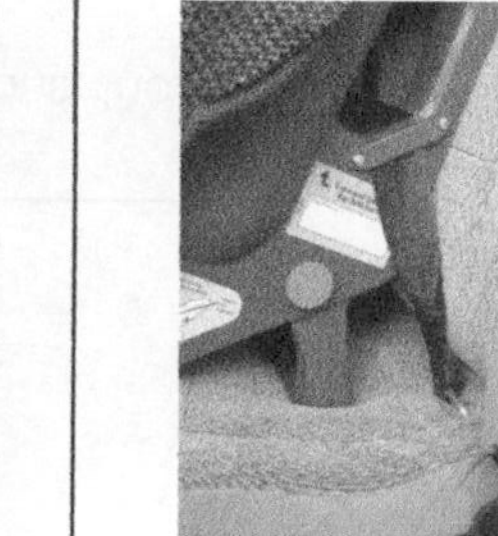

Figure 17.18 (NHTSA)

Use and Installation of Tether Straps

A tether strap is an additional belt that anchors the child safety seat top to the vehicle. A tether reduces the forward movement and rotation of the child safety seat in a crash. A tether strap can be optional or factory installed on a child safety restraint. Most child safety seats made after September 1999 provide tether straps; if a tether strap did not come with your child safety seat, contact the manufacturer to determine if it can be ordered as an option.

Ready-to-Use Tether Anchors, Newer Carsa

Most new vehicles, beginning with models of year 2000 have ready-to-use tether anchors.

1. Locate the tether anchor by reading your vehicle's owner manual.
   Some may be easy to see (as on the rear shelf of a sedan) while others may be more difficult to locate, such as those underneath a rear seat or on the roof of some vehicles (Figure 17.19).
2. Attach the child safety seat tether strap and pull excess webbing (Figure 17.20).

Figure 17.19 (NHTSA)

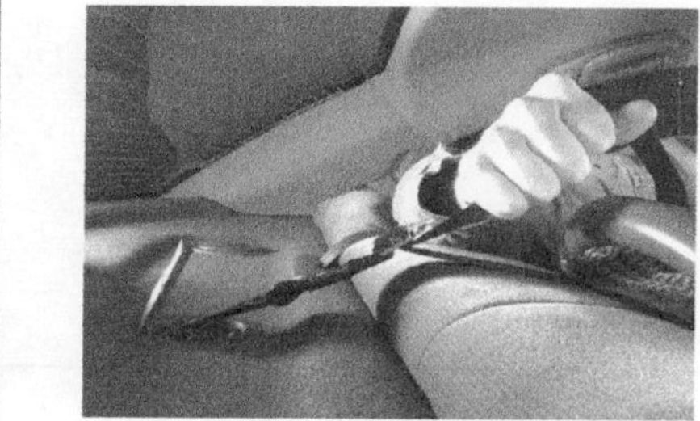

Figure 17.20 (NHTSA)

(continued)

Table 17.2 Installing the LATCH System (*Lower Anchors and Tethers for Children* Standard for Anchoring Vehicles) (Continued)

The LATCH system enables child safety seats to be installed without using seat belts

Tether Anchors in Older Vehicles[b]

1. If your vehicle does not have a ready-to-use tether anchor, read your vehicle's owner manual, which may include information on where to install tether anchors in your vehicle.
   Contact the vehicle dealer or manufacturer customer service line for tether anchor hardware and installation information. Use vehicle-specific parts whenever possible.
2. In some vehicle models, tether anchor hardware is easy to install by consumers themselves.
   In other vehicles, however, special equipment or complex procedures are necessary, so the dealership or a mechanic should do the installation.
   a. 1985 and older vehicles frequently require the drilling of a tether anchor point hole.
      If tether anchor hardware is not available from the vehicle manufacturer for older vehicles, use tether anchor hardware offered by child restraint manufacturers.
   b. 1986–1988 vehicles may have predrilled holes fitted with a plug or filler.
   c. 1989 and more recent model vehicles usually have threaded weld nuts, predrilled holes, or dimpled drill locations designated for tether anchor use.

Steps for installing a tether anchor

a. Identify and find the preferred anchor point.
b. Obtain the tether anchor kit.
c. If necessary, drill a tether anchor hole in the appropriate location.
d. Install the anchor hardware according to manufacturer's instructions.

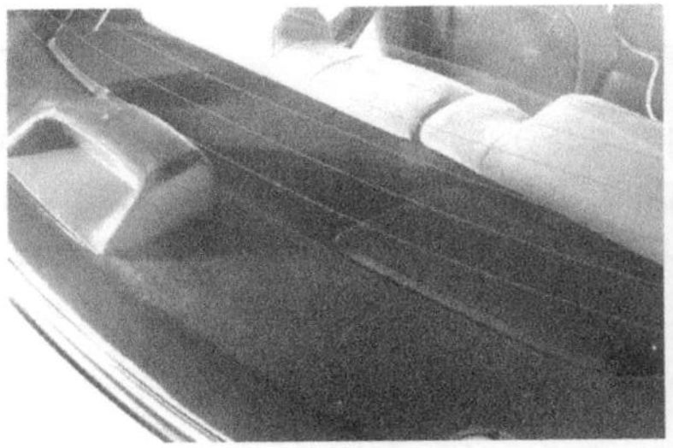

Figure 17.21 (NHTSA)

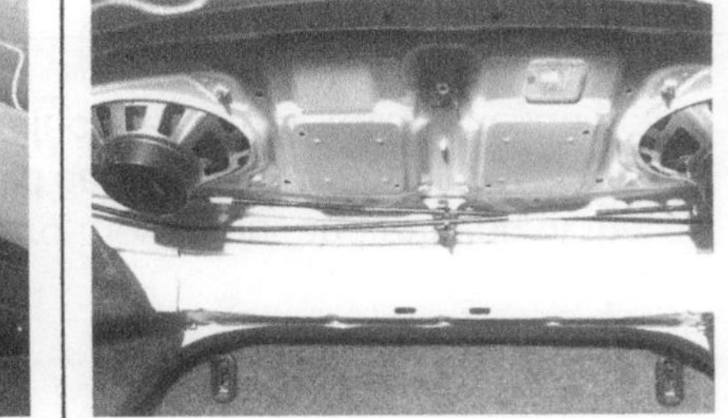

Figure 17.22 (NHTSA)

[a] Most new vehicles, beginning with models in year 2000 have ready-to-use tether anchors.

[b] The information presented in this section is obtained from Tethering Child Restraints, developed by Safe Ride News Publications, Inc.

BELT-POSITIONING BOOSTER SEATS

Although more than 90% of infants and toddlers use child safety seats, only 10%–20% of 4- to 8-year-old passengers are believed to use age-appropriate child restraints* (NHTSA, 2005). Yet, children who are shorter than 4 ft 10 in. (147 cm) and weigh less than 80 lb (36.3 kg) do not fit safely in adult lap and shoulder belts. Staunton et al. (2005) found that most 5- to 8-year-old children are placed in seat belts rather than belt-positioning booster seats.

Belt Positioning Booster Seats raise a child's sitting height to improve the fit with the seat belt system. These reduce the risk of injury in 4- to 7-year-old children by 59% (odds ratio) compared to those using seat belts alone. These reductions were particularly dramatic with injuries to the abdomen, neck or spine or back, and lower extremities (Durbin et al., 2003).

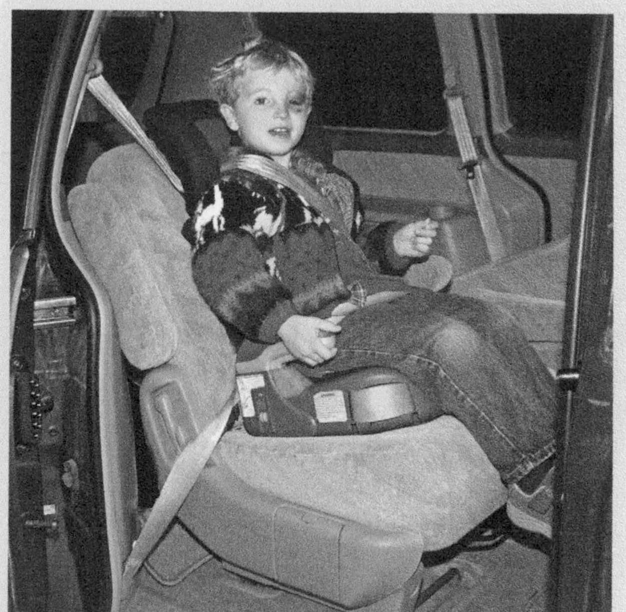

Figure 17.23 (Ursy Potter Photography)
Children should start using booster seats when they grow out of their child safety seats. When used properly, these raise the child so that the combined lap and shoulder belt fit correctly.

Shield booster seats, which have a plastic shield in front of the child, offer less protection and should not be used.

POSITION IN CAR

The safest place for children (and adults) to ride is in the back seat, whether or not the vehicle has air bags† (Berg et al., 2000; Braver et al., 1998, Durbin et al., 2005). Until about the age of 13 (when children can properly fit lap and shoulder belts), children should ride in the back seat.

The seat position is particularly important for infants in child safety seats. If infants must sit in the front row, they are better off facing forward than to the rear because this positions them somewhat farther from the air bag inflation point. In this situation, move the seat of the vehicle as far back as possible.

While properly secured safety seats afford more protection than that associated with sitting in the back seat, the combination of the two is by far the safest

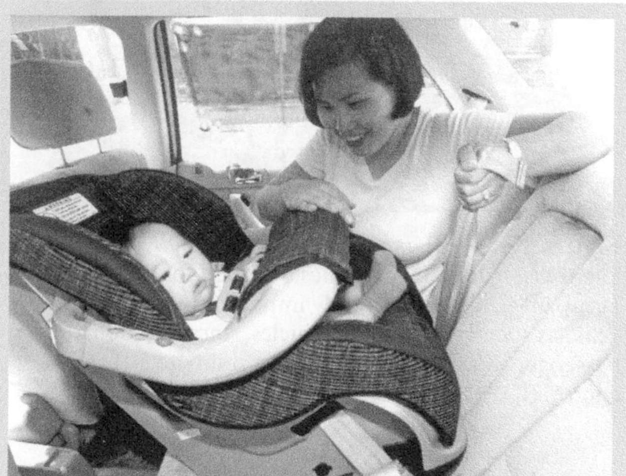

Figure 17.24 (NHTSA)
Child safety seats usually position children several inches forward in the seat. As a result, these children will be closer to the dashboard than adults, further increasing risk of injury.

If the vehicle does not have a rear seat, position children in the front seat facing forward, and as far back as possible from the air bag.

* Staunton et al. (2005) found in their Georgia survey that 94% of 5- to 8-year-olds were not in booster seats.

† While parents can install an on–off switch for air bags to enable children to ride in the front, children are 35% less likely to die while riding in the back, with or without an air bag.

way for children to travel (Berg et al., 2000, Durbin et al., 2005). Until the age of 13 to 15, risk doubled when children were not properly restrained compared with a 40% greater risk imparted by sitting in front (Durbin et al., 2005).

SAFETY FEATURES

SEAT BELTS

Adults who do not use seat belts are less likely to properly restrain their children; almost 40% of children riding with unbelted drivers were also unrestrained (Cody et al., 2002). Therefore, legislation that increases adult use of seat belts should correspondingly increase the use of restraints in children* (Dinh-Zarr et al., 2001).

Motor Vehicle Safety Guidelines

- Every person riding in a car or truck needs his or her own seat belt.

  Do not let passengers ride in storage areas or on other people's laps.
- Children always ride restrained with a car seat or seat belt and in the back seat.
- Infants should ride in rear-facing car seats until they are at least 20 lb (9 kg) and at least 1 year old.

  Do not put a rear-facing car seat in the front seat of a vehicle with an active passenger air bag.
- Children over 1 year old and between 20 lb (9 kg) and 40 lb (18 kg) should ride in forward-facing car seats.
- Children aged 4 to 8 and between 40 lb (18 kg) and 80 lb (36 kg) should ride in booster seats restrained with lap and shoulder belts.

  A regular seat belt will not fully protect a child this size in a crash.
- Children and adults over 80 lb (36 kg) should use a seat belt for every ride.

http://safekids.org

Figure 17.25 (Ursy Potter Photography)
Do not ride double in vehicles designed for one occupant. This is particularly important with vehicles such as all-terrain vehicles, since additional passengers may impair the driver's ability to steer and control (CPSC).

Berg et al. (2000) and others found that older children aged 12 to 14 are more likely to sit unrestrained in cars than younger children, perhaps because parents consider these children old enough to be responsible for securing themselves. Yet these adolescents are still learning safe behaviors and they may consider seat belts "uncool."

* There is a complex relationship between these factors. For example, the increased use of seat belts has not reduced the death rate as much as had been anticipated, given the known effectiveness of seat belts in preventing injuries, perhaps because unsafe drivers are particularly unlikely to use seat belts. See Dinh-Zarr et al. (2001) for a review of the literature on this topic.

AIR BAGS

Most new cars have air bags, which are designed to inflate during frontal crashes, for front-seat passengers.[†] These do not protect from side, rear, and rollover crashes.[‡] Air bags are also considered less effective than seat belts and are intended to supplement rather than replace seat-belt protection for passengers in frontal crashes[§] (Barry et al., 1999; McGwin et al., 2003).

While airbags have saved many lives, these may also in some situations increase risk in adults. When used with lap or shoulder belts, air bags help protect ***older*** children and adults in front seats, but less so than NHTSA originally predicted (Kent et al., 2005).

More sophisticated types of air bags that are not fully understood have been introduced in some vehicles.

- *Depowered air bags* are designed to inflate less forcefully in minor crashes.*
- *Dual and multistaged inflation air bags* aim to increase inflation pressure based on the severity of the crash. The system shuts off, suppresses, or uses less inflation force in situations such as low-speed crashes or if the occupant is out of position.

Frontal Air Bags are installed in the dashboard.

Side Impact Air Bags are intended to provide additional chest protection to adults in many side crashes.

Children seated close to a side air bag may be at risk of serious or fatal injury if the air bag deploys.

Always review manufacturer instructions.

http://nhtsa.dot.gov

Children in front

Children aged 12 and younger should sit in the back. If a child ***must*** ride in the front seat of a vehicle with an active air bag, consider installing an on–off switch. Regardless, children are safer in back.

Figure 17.26 (Ursy Potter Photography)

Figure 17.27 (Ursy Potter Photography)

The air bag warning on the vehicle visor (Figure 17.26 and Figure 17.27) states

- Death or serious injury can occur
- Children 12 and under can be killed by the air bag
- The BACK SEAT is the SAFEST place for children
- NEVER put a rear-facing child seat in the front
- Sit as far back as possible from the air bag
- ALWAYS use SEAT BELTS and CHILD RESTRAINTS

* These have been in most vehicles in the United States since the NHTSA allowed their use in 1997.

† Air bags are designed to inflate in moderate-to-severe frontal and near-frontal crashes when crash force resembles hitting a brick wall head-on at 10-15 mph (16–24 km) or a similar sized vehicle head-on at 20–30 mph (32–48 km).

‡ Unsecured occupants in particular are in danger in rollover crashes; unbelted occupants can be thrown about, striking steering wheels, windows, and other parts of the vehicle. Unbelted passengers are also more likely to be ejected and may be struck by their own or other vehicles.

§ Kent et al. (2005) reviews the literature of field performance of frontal air bags.

Figure 17.28 (Ursy Potter Photography)

Figure 17.29 (Ursy Potter Photography)

Helmets save lives. Without helmets, the risk of head injury doubles.

It is ***not safe*** for children and infants in child safety seats to sit facing rearward while in the front seat of vehicles with air bags. These may also seriously injure unbuckled children or adults who sit too close to the airbag or who are thrown forward. Because the back of a rear-facing child seat sits very close to the dashboard, air bags may also deploy so forcefully as to seriously injure young children and infants.

TYPE OF VEHICLES

The safety of vehicles vary considerably between types and between passengers, pedestrians, and occupants of the other vehicle. Important safety factors include the size and mass of the vehicles, padding, design features that distribute impact forces away from the passenger area, the design of seat belts, air bags, and other safety features.

The risk of children in compact extended-cab pickup trucks is three times that of children in other vehicles, largely because of the greater danger of sitting in the back seat. Children sitting in the back of these trucks are four times more likely to be injured than children sitting in the back of other vehicles (Winston et al., 2002). Some vehicles such as smaller sports utility vehicles (SUVs) are more likely to roll over (Daly et al., 2006; Rivara et al., 2003).

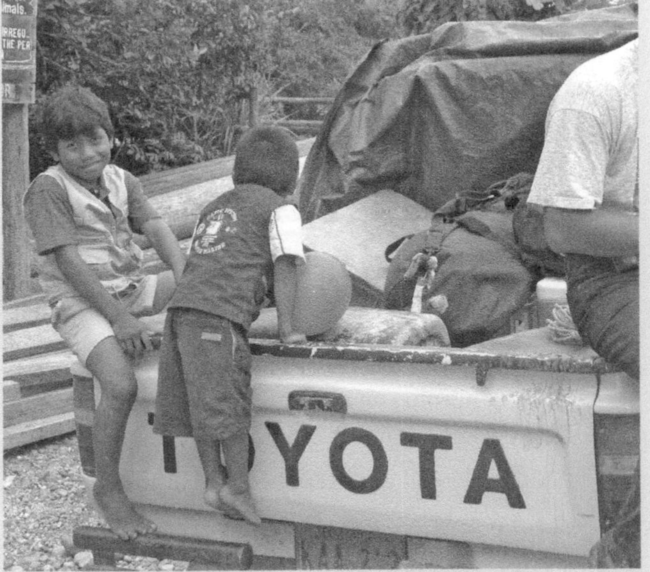

Figure 17.30 (Ursy Potter Photography)
Children traveling in the back of trucks are more likely to be injured in noncrash events such as with quick stops, and are more likely to sustain brain injuries and injuries involving multiple body areas (Agran et al., 1990). Compared with restrained interior occupants, the risk of death for those in the cargo is eight times higher (AAP, 2000).

* Further, pedestrians struck by LTVs have a three times higher risk of severe injuries than passenger vehicles and 2 to 3.4 (adjusted) risk of death. This may be due in part to the front-end design of the vehicle (Roudsari et al., 2004).

Danger is greatest when children ride unrestrained the cargo area (AAP, 2000). Children riding as passengers in light-terrain vehicles (LTVs) are more likely to sustain head, neck, and thorax injuries during quick stops that do not involve a crash* (Roudsari et al., 2004).

PREVENTION

HEAT-RELATED DEATHS

Each year, young children die of heat in vehicles. These are more likely to occur during summer months and may take place in as little as 15 minutes.

1. 27% of deaths resulted from children accessing unlocked vehicles, mostly at home. Some cars were inoperable or abandoned.
   - Almost four out of five of these children were male.
   - Half of these children were between 2 and 3 years old.
2. 73% of heat-related deaths took place after adults left children unattended in vehicles.
   - Most were under 1 year old. Three-fifths were boys.
   - In just over half of these deaths, the adults were not aware of or forgot about the children. At other times, the infants were intentionally left in safety seats to restrain them while the adult was occupied.
3. Sports utility vehicles, vans, and minivans apparently contributed to some deaths.
   - High and tinted windows made it more difficult to see the infants and heavy sliding doors and electric locks may have impeded children's attempts to leave.

Help prevent child deaths (Safe Kids, 2006):

1. Never leave a child in an unattended car, even with windows down.
2. On arrival, make sure all children leave the vehicle. Do not forget sleeping infants.
3. Always lock car doors and trunks, especially when parked in or near home.
4. Keep car keys out of reach and sight.
5. Teach children
   a. not to play in or around cars and
   b. how to disable driver door locks if they become trapped.
6. Help keep kids from getting into the trunk from inside the car by keeping the rear fold-down seats closed.
7. Vehicle
   a. Be cautious with child-resistant locks.
   b. Contact your automobile dealership about retrofitting the vehicle with a trunk release.
8. Pay attention to the temperature
   a. Before restraining children in a car, check the temperature of the child safety seat and safety belt buckles.
   b. Shade the seat of your parked car with a light cover and windows.
9. Call emergency immediately if your child is locked in a car.

Actual rates of these injuries likely differ between cultures. Guard and Gallagher (2005) note that while 10% of parents surveyed in the United States found it "acceptable" to leave young children alone in a vehicle, this behavior was rated second on a list of crimes, "unethical and immoral acts" in Japan.

Figure 17.31 (Ursy Potter Photography)

Figure 17.32 (Ursy Potter Photography)

Figure 17.33 (Ursy Potter Photography)

Figure 17.34 (Ursy Potter Photography)

Many children and adolescents in poverty are exposed to hazardous conditions as they struggle to get past the immediate difficulties in their day-to-day lives. Of the challenges we face, one is to find a way to protect all of our children.

CHILD AND ADOLESCENT DRIVERS

Each year about 85 deaths in the United States occur from motor vehicle crashes where with drivers are aged 7 to 14. Death rates vary considerably in different states* (Frisch et al., 2003).

It is particularly critical that adolescent drivers wear seat belts. They need driver education training that includes driving time and extensive practice with adult supervision to prepare them.

COMMUNITY EFFORTS

There are large variations in child-crash fatalities between states.† While legislation can increase the use of restraints, these efforts were most effective in communities that had prevention programs in place (Ekman et al., 2001).

---

* This analysis excluded crashes with farm equipment, all-terrain vehicles, motorcycles, dune buggies, and go-carts. There were dramatic differences between states; five-year rates were highest in South Dakota, North Dakota, Montana, Wyoming, and Kansas. The highest number of fatal crashes were in Texas, California, Georgia, and Florida.

† Frisch et al. (2003) commented, "despite very large populations of children, New York and Pennsylvania each had small number of crashes with rates only 5% of those in comparably sized Texas and Florida." They attribute such differences between states to differences in driving culture, parent's attitudes and law enforcement.

A review of the effectiveness of community-based interventions to reduce motor vehicle–related injuries and deaths by promoting proper use of child safety seats, increasing the use of safety belts, and reducing alcohol impaired driving is summarized in Table 17.3 (Anstadt, 2002).* In their review, a determination of "insufficient evidence to determine effectiveness" does not mean that the intervention does not work, but rather indicates that additional research is needed to determine whether the intervention is effective.

One out of four of all occupant deaths among children aged 0 to 14 years involve a drinking driver. More than two-thirds of these fatally injured children were riding with a drinking driver (Shults, 2004).

Table 17.3 The Effectiveness of Community Interventions to Reduce Motor Vehicle Injuries

Use of Child Safety Seats	(Anstadt, 2002)
Child Safety Seat Laws	Recommended (Strong evidence)
Community Wide Info & Enforcement	Recommended (Sufficient evidence)
Distribution & Education Campaigns	Recommended (Strong evidence)
Incentive & Education Programs	Recommended (Sufficient evidence)
Education-Only Programs	Insufficient evidence
Use of Safety Belts	
Safety Belt Laws	Recommended (Strong evidence)
Primary Enforcement Laws	Recommended (Strong evidence)
Enhanced Enforcement	Recommended (Strong evidence)
Reducing Alcohol-Impaired Driving	
08 Blood Alcohol Content (BAC) Laws	Recommended (Strong evidence)
Lower BAC Laws for Inexperienced Drivers	Recommended (Sufficient evidence)
Minimum Legal Drinking Age Laws	Recommended (Strong evidence)
Sobriety Checkpoints	Recommended (Strong evidence)
Intervention Training Programs for Servers	Recommended (Sufficient evidence)
Mass Media Campaigns	Recommended (Strong evidence)
School-Based Programs	
School-Based Instructional Programs	Recommended (Sufficient evidence)
Peer Organizations	Insufficient evidence
Social Norming Campaigns	Insufficient evidence
Designated Driver Programs	
Population-Wide Promotion Campaigns	Insufficient evidence
Incentive Programs	Insufficient evidence

* This review focus on the effectiveness of these interventions to findings of this review, published in the *American Journal of Preventative Medicine* was later summarized in a Community Guide. The full report is available at http://thecommunityguide.org and http://childergo.com/recc-children.htm

CONCLUSION

Motor vehicle injuries are the leading cause of death in children at every age after their first birthday and their injuries are more severe than those of adults. Yet, shocking numbers of children sit in the front seat, lack appropriate child safety seat, or their seats are not properly set up. Many children ride unrestrained, or ride in lap and shoulder belts that do not fit them properly. Such hazardous conditions cause many deaths that would easily be avoided.

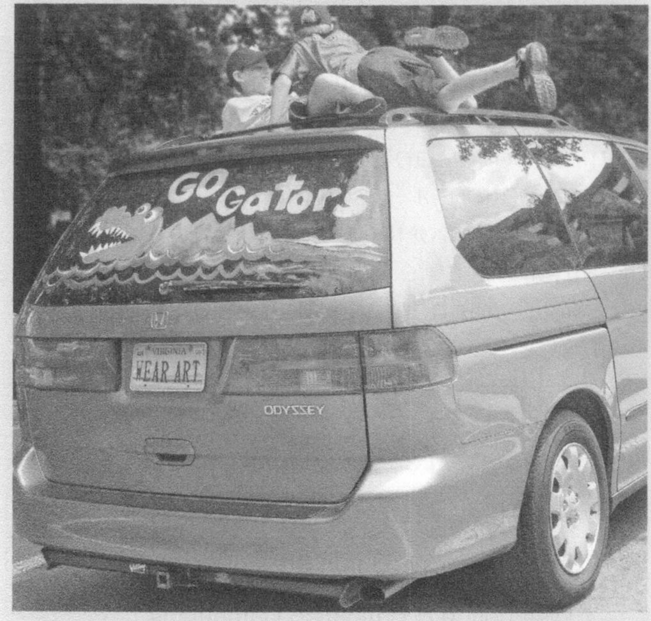

Figure 17.35 (Ursy Potter Photography)

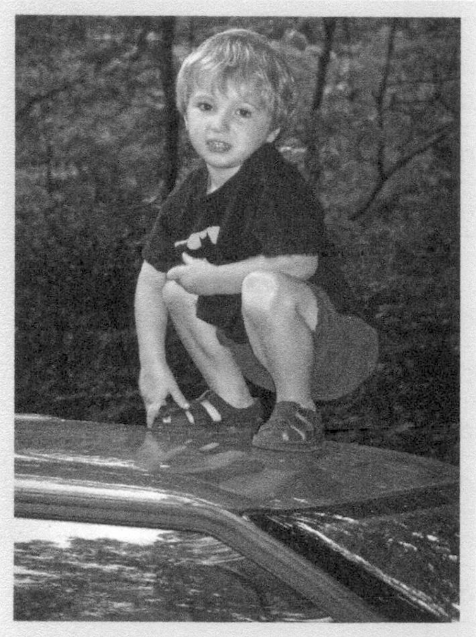

Figure 17.36 (Ursy Potter Photography)

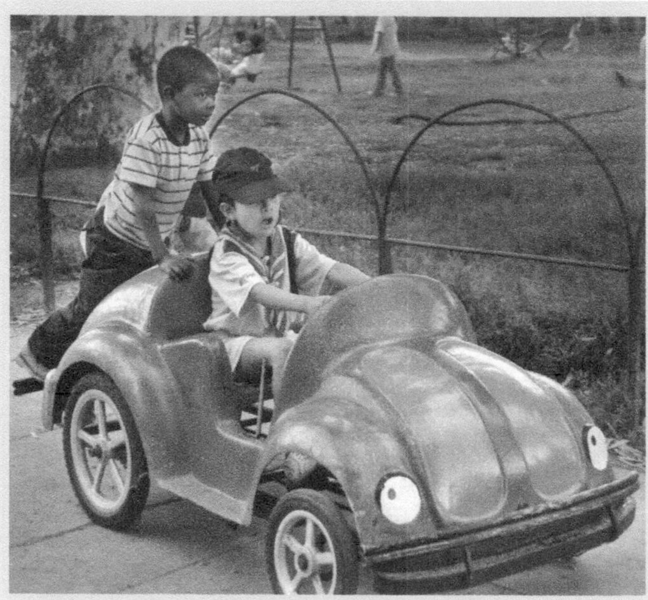

Figure 17.37 (Ursy Potter Photography)

REFERENCES

AAP (2000) Children in pickup trucks. American Academy of Pediatrics. Committee on Injury and Poison Prevention. *Pediatrics*, 10(4), 857–859.

AAP (2001) American Association of Pediatrics. School bus transportation of children with special health care needs. *Pediatrics*, 108(2), 516–518.

Agran, P.F., Winn, D.G., and Castillo, D.N. (1990) Pediatric injuries in the back of pickup trucks. *JAMA*, 264(6), 712–716.

Aitken, M.E., Graham, C.J., Killingsworth, J.B., Mullins, S.H., Parnell, D.N., and Dick, R.M. (2004) All-terrain vehicle injury in children: strategies for prevention. *Inj. Prev.*, 10(5), 303–307.

Anstadt, G. (2002) The guide to community preventive services: Reducing injuries to motor vehicle occupants; systematic reviews of evidence. Recommendations from the Task Force on Community Preventive Services, and expert commentary. *Am. J. Prev. Med.*, 22(4), 330.

Barry, S., Ginpil, S., and O'Neill, T.J. (1999) The effectiveness of air bags. *Accid. Anal. Prev.*, 31(6), 781–787.

Berg, M.D., Cook, L., Corneli, H.M., Vernon, D.D., and Dean, J.M. (2000) Effect of seating position and restraint use on injuries to children in motor vehicle crashes. *Pediatrics*, 105(4) Pt 1, 831–835.

Blank, D. (1999) Kids in the back seat: Brazil's strides in enforcing its new traffic law. *Inj. Prev.*, 5(1), 77.

Braver, E.R., Whitfield, R., and Ferguson, S.A. (1998) Seating positions and children's risk of dying in motor vehicle crashes. *Inj. Prev.*, 4(3), 181–187.

Brewer, R.D., Morris, P.D., Cole, T.B., Watkins, S., Patetta, M.J., and Popkin, C. (1994) The risk of dying in alcohol-related automobile crashes among habitual drunk drivers. *N. Engl. J. Med.*, 331, 513–537.

CDC (2000) Centers for Disease Control and Prevention. Motor-vehicle occupant fatalities and restraint use among children aged 4–8 years—United States, 1994–1998. *JAMA*, 283(17), 2233–2234.

CDC (2002a) Centers for Disease Control. Ten leading causes of death. Atlanta, Georgia: U.S. Department of Health and Human Services, Public Health Service. http://cdc.gov

CDC (2006) *Injury Fact Book*. National Center for Injury Prevention and Control. Centers for Disease Control and Prevention. Atlanta, November 2006. 12.1 ps. www.cdc.gov

Chen, I.G., Durbin, D.R., Elliott, M.R., Kallan, M.J., and Winston, F.K. (2005) Trip characteristics of vehicle crashes involving child passengers. *Inj. Prev.*, 11(4), 219–224.

Cody, B.E., Mickalie, A.D., Paul, H.P., and Coleila, J.M. (2002) *Child passengers at risk in America: A national study of restraint use*. Washington DC. Safe Kids campaign.

Daly, L.M., Kallan, M.J., Arbogast, K.B., and Durbin, D.R. (2006) Risk of injury to child passengers in sport utility vehicles. *Pediatrics*. 117(1), 9–14.

Dinh-Zarr, T.B., Sleet, D.A., Shults, R.A., Zaza, S., Elder, R.W., Nichols, J.L., Thompson, R.S., and Sosin, D.M. (2001) Reviews of evidence regarding interventions to increase the use of safety belts: Task Force on Community Preventive Services. *Am. J. Prev. Med.*, 21(4), 48–65.

Durbin, D.R., Kallan, M., Elliott, M., Arbogast, K.B., Cornejo, R.A., Winston, F.K. (2002) Risk of injury to restrained children from passenger air bags. *Annu. Proc. Assoc. Adv. Automot. Med.*, 46, 15–25.

Durbin, D.R., Elliott, M.R., and Winston, F.K. (2003) Belt-positioning booster seats and reduction in risk of injury among children in vehicle crashes. *JAMA*, 289(21), 2835–2840.

Durbin, D.R., Chen, I., Elliott, M., and Winston, F.K. (2004) Factors associated with front row seating of children in motor vehicle crashes. *Epidemiology*, 15(3), 345–349.

Durbin, D.R., Chen, I., Smith, R., Elliott, M.R., and Winston, F.K. (2005) Effects of seating position and appropriate restraint use on the risk of injury to children in motor vehicle crashes. *Pediatrics*, 115(3), e305–e309.

Ekman, R., Welander, G., Svanstrom, B.L., and Schelp, L. (2001) Long-term effects of legislation and local promotion of child restraint use in motor vehicles in Sweden. *Accid. Anal. Prev.*, 33(6), 793–797.

Frisch, L., Coate Johnston, S., Melhorn, K., Phillips Hill, C., and Boyce, M. (2003) In the hands of children: Fatal car, van, and truck crashes involving drivers aged 7 through 14 years. *Arch. Pediatr. Adolesc. Med.*, 157(10), 1032.

Guard, A. and Gallagher, S.S. (2005) Heat related deaths to young children in parked cars: An analysis of 171 fatalities in the United States, 1995–2002. *Inj. Prev.*, 11(1), 33–37.

Kent, R., Viano, D.C., and Crandall, J. (2005) The field performance of frontal air bags: a review of the literature. *Traffic Inj. Prev.*, 6(1), 1–23.

MADD (2004) Mothers against drunk driving. *Alcohol and Child Safety Laws*. Irving, TX: Mothers Against Drunk Driving. www.madd.org

McGwin, G. Jr., Metzger, J., Alonso, J.E., and Rue, L.W. III. (2003) The association between occupant restraint systems and risk of injury in frontal motor vehicle collisions. *J. Trauma*, 54(6), 1182–1187.

NHTSA (1996a) National Highway Traffic Safety Administration. Third Report to Congress: Effectiveness of Occupant Protection Systems and Their Use. US Department of Transportation.

NHTSA (1996b) National Highway Traffic Safety Administration. Police Accident Report Quality Assessment Project. Washington, D.C.: US Department of Transportation. DOT-HS-808-487.

NHTSA (2002a) National Highway Traffic Safety Administration. Are you using it right? Report DOT HS 809 245. US Department of Transportation.

NHTSA (2002b) National Highway Traffic Safety Administration. Transitioning to multiple imputations—A new method to impute missing blood alcohol concentration (BAC) values in FARS. Washington, D.C.: US Department of Transportation, National Center for Statistics and Analysis. DOT-HS-809-403.

NHTSA (2005) National Highway Traffic Safety Administration. Improving the safety of older-child passengers: A progress report on reducing deaths and injuries among 4- to 8-year-old child passengers. Washington, DC: US Department of Transportation.

NHTSA (2005) *Traffic Injury Facts, 2005 data.* National Highway Traffic Safety Administration. DOT HS 810 618. www.nhtsa.gov

Ostrom, M., Huelke, D.F., Waller, P.F., Eriksson, A., and Blow, F. (1992) Some biases in the alcohol investigative process in traffic fatalities. *Accid. Anal. Prev.*, 24, 539–545.

Quinlan, K.P., Brewer, R.D., Sleet, D.A., and Dellinger, A.M. (2000) Characteristics of child passenger deaths and injuries involving drinking drivers. *JAMA*, 283, 2249–2252.

Rivara, F.P., Cummings, P., and Mock, C. (2003) Injuries and death of children in rollover motor vehicle crashes in the United States. *Injury Prevention,* 1, 76–80.

Roudsari, B.S., Mock, C.N., Kaufman, R., Grossman, D., Henary, B.Y., and Crandall, J. (2004) Pedestrian crashes: Higher injury severity and mortality rate for light truck vehicles compared with passenger vehicles. *Inj. Prev.*, 10(3), 154–158.

Safe Kids (2006) National safe kids campaign. *Heat Entrapment.* http://usa.safekids.org

Shults, R.A., Elder, R.W., Sleet, D.A., Nichols, J.L., Alao, M.O., Carande-Kulis, V.G., Zaza, S., Sosin, D.M., and Thompson, R.S. (2001) Reviews of evidence regarding interventions to reduce alcohol-impaired driving. *Am. J. Prev. Med.*, 21(4) Suppl, 66–88.

Staunton, C., Davidson, S., Kegler, S., Dawson, L., Powell, K., and Dellinger, A. (2005) Critical gaps in child passenger safety practices, surveillance, and legislation: Georgia, 2001. *Pediatrics*, 115(2), 372–379.

Taft, C.H., Michalide, A.D., and Taft, A.R. (1999) *Child Passengers at Risk in America: A National Study of Car Seat Misuse.* Washington, D.C.: National SAFE KIDS Campaign.

Wells-Parker, E., Bangert-Drowns, R., McMillen, R., and Williams, M. (1995) Final results from a meta-analysis of remedial interventions with drink/drive offenders. *Addiction*, 90, 907–926.

Winston, F.K., Kallan, M.J., Elliott, M.R., Menon, R.A., and Durbin, D.R. (2002) Risk of injury to child passengers in compact extended-cab pickup trucks. *JAMA*, 287(9), 1147–1152.

Zador, P.L., Lund, A.K., Fields, M., and Weinberg, K. (1989) Fatal crash involvement and laws against alcohol-impaired driving. *J. Public Health Policy*, 10, 467–485.

CHAPTER 18

PREVENTING MUSCULOSKELETAL DISORDERS FOR YOUTH WORKING ON FARMS

THOMAS R. WATERS

TABLE OF CONTENTS

INTRODUCTION

Children and adolescents are essential to the farming workforce. Children begin helping at very young ages, performing physically demanding jobs designed for adults that may exceed their capabilities.

Children often lift and move materials, operate farm equipment, and perform other tasks that require considerable strength and coordination (Figure 18.1 and Figure 18.2). Nearly 2 million American youth, younger than age 20, lived or worked on a US farm in 1998 (Table 18.1).

Figure 18.1 (NIOSH) **Figure 18.2** (Ursy Potter Photography)
Children who work on farms perform physically demanding tasks that place them at risk of developing musculoskeletal disorders. The above tasks require the youth to perform heavy lifting and forceful muscle exertions while working in awkward overhead postures. The physical demands are significantly increased if the work is highly repetitive.

Farms are one of the most hazardous places for anyone to work. In the United States, about 8% of adult farm workers sustain injuries involving lost work-time each year (OSHA, 2001). Though we know less about injuries among farm children, they are clearly at considerable risk (Xiang et al., 1999; Allread et al., 2004).

Table 18.1 Ages of Children Working on US Farms

Age (Years)	Percent (%)	Number
<10	9	130,000
10–15	47	650,000
16–19	53	745,000

Source: Meyers and Hendricks (2001).

INJURIES AMONG CHILDREN

Farm children may sustain musculoskeletal injuries from acute events such as traumatic injuries or by repeated or sustained physical exertions.

TRAUMATIC INJURIES

Traumatic injuries typically result from sudden events. These include motor-vehicle accidents; slips, trips, or falls; and being struck or hit by animals or objects. Traumatic injuries may include lacerations, contusions, broken bones, or extremities caught in machinery.

MANY CHILD INJURIES ON FARMS ARE PREVENTABLE

One study concluded that half the injuries among farm children could have been avoided by following currently existing guidelines. Although most studies focus on farming, many youth also work in commercial or production agriculture, where serious injuries also occur. Unfortunately, there is a lack of formal reporting systems to capture information on the types and causes of injuries children incur while working on farms or in commercial enterprises.

MUSCULOSKELETAL DISORDERS RELATED TO WORK

Musculoskeletal disorders (MSDs) are caused by excessive physical demands or repetitive exertions (Table 18.2). These MSDs can affect the upper and lower back, upper extremities (hands, wrists, and elbows), shoulder, neck, and lower extremities (legs, ankles, and feet).

Table 18.2 Description of Musculoskeletal Disorders and Their Causes

What Are Musculoskeletal Disorders?	
Definition	Health conditions or disorders, which involve the muscles, nerves, ligaments, tendons, joints, cartilage, spinal discs, and other supporting structures of the body that are caused or exacerbated by a person's tasks and activities (NIOSH, 1997).
Risk Factors	• High repetition • Extended use of awkward postures • Long-duration static hold • Direct pressure on soft tissues of the body • Heavy loads • Exposure to vibration
Symptoms	Tingling, numbness, pain
Synonyms	Overuse injuries, repetitive motion injuries, repetitive strain injuries, cumulative trauma disorders

Farm workers are at risk of developing MSDs, particularly of the low back. In fact, farming ranks fifth among jobs at high risk of developing back pain. About 19% of farm workers experience work-related back pain that lasts a week or more.* This is more than double the rate among the general working population (9% for males) (Guo et al., 1995).

Characteristics of industrial work that contribute to injuries† (particularly back injuries) are similar to those often found in farming (Allread et al., 2004). Youth working on farms face these very same risks (Ehrmann-Feldman et al., 2002). Many farming jobs are physically demanding, often requiring lifting, pushing, pulling, and carrying of heavy bags, buckets, and other objects. Examples of farm jobs performed by youth that may lead to MSDs are listed in Table 18.3 (Bartels et al., 2000). Other examples of tasks performed by youth on farms are shown in Figures 18.5 through 18.8.

Figure 18.3 (NIOSH)
Physically demanding tasks on farms include pushing, pulling, lifting, and carrying heavy objects. Most children who work on farms develop strong muscles and bones through the continued use of their bodies over time, yet there can also be a tendency for them to push themselves to do as much as an adult without asking for assistance.

Teaching children to recognize signs and signals of overuse, as well as when and how to push their own physical limitations safely, can help prevent injuries. Supervisors must take children's reports of pain or other signs of overuse seriously, so appropriate rest or treatment can occur.

Children who work on farms care for large and small animals; clean animal stalls, pens, and corrals; load, unload, and stack hay; and drive tractors with trailers. Aside from the tractor-driving task, these activities are physically demanding. They involve heavy force exertions, as well as repetition and awkward movements, all of which are risk factors in the development of MSDs among adults (NIOSH, 1997).

* This does not include horticulture workers.

† These task requirements include the weight of objects lifted, frequency of lift, postures, and other aspects of work that are known to increase risk of injury.

Table 18.3 Jobs Performed by Youth on Farms that May Lead to MSDs

Work with Animals	• Feed and water livestock daily • Milk cows • Spread straw in animal pens • Bottle or bucket feed milk to calves or lambs • Move sows and pigs from pen to pen • Sort animals	
Operating Equipment	• Drive tractor with no specific implement attached • Operate a lawn mower • Drive tractor to plant or drill crops • Drive a tractor raking hay • Drive tractor to till the soil • Operate combine to harvest corn, soybeans, or wheat • Operate a skid loader to clean pens • Drive tractor and grain wagon to barn, and then unload grain into bin • Drive tractor to pull wagons of hay to barn • Drive tractor to hay baler • Drive truck around the farm	
Moving Materials (Manual Material Handling)	• Stacking bales of hay or straw on wagon • Place bales on an elevator off the wagon • Stack bales in the barn • Shovel, fork, or scrape manure to clean barn • Shovel grain or silage	
Tasks that Require Awkward Postures	• Pull and sort sweet corn • Weed vegetables with hand hoe • Pick up rocks	

Of course, not all physically demanding work is harmful. Physical activity builds bones and muscles and improves physical endurance. Parents and caregivers of children who do not live on farms encourage them to engage in gym and sports activities to develop similar levels of fitness! However, the way a child accomplishes a task and the matching of a child's capabilities with the demands of the task are important to avoid injury and possible permanent damage.

Children of different ages have specific developmental abilities that allow them to perform certain jobs with less risk of injury (Murphy and Hackett, 1997). Table 18.4 lists the various growth stages of children, their ages, developmental characteristics, types of injuries or fatalities that occur at each stage, prevention suggestions, and developmentally appropriate work tasks for each age group.

According to the report by Murphy and Hackett (1997):

> *The developmental characteristics, types of injuries or fatalities and preventive strategies for each age group are more scientifically grounded or experienced based than are the developmentally appropriate work tasks suggested. Task suggestions for the age groups may not exactly describe tasks or be suitable for all children: Each child is a unique individual and may not perfectly fit within any grouped criteria or classification.*
>
> *On the other hand, it is normal for parents to overestimate the skills and abilities of their own children. The developmentally appropriate work task suggestions for each age group represent the best opinions of several child development and farm safety experts.*

Parents and supervisors of youth working on farms need nonthreatening information on MSDs, risk factors and methods to decrease their children's exposure to risk factors. This will help them make wise decisions about the tasks their children perform that they may not have considered previously.

Excessive physically demanding work as a child may also increase the risk of latent, chronic musculoskeletal disorders, such as osteoarthritis, in adulthood. Although adults experience these disorders more often than children, the initial insult can occur during childhood. Figure 18.4 illustrates one theoretical model of this potential chronic injury pathway.

According to the theoretical model, repetitive exposure to large muscle exertions likely leads to significant increases in bone mineral density and bone stiffness. Mechanically,

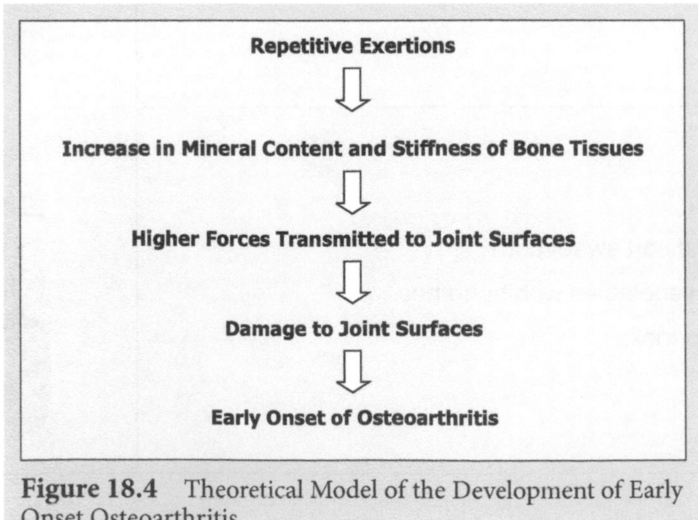

Figure 18.4 Theoretical Model of the Development of Early Onset Osteoarthritis.

stiffer bones likely would lead to increased transmission of forces through the bone to the joints. Over many years, these increased loadings may cause wear and tear or micotrauma to the musculoskeletal system, resulting in chronic early onset of chronic MSD symptoms such as osteoarthritis (Thelin et al., 2004).

Figure 18.5 (Ursy Potter Photography) **Figure 18.6** (Ursy Potter Photography)
Farm work involves many chores: planting, growing, harvesting, transporting, and often selling. Children in farming families help where and when they can, often with each part of the process. In the photo on the left, two young boys sort produce to sell at the market.

Although the work is not a physically heavy task, it requires repetitive motions of the small muscles of the fingers and hands (which fatigue quicker than larger muscle groups) and, in this case, the boys' posture during the process can be fatiguing. The photo on the right shows that even the youngest children are involved in the family business of farming and selling.

Figure 18.7 (Ursy Potter Photography)
This father has placed his little girl on the hood of his vehicle, so she is not "underfoot" of the llamas he has taken to a market. Even when the children are not actively engaged in the care of the animals, children in farming families are around the animals and daily work activities. Many parents cannot imagine doing their job and keeping an eye on their children at the same time, but farming families do so every day.

Figure 18.8 (Ursy Potter Photography)
Transporting animals to market can be dangerous work. Getting the animals from the pen or barn to the vehicle and then into the vehicles takes careful planning and firm, even handling of these large, sometimes resistant animals. It can be difficult for an adult, as well as a child.

This boy is handling nine pigs at once. Although it may work out just fine, if one or several of them move in an unanticipated direction, he could have difficulties. In fact, two pigs are moving in a different direction than the rest.

What types of farms are children injured on?

Livestock (2.17 per 100)

Crop operations (1.59 per 100)

Who is injured?

Children younger than 10 years of age are injured most frequently.

Boys are injured more often than girls.

The most common type of injury among youth is a sprain or strain.

http://www.fs4jk.org/

Table 18.4 Child Development and Appropriate Work Tasks from the Penn State Report

Child Development and Appropriate Work Tasks				
Growth Stage (in Years)	**Developmental Characteristics**	**Causes of Deaths or Injuries**	**Preventive Strategies**	**Developmentally Appropriate Work Tasks**
Birth–4 (Infant/ Toddler/ Preschooler)	• Rapid growth, beginning motor skills development • Has balance problems, slow reaction time • Is curious, exploring • Is fascinated by movement • Has illogical or "magic" thinking • Is very energetic, releases tension by playing, even when exhausted • Is self-centered but interested in group activities	• Falling from tractors or heights, such as ladders • Ingesting poisons • Being kicked or trampled by animals • Being run over by a tractor • Drowning in ponds or manure pits	• Never have a child as an extra rider. • Use strong physical barriers such as locks and fences around ponds and manure pits. Lock up chemicals. • Store ladders out of sight and reach. • Provide a fenced-in play area away from farming activities. • Provide maximum supervision at all times because of small children's poor coordination, high energy, and lack of fear.	• None, children this age should not be exposed to work hazards
5–9 (Preschooler/ Early Elementary School Age)	• Is learning to use small and large muscles—slow, steady growth stage • Has poor hand–eye coordination • Tries to master more complex skills • Operates with concrete facts, not capable of abstract ideas or thinking • Wishes to appear competent; seeks parental approval • Wishes to take on tasks without adult supervision • Is discovering that parents make mistakes, are human • Rarely follows through on a task—not yet ready for responsibility	• Slipping and falling from tractors, trucks, or heights • Entangling in augers, other machines • Suffocating in grain • Being kicked or trampled by animals	• Set rules. • Discuss safe behavior with children. • Assign and closely supervise chores. • Talk openly about types of injuries and consequences. • Never assign intense, physical chores—they can lead to exhaustion. • Play games (with adult supervision) that focus on farm safety issues. • Use Job Safety Analysis (JSA) tools.	• Tasks of short duration that do not require excellent hand–eye coordination • Projects with hand tools, not power tools • Help with watering plants and feeding small animals, such as pets or orphaned baby animals • Collect eggs

(continued)

Table 18.4 Child Development and Appropriate Work Tasks from the Penn State Report (Continued)

Child Development and Appropriate Work Tasks				
Growth Stage (in Years)	**Developmental Characteristics**	**Causes of Deaths or Injuries**	**Preventive Strategies**	**Developmentally Appropriate Work Tasks**
10–13 (Middle School Age/ Early Teen)	• Is growing at a steady rate—approaching puberty; boys grow more quickly than girls • Small muscles are developing rapidly • Has same coordination as adults but lapses of awkwardness are common • Has greater physical and mental skills • Desires peer and social acceptance • Wishes to try new skills without constant adult supervision • Signs of independence emerging • Success important for self-concept	• Becoming entangled with machinery • Hearing loss from exposure to noisy machinery • Injuring head or spine in motorcycle and all-terrain vehicle accidents • Extra rider falling from tractor or other equipment	• Potentially the most dangerous age because of constant risk taking and ease of distraction and clumsiness—never mistake a child's size for ability to do work! • Enroll child in bike safety classes; always require helmets. • Set clear and consistent rules; discuss consequences and rewards. • Provide specific education on farm hazard prevention. • Plan increases in chores and responsibilities. • Start with low-risk tasks; give more responsibility for follow-through with less supervision. • Use JSA.	• Hand raking, digging • Limited power tool use (with supervision); hand tools better • Operating lawn mower (push mower, flat surface, under supervision) or garden tractor • Handling and assisting with animals

Table 18.4 Child Development and Appropriate Work Tasks from the Penn State Report (Continued)

Child Development and Appropriate Work Tasks				
Growth Stage (in Years)	**Developmental Characteristics**	**Causes of Deaths or Injuries**	**Preventive Strategies**	**Developmentally Appropriate Work Tasks**
13–16 Adolescent/ Young Teenagers)	• Is growing rapidly and changing physically; can be an uneasy time • Girls growing faster than boys • Has moved from concrete thinking to abstract; enjoys mental activity • Can find solutions to own problems but still needs adult guidance • Feels need to be accepted by peers • Resists adult authority • Feels immortal	• Hearing loss from exposure to loud machinery • Head and spine injuries from motor-cycle or all-terrain vehicle accidents • Machinery rollover or roadway accident • Amputation due to power take-off (PTO) entan-glement	• Judge size, age, cognition, and judgment to assess maturity for tasks. • Be consistent with rules. • Provide education from peers with farm injuries. • Provide all-terrain vehicle training, protective gear. • Become involved in 4-H and FFA safety projects. • Use JSA.	• Still needs adult supervision but ready for more adult jobs such as equipment operation and maintenance • Gradually increase tasks as experience is gained • Manual handling of feed and feeding animals • Can operate a tractor over 20 PTO horsepower or connect and discon-nect parts to or from tractor at ages 14 and 15 after the completion of a 10 hour training program • Can assist with and operate (including stopping, adjusting, and feeding) the following after completing a 10 hour training program: corn picker, cotton picker, grain com-bine, hay mower, forage harvester, hay baler, potato digger, mobile pea viner, feed grinder, crop dryer, forage blower, auger conveyor, the unloading mechanism of a nongravity-type self-unloading wagon or trailer, power post-hole digger, power post driver or nonwalking rotary tiller

(continued)

Table 18.4 Child Development and Appropriate Work Tasks from the Penn State Report (Continued)

Child Development and Appropriate Work Tasks				
Growth Stage (in Years)	**Developmental Characteristics**	**Causes of Deaths or Injuries**	**Preventive Strategies**	**Developmentally Appropriate Work Tasks**
16–18 (Middle/Older Teenage)	• Awkwardness overcome, mastery of small and large muscles basically complete • Knows abilities, moving further away from family and into community as an independent person • Feels immortal • May act like child one day, adult the next • Rebellion, risk-taking, and aggressiveness are typical behaviors • Consistent treatment from adults important • Needs independence and identity • Has increased sense of adult responsibilities, thinking of future • May experiment with drugs or alcohol	• Same as adult risks: respiratory illness, hearing loss, muscle and bone injuries, rollover from tractor, machinery entangle ments • Additional risk if experimenting with or under the influence of drugs or alcohol	• Provide rules regarding drugs and alcohol; open communication. • Reward for accepting adult responsibilities. • Serve as role model—teach younger children farm safety. • Parents may still have cause for concern with recklessness and risk-taking and may work side-by-side with young adult until they are absolutely ready to work independently. • Use JSA.	• May be ready to work with tractors, self-propelled machinery, augers, elevators, and other farm equipment, but must earn this responsibility. Should be trained, educated and supervised at regular intervals

Source: Murphy, D. and Hackett, K. Children and Safety on the Farm. Ps6361, U.Ed.AGR97-25, 1997. Reprinted with permission.

Farmers and farm supervisors need to understand that MSDs are real and not "just part of doing business." They also need to learn how to recognize and avoid risk factors. Doing so would enable them to more safely structure tasks and assign jobs to children (Bartels et al., 2000).

NONWORK INJURIES

Figure 18.9 (Ursy Potter Photography)

Figure 18.10 (Ursy Potter Photography)

Figure 18.11 (Ursy Potter Photography)

Injury and death occur even when farm children are not working. Although little is known about nonwork injuries among children on farms, it is a serious problem in farming populations.

Accidental injuries occur to children who are not actively working on the farm. Often, they are merely playing near their working parents when the injury occurs.

Nonwork injuries, such as those listed in Table 18.5, occur more often among children from 1 to 6 years old and the victims tend to be relatives of the farm owners (Pickett et al., 2005). Hartling et al. (2004) reviewed the effectiveness of recommendations for preventing nonwork injuries to youth.

Table 18.5 Types of Nonwork Injuries, Accidents, and Deaths, along with Potential Solutions

Causes	Activities	Potential Solutions
• Bystander and passenger runovers • Drowning • Machinery entanglements • Falls from heights (often from haylofts) • Animal trauma	• Playing at the worksite • Being an extra rider on farm machinery or being a bystander • Recreational horseback riding	1. Keep children away from the worksite while adults work. 2. Provide alternate childcare so adults do not need to divide their attention between supervising children and doing their work. 3. Install passive physical barriers such as fencing such as around animals or in haylofts (Figure 18.14). 4. Store materials carefully, such as large items that might fall. 5. Institute guidelines for recreational riding of horses and all terrain vehicles.

Source: From Pickett, W., Brison, R.J., Berg, R.L., Zentner, J., Linneman, J., and Marlenga, B. (2005). *Inj. Prev.*, 11, 6–11, 2005.

Figure 18.12 (Ursy Potter Photography)
Many farming families do not have the luxury of daycare or hiring babysitters for their children while they work. Instead, their children accompany them. This means potential exposure to safety hazards associated with farm machinery or toxic chemicals such as pesticides.

Figure 18.13 (Ursy Potter Photography) **Figure 18.14** (Ursy Potter Photography)
Careful instruction of how to interact with animals and emphasizing that animals can be unpredictable in their behavior can help prevent accidental injuries. However, young children are distractible and do not fully understand the consequences of their actions (see chapters in Section B of this book). When possible, it is best to place barriers such as fences around livestock to prevent accidental injuries among younger children. This separates the child from the hazardous situation.

PERCEPTIONS DRIVE ACTIONS

Our perceptions drive our actions. Adults hire, supervise, and assign youth working on farms to jobs according to their perceived maturity, rather than their chronological age (Bartels et al., 2000). The young workers believe adults base their assignments on need (i.e., the task must be completed, and they are the only ones available to do it), rather than their capability or readiness to perform the task.

It is likely that adults consider both perceived maturity and need when making task assignments. However, parents and supervisors may not consider a young person's stature, strength, and stage of bone and muscular development when assigning their work. The adults may not have the knowledge or the background to understand why such considerations are important.

Table 18.6 lists the activities thought to cause the most serious MSDs in children during farm work.* Children and adolescents rarely report musculoskeletal disorders. However, they describe muscle aches and strains of the legs, arms, shoulder, back, or neck as everyday occurrences. Young workers often believe that "if it is not broken, you are fine."

Table 18.6 Causes of Serious Musculoskeletal Disorders in Children as Reported by Parents and Supervisors of Children Working on Farms

	Causes of Serious Child Musculoskeletal Disorders
Primary	• Lifting objects • Forking • Shoveling
Secondary	• Bending over while working • Sitting in an awkward position looking back at equipment from a tractor • Sitting in a cramped position, looking down at a combine header • Long hours of work

Source: From Bartels, S., Niederman, B., and Waters, T. *J. Agric. Saf. Health*, 6, 191, 2000.

Parents and supervisors generally do not believe special training is necessary for youth working on a farm; they believe youth learn best by observing. They also believe that parents and supervisors would ignore guidelines for assigning youth to work tasks, unless required by law, even if physicians made the recommendations (Bartels et al., 2000).

Such perceptions suggest parents do not understand:

- the nature of MSDs
- the potential long-term consequences of MSDs that begin in childhood
- the importance of training (for themselves and for the their children)
- the benefits of eliminating or reducing risk factors for MSDs for their children

Most parents and supervisors would not intentionally put young workers at risk. Instead, they lack information about the injuries and prevention strategies that apply to their children's lives. They may also be like many people and simply believe the injury reports are "exaggerated" or "do not apply to me and my family."

IDENTIFYING HIGH-RISK JOBS

Identify associations between tasks and injuries

One way to recognize high-risk jobs is to identify the various physical exertions and tasks within a job that are associated with MSDs. This is difficult when dealing with children because most information on children's performance is on standardized physical education tests (e.g., chin-ups, sit-ups, and push-ups) rather than work-related tasks.

* As reported by parents and supervisors.

To identify high-risk jobs and evaluate the child's capabilities to perform these demands, a specialized ergonomic tool developed by NIOSH researchers allows users to determine whether a specified work task is likely to exceed the strength capabilities of a youth.

Match task demands with child capabilities

Performance information for some age categories is also lacking. For example, US strength data is available for 3–10 year olds, but not for 11–20 year olds (Owings et al., 1975).

In order to best match specific task demands with the personal capabilities of children a task-related assessment method is needed, such as a strength prediction program for children.

Matching children's capabilities with the demands of their tasks requires an understanding of the "fit" between the child and equipment they use. The two data sets used in the United States to identify size characteristics of children are anthropometric data (infant–20 years, Snyder et al., 1977) and clinical growth charts (NCHS, 2003). Clinical growth charts include age-related percentiles for stature, weight, and body mass for boys and girls.

Table 18.7 Output Screen from NIOSH Static 2-D Biomechanical Model for Children

Researchers at the National Institute for Occupational Safety and Health (NIOSH) and the University of Wisconsin-Milwaukee used the available strength and anthropometric data to develop the NIOSH 2D Child Strength Program, a computerized two-dimensional strength prediction program[a] for use with young workers (Waters, 2003).[b]

This NIOSH computer program calculates the percent of young persons who will be sufficiently strong to perform a specific pushing, pulling, or lifting task by age and gender.

This Child Strength Prediction Program also estimates loads on the spine based on the administrators' responses regarding the child's age, gender, necessary body posture, weight, and direction of force (i.e., push, pull, or lift).

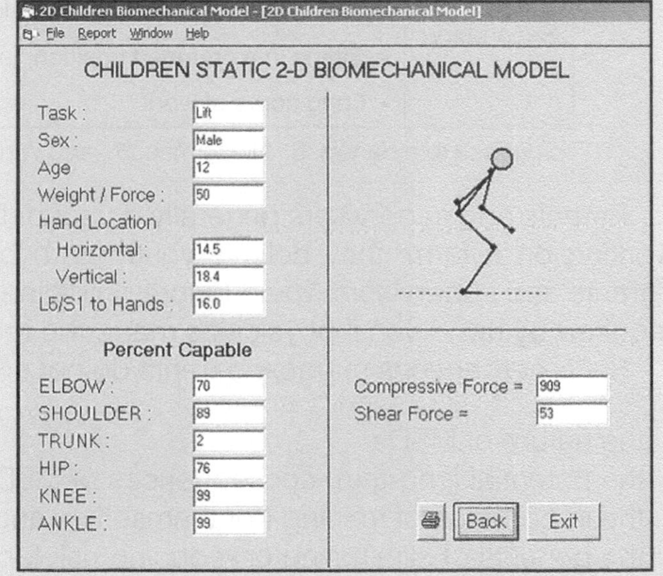

An example of this tool is on the right. In this example, between 70% and 99% of 12-year-old males have the knee, ankle, hip, shoulder, and elbow strength to lift 22.7 kg (50 lb) in the posture shown, but only 2% have the necessary trunk strength.

This program estimates that the task involves a compression of the spine and shear force of 412 kg (909 lb). The acceptable compression force for adult males is 350 kg (770 lb), so this is obviously far above an acceptable level for a child (Waters et al., 1993).

This software program is still under development and has not yet been validated against alternate strength norms for children. Unfortunately, we have not been able to identify any alternate strength norms for comparison. Even so, it is an excellent tool to evaluate the fit between a child's strength and the demands of tasks. The results tell you when to reengineer a task or when someone with greater strength or a team of workers may be required.

[a] Further information about the NIOSH 2D Child Strength Program is available at www.childergo.com/farmchildinjuries.htm

[b] Gaps in the data among older youth were extrapolated to estimate the continued increases in strength.

CURRENT GUIDELINES FOR CHILDREN WORKING IN AGRICULTURE

Guidelines provide information about ways to reduce risks for children working on farms. Agriculture safety and health experts from the National Children's Center in Marshfield, Wisconsin, published the North American Guidelines for Children's Agricultural Tasks (NAGCAT) (NCCRHS, 1999).* NAGCAT guidelines cover 62 jobs in seven major work areas, including animal care, general activities, haying operations, implement operations, manual labor, specialty production, and tractor fundamentals. Table 18.9 (A, B, and C), Table 18.10 (A, B, and C), and Figure 18.15 show three sample NAGCAT Guidelines. These provide guidance on tasks that involve manual lifting, hand picking vegetables, and operating a tractor.

NAGCAT Guidelines are easy to use. First, it provides information on the task. Then, it poses a series of questions about a child's ability to perform that specific task and suggests training needs and level of adult supervision needed for the task. Each "No" response triggers an explanation as to why that job may be hazardous. The guideline also includes specific age-based recommendations regarding supervision for children for each of the 62 farm tasks.

The guidelines are important and helpful for parents and supervisors. Adding data on the physical demands of farm work (as they become available) will make it even more so.

Table 18.8 Checklist Developed from the North American Guidelines for Children's Agricultural Tasks

Sample Checklist (NAGCAT)	
☐ Yes ☐ No	1. Can the child lift safely?
☐ Yes ☐ No	2. Has the child been trained to lift safely?
☐ Yes ☐ No	3. Can the child easily push up to 10% to 15% of his or her body weight?
☐ Yes ☐ No	4. Does the amount of material to be lifted weigh less than 10% to 15% of the child's body weight?
☐ Yes ☐ No	5. Does the child have to carry the object less than 10–15 yards?
☐ Yes ☐ No	6. Are the tools the right size for the child?
☐ Yes ☐ No	7. Has the child shown they can do the job safely 4–5 times under close supervision?
☐ Yes ☐ No	8. Can an adult supervise as recommended?

* North American Guidelines for Children's Agricultural Tasks can be downloaded at www.nagcat.org

Table 18.9A NAGCAT Guideline: Manual Labor Lifting

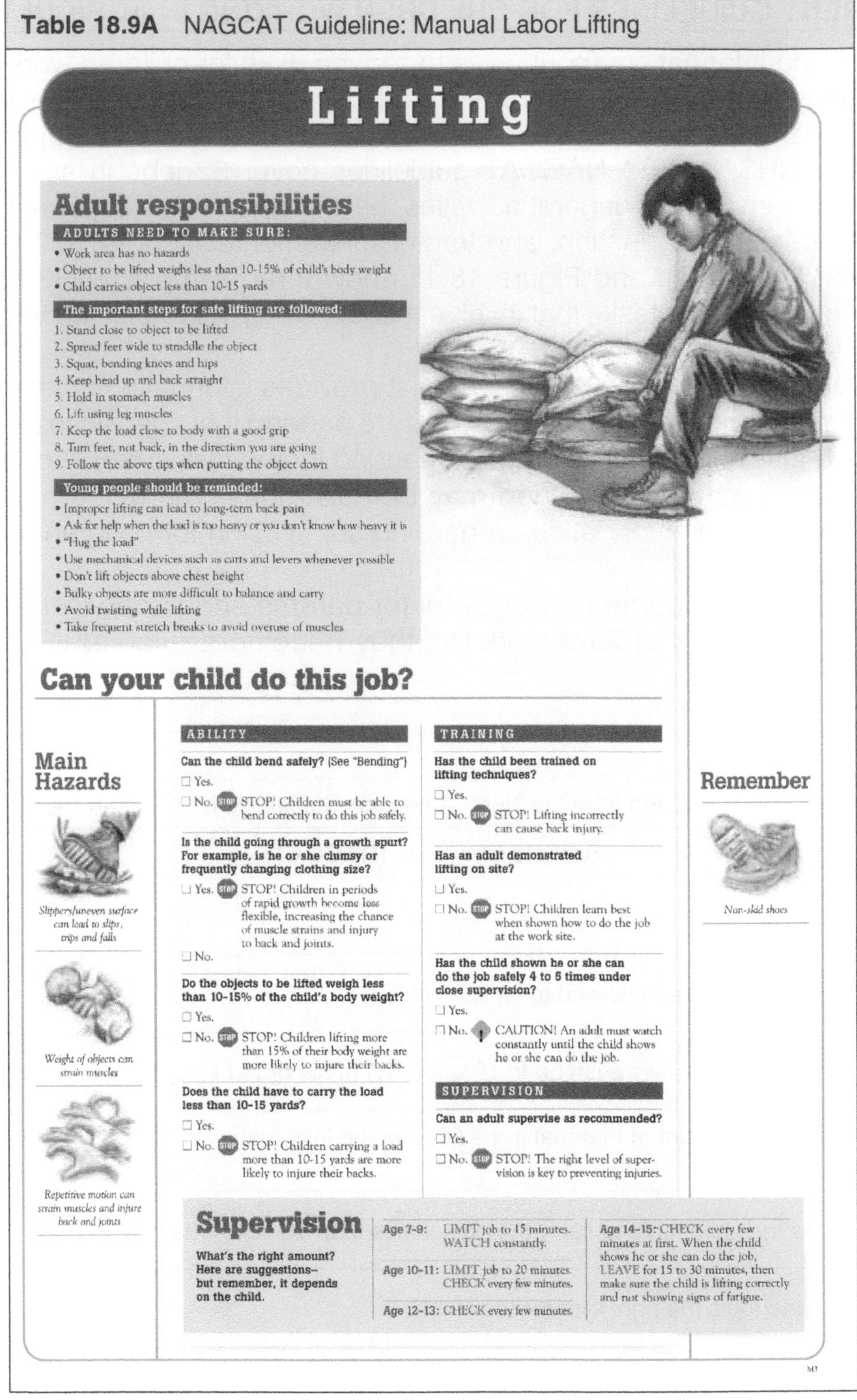

Lifting

Adult responsibilities

ADULTS NEED TO MAKE SURE:

- Work area has no hazards
- Object to be lifted weighs less than 10-15% of child's body weight
- Child carries object less than 10-15 yards

The important steps for safe lifting are followed:

1. Stand close to object to be lifted
2. Spread feet wide to straddle the object
3. Squat, bending knees and hips
4. Keep head up and back straight
5. Hold in stomach muscles
6. Lift using leg muscles
7. Keep the load close to body with a good grip
8. Turn feet, not back, in the direction you are going
9. Follow the above tips when putting the object down

Young people should be reminded:

- Improper lifting can lead to long-term back pain
- Ask for help when the load is too heavy or you don't know how heavy it is
- "Hug the load"
- Use mechanical devices such as carts and levers whenever possible
- Don't lift objects above chest height
- Bulky objects are more difficult to balance and carry
- Avoid twisting while lifting
- Take frequent stretch breaks to avoid overuse of muscles

Can your child do this job?

Main Hazards

Slippery/uneven surface can lead to slips, trips and falls

Weight of objects can strain muscles

Repetitive motion can strain muscles and injure back and joints

ABILITY

Can the child bend safely? (See "Bending")

☐ Yes.

☐ No. STOP! Children must be able to bend correctly to do this job safely.

Is the child going through a growth spurt? For example, is he or she clumsy or frequently changing clothing size?

☐ Yes. STOP! Children in periods of rapid growth become less flexible, increasing the chance of muscle strains and injury to back and joints.

☐ No.

Do the objects to be lifted weigh less than 10-15% of the child's body weight?

☐ Yes.

☐ No. STOP! Children lifting more than 15% of their body weight are more likely to injure their backs.

Does the child have to carry the load less than 10-15 yards?

☐ Yes.

☐ No. STOP! Children carrying a load more than 10-15 yards are more likely to injure their backs.

TRAINING

Has the child been trained on lifting techniques?

☐ Yes.

☐ No. STOP! Lifting incorrectly can cause back injury.

Has an adult demonstrated lifting on site?

☐ Yes.

☐ No. STOP! Children learn best when shown how to do the job at the work site.

Has the child shown he or she can do the job safely 4 to 5 times under close supervision?

☐ Yes.

☐ No. CAUTION! An adult must watch constantly until the child shows he or she can do the job.

SUPERVISION

Can an adult supervise as recommended?

☐ Yes.

☐ No. STOP! The right level of supervision is key to preventing injuries.

Remember

Non-skid shoes

Supervision

What's the right amount? Here are suggestions– but remember, it depends on the child.

Age 7-9: LIMIT job to 15 minutes. WATCH constantly.

Age 10-11: LIMIT job to 20 minutes. CHECK every few minutes.

Age 12-13: CHECK every few minutes.

Age 14-15: CHECK every few minutes at first. When the child shows he or she can do the job, LEAVE for 15 to 30 minutes, then make sure the child is lifting correctly and not showing signs of fatigue.

M3

Table 18.9B Can Your Child Do this Job?

Main Hazards	ABILITY		Remember
Slippery/uneven surface can lead to slips, trips and falls	**1. Can the child bend safety?** (See "Bending") ___Yes. ___ No. STOP STOP! Children must be able to bend correctly to do this job safety.	**2. Is the child going through a growth spurt? For example, is he or she clumsy or frequently changing clothing size?** ___ Yes. STOP STOP! Children in periods of rapid growth become less flexible, increasing the chance of muscle strain and injury to back and joints. ___ No.	*Non-skid shoes*
Weight can strain muscles	**3. Do the objects to be lifted weight less than 10-15% of his or her body weight?** ___Yes. ___ No. STOP STOP! Children lifting more than 15% of their body weight are more likely to injure their backs.	**4. Does the child have to carry the load less than 10-15 yards?** ___Yes. ___ No. STOP STOP! Children carrying a load more than 10-15 yards are more likely to injure their backs.	
	TRAINING		
Repetitive motion can strain muscles and injure back and joints	**5. Has the child been trained on lifting techniques?** ___ Yes. ___ No. STOP STOP! Lifting incorrectly can cause back injury.	**6. Has an adult demonstrated lifting on site?** ___ Yes. ___ No. STOP STOP! Children learn best when shown how to do the job at the work site.	
		SUPERVISION	
	7. Has the child shown he or she can do the job safely 4 to 5 times under close supervision? ___ Yes. ___ No. ! CAUTION! An adult must watch constantly until the child	**7. Can an adult supervise as recommended?** ___ Yes. ___ No. STOP STOP! The right level of supervision is key to preventing injuries.	

Table 18.9C What's the Right Amount of Supervision?	
Supervision	
What's the right amount? Here are suggestions—but remember, it depends on the child	
Age 7–9:	**LIMIT** job to 15 minutes. **WATCH** constantly.
Age 10–11:	**LIMIT** job to **20 minutes. CHECK** every few minutes.
Age 12–13:	**CHECK** every few minutes.
Age 14–15:	**CHECK** every few minutes at first. When the child shows he or she can do the job, **LEAVE** for 15–30 minutes, then make sure the child is lifting correctly and not showing signs of fatigue.

Table 18.10A NAGCAT Guideline: Hand Harvesting Vegetables

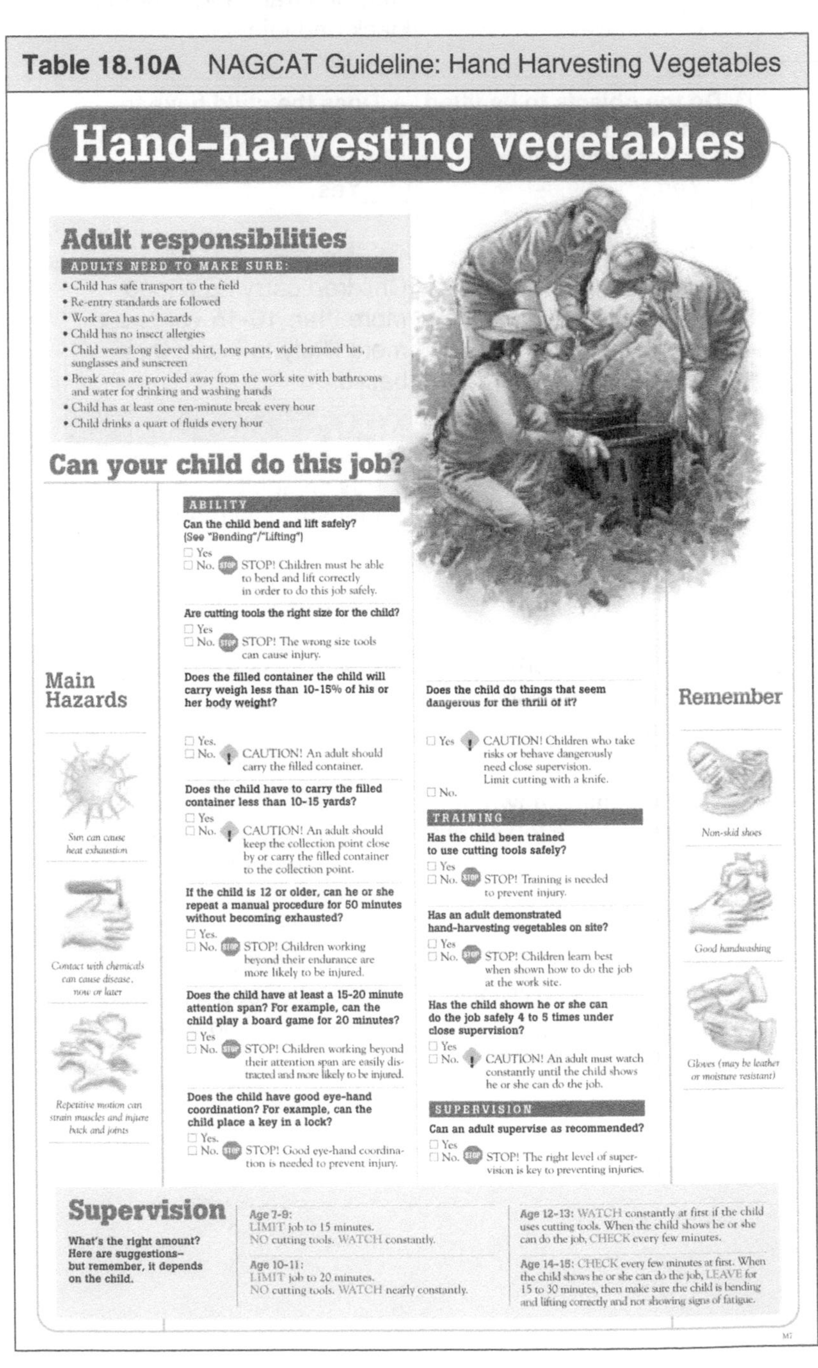

Hand-harvesting vegetables

Adult responsibilities

ADULTS NEED TO MAKE SURE:

- Child has safe transport to the field
- Re-entry standards are followed
- Work area has no hazards
- Child has no insect allergies
- Child wears long sleeved shirt, long pants, wide brimmed hat, sunglasses and sunscreen
- Break areas are provided away from the work site with bathrooms and water for drinking and washing hands
- Child has at least one ten-minute break every hour
- Child drinks a quart of fluids every hour

Can your child do this job?

ABILITY

Can the child bend and lift safely? (See "Bending"/"Lifting")
☐ Yes
☐ No. STOP! Children must be able to bend and lift correctly in order to do this job safely.

Are cutting tools the right size for the child?
☐ Yes
☐ No. STOP! The wrong size tools can cause injury.

Does the filled container the child will carry weigh less than 10-15% of his or her body weight?
☐ Yes.
☐ No. CAUTION! An adult should carry the filled container.

Does the child have to carry the filled container less than 10-15 yards?
☐ Yes
☐ No. CAUTION! An adult should keep the collection point close by or carry the filled container to the collection point.

If the child is 12 or older, can he or she repeat a manual procedure for 50 minutes without becoming exhausted?
☐ Yes.
☐ No. STOP! Children working beyond their endurance are more likely to be injured.

Does the child have at least a 15-20 minute attention span? For example, can the child play a board game for 20 minutes?
☐ Yes
☐ No. STOP! Children working beyond their attention span are easily distracted and more likely to be injured.

Does the child have good eye-hand coordination? For example, can the child place a key in a lock?
☐ Yes.
☐ No. STOP! Good eye-hand coordination is needed to prevent injury.

Does the child do things that seem dangerous for the thrill of it?
☐ Yes CAUTION! Children who take risks or behave dangerously need close supervision. Limit cutting with a knife.
☐ No.

TRAINING

Has the child been trained to use cutting tools safely?
☐ Yes
☐ No. STOP! Training is needed to prevent injury.

Has an adult demonstrated hand-harvesting vegetables on site?
☐ Yes
☐ No. STOP! Children learn best when shown how to do the job at the work site.

Has the child shown he or she can do the job safely 4 to 5 times under close supervision?
☐ Yes
☐ No. CAUTION! An adult must watch constantly until the child shows he or she can do the job.

SUPERVISION

Can an adult supervise as recommended?
☐ Yes
☐ No. STOP! The right level of supervision is key to preventing injuries.

Main Hazards

Sun can cause heat exhaustion

Contact with chemicals can cause disease, now or later

Repetitive motion can strain muscles and injure back and joints

Remember

Non-skid shoes

Good handwashing

Gloves (may be leather or moisture resistant)

Supervision

What's the right amount? Here are suggestions– but remember, it depends on the child.

Age 7-9: LIMIT job to 15 minutes. NO cutting tools. WATCH constantly.

Age 10-11: LIMIT job to 20 minutes. NO cutting tools. WATCH nearly constantly.

Age 12-13: WATCH constantly at first if the child uses cutting tools. When the child shows he or she can do the job, CHECK every few minutes.

Age 14-15: CHECK every few minutes at first. When the child shows he or she can do the job, LEAVE for 15 to 30 minutes, then make sure the child is bending and lifting correctly and not showing signs of fatigue.

M7

Table 18.10B Can Your Child Do this Job?

Main Hazards	ABILITY		Remember
Sun can cause heat exhaustion	**1. Can the child bend and lift safely?** (See "Bending/Lifting") ___Yes. ___ No. STOP STOP! Children must be able to bend and lift correctly to do this job safely.	**2. Are cutting tools the right size for the child?** ___Yes. ___ No. STOP STOP! The wrong size tools can cause injury.	*Non-skid shoes*
Contact with chemicals can cause disease, now or later	**3. Does the filled container the child will carry weigh less than 10-15% of the child's body weight?** ___ Yes. ___ No. ! CAUTION! An adult should carry the filled container.	**4. Does the child have to carry the filled container less than 10-15 yards?** ___ Yes. ___ No. ! CAUTION! An adult should keep the collection point close by or carry the filled container to the collection point.	*Good hand washing*
	5. If the child is 12 or older, can he or she repeat a manual task for 50 minutes without becoming exhausted? ___ Yes ___ No. STOP STOP! Children working beyond their endurance are more likely to be injured.	**6. Does the child have at least a 15-20 minute attention span? For example, can the child play a board game for 20 minutes?** ___ Yes ___ No. STOP STOP! Children working beyond their attention span are easily distracted and more likely to be injured.	*Gloves (may be leather or moisture resistant)*
Repetitive motion can strain muscles and injure back and joints	**7. Does the child have good eye-hand coordination? For example, can the child place a key in a lock?** ___ Yes ___ No. STOP STOP! Good eye-hand coordination is needed to prevent injury.	**8. Does the child do things that seem dangerous for the thrill of it?** ___ Yes. ! CAUTION! Children who take risks or behave dangerously need close supervision. Limit cutting with a knife. ___ No	
	TRAINING		
	9. Has the child been trained to use cutting tools safely? ___ Yes ___ No. STOP STOP! Training is needed to prevent injury.	**10. Has an adult demonstrated hand-harvesting vegetables on site?** ___ Yes ___ No. STOP STOP! Children learn best when shown how to do the job at the work site.	
		SUPERVISION	
	11. Has the child shown he or she can do the job safely 4 to 5 times under close supervision? ___ Yes ___ No. ! CAUTION! An adult must watch constantly until child shows he or she can do the job.	**12. Can an adult supervise as recommended?** ___ Yes ___ No. STOP STOP! The right level of supervision is key to preventing injuries.	

Table 18.10C What's the Right Amount of Supervision?	
Supervision	
What's the right amount? Here are suggestions—but remember, it depends on the child	
Age 7–9:	**LIMIT** job to 15 minutes. **NO** cutting tools. **WATCH** constantly.
Age 10–11:	**LIMIT** job to 20 minutes. **NO** cutting tools. **WATCH** nearly constantly.
Age 12–13:	**WATCH** constantly at first if the child uses cutting tools. When the child shows he or she can do the job, **CHECK** every few minutes.
Age 14–15:	**CHECK** every few minutes at first. When the child shows he or she can do the job, **LEAVE** for 15 to 30 minutes, then make sure the child is bending, lifting correctly and not showing signs of fatigue.

Tractor operation chart

CHILDREN SHOULD ONLY OPERATE WIDE-FRONT TRACTORS EQUIPPED WITH ROPS AND SEATBELTS. An adult should ensure that the child can reach all controls while wearing a seatbelt, that a pre-operations service check has been completed, and that no extra riders are allowed on the tractor. This guideline assumes that the child will be operating the tractor in daylight, under dry conditions, while not on a steep slope and with reasonable distance from ditches, trees and fences.

Refer to the specific guideline for recommended supervision

Size of tractor

Increased complexity of job	LAWN & GARDEN less than 20hp	SMALL 20hp to 70hp	MEDIUM/LARGE more than 70hp	ARTICULATED
OPERATING A FARM TRACTOR (no equipment attached)	12-13 years	12-13 years	14-15 years	16+ years
TRAILED IMPLEMENTS fieldwork	12-13 years	12-13 years	14-15 years	16+ years
3-POINT IMPLEMENTS fieldwork	12-13 years	14-15 years	14-15 years	16+ years
REMOTE HYDRAULICS fieldwork	14-15 years	14-15 years	14-15 years	16+ years
PTO-POWERED IMPLEMENTS fieldwork	14-15 years	14-15 years	14-15 years	16+ years
TRACTOR-MOUNTED FRONT-END LOADER	14-15 years	16+ years	16+ years	16+ years
WORKING IN AN ORCHARD	14-15 years	16+ years	16+ years	16+ years
WORKING INSIDE BUILDINGS	14-15 years	16+ years	16+ years	16+ years
DRIVING ON PUBLIC ROADS*	N/A	16+ years	16+ years	16+ years
PULLING OVERSIZE OR OVERWEIGHT LOAD	Due to increased hazard and complexity, these jobs should **not** be assigned to children.			
HITCHING TRACTOR TO MOVE STUCK OR IMMOVABLE OBJECTS				
SIMULTANEOUS USE OF MULTIPLE VEHICLES				
ADDITIONAL PERSONS WORKING ON A TRAILING IMPLEMENT				
PESTICIDE OR ANHYDROUS AMMONIA APPLICATION*				

* follow state/province laws

Figure 18.15 NAGCAT Guideline: Tractor Operation.

The most important question is whether using the NAGCAT helps to prevent injuries. So far, the evidence is weak, but positive (Pickett et al., 2003). The effort is strongest if three things happen in conjunction: (1) a safety specialist visits the farm; (2) the safety specialist gives the parents or supervisors the NAGCAT guides; and (3) the safety specialist provides information on basic child development. Most studies evaluate training and behavior modification rather than actual changes in the workplace itself. Stopping the injury prevention effort at parental and supervisor education may not be sufficient to attain behavioral and design changes on farms. Safety standard regulations may be necessary.

FUTURE RESEARCH

In 2002, NIOSH and the Great Lakes Center for Agricultural Safety and Health at The Ohio State University cosponsored a meeting to discuss research gaps and risk factors for MSDs among children and adolescents working in agriculture.*

Experts at the meeting made the following suggestions:

1. Evaluate jobs performed by youth to identify high-risk jobs and objectively determine appropriate job assignments for youth.
2. Develop and implement health and hazard surveillance systems for measuring and tracking the magnitude of risk for children and adolescents working in agriculture.
3. Develop and evaluate ergonomic interventions for reducing risk or MSDs for children and adolescents working in agriculture.

Table 18.11 summarizes important research gaps identified at the meeting.

Table 18.11 Important Gaps in Research on Children and Youth in Farms

Issues Related to Assessing High Risk Jobs	1. Develop an "Enterprise Classification" system and evaluate risk of MSD based on this classification (e.g., determine risk by region, agriculture sector, or size of enterprise). 2. Determine the number of exposed youth and what jobs they are doing in each commodity area. 3. Identify the hazards or physical work factors in each job or task and determine the number of hours worked per year. 4. Evaluate the effectiveness of different methods of risk assessment, including self-assessment, professional judgment, and objective quantitative methods. Use "health outcome" or "level of exposure" as a measure of risk. 5. Evaluate risk in unmechanized production (e.g., tool usage in manual labor).
Surveillance Issues	1. Develop a National Registry of musculoskeletal hazards and health outcomes (e.g., National Health and Hazard Exam). 2. Supplement existing surveillance systems (e.g., National Health Interview Survey, National Health and Nutrition Examination Survey, Behavioral Risk Factor Surveillance System, and prospective community-based surveys such as the Keokuk and Iowa Safe Farm surveys). 3. Conduct ad-hoc population-based health and hazard surveys, such as clinic- or school-based methods or face-to-face interviews. 4. Develop partnerships with those who know and are known by the population under study. 5. Conduct quality cross-sectional and longitudinal studies. 6. Develop and validate a consensus list of jobs and health outcomes.

* The proceedings from this meeting are available in a special NIOSH report (Waters and Wilkins, 2004).

Table 18.11 Important Gaps in Research on Children and Youth in Farms (Continued)

Intervention Issues	1. Develop more solutions. Local and federal government agencies should encourage private industry and academic–industry partnerships and develop vocational agriculture awards program for interventions at the high school or college level. 2. Develop improved methods of disseminating information. 3. Conduct studies that address liability, cultural, ethical, and economic barriers. 4. Encourage more intervention evaluations using randomized trials and quasiexperimental and blended evaluations. 5. Adopt successful models for implementation of solutions, such as the tobacco model for increasing awareness of interventions.
Etiological Issues	1. Conduct studies to assess physical, cognitive, and developmental capabilities. 2. Encourage development of studies to examine the dose–response relationships for MSDs to determine the magnitude of exposures and symptoms, including examination of multiple exposures (e.g., sports, 2nd job). 3. Increase development of instrument-based and laboratory-based work assessment systems to improve measurement of exposure and health outcomes. 4. Conduct population, clinical, and laboratory studies to evaluate the short-term impact of risk factors on MSDs, such as effects of different types of exposures on MSD risk and early indicators, such as bone density, stiffness, and pain. 5. Conduct population, clinical, and laboratory studies to evaluate the long-term impact of repeated exposure (e.g., study to compare health status of retired farmers compared with nonfarm workers, evaluation of the permanent effects of physical loading and include groups with maximal exposures, etc.).

Source: Waters, T.R. and Wilkins J.R., Conference Proceedings: Prevention of Musculoskeletal Disorders Among Youth and Adolescents Working in Agriculture, NIOSH, Cincinnati, OH. DHHS(NIOSH) Publication No. 2004-119, 2004.

CONCLUSIONS

Farming is physically demanding. The strength and endurance training that children get while working on farms can contribute to their overall health and fitness. However, excessive physical demands may lead to acute or chronic injuries and illnesses to the musculoskeletal system. The risk is especially great for the lower back, shoulders, and upper extremities.

Table 18.12 Ergonomic Interventions for Youth Working in Agriculture

1. Design equipment so workers with smaller body sizes can adjust controls.
2. Use backpacks, knapsacks, or straps to support equipment, such as sprayers and weed eaters.
3. Improve design of existing mechanical equipment and develop new equipment to eliminate bending and twisting during physically demanding farm work.
4. Use lightweight bags and wheeled carts for picking and transporting ground crops.
5. Use small tubs and baskets for carrying fruit and produce. The full tubs will weigh less.
6. Buy feed and other materials in smaller, lighter bags.
7. Match the task requirements with the child's physical and mental capabilities. Alter the task design or the number of individuals assigned to the task or develop alternate means of completing the task if the demands exceed the child's abilities.
8. Design work tools to fit the body dimensions of children.

The demands of the job must match the child's abilities. Parents, supervisors, and children need to understand the hazards and know how to minimize risks. Children themselves will benefit from information that teaches them to recognize "normal" soreness versus soreness that indicates a need to rest or recover.

Table 18.12 provides additional suggestions for youth working in agriculture. More specific examples of successful ergonomic interventions are listed in "Simple Solutions: Ergonomics for Farm Workers" (NIOSH, 2001).

Since perceptions drive action, the first target for intervention is the perceptions of farm workers and supervisors. They need information on the problems in order to understand and make conscientious decisions on assigning farm tasks to children. The most effective ergonomic interventions involve a collective effort that includes children, parents, supervisors, and safety experts.

Children, parents, and supervisors will have ideas about task design and safety that fit their surroundings and lifestyle. They are also more likely to use the information if they are active participants in their own health and welfare, rather than recipients of programs designed by "outsiders."

Regardless of the risk of MSDs, children and adolescents will work in agriculture and on farms with their families to gain extra income. However, they should be able to do so without exposing themselves to significant risk. Every effort should be made to assign children to jobs appropriately, according to their capabilities and to identify and eliminate jobs with high risk of injury.

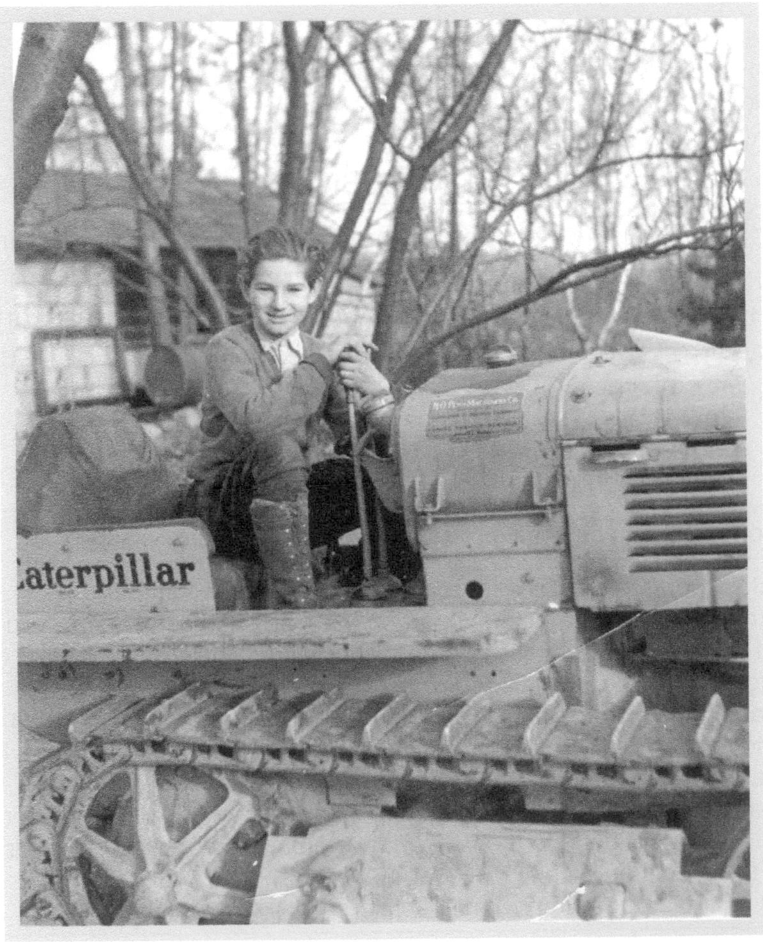

ACKNOWLEDGMENTS

A special thanks to the editors, Rani Lueder and Valerie Rice, for their diligent efforts and important contributions to the chapter. Also, thanks to Ursy Potter Photography for her wonderful photographs.

REFERENCES

Allread, W. G., Wilkins, J. R., Maronitis, A. B., and Waters, T. R. (2004). Using the lumbar motion monitor to assess the physical demands for children and adolescents working on farms. *Journal of Agricultural Safety and Health, 10*(4), 255–272.

Bartels, S., Niederman, B., and Waters, T. (2000). Job hazards for musculoskeletal disorders for youth working on farms. *Journal of Agricultural Safety and Health, 6*(3), 191–201.

Ehrmann-Fledman, D., Shrier, I., Rossigno, M., and Abenhaim, L. (2002). Work is a risk factor for adolescent musculoskeletal pain. *Journal of Occupational and Environmental Medicine, 44*(1), 957–961.

Guo, H., Tanaka, S., Cameron, L., Seligman, P., Behrens, J., Ger, J., Wild, D., and Putz-Anderson, V. (1995). Back pain among workers in the United States: National estimates and workers at high risk. *American Journal of Industrial Medicine, 28,* 591–602.

Hartling, L., Brison, R., Crumley, E., Klassen, T., and Pickett, W. (2004). A systematic review of interventions to prevent childhood farm injuries. *Pediatrics, 114*(4), 483–496.

Marlenga, B., Pickett, W., and Brison, R. (2003). Evaluation of the North American Guidelines for Children's Agricultural Tasks using a case series of injuries: Interim results. Presented at the American Public Health Association 131st Annual Meeting, November 18, 2003, San Francisco, CA. Abstract 68453.

Meyers, J. R. and Hendricks, K. J. (2001). Injuries among Youth on Farms in the United States, 1998. Department of Health and Human Services, National Institute for Occupational Safety and Health [DHHS (NIOSH) Publication No. 97-141]. Cincinnati, OH.

Murphy, D. and Hackett, K. (1997). Children and Safety on the Farm. Ps6361, U.Ed.AGR97-25. For further information contact Penn State College of Agricultural Sciences, Publications Distribution Center, Pennsylvania State University, 112 Agricultural Administration Bldg., University Park, PA 16802 or Pennsylvania Safe Kids Coalition 1578 Interstate Drive, P.O. Box 68525, Harrisburg, PA 01706-8525.

NCCRHS (National Children's Center for Rural and Agricultural Health and Safety, Marshfield Clinic). (1999). *North American Guidelines for Children's Agricultural Tasks*. Marshfield, WI: Marshfield Clinic.1000 North Oak Av. Marshfield, WI 54449.

NCHS (National Center for Health Statistics). (2003). *Clinical Growth Charts*. www.cdc.gov/nchs/about/major/nhanes/growthcharts/clinical_charts.htm

NIOSH. (1997). Musculoskeletal Disorders and Workplace Factors: A Critical Review of Epidemiologic Evidence for Work-related Musculoskeletal Disorders of the Neck, Upper Extremity, and Low Back, B.P. Bernard (Ed.). DHHS (NIOSH) Publication No. 97-141. NIOSH, 4676 Columbia Parkway, Cincinnati, OH 45226.

NIOSH. (2001). Simple Solutions: Ergonomics for Farm Workers, S. Baron, C. Estill, A. Steege, and N. Lalich (Eds.). DHHS (NIOSH) Publication No. 2001-111. Department of Health and Human Services, National Institute for Occupational Safety and Health, 4676 Columbia Parkway, Cincinnati, OH 45226.

OSHA. (2001). US Department of Labor, Bureau of Labor Statistics. *Injury and Illness Tables.* Washington D.C. www.bls.gov/iif

Owings, C. L., Chaffin, D. B., Snyder, R. G., and Norcutt, R. H. (1975). Strength Characteristics of U.S. Children for Product Safety Design. Contract No. FDA-73-32, Prepared for Consumer Product Safety Commission, 5401 West Bard Avenue, Room 100, Bethesda, MD 20207.

Pickett, W., Brison, R. J., Berg, R. L., Zentner, J., Linneman, J., and Marlenga, B. (2005). Pediatric farm injuries involving non-working children injured by a farm work hazard: Five priorities for primary prevention. *Injury Prevention, 11*, 6–11.

Snyder, R. G., Schneider L. W., Owings, C. L., Reynolds, H. M., Golomb, D. H., and Schork, M. A. (1977). Anthropometry of infants, children and youths to age 18 for product safety design: SP-450. Contract No. CPSC-75-0068 by Highway Safety Research Institute. The University of Michigan. Published by the Society of Automotive Engineers, Inc. Warrendale, PA 15096.

Thelin, A., Vingard, E., and Holmberg, S. (2004). Osteoarthritis of the hip joint and farm work. *American Journal of Industrial Medicine, 45*, 202–209.

Waters, T. R., Putz-Anderson, V., Garg, A., and Fine, L. J. (1993). Revised NIOSH equation for the design and evaluation of manual lifting tasks. *Ergonomics, 36*(7), 749–776.

Waters, T. R. (2003). Two-dimensional biomechanical modeling for estimating strength of youth and adolescents for manual material handling tasks. Poster presented at 2003 Challenges in Agricultural Health and Safety Conference, San Francisco, CA, September 7–9, 2003.

Waters, T. R. and Wilkins J. R. (2004). Conference Proceedings: Prevention of Musculoskeletal Disorders among Youth and Adolescents Working in Agriculture. DHHS(NIOSH) Publication No. 2004-119. NIOSH, 4676 Columbia Parkway, Cincinnati, OH 45226.

Xiang, H., Stallones, L., Hariri, S., Darrahg, A., Chui, Y., and Gibbs, J. (1999). Back pain among persons working on small or family farms—eight Colorado counties, 1993–1996. *Morbidity and Mortality Weekly Reports, 48*(45), 301–304.

SECTION F

Children and School

CHAPTER 19

PRESCHOOL AND DAYCARE DESIGN

LORRAINE E. MAXWELL

TABLE OF CONTENTS

INTRODUCTION

Parents from every nation need assistance with caring for their preschool-age children. In situations where both parents work, the need is especially great. Approximately 13 million children in the United States under the age of 6 are in some form of nonparental care for at least some of the day (CPSC, 1999). This includes nearly 3.7 million children who attend day-care centers, Head Start, and nursery school programs.

We expect quality childcare to contribute to children's learning and social development, and caregivers to be nurturing and kind. Other important factors that affect the quality of childcare programs are staff training and experience, the director's leadership style and skills, caregiver-to-child ratios, the number of children in the center, and the physical environment of the classrooms and playrooms (Figure 19.1).

Although attractive surroundings appeal to us all, caregivers may underestimate the critical role that physical environments play on children's cognitive, social, physical, and emotional development. Stimulating and well-organized childcare settings help children develop their vocabularies, attention and memory skills, and their social interactions with peers (NICHD, 2000).

Figure 19.1 (Ursy Potter Photography) Day-care centers do not merely provide "paid babysitting." A good day care uses all of its resources to nurture children's physical, cognitive, and emotional growth. While the resources include an educated staff and educational materials, it also includes the design of the physical environment itself.

IMPACT OF THE PHYSICAL ENVIRONMENT

LEARNING

Infants and toddlers (0–3 years) learn by experiencing the environment through their senses. Piaget referred to this time as the sensorimotor period (Piaget, 1974). Children of this age learn through touch, vision, hearing, kinesthetics or equilibrium, hearing, taste, and smell. Day-care environments should reflect this and invite children to handle toys, look around, interact, and explore.

Four- to six-year-old children learn by repetition and practice (Piaget's preoperational period). Preschool environments for these children should encourage rehearsal, such as with child-size kitchens, grocery stores, and banks (Figure 19.2 and Figure 19.3). As they play, children unknowingly prepare themselves for both real life and school-related academic activities such as reading and mathematics. The design of the physical environment is critical during both of these stages (see also Chapter 2 on child development).

Figure 19.2 (Ursy Potter Phototgraphy) **Figure 19.3** (Valerie Rice)
Much of children's play is a rehearsal for life. Children mimic their parents and other caregivers, trying on different tasks and roles. Here they try cooking, serving food, taking food orders, and "giving change." They also mimic adult occupations they come in contact with, such as playing "dentist." It is all in fun, and they learn about themselves, as well as the tasks.

Childcare spaces are more than "containers" for social interaction, exploration, and learning (Gandini, 1998). Responsive environments encourage learning (e.g., round objects roll, round objects can not be stacked one on top of the other) and mastery (e.g., drawing, climbing to the top, getting a drink of water). Unfortunately, the reverse is also true. Poorly designed environments that lack age-appropriate materials can impede how children grow and develop. Thus, child environments are more than "containers" in which they learn (Gandini, 1998). Childcare spaces affect cognitive and physical development by supporting or inhibiting behaviors.

Activity centers are an excellent example of how design can facilitate behavior (Figure 19.4 and Figure 19.5). Childcare spaces can also promote children's development by scaling the environment to fit them (Figure 19.6 and Figure 19.7).

Figure 19.4 (Valerie Rice)
Activity centers permit children to engage in center-specific activities, such as a reading corner or a drawing area. Arranging centers to create singular spaces lets children engage in small group or one-on-one interactions with peers.

These types of interactions develop socio-emotional skills. Locating activity centers apart from circulation paths permits children to concentrate on their play activities. Low height furniture enables caregivers to see over the top and be involved with the children in the activity centers, while spending less time monitoring or organizing.

The child-size scaling of the furniture in these centers makes task accomplishment easier for children, lets them practice "grown-up" roles and encourages their feelings of mastery.

Figure 19.5 (Lorraine Maxwell)

Figure 19.6 (Lorraine Maxwell)

Figure 19.7 (Ursy Potter Photography)
Even the size of the furniture and shelves, as well as the placement of toys and activities influences children's development. Children in child-scaled environments begin complex play sooner than children in adult-scaled settings (DeLong and Moran, 1992).

HEALTH AND SAFETY

In addition to influencing children's behavior, ergonomic designs can promote children's and adult caregivers' health and safety. Obviously, toys and equipment must be easy to clean and disinfect to prevent spread of disease and illness. Although not as obvious, falls from furniture (or as the result of collisions) cause injuries in indoor childcare settings (Chang et al., 1989). Careful design of the space can make it easy for caregivers to supervise children, encourage development, and maintain safety.

At times, these goals seem to conflict with one another. Equipment designed for indoor climbing must both challenge the child and protect against falls. Spacing of furniture must help prevent collisions, while still defining alcoves for specific activities and small group interactions. Tradeoffs are often necessary when designing environments that stimulate development and promote safety (Figure 19.8).

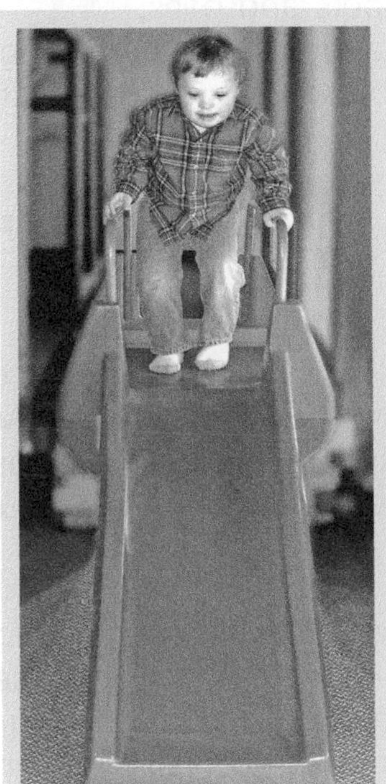

This indoor slide provides a challenge, yet protects children from injury.
It is on a soft, carpeted surface, has handrails and is low to the ground.

Figure 19.8 (Ursy Potter Photography)

Childcare environments must also protect adult caregivers. They need adult-sized furniture. For example, higher changing tables can reduce the loads on the back and lower extremities (Grant et al., 1995; Shimaoka et al., 1998).

ENVIRONMENTAL CONSIDERATIONS

Ergonomic environments fit their individual users. They enhance performance (in this case learning) while ensuring safety. Designing such environments requires attention to detail. Important aspects of the physical environment include storage, floors and carpeting, lighting, furniture, walls, windows, and room design, as well as age-appropriate toys and play materials.

A well-designed environment for children requires attention not only to the types of furniture, but also to their variety, dimensions, arrangement, construction, colors, fabrics, and ease of cleaning, storage, and repositioning. It includes light switches, doorknobs, water faucets, pots and pans, tools, closets, stairs, handrails—in short, it includes every aspect of the child's environment! Box 19.1 contains some questions designers must ask to assess how the design impacts the users.

BOX 19.1 DESIGN QUESTIONS FOR EVALUATING CHILDCARE ENVIRONMENTS

Design Questions

- Does the space support positive social interaction yet provide private space?
- Does its configuration lead to continuous interruptions from others?
- Can an adult easily observe children throughout the room, while permitting some independence?
- Can a child choose from a variety of materials to facilitate cognitive and fine-motor development?
- Does it encourage gross motor development without undue risk?
- Does the environment encourage a child's growing independence?

GENERAL DESIGN REQUIREMENTS

ACCESSIBILITY

Childcare centers and spaces within the centers must be accessible to individuals who use wheelchairs or have motor disabilities. Areas requiring particular attention include doorways, toilets, sinks, drinking fountains, classrooms, and outdoor play areas. Children with health, vision, or hearing impairments require additional provisions (AAP, 2002).

SAFETY

As with other buildings, designs must be structurally sound and protect against entry of mold, dust, and pests. All rooms in childcare centers must have an exit to the outside or a common hallway that leads out of the building. Corridors leading to exits must be free of obstructions. In addition, the minimum width for an egress path is 91 cm (36 in.) and doors must have a minimum clear width of 81 cm (32 in.). Emergency exits should be clearly marked and legible to care providers (AAP, 2002).

NOISE

Both children and caregivers benefit from controlling noise levels. Constant exposure to high levels of noise can be stressful to children and adults. Health risks include high blood pressure for children and adults, diminished pre-reading skills and motivation for preschool children, and impaired cognitive development in infants (Maxwell and Evans, 2000; Wachs, 1979; see Chapter 5 of this book).

Recommended noise levels in an unoccupied classroom for childcare are 30–35 dBA and 45–55 dBA in occupied spaces (ANSI, 2002). Noise levels are often much higher in childcare settings. Air handling and heating or cooling equipment as well as external conditions (road traffic and airport-related traffic) increase noise levels in unoccupied spaces. Use caution when designing and placing mechanical equipment for childcare spaces.

Window and door designs should minimize the transmission of unwanted noise from both outside sources and noisy sections such as multipurpose areas. Caregivers can help reduce noise levels in class by using a combination of hard and soft surfaces and by avoiding continuous background music.

LIGHTING

A combination of glare-free natural and artificial light should effectively illuminate activities. Exposure to sunlight provides beneficial Vitamin D, a sunlight prohormone that influences how our body uses calcium. The number and placement of windows should therefore expose the inhabitants to some sunlight, especially during the winter months, when children have fewer opportunities to play outside.

Eyestrain and headaches have been linked to inadequate artificial light. Poorly illuminated childcare spaces are often characterized by glare, too little light, or confusing shadows (CPS, 1996).

Some studies indicate that lighting influences academic performance. Others report that lighting affects mood and health. The California Energy Commission recommends* (CEC, 2003):

- ample and pleasant window views,
- reducing sources of glare,
- avoiding direct sun penetration, particularly from east or south facing windows (contributes to thermal discomfort and glare), and
- providing all windows with sun control (i.e., shades) in order to protect against glare or distraction.

CROWDING

Some countries let individual states, territories, or provinces regulate childcare group sizes and adult-to-child ratios. Other countries have national standards.

Infants require higher adult-to-child ratios and smaller group sizes than toddlers and preschoolers. Group size (per room) varies depending on the children's ages, but can range from 6–10 for infants up to 18 or 25 for the oldest preschool children. Adult-to-child ratios vary from 1 adult to 3 children for the youngest infants to 1 adult for every 7–10 children in the United States.

Minimum space requirements for playrooms and classrooms are also common. The United States requires at least 3.25 m^2 (35 sq. ft) but recommends 5 m^2 (50 sq. ft) per child. See Table 19.1 for a summary of standards in several countries.

* The above recommendations refer to elementary and secondary schools but the issues may be similar for childcare settings.

Table 19.1 International Standards for Preschool Environments

Standard	United States	Australia	The Netherlands	Calgary, Canada	Nova Scotia, Canada	Finland	South Korea
Maximum Group Size Appropriate group size for children to interact and learn social skills	6–8 infants (0–1 year) 16–12 toddlers (1–3 years) 14–20 preschool (3–5 years)	8 infants (0–1 year) 10 toddlers (2–3 years) 24 preschoolers (3–6 years)	12 infants (0–1 year) 16 toddlers and early preschoolers (1–4 years) 20 4–12 years	Not Available	10 infants 18 toddlers 25 preschoolers	12 toddlers (1–2 years) 21 preschoolers (3–6 years)	
Minimum Space per Child Enough space to accommodate children's activities including fine & gross motor activities	3.25 m^2 (35 sq. ft) all ages	3.25 m^2 (35 sq. ft) all ages	3 m^2 all ages	3 m^2 all ages	2.75 m^2 all ages	3.7 m^2 all ages	2.64 m^2—children under 3 years 1.98 m^2—children 3 years and older
Adult-to-Child Ratios Balance between children's developing autonomy, sense of belonging to a group, and interaction with an adult	1:4—infants 1:5 to 1:7—toddlers 1:7 to 1:10—preschoolers	1:4—infants 1:6—toddlers 1:12—preschoolers	1:4—0–1 years 1:5—1–2 years 1:6—2–3 years 1:8—3–4 years 1:10—4–12 years	Not Available	1:4—infants 1:6—toddlers 1:8—preschoolers	1:4—toddlers 1:7—preschoolers	1:5—infants (0–2 years) 1:7—toddlers (2 years) 1:20—preschoolers (3–5 years)
Maximum Noise Level Background noise at level to accommodate range of activities including play & sleep	35 db(A)	Not Available	External noise source 40 db (A) Reverberation—less than 1 second	Not Available	Not Available	Not Available	Not Available
Light Level Appropriate for all activities and children's vision abilities	300–460 Lux for general play areas 460–900 Lux for close work areas 50 Lux for sleeping	Not Available	Minimum 200 Lux on the work surface	Minimum 800 Lux on the work surface	Not Available	Not Available	Not Available

INDOOR AIR QUALITY

Children spend up to 90% of their time indoors in childcare centers. Poor indoor air quality is associated with nasal discharge, eye irritation, throat dryness, headache, irritability, and lethargy (Staes et al., 1994) (Figure 19.9). Table 19.2 contains guidelines for maintaining healthy air quality in day care or other classrooms (WEA, 2004).

Table 19.2 Guidelines for Maintaining Classroom Air Quality

Cleaning	• Use high-efficiency vacuums.
	• Hot water or steam "extraction" is the best for cleaning, with no strong chemicals or soaps.
	• Clean carpet at least quarterly.
	• Carpet should dry thoroughly—within 24 hours—after cleaning.
Classroom Behavior	
Classroom Behavior	• Maintain cleanable horizontal surfaces.
	• Do not allow food or beverages in rooms with carpeting.
	• Have children help by putting their chairs on the desk at the end of the day.
	• Wet-wipe surfaces weekly.
	• De-clutter shelves on the day the janitor is scheduled to dust.
	• Do not use a feather duster, it can disperse the dust.
	• Avoid clutter. Put items into plastic boxes that can be wet-wiped.
	• Avoid hanging items that collect dust: streamers, projects, papers, piñatas, etc. Clean them if you hang them.
	• Do not keep warm-blooded animals in classrooms, including hamsters and guinea pigs. If you do, clean the cages often and house them away from air vents.
	• Remove fleecy items that can harbor allergy triggers—old overstuffed furniture, area and throw rugs, pillows, blankets, or stuffed animals that can't be properly and regularly cleaned.
	• Avoid use of "stinky" dry-erase board markers and cleaners.
	• Avoid spray adhesives, contact cement, and volatile paints.
	• Use nontoxic water-based materials when possible.
	• Do not bring chemicals, paints, or sprays from home without clearing them with the maintenance staff.
	• Avoid room deodorizing sprays or plug-ins.
	• Do not use ozone machines in occupied areas.

(continued)

Table 19.2 Guidelines for Maintaining Classroom Air Quality (Continued)

Classroom Behavior	• Inventory supplies and materials in terms of indoor air quality: Consider if they are of low odor. Can they create dust or other particles? Do they harbor allergens?
	• Report water leaks, water stains, damp materials, or "musty" or "moldy" smells immediately.
	• Keep your room comfortable—learn how to operate your heating and cooling system for comfort and energy efficiency and communicate with the facility staff regarding mechanical systems for your classroom.
	• Help ensure students get adequate fresh air ventilation. Do not block air supply or exit grills.
	• Do not turn off ventilators—work with maintenance staff to fix noisy units, control temperatures, and control drafts.
	• If a classroom does not have mechanical ventilation, open windows or doors frequently to provide a quick "flush-out" of the stale air.
	• Monitor windows. They should not show condensation except on the very coldest of days. Condensation suggests either a moisture problem or not enough ventilation (or both!).
	• Notify maintenance if you smell odors or particle matter from other zones in the building: shops, science, laminator, locker room, graphics, custodial, storage areas, combustion equipment, kitchen, buses at the curb, etc.—air should move from "clean areas to dirty areas."
Administration	
Administration	• Educate teachers and children about indoor air quality and asthma and allergy triggers.
	• Request a ventilation system that supplies the state code minimum of 15 cubic feet per minute per person outside air at all times the school is occupied.
	• Remove stained ceiling tiles—they can harbor mold, and it is hard to tell if or when they get wet.
	• Install walk-off mats that provide "four good footsteps" at all outside entry doors.
	• Hallways should be hard surface, not carpet.
	• Assign an indoor air quality (IAQ) person so all IAQ (moisture, odor, mold, etc.) issues can be directed to one person.
	• Make sure all teachers and staff know who the IAQ person is.

Source: From the Washington Education Association website: http://www.wa.nea.org/iaq/tips.htm (Paraphrased with permission.).

Indoor air quality testing should comply with local ordinances and include carbon dioxide, lead, asbestos, and radon testing. Respiratory problems and irritability—for children and caregivers—can occur with high levels of carbon dioxide.

Public awareness in the United States of the dangers of lead has eradicated the use of lead-based paint. However, the original paint in some older facilities or home-based centers may be lead based and exposure can occur if the paint chips or flakes (Box 19.2). The presence of lead (in paint, dust, stoneware, and other sources) can result in long-term health and developmental problems including low IQ, learning disabilities, hyperactivity, and attention-deficit disorder. Basement space (only permitted for children older than 2 years) must be free of friable asbestos and radon in excess of 4 picocuries per liter of air (AAP, 2002).

BOX 19.2 DANGERS OF LEAD POISONING

Although public awareness of the dangers of lead poisoning in children is growing, children in the United States and other nations are still at risk. About one million children in the United States under the age of 6 have lead poisoning (1999).

The CDC (Center for Disease Control) and the WHO (World Health Organization) recognize that the problem is not confined to the United States. Recent research indicates that levels of lead in the blood that were previously thought to not pose a risk are now questioned. Blood lead concentrations below 10 mg/dL in children under the age of 5 are associated with adverse health effects, including low IQ scores (2003).

See:
www.humanics-es.com/lead.htm
www.nursingceu.com/NCEU/courses/lead

Manufacturers and designers are responsible for ensuring that furniture, floor coverings, window treatments, laminates, and paint do not contain toxic chemicals. For example, pressure-treated lumber is toxic to children because it is made with the preservative arsenic.* Now that it is recognized as hazardous for children, it is no longer widely used for building outdoor playground equipment, decks, and patios.

Other examples include wall-to-wall carpeting and particleboard. Newly installed carpeting and particleboard can release formaldehyde, as well as other volatile organic compounds (VOCs). Formaldehyde is a potential contributor to asthma and other respiratory and neuropsychological problems. Do not install carpeting or other flooring while children are in the space. Ventilate the space for at least 24 hours before the space is occupied.

Indoor air quality also includes temperature recommendations for day-care centers (see Table 19.3).

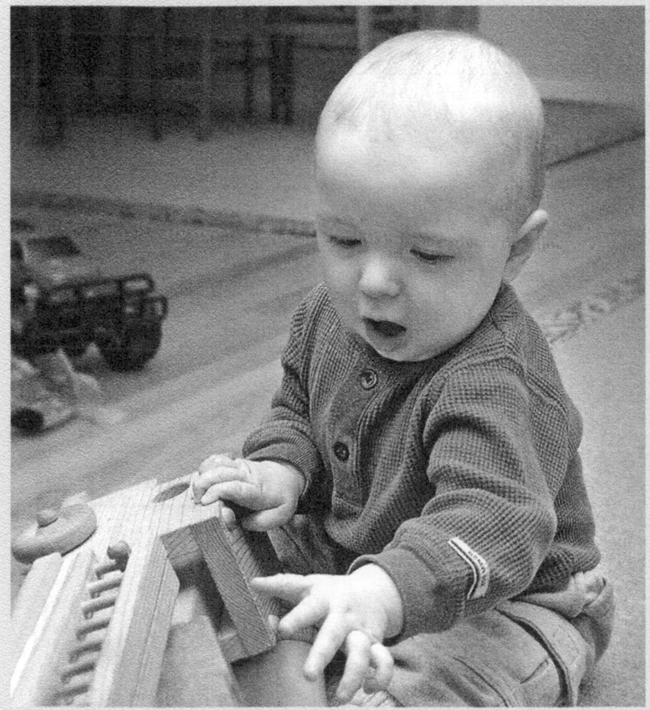

Figure 19.9 (Ursy Potter Photography) Children's breathing zones are closer to the floor than adults (Staes et al., 1994). Architects and builders must consider that the air quality for adults (who are taller) may be quite different from children. Air exchange should be a minimum of 0.42 cubic meters (15 cubic feet) per minute per person of outdoor air.

* Arsenic is injected into pressure-treated wood to protect it from invasion by insects, fungi and molds. However, it leaches from the wood. See chapter 8 of this text for more information on accidental poisoning.

ANTHROPOMETRIC STANDARDS

Children from birth to age 5 grow rapidly and vary considerably in body sizes and rates of growth. While Table 19.4 provides basic anthropometric dimensions of preschool age children, Chapter 3 covers the topic in more detail.

Table 19.3 Temperature Recommendations for Day Care Centers

Winter	Summer	Relative Humidity
18°C–24°C (65–75° F)	20°C–28°C (68–82° F)	30%–50%

Source: AAP (2002).

For more detailed information on child anthropometrics, please see Chapter 3 of this book.

Table 19.4 Selected Children's Anthropometric Standards

Age	Boys	Girls	
	Height		
0 year	56 cm (22 in.)	51 cm (20 in.) (Length for Infants)	
1 year	76 cm (30 in.)	71 cm (28 in.)	
2 years	86 cm (34 in.)	86 cm (34 in.)	
3 years	97 cm (38 in.)	97 cm (38 in.)	
4 years	102 cm (40 in.)	102 cm (40 in.)	
5 years	112 cm (44 in.)	107 cm (42 in.)	
	Boys and Girls		
	Standing Eye Level	Seated Eye Level	
1 year	76 cm (30 in.)	41 cm (16 in.)	
2 years	81 cm (32 in.)	43 cm (17 in.)	
3 years	86 cm (34 in.)	44 cm (17½ in.)	
4 years	94 cm (37 in.)	46 cm (18 in.)	
5 years	99 cm (39 in.)	50 cm (19½ in.)	
	Vertical Reach to Grip	Span (Arms)	Hand Width
1 year	86 cm (34 in.)	NA	5 cm (2⅛ in.)
2 years	97 cm (38 in.)	86 cm (34 in.)	6 cm (2¼ in.)
3 years	107 cm (42 in.)	97 cm (38 in.)	6 cm (2½ in.)
4 years	117 cm (46 in.)	102 cm (40 in.)	7 cm (2⅝ in.)
5 years	127 cm (50 in.)	107 cm (42 in.)	7 cm (2¾ in.)

Table 19.4 Selected Children's Anthropometric Standards (Continued)		
Age	**Boys**	**Girls**
	Seated Knee Height	Head Width
0 year	NA	10 cm (3¾ in)
1 year	20 cm (8 in.)	12 cm (4¾ in.)
2 years	23 cm (9 in.)	13 cm (5 in.)
3 years	28 cm (11 in.)	13 cm (5¼ in.)
4 years	30 cm (12 in.)	13 cm (5¼ in.)
5 years	33 cm (13 in.)	14 cm (5½ in.)

Source: From Ruth, L.C., *Design Standards for Children's Environments*, McGraw-Hill, New York, 1999.

DISABILITIES

Childcare facilities need to accommodate children with and without disabilities. Although this chapter will not cover designing for children with disabilities in great detail, some suggestions are offered.

In the United States, readers can consult the Universal Building Code and the Americans for Disabilities Act Accessibility Guidelines (ADAAG). Other countries have their own guidelines or regulations for consultation. The ADAAG does not have accessible (wheel chair) requirements (height of table and knee clearance) for tables for children under the age of 5. However, these require 76 × 122 cm (30 × 48 in.) of clear floor space parallel to a table or counter to enable children in wheelchairs to approach the worktop surface from the side (Ruth, 1999).*

Children with mobility impairments may use furniture for support. For example, a child who has difficulty standing up or sitting on the floor unassisted (Olds, 2000) might lean on a chair or table for support. Stable construction and proper placement are particularly important for such users.

Deaf children benefit from environments with surfaces that transmit vibrations, such as wooden platforms and lofts. This enhances their awareness of ambient sounds and movements in play spaces (Olds, 2000).

Changes in floor texture or color can help orient both visually impaired and sighted children. Such changes may delineate activity areas or signal an upcoming doorway. Visually impaired children can use the information to explore their environments and learn where things are and how they are shaped. Visually impaired children need play spaces that are free of glare. Quality lighting encourages their free, independent movement through childcare spaces.

It is important to keep consistent circulation paths because visually impaired children memorize special configurations. Large muscle equipment such as swings, tumblers, and rockers give visually impaired children support and help them move their bodies in different directions (Olds, 2000).

* For a complete set of anthropometric standards and discussion of accessibility guidelines, consult Linda C. Ruth (1999). *Design Standards for Children's Environments.*

CAREGIVERS

One of the most frequent musculoskeletal symptoms among adult childcare workers is lifting-related lower back pain. Since those who work with infants and toddlers do more lifting than those who work with preschoolers, the former are at greater risk.

Caregivers lift, carry, and hold children when consoling or feeding them. They bend over children while changing their diapers, feeding them at their low tables and even while playing with them (Grant et al., 1995). Raising changing tables and having children climb steps to have their diapers changed can help. Of course, caretakers have to supervise and assist children's climbing (Figure 19.10).

Adult-size, comfortable furniture is important for caretakers. The Netherlands requires that chairs and tables used for activities that take more than 4 minutes (e.g., eating with children) be adult height.

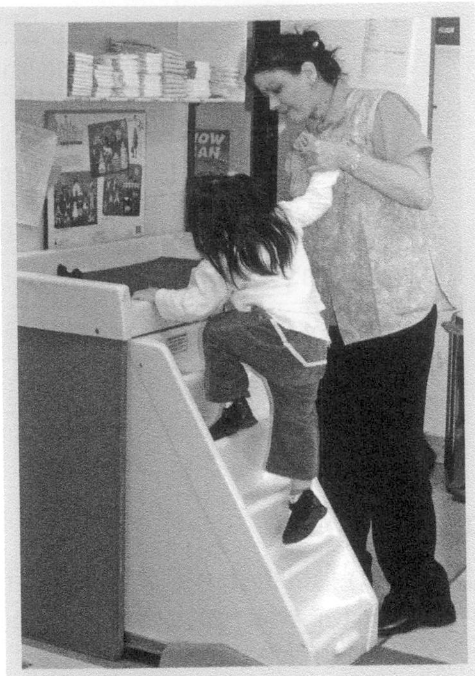

Figure 19.10 (Ursy Potter Photography)
Back pain is a common musculoskeletal symptom among caregivers of young children. Having a higher changing table means caregivers do not have to lean over the changing table while changing diapers, thus reducing their sustained back flexion. Steps up to the changing table for children who are able to climb with assistance reduce caregivers lifting requirements.

RECOMMENDATIONS FOR ADULTS:

- Provide adult-size chairs and tables for administrative work.
- Provide adult-height sinks for hand washing and rest breaks.
- Provide adult-sized seating with good back support for caregivers when they feed or cuddle children.
- Provide wide, cushioned seats to distribute the caretaker's weight over a larger portion of the buttocks (Grant et al., 1995).

DETAILED DESIGN ACCORDING TO AGE

As mentioned, children from birth to 3 years experience the world through their senses (Box 19.3). They touch and examine, smell, taste, listen, and watch! Infants and toddlers learn through experimentation, imitation, repetition, and trial and error (Lally et al., 1987). Childcare centers serving infants and toddlers must support all of these activities. For each age group, there will be differences in the design of toys, play materials, equipment, furniture and furniture arrangement, flooring, walls, doors, windows, storage, and lighting.

INFANT PLAY ITEMS

- Busy Boards
- Nesting Bowls
- Snap Together Beads
- Small Blocks
- Squeeze Toys
- Shape Sorters
- Foam Mats
- Tunnels (Crawling)
- Wedges or Bolsters
- Swings
- Ramps

(Bredekamp, 1997)

BOX 19.3 INFANT AND TODDLER ACTIVITIES (GREENMAN, 1988)

Infants:

- See, look, watch, inspect, hear, listen, smell, taste, eat, feel, touch, reach, grasp, hold, squeeze, pinch, drop, bang, tear, clap, take in and out, and find.
- Roll, sit up, turn, crawl, creep, swing, rock, coo, babble, imitate sounds, react to others, solicit from others, and experiment.

Toddlers:

- Walk (in, out, up, down, over, under, around, and through), climb (in, up, over, on top), slide, swing, bang, jump, tumble, stack, pile, nest, set up, collect, gather, fill, and paint.
- Take apart, put together, dump, examine, sort, match, select, order, carry, rearrange, put in, take out, hide, explore, investigate, imitate, and discover.
- Draw, smear, mix, pour, sift, splash, doll play, label, cuddle, follow directions, hug, kiss, wash, eat, dress themselves, and read labels.

INFANTS: AGES 0 TO 18 MONTHS

TOYS, PLAY MATERIALS, AND EQUIPMENT: This is a time of rapid development of gross and fine motor skills. Place toys and materials within the child's skill level to prevent frustration; at the same time, it should challenge them to promote interest and development (Figure 19.11). As children learn through their senses, they try almost everything to discover just how the world works. Toys should expose children to different sensations. The mouth has extensive nerve endings, and children of this age put nearly everything in their mouths.

Varying materials expose children to differing tactile sensations. Older infants enjoy playing in water or sand placed in a trough that is 30.5 cm (12 in.) high. Children prefer bright colors and gravitate to colorful toys and books.

While developing fine motor skills, children can begin to disassemble their toys! Therefore, avoid toys with small parts for infants.

For gross motor skills, toys that children can climb or move upon enhance equilibrium and kinesthetic sensations. Child-powered walkers are not recommended, as these cause more injuries than any other baby product (AAP, 2002) and are associated with delayed motor development in infants (Aronson, 2000). Playpens, jolly jumpers, and walkers are not permitted in childcare facilities in Nova Scotia, Canada (Day Care Regulations, Day Care Act Nova Scotia Reg. 2000). Experts debate both the safety and developmental enhancing qualities of jumpers.

> Infant care programs generally have 6–8 infants in the United States and elsewhere.

Cognitively, infants and toddlers are just learning they can impact their environment. Toys should respond to a child's actions while encouraging a range of play behaviors. Children will repeat actions with toys that have corresponding reactions, thus learning about cause and effect (Harrison, 1990).

Infants and toddlers are in a great period of discovery, and they want to learn about everything. They happily explore and manipulate their environment—so the environment must be safe as they will be everywhere, getting into everything.

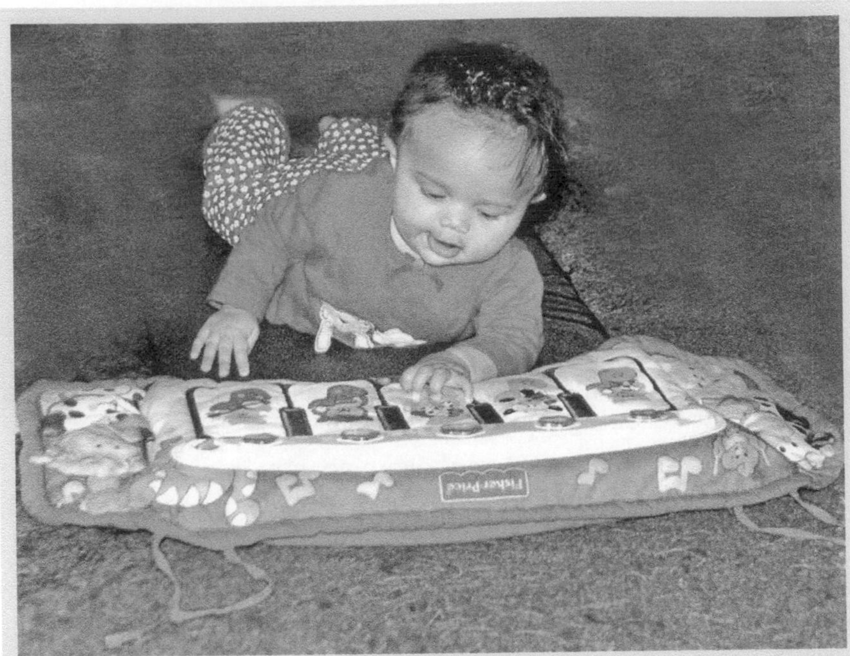

Figure 19.11 (Ursy Potter Photography)
A baby's interest often goes beyond what she can reach. By putting items within their reach, they are able to continue to explore their environment. This item stimulates vision, sound, touch, and proprioception.

FURNITURE: Infants need furniture that allows for sleep, diaper changing, and eating. Furniture designs and its arrangement can promote safety and social interactions.

Sleeping Separate sleep alcoves are ideal to keep nearby activities from disturbing sleeping infants. Canada requires separate sleeping rooms. Few day-care programs in the United States have separate sleeping rooms; for those that do, some states require that they be staffed at all times. Intercoms or windows into the space may not comply with these regulations.

Although most day-care programs schedule sleep times, not all children will sleep at the same time. Arrangements must permit caregivers to see which children are asleep and which are awake and active.

Each child requires a crib (full size or portable), yet cribs take up considerable floor space for children's play and crawling. Even so, stacking cribs are not recommended (AAP, 2002) and not permitted in some states to prevent falls. Spacing cribs a certain distance apart may decrease transmission of infectious diseases and the United States requires that cribs be at least 1 m (about 3 ft) apart. Canada requires placing cribs at least 46 cm (18 in.) apart.

> Caregiver Safety
>
> In the Netherlands mattresses must be 95–110 cm (37–43 in.) above the floor so adults do not have to bend to lift the child in and out of the bed.

Cribs in the United States must meet standards set by the Consumer Product Safety Commission (CPSC) and the American Society for Testing Materials (ASTM). Both list the standards mentioned in this chapter.

Crib construction is important. Obviously, cribs must be strong and sturdy. They should not have loose or protruding brackets, screws, or other hardware or have broken parts.

In some older cribs, children's heads could fit between the slats. Crib slats should be less than 6 cm (2 3/8 in.) apart* to prevent possible strangulations. Cribs with Plexiglas ends provide easy visual access from all sides. Head or footboard cutouts also offer easy visual access, but they must comply with standards for preventing head entrapment.

Infants can also slip between a mattress and the side of a crib and suffocate. Therefore, avoid gaps wider than two adult-size fingers between the mattress and the lowest side of the crib in its lowest setting (AAP, 2002). It is dangerous to place infants on waterbeds, sofas, soft mattresses, or other soft surfaces. Infants may suffocate when they are unable to raise themselves above the surface (AAP, 2002).

Crib "gyms" can be a safety hazard once a child is over 5 months of age. Five-month-olds can grab them, put parts in their mouths, or inadvertently wrap the strings around their necks (AAP, 2002).

Other sleeping arrangements for infants include cradles (for infants who are not able to pull themselves up), large beanbag chairs (surface must be nonabsorbent and easily cleanable), large baskets with pillows (the Reggio Emilia program† in Italy), hammocks, and sleeping mats. Each child should have at least two changes of personal bed linens in case of spills or other accidents.

Diaper changing The design of infants changing areas need to fit both the child and the caregiver, physically and socially. Changing time should not simply be a chore for the adult; it is an opportunity for the caregiver and child to interact through language and eye contact (Figure 19.12).

If the diaper changing station is not located in a toilet room with an exhaust system, the station should have its own exhaust system to reduce unpleasant odors. A constant-volume outside–inside exhaust grille located directly over changing tables and toddler training toilets is good (Barker, 2002) and an operable window nearby is ideal. A radiant heat panel suspended over the changing area will warm the child (Olds, 2000). Infants should never be changed on the floor (even with a mat) to protect others from possible spread of infectious diseases. Table 19.5 lists additional requirements for diaper changing areas.

Figure 19.12 (Ursy Potter Photography)
Diaper changing is one more opportunity for the caregiver and child to interact with each other. Making eye contact, talking and providing comfort can all be part of this interaction.

> The Calgary, Canada Environmental Health Dept. recommends foot, knee, or wrist-operated controls on sinks (2002).

* More about these recommendations by Hendricks can be seen at www.earlychildhood.com and www.cpsc.gov/cpscpub/pubs/5020.html

† The Reggio Emilia (Italy) approach emphasizes classroom projects that provide learning experiences to young children. A critical feature of this approach is that the physical environment is considered the child's third teacher (after parents and teachers). Therefore, the spaces should be well planned, welcoming, and reflect the culture.

RECOMMENDATIONS FOR DIAPER CHANGING AREAS:

- Provide railings so that the child cannot roll off the table (infants should not be strapped to the table).
- Use nonabsorbent surface material that is easy to clean.
- Locate necessary materials immediately adjacent to the changing table including clean diapers, a sealed and odor-controlled container for soiled diapers, cleaning supplies, lotion, and clean clothes. Caregivers should be able to keep one hand on the child at all times, even when reaching for objects.
- Locate the hand washing sink with a soap dispenser and towels near changing table (not used for food preparation—there must be a physical separation from food preparation area). Elbow-operated handles or foot pedals will help avoid contamination (Olds, 2000). Place the changing table so the caregiver can easily see (and be seen) by other children in the room.
- Set changing tables at adult height to help prevent back injuries for caregivers (AAP, 2002).

Table 19.5 Infant Diaper Changing Are

Changing table	91 H x 60 W x 107 cm L (36 H x 24 W x 42 in. L)
Pad	5–7 cm thick (2–3 in.)
Raised edge (around the pad, three sides)	15 cm (6 in.)
Diapering station (sink and changing table)	5.5–9 sq. m. (60–100 sq. ft.)
Personal storage units (one per child)	25 W x 35 D x 15 cm H (10 W x 14 D x 6 in. H)

Source: From Olds, A.R., *Child Care Design Guide*, McGraw-Hill, New York, 2000.

FEEDING: Caregivers who bottle feed infants need a comfortable place to sit. Often, when mothers come to the center to feed their infants, they come to breast feed. Providing a private space, comfortable chair, and pleasant surroundings will make the visit more pleasant. Position overhead lights so they do not shine in the eyes of supine infants.

When infants are ready for soft foods, they will need to sit. At first, they may need supported sitting. Some programs use infant car seats for this purpose. Caregivers can also use a high chair or hold an infant on their lap and place the food on a nearby table or counter. Newer high chairs may transition between supported and unsupported sitting (Figure 19.13).

Once children can sit unsupported, it is better to use child size chairs (with sides) and tables or provide child-height feeding chairs with trays. It is easier if trays are removable for cleaning and storage.

Older children enjoy sitting at tables in groups of three to four with an adult. This smaller ratio of caregivers to children allows caregivers to assist children and quickly intervene if a child begins to choke. Mealtime is another opportunity for verbal interaction, as well as modeling of eating and social behaviors. Table 19.6 shows recommendations for infant feeding.

Table 19.6 Infant Feeding Area

Table	30–35 cm high (12–14 in.)
Chair with sides	12–17 cm (5–6½ in.) seat height
Warming area	Sink, refrigerator, food warming appliances (microwave ovens are not recommended), storage for utensils, juices, crackers, etc.

Source: From Olds, A.R., *Child Care Design Guide*, McGraw-Hill, New York, 2000.

Figure 19.13 (Stokke LLC)
Child seating with a footrest decreases their "fidgeting," improves body stabilization, and increases reach distance (Hedge, et al. 2003).

SOCIAL INTERACTION: Infants need to be held and cuddled. Their growth and development depends on it. Rocking or gliding chairs add gentle movements (and stimulate inner ear canals for balance and equilibrium) as well as comfort while adults hold a child. Careful placement of chairs, watching crawling infants, and blocking access helps protect crawling infants from accidentally placing their fingers or other body parts beneath the casters of a rocking chair. Another alternative is to hold infants while sitting on low sofas that permit adults to interact easily with other children in their care.

Arranging adult-size and child-size furniture in various areas encourages adults and children to disperse throughout the room. This easy movement and resting throughout the area should foster more frequent and relaxed interactions.

Furniture placement should not block a caregiver's view of children. Instead, open viewing lets caregivers see and intervene with little difficulty.

SAFETY: Children under 2 years of age are more prone to sustaining injuries associated with furniture than older children (Mackenzie and Sherman, 1994). Young children fall from or against furniture. They use furniture to pull themselves up to standing and sometimes fall while "cruising" around the furniture.* They may also quickly climb onto furniture when caregivers are busy with other children or activities.

Furniture in preschools should be sturdy to support a child's weight. Rounded furniture edges protect children from bumps or scrapes. Retrofitted rubber "bumpers" can cover sharp table edges, but purchasing original furniture without sharp edges eliminates the possibility of the bumpers coming loose.

* "Cruising" is a term used to describe when a child cannot yet walk independently, instead they 'cruise' around furniture, holding on to it for support while they walk.

FLOORING: As infants begin to move around on their own (creep and crawl) they need unencumbered floor space. Warm floors (without drafts) that are soft, easily cleaned, and nonabrasive are best.

If carpets are used, these should have low static electricity. The edges of rugs can be a trip hazard. Securing the edges of area rugs to the floor helps keep infants and toddlers from tripping. Although many caregivers might not consider carpet-drying time when purchasing floor coverings, it is important since moisture encourages mold growth (see Table 19.2).

Vinyl tile (without asbestos) floors can be a good choice as they provide both resiliency and durability. These may be preferable to sheet vinyl because it is fairly easy to replace individual tiles.

Floors can provide a variety of tactile experiences, including more than one type of surface can contribute to a child's development. Floor-based tactile experiences can also include activities such as sitting on the floor while painting.* Both on- and off-the-floor activities influence floor-covering choices. For example, tile flooring and drains are good choices for water-play areas.

Carefully placed floor level changes also provide variety and opportunities to explore without tripping. Carpeted platforms 10 cm (4 in.) high and arranged on the periphery serve this purpose while providing new opportunities for adults to sit. Covered mattresses serve the same purpose.

Mirrors (shatterproof, tempered glass) and pictures can be adhered to the floor to provide visual feedback. This unique design adds complexity and provides infants with a new perspective. Of course, adhesions must not allow infants to dislodge the mirrors from the floor.

WALLS: Pictures and other items (e.g., activity boards) to stimulate vision are typically placed on walls. Unfortunately, even in preschool settings they are often at an adult height! It is best to place them low, so children can see them while scooting, crawling, sitting, or standing (Figure 19.14). Shatterproof (tempered glass) mirrors are especially enticing to infants who enjoy looking at themselves.

Wall hangings of various materials add a tactile element to the visual experience. Textured walls (nonabrasive) and carpeting on lower portions of the wall provide tactile information. Wall carpeting also reduces noise, if there are acoustical concerns.†

Wall coloring (and even picture placement) should not over stimulate (Figure 19.15). It is better to rotate pictures and hangings than to overwhelm infants with a barrage of colors, shapes, and textures. In terms of colors, the toys and play materials, displays of children's artwork, and people's clothing provide an adequate amount of complexity. Adding more can make an environment too busy.

A nice touch is to have wall coloring reflect the activity in the space. For example, the walls of a space for sleeping may have one color, while walls in a space for active play have another color. A space may have specific architectural features (e.g., an arched way) or photos depicting the activity for that area may adorn the wall.

Glare interferes with vision and is irritating for children and adults. Matte finishes help reduce glare (Olds, 2000). Pure white walls contribute to glare, while warm whites are easier on the eyes. Many playrooms use satin-type finishes that are easy to clean.

* See Chapter 6 for guidance regarding the impact of floor sitting on muscle development and posture. While on-the-floor sitting is fine for occasional sitting, having children consistently sit unsupported is not a good practice.

† Keep in mind that materials used for walls, ceilings, and flooring should contribute to an appropriate acoustical environment. External sound should be minimized but controlling internally generated sound is also important. For children with hearing impairments, it is especially important that speech is intelligible in the childcare space.

Figure 19.14 (Lorraine Maxwell) **Figure 19.15** (Valerie Rice)
It is important to place items low, so preschoolers can see and interact with them. There is a balance to maintain, though, as children can be over stimulated with too much visual input. In the classroom on the right children can look at cards and posters depicting shapes and colors that are placed on the floor and low on the wall. Even the rug has interesting shapes and colors. Yet, the busy flooring combined with wall decorations may be overstimulating for small children. In addition, the very high wall decorations are really more for visiting parents and teachers, as they are too high to benefit preschoolers.

In the room on the left (activity center 1 with cushions), children can sit and read a book or play with the objects that are on the low shelves and within their reach. The decor is interesting for children, yet not overstimulating. One way to maintain an interesting room without overstimulation is to limit the number of posters and placards in children's visual field, and rotate them on a regular basis.

DOORS: Doors provide visual, acoustic, and access separation between spaces. Although separation is often necessary, most children and adults are curious about what is on the other side of the door.

Door glazing can provide crawling infants with a view to the other side. Glazing at adult height lets adults see children who are crawling or sitting near the door and avoid hitting them when opening the door. If a large expanse of a door is glass (or glazed), some children may not realize it is a door at all. Placing decals on the door glazing will help people realize it is a door, so they do not accidentally walk into it.

Dutch doors work well in infant spaces (Figure 19.16). Doors accessible to children should have finger-pinch protection devices such as rubber gaskets or slowly closing doors (AAP 2002). Some spaces require solid doors, such as those for fire safety, storage areas, janitorial closets, and adult bathrooms.

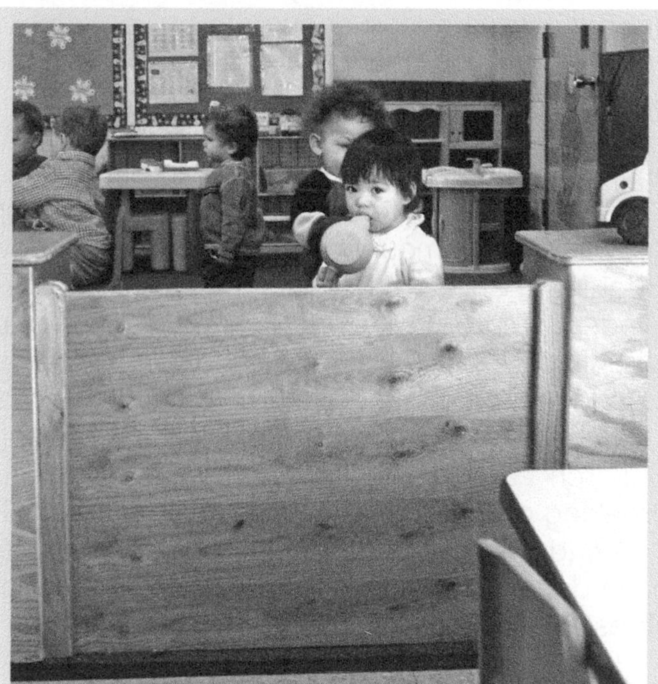

Figure 19.16 (Ursy Potter Photography)
Dutch doors work well in day cares. They permit light, sound, and visual connection between spaces while controlling infant and toddler acccess (Olds, 2000).

Door placement can influence the use of spaces. For example, placing a door in a corner of a room affords more wall space for activity areas, whereas placing a door in the middle of a long wall in a large rectangular room leaves the four corners available for activity areas.

If children need to be well aware of a particular door, it can be painted a different color so it stands out. On the other hand, if a space feels rather small, doors can be the same color as the wall so they tend to "disappear," making the room seem larger (Olds, 2000).

Crawling infants like to go through doors at their height. In fact, most small children enjoy going in and out of doors. If children must go into and out of a door independently, they must also be able to safely open and close that door independently and the opening must be large enough for the child's entire body. This sounds like common sense, but playhouse doors do not always permit such easy access and movement.

WINDOWS: Nearly everyone likes to be able to see outside at some time during the day. Adults like to check on the weather visually and children like to watch their natural world changing before their eyes (Figure 19.17).

For infants to see out of a window, the window needs to start about 30 cm (1 ft) above the floor. Of course, the view has to be worth seeing.

Windows do more than provide outside views, they bring natural light and ventilation to a classroom. Skylights also supply natural light. Operable windows provide full-spectrum light into spaces when open and let the teacher have some control over classroom ventilation. Infants and toddlers should not be able to reach windows or the ventilation adjusters. If operable windows are at child height, then limit ventilation adjusters to 15 cm (6 in.) or less (Olds, 2000) for safety reasons.

Screens are necessary if windows are to be open for ventilation. However, screens are not always strong enough to prevent a child from falling from a window. It is therefore important to use caution and control access to open windows with infants and toddlers who will not be fully aware of the dangers.

Window coverings (including coverings for skylights) permit caregivers to control the amount of natural light coming into the space, important for nap times! Various window coverings block different amounts of light, so one piece of information for selecting covering will be the amount of light it blocks. Window coverings with cords pose the danger of catching a child's neck in the cord, especially when the cords loop at the end. Venetian blinds (lead free) with wands and roller shades with a continuous beaded chain work well for childcare spaces. Drapes and vertical blinds may not be a good choice, as they usually extend to the floor and children can pull them down and injure themselves (Olds, 2000).

Shatter-resistant windows help ensure safety, especially with very young children. Children can unexpectedly hit windows as they do not yet understand the concept of breaking glass. They may also inadvertently fall against low windows.

Windows with deep reveals* help to reduce glare, provide protection from drafts, and have deep sills for placing plants and other displays. Rooms with windows on at least two sides also help to reduce glare.

Coupling window placement with door and furniture placement helps design environments that take advantage of wall and room spaces, as well as controlling room temperature. For example, in the northern hemisphere, south facing light is warm and north facing light is cooler. Tempered, double- or triple-thermopane glass that is tightly sealed will prevent drafts (Olds, 2000).

* A reveal is the wall opening around the window between the frame and the outer surface of the wall (Olds, 2000).

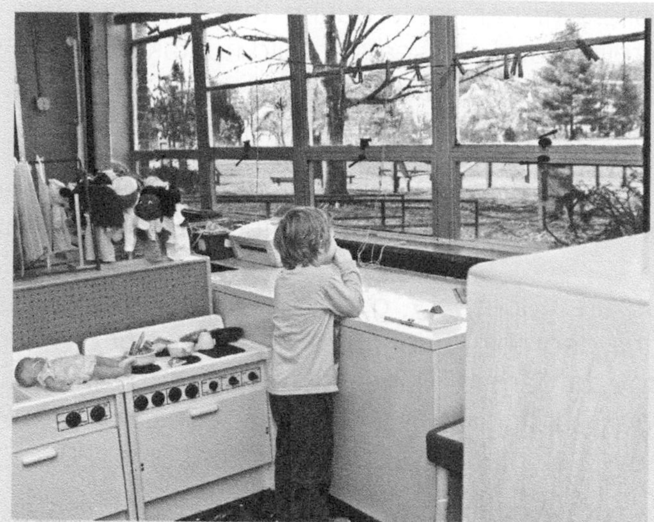

Figure 19.17 (Ursy Potter Photography)
Windows bring natural light into the classroom and an opportunity to teach children about their environment.
Surprisingly, architects often forget to consider the child's view when placing windows into an architectural drawing. One way to help trigger the thought is to include the child user and the landscaping in the architectural drawing. Children should never be able to reach the cords on blinds or adjust the windows themselves.

STORAGE: Convenient storage can be fun, encourage responsibility, and prevent floor clutter. Floor level bins in open shelves allow infants to get their own toys (Figure 19.18). Their size also limits the number of available items. Open storage and ample closet space lets caregivers stock additional toys and supplies. Locked storage—not accessible to children—is essential for storing hazardous items.

Every child needs personal storage space they can reach, for outer clothing, a change of clothes, bedding, and perhaps favorite toys from home. In most day cares and preschools, these open-face cabinets are known as "cubbies" (Figure 19.19).

Figure 19.18 (Ursy Potter Photography)
Storage can be fun, especially for infants. Infants like to take things out of containers and put them back in. They like to examine items and hear them drop onto a surface. Selecting bright colors and placing them within reach helps young children learn a sense of place, order, and responsibility.

Figure 19.19 (Lorraine Maxwell)
Children's "cubbies" provide spaces for their jackets, backpacks, shoes, boots, and even a few items for comfort, such as a stuffed toy from home. Once again, they should be child-size, so children can put away and retrieve their things by themselves.

If parents leave strollers or car seats during the day, they will need ample storage space for those also. Strollers and car seats should not take up play space. Ideally, these items will be stored outside of the class or play room.

LIGHTING: Even with windows, some artificial lighting will probably be needed. Full spectrum lighting is best and warm lighting is preferable to cool lighting even though it is more expensive.

Two types of lighting are necessary, general illumination and task lighting. It helps if caregivers can control the degree of brightness (not just on or off). Wall sconces and other indirect lighting can provide general illumination without the institutional effect of ceiling fluorescent lighting. Track lighting is nice for tasks (Greenman, 1988), and task lighting should not shine in the eyes of infants lying on the floor (Olds, 2000).

Some states within the United States do not permit table lamps as task lighting. Dresser lamps are good alternatives that also impart a "homey" feel.

Locate and construct light fixtures to avoid hazards to children or adults. This includes the risk of food contamination if a fixture breaks (CEHD, 2002). Caregivers should have easy access to changing bulbs. Once again, avoiding glare will benefit both children and adults.

TODDLERS—AGES 18 TO 36 MONTHS

In most childcare centers toddlers and infants stay in separate spaces. Some centers have transitioning programs, so younger toddlers (18–24 months) can continue to mix with older infants (12–18 months) until they "graduate" into a class of all toddlers (18–36 months). Spaces used for transitioning must meet the needs of both older infants and toddlers.

Toddlers still experience the world through their senses, but their ability to ambulate lets them explore more independently. Some items need to change as they become more independent, in order for their play space to be safe, challenging, and interesting. At each age, children need to explore and experiment and not get hurt in the process!

Caring for a room full of active toddlers is demanding. Therefore, the physical environment should minimize supervision issues and maximize opportunities for adults to provide emotional support and encourage exploration and independence.

PLAY ITEMS FOR TODDLERS

- Stairs
- Balance beams
- See saw
- Small climber
- Suspension bridge
- Push and pull toys
- Construction & Transportation toys
- Dolls, arts & crafts materials
- Items for pretend play
- Sturdy picture books
- Nesting, stacking, & sorting toys
- Rocking toys

Appropriate design enhances physical, cognitive, and psychological development. For example, low, easy to access supply areas let toddlers retrieve and put away their play supplies. Playing with toys that match their age and development helps children learn. Children gain a sense of accomplishment and independence when the environment supports their needs.

TOYS, PLAY MATERIALS, AND EQUIPMENT: The increased mobility and dexterity of toddlers introduces additional safety issues. For example, toddlers like to take things apart. Toys with small parts, easily removed by a curious child, are a choking hazard. To guard against choking, toy pieces and *manipulatives* should be at least 3.2 cm (1 ¼ in.) in diameter and 5.7 cm (2¼ in.) in length (AAP, 2002).

As with infants, different textures help toddlers learn about their environment through touch. This can include water, sand, pebbles, or various shapes of uncooked pasta. Play with textures can take place in a trough (commonly called a water or combined water and sand tables) 45.7 cm (18 in.) high. Optimally, they should have their own drain with a cleanout trap under a plumbed faucet (Olds, 2000). The floor surface around the water table should be easy to clean.

FURNITURE: Toddlers need open floor space for play, as well as furniture specific to their routine such as napping and eating. Cribs are not necessary for toddlers (although cribs are permissible); cots or mats suffice.

Sleeping Children who are in day care four (or more) hours a day should have their own cot or mat. Regulations on the use of cribs and mats provide guidance on sleep arrangements. For example, some states in the United States do not permit padded mats placed directly on the floor for children under the age of 3.

> *Manipulatives* are small play items, such as Legos, small blocks, play people and animals.

Separate alcoves or areas for sleeping are ideal, but when space is at a premium it is acceptable to place cots or mats throughout the playroom. Spacing of furniture for sleeping is subject to local codes, but it is important to provide an area away from the general play vicinity since some children sleep longer than others.

Most cots used in the United States are mesh vinyl with metal or plastic legs. Cots typically measure 61 cm by 107 cm (24 in. by 42 in.). Stacking cots preserves space. Stackable cots with foam mattresses (as opposed to the mesh vinyl surface) are another option (Olds, 2000). These cots are attractive and the foam mattresses may be more appropriate for children with allergies or asthma. Clearly, the type of bedding impacts storage needs, which includes children's personal storage.

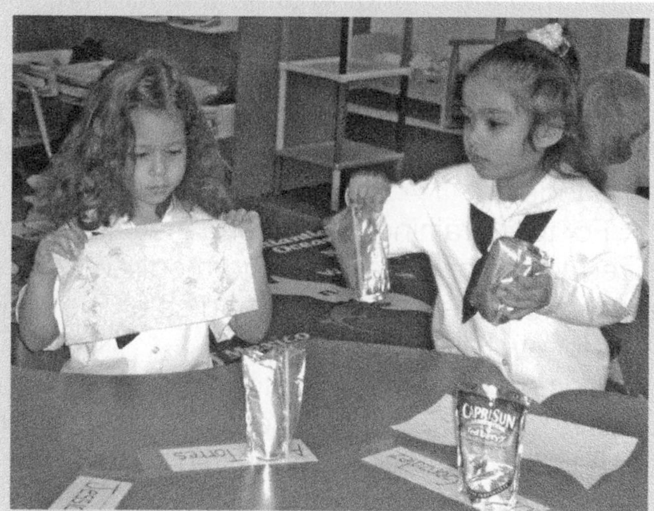

Figure 19.20 (Valerie Rice)
Young children take pride in eating and drinking independently. They consider it a privilege to pass out napkins and drink packs to classmates.
Small tables (seating 4 children with an adult caregiver) of various shapes (semicircle, square, rectangular, and hexagonal) are perfect for eating and other activities.

Table activities Toddlers enjoy eating on their own and should be encouraged to do so. Appropriate furniture helps toddlers accomplish this task (Figure 19.20).

Tables that combine to form a larger table create flexible space. Since many toddler activities do not involve table work, a flexible and less crowded room is helpful. If possible, the height of the tables should be adjustable between 40 and 45 cm high (16–18 in.) (Greenman, 1988). Occasionally, tables or counters attach to the wall for activities such as art. If these tabletops are 25 cm (10 in.) high, both infants and toddlers can use them comfortably (Olds, 2000).

Chairs should generally range from 20 to 25 cm high (8–10 in.). Chairs with good back support that permit toddlers' feet to rest firmly on the floor help encourage good postures (Figure 19.21).

Tables and chairs should be easy to clean and with rounded edges. Plastic chairs are less expensive, whereas wood chairs are more durable but do not stack as easily. Chairs should be light, so toddlers can move them.

Once again, adults need furniture that fits them. Adult-sized sofas or beanbag chairs can fit the caregiver comfortably, while they meet the cuddling needs of children.

Figure 19.21 (Lorraine Maxwell)
Sturdy child-size chairs provide support while children work. Chairs that stack add flexibility to the classroom as more floor space is available when needed. Yet it can be difficult to find comfortable chairs with good back support that are also stackable! Note the storage shelves in the background, they are uncluttered and at child level.

TOILETING: Toddler rooms need both changing tables and toddler-sized toilets. Requirements for changing tables are the same as for infants (see previous section) with the added caveat of accessibility. That is, toddlers should be able to climb onto the changing table with minimal supervision.

> In the Netherlands changing tables must be height adjustable to fit the needs of the caregiver.

An appropriately sized and stable step aid provides such accessibility. This eases the burden on caregivers who would otherwise have to lift each child onto the table. It also gives toddlers a sense of independence and accomplishment while they exercise their newly developing gross motor skills. It may also add some fun to an otherwise difficult time for a toddler.

For toilet training, the American Academy of Pediatrics (AAP) recommends either child-sized toilets or a step aid and modified toilet seat for adult-size toilets (AAP, 2002). Step aids must be easy to clean and safe. The AAP recommends a minimum of one toilet and sink for every 10 toddlers.

Since staff can come into contact with fecal matter and urine when children use "potty chairs," they are not sufficiently sanitary for preschools. Hands-off flushing is far superior for handling waste! Silent flushing toilets are nice because some toddlers fear the loud standard flush (Olds, 2000).

Good hygiene dictates placing hand-washing sinks immediately adjacent to changing and toileting areas. Toddlers using this area must have adult visual supervision at all times (Figure 19.22). See the infant section for ventilation and exhaust requirements for the toilet training area and bathroom.

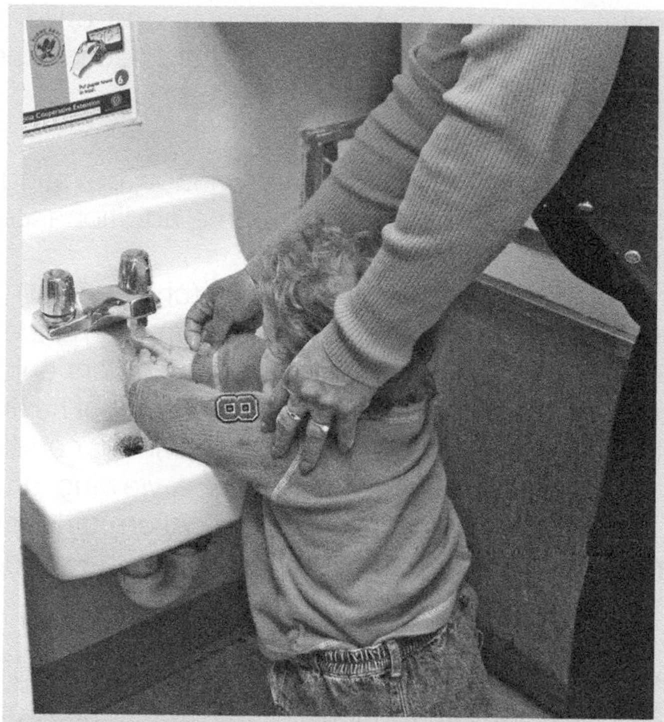

Figure 19.22 (Ursy Potter Photography)
Childsize sinks need to be located adjacent to the toileting area. Teachers can verbalize and posters can show the steps to hygienic hand washing.

FLOORING: Toddlers need floors they can crawl, sit, and walk on. Although they are more mobile than infants, they are just beginners when it comes to walking. They are unsure on their feet and occasionally fall. They sit and lie on the floor to play. Floors should be soft, warm, and free from drafts.

The floor coverings should be similar to the recommendations for infants. They need more open space to move around and practice their walking.

In areas where children engage in messy activities (water, paint, sand), tile floors with drains can be a good choice. Nonskid flooring materials helps avoid potentially dangerous falls when floors become wet.

WALLS: The requirements for walls in toddler spaces are similar to those for infants. One possible difference is the use of wall-mounted Plexiglas as easels for painting or drawing.

DOORS: Toddlers quickly learn the concepts of in and out, close and open. Doors become a source of fascination and unending delight. It will be important to provide them with opportunities to experiment with doors within a safe context.

Obviously, some doors cannot be "played with." For example, children should not have access to hazardous materials and should not exit the playroom space unaccompanied by an adult. Toilet areas in toddler rooms must have barriers to prevent children from playing in the toilets, for sanitary reasons and to prevent drowning.*

Door hardware is quite important. Toddlers easily operate hardware that meets ADA accessibility requirements. At this time, there is no specific solution to this problem but

* Children have drowned in toilets.

designers must carefully examine ADA requirements and child safety to make the best possible design choices (Barker, 2002). See the infant section for other issues related to doors in toddlers' spaces.

WINDOWS: Recommendations for windows and window coverings in the infant section of this chapter also apply to spaces for toddlers. Some additional considerations for toddler areas include cleaning and the use of lofts.

Toddlers get around more than infants and low (child-height) windows in high activity areas will require cleaning—often. However, this high-maintenance cleaning can be part of children's shared responsibility in the classroom. Most toddlers enjoy cleaning activities and the sense of responsibility. Teachers can eliminate the worry of having children use chemical cleaners by having them use balled-up newspapers or vinegar and water with paper towels.

If existing window heights do not permit outside viewing and the height cannot be changed, consider adding appropriately sized lofts. Alternately, place "practice stairs" for developing gross motor skills where a child can see out the window (but not fall out of or into the window).

STORAGE: Storage requirements for toddler spaces are similar to those for infants' spaces. Once again, toddler's new independence guides additions. For example, toddlers are interested in dressing themselves (they also require some assistance from caregivers); therefore, they need easy access to their outdoor clothing.

Toddler cubbies need separate spaces for boots and coats or snow suits, racks for drying mittens, and hooks at child height. Places for toddlers to sit and plenty of floor space will help while they dress themselves in the cubbies area. One of the first methods toddlers learn for putting on jackets requires them to place the jacket on the floor.* Sometimes toddlers spread their snowsuit on the floor and lay on top to put it on (Olds, 2000). Storage areas for a change of clothes and bedding may be separate from the cubbies. For example, it helps to locate additional clothing near changing areas.

Easy access to toy storage helps toddlers learn to put things away. Toddlers use repetition to learn when they practice the concepts of "in" and "out," as they put toys into and remove them from toy bins. Small bins work better than large ones since the latter may be too deep.

Too many toys can distract young children. Ample storage, accessible only to caregivers, permits adults to store some of the toys and materials. They can rotate toys, rather than having large numbers of toys available to children at any given time.

LIGHTING: Lighting issues for toddler spaces are similar to those for infant spaces.

PRESCHOOLERS—AGES 36 TO 60 MONTHS

Preschoolers are in what Piaget (1974) describes as the preoperational stage of cognitive development. This means they are not capable of logical thought (e.g., the ability to reason, associate one situation with another, or reverse an action). Although children in this stage depend less on sensory experiences for learning, they still benefit from a responsive environment. They also thrive on new information and continue to grow in independence and self-control. They play in social groups, are learning cooperation and social interaction skills, and may develop their first "best friend" (Figure 19.23).

* Toddlers sometimes lay their jacket upside down on the floor in front of them, with the jacket open (unzippered or unbuttoned) and the open side facing up. They then lean over and put one hand in each sleeve, stand up, and let the coat "fall into place" as it slides down their arms.

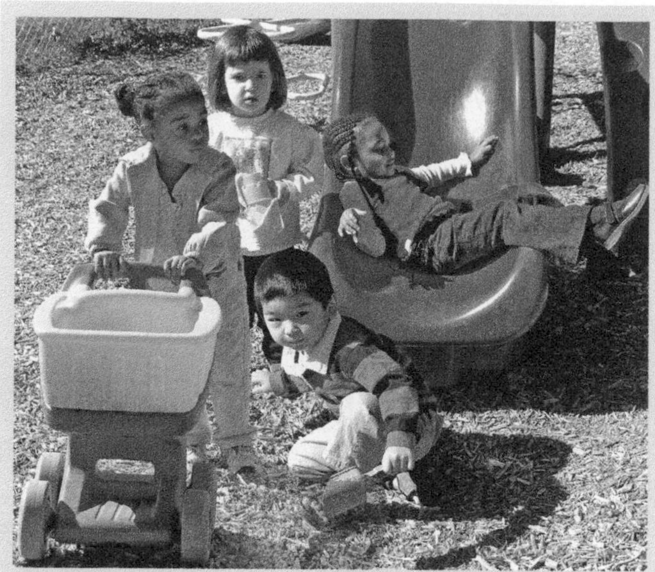

Figure 19.23 (Ursy Potter Photography)
Preschool children enjoy socializing as they learn to take turns, play together, and copy one another's antics.

Box 19.4 describes their physical skills.

BOX 19.4 PHYSICAL SKILLS OF PRESCHOOL AGE CHILDREN

In addition to all of the skills and interests of infants and toddlers, preschoolers can

- run, jump, hop, walk confidently,
- imitate others, participate in make-believe, listen to stories, sing songs,
- manage tools and implements, construct and build things from raw materials, and
- begin to play games with rules, follow directions, recognize gender differences.

TOYS, PLAY MATERIALS, AND EQUIPMENT: There is a large range in the abilities of preschoolers. They therefore need a wide variety of toys and play materials,* Since preschoolers are more independent (and can help get materials out as well as clean up), their classrooms tend to have more materials available to children than infant or toddler classrooms.

Activity centers are excellent for preschoolers. Challenges within each area can range from simple to complex as a means to individualize activities. Activity centers can have areas for messy play (paint, sand, and water), science, reading, writing, dramatic play, construction, music or listening, group activities, manipulative play, and quiet time.† Providing an area for quiet, restorative activities is essential to calm children and give them quieter areas where they can interact with others.

* See Bronson, M. (1995). *The Right Stuff for Children Birth to 8* for detailed suggestions of play materials for children in each age category discussed in this chapter.

† See Olds, A. (2000) *Child Care Design Guide*, pp. 350–369 for good descriptions of equipment and materials in each of these activity areas.

Play materials that encourage higher levels of activity include ramps, foam mats, wedges or bolsters, riding toys and wagons, balance beams, suspension bridges, and climbing structures (Figure 19.24). Grading challenges on climbing structures often means providing more than one way to climb or leave the structure, each with a different degree of difficulty. Preschoolers may use climbing structures inappropriately as they try their new skills (Barker, 2002). For example, a child might climb the structure correctly, but seek to leave it by jumping off in an unsafe way.

It is impossible to have the foresight into every potential misuse. However, designs should take potential misuse into account (when possible) and design to prevent or guard against it.

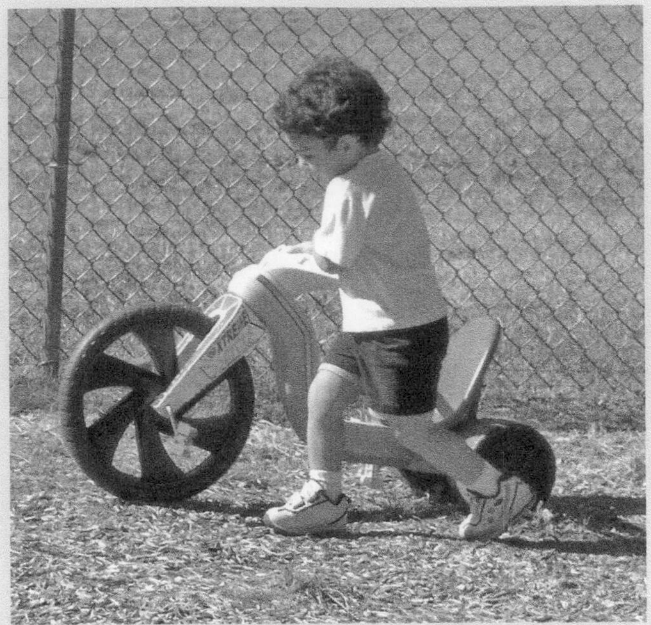

Figure 19.24 (Valerie Rice)
Toddlers need a variety of sizes of equipment that will challenge them to try new skills, yet permit them to build confidence slowly. This boy is having fun just pushing the tricycle and soon he will learn to ride it.

FURNITURE AND ARRANGEMENT: Just as they need a wide assortment of play materials, preschoolers also need a variety of options for interacting with materials. They can use counter tops and tables, chairs, wall-mounted or standalone easels, and stools. They will also choose to work on the floor at times (Figure 19.25).

Separate activity areas let children focus on particular interactions or skills. Tables, shelving units, and other furniture can form discreet activity areas. Rugs, markings on the floor, architectural features of the space, or changes in floor level can also organize spaces.

Interruptions make it difficult for some children to focus on an activity. Clearly defined circulation paths between activity areas minimize interruption of children's work and possibly lessen resulting negative interactions between children.

Activity areas also influence how children interact with each other. Areas that accommodate four to five children at a time might promote cooperative play whereas those designed for larger groups may discourage small group cooperation.

Soft seating and cushions provide comfortable areas, and rounded edges protect children from bumps and bruises. The more flexible the furniture is for producing different arrangements, the better. Children can use the same chairs and tables for play and for meals. Again, small tables that combine to form larger units provide flexibility.

Work surfaces with sloping tops may help prevent hunching. For example, preschool age children seated at desks with 15° sloping desktops and chairs that support feet, knees, and back have better posture and less tension of the low and mid-back (Marschall et al., 1995). Obviously, desks, chairs, and other furniture should fit the body sizes of its children users. Children's feet should rest firmly on the floor while seated. Provide table heights that enable children to be comfortable during sitting and standing activities.

Since preschoolers are just developing strength, their core (torso) muscles can tire, causing them to slouch. Provide work areas that allow them to shift through a range of good postures over their day.

Adjustable computer work surfaces allow caregivers to raise or lower them to the child user's size. Prolonged sitting and gazing at a monitor that is too high or too low may lead to neck, shoulder, or back discomfort among children. Although children of this age will probably not spend a great deal of time working at computers, correct environments promote prevention and education. This way, children learn good postural habits at an early age.

All furniture that supports activities needs to accommodate children with mobility impairments. Circulation paths in the classroom should also accommodate children with special needs.

Table 19.7 shows the recommended dimensions for tables and chairs for preschoolers.

Table 19.7 Preschool Furniture, ages 3–5 years old	
Tables	45–50 cm high (18–20 in.)
Water & Sand Table	45–50 cm high (18–20 in.)
Shelf Height	94–102 cm (37–40 in.) maximum
Chairs	25–30 cm high (10–12 in.) 29–32 cm wide (11½–12 ¾ in.) 24–28 cm depth (9½–11 in.)

Source: From Olds, A.R., *Child Care Design Guide*, McGraw-Hill. New York, 2000.

Preschoolers still take naps, so cots or mats remain necessary (see the above discussion on separate sleeping). Arranging cots so that they are from 60 cm (2 ft) (Australia) to 91 cm (3 ft) (US) apart is thought to reduce the transmission of infectious diseases (AAP, 2002; ACT-Australian Capital Territory, "Centre Based Children's Services" August 2000).

Separate areas let one child rest while others are active. This is especially helpful for a child who does not feel well or needs to sleep longer. Preschoolers' cots are typically 56 × 132 × 13 cm high (22 × 52 × 5 in.) (Olds, 2000). Store mats to avoid contacting one sleeping surface with another (AAP, 2002).

Figure 19.25 (www.SIS-USA-Inc.com)
Having furniture that teachers can easily move and adjust permits them to create new spaces that meet children's needs as they grow.

TOILETING: Preschoolers should have access to a toilet room that is separate from the playroom, without asking permission from caregivers. This promotes independence and self-control.

Recommendations for the ratio of toilets and sinks to children ranges from 1 of each for every 15 children (Capital Territory, Australia), to 1 toilet and 1 sink for every 10 children in the United States and Canada. The maximum height recommendation for toilets is 28 cm (11 in.) and 56 cm (22 in.) for sinks (AAP, 2002).

Children can use adult fixtures that are adapted for safety. For example, warm water is preferable to hot water.

Hygiene is part of their education; consider providing a mirror, toothbrushes, toothpaste, toilet paper and holder, soap dispenser, and paper towels. Other important elements are a floor drain, hamper for soiled clothes, trash basket, and shelf space at adult height for storage. All toileting areas need proper ventilation.

Preschoolers appreciate some privacy in the toilet room. This does not require separate bathrooms with closing doors; partitions between toilets provide some privacy. Preschoolers do not need gender separation in toilets.

FLOORING: Although preschoolers do not sit on the floor as often as infants and toddlers, some still prefer sitting on the floor and some activities lend themselves to floor work. A combination of carpeted and noncarpeted areas is appropriate. Easy to clean floors are excellent for this active and often messy group. Soft flooring is nice for quiet play, reading, and music. Carpeting helps reduce noise levels in the block areas, particularly when buildings tumble.

Simple carpet designs are preferred as complicated designs increase the complexity of the space. Carpeting with intricate or complicated designs can lead to tripping hazards for children with diminished depth perception (Bar and Galluzzo, 1999).

Visually impaired children will benefit from visual and tactile cues in the environment, especially when there is a change in floor level. Bright, contrasting colors or changes in type of flooring are good cues.

WALLS: A variety of displays helps to stimulate visual and cognitive development. Children or caregivers can make displays or caregivers can purchase them. Displaying children's work highlights their accomplishments and parents and children alike delight in the artwork. Displays can cover entire walls or only portions of a wall (see Table 19.8 for suggested coverings). Placing displays intended for children at their height enhances their ability to gaze at it, without straining their necks.

Table 19.8	Materials to Ready Wall Surfaces for Displays (Olds, 2000)
Cork	Available in sheets or tiles and in different textures, thicknesses, and color provides some acoustical insulation.
Homosote	Less expensive than cork, a recycled paper of which some varieties are fire rated; can be painted; needs periodic replacement, provides some acoustical insulation.
Forbo	Most expensive but durable and self-seals when tacks and staples are removed, nontoxic, resilient linoleum, available in sheets of a variety of colors, easy to install, excellent acoustical properties.

Walls can serve as displays (visual and tactile), separators between spaces, and acoustical controls. Materials that provide visual stimulation (besides standard paint) include wood, carpet, tile, stucco, and brick. The abrasive qualities of cinder block make it less desirable. A variety of wall textures provides complexity and can highlight specific architectural features. For example, carpet on a column may highlight it visually, protect children who might bump into it while playing, and provide a comfortable backrest for children sitting on the floor.

Preschoolers enjoy looking at themselves, just as younger children do. Consider placing tempered glass mirrors where children can easily see themselves. The wall description in the infant section includes suggestions for paint colors appropriate for childcare classes and playrooms.

DOORS: Preschoolers who access certain spaces on their own need to be able to operate the doors and door hardware. Doors to toilet rooms must be easy to open from inside and outside (should children need assistance) (AAP, 2002). Requirements for doors described in the infant and toddler sections also apply for preschooler spaces.

WINDOWS: Windows provide light, ventilation, and the opportunity to observe what is going on in the outside world. Consider all of these issues, including guarding against glare, when placing windows in preschool classrooms.

For preschoolers to comfortably see out of windows, start them at 61–76 cm (24–30 in.) above the floor (Olds, 2000). If possible, also provide at least one lower window since preschoolers spend time on the floor.

Comfortable seating for one or two children near a window creates restorative spaces where children can sit and observe interesting outdoor scenes. Of course, windows must face areas that provide such a scene. Lofts or raised floor areas can also lift children to a height to see out higher windows. The infant and toddler sections contain more detail on window safety and coverings.

Windows to interior spaces are important. Since preschoolers move through the entire childcare center more often than infants or toddlers, windows in corridors—that let them peer into other spaces (e.g., infant space, kitchen)— can be a valuable way of connecting children with other activities and spaces in the center (Olds, 2000). Clearly, these windows should be at child height.

STORAGE: Access to most toys and other play materials can promote independence, good habits, and cooperation. The larger variety of materials required by this age group translates to a need for more visible storage. Low, open shelving (that does not tip over) provides storage and delineates activity centers. After covering the backs of the shelving units with homosote, they can hold displays (Olds, 2000).

Shelving is preferable to deep storage bins (Figure 19.26). Shelving allows children to see what is available and caregivers often arrange materials on the shelves to teach concepts such as color, size, and categories. Shelving also makes it easier for preschoolers to participate in cleaning. Children cannot see materials in deep storage bins and are tempted to empty everything to see what is at the bottom. Small, clear bins can hold manipulatives and other small items on shelves. Closeted storage for more risky items (such as scissors) or for items to be rotated is necessary.

Olds (2000) recommends 1 linear foot per child for personal cubbies to hold coats and other personal items. It helps if the cubbies are large enough for the entire group to put on their outer clothes, there are benches to sit on while dressing and there is adequate space for cold or wet weather gear. An adjacent area to store children's art will assist both parents and caregivers. Having cubbies near the entry to the classroom prevents disruptions of other activities when leaving and retrieving children.

Figure 19.26 (Lorraine Maxwell)
Shelves for toddlers need to be low and open. This lets them easily see and remove items independently.

LIGHTING: Wall scones or track lighting can provide direct and indirect lighting, as well as spot lighting. Full-spectrum light fixtures with a dimmer permit caregivers to adjust light levels for specific activities, including naps.

CONCLUSIONS

Virtually every aspect of the childcare space can contribute to children's education and growth. A well designed, ergonomically appropriate childcare space benefits child development, helping children move through their sequential developmental patterns. On the other hand, designs that do not enhance children's cognitive, emotional, social, and physical development can actually hamper their achievements. An "ergonomically appropriate" design does not happen accidentally. Only through careful design can environments at once challenge growth while ensuring safety.

Figure 19.27 (www.SIS-USA-Inc.com)

REFERENCES

American National Standard Institute. (2002). ANSI512.60-2002. Acoustical Performance Criteria, Design Requirements and Guidelines for Schools. http://www.ansi.org/

Aronson, S.A. (2000). Updates on healthy eating and walkers. *Child Care Information Exchange, 134*, 30.

Australian Capital Territory (2000). "Centre Based Children's Services".

Bar, L. and Galluzzo, J. (1999). *The Accessible School: Universal Design for Educational Settings*. Berkeley, CA: MIG Communications.

Barker, G.A. (2002). Safety in childcare facilities. *Research Design Connections, 1*(2), 10.

Bredekamp, S. (1987). *Developmentally Appropriate Practice in Early Childhood Programs Serving Children from Birth Through Age 8*. Washington, D.C.: National Association for the Education of Young Children.

Bronson, M.B. (1995). *The Right Stuff for Children Birth to 8*. Washington, D.C.: National Association for the Education of Young Children.
Calgary Health Region (2002). *Policy for Social Care Facilities, Draft Policy*. Calgary, AB: Environmental Health.
California Energy Commission (2003). http://www.energy.ca.gov/
Canadian Pediatric Society. (1996). *Well Beings: A Guide to Promote the Physical Health, Safety and Emotional Well-being of Children in Child Care Centres and Family Day Care Homes*, 2nd edn. Ottawa, ON: Canadian Pediatric Society.
Caring for our children. National Health and Safety Performance Standards: Guidelines for Out-of-Home Care, 2nd edn. (2002). Washington, D.C.: American Academy of Pediatrics, the American Public Health Association, and the National Resource Center for Health and Safety in Child Care. http://nrc.uchsc.edu/CFOC/index.html
Chang, A., Lugg, M.M., and Nebedum, A. (1989). Injuries among preschool children enrolled in day-care centers. *Pediatrics, 83*(2), 272–277.
Consumer Product Safety Commission (1999). *Safety Hazards in Child Care Settings*. Washington, D.C.: U.S. Consumer Product Safety Commission, www.cpsc.gov
DeLong, A.J. and Moran, J. (1992). Effects of spatial scale on cognitive play in preschool children. Paper presented at International Conference on Child Day Care health: Science, Prevention, and Practice, Atlanta.
Gandini, L. (1998). Educational and caring spaces. In C. Edwards, L. Gandini, and G. Forman (Eds.), *The Hundred Languages of Children: The Reggio Emilia Approach-Advanced Reflections* (pp. 161–178). Greenwich, CT: Ablex Publishing Corp.
Grant, K.A., Habes, D.J., and Tepper, A.L. (1995). Work activities and musculoskeletal complaints among preschool workers. *Applied Ergonomics, 26*(6), 405–410.
Greenman, J. (1988). *Caring Spaces, Learning Places: Children's Environments that Work*. Redmond, WA: Exchange Press, Inc.
Harrison, L. (1990). *Planning Appropriate Learning Environments for Children Under 3*. Australian Early Childhood Association.
Hedge, A., Ruchika, J., Jagdeo, J., Ruder, M., and Akaogi, M. (2003). Effects of chair design on toddler behavior. Proceedings of the Human Factors and Ergonomics Society 47th Annual Meeting, October 13–17, Denver, CO, pp. 937–941.
Hendricks, C. (2002). How safe is your classroom? Identifying hazards before accidents happen. *Early Childhood News*. Monterey, CA.
Lally, J.R., Provence, S., Szanton, E., and Weissbourd, B. (1987). Developmentally appropriate care for children from birth to age 3. In S. Bredekamp (Ed.), *Developmentally Appropriate Practice in Early Childhood Programs Serving Children from Birth Through Age 8* (pp. 17–33). Washington, D.C.: National Association for the Education of Young Children.
Mackenzie, S.G. and Sherman, G.J. (1994). Day-care injuries in the data base of the Canadian hospitals injury reporting and prevention program. *Pediatrics, 94*, 1042–1043.
Marschall, M., Harrington, A.C., and Steele, J.R. (1995). Effect of work station design on sitting posture in young children. *Ergonomics, 38*(9), 1932–1940.
Maxwell, L.E. and Evans, G.W. (2000). The effects of noise on pre-school children's pre-reading skills. *Journal of Environmental Psychology, 20*, 91–97.
NICHD Early Child Care Research Network (2000). Characteristics and quality of care for toddlers and preschoolers. *Journal of Applied Developmental Science, 4*, 116–135.
Olds, A.R. (2000). *Child Care Design Guide*. New York: McGraw-Hill.
Piaget, J. (1974). *The Origins of Intelligence in Children*. New York: International Universities Press, Inc.
Ruth L.C. (1999). *Design Standards for Children's Environments*. New York: McGraw-Hill.
Shimaoka, M., Hiruta, S., Ono, Y., Nonaka, H., Hjelm, E.W., and Hagberg, M. (1998). A comparative study of physical work load in Japanese and Swedish nursery school teachers. *European Journal of Applied Physiology, 77*, 10–18.
Staes, C., Balk, S., Ford, K., Passantino, R.J., and Torrice, A. (1994). Environmental factors to consider when designing and maintaining a child's day-care environment. *Pediatrics, 94*, 1048–1050.
Wachs, T. (1979). Proximal experience and early cognitive intellectual development: The physical environment. *Merrill-Palmer Quarterly, 25*, 3–24.
Washington Education Association. http://www.wa.nea.org/iaq/tips.htm.

ADDITIONAL REFERENCES

Indoor Air Quality for Schools

EPA: http://www.epa.gov/iaq/schools
Northwest Clean Air Agency: http://www.nwcleanair.org/links/#iaq
WSU Extension Energy Programs: http://www.energy.wsu.edu/projects/building/iaq.cfm

CHAPTER 20

CHILDREN AND HANDWRITING ERGONOMICS

CINDY BURT AND MARY BENBOW

TABLE OF CONTENTS

INTRODUCTION

Many of us take children's ability to learn handwriting for granted. Yet handwriting is physically and intellectually demanding. So demanding that between 10% and 20% of school-age children find it difficult to write (Hamstra et al., 1993).

Postural instability, paper and pencil positioning, and limited gripping ability are correlated with poor handwriting performance (Amundson and Weil, 1996; Parush et al., 1998; Jacobs et al., 2002). Debates from the nineteenth century about "proper" desk heights and inclinations (Warren, 1846) continue to this day.

Creating effective child environments for writing requires more than simply supplying a place for children to copy letters. Effective learning environments must be directive, supportive, and intriguing for children as they develop this new way to communicate. Children need appropriate writing tasks and tools and they must be developmentally ready to write.

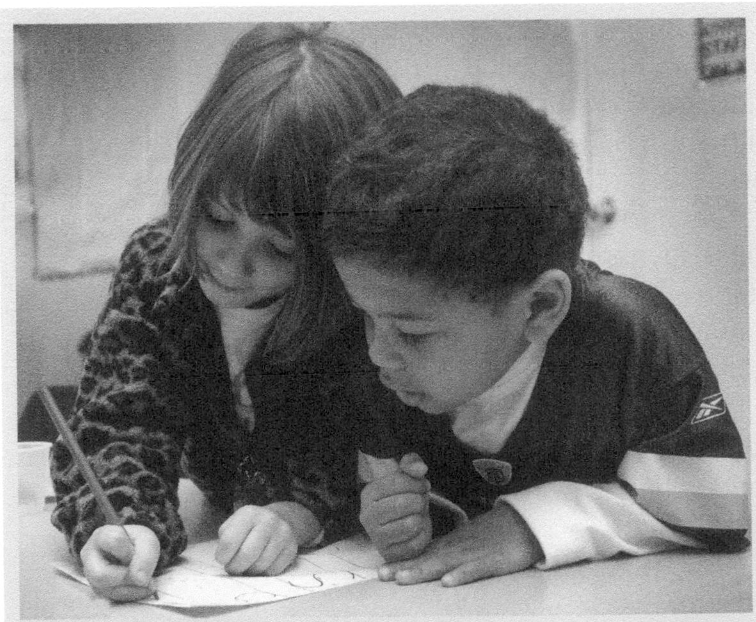

Figure 20.1 (Ursy Potter Photography)
Creating effective child environments for writing requires more than simply supplying a place for children to copy letters. Children must be developmentally ready to write and have the right tools and support systems.

BACKGROUND

Our ability to use scribing tools to communicate sets us apart from other animals. It took thousands of years for humans to develop the precision needed to use crayons, pens, or pencils. Writing requires interaction and coordination of multiple physical, perceptual, and cognitive skills.

POSTURAL STABILITY

> Writing tool manipulation requires postural stability for distal control.

Postural muscles provide stability. Also known as "core muscles," they support upright postures that withstand gravity. These short, strong muscles lie close to the bones and deep within the body to control the neck, spine, shoulder, and pelvic girdles.

Shoulder and hip girdle stability enables children to control extremities from a stable base of support. The importance of this postural stability cannot be overstated. Children must be able to sit or stand in a range of stable postures before they can read and write effectively (Figure 20.2 and Figure 20.3).

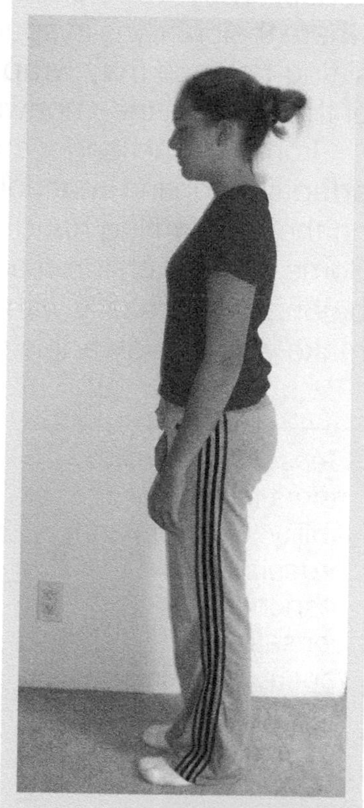

Figure 20.2 (Mary Benbow)

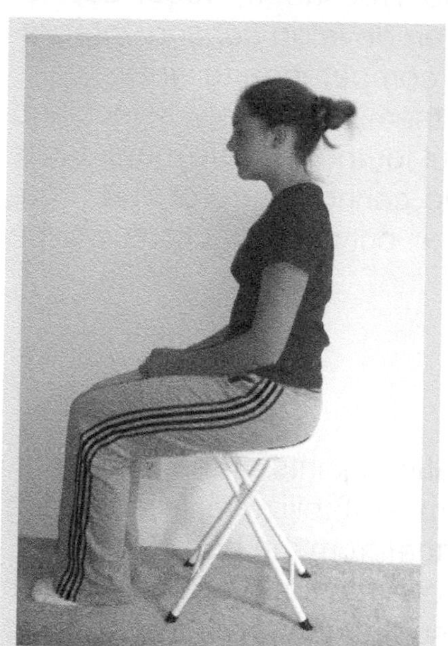

Figure 20.3 (Mary Benbow)

Unfortunately, postural problems are increasingly common among children. Many developmental specialists suggest that today's infant rearing practices impede the development and strength of postural muscles and contribute to postural deficits (see Chapter 6 for additional information).

Spinal and neck extension and shoulder and pelvic girdle control begin to develop during floor time when babies play and explore on their tummies. Babies develop holding strength and balance by gaining control of neck and upper back muscles, incrementally lifting their head and chest into higher postures. They begin to stabilize postures and maintain balance by shifting weights to free arms and hands for support and exploration. This normal progression to upright sitting and standing postures prepares them for manipulating writing instruments and other tools.

In the 1990s, as fears about sudden infant death syndrome (SIDS) became widespread, parents were cautioned to avoid SIDS by positioning babies on their backs while sleeping. Many parents overinterpreted this warning to mean their babies should never be prone, even during daytime play periods. Yet, this "prone on tummy" stage is crucial to develop postural stability and balance, and to learn to shift weight and integrate the two sides of the body. Children who bypass this stage often miss motor milestones. This can delay their development.

> *Adaptive righting reactions* enable very small children to balance themselves, and to sustain and change postures. Children make these body adjustments when they sense being out of balance.

Children develop motor competence by exploring. Plastic "exosketetal" devices such as recliners, jumpers, walkers, strollers, and joggers that aim to support and control children can inadvertently deprive them of valuable hours of motor exploration. Their use can be especially detrimental for children with low muscle tone by impairing their development of core strength and motor competence.

Children with low postural stability and balance often have difficulty functioning at school. They wobble and sway when sitting at a desk. Others fall from chairs, not sensing where their body is in space while concentrating on teacher instructions. Children with poor sitting balance may wrap feet around the legs of their chair or lean against the front edge of the desk on their nonwriting arms to brace themselves.

Ironically, the longer children with posture and balance problems sit, the more challenging adaptive righting reactions or responses become. Many of these children lay their head on their nonwriting hand or lean forward to prop themselves against the edge of the desk. Some of these children actually write more effectively while standing beside the desk. They balance more easily with their feet on the floor, which stabilizes them and enables them to make whole body adjustments.

Children aged 4 to 6 demonstrate *postural sway* when engaged in challenging grasping movements (Pare and Dugas, 1999; Shumway-Cook and Woollacott, 1985; Laszlo, 1990). This is the age when children develop *sensory integration strategies* required for motor competence to write.

Sensory integration refers to children's ability to understand, organize, and respond to sensations of touch, balance and equilibrium, vision, taste, and smell.

Sensory integration prepares them to actively select and organize the sensations most useful for the specific motor tasks required for writing tasks. If a child cannot interpret and organize these sensory messages, their outward reactions will be different from desired or expected.

HAND MOBILITY

Hand dexterity requires smooth interactions between the various joints of the shoulders, arms, and hands. Joints that are closer to the body (*proximal*) must stabilize while facilitating movements of joints that are further from the body (*distal*).

Elbow motions help position the hand in space. *Forearm supination* enables the palm to rotate up facing the sky and *pronation* enables the palm to rotate down facing the floor.

Children need active elbow motion in all available directions (flexion, extension, supination, and pronation) to position forearms at midrange. This enables them to hold and manipulate writing tools at comfortable angles and distances. Children who cannot position and hold their forearms at midrange often revert to immature postures (pronation) for better control. For example, children need shoulder stability to control movement of elbows and wrists—and they need stable wrists to control their thumb and fingers while writing.

School children with limitations in palm-up postures hold writing hands parallel to their desks while coloring or writing, rather than with partially rotated forearm positions that allow them to see and monitor their work. Insufficient forearm rotation will also limit their ability to actively flex their wrists downward while coloring or writing, reducing both quality and speed of output (Figure 20.4).

A stable and extended wrist helps children separate the thumb from the palm of the hand. Thumb separation is necessary for a *three-finger pulp* or *tripod grip* (Figure 20.5).

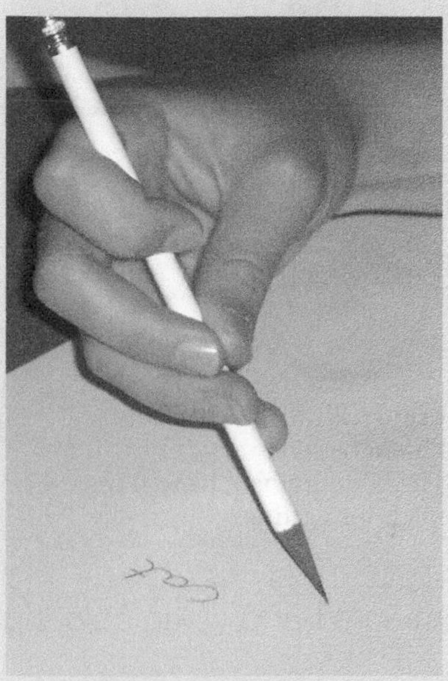

Figure 20.4 (Mary Benbow)
Index grip.

Motor control of the upper extremities requires postural stability. Children with poor wrist stability compensate by flexing their wrist joints to stabilize it (Figure 20.7), pressing "bone on bone." This posture reduces both finger strength and dexterity needed to manipulate writing instruments.

If the child needs to flex his wrist to stabilize it, he cannot use active wrist flexion for downstroking when making letters. They compensate by using larger and more proximal joints and sacrifice precision of movement.

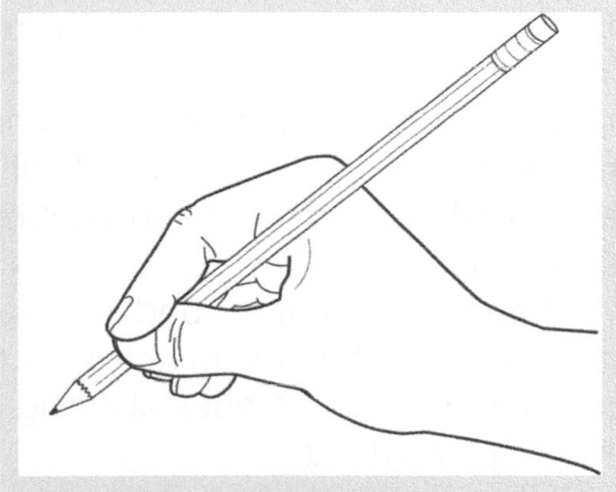

Figure 20.5 (Mary Benbow)
Tripod grip: Static to dynamic.

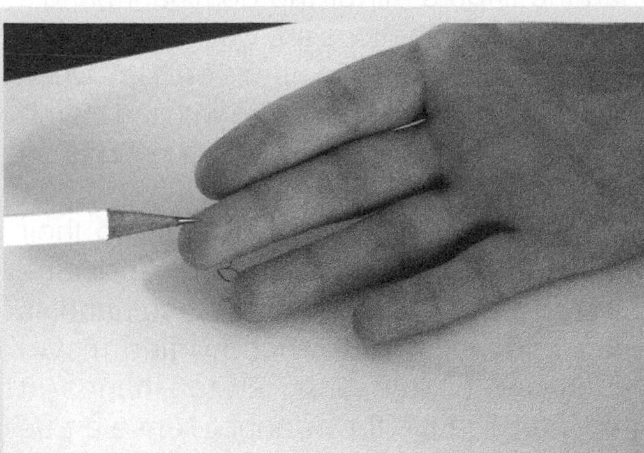

Figure 20.6 (Mary Benbow)
The finger pulp is the padded part of fingers near the tip with fingerprints.

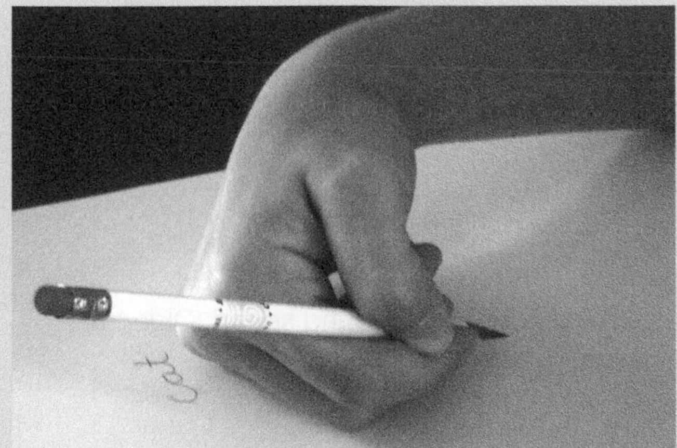

Figure 20.7 (Mary Benbow)
Wrist extension is necessary to position the thumb for efficient writing. Wrist flexion does not allow proper thumb posture for efficient writing.

HAND DOMINANCE

> *Hand dominance.*
> The hand most often used to eat, finger paint, or pick up objects is usually the hand preferred for writing.

Babies begin to differentiate hand movements within the first year of life. Between 2 and 3 years of age, they use opposing or asymmetric hand and arm movements to complete highly differentiated activities such as cutting with scissors (Guiard, 1987).

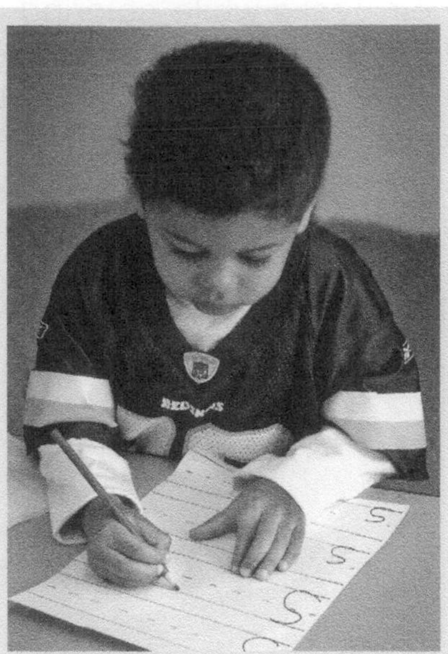

Figure 20.8 (Ursy Potter Photography)
The nondominant hand frames the space for writing.

The development of hand dominance requires *bilateral integration* of the body sides, which enables a smooth interaction of the two hands. The "dominant" hand completes skilled movements as the recessive hand assists. The recessive hand assists movement for the dominant hand by "framing" the space in which the dominant hand moves (Figure 20.8). Movements of the dominant hand are faster and more refined and controlled than the nondominant hand.

Bilateral integration and development of hand dominance are vital for writing activities. Children with immature body integration neglect stabilizing their paper with the nondominant hand when coloring, drawing, or writing.

If the dominant hand does not assume its leadership role, the recessive hand will not develop an assisting role. Competition between the sides of the body will continue and cause motor overflow (also called *synkinesis*). The nondominant fingers exhibit synkinesis when the fingers mirror the dominant hand's movements. Finger motions are visible to a lesser degree. Synkinesis can sometimes be seen when children try to touch each fingertip with the thumb on one hand. Watch for overflow movements on the other hand.

> Never force selection of hand dominance. Direct children who switch hands to complete similar tasks simultaneously with both hands at midline. This will help the child determine which hand is most proficient.
>
> Spinning two identical tops simultaneously will show which hand has the best control, force, and speed. Older children can write the numbers 1–10 with both hands on either side of a vertical line on a chalkboard. Observe and help the child identify which hand leads.

Children typically shift their weight to their nonwriting side as they write. When postural instability restricts shifting, their writing hand is not freed and which slows handwriting movements. These children often switch hands at their midline because they cannot sense a true dominant hand.

VISUAL SKILLS AND WRITING

Visual problems affect school performance. When a child's school performance lags behind their verbal skills, have their vision checked.

Writing requires visual coordination. Children need coordinated central and peripheral visual systems to write using consistent letter size, even spacing, and correct orientation on writing surface.

It is not surprising that children with visual problems often find it difficult to learn to write. Yet school professionals may not detect such dysfunctions among schoolchildren. One indicator is when school performance (such as reading and writing) lags behind verbal ability.

Primary visual skills that are necessary to read and write include ocular mobility, binocularity, and accommodation. Ocular mobility (eye-movement control) governs the speed and accuracy of visual inspection and scanning. Poor ocular mobility is fatiguing and compromises learning and comprehension.

Near-point activities such as reading and writing can reveal symptoms of poor ocular mobility. Uphill and downhill handwriting involving inconsistent size and spacing of letters graphically displays the poor coordination of peripheral and/or foveal (central visual field) visual systems in children who experience poor ocular mobility (Figure 20.10).

Figure 20.9 (Valerie Rice) Compensating for impaired binocularity.

Binocularity is the ability of the two eyes to coordinate and fuse visual images into one (teaming). Impaired binocularity can affect spatial orientation, spatial relationships, and depth perception.

Children with impaired binocularity may close or cover one eye, squint, or blink. They may compensate by resting their head on their hand or arm to write or turning writing paper at extreme angles to write uphill or downhill (Figure 20.9).

Accommodation allows the eyes to shift from near (desk level) to far point (such as the chalkboard) quickly, smoothly, and without losing one's place or exhausting the visual system. Copying letters at functional speeds requires accommodative flexibility to change viewing distances as children shift their gaze from the sample letters to their writing surface. Children also use accommodative flexibility when they progressively lower their gaze down the page when they write.

Visual perception affects movement. Children need spatial perception to determine their position and their relationship to other objects in space. Depth perception provides the child with awareness of how far away something is, helping them move in space.

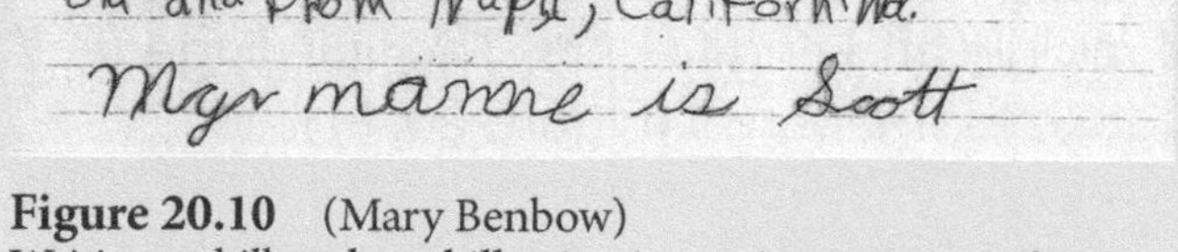

Figure 20.10 (Mary Benbow) Writing uphill or downhill using inconsistent size results from impaired binocularity.

Chapter 4 provides a more detailed description of how visual problems can affect a child's performance.

GRIPPING SKILLS

Children change their *writing grips* (the way they hold their writing tool) as they grow and develop (Rosenbloom and Horton, 1971). Grips vary in stability, mobility, and sensory motor control. The earlier children begin to write, the more likely they are to use awkward grips.

> *Sensory feedback* helps children regulate grip pressure and movement when writing.

Some writing grips are inefficient and may harm the joints. Fortunately, most grip patterns mature as children develop and integrate sensory and motor skills.

> Grip control affects hand placement when writing. Children with immature grips hold writing tools near the middle or top of the shaft. As grips mature, children hold the writing device near the writing tip using the pulps of the fingers.

An individual child can use a combination of grips, skipping some, as they progress from immature static grips to more mature dynamic grips. There are many variations of each grip, but they generally fit into categories. A discussion of grips follows the developmental sequence, with an adaptation of the mature tripod grip presented last.

INDEX GRIP: Children who start to write before their hand dominance is well developed often adopt an immature grip pattern called the *index writing grip*. Index grips are not very functional for children once written work progresses beyond circling correct responses or filling in blanks.

> The ulnar styloid is the pointed bone evident on the outside (little finger side) of the wrist.

The flexed joints of the index finger cradle and stabilize the upper shaft of the pencil proximally (Figure 20.11). The thumb clutches the mid shaft of the pencil. The little finger braces the far end of the pencil. The wrists extend (rather than flex) and stabilize on the *ulnar styloid process* against the desk surface.

Although children can visually monitor their hands as they write, this grip limits *tactile* and *proprioceptive input* from the fingers. It also causes children to rotate their elbow beyond its midrange, limiting its use while writing. This writing grip causes the three skilled fingers to stabilize (rather than move) the pencil.

The position of the flexed wrist interferes with finger movement. Downstrokes of the letters require flexing of the wrist and thumb in combination with the flexion of middle and ring fingers. The little finger stabilizes the pencil (Figure 20.11).

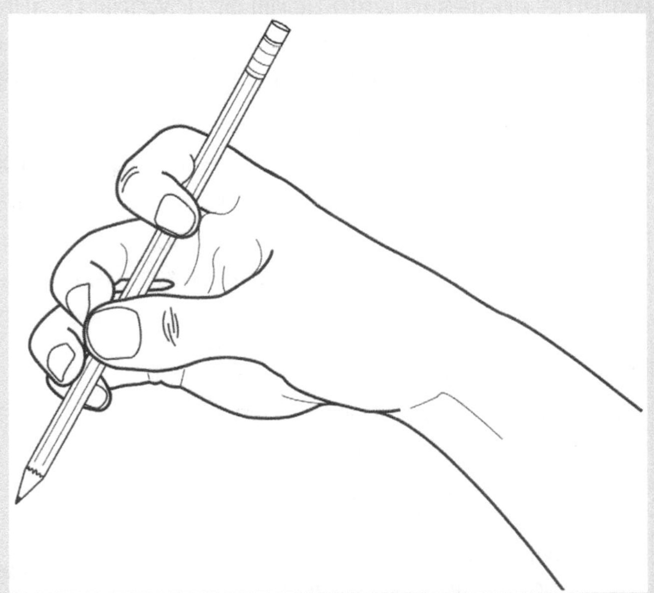

Figure 20.11 (Mary Benbow)
Children who are expected to write before establishing dominance often use the index grip.

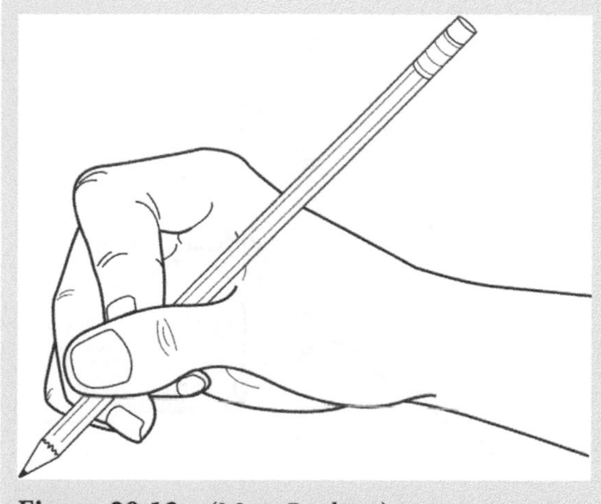

Figure 20.12 (Mary Benbow)
Thumb wrap grip.

Proprioceptive, kinesthetic, and *tactile feedback* provides children with information about the position and movements of the fingers from joints and muscles and skin.

During letter downstrokes, the distal shaft of the pencil (part of pencil near the writing tip) passively pushes the little finger down. The little finger extends with the thumb joints and wrist to make upward strokes. *Wrist extension, ulnar deviation*, and thumb *hyperextension* combine with extension of the little finger for rotation over the tops of letters.

"LOCKED" GRIP WITH THUMB WRAP: Children who need greater tactile input or struggle to develop muscle control must often "lock" their grip by wrapping the thumb around the pencil in order to hold it (Figure 20.12). Writing with the *locked grip* is slow in speed, crude, and limits sensory input needed to control grip pressure and force against the writing surface.

This grip is also called the *white knuckle grip* because of the amount of pressure children exert both on the paper, sometimes ripping it with their pencil, and on their fingers and joints.

During *thumb wrap grips*, the thumb presses the tip of the index finger downward against the pencil shaft. This pressure forces the pencil against the radial (little finger) side of the middle finger above or below the *distal interphalangeal joint* (DIP) (Figure 20.13).

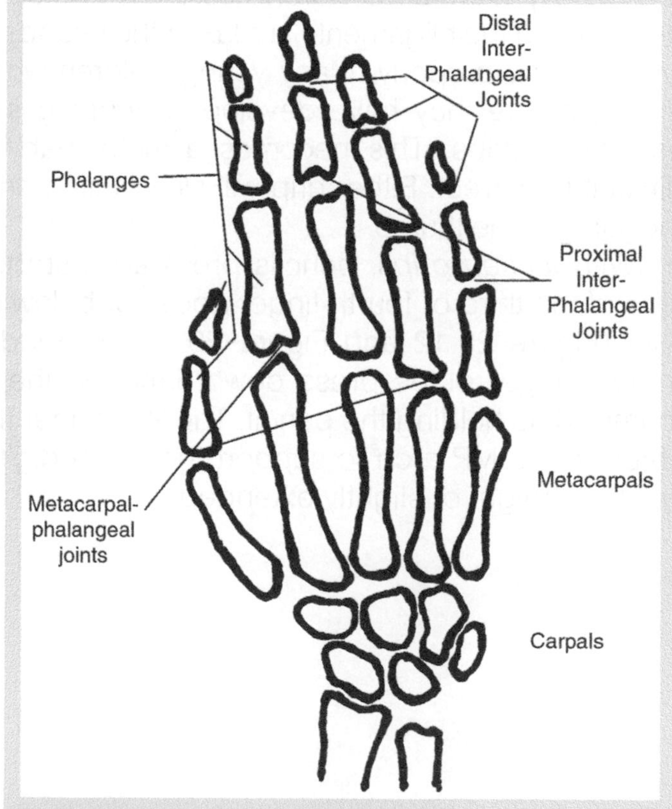

Figure 20.13 (Cindy Burt)
Joints of the hand.

Children writing for prolonged periods with "white knuckle grips" can injure their thumb (metacarpal phalangeal) joint. Children who have problems will have very weak pinch. When holding a ball or glass, their thumb will point downward and away from the index finger (Figure 20.14). To test, ask the child to pinch a piece of stiff paper between the thumb and index finger pulps tightly so you can't pull it away. If pinch is weak, the paper will slip out easily.

The heavy pressure on the shaft of the pencil forces the *proximal interphalangeal joint* (PIP) of the index finger into *hyperflexion*. This allows the DIP finger joints to hyper-extend.

Since the hand is held flat, the ring and little fingers must flex to support the middle finger. The ring and little fingers flex and extend to produce the writing strokes. The wrist is straight or slightly extended with the forearm slightly rotated upward.

This "locked" grip mechanically restricts the fingers from manipulating the pencil to make strokes. With this grip, thumb pressure stresses the metacarpophalangeal (MP) joint of the thumb. Persistent use can cause pain and degenerative disease over time.

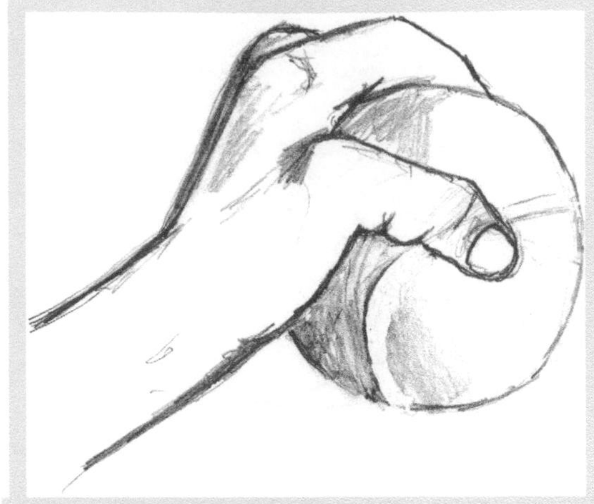

Figure 20.14 (Mary Benbow)
Grip following ulnar collateral ligament partial tear of the thumb MP joint.

If the ligaments of the hands are very lax the *adductor* muscle of the thumb rotates the thumb metacarpal into full supination. The combined stress may partially tear the ulnar collateral ligament at the MP joint of the thumb (Figure 20.14).

"LOCKED" GRIP WITH THUMB TUCK: While some children wrap their thumbs around the pencil to stabilize it, other children elect to "lock" their grips while tucking their thumbs under (rather than over the index and middle fingers). Both grips are equally harmful. Children use them to stabilize the pencil when their ligaments are lax or their hands are too weak to grip properly. Many young children who begin writing before they have developed gripping skills will use these grips. This becomes a motor habit that is difficult to correct. Either grip will slow writing and harm the joints of the thumb.

With *locked grips*, pencils press against the radial side of the third or fourth finger above or below the DIP joint (Figure 20.12 and Figure 20.15). The index and middle finger pulps press down firmly on the tucked thumb while holding the pencil. The little finger flexes to stabilize the MP arch to support the radial digits.* The wrist is straight or slightly extended.

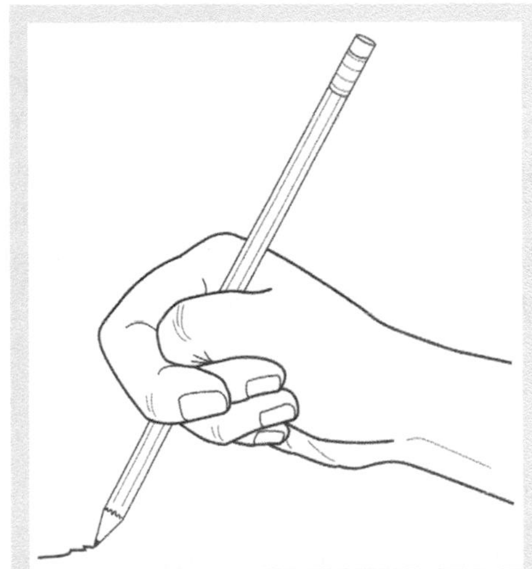

Figure 20.15 (Mary Benbow)
Locked grip with thumb tuck.

* The radial digits are the fingers located on the same side as the long bone in the forearm known as the radius, on the "thumb side" of the hand.

LATERAL PINCH GRIP: During *lateral pinch grips*, the adductor muscle of the thumb stabilizes rather than manipulates the pencil. This can result in thumb pain at the MP joint of the thumb.

During this grip, the proximal section of the thumb (section closest to the palm) presses against the side of the index finger to stabilize the pencil shaft (Figure 20.16).

The thumb-index web space is partially closed while the fingers are adducted and loosely flexed into the palm. The forearm is pronated on the desk surface.

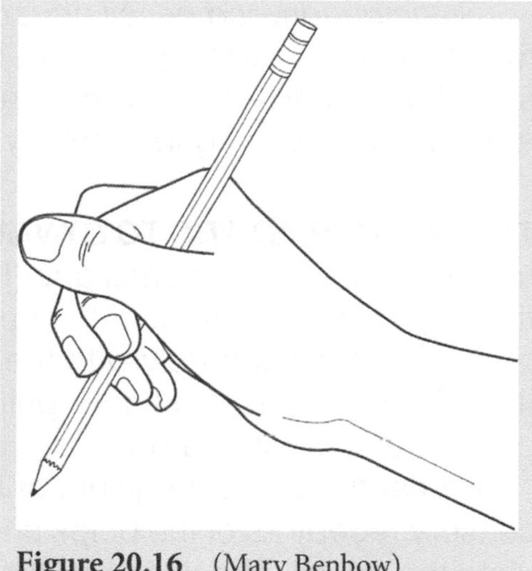

Figure 20.16 (Mary Benbow)
Lateral pinch grip.

Figure 20.17 (Mary Benbow)
Limited elbow motion restricts forearm rotation.

The pencil moves by flexing and extending the joints of either the index or middle finger or all four fingers working as a unit. Pressed against the index finger, the pencil mechanically blocks rotation at the index MP joint and impedes the rounding of strokes over the tops of letters.

The pronated posture of the forearm restricts coordinated movement of the elbow, wrist, and hand when writing (Figure 20.17).

QUADRUPOD GRIP: STATIC TO DYNAMIC: Children progressing from an immature grip to a mature grip will often begin with a *static quadrupod grip*. The pulps of the thumb, index, middle, and ring fingers hold the pencil securely as the thumb pulp stabilizes the pencil against the radial side of the ring finger (Figure 20.18). The little finger flexes to stabilize the MP arch and support the other fingers.

The wrist is extended 25° to 35°, allowing subtle flexion and extension interplay with the fingers and thumb. Positioning of the forearm in 50° to 60° of pronation allows elbow rotation for rounding over the tops of letters when writing.

The pencil, held in the open index-thumb web space, is positioned for writing that is more dynamic. This is necessary to print smaller letters within lines. The MP joints of the hand can laterally rotate to round over the tops of letters, eliminating the need to use more proximal, less accurate, and slower joints such as the wrist or elbow.

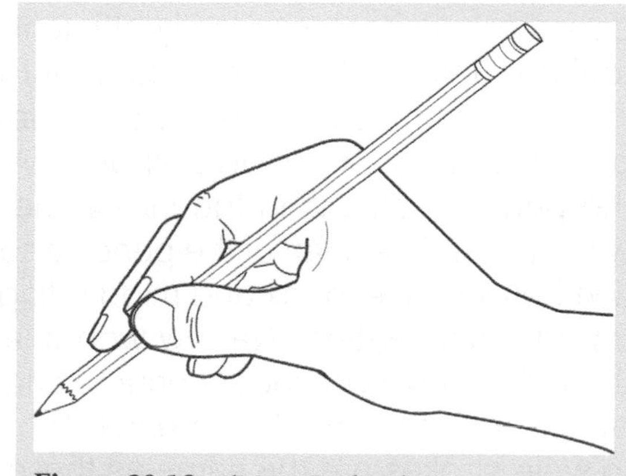

Figure 20.18 (Mary Benbow)
Quadropod grip: Static to dynamic.

Static quadrupod pencil grips often help children develop more fluid, mobile, and comfortable dynamic writing patterns. If the *quadrupod grip* does not become dynamic, writing speed and quality are adversely affected. However, the *quadrupod grip* usually becomes a dynamic, efficient grip (Benbow, 1990).

TRIPOD GRIP: STATIC TO DYNAMIC: The *three-digit tripod grip* is considered the most mature and efficient pencil grip (Figure 20.19). The MP arch, creating a balanced three-finger (chuck) grip to hold the pencil. The pulps of the thumb, index, and middle finger hold the pencil securely with the thumb stabilizing the pencil against the radial side of the middle finger. The finger pulps act like a "chuck" to hold the pencil securely like a chuck holds a bit in a drill.

Partial flexion of the joints of the thumb, index, and middle fingers allow the dense sensory receptors of the pulps and joints to regulate grip pressure and effectively manipulate the pencil. The ring and little fingers flex to stabilize the MP arch, support the third finger, and shift control to the skilled side of the hand.

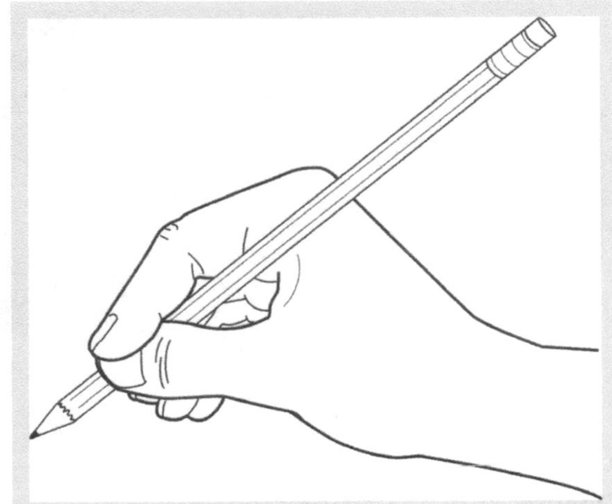

Figure 20.19 (Mary Benbow)
Tripod grip: Static to dynamic.

The full expansion of the thumb-index web space allows dynamic movement as children move from widely ruled to narrowly ruled paper where the spacing size between the lines is smaller. Practice paper for younger children is usually a 0.5 in. (1.27 cm) rule with a dotted dividing line. Most notebooks are ruled at 3/8 in. (0.95 cm) for older students.

The wrist extends about 20°, allowing for a subtle interplay of wrist and finger movements. If the grip remains static, the proximal upper extremity joints will produce strokes at the expense of speed.

ADAPTED TRIPOD GRIP: Children who find writing uncomfortable or overly fatiguing can benefit by using the *adapted tripod grip*. The *adapted tripod grip* is the adaptation of choice when the acquired pencil grip is not motorically efficient or is stressful to the thumb and finger joints (Figure 20.20).

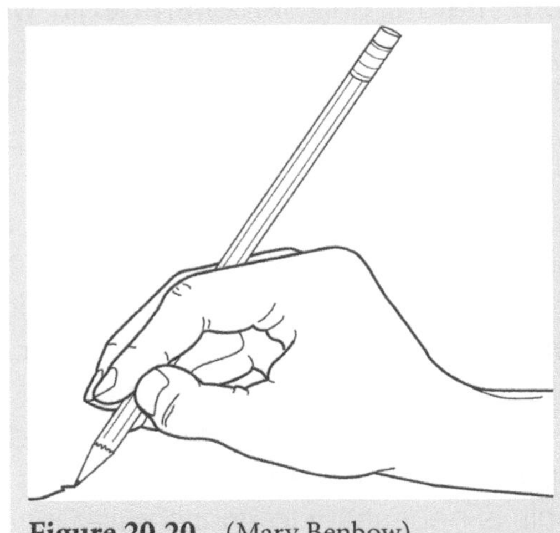

Figure 20.20 (Mary Benbow)
Adapted tripod grip.

The thumb pulp holds the *distal* (lead) end of the pencil against the middle finger, while stabilizing the pencil shaft within the index and middle finger web space. The shaft of the pencil wedges a separation between the index and middle fingers for active tripod manipulating. Narrow web spacing between the index and long finger secures and stabilizes the pencil shaft to control the pencil. The muscles used to write strokes are the same as those used with the *dynamic tripod grip*.

INDICATION FOR A SPECIALIZED HAND SKILLS EVALUATION

Children who have difficulty progressing to mature pencil grips may need a specialized hand skill evaluation. Pediatric occupational therapists can work with educators and parents to identify problems and develop a writing intervention program. Signs that may indicate a need for such an evaluation include (see also Appendix A):

1. Limited forearm rotation
2. Limited wrist extension with stabilization
3. Inability to write with an open thumb-index web space
4. Weak or unstable *arches* of the hands
5. Inability to separate the motor functions of the two sides of the hands
6. Poor postural stability
7. Weak muscles of the trunk and shoulder girdle

DEVELOPING MATURE WRITING GRIP

READINESS

Some children (particularly girls) begin "writing" as early as 2½. At this age, they usually have insufficient hand strength, stability, and dominance. They are usually unable to name letters and form words. Although fascinated with writing tools, they are not ready for formal training.

The best time to begin printing is in the latter half of kindergarten. Perceptual–motor integration of the diagonal, which is required to print 10 capital letters, develops at this time. Children between the ages of 5 and 6 also have the basic motor skills, attention, and recall needed to learn to form letters.

The beginning of second grade is optimal for introducing cursive writing. This is when children are better able to chain movement sequences and combine letters into words.

No matter what age the child is, both proximal postural stability and dynamic movement are necessary to develop hand dexterity and coordination while manipulating writing tools. Specific hand functions required for handwriting include:

> Ten capital letters that require use of diagonal:
>
> **A K M N R V W X Y Z**
>
> Diagonal perception is needed to print these capital letters.

1. Ability to open and stabilize the thumb-index web space
2. Ability to stabilize arches of the hand
3. Motor separation of the two sides of the hand
4. Ability to coordinate precise hand movements

WHAT IS A MATURE WRITING GRIP?

Mature grips use the fingertips or pulps to grip the pencil near the writing tip (distal end) to promote optimal hand and wrist mobility. This grip is often called the *distal digital grip*. Both quadruped and tripod grips are distal digital grips.

> *Sensory feedback* helps children regulate grip and movement when writing. *Tactile* feedback helps them hold pencils using proper force. *Kinesthetic feedback* provides ongoing information related to movement to direct writing tools.

The *distal digital pen grip* facilitates active finger gripping and movement by utilizing the finger pulps, which have the highest density of sensory and motor receptors (Johansson and Westling, 1987; Moberg, 1983). These receptors send information to the brain to help regulate grip force and direct writing movements for fluid and dynamic writing motions.

The *lateral pinch grip* (Figure 20.21) positions pencils against the side of the index finger when writing. This provides considerably less sensory feedback for gripping and movement than provided by the finger pads.

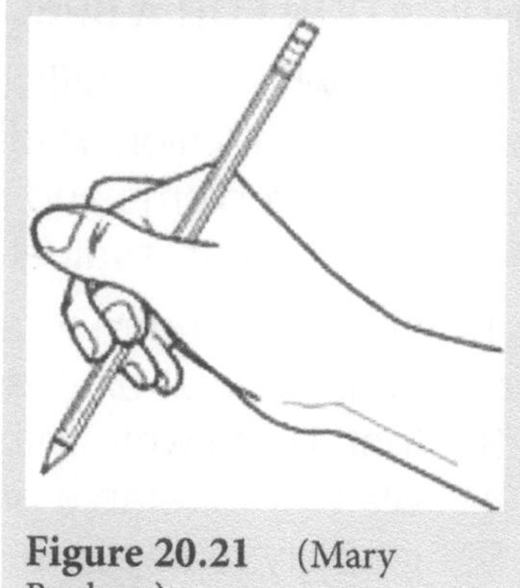

Figure 20.21 (Mary Benbow)
Lateral pinch grip.

Holding a pencil in a power configuration (a *closed-thumb web or thumb wrap grip*, Figure 20.22) also compromises input necessary to regulate grip force and promote movement (Long et al., 1970).

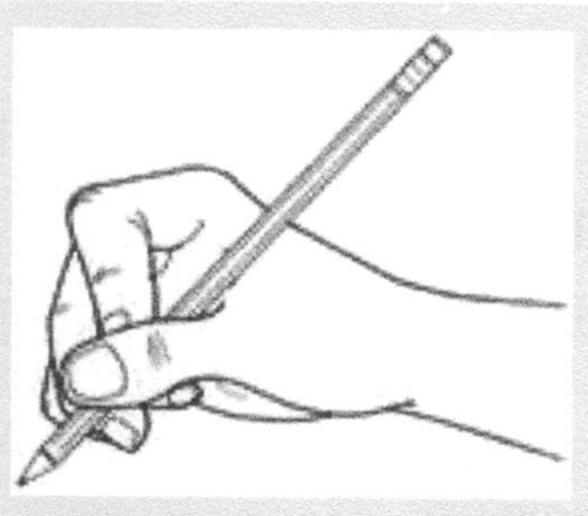

Figure 20.22 (Mary Benbow)
Thumb wrap grip.

> Writing requires up and downstrokes of the pen. The finger joints must alternately flex and extend to make the necessary "needle threading type movements."

Distal digital finger grips make it easier to manipulate pencils while forming letters. Gripping with the finger pads near the lead end of the pencil allows the small finger joints to flex and extend smoothly when writing (Figure 20.23).

When the pencil rests in the open web of the thumb-index web space, the intrinsic muscles of the palm control grip strength and allow for precise movement (Long et al., 1970). *Open-web thumb grips* also provide better sensory feedback for developing movement patterns (Figure 20.23). These grips promote smooth, stress-free, and efficient movements needed during extended writing (Figure 20.24). Encourage children to hold pencils on the shaft close to the writing tip with an open-web thumb grip.

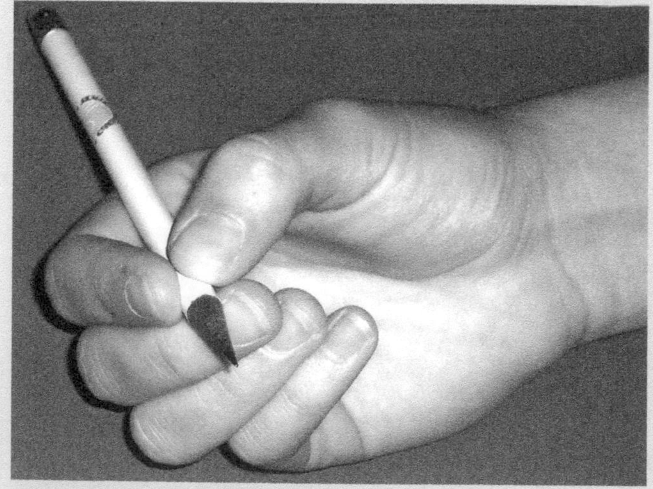

Figure 20.23 (Mary Benbow)
The distal digital finger grip is an example of an open-web thump grip.

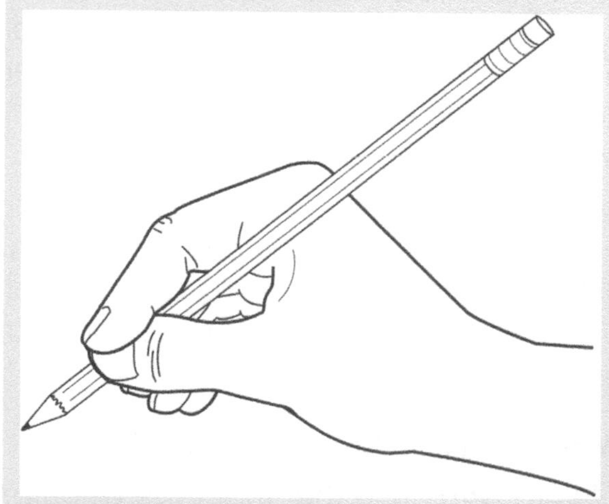

Figure 20.24 (Mary Benbow)
The dynamic tripod grip utilizes an open-web thumb grip.

The following table describes the various grip patterns and reasons for recommending or not recommending each grip.

Table 20.1 Grip Patterns

Grip	Description	Picture	Recom-mended	Reasons
Index Grip	Also called closed-thumb web grip, immature grip, and proximal grip. Index finger cradles upper shaft of pencil, thumb clutches middle of pencil with little finger bracing pencil near the writing tip. Wrist flexed. Hand rests on ulnar styloid.		No	1. Limited sensory input from fingers 2. Fingers stabilize rather than move the pencil 3. Immature grip pattern
Locked Grip With Thumb Wrap	Also called proximal grip, power grip, white knuckle grip, closed-thumb web grip, nonfunctional grip. Thumb presses tip of index finger with strong force against pencil shaft. Pencil held so tightly knuckles turn white.		No	1. Poor ability to manipulate pencil 2. Can result in thumb joint pain 3. Immature grip pattern
Locked Grip With Thumb Tuck	Also called proximal grip, power grip, closed-thumb web grip, and nonfunctional grip. Thumb tucked under index and middle fingers. Index finger pressed against pencil shaft with strong force. Wrist straight or slightly extended.		No	1. Poor ability to manipulate pencil 2. Potential for thumb joint pain 3. Immature grip pattern
Lateral Pinch Grip	Also called closed-thumb web grip, proximal grip, and nonfunctional grip. Thumb used to push index finger against pencil shaft to stabilize it. Pencil moved with joints of fingers.		No	1. Limits pencil motion, especially when rounding over tops of letters 2. Immature grip pattern
Quadrupod Grip	Also called open-thumb web grip, distal digital grip, and functional grip. Pulps of thumb, index, and middle fingers stabilize pencil shaft near writing tip against ring finger. Coordinated wrist and finger motion present.		Yes	1. Able to manipulate pencil. Can begin as static grip, but usually develops into dynamic grip 2. Considered mature grip pattern

(continued)

Table 20.1 Grip Patterns (Continued)

Grip	Description	Picture	Recommended	Reasons
Tripod Grip	Also called three jaw chuck grip, open-thumb web grip, three-finger pulp grip, distal digital grip, and functional grip. Balanced three-finger (chuck) grip used to hold pencil near writing tip with finger pulps. Coordinated wrist and finger motion present.		Yes	1. Most efficient writing grip 2. Can begin as static grip, but usually develops into dynamic grip 3. Considered mature grip pattern
Adapted Tripod Grip	Also called closed-thumb web grip, adapted distal digital grip, and functional grip. Thumb pulp used to stabilize writing end of pencil against middle finger. Pencil shaft stabilized between index and middle finger.		Yes	1. Provides secure writing grip when standard pencil grip is difficult due to weakness or joint problems 2. Considered a viable adaptation of a mature grip

THE CHALLENGES OF HANDWRITING ERGONOMICS

Many children have problems achieving handwriting competence. Writing is a complex activity requiring a multitude of physical, cognitive, perceptual, and visual skills. The manipulation of a writing tool as an extension of the hand requires postural support, counterbalance, and eye–hand coordination.

Handwriting, long considered a rite of passage for children, can be a daunting task for a child to master. Parents and educators can help children master handwriting skills by creating supportive environments to maximize physical, cognitive, perceptual, and visual skill development. The following section provides specific recommendations that can help children develop the skills required for handwriting.

POSTURAL STABILITY AND CONTROL

1. Promote the development of prone postures in babies and toddlers with floor time activities and active exploration of the environment.
   a. Avoid propping infants with boosters and pillows in semiupright positions. Provide exploratory playtime on stomach.
   b. Delay and limit use of baby bouncers and walkers. These restrict exploration and are hard on the hip and ankle joints.
2. Encourage standing rather than prolonged sitting when teaching children with low postural stability to write. Standing promotes postural stability and development of proximal strength.
3. Support feet to promote upright sitting posture. Foot sensation triggers the spinal erector reflex. Without foot contact, the child will need to consciously think about and repeatedly correct their posture in order to sit upright or lean against desk surface.

4. Use vertical surfaces for writing activities with young children to promote wrist stabilization and postural control (Yakimishyn and Magill-Evans, 2002) (Figure 20.25). Some children are not developmentally ready to work on horizontal surfaces until they are about 7 years old. If the wrist cannot maintain extension while the fingers are manipulating the pencil or crayon, the child should continue to work on inclined or vertical surfaces.

   Children working on vertical surfaces above eye level will automatically position the wrist in its proper extended position to function. This facilitates the wrist extension and thumb abduction synergy needed for functional grips (Figure 20.26).

   a. Vertical and inclined surfaces promote upright postures and scapular stability needed for writing. Use blackboards or paper taped to walls or doors.
   b. Help children develop proximal stability needed for writing. The more vertical the surface, the better.
   c. Vertical and inclined surfaces facilitate wrist positioning and stabilization needed to write.
   d. Wrist extension facilitates thumb separation and rotation to be able to oppose the thumb for pinch and grip activities.
   e. After wrist extension and thumb abduction synergy is mature, progress writing activities to less inclined surfaces. Bendix (1986) and Sciascia and De Faenza (2003) demonstrated that inclined surfaces between 22° and 45° improve reading posture, but are not beneficial for writing.
   f. An old bifold door can be fastened to a table and used as a large group activity easel (see Figure 20.28) (Olsen, 2001). Bifold doors are designed for use in closets, pantries, or in some cases as folding doors between rooms. The hinge allows them to fold up for opening. The hinged ends of the doors can

Figure 20.25 (Mary Benbow)
Writing on vertical surfaces promotes upright postures and scapular stability.

Figure 20.26 (Mary Benbow)
Writing on vertical surfaces above eye level positions the wrist in extension and increases ability to grip.

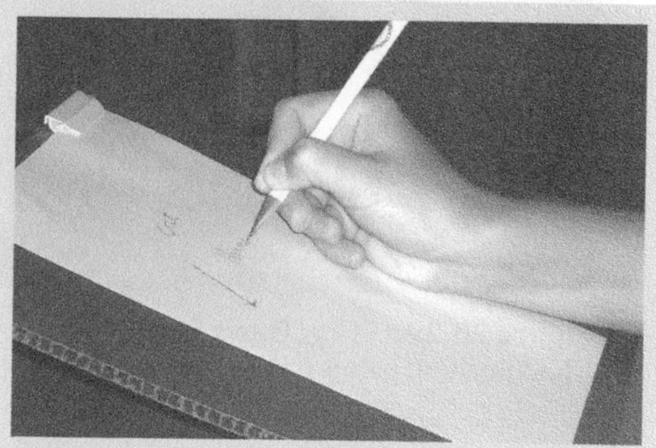

Figure 20.27 (Mary Benbow)
Inclined surfaces facilitate wrist extension and stabilization.

be placed at an angle to form an easel. Additional hinges or ledges can be used to secure the outside ends of the bifold door to the table.

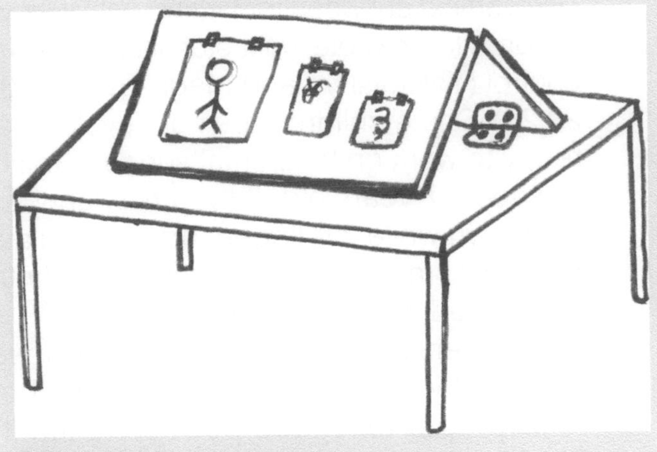
Figure 20.28 (Cindy Burt)
Bifold door used as an easel.

5. Bring posture to awareness and make the activities fun.
    a. Teach arm and hand warm-ups such as pushing palms together, pulling hands apart, hugging yourself tightly, and "flapping your wings" (Olsen, 2001).
    b. Ensure that elbows are below wrist level to avoid placing unnecessary pressure on the carpal tunnel (Figure 20.29).
    c. Teach children shoulder awareness by raising the shoulders, pulling them back, and letting them down.

Figure 20.29 (Ursy Potter Photography)
Child pushing palms together correctly.

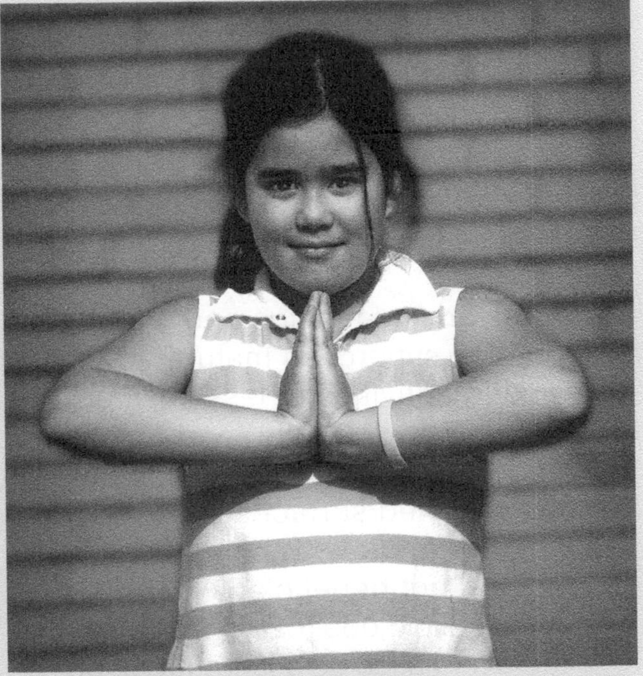
Figure 20.30 (Ursy Potter Photography)
Child pushing palms together incorrectly.

d. Use the posture stomp to help children sit correctly (see Figure 20.31).

While sitting, show them how to stomp feet and raise arms in the air. Their feet will be on the floor and parallel in front of them. Such arm movements straighten their trunks (Olsen, 2001).

Figure 20.31 (Ursy Potter Photography)
The posture stomp is a fun way to increase sitting balance.

WRITING READINESS

1. Make activities fun.
2. Western writing progresses from left to right and begins at the top of the page. Manuscript letters and numbers also initiate from the top and move downward. Consider these concepts when teaching children skills needed to write. Start writing at the top left margin of the paper. Design games and activities that incorporate these concepts before writing training.
3. Teach the 10 number symbols first.
4. Introduce manuscript capital letters after the 10 number symbols. Manuscript capital letters are bold, familiar, and easy to recognize. They are all initiated from the top at equal heights.
5. Follow developmental principles. Children learn forms in a predictable order (Gesell, 1940). Letters with vertical and horizontal lines are easier to write than letters with diagonal lines.

   Begin with letters using horizontal and vertical lines. Wait to teach the 10 capital letters containing diagonals until children master perceptual–motor development of the diagonal (usually around age 5 to 5½) (Beery, 1982).

A K M N R V W X Y Z

Ten capital letters with diagonals.

6. Our cultural system of reading and writing is biased for motor preference of the right-handed, right-eyed population. Reading and writing are the only skills that move in one dimension and one direction—from the body midline into right space.

   Left-handed children have more difficulty learning to write because they must initiate movement from the periphery and move toward their midline. This unnatural movement pattern makes learning to write more difficult. Teachers and parents should be sensitive to this difficulty.

   Some left-handed children develop “hooked wrists” when they write. Writing on a chalkboard above eye level will help correct wrist and finger movement for more efficient wrist–finger interplay for speed.

   When modeling letter patterns for left-handed writers, the teacher should use her left hand so the child can imitate the movements.

EFFICIENT GRIPS

Children, especially girls, experiment with "writing" tools at a young age, frequently before the age of 3. The child who starts "writing" early may not have the motor skills needed to stabilize and guide writing devices in a functional dynamic grip. When children insist on writing before they are physically ready, they can develop awkward or immature grip patterns that can be difficult to change. The following techniques can be used to avoid problems.

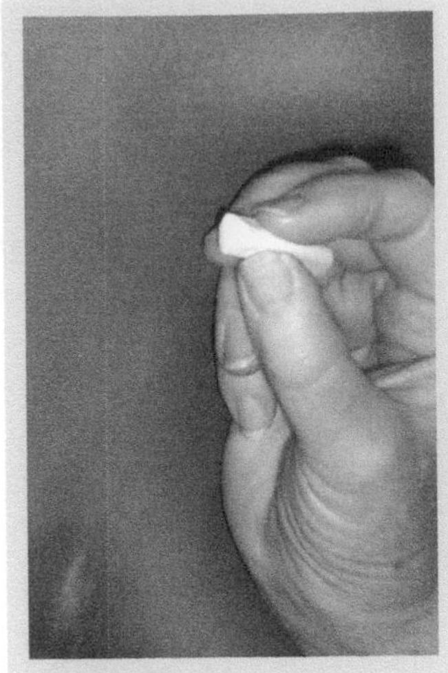

Figure 20.32 (Mary Benbow)
Crayon in three jaw chuck grip.

1. Instruct beginning writers to use gross motor movements and write on vertical surfaces using large free-flowing patterns.
2. Use chalk or crayons broken into ½ to 1 in. pieces. Large diameter, short pieces promote use of a tripod or three jaw chuck fingertip grip (Yakimishyn and Magill-Evans, 2002) (Figure 20.32).
3. Teach the child to hold pencils in a dynamic tripod with an open-thumb web space. Grip pencils close to the distal end near the writing tip (Figure 20.33).

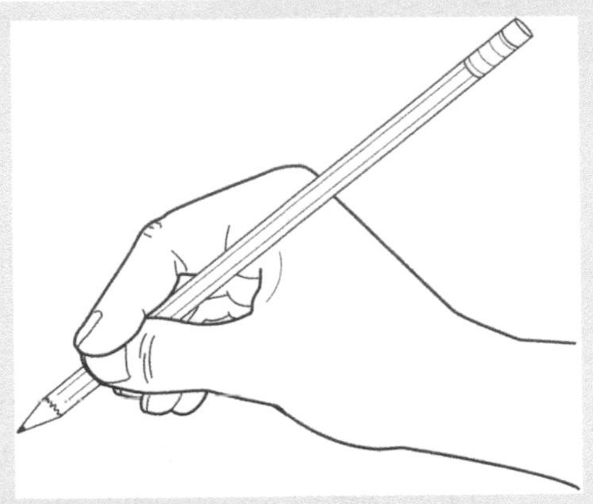

Figure 20.33 (Mary Benbow)
Dynamic tripod grip.

The following techniques can help children grip properly (Olsen, 2001).

A-OK			
A-OK	Step A	Make an A-OK sign with the writing hand	
	Step B	Drop the fingers	
	Step C	Open the A-OK and pinch the pencil	

Flip the Pencil Trick		
Flip the Pencil Trick	Step A	Place the pencil on a table pointing way from you. Pinch the pencil where you should hold it (on the paint where the paint meets the wood) and pick it up. 
	Step B	Hold the eraser and twirl it around. 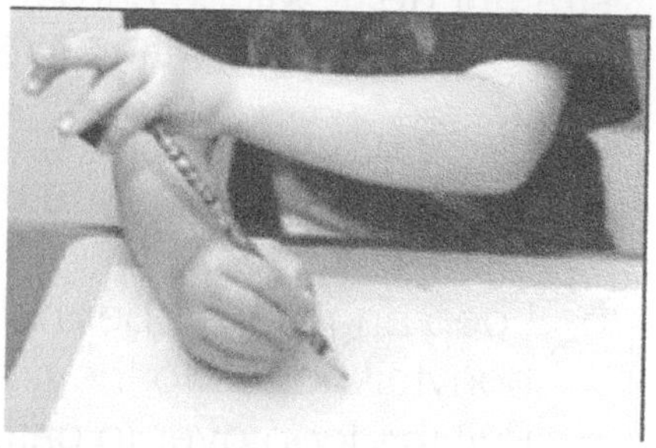
Pencil Driving Tip		
Pencil Driving Tip	Step A	Name the fingers. Thumb is Dad, index finger is Mom, and the remaining fingers are brothers, sisters, pets, or friends. 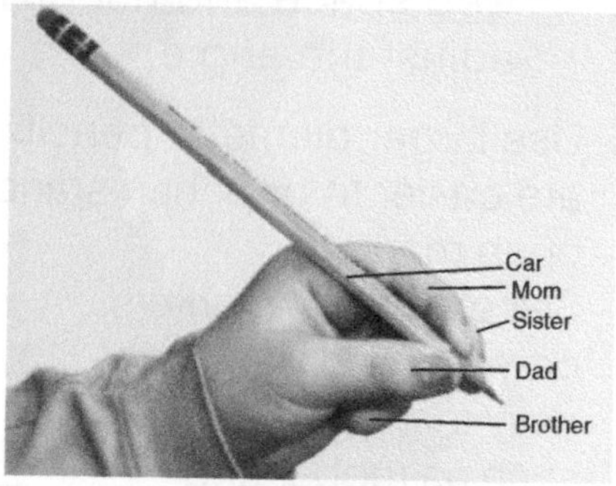
	Step B	Say the pencil is the car. Like a real car, Mom and Dad sit in front and everyone else in back. For safe driving, Mom and Dad face forward (toward front of the pencil). Dad shouldn't sit on Mom's lap (thumb on top of index finger) and Mom shouldn't sit on Dad's lap (index finger over thumb). If the child uses an overlapped or tucked-in thumb posture, remind him or her that no one can sit on another's lap while driving (Bradshaw and Daniels in Olsen, 2001). 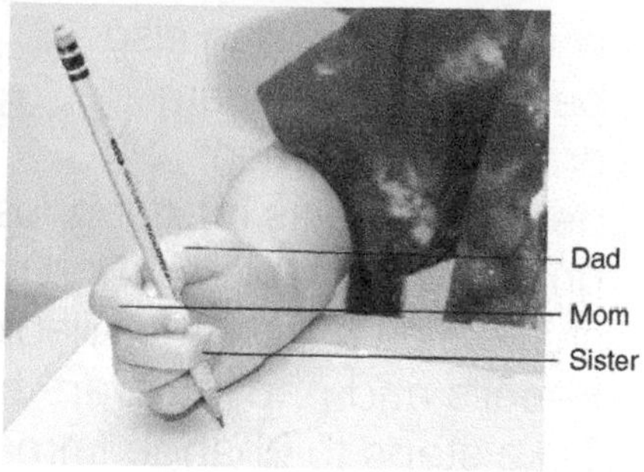

4. Use positioning devices such as the Pencil Pal (Figure 20.34) to provide a higher point of stability to reduce DIP (distal joint of the index finger) hyperextension.
5. Rubber bands or ponytail holders can be used to check the angle of the pencil. If the pencil is standing straight up, it will be hard to hold and cause tension in the fingertips.
   a. Place a rubber band or ponytail holder around the child's wrist.
   b. Loop another rubber band or ponytail holder to the first one.
   c. Pull the loop over to catch the pencil eraser.
   d. Make sure the pencil is positioned at a good writing angle in the thumb web space and not pulled too tight. If so, use longer rubber bands to adjust the angle.

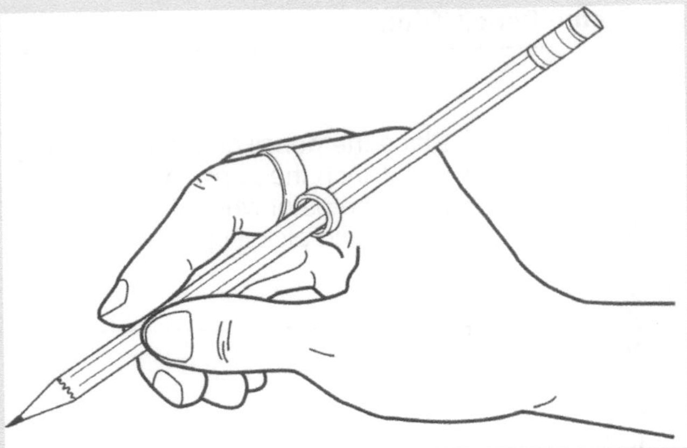

Figure 20.34 (Mary Benbow)
The Pencil Pal.

6. Use larger diameter pencils, markers, or crayons. Larger diameter writing devices are easier to grip than standard size devices. Many are readily available on the open market.
7. Select pencils, markers, crayons, and pens with easy to grip surfaces such as rubber and foam. The surface of writing devices can also be padded with foam or finger placement devices such as rubber grips (Figure 20.35) to assist in gripping. Foam cylinders and grips are available through many educational or medical and rehabilitation supply stores. Foam tape is available at hardware and office supply stores.

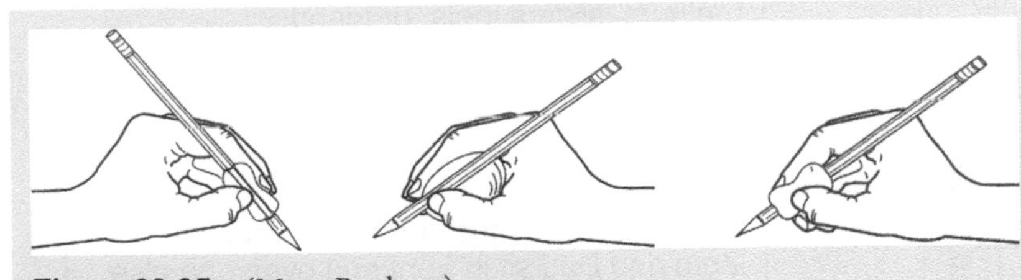

Figure 20.35 (Mary Benbow)
Finger-positioning devices to aid pencil grip.

8. Pencils and pens that are heavier near the writing tip are easier to use.
9. Take steps to change immature pencil grips that are inefficient or harmful. Try to get the child to use the adapted tripod grip (Figure 20.36) where the pencil stabilizes in the web space between the index and long finger. This grip is easier to adopt than the dynamic tripod grip (Figure 20.33), but still utilizes the same hand muscles to manipulate the pencil.
10. When teaching new grips, follow a developmental progression of scribble, free-flowing shapes, and repeated patterns. Put the child in control, letting them alternate between the inefficient grip and the new grip. Allow the child to shift back to the familiar grip when the adapted grip begins to "bug" them. When ready, the child can

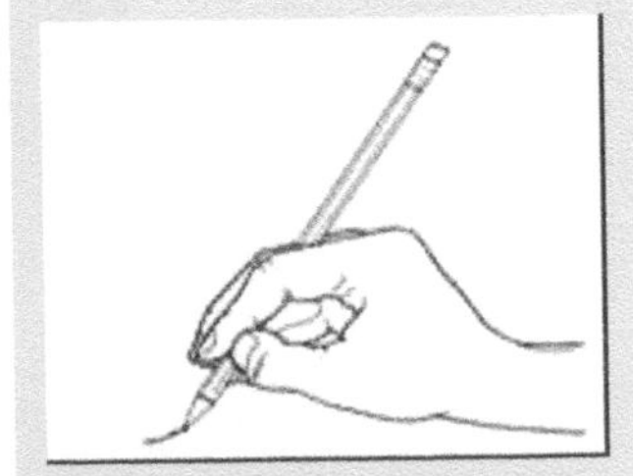

Figure 20.36 (Mary Benbow)
Adapted tripod grip.

return to the new grip. Over time, the new grip will become more comfortable and efficient, and the child will use it more consistently with less stress.

11. Monitor left-handed writers to promote a relaxed grip to avoid discomfort and development of the "left-handed hook" writing position.

PAPER AND WRITING TOOLS SELECTION

Paper and tool design can affect writing quality, comfort, and speed. Suggestions for selecting paper and writing tools include:

1. Experiment with different types of paper to find the type that works best for each child. Some prefer unlined paper with a highlighted baseline on which to locate and orient letters.

   Make sure the width of the writing spaces between the lines is appropriate for the skill level of the child and the type of writing device used. Markers and crayons require higher vertically spaced lines than pencils.

   When using blunt writing tips such as magic markers, make sure lines are separated by a 1 to 1½ in. (2.5–3.75 cm) vertical space. When writing with a #2 pencil, the rule or vertical space should be about 3/8 in. (1 cm) rule (typical notebook paper). Rule that is more than ½ in. (1.3 cm) will be difficult and uncomfortable to use with a #2 pencil.

   Grip excursion is important to consider when selecting the writing device and line rule. Children who have difficulty making long, smooth strokes will find large ruled paper harder to use. Measure the length of the child's stroke with the writing device to determine maximum line rule.

> *Grip excursion* is the distance of the stroke. Measure the length of the line drawn by the pencil with the fingers moving from full extension to full flexion in downstroking, or with fingers moving from full flexion to full extension in upstroking.

2. New writers who are beginning to use lowercase letters write more effectively on ½ in. (1.3 cm) divided, lined paper. Highlight the baseline of each writing space. This helps to orient the child to reduce line and spatial confusion.
3. Experiment with a variety of writing tools and let children select the one that feels the most comfortable. Options include ballpoint pens, markers, calligraphy pens, and varying hardness of lead in pencils.
4. Evaluate pens with a variety of liquid and solid inks. Liquid inks flow more readily than solid inks and require less pressure.
5. Select high contrast paper and marking ink so children can clearly see their written work.
6. Avoid use of ballpoint pens and markers that smudge. This inhibits all writers, especially left-handers.

Figure 20.37 (Ursy Potter Photography)
Experiment with a variety of writing tools. Let children select the one that feels the most comfortable.

MANAGEMENT OF THE WRITING PAPER

1. Set table height for writing 2 in. (5 cm) above the elbow with the child's heels touching the floor. This supports the forearm at an optimal writing angle.
2. Angle the slant of the paper parallel to the writing arm when the hands are relaxed and together at midline on the tabletop. Slanted paper facilitates use of rapid wrist flexion in downstroking of letters in manuscript and cursive writing.
3. Give children with postural instability narrow strips of paper for writing tasks.

   Standard writing paper folded into narrower 4 in. (10 cm) widths helps children work close to midline where stability and control is greatest (Figure 20.38). The narrower strips of paper require stabilizing from the nondominant hand to keep them from moving around the desk when writing. Stabilizing the paper will encourage the child to use the nondominant hand for postural support and to frame the writing space.
4. Encourage children to keep their writing hand below the writing line for all writing tasks. This positions the wrist for accurate and rapid downstroking. It also prevents the "left-handed hook" writing position.
5. Teach children to advance paper upward on the desk as writing progresses downward. This prevents children from bending their necks down to see the bottom of the page.

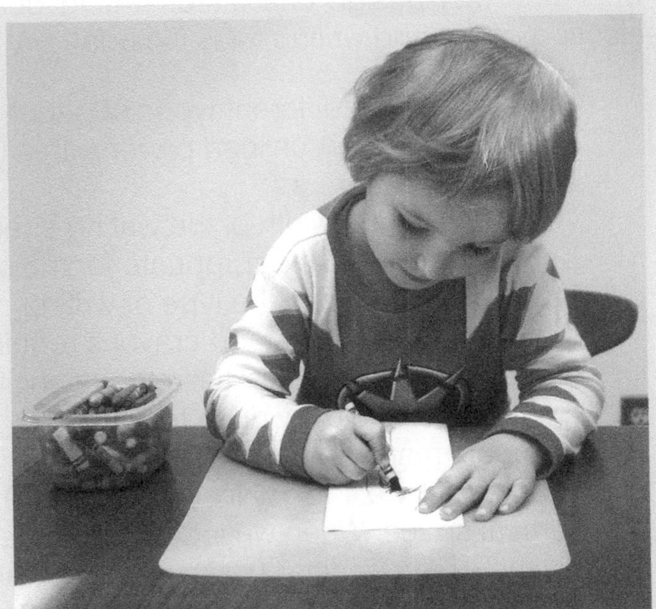

Figure 20.38 (Ursy Potter Photography)
Stabilizing the narrow strips of paper with the nondominant hand helps children balance.

FURNITURE DESIGN

1. Provide children with chairs that support their low back curve when they sit.
2. Shoulder mobility and open shoulder postures are important, especially for children developing lifelong habits and postures. Select chairs with backrests that stop just below the inferior (lower) angle of the scapula (shoulder blade) (Cotton, 1904; Zacharkow, 1988). Adjust seat height so heels contact the floor for weight shifting and balance as the writing arm moves horizontally across the paper.
3. Consider chairs with adjustable seat tilt. Forward sloping seats (10°–15° tilt) can help reduce slouching and forward head postures (Bendix, 1984; Bendix and Hagberg, 1986; Mandal, 1982, 1984; Marschall et al., 1995; Aagaard-Hensen and Storr-Paulsen, 1995; Taylour and Crawford, 1996).
4. Avoid seats that have rounded or curved designs that encourage slouched postures. Seats with distinct angles between the base of the spine and the thighs are better choices.

5. Children need back support when writing on desk surfaces. Teach them to sit with their backs supported against the backrest.
6. The space or distance between the desk and chair should adjust to promote upright postures. In most cases, the edge of the desk should overlap the seat.

   Make sure there is enough space between the chair and desk to allow the trunk to rotate as the writing hand moves horizontally away from the midline of the body.

Figure 20.39 (Ursy Potter Photography)
Writing on high surfaces raises shoulders, leads to fatigue, and reduces hand dexterity.

7. Desktops that are about 2 in. (5 cm) above seated elbow level provide forearm support while maintaining a relaxed shoulder position when writing. When the writing surface is higher, the shoulder and elbows are elevated and forced into overly abducted positions. This tires the shoulders and reduces dexterous use of the hands.

   When the desk level is too low, children will slouch into a forward, rounded posture with the neck flexed. Children sitting in this position often need to support their head with the nonwriting hand. This causes the hand to lose its role in stabilizing and adjusting the writing paper.
8. Moveable, slanted reading surfaces can help facilitate mobile and upright postures. Adjustable surfaces can help prevent forward leaning and awkward head, neck, and trunk postures when reading. However, it is more comfortable to write on a horizontal surface. Place papers used for writing on a horizontal surface in front of a moveable angled reading platform (Bendix, 1984; Bendix and Hagberg, 1986, Sciascia and De Faenza, 2003).
   a. Horizontally positioned 3 in. (7.6 cm) wide binder notebooks can serve as low-cost sloped reading surfaces if reading stands and document holders are not available.
   b. More vertical writing surfaces are appropriate if wrist and postural stabilization are primary issues that must be addressed.
9. Concave "cutout" work surfaces that accommodate to the anterior trunk help facilitate trunk control and provide forearm support (Shen et al., 2003). These types of desks can be useful for children with postural instability.
10. Provide children in kindergarten and elementary schools with a variety of vertical surfaces for writing. Chalkboards, upright easels, and bifold doors can help develop functional writing grips, especially when used by children with weak wrist extension strength.

IMPLEMENTING A KINESTHETIC WRITING PROGRAM

Cursive handwriting is a *kinesthetic* skill. The most efficient way to teach handwriting is using kinesthetic teaching strategies. Following kinesthetic steps to master handwriting is an effective way to develop writing skill to its highest level. From the first session, kinesthetic skills are taught, practiced, integrated, and evaluated as each letter is mastered.

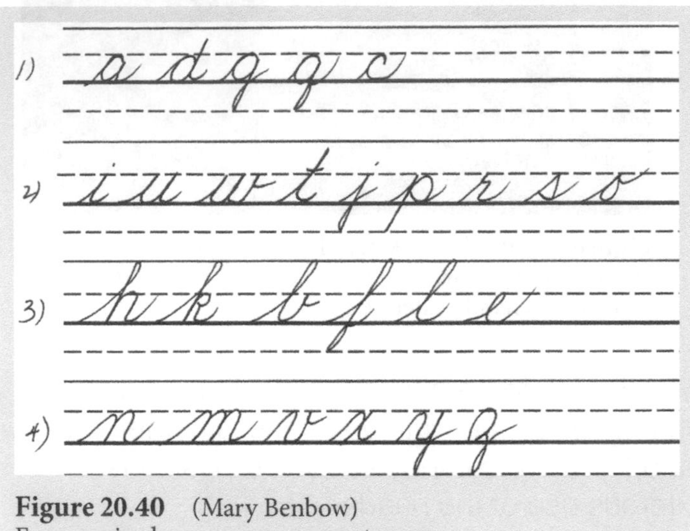

Figure 20.40 (Mary Benbow)
Four cursive lowercase movement groups.

The 26 lowercase letters reduce to four movement groups, determined by their movement pattern. The four groups should be visually associated with common objects in the child's world to facilitate learning. The child must learn the initial letter in each group to the automatic kinesthetic level so it can be modified for the remaining letters in the group (Benbow, 1990).

> *Kinesthesia* is the perception of body position, movement, weight, and force in space. It is enhanced by repeating movements with vision occluded (such as while blindfolded) until they become automatic.

INSTRUCTOR'S GUIDE

Draw a 2 ft model of the letter on the chalkboard carefully positioned within a three-unit divided line. Use the model to teach the following:

1. Indicate the starting point with a dot to left of the letter position on the writing line.
2. Analyze and describe the letter position within the ½ space vertical line units with the children.
3. Present letters in the groups identified in Figure 20.40. These groups are determined by their movement pattern, specifically the lead-in stroke used to make the letter. Discuss lead-in group assignment and reinforce visualizing all of the movements required to make the letter.

    a. Group 1: Have children visualize clock face to facilitate memory of lead-in stroke (*a*, *d*, *g*, *q*, *c*).
    b. Group 2: Have children visualize a kite string for the right diagonal stroke (*i*, *u*, *w*, *t*, *j*, *p*, *r*, *s*, *o*).
    c. Group 3: Ask the child to visualize a stunt plane moving in an upward right diagonal, looping over the top to the right, and descending with a downstroke to the writing line (*h*, *k*, *b*, *f*, *l*, *e*).
    d. Group 4: Ask children to visualize hills and valleys with this group (*n*, *m*, *v*, *x*, *y*, *z*).

4. Demonstrate while verbalizing the entire stroke progression using consistent terminology. For example, with group 1, tell children to climb over the clock to one o'clock, back tract to nine o'clock, round down to six o'clock, and tilt up to one o'clock. Retrace to writing line to release.

Teach each child to "talk to your hand" (the motor plan) and make your writing hand do exactly what you are telling it to do. Repeat verbalizing as you move your chalk along the "lace" of the letter. The "lace" of the letter is the visible progression of the stroke pattern left by the pencil's lead.

STUDENT MOTOR LEARNING

Instructors can encourage motor learning among the students as follows:

1. Have students point their index and long fingers while holding down the ring and little fingers with the thumb of their writing hand. This helps develop a sense of motoric separation of the two sides of the hand (Figure 20.41).
2. Have the children join the instructor in reciting aloud the movement pattern.
3. During the recitation, have the children "trace" the chalkboard model in the air. Do this several times using isolated shoulder movements (the shoulder moves with the elbow, wrist, and fingers held still). Try to get the children to use only their shoulders for the "tracing." The rich supply of movement receptors in the shoulder joint provide feedback to the brain and speed learning of new movement patterns.
4. As children trace the letter in the air, have them visualize the stroke progression, and pay attention to the "feel" of the movement.
5. Once the student is confident in visualizing and verbalizing the motor pattern, have them rest their elbows on the desk, close their eyes, visualize, and verbally self guide their pointed fingers to write the letter in the air using elbow and wrist movements.

   The instructor should assess the child's verbal and movement plan before letting them try the new letter on paper.

Figure 20.41 (Mary Benbow) Kinesthetic letter tracing in the air.

After the instructor is satisfied with the above steps, students should check their own learning using half-inch divided line paper. As they verbalize the motor plan aloud, have them write the new letter 10 times within the proper line segments. On completion, each child evaluates their own work and circles all correct letters. Have each child select its best letter for instructor verification. Using their new kinesthetic model, they should write the letter 15 more times.

Once the motor pattern is consistent, they should connect the letter to previously mastered letters in short words. When they can write the new letter correctly using visualization to kinesthesia, they will be eager to try the subsequent letter in the motor group. It will come naturally and easily because the basic motor pattern is kinesthetic.

It is estimated that up to 95% of letters on a typical page of cursive writing are in lowercase. Teach lower-case letters to the kinesthetic or automatic level at a functional speed before introducing the capitals. Note that this is the reverse of the learning sequence for the lowercase manuscript letters. Children typically learn block capital letters before lowercase block letters.

After students become proficient on cursive writing of lowercase letters, teach them capital letters approximately 2 to 3 months later. Group capital letters by motor patterns to help them learn more efficiently. There are 10 motor patterns. Use the lead-in names to facilitate memory of letters just as used in teaching the lowercase cursive letters (Figure 20.42).

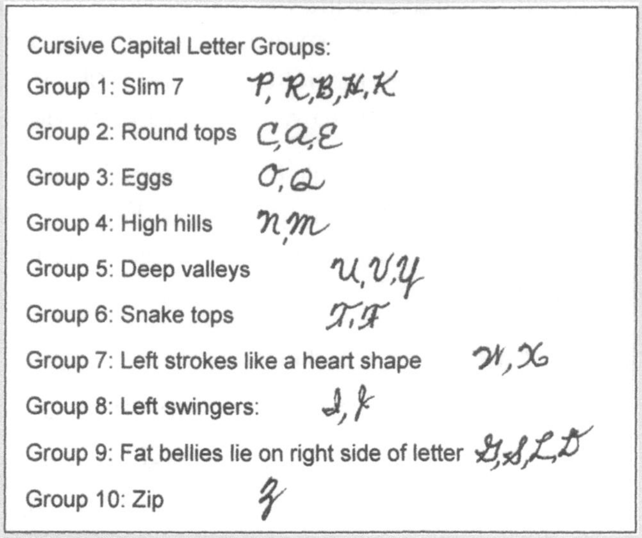

Cursive Capital Letter Groups:

Group 1: Slim 7 P, R, B, H, K

Group 2: Round tops C, A, E

Group 3: Eggs O, Q

Group 4: High hills N, M

Group 5: Deep valleys U, V, Y

Group 6: Snake tops T, F

Group 7: Left strokes like a heart shape W, X

Group 8: Left swingers: I, J

Group 9: Fat bellies lie on right side of letter G, S, L, D

Group 10: Zip Z

Figure 20.42 (Mary Benbow)
Cursive capital letter groups.

SUMMARY

Handwriting is a complex and demanding task that we often take for granted. Since handwriting ability can influence future academic success, children need to develop competence.

Successful learning environments are supportive, directive, intriguing, and most of all, fun. Designing the learning progression according to children's developmental progression will give children the tools they need to become proficient at writing.

Flexibility and attentiveness are necessary to recognize and meet the developmental level of each child cognitively, emotionally, and physically. Developmental readiness is necessary for writing skill acquisition. Postural stability, fine motor control, visual coordination, and cognitive perceptual skills are all required to write effectively. Planning the learning process to sequentially build these skills will provide children with a base from which to begin and progress through scribbling, coloring, printing, and cursive writing.

Figure 20.43 (Ursy Potter Photography)

Physically, it is important that furniture and writing tools fit the child, with adjustable furniture being optimal to meet the needs of a variety of growing children. Furniture that fits helps to position the child so writing occurs with the least amount of effort. Specialized writing tools can help during learning or to remediate poorly learned skills. A variety of devices are available to assist with the process.

Kinesthetic and visualization handwriting techniques result in children learning faster, being more accurate writing individual letters, and ultimately in better handwriting. Although the techniques may appear "too easy" or too "mind-oriented," their use is a foolproof way to help student's master handwriting.

Paying attention to the details of teaching handwriting will pay great dividends. After all, writing and printing are skills that are necessary for a lifetime.

APPENDIX A: WRITING PROBLEM INDICATORS

Skills	Problem Indicators
Sitting Balance and Endurance	Slumps forward on desk for support
	Supports head on hand
	Slumps sideways and falls from chair
	Wraps feet and ankles around legs of chair for support
	Sits on one hip and then the other
Shoulder Stability	Scapulae "wings" out during movements
	Scrunches up shoulders
	Leans on forearms when writing
	Writes low and small on easel or chalkboard
Arm Mobility	Holds writing hand parallel to desk rather than in partially rotated forearm position to monitor work
	Unable to make downstrokes with wrists when coloring or writing
	Writes with wrist in flexed position (bent downward)
Vision–Perception–Coordination	Writes with inconsistent letter size and spacing
	Turns head to view with one eye only
	Closes one eye or squints when writing
	Turns paper at extreme angles to write uphill
	Unable to stay within lines to write and color
	Reading and writing lag behind verbal skills

(continued)

Appendix A: Writing Problem Indicators (Continued)	
Skills	**Problem Indicators**
Dominance	Unable to use contralateral side to stabilize trunk when writing
	Ignores nondominant hand when doing one-handed activity
	Switches hands at midline when writing
Grip	Holds writing device too high or too low for easy control
	Laps thumb over fingers to hold pencil
	Tucks thumb under fingers when writing
	Complains of sore hands when writing
	Holds and presses so hard that knuckles turn white
	Presses hard (breaks lead, tears paper)

APPENDIX B: GLOSSARY

Accommodation: Ability of the eyes to efficiently shift from near to far point.

Abductor muscle: Muscle that moves a limb or digit away from the body.

Abduct: Movement away from the body.

Adductor muscle: Muscle that moves a limb or digit toward the body.

Adduct: Movement toward the body.

Adaptive righting reactions: Body adjustments that children make to balance themselves, and to change and sustain postures.

Arches of the hand: The hand is not flat. It is composed of three arches that enable hand function. The arches are important for handwriting and other fine motor activities because they provide a balance between stability and mobility for grasping.

Bilateral integration: The brain function that enables coordination of the two sides of the body.

Binocularity: Ability of the two eyes to coordinate and fuse visual images into a single image.

Distal interphalangeal joint (DIP): The joint at the distal end of a phalange or finger (closest to the fingernail).

Extension: The act of straightening a joint; movement which brings the members of a limb into or toward a straight position.

Flexion: The act of bending a joint or of a joint being in a bent position.

Foveal vision: Central part of the visual field used for focused vision with attention.

Grip: Finger and hand position and technique used to hold a writing device when writing.

Hyperextension: Extreme or extensive extension of a joint beyond the straight position (more than 180°).

Hyperflexion: Extreme or extensive flexion of a joint beyond the normal bent position.

Kinesthetic feedback: Sensory feedback information related to the movement of direction of movement of joints.

Metacarpophalangeal (MCP or MP) joints: The knuckle joints of the hand; the joints between the wrist and the fingers.

Oppose (or opposition): This is a movement of the hand in which the thumb moves across the hand toward the ulnar (little finger) side of the hand or the little finger.

Ocular mobility: Eye-movement control which governs the speed and accuracy of visual inspection and scanning.

Peripheral vision: Vision outside the central part of the visual field used for nonfocused vision.

Proximal interphalangeal joint (PIP): The joint at the proximal end of a finger (closest to the knuckles).

Postural muscles: Core muscles that support upright postures.

Postural sway: Natural movement of the child's center of gravity within a base of support.

Pronation: The rotation of the forearm to position the palm downward.

Proprioceptive feedback: Sensory feedback related to body or limb position in space.

Proximal joints: Joints that are close to the center or core of the body, as opposed to distal joints, which are further away from the center of the body.

Pulp: The padded part of the finger or thumb near the tip with the fingerprints.

Sensory feedback: Awareness of tactile, proprioceptive, kinesthetic, visual, auditory, gustatory, and olfactory stimuli.

Sensory integration: Ability to understand, organize, and respond to sensations of touch, balance, equilibrium, vision, taste, and smell.

Spinal erector reflex: A reflex present at birth that assists in development of upright posture.

Supination: The rotation of the forearm to turn the palm upward.

Synkinesis: Associated movements or unintentional movements on the nonperforming side accompanying a volitional movement with the dominant hand.

Tactile feedback: Sensory awareness related to touch, pressure, temperature, pain, and vibration.

Ulnar deviation: Hand shifts toward the little finger side of the wrist.

Ulnar styloid process: The sharp bone evident on the little finger side of the wrist.

REFERENCES

Aagaard-Hansen, J. and Storr-Paulsen, A. (1995). A comparative study of three different kinds of school furniture. *Ergonomics, 38,* 1025–1035.

Amundson, S. J. and Weil, M. (1996). Prewriting and handwriting skills. In: J. Case-Smith, A. S. Allen, and P. N. Pratt (Eds.), *Occupational Therapy for Children,* 3rd edn. St. Louis, MO: Mosby-Year Book, pp. 524–541.

Beery, K. (1982). The Development Test of Visual-Motor Integration. Cleveland, OH: Modern Curriculum Press.

Benbow, M. (1990). *Loops and Other Groups. A Kinesthetic Writing System.* Tucson, AZ: Therapy Skill Builders.

Bendix, T. (1986). Sitting postures—A review of biomechanic and ergonomic aspects. *Manual Medicine, 2,* 77–81.

Bendix, T. and Hagberg, M. (1984). Trunk posture and load on the trapezius muscle whilst sitting at sloping desks. *Ergonomics, 27,* 873–882.

Cotton, F. J. (1904). School-furniture for Boston schools. *American Physical Education Review, 9,* 267–284.

Gesell, A. (1940). *The First Four Years of Life.* New York: Harper.

Guiard, Y. (1987). Asymmetric division of labor in human skilled bimanual action: The kinematic chain as model. *Journal of Motor Behavior, 19,* 486–517.

Hamstra-Bletz, L. and Blote, A.N. (1993). A longitudinal study on dysgraphic handwriting in primary school. *Journal of Learning Disabilities, 26,* 689–699.

Jacobs, K., Bhasin, G., Bustamante, L., Buxton, J. C., Chiang, H., Greene, D., Ford, S., Frank, R., Moczydlowski, E., Naidu, J., Nielsen, E., Savini, A., Singer, B., and Wieck, A. (2002). Everything you should know about ergonomics and youths, but were afraid to ask. *OT Practice, May 27,* 11–19.

Johansson, R. S. and Westling, G. (1987). Signals in tactile afferents from the fingers eliciting adaptive motor responses during precision grip. *Experimental Brain Research, 66,* 141–154.

Laszlo, J. I. (1990). Grasp stability during manipulative actions. In: C. A. Hauert (Ed.), *Developmental Psychology.* North-Holland: Elsevier, pp. 273–309.

Long, C., Conrad, M. S., Hall, E. A., and Furler, M. S. (1970). Intrinsic–extrinsic muscle control of the hand in power and precision handling. *Journal of Bone and Joint Surgery, 52A,* 853–867.

Mandal, A. C. (1982). The correct height of school furniture. *Human Factors, 24,* 257–269.

Mandal, A. C. (1984). The correct height of school furniture. *Physiotherapy, 70(2),* 48–53.

Marschall, M., Harrington, A. C., and Steele, J. R. (1995). Effects of workstation design on sitting posture in young children. *Ergonomics, 38,* 1932–1940.

Moberg, E. (1983). The role of cutaneous afferents in position sense, kinaesthesia and motor function of the hand. *Hand, 106,* 1–19.

Olsen, J. Z. (2001). *Handwriting without Tears Teacher's Guide,* 8th edn. Potomac, MD: Jan Z. Olsen.

Pare, M. and Dugas, C. (1999). Developmental changes in prehension during childhood. *Experimental Brain Research, 125,* 239–247.

Parush, S., Levanon-Erez, N., and Weintraub, N. (1998). Ergonomic factors influencing handwriting performance. *Work, 11,* 295–305.

Rosenbloom, L. and Horton, M. E. (1971). The maturation of fine prehension in young children. *Developmental Medicine and Child Neurology, 13,* 3–8.

Sciascia, G. and De Faenza, M. D. (2003). School ergonomics: Seated position. *Italian Rehabilitation, Orthopedics, 2.* www.riabilitazionaitala.it/orthoped2/. Viewed 04/05/06.

Shen, I., Kang, S., and Wu, C. (2003). Comparing the effect of different designs of desks with regard to motor accuracy in writing performance of students with cerebral palsy. *Applied Ergonomics, 34,* 141–142.

Shumway-Cook, A. and Woollacott, M. H. (1985). The growth of stability; postural control from a developmental perspective. *Journal of Motivational Behavior, 17,* 131–147.

Taylour, J. A. and Crawford, J. (1996). The potential use and measurement of alternative work stations in UK schools. In: S. A. Robinson (Ed.), *Contemporary Ergonomics.* London: Taylor & Francis, pp. 464–469.

Warren, J. C. (1846). *Physical Education and the Preservation of Health.* Boston, MA: Ticknor.

Yakimishyn, J. E. and Magill-Evans, J. (2002). Comparisons among tools, surface orientation and pencil grasp for children 23 months of age. *American Journal of Occupational Therapy, 56,* 564–572.

Zacharkow, D. (1988). *Posture: Sitting, Standing, Chair Design and Exercise.* Springfield, IL: Charles C. Thomas.

CHAPTER 21

SCHOOL FURNITURE FOR CHILDREN

ALAN HEDGE AND RANI LUEDER

TABLE OF CONTENTS

INTRODUCTION

Classroom furniture affects children's postures, comfort, health, and ability to learn. Research indicates that many schoolchildren sit in furniture that does not fit them properly. Further, schoolchildren who sit in awkward postures for long durations are prone to experiencing back and neck discomfort and other musculoskeletal symptoms that may worsen with time and even progress into adulthood.

Yet common assumptions about what is ergonomically "proper" for adults may or may not be appropriate for children. To date there has been little systematic work to develop ergonomic guidelines for children that can accommodate the age spectrum of children and adolescents.

This chapter reviews ergonomic design considerations for classroom furniture and summarizes worldwide ergonomics research into the design of comfortable school furniture.*

Figure 21.1 (SIS-USA-Inc.com)
Research indicates that classroom furniture affects sitting postures and the corresponding risk of discomfort and pain. Childhood is a critical time for learning postural habits that continue through a lifetime.

Schoolchildren's furniture often reflects national differences. For example, while US schoolchildren typically sit at desks that are set at or below seated elbow height (Figure 21.2), Danish children may sit on high seats at shared desks (Figure 21.3).

Although the furniture may be different, the range of seated postures is often similar in different countries. Children's postures are often at odds with ergonomic guidelines for supporting natural sitting postures.

* The focus of this chapter is on how the design of conventional classrooms can promote health, comfort, and effective learning environments. The design of laboratory furniture or the design of furniture for teachers is beyond the scope of this chapter.

Figure 21.2 (Ursy Potter Photography) A typical elementary school group in a US school, with fixed-height chairs and a flat fixed-height table.

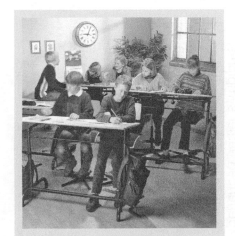

Figure 21.3 (SIS-USA-Inc.com) Danish classroom furniture provides features considered unusual in the United States, such as sit-to-stand desks, sit-to-perch chairs, tilting tables, and foot supports.

Schoolchildren in the last century and today

In the late nineteenth century, attendance at schools (as we know them today) became mandatory in many countries. Back then, classrooms were stark environments and school furniture was solid and designed for longevity rather than comfort. Schoolchildren often sat on bench seats attached to fixed height desks accommodating two students sitting side by side. Some writing surfaces were partly sloped to enable children to read and write on slate. Forward-facing seat arrangements allowed children to see the teacher and blackboard and the teacher to see the children.

By the middle of the twentieth century, students commonly sat on individual chairs with hard seats and backs at desks that typically provided in-desk or under-desk storage for books, paper, pencils, and pens. In the 1950s, it was common to find children staying in class while teachers moved from room to room to teach their lessons.

By the 1960s, children began to move from class to class. School furniture became lighter and movable so that teachers could reconfigure classrooms for different activities.

As we enter the twenty-first century, technological and social changes encourage us to rethink about the classroom furniture design. Although many of our educational tools and methods are vastly different from a century ago, the basic physical demands of the classroom have not changed substantially. On a typical schoolday, children spend much of their time sitting in class while listening to teachers, reading, writing, drawing, or working at a computer. As in the past, price and robustness assume a higher priority than ergonomic considerations when purchasing classroom furniture.

(Centers for Disease Control)

SCHOOL FURNITURE AND CHILDREN: MUSCULOSKELETAL SYMPTOMS*

A growing body of research indicates that schoolchildren report high rates of discomfort, particularly in the neck and back.† Such musculoskeletal symptoms often impair children's comfort and long-term health.

Of course, school furniture is not the only cause of pain and discomfort. However, research has shown that awkward and constrained sitting postures and poorly designed classroom furniture are important contributors to children's musculoskeletal discomfort.

Table 21.1 Back Pain Risk Factors among Schoolchildren (see Chapter 7, Table 7.3)

Classroom	• Extended sitting in static postures • Prolonged work on the computers • Finding their chair uncomfortable
Other	• Inadequate storage • Overly heavy book bags • Intensive competitive sports activities

These considerations are particularly important given that childhood is a critical time to learn and develop good postural habits that they might practice through their lifetime.

Schoolchild's discomfort and pain

1. *is common;*
   Schoolchildren report high rates of discomfort and pain (especially. adolescents), which compares with the rates of back pain in adults.
2. *adversely affects their ability to function;*
3. *often progresses through childhood and adolescence;*
   As children become adolescents, their symptoms often become more frequent and progressively more intense and local discomfort affecting specific body parts often becomes increasingly generalized.
4. *may continue into adulthood;*
   Back pain in children increases their risk of back pain as adults. Research suggests the strongest predictor of future back pain among children is, having experienced earlier symptoms.
5. *can interfere with learning.*
   Static postures may introduce new sources of distraction; discomfort can interfere with a child's ability to learn.

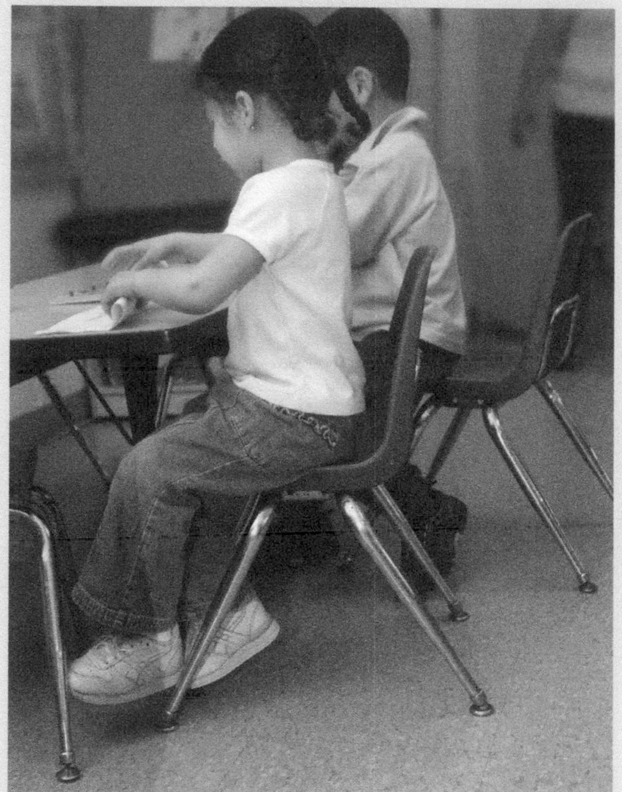

Figure 21.4 (Ursy Potter Photography)
Postures children adopt in their early school years expose them to risk factors that set the stage for long-term habits. Here a young child sits with her feet dangling and her back unsupported, while the boy sits on the front edge of his seat so his feet can touch the floor but his back also is unsupported by his chair.

* Gardner and Kelly (2005) provides an excellent review of the research on back pain among children. Other good reviews include Balague et al. (2003), Balague and Nordin (1992), Balague et al. (1999), Barrero and Hedge (2002), Jones and Macfarlane (2005), King (1999), Murphy et al. (2004), Steele et al. (2006) and Trevelyan and Legg (2006).

† Please see Table 6.2 and Table 6.3 of Chapter 6 for more details about musculoskeletal development in children and adolescents. Table 6.6 of that chapter summarizes many of the studies about back pain in schoolchildren.

AGE-SPECIFIC DEVELOPMENTAL CONSIDERATIONS

When designing school furniture, it may be useful to consider how children develop and mature to incorporate features that accommodate a wide range of ages in good postures.

Such considerations include*

1. *Growth patterns differ by age* (Bass et al., 1999)

    - Younger children tend to grow more in their extremities.

      Furniture design should account for their changing needs in seat height and seat depth.
    - Adolescent growth (following puberty) is largely at the spine† (Bass et al., 1999).

      In the early stages, the spine is growing quickly, adding length without adding mass; in the mid-stage, increasing volume; and in the later stages, increasing in density (see Chapter 6).

      The backrest support should accommodate such age-related differences. Postural support is more important at certain ages, particularly early adolescence.
    - Young children lack lumbar curves.

      Children and adolescents have different back shapes and therefore require different backrest profiles from each other.

      Children do not develop what we associate with a "normal" low-back (lumbar) curve until their early teens. The lumbar (low back) curve does not develop to maturity until adolescence.

      Teenagers need greater backrest support that reflects these changing back-profiles.

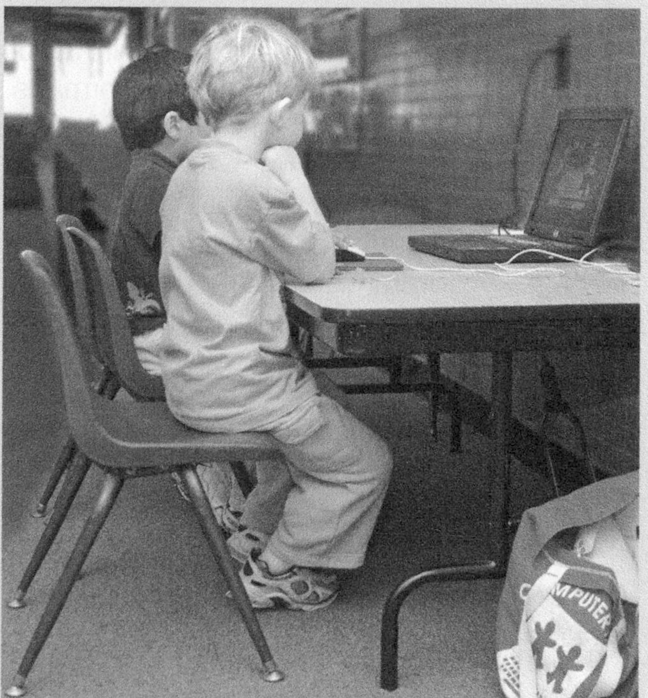

Figure 21.5 (Ursy Potter Photography)
Younger children tend to grow more in their extremities, while adolescent growth (following puberty) is largely in the spine.

2. *For all children and adolescents*

    - Furniture design should accommodate free movements for all while discouraging unnatural postures and postures at extreme positions. However, postural support is particularly important at certain ages.

      Although the ideal posture may change somewhat as children mature,‡ they are also developing postural habits that, if correct, should help protect them when they are adults.

      It can be difficult for adults to "unlearn" bad postural habits acquired in childhood.

* Please see Chapter 6 about musculoskeletal development in children and adolescents and age-specific risks. Table 6.3 reviews the research on children's developmental stages of the spine.

† Spear (2002) notes during puberty, teenagers gain about 15% of their final stature and 45% of their skeletal mass.

‡ Please see Table 6.3 about children's and adolescent's physical development.

3. *Some children need particular postural support*
   - The end growth plates of young children's bones are developing and are readily subject to injury, increasing the risk of awkward postures.*
   - Children's optimum viewing distance and viewing angles change through childhood.†
   - During childhood, the vertebrae at the neck (cervical vertebrae at C3–C7) tend to be more wedge shaped‡ until around their 7th or 8th year when they begin to differentiate and square off (Dormans, 2002).

     Before adolescence, about 20% of cervical spines are kyphotic; unlike in adults, this is considered normal.

     Such changes suggest that the position of the child's neck at the computer may deserve particular attention.
   - As children (particularly girls), grow and move through adolescence, they are more likely to experience back and neck pain than boys.

Figure 21.6 (Valerie Rice) Children working at computer tables rarely are able to read text on their computer screen at eye level. Their sightlines are worsened by its placement on the CPU of the computer (above).

This is cause for concern, because children's visual systems are still developing. These age-related changes in vision may affect their effective viewing-angles and distances at the computer.§

* Bass and Bruce (2000) note "in children, most ligaments are attached to the epiphyses (growing ends of bones) above and below. This anatomical arrangement concentrates forces on the physis (horizontal growth plate). While low velocity trauma can result in ligamentous injury, high velocity trauma is more likely to result in physeal disruption, because forces applied rapidly do not allow time for ligaments to stretch".

† Please see Chapter 4 of this book on children and vision.

‡ Due to the shallow angle of the facet joints of vertebrae.

§ See Chapter 4 on children's vision for a thorough review of this topic.

FIT BETWEEN CHILDREN AND THEIR CLASSROOM FURNITURE

Major mismatches in "fit" are common between schoolchildren and their classroom furniture dimensions:*

1. Poor sitting postures are associated with discomfort and with inappropriate "furniture fit" (Panagiotopoulou et al., 2004)
   a. Mismatches between seat depth and thigh length cause seat discomfort (Evans et al., 1992)
   b. Mismatches between seated elbow and desk height† contribute to neck and shoulder pain (Evans et al., 1992)
   c. The degree of mismatch between the sizes of adolescent children and their school furniture is associated with adolescent low back pain (Milanese and Grimmer, 2004)
   d The tallest students report the highest rates of low back pain (Milanese and Grimmer, 2004)
   e. Poor furniture fit contributes to high-risk sitting postures in elementary school children (Oates et al., 1998)
2. Long-term sitting in school furniture is associated with reports of musculoskeletal discomfort and pain (Fallon and Jameson, 1996)
3. Some body dimensions are better able to predict chair design requirements
   a. Lower leg-length (popliteal height) is more appropriate than child stature for defining chair sizing (Molenbroek and Ramaekers, 1996; Molenbroek et al., 2003)
4. Boys and girls differ in "fit" (Jeong and Park, 1990; Parcells et al., 1999)
   a. Boys taller than 126 cm (50 in.) in stature require higher desks and chairs than girls of the same stature
   b. Girls taller than 120 cm (47 in.) require deeper and wider seats than boys of the same height

* See Bruynel and McEwan (1985); Legg et al. (2003); Panagiotopoulou et al. (2004); and Parcells et al. (1999).

† The other furniture/occupant measures were significantly related to postural discomfort.

SITTING AND CLASSROOM FURNITURE

1. FIRST PRINCIPLES OF CLASSROOM SITTING POSTURES

Some school furniture considerations

1. Discourage prolonged static sitting postures.
   Allow for sitting in a range of neutral sitting postures. Older children particularly need postural support, especially after puberty.
2. Provide desks that accommodate both computer and paper-based tasks.
   Slanted work surfaces can promote good postures while reading, but may be less useful when writing and using computers.
3. Check that the computer screen is not too high for the child because this can cause neck and back discomfort (also see Chapter 4, Vision).
4. Provide computer tables that are deep enough to set the computer screen behind the keyboard and in front of the child.
5. If the child uses a notebook computer for more than an hour set this up with a separate keyboard and mouse.
   An additional computer may introduce new problems if it causes their neck to crane upward.
6. Provide computer tables wide enough for a keyboard and mouse or a height-adjustable negative slope keyboard tray, and for left- or right-handed use and for documents.
   Use a document holder where possible that fits the task (e.g., allows writing on documents).
7. Position the computer screen to avoid reflections and glare.
8. Consider ways to accommodate differently abled users.
9. If children are engaged in collaborative activities, provide furniture that accommodates team activities.
10. If the furniture is adjustable, make sure that the children know how to adjust their furniture appropriately.
    Look for easy-to-use adjustment controls, or adaptable furniture

Figure 21.7 (Ursy Potter Photography)
Avoid excessively high monitors that require children to look up while they look at screen text (see Chapter 4 on Vision).

Figure 21.8 (SIS-USA)
Reading postures may be more comfortable at sloped desks or with a reading stand used on a flat desk to angle a book. Sloping surfaces do not necessarily help with the use of a laptop computer, or while writing.

Adjustable surfaces that allow different height to schoolchildren to alternate between sitting and standing, and between working at sloped and tilted work surfaces provide them with alternative postures that ease the demands of static sitting. Alternatively, students can be encouraged to periodically stand up and move around.

Figure 21.9 (Ursy Potter Photography) **Figure 21.10** (Ursy Potter Photography)
Vision and tasks affect posture. Consequently, lighting, school furniture, training, and the task interact and shape children's postures in important ways. Children (and adults) often adopt unnatural postures in order to see what they are doing. In doing so, children become accustomed to and develop a sense of familiarity with postures that contribute to musculoskeletal disorders as they become adolescents and adults.

Figure 21.11 (SIS-USA-Inc.com)
Vision and tasks affect posture, in that children (and adults) will often adopt a particular posture to be able to see clearly without even recognizing that they are leaning or hunched. Their focus is on being able to see and they only recognize the posture as fatiguing when they begin to feel stiff or sore.

CHILDREN AND CLASSROOM FURNITURE: IS THERE A MATCH?

Research studies conducted worldwide show that mismatches between the dimensions of schoolchildren and the dimensions of classroom furniture are commonplace (Table 21.2). Such mismatches result in poor seated and work postures and in an increase in musculoskeletal discomfort among children.

Table 21.2 Research Studies of the Dimensions of School Furniture and Schoolchildren

	Citation	Subjects	Findings	Conclusions
Australia	Evans et al. (1992)	• Surveyed 224 students (aged 12, 14, and 16) at four schools about activities and symptoms • Compared furniture dimensions with child sizes	• 55% schoolchildren reported pain and discomfort • However, anthropometric differences did not fully account for the students' discomfort • Class furniture was inappropriately sized for four out of five children	• Mismatches between seat depth and thigh length cause seat discomfort • Mismatches between seated elbow and desk height[a] contribute to neck and shoulder pain
Australia	Milanese and Grimmer (2004)	• Surveyed 1269 adolescent Australian students • Examined the fit between sizes of individual children and their school furniture[b] • Compared the reported back-symptoms with mismatches between children and their school furniture	• The smallest students had the "best fit" with the school furniture • The tallest students reported higher rates of low back pain	• The degree of mismatch between the sizes of children and their school furniture is associated with adolescent low back pain

Greece	Panagiotopoulou et al. (2004)	• 118 randomly selected elementary schoolchildren in Greece This included 10 boys and 10 girls from each of the 2nd, 4th, and 6th grades at three primary schools. • Compared the fit of older and newer furniture provided to children in grades 1–3 and 4–6	***For the 2nd grade children*** • The chairs were too high for 95% of the schoolchildren[c] • All seats were too deep for the children (35 cm/13.8 in. depth) • None of students' body size measurements matched their chair dimensions Only 3% of old desks and 12% of new desks fit the children when they sat on the new chair ***For the 4th grade children*** • Both old and new desks were too high for most of the 4th, 5th, and 6th grade students[d] • Old and new chairs were too deep for 70% and too high for two-thirds of these children[e] • Old-style desks were too high for all children Even the new chair-and-desk combination was too high for 93% of children ***For the 6th grade children*** • These children had a somewhat better fit with their furniture • Seat depth (35 cm) fitted 71.7% of children; 23% found this too big and 5% too small • Only 7% fit the old desks (71 cm), 27% fit the new desks (66 cm), and 3% fit the new high-school desks (74 cm)[f] ***General findings*** • 15% of 4th-grade children and 23.3% of 6th-grade children reported suffering from back pain associated with sitting at school • 8.3% of 2nd graders, 10% of 4th graders, and 15% of 6th graders reported leg pain from sitting at school	• Major mismatches between the anthropometric measures and classroom furniture dimensions • Desks and chairs were too high for all 2nd and 4th grade students and most 6th graders • Although the fit of the desk–chair combination was better for 6th grade than younger students • Sitting is associated with musculoskeletal discomfort and it progresses with the age of the child with 16.7% of 2nd grade children • Comfort ratings of school desks and chairs declined with the grade of the child, as did the ratings of comfortable sitting for writing, reading, and listening in class (Table 21.6) • Table 21.7 lists recommended dimensions for school chairs and desks for children in Greece

(continued)

Table 21.2 Research Studies of the Dimensions of School Furniture and Schoolchildren (Continued)

	Citation	Subjects	Findings	Conclusions
Holland	Molenbroek and Ramaekers (1996)	• Delft University of Technology developed a size system for school chairs-and-tables for the Dutch Standard Committee on School Furniture	• Used dimensions from ChildData (UK) and from Peoplesize software (UK) as the basis for designs	• Popliteal height is a far better key dimension than the body height for defining furniture sizing
	Molenbroek et al. (2003)	• Compared body sizes data of Dutch children with data from German and British national standards	• Statistically estimated the fit of a draft European school furniture design standard (CEN/TC 207 prEN 1729 1, 2, 2006a,b). They found it would not fit Dutch schoolchildren	• Unless sizing is based on popliteal height rather than stature, draft European standard will not fit Dutch schoolchildren
Hong Kong	Evans et al. (1988)	• Evaluated the anthropometric data for 684 schoolchildren in Hong Kong for the design of school furniture	• They found that current furniture sizes, based on British standard BS 5873 that specifies five sizes, were not appropriate for the Hong Kong schoolchild population	• Five sizes of chair and table combinations can accommodate 6–18-year-old Hong Kong schoolchildren (see Table 21.5)
Iran	Mououdi and Choobineh (1997)	• Collected 17 anthropometric measurements from 1758 randomly selected schoolchildren aged 6 to 11 years in Mazandaran province	• Anthropometric data significantly different between sexes	• Anthropometric data can be used for the design of school furniture
	Habibi and Hajsalahi (2003) Habibi and Mirzaei (2004)	• Anthropometric survey of 17 body dimensions of male and female Iranian primary school children aged 7 to 11 years in the city of Esfahan	• 71% of these schoolchildren reported back pain • 42% reported problems related to classroom space • Students' mean height was 1305 mm and their mean weight was 29 kg[9]	• Argued for using 17 body size dimensions in the design of school furniture and equipment in order to minimize musculoskeletal, visual, and circulatory problems

Ireland	Fallon and Jameson (1996)	• Assessed primary-school furniture in the western region of Ireland • Used an anthropometric survey, a posture analysis, and a subjective comfort evaluation	• Current Irish design standards that specify dimensions for primary-school furniture do not reflect male and female students' body sizes • They conclude that pupil's postures while using their school furniture was not necessarily a serious threat to their health and welfare	• Pupils reported significant back, neck, and leg discomfort when using their school furniture for long durations
Korea	Jeong and Park (1990)	• Compared body sizes of 1248 Korean schoolchildren aged 6 to 17 years with their school furniture dimensions	• Stature was highly related to body dimension for school furniture design • There were significant sex differences in relationships between stature and the body dimensions	• Boys above 126 cm in stature required higher desks and chairs than girls of the same stature • Girls taller than 120 cm required greater seat depth and breadth of chair than boys of the same height
Mexico	Prado-Leon et al. (2001)	• Studied 50 anthropometric dimensions on 4758 Mexican primary school children, boys and girls aged 6 to 11 in the city of Guadalajara • Compared these measurements to those of American, Cuban, and Mexican children	• Mexican children in this part of Mexico have different body dimensions than do American, Cuban, and other Mexican children, possibly due to ethnic differences and temporal differences when studies were conducted	• It requires 50 anthropometric dimensions to properly design the school furniture; fittings and equipment to minimize risk of musculoskeletal, visual, and circulatory problems
New Zealand	Bruynel and McEwan (1985)	• 576 13- and 14-year-old schoolchildren	• 30% children reported back pain • Furniture was generally too low	• Attributed student's back pain can to a mismatch between the children's body sizes and school chair sizes
	Legg et al. (2003)	• 189 students in three New Zealand secondary (high) schools	• A high level of mismatch between school furniture and anthropometric dimensions	• There is a considerable mismatch between the sizes of children and their school furniture (Table 21.8)

(continued)

Table 21.2 Research Studies of the Dimensions of School Furniture and Schoolchildren (Continued)

	Citation	Subjects	Findings	Conclusions
Taiwan	Lin and Kang (2000)	• Applied anthropometric data of Taiwanese people to the design of high-school classroom equipment	• Compared advantages and disadvantages of six current desks and chairs used in Taiwan high schools	• Summarizes the ideal desk and chair design dimensions for Taiwanese High School children
Turkey	Kayis (1991)	• Detailed anthropometric survey of 3584 primary school Turkish children	• Based upon the dimensions of the children, recommended that five ranges of furniture size would cover the needs of children aged 6 to 13 and above However, for economic reasons, two and three size ranges of size should be adequate	
United States	Parcells et al. (1999)	• Compared 6th and 8th grade students' body sizes[h] with their school furniture dimensions • Subjects were 74 students (37 male and 37 female) in 6th through the 8th grade during physical education classes • Compared three chair styles and three desk styles	• Fewer than 20% of students sat at acceptable chair and desk combinations • Most students had desks that were too high and sat on chairs with seats that were too high or too low • After statistically controlling for body stature, they found that girls were less likely to fit their chairs than boys	• There was a considerable mismatch between the student body dimensions and furniture dimensions

[a] The other furniture/occupant measures were significantly related to postural discomfort.
[b] Children were divided into four size groups. They compared the smallest size with the largest groups with their school furniture dimensions.
[c] The seat pan was 35 cm (13.8 in.) above the floor.
[d] The old desks were 71 cm (28 in.) high; the new desks were 66 cm (26 in.), and 74 cm (29 in.) high.
[e] The old seats were 42 cm (16.5 in.) high. The new seats were 39 cm (15.4 in.) high.
[f] The old desks were 71 cm (28 in.) high; the new desks were 66 cm (26 in.), and the new high-school desks were 74 cm (29 in.) high.
[g] They also compared anthropometric measurements of Iranian children with children from American, Cuban, Mexican, and children of other nationalities.
[h] They measured students' elbow height, shoulder height, upper-arm length, knee height, popliteal height, buttock–popliteal length, and stature.

2. STUDENT/SCHOOL FURNITURE FIT

Anthropometric dimensions of schoolchildren vary by ethnicity, gender, and age. Some researchers suggest that designers require at least 50 different body-size (anthropometric) dimensions to develop appropriate furniture (Prado-Leon et al., 2001). Others strive for a simpler approach that defines a relatively small number of size ranges to provide a tolerable fit for a majority of students, such as the five sizes recommended in the British standard for educational furniture (British Standards Institution BS 5873, 1980).

One study of Hong Kong schoolchildren found that gender differences in dimensions were not significant until the age 14, after which the boys grew more than the girls (Evans et al., 1988). They also found that children's dimensions did not match furniture dimensions that were conformed to British Standard BS 5873 on Educational Furniture;* in some cases, there were discrepancies of over 6 cm (2.4 in.) (see Table 21.3).

Table 21.3 Discrepancy in Furniture Sizing Based on British Standard 5873 and Anthropometric Body Size Dimensions of Hong Kong Schoolchildren

Dimension	Size 1 (6–8 y)	Size 2 (8–10 y)	Size 3 (10–12 y)	Size 4 (12–14 y)	Size 5 (14–18 y)
Seat height (HK)	29.5 cm	32.0 cm	35.0 cm	37.5 cm	40.0 cm
Seat height (BS)	26.0 cm	30.0 cm	34.0 cm	38.0 cm	42.0 cm
Difference (BS–HK)	*–3.5 cm*	*–2 cm*	*–1.0 cm*	*+0.5 cm*	*+2 cm*
Table height (HK)	49.0 cm	52.0 cm	57.0 cm	61.0 cm	67.0 cm
Table height (BS)	46.0 cm	54.0 cm	58.0 cm	64.0 cm	70.0 cm
Difference (BS–HK)	*–3.0 cm*	*+2.0 cm*	*+1.0 cm*	*+3.0 cm*	*+3.0 cm*

Source: Evans, W.A., Courtney, A.J., and Fok, K.F. *Appl. Ergon.*, 19, 122, 1988.

Defining appropriate size ranges is complicated by the fact that children do not grow at uniform rates. There is rapid growth between ages 6 and 14 and the degree of dimensional mismatch can be striking in adolescents around puberty and older than 14 years (Evans et al., 1988).

* British Standards Institution BS 5873 "Educational furniture. Specification for strength and stability of chairs for educational institutions." These may be purchased from the British Standards Institution at www.bsiglobal.com or by telephone: 01442 230442.

Ideally, school furniture design should be adjustable so as to accommodate the size diversity seen in classrooms in ethnically mixed countries, such as the United States. However, adjustable furniture is the exception in most classrooms. Research shows that even in countries with fairly homogenous populations, such as Greece, comfort ratings for school-furniture decline considerably with student's age; by the 6th grade, only 1 in 10 students say that they have a good chair—and 1 in 3 consider their desk comfortable (Panagiotopoulou et al., 2004; see Table 21.4).

Table 21.4 Classroom Comfort Ratings, Greek Schoolchildren

Schoolchild Comfort Ratings	Grade Level		
	2	4	6
Good chair	63.3%	35%	11.7%
Comfortable seat height	63.3%	45%	20%
Good desk	65%	36.7%	21.7%
Comfortable desk height	70%	45%	35%
Comfortable writing	75%	56.7%	45%
Comfortable reading	80%	76.7%	53.3%
Comfortable listening	83%	78.3%	43.3%

Source: Panagiotopoulou, G., Christoulas, K., Papanckolaou, A., and Mandroukas, K., *Appl. Ergon.,* 35, 121, 2004.

On the basis of their findings, Panagiotopoulou et al. (2004) proposed these school furniture dimensions to fit most Greek schoolchildren (see Table 21.5).

Table 21.5 Recommended Furniture for Schoolchildren in Greece

	Grade Level		
Dimension	2	4	6
Seat depth (% fit)	29 cm (90%)	32 cm (60%)	Seat height + 20 cm (80%)
Seat height (% fit)	31 cm (92%)	34 cm (57%)	Seat height + 22 cm (73%)
Desk height (% fit)	34 cm (83%)	37 cm (55%)	Seat height + 23 cm (68%)

Source: Panagiotopoulou, G., Christoulas, K., Papanckolaou, A., and Mandroukas, K., *Appl. Ergon.,* 35, 121, 2004.

In a study of New Zealand schoolchildren, Legg et al. (2003) found schoolchildren's body dimensions did not fit the furniture they used* (Table 21.6).

Table 21.6 Student Dimensions and School Furniture "Fit"

Seat Height versus Lower Leg Length (Popliteal Height)		**Conclusion**
Seat Height	• 40.7 (34.4–48) cm	• 96% students sat at seats that were too high for them • No seats too low
Popliteal Height	• 45.7 cm (boys school) • 41.3 cm (girls school) • 44 cm (coeducational school)	
Seat Depth Versus Upper-Leg Length (Buttock–Popliteal Length)		
Seat Depth	• 43.7 cm (boys school) • 33.5 cm (girls school) • 39 cm (coeducational school)	• 54% students had a mismatch • 48% seats were too shallow • 6% seats were too deep
Buttock–Popliteal Length	• 48 (40.2–57.7) cm • SD 3.2, range 40.2–57.7 cm	
Desk Clearance versus Knee Height		
Desk Clearance	• 69.5 cm (boys school) • 69 cm (girls school) • 65 cm (coeducational school)	• No mismatches All desk–chair combinations provided adequate knee clearance.
Knee Height	• 51.4 cm • SD 3.4, 44.2–59.7 cm	

Source: Legg, S.J., Pajo, K., Sullman, M., Marfell-Jones, M., Mismatch between classroom furniture dimensions and student anthropometric characteristics in three New Zealand secondary schools, *Proceedings of the 15th Congress of the International Ergonomics Association, Ergonomics for Children in Educational Environments Symposium,* 6, 395–397, Seoul, Korea, 24–29 August 2003.

Unfortunately, the research studies that have been conducted to date have generally provided little guidance on how to change the school furniture design so that this can be more effective in promoting neutral postures.

* Legg et al. (2003) measured the body sizes (anthropometric dimensions) of children aged 13 to 14, 15 to 16, and 17 to 18, in three schools (all boys, all girls, and coeducational).

CLASSROOM DESK: DESIGN ALTERNATIVES

More than 100 years ago, Mosher (1899) designed a chair and a desk that independently adjusted in height. The chair also had a curved backrest to support the child's back. He recommended adjusting the chair height so that student could sit with feet firmly on the floor, and adjusting the height of the desk to provide a comfortable working arm position. The surface of the desk was slightly sloped for more comfortable writing and the desk had a rotating and movable reading platform to orient reading materials in a more vertical position so that materials could be read with the head in a more neutral posture (Figure 21.13).

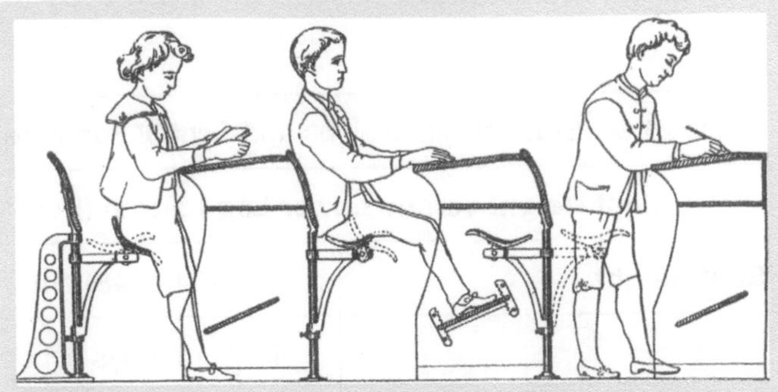

Figure 21.12 This sit–stand school furniture patent from over a century ago (Schindler, 1892) looks far superior to designs we commonly see in today's schools. It enables schoolchildren to assume a range of good sitting-and-standing postures in class. These design options also accommodate a range of child user sizes with one design.

Figure 21.13 An adjustable-height chair-and-desk design. (From Mosher, E.M., *Educ. Rev.*, 18, 9, 1899.)

This design was far ahead of its time. Indeed, even today few schoolchildren sit in furniture that is as adjustable as Mosher's chair-and-desk arrangement.

Recently, there has been a surge of interest in how to improve the fit between furniture and its users, particularly when using computers. In this section of the chapter, we will review research on the design of classroom furniture, including the desk, chair, desk and chair combinations, and computer furniture.

In the United States, a number of classroom desk designs are commonly available in schools. These designs typically feature desktops that are either plastic or laminate (plastic is generally more durable and expensive).

The various classroom desk designs differ in advantages and limitations. Table 21.7 reviews common desk design alternatives.

Table 21.7 Common Types of US Classroom Furniture (images courtesy of Virco)

Style		Features	Advantages	Disadvantages
Desks				
Adjustable Height Open Front Desk with Metal Book Box		1. A rectangular top with under-desk storage provided by a plastic, wire-cage, or metal book box 2. Desk can be of fixed height or adjustable a. Fixed-height desks are often 74 cm (29 in.) b. Adjustable desks ranges are often 56–74 cm (22–29 in.) or 76 cm (30 in.) in 2.5cm (1 in.) increments 3. The desk area is usually 46 cm (18 in.) deep x 61 cm (24 in.) wide 4. Double desks enable two students to sit side-by-side This is often sized 46 cm x 102 cm (18 in. x 48 in.)	1. This design is very popular in US grade schools 2. Available as single or double desks This may save floor space and facilitate child teams 3. Accommodates left- and right-handed child users 4. Provides a somewhat larger work area 5. Front storage may sometimes add more security	1. Storage access only available from the front. This design may contribute to awkward postures when child users attempt to access materials while seated
Adjustable Height Lift Lid Desk with Metal Book Box		1. Desk lids can be lifted to access to a storage book box or bin 2. Dimensions and height adjustment options are similar to the Open Front design (above)	1. May be the common desk design in US grade schools 2. May reduce awkward postures when reaching for storage materials while seated in the desk 3. The desk might be retrofitted for specialized work surface requirements	1. Child users must remove work surface items before accessing desk storage. 2. This may contribute to awkward postures, and obstruct passageways 3. May in some cases obstruct vision while lifting the desk flap

(continued)

Table 21.7 Common Types of US Classroom Furniture (images courtesy of Virco) (Continued)

Style	Features	Advantages	Disadvantages
Chair Desks			
Tablet Arm Chair Desk with Metal Frame Book Box	1. Desktop is attached to the chair, with book storage beneath 2. There are many chair desk designs 3. Chair heights are typically 39 cm (15½ in.) or 44 cm (17½ in.) with desktop areas that are 50 cm (19½ in.) wide a. Right-side desks are 44 cm (17½ in.) deep b. Left-side desks are 34 cm (13½ in.) deep 4. Often designed with an L-shaped "tablet arm" that provides an armrest as well as a writing surface. Sometimes has a trapezoid-shaped top	1. These designs are popular in high schools 2. Writing surface moves location with chair 3. Access to storage without moving desktop items	1. The right-sided tablet is inconvenient for left-handed students and vice versa 2. Fixed-height writing surface 3. Chair access only on one side 4. Fixed depth between tablet and back of chair Consequently, posture is different for children who are smaller or with greater body girths
Sled Chair Desk with Metal Frame Book Box	1. This desk is sometimes attached to the chair in a sled arrangement	1. Fits both right- and left-handed students. This desk provides the same amount of work area for right- and left-handed students 2. Right- and left-side access to chair	1. Fixed depth between tablet and back of chair. Consequently, postures differ among children who are smaller or with greater girths 2. Fixed-height writing surface

Combo Desks				
Combo Desk with Metal Frame Book Box		1. The desk and chair unit has two legs on the chair and two coming off the desk 2. The design can accommodate more size options 3. Common chair heights are 34, 39, and 44 cm (13½, 15½, and 17½ in.)	1. Combo desks tend to feature larger writing surfaces and allow more room for body girth 2. Can be equally suitable for right-or left-handed students	1. Tablet is only on one side so inconvenient for a left-handed student to use a right-handed tablet and vice versa 2. Fixed-height writing surface 3. Chair access only on one side 4. Fixed depth between tablet and back of chair, so posture is different for smaller and greater body girths

"ERGONOMIC" CLASSROOM FURNITURE DESIGNS

In the 1980s, a Danish researcher proposed that schools replace conventional flat desks and low chairs with higher chairs with forward-tilting seat pans and with slope-adjustable table surfaces (Mandal, 1982, 1984, 1997). Mandal argued that this design would reduce forward bending of the low back and neck (lumbar and cervical flexion), improve the abdominal angle, and allow children to work in more neutral body postures. He provided extensive photographic and experimental support for his design ideas. By the early 1980s, some 5000 schoolchildren worked at this type of furniture in Denmark.

Studies that evaluated the effects of different types of furniture suggested by Mandal generally have found favorable results (Linton et al., 1994; Aagaard-Hensen and Storr-Paulsen, 1995; Taylour and Crawford, 1996).

However, simply providing the furniture does not guarantee that schoolchildren will sit in good postures. Linton et al. (1994) found that 4th graders (10 years old) did not automatically sit "properly," even in "ergonomic" furniture, which highlights the need to provide children with proper instructions and adjustment.

Sloping desks that adjust—together with proper training—may improve posture for paper-based classroom tasks, such as writing, drawing, or reading from a book. Yet such adjustability is not likely to provide the same benefits for computer work, which is increasingly a component of classroom instruction. Alternative furniture designs are needed to accommodate this technology (Barrero and Hedge, 2002). Results from studies of alternative "ergonomic" classroom furniture designs are summarized in Table 21.8.

Figure 21.14 (SIS-USA-Inc.com) Classroom with a mix of flat and sloped surfaces. While well-designed furniture has been found to improve schoolchildren's postures, it is not sufficient to guarantee that they will consistently do so.

Table 21.8 Summary of Studies of "Ergonomic" Classroom Furniture Designs

Source	Analysis	Testing	Conclusions
Aagaard-Hensen and Storr-Paulsen (1995)	• Compared Danish children (7 to 11 years) working at three workstation designs	• **Workstation A:** A flat work-surface and a level seat pan that conformed to the ISO (1979) design principles • **Workstation B:** 15 cm higher than Workstation A and had adjustable table height, adjustable tabletop angle (0° and 20°), and a forward sloping seat • **Workstation C:** Slightly higher than Workstation A, had a fixed table height, an adjustable tabletop angle like Workstation B, and a forward slanting seat (3°)	• Children preferred the table height, chair height, reading position, backrest, and overall design of Workstation B • They gave overwhelmingly positive reactions to the adjustable tabletop independent of the table height
Knight and Noyes (1999)	• Studied the reactions of British 9- and 10-year-olds to two different types of furniture designs • They measured children's sitting behavior (foot placement on the floor relative to chair) and their time on-task and off-task	• **Design A** consisted of a standard polypropylene seat with a uniformly sloping, flexible back and a backward sloping seat • **Design B** had a seat that sloped slightly back but had a deeper and shallower curve than the conventional seating and a protruding back support located halfway up the rigid chair back	• 2% increase in on-task behavior for design B However, feelings toward the alternative design were polarized; children either disliked it the most or liked it the best • Observed nonstandard sitting when the popliteal and seat height difference exceeded 5 cm, when the seat depth and upper-leg length did not match, and when there were discrepancies in elbow–floor/table height. The child's stature did not predict nonstandard sitting and individual body dimensions were better indicators of furniture fit
Laeser et al. (1998)	• To explore requirements for computer tables, tested the reactions of 30 6th graders and 28 8th graders to working with a computer • Used the RULA method (McAtamney and Corlett, 1993) to evaluate children's posture, task performance for typing and mousing was measured • Children were videotaped to determine the amount of engaged behavior and students were asked about their preferences	• **Workstation A:** Conventional keyboard and mouse when this was placed on a flat desktop or on a downward sloping (~11° declination angle front-to-back) • **Workstation B:** Keyboard tray system with a separate flat, position-adjustable mouse platform about 4 cm over the numeric keypad of the keyboard	• Children's seated posture improved (i.e., lower RULA scores) when using the adjustable tiltdown keyboard tray system • Benefits were greatest for upper arm and lower arm postures during mousing and wrist postures during keyboarding • Typing performance was slightly slower (<1% slowing of typing speed) with the tiltdown tray arrangement but mousing performance slightly improved (3% improvement in mousing accuracy) • Overall, 40% of children chose the tiltdown keyboard tray system as easiest 38% preferred this work arrangement 33% found this to be more comfortable than the conventional desktop arrangement. Other children expressed no strong preferences

(continued)

Table 21.8 Summary of Studies of "Ergonomic" Classroom Furniture Designs (Continued)

Source	Analysis	Testing	Conclusions
Marschall et al. (1995)	• Australian Researchers quantified the benefits of giving the children a short training session on positive sitting • Children (mean age: 4.7 years) performing tracing tasks	**Compared the effects on these stations** • **Type A:** A traditional workstation with a level desktop • **Type B:** 5° backward sloping seat and an "ergonomic workstation" with a 15° sloping desktop and a 15° forward sloping seat	• They found that positive sitting behaviors increased between 28% and 46%, but this effect gradually decreased after instructions had ceased (back to 40%), showing poorer recall, and retention of the instructions • They found significantly less latissimus dorsi (LD) activity at the "ergonomic" workstation, a significantly greater hip angle and less neck flexion, which confirmed Mandal's original observations and the children expressed a preference for the ergonomic workstation • However, there were no differences in erector spine (ES) or superior trapezius (ST) activity between the two workstations and no tracing performance differences
Taylour and Crawford (1996)	• Compared the effects of three furniture designs on the sitting behavior of 11-year-old British children • Scored children's sitting behavior as positive or negative according to postural variables, such as being slumped, having the neck flexed, sitting erect, etc.	The three designs were: • **Type A:** A polypropylene chair with separate seat and backrest and steel frame, and a conventional four-legged double desk with a horizontal laminate top • **Type B:** A conversion of Type A with a raised wedged-shaped seat to alter the slope and a forward sloping work surface mounted to the original desk • **Type C:** An "ergonomic" workstation with adjustments for the angled work surface and a raised chair with a forward sloping seat at the front	• There were significant differences in sitting behaviors across the three types of furniture • Positive sitting behavior was observed for only 28.5% of the time with Type A furniture, but for 48% and 50% of the time with Types B and C • Sitting hunched over the desk was the most frequent negative sitting behavior, followed by sitting with a flexed neck • Children's survey responses about their sitting comfort, ease of use, and pain or discomfort generally agreed with the positive sitting behavior results for each chair

COMPUTER FURNITURE WORKSTATIONS

The introduction of computers in school classrooms represents the greatest technological change to affect the design of school furniture in our history. Yet, research also indicates that schoolchildren must often use computers on inappropriately designed school furniture, causing poor postures that may increase the risk of developing musculoskeletal disorders.

Research on elementary schoolchildren* using desktop computers showed that many children study in high-risk postures (Oates et al., 1998).

Figure 21.15 (Alan Hedge) Space efficient furniture cluster for a school computer laboratory.

These were caused by

- inclined keyboards on overly high surfaces that led to deviated wrists and arm postures;
- overly high computer monitors that aggravated neck postures; and
- chairs that were so high that many children's legs dangled.

Other research found that 60% of children report postural discomfort when using laptop computers. Further, these discomforts increase with time at the computer (Harris and Straker, 2000a,b).

Manufacturers have produced various furniture designs for school computer-laboratories that are space efficient (see Figure 21.15). As computer technology changes, especially the move to flat-panel displays, new furniture designs that allow for greater adjustability by its child users, are space efficient and provide more flexibility in layout of classrooms. Some designs incorporate ergonomic features, such as separate surfaces for a keyboard and mouse, and adjustable footrests (Figure 21.16).

30" wide desk, footrest extended,
K-6 chair has standard glides.
30" wide desk, footrest retracted.
7-12 chair has optional casters.

Figure 21.16 Style C "Edgar" computer furniture accommodates children of different sizes with adjustable-height chair and movable foot supports.

---

* These elementary school children were between 8.5 and 11.5 years old and in the 3rd through 5th grades (Oates et al., 1998).

Figure 21.17 (Alan Hedge)
The "Smart Desk Train" design for computer classrooms accommodates adjustable-height LCD screens and a keyboard and mouse.

In recent years, the United States has devoted considerable resources in purchasing computer technologies and networking infrastructures for schools. However, improving the physical design of the classroom furniture to better accommodate computers usually has been a low priority and there is a marked lack of attention to ergonomic issues in schools. Many schools do not have a sufficient budget to acquire more ergonomic classroom furniture. While classroom computer work is typically of short duration, this might mitigate any adverse effects of poor posture.

SCHOOL FURNITURE STANDARDS

There is a lack of standardization for the design of classroom furniture for educational settings in many countries, with the exception of Europe where the development of school furniture standards has been most actively pursued. There is a European standard that gives very general requirements (see inset) and an Austrian standard.

Currently, there is also discussion of a European draft educational furniture standard prEN 1729. Given the anthropometric diversity of many countries, the development of an international standard could be of benefit.

Table 21.9 European Standard: Chairs and Tables for Educational Institutions (CEN,[a] 2003)

The European Standard for Chairs and Tables for Educational Institutions specifies requirements for functional dimensions, adjustability, and markings between three kinds of educational furniture:

1. Chairs with seat pans that tilt between –5° and +5° and, associated tables.
2. Chairs and tables for high chairs with double-sloped seats.

   These enable child users to sit in two ways:

   a. Upright, with their thighs tilting forward; and also
   b. Leaning back with feet supported on a footrest or the floor
3. Standing height tables.

   Tables to accommodate child users while sitting

It does not apply to special purpose furniture (such as for laboratories) or schoolteachers.

They define eight dimensional categories (sizemarks 0–7). School-furniture manufacturers must label their products to indicate the sizes that their products meet.

[a] European Committee for Standardization CEN/TC207/WGS. PrEN 1729. Chairs and tables for educational institutions. Part 1: Functional dimensions. Furniture Industry Research Association / FIRA International. Hertfordshire, UK. www.fira.co.uk
Austria is currently the exception; this nation requires that school furniture meet the requirements of Austrian Standard ÖNORM A 1650 at this time.

CONCLUSIONS

More research is needed that systematically reviews the potential impact of ergonomic classroom furniture on schoolchildren's educational performance and well-being. To date, the term "ergonomic" has often been ill-defined in the research and loosely applied to alternative furniture designs. We lack a standard that defines what constitutes an "ergonomic furniture design."

The research to date has provided mixed results:

- In one study, children rated their "ergonomic" classroom furniture as more comfortable and also reported less short-term discomfort than another group that used conventional classroom furniture; even so, this study failed to demonstrate significant differences in posture (Linton et al., 1994).
- A long-term follow-up found no differences in back pain rates and physical symptoms such as scoliosis between two groups of schoolchildren who used different furniture designs for 4 to 5 years (Troussier et al., 1999).
- Another study developed an "ergonomic" desk design for Brazilian pre-schoolchildren, but did not report empirical findings related to this new design (Paschoarelli and da Silva, 2000).

The design of classroom furniture must fit its student child users, accommodate the necessary activity (e.g., reading, drawing, writing, paper-based or computer-based tasks at personal computers or laptops), and help them learn a new postural language that they can transfer to their work life as they become adults. We are still on the path to learning how to do so.

ACKNOWLEDGMENTS

We gratefully acknowledge Stephanie Ishack's assistance in gathering some of the literature in this chapter.

REFERENCES

Aagaard-Hensen, J. and Storr-Paulsen, A. (1995) A comparative study of three different kinds of school furniture. *Ergonomics*, 38, 1025–1035.

Balague, F., Dudler, J., and Nordin, M. (2003) Low-back pain in children. *Lancet,* 361(9367), 1403–1404.

Balague, F. and Nordin, M. (1992) Back pain in children and teenagers. *Baillieres Clin. Rheumatol.*, 6(3), 575–593.

Balague, F., Troussier, B., and Salminen, J.J. (1999) Non-specific low back pain in children and adolescents: Risk factors. *Eur. Spine J.*, 8(6), 429–438.

Barrero, M. and Hedge, A. (2002) Computer environments for children: A review of design issues. *Work*, 18, 227–237.

Bass, A.A. and Bruce, C.E. (2000) The persistently irritable joint in childhood: An orthopaedic perspective. *Eur. J. Radiol.*, 33, 135–148.

Bass, S., Delmas, P.D., Pearce, G., Hendrich, E., Tabensky, A., and Seeman, E. (1999) The differing tempo of growth in bone size, mass, and density in girls is region-specific. *J. Clin. Invest.*, 104(6), 795–804.

Bruynel, L. and McEwan, G.M. (1985) Anthropometric data of students in relation to their school furniture. *NZ. J. Physiother.*, December, 7–11.

BSI (1980) British Standards Institution BS 5873—Part1: Educational furniture. Specification for functional dimensions, identification, and finish of chairs and tables for educational institutions.

CEN/TC 207-Standards—prEN 1729-1 (2006a) Furniture—Chairs and tables for educational institutions—Part 1: Functional dimensions, Under Approval.

CEN/TC 207-Standards—prEN 1729-2 (2006b) Furniture—Chairs and tables for educational institutions—Part 2: Safety requirements and test methods, Under Approval.

Dormans, J.P. (2002) Evaluation of children with suspected cervical spine injury. *J. Bone Joint Surg. Am.*, 84-A1, 124–132.

Evans, O., Collins, B., and Stewart, A. (1992) Is school furniture responsible for student sitting discomfort? Unlocking potential for the future productivity and quality of life. In Edited by E. Hoffmann and O. Evans (Eds.), Proceedings of the 28th Annual Conference of the Ergonomics Society of Australia Inc, Melbourne, Australia, 2–4 December 1992. Ergonomics Society of Australia Inc, Downer, ACT, Australia, pp. 31–37.

Evans, W.A., Courtney, A.J., and Fok, K.F. (1988) The design of school furniture for Hong Kong schoolchildren. An anthropometric case study. *Appl. Ergon.*, 19(2), 122–134.

Fallon, E.F. and Jameson, C.M. (1996) An ergonomic assessment of the appropriateness of primary school furniture in Ireland. In A.F. Ozok and G.Salvendy (Eds.), *Advances in Applied Ergonomics*. West Lafayette, IN: USA Publishing, pp. 770–773.

Feldman, D.E., Shrier, I., Rossignol, M., and Abenhaim, L. (2002) Risk factors for the development of neck and upper limb pain in adolescents. *Spine,* 27(5), 523–528.

Gardner, A. and Kelly, L. (2005) Back pain in children and young people. BackCare, London. Available at www.backcare.org.uk/catalog/detail.php?BID=20&ID=2938

Habibi, E. and Hajsalahi, A. (2003) An ergonomic and anthropometrics study of Iranian primary school children and classrooms. In H. Luczak and K.J. Zink (Eds.), *Human Factors in Organizational Design and Management—VII. Re-Designing Work and Macroergonomics—Future Perspectives and Challenges*. Santa Monica, CA: IEA Press, pp. 823–830.

Habibi, E. and Mirzaei, M. (2004) An ergonomic and anthropometric study to facilitate design of classroom furniture for Iranian primary school children. *Ergonomics SA*, 16(1), 15–20.

Harris, C. and Straker, L. (2000a) Survey of physical ergonomics issues associated with school children's use of laptop computers. *Int. J. Ind. Ergon.*, 26(3), 337–346.

Harris, C. and Straker, L. (2000b) Survey of physical ergonomics issues associated with school children's use of laptop computers. *Int. J. Ind. Ergon.*, 26, 389–398.

Jeong, B.Y. and Park, K.S. (1990) Sex differences in anthropometry for school furniture design. *Ergonomics*, 33(12), 1511–1521.

Jones, G.T. and Macfarlane, G.J. (2005) Epidemiology of low back pain in children and adolescents. *Arch. Dis. Child.*, 90(3), 312–316.

Kayis, B. (1991) Why ergonomic designs in schools? Ergonomics and human environments. In V. Popovic and M. Walker (Eds.), Proceedings of the 27th Annual Conference of the Ergonomics Society of Australia, Coolum, Australia, 1–4 December 1991. Ergonomics Society of Australia Inc., Downer, ACT, Australia, pp. 95–103.

King, H.A. (1999) Back pain in children. *Orthop. Clin. North Am.*, 30(3), 467–474.

Knight, G. and Noyes, J. (1999). Children's behaviour and the design of school furniture. *Ergonomics*, 42, 747–760.

Laeser, K.L., Maxwell, L.E., and Hedge, A. (1998) The effects of computer workstation design on student posture. *J. Res. Comput. Educ.*, 31, 173–188.

Legg, S.J., Pajo, K., Sullman, M., and Marfell-Jones, M. (2003) Mismatch between classroom furniture dimensions and student anthropometric characteristics in three New Zealand secondary schools. In Proceedings of the 15th Congress of the International Ergonomics Association, Ergonomics for Children in Educational Environments Symposium, 6, 395–397, Seoul, Korea, 24–29 August 2003.

Lin, R. and Kang, Y.Y. (2000) Ergonomic design for senior high school furniture in Taiwan, ergonomics for the new millennium. In Proceedings of the 14th Triennial Congress of the International Ergonomics Association and the 44th Annual Meeting of the Human Factors and Ergonomics Society, San Diego, California, USA, July 29–August 4, 2000. Human Factors and Ergonomics Society, Santa Monica, CA, Vol. 6, pp. 39–42.

Linton, S.J., Hellsing, A.L., Halme, T., and Akerstedt, K. (1994) The effects of ergonomically designed school furniture on pupils' attitudes, symptoms and behaviour. *Appl. Ergon.*, 25, 299–304.

Mandal, A.C. (1982) The correct height of school furniture. *Hum. Factors*, 24, 257–269.

Mandal, A.C. (1984) The correct height of school furniture. *Physiotherapy*, 70(2), 48–53.

Mandal, A.C. (1997) Changing standards for school furniture. *Ergon. Des.*, 5, 28–31.

Marschall, M., Harrington, A.C., and Steele, J.R. (1995) Effect of workstation design on sitting posture in young children. *Ergonomics*, 38, 1932–1940.

McAtamney, L. and Corlett, E.N. (1993) RULA: A survey method for the investigation of work-related upper limb disorders. *Appl. Ergon.*, 24, 91–99.

Milanese, S. and Grimmer, K. (2004) School furniture and the user population: An anthropometric perspective. *Ergonomics*, 47(4), 416–426.

Molenbroek, J. and Ramaekers, Y. (1996) Anthropometric design of a size system for school furniture. In S.A. Robertson (Ed.), *Contemporary Ergonomics 1996*. London: Taylor & Francis, pp. 130–135.

Molenbroek, J.F.M., Kroon-Ramaekers, Y.M.T., and Snijders, C.J. (2003) Revision of the design of a standard for the dimensions of school furniture. *Ergonomics*, 46(7), 681–694.

Mosher, E.M. (1899) Hygienic desks for school children. *Educ. Rev.*, 18, 9–14.

Mououdi, M.A. and Choobineh, A.R. (1997) Static anthropometric characteristics of students age range 6–11 in Mazandaran Province/Iran and school furniture design based on ergonomics principles. *Appl. Ergon.*, 28(2), 145–147.

Murphy, S., Buckle, P., and Stubbs, D. (2004) Classroom posture and self-reported back and neck pain in schoolchildren. *Appl. Ergon.*, 35(2), 113–120.

Oates, S., Evans, G., and Hedge, A. (1998) A preliminary ergonomic and postural assessment of computer work settings in American elementary schools. *Computers in the Schools*, 14, 55–63.

Panagiotopoulou, G., Christoulas, K., Papanckolaou, A., and Mandroukas, K. (2004) Classroom furniture dimensions and anthropometric measures in primary school. *Appl. Ergon.*, 35(2), 121–128.

Parcells, C., Stommel, M., and Hubbard, R.P. (1999) Mismatch of classroom furniture and student body dimensions. *J. Adolesc. Health*, 24, 265–273.

Paschoarelli, L.C. and Da Silva, J.C.P. (2000) Ergonomic research applied in the design of pre-school furniture: The Mobipresc 3.6. In Proceedings of the 14th Triennial Congress of the International Ergonomics Association and 44th Annual Meeting of the Human Factors and Ergonomics Society, Vol. 2, pp. 24–27.

Prado-Leon, L.R., Avila-Chaurand, R., and Gonzalez-Munoz, E.L. (2001) Anthropometric study of Mexican primary school children. *Appl. Ergon.*, 32(4), 339–345.

Schindler, C.C.A. (1892) United States Patent US0483266 A1. Issued September 27, 1892. United States Patent Office.

Spear, B.A. (2002) Adolescent growth and development. *J. Am. Diet. Assoc.,* March 2002, S23–S29.

Steele, E.J., Dawson, A.P., and Hiller, J.E. (2006) School-based interventions for spinal pain: A systematic review. *Spine*, 31(2), 226–233.

Taylour, J.A. and Crawford, J. (1996) The potential use and measurement of alternative work stations in UK schools. In S.A. Robertson (Ed.), *Contemporary Ergonomics*. London: Taylor & Francis,, pp. 464–469.

Trevelyan, F.C. and Legg, S.J. (2006) Back pain in school children—Where to from here? *Appl. Ergon.*, 37(1), 45–54.

Troussier, B., Tesniere, C., Fauconnier, J., Grison, J., Juvin, R., and Phelip, X. (1999) Comparative study of two different kinds of school furniture among children. *Ergonomics*, 42, 516–526.

FURTHER REFERENCES

Arikoski, P., Komulainen, J., Voutilainen, R., Kroger, L., and Kroger, H. (2002) Lumbar bone mineral density in normal subjects aged 3–6 years: A prospective study. *Acta Paediatr.*, 91(3), 287–291.

Balague, F., Nordin, M., Skovron, M.L., Dutoit, G., Yee, A., and Waldburger, M. (1994) Nonspecific low-back pain among schoolchildren: A field survey with analysis of some associated factors. *J. Spinal Disord.*, 7(5), 374–379.

Bejia, I., Abid, N., Ben Salem, K., Letaief, M., Younes, M., Touzi, M., and Bergaoui, N. (2005) Low back pain in a cohort of 622 Tunisian schoolchildren and adolescents: An epidemiological study. *Eur. Spine J.*, 14(4), 331–336.

Bendix, T. and Hagberg, M. (1984) Trunk posture and load on the trapezius muscle whilst sitting at sloping desks. *Ergonomics*, 27(8), 873–882.

Biro, F.M., Khoury, P., and Morrison, J.A. (2006) Influence of obesity on timing of puberty. *Int. J. Androl.*, 29(1), 272–277.

Brattberg, G. (2004) Do pain problems in young schoolchildren persist into early adulthood. A 13-year follow-up. *Eur. J. Pain*, 8(3), 187–199.

Burton, A.K., Clarke, R.D., McClune, T.D., and Tillotson, K.M. (1996) The natural history of low back pain in adolescents. *Spine*, 21(20), 2323–2328.

Cakmak, A., Yucel, B., Ozyalcn, S.N., Bayraktar, B., Ural, H.I., Duruoz, M.T., and Genc, A. (2004) The frequency and vassociated factors of low back pain among a younger population in Turkey. *Spine*, 29(14), 1567–1572.

CDC (2000) Centers for Disease Control and Prevention, National Center for Health Statistics. CDC Growth Charts: United States. www.cdc.gov/growthcharts. May 30, 2000.

CDC (2005) Centers for Disease Control and Prevention. Health, United States, 2005 with Chartbook on trends in the Health of Americans. National Center for Health Statistics. Health, United States, 2005.

Diepenmaat, A.C., van der Wal, M.F., de Vet, H.C., and Hirasing, R.A. (2006) Neck/shoulder, low back, and arm pain in relation to computer use, physical activity, stress, and depression among Dutch adolescents. *Pediatrics*, 117(2), 412–416.

El-Metwally, A., Salminen, J.J., Auvinen, A., Kautiainen, H., and Mikkelsson, M. (2004) Prognosis of nonspecific musculoskeletal pain in preadolescents: A prospective 4-year follow-up study till adolescence. *Pain*, 110(3), 550–559.

Feldman, D.E., Shrier, I., Rossignol, M., and Abenhaim, L. (2001) Risk factors for the development of low back pain in adolescence. *Am. J. Epidemiol.*, 154(1), 30–36.

Grimmer, K. and Williams, M. (2000) Gender-age environmental associates of adolescent low back pain. *Appl. Ergon.*, 31(4), 343–360.

Gunzburg, R., Balague, F., Nordin, M., Szpalski, M., Duyck, D., Bull, D., and Melot, C. (1999) Low back pain in a population of schoolchildren. *Eur. Spine J.*, 8(6), 439–443.

Harreby, M., Nygaard, B., Jessen, T., Larsen, E., Storr-Paulsen, A., Lindahl, A., Fisker, I., and Laegaard, E. (1999) Risk factors for low back pain in a cohort of 1389 Danish school children: An epidemiologic study. *Eur. Spine J.*, 8(6), 444–450.

Hedley, A.A., Ogden, C.L., Johnson, C.L., Carroll, M.D., Curtin, L.R., and Flegal, K.M. (2004) Overweight and obesity among US children, adolescents, and adults, 1999–2002. *JAMA,* 291(23), 2847–2850.

Hewes, G.W. (1957) The anthropology of posture. *Sci. Am.*, February, 123–132.

Jacobs, K. and Baker, N.A. (2002) The association between children's computer use and musculoskeletal discomfort. *Work*, 18(3), 221–226.

Jones, G.T., Watson, K.D., Silman, A.J., Symmons, D.P., and Macfarlane, G.J. (2003) Predictors of low back pain in British schoolchildren: A population-based prospective cohort study. *Pediatrics*, 111(4) Pt 1, 822–828.

Jones, M.A., Stratton, G., Reilly, T., and Unnithan, V.B. (2004) A school-based survey of recurrent nonspecific low-back pain prevalence and consequences in children. *Health Educ. Res.*, 19(3), 284–289.

Juul, A., Teilmann, G., Scheike, T., Hertel, N.T., Holm, K., Laursen, E.M., Main, K.M., and Skakkebaek, N.E. (2006) Pubertal development in Danish children: Comparison of recent European and US data. *Int. J. Androl.*, 29(1), 247–255.

Kennedy, E., Cullen, B., Abbott, J.H., Woodley, S., and Mercer, W. (2004) Palpation of the iliolumbar ligament. *NZ J. Physiother.*, 32(2) 76–80. http://nzsp.org.nz/index02/index_Welcome.htm

Kjaer, P. (2004) Low back pain in relation to lumbar spine abnormalities as identified by magnetic resonance imaging (Report 1). PhD Thesis. Back Research Center. University of Southern Denmark.

Knusel, O. and Jelk, W. (1994) Pezzi-balls and ergonomic furniture in the classroom. Results of a prospective longitudinal study [Sitzballe und ergonomisches Mobiliar im Schumzimmer. Resultate einer prospektiven kontrollierten Langzeitstudie] *Schweiz. Rundsch. Med. Prax.*, 83(14), 407–413.

Koskelo, R. (2001) Function responses upper secondary school students considering the saddle chair and the adjustable table. Department of Physiology, University of Kuopio, 1–24.

Kovacs, F.M., Gestoso, M., Gil del Real, M.T., Lopez, J., Mufraggi, N., and Mendez, J.I. (2003) Risk factors for non-specific low back pain in schoolchildren and their parents: A population based study. *Pain*, 103(3), 259–268.

Kristjansdottir, G. and Rhee, H. (2002) Risk factors of back pain frequency in schoolchildren: A search for explanations to a public health problem. *Acta Paediatr.*, 91(7), 849–854.

Leboeuf-Yde, C. and Kyvik, K.O. (1998) At what age does low back pain become a common problem? A study of 29,424 individuals aged 12–41 years. *Spine*, 23(2), 228–234.

Leboeuf-Yde, C., Wedderkopp, N., Andersen, L.B., Froberg, K., and Hansen, H.S. (2002) Back pain reporting in children and adolescents: The impact of parents' educational level. *J. Manipulative Physiol. Ther.*, 25(4), 216–220.

Limon, S., Valinsky, L.J., and Ben-Shalom, Y. (2004) Children at risk: Risk factors for low back pain in the elementary school environment. *Spine*, 29(6), 697–702.

McMillan, N. (1996) Ergonomics in schools. Contemporary ergonomics in New Zealand—1996. In F. Darby and P. Turner (Eds.), Proceedings of the 7th Conference of the New Zealand Ergonomics Society, Wellington, August 2–3, 1996. New Zealand Ergonomics Society, Palmerston North, New Zealand, pp. 87–93.

Mikkelsson, L.O, Nupponen, H., Kaprio, J., Kautiainen, H., Mikkelsson, M., and Kujala, U.M. (2006) Adolescent flexibility, endurance strength, and physical activity as predictors of adult tension neck, low back pain, and knee injury: a 25 year follow up study. *Br. J. Sports Med.*, 40(2), 107–113.

Mikkelsson, M., Sourander, A., Salminen, J.J., Kautiainen, H., and Piha, J. (1999) Widespread pain and neck pain in schoolchildren. A prospective one-year follow-up study. *Acta Paediatr.*, 88(10), 1119–1124.

NCHS (2003) US Department of Health and Human Services. Prevalence of overweight among children and adolescents: United States, 1999–2002. www.cdc.gov/nchs/products/pubs/pubd/hestats/overwght99.htm

Ogden, C.L., Flegal, K.M., Carroll, M.D., and Johnson, C.L. (2002) Prevalence and trends in overweight among US children and adolescents, 1999–2000. *JAMA*, 288, 1728–1732.

Olsen, T.L., Anderson, R.L., Dearwater, S.R., Kriska, A.M., Cauley, J.A., Aaron, D.J., and LaPorte, R.E. (1992) The epidemiology of low back pain in an adolescent population. *Am. J. Public Health*, 82(4), 606–608.

Orbak, Z. (2005) Does handedness and altitude affect age at menarche? *J. Trop. Pediatr.*, 51(4), 216–218.

Petersen, S., Brulin, C., and Bergstrom, E. (2006) Recurrent pain symptoms in young schoolchildren are often multiple. *Pain*, 121(1–2), 145–150.

Reilly, J.J. and Dorosty, A.R. (1999) Epidemic of obesity in UK children. *Lancet*, 354(9193), 1874–1875.

Reilly, J.J., Dorosty, A.R., and Emmett, P.M. (1999) Prevalence of overweight and obesity in British children: Cohort study. *BMJ*, 319(7216), 1039.

Sairyo, K., Katoh, S., Sakamaki, T., Inoue, M., Komatsubara, S., Ogawa, T., Sano, T., Goel, V.K., and Yasui, N. (2004) Vertebral forward slippage in immature lumbar spine occurs following epiphyseal separation and its occurrence is unrelated to disc degeneration: Is the pediatric spondylolisthesis a physis stress fracture of vertebral body? *Spine*, 29(5), 524–527.

Sairyo, K., Katoh, S., Ikata, T., Fujii, K., Kajiura, K., and Goel, V.K. (2001) Development of spondylolytic olisthesis in adolescents. *Spine J.*, 1(3), 171–175.

Salminen, J.J., Erkintalo, M.O., Pentti, J., Oksanen, A., and Kormano, M.J. (1999) Recurrent low back pain and early disc degeneration in the young. *Spine*, 24(13), 1316–1321.

SchoolOutfitters, Good things to know about school desks, www.schooloutfitters.com/catalog/CAT27_gttk.php, accessed on January 20, 2005.

Sheir-Neiss, G.I., Kruse, R.W., Rahman, T., Jacobson, L.P., and Pelli, J.A. (2003) The association of backpack use and back pain in adolescents. *Spine*, 28(9), 922–930.

Shinn, J., Romaine, K.A., Casimano, T., and Jacobs, K. (2002) The effectiveness of ergonomic intervention in the classroom. *Work*, 18(1), 67–73.

Smith, S.A. and Norris, B.J. (2004) Changes in the body size of UK and US children over the past three decades. *Ergonomics*, 47(11), 1195–1207.

Smith-Zuzovsky, N. and Exner, C.E. (2004) Effect of seated positioning quality of typical 6- and 7-year-old children's object manipulation skills. *Am. J. Occup. Ther.*, 58(4), 380–388.

Stahl, M., Mikkelsson, M., Kautiainen, H., Hakkinen, A., Ylinen, J., and Salminen, J.J. (2004) Neck pain in adolescence. A 4-year follow-up of pain-free preadolescents. *Pain*, 110(1–2), 427–431.

Storr-Paulsen, A. and Aagaard-Hensen, J. (1994) The working positions of schoolchildren. *Appl. Ergon.*, 25, 63–64.

Szpalski, M., Gunzburg, R., Balague, F., Nordin, M., and Melot, C. (2002) A 2-year prospective longitudinal study on low back pain in primary school children. *Eur. Spine J.*, 11(5), 459–464.

Tertti, M.O., Salminen, J.J., Paajanen, H.E., Terho, P.H., and Kormano, M.J. (1991) Low-back pain and disk degeneration in children: A case-control MR imaging study. *Radiology*, 180(2), 503–507.

Trentham-Dietz, A., Nichols, H.B., Remington, P.L., Yanke, L., Hampton, J.M., Newcomb, P.A., and Love, R.R. (2005) Correlates of age at menarche among sixth grade students in Wisconsin. *WMJ*, 104(7), 65–69.

Watson, K.D., Papageorgiou, A.C., Jones, G.T., Taylor, S., Symmons, D.P., Silman, A.J., and Macfarlane, G.J. (2002) Low back pain in schoolchildren: Occurrence and characteristics. *Pain*, 97(1–2), 87–92.

Watson, K.D., Papageorgiou, A.C., Jones, G.T., Taylor, S., Symmons, D.P., Silman, A.J., and Macfarlane, G.J. (2003) Low back pain in schoolchildren: The role of mechanical and psychosocial factors. *Arch. Dis. Child.*, 88(1), 12–17.

Wedderkopp, N., Andersen, L.B., Froberg, K., and Leboeuf-Yde, C. (2005) Back pain reporting in young girls appears to be puberty-related. *BMC Musculoskelet. Disord.*, 6, 52.

Wedderkopp, N., Leboeuf-Yde, C., Andersen, L.B., Froberg, K., and Hansen, H.S. (2001) Back pain reporting pattern in a Danish population-based sample of children and adolescents. *Spine*, 26(17), 1879–1883.

CHAPTER 22

CHILD-FRIENDLY USER INTERFACES IN THE DIGITAL WORLD

LIBBY HANNA

TABLE OF CONTENTS

INTRODUCTION

Today's children have access to a huge range of electronic media. The use of computers and Internet is growing quickly, video games are popular, and television is everywhere. Immersive virtual reality is becoming available to children, and they are discovering new ways of communicating via instant messaging with computers and mobile phones.

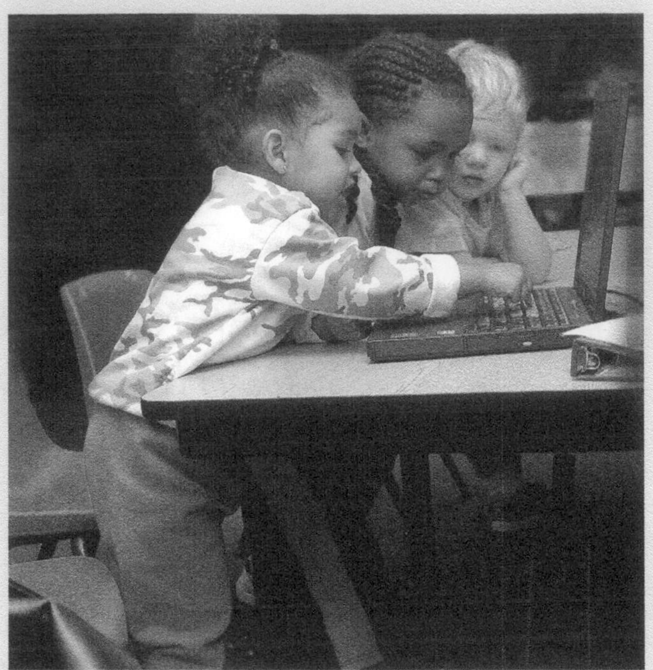

Figure 22.1 (Ursy Potter Photography)
Young children collaborating in the classroom as they explore how to work a computer.

These experiences can be positive or negative, both in the content that children see and hear, and in the flow of the interaction. For example, although video games can positively affect cognitive skills, there are also links between violent video games and violent behavior. Too much television may contribute to physical and behavioral problems, but well-designed television programming can enhance children's learning. Advertising is pervasive in media for children and needs to be dealt with responsibly.

Boys' and girls' different preferences also affect media content and design, and should be taken into account. What appeals to one gender and supports development may not work successfully for the other gender. On the other hand, strongly gender-stereotyped material may increase negative self-images and encourage gender divisiveness.

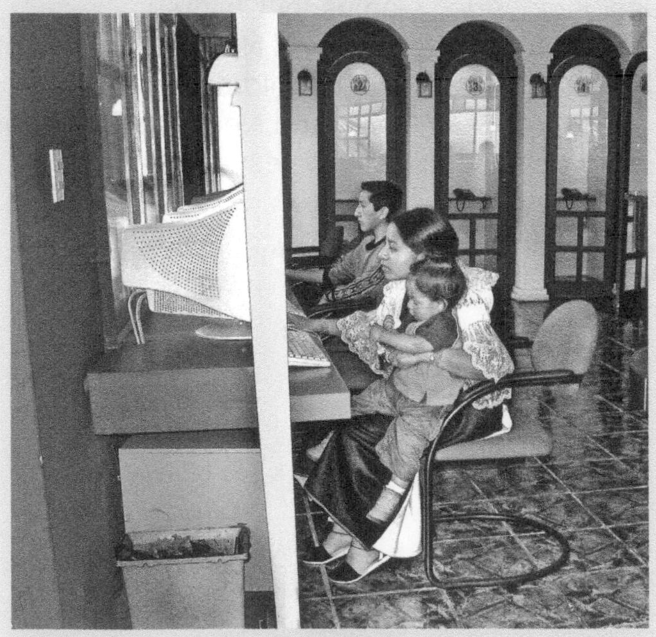

Figure 22.2 (Ursy Potter Photography)
A child and mother amusing themselves at an Internet café in Ecuador.

Children, like adults, need ergonomic user interfaces in electronic media. Ergonomic design guidelines can set the stage for children's initial interactions with a product, enabling them to use products intuitively and fluidly. The products they interact with should be easy to use and geared to their particular developmental stages in hand–eye coordination and cognitive skills.

Children also need media content that offers opportunities for positive growth. Ergonomic content guidelines can stimulate and nurture development by providing electronic media that will enrich children's lives.

DESIGNING USER INTERFACES FOR CHILDREN

User interfaces (UIs) consist of all the design elements and context that children interact with while using electronic media. UIs incorporate icons (clickable images that allow users to access various computer functions, such as the standard Windows "Close" icon (☒)). UIs also include menus or lists of items that users can choose from as well as explanatory text. UIs contain sound, animation, and other elements that guide attention and support interaction. (UIs also include input devices and hardware buttons and knobs—please see Chapter 18, Home Computing, for discussion of hardware ergonomics.)

Children of different ages approach such UIs in vastly different ways (Bruckman and Bandlow, 2003). Very young children cannot comprehend text and need large icons and targets. Children entering school are more adept at using input devices, but still need specially designed interfaces with familiar imagery and audio support for on-screen text. Teenagers have their own interests and abilities, and often benefit from customized UIs rather than a standard grownup look-and-feel.

Of course, the skills and interests of individual children will vary greatly. This chapter makes these distinctions in order to describe general (rather than absolute) differences between age groups.

DESIGNING FOR VERY YOUNG CHILDREN (UNDER 3 YEARS)

Between the ages of 6 months (when they are able to sit up and reach out for things) to about 2½ years, children learn the most from both physical contact with objects and social and emotional interactions with people.

Some experts believe computers exert a negative influence on children of this age range, i.e., directing children's attention to visual displays may discourage sensory and motor explorations of their world as well as interpersonal interactions (Elkind, 1998). Yet very young children are enticed by computers. They like seeing bright shapes and movements, hearing intriguing sounds, and pushing buttons. Young children also learn by other's examples—and they see siblings and parents captivated by computers for hours.

Software designed for very young children should provide opportunities to share computer experiences safely and without the inevitable frustration of using programs designed for older children. If used with affectionate and communicative caregivers, such programs offer experiences that resemble looking at picture books or working on puzzles together. Appropriate software products for this age range are called "lapware," because they are designed to be used by a child sitting on an adult's lap (Figure 22.3).

Figure 22.3 (Ursy Potter Photography)
Infants and toddlers can enjoy making things happen on the computer while accompanied by a caregiver.

Adults are also involved at this age because infants and toddlers can rarely use computer mice. They usually lack the eye–hand coordination needed to move mice and follow pointers on the screen until they reach their third year (Strommen et al., 1996). They also have difficulty finding specific keys on the keyboard.

Young children may focus on a particularly noticeable image (e.g., a bright red circle) while ignoring other areas of the computer screen. Their attention span is short, so they may want to move on to something else after listening to only a few bars of a song or hearing a few words in a rhyme.

Children under 3 years are learning how to control their world—they repeat the same actions over and over, partly to convince themselves they really are making something happen. A feeling of mastery comes with successful use of well-designed software, just as it does with busy boxes or pull toys. Table 22.1 offers guidelines for designing and evaluating effective software for children under 3.

Table 22.1 UI Design Guidelines for Very Young Children

Guideline	Rationale
1. ***Respond immediately*** Offer instantaneous reactions for children's actions and allow interruption with new actions.	• Children this age need immediate responses to their actions to understand the cause and effect relationship. • If they hit a key and start a song, they should be able to hit another key (or the same key again) and start something else and not have to wait for the song to finish.
2. ***Avoid text*** Text is for adults; therefore keep it out of the way so it will not detract attention from the visuals for the primary user.	• Children under 3 see text as additional objects onscreen, and may expect to be able to interact with it like other targets if it attracts their attention. • Consider picture books for the very young as an example—relatively small text on the bottom or top of the page apart from the illustrations.
3. ***Keep it simple*** Use visuals that are cleanly separated in space and offer simple cause-and-effect interaction.	• Help children understand what to pay attention to by isolating targets and using saliency cues (brightness of colors, size, and form of shapes) to draw focus. • Allow children to make things happen through simple interaction—they are not ready to follow instructions or navigate.
4. ***Use audio*** Attract attention and give feedback with sounds, and use verbal instructions along with or in place of written ones.	• Sounds that accompany visual effects strengthen feedback and invite interaction. Simple spoken instructions like "look at the red balloon" or even just "balloon" can focus attention and foster language development. • Do not overdo it—too much noise can be overwhelming and distressful to very young children.
5. ***Allow "any key" interaction*** Children should be able to make something happen by pressing any key on the keyboard.	• Children in this age range generally cannot use input devices well enough to successfully point to and hit targets onscreen. • Some programs for this age use sequential key presses to activate targets and then move on to a new target. Other programs randomize what will happen with each key press.

DESIGNING FOR 3- TO 5-YEAR-OLD CHILDREN

Three- to five-year-old children are brimming with curiosity as they make sense of what they see around them. They spend a good deal of time figuring out what is real (and what is not) by throwing themselves into pretend play.

Computer programs enable children to explore fantasy worlds and meet fantasy characters that become "friends" while they play on the computer. Software can also help children become ready for school in much the same way as coloring or activity books. These programs let children practice their newfound skills in different venues, enabling them to actively learn and experiment (Figure 22.4).

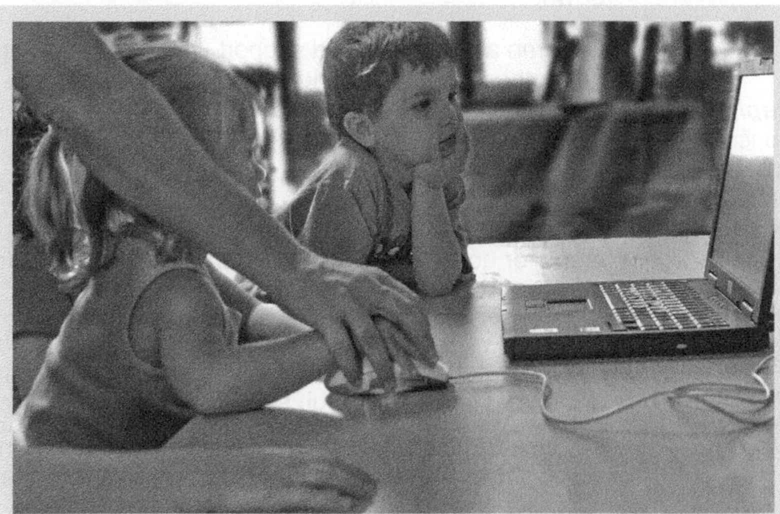

Figure 22.4 (Ursy Potter Photography)
Children around 3 years of age can learn to use a computer mouse and begin to exponentially expand their potential activities on the computer.

Children between the ages of 3 and 5 can typically use computer mice to move the pointer to targets and activate these targets with a mouse click. Even so, they move the mouse erratically, often causing them to overshoot targets. They have trouble doing two things at the same time, such as holding down the mouse button while moving the mouse in order to drag an object (Joiner et al., 1998). Their attention span is more developed, but they are still impatient and need immediate feedback for their actions.

This is the age when challenge and complexity first become important to design. Children under 3 are learning simple interaction with the computer. Children between the ages of 3 and 5 want to accomplish something with the program. Even so, open-ended programs such as painting or storybook programs can offer just as much challenge as skill-building programs with explicit goals.

Open-ended programs can support children's implicit goals of creating a beautiful or exciting picture, discovering hidden surprises in a story, or familiarizing themselves with letter shapes and sounds. Programs can also challenge children by offering increasingly complex tools and graphics.

Table 22.2 provides software guidelines for children aged 3 to 5. Box 22.1 lists Internet concerns and Box 22.2 gives a "best practice" example.

Table 22.2 UI Design Guidelines for 3- to 5-Year-Old Children	
Guideline	**Rationale**
1. Exaggerate the UI to support newly emerging computer skills Offer large targets, wait for responses, and include lots of visual and audio clues and feedback to help identify UI elements.	• These children are just learning computer skills. They may be haphazard at navigating, and may have difficulty pinpointing exact targets. • Use visual and auditory rollover effects so children can identify targets, and give clear feedback for mouse clicks. • Avoid drag-and-drop interaction, instead use a "sticky drag" method where the first click attaches the object to the cursor. Then the object can be moved around the screen without holding the mouse button down.
2. Use familiar metaphors Offer recognizable icons and navigational metaphors that children will readily understand.	• Children at this age need support when generalizing from one situation to another. They will not know what unfamiliar icons mean and are still struggling with left and right. • Navigational elements onscreen should use images that children this age see every day: stop signs, arrows, doorways, pointing hands, and beckoning faces. • Icons for activities and features should enable children to recognize the functionality of the feature.
3. Allow independence Children should be able to work on their own and choose the activities they want to do.	• Children want to be in control and become easily frustrated or disenchanted when they need help to finish the activity or when they are forced to work through specific activities in a predetermined order. • Designs for this age should expect children to be on their own at the computer and allow children to choose activities that reflect their own interests and abilities.
4. Appropriate challenge or novelty Include a range of difficulties or complexities so younger children can still enjoy the program, while increasing the challenge as they progress and become more familiar with the activities.	• There is huge variation in abilities and world knowledge in this age range. • Appropriate designs must offer access to the very young and the very new to computers, while including enough difficulty levels or complexity to intrigue more experienced children. • Even open-ended creativity applications can support a range of abilities by offering layers of tools or features or more unusual stimuli and information.
5. Use on-screen characters Engage children with social interaction and give a visual focus to audio instructions.	• Much of children's learning occurs in a social context. • These children respond eagerly to animated characters onscreen and absorb information much better when given a face to look at rather than listening to a disembodied voice. • These characters can be used to point out crucial UI elements on the screen to further illustrate instructions, and can offer help and feedback when children make mistakes.
6. Keep to a high standard of quality Children benefit from first-rate visual and audio quality and technical excellence as much as any other user.	• Technical problems with children's products severely disrupt interaction—this audience has no idea why the computer suddenly stopped working or why they cannot make a character stop talking, and they may quickly dissolve into tears. • Children deserve best efforts in both technical and aesthetic fields. Clearly this guideline applies to all ages, but is mentioned here as an age where quality sometimes gets overlooked.

BOX 22.1 INTERNET CONCERNS FOR YOUNG CHILDREN

On the Internet, children between 3 and 5 can interact with simple screens much as they do in CD-ROM programs. However, navigation through links from one Web site to another can quickly become overwhelming. These children will have difficulty remembering how they have traveled when they want to get back. They almost never use the Back button or any other browser tools to navigate (Ashmore, 2002).

Children in this age range should be accompanied by adults when on the Internet. Otherwise they may get lost, frustrated, or stumble on to inappropriate material. For this age, "inappropriate material" refers both to content that is offensive and content designed for older children.

BOX 22.2 EXAMPLE OF BEST PRACTICES FOR 3- TO 5-YEAR-OLD CHILDREN: WWW.BEMBOSZOO.COM

One example of Web design that works well for children in the 3–5 age range is www.bemboszoo.com,* created to publicize the book *Bembo's Zoo: An Animal ABC Book*.† The interaction on the site and the simple navigation make it ideal for young children.

Once users have clicked into the Flash activity from the home Web page, they see an array of the alphabet, shown in Figure 22.5. No further navigation is required from here to enjoy the Web site.

Figure 22.5 www.bemboszoo.com displays a simple Web interface that young children can easily interact with. (Roberto de Vicq de Cumptich and Mucca Design Corp)

The letters animate to draw attention and show that they are "clickable." As children roll their cursor over each letter, it rotates and plays a sound file of an animal noise. This combination of animation and audio gives children feedback about where they are onscreen and what they should click.

Once children click a letter, the screen fades to the word for that letter, and then the letters of the word animate and form the shape of an animal. Figure 22.6 and Figure 22.7 show the partial transformation and the completed animation for the letter "N" and the word "Newt." After the animation is complete, the screen fades back to the alphabet array. Children can click on a different letter or the same letter again and watch more transformations. The animations are intricate and entertaining and easily hold attention for their brief duration.

* Note: Because of the transient nature of the Web, these and the other Web sites mentioned in this chapter may be altered or unavailable when this book comes to press.

† The site is highly rated by the Blue Web'n library of outstanding Web sites at www.kn.com/wired/bluewebn/index.html

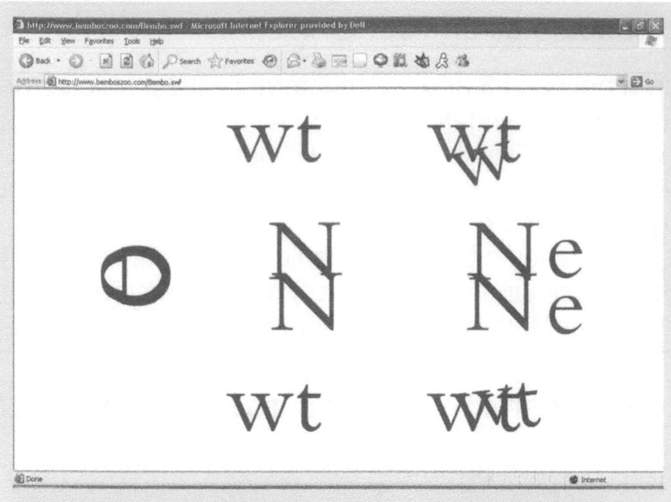

Figure 22.6 Clicking on a letter in www.bemboszoo.com starts an animation with letters, here for the letter N. (Roberto de Vicq de Cumptich and Mucca Design Corp)

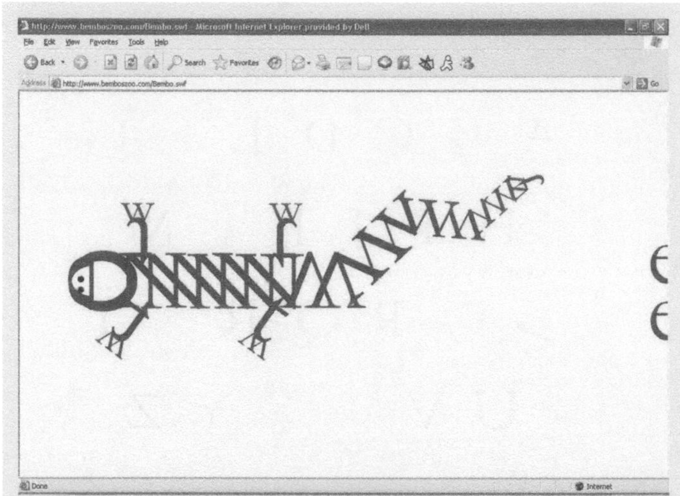

Figure 22.7 At the end of the animation, the letters create an animal and the display goes back to the full alphabet, ready for children to click on a new letter. (Roberto de Vicq de Cumptich and Mucca Design Corp)

DESIGNING FOR 6- TO 8-YEAR-OLD CHILDREN

Six- to eight-year-old children are beginning to develop higher-level social and academic skills. Their eye–hand coordination and fine motor skills have progressed, and they can better control their movements and orient themselves in time and space.

Although they now navigate computer screens with ease, they still reach targets more slowly than older children and adults (Jones, 1989). With practice, they learn to "click-and-drag" and use the right mouse button.

Most children in this age range are beginning readers and tend to avoid reading words on the screen. They often ignore instructions for computer games, and instead try to figure it out on their own. They have pride in their ability to do things "like a grown-up" and enjoy showing others how to play a game they already know.

Children between 6 and 8 like structured games and skill-based activities, as well as open-ended creativity programs and rich fantasy games. Their thought and play become more rule-based, and they enjoy having concrete goals.

Successful designs help children understand the objectives of the game. Table 22.3 contains UI design guidelines for 6- to 8-year-olds, while Box 22.3 shows a "best practice" example.

Table 22.3 UI Design Guidelines for 6- to 8-Year-Old Children

Guideline	Rationale
1. ***Use text sparingly*** Keep it easy to read and support it with audio.	• Children at this age do not want to have to struggle to read instructions on the screen. They would rather start on their own and see if they can figure it out. They respond better to spoken instructions (offered in a way that children can hear them again if needed). Text onscreen can be effective in one- or two-word signs that children quickly identify, like "Stop," "Go," or "Exit."
2. ***Make the UI intuitive for this age*** Use common game strategies and recognizable environments and metaphors to ease navigation.	• These children, even more than younger children, want to be independent on the computer. The design should be as intuitive as possible so children recognize at a glance how to interact with the program. Children this age can identify with navigation metaphors from a broader pool than before: space travel, road maps, and cityscapes, in addition to familiar environments like neighborhoods and backyards.
3. ***Make goals obvious*** Challenge should come from the difficulty in reaching the goal, not in figuring out what they are supposed to do.	• Make sure children know what they are supposed to be working toward, or what the purpose of the activity is, even if it is as simple as coloring a picture. No one wants to feel stupid or unsure of themselves when first exploring a program.
4. ***Motivate children with appropriate challenges and choices*** Expect a certain skill level and offer activities just one step ahead, while allowing child control.	• Children are most motivated to keep working and learning on tasks that are just barely out of reach—offering a feeling of almost making it with each successive effort. This can happen in creativity programs that offer a range of sophisticated tools and templates, in information programs that offer new perspectives and unusual facts, as well as in games that offer levels and sublevels of difficulty. Also, children learn better if they choose their own path through activities than if they have to play an uninteresting game in order to get to a more rewarding one.
5. ***Keep giving lots of feedback*** Offer both visual and audio feedback for rollover and click interaction.	• Children this age need just as much immediate feedback about what they are doing as younger children. They tend to click repeatedly if something does not happen right away (and often find themselves launched further into the program or an activity than intended), or they only click on something that gives a visual rollover clue that it's "hot."

BOX 22.3 EXAMPLE OF BEST PRACTICES FOR 6- TO 8-YEAR-OLD CHILDREN: LOGICAL JOURNEY OF THE ZOOMBINIS

Logical Journey of the Zoombinis offers a variety of mathematical logic puzzles that are undertaken while moving a group of characters (the Zoombinis) from one location to another. This program entrances both children and adults and can keep them involved for hours. Although the official age range for the software is 8 and up, it is easily accessible for 6-year-olds and illustrates many of the guidelines in Table 22.3.

There is little or no text onscreen and there is abundant feedback in the form of rollover animation, audio, and animated highlighting of relevant UI elements. The goals of each activity are very clear.

For example, at one point in the Zoombinis' travel, they meet a tree troll, shown in Figure 22.8. The tree troll stands in their way and shouts, "Make me a pizza!" Children choose combinations of toppings and send a pizza to the troll. He then indicates whether he dislikes one of the toppings (saying "Something must go!" and throwing it into the pit in the front) or wants more toppings ("More toppings!," tossing it back to his stump). As children progress they encounter increasing difficulty involving additional trolls or foodstuffs.

Figure 22.8 In this activity in Logical Journey of the Zoombinis, children test hypotheses to figure out what combination of pizza toppings the tree troll will like. (©1996 Riverdeep Interactive Learning Limited. All rights reserved.)

Challenge builds gradually throughout the program while allowing chances for success at intermittent steps, and the metaphor of traveling safely through a world of obstacles and dangers is extremely motivating. The program is filled with humor that appeals to both children and adults. Independent choices are offered throughout to allow children to feel in control and keep different abilities and interests satisfied. At the very start, children can choose physical characteristics (such as hairstyle or number of eyes) for the Zoombinis to suit their tastes. However, their actions have consequences later as they have to sort and match the Zoombinis on these characteristics. There is also a practice mode where children can play in any puzzle they like.

Figure 22.9 In another Zoombinis activity, highlighting of the lever on the bottom left of the screen shows children what to do next once they have matched the images. (©1996 Riverdeep Interactive Learning Limited. All rights reserved.)

Another example of successful UI design is shown in Figure 22.9. In this puzzle, children must pick a "crystal" on the right side of the screen that matches one of the Zoombinis on the left. Children click directly on the objects (the crystals or the Zoombinis), the objects "stick" to their cursor, and then as they move around the screen rollover highlighting shows where to place the objects. Once the objects are positioned, the lever on the bottom left begins to flash, indicating that it is the next step.

Logical Journey of the Zoombinis not only includes excellent UI design but also has exemplary educational design. Educational aspects of the program are intrinsic to the interaction. In every activity, children are making choices, testing hypotheses, and trying to figure out the correct answer. Learning is not a separate activity tested by a question-and-answer drill, but an integral part of the process of play.

DESIGNING FOR 9- TO 11-YEAR-OLD CHILDREN

Children between 9 and 11 years can become proficient on the computer and may be learning to type and use a computer for schoolwork. Depending on exposure, they may be familiar with many computer metaphors and so navigate with ease through program menus and Internet sites. They are comfortable reading on screen, but they probably will not read much if something else interests them more.

Many in this age range are competent with a mouse (Scaife and Bond, 1991). Children who are more sophisticated on the computer use right mouse buttons and keyboard commands as shortcuts. They are increasingly curious about tips and tricks for using the computer and playing games.

Children from 9 to 11 scan the screen efficiently to find what they are looking for. They are relatively unself-conscious about experimenting with new content. They do not hesitate to click on something that looks like it might be right.

However, they can be misled by unfamiliar icons and unclear instructions. They are still rooted in concrete thinking, and overly abstract content and navigational metaphors bewilder them. They may attribute more intelligence to the computer than it really possesses (Van Duuren et al., 1998). They tend to prefer realistic images to cartoon and are increasingly attracted to products designed for teens and adults.

Table 22.4 offers UI guidelines for 9- to 11-year-olds. Box 22.4 lists Internet concerns while Box 22.5 reviews Internet filtering devices.

Figure 22.10 (Hanna Research and Consulting, LLC) Girls are beginning to catch up to boys on the amount of time they spend on the computer, and like to socialize with chat and e-mail online.

Table 22.4 UI Design Guidelines for 9- to 11-Year-Old Children

Guideline	Rationale
1. ***Offer a sophisticated "kid" look*** Keep the UI kid friendly but do not dumb it down.	• Children this age are not quite ready to be adults, but want to avoid "little kid" stuff at all costs. They are often attracted to things just above their age range—a preteen or teenage look or style. They prefer cool and sharp over cute and warm. • However, they still need a relatively clean design that supports text with images and offers lots of feedback.
2. ***Use the same guidelines for usable interfaces that apply to anyone*** For example, use an "esthetic and minimalist" design, allow user control and freedom, and follow real-world and platform conventions.	• Much of appropriate design for this age group is appropriate design for all users. These children are competent enough to use adult word-processing and spreadsheet programs and will benefit from the same guidelines that make these programs usable. • Follow standard usability heuristics: give appropriate feedback, speak the user's language, allow user control, use consistent conventions, prevent user errors, keep options visible, allow for flexibility, use a clean design, support error recovery, and provide help (Nielson, 1994).

(continued)

Table 22.4 UI Design Guidelines for 9- to 11-Year-Old Children (Continued)

Guideline	Rationale
3. ***Stay concrete*** Use metaphors that are not too far from reality and present information in nonabstract terms.	• These children are still in a concrete stage of cognitive development and struggle with abstract thought. • For example, they may notice a search tool on a Web site and assume it is the same as a general Internet search. It is hard for them to understand that "search" can have the same functionality but not the same scope across sites.
4. ***Take gender differences into account*** If designing specifically for one gender, make sure to appeal to those interests. If trying to be gender neutral, make sure the design will not put off one or the other gender (Figure 22.10).	• Children in this age range are more likely than younger children to choose computer activities (when it is left up to them) that are attractive to their gender. • Boys choose action and sports games as their favorite computer activities. Girls choose creativity or productivity applications, and e-mail and chat on the Internet. Gender-neutral programs need to avoid appearing too "girlish" to keep boys interested, and need to avoid themes of dominance and aggression that are unappealing to girls.

BOX 22.4 INTERNET CONCERNS FOR SCHOOL-AGE CHILDREN

Schools and homes are embracing the Internet for education, with good reason. There are vast stores of knowledge available online, much of it free, and traditional resources like encyclopedias are more current and informative online than in printed editions. Children in the 9- to 11-year age range are increasingly using the Internet both to complete school assignments and to explore what is available online. Once children are competent on computers, they can be equally competent at navigating on the Internet, quickly learning to recognize hyperlinks, and use their browser Back button.

The social and community aspects of the Internet are very attractive to children. Both boys and girls enjoy instant messaging and e-mail with friends and group discussions in chat rooms and bulletin boards. (However, children will not usually enter into these activities until they can type proficiently. Many schools do not begin any kind of formal typing instruction until 9 or 10 years of age.) They are eager to join Web sites that offer a "club" experience and allow them to post creations and share with others (Druin and Platt, 1998).

With this freedom and sophistication, however, come the risks of wandering into dangerous territory. The Internet is full of opinion and contains content designed to appeal to both the high and the low of human instincts. Children are not good at evaluating the quality of information in Web pages (Hsieh-Yee, 2001).

There are plenty of Web sites that are designed for children and try to keep children within bounds by offering a large variety of activities within the site. Some of these Web sites open links to other sites within their own frame, so children still have access to pages within the original site. However, the very nature of surfing the Internet leads children to roam and eventually they may stumble, either by chance or out of curiosity, on inappropriate material. The best defenses against these risks are monitoring of children's online activity and teaching children safe and smart online behaviors.

BOX 22.5 FILTERING SOFTWARE

Many parents and educators are turning to Web-filtering or blocking software to try to protect children from offensive material. These programs work by screening sites for particular keywords and then either not opening the site or forcing children to get permission to access the site. The programs can also restrict access to chat rooms, control information children give out on Web sites, and filter out specific words in e-mail and instant messages both coming in and going out.

Filtering software has come under fire both for its limited effectiveness and for "false alarms"—blocking access to sites that users value and find inoffensive. CNET reviewed three top Web-filtering programs and found that even the most restrictive program blocked only half the offensive sites they tested as well as several innocuous sites (Tynan, 2001). Recent parental controls provided by two top Internet Service Providers restricted access to Google, a search engine that teachers frequently recommend to their students (Tynan, 2002).

Two approaches may protect children more effectively than filtering software:

1. Monitor children's activity:
   Parents and educators can stay aware of what children are doing online either by direct supervision, monitoring services, or by simply using a browser's History cache. If children know that caregivers are going to be looking at what they do online, they may be less likely to explore forbidden territories.
2. Teach children to be responsible and savvy Internet users by teaching them to:
   a. recognize reputable sites versus disreputable ones,
   b. identify links for advertising,
   c. know when and how it may be appropriate to give out information over the Internet, and
   d. how to use e-mail and instant messaging appropriately.*

DESIGNING FOR TEENAGERS 12 YEARS OLD AND UP

Wherever possible, teenagers try to integrate technology into all aspects of their lives. Teenagers who have easy access to computers and the Internet do the bulk of their schoolwork on the computer, shop and talk with friends online, and expect instant access to information. They are proficient at the computer, comfortable with typing, familiar with most computer conventions, and can extricate themselves from common computer glitches (Figure 22.11).

Figure 22.11 (SIS-USA-Inc.com) Teenagers spend increasing proportions of time in front of computers, both for schoolwork and for social and entertainment purposes.

Teenagers in cultures where mobile phones are ubiquitous have spurred the development of text messaging, Internet access, and video capabilities on wireless phone services. Their speed and adaptability with dial pads and small screen displays can quickly outpace adults (Figures 22.12 and 22.13).

* There are several reputable Web sites that offer suggestions for parents and teachers. The American Libraries Association has a section of their Web site devoted to children's online safety.
(www.ala.org/parentspage/greatsites/guide.html), as does the U.S. Department of Education.
(www.ed.gov/pubs/parents/internet/) and The Children's Partnership (www.childrenspartnership.org/).

Figure 22.12 (Ursy Potter Photography)
Text messaging is a popular way to stay in touch with friends and family. Adolescents (and others) can use it almost anytime, anywhere. However, the ability to use it in a variety of situations introduces design challenges. For example: the display should be equally easy to read in bright sunshine or at dusk; the device should be rugged enough to withstand being dropped while used during physical activities; and it helps if it is sufficiently water resistant for use in the rain.

Teenagers can work logically with many factors at once, can be introspective, and can deal with the possible as well as the real. They are exploring identity as well as how to express their unique voices. Design for this age range focuses more on content—what kind of information and activities can support the particular needs of teens—rather than on age-specific UI elements.

The Internet has special value for this age range. Teenagers need to experiment with how they present themselves to others and benefit from online activities that enable them to change how they appear to others with a click of the mouse (Hanlon, 2001). The Internet also offers both large and small communities where teens can find peers.

greengoblin32: *whats shakin*
TokyoNinja7: not much u?
greengoblin32: *im sitten round*
TokyoNinja7: me too!
greengoblin32: *ssssswwwwweeeeetttttttttt*
TokyoNinja7: yyyeeeeeaahhhhh
greengoblin32: *woooohoooo*
TokyoNinja7: weeeeheeeee
greengoblin32: *yippeeekiyay*
TokyoNinja7: hooohooohooo
greengoblin32: *wooooooooooooooooooooooohoooooooooooooooooooooooooo*
greengoblin32: *hoot!*
TokyoNinja7: rooof rooof roof
greengoblin32: *mooooooooooooooo*
greengoblin32: *kakaaaaaaaaaaaaaaaaa!*
TokyoNinja7: wallla wallla walla
greengoblin32: *ok its done*
TokyoNinja7: ok

Figure 22.13 (Hanna Research and Consulting, LLC)
Instant message conversation between two 14-year-old boys. Screen names have been changed to protect identity.

The explosion of instant messaging (via computers and mobile phones) in this age group in many cultures demonstrates their need to socialize with peers and explore interpersonal relationships. Instant messaging services allow teens to virtually hang out together while being physically apart. Instant messaging may reduce the feelings of loneliness or isolation that some teenagers reported with earlier technology (Subrahmanyam et al., 2000). Now they can have online experiences with familiar "buddies," rather than strangers in online chat rooms and multiuser domains.

Table 22.5 suggests UI design guidelines for 12-year-olds through teenagers and Box 22.6 provides a "best practice" example for this age range.

Table 22.5 UI Design Guidelines for 12-Year-Olds and Up

Guideline	Rationale
1. Use a unique look-and-feel Respect this age group with a style that is "non-childish" while not going overboard with a straight business look for adults.	• Teenagers want their own look that differs from adults and children and from preceding generations. • Unique styles are part of exploring identity and self-expression. Appealing designs for this age will generally be artistic, creative, trendy, and noticeably different from generic looks. Consider allowing teenage users to personalize the look with choices of colors and themes.
2. Keep to standard UI guidelines for navigation and organization Offer a usable design by adhering to established recommendations for computer–human interactions.	• While teenagers are still going through stages of physical, social–emotional, and cognitive development, their perceptual development is essentially complete. • Familiarity with computers is also on par with adults, so standard UI design guidelines are appropriate. • Follow traditional usability heuristics: give appropriate feedback, speak the user's language, allow user control, use consistent conventions, prevent user errors, keep options visible, allow for flexibility, use a clean design, support error recovery, and provide help (Nielson, 1994).
3. Offer content that supports a developing identity Include social interaction, self-expression, safe exploration, and a sense of community.	• Teenagers need to be heard—as they grow to understand their place in the world they need to work out their own opinions and get feedback from their peers. • They are searching for companionship in new ways, looking for others who "think like me" instead of previously choosing friends who "look like me" or "like the same things I do." • Online environments can support and encourage teenagers by establishing safe places to converse with others and join a community.
4. Be responsible Offer positive role models and activities, promote development of values, and be honest about advertising and sponsorship.	• Teenagers want to do the right thing, but on their own terms and following their own beliefs about what is right. They are increasingly looking for outside perspectives on morality and values, and abhor hypocrisy. • Web sites for this age can attract and nurture teens by having a prosocial and exemplary message, while respecting the intelligence and independence of users.

BOX 22.6 EXAMPLE OF BEST PRACTICES FOR TEENS: WWW.SHiNE.COM

Most of today's computer products geared specifically toward teenagers are on the Internet. There are few CD-ROM products developed exclusively for the teen market. On the Internet, there are teen magazine Web sites, teen health sites, teen entertainment sites, teen advice sites, and a wealth of sites hosted by individual teens for other teens. One Web site for teens that embodies the guidelines listed above is www.SHiNE.com.*

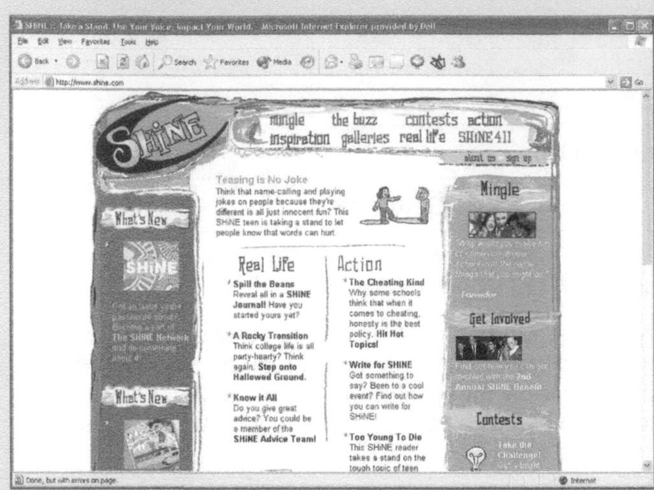

Figure 22.14 (SHiNE, Inc.)
The home page of www.SHiNE.com attracts teenagers with a distinctive style and activities for socializing, making statements, and getting involved in a cause.

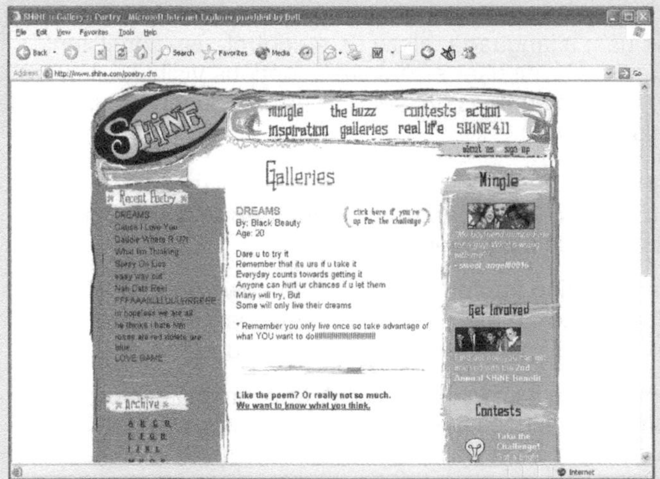

Figure 22.15 (SHiNE, Inc.)
The Poetry Gallery at www.SHiNE.com is a place where teenagers can anonymously express themselves online.

SHiNE.com is the Web site for a national nonprofit organization that works with teens to promote social harmony and nonviolence. They sponsor community events like "teen town halls," and offer local SHiNE clubs that teens can join to work with others for the cause.

The Web site has an eclectic design decidedly different from adult news-headline pages or magazine-style designs, while still appearing sophisticated and artistic. The layout is clean, without an overwhelming list of links, and different areas of the Web site are delineated with graphics as well as text. The offerings on the home page speak directly to teens: there are chat groups, bulletin boards for advice, galleries for artwork, and several ways to get involved in online and local communities (Figure 22.14). The message is one of empowerment and promoting independence and activism in socially positive ways.

Teenagers can choose to join the online community by registering or they can remain anonymous and still contribute to the Web site. Their ability to remain anonymous is an important aspect of their online exploration—teens often change user names and other descriptive information to experiment with their public identity. In the SHiNE.com poetry gallery shown in Figure 22.15, teens can share their thoughts and feelings in anonymous poems while staying safely private.

* This site appears in the "Most Popular" list within the Teenager category of the Yahoo! Directory: http://dir.yahoo.com/Society_and_Culture/Culture_and_Groups/Teenagers/

DESIGNING VIDEO GAMES FOR CHILDREN

The term video game, as used here, refers to console-based games such as Nintendo's GameCube, Sony's PlayStation, and Microsoft's Xbox. There is a great deal of overlap between computer games and video games; for example, you can often find games with the same titles in both formats.

However, the different input devices used with console-based games (digital game pad controllers, joysticks, etc.) have created unique game styles for console-based video games. Many games use simple controllers that restrict the action to only a few moves (for example, jump, leap, and turn) with extra moves added in with combinations of button presses.

The best video games appeal to our human nature. They are exhilarating, challenging, and humorous. Even adults spend hours trying to reach the next level of the game. Very young children can enter into the world of video games on early levels and gain expertise quickly. Although they may not grasp high-level goals or attend to all the cues in the game (Blumberg, 1998), they still enjoy the basic thrill of running and jumping to the next tree branch.

Boys between 8 and 14 years are the core market for video games. Game manufacturers acknowledge that they focus product designs on boys in this age range (Subrahmanyam et al., 2001). With growing concerns about teenage violence, especially among boys, appropriate design in video games must take into account research on both positive and negative effects of playing these games.

RESEARCH ON POSITIVE AND NEGATIVE EFFECTS OF VIDEO GAMES

The largest research projects to date on video games have focused on the relationship between violent content in the game (defined as acts of intentional physical assault including cartoon and fantasy violence) and aggressive behavior in the game player. Most of these studies, including a new one from Japan, have found that aggressive behavior increased in children (of both genders) and adults after playing violent games (Anderson and Bushman, 2001; Gentile et al., 2003; Subrahmanyam et al., 2001; Ihori et al., 2003) (Figure 22.16).

Playing violent games increases physiological arousal and aggressive thoughts, feelings, and actions. It also decreases prosocial behavior. Such findings have come from experimental studies that controlled how much and what kinds of games were played as well as observations of children's behavior after playing violent games. Exposure to violence in television has similar effects on children (see subsequent discussion on television violence).

Figure 22.16 (©2003 by King Features Syndicate, Inc. Reprinted with special permission of King Features Syndicate)

Video game images and animation are becoming increasingly realistic and taking on more aspects of virtual reality, providing a truly immersive feel. The realism and terror of such violence heightens concerns by parents and educators. In the past, an overall lower effect of violence from Japanese video games and television was hypothesized to be related to different portrayals of violence in Japanese media than in the United States (Iowa et al., 1981). Japanese violence was generally perpetrated by the "bad guys" and intended to elicit sympathy in viewers, versus the "justifiable" violence shown in US media. Although the Japanese still prefer role-playing fantasy games over fighting games, their exposure to US-style violence is growing.

Children tend to prefer violent games. Yet parents are often unaware of what their children are playing. One survey found that when parents were asked about their child's favorite game, the parents tended to name incorrect games or not be able to name any game at all. When they named the wrong game, parents consistently named nonviolent games in place of the violent games that the child actually preferred (Funk et al., 1999).

Another negative effect of video games relates to their potential to promote child inactivity. Children's risk of obesity increases with the cumulative amount of "screen time," including video games, computers, and television (Arluk et al., 2003).

Other physical effects include nausea when using 3D displays, palm blisters, and other discomforts from excessive use of game pad joysticks, and increased risk of seizures for children with photosensitive epilepsy (Subrahmanyam et al., 2000). Please see Chapter 18, Home Computing, in this volume for more discussion of input devices and physical setup of computer workstations.

On the other hand, there are documented positive effects of video games. Children who play video games demonstrate improvements in visual intelligence skills such as mental rotation of objects and tracking moving objects, eye–hand coordination and reaction time, and overall intelligence scores (Subrahmanyam et al., 2000; Van Schie and Wiegman, 1997; McSwegin et al., 1988).

Figure 22.17 (Ursy Potter Photography)
Contrary to popular opinion that video games isolate children, boys often play together and use games as something to talk about. Boys compare how far they have managed to get in the game, give each other hints on strategies and share codes or "cheats."

Playing video games in general (when not isolated to violent games) does not appear to detract from other activities or reduce children's social interaction with others. Some researchers hypothesize that recent increases in performance IQ scores across the United States are linked to the rise of video games (Subrahmanyam et al., 2001).

Guidelines for video games in Table 22.6 are intended to foster positive effects of video games while avoiding negative ones. For example, there are ways to include game violence without harm. Box 22.7 describes recent research on virtual reality devices and children.

Table 22.6 Design Guidelines for Video Games for Children

Guideline	Rationale
1. ***Use good game design*** Make the game easy to start but hard to master, include variety for replayability, balance forces, and make a compelling story.	• General game design principles (Crawford, 1982; Howland, n.d.) apply to children too. • These include smooth learning curves, so children can start quickly at easy levels and progress evenly, but stay engaged while they work for ever-harder goals. • The game should also include the "illusion of winnability," so children do not get frustrated and give up thinking they will never make it. • Appealing characters are important, especially for young children. Good games include conflict but do not have to be violent.
2. ***Follow age-specific UI guidelines for on-screen elements*** Know the target age and design appropriately.	• Children need the same support in the UI for video games as they do for computer software. • Depending on the age of the children who will be playing the game, follow guidelines suggested in the previous section of this chapter to design for recognizability, familiarity, and immediate comprehension.
3. ***Design for healthy video game playing*** Avoid interaction that would require extreme or injurious movement and keep to higher flicker frequencies.	• Children have been known to injure hand tissues while playing video games—there was one popular game that children played by using their palm to quickly rotate the controller, and ended up with abrasions. • Flicker frequencies between 5 and 30 flashes per second (Hz) may cause seizures in children with photosensitive epilepsy. • Although it is difficult to rule out all possible injuries that children may incur while playing, known issues should be avoided.
4. ***Design nonviolent games*** Exclude cartoon and fantasy violence as well as realistic violence for young children.	• Long-term effects of violent media, including video games, are especially negative for children under 6 years of age. • Although violent games are appealing, designs for young children should take into account the impressionability of children and their difficulties distinguishing reality and fantasy. • Game content can instead focus on individual challenge and conflict that is overcome with nonviolent means.
5. ***Be judicious and responsible*** If violence is included for older children, lessen the negative impact.	• Children over 6 years of age, and adults, also show negative effects from video game violence. • If using violence in a game for children, lessen the negative impact by making sure violence does not lead to overly positive outcomes for the most aggressive characters. • Avoid very graphic and realistic violence. Describe and label content (using content-based ratings or codes instead of age-based ratings) so parents and children can evaluate the product before buying and know what they are getting into.

BOX 22.7 RESEARCH ON VIRTUAL REALITY AND CHILDREN

Virtual reality is a new immersive technology, using head-mounted displays (ranging from simple goggles to large helmets with projection screens and speakers) and electronic wands or gloves that enable users to interact with virtual environments as if walking through rooms and manipulating objects. The equipment is expensive and has yet to be universally adapted to smaller heads for general use. However, virtual reality may be the wave of the future in electronic media (especially 3-D gaming) and is starting to trickle down to children.

One study comparing children (7- to 9-year-olds) and adults found that children were more sensitive to field-of-view restrictions in head-mounted displays. In a field-of-view 30° high and 22° wide, children had more difficulty navigating and manipulating objects and experienced greater disorientation than in less restrictive fields (and much more so than the adults). They did not retain any route knowledge of how they had navigated during the experiment (McCreary and Williges, 1998).

Use of lower-quality images (by stretching pixels in order to increase field-of-view) may detract children's ability to identify objects in the virtual environment. In another study comparing students from the 4th grade to the 12th grade, the youngest children had significantly more difficulty controlling a wand while using a head-mounted display, often having to resort to two hands. However, they did not report more nausea or discomfort than older children did. Younger children reported a greater feeling of "presence" (being in the environment) than the older students and very high enjoyment of the virtual experience (Taylor, 1997).

Virtual reality has shown benefits in therapeutic settings, where immersive environments can make use of multisensory feedback for children with disabilities (Passig and Eden, 2001; Rizzo et al., 2000) and effectively distract and soothe children undergoing medical procedures (Schneider and Workman, 1999; Holden et al., 1999).

Virtual reality is appealing to children and offers an effective immersive experience. However, the design of virtual experiences will need to be adapted to children's developing perceptual skills for them to have successful experiences. It will be important that the equipment is sized correctly for children and the field-of-view is large and uncluttered, with easily distinguishable objects. New products and future research will illuminate further guidelines for virtual reality and potential negative effects on children's sensory development, if any.

DESIGNING TELEVISION FOR CHILDREN

Children do not watch television passively. As they watch, they actively learn, develop values, and make decisions. They are most attracted to fairly novel content that is not too complicated and fits in with what they already know (Bickham et al., 2001).

Children learn more when the story line is straightforward and educational content is clear and closely tied to the story line. They have difficulty understanding complicated stories that explore unfamiliar stories and characters (Fisch, 2000).

Young children under age 5 prefer children's programming, cartoons, and fantasy shows that are slower paced with lots of music and art (Comstock and Scharrer, 2001). Older children prefer real-world shows and heroes. Older children are also more aware of—and critical of—the quality of the production (animation, cinematography, sound recording, etc.). All children are attracted to scary content, which may actually relieve their anxiety if the story ends happily (Valkenburg and Cantor, 2000).

Most research on the impact of television on children has focused on the negative effects of content such as news programs and violence. However, other research has consistently demonstrated positive effects from educational/informational children's programming. These findings underscore that quality television programs can and do support children's development.

RESEARCH ON POSITIVE AND NEGATIVE EFFECTS OF TELEVISION

Research indicates that television violence promotes aggressive behavior. This correlation between media violence and aggression is only slightly smaller than that between smoking and lung cancer (Bushman and Huesmann, 2001).

TV violence in cartoons and action shows has the greatest impact on children under 6. Violence is especially detrimental if children see heroic characters rewarded or succeed by their violent acts.

Viewing violent programs is not cathartic—it does not relieve violent impulses. Rather, it demonstrates violent behavior and beliefs that children may later imitate; it cues and primes aggressive thoughts and feelings that children may then act upon; and it desensitizes children in their reactions to violence.

Children may also be negatively affected by other adult programs, such as the news. Watching the news seems to exaggerate their perception of crime in the larger world. The news is more likely to frighten children aged 9 to 12 than children aged 5 to 8 (Smith and Wilson, 2002). Of course, 9- to 12-year-olds are also more likely to understand the news than younger children. They are more disturbed by crime stories than younger children, who in turn are more frightened by disasters.

Very recent research has linked greater amounts of early television watching (at ages 1 and 3) with an increased risk of attention problems at age 7 (Christakis et al., 2004). Childhood television viewing is also associated with teenage smoking, drinking, and sexual behavior (Gidwani et al., 2002; Villani, 2001).

The American Academy of Pediatrics now urges pediatricians to evaluate how child patients use the media and to educate parents about these negative effects (AAP, 2001). At routine well-child visits, many doctors take a "history" of television watching, and recommend no more than 1 to 2 hours a day of watching carefully selected programs (and none for children under 2 years of age).

The positive effects of television are equally well documented with long-term research. Children who watch educational programming like *Sesame Street* at 2 and 3 years of age have higher academic scores at 5 years of age and higher grades at 15 to 19 years of age than children who watch other programming (Bickham et al., 2001; Fisch et al., 1999.) This is especially true for boys.

Children show benefits in cognitive development from watching *Blue's Clues* (Anderson et al., 2000), and gains in vocabulary, counting skills, imaginative play, and awareness of manners after watching *Barney and Friends* (Singer and Singer, 1998). Children show the most benefits from these shows (and sometimes only show benefits) when parents or teachers explain and reinforce the concepts being presented.

One study of 5-year-olds found most preferred commercial to noncommercial television shows, and they were drawn to sequences that contained conflict and

Figure 22.18 (Ursy Potter Photography)
Boys tend to show the most benefit from educational/information television programming and the most harm watching televised violence.

aggression (Hake, 2001). Children may prefer their own "television counterculture" over adult media or they may be attracted to forbidden fruits, desiring what their parents are trying to guard against. Restricted labels and age-based ratings for both television and video games tend to attract children to what is supposed to be for teens or adults (Bushman and Huesmann, 2001).

On the other hand, content codes which describe the kind or amount of negative elements in a show (for example, "V" for violence, "FV" for fantasy violence, etc.) do not attract viewers and appear to be more effective at allowing parents to judge the kinds of shows their children should watch. Parents and educators can also teach children to be critical viewers, which may ameliorate harmful effects. Adults can point out theatrics and special effects and help children to understand what is going on within the story line.

One of the most important themes of research on children and television is the role of adults. Children are least affected by negative content, and learn the most from positive content, when adults are actively watching with them. This has been established by US and Japanese researchers (Kodaira, 1990) and is true for teenagers as well as young children (Collins et al., 2003). Designers of content for children should encourage adults and children to watch shows together by including elements that will appeal to both.

See Table 22.7 for guidelines for designing televised media for children and Box 22.8 for an example of "best practices."

Table 22.7 Design Guidelines for Television for Children

Guideline	Rationale
1. Involve children with moderate novelty and comprehensible narrative	• Children pay the most attention to, and learn the most from, content that is just on the edge of what is familiar and comprehensible. • Elements in the story should be novel enough to be interesting without being so strange that children cannot figure it out. The story line should be easily followed so children can absorb concepts and anticipate next steps.
2. Appeal to affective needs by resolving issues of conflict and fear	• Watching television involves emotion as well as cognition. • Children (and adults) are especially attracted to themes of conflict and fear—these touch basic instincts in human nature and are issues that humans constantly struggle with. Presenting these issues in televised content with strong resolutions can help children deal with their emotions and learn effective responses.
3. Appeal to adults with occasional adult-directed humor or intrigue	• Watching television with an adult helps children gain the most positive benefits from the experience and best protects them from harmful effects. • Content that appeals to adults as well as children will encourage caregivers to watch with their children.
4. Avoid violence, including cartoon and fantasy violence, in shows for young children	• The negative effects of watching violence on TV include increased aggressive behavior and desensitization to violence. These effects are particularly potent and long term for young children. • Content designed for young children should avoid depicting physical violence or attempts at physical violence, and especially depictions where the violence results in a positive outcome for the aggressor.
5. Be judicious and responsible about using violence for older children	• As recommended for video games, avoid depicting violence as the way the hero wins. • Avoid overly realistic and intense violence, which has the most potential for physiological arousal and desensitizing effects. • Accurately code and label violent content (using content codes instead of age-based ratings) so viewers are aware of what they might see.

BOX 22.8 EXAMPLE OF BEST PRACTICES IN TELEVISION FOR YOUNG CHILDREN: SESAME STREET

Sesame Street is an example of great television for children, both for its proven value as an educational tool and for its enduring appeal to generations of children. *Sesame Street* offers fantasy characters for young children like Big Bird and Oscar the Grouch, and diverse real people so that children from many different backgrounds can relate to them. The pace is lively and keeps both children and adults entertained. There is plenty of repetition for young children to enjoy while helping them learn. There is a great deal of humor both children and adults appreciate.

Figure 22.19 "Elmo's World" on Sesame Street offers an example of educational television that delights and supports young children. (All Rights Reserved. Used with permission.)*

For example, the show now includes a segment called "Elmo's World." Designed to appeal to very young children, Elmo is presented as a young child himself, with all of a young child's worries and feelings of inadequacy. In Elmo's World, he often plays with a real baby, giving the baby a task that Elmo has mastered, but the baby has not. Watching the baby try to eat spaghetti with a fork is funny for everyone.

Sesame Street also presents interpersonal conflict (witness Oscar or the famous fights between Grover and a customer at a restaurant) and real child fears that are dealt with warmly and resolutely. Production quality is high and incorporates innovative film techniques and combinations of animation and film, along with the familiar and artistic puppetry of the Muppets. Guest artists have included highly respected and talented artists, entertainers, and government leaders. All these features encourage families to watch together.

GENDER DIFFERENCES IN INTEREST AND INTERACTION

There are definite differences between boys and girls and what they like in media. Designs for children should not necessarily strive to be gender neutral. Designs that support the particular interests and interaction styles of each gender may be the most involving and rewarding. As long as products avoid incorrect or negative stereotypes and offer positive examples and role models for children, immersion in a "girls' world" onscreen or online is simply another way of exploring what it means to be a girl (Figure 22.20).

Figure 22.20 (Photo courtesy of Valerie Rice)
Girls find just as much reward and satisfaction in playing games with female heroes and feminine themes as boys do with fast-paced action games.

In the United States, boys spend up to three times as much time as girls playing electronic games, primarily action games (Wright et al., 2001). Boys are more likely to socialize with friends or family members around computers than girls do, probably because they often play computer games with friends, brothers, or fathers (Orleans and Laney, 2000). Girls spend more time on word-processing and creativity software than boys do (Kafai and Sutton, 1999).

Boys show stronger preferences for masculine versions of computer activities (for example, "pirate" versus "princess") than girls show for feminine versions (Joiner, 1998). This fits with a general tendency for boys to show stronger preferences for gender-stereotyped materials and to go out of their way to avoid female-stereotyped designs. Elementary- and middle-school boys view computers as a male-stereotyped activity and use computers more than girls do (Whitley Jr., 1997).

However, there may be decreasing differences in the amount of time spent on the computer and attitudes about computers. With improvements in Internet access and new communication platforms like instant messaging, girls are finding more things to interest them on the computer. Girls aged 10 and up engage in more social activities on the Internet than boys (Subrahmanyam et al., 2000).

Differences in television preferences support the idea that girls are more interested in social interaction and relationships. Young children watch child-directed shows like cartoons and educational/informational programs about equally. When they shift to general audience programming, girls move to situation comedies and relationship dramas, while boys switch to sports and action shows. Girls are more likely than boys to continue to watch educational shows as they get older, while boys are more likely than girls to continue to watch cartoons (Wright et al., 2001). Girls are also more likely to watch an entire show and pay more attention to the narrative than boys (Valkenburg and Cantor, 2000).

The guidelines in Table 22.8 offer some general ideas for how to appeal to boys versus girls and support diverging needs. There are also suggestions for designing gender-neutral products when aiming to please all children.

Table 22.8 Design Guidelines for Gender Appeal

Guideline	Rationale
1. ***For girls, offer interpersonal interaction and narrative***	• Girls are drawn to content that deals with social interaction and social relationships. • They like strong story lines. They are less concerned with the look of the design than boys are, and are not necessarily turned off by masculine themes. • However, they need to find the content they are interested in order to stay involved.
2. ***For boys, offer action and a nonfeminine look-and-feel***	• Boys avoid designs that look like they might be intended for girls (anything in pink). • They want designs that clearly offer masculine activities and information. They like fast-paced action games and themes of dominance.
3. ***For both boys and girls use "gender-neutral" designs*** Avoid an obvious masculine or feminine look, or include a wide variety of styles.	• A feminine design will turn off boys, and a masculine design may indicate to girls that they will not find anything interesting in the product. • Strive for a balance on the feminine/masculine continuum so that children are not quite sure if it is supposed to be for boys or girls. • Alternatively, offer variety, including some design elements that are clearly masculine, some clearly feminine, and some in between, so that all children can find something to identify with.

ADVERTISING AND CHILDREN

Advertising is intertwined with media in most modern cultures—there is almost no way to escape it. Stand-alone CD-ROM computer products and video games do not include advertising within the game, but many are tied to other merchandise and are an additional way to promote a brand. The Internet is full of advertising and many Web sites exist solely as an advertisement for a company or brand.

Some countries have laws against direct advertising to children (notably Sweden and Norway, with other countries that restrict children's television advertising including Greece, Denmark, Belgium, Taiwan, New Zealand, and India, among others). However, the line between children's television programming and advertising is often blurred.

By their second year, children begin to connect television advertising with products they see in the store, and ask parents to purchase advertised products (Valkenburg and Cantor, 2001). Children under 5 have trouble discriminating between commercials and regular television, and do not understand the persuasive intent of commercials until they are 7 or 8 years old.

Advertising is less regulated on the Internet than on television. Most Web sites designed for children either contain advertisements or are commercial sites specifically designed to advertise products. Most children under 11 years of age do not recognize the persuasive intent of commercial Web sites (Henke, 1999).

Children aged 6 to 12 click on online advertisements much more than adults do, because they do not see a difference between the content and the advertisement. They click on a favorite character they see in an ad and go to a new site full of games and activities, designed to keep them there. They do not notice or understand the small "advertisement" or "ad bug" label next to ads (Nielson, 2002). Children are eager to sign up for contests and clubs on Web sites and are rarely aware of how their personal information may be used by advertisers (Austin and Reed, 1999).

Growing concern over children unwittingly giving out personal information on the Internet in the United States led to the 1998 Children's Online Privacy Protection Act. Web sites that collect information from children under 13 are now required to have "verifiable parental consent," usually meaning a signature by mail or fax, a telephone call, or a credit card number (FTC, 1999). However, this does not address the pervasiveness of advertising in general and how it may deceive children online.

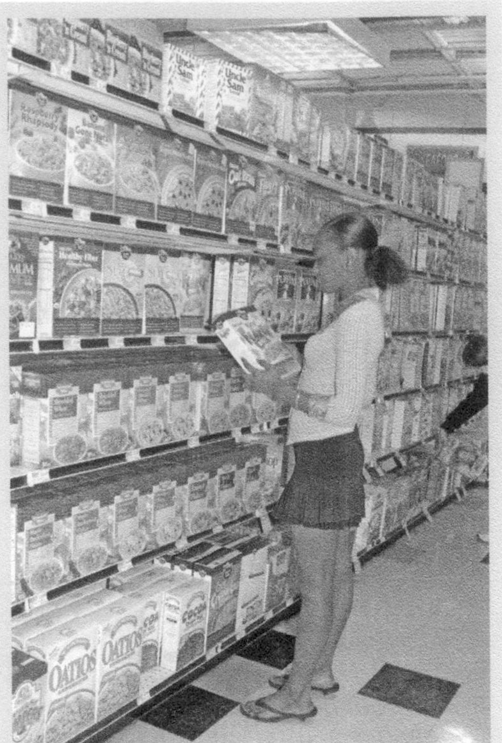

Figure 22.21 (Ursy Potter Photography) Advertising is more effective with children than adults, especially when it focuses on a "premium" offer (the toy in the cereal box). Interestingly, Taiwan specifically restricts this kind of ad. Ads for children are almost exclusively for toys, cereal, candy, and fast food (Kunkel, 2001).

On the other hand, much creativity and innovation on the Internet can be found on commercial sites. At www.barbie.com, girls can play in activities that mirror much of their real-world play with dolls. The fantasy and dress-up activities on www.barbie.com support girls' needs and interests and are hard to find on other sites.

Although persuasive intent certainly exists on the site (girls have many opportunities to view things for sale and receive personalized shopping suggestions based on what they click on), it feels direct rather than deceptive. Clicking on the flashing star that says, "See more!" takes girls to pictures of more Barbie paraphernalia, not to a different Web site with a different purpose. Girls who visit www.barbie.com are likely to be Barbie owners and have a legitimate interest in what is being offered.

The Web sites used as examples of best practices in this chapter do not contain advertising. Few Web sites appropriately handle advertising for children. The goal of responsible advertising for children should be to control children's exposure and clearly delineate advertisements from other content. Adults need to work with children to increase perception and understanding of advertising, helping children learn what it means to be a smart consumer. Table 22.9 contains design guidelines for advertising that focuses on children.

Table 22.9 Design Guidelines for Incorporating Advertising for Children

Guideline	Rationale
1. Help teach children to be savvy about advertising Encourage caregivers to watch television and browse the Internet with children and foster discussion about advertising.	• Children have difficulty distinguishing advertising from content in a variety of media, and do not understand the intent of advertising. • Children can be taught to discriminate advertising by discussing the story line of a show and how the advertising does not relate, or how the different graphics of an online ad have nothing to with the rest of the site. • Caregivers and children can talk about how pictures in advertisements may be different from what really goes on around them. • Responsible advertising and teaching children to be "savvy consumers" can promote positive media experiences.
2. On the Internet, exaggerate the distinction between content and advertising Place advertising on a separate "sponsors" page or in a side panel that is easily distinguished from the other activities on the page and does not distract from them.	• Children are not aware of the conventions of adult-style banner ads nor do they notice text or character labels that distinguish advertising. • Having a separate page for advertising might help children better understand that offers of food, toys, or links to television shows are different from the other activities on the site. • Links on other pages could invite children to visit "sponsors" and then show all the possibilities in comparison to one another.
3. On the Internet, help children keep track of where they are going and offer ways to get back	• Children may not realize they have gone to a new Web site when they click on advertising. They can easily get confused and not be able to get back to the place they want to be. • Advertising should take children to a game or product information that stays within the original site, so children will not lose their way. • Offer obvious links back from the sponsor's Web site.

CONCLUSIONS

Overall, there are two important concepts in ergonomic and developmentally appropriate design across these different kinds of media. First, the UI should be familiar and age appropriate. Children need to recognize UI elements in order to find a product easy to use and they can only recognize what is familiar to them. Designers need to understand what different ages find familiar and comprehensible, and evaluate designs with children of the target age to make sure children will find them usable and satisfying.

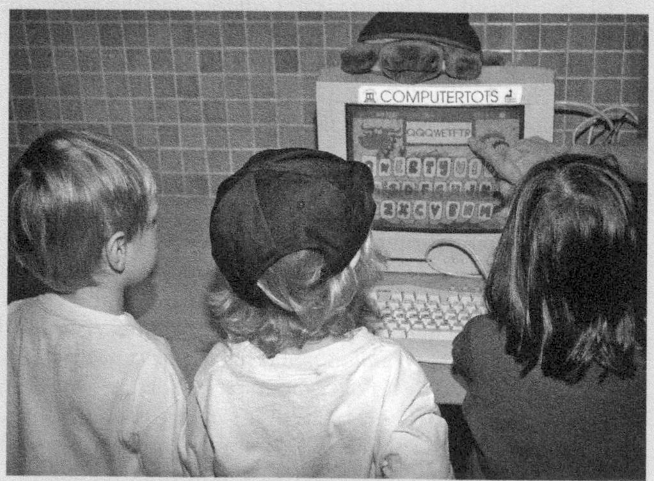

Figure 22.22 (Ursy Potter Photography)

The second concept is moderate novelty or increasing challenge and complexity of content. To be most attracted to a product, to find it most compelling and rewarding, and to learn the most from it, children need to find the content approachable yet stimulating. They need to understand enough to find it accessible, but they also need to encounter challenging situations.

The tools children use to interact with the content—the UI—need to be familiar and simple. The content itself needs to offer novelty and complexity. Finding this difficult balance is the key to quality media products for children. The guidelines in this chapter can help designers toward this balance, and toward excellence in product design.

ACKNOWLEDGMENTS

Many thanks to the stewardship and wealth of resources from the editors of this volume, Rani Lueder and Valerie Rice. I would also like to thank my friends and colleagues Kirsten Risden, Denise Neapolitan, Diane Brown, and Kristin Alexander for letting me bounce ideas off them and responding with openness and intelligence. Kudos to Mary Czerwinski who has always encouraged researching and publicizing children's issues and to Dennis Wixon and Edie Adams for promoting excellence in user research. As always, I am incredibly grateful to Don Fleming, my personal editor and writing coach.

REFERENCES

AAP (2001). American Academy of Pediatrics Committee on Public Education: Media violence. *Pediatrics, 108*(5), 1222–1226.

Anderson, C. A. and Bushman, B. J. (2001). Effects of violent video games on aggressive behavior, aggressive cognition, aggressive affect, physiological arousal, and prosocial behavior: A meta-analytic review of the scientific literature. *Psychological Science, 12*(5), 353–359.

Anderson, D. R., Bryant, J., Wilder, A., Santomero, A., Williams, M., and Crawley, A. M. (2000). Researching Blue's Clues: Viewing behavior and impact. *Media Psychology, 2*(2), 179–194.

Arluk, S. L., Swain, D. P., and Dowling, E. A. (2003). Childhood obesity's relationship to time spent in sedentary behavior. *Military Medicine, 168*(7), 583–586.

Ashmore, L. (2002). Web site usability and the theory of multiple intelligences. www.ekidsresearch.com/dissertation.html

Austin, M. J. and Reed, M. L. (1999). Targeting children online: Internet advertising ethics issues. *Journal of Consumer Marketing, 16*(6), 590–602.

Bickham, D. S., Wright, J. C., and Huston, A. C. (2001). The educational influences of television. In D. G. Singer and J. L. Singer (Eds.), *Handbook of Children and the Media* (pp. 101–119). Thousand Oaks, CA: Sage Publications, Inc.

Blumberg, F. C. (1998). Developmental differences at play: Children's selective attention and performance in video games. *Journal of Applied Developmental Psychology, 19*(4), 615–624.

Bruckman, A. and Bandlow, A. (2003). Human-computer interaction for kids. In J. Jacko and A. Sears (Eds.), *The Human-Computer Interaction Handbook* (pp. 428–440). Mahwah, NJ: Lawrence Erlbaum Associates, Inc.

Bushman, B. J. and Huesmann, L. R. (2001). Televised violence and aggression. In D. G. Singer and J. L. Singer (Eds.), *Handbook of Children and the Media* (pp. 223–254). Thousand Oaks, CA: Sage Publications, Inc.

Christakis, D. A., Zimmerman, F. J., DiGiuseppe, D. L. and McCarty, C. A. (2004). Early television exposure and subsequent attentional problems in children. *Pediatrics, 113*(4), 708–713.

Collins, R. L., Elliott, M. N., Berry, S. H., Kanouse, D. E., and Hunter, S. B. (2003). Entertainment television as a healthy sex educator: The impact of condom-efficacy information in an episode of *Friends. Pediatrics, 112*, 1115–1121.

Comstock, G. and Scharrer, A. (2001). The use of television and other film-related media. In D. G. Singer and J. L. Singer (Eds.), *Handbook of Children and the Media* (pp. 47–72). Thousand Oaks, CA: Sage Publications, Inc.

Crawford, C. (1982). The art of computer game design. www.vancouver.wsu.edu/fac/peabody/game-book/Coverpage.html

Druin, A. and Platt, M. (1998). Children's online environments. In C. Forsythe and E. Grose (Eds.), *Human Factors and Web Development* (pp. 75–85). Mahwah, NJ: Lawrence Erlbaum Associates, Inc.

Elkind, D. (1998). Computers for infants and young children. *Child Care Information Exchange, 123*, 44–46.

FTC (1999). Federal Trade Commission: How to comply with the Children's Online Privacy Protection Rule. www.ftc.gov/bcp/conline/pubs/buspubs/coppa.htm

Fisch, S. M. (2000). A capacity model of children's comprehension of educational content on television. *Media Psychology, 2*, 63–91.

Fisch, S., Truglio, R. T., and Cole, C. F. (1999). The impact of Sesame Street on preschool children: A review and synthesis of 30 years' research. *Media Psychology, 1,* 165–190.

Funk, J., Hagan, J., and Schimming, J. (1999). Children and electronic games: A comparison of parents' and children's perceptions of children's habits and preferences in a United States sample. *Psychological Reports, 85,* 883–888.

Gentile, D. A., Lynch, P. J., Linder, J. R., and Walsh, D. A. (2004). The effects of violent video game habits on adolescent hostility, aggressive behaviors, and school performance. *Journal of Adolescence, 27,* 2–22.

Gidwani, P. P., Sobol, A., DeJong, W., Perrin, J. M., and Gortmaker, S. L. (2002). Television viewing and initiation of smoking among youth. *Pediatrics, 110,* 505–508.

Hake, K. (2001). Five-year-olds' fascination for television: A comparative study. *Childhood: A Global Journal of Child Research, 8*(4), 423–441.

Hanlon, J. (2001). Disembodied intimacies: Identity and relationship on the Internet. *Psychoanalytic Psychology, 18*(3), 566–571.

Henke, L. L. (1999). Children, advertising, and the Internet: An exploratory study. In D. W. Schumann and E. Thorson (Eds.), *Advertising and the World Wide Web* (pp. 73–80). Mahwah, NJ: Lawrence Erlbaum Associates, Inc.

Holden, G., Bearison, D. J., Rode, D. C., Rosenberg, G., and Fishman, M. (1999). Evaluating the effects of a virtual environment (STARBRIGHT World) with hospitalized children. *Research on Social Work Practice, 9*(3), 365–382.

Howland, G. (n.d.) Game design: The essence of computer games. www.lupinegames.com/articles/essgames.htm

Hsieh-Yee, I. (2001). Research on Web search behavior. *Library and Information Science Research, 23,* 167–185.

Ihori, N., Sakamoto, A., Kobayashi, K., and Kimura, F. (2003, August). Does video game use grow children's aggressiveness?: Results from a panel study. Paper presentation at the 34th Annual Conference of the International Simulation and Gaming Association, Kazusa Academia Park, Chiba, Japan.

Iowa, S., Pool, I.S., and Hagiwara, S. (1981). Japanese and U.S. media: Some cross cultural insights into TV violence. *Journal of Communication, 31,* 28–36.

Joiner, R., Messer, D., Light, P., and Littleton, K. (1998). It is best to point for young children: A comparison of children's pointing and dragging. *Computers in Human Behavior, 14*(3), 513–529.

Joiner, R. W. (1998). The effect of gender on children's software preferences. *Journal of Computer Assisted Learning, 14*(3), 195–198.

Jones, T. (1989). Effect of computer-pointing devices on children's processing rate. *Perceptual and Motor Skills, 69,* 1259–1263.

Kafai, Y. B. and Sutton, S. (1999). Elementary school students' computer and Internet use at home: Current trends and issues. *Journal of Educational Computing Research, 21*(3), 345–362.

Kodaira, S. I. (1990). The development of programs for young children in Japan. *Journal of Educational Television,* 16(3),127–150.

Kunkel, D. (2001). Children and television advertising. In D. G. Singer and J. L. Singer (Eds.), *Handbook of Children and the Media* (pp. 375–393). Thousand Oaks, CA: Sage Publications, Inc.

McCreary, F. A. and Williges, R. C. (1998). Effects of age and field-of-view on spatial learning in an immersive virtual environment. Proceedings of the Human Factors and Ergonomics Society 42nd Annual Meeting (pp. 1491–1495). Santa Monica, CA: Human Factors Society.

McSwegin, P. J., Pemberton, C., and O'Banion, N. (1988). The effects of controlled video game playing on the eye–hand coordination and reaction time of children. In J. E. Clark and J. H. Humphrey (Eds.), *Advances in Motor Development Research, Vol. 2* (pp. 97–102). New York: AMS Press, Inc.

Nielson, J. (1994). Ten Usability Heuristics. www.useit.com/papers/heuristic/heuristic_list.html

Nielson, J. (2002). Jakob Nielson's Alertbox, April 14, 2002: Kids' Corner: Web site Usability for Children. www.useit.com/alertbox/20020414.html

Orleans, M. and Laney, M. C. (2000). Children's computer use in the home. *Social Science Computer Review, 18*(1), 56–72.

Passig, D. and Eden, S. (2001). Virtual reality as a tool for improving spatial rotation among deaf and hard-of-hearing children. *CyberPsychology and Behavior, 4*(6), 681–686.

Rizzo, A. A., Buckwalter, J. G., Bowerly, T., van der Zaag, C., Humphrey, L., Neumann, U., Chua, C., Kyriakakis, C., van Rooyen, A., and Sisemore, D. (2000). The virtual classroom: A virtual reality environment for the assessment and rehabilitation of attention deficits. *CyberPsychology and Behavior, 3*(3), 483–499.

Scaife, M. and Bond, R. (1991). Developmental changes in children's use of computer input devices. *Early Childhood Development and Care, 69,* 19–38.

Schneider, S. M. and Workman, M. L. (1999). Effects of virtual reality on symptom distress in children receiving chemotherapy. *CyberPsychology and Behavior, 2*(2), 125–134.

Singer, J. L. and Singer, D. G. (1998). Barney & Friends as entertainment and education: Evaluating the quality and effectiveness of a television series for preschool children. In J. K. Asamen and G. L. Berry (Eds.), *Research Paradigms, Television, and Social Behavior* (pp. 305–367). Thousand Oaks, CA: Sage Publications, Inc.

Smith, S. L. and Wilson, B. J. (2002). Children's comprehension of and fear reactions to television news. *Media Psychology, 4*(1), 1–26.

Strommen, E. F., Revelle, G. L., Medoff, L. M., and Razavi, S. (1996). Slow and steady wins the race? Three-year-old children and pointing device use. *Behaviour and Information Technology, 15*(1), 57–64.

Subrahmanyam, K., Greenfield, P., Kraut, R., and Gross, E. (2001). The impact of computer use on children's and adolescents' development. *Applied Developmental Psychology, 22*, 7–30.

Subrahmanyam, K., Kraut, R. E., Greenfield, P. M., and Gross, E. F. (2000). The impact of home computer use on children's activities and development. *The Future of Children, 10*(2), 123–144.

Taylor, W. (1997, March). Student responses to their immersion in a virtual environment. Paper presentation at the Annual Meeting of the American Educational Research Association, Chicago, IL.

Tynan, D. (2001, June 12). Web watchers: CNET reviews 3 online filtering programs. www.cnet.com/software/0-352110-8-6135181-1.html

Tynan, D. (2002, October 24). MSN 8.0. www.cnet.com/Internet/0-3762-8-20576873-2.html?tag=st.is.3762-8-20576873-1.arrow.3762-8-20576873-2

Valkenburg, P. M. and Cantor, J. (2000). Children's likes and dislikes of entertainment programs. In D. Zillmann and P. Vorderer (Eds.), *Media Entertainment: The Psychology of Its Appeal* (pp. 135–152). Mahwah, NJ: Lawrence Erlbaum Associates, Inc.

Valkenburg, P. M. and Cantor, J. (2001). The development of a child into a consumer. *Applied Developmental Psychology, 22*, 61–72.

Van Duuren, M., Dossett, B., and Robinson, D. (1998). Gauging children's understanding of artificially intelligent objects: A presentation of "counterfactuals." *International Journal of Behavioral Development, 22*(4), 871–889.

Van Schie, E. G. M. and Wiegman, O. (1997). Children and video games: Leisure activities, aggression, social integration, and school performance. *Journal of Applied Social Psychology, 27*(13), 1175–1194.

Villani, S. (2001). Impact of media on children and adolescents: A 10-year review of the research. *Journal of the American Academy of Child & Adolescent Psychiatry, 40*(4), 392–401.

Whitley, Jr. B. E. (1997). Gender differences in computer-related attitudes and behavior: A meta-analysis. *Computers in Human Behavior, 13*(1), 1–22.

Wright, J. C., Huston, A. C., Vandewater, E. A., Bickham, D. S., Scantlin, R. M., Kotler, J. A., Caplovitz, A. G., Lee, J. H., Hofferth, S., and Finkelstein, J. (2001). American children's use of electronic media in 1997: A national survey. *Applied Developmental Psychology, 22*, 31–47.

CHAPTER 23

INFORMATION AND COMMUNICATION TECHNOLOGY IN SCHOOLS

CLARE POLLOCK AND LEON STRAKER

TABLE OF CONTENTS

Use the internet to research the events leading up to the start of World War I with particular emphasis on the impact of world events on local events in (town). The assignment should be word processed and illustrated. Illustrations may be downloaded or scanned...

(Ninth grade homework assignment)

INTRODUCTION

This homework assignment (above) demonstrates the level of technological sophistication we expect of children in industrially advanced countries. We expect children to use a variety of ***information*** *and* ***communication technologies (ICTs)***, and we expect teachers to integrate these technologies into their curriculum.

This chapter reviews the ergonomic implications of using ICTs in schools. It also provides guidelines for implementing and using ICTs to promote child learning and well-being.

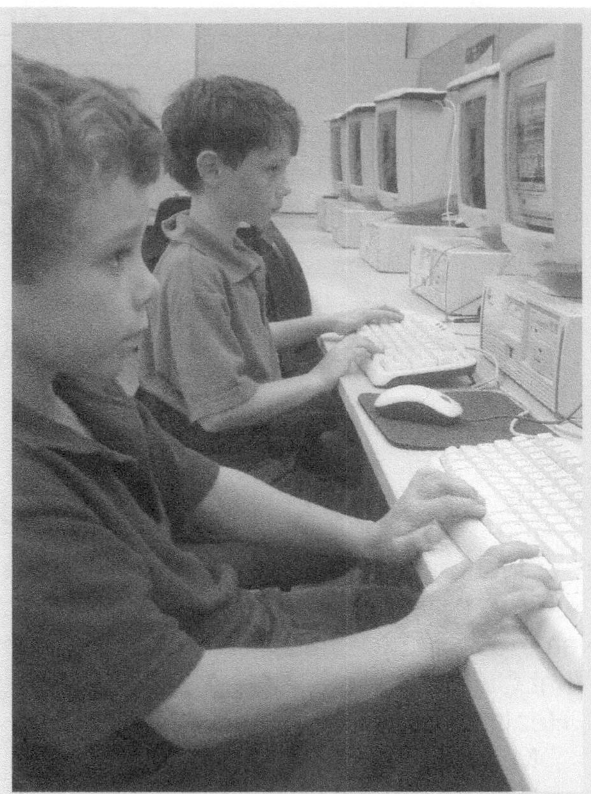

Figure 23.1 (Leon Straker)

> *"If by a miracle of mechanical ingenuity a book could be so arranged that only to him who had done what was directed on page one would page two become visible and so on much that now requires personal instruction could be managed by print"*
>
> (Thorndike, 1912)

CURRENT STATE OF ICT USE IN SCHOOLS

Microcomputers were introduced into schools in the 1970s and 1980s* as they became widely available on the market. Today, desktop computers are often available in specialized computer laboratories or as part of normal classroom activities (Figure 23.2a and Figure 23.2b). Some schools make laptop computers widely available in classrooms (see Figure 23.2c).

> In this chapter **Information and Communication Technologies (ICTs)** refers to desktop and laptop computers.
>
> **ICTs** also include palm-top computers, mobile phones, and electronic games.

Many developed countries spend great sums to implement and integrate computers into schools.† Even so, poorer children can have less access to computers at home and school than those in more affluent areas. This 'digital divide' may lead to unequal opportunities to learn and develop skills needed for employment.

When implementing technology into schools, a number of factors deserve consideration in addition to its impact on learning. These include the effect of ICTs on children's

* Initially, schools' computers were restricted to mathematics or computer training classes. However, proponents advocated widespread use of computers to accelerate children's cognitive development (Papert, 1980). Other incentives for using ICTs in class include reduced cost, the internet and wide-reaching potential of computer systems.

Some educators see technology as a solution to educating the burgeoning public school population in the United States, as they can educate more children with fewer personnel resources (teachers) (Januszewski, 2001).

† For example, Singapore aims to have 30% of teaching activities to use technology, while Queensland plans to have ICT used in 40% of learning areas. In 2003, the United States budget for state grants for educational technology was more than $700 million (USD).

Figure 23.2a (Valerie Rice)

Figure 23.2b (Keith Lightbody, Western Australian Department of Education)

Figure 23.2c (Keith Lightbody, Western Australian Department of Education)

Examples of current ICT use in schools:

a. Computer corner in traditional classroom
Note that the table, keyboard, and VDT are too high for the students. Note also the windows directly behind the computers subjecting them to direct glare and possible visual fatigue.
b. Computer laboratory with older desktop equipment
Note that the equipment is not adaptable.
c. Laptop-based classroom showing high use of computers.

health and safety, such as whether sustained use of computers may expose them to risk factors that contribute to overuse injuries.

NAMING OF PARTS: WHO IS PART OF THE SYSTEM?

The science of ergonomics is based on a systems-approach that can also be applied to implementing school technologies. Figure 23.3 illustrates a systems model of ICT in education that incorporates the community, home, and school.

In this model, the "community" represents the larger geo-social-political environment that influences attitudes and use of computers at school and home. For example, local business and parent groups are important because they provide most of the ICT funding in US public schools (NCES, 2000).*

Communities differ in the resources available to them (Figure 23.4). Schools

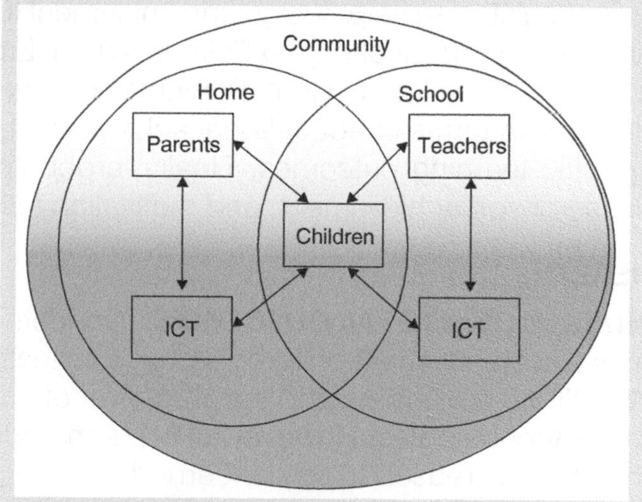

Figure 23.3 (Leon Straker)
A systems model of information and communication technology use at school and home.

* Based on a 1999 survey by the United States Department of Education.

and households also vary in their access to computers—how many, what type, and how often these are replaced. They also differ in how adults (teachers, parents, and caregivers) and children use them.

Home environments affect how and where children use computers. For example, the number of family members who share ICTs and the child's perception of what family members know about and perceive computers influences how they interact with them (Attewell and Battle, 1999; Colley et al., 1994) (see Chapter 17).

The three central parts in the school-based-ICT system are the children, teachers, and ICTs (hardware and software). The child interactions refer not only to an interaction between a child and teacher, or a child and ICT, but also to other children, as peer interactions impact the social and educational outcomes of ICT use (Heft and Swaminathan, 2002).

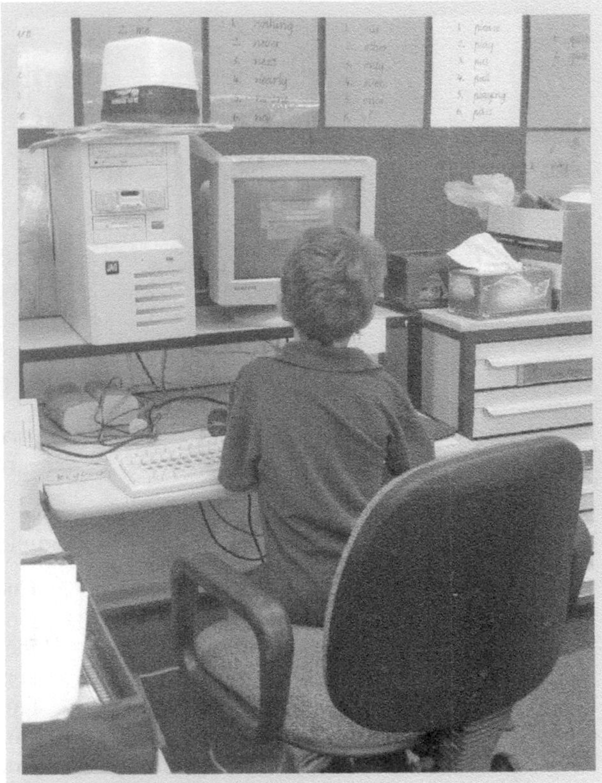

Figure 23.4 (Leon Straker)

Teachers and school administrators influence the educational and technical support and the equipment made available to child users. They also influence the culture of technology use in the school (Cook and Finlayson, 1999; Lankshear and Snyder, 2000).

The above is a brief overview of the systems involved in IT integration into schools. "Managing the System" in this chapter provides guidelines for taking advantage of new technologies as a system.

WHAT CAN ICT USE ACHIEVE?

Ergonomists can help design and implement ICT systems that enhance health, productivity, and well-being (Straker and Pollock, 2005). Ergonomics can enhance health by preventing musculoskeletal and visual problems and promoting physical activity. It can promote productivity in terms of educational achievement, both in general cognitive skills and through specific learning outcomes. Finally, ergonomics can promote the child self-confidence, sense of self-achievement, and motivation to learn.

HEALTH

"**Musculoskeletal**" refers to the muscles, bones and joints of the human body.

MUSCULOSKELETAL OUTCOMES: Concern is growing that ICTs may adversely affect children's physical health (McGrane, 2001). We still do not understand the long-term effects on children using these technologies. However, short-term findings on symptoms among children suggest that there is reason to be concerned.

As with adults, children who use computers often report discomfort. Sixty percent of 10- to 17-year-olds report discomfort from using laptop computers at school (Harris and Straker, 2000). Although most children report mild discomfort, some experience extreme discomfort or pain. For example, in self-reports by 11- to 12-year-olds at a technology-rich school,

5%–10% said their discomfort was severe enough to justify a medical referral (Straker, 2001).

Children report discomfort in the neck and upper limb areas most often, followed by the lower back (Figure 23.5 through Figure 23.7) (Balague et al., 1999; Straker, 2001; Hakala et al., 2002). This pattern of discomfort is similar to that of adults, and underscores the concern that children may develop longer lasting musculoskeletal disorders.

Children may also be vulnerable to longer-term problems because their muscles and bones are still developing while they are experiencing these symptoms. As they develop, the pull of a child's muscles and tendons influences their bone growth (see Chapter 6).

This suggests that using computers may expose children to physical demands that contribute to their early development of health disorders. Clearly, the postural habits they develop as children are likely to carry through adulthood.

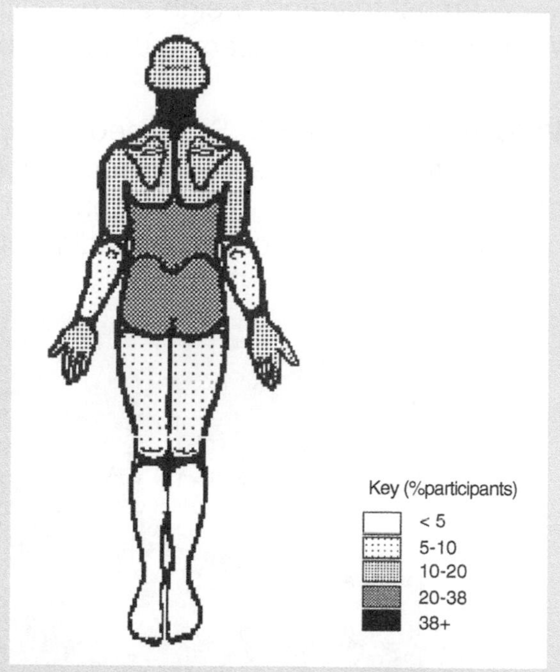

Figure 23.5 This body map shows where children report discomfort when using computers: 49% reported neck pain, while 27% reported pain in the back, and 16% reported pain in the shoulders.

Figure 23.6 (SIS-USA-Inc.com) **Figure 23.7** (SIS-USA-Inc.com)

Schools that are incorporating technology must consider a range of factors that affect children's risk of musculoskeletal discomfort and visual strain. These include:

- the technology itself, in terms of hardware and software
- the supporting equipment, including furniture and classroom space (see Chapter 21 of this text)
- education for students and staff on how to use the technology, structure their environment, and prevent musculoskeletal and visual strain

It may not be sufficient to simply provide education; students need to develop an understanding of how physical fitness affects posture, how their posture influences their development, and how both affect their performance!

We must therefore weigh the benefits of technologies against the potential health risks. We must teach child users to use technology wisely.

VISUAL OUTCOMES: Sustained computer work is visually demanding, requiring the children to focus on nearby objects for long periods. This can cause visual stress. Traditional classroom teaching uses a variety of media with a range of visual demands (e.g., using near vision to read books vs. reading blackboards from a distance).

Figure 23.8 (Leon Straker)
Near vision dominates when working on the computer. Excessive use of near vision may strain the visual system and lead to changes to the eyeball that creates myopia (near-sightedness) (Tan et al., 2000) (see Chapter 4 on child vision).

PHYSICAL ACTIVITY: Moderate and vigorous physical activity is declining in children, prompting health authorities around the world to set targets for children's physical activity levels. For example, the goals of the US Healthy People Plan include the recommendations for children's activity shown in Table 23.1 (Healthy People, 2010).

Physical activity promotes normal musculoskeletal development in children and adolescents (Bauman et al., 2002). It also reduces the risk of coronary heart disease, hypertension, obesity, osteoporosis, Type II diabetes, colon cancer, and depression for adults (Pate et al., 2002; Janz, 2002). Regular physical activity helps children to develop attitudes and habits of daily living that may encourage them to continue to exercise throughout adulthood (Cavill et al., 2001).

Table 23.1 US Children's Level of Physical Activity and Goals (Healthy People, 2010)

Activity Frequency	1999	Goal
30 min, 5 or more days/week	27%	35%
20 min, 3 or more days/week	65%	85%

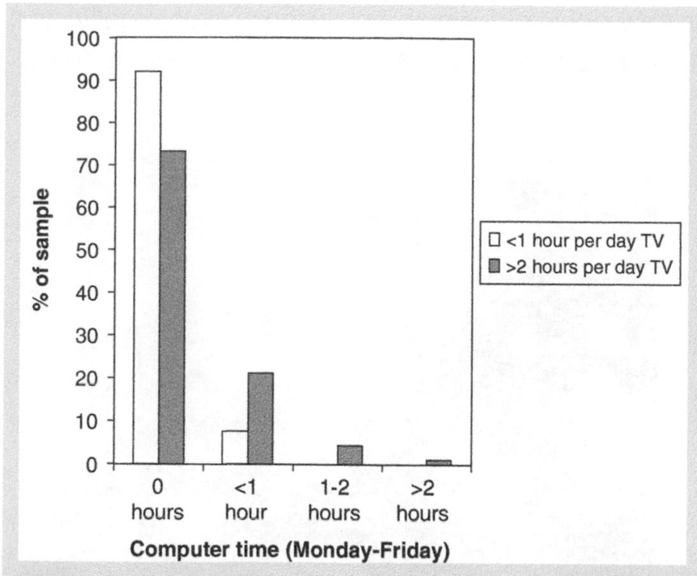

Figure 23.9 (Leon Straker)
Children who watch more TV (dark columns) also tend to spend more time on computers (1600 5-year-old children).

Excessive television watching appears to discourage physical activity. Further, there is growing concern that increased ICT use will exacerbate the problem (Grund et al., 2001) because children who spend the most time watching television are also the heaviest users of computers at home (Straker, 2001).*

PRODUCTIVITY

Research on the impact of computers on health is sobering. Yet many parents and teachers believe that ICT benefits children's cognitive development and learning.

EDUCATIONAL OUTCOMES: People who advocated microcomputers in schools in the 1970s and 1980s believed that early and structured use of programming at school and home would promote children's development of math and writing skills (Papert, 1980; Noss, 1988; Clark, 1985). However, research findings on the relationship between ICT and children's achievement has not been consistent; some studies show a relationship between ICTs and learning while others show no relationship.

In one of these studies, children with unlimited access to advanced technology at school also showed improvements in learning. Teaching improved as well during this 10-year study; teachers became more motivated and developed more advanced learning approaches (Apple, 1995).

Another more recent study also found that ICTs helped children learn.† Their findings are particularly impressive given that relatively low use of ICT improved children's performance (Harrison et al., 2002).

In contrast, BECTA (2000) found that schools that had more ICT resources did not get higher achievement in English and Math from their children. Another found no correlation or a negative correlation between the amount of time spent using school computers and English and Math scores (Moseley et al., 2001).

Such conflicting findings may have resulted from different approaches used by researchers. For example, researchers who measured the total time children spend using computers that included game playing found a negative effect of ICT on learning (Moseley et al., 2001), while others who tracked time spent on educational activities found a positive effect (Harrison et al., 2002).

---

* See Figure 4 and Chapter 6 for more on physical activity in children.

† In this study, children in schools across England (ages 10–11, 13–14, and 15–16) participated. In all cases, there were greater improvements in English, Math, and Science among high users of ICT than among low users of ICT.

So, while it appears that using computers for work in class helps children scholastically, and playing games has a negative (or no) effect, we need more research before we can be sure.

CONFIDENCE AND ATTITUDES

Technology will become an increasingly important part of most children's lives, as advances in technology inevitably lead to new applications for both work and play.

Learning is more fun when we feel confident and positive. Children who use computers are less anxious (Todman and Lawrence, 1992) and more positive towards computers and technology (Moseley et al., 2001; Somekh et al., 2002) (Figure 23.10).

Yet, exposure to computers does not necessarily reduce anxiety, rather, children's perception of their own ability and knowledge of computers are better predictors of their levels of anxiety (Anderson, 1996). Children may spend many hours playing computer games, yet feel baffled by new programs.

Negative experiences with computers may heighten anxiety, thus reducing their interest in using computers in the future (Gos, 1996). This suggests that the quality of experience affects attitudes more than the amount of exposure to computers.

Children have more positive experiences with computers when their teachers are confident and competent in these technologies. Teachers' anxieties about computers limit their ability to use them effectively (Van-Braak, 2001; Atkins and Vasu, 2000).

Figure 23.10 (Leon Straker)
Carefully introducing ICT into schools can promote positive attitudes towards computers. This requires a consistent emphasis and incremental education that promotes positive attitudes and confidence among children and teachers.

Computer use can also increase students' feelings of competency and self worth, as they improve their skills and "move up" through programs. Contrary to popular belief, computer use for learning and schoolwork is not always a solitary experience; it can involve interaction between students, promoting socialization, cooperative learning, and teamwork.

Boys Versus Girls

Boys and girls have different attitudes towards computers; girls perceive that computers are less interesting and useful than do boys (e.g., Colley et al., 1994). These differences appear to diminish as girls become familiar with computers and gain access to girl-friendly software (Arrowsmith, 2001).

Many computer-literate teachers champion school technologies that ultimately only they use. Teachers tend to resist technologies that do not provide them with the technical assistance and training for their use (Rogers, 1995).

Unfortunately, many teachers perceive that computers require expertise and rely on *early innovators* to "champion" and assume responsibility for school technologies. This may cause schools to emphasize ICT skills training classes for students, instead of integrating ICTs into teaching and learning activities throughout the curriculum (Somekh et al., 2002).

Carefully introducing ICT into schools can promote positive attitudes towards computers. This requires a consistent emphasis and incremental education that promotes positive attitudes and confidence among children and teachers.

MANAGING THE SYSTEM

The section, "Naming of Parts: Who is Part of the System?," of this chapter reviewed some of the different elements that affect outcomes of implementing technologies. The following section provides guidelines for each part of the system.

COMMUNITY

HOW IS THE ICT FUNDING SPENT?:

Figure 23.11 (Leon Straker)
Social and political environments can powerfully influence school ICT resources and support. Overzealous advocacy of ICTs may waste technology budgets. In the process, other important projects may flounder through lack of financial or political support.

Of course, the reverse can also be true; insufficient funding for technologies can be detrimental. Educational priorities involve careful weighing up of such trade-offs.

WILL ICTS MEET EXPECTATIONS?: The implementation of ICTs may also fail by not meeting expectations. This is particularly so when teachers lack the motivation to take full advantage of these technologies.

Professional development programs for teachers and centers of excellence might help avoid such attitudes. For example, the Education Department of Queensland, Australia earmarked certain schools as Centers of Excellence. These centers serve as examples of successful integration of technology. They provide an educational resource for teachers in other areas.

WILL ICT GET ADEQUATE SUPPORT?: Local communities can help schools benefit from ICT by providing financial and other kinds of support. Many schools in the United States receive hardware, software, or technical assistance from their communities, supplementing government funding (NCES, 2000).

Community support for ICT systems should

1. provide financial and technical support for ICT to schools and
2. respect the education priorities set by teachers for the use of ICT in specific communities and groups of children.

SCHOOL

The wide-scale use of ICT throughout the curriculum may cause children to work long hours on computers at school and at home, putting them at risk for musculoskeletal overuse symptoms. Teachers and parents should consider the impact of children's combined use of computers for schoolwork and for leisure (Figure 23.12).

As a rule, children tend to use computers more at home than at school (Harrison et al., 2002). These computers also tend to be more technologically advanced. However, the opposite may be true in underprivileged areas.

Implementing school ICT often progresses through three stages:

1. Purchasing hardware and software
2. Providing ICT skills classes for children (and teachers)
3. Integrating ICT into the curriculum

The final stage has the greatest effect on educational outcomes (Somekh et al., 2002).

Communications between teachers, parents, and students can help balance how ICTs are used at home and school. For example, computer classes should reflect the knowledge and skills of children.*

Guidelines for proper use of ICTs at school:

1. Consider ICT resources for children within the community before defining school budgets and training for technology.
2. Train and help teachers evaluate how and when to implement ICTs. At the same time, respect individual teacher's decisions about how they will use ICTs in their class. Help teachers make the transition from their traditional role as instructors to roles more akin to guides.
3. Provide technical assistance to teachers who need it.
4. Avoid concentrating all ICT expertise in one or two teachers in a school.
5. Evaluate the need for ICT skills classes, particularly when children are already competent with computers.
6. Teach all students to touch-type.
7. Teach children and teachers about the importance of changing physical activities over the day, how to set up their workstations and correct postures for computer use.

Figure 23.12 (Leon Straker)
Approaches for reducing schoolchildren's risk of discomfort include their learning to touch-type and regularly changing their posture. Children who touch-type spend less time doing computer-based schoolwork and reduce neck, arm, and hand strain (Brandis and Straker, 2002).

Alternating activities throughout the day encourages changes in postures. Children must also learn about proper sitting postures at the computer and the benefits of movement and breaks away from the computer.†

* See Chapter 17 of this book on home use of computers and Chapter 13 for Usability.

† See Chapter 6 for information on providing exercises and activities to counteract postures that children use consistently throughout the day.

TEACHERS

Most teachers want to teach more effectively, but they may be unsure about how they can take advantage of ICTs to improve their curriculum. Since unstructured and undirected use of ICT can negatively affect children's education, avoid using ICTs to reward good behavior or to distract/babysit hard-to-manage children.

Each ICT use in class should have a purpose. Although this purpose is usually to learn classroom material, it may also be appropriate to focus on social skills. For example, playing games on a computer may help promote socialization and class discussion (Orleans and Laney, 2000).

How teachers can support children's use of ICTs:

1. Recognize that ICTs can and should change the nature of teaching from telling/giving information, to helping children discover information for themselves
2. Actively seek opportunities to learn what ICTs can add to children's learning experiences and what they can not do
3. Evaluate children's access to and familiarity with technologies at school and at home
4. Ensure that children's school and home assignments reflect their ICT abilities and access
5. Take advantage of children's skills to motivate them to establish learning goals for themselves, for classmates and teachers. Encourage technologically sophisticated students to teach and help others
6. Balance the use of ICTs to benefit education while avoiding overuse

Integrating ICTs into the curriculum changes the nature of child–teacher relationships. Children may be more skilled at the computer than the teacher, eroding traditional teacher–student roles. Teachers may find this situation daunting, particularly if they are already anxious about using computers.

ICT fosters a more independent and exploratory kind of learning. Teachers need to feel comfortable moving away from traditional instructor roles towards child–teacher relationships that involve working together towards shared goals.

PARENTS

Parents can help their children to use computers at school more effectively by providing them with a computer and internet connection at home. This helps them develop skills and confidence they can share with their peers (Figure 23.13).

Figure 23.13 (Valerie Rice)
Children who use computers at home achieve more at school. They get better grades and math and English scores, even after accounting for socioeconomic status, race, gender, family structure, and social/educational experiences (Attewell and Battle, 1999).

These findings challenge communities to ensure that all children can use computers outside of school. Parents can encourage children to use computers at home by helping them look up answers to their questions on the Web (e.g., "what's the name of this shell?") and by playing computer games with them.

Parents can also help schools receive funding and support. They can donate materials, seek donations from local industries, or petition the school board for additional resources. As mentioned earlier, the children who use computers the most also watch more television and videos. Parents can monitor and limit the amount of time their children spend in sedentary activities. They should encourage regular, moderate, and vigorous activity. They can encourage physical activities by engaging in them as a family and by serving as good models by being physically active themselves.

> ***How parents can support their child's use of ICTs:***
>
> Provide them with a computer outside of school.
>
> If a computer is not available at home, consider using community resources such as public libraries.
>
> Become involved in how your child uses the computer.
>
> Monitor and limit how much time they spend in sedentary activities such as television and video games.
>
> Encourage children to engage in a variety of physical activities outside of school.
>
> Support your school's fund raising efforts to provide ICT resources for children in school.

ICT

ICT resources for teaching include:

1. Information (the internet and electronic encyclopedias)
2. Communication (e-mail and chat rooms)
3. Document creation (word processing and graphics packages)
4. Specific interactive educational packages (e.g., spelling drill games and math adventure puzzles).

The type of ICT resources teachers use depends on school and teacher goals for ICT use, curriculum demands, funding, cycles of ICT development, and gender needs. Cycles of ICT development refer to the availability of new equipment and software (and updates).

Schools must plan implementations carefully. Technology advances at a rapid pace and school equipment can quickly become antiquated and no longer work with newer systems.

Newer technologies can benefit children. For example, faster processing and internet speeds may hold children's attention better.

> ***When using ICTs at schools:***
>
> 1. Match hardware and software purchases with school and curriculum goals.
> 2. Plan resources for emerging technologies carefully.
> 3. Evaluate whether the technologies will improve performance and reduce physical and mental stress.
> 4. Plan for continued updates and replacement costs.
> 5. Include software programs for boys and girls, collectively and separately.
> 6. Specifically introduce girl-oriented resources.

Similarly, new input and output devices, such as wall-mounted displays and voice and gesture input, may provide more variety in the physical demands than today's monitor–mouse–keyboard systems. However, newer applications require faster computers and more memory. Putting new applications on older hardware leads to poor performance and reliability, resulting in frustration for the child using it!

Boys and girls have historically differed in their attitudes toward ICT. Girls were less likely to enter ICT-related jobs than boys (Newton and Beck, 1993) and girls were less satisfied with their access to technologies at school (Beynon, 1993).

Boys and girls also tend to use ICT differently. Girls use technology to communicate (e-mail and chat rooms), while boys play games. Even so, as mentioned in the section, "What can ICT use achieve?," providing girls with ICTs that reflect their interests may help reduce this gender gap.

SCHOOL WORKSPACES

"School workspace" refers to the design of classrooms, while "workstation" describes the child's work area, such as the desk, chair, computer, and keyboard. Obviously, the design of school workspaces and workstations must accommodate child users.

Workstations that do not fit children may cause discomfort (Barrero and Hedge, 2002). Adaptable workstations are important as children's sizes vary considerably within an age range.

The child user's schoolwork also affects the best workstation design. Pencil and paper tasks place different physiological and biomechanical demands on children than computer tasks (Straker et al., 2002). Figure 23.14 shows different patterns of muscle work involved when using computers and reading books.

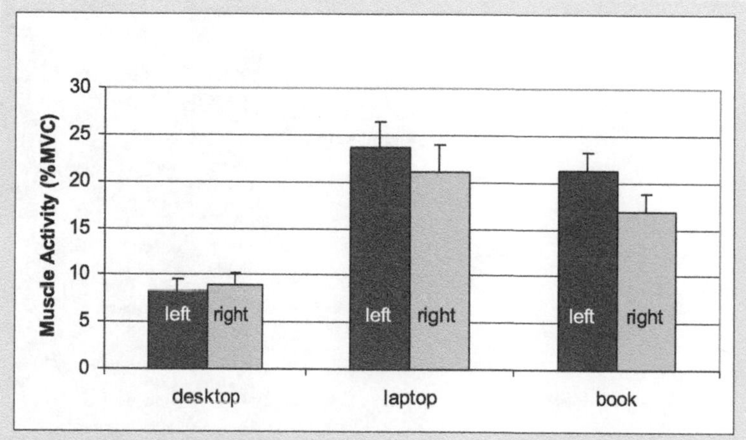

Figure 23.14 (Leon Straker)
Comparison of neck muscle activity during book reading, and reading from laptop and desktop computers.

Working at a computer poses new and different demands on children compared with traditional schoolwork. Traditional workstations allow for little postural variety when working on a computer, which increases the risk of musculoskeletal discomfort (Figure 23.15).

Therefore, traditional workstations are often not suitable for computer work unless they "fit" the child, accommodate their tasks, and allow for a range of acceptable positions (see Chapter 21 on workstation design.)

Currently, many schools have computer resource rooms that children use intermittently over the day. As schools push to integrate technologies into the curriculum, computers are appearing on students' desks. The chair and desk height may not be appropriate for keying. The desk height, combined with the monitor size, may place the monitor too high for relaxed viewing. Together with unsuitable lighting for the task, this can cause eyestrain and arm, shoulder, and neck discomfort.

Workstation layouts should foster appropriate social groupings and interactions. In short, the classroom of the future may look very different from the current classroom.

Classroom guidelines:

1. Design child workspaces and workstations for both paper-based and computer work.
2. Design teachers' workspaces and workstations for both also. Adjustable equipment is best.
3. Train teachers and students in proper positioning of their workstation equipment.
4. Instruct teachers and students to change postures periodically while they work on computers.
5. Design workstations to enable postural changes.
6. Design workspaces that allow children to work cooperatively with their peers and teachers.
7. Avoid glare and excessive screen reflections of computer monitors.
8. Teach children and teachers to vary their focus (far and near) periodically when working on computers.

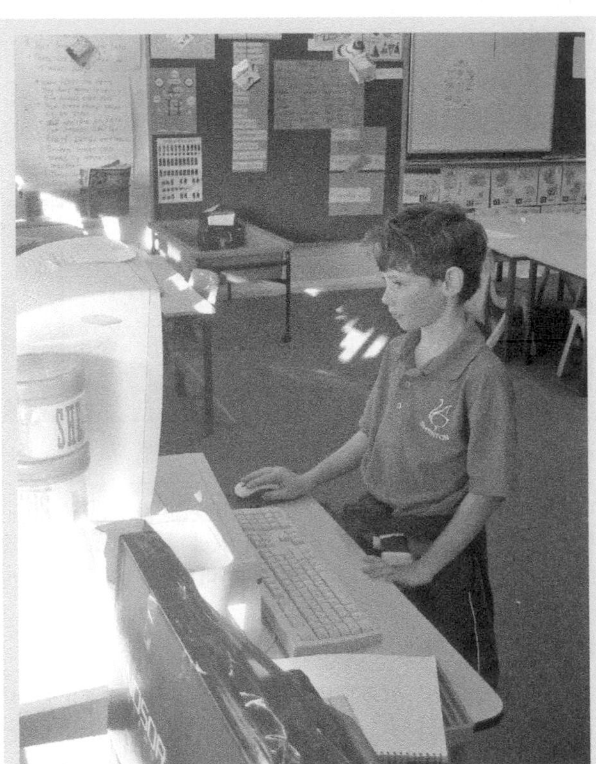

Figure 23.15 (Leon Straker)
Workspaces (and workstations) must "work" for how we use computers now and how we expect to use them in the future.

CHILDREN

As mentioned, children may be more proficient with ICTs than their parents or teachers. This may intimidate the teacher, parent, or other students. Yet it also provides children the opportunity to transfer experiences learned at home and at school.

It also enables them to assume the role of "instructor" or "assistant" when their knowledge surpasses their teachers' or peers'. If the student and teacher handle this properly, it can also help the child develop self-assurance and competence (Figure 23.16).

Recommendations for children:

1. Have fun using ICT at school.
2. Share experiences and expertise with friends and teachers.
3. Allow everyone equal access to ICT.
4. Learn to touch-type.
5. Remember to frequently change your position, take breaks and be physically active to help prevent discomfort.

When a school relies on one or two male teachers to be responsible for ICT, girls lack adequate role models for being involved with ICT at school. Because boys show more interest in computers, they may also clamor to use computers, keeping girls from their share of time on the computer.

Providing girls with encouragement and interesting tasks will help them realize they deserve computer time just as boys do. Obviously, there should be no gender bias in sharing time on the computer (Beynon, 1993).

Learning to touch-type is useful. Despite advances in voice recognition and novel input technologies, the keyboard is likely to remain a dominant input device for text.

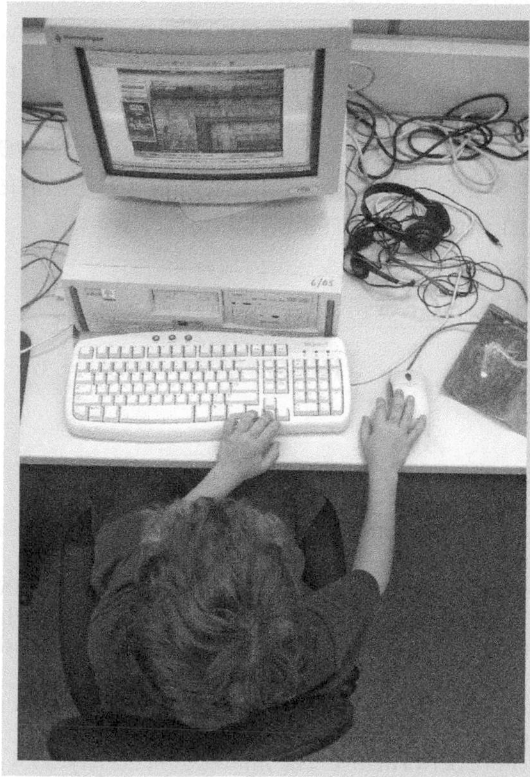

Figure 23.16 (Leon Straker) Children tend to view home-related ICT activities as 'games' and school-related ICT activities as 'schoolwork' (Somekh et al., 2002). Yet both can be fun. Teachers must challenge themselves to provide interactions that are as stimulating at school as those children enjoy in their leisure time.

FUTURE DIRECTIONS AND CONCLUSIONS

Few schools have effectively integrated technologies into all areas of learning due to financial and personnel constraints. Finances are often insufficient.

Teachers may be uncertain about their own ICT abilities and unsure of how to take advantage of new technologies. They do not always get the ICT education they need to feel comfortable and easily integrate ICT into their classrooms. Most school ICT use remains in special ICT classrooms, focused on skill acquisition instead of being part of "every-day learning."

A more comprehensive use of ICT requires that teachers change their philosophical approaches in teaching. Thus, professional development for teachers is as important as providing hardware and software in classrooms. School managers must balance financial and human resources within their technology infrastructure.

When used effectively, ICTs can promote educational achievement. It can allow for self-paced learning that enables students to progress at their own rate. It can bring information into the classroom that has never been available so quickly and elegantly.

Yet introducing ICT is not enough. Teacher and student competencies, assignments using ICT, design of workspaces and workplaces, and purchases of hardware and software take careful planning.

New technologies will continue to emerge as products aimed at adult corporate users filter down to educational use by children. These may include pocket-sized computers that incorporate digital cameras and pointing devices and are speech-activated. They will undoubtedly have wireless connectivity throughout the school. Similar access might be available while commuting and at home. There may be multiple display options with inexpensive thin, flat panels covering desks, large digital whiteboard walls in classrooms, and even community access to digital screens in public places. No matter how these new technologies play out, our children's lives will be increasingly connected and information will continue to be more complex.

We now have information on how ICT affects health, productivity, and well-being for children (as well as parents, teachers, and the broader community). It is time to put the knowledge to work as we integrate ICT into our children's schools.

RESOURCES

www.becta.org.uk
www.ed.gov/Technology

REFERENCES

Anderson, A.A. (1996). Predictors of computer anxiety and performance in information systems. *Computers in Human Behavior*, 12, 61–77.

Apple (1995). Changing the conversation about teaching, learning and technology: A 10-year report on the Apple classrooms of tomorrow project. Available at http://www.apple.com/education/k12/leadership/acot/library.html

Arrowsmith, K.M. (2001). The occurrence of flow in year 6 children's use of computers. Unpublished dissertation. School of Psychology, Curtin University of Technology, Perth, Australia.

Atkins, N. and Vasu, E. (2000). Measuring knowledge of technology usage and stages of concern about computing: A study of middle school teachers. *Journal of Technology and Teacher Education*, 8(4), 279–302.

Attewell, P. and Battle, J. (1999). Home computers and school performance. *The Information Society*, 15, 1–10.

Balague, F., Troussier, B., and Salminen, J.J. (1999). Non-specific low back pain in children and adolescents: Risk factors. *European Spine Journal*, 8(6), 429–438.

Barrero, M. and Hedge, A. (2002). Computer environments for children: A review of design issues. *Work*, 18(3), 227–238.

Bauman, A., Bellew, B., Vita, P., Brown, W., and Owen, N. (2002). Getting Australia active: Towards better practice for the promotion of physical activity. Melbourne, National Public Health Partnership.

BECTA. (2000). A preliminary report for the DFEE on the relationship between ICT and primary school standards. Coventry: British Educational Communications and Technology Agency (BECTA) www.becta.org.uk/research/research.cfm?section=1&id=538

Beynon, J. (1993). Computers, dominant boys and invisible girls: or 'Hannah it's not a toaster it's a computer'. In J. Beynon and H. Mackay (Eds.), *Computers into Classrooms: More Questions than Answers*. London: Falmer, pp. 160–189.

Brandis, H and Straker, L. (2002). The effect of touch typing skill on discomfort and performance in office workers. In A. Thatcher (Ed.), *Proceedings of CybErg 2002. The Third International Conference on Ergonomics in Cyberspace*. Johannesburg: International Ergonomics Association Press.

Cavill, N., Biddle, S., and Sallis, J.F. (2001). Health enhancing physical activity for young people: Statement of the United Kingdom expert consensus conference. *Pediatric Exercise Science*, 13(1), 12–25.

Clark, M. (1985). Young writers and the computer. In D. Chandler and S. Marcus (Eds.), *Computers and Literacy*. Milton Keynes: Open University Press.

Colley, A.M., Gale, M.T., and Harris, T.A. (1994). Effects of gender identity and experience on computer attitude components. *Journal of Educational Computing Research*, 10(2), 129–137.

Cook, D. and Finlayson, H. (1999). *Interactive Children, Communicative Teaching*. Buckingham: Open University Press.

Gos, M.W. (1996). Computer anxiety and computer experience: A new look at an old relationship. *The Clearing House*, 69, 271–276.

Grund, A., Krause, H., Siewers, M., Rieckert, H., and Muller, M.J. (2001). Is TV viewing an index of physical activity and fitness in overweight and normal weight children? *Public Health Nutrition, 4(6), 1245–1251.*

Hakala, P., Rimpela, A., Salminen, J.J., Virtanen, S.M., and Rimpela, M. (2002). Back, neck, and shoulder pain in Finnish adolescents: National cross sectional surveys. *British Medical Journal, 325* (5 October), 1–4.

Harris, C. and Straker, L. (2000). Survey of physical ergonomics issues associated with schoolchildren's use of laptop computers. *International Journal of Industrial Ergonomics, 26*, 337–347.

Harrison, C., Comber, C., Fisher, T., Haw, K., Lewin, C., Lunzer, E., McFarlane, A., Mavers, D., Scrimshaw, P., Somekh, B., and Watling, R. (2002). ImpaCT2: The Impact of Information and Communication Technologies on Pupil Learning and Attainment. A report to the Department for Education and Skills. ICT in School Research and Evaluation Series—No. 9. London: British Educational Communications and Technology Agency.

Heft, T.M. and Swaminathan, S. (2002). The effects of computers on the social behavior of preschoolers, *Journal of Research in Childhood Education*, 16, 162–174.

Januszewski, A. (2001). *Educational Technology: The Development of a Concept*. Englewood, CO: Libraries Unlimited.

Janz, K. (2002). Physical activity and bone development during childhood and adolescence. Implications for the prevention of osteoporosis. *Minerva Pediatrica, 54*(2), 93–104.

Laeser, K., Maxwell, L., and Hedge, A. (1998). The effect of computer workstation design on student posture. *Journal of Research on Computing in Education, 31*(2), 173–188.

Lankshear, C. and Snyder, I. (2000). *Teachers and Technoliteracy: Managing Literacy, Technology and Learning in Schools*. St Leonards, Australia: Allen & Unwin.

McGrane, S. (2001). Creating a generation of slouchers, *The New York Times*, 4 January, section G page 1.

Moseley, D., Mearns, N., and Tse, H. (2001). Using computers at home and in the primary school: Where is the value added. *Educational and Child Psychology*, 18(3), 31–46.

NCES (2000). National Center for Education Statistics. *Internet Access in U.S. Public Schools and Classrooms: 1994–1999*. U.S. Dept. of Education, Office of Educational Research and Improvement Report NCES 2000-086.

Newton, P. and Beck, B. (1993). Computing: An ideal occupation for women? In J. Beynon and H. Mackay (Eds.), *Computers into Classrooms: More Questions than Answers*. London: Falmer.

Noss, R. (1988). Geometrical thinking and LOGO: Do girls have more to gain? In C. Hoyles (Ed.), *Girls and Computers: General Issues and Case Studies of LOGO in the Mathematics Classroom*. Institute of Education, University of London.

Oates, S., Evans, G.W., and Hedge, A. (1998). An anthropometric and postural risk assessment of children's school computer work environments, *Computers in the Schools, 14*(3/4), 55–63.

Orleans, M. and Laney, M.C. (2000). Children's computer use in the home. *Social Science Computer Review, 18*, 56–72.

Papert, S. (1980). *Mindstorms: Children, Computers and Powerful Ideas*. New York: Basic Books.

Pate, R., Freedson, P., Sallis, J., Taylor, W., Sirard, J., Trost, S., and Dowda, D. (2002). Compliance with physical activity guidelines: Prevalence in a population of children and youth, *Annals of Epidemiology, 12*(5), 303–308.

Rogers, E. (1995). *Diffusion of Innovations*, 4th edn. New York: The Free Press.

Somekh, B., Lewin, C., Mavers, D., Fisher, T., Harrison, C., Haw, K., Lunzer, E., McFarlane, A., and Scrimshaw, P. (2002). ImpaCT2: Pupils' and Teachers' Perceptions of ICT in the Home, School and Community. A report to the Department for Education and Skills. ICT in School Research and Evaluation Series—No. 7. London: British Educational Communications and Technology Agency.

Straker, L. (2001). Are children at more risk of developing musculoskeletal disorders from working with computers or with paper? In A.C. Bittner, P.C. Champney, and S.J. Morrissey (Eds.), *Advances in Occupational Ergonomics and Safety*. Amsterdam: IOS Press.

Straker, L., Briggs, A., and Greig, A. (2002). The effect of individually adjusted workstations on upper quadrant posture and muscle activity in school children. *Work*, 18(3), 239–248.
Straker, L. and Pollock, C. (2005). Optimising the interaction of children with information and communication technologies. *Ergonomics*, 48(5), 506–521.
Tan, N.W., Saw, S.M., Lam, D.S., Cheng, H.M., Rajan, U., and Chew, S.J. (2000). Temporal variations in myopia progression in Singaporean children within an academic year. *Optometry & Vision Science*, 77(9), 465–472.
Thorndike, E.L. (1912). *Education*. New York: Macmillan.
Todman, J. and Lawrenson, H. (1992). Computer anxiety in primary schoolchildren and university students. *British Educational Research Journal*, 18, 63–72.
Van-Braak, J. (2001). Individual characteristics influencing teacher's class use of computers. *Journal of Educational Computing Research*, 25(2), 141–157.
Zandvliet, D. and Straker, L. (2001). Physical and psychosocial aspects of the learning environment in information technology rich classrooms. *Ergonomics*, 44(9), 838–857.

CHAPTER 24

RETHINKING SCHOOL DESIGN: NEW DIRECTIONS IN CHILD LEARNING

DAVID GULLAND AND JEFF PHILLIPS

TABLE OF CONTENTS

OVERVIEW

New directions in learning are having a major impact on the design of schools for learners of all ages. A standard design and list of rooms is no longer sufficient criteria for designing a school. A standard set of drawings will not apply to all schools, just as not all students can learn in the same place, at the same pace, and from the same set of standardized texts.

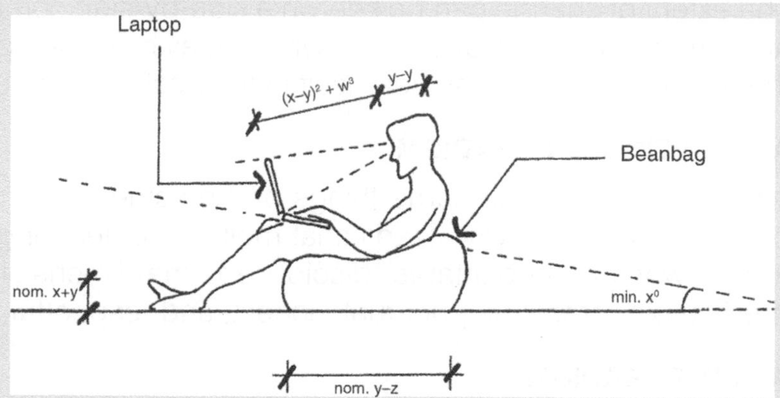

Figure 24.1 (Drawing courtesy of Hassell)
Designers need to understand how students structure their quest, that is, how students learn should determine the physical structure of a school.

Given an appropriate environment, students will construct their learning as a quest for meaning of the world around them.

The educational debate is in constant motion and in a discontinuous state, surging, halting, shifting, and swinging between views and reacting to new knowledge. Often the decisions about the design of a school are made at various points within an ongoing debate, rather than at the conclusion of the debate. Beliefs about what students should learn and how they should learn are constantly under challenge. The ensuing debate challenges traditional patterns of both education and facilities design.

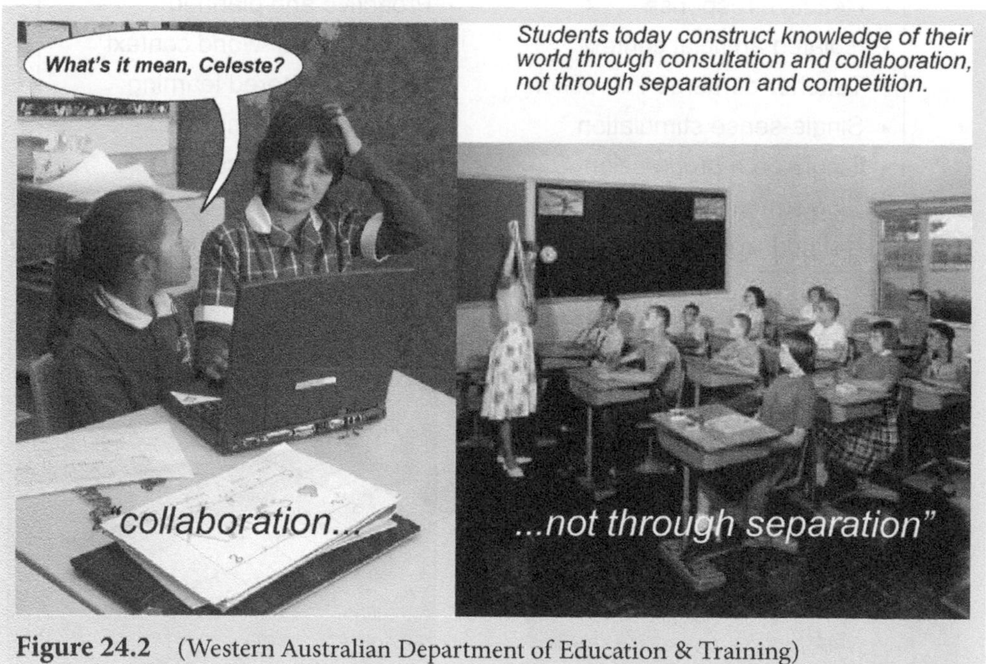

Figure 24.2 (Western Australian Department of Education & Training)

THE MODEL AND PARADIGM HAS CHANGED

The extent of change can be seen in a side-by-side comparison of the traditional paradigm and a model with new directions. While it may be easier to remain in a comfort zone between the two models, this may not result in the best design decisions for the learner.

A TRADITIONAL PARADIGM

All students learn the same things at the same time from the same people at the same place in the same way. Traditional methods of learning teach the student to be reliable, a hard worker, accountable, disciplined, one-dimensional expert, organized, productive, predictable, practical, punctual, honest, and respectful.

A NEW PARADIGM

Each student learns different things from different people at different times in different places in different ways. New directions in learning also teach the student to be a trail-blazer, motivator, knowledgeable, curious—with a desire to learn, a risk-taker, creative, honest, intuitive, nurturing, humble, flexible, to work smart, and to be a strong communicator and collaborator.

Table 24.1 How the Paradigm Has Changed

Traditional School	New Directions
• Information delivery	• Information exchange
• Passive learning	• Active, inquiry-based learning
• Factual knowledge based	• Critical thinking
• Decisions by others	• Informed decision-making
• Reactive response	• Proactive and planned
• Isolated artificial context	• Authentic real-world context
• Teacher-centered instruction	• Student-centered learning
• Single-sense stimulation	• Multisensory stimulation
• Single-path progression	• Multipath progression
• Single media	• Multimedia
• Isolated work practices	• Collaborative work practices

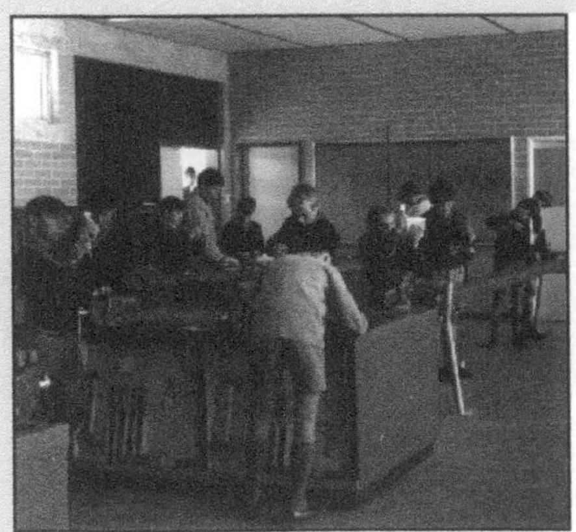

All boys work on identical tasks at the same time with the same materials and tools.

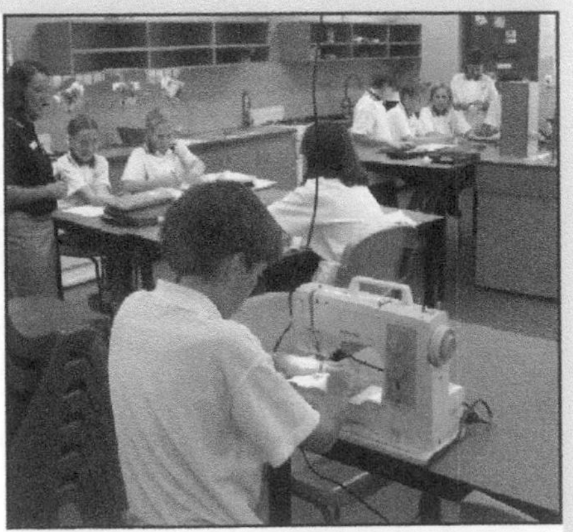

Mixed gender group of students working on individual tasks and in groups.

Figure 24.3 (Western Australian Department of Education & Training)

NO LONGER CAN ONE SIZE FIT ALL

The primary basis for the design of the nineteenth century schoolhouse was accommodating large numbers of students. The reach of the teacher's voice determined classroom size. The architectural form of the room was based on maintaining control and discipline of the large numbers of students.

By essentially reproducing the same classroom throughout the buildings structure, students and teachers were immediately familiar with how to act and where to go. Their measure of design success was how quickly students could change classes and settle into the classroom routine at the sound of the bell. Designers did not consider the impact of the physical learning environment on the learner and students were treated as a single homogeneous mass. The same designs exist in most of today's large elementary and high schools.

IMPACT OF PHYSICAL ENVIRONMENT ON LEARNING

We now know that the physical learning environment impacts the learner. The quality of the learning environment can enhance learning outcomes, as learners respond positively and joyously to stimulating spaces (Lackney, 2002). Poor designs create barriers to learning by physically isolating students from each other and giving no sense of belonging, ownership, or engagement with the space. A major paradigm shift for school designers is that schools once designed to keep people apart are now designed to bring people together.

The relationship between the process of learning and the quality of the physical learning environment is at the center of a new culture of school design. The learner's needs are paramount (Jilk et al., 2004). This supportive learning environment involves questioning traditional concepts and cultures and creating new ones.

Connection to outside areas.

Flexible classroom environment able to accomodate a range of activities.

Figure 24.4 (Photos courtesy of Hassell)

NEW DIRECTIONS

RECOGNITION OF LEARNER'S NEEDS

In identifying the learner's needs, school designers need to understand how students learn and what students should learn at each stage of development.

LEARNING CONCEPTS

Research on learning styles, brain-based learning, and multiple intelligences form the basis for concepts such as lifelong learning, student-centered learning, project-based learning, and just-in-time learning.

LEARNING CULTURE

While each learning concept is a major field of study, collectively they create a new culture of learning. Diversity in intellectual development, activities, and outcomes characterizes the new learning culture. Developmental diversities exist between individual students and also within each student as they grow and mature. A student may engage in a variety of formats for learning, depending on the topic, the environment, and their own developmental stage.

LEARNING CLIMATE

One of the major barriers to establishing a new culture within the physical learning environment is the difficulty that comes with accepting and adapting to change. Change involves added effort. Developing a climate where learning can flourish means designers must think more about the functions associated with a space, as well as the relationships between these functions and their impact on learning.

Figure 24.5 (Western Australian Department of Education & Training)
Students engaged in learning.

DESIGN ELEMENTS FOR A NEW DIRECTION

FOUR ELEMENTS FOR DESIGN

Educators and architects consider four elements when designing a new school: curriculum, pedagogy, organization, and technology. Each has undergone a major transformation in the last few decades, strongly influenced by ongoing research into how students learn.

Table 24.2 Four Elements for Design	
Curriculum	Prescribed content replaced with learning outcomes
Pedagogy	From learning by rote to just-in-time learning
Organization	From one-size-fits-all to individual programs
Technology	From limited access to ever-present access

Performance criteria describe the conditions that must be met to support the learning and teaching program for each element. The criteria identify the essential features for twenty-first century school design.

FEATURES OF A TWENTY-FIRST CENTURY SCHOOL

1. Ever-present computing with wireless networking
2. Small signature schools and school-within schools
3. Technology-intensive teaching and learning styles
4. Many areas with unprogrammed uses
5. Traditional classroom as sole center of learning de-emphasized
6. Lunch taken in a food court and not on fixed schedule
7. Common areas shared with community
8. Imaginative furniture design and variety of furniture configurations
9. Team-teaching, interdisciplinary curricula, nonchronological grouping
10. Emphasis on community service learning and school-to-career programs
11. Significant off-campus learning
12. Students creating products for business as part of their education
13. Distance learning electronic studios replace computer laboratories
14. More hi-tech production facilities with special emphasis on business, communication, and arts
15. Offer parent and community education programs in schools
16. New global partnerships with other schools and universities

During break a first grader asks her sixth grade mentor to help her email her mom to ask if she can play at a friends house after school.

Team teachers stand back to let students get on with learning.

A variety of furniture configurations support students engaged in different tasks.

A first grader learns about technology from his sixth grade mentor.

Figure 24.6 (Western Australian Department of Education & Training) Technology is "ever present".

SOCIETY'S NEEDS AND COMMUNITY EXPECTATIONS

While the design of new schools can reflect new models, most school systems have a large number of schools operating within traditional models and at all stages in between. Also, many schools have similar physical elements, but differ in terms of the needs and expectations of their communities. Obviously, today's needs and expectations both influence today's design, but they will also influence designs in the future, just as yesterday's experiences still have an impact on designs today. Designs need to be adaptable enough to include the concerns of future communities, rather than leaving them with buildings that meet needs for another time, long past.

The differences between an industrial society and an electronic society also demonstrate the contrast between today's "learning society" and that of past generations.

There is a quite a difference between being educated to learn skills to earn a living and being educated to learn skills to make a life.

A teacher helps one student. while classmates help each other.

Taking the learning outdoors stimulates the brain and energizes the body.

Participating in healthy and enjoyable activities is an essential element of every child's schooling today.

Figure 24.7 (Western Australian Department of Education & Training)
Students learn in different ways and places. In an industrial society, learning reflects the need for a society of producers, using vast quantities of natural resources. It is unsustainable economically, environmentally, and ecologically. In an electronic community, learning arises from the connectivity of ideas and information, through e-business and e-learning. It is often changing directions and never coming to a final outcome. Learning opportunities are discontinuous, expanding exponentially, and empowering.

CLASSROOM

OR... WHAT COULD A CLASSROOM LOOK LIKE IN THE TWENTY-FIRST CENTURY?

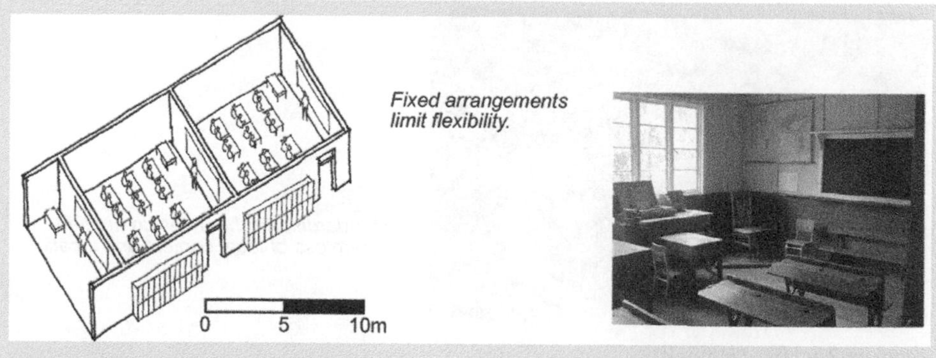

Figure 24.8 (Western Australian Department of Education & Training)

Table 24.3 From the "Old" to the "New"

Traditional Design	New Directions
• Fixed arrangement of spaces, user has little ability to change the layout	• Empowerment as the users can arrange their own environment
• Few alternatives for a range of different activities	• Flexibility, rearranging space permits different activities
• Little or no link to external space: rigid, self-contained learning boxes	• Strong link to external areas, as an extension of the learning environment
• Technology access in permanent and fixed locations	• Ever-present technology access, such as the wireless LAN—"anytime, anywhere"
• Fixed arrangement of space with few options for spontaneous gathering of groups	• Flexible display space and gathering points that modify quickly and easily
• Space emphasizes learning style of didactic, "sage on stage" approach	• Physical space accommodates a range of learning styles
• Little ability for interaction between different class groups	• Many options for different size groups to work together

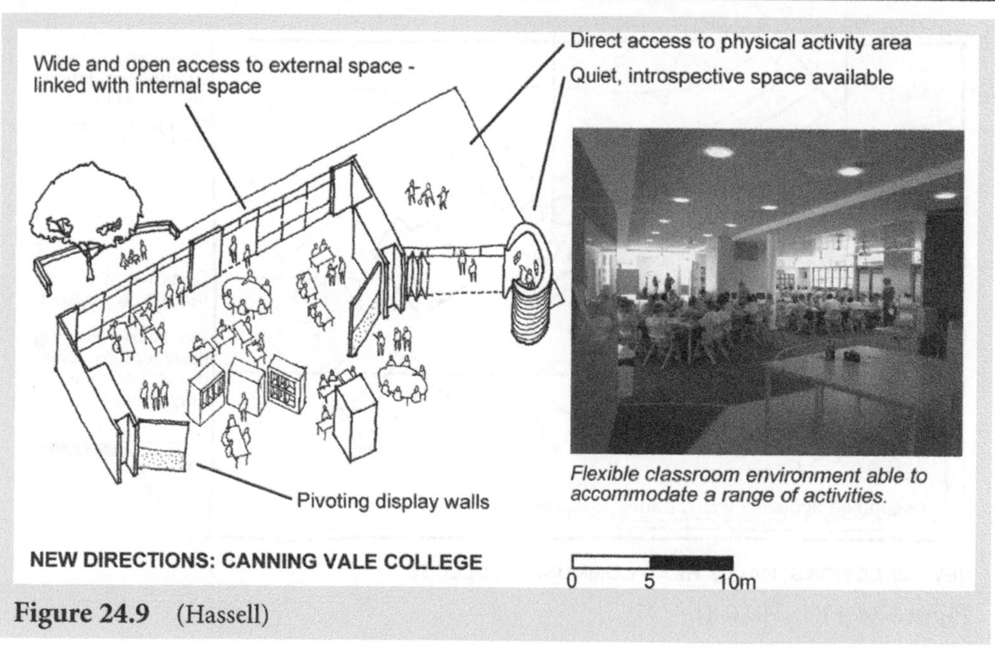

Figure 24.9 (Hassell)

MASTER PLAN

OR... DOES THE SCHOOL RESPOND TO ITS CONTEXT?

Standardised school design applied regardless of local context and location.

Figure 24.10 (Hassell)
Aerial view of a traditional school and neighborhood.

Table 24.4 Planning—for Yesterday or Tomorrow?

Traditional Design	New Directions
• The same standard modular plan in many sites	• Each site has a unique plan to suit their conditions
• Modular approach can lead to linking spaces occurring as leftover elements	• Hierarchy and linking of space is part of the plan: e.g., public/private, community/intimate
• Standard plan with fixed relative positions of buildings and facilities	• Opportunities for sharing facilities between school and community affect layout
• "Faculty-based" planning separates curriculum areas into separate fiefdoms	• Different curriculum areas and facilities linked together to increase cooperation and flexibility
• Standard building type has little flexibility to manage local micro/macro climate	• Building form complements the local micro/macro climate
• Separate, single purpose facilities limit cross-curricular learning opportunities	• Flexibility achieved through linking of buildings and designing multiple use facilities
• Standard design offers little opportunity to create local community ownership of the school	• Local community participates in the design process to create unique school identify and spirit

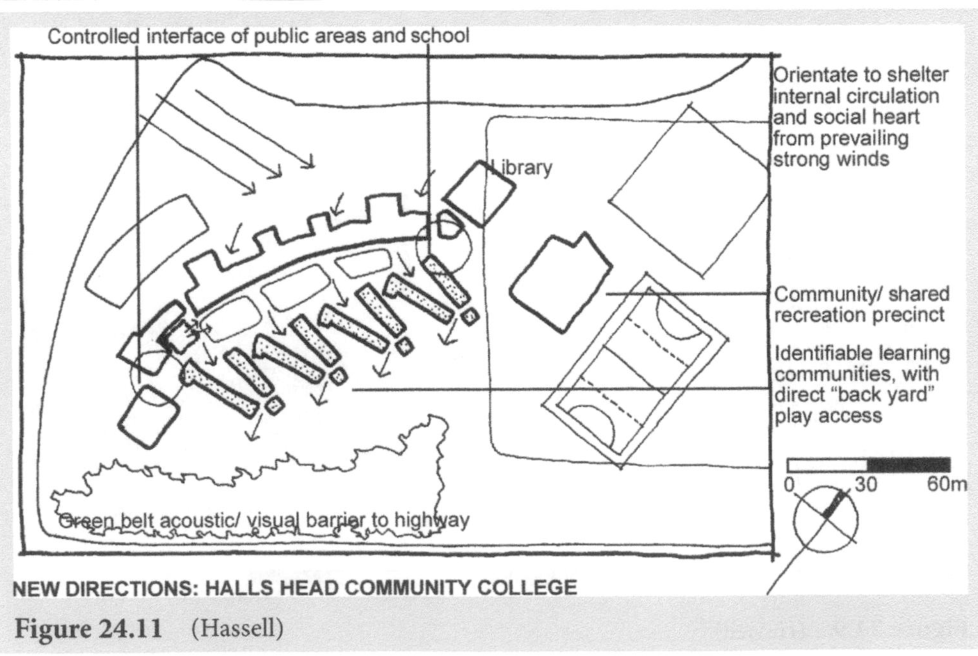

NEW DIRECTIONS: HALLS HEAD COMMUNITY COLLEGE

Figure 24.11 (Hassell)

COMMUNITY

OR... IS THE SCHOOL ENGAGED WITH ITS COMMUNITY?

Traditional: School with minimal connection to its surrounding context.

Figure 24.12 (Hassell)
Aerial view of traditional school and community.

Table 24.5 Connecting to the Community

Traditional Design	New Directions
• Different authorities and groups want separate but similar projects	• Consultation between groups to find opportunities for sharing facilities
• Minimal consultation with community on school design	• School and community participate in the design process
• Relative positions of buildings and facilities follow standard plan	• Economies of scale through combining budgets for larger, shared facilities
• School competes with other facilities for resources and patronage	• School community has direct access to larger, higher quality facilities
• "Intruder" mentality towards public, and student disassociation from "real world"	• Greater community ownership of school, with the benefits of more people watching out for each other
• Barrier-type perimeter to school with "lock down and lock out" approach	• Encourage controlled community and public access into appropriate areas
• Perception that learning and school has relevance only for school students	• More opportunity to engage students with community, and the community with learning

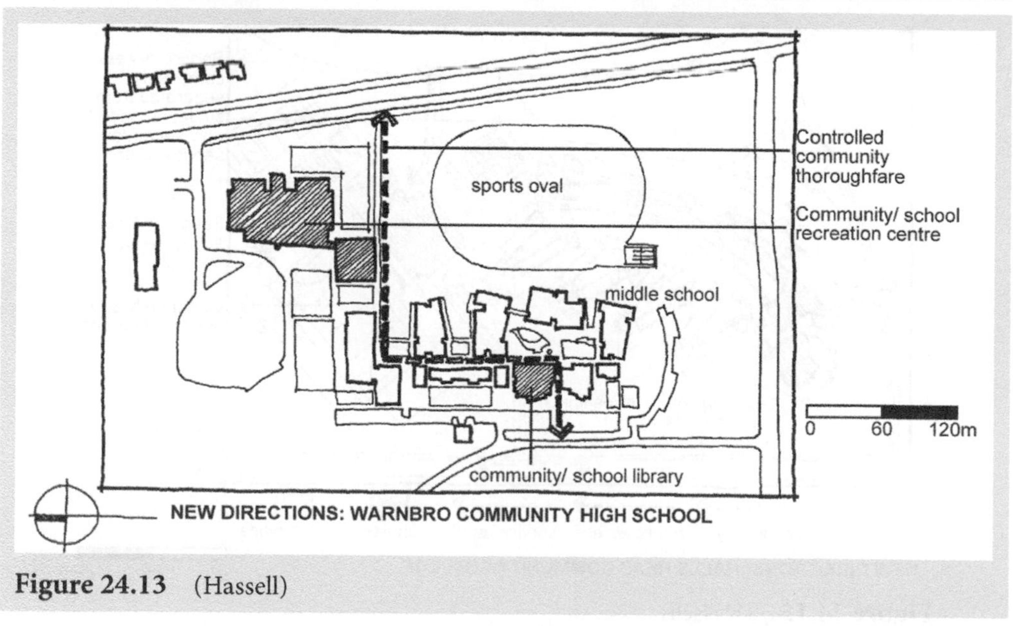

Figure 24.13 (Hassell)

CIRCULATION

OR... HAVE ALL ASPECTS OF CIRCULATION BEEN CONSIDERED?

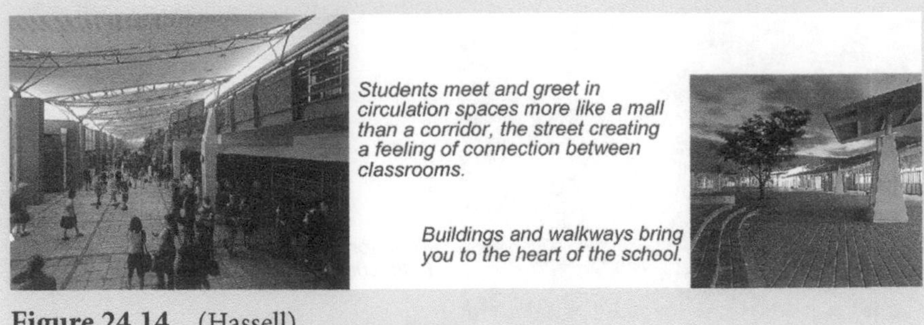

Students meet and greet in circulation spaces more like a mall than a corridor, the street creating a feeling of connection between classrooms.
Buildings and walkways bring you to the heart of the school.

Figure 24.14 (Hassell)

KEY GUIDELINES

- Consider all modes of circulation: pedestrian, bicycle, service, vehicles
- Consider safety and supervision issues
- Separate student movement and vehicular traffic
- Ensure equal access for all, accommodating disabled persons
- Make way-finding through the school obvious, intuitive, and stimulating
- Create a hierarchy of circulation patterns: primary, secondary, incidental
- Identify staff, private, and servicing movement patterns
- Consider having circulation routes contain more functions than just movement, e.g., socializing, playing
- Make circulation routes physically comfortable, provide shelter and safety

CASE STUDY EXAMPLES

- Internal focus to school "heart" combines circulation, socializing, physical and visual comfort, and safety through pathways, courtyards, shelter, and landscaping
- Primary circulation allows six students abreast under shelter with no outer columns
- Primary circulation adjoins continuous learning "street" of a range of activities
- Landscaping includes central court for internal focus and clear supervision lines
- Avoidance of "island" circulation, confusing/hard to supervise alleys
- Principles of way-finding clarify routes and separation of shared/private areas

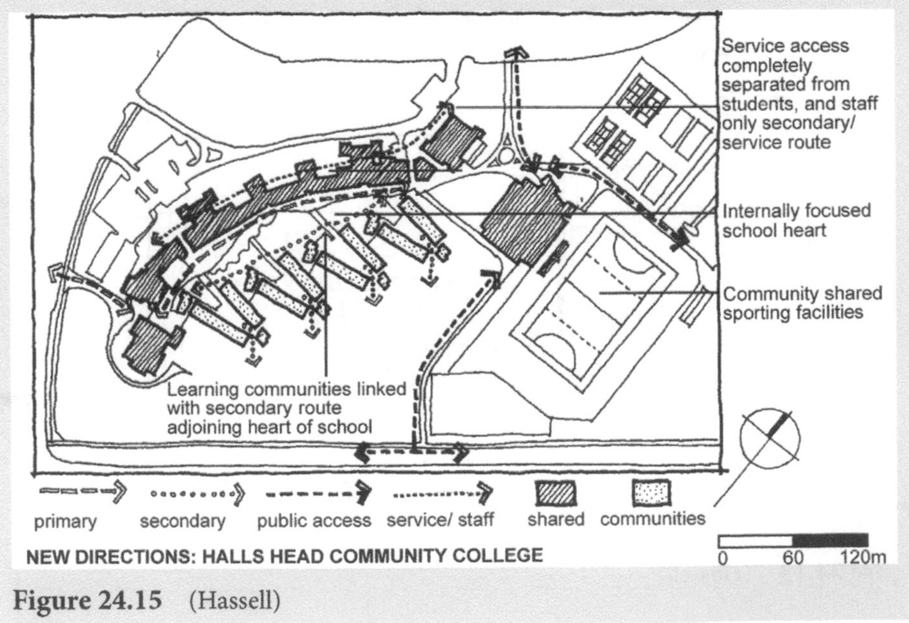

Figure 24.15 (Hassell)

ENVIRONMENT

OR... IS THE SCHOOL ENVIRONMENTALLY RESPONSIBLE AND COMFORTABLE?

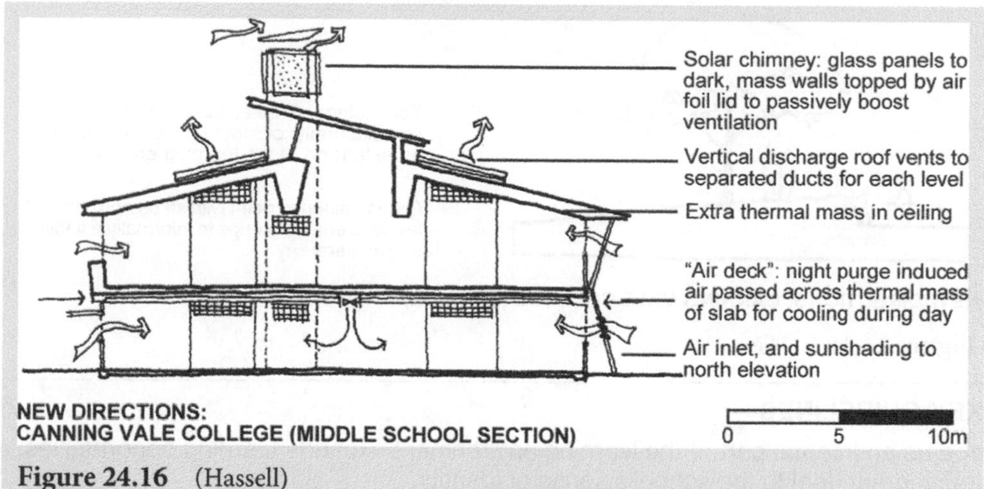

Figure 24.16 (Hassell)

SOME KEY GUIDELINES

Environmentally sustainable design (ESD) principles improve user comfort, running cost, and have the potential to act as "real world" learning opportunities.
ESD must consider a wide range of interrelated environmental issues:

- Thermal performance, including analysis of building envelope
- Lighting levels appropriate to activities, daylighting
- Natural ventilation (e.g., minimum open area 5% of floor area) and air movement
- Sunshading (particularly to windows and east/west walls) and building orientation
- Acoustic performance: sound transmission/insulation, reverberation, speech intelligibility (see Chapter 5 in this text)
- Consider the environmental impact of energy used in production of building materials
- Design response to micro- and macroclimate, such as prevailing winds, noise sources, etc.

CASE STUDY EXAMPLES

- Students control solar wall panels using venturi effect to heat/aid ventilation
- Night cooling for climates with high diurnal range, e.g., with "air deck" system
- Sun-assisted ventilation, such as a solar chimney
- Geothermal systems, stabilizing internal environments by using below ground conditions
- Wind scoops to capture cooling coastal breezes

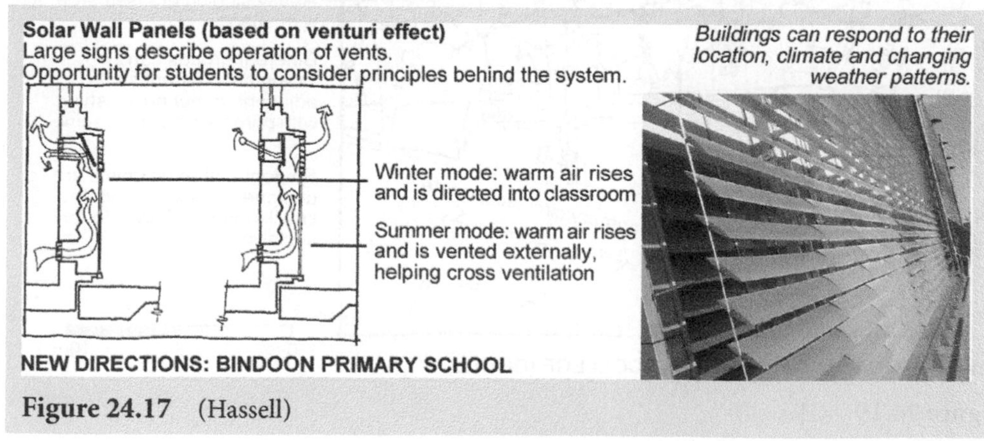

Solar Wall Panels (based on venturi effect)
Large signs describe operation of vents.
Opportunity for students to consider principles behind the system.
Buildings can respond to their location, climate and changing weather patterns.
Winter mode: warm air rises and is directed into classroom
Summer mode: warm air rises and is vented externally, helping cross ventilation
NEW DIRECTIONS: BINDOON PRIMARY SCHOOL

Figure 24.17 (Hassell)

LANDSCAPE

OR... IS THE LANDSCAPE INTEGRATED INTO THE LEARNING ENVIRONMENT?

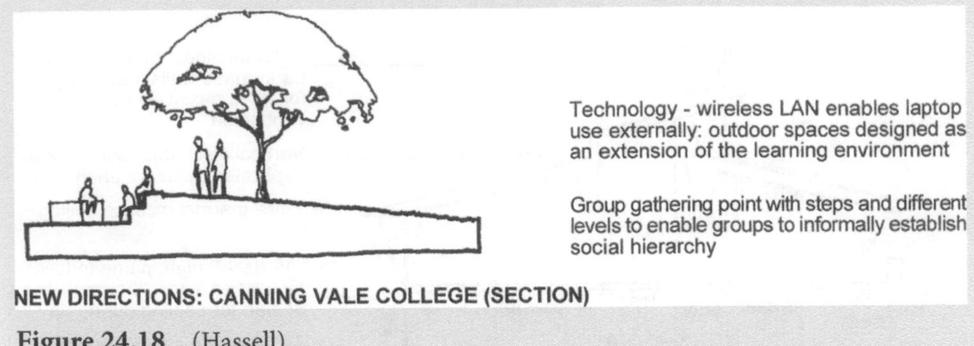

Technology - wireless LAN enables laptop use externally: outdoor spaces designed as an extension of the learning environment
Group gathering point with steps and different levels to enable groups to informally establish social hierarchy
NEW DIRECTIONS: CANNING VALE COLLEGE (SECTION)

Figure 24.18 (Hassell)

SOME KEY GUIDELINES

Landscape as an integral part of the learning environment expands learning opportunities, and contributes a great deal to the school's sense of identity.
Consider a wide range of interrelated landscape issues including:

- Select plants for safety, maintainability, variety, and minimal resource waste
- Hard and soft landscape elements create a range of spaces for a range of activities
- Landscape as an active ESD strategy, e.g., wind/acoustic buffers, sunshading, etc.
- Landscape to reinforce key circulation patterns and to aid way-finding
- Direct access to recreational landscape: particularly important for adolescent boys
- Consideration of building to landscape interface, encourage access and links
- Design for learning opportunities and stimulation of other senses (smell, touch)

CASE STUDY EXAMPLES

- Preservation/enhancement of two distinct natural environments on site as a learning resource
- Integration of natural environment and data streaming technology for an art project
- Socializing and pause points able to be "owned" by different groups
- Tree types/locations aid winter daylighting while maximizing summer shading
- Open storm water ditch turned into "living stream" in consultation with local community

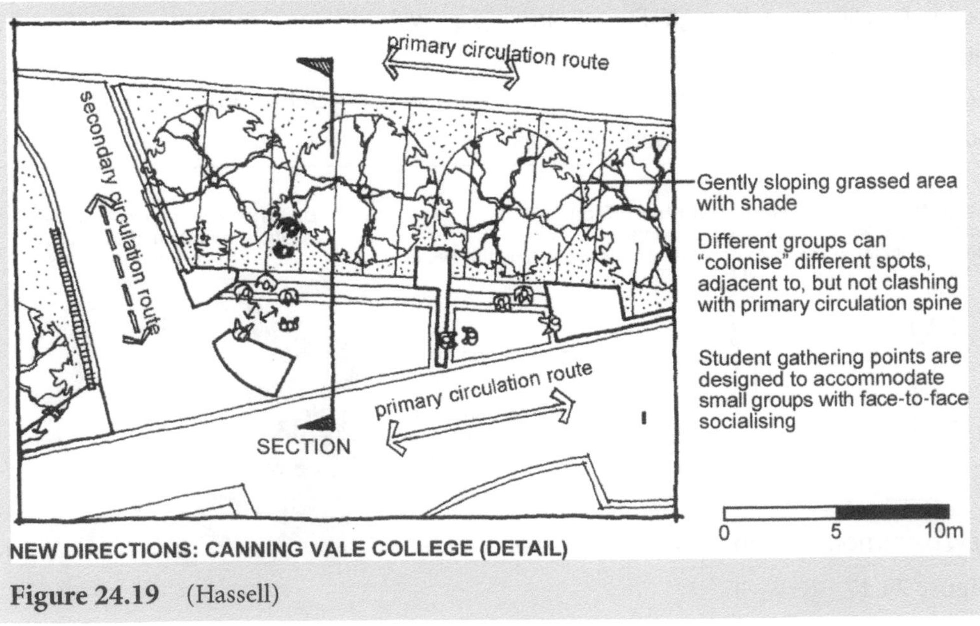

Gently sloping grassed area with shade

Student gathering points are designed to accommodate small groups with face-to-face socialising

Figure 24.19 (Hassell)

STORAGE

OR... HOW DO STUDENTS ACCESS THE RESOURCES?

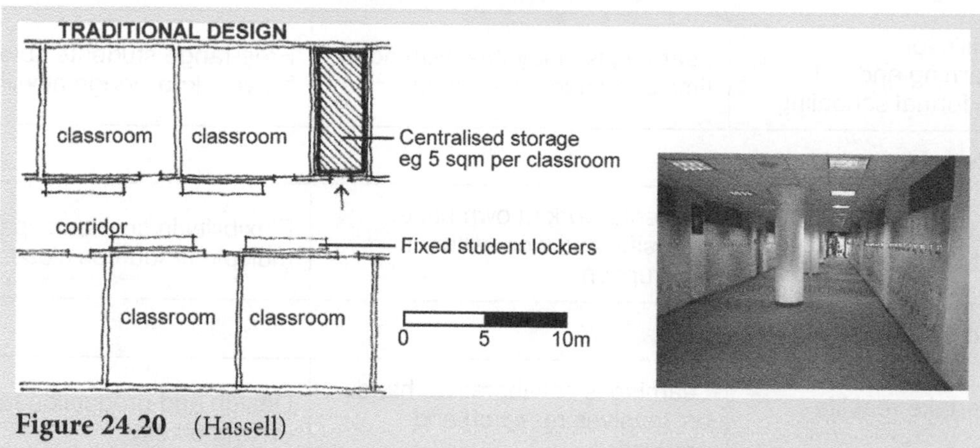

Figure 24.20 (Hassell)

Table 24.6 Access and Storage

Traditional Design	New Directions
• Storage is separate, secure, and not accessible by students without permission	• Integrated storage, easily accessed by students
• Instructor collects materials from central store for entire class before lesson	• Storage immediately adjacent to learning activities
• Standard, generic allowance of area per student, regardless of function of area	• Dispersed storage areas as part of open plan "learning neighborhoods"
• One standard type of storeroom to fit all needs	• Different types of stores, e.g., open "nooks," secure laptop trolley recharging areas
• Storeroom is a single purpose space to standard specifications	• Flexibility of spaces, e.g., storage "nooks" can become part of the learning space
• Student lockers are in fixed location often blocking visual links into learning spaces	• Mobile storage units that may be used for display or dividing up space
• Lack of storage for student "work in progress" and project work	• Student accessible storage for "work in progress" and project work

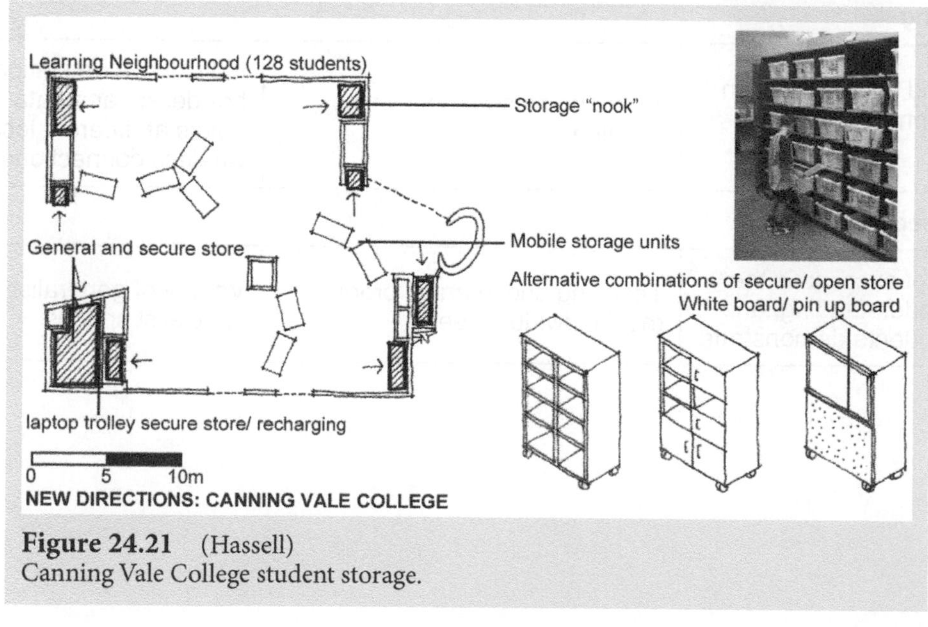

Figure 24.21 (Hassell)
Canning Vale College student storage.

Learning Concepts →	Learning Culture →	Learning Climate
Lifelong learning		
Developing skills for continuous learning and not dependent on formal schooling	Learning is enjoyable, with no time and space boundaries	Free-range students able to harvest knowledge at will
Student-centered learning		
Students have greater choice learning: what, how, and when	Students work at own pace, diversify, and work without interruption	Flexibility in time and space with multiple resources close at hand
Project-based learning		
Students undertake real life projects	Learning is collaborative, hands-on, involves research and production	Design and production facilities on and off site
Problem-based learning		
Curriculum includes student as active problem solver	Basis of learning is inquiring, cooperation, and reflection	Individual space, on-line access, small group work
Just-in-time learning		
Student motivation for learning is their own natural curiosity	Individual responses drive learning from different sources	Instant access to resources and adaptable spaces
Brain-based learning		
Conditions support natural learning processes and patterns	Learning involves the whole person, e.g., their entire physiology	Changing and stimulating environment with active and passive spaces
Collaborative learning		
Learning is interactive between peers	Location and technology connects learners	Small group areas, social chat spaces
Online learning		
Information and communication technology connects students and teachers	Time and place do not limit learning	Individual workstations, hot-desks, and data connection points at different locations, and wireless connections
Outcome-based learning		
The measure of success is the knowledge, understanding, and values that students demonstrate	Teaching and learning programs meet individual needs	Variety of generalist and specialist spaces

REFERENCES AND FURTHER INFORMATION

SCHOOLS FOR THE FUTURE

A school for the future should not only provide excellent educational facilities, but also serve as a center for the local community. There is no design blueprint for achieving this; a variety of models will emerge. This *Schools for the Future* Web page aims to provide information and inspiration to those contributing to school building projects. The site changes regularly to respond to new information and comments.
www.teachernet.gov.uk/management/resourcesfinanceandbuilding/schoolbuildings/sbschoolsforthefuture

THE CLASSROOM OF THE FUTURE™ (COTF)

The COTF program is helping to bridge the gap between America's classrooms and the expertise of NASA scientists. NASA scientists are responsible for advancing the frontiers of knowledge in virtually every field of science over the last 40 years.
www.cotf.edu/
It is a classroom, but not as we know it.
The Education Secretary, Estelle Morris, has unveiled a vision of a UK school of the future making extensive use of computer technology and classroom assistants.
http://news.bbc.co.uk/1/hi/education/1749817.stm

LEARNING AND TEACHING

UK SCHOOL BUILDING AND DESIGN UNIT

The Web site of the Schools Building and Design Unit of the UK Department for Education and Skills promotes best practices in design, use, and management of school facilities.
www.teachernet.gov.uk/management/resourcesfinanceandbuilding/schoolbuildings

WHAT IS STUDENT-CENTERED LEARNING?

www.bath.ac.uk/dacs/cdntl/e-Learning/Website/StudentCentred.html

EDUCATION IN AUSTRALIA

Online is a service that supports and promotes the benefits of the Internet for learning, education, and training in Australia. It shows Australian curricula and its tools are free to Australian educators. The funding sources are those responsible for education in Australia—all Australian governments.
www.edna.edu.au/

TECHNOLOGY AND LEARNING

Teaching & Learning Magazine produces techLEARNING.com. Their production includes contributions from hundreds of K-12 teachers, administrators, and other experts in the field.
www.techlearning.com/

CURRICULUM FRAMEWORK

The Curriculum Framework represents a major step in the reform of school curriculum in Western Australia. Their philosophies are that learning is continuous and the essential purpose of schooling is to improve the learning of all students. The Framework establishes

learning outcomes for students from kindergarten to year 12, regardless of who they are, where they are from, or which school they attend. Rather than being prescriptive, the Curriculum Framework will guide schools to develop and implement their teaching and learning programs according to the needs and characteristics of their students.
www.curriculum.wa.edu.au/

THE PHYSICAL LEARNING ENVIRONMENT

SCHOOL CONSTRUCTION NEWS

Online source for design, construction, management, and operation of schools.
www.schoolconstructionnews.com

DESIGNSHARE IS A GLOBAL FORUM ON INNOVATIVE SCHOOL DESIGN

DesignShare is the central address for the very best in educational facilities and their impact on the learning process. DesignShare provides an invaluable service as a facilitator of ideas and resources about best practices and innovation in schools from early childhood through the university level.
www.designshare.com

IMPACT OF BUILDING DESIGN ON EDUCATION

OECD study highlights the impact of building design in education.
http://www1.oecd.org/media/publish/pb01-11a.htm

THE EDUCATION DESIGN INSTITUTE

The Educational Design Institute, in collaboration with students, parents, educators, school administrators, school boards, and communities, seeks to promote and encourage the creation of safe, accessible, flexible, and developmentally appropriate learner-centered environments. The goal of these environments is to challenge and motivate student's minds, strengthen their bodies, and commit them to their communities.
www.edi.msstate.edu/

US DEPARTMENT OF EDUCATION

www.ed.gov/inits/construction/commguide.pdf

NATIONAL CLEARINGHOUSE FOR EDUCATIONAL FACILITIES

The National Clearinghouse for Educational Facilities (NCEF) is a free public service that provides information on planning, designing, funding, building, improving, and maintaining schools. Funding comes from the U.S. Department of Education. The Clearinghouse works with the U.S. Department of Education's Educational Resources Information Center (ERIC).
www.edfacilities.org/

COUNCIL OF EDUCATIONAL PLANNERS INTERNATIONAL (CEFPI)

CEFPI is a unique organization with international membership. It brings together educationalists, architects, facility planners, builders, information technologists, and other groups in the community. Their aim is to create appropriate educational facilities for all levels of learning in the twenty-first century. This organization is an advocate for excellence in student learning environments.

They acknowledge (and act on) the integral relationship between the quality of educational facilities and the level of student achievement.
www.cefpi.org/

AUSTRALASIA REGION OF CEFPI

The Australasia Region of the CEFPI is the first formed outside North America. The Australasia Region is a forum for professional development and advocacy of the importance of education facilities. Their aim is to facilitate creative and responsible planning, design construction, and renovation of schools to provide the most effective learning environment for students of all ages.
www.cefpi.org (look for the link to Australasia region).

SECTION G

Children and Public Places

CHAPTER 25

DESIGNING CITIES AND NEIGHBORHOODS FOR CHILDREN

RANI LUEDER

TABLE OF CONTENTS

OVERVIEW

Young children's neighborhoods* are central to their development, second only to their families. As children become older, more independent, and their territory expands, their immediate neighborhood tends to become less critical.
Effectively designed environments are more than functional. These may

- support child-directed and developmentally appropriate activities,
- promote children's health by continuing to shape their eating habits and physical activities as they mature,
- affect the perceived and actual safety of their environments, and
- provide children with opportunities to learn how to interact with others and develop a sense of self.

This chapter reviews environmental design considerations that may create safe, healthful, and effective neighborhoods for children.†

Figure 25.1 (Ursy Potter Photography)
In contrast to many children's highly controlled play under adult supervision, playing in streets encourage interaction with peers, spontaneity, playing many different games, and allowing children to become familiar with their environment (Brougere, 1991).

* This chapter uses the word ***neighborhood*** loosely to describe the general physical boundaries of a community that interact, are in some ways interdependent, and share a sense of identity. (*s.f.* McMillan and George, 1986). Such neighborhoods are generally part of a city, suburb, or large rural area.

† Excellent reviews of this research are available from Evans et al. (2000, 2001, 2003); Leventhal and Brooks-Gunn (2000); Stokols and Altman (1991); and Frank et al. (2003).

CHILDREN AND PLAY*

It is unfortunate that children can't design their outdoor play environments. ... their creations would be completely different from the areas called playgrounds that most adults design for them. Outdoor spaces designed by children would not only be fully naturalized with plants, trees, flowers, water, direct sand, mud, animals, and insects, but also would be rich with a wide variety of play opportunities of every imaginable type... they would be rich developmentally appropriate learning environments where children would want to stay all day.

(Randy White and Vicki Stoecklin)

Figure 25.2 (Ursy Potter Photography) Young people experience their world in an *environment of relationships*, of critical actions and interactions that continue to shape their physical, emotional, social, cognitive, behavioral, and moral development (NSCDC, 2004).

Children's play helps them develop physically, mentally, and creatively. Through play, children learn about their surroundings, their social relationships, and their place in the world. They need environments that continue to support and challenge them to progress to the next stage of development[†] (CPSC, 2002).

- Play helps them develop *intellectually* through creativity, discovery, language and verbal exchanges, problem solving, and abstract thought.
- Play fosters *emotional* development as children experience happiness, spontaneity, humor, and control over their environments. They learn to be aware of their emotions, sensitive to others and develop a sense of self.
- Play helps children develop *socially* by learning appropriate behaviors with others.[‡]

Figure 25.3 (Ursy Potter Photography) Children's activities are varied and change with age; the most effective environments provide them with a range of choices (Sallis and Glanz, 2006). Locate children's areas and public restrooms near staffed areas to make it easier to observe them (CPTED, 2005).

Yet research from various countries indicates that children today are considerably more restricted than previous generations[§] (Chawla, 2005; Cobb and Danby, 2005; Hillman et al., 1991; Karsten and van Vliet, 2006; Pooley et al, 2005). They are afforded less independence, more oversight, and have fewer opportunities for free play and physical activity.

* Additional information on children and play is provided in Chapter 2 (Child Development); Chapter 7 (Children and Exercise); Chapter 8 (Children and Injuries); Chapter 11 (Products for Children); Chapter 12 (Children and Toys); and Chapter 28 (Children and Playgrounds).

† Valentine and McKendrick (1997) provide an excellent review of the many dimensions of play and its essential role in child development.

‡ Play spaces for children also encourage their use by the family. Facilities provide cues as to the uses of the space. Such actions promote the community as well as safety (CPTED, 2005).

§ Researchers often refer to children's freedom to make their own decisions as *governance* (Bartlett, 2005).

> *United Nations Rights of the Child*
>
> 1. ...recognize the right of the child to rest and leisure, to engage in play and recreational activities appropriate to the age of the child, and to participate freely in cultural life and the arts
> 2. ...respect and promote the right of the child to participate fully in cultural and artistic life and shall encourage the provision of appropriate and equal opportunities for cultural, artistic, recreational, and leisure activity.
>
> *United Nations Convention on the Rights of the Child (Article 31)* (www.ohchr.org)

Figure 25.4 (Ursy Potter Photography)
Children need challenges, but with these challenges come risk. As communities focus on eliminating risk in children, they also remove the challenges that are important for their development.

For example, one British study found that while 80% of 7- and 8-year-old children in 1971 walked to school, fewer than 10% did so two decades later (Hillman et al., 1991). Such findings are a cause for concern over the potential implications on play-related behaviors that are necessary for children's development.

CHILDREN IN STREETS

Being struck by cars represents a leading cause of child injuries and fatalities. Young children have difficulty evaluating distances and vehicle speeds and recognizing when it is safe to cross a street, particularly in their first 9 or 10 years. They are impulsive and act before they think—and still do not recognize the consequences of their actions (TCE, 2006).

Parents may also overestimate their child's abilities; many elementary school children are still unable to understand traffic signals or foresee driver behaviors.* Their smaller size makes them difficult for drivers to see, especially if they are standing between parked cars on the side of the road (TCE, 2006). Danger is greatest where there is heavy or fast

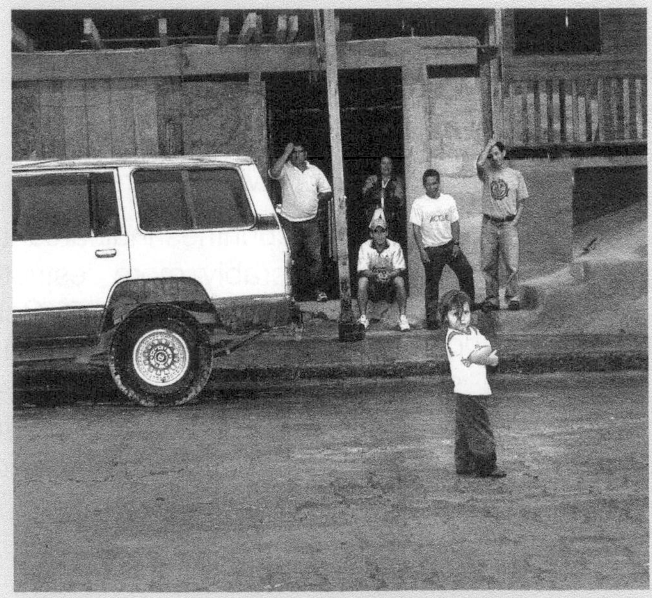

Figure 25.5 (Ursy Potter Photography)
Children are most often hit by cars when playing near home. They often run into the street in the middle of the block, where drivers do not expect them (TCE, 2006).

* Drivers and child pedestrians may each incorrectly assume the other will yield to them.

traffic, where visibility is limited, or where drivers are distracted and fail to notice pedestrians and cyclists on the road (DSTI, 1998).

Children require environments that allow them to move freely in protected surroundings. Environments can also protect children by promoting safe driver behaviors (DSTI, 1998).

One important way the built environment can reduce the risk of injury to children is by slowing traffic down* (*s.f.*, Ewing, 1999). A New Zealand study found that at the busiest sites the risk of injury to children was 14 times greater than at the least busy sites. High-density curb parking was also associated with an eightfold risk (Roberts et al, 1995).

This underscores that environmental interventions such as restricting curb parking, stoplights and stop signs, speed bumps, warnings, and more traffic cops may significantly reduce risk to children (Ewing, 1999; Roberts et al., 1995). The effectiveness of any interventions will vary; characteristics of streets, neighborhoods, and traffic varies, and the environment should reflect these differences (Ewing, 1999)

The design of residential driveways may also put children at risk. Children who live in homes where the play area is not separated from the driveway by a fence or a gate were 3.5 times more likely to be injured than other children. Living in homes with shared driveways also more than tripled the risk of child injuries† (Norton and Jackson, 1995).

Child crash injury facts:

1. Most preschooler injuries are near their home or on their own street.
2. Most crashes involving preschool children happen between 3 and 6 p.m.
3. Most crashes involving preschoolers occur in fair and warm weather.
4. Preschool boys are twice as likely to be injured as preschool girls.
5. One study in New York city found that most crashes involving children occurred while drivers were driving straight on the road (DiMaggio and Durkin, 2002).

   Younger children (aged 5–9) were more likely to be hit at the middle of the block during the day, while adolescents were more likely to be hit at night and at intersections.

   Motor vehicle accidents involving children under the age usually took place in driveways, rather than streets.

Note: Source for the first four points in this text box: The National Highways Traffic Safety Administration. Source for the last point in the text box: DiMaggio and Durkin (2002).

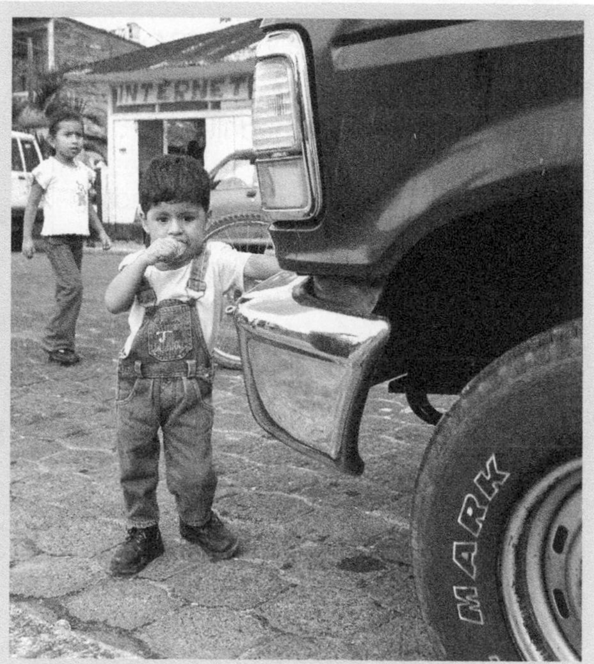

Figure 25.6 (Ursy Potter Photography)
Young children's smaller size makes them difficult for drivers to see, especially if they are standing between parked cars on the side of the road.

* Ewing (1999) provides an excellent review of efforts in different countries to slow traffic (*traffic calming*).

† The typical driveway-related child pedestrian injury involved a 2-year-old child hit by a car driven by a friend or relative as they were reversing out of the driveway (Norton and Jackson, 1995).

Common crash situations

- Child darts out into street at corner or mid-block
- Vehicle turns into the path of a child
- Child hidden by an ice cream truck
- Child hidden by bus and driver does not stop
- Vehicle backing up on a roadway, driveway, or parking lot

National Highway Safety Administration: NHTSA.gov

Figure 25.7 (Ursy Potter Photography)
This adolescent runs after a ball under the car without verifying that it will remain stationary. Each year, many children are injured or killed when they run into the path of a car. Children who sustain traffic injuries are more likely to incur major and minor head trauma than all other types of injuries (Durkin et al., 1999).

Rules of school bus safety (NHTSA)

1. Be at the bus stop at least 5 minutes before the bus is scheduled to arrive.
2. When the bus approaches, stand at least three giant steps (6 ft/183 cm) away from the edge of the road, and line up away from the street.
3. Wait until the bus stops, the door opens, and the driver says that it is okay before stepping on the bus.
4. When crossing the street in front of the bus, walk on the sidewalk or side of the road to a point at least five giant steps (10 ft/305 cm) ahead of the bus before you cross. Be sure the bus driver can see you, and you can see the driver.
5. When exiting the bus, be careful that clothing with drawstrings and book bags with straps are not caught in the handrail or door.
6. Never walk behind the bus.
7. Walk at least three giant steps away from the side of the bus any time you are walking beside the bus.
8. If you drop something near or under the bus, tell the bus driver. Never try to pick it up. The driver may not see you and begin to drive away.

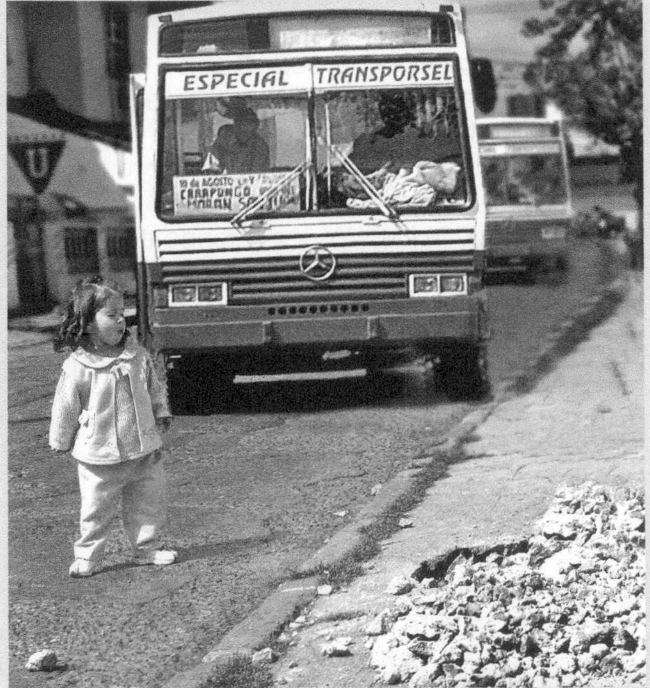

Figure 25.8 (Ursy Potter Photography)
Young children have difficulty evaluating distances and vehicle speeds and recognizing when it is safe to cross a street, particularly in their first 9 or 10 years. They are impulsive and act before they think, and still do not recognize the consequences of their actions (TCE, 2006).

COMPLETE STREETS

There has been a growing concern over urban developments that emphasize the convenient movement of vehicles over other needs in the community. The impact is profound in the United States; injury rates and fatalities are much higher in the United States than in pedestrian-friendly Western European countries such as Germany and the Netherlands.

This has generated interest in the ***Complete Street*** movement in the United States, which focuses on the comfort, safety, and functional requirements of pedestrians and cyclists in addition to motorists—particularly in *walkable neighborhoods* that promote health and community (Sallis and Glanz, 2006).

The streets of our cities and towns ought to be for everyone, whether young or old, motorist or bicyclist, walker or wheelchair user, bus rider or shopkeeper. Too many of our streets are designed only for speeding cars.... They are unsafe for people on foot or bike—and unpleasant for everybody.

www.completestreets.org

In the first half of the last century, communities were, by necessity, designed around people's need to walk to school, shop, and work. Such "traditional" neighborhoods were interconnected, complex, and typically involved moderate to high-density populations. Areas and buildings were used in a variety of ways. Streets were laid in interconnected grid patterns that facilitated pedestrian movements.

In recent decades, communities expanded to suburbs that emphasized automobile travel. Zoning codes separated ways of using land that were too far to walk. Low-traffic residential streets fed into multilane, high-speed streets and freeways that are dangerous for pedestrians and discourage walking.

(Sallis and Glanz, 2006)

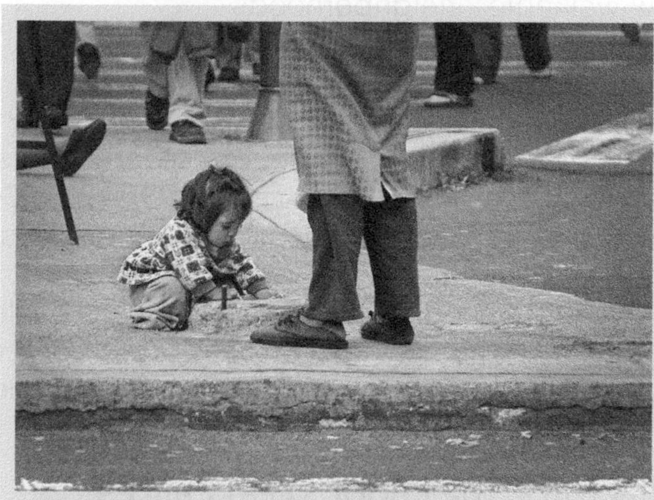

Figure 25.9 (Ursy Potter Photography)

WALKABLE NEIGHBORHOODS

Older cities and neighborhoods were once designed to increase the convenience of pedestrians by interconnecting commercial and residential areas. In contrast, urban planning has become more closely associated with urban sprawl, designed to increase the convenience of motor vehicle traffic. Work, home, school, and stores were separated by miles, and this discouraged walking.

This shift in developmental planning is cause for concern due to the adverse health consequences of sedentary lifestyles and the adverse effect on community life.

Yet research suggests that neighborhood design features can promote or discourage physical activity. These include the presence of sidewalks, street-crossing aids, bike lanes,

Improving public health aspects of the built environment that would ultimately help decrease obesity, cardiovascular disease, and asthma while improve mental health are potential benefits of living in sustainable communities.

National Institute of Environmental Health Sciences (NIEHS): www.niehs.nih.gov

closeness of facilities, density of housing, availability of public transit, pedestrian and bicycle facilities, community and recreational facilities, public parks, hiking trails, and open areas. Further, lack of sidewalks, long distances to schools, and the need to cross busy streets discourage walking and biking to school.

Figure 25.10 (Ursy Potter Photography)
This family must walk single file because there is no sidewalk and the street is ill suited for pedestrian travel. More sidewalks and improved street crossings have been found to encourage walking to school.

Likewise, heavy traffic reduces the likelihood that children will walk and may thus keep children indoors and sedentary. Residents of low-walkable neighborhoods spend more time driving than walking, so community design will likely similarly affect children's weight. People who perceive that there is heavy traffic and those who must cross several roads to get somewhere were also associated with being less active (Sallis and Glanz, 2006). Research consistently indicates that boys are more active than girls, suggesting special efforts may be needed to involve them (Sallis et al., 1999).

Neighborhoods that are perceived as unsafe were independently associated with overweight in first-graders than those who considered their neighborhood more safe* (Lumeng et al., 2006). Without safe places to play near home, children may spend more time being inactive indoors.

KIDS WALK-TO-SCHOOL SAFETY RULES

Walkable principles

- Encourage children to walk in groups
- Organize responsible adults to accompany children walking
- Teach and practice safe pedestrian behavior on the walk
- Empower neighborhoods to identify and create safe routes to school
- Collaborate with local police to enforce speeds in and around the neighborhood and school
- Provide crossing guards for children on their way to and from school

* This effect remained after controlling other factors. Children were considered overweight if they exceeded the 95th percentile body mass index (BMI) for the child's age, sex, and height (Lumeng et al., 2006).

Figure 25.11 (Edwin P. Ewing, Jr.)
One of the first "Walk to School Programs" in 1988. These efforts aim to promote physical activity for children and adults in neighborhoods, reduce traffic emissions, and prevent crime by increasing neighborhood interactions.

Table 25.1 Teach Older Children to Cross the Street Safely[a] (NHTSA)

1. *Supervise Preschoolers at all Times*	• Do not allow preschoolers to cross the street alone • Teach them who can help them cross the street safely
2. *Teach by Explaining*	*Explain to your child the safe way you cross a street* • Say: "When I cross a street, I always stop at the curb. • I look for cars. I look left for any traffic coming, and then I look right for traffic coming that way. Then I look left again. • When it is clear, I cross the street and keep looking left and right."
3. *Teach by Example*	*When you cross a street with your child, always:* • Stop at the curb • Look LEFT-RIGHT-LEFT for traffic in all directions • Stop to look around parked cars or other objects that block the view of traffic • Make eye contact with drivers to make sure they see you • Cross when it is clear • Keep looking for cars as you cross
4. *Encourage your Child*	• As you both safely cross the street together, praise the child for copying your safe actions or words • Practice what you teach at all times

[a] up to the age of about 10, young children still need to be supervised around traffic.
For more on this and related information, see the National Highways Transportation Safety Administration Web site: www.nhtsa.doc.gov.

BICYCLING

Figure 25.12 (Ursy Potter Photography)
Many bicycle injuries are associated with not wearing helmets, motor vehicle incidents, and unsafe riding environments. Communities can reduce the risk of such injuries by ensuring that roads are properly designed, maintained, and that there are separate paths for biking (Thompson and Rivara, 2001).

Figure 25.13 (Ursy Potter Photography)
Helmets are the most effective way to prevent head injuries and bicycle-related fatalities. These reduced the risk of head and brain injury by 69%–74% for all age groups, whether or not these involved motor vehicles* (Thompson et al., 1996; see Chapter 11 for more on helmets).

Figure 25.14 (Ursy Potter Photography)
Children are particularly vulnerable as their abilities to deal with traffic evolve with age and remain severely limited in the first 9 or 10 years of their life. It is not enough to instruct children how to behave. They are at risk in any situation that involves heavy or fast traffic, limited visibility, or when drivers focus their attention on other vehicles and tend to forget about pedestrians and cyclists.

* It should be noted that many sources report that helmets are 85% to 88% effective in mitigating head and brain injuries in bicycle incidents, based on an earlier study by Thompson et al. (1989). However, these earlier findings were less well controlled than their subsequent study.

Table 25.2 Talking Points on Biking for Parents (NHTSA)	
Be Smart	• Always wear an approved bicycle helmet that meets current crash safety standards, and make sure it is properly fitted • Road rash and broken bones heal; brain damage is often permanent • Consider what you are investing in your college education • Consider your future with a head injury • Wearing a bicycle helmet is common sense
Be Thorough	• Check and maintain your bicycle regularly • Keep it clean, and make sure all nuts and bolts are tight • Your brakes must work • If you prefer not to do your own bicycle maintenance, most bicycle retailers offer checkups for a modest fee
Be Seen	• Fluorescent and brightly colored clothing and helmets help make you more visible • Reflective clothing makes you easier to see at night • If you choose to ride at night, remember that a headlight, taillight, and reflectors are vital to your safety and are required by law in most states
Be Safe	• The most important part of safe bicycling is predictability • When you are on a bicycle, you are part of vehicular traffic and must obey the same laws • Ride your bicycle with traffic, not against it • Riding on the left-hand side of the road, which is illegal, is a common cause of crashes. It puts you in an unexpected position for motorists, pedestrians, and other cyclists • Always scan your environment when approaching intersections, and obey stop signs and traffic signals
Be Aware	• Motorists often find it difficult to see bicyclists • Never assume that they see you • Try to be visible by communicating with hand signals and establishing eye contact • If you are in doubt about a motorist's intentions, be prepared to yield in the event that the motorist does not see you • Control your speed, and watch for approaching vehicles turning left into your path • Scan the road for hazards such as potholes and drainage grates • Identify hazards in time to avoid them without swerving into traffic
National Highway Safety Administration (NHTSA): www.nhtsa.dot.gov	

SAFE COMMUNITIES

CRIME PREVENTION THROUGH ENVIRONMENTAL DESIGN

Crime Prevention through Environmental Design (CPTED) is a multidisciplinary design-centered approach that aims to prevent behaviors that lead to crime.

Design can reinforce the shared dimensions of a community that looks out for each other and feels a sense of responsibility for others, such as by defining areas for the community with different degrees of public and private space to encourage a sense of ownership, participation, and surveillance.

Figure 25. 15 (Ursy Potter Photography)
This area does not follow the principles of *Crime Prevention Through Environmental Design* because the heavy landscaping and barriers isolate the boy from the sight of residents nearby.

Britain and the Netherlands have designed neighborhoods that strictly control or block street traffic or that orient homes toward landscaped common areas in order to protect children and pedestrians and reinforce a sense of community without isolating them from surrounding neighborhoods (Marcus and Sarkissian, 1988).

Some community designs found a compromise between the needs of drivers and pedestrians by segregating traditional streets. In the 1960s, the Netherlands instituted a form of *shared streets* (referred to as "*Woonerfs*"), where children and adults are free to use the entire street to walk, bike, and play. Car drivers, on the other hand, are considered guests and must proceed with extreme caution (Gill, 2006; Hubsmith, 2006). The designs of these streets (through paving, landscaping, and orientation) clearly demonstrate that the street is primarily for pedestrians. One of the benefits of this design is that it is safer as it causes traffic to slow to a walking speed (Marcus and Sarkissian, 1988).

- Jane Jacob's (1961) book *The Death and Life of Great American Cities* argued that "urban renewal" policies were destroying the complex fabric of inner city communities, particularly by creating isolated and unnatural commercial and industrial areas.
  She wrote that designing smaller scale neighborhoods that reflect the complex interacting nature of a community might promote safety.
- Oscar Newman's (1972) book *Defensible Space* described how building design features could discourage a sense of identity, social control, and ownership.
- C. Ray Jeffery's (1971) book *Crime Prevention through Environmental Design* provided a model for using environmental design to prevent behaviors that lead to crime.

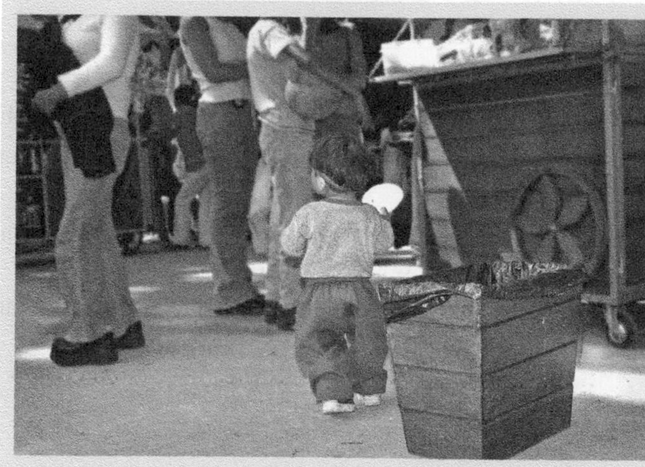

Figure 25.16 (Ursy Potter Photography) Areas that limit visibility are associated with greater risk.

Table 25.3 Principles of Crime Prevention through Environmental Design (CPTED, 2005)

Natural Access Control	• Using the placement, design, and orientation of entrances and exits, lighting, landscaping, and barriers to define and control public and private space to discourage behaviors associated with crime.	• Define entrances to the site and each parking lot with landscaping, architectural design, or symbolic gateways. • Block off dead-end places with fences or gates. • Use devices that automatically lock upon closing on common building entrances. • Provide good illumination in hallways. • Allow no more than four apartments to share the same entrance; individual entrances are recommended. • Centrally locate elevators and stairwells where many users can watch them. • Limit access to buildings to only one or two points. • If pay telephones must be installed in higher crime areas, use rotary dial phones that can only make outgoing calls.
Natural Surveillance	• The physical design of the environment encourages positive social interactions in public areas and discourages criminal behaviors by creating a perception that others can see them and that there is limited opportunity for escape.	• Design buildings so that exterior doors are visible from the street or by neighbors. • Use good lighting at all doors that open to the outside. Install windows on all four facades of buildings to allow good surveillance. • Assign parking spaces to residents. Locate the spaces next to the resident's unit, but not marked with their unit number. This makes unauthorized parking easier to identify and less likely to happen. • Designate visitor parking. • Make parking areas visible from windows and doors. • Adequately illuminate parking areas and pedestrian walkways. • Position recreation areas (pools, tennis courts, clubhouses) to be visible from many of the units' windows and doors. • Screen or conceal dumpsters, but avoid creating blind spots or hiding places. • Build elevators and stairwells in locations that are clearly visible from windows and doors. • Allow shrubbery to be no more than 3 ft high for clear visibility in vulnerable areas. • Site buildings so that the windows and doors of one unit are visible from another (although not directly opposite). • Construct elevators and stairwells to be open, well lighted, and not enclosed behind solid walls. • Place playgrounds where they are clearly visible from units, but not next to parking lots or streets.

(continued)

Table 25.3 Principles of Crime Prevention through Environmental Design (CPTED, 2005) (Continued)

Territoriality	• Redefining space into public, semi-public, and private space with the design of buildings, pathways, barriers, signs, lighting, and landscape. This promotes a sense of identity and ownership that encourages proper maintenance, watching over, and challenging people who invade their territory. It encourages the identification of suspicious outsiders.	• Define property lines with landscaping or decorative fencing. • Use low shrubbery and fences to allow visibility from the street. • Accentuate building entrances with architectural elements, lighting, and landscaping. • Clearly identify all buildings and residential units using street numbers that are a minimum of 3 in wide and well lit at night. • Try to locate individual locking mailboxes next to the appropriate units.
Maintenance	• An emphasis on maintaining property may discourage crime by reflecting a commitment to community and ownership.	• Maintain all common areas to very high standards, including entrances, esplanades, and right-of-ways. • Prune trees and shrubs back from windows, doors, and walkways. • Use and maintain exterior lighting. • Strictly enforce rules regarding junk vehicles and inappropriate outdoor storage. • Disregard above rules that make a site appear not cared for and less secure.

WAYFINDING

Children are explorers by nature and can become easily lost. Although children vary in their special knowledge of their environment, their wayfinding abilities improve with age.*

Figure 25.18 (Ursy Potter Photography) A mother shows her son the beginning of a simple cognitive map depicting the relationship between important landmarks.

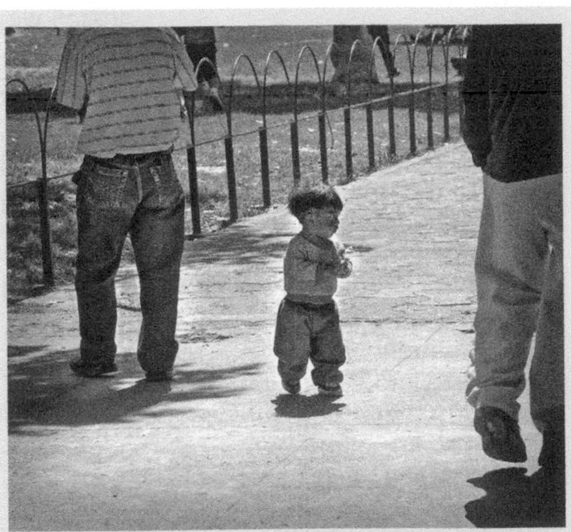

Figure 25.17 (Ursy Potter Photography) Most parents in the United States express confidence with their communities. 84% of parents believe that their children are safe in their neighborhood; 88% consider their children safe at school.

* Please see Chapter 26 for a more detailed review of children and wayfinding.

Differences are most pronounced between 10 or 12 years of age and in younger children. Yet 5-year-old children are able to demonstrate spatial abilities and children can be taught to improve their ability (Lehnung et al., 2003; see also Chapter 26 on wayfinding in children).

Figure 25.19 (Ursy Potter Photography)
Parents can help children expand their "home range" by teaching them to identify specific landmarks they can use to orient themselves from every direction. Important landmarks are unique, distinct, easy to identify, convey orientation and distance to help them localize their orientation when they find themselves in unfamiliar territory.

This ice cream cart would make a poor landmark because it does not maintain a consistent orientation (see Chapter 26 on wayfinding for children).

PUBLIC HAZARDS

Boys are more likely than girls to sustain fall-related injuries. While preschool children are more likely to fall from windows, older boys are more prone to falling from hazardous play areas (AAP, 2001; see also Chapter 8 on preventing fall injuries in children).

Children have age-specific climbing behaviors.*

- Until their sixth year, children are just starting to learn to climb yet lack strength to climb high objects. They need to be protected from truly dangerous and extremely high objects.
- Many 6- to 9-year-olds like to play unsupervised.
- Children aged 10–14 want more challenging climbing activities, such as on climbing walls and rocks.

Figure 25.20 (Ursy Potter Photography)
Falls are the most common cause of childhood injury.† Most fall-related fatalities occur primarily when children fall from great heights (more than two stories/6.7 m [22 ft]) or when children's head hit hard surfaces (AAP, 2001).

Falls from heights are more common in urban areas where children live in low-income multistory housing. These are often in the summer when windows are open and children play on fire escapes, balconies, and roofs (Lallier et al., 1999).

* For more information, please read the excellent reviews on children and climbing skills by van Herrewegen et al. (2004). These can be downloaded for free at the Web site of ANEC (www.anec.org), the European consumer voice in standardization.

† Please see Chapter 8 of this text on injuries in children for more on this topic. The American Academy of Pediatrics (AAP, 2005) and Lallier et al. (1999) also provide excellent reviews of the research on fall-related injuries in children. It should be noted that while falls are the leading cause of injury in children, motor vehicle accidents are the leading cause of childhood fatalities.

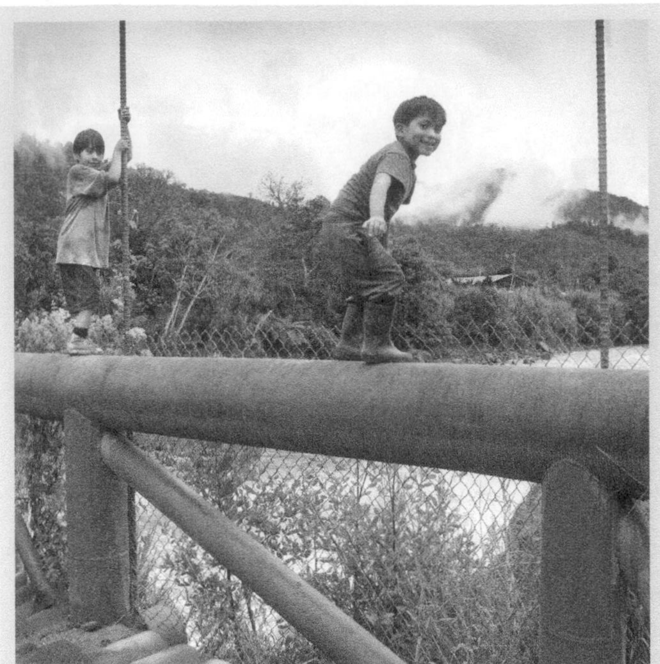

Figure 25.21 (Ursy Potter Photography)
Children climb on anything. They also fall (van Herrewegen et al. (2004) www.anec.org).

While environmental interventions are necessary, such as fences, window guards, and other barriers that can reduce such injuries (Pressley and Barlow, 2005), children can also be trained to develop age-appropriate climbing skills (van Herrewegen et al., 2004).

Figure 25.22 (CPSC)
Chain link fences with 1¼ in. mesh discourage young children from climbing.

PUBLIC BUILDINGS

SIZING PLACES TO CHILDREN

Public buildings and environments are designed by adults and usually aimed to accommodate these adults. Children are often forgotten when designing public services such as telephone booths and water fountains.

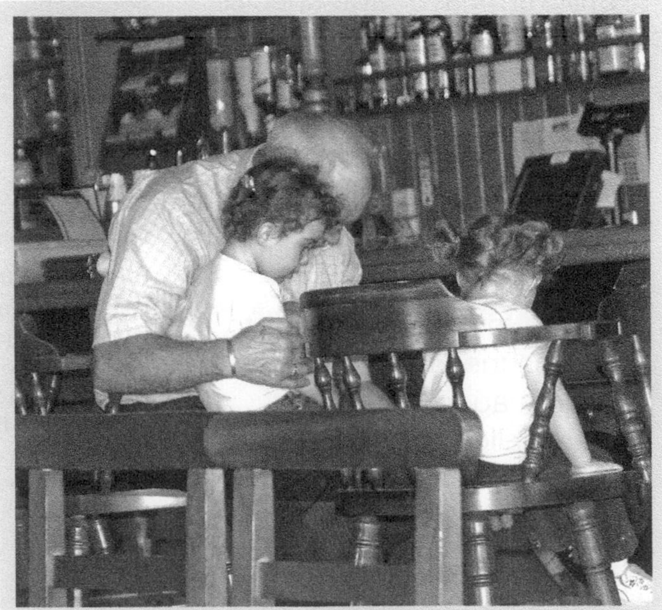

Figure 25.23 (Ursy Potter Photography)

Figure 25.24 (Valerie Rice)
This is a common sight in public restrooms; the mother and child sustain awkward and uncomfortable postures as they try to wash her hands. Providing lower adjacent sinks might accommodate both.

Figure 25.25 (Valerie Rice)
Daniel has difficulty seeing and reaching inside his mailbox. A longer and more horizontal distribution of mailboxes would have been easier for children with high boxes and adults with lower positions.

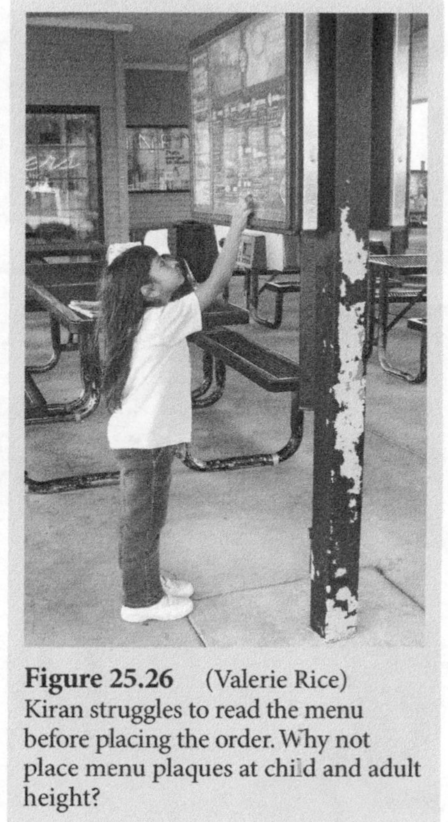

Figure 25.26 (Valerie Rice)
Kiran struggles to read the menu before placing the order. Why not place menu plaques at child and adult height?

Figure 25.27 (Ursy Potter Photography) **Figure 25.28** (Ursy Potter Photography)
Placing recycling and trashcans at child height enables children to use these at a young age, but also encourages children to poke around inside.

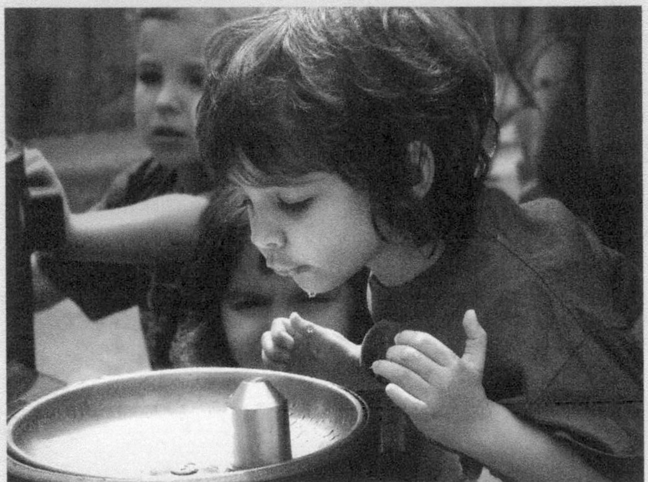

Figure 25.29 (Ursy Potter Photography)
Lower water fountains are one of many examples of universal designs that benefit children as well as wheelchair users. While the optimum height will vary, some general guidelines may often apply:

- Water fountains should be free of sharp corners, edges, and projections.
- The controls should be operable from the front and side.
- The spout should be no more than 29½ in. (75 cm) above the floor.
- A knee clearance of 16 in. (40.6 cm), 18 in. (45.7 cm) deep and 30 in. (76 cm) wide, should be provided.

ATBCB (1998) ADA Accessibility Guidelines Building Elements Designed for Children's Use.

ESCALATORS

CPSC estimates that 6000 people experience injuries from escalators that required hospital emergency room treatments each year in the United States. Most of these injuries (75%) were caused by falls; another 20% took place when hands, feet, or shoes were trapped in escalators (CPSC, 2004).

Figure 25.30 (Ursy Potter Photography)

The United States voluntary guideline for escalators requires the following (ASME/ANSI)*:

- The emergency shutoff buttons must be at the top and bottom of each escalator. The button should be on the right side of the escalator when facing the stairs.
- Sidewalls must be made of low-friction material so soft-soled shoes cannot be caught easily.
- "Skirt obstruction devices" must be at the top and bottom of the escalator. These automatically shut off the escalator upon sensing the presence of a foreign object.
- Side clearance at the edges of steps be no more than 3/16 in.
- Warning signs should be placed on escalators reminding parents to hold children's hands and face forward.
- Each step should have painted footprints or brightly colored borders.

Figure 25.31 (Ursy Potter Photography)

Escalators can trap loose shoelaces, drawstrings, scarves, and mittens.

- If your child's clothing still has drawstrings, remove them.
- Always hold children's hands on escalators.
- Do not permit children to sit or play on the steps.
- Do not bring children onto escalators in strollers, walkers, or carts.
- Always face forward and hold the handrail.
- Avoid the edges of steps where entrapment can occur.
- Learn where the emergency shutoff buttons are in case you need to stop the escalator.

* ASME is the American Society of Mechanical Engineers (ASME). ANSI is the American National Standards Institute (ANSI).

Figure 25.32 (Ursy Potter Photography)
Many escalators, such as this one, display a distracting complex of warnings, instructions, and controls that may or may not be accessible to the user. This arrangement may cause them not to see or notice the bottom warning and instruction.

Figure 25.33 (Ursy Potter Photography)

Figure 25.34 (Ursy Potter Photography)

Elevator warnings. Above left, safety rules for safe use of escalators. Above right, a warning to parents to hold the handrail and their child's hand while avoiding the sides.

MOVEABLE PUBLIC FURNITURE

Table 25.4 Furniture in Public Buildings

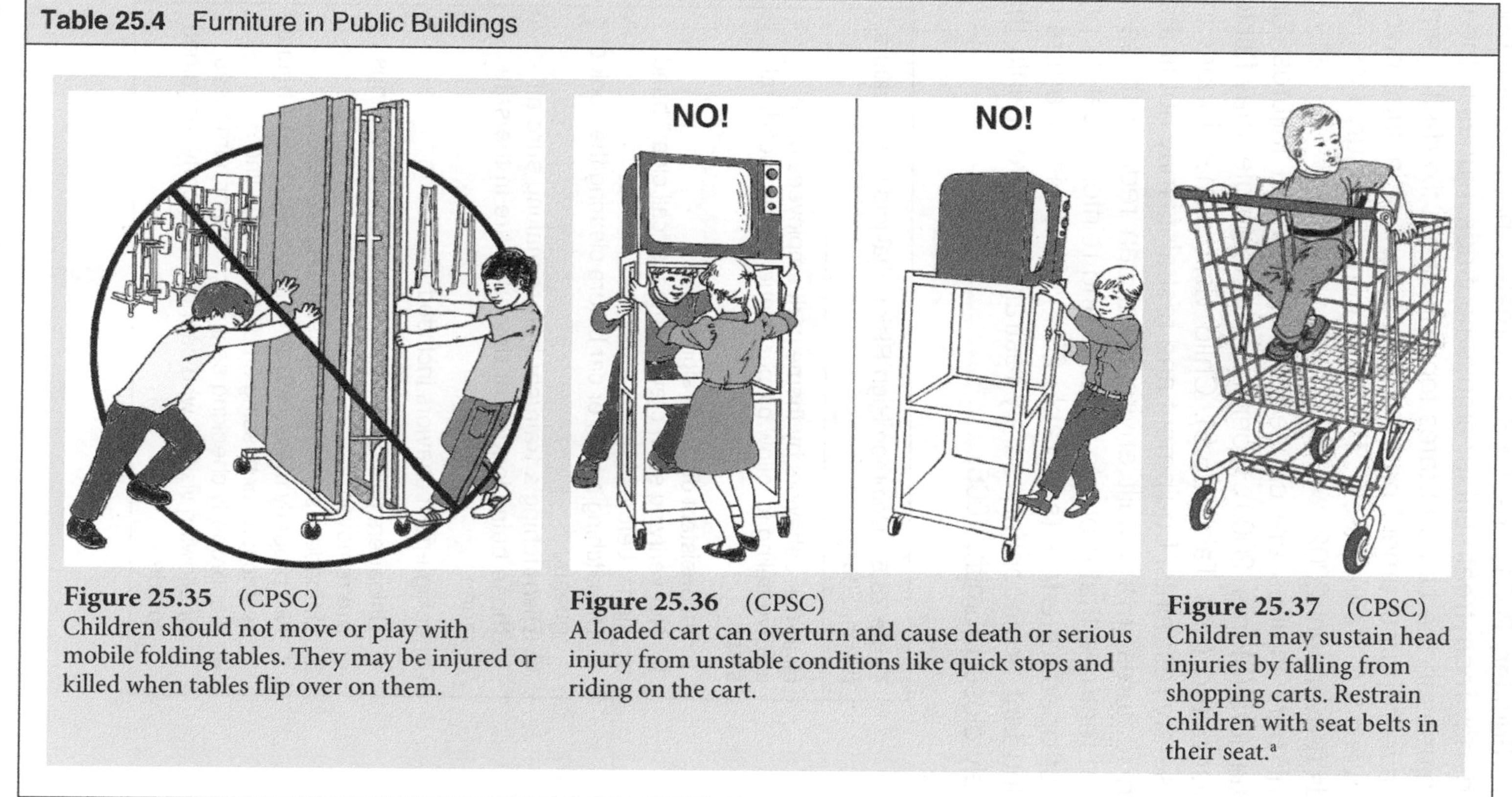

Figure 25.35 (CPSC)
Children should not move or play with mobile folding tables. They may be injured or killed when tables flip over on them.

Figure 25.36 (CPSC)
A loaded cart can overturn and cause death or serious injury from unstable conditions like quick stops and riding on the cart.

Figure 25.37 (CPSC)
Children may sustain head injuries by falling from shopping carts. Restrain children with seat belts in their seat.[a]

[a] See Chapter 8 on preventing injuries in children for more on shopping cart injuries.

AMUSEMENT PARKS

Parents know very little about ride safety in amusement parks. They are likely to focus on their own enjoyment rather than safety and often have a childlike trust that amusement parks are safe (Fackler, 2001). Yet CPSC estimates that in 2003, 6900 visitors received hospital emergency room treatment for amusement ride injuries in the United States.* Four out of five of injuries in the United States took place at "fixed-site" amusement parks; the rest were at traveling amusement parks. Five percent of injuries involved children under 5 years; 30% were 5 to 14 years-old. The most common injury (40%) was of the shoulder, arms, or hands (Levenson, 2002).

Amusement park injuries are caused by mechanical failures, operator errors, and consumer behaviors (Morris, 2000). Operators, for example, may fail to follow procedures or become distracted (see Table 25.5). Child visitors may ignore instructions or try to override their safety restraints. Equipment and restraints may not fit them properly.

Yet design features in the built environment can reduce these risks. For example, physical design features that promote visibility and reinforce safety practices can reduce the potential for operator errors (see Table 25.7). Child visitor errors may be reduced by providing redundant or secondary safety restraints and restraints that indicate whether they are actively closed (Smith, 2005).

Table 25.5 Common High-Risk Behaviors by Operators (Lyon, 2001)

High-risk behaviors by theme park employees often involve failing to follow proper sequence of procedural steps:

1. Inconsistent dispatch signals
2. Not verifying each crew member's "call clear" hand signal before acting
3. Dispatching a train or car before clearing the dock of guests
4. Dispatching a train/car before making sure all those being loaded or unloaded are in the safe zone

Other high-risk behaviors included:

1. Unloaders failing to ensure visitors and belongings are removed
2. Unloaders standing in "no standing" zones
3. Not properly measuring or enforcing children's height restrictions or missing small children in line
4. Not properly checking seat restraints or harnesses or allowing visitors who cannot be properly restrained to ride

---

* Amusement park fatalities are considerably less common than injuries. CPSC estimates about 4.4 nonworker deaths a year in the United States (Levenson, 2002).

These estimated injury rates in 2003 are an improvement over those in observed in 2001; in that year, 8313 visitors required hospital emergency room treatment for amusement ride injuries. Braksiek and Roberts (2002) argue that actual rates in the United States are likely to be considerably higher because of weak oversight by the states and limitations in injury recording practices.

See their study for a comparison of amusement park injuries in the United States and other countries.

Table 25.6 High-Risk Behaviors at Amusement Rides

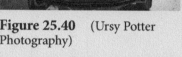

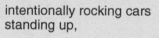

Figure 25.38 (Ursy Potter Photography)

Figure 25.39 (Ursy Potter Photography)

In the United States, inflatable amusement attraction injuries have increased dramatically.[a] Risk factors for injury with inflatable rides include (Morris, 2001):

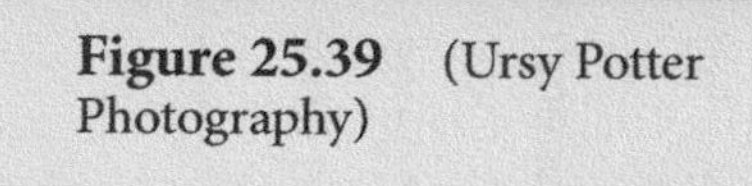
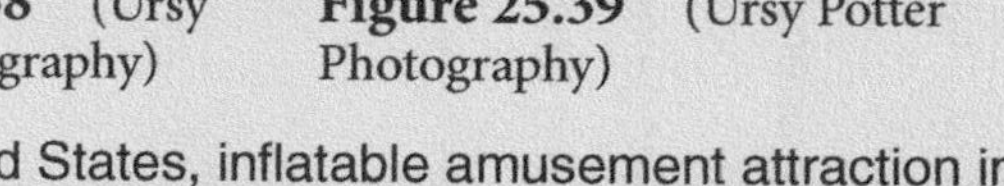

- Collision when several children are jumping or sliding at the same time, particularly between a large and smaller child
- Falling from a jumping balloon or inflatable slide on a hard object
- Catching a foot or appendage while jumping
- Jumping or sliding head first or trying to perform flips may injure head or neck
- Unexpected wind gusts and improperly secured inflatable devices throw children on the ground or against objects
- Fan power loss causes slides to rapidly deflate

Figure 25.40 (Ursy Potter Photography)

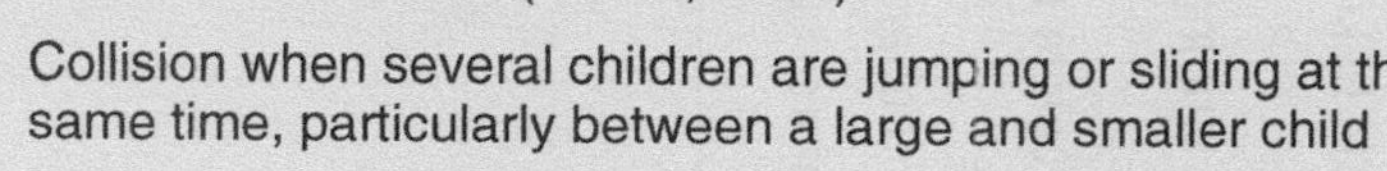

High-risk behaviors in children include (Morris, 2001):

- intentionally rocking cars standing up,
- defeating safety restraints, and
- sitting improperly, such as sideways or with feet above lap bar.

Figure 25.41 (Ursy Potter Photography)

There is no safe way to strap toddlers and preschoolers.

- Carousels often provide safety restraints that loosely buckle around a child. Although these may prevent serious head injuries, they may also introduce new risk factors. If the child falls, they may become entangled and drag them (Fackler, 2001).
- Parents should ride with small children, standing next to the horse to keep the child from falling. The built environment must also support these parents in this unstable environment.

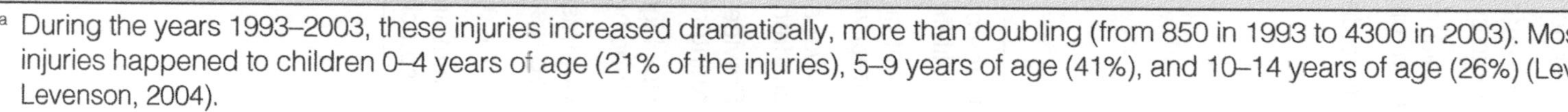

[a] During the years 1993–2003, these injuries increased dramatically, more than doubling (from 850 in 1993 to 4300 in 2003). Most of these injuries happened to children 0–4 years of age (21% of the injuries), 5–9 years of age (41%), and 10–14 years of age (26%) (Levenson, 2002; Levenson, 2004).

Table 25.7 "Best Practices" to Prevent Theme Park Injuries (Lyon, 2001)

	Design Solutions	Work Practice Solutions
Attention	• Provide shade in wait lines, loading dock and lift platforms, and control stations to reduce worker fatigue and heat stress	
Communication Errors	• Position control panels so the lead and secondary control operators face each other • Automatically stop background music and sound effects when audio system is on • Position convex mirrors to enable operators to see blind spots (primarily for flat rides) • Recorded stories in visitor lines tell visitors what to do	• Provide lead control operators hands-free operation with headset microphones • Do not use any hand signals until clearing visitors and closing exit gates • Lead operators verbally confirm status over the audio system with hands off controls • Practice universal hand signal sequences at all parks to avoid inconsistencies • Use PA to signal "all clear" to crew before dispatch • In emergencies, employees whistle in children's rides
Procedural Errors	• Place child height markers in stalls and waiting lines to facilitate verifying if a child is tall enough to ride • Paint numbers on loading stall floors to communicate more effectively • Lower air gates to 20 cm (8 in.) from the floor to prevent small children from crawling under gates • Provide air gates or turnstiles at exits to keep visitors from reentering • Use highly visible yellow and black striping to define "no standing zones" • Contrast lapbars and car headrest colors to help operators see restraint positions • Provide "safe docking zones" behind chains or air gates • Paint areas to identify where operators must stand while dispatching and returning rides	• Provide ride operators with elastic wristband key chains to ensure they do not leave keys unattended on the control panel • Operator provides crossing signals before rather than during crossing

STADIUMS

Bleachers are structures that provide tiered or stepped seating, usually without backrests. In 1999, about 4910 children under the age of 15 required emergency room treatment from falling from or through bleachers (CPSC, 2001; see also Chapter 8 of this text for more on this topic).

Causes of child injuries (CPSC, 2001)

1. Missing guardrails (sides or back)
2. Openings between guardrails, seats, and footboards large enough for a child to fit through
3. Bleachers collapse after improperly being opened, closed, or maintained
4. Broken or missing supports such as handrails, aisles, and nonskid surfaces
5. Older bleachers requiring retrofitting

Figure 25.42 Manually operated telescopic bleachers (above) may collapse if not operated properly (CPSC, 2001).

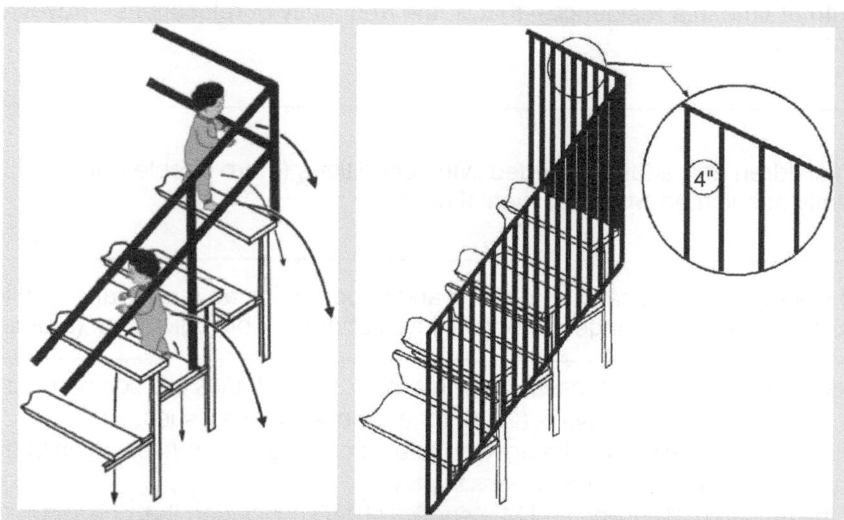

Figure 25.43 **Figure 25.44**
Children are at risk of falling if they can pass under or through a guardrail (above). Children should not be able to pass through 4 in. (10 cm) openings. Guardrail designs should also discourage children from climbing on them (CPSC, 2001; see Chapter 8 of this text).

RESIDENTIAL ENVIRONMENTS

Oscar Newman's pivotal book about creating defensible space (Newman, 1972) described how building design features such as building height, the size of the complex, the number of occupants sharing an entrance, and its layout could promote or deter crime. For example, he described how larger high-rise buildings with many people sharing entrances are depersonalizing, difficult to monitor, and promote crime. Walkways with public and anonymous spaces that discourage a sense of identity, social control, and ownership adversely affect security (Newman, 1972) and mental health (Evans, 2003).

Table 25.8 Residence Design and Mental Health (based on reviews by Evans (2003) and Evans et al. (2003))[a]

Number of Building Floors	• Living in multidwelling housing is consistently associated with more mental health problems than those living in low-rise or separate houses for children and adults. These effects are stronger in low-income residents.[b] ◦ Possible factors include the impact of the design on social isolation, a lack of play spaces for children, fear of crime, or perceived social stigma.[c] ◦ Design features of multidwelling housing influences that adversely affect residents' perception of social support include the use of long corridors—layouts that discourage interactions. • People who live in single-family detached homes typically have the fewest mental health problems.
Building Floor	• Residents of higher floors tend to have poorer mental health than those living on lower floors. One possible factor: it was not clear whether residents living in some floors tend to live longer than those in others do.
Quality	• Higher quality housing is consistently related to improved mental health.[d]
Relocation	• The length of time in a residence—that is, the frequency of relocations—adversely affects mental health.
Crowding	• For both children and adults, crowded living conditions (more people living in a room) is consistently associated with poor mental health.[e]
Noise	• Much of the research indicates that traffic and airport noise adversely affect children's mental health. Effects may be stronger among children at greater biological risk (prematurity).

[a] While there is a considerable body of research on this topic, most studies are confounded in one way or another to some degree. Please see Evans (2003) for a review of these issues.

[b] See also the review by Wells (2000) on the adverse affects of living in high-rises on children's health, well-being, behavioral problems, and restrictions in play.

[c] In their review of the research, Valentine and McKendrick (1997) concluded that children in high-rise buildings consistently spend more time indoors than children living in low-rise buildings, but their behavior is also more restricted because the dense population affords them less privacy and discourages being noisy.

[d] Inner-city adolescents who see their environments less positively also experience more difficult relationships with family and friends, impaired cognitive functioning, and delinquency (Obasanjo, 1998). Deteriorating buildings and the presence of trash are routinely associated with perceived lack of safety and fear (Austin et al., 2002).

[e] Overcrowding appears to affect children's health, education, childhood growth, and development, although we need better-controlled research to make absolute conclusions. Such effects on children can be direct and indirect. For example, children's schooling may be hindered directly by a lack of space for homework as well as indirectly, such as by illness caused by overcrowding (ODPM, 2004).

> *"Tall, large structures, long interior corridors, lack of small group spaces and poor visual surveillance... interfere with territorial control and feelings of ownership and are associated with both actual crime levels and fear of crime"*
> (Evans, 2005)

Environments that impart a sense of control over their environments are associated with positive mental health. Children consistently exposed to noise and crowded living conditions, for example, have been found to develop a sense of learned helplessness that may (Evans, 2003) generalize to other areas of their life and exert cumulative and long-term effects.

> *"Homes should not be either large condominiums with lots of apartments or single family cottages. In large condominiums, you are afraid of the others and stay shut up in the house; in a cottage, you do not need others and end up being just as lonely.*
> (Tonucci and Rissotto, 2001), quoting a young girl participant

Table 25.9 Creating Defensible Spaces

Figure 25.45

Figure 25.46

Before (above left) and after (above right) examples of *Defensible Space*. The photo on the top right demonstrates environmental changes directed toward better defining public, semipublic, and private spaces in order to promote residents' sense of control and ownership of their streets, grounds, and nearby areas (From Newman, O., *Creating Defensible Space*, US Dept. of Housing and Urban Development, Office of Policy Development and Research, Institute for Community Design Analysis, Washington, D.C., 1972. With permission.)

CHILDREN WITH DISABILITIES

There has been a growing recognition that providing different accommodations for people with special needs isolates them and emphasizes their differences. ***Universal Design*** is the design of places and products to fit the broadest range of people and applications.

Effective examples of *Universal Design* accommodate everyone. For example, a ramped curb may enable wheelchair users to cross a street intersection, but it also helps mothers with baby strollers, travelers rolling their luggage, and shoppers with vision obstructed by grocery bags.

This shift has largely reflected the World Health Organization's (WHO) new system for classifying disabilities, which assumes that people fall along a continuum of degrees of abilities in the population and each person's functional abilities change dynamically over time, reflecting their experiences and life situation. Implicit in this is that the person's disabilities largely relate to their social and financial circumstances.

Figure 25.47 (Valerie Rice)
This playground ramp aims to accommodate children with wheelchairs. Yet while the ramp enables children and parents in wheelchairs to get a little closer to the action and watch children play, the playground itself provides few opportunities for these children to play with others.

Play spaces that fail to reflect principles of *Universal Design* can be less functional and emphasize children's differences over similarities (see Chapter 28 on playgrounds and international playground accessibility standards).

Children with disabilities have critical requirements in the design of public places. Effective designs for these users will provide them with access, help them feel secure and safe, and enhance their life experiences.

While a comprehensive review of this important topic is beyond the scope of this chapter, readers may find additional information from numerous resources, including these:

- ***The Access Board*** is an independent federal agency devoted to accessibility for people with disabilities.*
- Its guidelines includes *"Building elements designed for children's use"* (1998); *"Play areas"* (2000); and "*Recreation facilities*" (2002). The latter includes topics such as amusement rides, sporting activities, and swimming pools. (www.access-board.gov)
- The ***Japanese Society for Rehabilitation of Persons with Disabilities*** (JSRPD) has extensive content on international activities and resources for people with disabilities. (www.dinf.ne.jp)
- The Humanics ErgoSystems Web site provides research and resources on ergonomics for children, people with disabilities, and on the design of public places.

Research cited in this chapter that is public domain is available at these links:

www.humanics-es.com/recc-children.htm
www.humanics-es.com/recc-design.htm
www.humanics-es.com/recc-disabled.htm

* The *Access Board* interacts and coordinates with U.S. Federal agencies in order to represent the public, particularly people with disabilities. Half of its members are from federal agencies. The other half are members of the public appointed by the President, of which most have a disability.

Table 25.10 Principles of Universal Design (CUD, 1997)

Principles	Reason	Guidelines	Examples
1. Equitable Use	The design is useful and marketable to people with diverse abilities.	1a. Provide the same means of use for all users 1b. Identical whenever possible; equivalent when not 1c. Avoid segregating or stigmatizing any users 1d. Provisions for privacy, security, and safety should be equally available to all users 1e. Make the design appealing to all users	• Play areas in natural environments that stimulate senses (sound, smell, light, water) These areas must be safe, free of barriers, and include solid paths and common areas with benches next to wheelchairs.
2. Flexibility in Use	The design accommodates a wide range of individual preferences and abilities.	2a. Provide choice in methods of use 2b. Accommodate right- or left-handed access and use 2c. Facilitate the user's accuracy and precision 2d. Provide adaptability to the user's pace	• Child museums that allow children to activate exhibits in different ways • Thicker pens that require less grip force and writing skill
3. Simple and Intuitive Use	Use of the design is easy to understand, regardless of the user's experience, knowledge, language skills, or current concentration level.	3a. Eliminate unnecessary complexity 3b. Be consistent with user expectations and intuition 3c. Accommodate a wide range of literacy and language skills 3d. Arrange information consistent with its importance 3e. Provide effective prompting and feedback during and after task completion	• Large and high-contrast buttons with graphic symbols that depict its function • ISO international signage • Intelligent public displays that adapt content options to reflect the child's prior commands
4. Perceptible Information	The design communicates necessary information effectively to the user, regardless of ambient conditions or the user's sensory abilities.	4a. Use different modes (pictorial, verbal, tactile) for redundant presentation of essential information. 4b. Provide adequate contrast between essential information and its surroundings. 4c. Maximize "legibility" of essential information. 4d. Differentiate elements in ways that can be described (i.e., make it easy to give instructions or directions) 4e. Provide compatibility with a variety of techniques or devices used by people with sensory limitations	• Stop lights that emit directional sounds (500–1000 Hz) as well as lights to indicate when it is safe to cross. Buttons include raised tactile directional arrows[a] • Elevator controls indicate floors in Braille and numbers that are tactile Once activated, these provide kinesthetic and auditory "click" feedback • In the event of fire, public buildings flash very bright lights as well as sounding alarms.

(continued)

Table 25.10 Principles of Universal Design (CUD, 1997) (Continued)

Principles	Reason	Guidelines	Examples
5. Tolerance for Error	The design minimizes hazards and the adverse consequences of accidental or unintended actions.	5a. Arrange elements to minimize hazards and errors: most used elements, most accessible; hazardous elements eliminated, isolated, or shielded 5b. Provide warnings of hazards and errors 5c. Provide fail-safe features 5d. Discourage unconscious action in tasks that require vigilance	• Upon detecting any pressure, elevator door sensors stop shutting doors • Emergency shut off switches are prominent and easily activated
6. Low Physical Effort	The design can be used efficiently and comfortably with a minimum of fatigue.	6a. Allow user to maintain a neutral body position 6b. Use reasonable operating forces 6c. Minimize repetitive actions 6d. Minimize sustained physical effort	• Doors in public buildings open electronically and by pushing a horizontal bar • Door handles open if users place the weight of their full arm on it (without grasping) • Theater seats in standard seating areas with added leg room for people with leg braces
7. Size and Space for Approach and Use	Appropriate size and space is provided for approach, reach, manipulation, and use regardless of user's body size, posture, or mobility.	7a. Provide a clear line of sight to important elements for any seated or standing user 7b. Make reach to all components comfortable for any seated or standing user 7c. Accommodate variations in hand and grip size 7d. Provide adequate space for the use of assistive devices or personal assistance	• Automatic water faucets in public restrooms activate when users place hands in position • Text displays such as menus set at both wheelchair and standing height
Copyright ©1997 NC State University, *The Center for Universal Design*. www.design.ncsu.edu/cud			Examples from Rani Lueder

[a] Bertzen and Tabor (1998) provided a useful review of worldwide accessible pedestrian signals for the US Access Board. (www.access-board.gov)

COMMUNITY

Each neighborhood is unique, and its design should reflect and accommodate differences that support the values of the community. Even very poor communities have positive and negative design features. It is important to understand the implications of these differences before starting renovations (Coughy et al., 1999).

Figure 25.48 (Ursy Potter Photography)
Recently there has been a greater recognition of the potential benefits of designing communities with interconnected mixed-use areas that can promote physical activities (Sallis and Glanz, 2006). These efforts have been hindered by restrictions that often requires changes in zoning codes and regulations.

IMPOVERISHED NEIGHBORHOODS

The single most consistent effect of the environment on children and adolescents is from the socioeconomic status of the neighborhood. Its effects are cumulative, progressing through childhood and adolescence into adulthood.

Important differences in these communities are its resources, which for example affect the availability of parks, sports areas, trails, community centers and libraries, and the quality and availability of childcare, medical, and other resources. During adolescence, boys seem to be more affected than are girls of the same age (Leventhal and Brooks-Gunn, 2000).

Figure 25.49 (Ursy Potter Photography)
Parents with lower incomes feel they have less support from their neighborhood,* and they also feel less safe in their neighborhood and that their children are less safe in school. Children in lower income families are also rated lower health and more likely to experience chronic physical or mental health problems (DHHS, 2005).

* These parents were less likely to feel that people in their neighborhood help each other out, watch out for each others' children, and can be trusted to help their child if they are hurt (DHHS, 2005).

Research indicates that children who live in neighborhoods that are more affluent consistently demonstrate better school readiness, verbal ability and reading, and other achievements and improved behavioral and emotional well-being* (Leventhal and Brooks-Gunn, 2000). A leading survey of parents about their children's health in the United States found children of lower incomes are less likely to attend preschool,† participate in outside activities outside of school‡ (DHHS, 2005).

CHILDREN IN NATURAL ENVIRONMENTS

A growing body of research suggests that natural environments can have a profound effect on children's well-being. Children living in places with more nature and "restorative" places demonstrate improved cognitive functions, ability to focus, and motor coordination (Wells, 2000).

One study that compared people who were arbitrarily assigned to different buildings found that those who happened to have more trees near their building reported significantly better relationships with their neighbors, more visitors, liked where they lived more, and felt safer than people in identical buildings with fewer trees (Sullivan and Kuo, 1996). People who live in greener surroundings also report less fear and aggressive and violent behavior (Kuo and Sullivan, 2001).

At this point, the reader may be experiencing cognitive dissonance. While the CPTED movement emphasizes clearing vegetation to prevent crime, the natural environments movement emphasizes reinstating vegetation.

Figure 25.50 (Ursy Potter Photography) Nature may have long-term benefits for children. Children who experienced the greatest increase in the "restorativeness" of natural elements tended to have the greatest ability to focus attention several months after moving to a new home (Wells, 2000).

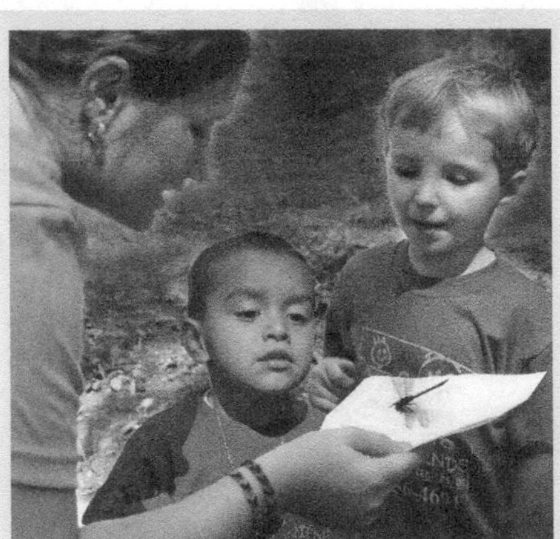

Figure 25.51 (Ursy Potter Photography) Green inner-city neighborhood spaces are twice as likely to be supervised by an adult as children in similar but barren spaces. Spaces with trees also support more creative play behaviors (Taylor et al., 1998).

Figure 25.52 (Ursy Potter Photography)

* This difference was most pronounced among European-American children and adolescents (Leventhal and Brooks-Gunn, 2000).

† These preschool programs include preschool, nursery school, Head Start or Early State programs, or kindergarten (DHHS, 2005).

‡ Specifically, they found that children of minorities and lower incomes are less likely to participate in sports teams, lessons, Scouts, religious groups, and Boys and Girls' clubs outside of school (DHHS, 2005).

While dense vegetation such as large shrubs, underbrush, and dense forests may enable criminals to conceal their behaviors, not all vegetation blocks views. Widely spaced, high-canopy trees have little effect on visibility, other visibility-preserving forms of vegetation do not promote crime, and there is research to suggest that it may even discourage crime by increasing actual and implied surveillance when more people participate in their environment. It can also provide greater perceived and actual safety by more effectively defining territory and promoting identity, and by reducing mental fatigue (Kuo and Sullivan, 2001).

CHILD PARTICIPATION IN URBAN DESIGN

In recent years, there has been a heightened attention on whether and how to allow children to participate in the design of their own environments.

Reasons for this interest include the potential benefits of incorporating input from children in the design process. For example, while urban planning has typically centered on the need to protect children from perceived or actual harm, children and adolescents have often expressed an interest in designing public places that provide them with greater opportunities for independent action and places that encourage meeting and socializing with their peers (Simpson, 1997). These needs will vary between children of different ages.

Many maintain that children have the right to be heard and as they mature, their input should be provided greater weight. Such participation should of course reflect age-related and developmental implications. Children learn decision-making skills as they mature and become better able to think objectively and act independently. Important skills include the ability to process emotions, interpret expressions, evaluate alternatives, and make value judgments. As their decision-making improves, so should their role in the participation process (Simpson, 2003). Others remind us of the great personal costs of children living in many communities who are afforded considerably less freedom and opportunities to explore than previous generations (Tonucci and Rissotto, 2001).

The "*Children's City*" project involved the participation of children in 40 cities in Italy in planning the structure of social and urban environments. Groups of 6-to 10-year-old children working with an adult monitor to "advise" city administrators and planners about their point of view on topics that affect city life.

An example of a suggestion from a little girl was to widen the street pavements, which discouraged family outings because they were too narrow for the family to walk together. Another wanted the designs to make it easier for her grandmother to take her on walks.

This approach has since expanded to cities in Spain and Argentina (Tonucci and Rissotto, 2001).

Figure 25.53 (Ursy Potter Photography)

Figure 25.54 (Ursy Potter Photography)
One argument for listening to children's views of the city is that changes in urban environments come at a high cost for urban children, many of whom spend most of their day confined in enclosed spaces while being carefully controlled by adults. Many of these children are afforded little independence or opportunity to go outside and play with friends or explore new places (Tonucci and Rissotto, 2001).

UNICEF Principles for Child Friendly City.
Enable every young citizen to:

- Influence decisions about their city
- Express their opinion on the city they want
- Participate in family, community, and social life
- Receive basic services such as health care and education
- Drink safe water and have access to proper sanitation
- Be protected from exploitation, violence, and abuse
- Walk safely in the streets on their own
- Meet friends and play
- Have green spaces for plants and animals
- Live in an unpolluted environment
- Participate in cultural and social events
- Be an equal citizen of their city with access to every service regardless of ethnic origin, religion, income, gender, or disability

UNICEF Principles for Child Friendly Cities
www.childfriendlycities.org

COMMUNITY-BASED CHILD INJURY PREVENTION PROGRAMS

Community efforts to develop programs aimed at preventing child injuries have become accepted practice. These efforts sometimes arose from recognition that safety interventions such as bicycle helmets and child safety restraints were less effective than had been anticipated and required additional support at the community level.* Such efforts might include, for example, community education programs, safety festivals, targeted counseling of parents or training of nurses, environmental repairs of unsafe public areas, and distribution of printed materials that educate about child injuries and prevention.

In their review of the research on these efforts, Spinks et al. (2004) reported that they could find only nine studies that evaluated community-based intervention programs. Five of these nine studies found the interventions significantly reduced injuries among children. Yet more research is needed to help us better understand which interventions will be more effective.

- The environments provide a context for the influence of relationships on healthy human development.
- They incorporate the quality that best promote competence and well-being, individualized responsiveness, mutual action and interaction, and emotional connection to another being.
- Relationships engage children in the human community in ways that help them define who they are, what they can become, and why they are important to other people.

* The efforts of the United States National Safe Kids and Safe Kids Worldwide Campaigns, the World Health Organization (WHO) Safe Communities, Canadian Safe Communities Foundation, and the Beterem National Centre for Children's Health and Safety in Israel are noteworthy (Spinks et al., 2004).

SMART GROWTH PRINCIPLES

Smart Growth Principles for Child-Friendly Communities[a]	
1. Mix land uses	Environments that support a mix of living and commercial activities in close proximity facilitate the development of alternatives to driving, such as walking or biking. These complex environments provide a commercial base that supports public transit while fostering a sense of community and safety.
2. Compact building designs	Compact building designs (e.g., high rises) may enable efficient use of land and resources that may then be used to preserve open spaces and provide safe places for children to learn, play, and interact.
3. Create housing alternatives	Providing quality housing that accommodates a broad range of income levels and commuting patterns within a community can promote more alternatives in transportation, education, and other kinds of services that benefit all children in the community.
4. Walkable neighborhoods	Walkable communities providing safe and hospitable areas for pedestrians, bike riders, and public transit riders in communities that integrate housing, commercial business and services can foster safe communities and facilitate the development of alternative forms of transportation, schools, and community centers such as libraries, parks, and activity centers.
5. Distinctive communities	Development and construction standards that reflect the values and cultures of its residents create livable and child-friendly communities.
6. Preserve open space and natural environments	Preserving natural areas in and around neighborhoods preserve the environment and improves the quality of life in all communities.
Adapted from Smart Growth Online, a service of the Smart Growth Network. www.smartgrowth.org	

[a] See also ICMA (2002, 2003).

CONCLUSION

Young children's neighborhoods are central to their development. Research demonstrates that the design of the built environment can help promote children's physical, emotional, intellectual, behavioral, and moral development. It can foster a sense of identity and community and encourage them to become active participants in the effectiveness of their neighborhood. It can reduce the risk of injuries, help discourage behaviors that can lead to crime, promote health and a sense of well-being.

Behavioral interventions are often most effective when they are multifaceted in approach that include a range of strategies that include environmental design (Cross and Hall, 2005).

Figure 25.55 (Ursy Potter Photography)

REFERENCES

AAP (2001). American Academy of Pediatrics Policy Statement reaffirmed in 2005. Falls from heights: Windows, roofs, and balconies. *Pediatrics*, 107(5), 1188–1191.

Anderson, P.M. and Butcher, K.E. (2006). Childhood obesity: Trends and potential causes. *Future Child*, 16(1), 19–45.

Arndt, S.R., Cargill, R.S., and Hammoud, P.E. (2004). Head accelerations experienced during everyday activities and while riding roller coasters. Proceedings of the Human Factors and Ergonomics Society 48th Annual Meeting. 1973–1977.

ATBCB (1998). Architectural and Transportation Barriers Compliance Board. ADA Accessibility Guidelines for Building Elements Designed for Children's Use. 36CFR Part 1191 (Docket No. 94-2) Architectural and Transportation Barriers Compliance Board: www.access-board.gov

Austin, D.M., Furr, L.A., and Spine, M. (2002). The effects of neighborhood conditions on perceptions of safety. *Journal of Criminal Justice*, 30, 417–427.

Bartlett, S. (2005). Good governance: Making age a part of the equation—and introduction. *Children, Youth and Environments*, 15(2), 1–17.

Bentzen, B.L. and Tabor, L.S. (1998). *Accessible Pedestrian Signals*. Washington, D.C.: US Access Board. 40 pp. www.access-board.gov

Braksiek, R.J. and Roberts, D.J. (2002). Amusement park injuries and deaths. *Annals Emergency Medicine*, 39(1), 65–72.

Brougere, G. (1991). Espace de jeu et espace public. *Architecture & Comportement*. 7, 165–177. (Cited in Tonucci and Rissotto (2001), *op. cit.*)

Brownson, R.C., Boehmer, T.K., and Luke, D.A. (2005). Declining rates of physical activity in the United States: What are the contributors? *Annual Review of Public Health*, 26, 421–443.

Chawla, L., Blanchet-Cohen, N., Cosco, N., Driskell, D., Kruger, J., Malone, K., Moore, R., and Percy-Smith, B. (2005). Don't just listen—do something! Lessons learned about governance from the growing up in cities project. *Children, Youth, and Environments*, 15(2), 53–88.

Choudhary, R., Sharma, A., Agarwal, K.S., Kumar, A., Sreenivas, V., and Puliyel, J.M. (2002). Building for the future: Influence of housing on intelligence quotients of children in an urban slum. *Health Policy Plan*, 17(4), 420–424.

Cobb, C. and Danby, F. (2005). Governance of children's everyday spaces. *Australian Journal of Early Childhood*, 30(1), 14–20.

Cook, T.D., Herman, M.R., Phillips, M., and Settersten, R.A. Jr. (2002). Some ways in which neighborhoods, nuclear families, friendship groups, and schools jointly affect changes in early adolescent development. *Child Development*, 73(4), 1283–1309.

Coughy, M.O., O'Campo, P., and Brodsky, A.F. (1999). Neighborhoods, families, and children: Implications for policy and practice. *Journal of Community Psychology*, 27(5), 615–633.

CPSC (2001). US Consumer Product Safety Commission. Guidelines for Retrofitting Bleachers. Publication 330-000011. Washington, D.C. 16pp. www.cpsc.gov

CPSC (2002). US Consumer Product Safety Commission. Age Determination Guidelines: Relating Children's Ages to Toy Characteristics and Play Behavior. Based on work performed under contract by Play Today. Smith, T.P. (Ed.). 269pp. www.cpsc.gov

CPSC (2004). US Consumer Product Safety Commission. *Escalator Safety*. CPSC Document #5111. www.cpsc.gov

CPTED (2005). Crime Prevention through Environmental Design Committee. CPTED Guidelines. Safer by Design: Creating a Safer Environment in Virginia. Richmond, VA: Virginia Crime Prevention Association. http://vcpa.org

CUD (1997). Center for Universal Design. The Principles of Universal Design. Version 2.0. © NC State University. The Center for Universal Design. www.design.ncsu.edu/cud

Cummins, S.K. and Jackson, R.J. (2001). The built environment and children's health. *Pediatric Clinics of North America*, 48(5), 1241–1252.

DHHS (2005). US Department of Health and Human Services. The National Survey of Children's Health 2003. Rockville, MD: US Department of Health and Human Services, Health Resources and Services Administration, Maternal and Child Health Bureau, 2005. www.mchb.hrsa.gov

DiMaggio, C. and Durkin, M. (2002). Child pedestrian injury in an urban setting: Descriptive epidemiology. *Pediatric Pedestrian Injuries*, 9(1), 54–62.

DSTI (1998). Directorate for Science, Technology, and Industry. Safety of Vulnerable Road Users. Scientific Expert Group on the Safety of Vulnerable Road Users (RS7) DSTI/DOT/RTR (98) Final. Organization for Economic Co-operation and Development. Paris, 229pp.

Durkin, M.S., Laraque, D., Lubman, I., and Barlow, B. (1999). Epidemiology and prevention of traffic injuries to urban children and adolescents. *Pediatrics*, 103(6), e74.

Evans, G.W., Saltzman, H., and Cooperman, J. (2001). Housing quality and children's socioemotional health. *Environment and Behavior*, 33, 389–399.

Evans, G.W., Wells, N.M., Chan, E., and Saltzman, H. (2000). Housing and mental health. *Journal of Consulting and Clinical Psychology*, 68, 526–530.

Evans, G.W. (2003). The built environment and mental health. *Journal of Urban Health: Bulletin of the New York Academy of Medicine*, 80(4), 536–555.

Evans, G.W., Wells, N.M., and Moch, A. (2003). Housing and mental health: A review of the evidence and a methodological and conceptual critique. *Journal of Social Issues*, 59(3), 475–500.

Ewing, R.H. (1999). Traffic Calming: State of the Practice. Publication No. IR-098. Sponsored by the Federal Highway Administration: Office of Safety Research and Development. Washington, D.C.: Institute of Transportation Engineers. www.ite.org

Exponent* (2002). Investigation of Amusement Park and Roller Coaster Injury Likelihood and Severity. Report DC18967.000 A0T0 0802 FR01. Failure Analysis Associates. 66pp. http://emerson-associates.com

Fackler, K. (2001). Child Safety Techniques for Prevention of Amusement Ride Accidents. La Jolla, CA: Saferparks. 22pp. www.saferparks.org

Frank, L., Engelke, P., and Thomas, T. (2003). Health and community design: The impact of the built environment on physical activity. Washington, D.C.: Island Press.

Frank, L.D. and Engelke, P. (2001). How Land Use and Transportation Systems Impact Public Health. A Literature Review of the Relationship between Physical Activity and Built Form. Active Community Environments (ACEs) Working paper #1. Schmid, T.L and Killingsworth, R.E. (Project Officers) Division of Nutrition and Physical Activity. Division of Physical Activity and Health Branch, Centers for Disease Control.

Gill, T. (2006). Home zones in the UK: History, policy, and impact on children. *Children, Youth, and Environments*, 16(1), 90–103.

Hillman, M., Adams, J., and Whitelegg, J. (1991). *One False Move . . . : A Study of Children's Independent Mobility*. Policy Studies Institute.

Holaday, B., Swan, J.H., and Turner-Henson, A. (1997). Images of the neighborhood and activity patterns of chronically ill school-age children. *Environment and Behavior*, 29(3), 348–374.

Hubsmith, D.A. (2006). Safe routes to school in the United States. Children, Youth, and Environments, 16(1), 168–190.

ICMA (2002). Getting to Smart Growth I: 100 More Policies for Implementation. Smart Growth Network and the International City/County Management Association. Washington, D.C.: ICMA.

ICMA (2003). Getting to Smart Growth II: 100 More Policies for Implementation. Smart Growth Network and the International City/County Management Association. 2003. Washington, D.C.: ICMA.

Istre, G.R., McCoy, M.A., Stowe, M., Davies, K., Zane, D., Anderson, R.J., and Wiebe, R. (2003). Childhood injuries due to falls from apartment balconies and windows. *Injury Prevention*, 9(4), 349–352.

Jacobs, J. (1961, 1993) *The Death and Life of Great American Cities*. New York: Random House Publishing.

Jeffery, C.R. (1971). *Crime Prevention through Environmental Design*. Sage Publications. 290pp.

Karsten, L. and van Vliet, W. (2006). Children in the city: Reclaiming the street. *Children, Youth, and Environments*, 16(1), 151–167.

Kerr, J., Rosenberg, D., Sallis, J.F., Saelens, B.E., Frank, L.D., and Conway, T.L. (2006). Active commuting to school: Associations with environment and parental concerns. *Medicine and Science in Sports and Exercise*, 38(4), 787–794.

Kuo, F.E. and Sullivan, W.C. (2001). Environment and crime in the inner city. Does vegetation reduce crime? *Environment and Behavior*, 33(3), 343–367.

Lallier, M., Bouchard, S., St-Vil, D., DuPont, J., and Tucci, M. (1999). Falls from heights among children: A retrospective review. *Journal of Pediatric Surgery*, 34(7), 1060–1063.

Lehnung, M., Leplow, B., Haaland, V.O., Mehdorn, M., and Ferstl, R. (2003). Pointing accuracy in children is dependent on age, sex, and experience. *Journal of Environmental Psychology*, 23, 419–425.

Levenson, M.S. (2002). Amusement Ride-related Injuries and Deaths in the United States: 2002 Update. Bethesda, MD: US Consumer Product Safety Commission. Directorate Epidemiology. 30pp.

Levenson, M.S. (2004). Amusement Ride-related Injuries and Deaths in the United States: 2004 Update. Bethesda, MD: US Consumer Product Safety Commission. Directorate Epidemiology. 16pp.

Leventhal, T. and Brooks-Gunn, J. (2000). The neighborhoods they live in: The effects of neighborhood residence on child and adolescent outcomes. *Psychological Bulletin*, 126(1), 309–337.

Lobstein, T. (2004). The prevention of obesity in children. *Pediatric Endocrinology Reviews*, 1 (Suppl. 3), 471–475.

Lumeng, J.C., Appugliese, D.A., Cabral, H.J., Bradley, R.H., and Zuckerman, B. (2006). Neighborhood safety and overweight status in children. *Archives of Pediatrics and Adolescent Medicine*, 160(1), 25–31.

Lyon, B.K. (2001). Behavior sampling of theme park ride operators. Professional Safety, a Publication of the American Society of Safety Engineers. 46(10) October, 35–43.

Malina, R.M. (2004). Motor development during infancy and early childhood: Overview and suggested directions for research. *International Journal of Sport and Health Science*, 2, 50–66.

Marcus, C.C. and Sarkissian, W. (1988). *Housing as if People Mattered: Site Design Guidelines for the Planning of Medium-Density Family Housing*. California series in urban development. University of California Press. 280pp.

McCulloch, A. and Joshi, H.E. (2001). Neighbourhood and family influences on the cognitive ability of children in the British National Child Development Study. *Social Science and Medicine*, 53(5), 579–591.

McMillan, D.W. and George, D.M.C. (1986). Sense of community: A definition and theory. *Journal of Community Psychology*, 14, 6–24.

* Exponent is a company. Authors were not named in the report, which is available on their Web site.

Morris, C.C. (2000). Amusement Ride-related Injuries and Deaths in the United States: 1987–1999. Bethesda, MD: US Consumer Product Safety Commission. Directorate for Epidemiology. 33pp. www.cpsc.gov

Newman, O. (1972, reprinted in 1996) *Creating Defensible Space*. Washington, D.C.: US Dept. of Housing and Urban Development, Office of Policy Development and Research, Institute for Community Design Analysis. www.huduser.org

Norton, R. and Jackson, R. (1995). Driveway-related child pedestrian injuries: A case-controlled study. *Pediatrics*, 95(3), 405–409.

NSCDC (2004). National Scientific Council on the Developing Child. Young Children Develop in an Environment of Relationships. Council Working Paper Series. Working paper #1. Summer. 16pp.

Obasanjo, O.O. (1998). The impact of the physical environment on adolescent functioning in the inner-city. Doctoral dissertation. Ann Arbor: University of Michigan.

ODPM (2004). Office of the Deputy Prime Minister. The Impact of Overcrowding on Health and Education: A Review of Evidence and Literature. London: Office of the Deputy Prime Minister. United Kingdom. www.odpm.gov.uk

Pooley, C.G., Turnbull, J., and Adams, M. (2005). The journey to school in Britain since the 1940s: Continuity and change. *Area*, 37(1), 43–53.

Pressley, J.C. and Barlow, B. (2005). Child and adolescent injury as a result of falls from buildings and structures. *Injury Prevention*, 11, 267–273.

Roberts, I., Norton, R., Jackson, R., Dunn, R., and Hassall, I. (1995). Effect of environmental factors on risk of injury of child pedestrians by motor vehicles: A case control study. *British Medical Journal*, 310, 91–94.

Sallis, J.F. and Glanz, K. (2006). The role of built environments in physical activity, eating, and obesity in childhood. *Future Child*, 16(1), 89–108.

Sallis, J.F., Prochaska, J.J., and Taylor, W.C. (2000). A review of correlates of physical activity of children and adolescents. *Medicine and Science in Sports and Exercise*, 32(5), 963–975.

Sapolsky, R. (2005). Sick of poverty. *Scientific American*. December 2005. 92–99.

Simpson, B. (1997). Towards the participation of children and young people in urban planning and design. *Urban Studies*, 34(5–6), 907–925.

Simpson, C. (2003). Children and research participation: Who makes what decisions? *Health Law Reviews*, 11(2), 20–29.

Smith, T.P. (2005). Human Factors Review of Restraint Failures on Mobile Amusement Rides. US Washington D.C.: Consumer Product Safety Commission. Division of Human Factors. 34pp. www.cpsc.gov

Spinks, A., Turner, C., McClure, R., and Nixon, J. (2004). Community-based prevention programs targeting all injuries for children. *Injury Prevention*, 10, 180–185.

Stokols, D. and Altman, I., Eds. (1991). *Handbook of Environmental Psychology*. Krieger Publising Company. Melbourne, FL. 1654pp.

Sullivan, W.C. and Kuo, F.E. (1996). Do trees strengthen urban communities, reduce domestic violence? Urban and Community Forestry Assistance Program Technology Bulletin No. 4. Atlanta, GA: USDA Forest Service, Southern Region, Athens, GA.

Taylor, A.F., Wiley, A., Kuo, F.E., and Sullivan, W.C. (1998). Growing up in the inner city: Green spaces as places grow. *Environment and Behavior*, 30, 3–27.

TCE (2006). Texas Cooperative Extension, Texas A&M University System. FDRM. Steps to Safety. With the Texas Department of Transportation. 2pp. http://txtownsafety.tamu.edu

Thompson, D.C., Rivara, F.P., and Thompson, R.S. (1989). A case-control study of the effectiveness of bicycle safety helmets. *New England Journal of Medicine*, 320(21), 1361–1367.

Thompson, D.C., Rivara, F.P., and Thompson, R.S. (1996). Effectiveness of bicycle safety helmets in preventing head injuries: A case-control study. *Journal of the American Medical Association*, 276(24), 1968–1973.

Thompson, M.J. and Rivara, F.P. (2001). Bicycle-related injuries. *American Family Physician*, 63(10), 2007–2014.

Tonucci, F. and Rissotto, A. (2001). Why do we need children's participation? The importance of children's participation in changing the city. *Journal of Community & Applied Social Psychology*, 11, 407–419.

Troiano, R.P. and Flegal, K.M. (1998). Overweight children and adolescents: Description, epidemiology, and demographics. *Pediatrics*, 101(3), 497–504.

van Herrewegen, J., Molenbroek, J., and Goossens, H. (2004). Children's Climbing Skills. R&T Project to Identify What Projects Children Can Climb on and How They Use Support Points While Climbing These Products. Part 1. ANEC: European Consumer Voice in Standardization. http://anec.org

Wakefield, J. (2004). Fighting obesity through the built environment. *Environmental Health Perspectives*, 112(11), A616–618.

Wells, N.M. (2000). At home with nature: The effect of greenness on children's cognitive function. *Environment and Behavior*, 32(6), 775–795.

CHAPTER 26

CHILDREN AND WAYFINDING

GARY L. ALLEN, RANI LUEDER, AND VALERIE RICE

TABLE OF CONTENTS

INTRODUCTION

Wayfinding refers to behavior that gets us to and from specific destinations successfully. As adults, we rarely think about what it takes to find our way; we usually travel in familiar environments, and our travel skills are well practiced.

Even so, adults with parental or supervisory responsibilities understand intuitively that children may become lost. It is therefore both prudent and practical for professionals and parents to consider wayfinding principles in the design of places and spaces used by children and when identifying methods to enhance children's wayfinding skills. Both aid children's wayfinding and prevent disorientation.

Any evaluation of wayfinding interventions should consider these facts about children:

1. **Children are explorers by nature.**
   Some children are more daring than others, but they all want to travel and explore their world. Their knowledge about their environment, that is, their "home range," increases steadily from kindergarten to their early teens (Cornell and Heth, 1996).

2. **Children's wayfinding skills improve dramatically with age and practice.** Young children's motivation to explore often exceeds their ability to stay oriented as they travel (Ambrose, 2000; Blades and Spencer, 1988). Yet, by their 12th year, their wayfinding skills commonly resemble those of adults—provided neither the adults nor children received specialized training in orienting themselves. Self-directed or independent travel is most important in acquiring wayfinding skills; being a passenger does not help much (Rissotto and Tonucci, 2002).
3. **Individual children vary greatly in terms of their wayfinding abilities.** It is possible to identify groups of "high spatial ability" children in much the same way as with adults (Fenner et al., 2000). Even so, there are many different types of spatial abilities; the individual who is outstanding with puzzles, diagrams, and geometry homework might not be skilled at the kind of navigation abilities that are the focus of this chapter.

   We should also note that although males tend to do better on tasks that involve mental rotation of figures (a standard test of "spatial ability") than do females, this ability is not usually important in wayfinding. Further, regardless of a child's level of ability, he or she can benefit from an effective intervention.

Any consideration of interventions requires an understanding of the problem. Thus, it is useful first to understand the types of wayfinding tasks and the abilities at our disposal to accomplish these tasks before considering the potential utility of interventions.

THE THREE TYPES OF WAYFINDING*

WAYS AND MEANS: HOW CHILDREN FIND THEIR WAY

Commuting involves repeated travel between destinations, as when children commute to and from school every day. In the end (as the user returns), the origin becomes the destination. Typically, as back-and-forth travel becomes familiar, commuting requires relatively little attention or thought.

Exploring is travel for the sake of seeing what is out there. Children of all ages explore environments on their own by foot, bicycle, skates, subway, and other means available to them.

As with *commuting*, the ultimate goal is to return *home*; the starting point typically becomes the final destination. Unlike *commuting*, *exploring* usually requires some amount of concentration in order to stay oriented relative to their *home range*.

Questing is travel to new destinations that the traveler knows to exist but has not visited. They use *maps* or *verbal descriptions* to find their destination.

Adults are accustomed to this type of travel. They may set out for a destination using information they learned from the *Yellow Pages* or the Internet.

This sort of *questing* is uncommon in young children, but improves slowly with age as their map-reading and verbal comprehension skills develop. Thus, *questing* involves a combination of comprehension, planning, and navigational skills; it is obviously an activity that demands attention and thought.

CHILDREN'S WAYFINDING ABILITIES

Children use a variety of abilities to find their way. Four of the following six types of wayfinding are based on direct experience. Two require assistance, such as with maps and verbal descriptions. These are ordered by increasing sophistication.

* For more information about these types of wayfinding, see Allen (1999).

Table 26.1 Children's Wayfinding Abilities

	What Is It?	Example	Description
Inborn Search Tendencies	• Wayfinding based on instinct or inborn tendencies. • Unconscious reenacting of behavior patterns in search of a destination.	A child may repeatedly circle about by walking away and then reversing back (in some cases).	• What appears to be aimless wandering sometimes involves a pattern of movements that increases the likelihood of finding a particular place, even without the traveler's awareness (Heth and Cornell, 1985). • While effective for animals, it is crude and unpredictable in both children and adults. There is no obvious way to base an intervention on such a tendency. • However, consideration of such behavior is an important factor in search and rescue efforts involving lost children (Heth and Cornell, 1998).
Path Integration	We remember our movements in order to keep track of where we started from (Loomis et al., 1999).	After walking an L-shaped path away from a place, a child senses a more direct shortcut back through unfamiliar territory.	• The fundamentals of path integration are present in infancy (Newcombe and Huttenlocher, 2000), but even challenge adults when integrating complicated routes (Loomis et al., 1999). • Although path integration is aided by inborn processes that do not require much conscious effort, it is greatly enhanced by attention and planning—which improves over the course of childhood. • Although children can be coached or taught to develop these skills, this form of intervention has inherent limitations; path integration tends to become less and less precise as distance increases. Thus, it may be a more successful strategy for staying orientated in smaller areas such as supermarkets or shopping malls than in larger areas such as theme parks or recreational areas.

(continued)

Table 26.1 Children's Wayfinding Abilities (Continued)

	What Is It?	Example	Description
Route Following	Following a route by taking actions in response to particular sights, sounds, tactile, or kinesthetic cues (i.e., landmarks).	A child rides her bicycle to school by turning right at the stoplight, left at the big yellow church, left at the fire hydrant, and right at the smell of fresh bread at the bakery.	• We often use route following in our day-to-day lives. It is the primary method people use when commuting and questing. • Route following skills can be readily coached or taught. It requires attention and effort to initially associate landmarks with actions. However, with repeated experience, even young children can learn to automatically follow a route. • Although the capacity for basic association learning is present in infancy, our ability to remember and associate sequences of landmarks with actions develops somewhat later. • The ability to distinguish between good and poor landmarks (those that are not consistent or may not be unique) emerges even later in childhood, years after the capacity for route learning per se (Allen et al., 1979) (Figure 26.1). • Elementary-age children are much better at following prescribed routes than learning one on their own.
Cognitive Maps	Knowledge about spatial relations between places that enables people to reach a destination (O'Keefe and Nadel, 1978).	A child who has made the trip from his own house to school and the trip from his own house to a friend's house infers how to get from school to his friend's house.	• In addition to providing a storehouse of information, cognitive maps permit people to infer how to get from place to place without making that specific trip beforehand. • Cognitive maps provide people with a powerful and flexible means of navigating that they can use in commuting, exploring, and questing. • Coaching or teaching cognitive mapping skills can be appealing, but require an investment of attention and thought and the emergence of skills in drawing logical inferences. • Basic cognitive mapping may emerge along with independent locomotion in late infancy or toddlerhood (Newcombe and Huttenlocher, 2000), but improves dramatically over the course of childhood as children learn to orient using distant landmarks in the environment (Anooshian and Kromer, 1986).

Figure 26.1 (Kit Lueder)
A good landmark should:

1. stand out from objects in the background rather than blending in,
2. appear at points where the traveler must choose which way to go,
3. be reliable and appear much the same for every trip,
4. be distinct within their environment, that is, no other objects closely resemble that object along the route.

COGNITIVE INTERVENTIONS TO FOSTER WAYFINDING SKILLS

Human performance is always the outcome of an individual's capabilities expressed within a specific task environment. Performance can be improved by enhancing the individual's capabilities or by altering the task environment (Figure 26.2). This section describes cognitive interventions designed to improve children's wayfinding behavior by enhancing their capabilities.

Each of these cognitive interventions involves an adult or skilled older child teaching an inexperienced child to improve his or her wayfinding skills. Just as with reading and math, children develop these skills by practicing specific tasks* that they will later generalize to other situations.

Table 26.1 summarizes the various types of cognitive interventions. The ages listed in the table are approximate and apply only in the context of the "take-home" messages in the text.

* For example, a task may involve finding one's way from home to school and back.

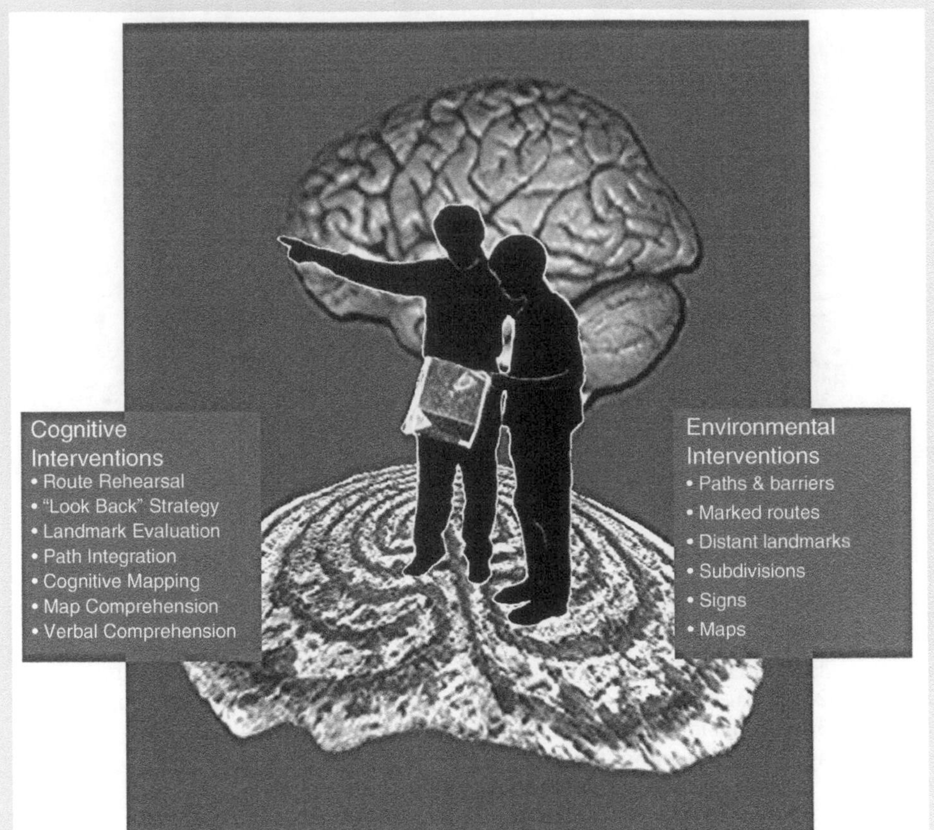

Cognitive
Interventions
• Route Rehearsal
• "Look Back" Strategy
• Landmark Evaluation
• Path Integration
• Cognitive Mapping
• Map Comprehension
• Verbal Comprehension
Environmental
Interventions
• Paths & barriers
• Marked routes
• Distant landmarks
• Subdivisions
• Signs
• Maps

Figure 26.2 (Figure courtesy of Adam Hutcheson)
You can improve a child's wayfinding skills by teaching them specific techniques of wayfinding and having them practice (cognitive interventions). Of course, a child must have sufficient cognitive skills to benefit from the instruction, so instructions must be tailored to the child's ability level.

If a number of children of different ages and abilities use an area, it is helpful to design the environment to "guide" the children's movements or offer child-friendly assistance for wayfinding (environmental interventions).

Table 26.2 Cognitive Interventions: Fostering Skills to Help Children Find Their Way

	What Is It?	Example	Description
Route Rehearsal	Repeated rehearsal to help children remember, as in the adage "practice makes perfect."	Adults escort children repeatedly along a specific route, pointing out the same good landmarks on each trip while repeating actions to take at each landmark (Figure 26.3).	• This intervention is widely used for a variety of memory tasks. In principle, route rehearsal takes advantage of the fact that children's recognition memory excels at a very early age. Its goal is usually to provide wayfinding competency with specific routes rather than to provide a generalizable skill. • Landmarks are important aspects of route rehearsal (see Figure 26.9 and Figure 26.10). Landmark recognition is a strategy that involves children making their way from one visible landmark to another, building on an existing strength (Cornell et al., 1994). • Route rehearsal plays a prominent role in many orientation-and-mobility training programs for visually impaired children and adults. Although such programs vary widely, some approaches incorporate the technique of tracking (that is, pointing out important items and reviewing progress along the way) to help them establish a return route. • It is also a useful way to provide very basic wayfinding skills to mildly to moderately mentally retarded or developmentally delayed individuals (Golledge et al., 1980).
"Look Back"	Looking back periodically to see how the path looks in the opposite direction while learning a route (Cornell et al., 1992).	Adults (or an older child) help by having children turn around and look at the route they have just taken, making sure to point out how a landmark and/or route looks from each direction (Figure 26.4).	• We often forget the path looks very different on the return trip from the way it looked on the outbound trip. In fact, young children tend to think of the return trip as distinct from the trip to a destination. For them, commuting initially involves learning two routes and the "look back" strategy enhances this learning process. • Clearly, route rehearsal combined with the "look back" strategy is a potent combination that can enhance route learning and can provide the basis for successful commuting and exploration. • In principle, "look back" should aid children of all ages, but practically speaking, it may provide little to no benefit to children under the age of 6 years because of their memory limitations.
Landmark Evaluation	Identifying various landmarks along the route.	Adults call children's attention to on-path and distant landmarks during travel, while emphasizing the features of good landmarks (Figures 26.9 and 26.10).	• Children can make use of good landmarks before the age at which they can actually *select* good landmarks on their own (Allen et al., 1979; Cornell et al., 1989). • The development of landmark evaluation skills requires some wayfinding experience.

(continued)

Table 26.2 Cognitive Interventions: Fostering Skills to Help Children Find Their Way (Continued)

	What Is It?	Example	Description
Path Integration	Children practice path integration along very simple routes to help them begin to develop the concept.	Show children how to move from one location to another, and then identify a shortcut back to the origin by having them walk various paths in an open field or a park with the point of origin clearly visible.	• Children appear to be capable of simple path integration over short distances as early as the preschool years, but this process is so error-prone and distance-sensitive that it makes a poor candidate for a general wayfinding strategy. Nonetheless, there is no reason to believe that it cannot be enhanced through tutoring. • The most effective form of practice is to show children how keeping track of their own movements can help them find a shortcut back. This is done by having them walk various paths with the point of origin clearly indicated. The idea would then be to transfer the same principle of taking a shortcut to a setting in which the starting point would not be visible from the ending point. • Typically, children should not attempt this practice until they reach about their eighth year and have developed independent wayfinding skills on (at least) a neighborhood scale.
Cognitive Mapping	Teach children to conceptualize relationships between objects in the environment.	• **Familiar environment** Help the child identify distinct elements (landmarks) in a familiar environment such as the street or neighborhood. Then draw a simple, relational map. • **Unfamiliar environment** Point out distinctive landmarks and how these change while moving through environments. • Such landmarks may include large, distinctive buildings in urban settings or in large amusement parks, or far off clocks in enclosed environments such as shopping malls or sports arenas (Figures 26.9 and 26.10).	• Cognitive maps have clear benefits. This flexible knowledge about distance and directions between places allows travelers to perform tasks that would otherwise be very difficult or impossible (e.g., inventing new routes and shortcuts between previously experienced locations). • There is reason to believe that practice can improve cognitive mapping skills, but very few children get tutoring in such ability. Instead, children piece together an overall picture of where places are in relation to each other through their independent travel experience (Cornell et al., 2001). • In general, adults can help children develop this skill by directing their attention to distant landmarks in different directions. • Research has not established the lower age limit on cognitive mapping skills. Evidence suggests these skills are not highly developed in children younger than 8, but improve steadily with age and experience. • As mentioned previously, cognitive mapping aided by vision is a challenge for young children, so it is clear that cognitive mapping without vision would be extremely challenging without adultlike cognitive skill, for example before age 12 years (Blades et al., 2002). For this reason, it is unlikely that cognitive mapping interventions would be useful with developmentally delayed individuals unless the environment was uncomplicated and the number of locations extremely limited.[a] Some orientation-and-mobility training programs for visually impaired children and adults involve specific instructions for the traveler to form a cognitive map of the environment. This is a more difficult strategy than route rehearsal in terms of cognitive interventions (although the two are not mutually exclusive).

[a] Obviously this depends on the individual's cognitive abilities.

Comprehending Maps	Have children explain or use maps while providing feedback as to the accuracy of the child's interpretation.	Beginning practice might involve teaching the child how a map of the home or neighborhood corresponds with that environment. More advanced practice might involve maps at the zoo or amusement park (Figures 26.13 and 26.14).	• Some people consider map reading innate, while others believe children learn the skill after infancy as they develop symbolic thought (Blaut, 1997; Liben and Downs, 1989). • In general, children do not consistently understand models or maps as wayfinding aids until they are about 4 years old. Even after their fourth year, they may misinterpret features of a map and have difficulty in understanding scale (DeLoache et al., 1997; Spencer et al., 1989). • In order to use maps for wayfinding, children must understand interrelationships between themselves, their maps and environments—an activity that many adults find difficult. • Helping in this learning process are standards for geographic education that are now part of many public school curricula. • Once familiar with map-reading conventions, children can use maps to facilitate route following and cognitive mapping activities. • Skill in comprehending maps improves with age and with relevant experience (Liben and Downs, 1989; Spencer et al., 1989). • With respect to wayfinding tasks, the ability to read a map makes possible an entirely new type of task, namely, questing, which involves setting out to destinations known only through symbolic means such as maps and verbal descriptions.
Comprehending Verbal Directions and Descriptions	Provide children with verbal directions; have them find their way from those directions.	Give children verbal route directions that involve sequences of actions that refer to sets of spatial relations and/or perceptions, such as "turn left here (action), after three blocks (special relation), turn right (action) on a narrow street (perception)" (Allen, 2000).	• As with actually walking the route, rehearsal of route directions is one of the best ways to get the information into memory. Verbal directions and descriptions are also the most common way of conveying wayfinding information. • Descriptions of environments can be a useful way of providing general information about the lay of the land, as opposed to a specific account of how to get from one point to another. In some ways, such descriptions are the verbal equivalent of maps, providing interlocation information that can be used for a variety of purposes that involve novel routes between places. • Children need to understand directional and spatial terms, such as distances, before they can properly understand and produce verbal instructions. Of course, understanding always precedes production, developmentally speaking. • Verbal comprehension skills begin to develop in late infancy. Children's understanding of spoken language advances rapidly between ages 2 and 3. During toddlerhood, they learn directional terms such as up and down. The terms left and right typically come a little later, around the ages of 4 and 5. Young children find it difficult to rapidly remember directional information because of working memory limitations, but the ability improves throughout childhood (Newcombe and Huttenlocher, 2000).

(continued)

Table 26.2 Cognitive Interventions: Fostering Skills to Help Children Find Their Way (Continued)

	What Is It?	Example	Description
Comprehending Written Directions	Provide children with written directions; have them find their way from those directions.	Games such as a treasure hunt, where children follow written directions to find hidden treasure: "X marks the spot."	• Using written rather than oral descriptions of routes can introduce its own set of difficulties. Travelers must simultaneously read text, understand content, remember what it is said in the proper sequence, and act on the information. • A child's reading comprehension skills greatly affects the child's ability to use written instructions. Nevertheless, once verbal comprehension skills are in place, children can use them to facilitate route following and cognitive mapping. Consequently, they can support the performance of all three wayfinding tasks (commuting, exploring, and questing). However, as with map comprehension skills, understanding language (verbal and/or written) is essential in questing, which involves seeking a previously unvisited destination.
Frame of Reference	Help children to understand locations and directions in relation to: • themselves (body-based frame of reference) • known areas (neighborhood or city within a state) • concepts (east, west, north, south)	Use a compass to orient oneself while camping.	• Understanding this concept is important in being able to understand and use verbal directions and descriptions. • Using a body-based frame of reference is a reasonable strategy, but some children and adults have some degree of left–right confusion. Such confusion, coupled with having to remember a long list of turns, interferes with a number of issues in wayfinding, including map reading and cognitive mapping. • Using cardinal directions (north–south–east–west) as a frame of reference is an option used by some individuals. However, it is challenging for children under the age of 12 years (as well as some adults) to apply these directions to actual movements in the environment (Ondracek and Allen, 2000). • Overall, a body-based frame of reference is probably the best with young children, especially when giving route directions. In contrast, cardinal directions may be more effective when providing an overall "lay of the land" description for older children.

Figure 26.3 (Valerie Rice) **Figure 26.4** (Valerie Rice)
Elia practices "route rehearsal" with his sister Kiran. As they walk to a friend's house, he points out distinct landmarks. The fire hydrant is distinct because it is the only hydrant on the street and its bright red color stands out. He tells her that this is where she should turn and then points the way back, showing her the route they traveled and how it will look on their return.

Routes with more than four or five choices can be difficult for preschool or kindergarten children to learn and more importantly, remember. Young children are in particular need of more rehearsal with longer routes.

This has obvious implications for children under the age of 6 years, whose working memory spans are generally shorter than those of older children and adults. For children under age 8 years, landmarks should be on or near the path rather than distant (Cornell et al., 1989).

Table 26.3 Cognitive Interventions for Children's Wayfinding

Intervention	Ability	Recommended Age
Route rehearsal	Route following	All ages
"Look back" strategy	Route following	6 years+
Landmark evaluation practice	Route following; Cognitive mapping	8 years+
Path integration practice	Path integration	8 years+
Cognitive mapping practice	Cognitive mapping	8 years+
Map comprehension practice	Route following; Cognitive mapping	4 years+ (small area) 8 years+ (large area)
Verbal description practice	Route following; Cognitive mapping	4 years+ (small area) 8 years+ (large area)

ENVIRONMENTAL INTERVENTIONS

This section addresses environmental design strategies to influence children's wayfinding. Naturally, this list includes a wide variety of interventions, some of which help to alter the structure of the environment itself and some of which simply inform the traveler of the structure, as it exists.

Table 26.4 summarizes environmental interventions. As with cognitive interventions, the ages listed are approximate and should be interpreted only in the context of the various caveats and conditions mentioned in the text (Table 26.5).

Table 26.4 Environmental Interventions: Design Features that Help Children Find Their Way

	What Is It?	Example	Description
Paths and Barriers	Paths and barriers clearly direct a child's movement along a route. • **Paths** Clearly marked walkways for a child to follow • **Barriers** Obstacles designed to prevent children's passage	Trails in national parks provide access to terrain that otherwise would be inaccessible to most visitors, but they are also designed to keep travelers out of sensitive natural areas.	• Paths make travel easier; barriers make travel more difficult. Both paths and barriers serve a dual purpose. Not only do they facilitate travel per se but they also direct or constrain travel. For example, sidewalks are designed to provide a network of paths for getting around in a city, while simultaneously keeping pedestrians out of street and off private property. • Paths and barriers can be particularly effective in constraining the wayfinding behavior of children (Figures 26.5 and 26.6). • Fewer paths and more barriers, such as in a museum or park, can help keep children oriented. • Commercial, entertainment, and educational enterprises often have something other than efficient wayfinding behavior in mind with their paths and barriers: ◦ Paths are often designed to put potential consumers in proximity with products. For example, exits from zoos or theme park attractions can force travelers through a gift shop or café area. The effect on children is quite predictable. ◦ Sometimes the desired effect is to **not** orient travelers. For example, winding pathways and circuitous routes in large amusement parks can make it difficult for visitors to remain oriented. Consequently, some settings seem larger and grander than the space they occupy. • Before using paths and barriers as environmental interventions, there needs to be careful consideration given to their impact on mobility-impaired travelers of all ages: ◦ Accessibility is a right protected by law. However, other issues such as physical and emotional fatigue are also salient concerns. In recognition of such factors, theme parks and other large tourist attractions frequently include direct access shortcuts for visitors whose mobility is impaired. Making such accommodations known to children and adults can be a challenge, frequently accomplished through printed literature and on-site signs (Robertson and Dunne, 1998).

Paths	A clearly marked track or course for a child to follow.	Neighborhood playgrounds often have paths without barriers.	• Paths affect wayfinding even without barriers. However, before placing paths without barriers in a setting, it is vitally important to consider patterns of pedestrian traffic in the overall environment. Pedestrians, including children, tend to ignore sidewalks in favor of making their own paths when the sidewalks are irrelevant to efficient travel. • In general, wayfinding tends to be easier in environments that feature routes with fewer turns, shorter paths, approximately equal distances between turns and turns of equal angles. • Certain factors can enhance the effectiveness of paths, such as attractive and interesting features along the path (e.g., colorful "street furniture" such as lampposts and benches). What children consider "interesting" obviously changes with their age. • Long pathways that wind around or have irregular intersections make orientation more difficult. This is particularly true for visually impaired individuals. Visually impaired travelers can walk for hundreds of meters on a gradual curve while under a false impression that their route has been straight.
Barriers	Using obstacles to block a child's path.	Gates, fences, or a simple chain across a path used to block a path.	• Barriers are effective only when they successfully constrain access. Often, fences and walls that are completely effective with young children (because of physical limitations) and adults (because of social convention) become "playground equipment" for older, more physically capable children (Figure 26.5 through Figure 26.7). • Environmental designers must take into consideration the specific purpose and the designated target group for barriers in order for the structures to have maximum effectiveness in constraining wayfinding behavior. For example, in many cases, barriers block physical access while allowing for visual access (e.g., fences, short walls, and hedges at zoos and gardens). This calls for a different type of barrier than those designed to limit physical access such as with taller fences, walls, and hedges.

(continued)

Table 26.4 Environmental Interventions: Design Features that Help Children Find Their Way (Continued)

	What Is It?	Example	Description
Marked Routes	Identifying symbols placed on a course or way taken in traveling, in order to help direct travelers.	Small swimming fish embedded in pavement tiles or placed on "path signs" that point the way to the aquarium.	• Routes can be marked with landmarks intermittently or with a continuous trail. Continuous trail landmarks are typically smaller human-made markers that may be placed on or beside features of the path that stand out from the background. • In some cases, landmarks to a specific destination are labeled or marked with a particular symbol: ◦ Because this method requires the traveler to remember only a single idea, it is particularly effective for young children. Of course, pictures can enhance the effectiveness of this intervention by reinforcing memory for the destination. • An entire route can be marked from beginning to end (see Figure 26.8). This is occasionally done in complex environments such as hospitals (showing the way to departments such as radiation or physical therapy), airports (showing the way to baggage claim), and amusement parks (showing the way to special attractions). • Routes can be marked for visually impaired travelers with a variety of tactile and auditory cues (Robertson and Dunne, 1998). For example, routes can be marked with local landmarks containing textured cues or Braille messages. Alternatively, continuous trails can be marked with textures such as carpet or artificial turf that can easily convey to the traveler whether he or she is on a desired path. • Overall, there are two drawbacks associated with this intervention. First, there are the logistics of marking a trail; it is simply not possible in all settings. Second, the effectiveness of this intervention diminishes rapidly as the number of marked trails grows.
Distant landmarks	A prominent object in the distance that guides travelers.	• Cinderella's Castle in Disneyland • Scoreboards on football fields • Distinctive attractions such as a merry-go-round in a shopping mall	• Designers can easily incorporate existing structures that define a region to serve as distant landmarks in the design phase of an environment, but these features can also be added to settings. Large, prominent landmarks make it easier for travelers to "get the lay of the land" and maintain orientation (Figure 26.9 and Figure 26.10). • Distant landmarks can be located either near the center of a region or at an extreme boundary of an environment. Because younger children tend not to pay attention to distant features, a large and centrally located landmark is often more effective as an aid to wayfinding and orientation. • Although symmetry is the rule in building large structures, strict symmetry limits the usefulness of distant landmarks for information about relative direction. For example, a large cylindrical building in the middle of an area is useful for information on distance (e.g., 'Do I go closer to or farther away from the building?') but not for direction (e.g., 'Am I on the right or wrong side of the building?'). Something as simple as placing a single business logo on the building remedies this situation.

Subdivisions	Splitting an area into smaller sections or regions.	Subdividing these neighborhood areas into sections: • A baseball field for soccer or flying a kite • A fenced playground for young children • Sheltered picnic areas	• Research indicates that the way children and adults think about an area is affected by how that area is subdivided into regions. • Places within the same region are reliably judged to be closer together than are places in different regions even when the distances are the same (Allen, 1981; Newcombe and Huttenlocher, 2000). • Subdivisions facilitate cognitive mapping by making it easier to learn the spatial relations among different environmental regions. • Designers typically use barriers to create subdivisions. For example, theme park designers create the illusion of completely differed "worlds" in areas just a few yards apart by using barriers. Subdivisions can also be created using interventions such as landscaping, color-coding, and furnishings (e.g., playground equipment or picnic tables). • Even preschool children can benefit from simple subdivision schemes involving four or fewer regions in which they can walk from region to region in just a few minutes. • For visually impaired wayfinders, textural cues on the walking surface can convey a sense of subdivision or at least mark boundaries between different areas.
Signs	Generally a board or placard indicating a location or direction.	Street signs (Figure 26.11)	• Signs are the standard design intervention to assist route following in public environments. These are typically placed at choice points. Ideally, they should also appear intermittently during long route segments to provide reassurance that the traveler is still on his or her way to the destination. • In order to be effective, signs must be placed appropriately and be easily seen and understood. • Effective signage for children (as seen in schools, museums, and amusement parks) tends to be colorful, easily visible at a child's height, and easily understood.
Signs in Built Environments	Directional indicators located on boards or placards.	Signs in amusement parks shaped like arrows that depict attractions while pointing the way to the attraction.	• Signs in built environments often include direction indicators, meaning that they contain arrows or are themselves shaped like a pointer (see Figure 26.12). • These indicators are usually easy to understand, especially in uncomplicated environments where the traveler's choices are to straight, turn left, or turn right. However, in some situations (e.g., roads that turn at unusual angles), such signs can even confuse adults. • Preschool children are typically not skilled in interpreting direction indicators on signs; either they fail to interpret the sign as a pointer or they mistakenly interpret the arrow as pointing to something immediately visible in the environment.

(continued)

Table 26.4 Environmental Interventions: Design Features that Help Children Find Their Way (Continued)

	What Is It?	Example	Description
Signs in Braille	Directional indicators placed on boards or placards that can be felt and read using fingertip tactile sensation.	City signs describing architecture, statues, and items of interest. These often do not provide directional information.	• Signs in Braille are a common feature to assist visually impaired travelers in obtaining information about the environment. • Braille does not consistently appear on directional signs with arrows or on signs that serve as directional indicators, perhaps because such signs are often raised in elevation to be visible at a distance, putting them out of reach for blind wayfinders.
Talking Signs	Travelers press a button at a landmark, or a sensor triggers when a traveler passes a particular place, producing auditory messages from "talking signs". The auditory signal is often sent to an electronic receiver carried by the traveler.	Talking "signs" or directional information may assist wayfinders of all ages and are particularly useful for visually impaired travelers.	• In their least sophisticated form, "talking signs" consist of a recorded message that names a landmark or place—sometimes referred to as "verbal landmarks" (Bentzen and Mitchell, 1995). • In a more sophisticated form, "talking signs" include recorded information about which way to turn at a particular point in order to approach a desired destination. • In their most sophisticated form, talking signs involve the interface of a Global Positioning System (GPS), which monitors the traveler's locations and a Geographic Information System (GIS), containing information about the location of key environmental features (Jacobson and Kitchin, 1997; Loomis et al., 1998). • The effectiveness of "talking signs" with children hinges on the children's ability to understand the wayfinding information provided. • Understanding "verbal landmarks" is not a problem, but naming a landmark per se does not help the traveler in unfamiliar settings. • "Talking" signs that include route information are potentially of greater use. However, because of general limits on short-term memory, messages must brief and to the point for children to comprehend and use the information. • Vocabulary (and directional concepts) must be age-appropriate. • It is likely that standard "talking sign" technology would be of limited effectiveness for children under age 8.

Maps	A regional representation, typically on a flat surface.	Road maps. Maps of city areas, subway systems, and built environments such as amusement parks and zoos (Figure 26.13).	• Designers often include maps in complex environments to assist wayfinders and most frequently; these are fixed-location "you-are-here" maps. Unfortunately, a large proportion of these maps have alignment flaws that make their use problematic. • Large proportions of "you-are-here" maps are situated so the environment portrayed on the posted map is misaligned with the map-reader's view of the environment (Levine et al., 1984). Consequently, the map-reader often heads off in the wrong direction. This misalignment occurs in airports, hospitals, schools, and urban areas frequented by tourists. It affects adults and children alike. • An effective "you-are-here" map aligns with the environment it portrays and has the position and orientation of the observer clearly indicated.
Posted Maps	A regional representation, placed vertically on a board, similar to a poster.	• Subway maps • Maps for zoos or amusement parks • Trail maps at a state park	• Posted maps can mark a route to specific destinations (route maps) or provide the general lay of the land (survey maps). Route maps can be a relatively simple, "string" of destinations, similar to a subway or other transport map. Pictures of landmarks and destinations help (Figure 26.13 and Figure 26.15). • Elementary school-age children should have little difficulty interpreting route maps. • The disadvantage to giving a child a simple route map is that it pertains only to the string of destinations on the path. That is, if the child "gets off the map" and loses contact with the pictured landmark sequence, they could have serious difficulties finding their way back.
Survey Maps	A map designed to show a particular regional area in its entirety.	An overhead-perspective map, such as a road map, is a typical example.	• Survey maps come in a variety of styles. Children are typically fascinated by road maps, but they rarely show any real skill in using them for wayfinding until about 10 years of age. • The chief problem with overhead-perspective maps is that they provide very little information about how things appear to the traveler and how things appear to the traveler is of great salience to child wayfinders.
Oblique-perspective Map	A map that depicts how environmental features appear from a particular perspective.	A theme park map that combines an overhead perspective (top or aerial view) with the oblique perspective (pictures of amusement park attractions).	• Oblique-perspective maps can have certain advantages for children (Liben and Yekel, 1996). For example, it combines items of interest for them (the attractions) with directional information. • It can be a great teaching experience in wayfinding. For many families navigating through such parks, the child views the pictures of the attractions on the map, selects the target location and the accompanying adult uses the overview information to navigate to the attraction. • Difficulties in map interpretation arise when the traveler's view of pictured features does not correspond with the view portrayed on the map. This applies to adults, but especially to children. • Another potential problem is violations of scale and distortions of distance. For example, the roller coaster that looms so large on the map is typically relatively much, much smaller when seen within its actual environmental setting, a fact that complicates wayfinding and orientation.

(continued)

Table 26.4 Environmental Interventions: Design Features that Help Children Find Their Way (Continued)

	What Is It?	Example	Description
Talking Maps	"Maps" that provide auditory information about a route or survey at a fixed location.	Such maps may range from simple tape-recorded messages delivered at fixed locations to complex GPS[a]/GIS interfaces.	• Particularly useful for visually impaired travelers. • Verbal route instructions are best when they refer to prominent landmarks and involve easily understood language. • With survey-type verbal descriptions, it is critical to select distinct distal landmarks to frame the environment and explain the relationships between locations and landmarks in spatial terms. • "Talking maps" are most useful when the verbal information is interpreted in terms of the special relationships of the traveler. This may be accomplished by pushing buttons on a map when the buttons are arranged to represent landmark locations in the environment. • In general, verbal route directions and survey descriptions are fine if there are only a few locations and landmarks, but a larger number would quickly overload the immediate memory capability of children, even older children. Verbal instructions and descriptions by themselves would be generally ineffective with young children (under the age of 8) and still rather challenging for older children.
Audio Recordings	Prerecorded auditory information that wayfind-ers listen to as they move around an environment.	Historical sites and museums provide head-phones for tourists. Pushing a button at that location activates an auditory message that describes items in the visual field and where the visitor should go next.	• An alternative to fixed "talking maps," the wayfinder listens as he or she travels. These presentations typically describe a panorama from the observer's vantage point. Such descriptions can be very useful for route following. • Just as with route maps, the traveler may become disorientated if he or she accidentally gets off-route. • The success of this technique hinges on the communications skills incorporated in the description and the comprehension skills of the listener. In general, the attentional and verbal comprehension demands of this technique make it best for children over 8 years of age, and even then in well-structured environments.

[a] GPS stands for "Global Positioning Systems." GIS represents "Geographical Information Systems." These systems provide global position information from satellite data (see section on "Assisted Wayfinding for Children in the Future").

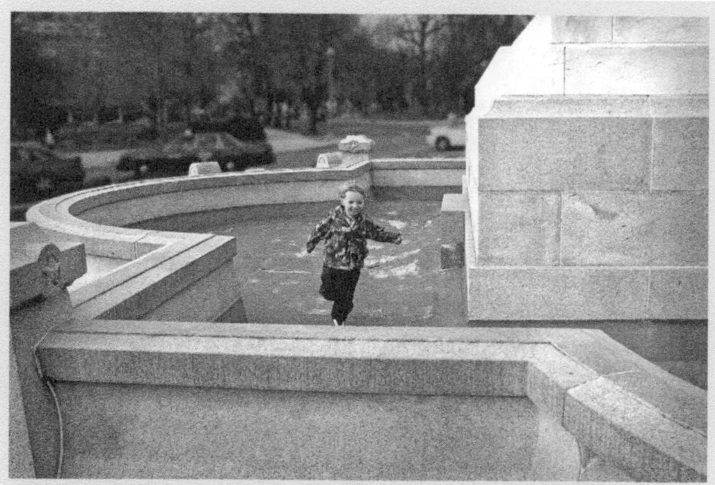

Figure 26.5 (Ursy Potter Photography)
Low barriers can channel movement along pathways while providing visual access to the surroundings. If planners believe the areas will have to channel large volumes of people, they may want to add directional signs (or other mechanisms) to keep people flowing in one direction.

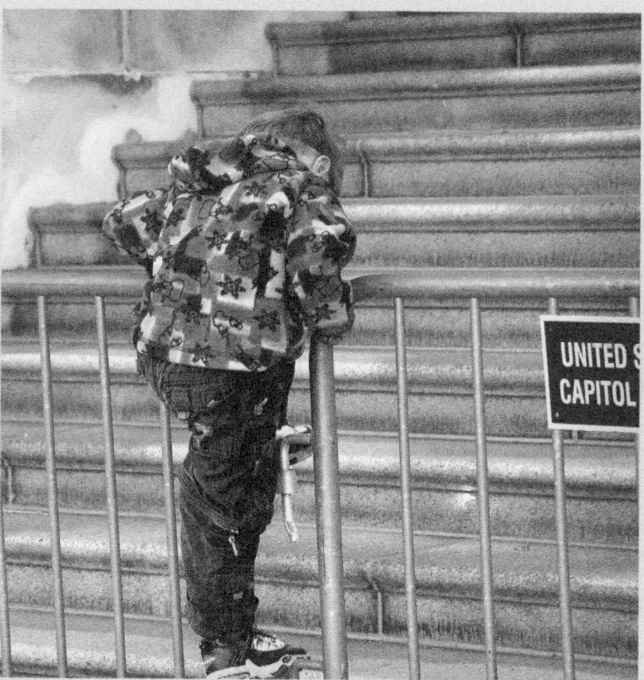

Figure 26.6 (Valerie Rice) **Figure 26.7** (Ursy Potter Photography)
Barriers are only effective if they actually prevent entry! Some barriers work well for very young children who cannot physically climb them, but they become fun and challenging structures for slightly older children. When designing barriers, it is important to think of the age ranges that will confront them and perhaps limit horizontal elements that can serve as ladders.

Figure 26.8 (Gary Allen)
A broken black line painted on the ground illustrates the use of a continuous trail. It takes the child visitor from one display to another, allowing them to spend as much (or as little) time at each one as they choose.

Figure 26.9 (Ursy Potter Photography)

Figure 26.10 (Kit Lueder)

- Distant landmarks are critically useful in maintaining orientation in complex urban settings.
- Each landmark requires a choice. The difficulty of a route is a function of how many choices the traveler must make; the more choices to make, the greater the potential for mistakes. This is particularly so in the middle portions of a route when the child is some distance from both the point of origin and the destination.
- Adults who select landmarks for children must also keep in mind the child's eye view of the world. Environmental features that are clearly visible to adults may not be visible to a child.

Figure 26.11 (Valerie Rice)
City or National Park trails often place markers intermittently and at intersections. These point travelers along specific paths and often include additional information, such as the length of the trail in miles. This sign informs travelers of the difficulty level of the path and whether it is paved for wheelchair travelers.

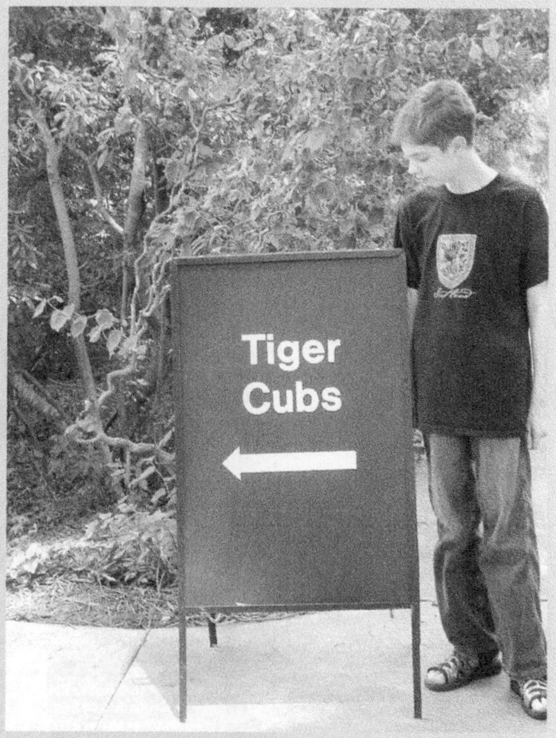

Figure 26.12 (Gary Allen)
Signs in built environments, such as amusement parks or zoos, often include direction indicators; that is, they contain arrows or are themselves shaped like a pointer. Designers try to keep these signs relatively uncomplicated.

Figure 26.13 (Valerie Rice) **Figure 26.14** (Ursy Potter Photography)
Pictorials help young children who cannot read to understand the region and the basic route structure. Even if they are unable to follow the route independently, it is good practice for them to try. For example, adults can point out the picture of the buffalo on the regional map, help them trace the route with their finger (kinesthetic memory), guide them along the route pointing out additional directional signs, and show them the matching buffalo picture at their destination (the buffalo enclosure).

Figure 26.15 (Ursy Potter Photography)
An aligned "you-are-here" map, such as this one, matches the observer's view of the map and his or her view of the environment. Even a young child can point to displayed icons in maps. However, usually only older children can actually use such maps for wayfinding.

Table 26.5 Environmental Interventions For Children's Wayfinding

Intervention	Ability	Recommended Age
Barriers and paths	Route following	All ages
Marked routes	Route following	Preschool+
Distant landmarks	Route following, path integration, and cognitive mapping	8 years+
Subdivisions	Cognitive mapping	Preschool+
Signs	Route following	6 years+
Maps	Route following Cognitive mapping	6 years+ (small area) 10 years+ (large area)

ASSISTED WAYFINDING FOR CHILDREN IN THE FUTURE

New research on how children of various ages find their way and use new technologies show great promise for wayfinding applications in the future. The components are already in place for a revolution in assisted wayfinding. The key ingredients in the revolution are the wedding of GPS with GIS. This technology will guide mobile entities (travelers on foot or in vehicles) to specific destinations.

Obviously, destinations must be predefined, indexed, and stored in existing databases. The same technologies that guide self-propelled munitions to military targets and car rental customers to their business appointments can guide the wayfinding efforts of individuals of all ages to their chosen destinations. Such systems can particularly benefit special populations. For example, it could inform individuals with visual and/or mobility limitations about physical impediments and access points (Jacobson and Kitchin, 1997; Loomis et al., 1998; Matthews and Vujakovic, 1995). Individuals with head injuries (i.e., traumatic brain injury) or with Alzheimer's disease who may have difficulty finding their way in familiar environments might also benefit from such systems.

A second technology that can revolutionize assisted wayfinding involves virtual environments (VE) (Durlach et al., 2000). This technology can digitally recreate and display any environment on head-mounted displays, enabling users to coordinate head and body movements with virtual views of places of interest. Although simple, desktop virtual tours are not unusual, movement within immersive environments may eventually become commonplace. What a way to learn, by practicing in "nearly real" environments!

While parents will appreciate a safe learning environment where they can teach their children how to travel, it may also provide children with a means of finding their way safely and without parental supervision! Each new technology brings both welcome innovations and worrisome challenges.

Two underexploited uses of VE technologies for wayfinding are **environment-to-map morphing** and **intelligent virtual guides**. Environment-to-map morphing involves the gradual visible transformation from a view of the world to a location on a map. It may also convey information about current locations relative to unseen destinations. Virtual guides combine GIS, GPI, and VE and provide a digitally based "person" who can review data concerning current location and surrounding landmarks and offer wayfinding advice. This system will require a portable visual display device and an interface of GIS and GPS, all of which currently exist.

Although the cognitive and developmental questions remain, there is already a great deal known about the effectiveness of various "virtual tutors." The notion of a child, guided by a nurturing, engaging, and unfailing virtual companion to any location on earth registered on a GIS database provides an intriguing (and perhaps unsettling) image.

CONCLUSION

Parents and guardians used to take sole wayfinding responsibility for their children. Now, children often must become competent independent wayfinders as they make their way to and from school, to after-school care and on recreational excursions themselves. Parents and adult mentors can implement a number of cognitive interventions to improve the wayfinding abilities of children in their charge. These interventions vary in their cognitive demands and thus in their age-appropriateness. The wise adult considers such developmental constraints when attempting to help children become better wayfinders.

How people find their way in new environments has always been a part of environmental design. It has also been a challenge, as designers try to help local travelers get to locations quickly and tourists visit attractions without getting lost. The need to assist travelers in wayfinding has grown as our environments are more complex, we spend more on traveling and we seek to meet the needs of an ever-changing population. The environmental interventions mentioned in this chapter are not novel ideas for environmental designers. However, the implications of each intervention for children's wayfinding may indeed provide new considerations for designers, particularly as they endeavor to create buildings and environments that are effective and age-friendly.

Figure 26.16 (Ursy Potter Photography)

REFERENCES

Allen, G. L. (1981). A developmental perspective on the effects of "subdividing" macrospatial experience. *Journal of Experimental Psychology: Human Learning and Memory, 7,* 120–132.

Allen, G. L. (1999). Spatial abilities, cognitive maps and wayfinding: Bases for individual differences in spatial cognition and behavior. In R. Golledge (Ed.), *Wayfinding Behavior: Cognitive Maps and Other Spatial Processes* (pp. 46–80). Baltimore, MD: Johns Hopkins University Press.

Allen, G. L. (2000). Principles and practices in communicating route knowledge. *Applied Cognitive Psychology, 14,* 333–359.

Allen, G. L., Kirasic, K. C., Siegel, A. W., and Herman, J. F. (1979). Developmental issues in cognitive mapping: The selection and utilization of environmental landmarks. *Child Development, 50,* 1062–1070.

Ambrose, G. V. (2000). Sighted children's knowledge of environmental concepts and ability to orient in an unfamiliar residential environment. *Journal of Visual Impairment and Blindness, 94,* 509–522.

Anooshian, L. J. and Kromer, M. K. (1986). Children's spatial knowledge of their school campus. *Developmental Psychology, 22,* 854–860.

Bentzen, B. L. and Mitchell, P. A. (1995). Audible signage as a wayfinding aid: Verbal landmark versus talking signs. *Journal of Visual Impairment and Blindness, 89,* 494–506.

Blades, M., Lippa, Y., Golledge, R., Jacobson, R. D., and Kitchin, R. (2002). Wayfinding by people with visual impairments: The effect of spatial tasks on the ability to learn a novel route. *Journal of Visual Impairment and Blindness, 96,* 407–419.

Blades, M. and Spencer, C. (1988). How do children find their way through familiar and unfamiliar environments? *Environmental Education and Information, 7,* 1–14.

Blaut, J. M. (1997). Piagetian pessimism and mapping abilities of young children. A rejoinder to Liben and Downs. *Annals of the Association of American Geographers, 87,* 168–177.

Cornell, E. H., Hadley, D. C., Sterling T. M., Chan, M. A., and Boechler, P. (2001). Adventure as a stimulus for cognitive development. *Journal of Environmental Psychology, 21,* 219–231.

Cornell, E. H. and Heth, C. D. (1996). Distance traveled during urban and suburban walks led by 3- to 12-year-olds: Tables for search managers. *Response! The Journal of the National Association for Search and Rescue, 15,* 6–9.

Cornell, E. H., Heth, C. D., and Alberts, D. M. (1994). Place recognition and wayfinding by children and adults. *Memory and Cognition, 22,* 537–542.

Cornell, E. H., Heth, C. D., and Broda, L. S. (1989). Children's wayfinding: Response to instructions to use environmental landmarks. *Developmental Psychology, 25,* 755–764.

Cornell, E. H., Heth, C. D., and Rowat, L. S. (1992). Wayfinding by children and adults: Response to instructions to use look-back and retrace strategies. *Developmental Psychology, 28,* 328–336.

DeLoache, J. S., Miller, K., Rosengren, K., and Bryant, N. (1997). The credible shrinking room: Very young children's performance with symbolic and nonsymbolic relations. *Psychological Science, 8,* 308–313.

Durlach, N., Allen, G., Darken, R., Garnett, R. L., Loomis, J., Templeman, J., and von Wiegand, T. E. (2000). Virtual environments and the enhancement of spatial behavior: Towards a comprehensive research agenda. *Presence: Virtual Environments and Teleoperators, 9,* 593–615.

Fenner, J., Heathcote, D., and Jerrams-Smith, J. (2000). The development of wayfinding competency: Asymmetrical effects of visuo-spatial and verbal ability. *Journal of Environmental Psychology, 20,* 165–175.

Golledge, R. G., Parnicky, J. J., and Rayner, J. N. (1980). Procedures for defining and analyzing cognitive maps of the mildly and moderately mentally retarded. In R. Golledge, J. Parnicky, and J. Rayner (Eds.), *The Spatial Competence of Selected Populations* (pp. 87–114). Columbus, OH: Ohio State University Research Foundation (Project 761142/711173).

Heth, C. D. and Cornell, E. H. (1985). A comparative description of representation and process during search. In H. Wellman (Ed.), *Children's Searching: The Development of Search Skill and Spatial Representation* (pp. 215–250). Hillsdale, NJ: Erlbaum.

Heth, C. D. and Cornell, E. H. (1998). Characteristics of travel by persons lost in Albertan wilderness areas. *Journal of Environmental Psychology, 18,* 223–235.

Jacobson, R. D., and Kitchin, R. M. (1997). GIS and people with visual impairments or blindness: Exploring the potential for education, orientation and navigation. *Transactions in Geographic Information Systems, 2,* 315–332.

Levine, M., Marchon, I., and Hanley, G. (1984). The placement and misplacement of "you-are-here" maps. *Environment and Behavior, 16,* 139–157.

Liben, L. S. and Downs, R. M. (1989). Understanding maps as symbols: The development of map concepts in children. In H. W. Reese (Ed.), *Advances in Child Development and Behavior* (Vol. 22, pp. 145–201). New York: Academic Press.

Liben, L. S. and Yekel, C. A. (1996). Preschoolers' understanding of plan and oblique maps: The role of geometric and representational correspondence. *Child Development, 67,* 2780–2796.

Loomis, J. M., Golledge, R. G., and Klatzky, R. L. (1998). Navigation system for the blind: Auditory display modes and guidance. *Presence: Virtual Environment and Teleoperators, 7,* 193–203.

Loomis, J. M., Klatzky, R. L., Golledge, R. G., and Philbeck, J. W. (1999). Human navigation by path integration. In R. Golledge (Ed.), *Wayfinding Behavior: Cognitive Mapping and Other Spatial Processes* (pp. 125–151). Baltimore, MD: Johns Hopkins University Press.

Matthews, M. H. and Vujakovic, P. (1995). Private worlds and public places—mapping the environmental values of wheelchair users. *Environment and Planning A, 27,* 1069–1083.

Newcombe, N. S. and Huttenlocher, J. (2000). *Making Space: The Development of Spatial Representation and Reasoning.* Cambridge, MA: MIT Press.

O'Keefe, J. and Nadel, L. (1978). *The Hippocampus as a Cognitive Map.* Oxford: Clarendon Press.

Ondracek, P. J. and Allen, G. L. (2000). Children's acquisition of spatial knowledge from verbal descriptions. *Spatial Cognition and Computation, 2,* 1–30.

Rissotto, A. and Tonucci, F. (2002). Freedom of movement and environmental knowledge in elementary school children. *Journal of Environmental Psychology, 22,* 65–77.

Robertson, B. S. and Dunne, C. H. (1998). Wayfinding for visually impaired users of public buildings. *Journal of Visual Impairment and Blindness, 92,* 349–355.

Spencer, C., Blades, M., and Morsley, K. (1989). *The Child in the Physical Environment.* Chichester: Wiley.

CHAPTER 27

*DESIGNING MUSEUM EXPERIENCES FOR CHILDREN**

JEFF KENNEDY AND MARJORIE PRAGER

TABLE OF CONTENTS

* All photographs in this chapter were taken by Kevin Burke at exhibitions designed by Jeff Kennedy Associates.

INTRODUCTION

Who visits museums? At the turn of the twenty-first century, the answer is—families with children.

Figure 27.1 (Kevin Burke)
Adler Planetarium & Astronomy Museum, Chicago, Illinois.

The last 25 years has seen a boom in museum experiences for children. In the United States alone, the number of children's museums has jumped from 38 to close to 300.* And museums of all kinds—science centers, art and history museums, zoos, aquariums, and nature centers—have come to realize that in order to survive and thrive, they must offer experiences that engage and entertain children. Whether they visit in family groups or on school trips, the future of museums depends on children.

WHY DO PEOPLE LIKE TAKING CHILDREN TO MUSEUMS? HOW DOES THIS INFORM DESIGN?

Parents and caregivers like to visit places where they can spend time with their children.† They like active rather than passive places that are fun, pleasant, and safe, and where they can socialize with other adults. Museums rank along with great natural settings, theme parks, and shopping malls as parents' favorite places to take their kids.

What does this mean for design? It means it is critical to design environments in which both children and adults feel comfortable playing active roles. Museums conducive to socializing and active learning *do* feel fun, pleasant, and safe.

* Association of Children's Museums.

† Museum staffs conduct surveys and interviews to learn why parents and caregivers bring their children to museums and what things they like best or would like to add.

CHILDREN DO NOT COME ALONE

Children do not come alone. Design for parents, caregivers, and children

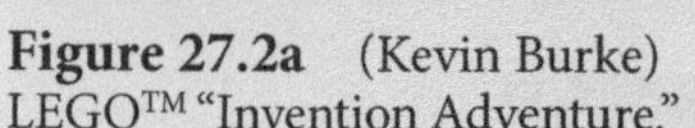

Figure 27.2a (Kevin Burke)
LEGO™ "Invention Adventure."

Figure 27.2b (Kevin Burke)
Stepping Stones Museum for Children.

Children and adults often play together in museums.
The experiences they share give adults insight into ways children develop and learn.

Museum spaces for small children often include labels and guidelines that illuminate aspects of early childhood behavior and development. Sometimes the labels suggest ways to extend the learning experience at home.

Staff and volunteers at many children's museums report that parents interact with each other. They share concerns, volunteer advice, and comment on the universality of behaviors common to their offspring.

Visitors value safety, comfort, and amenities, whether they visit once or return week after week. Museum designs therefore must attend to the needs of adults and children alike.

Figure 27.3a and Figure 27.3b (Kevin Burke)
Stepping Stones Museum for Children, Norwalk, CT. Increasingly, science centers are adding spaces with appeal for children aged 5 and younger while maintaining their offerings for the traditional children's museum target age range of 5- to 8-year-olds. This emphasis reflects changes in the population; many families now have children in this 3 to 5 age group.

Figure 27.3c (Kevin Burke)
The Kids Room, Maryland Science Center.

Many museums offer spaces specifically for toddlers. These spaces differ from those for older children. Toddlers are smaller and less agile. They enjoy full body activities and simple and direct interactions.

Parents need comfortable seating where they can watch their children play and socialize.

Figure 27.4a (Kevin Burke)
Stepping Stones Museum for Children, Norwalk, CT.

Bench type seating with adequate cushioning is often a good choice.

Figure 27.4b (Kevin Burke)
The Kids Room, Maryland Science Center.

Provide soft, cleanable floor coverings if you expect parents or caretakers to sit on the floor while sharing activities.

Adults with mobility impairments use benches with backs (and arms) as a means of support when getting up or sitting down.

Back support is welcome for all adults on a prolonged outing with their children. Parents with small children may also have babies to hold or nurse. Parents especially value benches with arms in spaces designed specifically for small children.

STAFFING, TRAINING, MAINTENANCE

STAFFING AS A PARAMETER FOR DESIGN

Children's museum spaces need ample staff, especially during busy periods, but in reality, adequate staffing is not always present.

The number of staff influences the design and less staff often means

- fewer full body activities (e.g., climbing) and
- open spaces with little clutter.

Designers should ask museum managers to provide a staffing plan before the design begins.

TRAINING

Adults and children notice when staff interact with them and are knowledgeable. Visits are more fun when the staff have good customer relations skills and can explain exhibits easily.

The design plan should include ongoing training in how to make the most of each exhibition. Floor staff positions have the highest turnover rate in museums—and institutional memory leaves with them. Ongoing training keeps employees up-to-date and communicates their importance to the success of the museum.

MAINTENANCE

Children's galleries often incorporate building activities, such as blocks, puzzles, computer pieces, and dress-up areas. Staff must keep these areas clean and neat by picking up pieces during the course of a busy day and at the end of the day. Items on the floor become stumbling hazards and impediments for wheel chair users and people with low vision.

Untidy areas also create a sense of disorderliness and lack of control. They do not feel safe and comfortable. Broken exhibits create a sense that museums do not put the visitors' experiences first. If maintenance staffing is minimal, components need to be simple and built to last.

As visitors flow through museums, they break and lose items. Designs need to be flexible, allowing replacement of components rather than entire exhibits to assist with ongoing renovation.

"HANDS-ON" DESIGN—*PLEASE TOUCH THAT!*

Figure 27.5 (Kevin Burke)
Stepping Stones Museum for Children, Norwalk, CT.

Younger children learn through play. The best hands-on experiences are open-ended and provide direct contact with elements that children manipulate, move, and control.

However, hands-on exhibits often incorporate knobs, push buttons, hand-cranks, and other controls. These present greater design challenges.

CONTROLS

Children know to spin or turn a wheel when they see it. Young children like to explore and will grab wheels to see what they do and what they might control! Exhibit designs should enable children to see the connection between controls and resulting actions.

The size of the control wheels should be appropriate for the task (e.g., larger diameters for more power). Larger mechanisms are easy for everyone. When in doubt, err on the side of larger controls. Controls such as wheels, knobs, and levers, which are too small, require fine motor control that young children and children with motor impairments may find difficult to manage.

Figure 27.6 (Kevin Burke)
Museum of Life & Science, Durham, NC.
Controls should be as simple as possible. Ideally, each hands-on component includes only one control mechanism.

Safety is critical. Handles and cranks may potentially transfer force back to the user. Safety mechanisms must be part of the design (e.g., one-way clutches). Avoid wheels and handles with spokes, which can be unsafe when spinning.

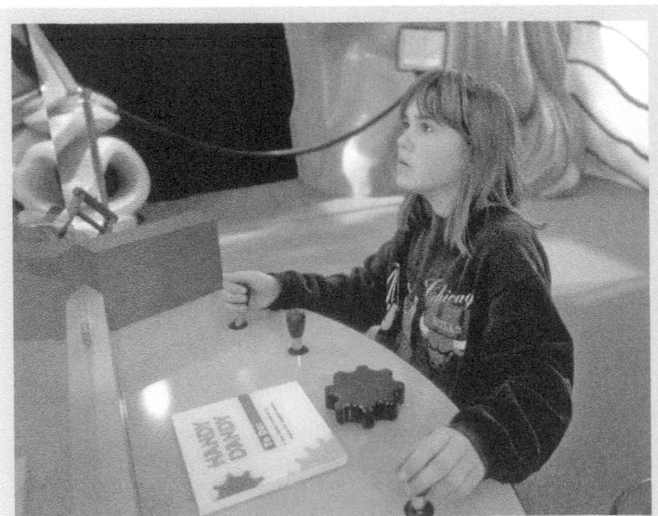

Figure 27.7 (Kevin Burke)
Health World Children's Museum, Barrington, Illinois.
Provide knurling, indenting, or other grip-enhancing treatments to controls (such as at the outer perimeter of a wheel) to increase children's ability to grip without sliding.

MANIPULATIVES

Many hands-on activities for children involve direct interaction with physical elements. They manipulate balls, building blocks, and parts that they combine in a variety of ways.

These are known as ***manipulatives***. Manipulatives are small items that children can easily maneuver with their small hands.

Design exhibits that include manipulatives in a way that promotes keeping pieces where they belong. Parts bins or containers in the activity area help, as will hidden storage for easy restocking. Adult visitors often help with this task, but museum staff must periodically pick up, especially on busy days. Parts on the floor pose tripping hazards, especially when they scatter to places where people do not expect to see them.

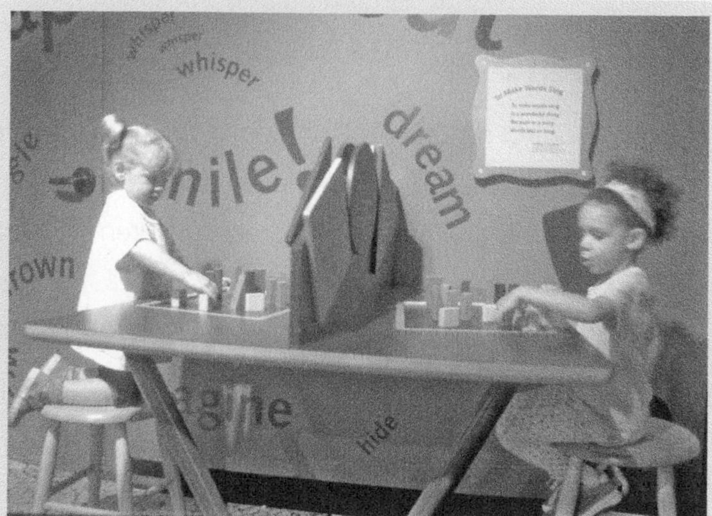

Figure 27.8 (Kevin Burke)
Stepping Stones Museum for Children, Norwalk, CT.

Less complex tasks are better for multiple-age users. Activities that require more controls or several steps with different controls are too complex for many children. One exception to this rule is when two or more users participate together, such as in playing a game.

Parts and pieces

What is the "right" number of pieces? Too many pieces can be a management problem; too few can frustrate children, especially on busy days when there are not enough pieces or parts to go around.

Trial and error usually determines the perfect balance, as the right number depends on the number of visitors that typically play at the exhibit. Communicate regularly with front-line floor staff about their experiences on the space to evaluate the design and restocking.

Figure 27.9 (Kevin Burke)
Stepping Stones Museum for Children, Norwalk, CT.

Replenishment. Choose materials and components that are widely available. For example, design a ball activity around a golf ball, squash ball, or tennis ball, rather than a special ball that the staff must order.

Always choose pieces that are too large to fit into a child's mouth, avoiding potential choking hazards.

Design Guidelines for Controls and Manipulatives

- Create simple designs
- Use controls that children intuitively understand, such as wheels
- Design control sizes to reflect the activity (larger wheels when tasks require more effort)
- Design controls large enough for everyone
- Design controls that are easy to grasp; children's fingers should not slip in an effort to move them
- Incorporate safety features
- Design for easy storage, easy replacement, and ease of upkeep
- Use contrasting colors for pieces to facilitate seeing parts on the floor
- Use readily available components
- Ensure that the pieces are easy to clean
- Avoid very small pieces that can fit in a child's mouth

DESIGNING THE ENVIRONMENT

Start with an exhibition plan

An exhibition plan considers the overall relationship of floor spaces and the exhibits within them, with an eye to the rhythms and flows of audiences, exhibit adjacencies, hot and cool spots, and shared service areas. Managers and designers should jointly develop and review the plan.

COMFORT

Children are the target audience, but adults bring the children. Since parents and caregivers can spend long stretches of time watching their children play, they will need seating and resting areas. Comfort becomes even more important during "worst-case" conditions—busy weekends with crowds of visitors.

"Comfort" refers to more than adequate seating and reasonable table and counter heights. *Sightlines* should allow parents to keep an eye on

Figure 27.10 (Kevin Burke)
The Kids Room, Maryland Science Center.

children throughout a gallery and easily view displays while sitting on the floor. Clear, well-placed building *signage* makes it easy to find one's way around. Family-friendly restrooms should be close by and large enough to accommodate busy-day traffic (see Chapter 26 on Wayfinding).

SOUND

Acoustic overload can make even visually attractive environments unappealing, especially for adults. The greatest source of sound in children's museums is the gleeful but shrill chorus of voices of children at play.

Environmental control is critical. Typically, well-padded carpeting and acoustic ceiling tiles dampen ambient sound. Some museums also choose to use sound-absorbing panels on walls, especially above the "touch" zone (the area children can reach). Well-made panels of this type are expensive, but can significantly reduce sound. Installing damping and insulation between floors, in walls, and in exhibit sound booths is less expensive during initial construction. It is more difficult and costly to retrofit.

Children and adults enjoy creating sound. Percussive activities (drums, xylophones, etc.) are particularly popular—but *loud;* so include them with caution. Sound may carry and make hearing difficult in other exhibits.

Figure 27.11a and Figure 27.11b (Kevin Burke)
"Making America's Music" Traveling Exhibition, Boston Children's Museum.
Specially designed spaces with acoustic protection can mitigate sound problems. Alternatives to true percussion instruments include electronic versions that allow designers to specify noise, by reducing or controlling the speaker systems.

LIGHTING

Overall light levels. Museum visitors prefer lighter and brighter spaces. Visitors often comment when an environment feels bright and cheery or seems too dark. Rarely do they verbalize appreciation of the subdued lighting of a darker space or remark that a well-lit space is too bright. They mention uncomfortable lighting, such as from direct natural light or glare that interferes with seeing exhibits.

Avoid stark changes in lighting. Passing from dark to brightly lit spaces requires time to visually adapt—a transition space should connect the two. Going directly from dark into a bright space causes discomfort. Use the magic of darkness sparingly and thoughtfully as a theatrical exhibition tool.

Light sources. Low-voltage fixtures or lamps tend to "hotspot," making it particularly difficult to read labels. They can create uncomfortably high contrast between disparate colors and finishes, as well as severe shadows.

Budget-conscious museums often ask designers to specify low-voltage fixtures. However, it takes many low-voltage fixtures to achieve appropriate illumination and cost savings are unlikely. If possible, a lighting-design professional should create the lighting plan, especially if it includes low-voltage fixtures.

Designers must aim fixtures to prevent glare or shadows. Both can make interacting with an exhibit component difficult or virtually impossible. As exhibits change over time, the lighting plan needs ongoing review. Fixtures will need replacing and refocusing. Reflection from a glass or plastic case is a classic glare problem, often extremely difficult to resolve because of the very broad range of viewing angles of adults and children. The best way to study and resolve this important design issue is with quick mock-ups in the design studio.

Fluorescent lighting, while not especially popular with museum designers, is often effective in children's spaces, especially when full spectrum lamps and incandescent fixtures accent lighting. This approach provides a good level of ambient illumination at a reasonable cost with minimal shadows and glare.

Color. Color brings appeal and excitement into children's spaces. Primary colors need not be the basis for color choices. Use a color palette with appeal for adults as well as children, so everyone will want to spend more time there.

Figure 27.12a and Figure 27.12b (Kevin Burke)
Stepping Stones Museum for Children, Norwalk, Connecticut.
A range of vibrant colors and thoughtful placement of objects can make an area attractive and stimulating for both children and adults.

SEATING

Adults often tire during their visit. While they may spend time on the floor with their children, they also need seats with good support.

> ***You can never have too much seating.***

Bench seating with pads and backrests let parents sit while they watch their children. Benches can be comfortable and easy to maintain. They promote social interaction among adults and permit visitors to move them to meet their viewing needs.

Benches that are not against a wall should have backs. Strength of construction and stability are important as people with mobility impairments use the backs or arms of the benches to help themselves when they sit down or get up.

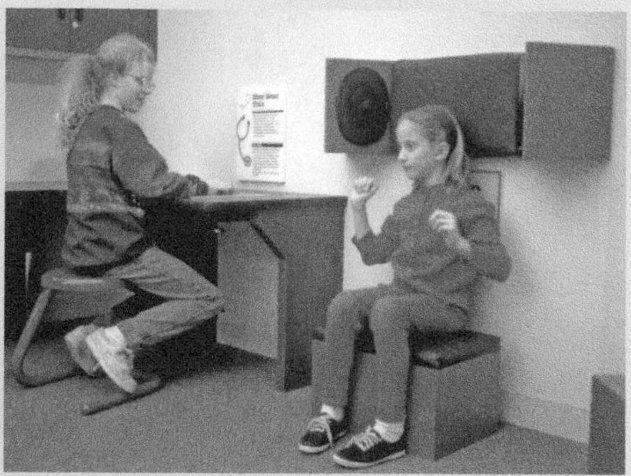

Figure 27.13a (Kevin Burke)
HealthWorks! Kids Museum, South Bend, Indiana.

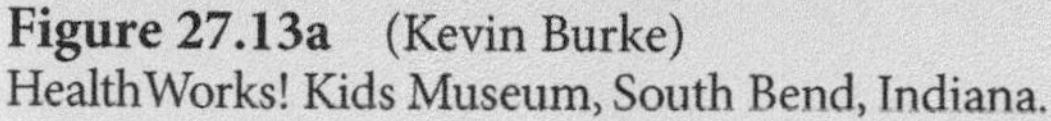

Figure 27.13b (Kevin Burke)
Health World Children's Museum, Barrington, IL.

Tabletop and computer activities often need stools or chairs for children and adults.
Not all furniture should be child sized (and children come in a variety of sizes themselves).

SIGHTLINES AND ACCESS

Parents and caregivers like to see their children at all times. This requires minimizing interior partitions and other large visual obstructions. Multilevel climbing activities should return children to their starting point (or very near) and keep them within easy view at all times.

Signs should clearly indicate where exhibit exits and entrances are. If this is not possible, either make them large enough for adults to accompany their children or anticipate the need for additional staffing to guide parents on busy days and respond to their concerns.

Figure 27.14a and Figure 27.14b (Kevin Burke)
Stepping Stones Museum for Children, Norwalk, CT.

Climbing structures should be readily accessible by adults in case of an emergency or accident. Crawl-through activities and tunnels should be relatively short and not too dark.

Figure 27.14c (Kevin Burke)
The Kids Room, Maryland Science Center.
A clearly marked entry invites visitors to come in and explore.

AIR QUALITY

Well-designed air-handling systems keep facilities comfortable with an even temperature and fresh-smelling air. As with all other aspects of facility design, design the heating, ventilation, and air conditioning (HVAC) for peak load conditions.

Water exhibits require special HVAC design. They can become breeding grounds for fungus and bacteria that pose a health risk and create an unpleasant smell that can easily permeate the area or entire museum. The HVAC system can prevent these problems.

Theater spaces may also need air supply and return if accommodating more than 20 people at a time. This is especially true when they are largely enclosed and have their own ceilings.

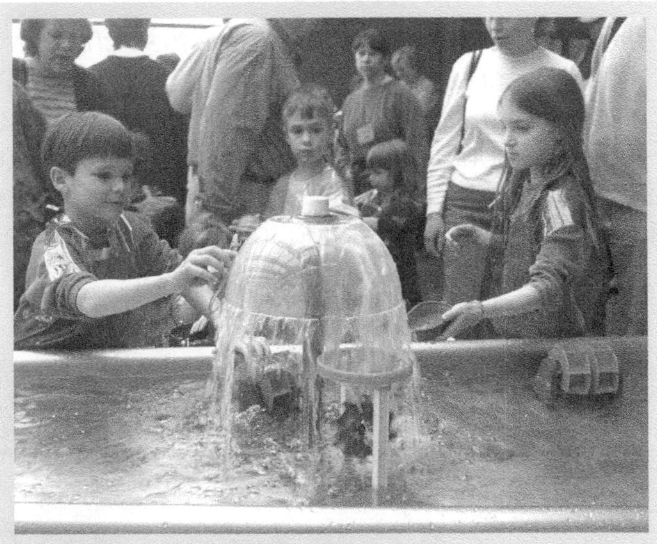

Figure 27.15 (Kevin Burke)
Stepping Stones Museum for Children, Norwalk, CT.

The Exhibition Plan

The *exhibition plan* is a periodic review of the overall relationship of floor spaces and the exhibitions with an eye to the rhythms and flows of audiences, exhibition adjacencies, hot and cool spots, and shared service areas. Managers and designers should jointly develop and review the plan.

SAFETY

Safety is the most important aspect of design.

Staff who perform the morning walkthroughs should check for potential safety hazards such as broken elements with sharp edges, drumsticks without a rubber tip, or cabinet doors left open.

Strategically placed checklists help staff identify problems. Broken or dangerous elements should undergo quick on-site repair or removal. Having a repair room near the exhibit floor is a wonderful safety asset.

General design guidelines for the environment

1. Design for everyone's comfort
   a. Consider abilities and limitations of children and adults
   b. Provide elements of each exhibit that appeal to all ages
2. Design for ease of movement
   a. Open spaces permit easy movement and visibility of children
   b. Frequent, clear signs show the way
3. Design so that people can hear each other and the sounds of the exhibit
   a. Design to keep ambient sound levels low—use sound dampening throughout
   b. Use spaces to "trap" sound
   c. Substitute mechanical for "real life" sounds when the latter are loud and carry
4. Design bright, light spaces
   a. Use transitional lighting between exhibits in which the light levels change
   b. Avoid glare
   c. Use dark spaces sparingly and only to create a "special effect"
5. Include places to refresh
   a. To rest and listen to a quiet story
   b. To eat
   c. Design for easy and quick access to restrooms
   d. Include comfortable seating with back support for adults
6. Plan for air quality

Figure 27.16 (Kevin Burke)
HealthWorks! Kids Museum, South Bend, Indiana.
Safe climbing components can include low walls with handholds that challenge children to move horizontally from one end to another along a ledge only a few inches above the ground. Walls should offer no possibility of climbing vertically without careful supervision.

CLIMBING

Children will climb anything they can, especially if it appears that climbing is part of the intent of the design. While this is often not the design intent, it is the interpretation of the user. Many adults allow children to climb on virtually everything in a space for children, perhaps mistakenly assuming it is acceptable. This leaves the task of avoiding unsafe climbing to designers and museum staff.

Characteristics that seem to invite climbing include step-like elements, footholds, limb-like elements, or any form of ladder. Of course, if a smaller child cannot see an exhibit, they will immediately look for a way to raise themselves higher, such as on a ledge or bench. Children can misinterpret vertical panel edges with 6 in. (15 cm) horizontal bars spaced 12 in. (30.5 cm) apart from top to bottom, as a ladder.

Multilevel climbing structures introduce the potential for stepping on children's fingers as they climb ladders and other scaling structures. Finger injuries are one of the most frequent injuries in facilities for children (see below). Injuries of this nature increase dramatically if the activity is available to children of all ages.

Younger children (3- and 4-year-olds) move more slowly (i.e., cannot get out of the way fast enough) and are somewhat less aware of others than are the older children (7- and 8-year-olds). Many facilities, recognizing that such structures are available at playgrounds, opt to eliminate climbing activities, especially when there are limitations in space and staff.

INJURIES

FINGER INJURIES: Finger-pinching injuries are common among children visiting exhibits. Doors with hinges and pivoting control levers are particularly hazardous. Doors or other hinged elements can catch fingers between two closing surfaces or pinch them in a space on the hinge side. Large and pivoting control levers with hard stops can catch fingers between the lever and stopping element. Avoid designs that can pinch fingers.

Many children develop finger injuries when they get stuck in a hole. This can lead to more severe injuries should children lose their balance or be bumped by another child while a finger is stuck. To avoid this, all holes should be smaller—or considerably larger—than a child's finger, and back perforated metal panels (with round perforations) with solid materials. The backing should be so close that a small finger cannot come between the two.

EYE INJURIES: Eye injuries are painful and can be quite serious. To reduce the risk of eye injury, avoid activities that involve small projectiles and stick-like components. If such elements are needed (e.g., a conducting baton or drumstick), making them oversize in diameter, with large radius rounded-over ends, will help. In the case of a drumstick or xylophone mallet, cover the end with a soft rubber ball-type tip.

Figure 27.17a and Figure 27.17b (Kevin Burke)
"Making America's Music" Traveling Exhibit, Boston Children's Museum.
Virtual conducting and "musical fences" activities make good use of tethered, oversized batons.

ELECTRICITY: Obviously, electrical components need to comply with applicable electrical codes.

Guidelines include:

- Use low voltages in areas that users might touch.
- If control of a component involving higher voltages is necessary, design an interface that lets children use a low-voltage relay to turn on–off the higher-voltage device.
- House electrical components and wiring in enclosures that require keys or special tools for access.
- Place light bulbs high and out of children's reach to avoid burns and breakage.

Other hazards include:

1. Elements that can become an unintended "weapon" (e.g., bat-like elements).
2. Elements that can become hot enough to the touch to cause pain (e.g., light bulbs inside metal housings).
3. Cords or cables that entangle children or their limbs (strangling or tripping danger).
4. Small pieces that may pose a choking hazard (diameters of 1.75 in. [4.5 cm] or less).
5. Access to the underside or inside of cabinets.
6. Sharp edges on cabinets, structures, and all other devices.
7. "Soften" or "ease" all outside corners of cabinets and tables at a child's head height by padding or rounding.

GERMS AND HEALTH

Design for quick and easy cleaning. For horizontal surfaces and elements that get a lot of use, use materials and finishes that are easy to clean with readily available cleaning products such as germicidal all-purpose cleaners.

Avoid items that go on children's heads, due to their potential for spreading head lice. This includes items used in role-play activities (hats, helmets, etc.), as well as audio headphones. Displays that include these items need daily cleaning. Replace hats and helmets regularly.

Figure 27.18 (Kevin Burke)
Stepping Stones Museum for Children, Norwalk, Connecticut.
Water invites touch. People may reach out to feel the water, even in areas that are not designed with this intent.

Figure 27.19 (Kevin Burke)
Stepping Stones Museum for Children, Norwalk, Connecticut.
Children of all ages love playing in water. As these areas are breeding grounds for germs, they need regular maintenance and germ-eradicating filtration systems.

If it is not feasible to install a filtration system, change the water frequently. This requires the location of the water supply and drains to be near by and easy to access, so staff can change the water quickly and efficiently.

Water exhibits require floor drains, sloping floor grades, and special flooring material to allow for drainage and cleaning. Rotating floor staffs are often responsible for the important day-to-day cleaning and maintenance of water features. Provide ongoing training for this maintenance, along with written rules and procedures.

General safety design guidelines

1. Design for safety first

a. Review each design plan and exhibit for safety concerns.

2. Check, check, and recheck

a. Train staff to scan for safety issues.
b. Provide safety checklists.

3. Pay attention to special areas such as

a. Climbing structures;
b. Hinged or pivoting components;
c. Stick-like elements;
d. Use of electricity; and
e. Water play.

UNIVERSAL DESIGN

The term *universal design* aims to enable all people to use them, rather than designing for particular groups such as people with disabilities.

The guiding principle is that everyone truly benefits from inclusive design, those with disabilities, children, elderly, and able bodied. For instance, captions on a video exhibit, while providing the narrative for deaf viewers, also offer a reading experience for a child just learning to read or a viewer not fluent in English. This section is written with the idea of inclusiveness at the forefront.

LABELS

The primary readers of labels and signs are adults. However, labels provide an important link between adults and children. Parents depend on labels to tell them what the things are or what to do. In some cases, the labels explain what their children are doing and why. Labels help adults feel comfortable coaching their children and providing guidance. Parents quickly gain "expertise" while their children busily experiment.

Design for size and reach. People of all ages attend children's museums and science centers and the physical elements need to "fit" each user. Often the chosen design approach is "one-size-fits-all."

However, adjustable-height tables and chairs are not the solution. They require adjustment after each use. Most adults do not want to make the effort and children definitely do not want to wait until adjustments are made.

One-size-fits-all works because visitors typically spend only a few minutes at a given activity. The variety of activities also keeps people moving among various postures and places; doing things on the floor, standing, crawling, and sitting.

Design dimensions are a compromise solution. For design purposes, the average size user in a children's museum is a 5-year-old boy, in a science museum it is a 12-year-old boy.

Suggested reading:

Norris, B.J. and Wilson, J.K. (1995) *Childata: The Handbook of Child Measurements and Capabilities—Data for Design Safety.* Department of Trade and Industry, London. ISBN 0 9522571 1 4.

Figure 27.20 (Kevin Burke)
"Making America's Music" Traveling Exhibition, Boston Children's Museum.

Type should be large (at least 24 point font size) so it is easy to read by a wide range of adults and is accessible to children just starting to read. Ample spacing between lines, high contrast with the background and, using fonts from the serif family facilitate reading. Even lighting and avoidance of glare will help.

LIGHTING

Rapid transitions from dark to light spaces cause discomfort for people with normal vision. For those with visual impairments, this can be a truly painful experience. Light-adapting transitional passages can alleviate this problem.

Labels are difficult to read if the background color is too light or too dark and if the lighting is too bright or "hot." For people with visual impairments, these conditions make a label illegible.

In dark environments (not especially recommended for children), labels are more easily read if the background is dark and the type is a highly contrasting light color (i.e., white type on a black background). Backlit graphics are the hardest to read of all, especially for people with vision impairments.

WHEELCHAIRS

Designing for wheelchair users benefits all visitors because exhibit spaces tend to open up, providing ample room for everyone to move about comfortably, even on crowded weekends.

Two basic rules of thumb are:

1. provide room for two wheelchairs to pass at all points (minimum 1.82 m [6 ft])
2. provide wheelchair turnaround space (minimum 1.82 m [6 ft] radius).

Wheelchair users need to pull up to exhibits to perform activities. Yet, wheelchairs come in a wide variety of sizes and configurations. This makes it difficult to arrive at a table height and under-cabinet foot clearance that meets everyone's needs. Pulling up sideways is not an acceptable alternative, except in extreme conditions. Wheelchair users need to get their feet far enough under an exhibit structure to reach controls or other elements requiring manipulation.

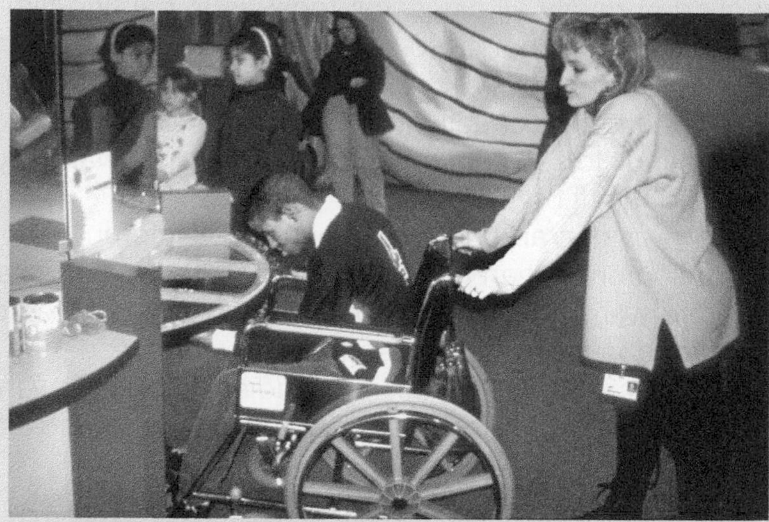

Figure 27.21 (Kevin Burke)
Health World Children's Museum, Barrington, Illinois.

Most children in wheelchairs sit higher than children on stools. This means that the table height for children on stools (or even standing up) is not ideal for wheelchairs.

If a tabletop activity is one in which reach is not important, then pulling up to the table (or other exhibit structure) to the point where the wheelchair arms make contact with the table or structure may be acceptable (as opposed to the arms going under the tabletop for a closer position).

Many hands-on activities require reach and many children using wheelchairs also have upper body impairments that constrain their reach. Building multilevel play structures is one answer.

CAPTIONING AND DESCRIPTIVE AUDIO

Figure 27.22 (Kevin Burke)
HealthWorks! Kids Museum, South Bend, Indiana.

Descriptive audio and captioning to multimedia components benefit all visitors.

The standard practice in nearly all museums today is to caption film and video productions for hard of hearing and deaf visitors. It is preferable for captioning to be "open," meaning that it is on screen at all times, rather than "closed," in which case it appears only when someone pushes a button or activates this feature. Drawbacks of "closed" captioning include that it is difficult to see caption buttons or requires making your way to the front of a group of viewers to turn it on.

Descriptive audio provides an additional soundtrack intercut with a film's voice track that describes the action on screen for visitors with sight impairments. Descriptive audio requires additional hardware, either in the form of a portable device or, in a theater-type setting, mounted headphones. For this reason, it is less common than captioning.

Well over half of a typical children's museum audience are below the reading age. This means that most captions are for accompanying adults to read and interpret for children.

CONCLUSION

Although children's museums are "for children," the best designs are for "children of all ages."

Museums for children are, in fact, *family museums.* The best museums make children and adults alike feel welcome and ready to explore. Design of museums for children becomes more complex and more eloquent, the more inclusive they become, engaging and inspiring visitors of all ages.

FURTHER READING

Belcher, M. (1992). *Exhibitions in Museums*. Published by Smithsonian Institution Press. Available from American Association of Museums, http://www.aam-us.org/bookstore, p. 248.

Kennedy, J. (1990). User *Friendly: Hands-On Exhibits That Work*. Published by the Association of Science-Technology Centers (ASTC). Available from ASTC, http://www.astc.org/pubs, 77pp.

McLean, K. (1993). *Planning for People in Museum Exhibitions*. Published by the Association of Science-Technology Centers. Available from ASTC, http://www.astc.org/pubs, 196pp.

Serrell, B. (1996). *Exhibit Labels: An Interpretive Approach*. Published by AltaMira Press. Available from American Association of Museums, http://www.aam-us.org/bookstore, p. 261.

CHAPTER 28

PLAYGROUND SAFETY AND ERGONOMICS

ALISON G. VREDENBURGH AND ILENE B. ZACKOWITZ

TABLE OF CONTENTS

Four-year-old Maria's grandma picks her up from preschool and takes her to a park for lunch. Maria plays on a geodesic climbing dome. After a while, her grandma calls and Maria races to her. As she does so, her finger catches on a bolt assembly that connects the arms of the dome. Maria tries to pull it out and lacerates her finger.

James, a 3-year-old boy, walks up the chute of a slide. His hand slips, he falls over the side and seriously injures his head. Typical of a child his age, he forgot that he had been instructed on proper use of the slide and had played there many times before. His guardian notes the sand around the play equipment was hard and there was not much of it. In this case, the compressed sand was 20 cm (8 in.) deep. The slide platform the boy fell from was 2.9 m (9 ft. 6 in.) off the ground.

How can we prevent such injuries? What steps can we take to protect our children on playgrounds? We often expect that spaces designed for children will protect them from possible harm. Guardians may fail to recognize the hidden hazards that lurk in familiar environments considered "child friendly." However, scenarios such as those described above happen more often than many people realize.

THE PROBLEM

Playgrounds enable children to develop physical and social skills in a fun and stimulating environment. This environment provides young children with "work" activities that help them develop skills such as coordination, figure ground (dimensions), and hand–eye coordination. During this skill-development phase, playground accidents and resulting injuries are possible.

User-environment fit is important for children as well as adults and is critical in the playground environment. Children need challenge, with minimal potential for injury. Improper playground design can contribute to injuries.

This chapter focuses on safety issues that will help reduce children's exposure to danger while they are using playgrounds. Specifically, this chapter reviews common playground hazards, child user requirements, and design guidelines.

CHILDREN AT RISK

Each year approximately 205,000 pre-school and elementary school children receive emergency room care for injuries on playground equipment in the United States. According to the US Consumer Product Safety Commission (CPSC), 76% of child injuries in the United States happen on public playground equipment while 23% occur on home playground equipment. A child is injured on a playground every 5 seconds in the United States alone.

In some cases, children have even died. In the United States, from January 1990 to August 2000, there were 147 deaths of children under age 15 involving playground equipment (CPSC). Seventy percent of those deaths occurred at home while 30% occurred on public playgrounds. Girls were involved in a slightly higher percentage of injuries (55%) than were boys (45%).

Injuries on playgrounds are an international concern. For example, the Child Accident Prevention Trust (CAPT) in the United Kingdom reported that in 1999 alone, over 34,500 children under the age of 15 were injured on public playgrounds (CAPT, 2004). The Canadian Hospitals Injury Reporting and Prevention Program (CHIRPP) is another body that tracks playground injury data. They indicated that in Canada, over 4200 children were injured on playgrounds in 1996 (CHIRPP, 1997). Table 28.1 depicts playground injuries in the United States.

Table 28.1 Child Playground Injuries in the United States by Age and Injury

Age	Most Common Type of Injury (%)	Most Frequent Equipment (%)
0–4	Head and face (49%)	Climbing equipment (40%) Slides (33%)
5–14	Arm and hand (49%)	Climbing equipment (56%) Swings (24%)

Injury data by type, rate, and age (Tinsworth and McDonald, 2001).

RECOMMENDATIONS

We can greatly reduce many of these injuries with properly designed, installed, and maintained equipment. We can avoid other injuries with appropriate supervision of children.

Playgrounds should be fun and interesting, yet safe places for play. This chapter reviews playground design, safety, and supervisory considerations for children between the ages of 2 and 12 years. This chapter also addresses age-appropriate capabilities and limitations of children.

Play facilities must meet the needs of *all* its child users, regardless of age, developmental level, and physical abilities. It should also enhance children's physical, intellectual, and social development. Because most public playgrounds attract a wide range of children with different backgrounds and abilities, supervision is fundamental and must also be considered in playground safety.

WHY KIDS PLAY

While play is as important to children's development today as in the past, other attractive activities compete for children's attention during their leisure time. Despite the importance of physical play for child health and development, several counterinfluences deprive today's children of vital physical play opportunities (Epstein, 2000). These include an emphasis on academics in school settings, eliminating school recess times, parental fears about perceived dangers on playgrounds, and competition for playtime from technology-related pastimes like video games and the Internet (Frost, 1999).

PHYSICAL AND COGNITIVE DEVELOPMENT

Sedentary lifestyles have contributed to obesity, poor physical fitness, and early signs of heart disease and diabetes among very young children (Ross and Gilbert, 1985). While playgrounds may have some inherent r isk, the relative risk of no exercise is considerable. Playgrounds give children of all ages the opportunity to challenge their bodies physically by climbing, running, and jumping. Use of overhead bars can influence in muscle development. These types of activities are important in today's society where kids are so tempted by technical diversions like computers, video games, and television. Adults can enhance a child's physical playground experience by offering challenges that are within the child's ability level.

BUILDING BLOCKS FOR FUTURE SKILLS

Children learn as they play. It helps them develop physical and cognitive abilities while nurturing social skills. Children will play more, and thus learn more in stimulating playgrounds.

For preschoolers and young adolescents, these play activities are essentially their "job." Playgrounds provide opportunities for cooperation, sharing, and taking turns, all essential social skills for school and for adult life (see Figure 28.1). Playgrounds with sandboxes and playhouses also provide opportunities for imaginary and pretend play. These are essential elements of early childhood development.

Figure 28.1 (Alison Vredenburgh)
Playgrounds enable children to develop physical and social skills.

DESIGNING PLAYGROUNDS

RISKS AND HAZARDS

Risk is inherent in play. It is important to balance safety with children's need for creativity, challenge, and novelty. Safe design should enable children to refine their cognitive and motor skills that help them function in everyday life and prepare them for tomorrow's challenges (Frost, 1999).

Rule #1. Increase challenge while reducing risk

Why not eliminate risk? Playgrounds wholly without challenge are also boring; children avoid them or use equipment in unintended ways. They seek experiences that develop skill sets and provide sensory input that helps them develop. For example, children who spin on the merry-go-round may need the sensory input to refine equilibrium and balance. Playgrounds must fit a variety of developmental stages for children of various ages.

We will never be able to completely eliminate accidents on playgrounds that children perceive as exciting and worthy of play. If play spaces are too "safe," children may even overcompensate and introduce other unintended risky behaviors.

The perception of risk can be designed into equipment. For example, high places where there is no chance of falling a long distance and low places such as a rope net bridge or a balance beam where challenge and dexterity are called for, appear risky but there is little chance of serious injury (Uzzell and Stone, 1990). However, potentially dangerous situations that children cannot anticipate or control are hazardous and should be eliminated.

Types of risks and hazards. Playgrounds sustain a broad range of accidents and injuries and require a varied response. Impact-absorbing surfaces can help protect children when falling from heights (a common playground hazard). However, surface materials are not the only issue; poor design, inadequate maintenance, and unintended use also result in accidents.

In addition to the hazards associated with children using age-inappropriate equipment and a lack of responsible adult supervision, a variety of general hazards are common to many types of playground equipment. Table 28.2 describes common playground hazards associated with injuries or fatalities.

Table 28.2 Playground Hazards Associated with Injuries or Fatalities[a]

Hazard	Description
Excessively High Equipment	Falls from tall equipment account for most serious injuries on playgrounds. Most approved surfacing materials cannot safely absorb children's falls from heights greater than 3 m (10 ft). We therefore recommend that equipment not be over 3 m (10 ft) high and that the surface protection is appropriate for the fall height of the equipment.
Inadequate Use of (fall) Zone Coverage	Provide effective resilient surfaces under and around all equipment so surfaces absorb impact energy when children fall. This resilient "fall zone" coupled with appropriate equipment height can reduce the incidence and severity of injury.
Lack of Guardrails	Any platform, deck, or walkway (including tops of slides) more than 76 cm (30 in.) above the ground should have a protective barrier to prevent falls.
Protrusions	Objects that project out from the equipment, such as nails, screws, bolts, and edges of broken equipment can cause cuts and bruises or entangle clothing and cause serious injury or death by strangulation. Such objects should be rounded or flush with the surface of the equipment.
Sharp Edges	Avoid sharp points, corners, or edges on all components of playground equipment that might puncture children's skin. Wooden and metal parts should be smooth or flush with the surface. Round out corners.
Head Entrapment	Openings that can entrap a child's head might be found in spaces between posts, ladder rungs, or deck levels. Therefore, bars should be less than 9 cm (3.5 in.) or greater than 23 cm (9 in.) apart.
Hard Swing Seats	Swings with heavy wooden or metal seats can seriously injure children who walk too close to moving swings. Use soft surfaces and wrap seats.
Pinch and Crush Points	Many moving structures contain parts in which a child can insert a finger or hand, resulting in a pinch or crush injury. These are common at the center of merry-go-rounds, swinging gates, and seesaws. Equipment barriers should block access to these areas.
Open S-Hooks	Ensure S-shaped hooks that attach swing seats to chains are completely closed. If the S is open, the seat may slip off the hook when a child swings high, causing the child to fall.
Tripping Hazards	Install anchoring devices for playground equipment below play surfaces to eliminate trip hazards. Other objects in the play environment that may cause children to trip include rocks and roots. Remove these objects when possible; if they cannot be eliminated, bury them or construct a barrier around them.

Source: From Jambor, T. and Palmer, S.D. (1990). *Playground Safety Manual.* Birmingham, AL: Injury Control and Research Center, University of Alabama at Birmingham

[a] These are listed above and also described in greater detail later in this chapter.

MAKING PLAYGROUNDS ACCESSIBLE

Playground designers can improve the accessibility of playgrounds for children with disabilities. The information in this section can be applied during the design, construction, and alteration of play areas for children ages 2 and over.* While this section outlines some provisions in this US standard, readers from other nations might refer to specific guidelines pertaining to their country. Table 28.3 lists the accessibility standards for several countries in Europe.

Table 28.3 Accessibility Standards in Europe

Country	Standard Number	Topic	Comments
European Standard	EN1176 and EN 1177	Playground equipment	Overall standard
	TC136	Sports, playground, and other recreational equipment	European standard subsections
	SC1	Playground equipment for children	
	SC1/WG3	Entrapment	
	SC1/WG4	Installation and maintenance	
	SC1/WG7	Individual types of equipment	
Austria	B 2607	Directives for planning	
Great Britain	BS 7188	Impact absorbing surfacing	Performance requirements and test methods
	BS EN 1177	Impact absorbing surfacing	Safety requirements and test methods
	BS EN 1176	Playground equipment	
	PAS 018	Playground equipment	
Germany	DIN 33942	Barrier-free playgrounds	Security requirements and test methods
Portugal	EN1176/1177		Adopted European standard as law
Spain	PNE 147102	Playground equipment	

Below are some guidelines from the Americans with Disabilities Act Accessibility Guidelines for Play areas (36 CFR Part 1191).

Clear floor or ground space. Provide clear space around play components that is at least 76 cm (30 in.) by 122 cm (8 in.). Clear space shall have a slope not steeper than 1:48 in all directions.

Accessible routes. Accessible routes must be located within the boundary of the play area and connect ground level and elevated play components, including their entry and exit points. Minimum clear width of 152 cm (60 in.) is required for ground level accessible routes in order for a wheelchair to turn.

* Guidelines do not exist to design play areas for children under age 2.

Play components. Locate each type of ground level play component and at least half of elevated play components on accessible routes.

Protrusions. Ensure that objects do not protrude into accessible routes at or below 203 cm (80 in.) above the ground.

Ramp slope and rise. Slopes for ground level accessible routes that border play areas may not exceed 1:16. Where a ramp is part of an elevated accessible route, the ramps are limited to 30 cm (12 in.) of change in elevation. Thus, a 3.5 m (12 ft) ramp would have a slope of 1:12, while a 4 m (14 ft) ramp would reduce the slope to 1:14.

Handrails. The top of gripping surfaces of handrails shall be 51 cm (20 in.) minimum to 71 cm (28 in.) maximum above the ramp surface. Handrails are not required for ramps within ground level use zones. Handrail extensions at the top and bottom of ramps are not required to avoid any potential protrusion into circulation paths.

Children's reach ranges. When designing play components with manipulative or interactive features, consider appropriate reach ranges for children seated in wheelchairs. Table 28.4 lists recommended ranges for forward and side reaches. Climbers on ground level accessible routes should have some climbing rings in these ranges.

Table 28.4 Appropriate Reach Ranges for Children Seated in Wheelchairs

Forward or Side Reach	Ages 3 and 4	Ages 5 through 8	Ages 9 through 12
High (maximum)	36 in. (915 mm)	40 in. (1015 mm)	44 in. (1120 mm)
Low (minimum)	20 in. (510 mm)	18 in. (445 mm)	16 in. (405 mm)

Ground surfaces. Measure the amount of effort it takes to propel a wheelchair straight ahead and while turning with a force wheel on a rehabilitation wheelchair. To meet the American standard (ASTM 1951), these forces must be less than those to propel a wheelchair up a ramp with a 1:14 slope.

Rule #2. Don't forget the intended users!

AGE APPROPRIATENESS

Playgrounds should allow children to progressively develop and test their skills by providing a series of graduated challenges. These challenges should be appropriate for the child user's age-related abilities and limitations. Adult supervision is necessary because all playgrounds should provide challenge and because children use equipment in unanticipated ways (CPSC, 1997).

When designing new playgrounds, it is recommended that there are separate areas for younger children with appropriately sized equipment and activities that will best serve their less advanced developmental level (NPPS, 1996). Delineate separate age-specific play areas with conspicuous signs that indicate appropriate ages for that area. Play areas for children younger than 5 years old may include areas to crawl, low platforms with multiple accesses, flexible spring rockers, sand areas, and slides no taller than 122 cm (4 ft high). Developmentally appropriate play areas for children older than 5 years may include rope or chain climbers, horizontal bars, swings, merry-go-rounds, seesaws, and slides (NPPS, 1996).

Physical and cognitive limitations of children. A key factor in many playground injuries is that children play on equipment not designed for their age group. Therefore, the steps or railings may be too far apart or require additional dexterity and coordination outside the range of typical capabilities for younger children. Most such injuries involve children under 4 years playing on equipment designed for children aged 5 to 12 (Tinsworth and McDonald, 2001).

Preschool and school-aged children differ dramatically in physical size and ability as well as in their cognitive and social skills. Preschool and young children's cognitive and physical abilities particularly depend on their level of development. Physically, they often need adult help. Young children need clear and simple directions to perform most activities due to their cognitive limitations.

Although they may be able to describe basic concepts like time and distance, they cannot make absolute or relative judgments at this age (Zackowitz and Vredenburgh, 1998). Furthermore, they have little or no understanding of their mortality or the mortality of others. What seems risky or dangerous to an adult does not seem so to a young child (Turner and Helms, 1995). They do not yet use the type of "if…then" analysis that leads people to understand the consequences of their behavior.

Young children do little monitoring of their own memory and comprehension. Children do not see the whole picture. They lack the experience, background knowledge, and ability to understand the world and are sometimes unable to carefully explore their surroundings. Since people make decisions based on prior experiences, young children may make illogical choices because existing knowledge is limited (Turner and Helms, 1995).

From a design perspective, separate playground sections by the users' age ranges. Most children younger than 5 years are smaller, weaker, less coordinated, and have a higher center of gravity than children older than 5 years. Therefore, they need smaller steps, crawl spaces, and railings with smaller grips positioned appropriately on platforms. Generally, equipment for children of age 5 and under should be lower to the ground (NPPS, 1996).

SUPERVISION

Because children demonstrate varying physical and cognitive capabilities and limitations, they need adequate supervision to ensure safe playground play. It is estimated that at least 40% of all playground injuries are related in some way to inadequate supervision (NPPS, 1996). Supervisors should guide children to play on age-appropriate equipment; equipment designed for children aged 5–12 is too big for preschoolers. Young children may lack the coordination to negotiate a large piece of equipment alone. However, playgrounds are full of attractive "toys" that children are often anxious to explore. Since they are not aware of their own physical and cognitive limits, adult supervision is an essential part of a safe play environment (Vredenburgh and Zackowitz, 2002).

In addition to a reduction in accidents resulting from misuse of equipment, social safety is less of an issue if adults are present. For example, if a diligent adult is present and observing the activities occurring on the playground, adults with bad intentions may be discouraged from victimizing children. The presence of adults will reduce accidents and injuries caused by bullying or other antisocial behaviors. Not only do adults provide supervision for preventing accidents, they are better equipped to handle emergencies that might occur (Uzzell and Stone, 1990).

MAINTENANCE AND INSPECTIONS

Maintenance is extremely important in preventing playground injuries. Even the best-designed and assembled equipment will weaken after repeated use. Equipment that needs repair can be hazardous and should be replaced or repaired immediately. Should such accidents occur, the owner (city) might be found negligent.

Poor maintenance is a common cause of playground injuries. City park and recreation departments typically have limited resources to spend on park equipment and maintenance. This lack of funds may result in broken and ill-maintained park equipment and ground surfacing. As part of a maintenance program, inspect playground equipment for potential hazards and deterioration. Check playground areas for broken glass or other dangerous debris.

Depending on playground location, weather conditions, and frequency of use, maintenance is necessary to insure adequate depth of the surfacing materials (CPSC,1997). Loose-fill materials typically have a lower initial cost but tend to have higher maintenance needs. In high-use areas, these materials may need to be raked frequently or tilled periodically to loosen the compaction (NPPS, 1996). Markers can be attached to equipment support posts that indicate the level of loose-fill protective surfacing around the equipment. Such markers will assist maintenance workers in determining when replenishment is necessary. Typical maintenance frequencies are presented in Table 28.5 (36 CFR 1191).

Table 28.5 Typical Maintenance Frequencies for Protective Surfaces

Maintenance Activity	Loose Fill	Engineered Wood Fiber	Rubber
Inspection	Daily to weekly	Daily to weekly	Weekly
Rake and level	Daily to weekly	Weekly to as required	Not required
Top off	1–3 years	3 years	Not required

Equipment maintenance inspections must comply with manufacturer recommendations and CPSC guidelines. Trained individuals should systematically perform inspections that emphasize moving parts and other components expected to wear from use. Items that should be checked regularly include depth of loose-fill materials, presence of playground trash, and equipment damage or vandalism. Annual or quarterly inspections should include replacing loose-fill material, missing equipment parts, equipment rust, and play area drainage (NPPS, 1996).

Consider using checklists to assure equipment is safe and well maintained. Inspections should also encompass less obvious hazards such as protruding nuts and bolts. These protrusions can catch fingers and hook on hooded clothing, possibly resulting in lacerations or strangulation.

Inspect for loose handrails as these items typically cannot be identified through visual inspection. Check for potential hazards while getting on the equipment. Repair such hazards or defects as soon as possible. Document all inspections and repairs and indicate the date that the inspection or repair took place. If an injury results in litigation, maintenance records may be important discovery materials (Vredenburgh and Zackowitz, 2002).

To address the issue of playground safety, the US Consumer Product Safety Commission (CPSC) has issued the *Handbook for Public Playground Safety* (1997) which provides guidelines for designing playgrounds and popular types of playground equipment.* While the CPSC handbook is not a mandatory standard, several states have codified the guidelines in the CPSC handbook. For example, in California, Senate Bill 2733, chaptered 9/21/90, calls for adopting regulations that are at least as protective as the CPSC handbook, and restricts state funding for new playgrounds unless they conform to the provision of SB 2733. The CPSC has also devised a safety checklist to ensure that general safety guidelines are followed (see Appendix A).

SPECIFIC DESIGN GUIDELINES

RULE #3. DESIGN OUT AS MANY HAZARDS AS POSSIBLE

DESIGN ISSUES: Playground design considerations include:

a. Can all potential users access play structures?
b. Might young children entrap their heads between components of equipment?
c. What factors should be evaluated when selecting and installing equipment?

ACCESS: Potential users must be able to safely access the playgrounds (Uzzell and Stone, 1990).

a. Place pedestrian walkways to avoid busy intersections where drivers are not required to give children the right of way.
b. Consider surrounding the playground with a fence or shrub barrier that prevents children from inadvertently running into the street (CPSC, 1997).
c. Fences and other barriers can help supervisors and parents keep track of children by preventing them from leaving the playground.

HEAD ENTRAPMENT: To prevent head entrapment hazards, ensure that no piece of equipment or its components form openings that might trap a child's head. This typically occurs when children insert their head through an opening in one orientation and then turn to a different orientation such that they cannot withdraw. This is especially true for younger children who may not have the cognitive skills necessary to release themselves from an entrapment or who panic upon realizing they are stuck. Generally, a space poses a head entrapment hazard if two opposing surfaces (such as two bars) are more than 9 cm (3.5 in.) and less than 23 cm (9 in.) apart.

SELECTION AND ASSEMBLY OF EQUIPMENT: Equipment and activities that promote skill development such as balance, coordination, muscular strength, and upper body development necessarily involve elements of risk. Total elimination of risk is impossible and mitigating risk may restrict necessary developmental challenges. Instead, select high-quality equipment and verify that it is properly assembled. This ensures structural integrity and stability and general safety. Follow manufacturer's assembly instructions for playground equipment installation closely. After assembly, inspect equipment before it is used (Uzzell

* This handbook presents safety information for playground equipment in the form of guidelines. It can be downloaded online at www.cpsc.gov/cpscpub/pubs/325.pdf

and Stone, 1990). As a precaution, maintain written information regarding assembly and installation in a permanent file.

Materials and construction. Consider several factors when selecting playground equipment. Analyze individual components of a play structure and consider the anchoring devices, platforms, stairs, and guardrails.

Individual pieces. When evaluating a playground, consider the safety of each individual piece of equipment as well as the layout of the entire play area. Locate moving parts and equipment that require open space for appropriate usage (e.g., merry-go-rounds and swings), on the perimeter or edge of the play area. This will reduce the risk of other children at play becoming obstacles during the use of this equipment.

Similarly, position slide exits so a swiftly moving child exiting the slide is less prone to collide with another child in the play area (CPSC, 1997). In addition, separate popular, heavily used equipment to avoid crowding in any one area.

Minimum use zones are areas to keep clear of obstructions in order to ensure that equipment is used safely (such as a slide exit). These zones should not overlap. The shortest journey between pieces of equipment—the desire lines—can be predicted and should not pass close to moving equipment (Uzzell and Stone, 1990). Facilitate supervision by eliminating visual barriers that prevent a clear line of sight between play areas (CPSC, 1997).

Anchoring devices. To eliminate tripping hazards from the playground, anchoring devices used to secure equipment to the ground should be installed so that the anchor is below ground level and under the surfacing material of a playground. This will also prevent injuries from occurring when children fall from a height and land on an anchoring device. Retaining walls are often used on playgrounds to help contain loose-fill surfacing materials. To prevent tripping accidents, retaining walls should be highly conspicuous and any change in elevation should be obvious. Bright paint is useful in making these barriers noticeable to playground users.

Platforms. Platforms are an important aspect of playground design; these provide children with an opportunity to rest and make decisions about the direction of their play activity. Platforms more than 2 m (6 ft) high should provide an intermediate standing surface (landing) where children can decide if they want to continue their ascent or choose an alternate means of descent. Playgrounds should not force a child into making risky choices that make them uncomfortable.

Stairs and ladders. To prevent entrapment, design stairways or rung ladders so that each step or rung is evenly spaced. Sometimes stairs are open between the treads. Close risers between steps that fall within the danger zone of greater than 9 cm (3.5 in.) and less than 23 cm (9 in.).

Rung ladders are typically used for both foot and hand support as they do not have handrails. To ensure that small hands are able to grasp rung ladder components, the diameter of the bars should be between 24 mm (0.95 in.) and 39 mm (1.55 in.). In fact, any device intended to be grasped by hand should have a cross-section diameter between 24 mm (0.95 in.) and 39 mm (1.55 in.). A diameter of 32 mm (1.25 in.) is ideal to benefit the weakest child in a setting, since grip strength is strongest in this power grip position.* Secure all rungs to prevent possible turning.

Provide all stepladders and stairways with continuous handrails at a vertical height between 56 cm (22 in.) and 66 cm (26 in.). These should extend on both sides the

* Of course, this depends on the hand size of the individual child.

entire length of the stairway or stepladder regardless of the height of equipment access. Handrails that serve only children older than 5 years can be as high as 97 cm (38 in.). In addition, all handrail diameters should fall between 14 mm (0.55) and 32 mm (1.25 in.).

Guardrails and barriers. Guardrails and protective barriers prevent inadvertent falls. Barriers should prevent children from overcoming the protective devices. Preschool children under 5 years require guardrails for surfaces elevated 51 cm (20 in.) or higher.

A full protective barrier should be in place for any surface over 76 cm (30 in.) elevation because it affords more protection from falls. For school-aged children over 5 years, guardrails should be in place for any surface over 76 cm (30 in.) elevation. A full protective barrier should be in place for any surface over 122 cm (48 in.) high. These barriers should completely enclose the raised platform they are guarding except for entrance and exit openings. Guardrails should also conform to the above to prevent head entrapment guidelines.

Guardrails should be high enough to keep the largest child from climbing over and the smallest from climbing under them. For children under 5 years, guardrails should be at least 74 cm (29 in.) high and no more than 58 cm (23 in.) above the platform they protect. For older children, guardrails should be at least 97 cm (38 in.) high and no more than 71 cm (28 in.) above the platform.

Equipment types. This section reviews features to look for when selecting and inspecting common types of playground equipment.

Climbing equipment. Because climbing equipment is more challenging than other playground structures, these are the leading source of fall accidents. Such devices include arch climbers, net or chain climbers, domes, balance beams, suspension bridges, and overhead hanging bars or rings. As children develop their skills and abilities, they progress to the more advanced equipment. Because many younger children have not yet developed the skills needed to master climbing equipment, such as upper body strength and coordination, they may have difficulty with more challenging climbing equipment.

Climbing equipment should enable children to descend as easily as they ascend the equipment. This may require an alternative way to exit, such as a slide or a set of steps. This is particularly important for equipment designed for preschool children, who develop skills necessary to "climb down" later than they learn to "climb up" such equipment.

Figure 28.2 (Alison Vredenburgh)
Nonrigid chain ladders are more challenging than traditional rigid ladders.

Climbers with nonrigid surfaces present a particular challenge to children because they do not offer steady or fixed support. Such equipment includes apparatus with chain, rope, or tire-climbing surfaces (see Figure 28.2). When providing access to platforms, these structures must be securely anchored at both ends. If secured to the ground, install the anchoring device below impact-absorbing ground surface material.

Children under 4 years of age should not use overhead horizontal ladders or rings; these require more upper body strength than is typical at younger ages (see Figure 28.3). Do not locate the first handhold on either end of an overhead climber directly above the platform or surface for mounting or dismounting; this minimizes the risk of children striking the platform during a fall if they miss their first handhold. The maximum height the hanging bar of overhead equipment is 152 cm (60 in.) above the ground for preschoolers and 213 cm (84 in.) for older children.

Figure 28.3 (Alison Vredenburgh) Overhead horizontal ladders require upper body strength.

Merry-go-rounds. Merry-go-rounds are a common example of rotating playground equipment. These require special consideration as users often mount and dismount while in motion. Ensure components of the equipment do not extend outside the radius of the base; this assures that protruding parts do not strike people standing nearby. The underside of rotating platforms should be at least 23 cm (9 in.) above the ground.

Provide secure handgrips that conform to guidelines. In addition, the merry-go-round should have a mechanism that keeps it from exceeding 4 m (13 ft) per second speeds. Most have holds that are relatively high in places, which encourage children to stand while they use the equipment.

Seesaws. Because standard fulcrum seesaws require two children to cooperate, they are typically not recommended for preschoolers who may not be able to use them successfully. Seesaws with spring fulcrums are safer for younger children as these minimize the risk of one end slamming to the ground when the other child dismounts. Providing both ends of the seesaw with stationary handholds also keeps children from sliding forward on the seat. Embeding shock-absorbing materials such as a rubber tire in the ground under each end of the seesaw reduces the shock of one end abruptly hitting the ground.

Slides. Children of all ages use slides in a variety of positions other than the traditional sitting with a feet-first orientation. Kids descend slides laying down head or feet first, on their knees or backward and they often climb up descent chutes. It is therefore important that slide designs comply with recommended guidelines.

Provide all slides with a platform that is sufficiently large to allow children to transition from mounting to seated positions. Freestanding slide platforms should typically be at least 56 cm (22 in.) deep. The platform should be at least as wide as the slide chute. Further, surround the platform with a protective barrier. Hoods can help channel users into a seated position before descent. Figure 28.4 depicts a typical freestanding straight slide.

The average slope of the chute surface should not exceed 30°. Slide chute sides should be at least 10 cm (4 in.) high to prevent children from falling off while sliding. Metal slides should be placed in shaded areas to prevent the sun from heating the surface and causing burns.

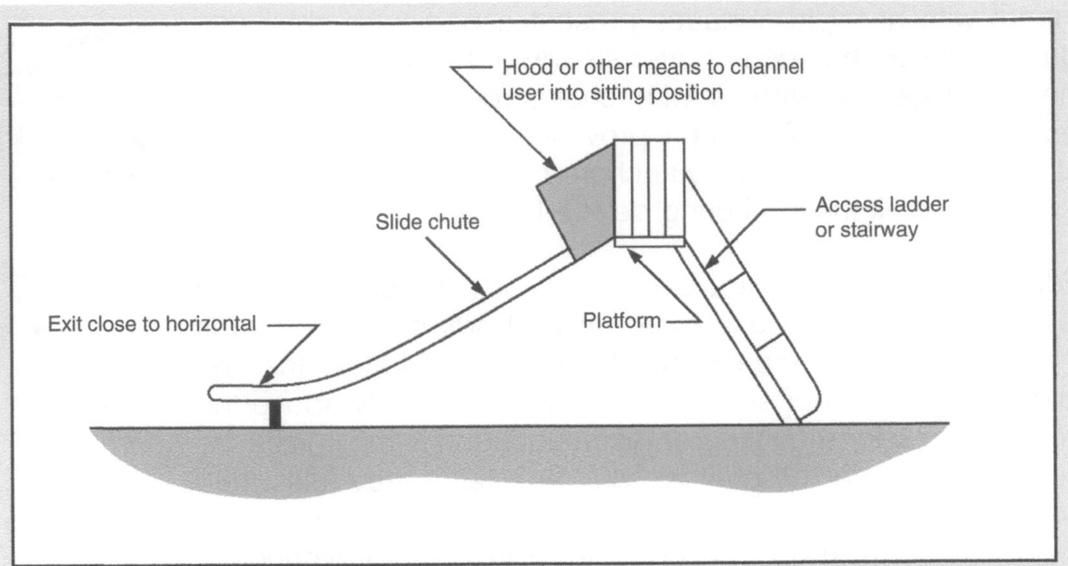

Figure 28.4 Typical freestanding straight slides. (From Consumer Product Safety Commission (1997). *Handbook for Public Playground Safety* (CPSC Publication No. 325). Washington, D.C.: U.S. Government Printing Office.)

Exit areas of slides should be close to horizontal, parallel to the ground, and be at least 28 cm (11 in.) long. Do not locate slide exits in a congested area of the playground. The exit area should be no more than 28 cm (11 in.) above the ground for slides lower than 1.2 m (4 ft) high. For higher slides, the exit area should be between 18 cm (7 in.) and 38 cm (15 in.) from the ground.

Figure 28.5 (Alison Vredenburgh)
Spring rockers are enjoyed by the youngest playground users. Close supervision is important.

Spring rockers. Preschool-aged children tend to enjoy the bouncing of spring rockers (see Figure 28.5). This equipment is typically not challenging enough for older children. Equip spring rockers with seats that accommodate the intended number of riders. Equip rockers with handgrips and footrests and design the spring to minimize the risk of a pinch injury.

Swings. Playground users of all ages enjoy swings. Aside from the typical position of sitting on the swings, younger children may swing on their stomachs while older children may try to swing while standing up (see Figure 28.6). Such distinctions are important to ensure safe play for all children.

Most swings operate on a single axis, as in traditional designs that only swing back and forth. Multidirectional axis swings, however, can swing in several directions. An example of this would be a tire swing suspended in a horizontal orientation using three suspension chains or cables connected to a single swivel mechanism that permits both rotation and a swinging motion in any axis. Do not suspend multiaxis swings from structures with other swings in the same bay or to composite structures.

Swings are usually attached to their supports with S-hooks. These hooks must be completely closed to prevent catching children's clothing and to prevent swings from

Figure 28.6 (Alison Vredenburgh)
Older children may use swings standing up.

coming off the hook. When in use, swings need a large area of clear space to avoid collisions. It is therefore best to have swings in a separate area of the playground to avoid injuring children who walk by.

To further reduce potential hazards, suspend no more than two swings from any single-frame structure. Separate these by at least 51 cm (20 in.) to keep multiple swings from colliding. Figure 28.7 depicts the minimum clearances for single-axis swings.

Swing seats should be made of soft materials like rubber. Avoid wood and metal as such hard swings may lead to head injury if they strike a child. Suspend swings intended for preschool use no higher than 30 cm (12 in.) off the ground. Swings for older children should be no more than 41 cm (16 in.) off the ground.

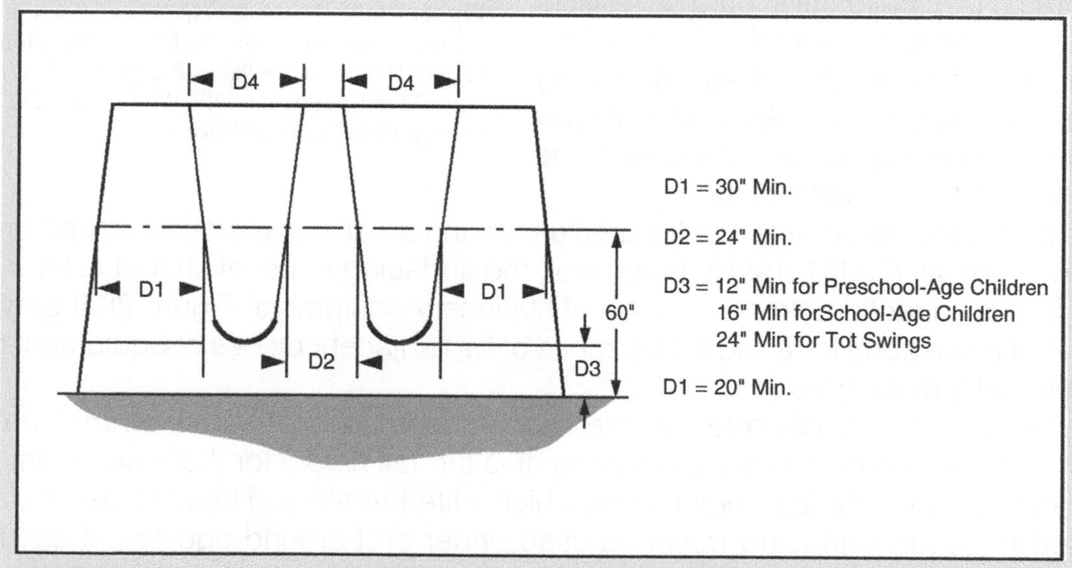

Figure 28.7 Minimum clearances for single-axis swings. (From Consumer Product Safety Commission (1997). *Handbook for Public Playground Safety* (CPSC Publication No. 325). Washington, D.C.: U.S. Government Printing Office.)

D1 = 76 cm (30 in.) minimum

D2 = 61 cm (24 in.) minimum

D3 = 30 cm (12 in.) minimum for preschool-age children

= 41 cm (16 in.) minimum for school-age children

= 61 cm (24 in.) minimum for toddler swings

D4 = 51 cm (20 in.) minimum

PLAYGROUND SURFACING

Once safe and properly designed playground equipment selections have been made, it is vital to carefully consider the playground surfacing material. Because falls account for the majority of playground injuries, surfacing is extremely important. Falls account for at least 75% of all playground accidents in the United Kingdom (CAPT, 2004), and 65% in Canada (CHIRPP, 1997). In the United States, 70% of all playground injuries result from falls to the ground surface (Tinsworth and McDonald, 2001) and about 80% of all playgrounds have unsuitable surfaces (NPPS, 1996).

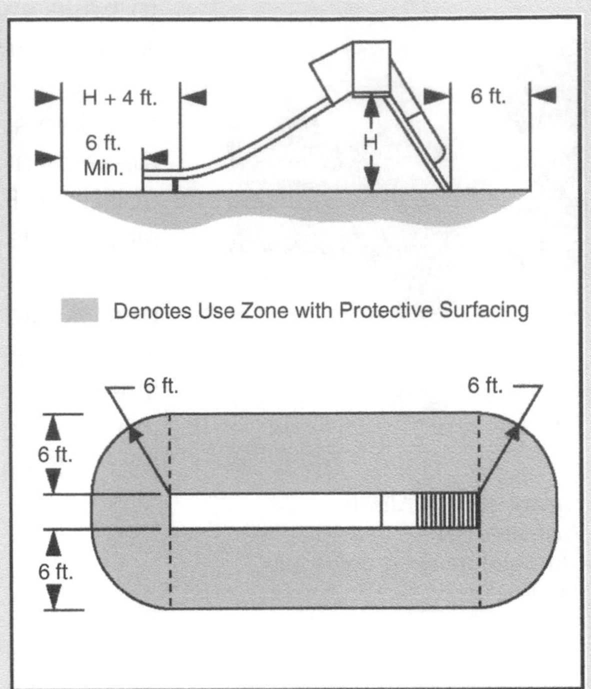

Figure 28.8 Example of a use zone for a slide. (From Consumer Product Safety Commission (1997). *Handbook for Public Playground Safety* (CPSC Publication No. 325). Washington, D.C.: U.S. Government Printing Office.)

Falls onto shock-absorbing surfaces are less likely to cause serious injury than falls onto hard surfaces. A fall from a 2.5 m (8-ft) high structure (the height of many slides) onto concrete or asphalt is equivalent to hitting a brick wall at 20 mph in a car (Ward, 1997). In contrast, if a child falls onto a resilient surface such as sand or mulch, the surface will deform upon impact and absorb energy from the fall (Hendricks, 1993). The deeper the shock-absorbing surface, the greater the potential to reduce severe injuries.

Use (fall) zones are defined as the area under and around playground equipment where children might fall (NPPS, 1996). In general, the surfacing material should extend at least 2 m (6 ft) in all directions from the edge of stationary equipment. Figure 28.8 provides an example of a use zone for a slide. Use zones of two adjacent pieces of equipment must be a minimum of 2 m (6 ft) apart.

The term *critical height* refers to the shock-absorbency of surfacing materials.* The highest play surface on a piece of equipment is the *fall height* for that equipment. Critical height approximates the fall height below which a life-threatening head injury would not be expected to occur. Surfacing materials used under and around equipment should have critical height values for at least the highest play surface on that piece of equipment.

Types of materials. Hard surfacing materials, such as asphalt or concrete, are unsuitable for using under and around playground equipment of any height. Avoid earth surfaces such as soil and hard packed dirt due to their poor shock-absorbing properties. Similarly, avoid grass and turf because wear and environmental conditions can reduce their effectiveness in absorbing shock during falls (Vredenburgh and Zackowitz, 2002). If located within use zones, ground surfaces shall comply with ASTM F1292 Standard Specification for Impact Attenuation of Surface Systems Under and Around Playground Equipment.†

* The technical definition of critical height is the maximum height from which an instrumented metal headform, upon impact, yields both a peak deceleration of no more than 200 times the acceleration due to gravity (200 G's) and a *head injury criteria* (HIC) value of no more than 1000 when tested in accordance with the procedure described in AST M F1292.

† You can purchase a copy of this ASTM standard from the ASTM Web site www.astm.com

There are two basic types of acceptable playground surfaces: loose-fill or unitary. *Unitary materials* are generally rubber mats or a combination of rubberlike materials held in place by a binder that may be poured in place at the playground. Some (but not all) unitary materials must be installed over a hard surface (CPSC, 1997). These mats are used to cover the entire area of some playgrounds. Unitary materials have varying degrees of absorbency; therefore, ask for critical heights for each material when selecting the appropriate material for a given playground.

Loose-fill materials include sand, gravel, shredded wood, and shredded tires and have acceptable shock-absorbing properties when installed and maintained at a sufficient depth (based on the critical height of the equipment). Avoid installing loose-fill materials over hard surfaces like concrete.

Table 28.6 lists critical heights for eight typical loose-fill materials when tested in an uncompressed state at depths of 15, 23, and 30 cm (6, 9, and 12 in.) and in a compressed state of 23 cm (9 in.). For example, if uncompressed wood chips were used at a depth of 15 cm (6 in.), the maximum fall height would be 2 m (7 ft). Note that as certain materials compress, the fall height decreases because the energy-absorbing potential is reduced. Therefore, these materials may need to be raked frequently or tilled periodically to loosen the compaction (NPPS, 1996). The National Program for Playground Safety generally recommends using 30 cm (12 in.) of uncompressed loose-fill material for equipment up to 2.5 m (8 ft) in height.

Table 28.6 Critical Heights of Loose-Fill Surfacing

Material	Uncompressed Depth			Compressed Depth
	6 in.	9 in.	12 in.	9 in.
Wood chips	7 ft	10 ft	11 ft	10 ft
Double-shredded bark mulch	6	10	11	7
Engineered wood fibers	6	7	>12	6
Fine sand	5	5	9	5
Coarse sand	5	5	6	4
Fine gravel	6	7	10	6
Medium gravel	5	5	6	5
Shredded tires[a]	10–12	NA	NA	NA

Source: From Consumer Product Safety Commission (1997). *Handbook for Public Playground Safety* (CPSC Publication No. 325). Washington, D.C.: U.S. Government Printing Office.

[a] Persons seeking to install shredded tires as a protective surface should request test data from the supplier showing the critical height of the material when tested in accordance with ASTM F1292.

CONCLUSIONS

The purpose of this chapter is to discuss ergonomic issues that help make our playgrounds safe places for children of all ages to develop critical physical and social skills. The key is to design the space with the capabilities and limitations of the intended users in mind, whether they are preschool or school-aged children. In addition, maintenance and supervision are imperative.

Risk-taking is a natural part of childhood, especially on playgrounds. This chapter illuminates design considerations that help reduce the inherent risk. The number and type of injuries greatly depends on what kind of play environment the child is provided. Equipment that is appropriately selected, installed, and maintained, as outlined in this chapter, will reduce the occurrence of certain injury types.

In addition to improving the playground environment, appropriate supervision and instruction on the playground are necessary. Children should be taught appropriate and safe playground behavior. Adult supervision will help ensure that children will not injure themselves by misusing playground equipment. Without addressing these factors, no playground, despite its design, will provide children with a safe play area.

APPENDIX A

General Maintenance Checklist (based on the CPSC Suggested General Maintenance Checklist, 1997).

Surfacing

____ The equipment has adequate protective surfacing under and around it and the surfacing materials have not deteriorated.
____ Loose-fill surfacing materials have no foreign object or debris.
____ Loose-fill surfacing materials are not compacted and do not have reduced depth in heavy use areas such as under swings or at slide exits.

General Hazards

____ There are no sharp points, corners, or edges on the equipment.
____ There are no missing or damaged protective caps or plugs.
____ There are no potential clothing entanglement hazards such as open S-hooks or protruding bolts.
____ There are no pinch or crush and shearing points or exposed moving parts.
____ There are no trip hazards, such as exposed footings or rocks and roots in the play area.

Deterioration of the Equipment

____ The equipment has no rust, rot, cracks, or splinters especially where it contacts the ground.
____ There are no broken or missing components on the equipment like guardrails, handrails, or protective barriers.
____ There are no damaged fences, benches, or signs on the playground.
____ All equipment is securely anchored.

Hardware

____ Are there fastening devices loose? Are connections worn?
____ Moving components, such as swing hangers or merry-go-round bearings, are not worn.

Drainage

____ The entire play area has satisfactory drainage, especially in heavy use areas.

General Upkeep of the Playground

____ The entire playground is free from debris and litter such as tree branches, glass, and cans.
____ There are adequate trash receptacles.

REFERENCES

Americans with Disabilities Act Accessibility Guidelines for Play Areas (November, 2000). 36 CFR Part 1191.

Canadian Hospitals Injury Reporting and Prevention Program (1997). *The CHIRPP Sampler*, Issue 12, November.

Child Accident Prevention Trust (2004). *Playground Accidents Fact Sheet*, November, CAPT: London.

Consumer Product Safety Commission (1997). *Handbook for Public Playground Safety* (CPSC Publication No. 325). Washington, D.C.: U.S. Government Printing Office.

Epstein, L.H. (2000). To get obese kids to exercise, turn off the TV. *Archives of Pediatrics and Adolescent Medicine*, 154, 220–226.

Frost, J.L. (1999). *Playground Safety*. Washington, D.C.: ERIC Clearinghouse on Teacher Education.

Hendricks, C.M. (1993). *Safer Playgrounds for Young Children*. Washington, D.C.: ERIC Clearinghouse on Teacher Education.

Jambor, T. and Palmer, S.D. (1990). *Playground Safety Manual*. Birmingham, AL: Injury Control and Research Center, University of Alabama at Birmingham.

Joint Powers Authority (2004). *Playground Safety Manual*. San Diego County Schools Joint Powers Authority.

National Program for Playground Safety (1996). Age appropriate design guidelines for playgrounds. University of Northern Iowa: NPPS.

Ross, J. and Gilbert, G. (1985). The national youth and fitness study. *Journal of Health, Physical Education, Recreation and Dance*, 56, 45–60.

Tinsworth, D. and McDonald, J. (2001). Special study: Injuries and deaths associated with children's playground equipment. Washington, D.C.: U.S. Consumer Product Safety Commission.

Turner, J.S. and Helms, D.B. (1995). *Lifespan Development*, 5th edn. New York: Harcourt Brace College Publishers.

Uzzell, D. and Stone, E. (1990). Children's playground safety guidelines. Department of Psychology, University of Surrey, Guilford, Surrey, UK.

Vredenburgh, A.G. and Zackowitz, I.B. (2002). Playground safety: A forensic human factors analysis of fall injuries. *Proceedings of the 54th Annual Meeting of the American Academy of Forensic Sciences*, Atlanta, GA, pp. 96–96.

Ward, A. (1987). Are playground injuries inevitable? *The Physician and Sports Medicine Journal*, 15(4), 162–168.

Zackowitz, I.B. and Vredenburgh, A.G. (1998). Preschoolers: A forensic human factors analysis. Proceedings of the 50th Annual Meeting of the American Academy of Forensic Science, San Francisco, CA, pp. 91–92.

AUTHOR BIOGRAPHIES

Gary L. Allen, PhD, is a developmental and cognitive psychologist at the University of South Carolina who for over 25 years has been doing basic research in children's and adults' spatial thinking and behavior. Since receiving his PhD from the University of Pittsburgh in 1979, he has collaborated with and learned from scholars in a variety of disciplines, including human factors, geography, education, linguistics, and rehabilitation psychology in efforts to better understand the nature of spatial knowledge and wayfinding skill. For more information, see /www.cas.sc.edu/psyc/facdocs/allen.html.

Dennis R. Ankrum is the president (and the only full-time employee) of Ankrum Associates, an office and Web site ergonomics consulting firm in Oak Park, Illinois. His research activity has focused on the interaction between vision and neck posture. He has published over 40 articles and book chapters on ergonomics. Ankrum is a standards junkie. He has been a member of many standards committees, which include ANSI HFES 100, the American National Standard for Human Factors Engineering of Visual Display Terminal Workstations, RP-1, the Illuminating Engineering Society standard on Office Lighting, RP-3, the Illuminating Engineering Society standard on Lighting for Educational Facilities, ASTM E34.85, the ASTM Standard Guide for the Integration of Ergonomics/Human Factors into New Occupational Systems (chairman), and BIFMA-G1-2001, the BIFMA Ergonomics Guideline for VDT (visual display terminal) Furniture Used in Office Work Spaces. In addition, he participated in the development of the National Safety Council's Z-365 standard on Control of Cumulative Trauma Disorders. He maintains the most popular office ergonomics site on the Internet (as per Google), www.office-ergo.com. His consulting Web site is www.AnkrumAssociates.com.

Mary Benbow has been researching and developing educational materials in hand skills and handwriting since 1976. She authored the innovative handwriting curriculum "*Loops and Other Groups*," a kinesthetic writing system, published in 1990. This system is used for learning by disabled and typically developing students in many English-speaking countries. Later, her techniques were adopted for a Hebrew writing system. She is best known in occupational therapy circles as an international lecturer for continuing education.

Cheryl L. Bennett, MSc, CSP, chairs a technical committee she organized for the International Ergonomics Association (IEA). The technical committee called Ergonomics for Children and Educational Environments (ECEE) has brought together interested researchers and practitioners from a growing number of countries around the world. The IEA/ECEE Web site is an important resource for researchers, teachers, students, and school administrators. She has organized a number of symposia focusing on children, schools, and ergonomics at international conferences and assisted in organizing a children and information technology symposium at The Center for Information Technology at Johns Hopkins University. She has developed ergonomics training modules for use in schools and is working to develop youth ergonomics strategies as a not-for-profit foundation to facilitate more education and research in this field.

Melissa Beran, as a research scientist in the Foreseeable Use Department at Interek-RAM, provides expertise in the area of foreseeable use as it relates to managing risks associated with consumer products. In order to better understand and predict the foreseeable use of consumer products, Ms. Beran facilitates child and caregiver studies at Interek-RAM and coordinates pediatric biomechanics data gathering. Ms. Beran's background is in psychology, counseling, child/family behavior research, child development, family studies, and elementary education. Currently, she is also an instructor for Interek-RAM's safety training series, which provides the customers with an improved understanding of the key product safety issues in today's marketplace.

Barbara H. Boucher, PT, PhD, OTR, has devoted her career to the health and well-being of children. She is an expert in the treatment of movement disorders and developmental problems. Her credentials include a baccalaureate degree in and license to practice occupational therapy, a master's degree in and license to practice physical therapy, and a doctor of philosophy in human development and family sciences. She has worked in public schools in four states and taught physical therapy to graduate students. Dr. Boucher continues to work with clients through her private practice, TherExtras, and Web site, www.therextras.com.

Tina Brown's background is in child psychology, child behavior research, child development, early childhood education, and statistical methods. As a former child development specialist at Interek-RAM, Tina Brown spent 7 years providing expertise in the area of foreseeable use, as it relates to managing the risks associated with consumer products. As part of this work, she conducted studies with children and caregivers to better understand and predict the foreseeable use of consumer products. She was also an instructor for the safety training series of Interek-RAM, which provides their customers with an improved understanding of the key product safety issues in today's marketplace. She is currently living in Michigan completing her master's degree in child development and putting her education and experience to use with her own 1-year-old daughter.

Cindy Burt, OTR/L, CPE, created and is manager of UCLA's ergonomics program. This program provides injury prevention services for 25,000 employees at UCLA. Before assuming this position, Cindy managed the UCLA Medical Center sports medicine/hand therapy program. Cindy is a licensed occupational therapist paneled by California Children's Services and a certified professional ergonomist. She has worked with both children and adults with physical injuries and disabilities and has written numerous articles and book chapters in these areas. Cindy is active in both occupational therapy and human factors/ergonomics associations. She developed UCLA's ergonomics Web site, www.ergonomics.ucla.edu.

Hsin-yu (Ariel) Chiang, MS, OT, is a doctoral student in occupational therapy at Boston University. She is a certified occupational therapist and medical technician in Taiwan. She worked at the Assistive Technology R&D Center as a research assistant for 1 year after her graduation from Chung Shan Medical University. Her research interests are assistive technology and ergonomics. She did her MA at Boston University, and published her study about middle-school students' backpack usage and its relationship to lower back pain. Currently, she is doing her doctoral study on the effectiveness of computer-based reading instruction with elementary and middle-school students.

Knut Inge Fostervold, PhD, is an associate professor at the University in Oslo, Norway, an adjunct associate professor at Lillehammer University in Lillehammer, Norway, and a licensed psychologist. Much of his research has focused on health, behavior, vision, and visual functions at work. Dr. Fostervold teaches and conducts research in the fields of ergonomics, cognition, vision and visual perception, and work and organizational psychology. His scientific research has been published in leading Norwegian and international journals. He served on the Board of the Nordic Ergonomics Society for several years and is currently president of the Norwegian Ergonomics Society.

David Gulland is an architect and principal of HASSELL, a design-based practice with offices throughout Australia, in the People's Republic of China, Hong Kong, and Thailand. HASSELL's areas of professional expertise are focused on architecture, interior design, landscape architecture, and urban design and planning. David has collaborated with the Western Australia Department of Education and Training as architect for a wide variety of schools and has been directly involved in the integration of educational philosophy with facility design for the past 15 years. For more information, see *http://www.hassell.com.au.*

Libby Hanna, PhD, principal of Hanna Research and Consulting, LLC, is a developmental psychologist who has been doing user research with children and adults on computer software and hardware products since 1996. She contracted with Microsoft to run usability testing for the Microsoft Kids Group from 1997–2000 and has written guidelines and articles for usability testing with children. She received her PhD from the University of Washington in Seattle, where she studied gender role development in very young children. For more information, see *www.hannaresearch.com.*

Alan Hedge, PhD, CPE, is a professor in the Department of Design and Environmental Analysis at Cornell University, where he teaches and conducts research in ergonomics. Dr. Hedge is widely published and has co-authored two books. He has been featured on numerous television and radio shows and his work is frequently cited in major newspapers and magazines. He is a fellow of the Human Factors and Ergonomics Society (USA), the Ergonomics Society (UK), and the International Ergonomics Association. In 2003, he received the Alexander J. Williams Jr. Design Award from the Human Factors and Ergonomics Society. He is a member of several other professional societies. He is a certified professional ergonomist and consults in ergonomics. His Cornell University Web site, http://ergo.human.cornell.edu/, contains extensive content on ergonomics for children.

Dr. Hal W. Hendrick is Emeritus Professor of human factors and ergonomics at the University of Southern California and Principal of Hendrick and Associates, a private consulting firm. Previously, he was executive director of USC's Institute of Safety and Systems Management and had chaired the university's human factors department. Hal is a former president of the Human Factors and Ergonomics Society (HFES), the International Ergonomics Association (IEA), and the Board of Certification in Professional Ergonomics (BCPE). He is a fellow of the IEA, HFES, American Psychological Association, and American Psychological Society. He has authored over 200 professional publications, including 3 books, and edited 10 books. He has served as an expert witness in numerous unintentional shooting cases and writes extensively about human factors of handgun training and design. He is a certified professional ergonomist and holds a PhD in industrial psychology and MS in human factors from Purdue University.

Karen Jacobs, EdD, OTR/L, CPE, FAOTA, is a past president and vice-president of the American Occupational Therapy Association (AOTA) and a 2005 recipient of a Fulbright Scholarship to the University of Akureyri in Akureyri, Iceland. Dr. Jacobs is a clinical professor and director of online post-professional occupational therapy programs at Boston University. She earned her EdD at the University of Massachusetts. Dr. Jacobs' research examines the interface between the environment and human capabilities. In particular, she examines the individual factors and environmental demands associated with increased risk of functional limitations among populations of university and middle school aged students, particularly in notebook computing and backpack use (see http://people.bu.edu/kjacobs). In addition to being an occupational therapist, Dr. Jacobs is a founding editor of the international journal *WORK: A Journal of Prevention, Assessment and Rehabilitation* (IOS Press, the Netherlands). Dr. Jacobs has a small business that provides consulting services in ergonomics and entrepreneurship.

Michael J. Kalsher, PhD (Virginia Tech, 1988), is associate professor of psychology and cognitive science at Rensselaer Polytechnic Institute. Kalsher served as Chair of the Department of Cognitive Science at Rensselaer from 1997 to 2002. He is a member of the American Psychological Society, the American Psychological Association, and the Human Factors and Ergonomics Society for which he currently serves as Chair of the Arnold M. Small Safety Lecture Series.

Jeff Kennedy is founder and principal of a US exhibition planning and design firm Jeff Kennedy Associates, Inc., known for innovative development and design for children's museums, science centers, history museums, and other informal learning environments. Jeff Kennedy is a recognized leader in universal design, an approach to exhibition planning and design that seeks to make exhibits accessible to the broadest possible public audience, including people with physical, sensory, or developmental impairments. His work in this area began with the book *User Friendly: Hands-On Exhibits That Work*, funded through a grant from the Institute for Museum and Library Services (IMLS). Reprinted several times since its original publication in 1990, the book has become a standard reference throughout the exhibition design industry. Their Web site is www.jeffkennedyassociates.com.

Renee Lockhart, MS, OTR/L, is an occupational therapist, currently providing outpatient services to adolescents and adults with acquired brain injury. Ms. Lockhart earned her MS in occupational therapy at Boston University and her BA in psychology at Boston College. She extensively studied back pain in adolescents caused due to backpack use and completed her thesis entitled "Middle school children's participation in activities and the effects of pain from backpack use on participation." Her work is published in *WORK: A Journal of Prevention, Assessment and Rehabilitation* (IOS Press: the Netherlands).

Lorraine E. Maxwell is an environmental psychologist and is currently an associate professor in the Department of Design and Environmental Analysis at Cornell University. She received her PhD in psychology from the Graduate Center of the City University of New York and holds a masters in city and regional planning from Rutgers University. Her research is in the field of children's environments including schools, childcare facilities, playgrounds, museums, and the home. Dr. Maxwell is interested in the effects of the physical environment on children's development and well-being and has studied the effects of chronic noise and high density on children's social and cognitive development. She is currently investigating attributes of the physical environment related to young children's development of self-identity and competency. Before coming to Cornell, Dr. Maxwell was an associate in an architectural firm where she worked as a facilities planner/programmer for 16 years.

Beverley Norris, PhD, is a principal research fellow of the University of Nottingham and projects manager of the Human Factors Research Group. She has over 15 years experience of ergonomics research, specializing in anthropometry, ergonomics, and design and product safety research. Her PhD was on methods to support designers in applying ergonomics to the design of safer products. Together with colleagues at Nottingham, Beverley produced the Childata/Adultdata/Older Adultdata series of publications, a United Kingdom government initiative to make anthropometric data more accessible to designers.

Mary Colleen O'Hara is a communications professional with a background in writing, editing, and public relations. Ms. O'Hara works as a freelance writer and reviews manuscripts for the international interdisciplinary journal, *WORK: A Journal of Prevention, Assessment and Rehabilitation.* She teaches English for speakers of other languages (ESOL) at Brookhaven National Laboratory in Long Island, New York. She earned her MA in English at Stony Brook University in Stony Brook, New York, and her BA in English from Saint Anselm College in Manchester, New Hampshire.

Jake Pauls, a certified professional ergonomist, is a Canadian now working in the United States with 38 years of international experience in research, codes and standards development, public health advocacy, and consulting. His subject matter expertise is in ergonomics, architectural and engineering aspects of movement of people, individually and in crowds, especially on stairways. He currently serves on 12 national standards and codes committees in the United States, where he is known for bridging among ergonomics, public health, and the development of codes and standards for building usability, safety, design, construction, and management.

Jeff Phillips is the principal consultant for research and development in Facility Planning for the Department of Education and Training in Western Australia. As an educator with over 30 years of experience, he has held senior administration and teaching positions in Australian and international schools. His current research involves an analysis of innovation and change in educational delivery and facilities provision throughout the world and identifying good practice in school building design. He has developed tools and methodology for postoccupancy evaluation, particularly in relation to assessing the educational adequacy of facilities.

Clare Pollock is a professor and head of school in the School of Psychology, Curtin University of Technology. She teaches human factors and psychology to undergraduate and postgraduate students. Professor Pollock supervises research students and conducts research into behavioral aspects of safety and people's interaction with technology. She has been a chief investigator on many research grants, has published over 80 scholarly works, and is an editorial board member on an international journal. She has been the scientific chair for two international conferences and has been an invited keynote speaker at national and international conferences. Clare is a member of the Human Factors and Ergonomics Society, the Human Factors and Ergonomics Society of Australia, and the Australian Aviation Psychology Association.

Marjorie Prager is an exhibit developer and project manager for many museum projects for children and families at Jeff Kennedy Associates, Inc., an exhibition planning and design firm in the United States. She is also a writer and producer of media. Jeff Kennedy Associates' Web site is www.jeffkjennedyassociates.com.

Stuart A. Smith, PhD, is a research fellow of the University of Nottingham, and senior human factors consultant of the Human Factors Research Group. Stuart has worked in a range of human factors topic areas, including human error, human modeling, and general occupational ergonomics. He has considerable experience in anthropometry and product safety research, including UK government publications on anthropometry (he was lead author of the *Older Adultdata Anthropometry Handbook*), strength capabilities with regard to product design, and work to assess the continuing validity of the anthropometric data contained in the "Childata" publication through the assessment of secular growth trends.

Robin Springer founded Computer Talk, a full-service assistive technology consulting firm, in 1995 to address the growing numbers of individuals with musculoskeletal injuries who required assistive technologies to remain productive. Her objective is to improve technological accessibility and productivity, regardless of physical ability. She performs needs–assessments and configures and implements technology plans for child and adult users at all points on the spectrum of abilities. Springer is a columnist for Speech Technology Magazine and advocate for individuals of varying abilities. She is currently conducting the second phase of a longitudinal study design about individuals with disabilities and their involvement in advanced degree education. Springer frequently contributes to publications and lectures about adaptive technologies and the disability paradigm. Through Springer's approaches, virtually anyone can participate in the world of technological advances—with confidence and ability. For more information, see *www.comptalk.com*.

Leon Straker is a professor of physiotherapy and director of research at the School of Physiotherapy, Curtin University of Technology, Perth, Australia. He teaches ergonomics and research methods to undergraduates and supervises doctoral research. His research over the last 20 years emphasized the prevention of work-related musculoskeletal disorders, starting with a government taskforce on repetitive strain injury. He published over

200 scholarly works including over 20 books/book chapters and 50 journal articles. He was chief investigator on grants of over AU$2.2 million. He served as an editorial board member for two leading journals, is currently an editor of *Ergonomics*, founded the international virtual ergonomics conference CybErg, and chaired its first two conferences. He was awarded the Stephen Pheasant Memorial Award for physical ergonomics research in 1999. He is a sought-after speaker and presented invited addresses in North America, Europe, Asia, and Africa. His ITKids Web site is http://itkids.curtin.edu.au.

Brenda A. Torres was a safety engineer at Intertek Risk Assessment & Management (Interek-RAM) from 1997 to 2004. Ms. Torres used child human factors, from developmental psychology, anatomy, physiology, and strength biomechanics, to evaluate consumer products for hazards. Her work for McDonald's Corporation, the LEGO Group, Gerber, Kolcraft, and other clients involved the design review of toys and juvenile products. She is a mother of two children, a girl and a boy. Ms. Torres received her MS from the Center for Ergonomics, University of Michigan, Ann Arbor. Ms Torres is a member of Human Factors and Ergonomics Society and the International Consumer Products Health and Safety Organization. She continues to do human factors work in consumer product safety for S.C. Johnson & Son.

COL Nancy L. Vause, PhD, Director of Strategic Partnerships Office for the US Army Medical Research and Materiel command brings theoretical and real-world field experience to psychoacoustic research. She examines the effects of noise and reverberation on hearing, localization, and communication. Her research efforts provide guidelines to design and optimize military equipment and audio interfaces (helmets, hearing protection, landmine detectors, and radio headsets). Her interest in classroom acoustics began before joining the army, with a special project at University of Memphis and serving as parent liaison in the Hearing Impaired Program, Memphis Special Education Program. As a military audiologist, she often provides hearing screening and acoustic consultation to post-day-care centers and DOD school systems. COL Vause received a PhD from Vanderbilt University and during her off-duty time teaches a distance-learning audiology doctoral course in hearing conservation for Central Michigan University.

Alison Vredenburgh, PhD, CPE, is the principal of Vredenburgh & Associates, Inc., a research and consulting firm specializing in human factors and safety. She performs industrial–organizational consulting in areas including employee injury prevention and the Americans with Disabilities Act. She regularly publishes in the areas of human factors, safety, and ergonomics. She has served as a forensic expert for personal injury and product liability cases including playground accidents. Before taking up consulting, she worked in the aerospace industry.

Dr. Thomas R. Waters is a certified professional ergonomist and holds advanced degrees in engineering science and biomechanics from the University of Cincinnati. As a researcher at the National Institute for Occupational Safety and Health for the past 17 years, Dr. Waters has conducted a wide range of ergonomics research and has published more than 40 peer-reviewed articles and chapters on manual material handling and prevention of low back disorders. He holds adjunct faculty positions in the Department of Environmental Health and Department of Mechanical, Industrial, and Nuclear Engineering at the University of Cincinnati and in the Systems and Industrial Engineering Department at the Ohio State University.

Dr. Waters is recognized internationally for his work on the revised NIOSH lifting equation. His primary research interests include occupational biomechanics, work physiology, low back injury prevention, and ergonomic risk assessment. Dr. Waters is the primary investigator for a series of studies examining risk of work-related musculoskeletal disorders for youth working in agriculture.

David Werner is cofounder and director of HealthWrights (Workgroup for People's Health and Rights) and a visiting professor at Boston University International School of Public Health. A biologist and educator by training, he has worked for the past 40 years in village health care, community-based rehabilitation, and child-to-child health initiatives in the Third World, mainly Mexico. He authored and illustrated the handbooks *Where There Is No Doctor* (now in over 90 languages), *Helping Health Workers Learn, Disabled Village Children*, and *Nothing About Us Without Us: Developing Innovative Technologies For, By and With Disabled Persons*. David is active in exploring the local-to-global social, political, and economic factors that affect the disadvantaged people's health and lives. In 1998, he coauthored with David Sanders *Questioning the Solution: the Politics of Primary Health Care and Child Survival*, which explores the root causes of poor health and possibilities for a healthier, more equitable way forward. He is currently an advisor and contributor to the Web-based "Politics of Health Knowledge Network" (www.politicsofhealth.org). All David Werner's books, in English and Spanish, plus many papers and newsletters on community-based rehabilitation and disability-related technology, are fully and freely accessible on www.healthwrights.org.

Michael S. Wogalter, PhD, is a professor of psychology, and currently is the coordinator of the ergonomics and experimental psychology PhD graduate program at North Carolina State University. He is also the director of the Cognitive Ergonomics Laboratory in the psychology department at North Carolina State University. Before arriving in Raleigh, North Carolina, in 1992, he held faculty appointments at the University of Richmond (1986–1989) and Rensselaer Polytechnic Institute (1989–1992). He received a BA in psychology from the University of Virginia (1978), an MA in human experimental psychology from the University of South Florida (1982), and a PhD in human factors psychology from Rice University (1986). Most of his research focuses on issues associated with warning effectiveness, with about two-thirds of his 300 peer-reviewed publications, including four of his six books are concerned with this topic. He is on the editorial boards of several scholarly scientific journals including *Applied Ergonomics, Ergonomics, Journal of Safety Research and Theoretical Issues in Ergonomics Science*. He has held a number of offices and roles in the Human Factors and Ergonomics Society (HFES) such as Chair and Program Chair of both the Safety Technical Group and the Forensic Professional Group, and having served as Secretary–Treasurer, and Executive Council Member. He is a fellow of HFES. He consults in various legal cases across the United States as an expert witness concerning issues associated with human factors/ergonomics, hazard perception, communication-human information processing, and warnings.

Ilene B. Zackowitz, PhD, CPE, performs industrial–organizational consulting as well as litigation consulting in the areas of human factors, ergonomics, and safety. She also conducts and publishes original research in these areas. She has assisted numerous organizations in developing and implementing proactive ergonomic plans.

Photographers

Ursy Potter is founder of Ursy Potter Photography, a specialized custom photography and animation studio based in McLean, Virginia. Her clients range from local schools to multinational organizations. She travels extensively, often to South America, Europe, and elsewhere for her work. She is also a well-known and award-winning artist whose work has been exhibited at the Renwick Museum, at the Folger Shakespeare Library (several times), in over 200 established art galleries, and in many publications. More details about her background and activities are available at her Web site, www.ursypotter.com

Kevin Burke, based in Cambridge, Massachusetts, specializes in social description photography. His personal projects reflect long-standing interests in calypso music and the history of license plates.

INDEX

Note: Page numbers in italic indicate information contained in illustrations, additional information may also be contained within the text on that page number.

B

C

D

E

F

G

I

J

K

L

N

Q

R

T

Z

Printed and bound by CPI Group (UK) Ltd, Croydon, CR0 4YY

23/10/2024

01778259-0020